T0134983

Advances in Intelligent Systems and Computing

Volume 1026

The series "Advances in Intelligent Systems and Computing" contains publications on theory, applications, and design methods of Intelligent Systems and Intelligent Computing. Virtually all disciplines such as engineering, natural sciences, computer and information science, ICT, economics, business, e-commerce, environment, healthcare, life science are covered. The list of topics spans all the areas of modern intelligent systems and computing such as: computational intelligence, soft computing including neural networks, fuzzy systems, evolutionary computing and the fusion of these paradigms, social intelligence, ambient intelligence, computational neuroscience, artificial life, virtual worlds and society, cognitive science and systems, Perception and Vision, DNA and immune based systems, self-organizing and adaptive systems, e-Learning and teaching, human-centered and human-centric computing, recommender systems, intelligent control, robotics and mechatronics including human-machine teaming, knowledge-based paradigms, learning paradigms, machine ethics, intelligent data analysis, knowledge management, intelligent agents, intelligent decision making and support, intelligent network security, trust management, interactive entertainment, Web intelligence and multimedia.

The publications within "Advances in Intelligent Systems and Computing" are primarily proceedings of important conferences, symposia and congresses. They cover significant recent developments in the field, both of a foundational and applicable character. An important characteristic feature of the series is the short publication time and world-wide distribution. This permits a rapid and broad dissemination of research results.

**** Indexing: The books of this series are submitted to ISI Proceedings, EI-Compendex, DBLP, SCOPUS, Google Scholar and Springerlink ****

More information about this series at http://www.springer.com/series/11156

Tareq Ahram · Waldemar Karwowski ·
Stefan Pickl · Redha Taiar
Editors

Human Systems Engineering and Design II

Proceedings of the 2nd International Conference on Human Systems Engineering and Design (IHSED2019): Future Trends and Applications, September 16–18, 2019, Universität der Bundeswehr München, Munich, Germany

Editors
Tareq Ahram
Institute for Advanced Systems Engineering
University of Central Florida
Orlando, FL, USA

Stefan Pickl
Department of Computer Science
Universität der Bundeswehr München
Neubiberg, Germany

Waldemar Karwowski
University of Central Florida
Orlando, FL, USA

Redha Taiar
Université de Reims Champagne Ardenne
Reims Cedex 2, France

ISSN 2194-5357 ISSN 2194-5365 (electronic)
Advances in Intelligent Systems and Computing
ISBN 978-3-030-27927-1 ISBN 978-3-030-27928-8 (eBook)
https://doi.org/10.1007/978-3-030-27928-8

This Springer imprint is published by the registered company Springer Nature Switzerland AG
The registered company address is: Gewerbestrasse 11, 6330 Cham, Switzerland

Preface

This volume, entitled *Human Systems Engineering and Design*, aims to provide a global forum for presenting and discussing novel design and systems engineering approaches, tools, methodologies, techniques, and solutions for integrating people, concepts, trends, and applications in all areas of human endeavor in industry, economy, government, and education. Such applications include, but are not limited to, energy, transportation, urbanization, and infrastructure development, digital manufacturing, social development, human health, sustainability, a new generation of service systems, as well as safety, risk assessment, health care, and cybersecurity in both civilian and military contexts. Indeed, rapid progress in developments in cognitive computing, modeling, and simulation, as well as smart sensor technology, will have a profound effect on the principles of human systems engineering and design at both the individual and societal levels in the near future.

This book focuses on advancing the theory and applications for integrating human requirements as part of an overall system and product solution, by adopting a human-centered design approach that utilizes and expands on the current knowledge of systems engineering supported by cognitive software and engineering, data analytics, simulation and modeling, and next-generation visualizations. This interdisciplinary approach will also expand the boundaries of the current state of the art by investigating the pervasive complexity that underlies the most profound design problems facing contemporary society today.

This book also presents many innovative studies of systems engineering and design with a particular emphasis on the development of technology throughout the lifecycle development process, including the consideration of user experience in the design of human interfaces for virtual, augmented, and mixed reality applications.

Reflecting on the above-outlined perspective, the papers contained in this volume are organized into eight unique research tracks with a total of eleven sections, including:

Section 1 Human-centered Design
Section 2 Human–Robot Interaction
Section 3 Transportation Design and Autonomous Driving

Section 4	Human-centered Design for Health Care
Section 5	User Experience and Virtual Environments
Section 6	Systems Design and Human Diversity
Section 7	Safety Engineering and Systems Complexity
Section 8	Sports Design and Sports Medicine
Section 9	Biomechanics, Health Management, and Rehabilitation
Section 10	Human Cyber-physical Systems Interactions
Section 11	Business Analytics, Design, and Technology

We would like to extend our sincere thanks to Universität der Bundeswehr München for their support. Our appreciation also goes to the members of the Scientific Program Advisory Board who have reviewed the accepted papers that are presented in this volume.

We hope that this book, which presents the current state of the art in human systems engineering and design, will be a valuable source of both theoretical and applied knowledge enabling the human-centered design and applications of a variety of products, services, and systems for their safe, effective, and pleasurable use by people around the world.

September 2019

Tareq Ahram
Waldemar Karwowski
Stefan Pickl
Redha Taiar

Contents

Human-Robot Interaction

Transportation Design and Autonomous Driving

Human-Centered Design

Interaction Design for the Dissemination and Sharing of Knowledge

Elisabetta Cianfanelli(✉) and Margherita Tufarelli

DIDA Department of Architecture, University of Florence, Design Campus, Florence, Italy
{elisabetta.cianfanelli,margherita.tufarelli}@unifi.it

Abstract. In this paper we propose the concept of a numbers of interactive device that can guide the user on a learning journey. The design of the material and immaterial device is developed on an analysis carried out in particular contexts and on the collection and evaluation of quantitative and qualitative data that led to the design of knowledge device concepts.

Keywords: Device · Knowledge transmission · Systems engineering

1 Introduction

Nowadays all information can be found easily and quickly, as Francesco Morace writes "the paradox of the information society is to know more and more and to understand less and less, replacing the deepening with the interactivity allowed by the permanent connection" [1]. It is an incremental phenomenon as it is supported by the evolving technological advances and by the implementation of calculation speed that has improved the performance of everyday objects.

Being always *connected* seems to have led to the generation of a network in every daily activity; a network in which new forms of communication coexist simultaneously in both a real and digital environment, creating "new landscape of communication" and "new learning environments" [2].

Moreover, this enormous spread of the Internet in multiple scenarios of everyday life has led to an expansion of the possibilities of communication, overcoming the barriers and limitations of time and space with the consequence of an increase in speed and a reduction in costs. This contribution addresses to the relationship between information and knowledge with a particular focus on what the latter has become in the digital age and what maybe the risks and opportunities of this contamination process.

Starting from these considerations, the research proposes three concepts of technological devices designed for a learning-by-doing process. Those products are not designed as specialized products, on the contrary, they have the aim to increase the knowledge transmission associated with particular activities in order to create a close relationship between users in the real and virtual world.

As always happens, the use of something can imply the possible abuse of it; therefore following the diffusion of new media (and of the Internet in general), we are witnessing an increase in psychopathological phenomena linked to excessive or

T. Ahram et al. (Eds.): IHSED 2019, AISC 1026, pp. 3–7, 2020.
https://doi.org/10.1007/978-3-030-27928-8_1

inadequate use of the Internet which manifests with symptoms similar to those dependent on psychoactive substances [3]. The massive use of the Internet can lead to serious psychological disorders that are not viruses or bacteria [4], but pathological conditions of the mind caused by the web or, rather, by the relationship with the tools that help being on the net. A set of pathologies that lead people to dependence phenomena generally indicated by the term "addiction" [5]. However, it should be pointed out that the mis-adaptive use of the Internet constitutes, especially where predisposing psychopathological factors are present, a real risk for the mental health of the individual. Internet abuse and addiction represent a really dangerous threats.

2 Approach and Methods

The design process was user-centered and carried out by steps. The first step involved a desk research and the development of a preliminary project framework. Then, in the second phase the team developed two questionnaires specifically designed to understand whether the concept hypotheses had a real response. the goal was to transform knowledge transmission from a competitive to a collaborative and empathic experience. Projects are designed for a generation that is used to interacting in open networks where knowledge is exchanged and does not accumulate.

Feedbacks obtained have detected a large amount of data that were used in the project fase to built personas profiles and user scenarios. In all the projects knowledge is not considered as a power to be used for personal success, but as an expression of shared responsibility aimed at the well-being of humanity.

3 Selected Concepts

Devices were designed as a physical relationship between tangible and intangible media and were developed on the definition of specific user scenarios dedicated to identified fields of knowledge. The aim is to connect people with same interests and try to involve others, stimulating the user's curiosity and looking for a continuous deepening of knowledge [1, 6].

Smart Tourist Device. The first concept is for city tourism: the device uses geo-localization to connect with digital archives aggregator that are ready online (e.g. Europeana). The device connects directly to the archive using their API (application programming interface) and is able to give informations about monuments and other things in the city to te user. The value of this concept is in its simplicity but effectiveness, in particular for the interesting use of digital archive materials (Fig. 1).

A Device for Learning Italian Dialects. Ciarlo is a bone conduction headset designed to spread and improve the knowledge of Italian dialects. The device main function is: Simultaneous translation of the terms and proverbs in dialect pronounced during a conversation. Touching the sensor once it is possible to activate the simultaneous translation mode so that all the words spoken within a 2-m radius are listened to and translated in real time. By double-tapping the touch sensor it is possible to activate the

Fig. 1. The smart tourist device

voice command, so the device receives instructions from the user. By sliding a finger on the touch sensor you can increase or decrease the volume of the speaker. The device has a control interface app with which it is possible to manage all Ciarlo functions and check the battery charge status (Fig. 2).

Amor Roma a Device for Explore the Underground Rome. This project aims at promoting and spreading Roman archaeological sites. The concept was developed on two levels: an app/game that exploits the potential of the network, reaching even those who are not in Rome. The device has also the function of audio guide within the archaeological sites themselves, to ensure that the user is more autonomous and involved in the visit. In this product system we find an additional element that uses game as a learning tool. The user, through the solution of various puzzles. The game includes extra downloadable content on site. Whoever benefits from the application delights in a playful activity, and at the same time becomes an active promoter facilitating the understanding of those archeological sites, encouraging more virtuous social behaviors. The device designed as an aid to the visit is given in the underground sites as a support to the visit of the site. It is an information container, which can be connected to the smartphone via NFC technology acing as an audio guide. While for the user who previously downloaded the application on his smartphone, this device will communicate curiosity and insights about the archaeological site he/she is visiting, through images in augmented reality (Fig. 3).

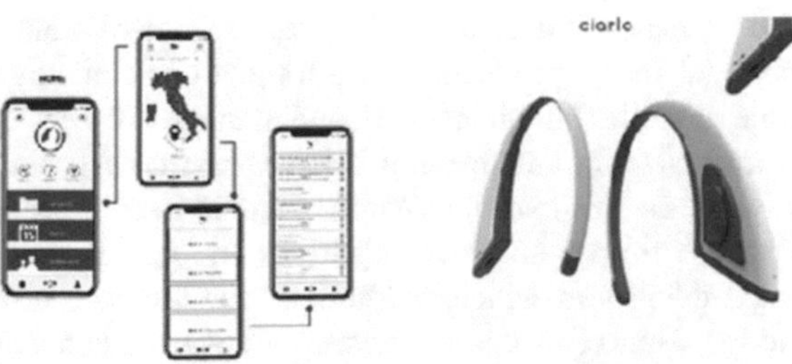

Fig. 2. A device for learning italian dialects: Ciarlo

Fig. 3. Amor Roma device and App

Learning New Languages with Herobo. This is a device that wants to change the way people study new languages focusing on the connection and collaboration of its users from all over the globe. The analysis conducted revealed the desire for interaction with a native speaker and the need to practice not only through reading and speaking but also with writing, digitally and not. The device, thus, acts as a physical "avatar", with which friends from any part of the globe can get in touch with the user and communicate in real time with simple and universal language through smilies displayed on a small screen (Fig. 4).

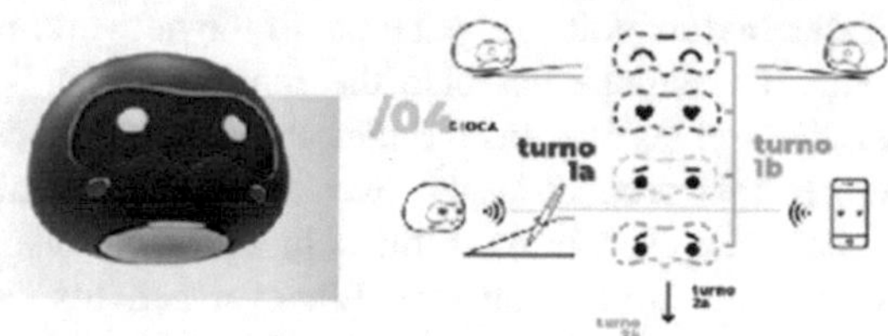

Fig. 4. Herobo, the device for learning a new language

4 Conclusions

All the device proposed in this paper have the aim to allow people to elaborate new social models starting from the assumption that knowledge is a complex concept with different meanings and different entities and its process of transmission is changing and not at all obvious. For a long time the reproduction of knowledge was based on the relationship between teacher and student. The research also wants to represent the passage from functional mono products to multi-functional, intelligent products that can perform multiple activities simultaneously and manage different opportunities. The devices wants to emphasize the relationship between virtual and material through the design culture. In synthesis, this design research intended to overturned the value given to the term "addiction": from a space of negativity to one of positivity through the design of specialized devices in which "addiction" can led to a profound interest in specialized knowledge. Hence, in this paper we propose the concept of a number of interactive devices that can guide the user on a learning journey.

Acknowledgment. All the project were developed by students during the course "Product Advanced Design" at University of Florence Design Campus. Precisely the device for city tourist was designed by Alex Caciatore, Alessio Martelli, Lucrezia Gori, Gaspare Tumbarello, XueYang e YingLiu. Ciarlo; the device for learning italian dialects was designed by Ghanem Brahmi, Letizia Capaccio, davide Di Bella, Booshra Nasri. Amor Roma was designed by Filomena Damiano, Camilla Storti, Giulia Pistoresi, Giorgia Marinelli, Shi Bin. Finally Herobo was designed by Maria Claudia Coppola, Andrea Capruzzi, Zhu Yue, Hu Yashan, Sepehr Mohajerani.

References

1. Morace, F.: Futuro + Umano. quello che l'intelligenza artificiale non potrà mai darci, Egea, Milano (2018)
2. Nixon, S.: Advertising Cultures: Gender, Commerce, Creativity. Sage (2003)
3. Miller, M.S.: Technology's interference in everyday life. University Park, PA, 10 December 2014
4. McCormack, M.: The role of smartphones and technology in sexual and romantic lives, Durham University, June 2015
5. Young, K.S.: Internet addiction: the emergence of a new clinical disorder. Cyberpsychology Behav. **1**(3), 237–244 (1998)
6. Munari, B.: Da cosa nasce cosa (1981)

Kaleidoscope of User Involvement – Product Development Methods in an Interdisciplinary Context

Anne Wallisch[1(✉)], Olga Sankowski[2], Dieter Krause[2], and Kristin Paetzold[1]

[1] Universität der Bundeswehr München, Werner-Heisenberg-Weg 39, 85577 Neubiberg, Germany
{Anne.Wallisch,Kristin.Paetzold}@unibw.de

[2] Technische Universität Hamburg, Am Schwarzenberg-Campus 1, 21073 Hamburg, Germany
{O.Sankowski,Krause}@tuhh.de

Abstract. Involving user perspectives into product development activities means involving empiric science methods into engineering sciences. On the one hand, human-centered research methods should assist engineering designers in problem definition, solution generation, decision-making and evaluation aligned to the users. On the other hand, different research logics and approaches of problem solving can be an obstacle to the application of methods from other disciplines. The article presents a terminology aimed at supporting interdisciplinary understanding of methods and its underlying conceptualizations.

Keywords: Design methodology · Interdisciplinary design · User-centered design · User involvement · User research methods

1 Introduction

Generally spoken, products are developed to support human striving for performance, self-determination, independence and satisfaction. Users, as experts in terms of their life conditions, contribute the validation of requirements, usability and user experience. As user-driven design activities increased in product development, so did interdependencies with adjacent as well as non-technical disciplines in design research. Currently, influences from various disciplines characterize a conceptual landscape in which the analytical boundaries become blurred and a lot of approaches and methods with similar goals, e.g. to increase the focus on user needs or usability, to reduce the risk of product failure, or to support the innovation process, exist. Thus, designers often cannot tell the difference between the various approaches existing and do not have an overview of methods and tools that can be used to gain a certain output or information.

This paper aims at contributing basics to a systematic exchange between scientific disciplines as well as between science and practice. In Sect. 2, relevant aspects of user-centered design research are introduced. Starting from this, the research approach chosen is presented in Sect. 3, the main findings in Sect. 4. In Sect. 5 a terminology

T. Ahram et al. (Eds.): IHSED 2019, AISC 1026, pp. 8–14, 2020.
https://doi.org/10.1007/978-3-030-27928-8_2

integrating relevant user-centered research concepts from various disciplines is synthesized. Eventually, conclusions for future user-centered design research are drawn.

2 Related Work

Although much work is done emphasizing that frequent and intimate involvement of user perspectives is important for improving product concepts, innovation capabilities, and a more complete assessment of user input requirements [e.g. 1, 2] only little consensus on what involving users in design tasks means had been achieved [3, 4]. Following Sanders and Stappers [2], user-centered design is referred to as having experts, i.e. designers or researchers, who are trained to observe and/or interview potential users and who include selected groups of users to perform different predelimited tasks, e.g. test prototypes, create personas or evaluate concepts. Other authors put user involvement on a level with need identification [e.g. 5, 6]. The ISO Standard 9241 [7] does not distinguish between human-centered design and user-centered design, further includes user experience and refers to human-computer interaction.

Previous attempts to collect and combine all methods are usually limited, the expected benefit and the potential risk of a method application compared to other methods is only evaluated in an extremely reduced form [8]: When and under which conditions the application of the method is goal-oriented and which results can be expected often remains unanswered [9]. Several researchers proposed frameworks to clean up activities and approaches of this research field [e.g. 10, 11] without referring to each other. Reviewing literature [e.g. 12, 13], the impression is given that sometimes involving users is a goal in itself [14] and many research papers do not even attempt to validate their findings [15]. In the field of user-centered design, different elements of methods embody different types of knowledge and thus require different approaches to validation drawings on both technical and social sciences [16]. Understanding the overall procedure and the elements each step contributes to the result of a method is an important part of applying a method. It is an important precondition for customizing methods in sensible ways, avoiding the risk of leaving out parts that are required for the overall result [17]. Overall, the need for systematic approaches to support designers in addressing both the user's demands and their expertise in the solution finding is to be stated.

3 Research Approach

For analyzing the user-centered design practice in the context of product development, conference proceedings serve as a fruitful data source. The wide range of conference contributions enriches the subject by insights of different disciplines; the homogeneous subthemes ensure the comparability of the single papers. All proceedings of the International Conference on Engineering Design (ICED) of 2013, 2015 and 2017 have been chosen as database. A keyword analysis of the terms user experience/UX, user-oriented/-centred/-centered/UCD and user integration/-participation/-involvement/-orientation/-involvement identified 121 papers. The search terms were chosen due to their synonym use. After eliminating footnotes, references and acknowledgements without reference to the subject, the sample finally comprises 98 proceedings. In a first step, this sample was analyzed towards the methods applied to involve user perspectives and the disciplines being referred to. From this content analysis, a terminology being used in the context of user-centered design has been condensed. In a next step, this terminology was supplemented by corresponding interdisciplinary definitions, which than were examined for similarities and differences. In order to create a basis for communication between scientific disciplines as well as between science and practice, finally, a uniform positioning of relevant terms was condensed.

4 Findings and Analysis

The content analysis revealed that methods from empirical sciences have found their way into product development. Alas without being described exactly enough. The explanatory power of a method seems quite limited by rather vague definitions of analytical concepts and targets. Some researchers only use one specific term to describe their approach, e.g. user participation or - integration, some only refer to a specific approach, e.g. inclusive or emotional design, some name no such thing and just use some kind of user oriented technique. The impression is given that user-centered design itself; acting on different levels with different elements, lead to the disciplinary and theoretical fragmentation that hinders a reliable total involvement of user perspectives.

Methods are often created for a particular context and purpose that need to be understood for applying the method correctly. The intended use of several user-centered methods must be articulated in such way that designers are quite informed about the suitability of a particular method in a specific context. By classifying methods into different categories, these can be compared in terms of expected benefits and required resources. An indispensable prerequisite for this, however, is that the terminology used is understood consistently and systematically applied. However, based on our analysis, we cannot confirm this as being state of the art. In contrast, we revealed that the concept ‘method’ itself is perceived very flexibly by different researchers and the connection of the analytical concepts ‘method’, ‘tool’ and ‘design approach’ is hardly considered. On the one hand, we found that most concepts, activities and approaches of involving user perspectives into development tasks exist on multiple levels. On the other hand, some authors lack a detailed description of the process performed.

From our content analysis, we identified sociology, psychology, market research, and human-computer interaction as being the main influencing disciplines for user-centered design. Sociology, psychology and market research are empirical sciences, aimed at describing and explaining specific observable phenomena. Descriptions are collected data about certain phenomena (e.g. experience and behaviour) and the conditions under which it occurs (regularities). An explanation is given when the conditions or even the causes of a phenomenon can be identified [18]. The decisive criterion for the correctness of corresponding assumptions and theories is their empirical verifiability. This is why the principles and rules for the collection, evaluation and interpretation of empirical data play such an important role. The interpretation here is already the synthesis, the joining of sense elements to a perceptual whole [19]. Without this theoretical integration, an explicit assignment to the phases of the development process cannot be made unambiguously. Methods to holistically involve user perspectives in the product development process include all variants of surveys, observations, tests and non-reactive procedures that are required for operationalizing the required user insights, for interaction with users, for user sampling, for data collection, for data analysis and finally for validation of the gained user perspectives and conversion into technical parameters.

5 Synthesis of Results

A greater clarity about what various user-centered design approaches, user research methods, guidelines and tools offer would aid clear communication among researchers and design practitioners, which in turn is fundamental for selecting the appropriate user research methods for specific development tasks. Furthermore, involving user perspectives in design activities requires an interdisciplinary communication basis, which ensures the common understanding of relevant terms.

To support this, a generic terminology is proposed here and illustrated in Fig. 1. The terms are allocated in a hierarchy of five levels; degree of concreteness is increasing from top to bottom level. Within each level, there is no strong hierarchy between the terms; some can be used synonymously to each other, some are distinct types of the same thing and can be used for classification. However, in order to name the levels one 'main term' was chosen for each, which we find to be most generic and most widely accepted over the different research areas. We omitted of connecting the various terms on from different levels in the generic overview, since every term can include all of the below or none of them, e.g. a design process can include a methodology, which, on the other side, contains various types of methods and activities.

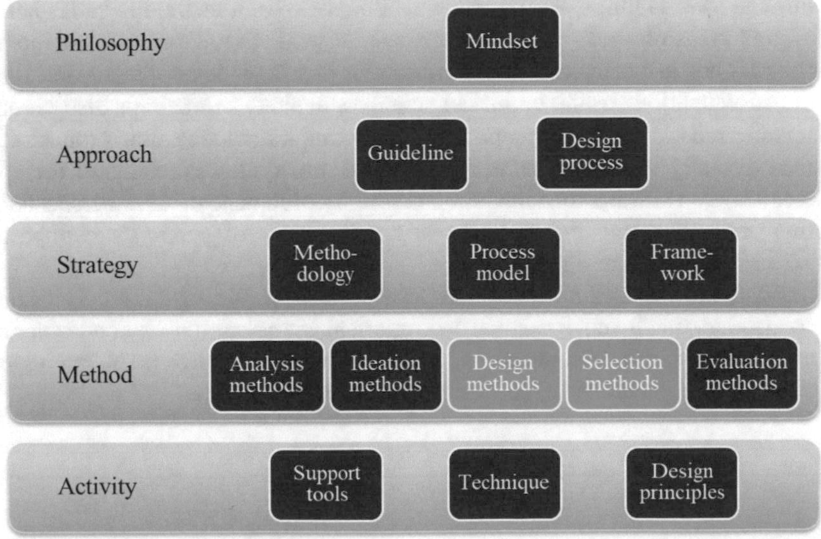

Fig. 1. Terminology relevant for user-centered design/user involvement

On top level, *Philosophy*, the main goal, idea or underlying mindset of the whole process/approach is described. Usually, all types of approaches and strategies are done to achieve a higher goal, e.g. reduce cost, increase quality, or come up with a new and innovative product. The *Philosophy* level stands for the reason *why* we do something. The *Approach* level on the other side represents *what* to do in a very rough way. This can be described either through a sequence of process steps, i.e. design process, or in a list of instructions, i.e. guideline. An approach is also often used to describe *what* has been done in a specific context.

Next two levels, *Strategy* and *Method*, both include general and abstract instructions and rules on *how* something should be done. A method always has a necessary input, a defined output and a distinct sequence of process steps that have to be performed quite strictly, whereas a strategy can include various methods and activities that can be done in parallel or sequential. Here the different terms can be used quite synonymously. On the *Method* level, however, not synonyms, but different types of methods are listed. Analysis methods are used to understand the user, the task and or the context. Ideation or creativity methods are helpful to come up with new/more ideas. Design methods are used for the creation and elaboration of the actual artefact. Selection methods can be used to reduce the amount of a considered set of e.g. ideas, solution, and options and are therefore helpful in making decisions. Evaluation methods are necessary for validation of the product's performance and quality.

The relevance of the different types of methods varies according to the research area. Ideation and design methods are more relevant for design research, while analysis methods are of particular importance in sociology and selection methods for decisions are of great importance for market research in addition to analysis methods. Within each research area, there is often no conceptual distinction between different types of methods. They are simply referred to as 'method' and stand side by side on an equal footing. Existing methods can be clearly classified in different types or classes. In an interdisciplinary context, however, it is not mandatory that they are only suitable for this purpose. The understanding of what a method is varies between the different research areas, measured against its goal. Empirical analysis methods, for example, are not only used in the analysis phase of development, but also in ideation or evaluation. In the empirical sciences, evaluation methods also belong in the analysis, related to the corresponding object of knowledge. For product development, especially in requirements management, the synthesis results of the empirical sciences are relevant for analysis. For the involvement of user perspectives in particular, selection and design methods are therefore of subordinate importance (marked light blue in the figure).

Finally, the lowest and most concrete level is *Activity*. It is also a representation of *what* to do, but much more specific to a particular application or context than the *Approach* level. It includes not only specific techniques and principles, but also supporting tools, such as artefacts, software, and templates.

6 Conclusion

Going thematically and methodically beyond the boundaries of the own field requires an attitude of (self-)critical reflection on the potentials and limits of the own as well as other discipline(s); it takes into account their historical growing and cultural embedding. Increasing complexity and technological advances keep disciplines constantly changing. The article aims facilitating methodological transfer by the analysis and relational explanation of essential terms, which are necessary for the interdisciplinary description of different methodical targets and theoretical concepts describing aspects of users in terms of their motivations, life goals, behavior, ethical values, desires and psychological needs. The analytical groundwork to support interdisciplinary understanding is deduced here. This groundwork can support the systematic selection of methods for user involvement based on the user insights required to address specific aspects of design accordingly. The revealed terminology condenses in a classification scheme based on theoretical arguments mostly. Further research need to be pursued to empirically prove and validate this scheme and, of course, it has to be extended in terms of disciplines and terms.

References

1. Kim, J., Wilemon, D.: Focusing the fuzzy front-end in new product development. R&D Manag. **32**, 269–279 (2002)
2. Sanders, E.B.-N., Stappers, P.J.: Co-creation and the new landscapes of design. Co-Design **4**, 5–18 (2008)
3. Cavaye, A.L.M.: User participation in system development revisited. Inf. Manag. **28**, 311–323 (1995)
4. Wallisch, A., Sankowski, O., Krause, D., Paetzold, K.: Overcoming fuzzy design practice: revealing potentials of user-centered design research and methodological concepts related to user involvement. In: 25th ICE/IEEE International Technology Management Conference, Nizza, France (2019)
5. Geyer, F., Lehnen, J., Herstatt, C.: Customer Need Identification Methods in New Product Development: What Works "Best"? (2018). https://doi.org/10.1142/S0219877018500086
6. Kujala, S.: Effective user involvement in product development by improving the analysis of user needs. Behav. Inf. Technol. **27**, 457–473 (2008)
7. ISO 9241-11:2018: Ergonomics of human-system interaction – Part 11: Usability: Definitions and concepts (2018)
8. Zogaj, S., Bretschneider, U.: Customer integration in new product development: a literature review concerning the appropriateness of different customer integration methods to attain customer knowledge (2012). http://dx.doi.org/10.2139/ssrn.2485240
9. Olsen, T.O., Welo, T.: Maximizing product innovation through adaptive application of user-centered methods for defining costumer value. J. Technol. Innov. **6**, 172–192 (2011)
10. Kremer, S., Lindemann, U.: A framework for understanding, communicating and evaluating user experience potentials. In: 20th ICED International Conference on Engineering Design, Milano, Italy (2015)
11. Bae, J., Cho, K., Kim, C.: Developing a framework of new mixed method, social networking services group diary and its application in practice. In: 20th ICED International Conference on Engineering Design, Milano, Italy (2015)

12. Prahalad, C.K., Ramaswamy, V.: Co-creation experiences: the next practice in value creation. J. Interact. Mark. **18**, 5–14 (2004)
13. Verschuere, B., Brandsen, T., Pestoff, V.: Co-production: the state of the art in research and the future agenda. Voluntas **23**, 1083–1101 (2012)
14. Voorberg, W.H., Bekkers, V.J.J.M., Tummers, L.G.: A systematic review of co-creation and co-production: embarking on the social innovation journey. Public Manag. Rev. **17**, 1333–1357 (2015)
15. Blessing, L.T.M., Chakrabarti, A.: DRM - A Design Research Methodology. Springer, London (2009)
16. Eckert, C.M., Clarckson, P.J., Stacey, M.K.: The spiral of applied research: a methodological view on integrated design research. In: 14th ICED International Conference on Engineering Design, Stockholm, Sweden (2003)
17. Gericke, K., Eckert, C.M., Stacy, M.K.: What do we need to say about a design method? In: 21th ICED International Conference on Engineering Design, Vancouver, Canada (2017)
18. Gerring, R.J.: Psychology and Life, 20th edn. Pearson International, London (2012)
19. Wundt, W.: An Introduction to Psychology. George Allen & Unwin Limited, London (1924)

Adaptive Augmented Reality User Interfaces Using Face Recognition for Smart Home Control

Bernardo Marques(✉), Paulo Dias, João Alves, and Beatriz Sousa Santos

DETI - IEETA, University of Aveiro, Aveiro, Portugal
bernardo.marques@ua.pt

Abstract. Augmented Reality (AR) offers the possibility to present information depending on the context/location: pointing at a given appliance will automatically present relevant interfaces. In this work, we explore Pervasive AR in the context of the Smart Home: users may select the control of a given appliance by pointing at it, simplifying the interaction process. This idea is then extended to consider user profile in the loop: face recognition is used to recognize a given user of and adapt the interface not only to location but also to specific users needs. A preliminary study was conducted to evaluate and validate the concept.

Keywords: Pervasive Augmented Reality · Human-centered interfaces · Adaptive user interfaces · Face recognition · Smart home

1 Introduction

With the evolution of technology, control and monitor of Smart Homes though digital devices is becoming a reality. As people's expectations of what technology can do for them are changing, new interaction paradigms are fundamental, as the vision of what a Smart Home entails is continuously evolving [1]. Augmented Reality (AR) appears as an opportunity to provide powerful and flexible user-interface for these intelligent environments [2].

Despite recent success of AR, most solutions present information with limited interaction rather than considering an integrated human-centered approach [3]. AR is starting to be extended to Pervasive AR: a continuous and pervasive user interface that augments the physical world with digital information registered in 3D, while being aware of and responsive to the user's context [4]. Continuous AR experiences raise challenges about how to display information (which type of information is distracting or relevant) and how to properly interact with that information. Clearly, this requires personalization and adaptation of the content.

Literature suggests user adaptive interfaces provide excellent opportunities for improving user experience based on context [5]. Hence, it is viable to extend this to AR, delivering content according to user profile, surrounding environment and device being used.

T. Ahram et al. (Eds.): IHSED 2019, AISC 1026, pp. 15–19, 2020.
https://doi.org/10.1007/978-3-030-27928-8_3

2 Uninterrupted Augmented Reality Experiences

The work presented is based on an initial application to create and customize uninterrupted AR experiences in a Smart Home. The application enables remote control of appliances by pointing at them through location-based interfaces. Depending on the available device, 1 of 3 different methods of interaction are presented: 2D standard interfaces, marker-based AR interfaces and continuous AR interfaces. The appliances controlled in this setup were a Philips Hue set, enabling control of several aspects of lights (Fig. 1 – (Left)), such as state (on/off) of specific rooms, brightness, color selection. A boiler was also used (Fig. 1 - (Right)), to control hot water, state (on/off), temperature regulation, and scheduling along the week period (enabling to program automatic turn on and off the boiler). The application was developed using the Unity Game engine, C# scripts and the Vuforia Library.

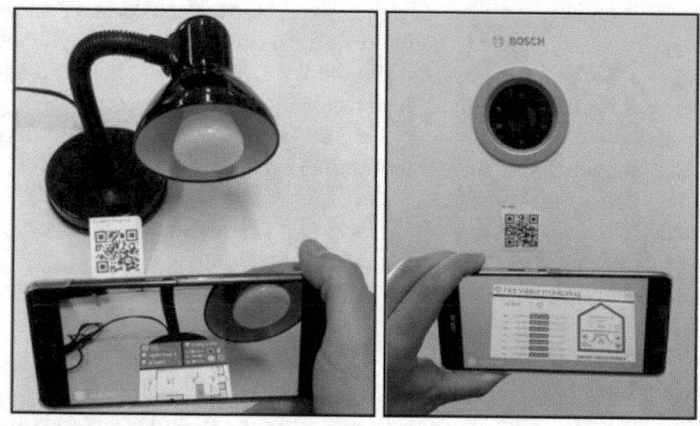

Fig. 1. Augmented Reality interfaces able to control lights (Left) and a boiler (Right).

3 Adaptive UI Using Face Recognition

The application was then evolved to mirror the conceptual framework of Fig. 3, composed of four interconnected modules, enabling the use of adaptive interfaces, based on face recognition authentication (Fig. 2).

The main modules are as follow:

The **Authentication Module** authenticates a user through credentials (traditional login/password) or face recognition (uses a photo taken with the camera of the device being used). The OpenCV asset for Unity was adapted to allow the identification of multiple users from a set of existing profiles with previously taken photos, based on the Eigen-Faces algorithm. Different user profiles with specific permissions exist to manage and control the Smart Home. New user profiles can also be added, edited or removed. Only a user with top-level permission on the hierarchy can perform such task.

Fig. 2. Adaptive user interfaces: authentication process (Left), available options after authentication is conducted with success (Right).

The **Pose Estimation Module** analyses the surrounding environment (6DOF Data Collection), which is then used by the content filter to generate specific adaptive interfaces.

The **Adaptive Module** uses a content filter and selects which interfaces to display according to user profile and location in the Smart Home. Then, the interface generator groups all models (text, images, 3D objects) necessary to create each interface. Based on the device being used, the application automatically displays the suitable interface.

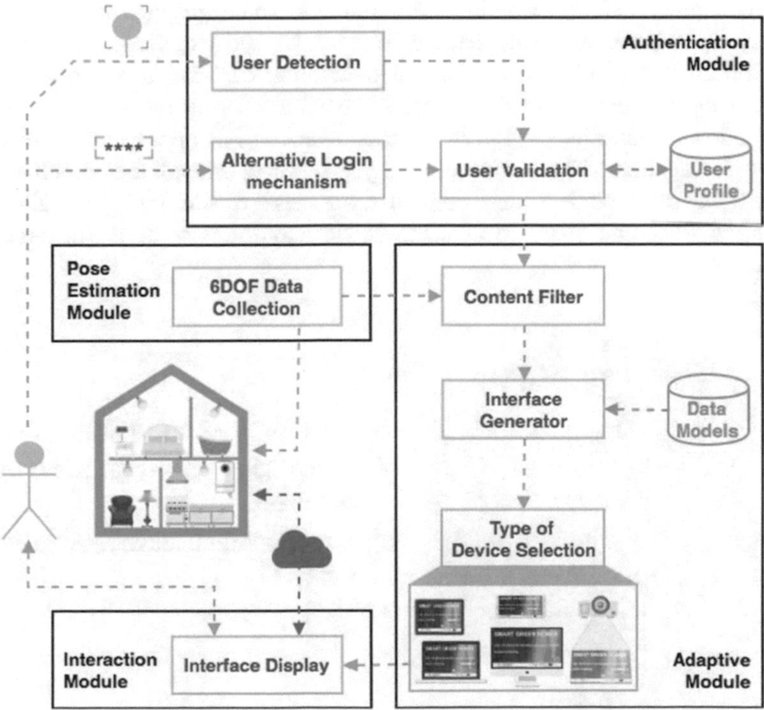

Fig. 3. Conceptual framework for context adaptive interfaces based on User Recognition and Identification.

Finally, the **Interaction Module** enables the user to manage and interact with the appliances in the Smart Home. The user is also able to select other type of interface.

4 Preliminary Results Using Face Recognition

A preliminary study was conducted with 5 participants (1 female and 4 males, ranging in age between 23 to 33 years old, simulating a family composed by two parents, plus three children) to evaluate the robustness of the face recognition authentication process. Participants were inserted in the system as new users, to create the training model. Each user was asked to log using face recognition ten times. Results show that 74% of the times, participants were successfully found and the remaining 26% of the times the application detected the wrong participant (Fig. 4).

The developed application opens up new possibilities for the way users interact and control the Smart Home concept moving forward, by pointing at specific localizations and appliances, while being confronted with adaptive interfaces based on the recognition of the user face. Adaptive interfaces can be useful for this, since they are able to provide the required information in the most suitable way, by taking into account current context of use, such as user related aspects (preferences and customization), aspects related to the technology (available interaction mechanisms) and environment context.

Virtual-generated objects and interfaces will continue to increasingly become more interactive through several modalities responding to voice, gestures and even allowing user interaction through touch, leading to an enhanced experience. Moreover, AR applications will become more persistent over time, enabling users to place virtual objects on top or next to physical objects, enabling sharing mechanisms with other users or allowing later interaction in another moment. AR applications will obtain a greater understanding of the surrounding physical space, through the combination with concepts like machine learning, context awareness or/and big data, allowing to immediately react and adapt to changes in the environment or to the presence of different users.

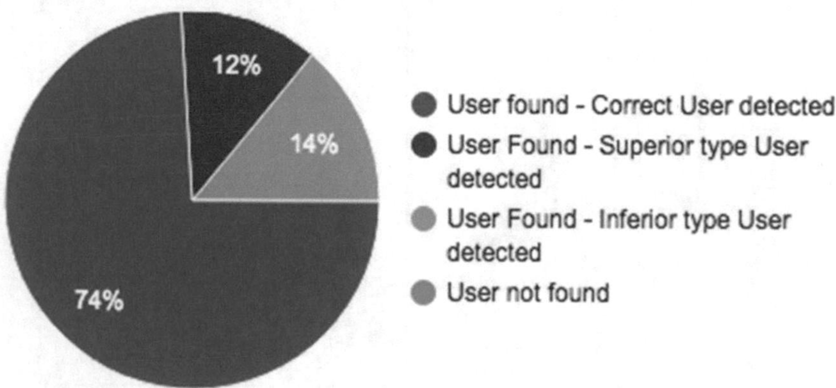

Fig. 4. Results associated to the use of the face recognition process to authenticate users in the application.

5 Concluding Remarks and Future Work

Through the use of Pervasive Augmented Reality, it is possible to upgrade from an application with a unique objective to a multi-purpose continuous experience based on the user face recognition, changing the way they interact with information and the surrounding environment. In the context of the Smart Home, considering a family, different members can have different levels or permissions. The interfaces can adjust its content, ranging from limited control over their rooms and shared-spaces, to full control over all the features of the application. Furthermore, based on the proposed framework, the interfaces can be adapted to other particular contexts, purposes and usages. Although at an early stage, the proposed framework seems promising, opening new possibilities for interaction and control of smart environments. By pointing at specific localizations and appliances, user tailored interfaces are presented in a specific device with minimum interaction. Future development includes the improvement of the application recognition robustness and a further validation. Additionally, we also want to explore emotion recognition, aiming to adapt the Smart Home environment (lightning settings, room temperature) to the user preferences and improve comfort and performance.

Acknowledgments. To all participants involved in the case studies, thanks for collaborating in providing essential data and relevant feedback. The present study was developed in the scope of the Smart Green Homes Project [POCI-01-0247-FEDER-007678], a co-promotion between Bosch Termotecnologia S.A. and the University of Aveiro. It is financed by Portugal 2020 under the Competitiveness and Internationalization Operational Program, and by the European Regional Development Fund. This study was also supported by IEETA - Institute of Electronics and Informatics Engineering of Aveiro, funded by National Funds through the FCT - Foundation for Science and Technology, in the context of the project UID/CEC/00127/2019.

References

1. Wilson, C., Hargreaves, T., Hauxwell-Baldwin, R.: Smart homes and their users: a systematic analysis and key challenges. Pers. Ubiquit. Comput. **19**(2), 463–476 (2015)
2. Kim, K., Member, S., Billinghurst, M., Member, S., Bruder, G., Duh, H.B., Member, S.: Revisiting trends in augmented reality research: a review of the 2nd decade of ISMAR (2008–2017). IEEE Trans. Vis. Comput. Graph. (2018)
3. Pedersen, I.: Radiating centers: augmented reality and human-centric design. In: Arts, Media and Humanities Proceedings - IEEE International Symposium on Mixed and Augmented Reality, ISMAR-AMH 2009, pp. 11–16 (2009)
4. Grubert, J., Zollmann, S.: Towards pervasive augmented reality: context awareness in augmented reality. IEEE Trans. Visual Comput. Graphics **23**(6), 1706–1724 (2017)
5. Abidin, R.Z., Arshad, H., Shukri, S.A.A.: Adaptive multimodal interaction in mobile augmented reality: a conceptual framework. In: ICAST 2017 (2017)

Comparison of Different Assembly Assistance Systems Under Ergonomic and Economic Aspects

Sven Bendzioch[1(✉)], Dominic Bläsing[2], and Sven Hinrichsen[1]

[1] Ostwestfalen-Lippe University of Applied Sciences and Arts, Campusallee 12, 32657 Lemgo, Germany
{Sven.Bendzioch,Sven.Hinrichsen}@th-owl.de

[2] University Greifswald, Franz-Mehring-Straße 47, 17489 Greifswald, Germany
Dominic.Blaesing@uni-greifswald.de

Abstract. More and more complex products are being fitted in small batches in manual assembly. Because of this, more information needs to be collected from employees and implemented in appropriate actions. At the same time, the informational design of assembly systems often shows deficits in operational practice. Manual assembly processes can be made more economical, reliable, and human-oriented with the help of informational assistance systems. Testing was carried out in the Laboratory for Industrial Engineering at the Ostwestfalen-Lippe University of Applied Sciences and Arts to verify this potential. Initial results on the use of augmented reality (AR) glasses in comparison to providing information in a paper-based format are presented.

Keywords: Worker assistance system · Manual assembly · Human-machine interaction · Informational complexity

1 Introduction

The introduction of informational assistance systems is becoming more and more important in light of new technological developments, existing deficits in information supply [1] and changing market requirements. As market requirements change, more complex products are assembled in a large number of varieties in small batches, or individually for specific customers [2]. Tasks involving a high workload, in particular, can result in great strain for employees and therefore assembly errors, making it impossible to achieve quantity and productivity targets [3]. In addition, providing the right information in line with requirements is highly relevant, especially in light of the lack of trained professionals and an aging workforce in western industrialized nations [4].

One unique challenge for companies is limiting the mental workload placed on assembly employees, while at the same time designing efficient assembly processes with high process capability. Introducing informational assembly assistance systems [5–7] is one potential solution. With the help of these systems, employees can receive the right information ("what") at the right time ("when") in the desired format ("how")

T. Ahram et al. (Eds.): IHSED 2019, AISC 1026, pp. 20–25, 2020.
https://doi.org/10.1007/978-3-030-27928-8_4

[8–10]. This allows relevant information to be collected and processed more quickly, reducing the mental workload placed on employees [5].

Initial empirical surveys [11] and laboratory testing [12–14] confirm the human-oriented and economical potential of informational assistance systems. These studies manly use Lego models with a small number of variations. Therefore, the studies only make it possible to draw limited conclusions regarding real assembly work involving a high level of product variability. In light of this, laboratory testing was completed at the Ostwestfalen-Lippe University of Applied Sciences and Arts using real industrial products. The goal of the testing is to assess the potential of different types of informational assembly assistance systems in comparison to providing information in a paper-based format. In particular, the tests were designed to investigate the influence of informational assistance systems on training time in consideration of products of different complexity levels. Five hypotheses were established for the testing. This paper will deal with the first three of these in relation to the use of augmented reality glasses as the selected assistance system type:

1. The use of an informational assistance system results in shorter training times in comparison to using paper-based instructions.
2. The use of an informational assistance system results in shorter execution times for the first assembly of a new product in comparison to using paper-based instructions.
3. The use of an informational assistance system results in a smaller number of picking and assembly errors in comparison to using of a paper-based instructions.
4. The use of an informational assistance system results in less mental workload in comparison to using paper-based instructions.
5. The use of an informational assistance system results in higher acceptance in comparison to using paper-based instructions.

2 Research Design

A 4 $\times$ 3 factorial research design with repeated measurements on the second factor was selected to investigate the hypotheses outlined above. Three different informational assistance systems were chosen – (1) augmented reality based assistance system, (2) tablet-based assistance system, (3) projection-based assistance system – and the control group got the information provided through a paper-based approach (4). Assembly complexity was implemented using three products of increasing complexity – (A) easy, (B) medium and (C) difficult. The information preparation (work plan) was carried out for all support media (1 to 4) with the same scope of information. This included parts lists, assembly hints, information of work materials and step-by-step instructions. Only the visualization of parts lists information was adjusted to the specific support medium. The respective container position was displayed in the work system for all informational assistance systems. The Vuzix M300 augmented reality glasses were used in the experiment. Using this output device, the work instructions were shown on a small display with the help of a photo app.

The three products to be assembled were pneumatic assemblies from a machine manufacturer that were used to simulate real laboratory applications. The complexity of the individual products and their classification into easy, medium and difficult complexity levels were determined using entropy dimension. In information theory, entropy is a measure for uncertainty before an event occurs [15], e.g. when selecting a component, and is used in the context of describing complexity (Table 1).

Table 1. Product degree of complexity.

	Product A	Product B	Product C
Number of parts	21	39	48
Number of different parts	7	20	27
Entropy	2.59 bit	3.83 bit	4.47 bit
Degree of complexity	Easy	Medium	Difficult

The following dependent variables were recorded to verify the hypotheses: Completion time, number of errors, mental workload and acceptance. The completion time is measured stepwise and overall. Each product is assembled four times in a row to take training time into consideration. The limit of four assembly processes (repetitions) is sufficient to detect training effects [16]. Assembly errors are recorded in an error list created in advance, to ensure they are described consistently. This error list includes the following features: (1) Picking errors: Incorrect tool picked, incorrect part picked and/or too many/too few parts picked (2) Assembly errors: Tool not used, mounted in incorrect position, component in incorrect orientation, component not installed, other assembly errors. Mental workload was evaluated after each assembly using the NASA-TLX (Task Load Index) [17]. For the measurement of user acceptance, a UTAUT questionnaire (Unified theory of acceptance and use of technology) [18] translated into German by Fischbach [19] is used, which has been extended by aspects of technology and process acceptance in the context of assistance systems in assembly.

The suitability of the test subjects for the study was determined by means of an eye test and a motor skills test using a reference model. After the completion of the eye test and the assembly of the reference model from Lego bricks, as well as a reference measurement of the NASA-TLX, the standardized assembly of the three products takes place. These are assembled four times in a row and in ascending complexity by the test person with the aid of a support medium. Acceptance is measured after the first assembly of product A and after the fourth assembly of product C. The examiner observed the study participant via cameras. The subjects were required to return assembly tools to their respective locations after use.

3 Initial Results: Paper Versus AR Instructions

During the initial testing phase, 16 persons (eight augmented reality based assistance system; eight provided with information through a paper-based system) completed the laboratory test. Study participants (14 male and 2 female) were between 21 and 37 years old *($M = 25.5$; $SD = 4.84$)*. The majority of the participants were students *(82.5%)*. SPSS 23 was used to evaluate the results. A variance analysis with repeated measurements was selected to test hypothesis 1. Mauchly's Test showed that the assumption of sphericity was violated *($\chi2(65) = 123.49, p < .001$)*. Therefore, the degree of freedom was adjusted using the Greenhouse-Geisser correction. The within-subjects factor was significant *($F(3.50, 49.04) = 234.43, p < .001$)*, indicating the differences between all completion times. Testing on the individual product level (easy, medium, difficult) had similar results ($F_{easy}(1.491, 20.88) = 46.46, p < .001$; $F_{medium}(1.73, 24.16) = 77.39$, $p < .001$; $F_{difficult}(1.30, 18.23) = 141.93, p < .001$). Significant differences in times within a single product, therefore, can be assumed. No significant difference was found between the support media (AR glasses, paper) both overall and in relation to individual products *($F(1, 14) = .371, p = .552$; $F_{easy}(1, 14) = .001, p = .978$ n.s.; $F_{medium}(1,14) = .684, p = .422$ n.s.; $F_{difficult}(1, 14) = .404, p = .535$ n.s.)*. Figure 1 shows the completion times for individual products depending on the medium used.

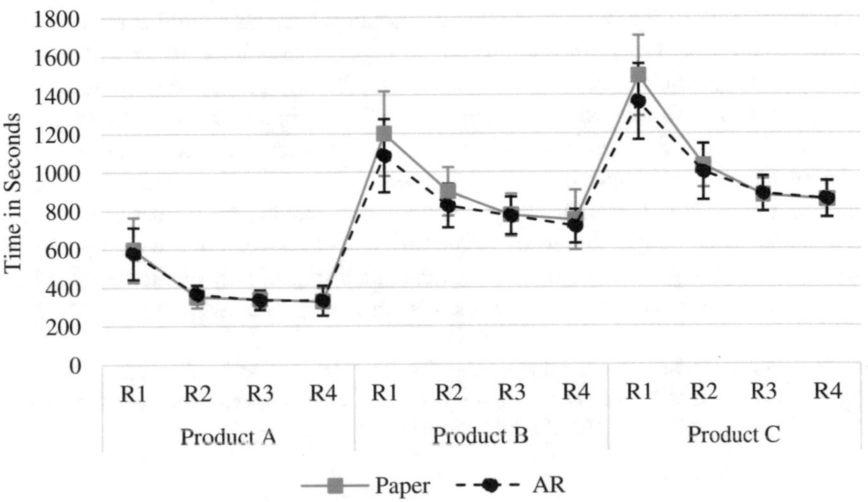

Fig. 1. Completion time (mean values and standard deviations) for the products and repetitions (R) in comparison between paper and augmented reality glasses.

The first completion times for the three products were compared to review hypothesis 2. Here, as well, we found no significant differences *($F(1, 14) = .956, p = .345$ n.s.)*. Errors in picking and assembly were investigated separately as indicators for hypothesis 3. Picking errors were investigated based on a violation of the normal distribution using a Mann-Whitney-U-Test of independent samples. A t-Test was used for assembly errors. There were no significant differences between the groups for both types of errors (picking errors: AR *($Mdn = 5.0$)* paper *($Mdn = 10.5$)*, $U = 14.0$, $z = -1.899, p = .059$ n.s; assembly errors: AR *($M = 10.25$ $SD = 7.96$)* paper *($M = 18.75$ $SD = 11.76$)*, $t(14) = -1.693, p = .113$ n.s.).

4 Discussion

Based on the available data, hypotheses 1, 2 and 3 must be rejected. No significant differences were found between the groups. The experimental manipulation of the complexity of the products was successful. This is indicated by the significant difference in completion times $(F(1.776, 24.86) = 269.0, p < .001)$. In addition, with respect to the mean completion times there are significant differences between assembly repetition 1 and 2 as well as 2 and 3 for product B and C. In product A, there are only significant differences with respect to the mean completion times for the first two assembly repetitions.

Based on the current small sample, it was not possible to prove any significant differences between the output systems (AR glasses, paper instructions) either in relation to completion times or errors. However, there is a tendency for using AR glasses to result in less errors and shorter completion times.

An assessment of the potential of using assistance systems should consider not only absolute completion times. In comparison to paper instructions, using AR glasses in initial assembly of the respective products resulted in an average time savings of 3.16% for product A, 9.51% for product B and 9.00% for product C. In assessing results, we must consider that information for the paper-based instructions was prepared according to ergonomic criteria. Often, paper-based assembly manuals have deficits in design in operational practice [1], so that using a paper-based manual from operational practice likely would have changed the results. The differences between the two output systems presented here (AR glasses, paper instructions) are also probably due to the fact that the respective container position in the work system is displayed when using AR glasses.

Acknowledgment. The authors acknowledge the financial support by the Federal Ministry of Education and Research of Germany (BMBF) and the European Social Fund (ESF) in the project Montexas4.0 (FKZ Grant no. 02L15A260 and 02L15A261).

References

1. Hinrichsen, S., Bendzioch, S.: How digital assistance systems improve work productivity in assembly. In: Nunes, I.L. (ed.) Advances in Human Factors and Systems Interaction. AHFE 2018, Advances in Intelligent Systems and Computing, vol. 781, pp. 332–342. Springer, Cham (2018)
2. Bächler, A., Bächler, L., Autenrieth, S., Kurtz, P., Heidenreich, T., Hörz, T., Krüll, G.: Entwicklung von Assistenzsystemen für manuelle Industrieprozesse. In: Rathmayer, S., Pongratz, H. (eds.) Proceedings of DeLFI Workshops 2015, pp. 56–63. CEUR-Workshop, München (2015)
3. Haller, E., Heer, O., Schiller, E.F.: Innovation in Organisation schafft Wettbewerbsvorteile – Im DaimlerChrysler-Werk Rastatt steht auch bei der A-Klasse-Produktion die Gruppenarbeit im Mittelpunkt. FB/IE **48**(1), 8–17 (1999)
4. Müller, R., Vette, M., Mailahn, O., Ginschel, A., Ball, J.: Innovative Produktionsassistenz für die Montage. In: wt Werkstattstechnik online **104**(9), 552–5650 (2014)

5. Bornewasser, M., Bläsing, D., Hinrichsen, S.: Informatorische Assistenzsysteme in der manuellen Montage: Ein nützliches Werkzeug zur Reduktion mentaler Beanspruchung? Zeitschrift für Arbeitswissenschaft **72**(4), 264–275 (2018)
6. Hold, P., Ranz, F., Sihn, W.: Konzeption eines MTM-basierten Bewertungsmodells für digitalen Assistenzbedarf in der cyberphysischen Montage. In: Schlick, C.M. (ed.): Megatrend Digitalisierung Potenziale der Arbeits- und Betriebsorganisation, pp. 295–322. GITO mbh Verlag, Berlin (2016)
7. Fast-Berglund, A., Fässberg, T., Hellman, F., Davidsson, A., Stahre, J.: Relations between complexity, quality and cognitive automation in mixed-model assembly. J. Manuf. Syst. **32** (3), 449–455 (2013)
8. Hollnagel, E.: Information and reasoning in intelligent decision support systems. Int. J. Man-Mach. Stud. **27**(5–6), 665–678 (1987)
9. Claeys, A., Hoedt, S., Soete, N., Van Landeghem, H., Cottyn, J.: Framework for evaluating cognitive support in mixed model assembly systems. In: Dolgui, A., Sasiadek, J., Zaremba, M. (eds.) 15th IFAC Symposium on Information Control Problems in Manufacturing (2015). IFAC-PapersOnLine **48**(3), 924–929
10. Hinrichsen, S., Riediger, D., Unrau, A.: Assistance systems in manual assembly. In: Villmer, F.-J., Padoano, E. (eds.) Production Engineering and Management. Proceedings 6th International Conference, Lemgo, pp. 3–14 (2016)
11. Kasselmann, S., Willeke, S.: Technologie Kompendium: Interaktive Assistenzsysteme. http://www.ipri-institute.com/fileadmin/pics/Projekt-Seiten/40ready/Technologie-Kompendium.pdf. Accessed 20 Dec 2018
12. Funk, M., Kosch, T., Schmidt, A.: Interactive worker assistance: comparing the effects of in-situ projection, head-mounted displays, tablet, and paper instructions. In: Proceedings of the 2016 ACM International Joint Conference on Pervasive and Ubiquitous Computing, UbiComp 2016, Germany, pp. 934–939. ACM, New York (2016)
13. Blattgerste J., Stenge, B., Renner, P., Pfeiffer, T., Essig, K.: Comparing conventional and augmented reality instructions for manual assembly tasks. In: Proceedings of the 10th International Conference on PErvasive Technologies Related to Assistive Environments, PETRA 2017, Greece, pp. 75–82. ACM, New York (2017)
14. Kosch, T., Kettner, K., Funk, M., Schmidt, A.: Comparing tactile, auditory, and visual assembly error-feedback for workers with cognitive impairment. In: Proceedings of the 18th International ACM SIGACCESS Conference on Computers and Accessibility, ASSETS 2016, USA, pp. 53–60. ACM, New York (2016)
15. Schlick, C., Bruder, R., Luczak, H.: Arbeitswissenschaft, 3rd edn. Springer, Heidelberg (2010)
16. Jeske, T.: Entwicklung einer Methode zur Prognose der Anlernzeit sensumotorischer Tätigkeiten. Industrial Engineering and Ergonomics Band 13. Shaker-Verlag, Aachen (2013)
17. Hart, S.G., Staveland, L.E.: Development of NASA TLX: results of empirical and theoretical research. In: Hancock, P.A., Meshkati, N. (eds.) Human mental workload, pp. 139–183. North Holland Press, Amsterdam (1988)
18. Venkatesh, V., Thong, J.Y., Xu, X.: Consumer acceptance and use of information technology: extending the unified theory of acceptance and use of technology. MIS Q. **36**(1), 157–178 (2012)
19. Fischbach, J.: Determinanten der Technologie- und Prozessakzeptanz im Kontext kooperativer Arbeit. Zeitschrift für Arbeitswissenschaft **73**(1), 35–44 (2019)

A Deep Learning Application for Detecting Facade Tile Degradation

Po-Hsiang Shih and Kuang-Hui Chi(✉)

Department of Electrical Engineering, National Yunlin University of Science and Technology, 123, University Road Section 3, Douliu, Yunlin County 640, Taiwan
{m10512021, chikh}@yuntech.edu.tw

Abstract. Facade tiles of buildings are likely to weaken, crack, or fall off due to aging or out of natural causes such as temperature variations during daytime and nighttime and earthquakes. Tile spalling of tall buildings often leads to accidents or even severe casualties. In view that a routine thorough inspection is costly, this study aims to develop a cost-effective means to detect facade tile degradation of tall buildings through machine learning. We leverage a drone to film outer walls of high-rise buildings at several dozens of sites, from which training data are produced for learning and validation. We resort to a convolutional neural network with deep learning capabilities that is trained with sufficient knowledge to identify hazardous conditions of cracked tiles in two or three levels. Core to our implementation is Jetson TX2—an embedded system—which is programmed in light of AlexNet over Keras and TensorFlow, open-source libraries for deep neural network programming. To heighten learning quality subject to limited amount of training data, image preprocessing involving gray-level transformation, thresholding, and morphological operations is introduced. Experimental results corroborate that our scheme achieves a correct classification rate of over 86%. Our development serves a moderate approach to deep learning in daily contexts, a practical scenario over which to inspire other applications.

Keywords: Machine learning · Deep learning · Convolutional neural network · Defect detection · Image processing · AlexNet

1 Introduction

Earthquakes hit Taiwan from time to time due to geological conditions in that the land is situated in the belt between the Eurasia Plate, Okinawa Plate, and Philippine Sea Plate producing constant collisions. Meanwhile, for decoration as well as water-proof purposes, the outer walls of buildings in Taiwan are often tiled. Façade tiles of buildings are likely to weaken, crack, or fall off due to aging or out of natural causes such as temperature variations during daytime and nighttime and earthquakes. Tile spalling of tall buildings may cause accidents or even lead to severe casualties. Although governments have spent budgets inspecting buildings outer walls for citizens, it is more obligatory for buildings residents to detect tile degradation regularly for

T. Ahram et al. (Eds.): IHSED 2019, AISC 1026, pp. 26–32, 2020.
https://doi.org/10.1007/978-3-030-27928-8_5

ensuring public safety. However, there are certain technical difficulties while inspecting tall buildings, a routine thorough examination is becoming costly.

As a remedy, this study aims to develop a cost-effective means to detect façade tile degradation of tall buildings. We leverage a drone to film outer walls of high-rise buildings. The captured images are then fed to an embedded system with certain intelligence to report assessment results to the user or ground staff for subsequent decision making. To this end, we resort to a machine-learning technique to approach a feasible solution. More specifically, we implement a convolutional neural network (CNN) with deep learning capabilities that is trained with sufficient knowledge to classify hazardous conditions of aging or cracked tiles in different levels. Core to our implementation is Jetson TX2 that is programmed with a deep learning engine for identifying flaws of photographed tiles in real-time. In use are TensorFlow and Keras, open-source software libraries for dataflow programming and deep learning, respectively.

The remainder of this paper is organized as follows. The next section provides a brief background on this study. Section 3 elaborates on the proposed approach. Section 4 accommodates performance results. Lastly, Sect. 5 concludes this work.

2 Background

As mentioned, façade tile spalling brings about non-negligible issues. Taiwan's regulations stipulate that, once such accidents occur, building residents, contractors, and management be fined $2000 to $10000 or 6-month to 2-year prison. Additionally, buildings aged over 15 years with more than 7 stories are considered potentially dangerous. Statistics indicate that there are several hundreds of thousands such buildings in Taipei. When it comes to nation-wide statistics, the number grows far more significantly, implying that handling of façade tile degradation warrants closer study.

As far as deep learning in the context of image recognition is concerned, there has been substantial research on classifying objects or target appearances from a large number of images through machines. Most traditional image recognition is achieved with OpenCV.[1] Thanks to recent advancement of graphics processing units (GPUs), image recognition has been further advanced by deep learning techniques for dealing with more involved situations and thereby has inspired a number of practical applications. For instance, FaceNet embedded with a deep CNN, performs a mapping from face images to Euclidean distances indicative of face similarity [4]. Accordingly, face verification and clustering can be done with ease. FaceNet gains representational efficiency by achieving state-of-the-art performance (99.63% accuracy on Labeled Faces in the Wild[2] and 95.12% on Youtube Faces database) using only 128-bytes per face.

[1] OpenCV, https://opencv.org/.

[2] Labeled Faces in the Wild is a database of face photographs designed for studying the problem of unconstrained face recognition. For more expository surveys, we refer the reader to http://vis-www.cs.umass.edu/lfw/.

Another example application lies in identifying space occupancies in parking lots, e.g., [5]. Such application adopted the *spatial transformer networks* [6] to tackle outdoor lighting variation and perspective distortion. Spatial transformer networks are a generalization of differentiable attention to any spatial transformation, allowing a neural network to learn how to perform spatial transformations on the input image for enhancing the geometric invariance of the model. Experimental results showed that the system reduces the error detection rate and reach an accuracy of 99.25%.

Literature review indicates that deep learning application of image recognition brings new opportunities [3]. In this light, we are concerned with a feasible solution to recognize images of outer walls of buildings by resolving degradation levels. We took into account well-known deep CNN models including LeNet-5, AlexNet [1], NiN [2], ZFNet, and GoogLeNet. Among others, we adopt AlexNet for its neatness, significance, and cost-effectiveness. Operating on input RGB images, AlexNet consists of five convolutional layers and three fully connected layers. Convolutional layers involve repeated feature extractions and max pooling, whereas the fully connected layers fit the model (weight, biases of neurons) for classification. AlexNet used the non-saturating ReLU (rectified linear unit) activation function, with advantage over tanh and sigmoid.

However, initial attempts reveal that our AlexNet-based implementation underperforms under first sets of training data (1200+ RGB images collected from the Internet and campus neighborhood.) Observing that system's performance is subject to the limited amount of training data, we investigate other means to strengthen our development. To this end, we introduce image preprocessing to offset limits of training data, as shall be covered in the next section.

3 The Proposed Approach

Investigation leads us to argue that machine learning alone may underperform if the amount of training data is limited. Joint use of image processing techniques may benefit the classification task, so we propose the following work-around treatment to detect façade tile degradation to a fuller extent, as depicted in Fig. 1, where the grayed box represents the introduction of image preprocessing to our embodiment. The introduced process aims to filter out noise or less significant information, so as to assist AlexNet in focusing on features like zigzag or irregular defects of captured images.

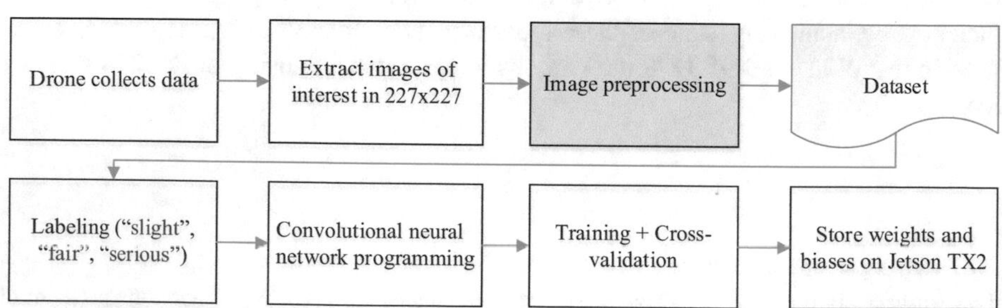

Fig. 1. Flowchart of the proposed approach.

To see what effects can result, Fig. 2 exemplifies before and after image preprocessing of a given input. It can be seen that, as compared with Fig. 2(a), (b) contains concise information highlighting cracks or defects signifying features in original images, implying clearer merits of such operations. Given an input image, preprocessing is carried out in 4 steps (gray-level transformation, image resizing, thresholding, opening operation) before output, as diagrammed in Fig. 3. Gray-level transformation is to simplify images because tile color does not make a feature in question. Image resizing is applied to ease AlexNet computations. Thresholding is to isolate objects or other relevant information in images. Opening operations, involving erosion and dilation, remove small objects from images for feature strengthening.

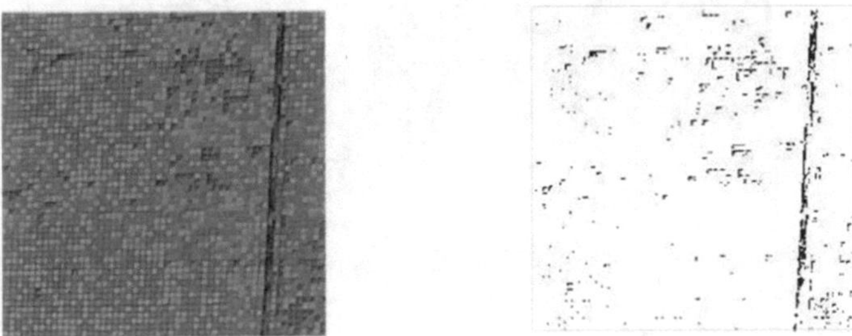

Fig. 2. Before and after image preprocessing.

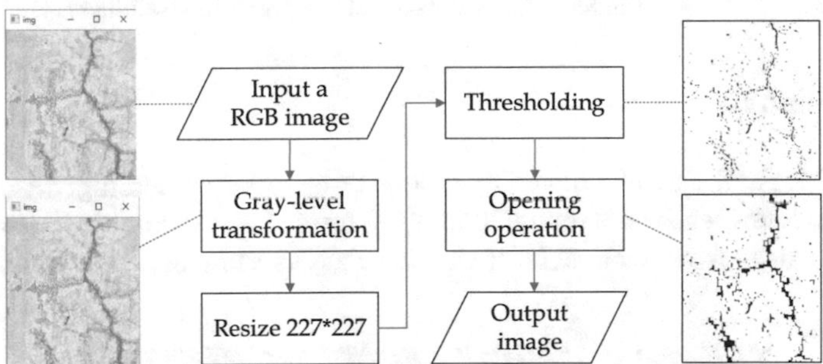

Fig. 3. Image preprocessing flowchart. For better visualization, images resulting from respective steps are illustrated alongside.

Image acquisition is done by utilizing an off-the-shelf DJI Phantom 3 SE quadcopter drone, which has excellent safety systems and is simple to fly using a smartphone or tablet. Offering up to 25-min flight time per battery, the drone comes with WiFi accessibility and dual-band satellite positioning, and is equipped with an one inch 20 Mega Pixel camera capable of shooting 4K super high-definition video and stills. High-resolution imaging is essential to our needs in that we mostly fly the drone to film outer walls aloft at distance of several meters away from the buildings. Such a distance is kept to avoid peripheral trees or other obstacles like window awnings. A high-definition camera is able to capture subtle cracks in façade tiles in course of flight, making images more likely to preserve important information.

As for training and cross-validation, Fig. 4 shows how different levels of tile degradation is learned and tested in our architecture. The training process reflects supervised learning, where training data are the shot images of buildings outer walls, each labeled with *slight*, *fair*, or *serious* according to its decay conditions. For pragmatic considerations, we first implemented AlexNet over a GPU-enabled PC (with Nvidia GeFore GTX970) and let the deep CNN model receive trainings to produce to which class each input image belongs. The trained model is then migrated to Jetson TX2 for field tests whose results shall be addressed in the next section.

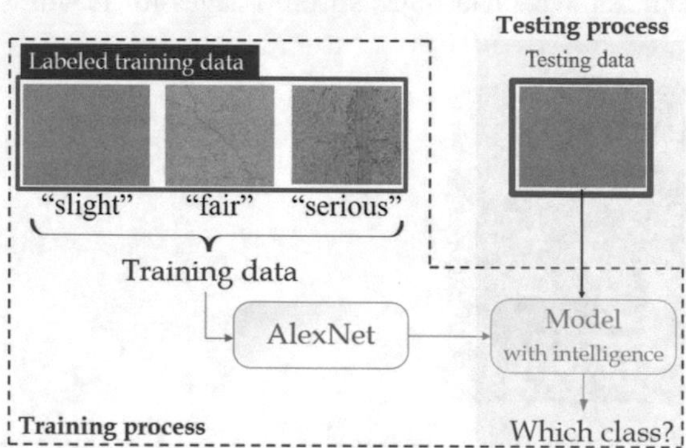

Fig. 4. Training process versus testing process. The dashed box encloses components of the training process. The trained model will undergo tests to prove its usefulness.

4 Preliminary Results

We proceed to put the trained model to various tests for benchmarking. Testing data are grouped into 5 sets, each containing 20 images, formed by randomly selecting a certain number of images from each class of training data as summarized in Fig. 5.

Classes	Test set 1	Test set 2	Test set 3	Test set 4	Test set 5
Slight	5	5	10	10	0
Fair	5	10	5	0	10
Serious	10	5	5	10	10

Fig. 5. Five test sets are generated for each experiment.

Provided that the learning rate η controls how fast or stably the system learns from training data and the fraction p of training data is used as the validation set, we have performance results in terms of classification success rate tabulated in Fig. 6. Experiments were conducted on-campus and off-campus, respectively, in consideration of 2 or 3 classes of decay conditions. It can be seen that the introduction of image pre-processing heightens model accuracy by over 46%, a marked improvement.

Scenario	w/ image preprocessing	w/o image preprocessing	improvement ratio
On-campus, 2 classes	97%	66%	46%
Off-campus. 2 classes	91%	33%	175%
On-campus, 3 classes	86%	44%	95%
Off-campus, 3 classes	50%	27%	85%

Fig. 6. Model accuracy ($\eta = 0.0004$; $p = 0.5$)

It is worth mentioning that off-campus experiments produced lower accuracy, even though image preprocessing had been applied. This is because, in contrast to campus buildings, off-campus buildings have a greater variety of tiles in different forms. Currently, data collected through our drone appears insufficient for this study. During this research, we were faced certain difficulty in acquiring enough images of interest in that we must seek resident permission to fly our drone to film outer walls of their buildings, so as not to raise privacy concerns. Such inconvenience limits data acquisition to a certain degree.

5 Conclusion

This study provided a means to detect facade tile degradation of tall buildings through machine learning. In the current stage of development, we leveraged a drone to film outer walls of high-rise buildings at several dozens of sites, from which training data were produced for learning and validation. We implemented a well-known deep CNN to identify hazardous conditions of cracked tiles in two or three levels. In order to improve learning quality subject to limited amount of training data, image preprocessing was introduced whose effectiveness was shown quantitatively. The next stage of development lies in extending our scheme to spot potentially dangerous areas of outer walls automatically in real-time and report assessments along with photographs shot during cruise back to ground staff for subsequent decision making. Featuring portability with ease to carry, our embodiment can be utilized to not only relieve the burden of labor-intensive work over long hours but also strengthen public safety as well as protection of individuals working aloft. This research shall be of value to alleviate the shortage of workforce or professionals in related occupations. Our development serves a moderate approach to deep learning in daily contexts, a practical scenario over which to inspire other applications.

Acknowledgments. This work was supported by the Ministry of Science and Technology, ROC, under grant MOST 107-2221-E-224-051.

References

1. Krizhevsky, A., Sutskever, I., Hinton, G.E.: ImageNet classification with deep convolutional neural networks. In: Proceedings 25th International Conference Neural Information Processing Systems, vol. 1, pp. 1097–1105 (2012)
2. Lin, M., Chen, Q., Yan, S.: Network in network. In: Proceedings International Conference Learning Representations (2013)
3. Schmidhuber, J.: Deep learning in neural networks: an overview. Neural Netw. **61**, 86–117 (2015)
4. Schroff, F., Kalenichenko, D., Philbin, J.: FaceNet: a unified embedding for face recognition and clustering. In: Proceedings 2015 IEEE Conference Computer Vision and Pattern Recognition (2015)
5. Vu, H.T., Huang, C.-C.: A multi-task convolutional neural network with spatial transform for parking space detection. In: Proceedings 2017 IEEE International Conference Image Processing (2017)
6. Jaderberg, M, Simonyan, K., Zisserman, A., Kavukcuoglu, K.: Spatial transformer networks. arXiv: 1506.02025, https://arxiv.org/abs/1506.02025 (2016)

Development of an Intelligent Pill Dispenser Based on an IoT-Approach

Nada Sahlab[1(✉)], Nasser Jazdi[1], Michael Weyrich[1], Peter Schmid[2], Florian Reichelt[2], Thomas Maier[2], Gerd Meyer-Philippi[3], Manfred Matschke[3], and Günther Kalka[3]

[1] Institute of Industrial Automation and Software Engineering (IAS), University of Stuttgart, Pfaffenwaldring 47, 70569 Stuttgart, Germany
{nada.sahlab,Nasser.Jazdi, Michael.Weyrich}@ias.uni-stuttgart.de

[2] Institute for Engineering Design and Industrial Design (IKTD), Department of Industrial Design Engineering, University of Stuttgart, Pfaffenwaldring 9, 70569 Stuttgart, Germany
{Peter.Schmid,Florian.Reichelt, Thomas.Maier}@iktd.uni-stuttgart.de

[3] Compware Medical GmbH, Robert-Bunsen-Straße 4, 64579 Gernsheim, Germany
{gmp,mm,gk}@compwaremedical.de

Abstract. This contribution describes a novel approach for developing a medication assistance system, namely a pill dispenser. The approach supports mobile and stationary use and applies IoT features for the system to be interconnected and remain synchronized. Based on a user survey and a literature review, functional requirements for the pill dispenser have been derived and use case scenarios conceptualized, followed by a rudimentary software and hardware development.

Keywords: Ambient Assisted Living · Internet of Things · Systems engineering requirements engineering · Pill dispenser · Raspberry Pi

1 Objective and Significance

Advancements in the medical field have led to prolonged life expectancies in industrial countries [1]. A prolonged life is not necessarily equal to a healthy one: The gradual biological regression associated with aging increases the likelihood of multimorbidity, i.e. the co-existence of two or more chronic conditions, and therefore the dependency on medication and on the health-care system. Due to declining sensory and cognitive abilities, the probability of adhering to daily medication decreases as the number of prescribed medication increases [2]. Non-adherence to medication among the elderly can lead to major health risks and require additional social and economic resources. Consequently, efficient solutions to medication non-adherence are necessary. Research in the Ambient Assisted Living, AAL, domain focuses on realizing technical assistance functions with the goal of preserving the personal autonomy while aging [1].

T. Ahram et al. (Eds.): IHSED 2019, AISC 1026, pp. 33–39, 2020.
https://doi.org/10.1007/978-3-030-27928-8_6

An assistance system for improving medication adherence falls under this category. Although a promising field, proposed AAL solutions have shortcomings. These include that most systems are closed ones with fixed infrastructures and system components thereby lacking interoperability, flexibility and outer connectivity [1, 3].

According to an online survey of 2000 participants aged 40 to 75 in Germany[1], users take 2.9 prescribed pills regularly on average. The majority either did not have a medication storage or dispensing unit for the pills or found the one they have impractical. Owned pill dispensers mostly lacked digital or connectivity features, while most users had a smartphone. Participants listed reminder functions and mobility support as high priorities. The older group, who mostly owned a pill dispenser, found safety and usability important while the younger group found digitization and automation aspects such as automatic pill dispensing and ordering to be efficient. Survey results show that there is a need for an efficient and modern medication assistance system. With the rise of the Internet of Things (IoT), more gadgets and sensors applied in daily activities become inter-connected as well as connected to a global infrastructure, allowing for more data exchange [2]. Furthermore, using remote computing capabilities enables the application of artificial intelligence algorithms on communicated data, therefore optimizing assistance functions and transforming them to user-centered, intelligent systems. This bridging of the physical world to the cyber world can enrich AAL systems with more flexibility and efficiency, as the available resources are no longer limited to predefined locations or infrastructures.

The objective of this research is to develop an intelligent, IoT-based pill dispenser to assist elderly with medication management and increase adherence.

2 Medication Assistance Systems

To proceed with our approach, first, the basic required functions for a medication adherence support system were investigated. Subsequently, criteria for an IoT-based approach were listed, upon which a literature review of currently available medication adherence systems was conducted and evaluated. As a result, functional requirements of our system were derived, based on which the system is currently being developed.

2.1 System Requirements

A medication assistance system should include the registration of medication and user data, a reminder function for the user, functions for the detection and registration of medication intake time as well as personal data management.

Furthermore, some additional aspects can be considered such as an escalation mechanism in case a reminder has not been responded to and long-term modeling of adherence to improve the promote an improved user behavior.

---

[1] https://www.cogitaris.de/.

2.2 IoT and AAL Requirements

IoT is said to overcome the shortcomings of available AAL-systems, such a system is characterized by the following requirements:

- IoT-Architecture: a heterogeneous landscape of remotely distributed physical and cyber components connected within a global network and exchanging data.
- Flexibility and Extensibility: components are loosely coupled and decoupled.
- Safety and Security: safety to users of the systems, security for data exchange.
- Intelligence and Adaptability: through using the computing capabilities of higher-level systems, data is processed to generate knowledge and make associations.
- Remote Services: based on the acquired knowledge, services from higher level systems are made available i.e. application interfaces to other systems and actors.

Besides IoT system requirements, AAL systems should be elderly appropriate, i.e include features corresponding to the needs of the elderly such as multimodal user-interaction and non-intrusiveness when acquiring medication intake data.

2.3 Evaluation of Available Systems

Based on IoT and AAL criteria listed above, nine medication assistance systems proposed in literature were evaluated on Table 1 with regard to the criteria fulfillment.

Table 1. Evaluation of medication assistance systems

Approach	IoT-architecture	Services	Flexibility	Extension	Intelligence	Elderly-appropriate
[4]	✓	✓	✓	×	×	×
[5]	×	×	✓	×	✓	✓
[6]	✓	✓	✓	✓	×	✓
[7]	✓	✓	×	×	✓	✓
[8]	×	✓	×	×	×	×
[9]	✓	✓	✓	✓	✓	×
[10]	✓	×	×	✓	×	×
[11]	✓	✓	✓	×	×	✓

[4] applies machine learning algorithms to detect medication intake through the wrist movement when opening a pill bottle cap while having a wearable device on while [5] also applies machine learning for the prediction of adherence using a smart pillbox and a web-based server monitoring system. [6] is a reminder enforcement system consisting of a wearable device and a web interface for communicating and storing data. [7] offers a comprehensive home-based platform consisting of an intelligent medicine box, specially coated pill films and a bio-patch as a system on a chip. The system uses RFID to open the pill box connects to a health-IoT cloud with interfaces to relevant actors and considers personalized health services, e-prescriptions, multimodal reminders as well as non-adherence control. [8] is also a comprehensive system for reminding, detecting medication intake as well as managing data. It detects

the skeletal tracking of patients when taking medications using Kinect and locally saves the results to inform caregivers of the patients' status. [9] is a theoretical system design approach presenting a full model for managing medication, computing adherence and intervening when necessary. It considers portable devices with personalized reminders, adherence monitoring as well as communication to healthcare professionals. [10] is a tablet application following a user-centered design for managing medication and tracking individual behavior. It supports adaptive reminding, mobility and connects the user to relevant actors. [11] is a system focusing on multimodal reminding through seamless integration in daily lives. It consists of an android smartwatch as a reminder and a cloud-based service to connect to other actors.

It is clear that only a few approaches propose a comprehensive system and either exhibit a high level of complexity, lack of flexibility or remain on theoretical basis. As for IoT features or system architectures, only a minority of consider IoT features and are either inflexible or invasive.

3 Intelligent Pill Dispenser

Derived System Requirements. Based on results from the survey, the literature review and interdisciplinary brainstorming sessions, we derived the following functional requirements for a pill dispenser as a medication adherence assistance system and mapped them in Table 2 to the IoT-requirements stated earlier:

Table 2. Mapping of IoT-requirements to the intelligent pill dispenser

Requirement	Intelligent pill dispenser
Architecture	A connected pill dispenser, phone app and cloud server exchanging medication, reminder and profile data as well as medication intake time
Services	Automatic pill ordering, interfaces to medical staff
Flexibility	Support for mobile and stationary use
Elderly-Appropriate	Automatic pill dispensing, multimodal reminding, pill filling assistance, non-invasiveness when detecting pill intake, multimodal reminding
Security	Authentication methods for secure cloud connection
Safety	Pill miss-use detection, escalation scenarios
Extensibility	Bluetooth interfaces for the acquisition of vital signs as a health gateway
Intelligence	Adaptive and personalized reminding

From the requirements, an architecture and use cases are conceptualized and described in the following section.

System Architecture. The proposed architecture consists of hardware as well as software components. Figure 1 shows the main components, which are the intelligent stationary pill dispenser storing one weeks' pills organized in daily boxes, a mobile pill dispenser, which houses one exchangeable daily pill box, a mobile app for personal data management and a cloud-based server for data storage and retrieval.

- The stationary pill dispenser manages pill boxes, timely and automatically dispenses pills and detects pill intake. It has a multimodal user interface and communicates with the cloud server to retrieve and update data.
- The mobile pill dispenser manages one exchangeable daily pill box and detects pill intake via box opening. It generates alarms and communicates with the app via Bluetooth. Exchanged data include box opening times and reminder settings.
- The mobile app scans the medication plan via its QR-code and automatically generates pill filling instructions and alarms for caregivers or users. Furthermore, the app allows the user to overview his medication and set his personalized alarms through preferred interfaces. It also exchanges and synchronizes data with the cloud server.
- The cloud server is responsible for ensuring overall system synchronization and provides interfaces with pharmacies for the automatic pill ordering as well as to medical professionals for updating the medication plan.

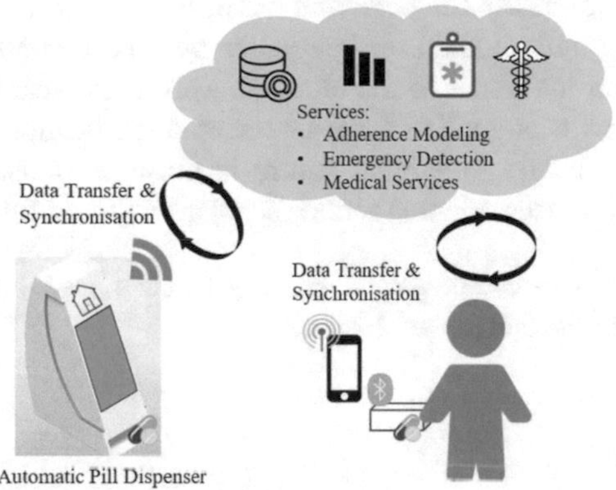

Fig. 1. System architecture

Use cases conceptualized so far are system set up, followed by medication intake and system synchronization. Emergency scenarios are also considered, which include an escalation mechanism in case of no response to the reminders as well as scenarios in case the mobile dispenser was lost or had no charge.

Prototypical Realization. The proposed system is undergoing prototypical implementation with first evaluation board constellation of the stationary pill dispenser, an Android-based mobile app and an experimental server based on a flow-based programming tool called Node-Red. Development of the mobile pill dispenser is underway. A Raspberry Pi and a set of commercially available electronics, shown on Fig. 2, control the stationary pill dispenser.

For the scope of this paper, the authors show the derivation of the system requirements, its architecture as well as the first steps of its prototypical realization. An extensive description of the software architecture or an overall system evaluation is planned for future publications.

Fig. 2. Constellation on evaluation board

4 Summary and Future Work

In this paper, a novel IoT-based approach for developing a mobile and stationary pill dispenser has been shown. As the development of the system is ongoing, future steps include implementing multimodal user-interaction by speech control and implementing secure communication interfaces to the cloud as well as to wearables. As a long-term plan, available data will be applied for user-oriented medication adherence modeling and context-based assistance, enabling future health-related use cases such as preventive diagnosis and a user-oriented recommendation system for healthy daily living.

Acknowledgements. This research project is funded by "Zentrales Innovations programm Mittelstand (ZIM)" with the duration of 2 years.

References

1. Wan, J., et al.: Internet of things for ambient assisted living: challenges and future opportunities. In: International Conference on Cyber-Enabled Distributed Computing and Knowledge Discovery (CyberC) (2017)
2. Ben Hmida, H., Braun, A.: Enabling an internet of things framework for ambient assisted living. In: Wichert, R., Mand, B. (eds.) Ambient Assisted Living. Advanced Technologies and Societal Change. Springer (2017)
3. Kunnappilly, A., et al.: Do We Need an Integrated Framework for Ambient Assisted Living? In: UCAmI (2016)
4. Kim, K., et al.: Algorithm and system for improving the medication adherence of tuberculosis patients. In: International Conference on Information and Communication Technology Convergence, pp. 914–916 (2018)
5. Hezarjaribi, N., et al.: A machine learning approach for medication adherence monitoring using body-worn sensors. In: Design, Automation & Test in Europe Conference & Exhibition (2016)
6. Rosner, D., Jurba, A.T., Dumitrof, A., Vasile, S.: PillBuzz-medication enforcement architecture for assisted living. In: RoEduNet Conference 13th Edition: Networking in Education and Research Joint Event RENAM 8th Conference, Chisinau, pp. 1–6 (2014)
7. Yang, G., et al.: A health-IoT platform based on the integration of intelligent packaging, unobtrusive bio-sensor, and intelligent medicine box. IEEE Trans. Ind. Inf. **10**, 1 (2014)

8. Moshnyaga, V.G., et al.: A medication adherence monitoring system for people with dementia. In: IEEE International Conference on Systems, Man, and Cybernetics (2016)
9. Varshney, U.: Smart medication management system and multiple interventions for medication adherence. Decis. Support Syst. **55**, 2 (2013)
10. Dalgaard, L.G., et al.: MediFrame: a tablet application to plan, inform, remind and sustain older adults' medication intake. In: IEEE International Conference on Healthcare Informatics, pp. 36–45 (2013)
11. Maglogiannis, I., et al.: Mobile reminder system for furthering patient adherence utilizing commodity smartwatch and Android devices. In: 4th International Conference on Wireless Mobile Communication and Healthcare (2014)

Types of Mimetics for the Design of Intelligent Technologies

Antero Karvonen(✉), Tuomo Kujala, and Pertti Saariluoma

Cognitive Science, Faculty of Information Technology, University of Jyväskylä, Jyväskylä, Finland
antero.karvonen@icloud.com, {tuomo.kujala, pertti.saariluoma}@jyu.fi

Abstract. Mimetic design means using a source in the natural or artificial worlds as an inspiration for technological solutions. It is based around the abstraction of the relevant operating principles in a source domain. This means that one must be able to identify the correct level of analysis and extract the relevant patterns. How this should be done is based on the type of source. From a mimetic perspective, if the design goal is intelligent technology, an obvious source of inspiration is human information processing, which we have called cognitive mimetics. This article offers some conceptual clarification on the nature of cognitive mimetics by contrasting it with biomimetics in the context of intelligent technology. We offer a two-part ontology for cognitive mimetics, suggest an approach and discuss possible implications for AI in general.

Keywords: Intelligent technology · Design methods · Design mimetics · AI

1 Introduction

A critical point in any design process is to get ideas for solutions. Often this is based on analogical thinking [1]. Mimetic design means using a source in the natural or artificial worlds as an inspiration for technological solutions. Designers may imitate existing solutions. In biomimetics they may imitate the biological structures found in nature. However, in creating intelligent technologies designers can use existing organizational and individual information processes as the source of ideas. Designing intelligent systems by utilizing existing human information processes as the source of solutions we have termed 'cognitive mimetics' [2–4]. Cognitive mimetics differs from typical and established biomimetics as it has different source of mimicking: human shared and individual cognitive processes, as well as the mental contents, representations, and constraints that establish the boundaries and forms it takes. It analyses how people carry out intelligent tasks today and uses this information in designing novel technological solutions.

The rationale for cognitive mimetics is clear. From a mimetic perspective, if the design goal is an intelligent artefact, then a natural source of inspiration is that which it is seeking to replace or support, and we can turn to human thinking for inspiration. However, it is not yet clear what this entails more exactly or how cognitive mimetics should methodologically speaking proceed. To take steps towards articulating these

T. Ahram et al. (Eds.): IHSED 2019, AISC 1026, pp. 40–46, 2020.
https://doi.org/10.1007/978-3-030-27928-8_7

issues, the purpose here is to explore the principles of mimetic design in general and to position cognitive mimetics in that conceptual space.

2 Mimetic Design

To solve the problems and to make life easier, people have for thousands of years developed new artefacts and new ways of working with them. It is perhaps no exaggeration to say that the world we inhabit is more man-made than it is natural [5]. Especially in modern engineering design, information systems and social system design have grown in importance [6–8]. Thanks to the intensity of modern innovative processes, reflection of design thinking has become central [5]. An important problem is formed by the origin of ideas. Design thinking begins with a problem which one cannot trivially solve. Thus, one must develop the solution with one's mind [9, 10]. Often the crucial hint is given by some cue or analogical formation [10–12]. When the origin of the cue is outside the actual design target domain, the designer is mimicking its source. The source may be a simple cue or a hint or it may be closer to the source [13]. Typically, there is an explicit mapping between the source and target domains. This is rarely an exact mapping, but of the aspects of the source which make it an effective solution. Thus, it is crucial that the mimesis occurs at the correct level to capture the working principles which enable the function [14].

The logical structure of mimetic design has certain necessary elements. To be an instance of mimetic design, there must be a **source domain**. The logical corollary to the source is the **target domain**. Furthermore, there is a process of interpretation or translation [15] between the source and target domain, which we can call **mimetic transfer**. Implicit here is the **designer** who can extract and implement design-goal-relevant information from a source. Important to note is that the process of interpretation is observer-relative given that designers with different backgrounds and knowledge observe different aspects in the source [16]. Thus, specifically identifying the source domain type and level enables the relevant analytical and empirical tools to identify the principles which make a solution work in the source. Here, we shall focus on the kinds of sources people can use for the design of intelligent technologies.

3 Types of Mimetics for Intelligent Technology

There should be no a priori limitations on what can act as a source of inspiration. However, given that technical artefacts are designed to achieve some function an obvious source the natural biological world. The field of biomimetics is well established. The number of patents granted for biomimetic solutions has exploded [17] and an analysis by Lepora and colleagues [18] encompassed approximately 18,000 publications in biomimetic research between 1995 and 2011 on an increasing growth track.

Intelligent technology is a major subfield in biomimetics as shown by the prominence of robotics and autonomous technology and their concern with control methods [18]. Other subfields include animal-based robot hardware; biomimetic actuators; biomaterials science; and structural bioengineering [18]. Bio-inspired computing is another field with direct applications to intelligent technology. Kar [19] reviewed bio-inspired algorithms including neural networks, genetic or evolutionary algorithms, and ant colony optimization among others. Biological neural networks were the early inspiration for artificial neural networks [20]. Deep Learning in neural networks took inspiration from Hubel and Wiesel's [21] analysis of the visual cortex [22]. Earlier advances were made by Hebb [22, 23], whose concern was to "present a theory of behavior that is based as far as possible on the physiology of the nervous system". Advances in intelligent technology have been made based on mimicking the *physiological basis* of information-processing or *human/animal behavior*. The examples share a common characteristic: they have not used the information contents and thinking *itself* as a source.

A major line of research and inspiration for intelligent systems has been the information processing in humans which we call cognitive mimetics [2–4]. Early AI was a collaborative effort between neuroscience, computer science and psychology [24] united under the topic of cognitive science. Mathematicians thinking was the inspiration behind the Turing machine [25]. Tree search, mimicking the way human searches for information [9, 26], has been used from the 1950s in AI solutions. Whereas Newell and Simon [27] focused on general cognitive abilities, a more domain-specific approach was taken by expert systems, which combine a knowledge base with an inference engine [28]. A recent article [29] argued for moving beyond both classical symbolic and neural networks methods by calling for machines that could learn and think like people by building in abilities such as causal modeling and intuitive physics and psychology, among others.

What is clear is that both bio- and cognitive mimetics have made significant contributions to the development of intelligent technology. Based on these examples alone, it is also clear that there is significant variation within them. Thus, from a mimetic perspective, we need more precise concepts to capture what is being mimicked in the source domain. A recent article identified seven analogy categories for biomimetic design: form; architecture; surface; material; function; process; and system [30]. However, the processing of information and its' contents are left implicit in the schema. Thus, for cognitive mimetics as a design method no similar schema is available. Clearly, there is a need for conceptual clarification on the sources for cognitive mimetics.

4 Cognition as a Source of Mimicking

Recall that the point of mimetic design is capture what makes the source an effective solution. When using human thinking as the source, the picture becomes complex. On the one hand, human behavior is structured by content-specific information processes and representations which organize work. In intelligent behavior, we think about specific things and represent them *as* something. On the other, those are based on various general cognitive abilities. It is obvious that all human behavior is predicated on some general abilities, articulations of which include cognitive architectures [31] and cognitive ontologies [32]. There the concern is on either task-specific cognitive modules or more general aspects of mind, like attention. However, it is clear that human expertise is in fact domain-specific: expertise in chess does not carry over to expertise in ship handling though both are on some level served by the same general abilities. Domain-specificity means that human thinking can be called content-specific [10, 33]. This is the first conceptual clarification for the sources of cognitive mimetics. The two levels of cognitive mimetics are complimentary, but stressing their difference makes the issues clearer.

Circling back to the question of effective solutions, are there reasons to favor one level over the other as a source? Given they are connected and design goals differ, no decisive answer should be given. However, if we use expert performance in a specific domain as a benchmark, it is clear going by the discussion before that the reasons are on the information level. Furthermore, technical solutions are usually domain-specific and starting with expert understanding makes it clearer what the technical system should do. Connections to general abilities can and should be drawn, but the disconnect between technical systems and humans causes the downward integration of information levels towards physical implementation to diverge regardless. For example, if a chess master or a ship captain makes a move or maneuver, our primary concern should be in an articulation of things like goals and reasons for the chosen action. Through empirical inquiry we may draw a rich description of the domain-specific information environment in which the action is embedded. Note that this inquiry is not trivial, because much of the information is tacit and unconscious [4]. The entry-point for cognitive mimetics are the representations which organize and structure information processes. Experts have a rich multi-leveled and domain-specific representation that accounts for their expertise. It is multi-leveled as much of the information-processing and contents are subconscious. It is domain-specific because it is populated by domain-specific mental contents. Capturing and articulating these is the first goal of cognitive mimetics as we see it. It is reasonable to approach the design of intelligent technology this way, because it allows flexibility on the implementation level. We may then connect the dots to logically necessary abilities and consider how those can be done by technical systems. Note that this approach may also afford a better integration of humans and technical systems.

Taking a wider perspective, the foundational ability on the information level is the capacity to take the shape of the demands of the environment, as shown by expertise [34]. This leads cognitive mimetics to slightly different problem formulations in terms of general goals for AI. For example, a super computer solves the problems stemming from the computational analysis of complex systems by increasing computational power, thus adding more energy and processing units to the system. In contrast, people solve problems by *lessening* the need for energy by having informational representations that organize work and mitigate the combinatorial explosion that has troubled AI systems in even moderately complex environments [35]. For cognitive mimetics, the formulation should be 'how is the system using information to narrow an information-processing task to minimize the size of the search space, energy use and time' and not 'how can we increase computational power so as to be able to solve this information-processing task within a specified time frame'. The optimization of programs over hardware is part and parcel of computer science and AI but illustrates the difference in perspective. Interestingly, this realization has also been recently articulated by Stuart Russell [35] who is the co-author of a very influential textbook on AI [36]. They argue for the rational approach in AI, which explicitly does not take direct inspiration from human intelligence. However, in the article, Russell [35] expresses doubts whether the standard method of "building calculatively rational agents and then speeding them up" will "enable the AI community to discover all of the design features needed for general intelligence". In the background is the insight that pressure towards optimality within finite resources leads to complexity on the "program level" [35]. In our terms, on the information level or more specifically its representations and contents.

5 Conclusion

The purpose of this article was to examine different types of mimetics for the design of intelligent technology. Mimetics can be classified into types based on the sources they use. From a mimetic perspective, if the design goal is intelligent technology, a viable source of inspiration is human information processing, which we have called cognitive mimetics. The basic idea has two parts. One is that mimetic design is based around the abstraction of the relevant operating principles in a source domain. This means that one must be able to identify the correct level of analysis and extract the relevant patterns. Since in our view major operating principles behind human intelligence are informational kinds, it follows that to use them in mimetic design, a viewpoint and a classification that can articulate those is needed. The categories available in biomimetic theory are not suitable because they have only tacitly implied information processes. We identified two major categories for cognitive mimetics: general abilities and domain-specific contents. Finally, we argued that there are practical and theoretical reasons for starting with the latter.

References

1. Holyoak, K.J., Thagard, P.: Mental Leaps: Analogy in Creative Thought. The MIT Press, Cambridge (1995)
2. Kujala, T., Saariluoma, P.: Cognitive mimetics for designing intelligent technologies. Adv. Hum. Comput. Interact. (2018)
3. Saariluoma, P., Kujala, T., Karvonen, A., Ahonen M.: Cognitive mimetics - main ideas. In: Proceedings on the International Conference on Artificial Intelligence (ICAI), pp. 202–206. The Steering Committee of the World Congress in Computer Science, Computer Engineering and Applied Computing (WorldComp) (2018)
4. Saariluoma, P., Karvonen, A., Wahlström, M., Happonen, K., Puustinen, R., Kujala, T.: Challenge of tacit knowledge in acquiring information in cognitive mimetics. In: Karwowski, W., Ahram, T. (eds.) Intelligent Human Systems Integration 2019. IHSI 2019. Advances in Intelligent Systems and Computing, vol. 903. Springer (2019)
5. Simon, H.A.: The Sciences of the Artificial. The MIT Press, Cambridge (1981)
6. Dym, C.L., Brown, D.C.: Engineering Design: Representation and Reasoning. Cambridge University Press, Cambridge (2012)
7. Pahl, G., Beitz, W., Feldhusen, J., Grote, K.H.: Engineering Design: A Systematic Approach. Springer, Berlin (2007)

8. Ulrich, K.T., Eppinger, S.D.: Product Design and Development. McGraw-Hill, New York (2011)
9. Newell, A., Simon, H.A.: Human Problem Solving. Prentice-Hall, Oxford (1972)
10. Saariluoma, P.: Chess Players' Thinking. Routledge, London (1995)
11. Duncker, K.: On problem-solving. Psychological Monographs, vol. 58, pp. 1–113 (1945)
12. Wertheimer, M.: Productive Thinking. Greenwood, Westport (1945)
13. Fayemi, P.E., Wanieck, K., Zollfrank, C., Maranzana, N., Aoussat, A.: Biomimetics: process, tools and practice. Bioinspiration Biomimetics **12**(1) (2017)
14. Drack, M., Limpinsel, M., de Bruyn, G., Nebelsick, J., Betz, O.: Towards a theoretical clarification of biomimetics using conceptual tools from engineering design. Bioinspiration Biomimetics **13** (2017)
15. Vincent, J.F., Bogatyreva, O.A., Bogatyrev, N.R., Bowyer, A., Pahl, A.K.: Biomimetics: its practice and theory. J. Roy. Soc. Interface **3**, 471–482 (2006)
16. Floridi, L.: The logic of design as a conceptual logic of information. Mind. Mach. **27**(3), 495–519 (2017)
17. Bonser, R.H.C.: Patented biologically-inspired technological innovations: a twenty year view. J. Bionic Eng. **3**(1), 39–41 (2006)
18. Lepora, N.F., Verschure, P., Prescott, T.J.: The state of the art in biomimetics. Bioinspiration Biomimetics **8** (2013)
19. Kar, A.K.: Bio inspired computing–a review of algorithms and scope of applications. Expert Syst. Appl. **59**, 20–32 (2016)
20. McCulloch, W.S., Pitts, W.: A logical calculus of the ideas immanent in nervous activity. Bull. Math. Biophys. **5**(4), 115–133 (1943)
21. Hubel, D.H., Wiesel, T.N.: Receptive fields and functional architecture of monkey striate cortex. J. Physiol. **195**(1), 215–243 (1968)
22. Schmidhuber, J.: Deep learning in neural networks: an overview. Neural Netw. **61**, 85–117 (2015)
23. Hebb, D.O.: The Organization of Behavior: A Neuropsychological Theory. Psychology Press (2005)
24. Hassabis, D., Kumaran, D., Summerfield, C., Botvinick, M.: Neuroscience-inspired artificial intelligence. Neuron **95**, 245–258 (2017)
25. Turing, A.M.: On computable numbers, with an application to the entscheidungsproblem. In: Proceedings of the London Mathematical Society, vol. 42, pp. 230–265, July 1936
26. De Groot, A.D.: Thought and Choice in Chess. Mounton, The Hague (1965)
27. Newell, A., Simon, H.A.: Computer science as empirical inquiry: symbols and search. Philos. Psychol. **407** (1975)
28. Buchanan, B.G., Davis, R., Feigenbaum, E.A.: Expert systems: a perspective from computer science. In: Ericsson, K.A, Hoffman, R.R., Kozbelt, A., Williams, A.M. (eds.) The Cambridge Handbook of Expertise and Expert Performance, 2nd edn., pp. 84–104. Cambridge University Press (2018)
29. Lake, B.M., Ullman, T.D., Tenenbaum, J.B., Gershman, S.J.: Building machines that learn and think like people. Behav. Brain Sci. **40** (2017)
30. Nagel, J., Schmidt, L., Born, W.: Establishing analogy categories for bio-inspired design. Designs **2**(4), 47 (2018)
31. Anderson, J.R., Bothell, D., Byrne, M.D., Douglass, S., Lebiere, C., Qin, Y.: An integrated theory of the mind. Psychol. Rev. **111**(4), 1036 (2004)
32. Price, C.J., Friston, K.J.: Functional ontologies for cognition: the systematic definition of structure and function. Cogn. Neuropsychol. **22**(3–4), 262–275 (2005)
33. Saariluoma, P.: Foundational Analysis: Presuppositions in Experimental Psychology. Routledge, London (1997)

34. Ericsson, K.A., Lehmann, A.C.: Expert and exceptional performance: evidence of maximal adaptation to task constraints. Annu. Rev. Psychol. **47**(1), 273–305 (1996)
35. Russell, S.: Rationality and intelligence: a brief update. In: Fundamental Issues of Artificial Intelligence, pp. 7–28. Springer (2016)
36. Russell, S.J., Norvig, P.: Artificial Intelligence: A Modern Approach. Pearson Education Limited (2016)

Implications of Mobility Service Diaries on Adaptive Mobility Platforms

Cindy Mayas(✉)

Technische Universität Ilmenau, Ilmenau, Germany
cindy.mayas@tu-ilmenau.de

Abstract. Mobility services include transport services and additional services, which are related to passenger journeys. This paper presents a study which combines the analysis of the varying mobility services in mobility service diaries with the journey characteristics duration, motivation and familiarity. Based on first results, hypotheses about the influences of the individual journey characteristics on the use of different mobility service types are derived. These results support the user-oriented development of adaptive mobility information platforms.

Keywords: Mobility services · Mobility service diaries · Adaptive systems

1 Introduction

Travelling is not only a sequence of transport services and movements but also a complex interplay of further individual services supporting a traveler's journey [1]. Therefore, passenger information changes from collective information on departures and routes to individual information on different mobility services. In particular, mobility information platforms integrate information on different primary transport services and secondary transport-related, ticketing, and comfort services for travelers [2]. In order to provide individual, adaptive information on the variety of services for travelers, the complexity of possible service combinations has to be reduced according to the travelers' needs.

Actual research on mobility platforms considers external context factors, such as congestion status, weather, and events to adapt mobility information [3]. However, individual context factors, such as motivation, company, and familiarity with the itinerary might also influence the choice of mobility services. The aim of this paper is to extract hypotheses for potential indicators from individual journey characteristics to improve the user-oriented development of adaptive mobility platforms for individual service information.

2 State of the Art

Internationally, travel surveys are carried out in order to analyze mobility behavior [4–8]. One part of these surveys are the so-called mobility diaries, which represent an individual documentation of ways for one or more days. The diaries collect data about

T. Ahram et al. (Eds.): IHSED 2019, AISC 1026, pp. 47–52, 2020.
https://doi.org/10.1007/978-3-030-27928-8_8

the purpose and the length of a way as well as the used means of transport. These studies focus on the quantification of mobility and the mobility market. In addition to the mobility diaries, demographic data is collected in inquiries. The combination of the demographic and mobility data enables the analysis of mobility user types [9]. However, data about further services while travelling is actually not included. This lack of data for systematic analysis of mobility services is addressed by the study described in this paper.

3 Mobility Service Diary Study

In contrast to the mobility diaries of national travel surveys, the method of "mobility service diaries" is developed to empirically identify indicators for the use of secondary mobility services, such as information, ticketing, entertainment or refreshment services. The method of mobility service diaries is based on the method of user diaries and modified for the context of user interaction with mobility services.

Method. The mobility service diary consists of two parts. The first part is a standardized questionnaire that collects demographic data and data about the journey. Table 1 shows the four key categories of journey characteristics - travel duration, motivation, companions, and familiarity.

Table 1. Variables and values for journey information

Variable	Values	Question type
Travel duration	- Less than 1 h - 1 h up to less than 3 h - 3 h up to less than 6 h - 6 h and more	Single response
Motivation	- Work - Shopping and errands - Free time - Vacation - Other	Multiple response
Companions	- One or two persons (at least 12 years old) - More than two person (at least 12 years old) - Children (less than 12 years old) - None	Multiple response
Familiarity	- I'm familiar with the whole journey - I'm familiar with the majority of the journey - I'm not familiar with the majority of the journey - I'm not familiar with the whole journey	Single response

The second part of the mobility service diary is the journey diary itself. In this explorative pre-study, the diary consists of open input fields to detect input categories for further studies about mobility services. The diary consists of four categories for each use of a mobility service – time, location type, service type, and touchpoint, shown in Table 2. For reasons of anonymization and protection of the private sphere, concrete names of providers, institutions or locations are not collected.

Table 2. Categories of mobility service diaries

Category name	Description	Examples
Time	Start time of a service use on the day of travel with an accuracy of about five minutes. Services, which are used before or after the day of travel, e.g. travel planning or billing, are documented without a concrete time	08:05, before day of travel, after day of travel
Location type	Generalized description of the place of service use without specifying the place	home, bus stop, airport, sharing station
Service type	General extent of the service used without further details	information, transport, food, drinks, hygiene, entertainment
Touchpoint	General description of the service provider	restaurant, bus, sanitary facilities, website

Study Design. The study is conducted as an explorative pre-study. The participation is voluntary and not related to any financial or material rewards. After the acquisition, 9 participants (4 male, 5 female) took part in the survey. The participants are between 20 and 40 years old. The documentation covers the journeys, which included at least one mobility service, of one up to three days within the range of one week.

4 Results and Hypothesizing

The participants documented 16 journeys with 142 uses of mobility services. The documented journeys include the following characteristics: short- and long-distance journeys, journeys for work and during free time, as well as familiar and unfamiliar journeys. The data analysis revealed the following five key service types [2]:

- Transport services: carriage of persons, e.g. by bus, train or plane,
- Transport-related services: carriage of luggage or preparing private or shared means of transport, e.g. refueling.
- Information services: requested individualized data about the planned or current itinerary, e.g. mobile applications or information desks.
- Ticketing services: sale or validation of reservations and tickets.
- Comfort services: services to ensure viability and well-being of the traveler, e.g. food and drinks, hygiene, or entertainment services.

In the next analysis step, hypotheses for the relation between the journey characteristics and the use of mobility services are derived.

The Longer the Journey Duration, the Higher the Extent of Use of Comfort Services. Overall, a linear relationship between the journey duration and the use of different mobility services can be assumed. However, Fig. 1 shows differences between the five key service types. Long-term journeys often contain a higher amount of transport services. Accordingly, the number of information services rises, too. In contrast, the use of ticketing services remains nearly constant, independently from the duration of the journey. One reason might be that most passengers prefer to buy all required tickets in advance of a trip, for instance at intermodal mobility platforms. The use of comfort services remains low for short-term journeys. However, the use of comfort services rises most on journeys, which take longer than 3 h. This higher comfort needs should be also considered by adaptive mobility platforms.

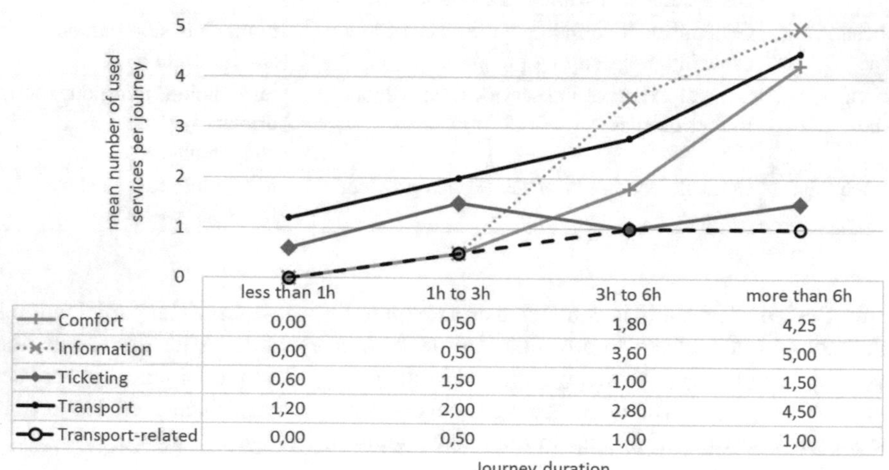

	less than 1h	1h to 3h	3h to 6h	more than 6h
Comfort	0,00	0,50	1,80	4,25
Information	0,00	0,50	3,60	5,00
Ticketing	0,60	1,50	1,00	1,50
Transport	1,20	2,00	2,80	4,50
Transport-related	0,00	0,50	1,00	1,00

Fig. 1. Usage of different service types per journey duration ($n_{journey} = 16$, $n_{service} = 142$)

Journeys with Work Motivation Require a Higher Number of Information Services than Journeys with Free Time Motivation. Figure 2 shows that free time journeys require in average less mobility services than work journeys. Only transport-related services, which depend on the means of transport, rise in free time journeys.

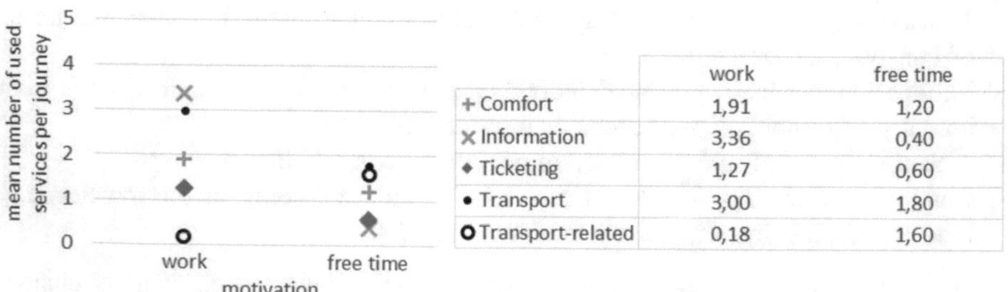

	work	free time
Comfort	1,91	1,20
Information	3,36	0,40
Ticketing	1,27	0,60
Transport	3,00	1,80
Transport-related	0,18	1,60

Fig. 2. Usage of different service types per motivation ($n_{journey} = 16$, $n_{service} = 142$)

In contrast, the number of used information services rises most for work journeys. This fact might be explained by the higher time pressure in working context, which results in higher information needs that should be addressed by adaptive mobility platforms.

The Lower the Journey Familiarity, the Higher the Extent of Use of Information and Comfort Services. Due to the low group size of the value "I'm not familiar with the whole journey", this value is grouped with the value "I'm not familiar with the majority of the journey." to the new value "mostly not familiar". Figure 3 shows that the number of service uses remains nearly constant between mostly not familiar journeys and mostly familiar journeys. In contrast, the use of services reduces for familiar journeys. Next to the transport services, information and comfort services show the highest reduction. Consequently, passengers on familiar journeys expect less detailed information about available additional services by adaptive mobility information platforms.

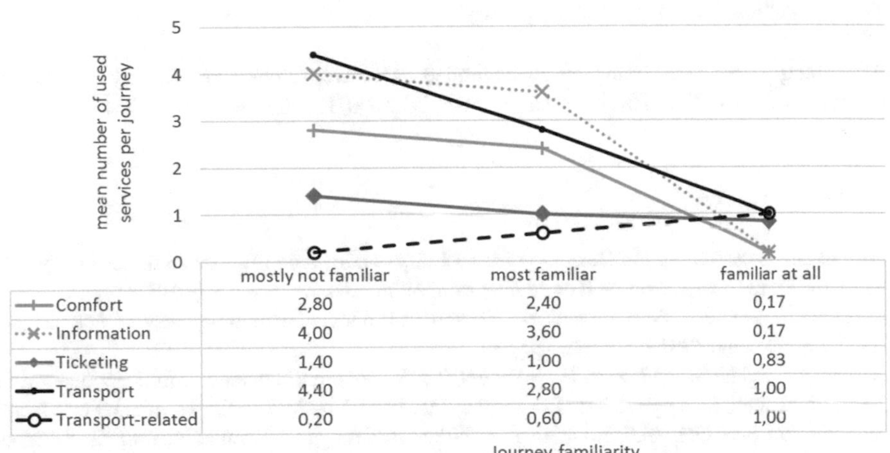

	mostly not familiar	most familiar	familiar at all
Comfort	2,80	2,40	0,17
Information	4,00	3,60	0,17
Ticketing	1,40	1,00	0,83
Transport	4,40	2,80	1,00
Transport-related	0,20	0,60	1,00

Fig. 3. Usage of different service types per journey familiarity ($n_{journey}$ = 16, $n_{service}$ = 142)

5 Discussion

The presented pre-study of mobility services has revealed hypotheses for further studies with larger sample sizes. Due to the low number of analyzed journeys some values had to be grouped to ensure a group size between 3 and 10. Data about the company while traveling was also collected, but 81 per cent of the analyzed journeys were unaccompanied. Therefore, the group sizes of the different accompanied journeys were too small for further analysis. Therefore, a larger sample size of at least 100 journeys is recommended in order to validate the hypotheses. In addition, the quality of the open-input entries varies between the participants. Therefore, standardized values has to be derived from the entries for further studies.

In addition, the duration of the journey should be collected in exact minutes, in order to analyze the correlation between journey duration and service use more detailed. The grouped values are too imprecise for further analysis. The journey characteristics should be extended with personal characteristics of the traveler, which might influence especially the need for information services.

6 Conclusion

The analysis shows that journey duration, motivation, and the familiarity with the travel itinerary influence the use of different types of mobility services while traveling. For instance, information services are most important for travel itineraries with a low familiarity or time-critical journeys for work. Comfort services are most important for long-term travelling and travel itineraries with a low familiarity.

The presented hypotheses provide indicators for the development of user-oriented adaptive mobility information platforms. Further mobility service diary studies with a larger sample size can support the improvement of current mobility information systems by detailed indicators for adaptivity.

Acknowledgments. Part of this work was funded by the German Federal Ministry of Transport and Digital Infrastructure (BMVI) grant number 19E16007F within the project DIMO-OMP.

References

1. Wienken, T., Krömker, H.: Designing for mobility experience - towards an understanding. In: Stopka, U. (ed.) Mobilität & Kommunikation. Winterwork, Borsdorf (2018)
2. Mayas, C., Steinert, T., Krömker, H.: Towards an integrated mobility service network. In: HCI International (2019, accepted paper)
3. Keller, C., Pöhland, R., Brunk, S., Schlegel, T.: An adaptive semantic mobile application for individual touristic exploration. In: Kurosu, M. (ed.) Human-Computer Interaction. Applications and Services. HCI 2014. Lecture Notes in Computer Science, vol. 8512. Springer, Cham (2014). https://doi.org/10.1007/978-3-319-07227-2_41
4. Department for Transport (eds.): Statistical release: National Travel survey England. https://assets.publishing.service.gov.uk/government/uploads/system/uploads/attachment_data/file/729521/national-travel-survey-2017.pdf. Accessed 01 Mar 2019
5. Institut national de la statistique et des études économiques (eds.): Enquête mobilité des personnes: une enquête sur les déplacements des personnes et leurs modes de transport. https://www.insee.fr/fr/information/3365007. Accessed 01 Mar 2019
6. Ecke, L., Chlond, B., Magdolen, M., Eisenmann, C., Hilgert, T., Vortisch, P.: Deutsches Mobilitätspanel (MOP) – Wissenschaftliche Begleitung und Auswertungen, Bericht 2017/2018: Alltagsmobilität und Fahrleistung. https://daten.clearingstelle-verkehr.de/192/162/Bericht_MOP_17_18.pdf. Accessed 01 Mar 2019
7. Infas Institut für angewandte Sozialwissenschaft GmbH (eds.): Mobilität in Deutschland. Kurzreport. http://www.mobilitaet-in-deutschland.de/pdf/infas_Mobilitaet_in_Deutschland_2017_Kurzreport.pdf. Accessed 01 Mar 2019
8. Federal Highway Administration: National Household Travel Survey. https://nhts.ornl.gov/. Accessed 01 Mar 2019
9. Kurosu, M. (ed.): HCI 2016. LNCS, vol. 9733. Springer, Cham (2016). https://doi.org/10.1007/978-3-319-39513-5

Design Process: The Importance of Its Implementation

Leticia Castillo(✉), David Cortés, and César Balderrama

Instituto de Arquitectura, Diseño y Arte (IADA), Departamento de Diseño, Universidad Autónoma de Ciudad Juárez (UACJ), Ciudad Juárez, Mexico
ldi.leticia.castillo@gmail.com

Abstract. Product design is a complex area that requires planning to successfully meet customer requirements. This complexity is proportional to the challenge in innovation and development, therefore the design acquires a new meaning. The lack of experience in design of both the companies and the designer should be guided in a direction in which the design is linked to concepts such as innovation and knowledge, in order to improve the benefits it provides. In this way, the design process becomes the structure of the development of a project with specific tasks from its initial stage, to the final stages of implementation.

Keywords: Design process · Product design · Product innovation · Iterativity · Convergence · Divergence

1 Introduction

The complexity rooted on Product design has been one of the factors in charge of setting the level within a company, even more so when concepts like innovation are continuingly emerging and evolving, so the utilization of tools and knowledge acquired has gotten essential.

One of the tools that has become more practical, for the capability of adaptation that it presents by allowing the selection of one among many others, is the design process, which will benefit the Product designer or the company, by means of having established phases to follow up during the project, encouraging to participate in an inquiry to identify the most relevant data for the project.

There are some qualities that have been identified as relevant to consider during the development of a design process, the iterativity plus the convergence and divergence that it presents. These concepts allow to take the project further by means of not closing it immediately after the last phase has been completed but to consider the data given by the results, for example, during the evaluation phase and return to previous stages to make it better.

T. Ahram et al. (Eds.): IHSED 2019, AISC 1026, pp. 53–58, 2020.
https://doi.org/10.1007/978-3-030-27928-8_9

2 Product Design

Product design is well known to be a discipline that is able to provide competitive advantages to companies from several factors like customer retention and company performance. When creating a new product, companies should create and sell these to appeal, like and motivate customers to buy them, so one of the principal purposes of a product designer will be to complement the quality of the user's life in any aspect [1].

The above, is accomplished through the multidisciplinarity of product design, where an analysis is able to show us the intervention of other areas to reach common goals, to have a successful product, functional, profitable that is able to make a remarkable differentiation from the competition, but that it is also able to hook the customer with its appearance and cost. In order to comply with these goals, the participation of several disciplines is optimal when developing a new product or improving and existing one, so this is where product design generates a convergence from discipline to discipline as seen on Fig. 1, that is adapted from Kim and Lee [2].

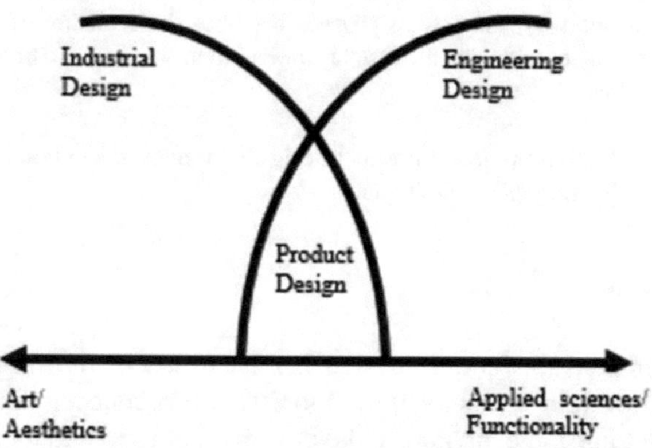

Fig. 1. Industrial Design (left) and Engineering Design (right) show their convergence towards meeting Product Design (center) in order to show the application of art/aesthetics (left) and applied sciences/functionality (right).

The incorporation of Industrial Design and Engineering Design as disciplines through Product Design is implied on Fig. 1, where it is possible to assure the integration of tools or processes that are capable to enrich the development of a product from art and aesthetics to applied sciences and functionality, where a previous immersion onto the product needs or specifications will lead to what is needed the most to make it a good product and avoid poorly designed products that will most likely fail cost wise, timewise and even with its principal purpose of being designed, whichever it may be [3].

So either visualizing a convergence more towards applying Industrial Design or Engineering Design to develop and design a product, the designer will be known as an object creator and we can see design itself as the axis from which factors like innovation can part [4].

Nowadays, innovation has become into an important factor for design, where the increase of its use is able to upgrade the competitive level, which we can see as creating a bond with knowledge [5].

Said bond is linked directly to the visualization where design stands along with the product, the process and the culture with the environment, without taking into consideration these concepts a certain approach would not be possible.

Figure 2 provides an explanation of the agents that are initially involved on a product development as established by the Argentine Industrial Union (UIA) [6].

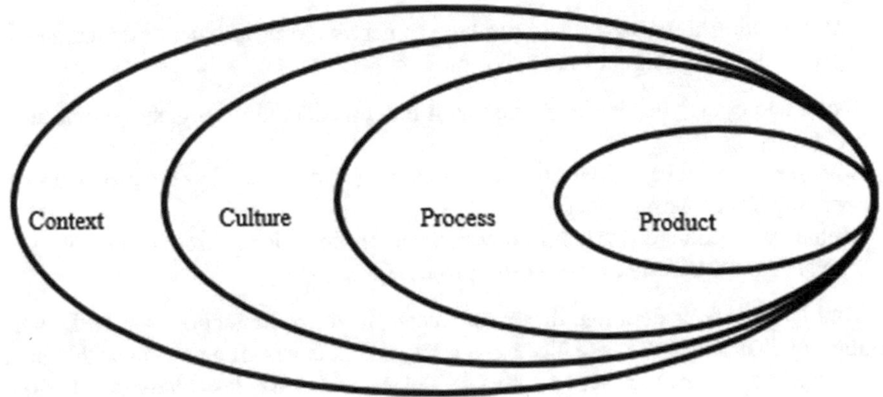

Fig. 2. Agents involving the development of a product, from *context (left)* to the *process (right)* that will take place to create de product, perform together in order to comply with the customer needs, as well as those of the company.

The context is determined as an agent of change for the design of a product, where through its analysis, the identification of new opportunities or ideas that are capable to perform and adapt to current trends is possible, either from economic, social and environmental factors to medium and long term referrals [6]. Also, the culture and the process are greatly involved on the design of a product, with which it is possible to first, target the market and then to establish the best process for its creation where having as an antecedent the context and the culture it will be easier to set a production plan according to the resources that are available. The involvement of these three agents, when implemented accordingly, will set the competitive level of an organization with factors like user approval and the fulfillment of costs and time.

3 Product Innovation

"Innovation can be defined as a continuous process within the companies that is addressed towards incremental improvements in product design" [7], it involves changes and improvements to technologies, products, processes and services that result in positive contributions for customers and other constituents of business organizations. An innovation is a creative new solution to the prevailing conditions and trends and fulfills the expressed and latent needs and wants of customers and stakeholders [8]. This means that innovation is the source from where companies are able to maintain its competitive level through new product development (NPD).

The creation of new products or solutions for existing ones is said to generate a value if they fit into a future of wellness for the users and the environment through innovation, where this wellness can be seen as a link between people, the planet and the profits.

The following three statements can help to get us closer to what designers need to consider while designing an innovative product [4]:

- Social and equity needs should be taken into consideration to create opportunities (people).
- Consider the load capacity of the ecosystem when implementing the production and consumption system (planet).
- Global value change is taken into consideration, in order to create an equity value for users and interested parts of the project (profit).

Taking into consideration these statements is a maximization for NPD where product innovation should take place as a strategy that is part of a process with which it is possible to get a product release into a safe place, where it is less likely that it will fail since product innovation makes it possible to consider concepts like why, what, who and how, same concepts that take the designer to more established tasks within the NPD like planning, developing, implementing and evaluating through a set of tools or steps in order to assure the best result for a NPD [8].

4 Design Process

The incorporation of a design process into a NPD has become part of the structure of a successful outcome since it suggests an analysis, a synthesis and an evaluation through its different phases where all of these become a whole in order to provide an optimal result. With this we could simplify the design process as the structure of a NPD.

Nowadays, there are multiple design processes that designers can use, still the main order can be identified from a called basic cycle of design, proposed by Roozenburg and Eekels in 1998 (see Fig. 3).

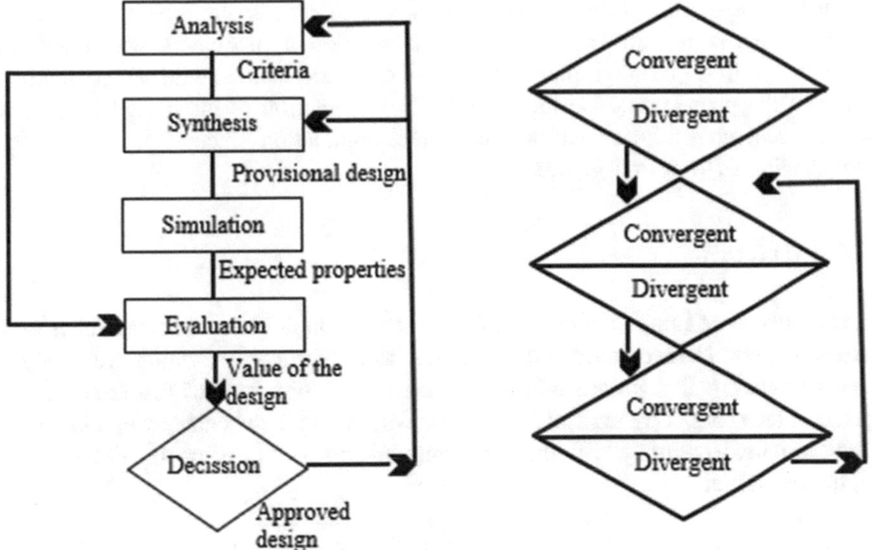

Fig. 3. The basic cycle of design by Eekels and Roozenburg characterized by its iterative quality and complemented by setting the convergence and divergence on its different phases.

As seen on Fig. 3 [9], the basic cycle of design integrates the phases that initiate to incorporate the whole of a NPD, from which it is able to part to more specific phases that will adjust according to the focus of the process, thus it will adapt to the needs of each project and organization. Also, the integration of the iterativity on this process opens a guideline for a well-established design process, where it allows to return to previous phases given the results on the current one.

One clear example could be the design process used by The James Dyson Foundation for their Dyson Supersonic™ hair dryer, since it is based by an iterative nature, which allowed the organization to prototype, test and improve as many times as they believe was necessary, this gave them a unique product [10].

Besides the creation of a unique product that will fulfill the needs of the user, we may identify some statements that are additional benefits of using a design process:

- It allows to maintain control, which allows to make a decision of what to communicate and to who.
- It is possible to administrate new resources towards a planned strategy.
- To order the development of the product and make sure to identify errors and opportunities that may be difficult to catch before the product has been manufactured.
- Economize resources and explode at a maximum the capacity of designers for exploration, attempting to work from something general to something particular.
- Start from an identified problem in order to get to original solutions or optimize an existing one.
- To visualize the process as a whole with anticipation, optimize resources and improve the strategies according to the detected needs [6].

Finally, the optimization of said benefits will be set also by the convergence and divergence that is indicated on Fig. 3, providing a more approachable way to apply the process by giving a more graphic explanation of where there should be an opening or closing of the project data, so the graphical representation of the design process will also represent an essential part for its correct application since it provides a better understanding of it at first glance.

5 Conclusion

The importance of implementing a design process though Product design relies on the contribution within innovation and knowledge to the companies, which eventually sets the differentiating factor with others by means of quality, use of resources and perception of the product by the end user, also it will establish the capacity and knowledge of the Designer to comply with the completion of complex projects that deliver a well-structured product.

References

1. Rodgers, P., Milton, A.: Product Design. Promopress, Barcelona (2011)
2. Kim, K.M., Lee, K.P.: Two types of design approaches regarding industrial design and engineering design in product design. In: 11th International Conference of Design, pp. 1795–1806. Faculty of Mechanical Engineering and Naval Architecture, University of Zagreb The Design Society, Glasgow, Dubrovnik (2010)
3. Maldonado, A., Balderrama, C., Pedrozo, J., García, J.: Diseño Axiomático: Fundamentos y aplicaciones. Libermex, Mexico City (2019)
4. Ferruzca, M., Fulco, D., Aceves, J., Gazano, G., Revueltas, J.: Aproximaciones conceptuales para entender el diseño en el siglo XXI Compilación. Universidad Autónoma Metropolitana, Mexico City (2016)
5. Pavón, J., León, G., Hidalgo, A.: La gestión de la innovación y la tecnología en las organizaciones. Pirámide, Madrid (2014)
6. Unión Industrial Argentina: Diseño de Productos: Una oportunidad para innovar. Instituto Nacional de Tecnología Industrial, San Martín (2012)
7. Fundación Prodintec: Diseño industrial guía metodológica Predica. Gráficas Rigel, Asturias (2006)
8. Rainey, D.: Product Innovation: Leading Change through Integrated Product Development. Cambridge University Press, Cambridge (2009)
9. Reinders, A., Carel, J., Brezet, H.: The Power of Design: Product Innovation in Sustainable Energy Technologies. Wiley, Sussex (2013)
10. The James Dyson Foundation. https://www.jamesdysonfoundation.com/content/dam/pdf/Standalone_DesignProcess.pdf

Bluetooth Tracking Approach for User Assistance Based in Sequential Patterns Analysis

Aitor Arribas Velasco(✉), John McGrory, and Damon Berry

School of Electrical and Electronic Engineering,
Technological University Dublin, TU Dublin, Kevin Street,
Dublin 8 D08 NF82, Ireland
{Aitor.ArribasVelasco, John.Mcgrory,
Damon.Berry}@dit.ie

Abstract. As a civilization, we are drowning in a raging torrent of data, of our own making, that is being harvested by our collective technologies and systems (e.g. Fitbit, phones). However, data itself is of no utility unless it is converted into beneficial knowledge. Design patterns have been shown to be a pragmatic solution to control and manage information flows and provide order and meaning to data within a given context. Assisting users within their daily activities has become a key aspect for modern Artificial Intelligence Systems. Nevertheless, although the GPS technologies work well for outside location, indoor positioning is still problematic, while is a vital awareness component in ambient assistance. This paper shows a preliminary Bluetooth tracking system with a focus on the user's transition between areas of interest. Our work aims to shed light on how the term design patterns can be applied for studying human behavior patterns in the smart environment.

Keywords: Bluetooth Low Energy · Design patterns · Smart environments · Presence detection

1 Introduction

A pattern is an entity that, when repeated in a sequence, contributes to guide or solve common reusable design issues [1]. They exist in different contexts such as nature, art and architecture, computer science and so on, allowing humans to mathematically model natural processes. For example; patterns, as observed in the natural environment, represent the optimization of a process [2].

Nowadays, technology has considerable impact on the use of indoor environments and the interaction of humans inside buildings such as houses or public spaces. Emerging technologies such as the Internet of Things (IoT) allow urban spaces to be designed in order to improve the experience of users while they are carrying out daily activities. IoT has therefore become an essential tool for enabling smart environments [3], with which systems can directly interact and assist users. Currently, IoT systems are developed and installed in a wide range of scenarios, such as smart homes or smart cities, in other words, enhancing the smart-living experience [4].

T. Ahram et al. (Eds.): IHSED 2019, AISC 1026, pp. 59–64, 2020.
https://doi.org/10.1007/978-3-030-27928-8_10

In line with both urban design and software design, the term design patterns has established the possibility of sharing and reusing general solutions to commonly occurring and repeatable problems within a given context. Thus, research has been carried out on these lines in order to optimize the design of new smart spaces [5]. However, a common reusable framework to identify and analyse the sequence of activities followed by users in the smart environment has not yet been defined [5].

Hence, we present a new design and implementation of a Bluetooth based tracking system, for both outdoors and indoors user's transitions between spaces, and not just focused on the location aspect. We combine the transition component to location and activity aspects to get an increase level of granularity related to the context. This paper relies on previous theoretical research published [6], which sought to lay the foundations for a technology-independent design pattern format.

Therefore, our work attempts to anonymously observe activity patterns and so provide guidance and insights into "where, what and who" a rule set based on the "design pattern" format and how this format could contribute to describe a general "understanding" of given cases in the smart environment domain, as well as allowing different processes to collaborate with each other. Applying in areas such as assistive technology and environmental monitoring [7].

2 Related Work

To illustrate the complexity of the environment Fig. 1 shows a domestic setting example. For ease, the different spaces are shown in a simplified version above the building. In this example, an Item (e.g. coffee, cooker), Activity (e.g. make coffee, cook in oven or cook) and Location (e.g. the location of the room from a map perspective) are highlighted, but many more aspects can exist.

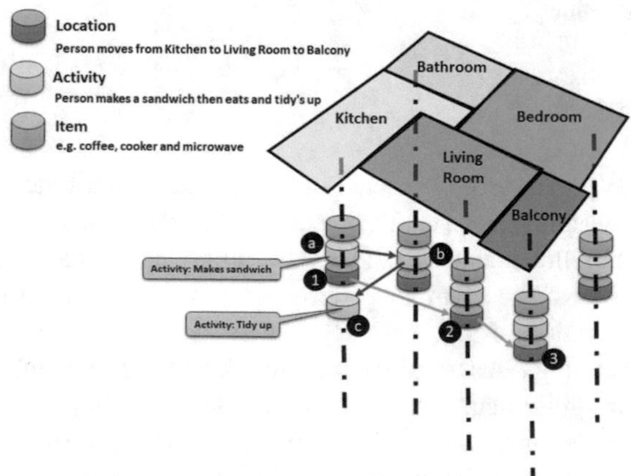

Fig. 1. Image showing vertices linking between nodes.

In Fig. 1 two scenarios are depicted illustrating transitions between locations aspect of each room. In this example, a person is moving from (1) Kitchen to (2) Living room to (3) Balcony. Hence, we are tracking their movements. Also shown on the same image are activity transitions aspects were a person is (a) making a sandwich, shown as a kitchen specific activity to (b) eating the sandwich, shown here as a living room (perhaps in front of TV) to (c) tidying up, which is a kitchen specific activity. These are two different node networks and the system can become very complex very quickly. The linking of nodes is not novel, the extracting of the context is.

However, although the term design patterns have been widely used to explain, predict or facilitate decision making, mathematical and computational models also have limitations. Furthermore, there is currently no pattern based system to cover smart spaces that is specifically designed for representing assistive technology problems in domestic, professional care and occupational therapy situations.

Our work focuses on how IoT technology can assist users with specific emphasis on providing context for particular user cases of daily activities, such as, independent living of elderly population. In particular, focusing on the transitions between locations within user activities in smart spaces, and understanding points in time when guidance may be needed.

3 Methodology

Design patterns and its related mathematical models have been used in a multitude of areas with different purposes, mainly to facilitate the work of designers [8]. From our current point of view, the study of design patterns is a view to use anticipation to optimize the decision-making based on previous experience. In other words, if I can see what is happening over the time and I can identify the pattern form of that sequence, then I can interact directly in the process in order to assist when an alteration occurs. This approach could also be applied to understanding the behaviour of users who are involved in different processes in the wide variety of smart public spaces, such as a home, museum or hospital.

3.1 Description of the Elements of the System

The core elements of our tracking system comprises two key components, an ARM-based embedded system, BBC Micro Bit, worn by the user and an array of detectors, and a set of Raspberry Pi 3, which are fixed in place at different areas of interest through the space. The first device, BBC Micro Bit, is being chosen due to its range of sensors and capability to work as a Bluetooth Low Energy (BTLE) peripheral [9]. These properties provide us with valuable data such as acceleration of the user, hence, being able to determine if the user is moving, standing upright, or he or she has fallen. On the other hand, the Raspberry Pis [10] which have on-board BLE transceivers, acted as anchor nodes receiving the RSSI signal from the BBC Micro Bit carried out by the user.

3.2 Description of the Implementation of the System

The system is described as follows, by following the steps represented in Fig. 2.

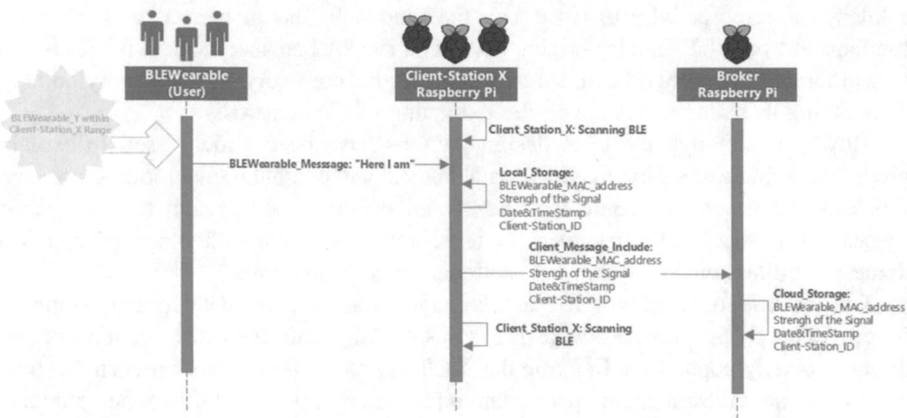

Fig. 2. First-part of the implemented system.

This project seeks to understand how citizens move in civic spaces. It necessarily involves gathering information which could disadvantage citizens if the information is not collected a manner that minimises data protection concerns. Therefore we have decided to adopt a policy of anonymisation at source where that is possible.

With the objective of preserving user anonymity, our system involves a technique called hash encoding that comprises of taking a string of characters such as "sesame 11:22:33:44:55:66" and turning it into a unique and very long number. Two of the interesting features of hash encoding is that each unique string always produces the same very long unique number/identifier, and also that the process is one way, once you have the number it is almost impossible to go backwards and get the original string. These two properties when taken together make the technique very useful for anonymization. The most popular hashing algorithm at the moment is the SHA algorithm and SHA-2 (the one selected for this study) is considered to be suitably secure.

It enables the movements of an individual user to be studied without recording mac address which could be related back to the identity of the user if the mac address was discovered independently. Furthermore, our source code will only record and store the hash encoded mac address of the BTLE peripherals used and described in the above section, BBC Micro Bit.

Hence, the BLEWearable device carried out by the user is constantly broadcasting a Bluetooth signal same as a lighthouse emits a white light. When the BLEWearable_Y is within a Client-Station_X range, this one, first, will locally store 4 data elements for every BLEWearable broadcast by a passers-by BBC Micro Bit. These four elements are: Hashed Mac Address, Received Signal Strength Indicator (RSSI), Timestamp (date and time), and the Client-Station_ID for the measurement. The RSSI parameter allows the system to estimate the distance to and, therefore, the current position of the user in a building. In relation to this last parameter, and in order to increase the precision of each measurement, we calculate the average of three values detected. In other words, each RSSI measurement recorded correspond to the average of three RSSI values scanned by the same station. We have implemented this feature, as multiple Client-Stations located in different rooms will pick up simultaneously the same BLEWeareble_Y, but this one, can only be in one place at the same time. Therefore, only an average greater that −80 dB will consider that an user is actually in that location. This value, −80 dB, is the result of multiple tests where standing 2.5 m away of the Client-Station-X we measured this average RSSI values, and so, allowing us to consider an user in a particular location.

Then, once a Client-Station_X has stored the previous data mentioned, this one will be sent to a data warehouse, or Broker. This last interaction allows us to gather all information from the multiple Client-Stations into a common database, as well as to back up the data recorded. Our system is being coded to handle multiple connections simultaneously, therefore, being able to track multiples peripherals at the same time.

4 Preliminary Results and Discussion

Our objective is not necessarily to reach conclusions but rather to prompt new ideas of how design-pattern techniques can be applied in the smart environment domain. Thus, our first scrip coded provides the following valuable information.

1. Number of transitions realized or edges.
2. Number of Client-stations or nodes.
3. For each node, the incoming and outgoing edges.
4. The number of times that a particular transition has been done. For example, this information will allow us to know how many times a user has transited between two particular locations.

This information could be used to identify possible pattern interactions between areas of interest and by, for example, in a museum, this information could provide insights into how the visitors move through the gallery, allowing our system to provide and suggest more personalized touring information, or save time from busy areas.

5 Conclusion and Future Work

A multitude of researches have shown how design patterns techniques can enhance the labour of designers, being of particular importance in computer science [11]. Thus, these entities are seen as a consolidated solution to a recurring problem. However, they are not popular in the recognition and analysis of human interaction in traditional householders. Our system attempts to identify those sequences within the known smart environment domain. Where if we describe each of the elements involved either in a process or in a sequence of activities, we begin to open the door to cross-compatibility and then to collaboration. This information will allow us to:

1. Identify similar behavioural patterns repeated over time by different users.
2. Assist users in real-time in case that an unexpected event or emergency occurs. For example, in a construction building context or assisting an elderly patient living on his/her own.
3. In public spaces, such as Museums, Airports or Hospitals, the behaviour pattern form of the users could contribute to managing patient waiting lines, thereby optimizing services and reducing patients' waiting time.

In addition, our research focuses on the transitions between spaces and how that could affect the original design and future proofing it through universal design concepts. Therefore, by exposing these Location Nodes, Activities and Transitions we can provide much more evidence-based education to the designers and installers of the future.

This research aims to provide an easy installation and configuration so over time evidence based information could be gathered to highlight the optimum design within the space. To conclude, in addition to recording and analyzing users transitions in a range of different smart spaces, our goal is to set a design patterns based platform able to predict and provide a basis for elderly assistance, decision making and guiding as well as understanding human behavior in the smart environment domain.

References

1. The Free Dictionary: Patterns. http://tinyurl.com/ybwug9lc. Accessed 02 Sept 2018
2. Ball, P.: Shapes: Nature's Patterns: A Tapestry in three Parts. Oxford University Press, Oxford
3. Cook, D.J., Das, S.K.: How smart are our environments? An updated look at the state of the art the role of physical components in smart environments. J. Pervasive Mob. Comput. **3**(2), 53–73 (2007)
4. Gomez, C., Chessa, S., Fleury, A., Roussos, G., Preuveneers, D.: Internet of things for enabling smart environments: a technology-centric perspective. J. Ambient Intell. Smart Environ. **11**(1), 23–43 (2019)
5. Vega-barbas, M., Pau, I., Augusto, J.C.: Interaction patterns for smart spaces: a confident interaction design solution for pervasive sensitive IoT services. IEEE Access **6**, 1126–1136 (2017)
6. Velasco, A.A., Mcgrory, J., Berry, D.: Patterns within patterns within the smart living experience, pp. 1–5 (2018)
7. Lea, D.: Christopher Alexander: an introduction for object-oriented designers. ACM SIGSOFT Softw. Eng. Notes **19**(1), 39–46 (1994)
8. Dearden, A., Finlay, J.: Pattern languages in HCI: a critical review. Hum.-Comput. Interact. **21**(1), 49–102 (2006)
9. BBC micro:bit MicroPython Documentation (2016)
10. Krishnan, R.: Introduction to Raspberry Pi. Control, no. April, p. 2014 (2014)
11. Ampatzoglou, A.: Research state of the art on GoF design patterns: a mapping study. J. Syst. Softw. **86**(7), 1945–1964 (2013)

Autonomous Learning Mediated by Digital Technology Processes in Higher Education: A Systematic Review

Washington Fierro-Saltos[1,2(✉)], Cecilia Sanz[2], Alejandra Zangara[2], Cesar Guevara[3], Hugo Arias-Flores[3], David Castillo-Salazar[2,4], José Varela-Aldás[4], Carlos Borja-Galeas[5,6], Richard Rivera[7], Jairo Hidalgo-Guijarro[8], and Marco Yandún-Velasteguí[8]

[1] Facultad de Ciencias de la Educación, Universidad Estatal de Bolívar, Guanujo, Ecuador
wfierro@ueb.edu.ec

[2] Instituto de Investigación en Informática LIDI- Facultad de Informática, Universidad Nacional de la Plata, Buenos Aires, Argentina
csanz@lidi.info.unlp.edu.ar, alejandra.zangara@gmail.com

[3] Mechatronics and Interactive Systems - MIST Research Center, Universidad Indoamérica, Ambato, Ecuador
{cesarguevara,hugoarias}@uti.edu.ec

[4] SISAu Research Group, Universidad Indoamérica, Ambato, Ecuador
{davidcastillo,josevarela}@uti.edu.ec

[5] Facultad de Arquitectura Artes y Diseño, Universidad Indoamérica, Ambato, Ecuador
carlosborja@uti.edu.ec

[6] Facultad de Diseño y Comunicación, Universidad de Palermo, Buenos Aires, Argentina

[7] Escuela de Formación de Tecnólogos, Escuela Politécnica Nacional, Quito, Ecuador
richard.rivera01@epn.edu.ec

[8] Grupo de Investigación GISAT, Universidad Politécnica Estatal del Carchi, Tulcan, Ecuador
{jairo.hidalgo,marco.yandun}@upec.edu.ec

Abstract. The concept of autonomous learning has been resignified in recent years as a result of the expansion of the different types of study. Online education in higher education institutions has become an effective option to increase and diversify opportunities for access and learning, however, high rates of dropout, reprisal and low averages still persist. academic performance. Recent research shows that the problem is accentuated because most students have difficulty self-regulating their own learning process autonomously. From this perspective, the purpose of the study was to examine and analyze, through a systematic review of the literature, on autonomous/self-regulated learning, theoretical models and determine which variables influence a learning process mediated by technology processes in the higher education. The findings indicate that: (1) autonomous learning is a synonym of self-regulation; (2) Pintrich's

T. Ahram et al. (Eds.): IHSED 2019, AISC 1026, pp. 65–71, 2020.
https://doi.org/10.1007/978-3-030-27928-8_11

self-regulatory model is the most used in digital contexts; and (3) the self-regulatory variables identified are wide and varied.

Keywords: Autonomous learning and self-regulation · Online education · Higher education

1 Introduction

The concept of autonomous or self-regulated learning has been resignified in recent years as a result of the expansion of the different modalities of presential, blended, distance and online study. Education mediated by digital technology processes in higher education institutions, has become an effective option to improve, increase and diversify access and learning opportunities. In addition, one of the fundamental characteristics of the educational process is the autonomy of the students, since they constitute the center of their own learning as active participants and with the capacities to learn by oneself. Although this concept is not new, nowadays in the universities they play a decisive role in the academic success of the students favoring the attainment of academic achievements and preventing the educational failure [1]. In the same line Kingsbury [2], highlights that Higher Education Institutions are responsible for motivating students to want to learn more autonomously, taking into account that knowledge is dynamic, distributed and complex, i.e. a university now it depends more and more on higher-order thinking skills, the ability to solve problems in interdisciplinary fields and knowing how to communicate, negotiate and collaborate effectively with others.

Nowadays there are already thousands of university students enrolled in online courses both formally, offered as part of an academic program to obtain a degree; as an informal level for professional development. However, and despite the high demand, recent studies reveal a significant index in educational failure and few students are able to successfully complete a learning process. Academic failure is shown in terms of low average academic performance, high dropout rates, repetition, student lag and low terminal efficiency rates. These causes are associated with factors such as the lack of teaching or technological support, lack of organizational support from the institution or lack of an adequate method of study and, above all, poor self-regulation skills on the part of students in this type of environment or study modality [1, 3, 4]. Following in this same line, research on self-regulation in the context of higher education mediated by technology is still very superficial and if a support of theoretical models, and more there are approaches to learning through courses on the web and internet. such as Massive and Open Online Courses (MOOCs) [5].

From this point of view, autonomous or self-regulated learning is a vital factor for the success of student learning, and it is still relevant for research on higher education learning, especially with the advent of educational technology and online learning environments [6]. Because of this perspective, this study presents the results of a systematic review which purpose is to examine autonomous and/or self-regulated learning, its theoretical models and determine which variables influence a process of process-mediated learning of digital technology in Higher Education. To answer this objective, three questions were raised that guided the investigation: RQ1. What is the relationship between autonomous learning and self-regulated learning (SRL)?; RQ2.

What are the main specific self-regulatory models that have had the greatest development and application in the field of higher education mediated by digital technology? and; RQ3. What are the main variables of self-regulation that influence the learning of students mediated by digital technology processes in Higher Education?

2 Method

To carry out this analysis, the method of systematic review of literature based on the Kitchenham guidelines was used [7]. It should be noted that, for this review, no analysis was carried out to determine the quality of the articles, given that the interest of the study is to include as many publications as possible, considering the contexts and questions that we investigate. The search for information was made in the most relevant and impacting scientific databases such as: Scopus, Web of Science, Science Direct, IEE Explorer and Springer. In addition, articles published between the years 2012 to 2019 and with inclusion criteria were considered as: Type Documents: Article, Conference Paper and Review; Source Type: Journals and Conference Proceedings; Language: English, Spanish; magazines of the discipline of Computers and Education. 195 items were retrieved according to the search criteria with the inclusion/exclusion elements. From these, we made a selection of articles, based on titles/abstracts and keywords. From this first set of articles, we exclude those that do not address self-regulation in Higher Education and eliminate duplicates. At the end of this process, we end with a total of 36 articles that were selected for the study.

3 Results and Discussion

3.1 RQ1. What Is the Relationship Between Autonomous Learning and Self-regulated Learning?

Different authors have defined autonomous learning in several contexts, many of them associate the concept of autonomous learning with self-regulation, self-directed learning and meta-cognition. In the analysis of these concepts, all of them have common and coinciding elements in which the ability of an apprentice to guide his own learning process, guided by objectives and different learning strategies, is highlighted. However, the concept of self-regulation in the academic context in a more comprehensive concept encompassing cognitive, affective, emotional, metacognitive processes, self-direction skills, self-control, and self-evaluation. From this perspective, autonomous learning, as well as self-regulated and self-directed learning, continue to represent three very similar theoretical areas, related and interdependent with each other, and that are only separated by disciplinary boundaries. These analogies are corroborated by important authors such as Persico and Steffens [8], who state that the concept of autonomous learning is very similar to self-directed learning (SDL). Similarly, Zimmerman [9] and Woolfolk [10] identify autonomous learning as a synonym for self-regulation (SRL); and, finally, Monereo and Barberá [11] and Dinsmore et al. [12], associate as synonyms the term autonomous learning with metacognition.

Although autonomous learning has a strong relationship with self-regulation, the position is to assume self-regulation of learning (SRL) as a broader and more inclusive principle for the research we are carrying out.

3.2 RQ2. What Are the Main Models of Specific Self-regulation that Have Had Greater Development and Application in the Field of Higher Education Mediated by Digital Technology?

The study of self-regulation of learning is based on the analysis of a set of variables and categories, which account for models that allow to describe, analyze and better understand the way in which a subject learns and takes control of learning by itself. From this context, literature studies report six theoretical models [13–15], among them the Model by Philip Winne and Allyson Hadwin; Zimmerman's cyclical model; the Pintrich model; the dual processing model of Boekaerts; the MASRL model of Efklides; and, the collaborative SRL model of Hadwin, Järvelä and Miller. These constructs are those that have had the most development and empirical evidence of application in the academic field of higher education. Regarding the prevalence of self-regulatory models used in digital contexts, it is observed that eleven studies adopted the Pintrich model in descriptive and/or quasi-experimental investigations; four studies were based on Zimmerman's cyclical model; two studies in the Winne and Hadwin model; a study in model Hadwin, Järvelä and Miller; and, the remaining eighteen studies, of which most are quasi-experimental studies, did not employ any theoretical model of self-regulation. From this context we can prove that Paul Pintrich's model is the one that has the greatest application in digital and virtual learning environments in Higher Education.

3.3 RQ3. What Are the Main Variables of Self-regulation that Influence the Learning of Students Mediated by Digital Technology Processes in Higher Education?

To identify the variables of self-regulation in contexts mediated by digital technology, a grouping of tools with a related function was carried out, in order to differentiate the variables of self-regulation taking into account that the environments are different; this is how we generate 3 types of categories. In the first category are the online learning systems associated with e-learning platforms (LMS, Blackboard), b-learning and m-learning, MOOCs, online forums, inverted classroom. In the second category are the online applications such as MyLA software, MetaTutor, ALEKS, CAF, CLICKER, Audacity, among others) and in the third category are social interaction systems such as the use of social networks, web2.0 technologies, Second Life, communications applications (instant messaging, twitter and even telephones to communicate with classmates).

From these scenarios, in the category of online learning systems, twenty-four variables or unit of analysis were identified in the different studies, these are: (1) Establishment and orientation of objectives and goals; (2) Time management; (3) Regulation of the effort, (4) Elaboration; (5) Critical thinking, (6) Search for help; (7) Habit of independent learning; (8) Academic self-efficacy and technology; (9) Metacognitive self-regulation,

(10) Intrinsic motivation, (11) Emotions (Test anxiety); (12) External scaffolding; (13) Regularity of studies; (14) Self-control; (15) Student persistence; (16) Study environment; (17) Commitment to tasks and problem solving, (18) Intrinsic understanding; (19) Performance orientation; (20) Metacognitive awareness; (21) Self-satisfaction; (22) Feedback at task level; (23) Level of Interaction (students -contents -professor); and (24) Academic skills (Reading comprehension, written communication, autonomous learning, use of Web 2.0 tools). In the second category of online applications, sixteen analysis variables were determined, comprising: (1) Goal setting; (2) Prior knowledge; (3) Intrinsic and extrinsic motivation; (4) Emotion (Anxiety before exams, - Boredom, Frustration); (5) Motivation in the value of the task, (6) Self-efficacy in the performance, (7) Metacognitive or critical thinking; (8) Online learning experiences; (9) Metacognitive reading strategies; (10) Time management in the study; (11) Learning control beliefs; (12) Peer learning; (13) Search for help; (14) Feedback and (15) Self-evaluation. Finally, in the category of social interaction systems, eight analytical variables were examined: (1) Control of attention; (2) Behavior; (3) Presence in social networks; (4) Self-efficacy; (5) Academic achievements; (6) Time management on the Internet and social networks, (7) Dialogue from informal learning spaces; (8) Search for academic help.

In general, it is observed that the analysis variables studied in the different investigations are wide and varied. In addition, among the most common SRL strategies that are supported by important theoretical models and that are also present in the educational processes mediated by technology in higher education are: Meta-cognition, time management, regulation of the effort, peer learning, elaboration, essay, organization, critical thinking and seeking help. These strategies of self-regulation are broadly coincident with the studies carried out by Broadbent and Poon [16], Panadero and Alonso-Tapia [17], Panadero, Jonsson and Botella [18].

4 Conclusions

Autonomous or self-regulated learning is a determining factor for the success of student learning in higher education, especially with the mediation of educational technologies. The complexity of online learning environments, the diversity of sources of information, the scarce orientation and attitudes toward study, the lack of teamwork and the difficulties in managing study time, demand high self-regulation of students, to improve academic performance, even though students may be "digital or millennial natives" and have access to technology, there is no guarantee that they will succeed or "the well-placed bits" term coined by Monereo [19]. In this systematic literature review, we analyze the relationship between autonomous learning and self-regulation, current theoretical models in digital contexts and determine variables that influence a process of learning mediated by digital technology processes in Higher Education. The results indicate that only a few researchers define a strong relationship of similarity between autonomous learning and self-regulation, six theoretical approaches to self-regulation and the Pintrich model have been identified, which has had the greatest application in both learning environments digital and virtual In addition, it is determined that the variables of self-regulation are broad, varied and in many cases common for different contexts and learning environments.

Acknowledgments. This work was supported by the State University of Bolivar, giving me the facilities to carry out the research thesis called "Strategies based on analytical learning for the study of the variables related to autonomous learning. Case study in Higher Education of Ecuador", as part of the Doctorate in Informatics at the National University of La Plata, Argentina.

References

1. Stoten, D.W.: Managing the transition: a case study of self-regulation in the learning of first-term business and management undergraduate students at an English university. Res. Post-Compuls. Educ. **20**(4), 445–459 (2015)
2. Kingsbury, M.: Encouraging independent learning. In: Fry, H., Ketteridge, S., Marshal, S. (eds.) A Handbook for Teaching and Learning in Higher Education: Enhancing Academic Practice, pp. 169–179 (2015)
3. Escanés, G., Herrero, V., Merlino, A., Ayllón, S.: Dropout in distance education: factors associated with choice of modality as the dropout conditions. Virtual. Educ. Sci. **5**, 45–55 (2014)
4. Miranda, M., Guzmán, J.: Analysis of the desertion of university students using data mining techniques. Univ. Educ. **10**(3), 61–68 (2017)
5. Berridi, R., Martínez, J.: Self-regulation strategies in virtual learning context. Educ. Profiles **39**(156), 89–102 (2017)
6. Schumacher, C., Ifenthaler, D.: Features students really expect from learning analytics. Comput. Hum. Behav. **78**, 397–407 (2018)
7. Kitchenham, B.: Procedures for performing systematic reviews. Keele University 24(TR/SE-0401), pp. 1–26 (2004)
8. Persico, D.; Steffens, K.: Self-regulated learning in technology enhanced learning environments. In: Proceedings of the STELLAR -TACONET Conference, pp. 115–126 (2017)
9. Zimmerman, B.J.: Achieving academic excellence: a self-regulatory perspective. Purs. Excellence Through Educ. 85–110 (2005)
10. Woolfolk, A.: Psicología Educativa. Pearson educación, London (2010)
11. Monereo, C.; Barbera, E.: Instructional design of learning strategies in non-formal educational environments. Obtenido de (2000). http://recursos.udgvirtual.udg.mx/biblioteca/bitstream/20050101/996/1/Lectura_10_Diseno_instruccional_de_las_estrategias_de_aprendizaje_en_escenarios_educativos.pdf. Accedido 14 de mayo de 2019
12. Dinsmore, D.L., Alexander, P.A., Loughlin, S.M.: Focusing the conceptual lens on metacognition, self-regulation, and self-regulated learning. Educ. Psychol. Rev. **20**(4), 391–409 (2008)
13. Panadero, E.: A review of self-regulated learning: six models and four directions for research. Front. Psychol. **8**(April), 1–28 (2017)
14. Zimmerman, B.J.: Self-regulated learning: theories, measures, and outcomes. Int. Encycl. Soc. Behav. Sci. 541–546 (2015)
15. Adam, N.L., Alzahri, F.B., Cik Soh, S., Abu Bakar, N., Kamal, N.A.M.: Selfregulated learning and online learning: a systematic review. In: Badioze, Z.H., et al. (eds.) Advances in Visual Informatics, IVIC 2017. LNCS, pp. 143–154. Springer, Cham (2017). https://doi.org/10.1007/978-3-319-70010-6_14
16. Broadbent, J., Poon, W.L.: Self-regulated learning strategies; academic achievement in online higher education learning environments: a systematic review. Internet High. Educ. **27**, 1–13 (2015)

17. Panadero, E., Alonso-tapia, J.: How do our students self-regulate? Review of Zimmerman's cyclical model of learning strategies. Ann. Psychol. **30**(2), 450–462 (2014)
18. Panadero, E., Jonsson, A., Botella, J.: Effects of self-assessment on self-regulated learning and self-efficacy: Four meta-analyses. Educ. Res. Rev. **22**, 74–98 (2017)
19. Monereo, C.: Digital competence: for what, who, where and how should it be taught. Classr. Mag. Educ. Innov. **181**, 9–12 (2009)

Human Aspects in Product and Service Development

Gabriela Unger Unruh(✉), Ana Maria Kaiser Cardoso, Kássia Renata da Silva Zanão, Thiago Augusto Aniceski Cezar, Roberta Ferrari de Sá, and Osíris Canciglieri Junior

Polytechnic School, Industrial and Systems Engineering Graduate Program (PPGEPS), Pontifical Catholic University of Paraná (PUCPR), Curitiba, Paraná, Brazil
gabriela.unruh@pucpr.br

Abstract. To create socially sustainable solutions, it is necessary to understand the human being in all its aspects and relations: personal, social and environmental; and to do so, a transdisciplinary work is needed. The objective of this article is, through bibliographic research, to identify the human aspects that should be considered in product and service development, the areas of knowledge related to each aspect and how to apply those aspects in projects, proposing a conceptual framework which allows to view these relationships and applications. As conclusion, future studies are suggested that may assist in the creation of sustainable solutions for complex systems.

Keywords: Product development · User-centered design · Emotional design · Sensorial design · Consumption

1 Introduction

Society is in transformation, opening possibilities for studies to understand the forces have driven consumption, improving the products quality and respect for the impact on the environment. In this way, its needed to understand what needs drive consumption for this new audience. "Needs are produced by internal psychological and cognitive processes, leading to choices within a potential market". The objective of this research is to understand how sensory and emotional aspects can influence the consumption behavior of society in relation to the current scenario and how these behavioral tendencies motivate product acquisition. This allowed to structure a conceptual framework for a better understanding of human interactions and their influence on product and service development. This article was divided into eight sections: (i) the human aspects; (ii) their influence on the interactions with products and services; (iii) the context and society; (iv) the human sensors as devices for interaction with the external world; (v) the importance in considering the emotions in the design process; (vi) the impacts caused by such aspects presented previously; (vii) conceptual framework that illustrates the interactions of presented aspects and how they influence the development of products and services; and (viii) the conclusion which indicates analyzed study opportunities.

T. Ahram et al. (Eds.): IHSED 2019, AISC 1026, pp. 72–78, 2020.
https://doi.org/10.1007/978-3-030-27928-8_12

2 Human Aspects

The human being consists of two main aspects: the body, visible part (flesh and bones), through which the sensory interactions occur, and the invisible psyche (soul, mind and spirit) where interactions of language and meaning occur. These aspects are essential to be understood in the process of developing products and services, so that the solution developed is appropriate for the people who will interact with them.

For Ergonomics [1], body aspects include anthropometry (human body measures); biomechanics (muscles, movement, handling, strength, posture); organism (neuromuscular function, metabolism); and senses (sight, touch, hearing, smell and taste). Ponciano et al. [2] presented aspects of the psyche including the cognitive system (mental processes including understanding, learning, and memory); motivation (impulses to achieve something); preferences (personal preferences generated by experiences); social behavior; emotion (perception, mood, affection, feeling, and opinion); individual differences, variability and changes (differences between people in physical, knowledge, skills and abilities). These aspects allow human beings to interact, as connection/social relationship is an essential human characteristic [3]. Then this article focus on the contextual, sensorial and emotional aspects.

3 Context and Society

Speaking about consumption, people live in a society where "having" stands out to "being" [4]. In a moment where consumption is justified as the pursuit of happiness and status, brands and designers impregnate their products of emotions, sensations, messages and sensory characteristics that imbue soul products and all this appeal makes people search for new objects that will give them a momentary happiness.

The term "Consumer Society" emerges in the period after Industrial Revolution where mass production generated lower costs, which led to lower prices, making buying accessible to the population. This practice generates an increase in profit through consumption, but carries with it high environmental damages that are felt worldwide.

If, in the post-industrial revolution period, where products have become accessible to a growing number of consumers and where the ideal of life has been consuming simply to satisfy needs often imposed by society, today realizes that the concept begins to reverse and the search for quality begins to surpass the concept of quantity and groups of people have consumed less, but better (with more quality) and for more life quality. Lipovetsky [5] calls lightness what today can also be called minimalist movement, essentialism, frugality, less consumption, etc. For him, this concept is not associated with the idea of deprivation or lack, but with mobility, virtual and respect for the environment. According to this principle, which is related to consuming the minimum and in the best possible way, accumulation and waste must be smaller, but without any type of deprivation, it is discarding what is no longer useful and, at the environmental level, understand that it is necessary to generate more than destroy to live [6]. Within this new scenario of consumption, it is necessary that all those involved in goods production and services understand this consumer and think about how to develop new products or adjust existing ones in order to arouse the right emotions in people and at the same time create economically attractive and sustainable solutions.

4 Human Sensors and Interaction

All experiences and perceptions are initially acquired through the senses, so the senses are devices for interaction with the external world. Almost all human understanding of the world happens through the senses. Vision is the predominant and most significant sensory system for human beings, since it is clearly and immediately perceived, causing an enchantment through colors, shapes, and design.

Hearing allows that certain sounds to be easily recognized and associated, has important influence in the emotional aspect through music [7]. Touch is the sense responsible for human contact with the world, it enables the increase of the perception of anyone, because it is often the touch that allows to possess the world. Humans instinctively feels the need to touch, grasp, hold an object. Smell evokes the most emotional memories and has the ability to influence people's moods, this sense has more direct access to the brain [8]. Thus, the smell of stimuli, quickly rescues an emotional memory that has been dormant in the unconscious. Palate binds and completes chemically with the smell, because the palate is awakened with the integration of other senses mainly the smell, which understands and recognizes as taste, reaching emotions and memories. In this context, people do not just seek to acquire products, they look for experiences that arouse their emotions and sensations.

5 Emotion

Humans are highly influenced by emotion. Memories, relationships and decisions are infused with emotion. It is central to human lives, which includes not only family and friends, but also the products and services with which they interact. Hence the role of emotions (positive and negative) in the man-product interaction and the benefits of design that evoke positive emotions have been widely discussed in the design literature [9]. Mainly taking into account that people do not react only to physical and mechanical properties of products, but especially to their symbolic and emotional properties. Given this, it is increasingly necessary for companies that develop products and services to better understand what the real emotions of their target audience are in front of what is offered to them. To this end, the link between Psychology and Design allowed the development of methodologies serve as a basis for the certification that the emotions that one wanted to provoke could, in fact, be obtained through projects. The field is heavily based on direct research with people, so the only way to certify that the project would really achieve its success, with a focus on emotion, is to bring designers closer to people and, therefore, to the activity of search.

Emotional design is an approach to designing products with additional emotional value, in order to extract prescribed emotions from the consumer. The products can evoke a diversity of positive emotions, for example pride, contentment, admiration, desire, relief or hope [10, 11]. Although they are all positive, these emotions are essentially different - both in terms of the conditions that cause them and in terms of their effects on human-product interaction. For example, while fascination encourages focused interaction, joy encourages a playful interaction. Desmet [12] has shown that there are at least 25 positive emotions that can be experienced in response to products. Although these positive emotions are all pleasurable, each one is different from the other in terms of feelings and how they influence people's thoughts and actions [13].

The methodology created by Jordan [14], on emotional design, proposes that products can bring four types of pleasures to users: (1) physiological (body and senses); (2) social (social relations); (3) psychological (mind); (4) ideological (people's values). An approach that focuses on translating consumers' feelings and emotions into concrete design parameters is known as Kansei Engineering [15], which defined how a product development methodology can translates feelings, impressions and emotions of customers and users into concrete design parameters [16]. It also focused his work on the way people deal with and use the information and influence of this process on the emotions, identifying three levels of processing: (1) visceral (direct perception); (2) behavioral (learned automatic responses); (3) reflective (conscious thought). These emotions are generated through cognition and the five senses: vision, hearing, taste, smell, and touch [17]. This is why it is important to understand how human sensory perception works, because each person's brain interprets the stimuli differently and takes into account three main aspects: their memory of past experiences, current state of attention and their genetic structure. All the technology, products, services and systems designed evoke emotions, and not considering these emotions in the design process is a missed opportunity at best.

6 Impact of Contextual, Sensorial and Emotional Aspects in Relation to Brand, Products and Services

All these contextual, sensorial and emotional aspects turn in experiences, which interferes in the relationship that people have with brands, that is, the expressions and representations of organizations, which include the organization itself, its products, image services and representation. Delivering unique and enjoyable brand experiences is essential for developing business loyalty, engaging consumers in a deep and lasting emotional connection with the brand. Thus, the need to establish an emotional bond with the consumer stimulated the emergence of sensory branding through the use of fragrances, sounds and textures. The concept of sensorial branding is defined as a system of communication about the brand, starting from the five sensorial perceptions of the human being [18]. As Pawaskar and Goel [19] refer, the relationship with a brand is stimulated by sensory branding, and allows the emotional factor to dominate rational thinking. By stimulating the five senses, brands that offer happiness, joy and affection lead to greater buying and loyalty attitudinal. The senses make bridges with emotions and memories, and for achieving this emotional connection requires multisensory brand experiences. The brand in its multisensory sense, always needs to keep the consumer in touch, put the individual in the center, show that the experience in which he was involved invites him always to enjoy a moment of pleasure, an immersion with the brand, to create a habit, and not just a passing experience.

7 Concept

Interdisciplinarity in social and environmental research has gained prominence [20], and is a fundamental aspect in the development of projects focused on the needs and well-being of people. People are complex and the product development process too, requiring interactions of various areas and activities of the company, generating a large amount of information [21]. In the aspects raised to date, several fields of knowledge can combine efforts to better understand human behaviors, therefore, a conceptual framework was created (Fig. 1) to illustrate the interactions of the presented aspects, which influence the development of products and services.

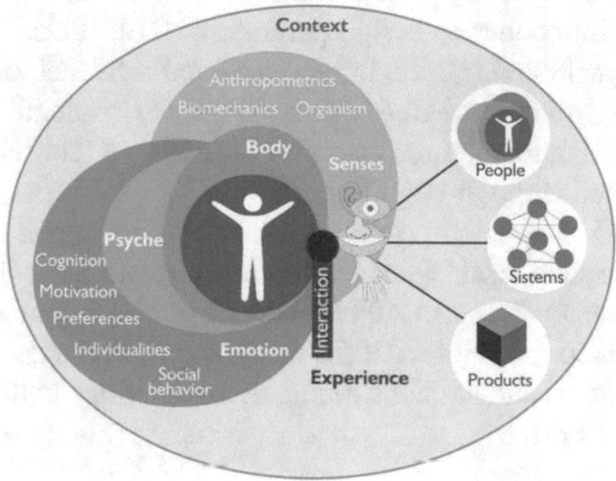

Fig. 1. Human aspects framework.

The first aspect addressed is the context, in which all human beings, interactions, relationships and products/services are inserted, because it is about social systems, environmental, historical, political and temporal aspects, so the context is what most influence human interactions. The context are the people (psyche and body), being the senses of the body where the interactions occur, information gathering and physical connection of people, which are processed in the psyche, where the perceptions, interpretations, emotions, experiences and reactions to interactions occur, which can also be manifested in the body after this processing. So the initial interaction between people, products/services and systems occurs through the senses, the central point of the framework. Already the interpretation and emotions from this interaction occur in the psyche and the manifestation of this occurs through psychological and physical reactions, related to the experience coming from this interaction. All of these aspects in an interdisciplinary way, should be considered in product and service projects, so that they fit people's needs and well-being.

8 Conclusion

This article identified human aspects and their influence on the interactions with products and services, presenting a conceptual framework, to visualize the aspects and where they are located in this interaction, generating a broad view of the human aspect and guidelines of these in product development. From this analysis, there are some study opportunities for product/service development. The first one is to understand consumer behavior in more depth, seeking to understand future motivations, contexts and interaction scenarios. Another opportunity is to study the application of Sensorial Design in product development to generate unique experiences, contributing to a differentiated positioning in the market. And the last opportunity was to identify ways to interpret people's emotions to aid in designing products that evoke positive emotions by checking the benefits of it. All of these opportunities should include interdisciplinary studies, especially engineering, design and psychology.

References

1. Iida, I.: Ergonomia: Projeto e Produção, 2nd edn. Edgar Blücher, São Paulo (2005)
2. Ponciano, L., Brasileiro, F., Andrade, N., Sampaio, L.: Considering human aspects on strategies for designing and managing distributed human computation. J. Internet Serv. Appl. **5**(10), (2014)
3. Liberman, M.D.: Social Why Our Brains are Wired to Connect. Crown Publishers, New York (2013)
4. Do Lago, F.G., Dos Reis, J.O.: Sociedade de consumidores na visão de Bauman e Drummond: interdiscursividade nas obras dos autores. Cadernos Zygmunt Bauman, vol. 6, no. 12 (2016)
5. Lipovetsky, G.: Da Leveza: Ruma a uma civilização sem peso. Amarilys, São Paulo (2016)
6. Carvalhal, A.: Viva o fim: almanaque de um novo mundo. Paralela, São Paulo (2018)
7. Lindstrom, M.B.: Brandsense: segredos sensoriais por trás das coisas que compramos. Campus, Porto Alegre (2012)
8. Braga, M.: Influência da música ambiente no comportamento do consumidor, vol. 06, pp. 05–12 (2012)
9. Crilly, N., Moultrie, J., Clarkson, P.J.: Seeing things: consumer response to the visual domain in product design. Des. Stud. **25**, 547–577 (2004)
10. Desmet, P.M.A.: Designing Emotions. Delft University of Technology, Netherlands (2002)
11. Desmet, P.M.A., Schifferstein, H.N.J.: Sources of positive and negative emotions in food experience. Appetite **50**, 290–301 (2008)
12. Desmet, P.M.A.: Faces of product pleasure: 25 positive emotions in human-product interactions. Int. J. Des. **6**(2), 1–29 (2012)
13. Frijda, N.H.: The Laws of Emotion. Lawrence Erlbaum Associates Publishers, London (2007)
14. Jordan, P.W.: Pleasure with products: human factors for body, mind and soul. In: Green, W.-S., Jordan, P.-W. (eds.) Human Factors in Product Design: Current Practice and Future Trends. Taylor & Francis, London (1999)
15. Schütte, S.: Towards a common approach in kansei engineering: a proposed model. In: Proceedings of the Conference. Interfejs Użytkownika - Kansei W Praktyce, Warszawa, pp. 8–17 (2007)

16. Sozo, V., Ogliari, A.: Stimulating design team creativity based on emotional values: a study on idea generation in the early stages of new product development processes. Int. J. Ind. Ergon. **70**, 38–50 (2019)
17. Lai, Y.C.: Emotion eliciting in affective design. In: International Conference on Engineering and Product Design Education. University of Twente, Netherlands (2014)
18. Drobysheva, E.A.: Sensory branding institutionalization. Life Sci. J. **11**(6), 522–524 (2014)
19. Pawaskar, P., Goel, D.M.: A conceptual model: multisensory marketing and destination branding. Proc. Econ. Financ. **11**, 255–267 (2014)
20. Rodela, R., Alasevic, D.: Crossing disciplinary boundaries in environmental research: interdisciplinary engagement across the Slovene research community. Sci. Total Environ. **574**, 1492–1501 (2017)
21. Pereira, J.A.: Modelo de desenvolvimento integrado de produto orientado para projeto de P&D do setor elétrico brasileiro. Ph.D. Thesis, Polytechnic School - Industrial and Systems Engineering Graduate Program (PPGEPS), PUCPR, Curitiba (2014)

Blueprint for a Priming Study to Identify Customer Needs in Social Media Reviews

Kristof Briele(✉), Alexander Krause, Max Ellerich, and Robert H. Schmitt

Laboratory for Machine Tools and Production Engineering, WZL of RWTH Aachen University, Campus-Boulevard 30, 52074 Aachen, Germany
{K.Briele,A.Krause,M.Ellerich, R.Schmitt}@wzl.rwth-aachen.de

Abstract. Unbiased customer reviews in social networks may hold the key for innovations in the saturated market of consumer goods. Customer reviews do not only offer information directly about the product, they also provide insights into the user's environment, customer habits and usage behaviour. These latent needs are stated objectively in reviews. This study aims to overcome the weakness of state-of-the-art machine-learning algorithms that can only extract explicitly stated needs. Key part of the study is the developed method to record and evaluate the reaction time of test subjects to analyse the association between a latent need category and a related word. As a result, we obtain word clusters that express an association with a latent need and a blueprint for upcoming studies that focus on the extraction and utilizing these needs. This knowledge can be used for further research in an automated need identification process and customer driven production.

Keywords: Priming study · Need identification · Innovation engineering

1 Introduction

With the increasing amount of unbiased data in social networks and product reviews, the companies' need in an automated evaluation of the data is growing. In order to design new and successful products, it is becoming important to consider the wishes and needs expressed by customers [1]. However, not only the expressed needs, but also the unconscious, latent needs are becoming especially interesting, as they provide the foundation for promising long-term innovations [2]. Since traditional survey methods might distort the result (e.g. online surveys with predefined questions and answers), the innovation performance can be improved by an enhancement of automated, computational analysis of written feedback [3–6].

The customer is free to choose his own words in reviews, thereby not only communicating his emotions, but also expressing his living conditions. One way of extracting unconscious needs from reviews is to analyse objective sentences that do not include emotions [7]. Work today focuses mainly on the used words in sentences [8] or on the customer's sentiments [9]. This research focuses on forming word groups from reviews, which can be used as a basis for a lexical analysis to extract latent needs. Key

T. Ahram et al. (Eds.): IHSED 2019, AISC 1026, pp. 79–84, 2020.
https://doi.org/10.1007/978-3-030-27928-8_13

part of this paper is the development of a study to assess the association between words and latent need categories.

The paper is structured into the four following parts: In Sect. 2, current methods to extract customer needs from written feedbacks are described. The model for the empirical study is explained in Sect. 3; the results are described in Sect. 4. The paper concludes with a discussion and further research.

2 Current Work

The freely accessible amount of data is available in unstructured form. In order to be able to use the data and generate information, scientists and companies try to extract customer needs using computers instead of traditional methods.

For this reason, Büschken and Allenby [8] present an algorithm to analyze reviews using the sentence structure. The work focuses on "understanding" the words used within the unstructured data. They are oriented towards the class of "topic" models. For each word within a sentence, a topic is searched. These topics refer to attributes of a product. Since they look at the probabilities of occurrence of used words and their topics, they can summarize all topics into a superordinate category. This allows them to show that entire reviews usually deal with only one parent category. Their results show that a topic approach provides better results in classifying and structuring sentences than bag-of-word approaches.

A way to extract opinions from written feedback is presented by Giatsoglou et al. [9]. Their approach consists of a sentiment analysis that includes a hybrid approach of lexica and word embedding model. The approach is realized with the help of a supervised training classifier. By combining both approaches, they achieve the same classification results as lexical and word embedding models, but they are significantly faster than the individual models.

Zhou et al. [4] present a two-layer model, which enables extraction of latent needs. The first layer emphasizes sentiment analysis with the goal of identifying explicit customer needs based on product attributes and common use cases from online product reviews, while the second layer focuses on implicit customer needs by explaining the semantic similarities and differences between different use cases. The model can only be used in extraordinary use-cases, since it translates explicitly expressed customer needs from ordinary use-cases into "latent" needs in extraordinary applications through semantic analysis.

Kang et al. [10] introduce a new sentiment analysis based on text-based hidden Markov model. They do not use a predefined lexicon but co-occurrence of words. They show in their work that they can achieve better results if no predefined sentiment words are present within a sentence.

The literature review shows that researchers and companies are making great efforts to extract customer needs from product reviews, which are quite successful for explicitly stated needs. Yet, their approaches lack evaluating the customers' underlying latent needs in a general manner and not already analyzing them concerning the product descriptions – it might bias and limit the findings.

3 Method

The approach of this paper is based on Reichard's nine latent need categories: security, belonging, respect, validity, self-realization, possession, efficiency, comfort and variety [11]. The idea is to evaluate to which degree the review addresses these categories by using a lexicon. If a certain amount of reviews is clustered, they can be further analyzed among each group. This approach might lead to clearer findings because it deals with structuring the highly unstructured data.

The lexicon itself consists of words with or with a product context and an association index for each category. To create the lexicon, two steps are necessary: Identifying the words and evaluating the strength of association.

To create groups of words, a brain writing method is used. It has the advantage that priming effects are reduced and ideas are not influenced by conversations with others. The participants have 10 min to write down as many words as possible to one of the latent categories. All words are collected and ordered for all nine categories. Subsequently, the existing words are extended with synonyms, opposite words and ambiguous words that describe both the need and a completely different issue, i.e. for the latent need 'Security' words like 'safety' (synonym), 'fear' (opposite) and 'lock' (ambiguous) are added (the second meaning of the word 'lock' is 'castle' in German). At the end, random words are added to each generic term that do not fit the category at all i.e. 'tree' for the same category. The result are word groups for each of the latent needs, consisting of 24–38 matching words and 14–26 not matching words. This is resulting in word clusters of 46–57 words for one latent need, i.e. a total of 9 latent needs and 468 targets.

A second study evaluates strength of association between the words and the latent need categories (Fig. 1). With the help of this study, the lexicon will be designed to extract latent needs in objective sentences.

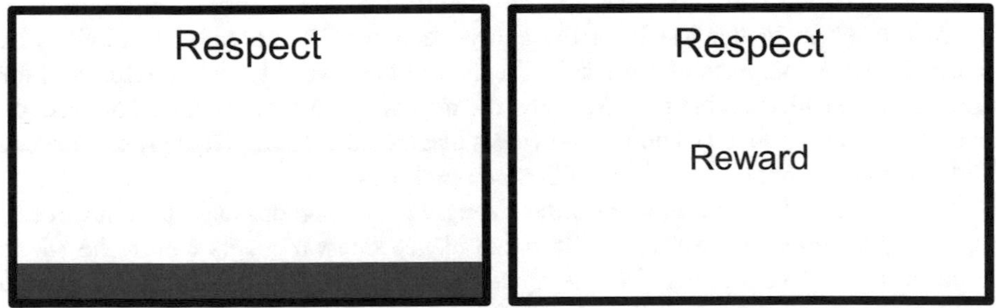

Fig. 1. Screenshots from the study: need category appears for 3 s, pressing is surpressed (left); subterm appears, reaction time is measured (right)

The study is divided into three blocks and each of these blocks consists of three latent needs. At the beginning of a block, the generic term is displayed in the middle of a white screen in black letters for 3 s, which is not annotated [12]. In these 3 s, a Bag-of-Word with associated words is formed in the mind.

After the 3 s, the subterm appears. The target word is also visible for 3 s. During this time, a click of the space bar is possible to signal an association. Due to the length of 3 s, effects are minimized that can be triggered by short reaction times (<1.5 s) described as affective priming [13, 14] but it also allows thinking and thus initiating a reaction that was not directly available in the brain [15]. The next target appears within the category of the generic term. Since response times to a stimulus are about 750 ms and wrong entries when changing the target should be eliminated, random pressing of the space bar within the first 500 ms are suppressed. The category changes after testing four target words.

The participant goes through all the words for each of the three categories of the first block. After a five minute break, the study is repeated for the other latent need categories.

The order of the categories and the stored targets are randomized to counteract any fatigue effects observed by Ahrens [16]. Since Much [12] showed that the word length had no influence on the reaction time, no further restriction is made. The total processing time of the study is between 30 to 40 min and 40 test subject take part.

The resulting table of the reaction times is presented in Table 1.

Table 1. Schematic representation of the results (translated into English)

Respect	Reaction time [s]	Belonging	Reaction time [s]	Security	Reaction time [s]
To reward	3,0108053	Partner	0,9672017	Top	3,0108053
To struggle	3,0108053	Marmalade	3,0108053	Future	0,7176012
Readiness	1,5912028	Union	1,9344034	Trust	1,289202
Circle	3,0108053	Bond	0,7800014	Insurance	1,0300016

4 Results

In order to evaluate the results of the study, each word is considered individually. Figure 2 shows examples of the results for the different word types. As expected, the words that are added during the first study, the matching words, have a median reaction time of 1.1–1.6 s (top left). Unmatched words that are added later in the process exceed the maximum reaction time of 3 s with some outliners.

Interestingly, the reaction times differ strongly for ambiguous and opposing words. Especially for opposing words, the 95% confidence interval covers almost the whole range of 1–3 s. Even though there is a strong association between the word and the latent need, the reaction time is high. The customer might use exactly these opposing words to write about a missing feature. When building a lexicon, this result has to be included: Not only the reaction time itself is an indicator of the association but also its polarity (given that the need categories are positive).

The second abnormality is the high reaction for ambiguous words. The median is in the same range as the one for matching words, but the range is about twice as big. Again, using only these reaction times to estimate the association, might distort the results. Words have multiple meanings; therefore, the lexicon should be – ideally – product and context-dependent.

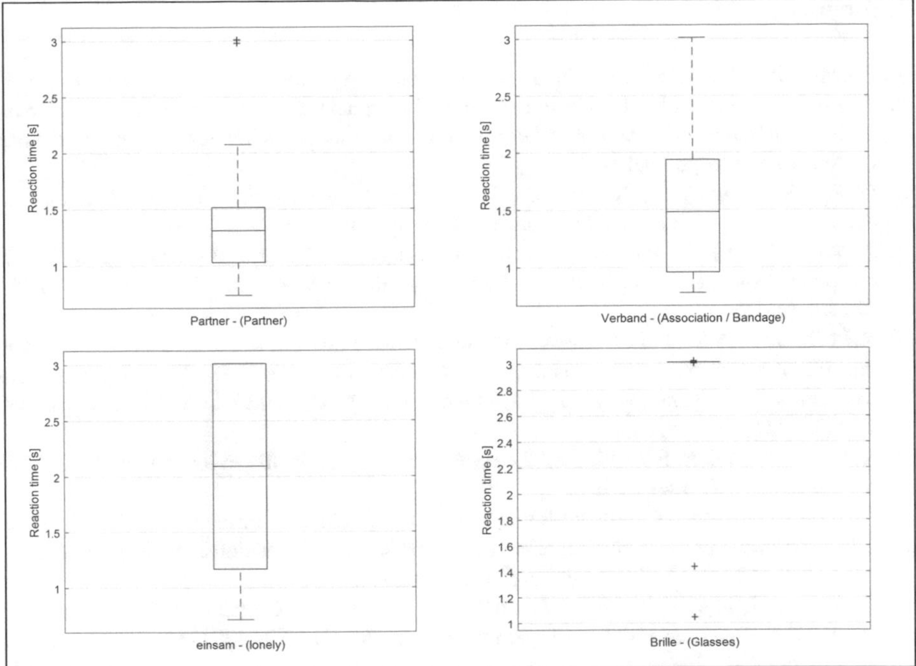

Fig. 2. Examplary representation of the different word categories to the latent need 'Belonging': synonymous word (top left), ambiguous word (top right), opposing word (bottom left), unmatched word (botting right)

5 Summary and Outlook

This study shows that latent needs can be addressed by many different words, which means that they are not addressed directly in texts. Synonymous terms have the fastest response time and thus the largest association, which confirms what has already been shown in the literature. The different reaction times of ambiguous and opposing words seem to indicate that there may be context-related dependencies. Apparently, priming effects have nevertheless occurred due to the randomization of the sequence, which has caused the different reaction times. Accordingly, when conducting such a study, one has to be careful what context can subconsciously be created with words. Conversely, this also means that the lexical basis cannot be generalised but must be adapted product-specifically in order to derive desired needs.

Acknowledgement. This paper results from the research project "Automated extraction of customer needs from reviews for the enhancement of innovation capability" (SCHM1856/82-1) of the Laboratory for Machine Tools and Product Engineering (WZL), RWTH Aachen University, Germany. The research project has been funded by the German National Science Foundation (DFG). The authors would like to express their gratitude to all parties involved.

References

1. Balazs, J.A., Velasquez, J.D.: Opinion mining and information fusion: a survey (2015)
2. Zogaj, S., Bretschneider, U.: Customer integration in new product development – a literatur review concerning the appropriateness of different customer integration methods to attain customer knowledge (2012)
3. Goffin, K., Lemke, F., Koners, U.: Identifying Hidden Needs: Creating Breakthrough Products, vol. 1. Palgrave Macmillan, Basingstoke (2010)
4. Zhou, F., Linsey, J., Jiao, R.J.: Latent customer needs elicitation by use case analogical reasoning from sentiment analysis of online product reviews. J. Mech. Design **137**(7), 071401 (2015)
5. Roberts, D.L., Piller, F.T., Lüttgens, D.: Mapping the impact of social media for innovation: the role of social media in explaining innovation performance in the PDMA comparative performance assessment study. J. Prod. Innov. Manag. **33**(3), 117–135 (2016). https://doi.org/10.1111/jpim.12341
6. Roberts, D.L., Piller, F.T.: Finding the right role for social media in innovation. MIT Sloan Manage. Rev. **57**(3), 41 (2016)
7. Aggarwal, C.C., Zhai, C. (eds.): Mining Text Data, 2012th edn. Springer, US, Boston (2012)
8. Büschken, J., Allenby, G.M.: Sentence-based text analysis for customer reviews. Mark. Sci. **35**(6), 953–975 (2016). https://doi.org/10.1287/mksc.2016.0993
9. Giatsoglou, M., Vozalis, M.G., Diamantaras, K., et al.: Sentiment analysis leveraging emotions and word embeddings. Expert Syst. Appl. **69**, 214–224 (2017). https://doi.org/10.1016/j.eswa.2016.10.043
10. Kang, M., Ahn, J., Lee, K.: Opinion mining using ensemble text hidden Markov models for text classification. Expert Syst. Appl. **94**, 218–227 (2018). https://doi.org/10.1016/j.eswa.2017.07.019
11. Reichardt, T.: Bedürfnisorientierte Marktstrukturanalyse für technische Innovationen: Eine empirische Untersuchung am Beispiel Mobile Commerce. Gabler Edition Wissenschaft. Betriebswirtschaftlicher Verlag Dr. Th. Gabler/GWV Fachverlage GmbH Wiesbaden, Wiesbaden (2008)
12. Musch, J., Elze, A., Klauer, K.C.: Gibt es Wortlängeneffekte in der evaluativen Entscheidungsaufgabe. Zeitschrift für Exp. Psychol. **45**, 109–119 (1998)
13. Eder, A.B., Erle, T.M.: Priming. In: Enzyklopädie der Psychologie (Bereich Sozialpsychologie) (2017)
14. Wentura, D., Degner, J.A.: Practical guide to sequential priming and related tasks. In: Gawronski, B., Payne, B.K. (eds.) Handbook of Implicit Social Cognition: Measurement, Theory, and Application. Guilford Press, New York (2010)
15. Karnath, H.-O., Ackermann, H. (eds.): Kognitive Neurowissenschaften: Mit … 28 Tabellen, 3., aktualisierte und erw. Aufl. Springer, Berlin (2012)
16. Ahrens, V.: Priming und Ermüdung (2014)

Risk Avoidance Through Reliable Attention Management at Control Room Workstations

Rico Ganßauge(✉), Annette Hoppe, Anna-Sophia Henke, and Norman Reßut

Chair of Ergonomics and Industrial Psychology, Brandenburg University of Technology Cottbus-Senftenberg, Siemens-Halske-Ring 14, 03046 Cottbus, Germany
{Rico.Ganssauge,Hoppe,Anna-Sophia.Henke, Norman.Ressut}@b-tu.de

Abstract. Nowadays the occupation in control rooms is becoming more and more important [1]. Working there requires highly responsible action. Control rooms often contain many screens [2], so important signals are potentially located out of the direct field of view. Detecting signals there is more difficult. The goal of the study is to identify signals' properties, which guide the operators' attention reliably. This is accomplished by capturing the attention within the direct field of view. Simultaneously stimuli are placed in the peripheral field of view on a background with visual ambient noise. Detection rate and -time are recorded. First results of the ongoing study are presented here. Later work-design indications are to be derived and put into practice, to foster a human-oriented design, resulting in a more secure work in control rooms. The German Research Foundation (DFG) funds the project (No. 358406233).

Keywords: Control room · Peripheral filed of view · Signal detection · Human-oriented design

1 Introduction

The occupation in control rooms plays an ever-increasing role for modern industries [1], especially in the light of a far-reaching change towards a more digitalized industrial production, often called by the catchphrase "Industry 4.0". Despite classical fields of application, as power stations and chemical plants, this work environment becomes more and more common in new fields, like healthcare [3]. Working there comes with many responsibilities as any mistake could possibly have far-reaching consequences. A significant amount of the work there is surveillance of the process, which usually is done with a large amount of screens [2]. This can have the implication that relevant information might be outside of the operators' direct field of view. The direct field, when having the head fixed and only moving the eyes, spans about 15° around the visual axis [4]. Relevant signals in there will be recognized properly most of the time. Outside of this area visual perception exists for about 90° horizontally each side and vertically about 45° to 70° up and down. This area is called the peripheral field of view. Most properties of the human perception decrease rapidly in there, e.g. the visual acuity

T. Ahram et al. (Eds.): IHSED 2019, AISC 1026, pp. 85–89, 2020.
https://doi.org/10.1007/978-3-030-27928-8_14

45° away from the line of sight is only 5% as accurate [5]. Colors are also mostly non-recognizable there [6], only strong brightness contrasts will be perceived [4]. Effects of lateral masking the signals by other visual information close by might also appear [7]. On the other hand, the perception of movements is quite good, which seems to be for evolutionary reasons (recognition of potential dangers) [8]. Another possibility for guiding attention might be the “Pop-Out-Effect” [9]: objects, which clearly differentiate from the background, can draw attention. The visual design on the control screens has a clear purpose: In case it is done improperly, the operators’ attention might not be guided appropriately. So necessary actions might not be taken and technological safety measures have to kick in [10]. This is not desirable; as it often involves a loss of production and potential dangers for men and environment. The knowledge described clearly shows the necessity for research in this area. The studies design is based on the possibilities for guiding attention described above.

2 Goal and Hypotheses

The aim of the study is to determine which signals in the peripheral field of view can draw the operators’ attention in the most reliable way while not disturbing him in any way. According to the theoretical background, a design was set up containing signals in different levels of brightness contrast and different blinking frequencies. Hypotheses derived are:

H1: High frequencies and brightness contrast will be recognized better than low frequencies and contrasts.

H2: The recognition will decrease with a higher angle in the peripheral field of view.

3 Research Methods

To reach the goal an experimental workplace in the laboratory was designed and implemented. The screen in the middle represents the direct field of view; two other screens on each side cover the peripheral field of view (Fig. 1).

Fig. 1. Experimental workplace at the laboratory (source: own creation)

To lock the attention in the direct field of view, the participants fulfill a problem-solving task there. For this purpose, the "Tower-of-London"-task from the PEBL-battery [11] was chosen. It is a typical task for problem solving in a psychological sense, containing complex thinking processes like anticipation, reasoning, planning and executing [12]. To create a constant challenge for the participant the level of difficulty alternates appropriately. This task stands as a proxy for solving a problem in a real control room and having all the operators' attention concentrated in just one part of the many screens used. On the screens for the peripheral field of view, we created a background-resembling flow diagram, which is often used in the chemical industry [13]. The graphical complexity is balanced by an equal distribution of equivalent elements over the screen. In reality, the control system creates a visual background noise. This is simulated by continuously changing the scales and numbers displayed. About one third of all scales will change at a given time in a continuous and logical manner. The signals to be detected appear as a bordering frame of a four-digit number that is as big as the recommended minimum size for proper visual recognition at the chosen distance of the screen [14]. Figure 2 shows the background and the signal as the black frame of 7.5 mm inner height and 16 mm inner width (for demonstration purposes indicated with an arrow).

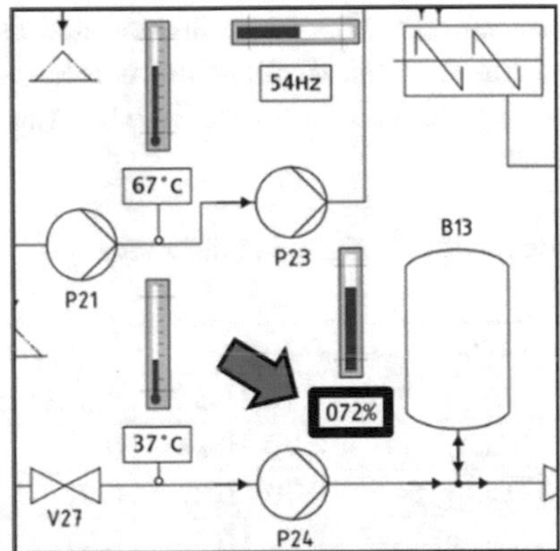

Fig. 2. The graphically complex background with a signal as a frame of a recommended minimum size number. (source: own creation)

The signals appear in four different frequencies of 0 Hz, 0.5 Hz, 2.5 Hz and 5 Hz. Four levels of contrast were chosen. The colors used range from light grey to black with two steps in between, chosen according to clearly distinguishable impressions and brightness contrasts with the surrounding white background. The used hexadecimal codes are #e1e1e1; #969696; #4b4b4b and #000000. This creates 16 combinations, which are shown on four different positions in the peripheral field of view: close to the direct field of view at 15°, at 40° and 65° and close to the border of the peripheral view at 90°. The signals will appear in a random manner and should be acknowledged by clicking a button on the central screen. Avoiding random gazing into the designated peripheral field of view is controlled by properly instructing the participant, a headrest to place the chin and an eye-tracking device. To maximize the primary variance caused by the experimental treatment, other influencing factors are carefully controlled. So for instance, the experiments take place in our laboratory, where climate and lighting conditions are on a constant level and loud noises are cancelled out, as relevant industrial standards for control rooms indicate [15].

4 Preliminary Results

For the first N = 12 participants we calculated the discrimination index A' according to Eq. (1) [16], which is recommended as a nonparametric alternative to the parametric d' index [17]. Values closer to 1 suggest a very good discrimination, while .5 or below indicates that a signal cannot be distinguished from noise. The equation includes calculating the ratio of correct recognitions of the signals, called "hits" (H) and compares it with the ratio of false alarms (F).

$$A' = \begin{cases} .5 + \frac{(H-F)(1+H-F)}{4H(1-F)} \; when \; H \geq F \\ .5 - \frac{(F-H)(1+F-H)}{4F(1-H)} \; when \; H < F \end{cases} \quad (1)$$

One participant obviously misunderstood the instructions somewhat and produced a very high number of false alarms. This case was counted as a missing value. It was replaced via single imputation with the mean of the whole dataset [18], which ranged from 0 to 11 false alarms (F), produced on 68 possible occasions. The calculated values of A' are shown in Table 1.

Table 1. Discrimination index A' for the signals in different angles, colors and frequencies used in the experiment

Angle	Color	Frequency			
		0 Hz	0,5 Hz	2,5 Hz	5 Hz
15°	Black	0.70	0.82	0.87	0.87
	Dark grey	0.70	0.87	0.87	0.87
	Medium grey	0.64	0.82	0.87	0.87
	Light grey	0.17	0.76	0.87	0.82
40°	Black	0.64	0.87	0.87	0.87
	Dark grey	0.70	0.82	0.82	0.87
	Medium grey	0.64	0.87	0.87	0.87
	Light grey	0.17	0.70	0.87	0.87
65°	Black	0.57	0.87	0.82	0.87
	Dark grey	0.41	0.87	0.87	0.87
	Medium grey	0.57	0.82	0.87	0.87
	Light grey	0.12	0.34	0.64	0.76
90°	Black	0.34	0.70	0.76	0.82
	Dark grey	0.34	0.70	0.87	0.82
	Medium grey	0.28	0.49	0.57	0.70
	Light grey	0.12	0.17	0.23	0.12

5 Discussion

The first results of this work in progress are still on a descriptive level, with no inferential statistical methods used yet. A sample size of N = 50 is estimated to give clear results and will be reached within the ongoing year. It seems likely that the results will support our hypotheses, as a trend towards better recognizable higher frequencies and brightness contrasts can be seen in the preliminary results. In the same manner, the recognition seems to get worse with higher angles in the peripheral field of view. Later analyses will focus on statistically validated differences to derive recommendations for a state-of-the-art visualization of signals. This will help to foster a more human-centered design in the safety-critical work environment of control rooms.

References

1. Andelfinger, V., Hänisch, T.: Industrie 4.0 – Wie Cyber-Physische Systeme die Arbeitswelt verändern. Springer, Wiesbaden (2017)
2. Kockrow, R.: Eye-Tracking Studien in Leitwarten – Evaluation einer 'Visuellen Komfortzone' für Operatortätigkeiten. Dissertationsschrift. Shaker, Aachen (2014)
3. Meinecke, S., Albat, D.: Gebündelt und transparent. Prozessorientierte Krankenhaus-Leitwarte erleichtert effiziente Ressourcenplanung. KU Gesundheitsmanagement, **5**(84), 56 – 58 (2015)
4. Schmauder, M., Spanner-Ulmer, B.: Ergonomie - Grundlagen zur Interaktion von Mensch, Technik und Organisation. Hanser, München (2014)
5. Schlick, C., Bruder, R., Luczak, H.: Arbeitswissenschaft. Springer, Berlin (2018)
6. Ditzinger, T.: Illusionen des Sehens. Springer, Heidelberg (2013)
7. Huckauf, A.: Zur Bedingungsanalyse lateraler Maskierung. Shaker, Aachen (1999)
8. Birbaumer, N., Schmidt, R.: Biologische Psychologie. Springer, Heidelberg (2010)
9. Vollrath, M.: Ingenieurpsychologie: Psychologische Grundlagen und Anwendungsgebiete. Kohlhammer, Stuttgart (2013)
10. IEC 61511-1:2016: Functional safety – Safety instrumented systems for the process industry sector - Part 1: Framework, definitions, system, hardware and application programming requirements
11. Mueller, S., Piper, B.: The psychology experiment building language (PEBL) and PEBL test battery. J. Neurosci. Methods **222**, 250–259 (2014)
12. Ward, G., Allport, A.: Planning and problem solving using the five disc tower of London task. Q. J. Exp. Psychol. A **50A**(1), 49–78 (1997)
13. ISO 10628-2:2012: Diagrams for the chemical and petrochemical industry - Part 2: Graphical symbols
14. DIN 1450:2013: Lettering: Legibility
15. ISO 11064-6:2005: Ergonomic design of control centres - Part 6: Environmental requirements for control centres
16. Stanislaw, H., Todorov, N.: Calculation of signal detection theory measures. Behav. Res. Methods Instrum Comput. **31**(1), 137–149 (1999)
17. Snodgrass, J., Corwin, J.: Pragmatics of measuring recognition memory: applications to dementia and amnesia. J. Exp. Psychol. Gen. **117**(1), 34–50 (1988)
18. Döring, N., Bortz, J.: Forschungsmethoden und Evaluation, p. 591. Springer, Berlin (2016)

Study on the Effect of Electronic Map Color Scheme on Operation Performance

Bei Zhange(✉), Tuoyang Zhou, and Yingwei Zhou

China Institute of Marine Technology and Economy, Xueyuan South Road, No. 70, Haidian District, Beijing 100081, China
zhangbeibj@126.com

Abstract. Electronic map is widely used in our daily life. The color matching of electronic map affects the recognition speed of users. In order to improve the user's recognition and search speed of the target, based on the actual needs of the user, this study proposes four color schemes for the existing electronic map, and carries out experimental research in the indoor light source environment. In the experiment, task performance index and subjective evaluation method are adopted to obtain the optimal scheme 3, followed by scheme 4. The conclusion of this study can be used in electronic map design.

Keywords: Electronic map · The color · Objective search

1 Introduction

At present, electronic maps have become the main form of people reading maps. Electronic maps are very different from paper maps in display [1]. Experimental studies have found that in order to improve the visual effect of the icon, the foreground and background colors of the screen should ensure a certain difference in brightness contrast and saturation [2]. Color visual performance depends on the specific match of the foreground and background colors as well as the brightness level. The effects of hue factors and brightness contrast factors on visual performance exist simultaneously [3].

Starting from the actual needs of electronic maps, and based on the existing electronic maps, by adjusting the brightness contrast and saturation of the background colors, four color schemes are proposed to investigate the differences in the operational performance of different electronic map color schemes under the environmental conditions of fluorescent lamps. This experiment provides an objective data basis for the optimization of electronic map color matching.

2 Experimental Design

2.1 Experimental Object

According to the current electronic map color matching, four kinds of color matching schemes are developed for comparative analysis. The RGB values of the four schemes are shown in Table 1 below (Table 2).

T. Ahram et al. (Eds.): IHSED 2019, AISC 1026, pp. 90–95, 2020.
https://doi.org/10.1007/978-3-030-27928-8_15

Table 1. Four color schemes in the experiment

Scheme	RGB value1	RGB value2	RGB value3
1	254.223.119	163.200.239	189.250.254
2	230.193.125	146.186.229	169.195.210
3	43.40.9	2.15.32	0.0.0
4	190.190.190	102.102.102	122.122.122

Table 2. Four color schemes in the experiment

1	2	3	4
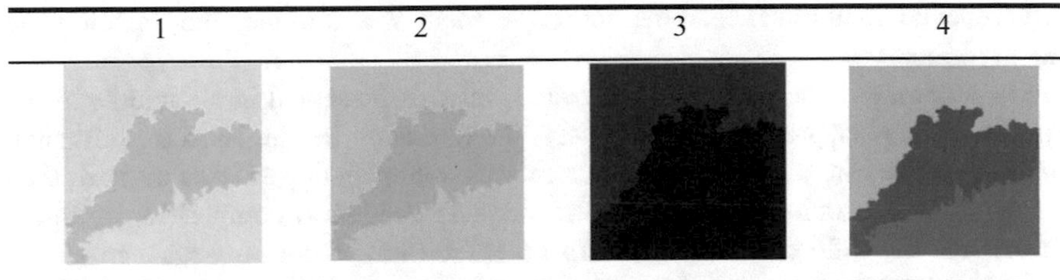			

2.2 Experimental Index and Participants

The experimental indicators are reaction time, task completion rate, and dual comparison score. The subjects were 26 male participants, all of whom had no color blindness or color weakness.

2.3 Experimental Materials and Tasks

The experimental materials are 4 color schemes. The target colors: red, blue, green, and yellow. See Fig. 1. The target consists of two parts: graphic and code. The graphic size is 30 px * 30 px, a total of 8 directions, Fig. 2, the code is 0001001-0001020.

The task of the participant is to search for the target of the 0001001-0001010 code in the interface as soon as possible. If a new target appears and the code is in this interval, the participant needs to click the target immediately.

Fig. 1. Four target color diagram

Fig. 2. Schematic diagram of eight target directions

2.4 Experiment Process

After the subjects entered the laboratory, they first adjusted the sitting position. Then, under the guidance of the experimenter, they conducted the experiment according to the experimental requirements. After the experiment was familiar, they entered the formal experiment.

In the experiment, the instruction language was first presented in the middle of the screen. After the user pressed the Enter key, the operation began. Each experimental task was completed in 40 s, and the next task was automatically entered in more than 40 s. For each electronic map scheme, the subjects needed to complete 8 judgment responses, and the different electronic map schemes were random in sequence.

After the experiment, enter the dual comparison task, the participants needed to compare the four color schemes under the three target quantity distribution maps of 32, 64, and 128 respectively.

3 Data Analysis

3.1 Performance Analysis of Clicking the New Target

(1) Correct Rate

Participants completed within 3 s after the new target appears, and the click is considered correct. The result is shown in Fig. 3.

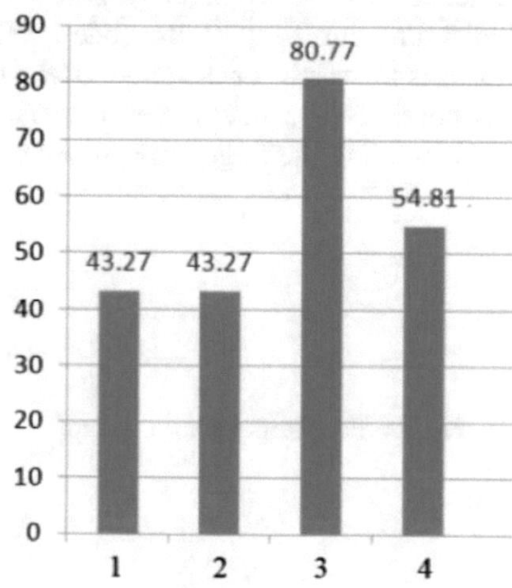

Fig. 3. Correct rate for different electronic map color schemes

The main effect of the electronic map color scheme was significant, F = 15.193, p = 0.000 < 0.001. Multiple comparative analysis found that scheme 3 was significantly higher than other schemes (p = 0.000 < 0.001), scheme 4, and scheme 2 were not significantly different from scheme 1.

(2) Reaction time

Calculate the average time of response for each click of the electronic map color scheme. The result is shown in Fig. 4.

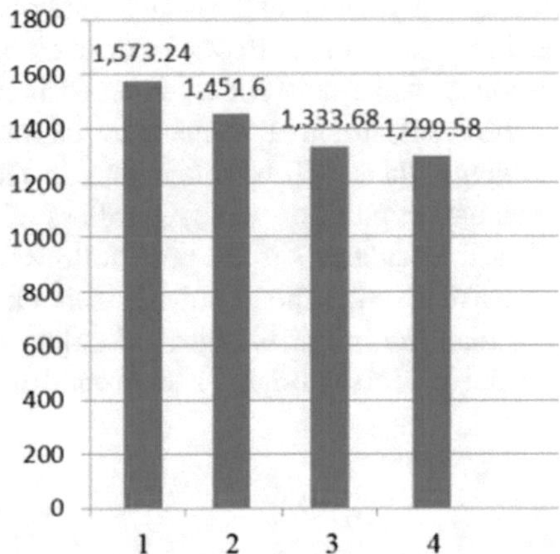

Fig. 4. Reaction time of click on different electronic map color schemes to new target

The main effect of the electronic map color matching scheme was significant, F = 2.658, p = 0.056 > 0.05. Multiple comparative analysis found that the fourth reaction was lower than that of the first one, and the marginal effect was significant (p = 0.061 > 0.05).

3.2 Dual Comparison Analysis

Calculate the number of selections under each color scheme, and the result is shown in Fig. 5.

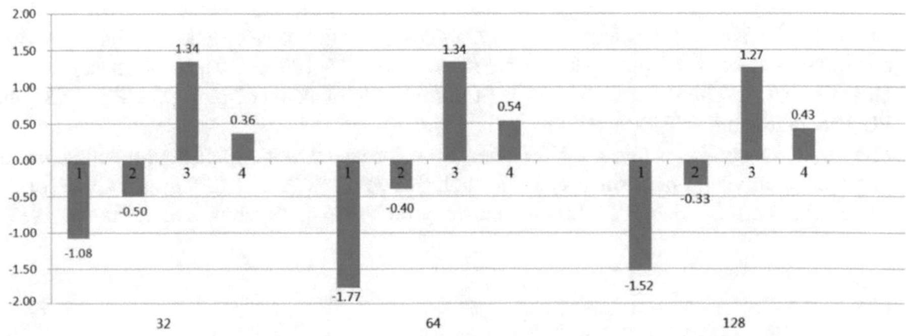

Fig. 5. Dual comparison score of map color scheme in three cases

In all three cases, the main effects of the electronic map color scheme were significant, F = 41.568, p = 0.000 < 0.001. Multiple comparative analysis found that protocol 3 was significantly higher than scenario 4 (p = 0.001 < 0.01), scenario 2 (p = 0.000 < 0.001), protocol one (p = 0.000 < 0.001), and scenario 4 was significantly higher than scenario 2 (p = 0.044 < 0.05).

4 Discussion

From the data analysis results, it is found that the correct rate ofscheme3 is significantly better than other schemes in terms of the correct rate of clicking the new target. From the perspective of the response of the new click, Scheme4 is significantly better than Scheme1. Because the participants found that the new target still needs to identify whether its code is in the range, there will be differences in different color schemes.

From the subjective evaluation point of view, regardless of the number of icons is 32, 64 or 128, it shows that the scheme 3 is the best, followed by the scheme 4.

This conclusion is consistent with the following scholars' opinion: "The dark background color is better than the bright background color, and the black (or dark) background is superior to the white (or bright) background in the display work efficiency [4–6]".

5 Conclusion

In the daily light source environment, among the four color schemes, scheme 3 is preferred, followed by scheme 4, and scheme 1 and scheme 2 are not recommended. Scheme 3 has a greater contrast with the foreground color, which is more conducive to search and recognition. Experiments show that electronic maps affect user performance and subjective feelings under different background color schemes.

Acknowledgments. This study was supported by the projects with the Grant No. JCKY2016206A001 and No. 41412040304.

References

1. Li, Z., He, L., Xie, P.: Research on color map design model of electronic map based on map perception theory. J. Mapp. Spat. Geogr. Inform. **10**, 187–189 (2017). (in Chinese)
2. Hua, G., Xi, J., Zhao, N.: Comparison of ergonomics of color display on CRT, LCD and PDP. J. Ergonomics. **7**(2), 5–9 (2001). (in Chinese)
3. Zhang, D., Zhang, Z., Yang, H.: VDT interface color visual work efficiency: the influence of hue factors on visual performance. J. Psychol. Sci. **31**(2), 328–331 (2008). (in Chinese)
4. Yang, G., Tang, B., Fang, Y.: Ergonomic study on VDU. J. Psychol. Sci. **1**, 40–46 (1989)

5. Santucci, G., Menu, J.P., Valot, C.: Visual acuity in color contrast on cathode ray tubes: role of luminance, hue, and saturation contrasts. J. Aviat. Space Environ. Med. **53**(5), 478 (1982)
6. Neri, D.F., Luria, S.M., Kobus, D.A.: The detection of various color combinations under different chromatic ambient illuminations. J. Aviat. Space Environ. Med. **57**(6), 555–560 (1986)

Configuration and Use of Pervasive Augmented Reality Interfaces in a Smart Home Context: A Prototype

Bernardo Marques[1(✉)], Paulo Dias[1], João Alves[1], Emanuel Fonseca[2], and Beatriz S. Santos[1]

[1] DETI - IEETA, University of Aveiro, Aveiro, Portugal
bernardo.marques@ua.pt
[2] Bosch Thermotechnology, Aveiro, Portugal

Abstract. The Augmented Reality (AR) is believed to be a trending mechanism for intelligent environments since it provides additional layers of information on top of the physical world. Pervasive AR extends this concept through context-awareness, 3D space positioning and tracking mechanisms, resulting in experiences without interruptions. This paper proposes a pervasive AR application to create and configure such experiences, allowing management and control of appliances in a Smart Home context Based on a virtual content re-calibration process, the application is able to place and highlight what in the smart home can be interacted with a high level of accuracy and resilience to changes in dynamic environments. We also present examples of other appliances that can be integrated into the management and control features. Finally, we propose possible scenarios besides the Smart Home where the application might also be used.

Keywords: Pervasive Augmented Reality · Context-sensitive computing · Information-display · Uninterrupted experiences · Smart Home

1 Introduction

As people's expectations of what technology can do for them are changing, the vision of what a Smart Home entails is continuously evolving as well [1]. The concept of the Smart Home is viewed as a way of better managing the demands of daily living through technology [2]. Augmented Reality (AR) is thought as an opportunity to provide powerful and flexible user interface for intelligent environments, for several years [3]. AR is able to overlay additional computer-generated content on specific positions or in the user's surrounding environment, in real time [4]. As such, there is a growing interest towards the use of this technology, since it has the potential to be used in our everyday lifestyles [5, 6]. Despite the recent success of AR applications, usually these serve a single purpose, being limited to specific markers or localizations [7]. Additionally, they are used for short amounts of time, not being able to deliver continuous, context-aware AR experiences. Yet, nowadays, the definition of AR is starting to be extended by the concept of Pervasive AR, using mechanisms capable of allowing continuous usage, in spite of changes and constrains on the user's context. In the near

T. Ahram et al. (Eds.): IHSED 2019, AISC 1026, pp. 96–102, 2020.
https://doi.org/10.1007/978-3-030-27928-8_16

future, AR will become more accessible, becoming a complementary part of our lives. Hence, it will be beneficial to map augmented interfaces on top of physical appliances to generate direct mapping [8], allowing to understand what we are looking at, resulting in real objects of the smart home to be augmented with information and controllable with different modalities. This paper explores the creation and configuration of uninterrupted AR experiences using the Tango motion-tracking platform combined with fiducial markers, aiming to be used as a mechanism to control appliances of a Smart Home. The goal was to control appliances by pointing at them using pop-up context aware AR interfaces.

2 Creating Uninterrupted AR Experiences

Creating uninterrupted AR experiences implies the capacity of placing and interacting with virtual content anywhere on top of the real-world environment. To explore this in the Smart Home context, we started by developing a mobile application, able to support configuration of augmented indoor spaces with several types of virtual content in a credible way (3D models, text, sound, video, etc.), as well as their exploration as an uninterrupted AR experience. The application is based on the use of the Tango platform motion-tracking capabilities combined with fiducial markers. It has been shown that this approach allows displaying virtual content with high accuracy in indoor scenarios (with an approximated error of 0.006 m), even in dynamic environments, where geometry and illumination changes may occur. This kind of approach also enables re-calibration of virtual content position and orientation, whenever needed, leading to smaller errors in pose estimation [9]. The resulting application provides the following main functionalities:

1 - Marker Detection
This option enables detection of QR codes. All markers must be the same size, to accurately determine their location and orientation in the environment, which is saved into a configuration file. Hence, there can be no repeated markers, since the application uses the information contained in the markers to locate their virtual position.

2 - Placement and Configuration of Virtual Content
This option allows the selection of a file with a previously obtained set of markers. Their position and orientation work as anchors to place virtual content in the surrounding environment. Thanks to the use of motion-tracking with markers, the mobile device can localize itself with a high accuracy, even if the mobile device is facing a surface without markers or has been shut down and re-started. After the detection of the first marker, the application understands its positioning in the environment, and allows to select previously loaded models from a list of 3D objects, interfaces, sounds, and videos. Next, the selected model can be placed on top of the real world. Additionally, the models can be rotated, scaled or moved to fit a specific position in the environment. In the end, the created configuration can be saved into a file. Furthermore, it is also possible to edit a configuration after it has been saved, in order to change or remove any of the existing content or add new one. Thus, based on the same set of markers, different virtual content can be placed, allowing the possibility of presenting different experiences in the same environment.

3 - Exploration Mode
This option allows selecting one configuration file and visualizing virtual content associated to it. This occurs after the detection of a first marker, allowing the application to understand its positioning in the environment. Depending on the content used, each virtual model may provide different interaction possibilities, namely: sound, animations, or pop-ups of additional menus or information. Moreover, the position of the virtual content can be re-calibrated when a marker is detected by the application.

3 A Prototype for Smart Home Management and Control

Creating Most appliances found in a Smart Home context are cumbersome to configure and use, and often require specific applications to be installed. Pervasive AR could enable an all-in-one solution for all appliances, while being more informative by providing mechanisms capable of allowing to understand what in the environment can be interacted with and presenting information aligned with the real-world environment. Also, it may turn high data volumes generated by smart appliances into usable and actionable information and aid in the interpretation of information in spatial context. In addition, Pervasive AR might help to reduce learning curves of new dwellers not acquainted with the environment. Furthermore, adaptive interfaces can be used to present specific content to dwellers according to their profile, which can also be customized according to the user needs/preferences, possibly leading to improved social interaction among dwellers in the Smart Home context. As such, and motivated by an ongoing research collaboration, we extended the application capable of creating uninterrupted AR experiences to allow remote management and control of appliances in a Smart Home context. Our goal was to help users understand what appliances can be interacted with, present their current status and functionalities, by overlaying additional information on top of them. We believe this approach makes management and control much more natural and can greatly ease user interaction.

We considered a small duplex apartment as our scenario and used as appliances a Philips Hue set (enabling the control of several aspects regarding lights, such as state (on/off) of specific rooms, brightness, color selection) and a Bosch boiler (enabling the control of hot water features: state (on/off), temperature regulation, and scheduling during the week period, enabling to automatically start and stop the boiler on specific moments of the day). Moreover, we added relevant videos, signs indicating for example danger or limited access to a specific room. To make the application cross-platform, allowing using it beyond mobile devices we developed user interfaces for other devices. Likewise, we enabled the interconnection among these devices to synchronize the state of all appliances. We started by evolving the application to support three different methods of interaction through different interfaces in different devices. All methods provide the previous mentioned functionalities allowing manage and control appliances of a Smart Home.

Method 1 - Standard 2D Interfaces

Method 1 presents standard 2D user interfaces, able to be used in several cross-platform devices, ranging from laptops, mobile devices, to projectors and others (Fig. 1). In this case, the user can choose from a set of different types of layouts, intended to facilitate the visualization of the information associated to the Smart Home appliances used.

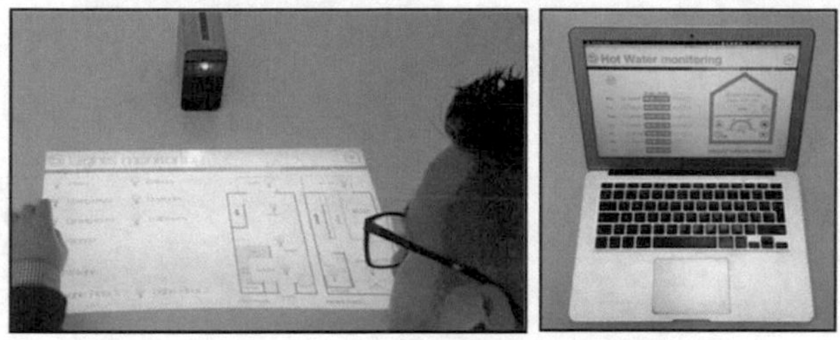

Fig. 1. Example of standard 2D interfaces to control lights (using a projector with touch recognition - left) and boiler (using a laptop - right) in the Smart Home

Method 2 - Conventional AR interfaces

Method 2 uses AR to display 2D interfaces on top of the physical world, based on marker detection (Fig. 2). This method is intended for mobile devices, due to their characteristics. Furthermore, in order to use this method, a previous step is required to locate the different markers in suitable places in the Smart Home environment. This step may be performed using functionalities 1 and 2, as previously described.

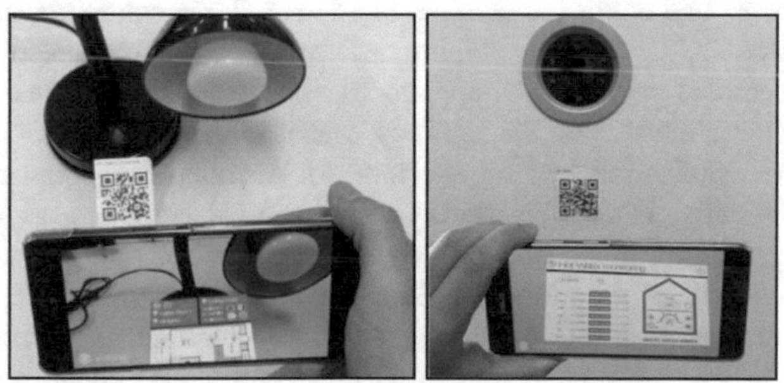

Fig. 2. AR interfaces using a mobile device to detect markers, allowing control lights (left) and boiler (right) in the Smart Home

Method 3 - Uninterrupted AR Interfaces

Method 3 uses the approach based on the Tango motion tracking and markers. Following the same mechanisms as before, in a configuration functionality the user can select and place the new interfaces near the appliances (Fig. 3). Later, it is possible to visualize the interfaces anywhere on top of the physical world in a subtle and unobtrusive way (presenting relevant and not distracting content), allowing to manage and control appliances in the Smart Home context (Fig. 3).

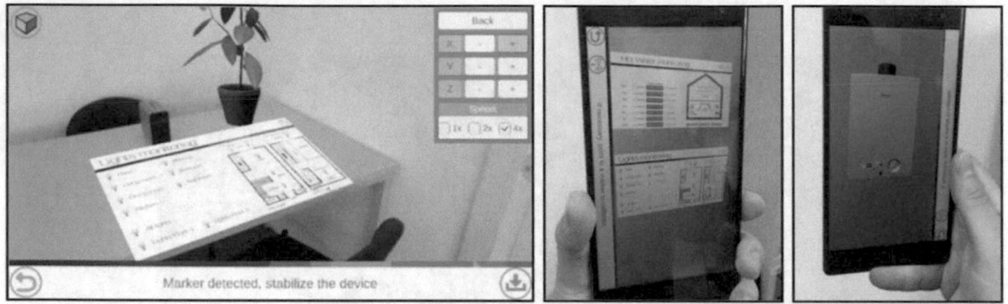

Fig. 3. Configuration of AR interfaces using motion-tracking and markers in the Smart Home (Left). Augmented Reality interfaces using motion-tracking and markers to control lights and boiler (center) and visualize a 3D representation of a real boiler in the environment (right) in the Smart Home.

4 Discussion

In this paper, we presented a Pervasive Augmented Reality application, allowing creation, configuration and use of uninterrupted AR experiences. The application focus on control of appliances in a Smart Home context by pointing at them using pop-up context aware AR interfaces. The developed application opens up new possibilities for the way we interact and control appliances in a Smart Home context, accordingly to the user interest/necessity. In a near future, AR applications may also allow greater awareness and control of additional appliances, such as: security systems cameras, air conditioning, lights, doors, windows, blinds, watering the garden, among others.

This type of application can be adequate to several professional or recreational activities. Specifically, it may be useful in professional activities, as warehouses or work support, providing personalized navigation aids to find specific products as well as present additional information regarding the product characteristics and also the existing stock in specific locations of the warehouse. Recreational activities are also a relevant example. Museums can use it to provide continuous enhanced experiences to the visitors by adding additional layers of information regarding current exhibits. Visitors can select specific areas of the museum that they are willing to see creating adaptable experiences. Using the same application, visitors can experience the same museum in different ways. Finally, other potential use is assistance in accommodation establishments, by helping to reduce learning curves of new dwellers not acquainted with new environments.

5 Conclusions and Future Work

Through the use of Pervasive Augmented Reality, it is possible to upgrade an application with a unique objective to a multi-purpose continuous experience, changing the way we interact with information and our surroundings. Combining the Tango platform motion-tracking capabilities with fiducial markers allowed developing a tool to create, configure and deploy uninterrupted AR experiences in indoor environments using three different types of interfaces to enable visualization, management and control of interactive content associated with specific appliances in a Smart Home context, accordingly to the interaction, localization and scenario characteristics.

Future developments will include integration of additional appliances in the environment, context awareness based on recognition and identification of users and thus enabling the adaptation of the interfaces to the user specifications and exploration of natural markers, through existing paintings or 3D-objects in the environment as re-calibration anchors to automatically align virtual content, avoiding the need for a more intrusive type of markers. Uninterrupted AR experiences raise challenges about how to display information (which type of information is distracting or relevant) and how to properly interact with that information. As such, an in-depth study will be conducted to explore AR for Smart Home control, allowing to test all the proposed features and evaluate user response and acceptance.

Acknowledgments. The present study was developed in the scope of the Smart Green Homes Project [POCI-01-0247-FEDER-007678], a co-promotion between Bosch Termotecnologia S.A. and the University of Aveiro. It is financed by Portugal 2020 under the Competitiveness and Internationalization Operational Program, and by the European Regional Development Fund. This study was also supported by IEETA - Institute of Electronics and Informatics Engineering of Aveiro, funded by National Funds through the FCT - Foundation for Science and Technology, in the context of the project UID/CEC/00127/2019.

References

1. Mennicken, S., Vermeulen, J., Huang, E.M.: From today's augmented houses to tomorrow's smart homes: new directions for home automation research. In: Conference on Pervasive and Ubiquitous Computing, UbiComp 2014, pp. 105–115 (2014)
2. Wilson, C., Hargreaves, T., Hauxwell-Baldwin, R.: Smart homes and their users: a systematic analysis and key challenges. Pers. Ubiquit. Comput. **19**(2), 463–476 (2015)
3. Macintyre, B., Mynatt, E.D.: Augmenting intelligent environments: augmented reality as an interface to intelligent environments. Interface, 93–95 (1998)
4. Azuma, R.T.: A survey of augmented reality. Presence: Teleop. Virt. Environ. **6**(4), 355–385 (1997)
5. Kim, K., et al.: Revisiting trends in augmented reality research: a review of the 2nd decade of ISMAR (2008–2017). IEEE Trans. Visual Comput. Graphics **24**(11), 2947–2962 (2018)
6. Van Krevelen, D.W.F., Poelman, R.: A survey of augmented reality technologies, applications and limitations. Int. J. Virtual Reality **9**(2), 1–20 (2010)

7. Ullah, A.M., Islam, M.R., Aktar, S.F., Hossain, S.K.A.: Remote-touch: augmented reality based marker tracking for smart home control. In: Conference on Computer and Information Technology, ICCIT 2012, pp. 473–477 (2012)
8. Heun, V., Hobin, J., Maes, P.: Reality editor: programming smarter objects. In: Conference on Pervasive and Ubiquitous Computing, UbiComp 2013, pp. 307–310 (2013)
9. Marques, B., Carvalho, R., Dias, P., Oliveira, M., Ferreira, C., Sousa Santos, B.: Evaluating and enhancing Google Tango localization in indoor environments using fiducial markers. In: ICARSC 2018, pp. 1–6 (2018)

Walkability in the Modern Arab Cities: An Assessment of Public Space Along Al-Qasba Canal and Lake Khaled in Sharjah

Mohamed El Amrousi and Mohamed Elhakeem(✉)

Abu Dhabi University, Abu Dhabi, United Arab Emirates
Mohamed.elhakeem@adu.ac.ae

Abstract. The new waterfront developments in Sharjah in the form of public gardens along lake Khaled and Al-Qasba water canal redefine public space in Sharjah with an identity that is more inclusive than privately developed more exclusive spaces and mega malls in Dubai. Such emerging landscapes and public spaces in Arab cities are necessary and when mitigated they represent an intermediate intervention of government and the private sector to improve the walkability of the city. This paper studies the strategies and effects of this assemblage of neo-Islamic buildings in Sharjah along its newly created landscapes and waterfronts. In addition, we used a 2D-hydrodynamic water surface model to simulate the flow pattern in the Sharjah canal and the lagoons.

Keywords: Sharjah · Lake Khaled · Al-Qasba Canal · Waterfronts and public space

1 Introduction

The city of Sharjah is one of the most popular residential cities in the United Arab Emirates because of its reasonable real-estate prices and proximity to Dubai. Its urban fabric gradually developed into a congested city where the stock of open spaces, public gardens and landscape has not been able to keep up with the population growth. The municipality of Sharjah developed several schemes in attempt to resolve traffic problems and to improve the walkability of the city. In the last decade, the border zone between Sharjah and Dubai has undergone a major gentrification process that revitalized Sharjah's waterfronts and created public spaces and pedestrian paths. The area around Lake Khaled and Sharjah's lagoons was redesigned to create new public spaces such as Al-Majaz garden and a new a water canal was dredged to connect between the lagoons improving water circulation (Fig. 1). The Central Souq of Sharjah represents part of an architectural trend to create an identity for the city, its form, decor and surrounding landscape best exemplify the role of neo-Islamic heritage along waterfronts that profoundly assure the community a sense of belonging in the modern city through a combination of cultural resonance and familiar everyday activity. Al-Qasba canal and its associated spaces and urban development included entertainment clusters, restaurants, bridges, pedestrian paths and 'Eye of the Emirates Wheel' where visitors can view the city from above. Sharjah's development of public spaces around Lake

T. Ahram et al. (Eds.): IHSED 2019, AISC 1026, pp. 103–108, 2020.
https://doi.org/10.1007/978-3-030-27928-8_17

Khaled and Al-Qasba represents a fresh approach to the provision of publicly accessible space that has been increasingly undertaken by the private sector highlighted by projects like Dubai's City Walk, La Mer and Bluewaters Island developed by Meraas. Privately developed public spaces are open to the public during certain hours and security prioritization makes space less desirable to certain social groups, as a result this may foster exclusion and threatens to concepts of inclusion [1]. Sharjah's urban renewal project and public spaces emerging along Lake Khaled and Al-Qasba are important as a counter-spaces to upscale urban facilities mitigated by private developers in Dubai such as mega malls, covered streets, waterfronts with jogging trails and tennis courts [2]. This paper studies the strategies and effects of the urban assemblage and public space in Sharjah shaped by neo-Islamic buildings along its newly created landscapes and water-fronts. In addition, we used a 2D-hydrodynamic water surface model to simulate the flow pattern in the Sharjah canal and the lagoons.

Fig. 1. Lake Khaled—Al-Majaz garden and Sharjah's lagoons.

Fig. 2. Al-Qasba and Sharjah's Central Souq (bazaar).

2 Urban Identity, Public Space and Waterfront Development

Sharjah's waterfront offers a genuinely new alternative rooted under a broader notion of regional expression and pan-Islamic architecture and ornament. The area around Lake Khaled and Al-Qasba (Fig. 2) highlights that when significant investment has allocated for improving sidewalks, crossing signals, and streetlights, create safe, comfortable public spaces in contrast to wide-open spaces that due to the hot arid climatic conditions of desert environments fail to attract pedestrians [3]. Al-Majaz garden its' pavilions the Central Souq, Al-Qasba and neo-Islamic buildings such as Al-Noor mosque shaping public space in Sharjah around lake Khaled. Their architectural style suggests that new forms of urbanism can be created in the shadow of historical and culturally resonant monuments. Sharjah's newly evolving public spaces created along the lagoons waterfronts and Al-Qasba canal when integrated within the urban context manifest the belief that through urban regeneration the city has the ability to gentrify its urban fabric to focus on more inclusive public space accessible to multi-ethnic socio-cultural groups. Al-Qasba canal and its Moorish-Hispanic arcaded buildings provide shade improved public space, path connectivity in addition to water circulation between the Sharjah lagoons. The positioning of architecture in relation to water and waterfronts has radically changed in the last decade to shape public space, and collaged forms and fragments of local tradition represent a narrative where the notion of authenticity and reference to the past is coupled by post-modernity and technological advancement [4]. Neither entirely the conservative vernacularism with whole-scale historical referencing, nor heavily founded on postmodernism Sharjah's waterfront landscapes represent a form of public spaces shaped by neo-regional urbanism [5]. In Gulf state cities waterfront regeneration project have resulted in creating leisure spaces shaped by cultural clusters that have become destinations for expatriate communities, especially landscapes that include water because they contribute to sense of place and identity [6]. Sharjah similar to most postindustrial cities developed a dense and congested urban fabric that suffered from lack of pedestrian connectivity [7]. However, the neo-Islamic buildings such as the Central Souq, Al-Noor mosque and the Moorish-Hispanic buildings of Al-Qasba shape pedestrian paths, foster walkability within the context of Sharjah's religious strategies and urban context. Al-Majaz garden and the neo-Islamic style of the buildings within it further cement Sharjah's urban identity and create spaces of visual interest. The Central Souq in Sharjah with its vaulted architecture, projecting wind-towers and stucco screen windows is an icon of the city reflecting on the role of architecture in weaving forms and fragments from a wide spectrum of Islamic arts to highlight the amalgamation of multi-ethnic communities within the city. Taken together this group of new monuments forms a complex urban whole that serves to reflect and deepen an emerging sense of identity that is built upon a similarly complex mix of multicultural ethnicities that make up the population of Sharjah.

Fig. 3. Al Noor mosque and neo-Islamic styled pavilions in Al-Majaz garden

The neo-Islamic style of the buildings created contemporary monuments and the public spaces associated with them together comprise a complex strategy that champions continuity in urban aggregation along the dimensions both of landscape and cultural identity. Al-Noor mosque (Fig. 3) with its clear reference to Ottoman architecture when coupled with the pavilions in Al-Majaz gardens and the public spaces around them are clear examples of how contemporary architecture can incorporate recognizably cultural assemblages signaling shared cultural virtue and Muslim identity. The congregational mosque its surrounding public space and the bazaar have always constituted a central node and place of gathering in Arab/Muslim cities [8]. These clearly signal a space of cultural and religious commonality within the overarching modernity of the Dubai-Sharjah metropole. The mosque's location in a garden overlooking the Khaled lagoon allows for complete views of its facades. In the design of Sharjah's Central Souq, Al-Noor mosque and Al-Qasba a model of meaningful urban development built around an assemblage of hybrid Muslim ornaments is represented to reinforce the diverse cultural identities of expatriate communities in Sharjah, while building a contingent sense of support for cultural and social activity [9]. Taken together this group of new monuments along Sharjah's waterfronts forms a complex urban whole that serves to reflect and deepen an emerging sense of identity that is built upon a similarly complex mix of multicultural ethnicities that make up the population of Sharjah.

3 Model Description

Due to the complex geometry of the Sharjah lagoons and Al-Qasba Canal, a 2D hydrodynamic model was used to investigate the flow pattern of the water in the lagoon. In this study, we chose the 2D Finite Element Surface Water Modelling System (FESWMS), which is part of the commercially available Surface water Modelling System (SMS) software package (version 12.1) developed by the Federal Highway Administration [10]. FESWMS solves the differential forms of the continuity and the momentum equations in the stream wise and transverse directions using the Galerkin method of weighted residuals providing water depth and depth-averaged velocity magnitude in x and y directions at each node in the grid [11].

The model inputs are the lagoons and canal bathymetry, Manning's coefficient $n = 0.025$, eddy viscosity $v = 1.0$ m^2/s, flow rate entering the lagoons from the sea from $Q = 50$ m^3/s, and flow depth of the water exiting the lagoon $y = 6$ m. Figure 4 shows the simulations of the velocity distribution and velocity vectors of the study reach. The simulation shows that the velocity is in the range of 0.04 to 0.4 m/s, with the highest values being in Al-Qasba Canal and the lowest values being in the lagoons. It can be seen from the figure that complex flow circulation patterns have been developed in the lagoons, which are desired because they help in controlling sediment deposition, aquatic growth, sulphide formation and improve water quality.

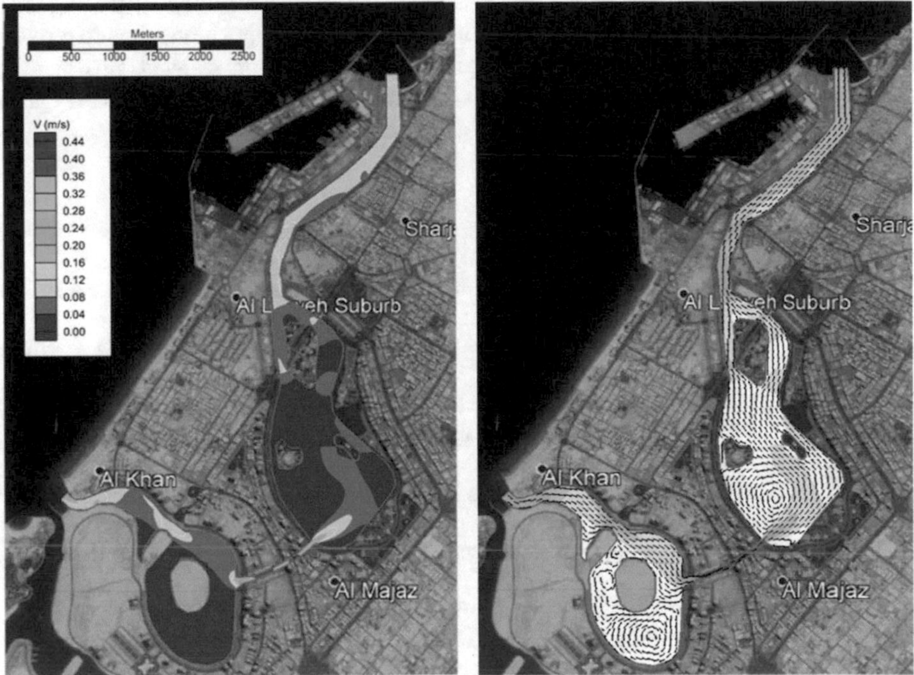

Fig. 4. Simulation of the flow velocity and flow circulation patterns in the Sharjah lagoons and Al-Qasba canal.

4 Conclusion

The prominent placement amidst open gardens and along the lagoon of the Sharjah neo-Islamic monumental buildings shape public space and collectively suggest a new urban image and socio-cultural space for a modern Muslim urban identity. Along with a series of restaurants, entertainment spaces, and office buildings, Ottoman Styled mosques and neo-Islamic styled cafes with eclectic references Sharjah's attempts to create public space amidst its densely populated urban fabric forms cultural connectivity and reference to State identity. Sharjah's Central Souq, Masjid al-Noor, and Al-Qasba use the most visually important components of Islamic ornaments, such as stepped crenellations, vaults, domes and pencil shaped minarets employed to their fullest extent in the articulate historically derived façade compositions to assert the cultural identity of public space around lake Khaled. They create a kind of shared communal space that connects communities via a highly eclectic space shaped by neo-Islamic architecture. The inclusivity and diversity of forms and fragments of Islamic architecture symbolized by the replication of traditional Islamic forms and fragments support larger issues of the perception, utilization, and visualization of public space in Sharjah. These spaces of gathering represent the pluralistic cultural understanding of spaces in the living city. We also studied through a 2D hydrodynamic model the flow patterns around the lagoons of Sharjah and the Al-Qasba Canal. The model predictions showed that complex flow circulation patterns have been developed in the lagoons, which are desired because they help in controlling sediment deposition, aquatic growth, sulphide formation and improve water quality. The simulation showed also that the velocity is in the range of 0.04 to 0.4 m/s, with the highest values being in Al-Qasba Canal and the lowest values being in the lagoons.

References

1. Nemeth, J., Schmidt, S.: The privatization of public space: modeling and measuring publicness. Environ. Plann. B: Plann. Des. **38**, 5–23 (2011)
2. Banerjee, T.: Public space, beyond invented streets and reinvented places. APA J. Winter **67**(1), 9 (2001)
3. Speck, J.: Walkable City: How Downtown Can Save America, One Step at a Time (2013)
4. Young, L.: Villages that never were: the museum village as a heritage genre. Int. J. Herit. Stud. **12**(4), 321–338 (2006)
5. El Amrousi, M., Biln, J.: Sharjah's Islamic urban identity and the living city. J. Islamic Archit. **1**(4), 190–198 (2011)
6. Ellin, N.: Canalscape: practicing integral urbanism in metropolitan Phoenix. J. Urban Des. **15**(4), 599–610 (2010)
7. Southworth, M.: Designing the walkable city. J. Urban Plann. Dev. **131**(4), 246 (2005)
8. Shamsuddin, S., Abdul Latiph, N.S., Sulaiman, A.B.: Waterfront regeneration as a sustainable approach to city development in Malaysia. WIT Trans. Ecol. Environ. **117**, 45–54 (2008). The Sustainable City V
9. Al-Asad, M.: Applications of geometry. In: Frishman, M., Khan, H. (eds.) The Mosque History, Architectural Development and Regional Diversity. Thames and Hudson, London (2002)
10. Froehlich, D.: User's Manual for FESWMS Flo2DH: Two-Dimensional Depth-Averaged Flow and Sediment Transport Model. Release 3 (2002)
11. Elhakeem, M., Papanicolaou, A.N., Wilson, C.G.: Implementing streambank erosion control measures in meandering streams: design procedure enhanced with numerical modelling. Int. J. River Basin Manag. **15**(3), 317–327 (2017)

Psychological Interpretation of Human Behavior to Atypical Architectural Shape

Young Lim Lee[1] and Yun Gil Lee[2(✉)]

[1] Department of Psychology and Psychotherapy, Dankook University, 119 Dandae-ro Dongnam-gu, Cheonan-si, Chungcheongnam-do, Republic of Korea
younglee13@dankook.ac.kr

[2] Department of Architecture, Hoseo University, 20 Hoseo-ro 79beon-gil, Baebang-eup, Asan-si, Chungcheongnam-do, Republic of Korea
yglee@hoseo.edu

Abstract. The aim of this study was to investigate the psychological effects of users' behaviors toward atypical architectural forms. The users' behaviors were interpreted from a psychological perspective that has been studied previously and is related to visual perception, active touch, equilibrium sense, poor shape perception, ecological psychology, canonical neurons, and affordances. The results theoretically stated that the users' different behaviors, when in atypical buildings, may have been caused by psychological actions. The ultimate purpose of this study was to develop a computerized tool to perform user simulations in the atypical architectural design process. The results of this study can be used as theoretical knowledge for developing advanced intelligence of the agent.

Keywords: Human factors · Human behavior · Psychological interpretation · Atypical architectural shape · Affordance

1 Introduction

The human factor is the most significant standard for judging the value of design alternatives. However, it is often overlooked in the design process, which places more weight on atypical shapes. Recently, atypically shaped buildings have appeared more frequently. Atypical buildings can induce new behaviors in users because architectural shapes and user behaviors are complementary and mutually restrictive. Architects must consider these factors when designing atypically shaped buildings for users' convenience and safety. Through preliminary research, the different behaviors that users exhibit in atypically shaped buildings were investigated, and several characteristics of human behaviors were found. These behaviors could have been related to the users' psychological reaction to the atypical forms. If the correlation between the users' behaviors and their psychological responses to a particular atypical form can be known, this correlation will predict users' behaviors toward that atypical form.

T. Ahram et al. (Eds.): IHSED 2019, AISC 1026, pp. 109–114, 2020.
https://doi.org/10.1007/978-3-030-27928-8_18

Table 1. Results of a human behavior survey using atypical architectural shapes. (1)

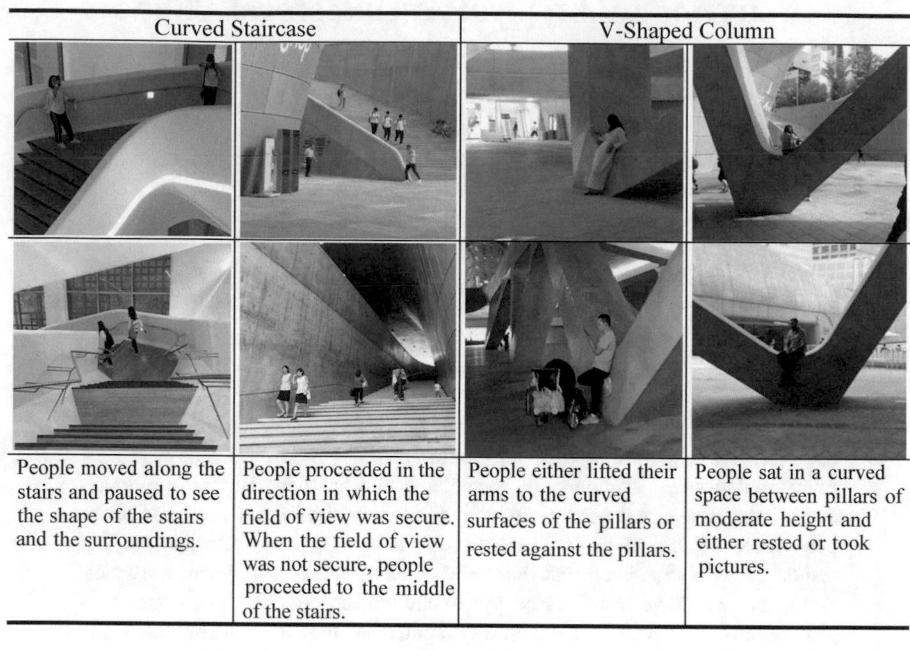

Curved Staircase		V-Shaped Column	
People moved along the stairs and paused to see the shape of the stairs and the surroundings.	People proceeded in the direction in which the field of view was secure. When the field of view was not secure, people proceeded to the middle of the stairs.	People either lifted their arms to the curved surfaces of the pillars or rested against the pillars.	People sat in a curved space between pillars of moderate height and either rested or took pictures.

Table 2. Results of a human behavior survey using atypical architectural shapes. (2)

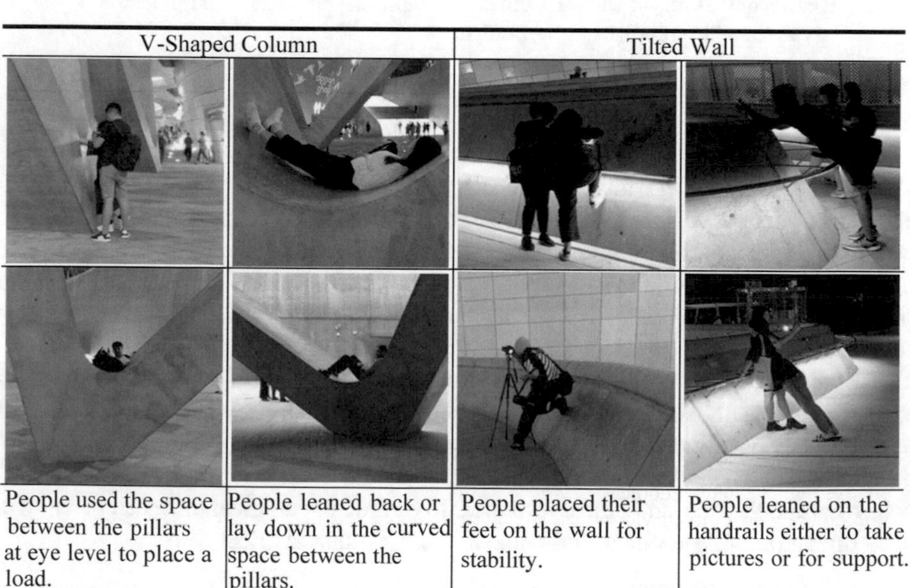

V-Shaped Column		Tilted Wall	
People used the space between the pillars at eye level to place a load.	People leaned back or lay down in the curved space between the pillars.	People placed their feet on the wall for stability.	People leaned on the handrails either to take pictures or for support.

The aim of this study was to investigate the psychological effects of users' behaviors toward atypical architectural forms. The users' behaviors that were seen toward the atypical architectural forms were interpreted from a psychological perspective that has been previously studied and is related to visual perception, active touch, equilibrium sense, poor shape perception, ecological psychology, canonical neurons, and affordances. The results theoretically show that the users' different behaviors toward an atypical building may have been caused by psychological actions. The ultimate purpose of this study was to develop a computerized tool to perform user simulations in the atypical architectural design process. The results of this study can be used as theoretical knowledge for developing advanced intelligence of the agent. This study helped identify the causes of the different behaviors that were found through field surveys and also provided a framework for further in-depth research.

2 Field Survey of Human Behaviors to Atypical Architecture

The previous study investigated the characteristics of human behavior in an atypical architectural space by analyzing users' behaviors when in a representative atypical building. The researchers visited two representative examples of Korea's atypical buildings, and they investigated what physical factors were exhibited by users in the field. Tables 1, 2, 3 and 4 summarize the results of the survey. Atypical architectural elements have also induced user movement through temporal stimulation, and users have conducted specific actions toward physical elements because of the physical characteristics (inclination, height, etc.) of the architecture.

Table 3. Results of a human behavior survey using typical architectural shapes. (3)

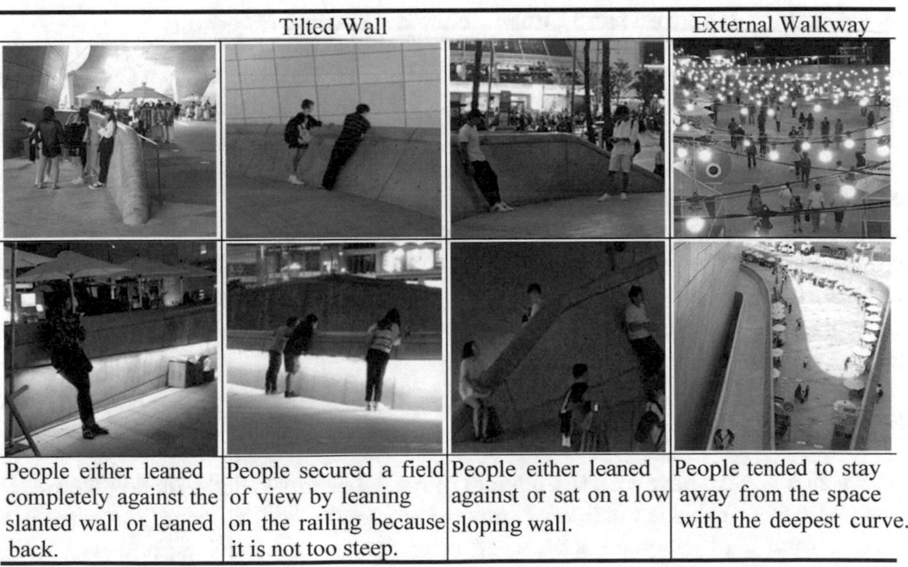

Tilted Wall			External Walkway
People either leaned completely against the slanted wall or leaned back.	People secured a field of view by leaning on the railing because it is not too steep.	People either leaned against or sat on a low sloping wall.	People tended to stay away from the space with the deepest curve.

Table 4. Results of a human behavior survey using atypical architectural shapes. (4)

Indoor Curved Staircase		

People entered the middle of the stairs and moved gradually toward the edge. People moved gradually toward the target point.	People leaned on the curve and moved to sit on the edge. People sometimes went to the edge to look down and grab the handrails.	People seemed to be moving toward a stable center; they were affected by the instability of the curve. Children jumped up the stairs one by one and moved toward the front of the stairs rather than out toward the surroundings.

3 Psychological Interpretation of Human Behavior Toward an Atypical Architectural Shape

3.1 Catching Handrails and Other Behavior Toward Handrails

Gibson [1] distinguished touching from being touched and touching, so called "active touch" provides a variety of information about the environment. He suggested that active touch tended to be related to the object being contacted. When you touch or grasp an object with your hand, the posture and movement of the hand naturally change depending on the shape of the object. The aesthetics of "active touch" are related to experiencing the positive or conspicuous form when touching or grasping objects. The texture of the surface and a graspable or dynamic form elicit the desire to touch. Due to the information that is coming from active touch senses and the desire to touch, the behaviors of grabbing the handrails of the stairs and going toward the rails may appear [1, 2].

3.2 Looking Down at the Railing

Freud stated that we have a certain amount of mental energy called a libido ("desire" in Latin), and if a forbidden action or impulse is suppressed, this energy appears in a different form. Not everyone is pursuing these dangerous behaviors; however, when evaluating situations where certain safety is ensured (the height of a staircase), some may look down [3].

3.3 Leaning on a Tilted Wall and a Railing

The tilted structure gives a dynamic impression, and perception of dynamic scenes possibly accompanied by your own movement can elicit the desire of exercising. Hence, it can trigger the urge to lean or move the body. Also, due to the oblique effect in which it is more difficult to perceive this orientation than in either the horizontal or the vertical orientations, it is possible to use haptic information to compensate when exploring the tilted structure [2, 4, 5].

3.4 Moving up to the Inside of the Curved Stairs

This behavior can be explained by Gibson's ecological psychology [6]. Humans are not able to perceive distance or depth correctly because retina images are two-dimensional. The traditional view of perception is that we internally represent the three-dimensional space using monocular and binocular cues. From this view, how successful our action is depends on how well we construct internal models about the environment. It has been found, however, that our internal representations do not always correspond to the visual space. In fact, perception of visual space is distorted. Nonetheless, our action is fairly correct. The ecological approach suggests that we can perform many actions without correct internal representations about space. Gibson believed that perceptual attributes which can be used immediately to control a particular behavior, are more important and necessary than spatial attributes such as size, distance, or form of objects in a three-dimensional space that are emphasized in the traditional approach. For instance, perceptual attributes needed for a person who wants to jump over a small creek are not the width of the creek itself, but whether the creek is wide enough to run over. To do this, we use optic flow produced when we move and the optic flow provides real-time information about where we are heading and how fast we are moving. Thus, the user might use optic flow when he/she moves in a curved staircase [6–9].

3.5 Children Jumping up the Stairs

This is related to the "affordance," which was introduced by Gibson [6]. The affordance characterizes the suitability of the environment to the observer, and so, depends on their current intentions and their capabilities. Perception is an act of intention, and when we perceive an object, both the physical property ("this object is something") and the meaning of the object ("how the object can be used") are perceived and two of them cannot be separated. Since the affordance includes the meaning of the object, it can be different depending on the observer. A stairway can be jump-able to a child, but walk-able to an adult. Therefore, the nature of the environment and the physical attributes of the actor are fused together to form a unique affordance. In other words, perception and action cannot be separable. The perception of the situation generates action and a new situation changed by the action generates another perception of new situation. Thus, the relation between perception and action is complementary and mutual constraint to achieve the goal-oriented movement [1, 6].

4 Conclusion and Discussion

In this study, we have analyzed certain psychological aspects of users' behavior in atypical spaces. In addition to the theoretical interpretation presented above, there is a possibility that a desire to rest simply may appear as an act of either sitting or leaning. It may also be that people behave in the same way in which they see others behaving. However, this is difficult to observe, and there is a limit to how many behaviors of other people can be observed by one person.

Through this study, we could summarize that human behavior can be expressed by psychological influence. To enhance human behavior simulations in unstructured spaces in the future, it is necessary to systematize and computerize those situations that can induce psychological behaviors and can be applied to a simulation system.

Acknowledgments. This work was supported by the National Research Foundation of Korea (NRF) grant funded by the Korea government (MSIT) (NRF-2018R1A2B6005827).

References

1. Gibson, J.J.: Observations on active touch. Psychol. Rev. **69**(6), 477–491 (1962)
2. Goldstrin, B.E.: Sensation and Perception, 9th edn. Cengage Learning, Mason (2013)
3. Schönhammer, R.: Einführung in die Wahrnehmungspsychologie: Sinne, Körper, Bewegung. UTB GmbH, Stuttgart (2013)
4. Arnheim, R.: A plea for visual thinking. Crit. Inq. **6**(3), 489–497 (1980)
5. Lee, Y., Crabtree, C.E., Norman, J.F., Bingham, G.P.: Poor shape perception is the reason reaches-to-grasp are visually guided online. Percept. Psychophys. **70**(6), 1032–1046 (2008)
6. Gibson, J.J.: The Ecological Approach to Visual Perception, classic edn. Psychology Press, London (2014)
7. Philbeck, J.W., Loomis, J.M.: Comparison of two indicators of perceived egocentric distance under full-cue and reduced-cue conditions. J. Exp. Psychol. Hum. Percept. Perform. **23**(1), 72–85 (1997)
8. Todd, J.T., Norman, J.F.: The visual perception of 3-D shape from multiple cues: are observers capable of perceiving metric shape? Percept. Psychophys. **65**(1), 31–47 (2003)
9. Land, M., Lee, D.: Where we look when we steer. Nature **369**, 742–744 (1994)

Green Ocean Strategy: Democratizing Business Knowledge for Sustainable Growth

Evangelos Markopoulos[1(✉)], Ines Selma Kirane[1], Clarissa Piper[1], and Hannu Vanharanta[2,3]

[1] HULT International Business School, Hult House East, 35 Commercial Rd, London E1 1LD, UK
evangelos.markopoulos@faculty.hult.edu, kinesselma01@gmail.com, clarissa.piper@gmail.com

[2] School of Technology and Innovations, University of Vaasa, Wolffintie 34, 65200 Vaasa, Finland
hannu@vanharanta.fi

[3] Faculty of Engineering Management, Poznan University of Technology, Poznan, Poland

Abstract. Sustainability should neither be dystopian, nor utopian, but better protopian. To achieve this, thinking green for the common good and benefit for all from all, does not only require thinking smart or green but also thinking together. This paper attempts to define a new set of organizational processes needed for a company to reach a continuous innovative and sustainable strategy by utilizing primarily its own human intellectual capital in an unbiased and democratic way. Such an approach can transmit an organization into a new management typology and leadership that can be defined as a Green Ocean strategy. The paper introduces the Green Ocean Strategy concept via a business transformation framework that can lead an organization from the Red to the Blue and from the Blue to the Green Oceans strategies. The framework is supported by phases, stages, processes, preconditions, and postconditions towards its effective adaptation.

Keywords: Strategy · Leadership · Knowledge · Management · Democracy · Transformation · Innovation · Green Ocean · Blue Ocean · Red Ocean

1 Introduction

Today's business environment undergoes a general market uniformization led by globalization. The ever-increasing information providence for environmental, economic and social sustainability and growth has brought organizations and customers to a more responsible and ethical business operations and product consumption. Business environment forces and markets are shifting competition globally towards sustainability, aiming to achieve higher degrees of innovation that can meet the societal, corporate and customer needs and expectations. Sustainability should frame equitable, viable and livable corporate activities and deliverables, while growth should target adding shared value to all stakeholders involved.

Since change is endemic and inevitable, every sustainable shift that can be transmitted to any stakeholder companies deal with, exponentially enhances each

T. Ahram et al. (Eds.): IHSED 2019, AISC 1026, pp. 115–125, 2020.
https://doi.org/10.1007/978-3-030-27928-8_19

achievement previously performed. Such changes impact Blue Oceans from the creation of uncontested markets [1] towards Green Oceans that can foresee and achieve the creation of sustainable market spaces for short, middle, and long-term profitability. As Blue Ocean Strategies seek to make competitors irrelevant by creating new customer value, Green Ocean Strategies can refine companies' mission and vision to compound sustainability over existing and profitable performances for a more adequate customer value creation.

Green Oceans Strategies can be seen as the logical, or natural, extension of Blue Ocean Strategies for companies aiming to reach and maintain their performance overtime, while being aligned with today's major social challenges. This can be achieved through the conversion strategies that take place in the 'turquoise canal', the transition passage from Blue to Green Oceans, in a journey that adapts democratic management practices for social, sustainable and shared value innovations.

Enacting a democratic management and leadership perspective in companies' operations management allow Specific, Measurable, Active, Realistic, and Time-bound (SMART) extroversion for internationalization in a globalized environment. The Company Democracy Model [2, 3], can be used and adjusted to form the base framework for driving an organization through the turquoise canal towards Green Oceans. The overall philosophy of the model is based on the principles of organizational sustainability driven by ethical values in a knowledge shared co-evolutionally y-type management and growth culture between the people, the organization and the society.

2 Towards Social Innovation

As business environments evolved through multiple disruptive waves over the past few centuries, management discipline turned into management science over the advancements in methods, tools and technologies evolved to manage volume, complexity, operations and above all the fast-changing innovations.

Management methods often rely on classical innovation, which comprises the process of finding new and more efficient solutions in the form of products, services or strategies. Such innovations can benefit directly and indirectly the stakeholders, but the concept has been evolved towards a wider and more social meaning.

Today companies direct their development strategies to create solutions that can solve social or environmental challenges whilst creating social value. This often requires the cooperation of different stakeholders, like businesses, governments, NGOs and individuals at various levels of engagement [4]. Social innovation extends to shared value innovation which derives from the collaborative (open) management ideology that values employees' potential to collaborate at any point of any initiative, though knowledge management for long term value creation [5]. The latest form of innovation, referred to as sustainable innovation, is a convergence of green and clean processes which start from the company's leadership, culture & capabilities and emerge in the new products or services portfolios [6]. Innovation in general becomes more demanding, not necessarily on the product's or service's functionality or performance, but more on the social impact the company must create through its business activities.

In the same sense, sustainability is defined with social factors but more than that with the utilization of environmentally friendly resources for higher profitability and long lasting operations.

The evolution of innovation tends to be a race on efficiency and sustainability mostly from the financial dimension. As the world gets interconnected at a tremendous pace, innovation extends its definition to cover as many dimensions and target groups it can possible touch. Therefore, the evolution of the classical innovation to the integration of the social, shared value and environmental dimensions define the desired sustainability needed to assure wide applicability, fast return on investment and of course strong competitive advantage.

3 The Wide Innovation

The evolution of the traditional innovation, also referred as closed innovation, to the open innovation on thinking outside the box, can be extended to the wide innovation concept characterized by innovation sustainability through its wide application. The width on such innovations varies from the utilization of the intellectual capital as the green fuel for their devolvement on achieving high and fast returns on investments with extensive sustainability.

Closed innovation is internally focused to the organizational boundaries and goals whereas open innovation includes externally focused elements in the organization's innovation model that can generate new products or explore new markets [7]. Wide innovation on the other hand extends the open innovation concept by targeting, and not only exploring new markets. Through rigorous innovation selection organizations move from research to development based on their sustainability strategy.

Open innovation can be related to the blue ocean strategy where new ideas explore new markets and extend to the green ocean where the effectiveness of the blue ocean success is sustained through a green user/client base (Fig. 1).

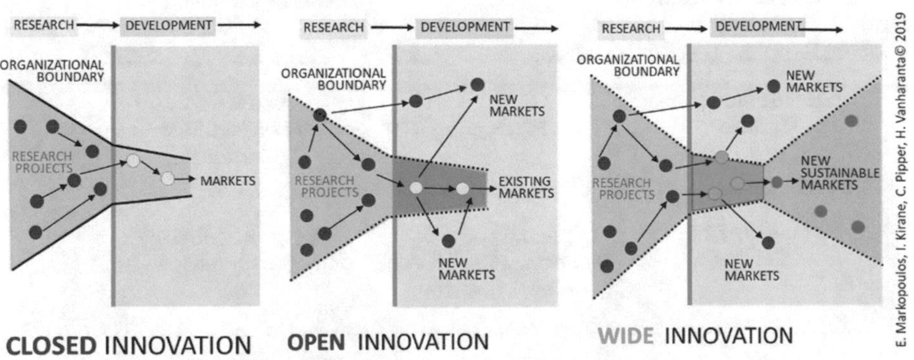

Fig. 1. Closed to wide innovation evolution.

Further characteristics of the wide innovation is that investments target ideas that require intellectual capital to generate web-based products and services that can bring shared value to the organization and the society by maximizing their target groups. High intellectual capital resources assure innovation maturity and technical quality while the web based innovations reduce production costs, logistics, transportation and are most economical and environmentally friendly. The environmental impact of software development, distribution and usage, especially over the web, is absolutely zero. Software products and services ideally define sustainable and green development.

4 The 3S Wide Innovation Matrix

The wide innovation elements provide innovation sustainability through social and shared value based on three human intellectual capital utilization dimensions which are the organizational inner change, and the organizational micro and macro environment influence. This triptych creates the 3S Wide Innovation Matrix.

The matrix combines the wide innovation elements with the utilization dimensions, creating an area of activity with nine cells. Each cell has a value of completeness and a value of importance. The completeness reflects the degree of fulfillment on the cell's target activities, while the degree of importance indicates the alignment of the cell with the corporate sustainability strategy. The depth and detail of both the target activities and the degree of criticality are set by each organization in particular based on the sustainable innovation strategy selected to execute. Figure 2 presents the Wide Innovation 3S Matrix with indicative values on the completeness and criticality factors. The maximum score an organization can obtain is 100 on each factor.

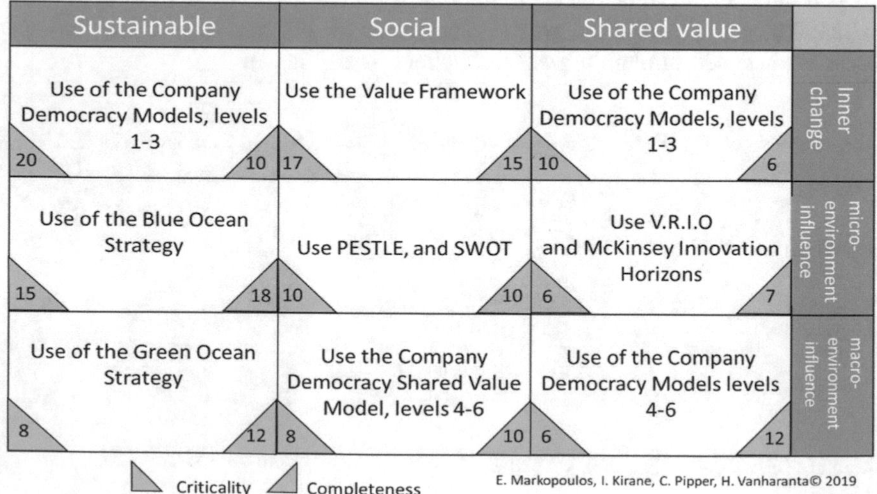

Fig. 2. The 3S Wide Innovation Matrix

Emphasis is given on the criticality dimension of the matrix. It is the degree of completion of the critical elements that directs the efforts on the organizations to reach high effectiveness and success. Each matrix element is represented with the vector WIe (Wide Innovation element) and the values of completeness (i) and criticality (j). (1)

$$\mathrm{WIe} = (i, j) \tag{1}$$

Every element has its score which is completeness against criticality (WIeS). (2)

$$\mathrm{WIeS} = (i/j), where\, j \neq 0 \tag{2}$$

The Wide Innovation score (WIS) an organization can achieve is calculated with the WIS formula. (3)

$$\mathrm{WIS} = \sum\nolimits_{i=1}^{9} i / \sum\nolimits_{j=1}^{9} j, where\, J \neq 0 \tag{3}$$

Wide Innovation initiatives with total score 1 are considered perfect in terms of goal settings and performance, while scores under 1 indicate incompleteness. However total scores over 1 indicate incomplete criticality targets. The total criticality of an initiative must be 100 spread over the matrix elements. It must be noted that the criticality values are subjective and set by the organization.

5 Green Ocean Strategy. Going Beyond Blue Ocean

Companies so far have been focusing on innovation to lead the competition while staying in it, or leave the competition by creating new markets. The existing Red Ocean Strategy is characterized by high competitiveness in an existing market, where boundaries are defined and accepted. The business' objective is to outperform their competitors and gain a greater market share. However, as the market gets overcrowded, potential growth gets restricted, prospects for profits decrease, and innovation targets short term goals, enough to give a temporary competitive advantage.

Diametrically opposed, the Blue Ocean describes a strategy which creates a new and uncontested marketspace. Blue ocean is led by open innovation initiatives to create and capture new demand making competition irrelevant [8]. Both strategies can be considered costly in resources and infrastructure in their attempt to create long term and profitable results. Emphasis is given on tangible results and on the market trends, while decisions are casement or management driven in each case. Both approaches are led by business development strategies either for short term in a saturated, red, market or long term in an uncontested, blue, market regardless the cost and effort needed.

These two extreme cases create room for a third strategy to be created that can work as an evolution of the Blue Ocean by maintain the Blue characteristics and build on them with human intellectual capital for sustainable and economic innovations mainly for the society, which is the actual market, and not the market as such.

As knowledge constantly flows within the organization, Green innovation via knowledge democratization can generate more challenging, rewarding, cost effective long-lasting opportunities. Democratic green innovation leads to Green Oceans where sustainability is neither dystopian nor utopian, but protopian. It encourages knowledge-based innovation through democratic cultures for collaborative thinking leading to profitable and sustainable innovation. By capturing and shifting the demand to new and social driven market spaces, the Green Ocean Strategy allows companies to turn their proactiveness into long-term competitiveness and sustainably. The Green Ocean Strategy is achieved via brain-driven, technology-oriented social innovations (Fig. 3).

6 The Road Map from Red Ocean to Green Ocean

Each of the three oceans can be considered as different market domain. The transition from one to the other can be done with an organizational transformation journey through process canals that create the infrastructure for such a transition, but also the organizational maturity to handle the market challenges and competition.

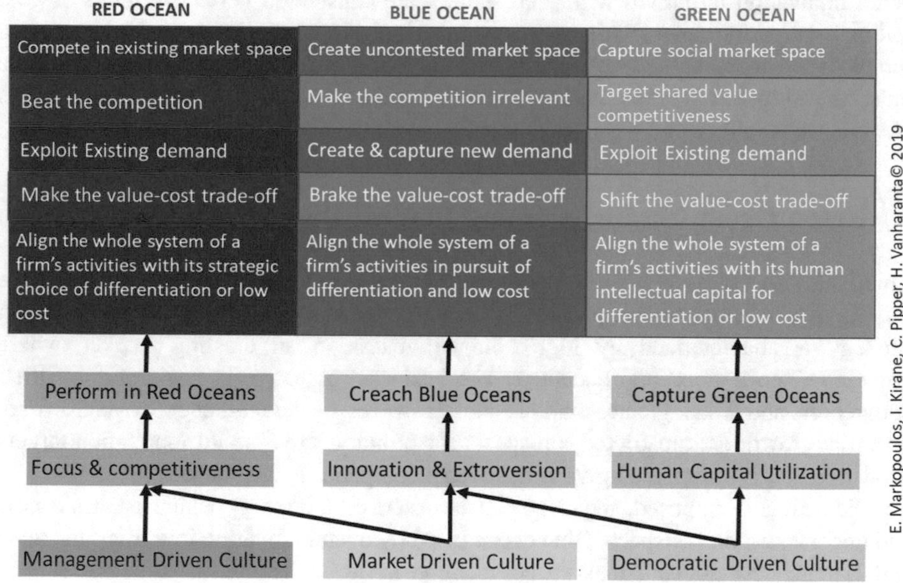

Fig. 3. Ocean strategies evolution and relationships

Companies competing in Red Oceans and seek transition to a Blue Ocean, need to shift their strategies and operations through the Purple Canal in order to reach a Blue Ocean. By using strategic management tools such as SWOT, PESTEL, VRIO, BCM and others, companies can find opportunities to enter an uncontested market which match their core competencies. Furthermore they can attempt to democratize their company structure and knowledge flow and undergo through operational changes that can ignite faster innovation and thereby differentiate from their competition through new markets creation by reaching Blue Oceans.

However Blue Oceans are not the end of the journey. Organizations willing the build on their existing activities and move towards more sustainable strategies must go through the Turquoise Canal to reach a Green Ocean. To analyze the transition strategies for this challenge and to select the proper strategy to go through this canal, companies can use the Wide Innovation 3S Matrix and the Company Democracy Model for effective identification, utilization and capitalization of their human intellectual capital.

This transformation is achieved with democratic generation in a knowledge-based culture for innovation where companies can adequately respond to the shifting demand of consumers within a given industry. This culture builds the self-awareness, self-questioning, self-improvement, and self-actualization operations through which managers can bring companies to the desired market responsiveness level needed to mirror contemporary changes in surrounding environments. Figure 4 presents the two organizational transformation canals towards green oceans.

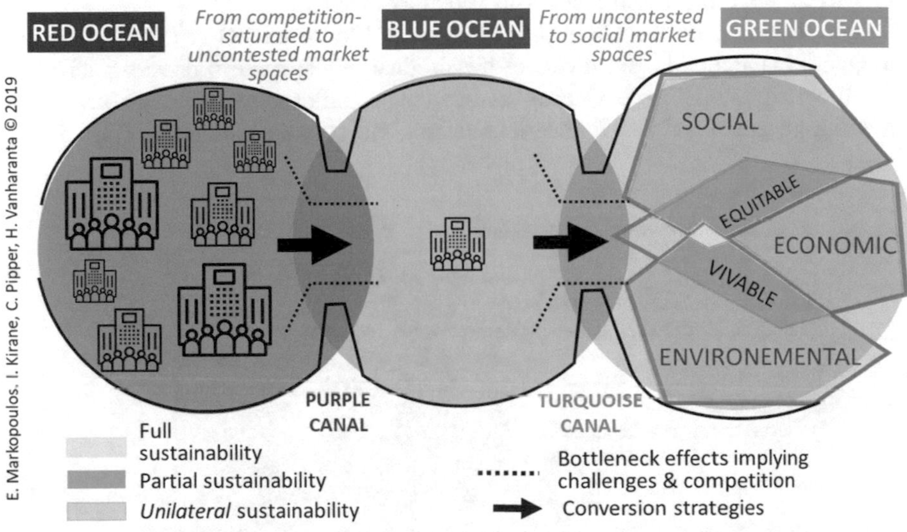

Fig. 4. The oceans transition process through the transformation canals.

7 Democratization of Knowledge for Green Ocean Strategy

The Green fuel required for an organization to reach Green Oceans through the turquoise canal is an everlasting resource that exists in every organization regardless its size, market or expertise. It is the knowledge of the employees which resides with an organization at no cost, enough to lead a green transformation. However the challenge is on the effective way of collecting, analyzing and utilizing such knowledge. Knowledge democratization is an approach to this challenge. No one knows where an idea can come from or what solutions can derive from the organization's human resources by treating employees as capital assets and not as white- or blue-collar workers.

By changing from within the corporate perspective of business actions, managers can aspire to drive their teams towards the green goals. Enacting the Company Democracy Model (CDM) for innovation creation and management, and by actively engaging the society, companies can successfully implement their conversion strategies in the Turquoise Canal.

The Company Democracy Model provides the actions to be performed in order to identify the extent of company democracy, based on the individual and collective evolution dimensions. The model adapts the ancient Hellenic wisdom from and the Delphic Maxims 'Know Thyself', 'Miden Agan' and 'Metron Ariston' [9] in a business context to build a co-evolutionary process framework. It is also aligned with the Co-Evolute methodology for innovation from internal corporate knowledge generation [10].

The classification of the knowledge is done through the use of ontologies to dynamically capture its creation, evolution, and behavior based on the employee's capability, maturity, competence and capacity [11]. This continuous approach turns organizational knowledge from tacit into explicit and generates the fuel for the organizations to move from Red Oceans to Blue Oceans [12].

The adaptation of the Company Democracy Model into Green Ocean Strategy framework maintains the six levels of the original model used to drive organizations from the Red to the Blue Oceans under a new philosophy based on knowledge screening for green and sustainable innovations and corporate strategies (Fig. 5).

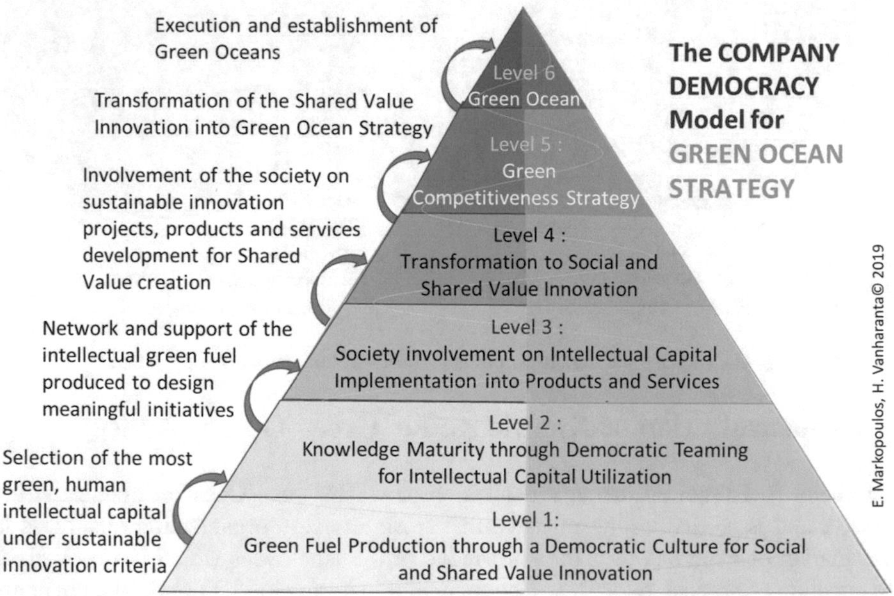

Fig. 5. Knowledge evolution for Green Oceans via the Company Democracy Model

In the model, companies evolve in a pyramid through an incremental progression. Starting from a retrospective/introspective phase, managers use democratic teaming processes to acquire the self-awareness needed to democratically re-organize their actions [13]. Companies move up the levels by expanding their democratic philosophy in their organization's structure and business actions.

8 A Co-evolutionary Spiralmid for Sustainable Innovations

The Company Democracy Model for Green Ocean creation is co-evolutionary based innovation management framework. The model is executed via a spiral process in which organizations create, share and utilize integrated knowledge from the society and its human resources (Fig. 6).

The first level of the model creates a democratic culture through which knowledge is generated and screened for sustainability and wide innovation characteristics. It is the stage where the 3S Matrix assess the quality of the knowledge for its transformation into green intellectual capital fuel to grow sustainable strategies and innovations.

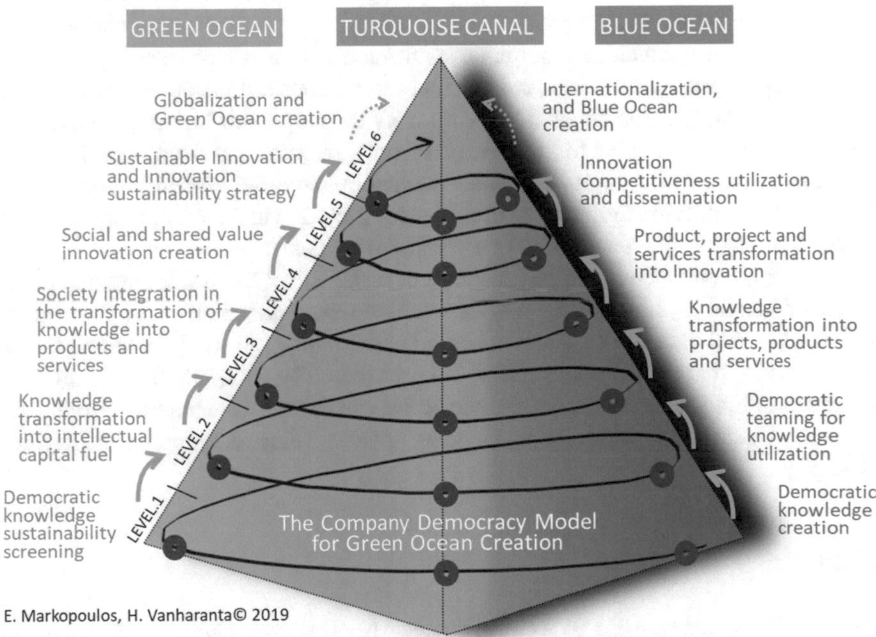

Fig. 6. Co-evolutionary spiral for dynamic sustainability & innovation management

On level two, the model supports the green thinkers of level 1 with the resources needed to validate their knowledge and transform it into practical products and projects designs. Level three integrates the society in the transformation of the green prototypes into actual products and services which are been tested, enhanced and evolved with the society needs, concerns and expectations.

The fourth level and based on the effectives of the deliverables of level three, and turns the social innovation into shared value innovation from which the benefits can be more direct and profitable to both the society and the economy. Level five utilizes the success of the previous levels by designing and executing an organizational innovation sustainability strategy that assures long lasting benefits and global awareness. Lastly level six is where Green oceans are identified based on the effectives of level five.

The global adaptation of sustainable innovation created from the organizational human intellectual capital and the society, is the epitome of the green ocean journey. Blue and Green oceans run in a symbiotic mode as the innovation of the Blue extents to the sustainability of the Green. However Green oceans can be reached without necessarily stopping in a Blue Ocean but by utilizing the innovation philosophy of the Blue directly into the Green innovation approach.

9 Conclusions

Knowledge can be the most valuable resource organizations have, however identifying this knowledge and utilizing it properly turns out to be quite a Herculian challenge. It is not the absence of knowledge from the organization's human recourses but mostly the lack on an appropriate democratic culture that can transform this valuable resource into clean, intellectual, fuel for organizational strategy development. Moving from data and information driven strategies to knowledge driven strategies and from open innovation to sustainable innovation requires qualitative and quantitative human intellectual capital. Blue oceans provide unique opportunities for growth and development which often requires the relative financial capital and time investments.

However Green Oceans can be reached under the Blue Ocean thinking and process with less financial capital investments and market uncertainty risks. The utilization of intellectual capital instead of monetary capital and the integration of the society instead of the trends is what differentiates the two approaches. The company democracy model and the transition journey through the turquoise canal can contribute on turning Blue Oceans to Green Oceans under a Y-theory shared value democratic management philosophy.

References

1. Kim, W.C., Mauborgne, R.: Blue Ocean Strategy: How to Create Uncontested Market Space and Make the Competition Irrelevant. Harvard Business School Press, Boston (2005)
2. Markopoulos, E., Vanharanta, H.: Democratic culture paradigm for organizational management and leadership strategies - the company democracy model. In: Proceedings of the 5th International Conference on Applied Human Factors and Ergonomics AHFE 2014 (2014)
3. Vanharanta, H., Markopoulos, E.: Creating a dynamic democratic company culture for leadership, innovation, and competitiveness. In: 3rd Hellenic-Russian Forum, 17 September 2013

4. World Economic Forum. http://www3.weforum.org/docs/WEF_Social_Innovation_Guide.pdf
5. Markopoulos, E., Vanharanta, H.: The company democracy model for the development of intellectual human capitalism for shared value. Procedia Manuf. **3**, 603–610 (2015)
6. Harvard Business Review. https://hbr.org/2007/10/harvard-business-ideacast-64-s
7. Krause, W., Schutte, C., du Preez, N.: Open innovation in South African small and medium-sized enterprises. In: 42, Proceedings, Cape Town, South Africa, 15–18 July 2012
8. Kim, W.C., Mauborgne, R.: Value innovation: a leap into the blue ocean. J. Bus. Strategy **26**(4), 22–28 (2005)
9. Parke, H., Wormell, D.: The Delphic Oracle, vol. 1, p. 389. Basil Blackwell, Oxford (1956)
10. Kantola, J., Vanharanta, H., Karwowski, W.: The evolute system: a co-evolutionary human resource development methodology. In: Karwowski, W. (ed.) The International Encyclopedia of Ergonomics and Human Factors. CRC Press, Boca Raton (2006)
11. Paajanen, P., Piirto, A., Kantola, J., Vanharanta, H.: FOLIUM - ontology for organizational knowledge creation. In: 10th World Multi-conference on Systemics, Cybernetics, and Informatics (2006)
12. Nonaka, I., Takeuchi, H.: The Knowledge-Creating Company: How Japanese Companies Create the Dynamics of Innovation. Oxford University Press, New York (1995)
13. Markopoulos, E., Vanharanta, H.: Project teaming in a democratic company context. Theor. Issues Ergon. Sci. **19**(6), 673–691 (2018)

Analysis of Correlation Between Surface Roughness of Aluminum Alloy and Human Psychological Perception

Wengqing Fu, Xiaozhou Zhou, and Chengqi Xue(✉)

School of Mechanical Engineering, Southeast University, Nanjing 211189, China
{220174232, zxz, ipd_xcq}@seu.edu.cn

Abstract. As modern products move toward embedded and flat design, the surface materials of products are becoming more and more important in the product experience. However, there has been a strong lack of studies to quantify and link surface materials of products to the human response. Here, we explore the correlation between perception to product surface properties of customers and controllable product surface properties. Analysis reveals five emotional dimensions related to perception of surface physical properties: "uncomfortable - comfortable", "cheap - luxury", "exotic - ordinary", "plain-beauitful" and "artificial - natural". Simultaneously, we deeply explored the relationship between these dimensions and surface roughness of the material. Through the "soft" judgment of psychological attributes and the "hard measurement" of the product surface properties, we have achieved a quantitative assessment of affective surface engineering. These help designers and engineers choose more active material embedded in the product.

Keywords: Affective surface engineering · Kansei · Quantitative

1 Introduction

When designing a product, it is important to understand the relationship between the surface properties of the material and the customer's psychology [1]. According to Choi [2], Materials should be considered for priority when designing due to buyers make decision to buy the products through the color and the materials.

In general, the texture of the materials can be classified by touch and vision perception [1] Chen et al. investigated that psychophysical and affective layers compose the tactile perception [3, 4]. Based on amounts of studies, it concludes that textures are consisted of three major perception dimensions: rough/smooth, hard/soft, and cold/warm [5].

Rosen et al. [6] demonstrated that Kansei adjectives about visual appearance of the texture can be linked to the roughness parameters, thus making design elements controllable by designers and possibly to use to specify product requirements.

The adjective clean expresses customer demands on: cleanability related to the surface properties of the material such as scratch proofness to withstand [7, 8]. ASME standard [9] connect the requirements of smooth to the texture average arithmetic amplitude (Ra).

T. Ahram et al. (Eds.): IHSED 2019, AISC 1026, pp. 126–131, 2020.
https://doi.org/10.1007/978-3-030-27928-8_20

However, only the requirement cleanability related to parameters is not enough. More psychological requirements of surface are not be explored systematicly. In the paper, we complete the affective layers and propose the emotional dimensions then connect these dimensions to the roughness parameters. Our study mainly aims at the aluminum alloy material texture surface.

The aluminum alloy material is divided into eight categories and classified according to the number [10].

In this study, we explored emotional dimentions of perception of surface physical properties and found the correlation between the emotional dimensions and the roughness paramenters. The results are further discussed to apply in the design and maximize the positive aspects of product experiences.

2 Methods and Experiments

2.1 Five Emotional Dimensions

In the experiment I, we explored the emotional dimensions of perception to product surface properties.

In this experiment, we selected twenty-five samples. The sample consists of four surface treatment of aluminum alloy products, including anodizing, drawing, sand-blasting and painting. Twenty participants aged 22–25 (mean age = 23.5 years; 10 men and 10 women) participated in the experiment on the same computer screen, so the only variable in this experiment was the different surface texture. The perceptual vocabulary in the questionnaire was collected from dozens of SCI articles [1, 2, 5, 11–13], and 28 pairs of bipolar adjectives were selected from 467 bipolar-adjective pairs. The questionnaire collects the user's emotional data by scoring, that is, adopts the SD method. Participants completed the experiment by scoring each material sample.

Data Analysis and Results. This experiment used SPSS for statistical analysis. Based on the descriptive statistics obtained from the experiments, we can draw the dimensions. According to the pearson correlation analysis and analysis of variance, five emotional bipolar-adjective pairs: "uncomfortable - comfortable", "cheap - luxury", "exotic - ordinary", "plain-beauitful" and "artificial - natural" were selected for experiment II.

2.2 Relationships Between Emotional Dimesions and Roughness

After the experiment I, we have acertained the emotional dimentions of the perception of surface. In experiment II, we will use five dimentions to explore the relationships between surface roughness parameters and perception of surface physical properties.

The questionnaire for this experiment consisted of 8 aluminum alloy samples and 5 pairs of bipolar words. The eight aluminum alloy samples were brushed and the drawing density was high to low. Its color and size are consistent. Thirty healthy participants (15 men and 15 women) aged 20–40 years underwent experiment II by SD. At the same time, we will use a photoelectric mirror to measure the roughness of these eight samples.

Data Reduction and Analysis. This experiment used SPSS for statistical analysis. The roughness data were matched with the perceptual data.

3 Results

The following four tables (Figs. 1, 2, 3, 4 and 5) show the relationship between the five emotional dimensions and the roughness paramenters of brushed aluminum alloy material surface.

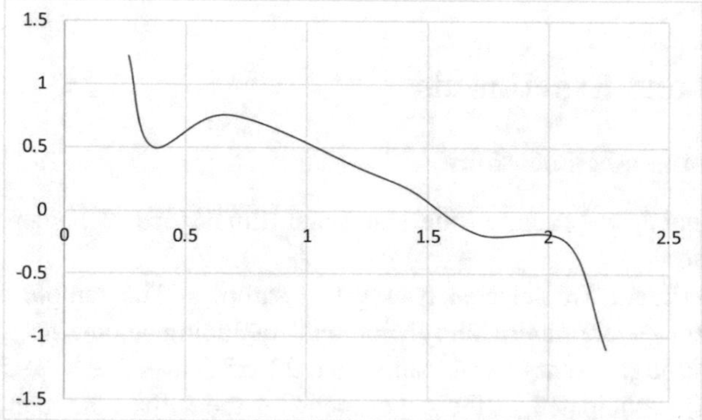

Fig. 1. The relationship between "uncomfortable - comfortable" and roughness

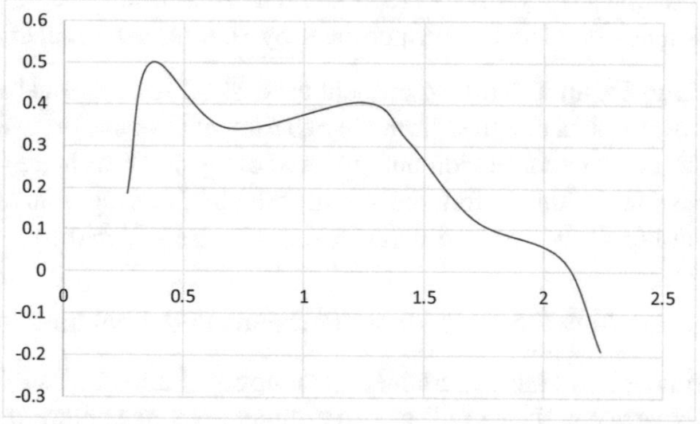

Fig. 2. The relationship between "cheap - luxury" and roughness

The ordinate represents the average value of five emotional dimensions on Likert method and the abscissa represents the roughness of eight samples.

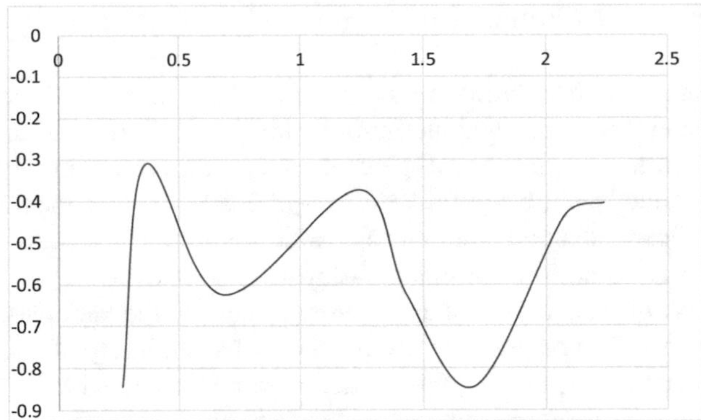

Fig. 3. The relationship between "exotic - ordinary" and roughness

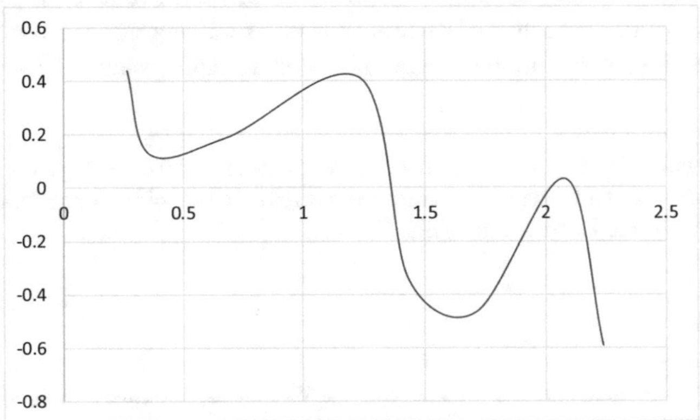

Fig. 4. The relationship between "artificial - natural" and roughness

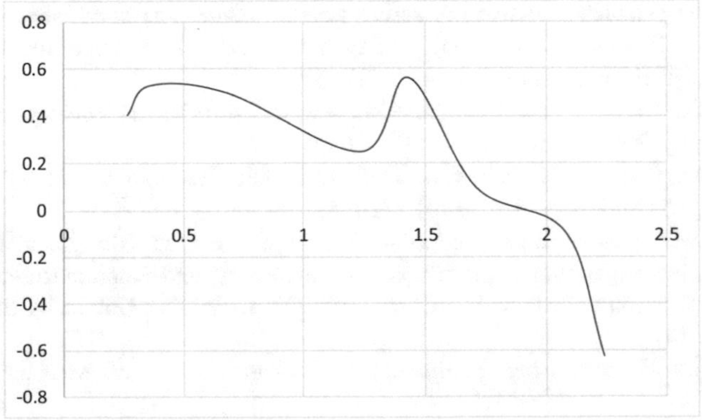

Fig. 5. The relationship between "plain-beauitful" and roughness

4 Discussion and Conclusion

In this study, we found the relationship between the five emotional dimensions and the roughness paramenters of brushed aluminum alloy. From Figs. 1, 2 and 5 above, the roughness will influence the visual perception of "uncomfortable – comfortable", "cheap – luxury" and "plain-beauitful". From Figs. 3 and 4 above, the roughness seems to have no significant effect on emotion of "exotic - ordinary" and "artificial - natural". The overall data is in the form of ups and downs.

The weakness of this study is that, in order to control the variables, the color and gloss are controlled. This may lead to the psychological similarity of the subjects in the process of the experiment, making the results more neutral. In addition to the brushed process there are many common processes worth exploring.

In general, it indicated that different density of brushed mental will influence the perception of customers. The positive meaning of the study is that we propose the emotional dimensions based on material texture surface and connect these dimensions to the roughness parameters. we have achieved a quantitative assessment of affective surface engineering. These not only helps consumers making purchase decisions easier, but also helps designers and engineers choose more active material embedded in the product.

Acknowledgments. This work was supported by Science and Technology on Avionics Integration Laboratory and Aeronautical Science Fund (No. 20185569008) and "the Fundamental Research Funds for the Central Universities".

References

1. Sakamoto, M., Watanabe, J.: Exploring tactile perceptual dimensions using materials associated with sensory vocabulary. J. Front. Psychol. **8**, 569 (2017)
2. Choi, J.: Material selection by the evaluation of diffuse interface of material perception and product personality. J. Int. J. Interact. Des. Manuf. **11**, 967–977 (2017)
3. Tiest, W.M.B.: Tactual perception of material properties. J. Vis. Res. **50**, 2775–2782 (2010)
4. Chen, X., Barnes, C.J., Childs, T.H.C., et al.: Materials' tactile testing and characterisation for consumer products'affective packaging design. J. Mater. Des. **30**, 4299–4310 (2009)
5. Okamoto, S., Nagano, H., Yamada, Y.: Psychophysical dimensions of tactile perception of textures. J. IEEE Trans. Hap. **6**, 81–93 (2013)
6. Rosen, B.G., Eriksson, L., Bergman, M.: Kansei, surfaces and perception engineering. J. Surf. Topog. Metrol. Prop. **4**, 033001 (2016)
7. Bergman, M., Rosen, B.G., Eriksson, L., et al.: Surface design methodology – challenge the steel. J. Phys.: Conf. Ser. **483**, 012013 (2014)
8. Bergman, M., Rosen, B.G., Eriksson, L., et al.: Surface design methodology—the cleanability investigation. In: The 5th Kanesi Engineering and Emotion Research. Linköping University Electronic Press, vol. 100, pp. 705–722. Linköping University Electronic Press, Sweden (2014)
9. ASME 2009 Bioprocessing Equipment. American Society of Mechanical Engineers, New York

10. Schelleng, R.D., Gilman, P.S., Jatkar, A.D., et al.: Research on mechanically alloyed aluminum-alloy products for aerospace applications. J. Metals. **36**, 24–25 (1984)
11. Zuo, H.: The selection of materials to match human sensory adaptation and aesthetic expectation in industrial design. J. Metu J. Fac. Archit. **27**, 301–319 (2010)
12. Guest, S., Dessirier, J.M., Mehrabyan, A., et al.: The development and validation of sensory and emotional scales of touch perception. J. Attent. Percept. Psychophys. **73**, 531–550 (2011)
13. Wang, W.M., Li, Z., Tian, Z.G., et al.: Extracting and summarizing affective features and responses from online product descriptions and reviews: a Kansei text mining approach. J. Eng. Appl. Artif. Intell. **73**, 149–162 (2018)

Human-Robot Interaction

Are We Ready for Human-Robot Collaboration at Work and in Our Everyday Lives? - An Exploratory Approach

Verena Wagner-Hartl(✉), Katharina Gleichauf, and Ramona Schmid

Campus Tuttlingen, Faculty Industrial Technologies, Furtwangen University, Kronenstraße 16, 78532 Tuttlingen, Germany
{verena.wagner-hartl,katharina.gleichauf, ramona.schmid}@hs-furtwangen.de

Abstract. An increase in human-robot collaboration is expected in the future. It can be assumed that this development will affect both, our working environment and our private lives. An exploratory study ($N = 116$) was conducted to get more insight in this important future field. The results of the study can help to understand how human-robot collaboration in the working environment and in our private lives is seen today, and should help to improve the acceptance of this important future field.

Keywords: Human-robot collaboration · Human factors · Working environment · Exploratory approach

1 Introduction

Nowadays, robots that interact with humans are no longer vision of the future. After industry, they are already part in our everyday life. Furthermore, an increase in human-robot collaboration is expected in the future [1]. It can be assumed that this development will affect both, our working environment and our private lives.

The idea of human-robot collaboration is to work hand in hand with a robot by sharing one workspace, interact and using the strengths of robots and humans both together. Through combining the specific skills of both, this kind of collaboration can lead to many advantages. In Germany, collaborating robots are already widespread in automotive or mechanical engineering industries. For example, they could relieve employees on their workplace by lifting heavy loads or increasing quality and precision of products [2].

In addition, in care human-robot collaboration has a great potential to cope with the challenges of demographic change. The use of robots could counteract social problems such as the increasing number of care-dependent people, the shortage of qualified nursing staff or the rising costs of healthcare [3]. For example, in Japan, a care assistant robot named "RIBA" was developed to support nursing staff by lifting bedridden people [4]. Other care robots or rather socially assistive robots are for example used in order to support cognitive training, affective and social aspects, for companionship and entertainment or in physiological therapy of elderly people as well as in rehabilitation,

T. Ahram et al. (Eds.): IHSED 2019, AISC 1026, pp. 135–141, 2020.
https://doi.org/10.1007/978-3-030-27928-8_21

to assist people to walk etc. [1, 3, 5]. Likewise, household assistant robots already have a place in our everyday lives. Besides supportive and informative activities, i.e. Fraunhofer's "Care-O-bot", a gentleman robot which offers the service of a butler, can further build a social bond to the user [6].

Aside from emotional interaction, appearance plays an important role in the acceptance of robots. According to Mori [7] robots that are more similar to humans are less accepted by humans. Up to a certain extent, the similarity of robot and human has an effect on the increase of acceptance. But, on the other hand, following Mori, if the similarity to humans gets too close, it gets to the point where the acceptance level of liking the robot completely collapses ("Uncanny Valley") [7].

The results of a special Eurobarometer survey [8] which was conducted in 2012 ($N = 26751$ participants of 27 Member States of the European Union) gave a first overview of the perception, acceptance and worries towards human-robot interaction. Following the results, more than two third of the participants "(…) have a positive view of robots" (ibid., p. 17). However, they reported mixed feelings about robots in the care section. So, more than half of the participants suggested that robots should not be used to care for people. Furthermore, 27% reported that the use of robots should be prohibited by law for the healthcare sector, 20% for leisure, 34% in education and 60% regarding the care of children, elderly and disabled persons. On the other hand, the participants feel more positive about a robot helping them with their work, especially for tasks that "(…) are to difficult or too dangerous for humans" (ibid, p. 32).

There is still uncertainly, however, how the increasing cooperation with robots will be accepted and in which areas human-robot collaboration is already desired today. Therefore, the aim of the exploratory study was to survey attitudes towards robots in the working environment, in private everyday life and in care. Furthermore, different tasks that can be supported by a robot should be evaluated regarding the acceptance of humans for cooperation with robots. In addition, it should be explored whether gender and age differences exist for the examined areas as well as the different tasks.

2 Materials and Method

2.1 Participants

Overall, 128 participants participated in the online study. Questionnaires from 12 participants (9.38%) were not included in the study because they had not finished the questionnaire. Therefore, the final sample consists of 116 participants who completed the whole questionnaire. Overall, 49 men and 67 women aged between 19 and 69 years ($M = 30.16$, $SD = 13.04$) participated in the online study. Participants were grouped into two age groups using a median split (younger: 24 years and younger, elderly: 25 years and older). 46.55% of the participants worked in a technical work field, 14.66% were from a social discipline, 12.93% from a commercial area and 2.59% from the craft sector. 23.27% were from other disciplines and areas like apprentices, trainees, academic staff, professors, higher grade of the civil service and/or did not describe their working field in more details.

2.2 Materials

The data of this study was collected with an online survey. The participants needed about 10 min to complete the questionnaire. The questionnaire addresses different aspects of participants' opinions about human-robot collaboration in different areas: at their working place, everyday lives and in the care sector. Participants' attitudes towards robots in their working environment as well as in the care sector were rated as follows: negative (−1) – neutral (0) – positive (+1). The different tasks at the working environment (loading and unloading tasks, hand things to someone, bring files and documents, assembly of parts and things at the work place, inform about current topics, others: open responses of the participants) as well as in the everyday live/household (clean-up tasks, organizational tasks, cleaning, cooking, homework assistance, babysitting, others: open responses of the participants) were assessed with a 5-point scale that ranges from very poor (1) to very good (5).

2.3 Analyses

The statistical analyses of the data were conducted using the software IBM SPSS Statistics. The analyses were based on a significance level of 5%. Open responses of the participants were analyzed using the method of qualitative content analyses by Mayring [9].

3 Results

3.1 Working Environment

The results of the study show that the participants' attitudes towards robots in their working environment are rather neutral to positive ($M = .21$, $SD = .81$). Following the results of an analyses of variance, a tendency towards significance is shown regarding age, $F(1, 105) = 2.75, p = .100, \eta^2_{part.} = .026$. Therefore, younger participants tend to see this more positively than elderly do. No significant effects can be shown for gender, $F(1, 105) = 2.23$, $p = .139$, $\eta^2_{part.} = .021$ or the interaction age x gender, $F(1, 105) = .13, p = .715, \eta^2_{part.} = .001$.

Following the results of an analyses of variance with repeated measures, significant differences regarding different tasks at the workplace and the need and acceptance of the help of a robot can be shown, $F_{HF}(3.57, 346.40) = 15.42, p \leq .0001, \eta^2_{part.} = .137$ (Fig. 1). Post-hoc analyses (Sidak) show that loading and unloading tasks were assessed as significant better regarding the need and acceptance of the help of a robot than all other tasks (hand things to someone, $p \leq .0001$; bring files and documents, $p = .002$; assembly of parts and things at the work place, $p \leq .0001$; inform about current topics, $p \leq .0001$). Furthermore, the task "bring files and documents" was assessed significant better than the task "inform about current topics" ($p = .004$). Age and gender differences did not reach the level of significance [age: $F(1, 97) = .28$, $p = .599$, $\eta^2_{part.} = .003$; gender: $F(1, 97) = .97$, $p = .327$, $\eta^2_{part.} = .010$; age x gender: $F(1, 97) = .84$, $p = .363$, $\eta^2_{part.} = .009$].

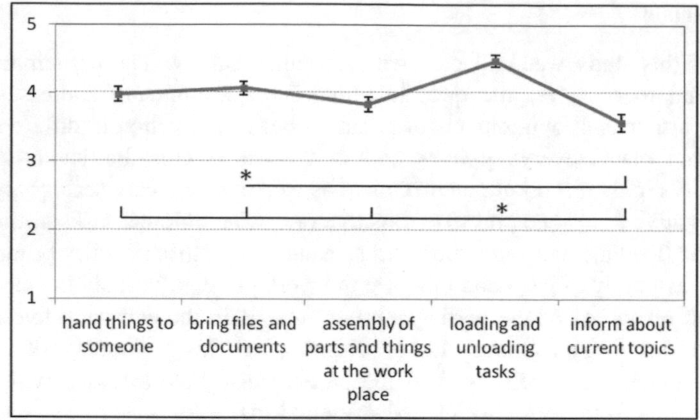

Note: 5-point scale ranges from very poor (1) to very good (5);
* ... p<.05, Sidak; I ... Standard error of mean

Fig. 1. Different tasks at the workplace: Need and acceptance of the help of a robot

Overall, 13 participants made further suggestions for the use of robots in their working environment. The most mentioned were: activities as support for nursing professions or hospital staff (e.g., patients lifting, serve food, sort medicaments, record vital signs and temperature; 4 participants), to make and bring coffee and/or soft drinks (3 participants) and to fill in and send documents (2 participants). Other suggestions that were only mentioned once were: to copy, sort files, storage tasks, query learning materials and schedule coordination.

3.2 Everyday Lives

In addition, the results show significant differences regarding the need and acceptance of the help of a robot for different tasks in the everyday live of the participants, especially regarding housekeeping tasks, $F_{HF}(4.54, 422.51) = 114.34$, $p \leq .0001$, $\eta^2_{part.} = .551$. Following the results of post-hoc tests (Sidak), significant differences can be shown between all six analyzed tasks (see Fig. 2).

Regarding age and gender differences an significant interaction age x gender can be shown, $F(1, 93) = 5.05$, $p = .027$, $\eta^2_{part.} = .052$. Following the results of post-hoc tests (Sidak), elderly men tend to assess the help of a robot in their everyday live poorer than elderly women ($p = .085$). Furthermore, elderly women assess it better than younger women ($p = .088$). All other effects did not reach the level of significance [age: $F(1, 93) = .01$, $p = .908$, $\eta^2_{part.} \leq .0001$; gender: $F(1, 93) = .08$, $p = .773$, $\eta^2_{part.} = .001$].

Overall, 15 participants made further suggestions for the use of robots in the private sector or rather everyday lives: washing and ironing laundry (3 participants), gardening, sweeping/cleaning of outdoor areas and snow shoveling (3 participants), to do one's shopping (2 participants) or to mow the lawn (2 participants). Other suggestions that were only mentioned once were: to repair things in the household, bring files and documents, receive parcels, carry heavy things (e.g., crate), clean windows or to vacuum the apartment.

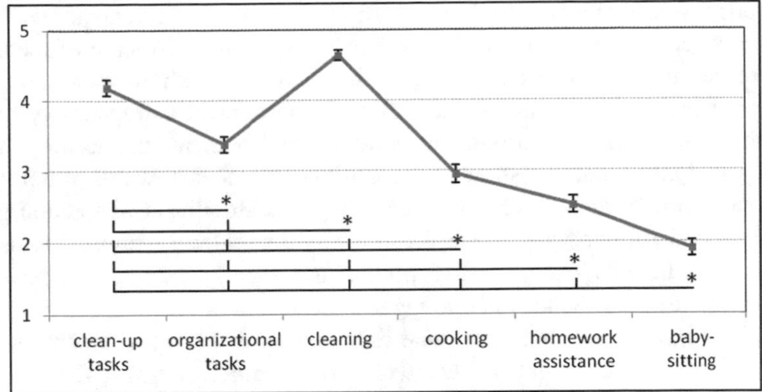

Note: 5-point scale ranges from very poor (1) to very good (5);
* ... p<.05, Sidak; I ... Standard error of mean

Fig. 2. Different tasks – everyday lives: Need and acceptance of the help of a robot

3.3 Care Sector

First, participants were asked about their attitudes towards robots in the care sector. Overall, they are rather neutral to negative ($M = -.12$, $SD = .74$). Gender or age differences cannot be shown [age: $F(1, 105) \leq .0001$, $p = .990$, $\eta^2_{part.} \leq .0001$; gender: $F(1, 105) = .61$, $p = .437$, $\eta^2_{part.} = .006$; age x gender: $F(1, 105) = .54$, $p = .464$, $\eta^2_{part.} = .005$].

Second, they were asked if they would let a robot take care of them. Mean values showed that again, this was rated rather neutral to negative by the participants ($M = -.43$, $SD = .71$). Furthermore, gender or age differences did not reach the level of significance [age: $F(1, 105) = .29$, $p = .594$, $\eta^2_{part.} = .003$; gender: $F(1, 105) = .70$, $p = .404$, $\eta^2_{part.} = .007$; age x gender: $F(1, 105) = .06$, $p = .815$, $\eta^2_{part.} = .001$].

4 Discussion

Overall, the study participants' attitude towards robots in their working environment was rather neutral to positive. All five rated tasks of the working environment (loading and unloading tasks, hand things to someone, bring files and documents, assembly of parts and things at the work place, inform about current topics) were assessed positively regarding the need and acceptance of a robot. The results show that loading and unloading tasks were the most preferred tasks for human-robot collaboration at the workplace. On the other hand, the task "inform about current topics", was the least preferred one of the participants.

Regarding everyday lives, tasks like cleaning and clean-up were very well accepted for human-robot collaboration by the participants. Otherwise, more complex tasks like cooking, or tasks related to childcare or child-rearing like homework assistance or babysitting were not well accepted to be executed by a robot. Furthermore, participants' attitude regarding robots in the care sector pointed in the same direction. Overall, their attitude towards robots in the care sector was rather neutral to negative. The same was shown when they were asked if they would let a robot take care of them.

Differences due to gender and age or the interaction of both, do exist, but can only be shown for two examined areas, the working environment and everyday lives, and not for the care sector. Regarding the working environment, the results show that younger participants tend to show more positive attitudes towards robots in their working environment then elderly do. Interestingly, a significant interaction gender x age can be found for everyday live-tasks in a way that elderly men tend to assess the help of a robot in their everyday live poorer than elderly women and furthermore, elderly women assess it better than younger women do.

This study has some limitations. One limitation is that the participants were only asked about their attitudes but did not work with a robot or use it for the different everyday-tasks (e.g., like a butler [6]), or experienced how it feels when a robot is used to take care of them or interacts with them (e.g., socially assistive robots, care assistant robots etc. [1, 3–5]). Further studies should provide the opportunity to analyze whether the attitude towards human-robot collaboration can be improved by providing real examples of collaborative robots for the different analyzed areas that can be tested and experienced by the participants. Another interesting point for future studies could be, to develop programs and interventions that support people to get used to interact with a robot.

To sum it up, the results of the presented exploratory study can help to understand how human-robot collaboration in the working environment and in our private everyday lives is seen today and should help to improve the acceptance of this important future field.

Acknowledgments. The authors would like to thank Tamara Weiß for her support in conducting the online study, and all participants who participated.

References

1. Ajoudani, A., Zanchettin, A.M., Ivaldi, S., Albu-Schäffer, A., Kosuge, K., Khatib, O.: Progress and prospects of the human–robot collaboration. Auton Robot **42**, 957–975 (2018)
2. Institut für angewandte Arbeitswissenschaft e.V. (ifaa): Mensch-Roboter-Kollaboration. [Human-robot collaboration]. ifaa – Zahlen, Daten, Fakten [ifaa – Figures, data, facts], pp. 1–4 (2017)
3. Olaronke, I., Oluwaseun, O., Rhoda, I.: State of the art: a study of human-robot interaction in healthcare. Int. J. Inform. Eng. Electron. Bus. **3**, 43–55 (2017)
4. Mukai, T., Hirano, S., Nakashima, H., Kato, Y., Sakaida, Y., Guo, S., Hosoe, S.: Development of a nursing-care assistant robot RIBA that can lift a human in its arms. In: The 2010 IEEE/RSJ International Conference on Intelligent Robots and Systems, pp. 5996–6001 (2010)
5. Abdi, J., Al-Hindawi, A., Ng, T., Vizcaychipi, M.P.: Scoping review on the use of socially assistive robot technology in elderly care. BMJ Open **8**, 1–20 (2018)
6. Kittmann, R., Fröhlich, T., Schäfer, J., Reiser, U., Weißhardt, F., Haug, A.: Let me introduce myself: I am Care-O-bot 4, a gentleman robot. In: Mensch und Computer 2015 Tagungsband, pp. 22–232. Oldenbourg Wissenschaftsverlag, Stuttgart (2015)

7. Mori, M.: The uncanny valley. IEEE Robot. Autom. Mag. **19**, 98–100 (2012)
8. European Commission: Public attitudes towards robots. Special Eurobarometer 382, report, 97 pp. (2012)
9. Mayring, P.: Qualitative Inhaltsanalyse: Grundlagen und Techniken [Qualitative content analysis: basics and techniques]. Beltz, Weinheim and Basel (2010)

Human-Robot Cooperation: Link Between Acceptance and Modes of Cooperation Chosen by Operator with a Robot

Adrian Couvent[1(✉)], Christophe Debain[1], Nicolas Tricot[1], and Fabien Coutarel[2]

[1] Université Clermont Auvergne, IRSTEA, UR, TSCF, Centre de Clermont Ferrand, 9 Avenue Blaise Pascal CS 20085, 63178 Aubière, France
{adrian.couvent, christophe.debain, nicolas.tricot}@irstea.fr

[2] Université Clermont Auvergne, ACTé, 63000 Clermont-Ferrand, France
fabien.coutarel@uca.fr

Abstract. Considering current developments, robotics allows users to choose the cooperation mode that facilitates the goals of each member of a team. This choice will influence the robot acceptance. In order to evaluate this, an indicator is proposed to evaluate the cooperation and applied during an experiment. At first glance, operators do not favour a cooperation mode. However, the robot autonomy level is linked with the evolution of the acceptance.

Keywords: User-centred systems · Human-systems integration · Systems engineering · Human-Machine systems · Usability and user experience

1 Introduction

Nowadays robots are more and more present and this in many fields. They are integrated in the industry for many years and have begun their development in other fields (agriculture, shipping, etc.). The capacities of robots, in particular their ability of adaptation to the environment, are constantly growing. Since the 2000s and even recently, they acquired a certain degree of intelligence, control and perception that allows them to evolve in complex working environments. This embedded intelligence on robots is also implemented to make them cooperative. Thus, the latest developments permit complex interactions between robots and humans through the convergence of automatic fields and of the human sciences. This allows imagining to configure and to reconfigure the robot behaviour in real time regarding its human partner to improve human-robot cooperation.

T. Ahram et al. (Eds.): IHSED 2019, AISC 1026, pp. 142–148, 2020.
https://doi.org/10.1007/978-3-030-27928-8_22

2 Background

Human-robot cooperation can be defined as cooperation between two agents. These agents are in cooperation if each of them strive towards goals although interfering with others and tries to manage such interferences in order to facilitate the activities of the other [1]. Different cooperation modes can be defined [2]: perceptive mode, mutual control mode, shared control mode, function delegation mode and fully autonomous mode. This work is focused on the shared control mode and the delegation mode. These different modes and models need different levels of autonomy. In robotics, autonomy is defined as the ability of a robot to detect its environment, plan for that environment and act on that environment [3]. Autonomy can be classified through the Levels Of Robot Autonomy (LORA) [4]. Considering this, autonomy levels depend on cooperation modes [2]. The autonomy level influences Human-Robot Interactions (HRI) and despite several existing metrics [5], none of them allows to identify cooperation mode and its performance.

In human-robot cooperation, different autonomy levels can influence a key human factor for a suitable integration of robotised or automated systems: the acceptance, [6] and [7]. This idea is widely developed with regard to autonomous cars [8]. The acceptance is defined as a progressive and complex phenomenon which can be apprehended with several approaches in order to evaluate and/or predict the adoption of a technology [9]. The acceptance can be described as a combination of attitudinal, intentional and behavioural acceptance [10]. Many models are proposed in literature to evaluate acceptance like the Technology Acceptance Model [11]. In 2003, a model was proposed to unify eight models in a professional and social context: Unified Theory of Acceptance and Use of Technology (UTAUT) [12]. This model has been extended to other context through the development of UTAUT2 [13].

Considering previous elements, acceptance is dependent on cooperation. The understanding of acceptance will improve human-robot cooperation as well as robot integration. So, is there a link between acceptance and cooperation modes? To ensure the best performance of the human-robot cooperation, its understanding is a key point for the robot. It is therefore relevant to ask modes of cooperation are favoured over others?

In the light of these elements, we hope that robotic allows users to choose cooperation with the best facilitation good and that this choice will improve the acceptance of the robot.

3 Indicator for Cooperation Mode

As autonomy levels depend on cooperation modes which influences Human-Robot Interaction [4]. A common metric in cooperation issues is the crypsis coefficient [14]. Therefore, its interest consists in the adaptation evaluation. Metrics as neglect tolerance or plan state represent the ability to be autonomous. None of them allows identifying cooperation mode and its performance. Considering this, an indicator (COoperative Action: Coa) has been designed (Table 2) to identify cooperation modes and autonomy levels. It is based on the Facilitation of Goals (FG) in the cooperation. This indicator represents the performance and the cooperation mode. The indicator sign represents if the robot executes the delegated function (Coa > 0) or if a shared control mode is needed (Coa < 0) considering the interactions in situation of interference.

Table 1. Definition of the indicator Coa

Type of cooperation	FG for operator	FG for robot	Coa definition
The operator adapts his behaviour and goes through the area with interferences	+++	+++	$Coa = 1$
The operator help the robot which is blocked with non-physical proxemics interaction	++	++	$Coa = 1 - \frac{Time_{robot\ blocked}}{Time_{in\ area\ with\ obstacles}}$
No area with obstacle encounter	+−	+	$Coa = 0$
The operator help the robot which is blocked with physical proxemics interaction	−	+	$Coa = -\frac{Time_{robot\ blocked}}{Time_{in\ area\ with\ obstacles}}$
The operator cannot help the robot, which is blocked and requires help from another agent	−	−	$Coa = -1$

If the Coa is equal to 0.7, the cooperation mode stays a function delegation mode because the robot continues to execute its function without physical operator interventions. If the Coa is equal to −0.7, the cooperation mode needs to change to a shared control mode with physical interventions of operator.

4 Field Test

In order to evaluate how users, cooperate and their acceptances progresses, we propose here to observe the human-robot cooperation thanks to an example of picking task.

This observation is achieved in different cases where the robot encounters difficulties to achieve its task. The operator chooses his cooperation mode considering if he physically interacts with robot (Coa < 0) or interacts without physical interaction (Coa > 0). In the first cooperation mode, the LORA is lower than in the second mode.

Considering these elements, an experiment has been developed to evaluate our hypotheses in the case of piece picking. For this purpose, an experimental platform has been designed.

4.1 Experimental Platform, Robot and Subjects

The experimental platform is composed of four shelves, identified with letters. On each shelf, there are six packages, identified with the letter of the shelf and a number according to their position on the shelf.

Sherpa by Norcan robot is used. The robot has three interfaces: a physical interface, a light interface and an audio interface. The physical interface is composed of two sets of four buttons, located at the front and at the rear. Each set is composed of three buttons to choose the mode: follow me, precede me mode or automated return. There is also a red emergency button. When the robot enters or exits a mode, it emits an audible signal and modifies its interface colour. The robot is able to follow (LORA = 4) or to precede (LORA = 3) someone thanks to rangefinders. The automated return defines its trajectory (LORA = 8). The robot can avoid obstacles or stop its movement, this function is full autonomous (LORA = 10). The cooperation model is the leader-assistant model with operators as leaders and the robot as the assistant.

The panel of subjects is composed of twenty-three voluntary people. They are not familiar with picking tasks and work in different fields.

4.2 Experiment

To begin, the operator answers a UTAUT2 questionnaire to evaluate his acceptance. After an explanation of the task, the operator picks up ten orders with a cart. Then the robot is introduced with a first order. Following this trial, the operator answers the UTAUT2 questionnaire. Then, the operator must pick up ten orders with the robot. To create interferences, four obstacles are introduced (Table 1). In the 2nd, 3rd, 4th and 5th orders, only one obstacle is used. In the 7th, 8th and 9th orders, two obstacles randomly chosen are used. At the experiment end, the operator answers the UTAUT2 questionnaire. During this experiment, all orders are recorded from a camera and robot data are recorded too.

Table 2. Obstacle to create interferences

Goals of the obstacle	Hide the operator	Reduce the space to manoeuvre	Narrow passage	Block the displacement of the robot
Schematisation				Obstacle on the ground
Interference severity	Low	Medium	Strong	Critical

5 Definition and Analysis

The Coa is computed after an analysis of the activity based on the viewing of videos and video processing on 20 subjects. The Table 3 shows the subjects distribution according to three levels of the Coa.

Table 3. Distribution of the Coa value on the twenty subjects

Coa	Number (percentage) of subjects
More than 0.05	9 (45%)
Between 0.05 and −0.05	2 (10%)
Less than −0.05	9 (45%)

Concerning the acceptance, its evolution between the three questionnaires is computed as the difference of the score between each questionnaire (2-1, 3-2 and 3-1).

In view of Table 3, no cooperation mode is favoured within the panel of subjects. This first result shows the capacity to identify the type of cooperation used by the operator with analysis of its activity. This can be automated directly on the robot.

Concerning the acceptance, its evolution and Coa values, a Multiple Correspondence Analysis (MCA) highlighted a link between the acceptance evolution (between the introduction and the end of experiment) and Coa values i.e. the cooperation mode.

The Fig. 1 shows that the acceptance has a positive evolution for the subjects majority who used a shared control mode (G1) contrary to subject who stayed in function delegation mode (G2). In regards of these results, cooperation modes have an impact on the acceptance evolution of operators.

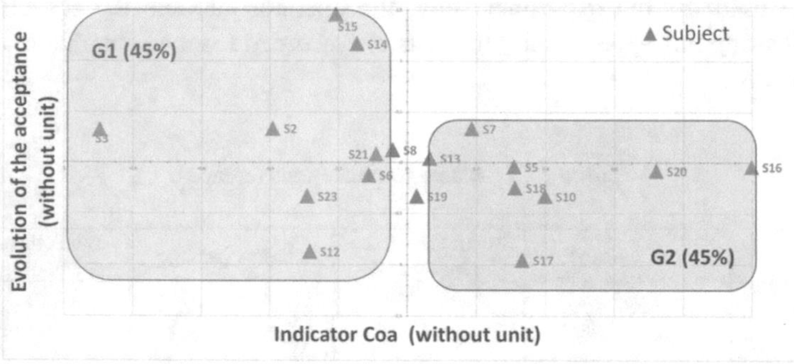

Fig. 1. Point cloud of subjects depending of Coa (abscissa) and the evolution of acceptance (ordinate)

6 Conclusion

In this paper, Coa indicator is proposed. It has been applied to evaluate cooperation type between robot and operator in picking task using a leader-assistant model. The acceptance has been evaluated using a UTAUT2 questionnaire before, during and after the cooperation with the robot.

Considering the results of the Table 3 and the dependence of autonomy levels to cooperation modes, as no cooperation modes are favoured between shared control mode and function delegation mode no autonomy levels are favoured too. In regards of the Fig. 1 results as the function delegation mode corresponds to a higher LORA than the shared control mode, the robot autonomy level has an impact on the acceptance and so the operator perception. Consequently, operators using robots in the function delegation mode with higher levels of autonomy will have a decreasing acceptance in case of interferences contrary to operators using robot in the shared control mode with a lower levels of autonomy.

So a link between human-robot cooperation mode, shared control mode or function delegation mode, and acceptance has been established. In the conditions of our experiment, no cooperation modes have been favoured. These elements are very important to understand and improve the user experience. It allows anticipating the acceptance evolution to improve the human-robot cooperation.

However, these results need to be nuanced because of the limited number of subjects, the limited number of cooperation modes proposed by the robot and their high acceptance. Nevertheless, these results allow imagining evaluating a level of cooperation with the observation of operator behaviour.

Acknowledgments. This research was financed by the French government IDEX-ISITE initiative 16-IDEX-0001 (CAP 20-25 with the support of the regional council Auvergne-Rhône-Alpes and the support with the European Union via the program FEDER).

References

1. Pacaux-Lemoine, M.-P., Simon, P., Popieul, J.-C.: Human-machine cooperation principles to support driving automation systems design. Presented at the Fast Zero 2015, Gothenburg, Sweden, p. 9 (2015)
2. Hoc, J.-M., Chauvin, C.: Cooperative implications of the allocation of functions to humans and machines. Working document (2011). http://jeanmichelhoc.free.fr/pdf/HocChau%202011.pdf
3. Beer, J.M., Fisk, A.D., Rogers, W.A.: Toward a psychological framework for levels of robot autonomy in human-robot interaction. Human Factors and Aging Laboratory, Atlanta, Technical report HFA-TR-1204 (2012)
4. Beer, J.M., Fisk, A.D., Rogers, W.A.: Toward a framework for levels of robot autonomy in human-robot interaction. J. Hum.-Robot Interact. **3**(2), 74–99 (2014)
5. Goodrich, M.A., Schultz, A.C.: Human-robot Interaction: a survey. Hum.-Comput. Interact. **1**(3), 203–275 (2007)
6. Lotz, V., Himmel, S., Ziefle, M.: You're my mate – acceptance factors for human-robot collaboration in industry. Presented at the International Conference on Competitive Manufacturing, Stellenbosch, South Africa, p. 10 (2019)
7. Beer, J.M., Prakash, A., Mitzner, T.L., Rogers, W.A.: Understanding robot acceptance. Human Factors and Aging Laboratory, Atlanta, Technical report HFA-TR-1103 (2011)
8. Nordhoff, S., de Winter, J., Kyriakidis, M., van Arem, B., Happee, R.: Acceptance of driverless vehicles: results from a large cross-national questionnaire study. J. Adv. Transp. **2018**, 1–22 (2018)
9. Bobillier Chaumon, M.-E.: L'acceptation située des technologies dans et par l'activité: premiers étayages pour une clinique de l'usage. Psychol. du Travail et des Organ. **22**(1), 4–21 (2016)

10. Quiguer, S.: Acceptabilité, acceptation et appropriation des Systèmes de Transport Intelligents: élaboration d'un canevas de co-conception multidimensionnelle orientée par l'activité. Ph.D. Thesis, Université de Rennes 2 (2013)
11. Bröhl, C., Nelles, J., Brandl, C., Mertens, A., Schlick, C.M.: TAM reloaded: a technology acceptance model for human-robot cooperation in production Systems. In: Stephanidis, C. (ed.) HCI International 2016 – Posters' Extended Abstracts, vol. 617, pp. 97–103. Springer International Publishing, Cham (2016)
12. Venkatesh, V., Morris, M.G., Davis, G.B., Davis, F.D.: User acceptance of information technology: toward a unified view. MIS Q. **27**(3), 425–478 (2003)
13. Venkatesh, V., Thong, J.Y.L., Xu, X.: Consumer acceptance and use of information technology: extending the unified theory of acceptance and use of technology. MIS Q. **36**, 157–178 (2012)
14. Damian, D., Hernandez-Arieta, A., Lungarella, M., Pfeifer, R.: An automated metrics set for mutual adaptation between human and robotic device. Presented at the IEEE International Conference on Rehabilitation Robotics, ICORR, pp. 139–146 (2009)

From HCI to HRI: About Users, Acceptance and Emotions

Tanja Heuer[1(✉)] and Jenny Stein[2]

[1] Department of Computer Science, Ostfalia University of Applied Sciences, 38302 Wolfenbüttel, Germany
ta.heuer@ostfalia.de

[2] Department of Electrical Engineering, Ostfalia University of Applied Sciences, 38302 Wolfenbüttel, Germany
je.stein@ostfalia.de

Abstract. HCI and HRI are two fields where a human being plays the main role. For this reason it is essential to investigate acceptance and its factors. As HRI is a relatively new and complex field, it is even more important to examine the wider range of acceptance coefficients. Therefore, we compare existing HRI acceptance models with existing HCI attitude models to point out missing factors in HRI. We could show that acceptance models leave out important factors although they should have been build upon HCI models and adapted according to extended influencing factors. Instead a new model is set up with "new" factors and "old" HCI elements such as cognitive components are left out. The cognitive component of attitude investigates the correlation of experience and situated knowledge.

Keywords: HRI · HCI · Mental models · Acceptance · User-centered · Design Interaction

1 Introduction

Human-computer interaction (HCI) as well as human-robot interaction (HRI) are focusing more and more on a user-centered design approach and acceptance. Several intersections and distinctions of HCI and HRI are already examined [2, 3]. Usability and user experience is a fundamental objective of research in HCI. In HRI research about acceptance and trust are just as important. HCI investigates emotional factors how software is used and perceived [1], HRI examines ethical and social implications [2] and acceptance factors [9, 10]. Technical devices in general are widely accepted and a life without smartphones, laptops and smart devices has become vital and will increase furthermore. Though, for robots as the next generation technology a lot of investigation is being put into ethical concerns and acceptance, whereas it is almost being deemed satisfactory and "normal" for HCI. Nobody discusses computerized call-centers where you do not talk to a person anymore. However for robots this looks slightly different. Expectations and perceptions differ a lot. Robots should not be seen as technical devices anymore, moreover they should be considered as a companion, behave like an autonomous being which opens a new dimension in human-machine-

T. Ahram et al. (Eds.): IHSED 2019, AISC 1026, pp. 149–153, 2020.
https://doi.org/10.1007/978-3-030-27928-8_23

interaction [4]. The robot is going to be the "executing" one and the reactions we receive differ a lot – in HCI it is a message box, in HRI it is about a verbal conversation, maybe with emotions and an "own" mind. Negative responses, for example an error message in HCI leads often to disappointment by the user whereas in HRI negative responses might be positive because the robot reacts the "right" way and give the user some interaction. There are a many differences but still it is all about the user and the possibilities how to interact with these devices and to accept them.

Perception of robot and its actions are important factors to investigate for acceptance in HRI [9], but it is left out why robots are perceived as they are and how this possibly can be improved. Therefore, the question we want to focus on in our paper is about how the mental models of HCI [1] can be transferred to HRI and how this can be included into the existing acceptance models to improve the interaction with and perception of social robots. Mental models investigate psychological and physiological aspects of information processing and these aspects should also gain importance in HRI. Both disciplines own a large set of methods and models regarding user-centered design and how the user itself and his data can be put into the focus of software development. Exchanging this tool-set open new perspectives for the research and improvement within both of these two disciplines.

In this paper we introduce mental models and acceptance models, the way they are used so far and how these models can be combined to terms of acceptance in human-robot interaction. First, we give an overview of acceptance and its factors in both disciplines. In a second step, we introduce mental models, compare both models and map out potential gaps in the acceptance model of HRI.

2 Acceptance Factors

2.1 Acceptance in HCI

HCI acceptance research focuses more on technology acceptance. The technology acceptance model (TAM), introduced in the 1980s, can be used to explain and forecost user acceptance and is determined by two key factors: perceived usefulness and perceived ease of use [16]. TAM focuses on how to increase users willingness to use information system, because in addition to task performance user acceptance is the one of the main factors in that are essential to a system's success [5]. Additionally, mental models as a method of psychology guideline the design of applications to enhance acceptance by addressing emotions and memory information [11, 12]. Mental models will be further discussed in Sect. 3.

2.2 Acceptance in HRI

In HRI, acceptance of interaction between human and robot is even more complex [16]. It is investigated how users perceive the robot and the interaction and how the interaction should take place. So far, two acceptance models describe the main factors for human-robot interaction [9, 10]. The acceptance model of Heerink et al. [9] sets up a model for the intention to use robots with factors and its correlations. Main factors are perceived adaptivity, sociability, usefulness, ease of use, enjoyment, trust and social influence. The second model of Shin et al. [10] simplifies the first model and leaves out factors like ease of use, trust and social presence of the robot. The focus on the model is on factors how the robot and its capabilities are perceived. It is a double sided communication which does not only need to work on the robots part but also from the perspective of the users.

Furthermore, Weiss et al. [17] introduced an USUS model to evaluate human-robot interaction. USUS addresses the main categories usability, social acceptance, user experience and societal impact where all categories can be seen as affective component of attitude and cognition and behavior is left out. Young et al. [15] emphasised the difference between interaction of human and robot (HRI) and human and technical device in general (HCI). As one important factor personal situated context (social and physical) is pointed out which is necessary to gain a successful interaction.

One important factor left out in all of the models is the connection of experienced and memory information itself. It is investigated how a robot should look like or how it should behave but why it is perceived like it is, is left out.

3 Mental Models

In general, there are two ways of understanding mental models. According to Johnson-Laird [11] mental models are representations that are formed and applied to a situation or problem, retrievable from the working memory. Especially logical reasonings are investigated in this context. Knauff et al. [12] mental models are functional models which are used to explain complex phenomena, retrievable from the long-term memory (production memory). To retrieve information, the working and the production memory is used and information is compared. The final action depends on the outcome of this comparison. One example therefore are icons for actions which are associated with the real object and the user knows "automatically" what it is for.

As the mental model of memories as theoretical method is difficult to adapt, we want to focus on the attitude component model [13, 14]. The attitude model describes the correlation between affective, behavioral and cognitive components of attitude. The affective component refers to emotions and feelings related to an object or situation. The behavioral component describes the way how we act or react towards the object or situation, often influenced by emotions. Cognition includes beliefs or attributes associated with an object or situation, e.g. background information.

One overall goal of the models is to design applications in a way, that it can be easily understood by rely on already known models. One example is the use of visual icons to make people automatically connect it with already known features. Users should not be forced to learn and adapt to new models with every new application [6].

For this reason we want to propose the approach of mental models, why acceptance models of HRI are not satisfactory and how mental models can be transferred for human-robot interaction. Figure 1 contrasts the acceptance model of Shin et al. [10] and components of the attitude model and its correlations [7, 8]. As it can be seen, the affective component is investigated quite detailed. With the analysis of perception regarding the robot and its capabilities, researchers gain a wide-ranging view.

Furthermore, the behaviour of the user can be analysed based upon reactions and responses towards the robot. What is left out in the acceptance model are the cognitive elements of the mental models. Even though, the field of robots is a relatively new field, users tend to compare and link new situations to already known objects. The way we perceive things or how we react, is always connected to a situated context such as social or cultural background, former experiences, associations and beliefs. For this reason, as emotions and companionship play an increasing role for human-robot interaction what and how information is linked in the memory and how perception is connected to daily routines and already known objects or situations.

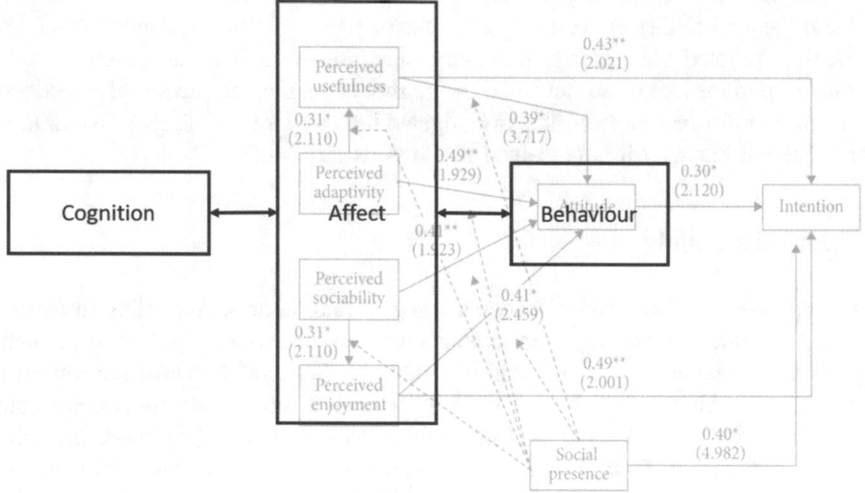

Fig. 1. Acceptance model [10], ABC model [15] and its correlations [8]

4 Discussion

As HRI still is a relatively new field and the interaction between human-robot and human-computer differs in the way, that human-robot interaction should be way more personal, emotional with the focus on companionship. To make people enjoy and accept that, factors for acceptance need to be adapted compared to HCI, especially the factors which makes these two fields different. Acceptance models should give a holistic view into essential factors. But as HRI can be seen as the extended form of HCI, methods do not need to be developed from scratch but existing models should be adapted to respective situations. As already highlighted by Young et al. [15] methods of HCI are not able to represent the whole process of human-robot interaction but need to be expanded. As it can be seen, the acceptance models created for HRI extended the technology acceptance model (TAM) but left out the factor of cognitive competences which might explain the perception of robot and its actions. As mental models and in particular cognitive sciences are already a substantial element of human-computer interaction, this should be also taken into account for human-robot interaction.

Acknowledgments. This work was supported by the Ministry for Science and Culture of Lower Saxony as part of the program "Gendered Configurations of Humans and Machines (KoMMa.G)".

References

1. Heinecke, A.M.: Mensch-Computer-Interaktion: Basiswissen für Entwickler und Gestalter. Springer, Heidelberg (2011)
2. Huang, W.: When HCI meets HRI: the intersection and distinction (2015)
3. Sucar, O.I., Aviles, S.H., Miranda-Palma, C.: From HCI to HRI-usability inspection in multimodal human-robot interactions. In: Proceedings of the 12th IEEE International Workshop on Robot and Human Interactive Communication, ROMAN 2003, pp. 37–41. IEEE, November 2003
4. Dautenhahn, Kerstin: Socially intelligent robots: dimensions of human–robot interaction. Philos. Trans. Roy. Soc. B: Biol. Sci. **362**(1480), 679–704 (2007)
5. Davis, F.D.: Perceived usefulness, perceived ease of use, and user acceptance of information technology. MIS Q. **13**(3), 319–340 (1989)
6. Lin, P., Abney, K., Bekey, G.A.: Robot Ethics: The Ethical and Social Implications of Robotics. The MIT Press, Cambridge (2014)
7. Anderson, J.: The Architecture of Cognition. Harvard University Press, Cambridge (1983)
8. van Harreveld, F., Nohlen, H.U., Schneider, I.K.: The ABC of ambivalence: affective, behavioral, and cognitive consequences of attitudinal conflict. In: Advances in Experimental Social Psychology, vol. 52, pp. 285–324. Academic Press (2015)
9. Heerink, M., Kröse, B., Evers, V., Wielinga, B.: Assessing acceptance of assistive social agent technology by older adults: the almere model. Int. J. Social Robot. **2**(4), 361–375 (2010). https://doi.org/10.1007/s12369-010-0068-5
10. Shin, D.H., Choo, H.: Modeling the acceptance of socially interactive robotics: social presence in human-robot interaction. Interact. Stud. **12**(3), 430–460 (2011). https://doi.org/10.1075/is.12.3.04shi
11. Johnson-Laird, P.N.: The history of mental models. In: Psychology of Reasoning, pp. 189–222. Psychology Press (2004)
12. Knauff, M.: Mentales modell. In: Wirtz, M.A. (Hrsg.), Dorsch – Lexikon der Psychologie. Zugriff am 05.11.2017 unter (2017). https://m.portal.hogrefe.com/dorsch/mentales-modell/
13. Rosenberg, M.J., Hovland, C.I.: Cognitive, affective, and behavioral components of attitude. In: Rosenberg, M.J., Hovland, C.I., McGuire, W.J., Abelson, R.P., Brehm, J.W. (eds.) Attitude Organization and Change: An Analysis of Consistency Among Attitude Components, pp. 1–14. Yale University Press, New Haven (1960)
14. Breckler, S.J.: Empirical validation of affect, behavior, and cognition as distinct components of attitude. J. Pers. Soc. Psychol. **47**(6), 1191–1205 (1984)
15. Young, J.E., Sung, J., Voida, A., Sharlin, E., Igarashi, T., Christensen, H.I., Grinter, R.E.: Evaluating human-robot interaction. Int. J. Social Robot. **3**(1), 53–67 (2011)
16. Venkatesh, V., Morris, M.G., Davis, G.B., Davis, F.D.: User acceptance of information technology: toward a unified view. MIS Q. **27**(3), 425–478 (2003)
17. Weiss, A., Bernhaupt, R., Lankes, M., Tscheligi, M.: The USUS evaluation framework for human-robot interaction. In: AISB2009: Proceedings of the Symposium on New Frontiers in Human-Robot Interaction, vol. 4, pp. 11–26, April 2009

Control of an Arm-Hand Prosthesis by Mental Commands and Blinking

José Varela-Aldás[1(✉)], David Castillo-Salazar[1,8], Carlos Borja-Galeas[2,7], Cesar Guevara[3], Hugo Arias-Flores[3], Washington Fierro-Saltos[4,8], Richard Rivera[5], Jairo Hidalgo-Guijarro[6], and Marco Yandún-Velasteguí[6]

[1] SISAu Research Group, Universidad Indoamérica, Ambato, Ecuador
{josevarela, dr.castillo}@uti.edu.ec

[2] Facultad de Arquitectura, Artes y Diseño, Universidad Indoamérica, Ambato, Ecuador
carlosborja@uti.edu.ec

[3] Mechatronics and Interactive Systems - MIST Research Center, Universidad Indoamérica, Ambato, Ecuador
{cesarguevara, hugoarias}@uti.edu.ec

[4] Facultad de Ciencias de la Educación, Universidad Estatal de Bolívar, Guaranda, Ecuador
washington.fierros@info.unlp.edu.ar

[5] Escuela de Formación de Tecnólogos, Escuela Politécnica Nacional, Quito, Ecuador
richard.rivera01@epn.edu.ec

[6] Grupo de Investigación GISAT, Universidad Politécnica Estatal del Carchi, Tulcan, Ecuador
{jairo.hidalgo, marco.yandun}@upec.edu.ec

[7] Facultad de Diseño y Comunicación, Universidad de Palermo, Buenos Aires, Argentina

[8] Facultad de Informática, Universidad Nacional de la Plata, Buenos Aires, Argentina

Abstract. Patients who lack upper and lower extremities have difficulties in carrying out their daily activities. The new technological advances have allowed the development of robotic applications to support people with disabilities, also, portable electroencephalographic (EEG) sensors are increasingly accessible and allow the development of new proposals which involve the mental control of electronic systems. This work presents the control by mental orders of an arm-hand prosthesis using low-cost devices, the objective is to command the arm using the user's attention and blinking, where the components are a brain signal sensor, a prosthesis, an Arduino board, six servomotors, and a computer. The developed program in Matlab allows controlling the arm by means of an attention level y blinking. The results show the functioning of the system through experimental tests and a usability test is applied, finally, the conclusions establish adequate coordination in the movements of the prosthesis and the patient indicate satisfaction with the proposal.

Keywords: Prosthesis · EEG · Attention · Blinking · Servomotor

T. Ahram et al. (Eds.): IHSED 2019, AISC 1026, pp. 154–159, 2020.
https://doi.org/10.1007/978-3-030-27928-8_24

1 Introduction

With the recent advances in robotics, the application of new technology in the service area is common, leaving the comfort of structured environments, as are the repetitive industrial processes, and appearing in domestic environments to solve problems of diverse nature [1]. Service robotics has allowed assistance in daily activities, such as cleaning homes, educational support, and medical assistance [2]; the latter involves considering the needs of patients in the characteristics of the system [3].

Modern prostheses are very attractive visually, and generate great expectations for future applications in medicine. The new demands of mobility in patients with disabilities has motivated to implement autonomous systems that provide mobility to prostheses of lower and upper limbs [4]. Usually, the patient controls the movements of the prosthesis; for this to be possible, the user must send the orders to the prosthesis, this has generated several proposals and the use of different sensors; the best thing is to perform the control without the help of another member of the human body, generating an independence in the user [5, 6].

Electroencephalographic sensors allow the acquisition of brain signals for interpretation and analysis [7]; the current market offers a wide variety of these devices and the advantages of modern technology offer comfort and portability. The most advanced sensors are able to distinguish multiple orders using several electrodes and intelligent algorithms, creating the possibility of connecting them with robotic prostheses that provide the movement [8]. The greatest limitation in the development of prosthesis controlled by brain signals is the high cost of these devices.

The insertion of low cost devices in the area of robotics has allowed the development of economic proposals to solve problems in the social sector. The use of the Arduino board in technological projects is something very common nowadays, due to the fact that it requires low budgets and provides flexibility [9]. Also, there are accessible cost sensors with feasibility and connectivity characteristics [10]. Developing technological proposals using low-cost devices that solve medical and social problems allows the economically vulnerable sector the opportunity to improve their living conditions and meet their personal needs.

This proposal presents the control of a prosthesis made up of the arm and the hand using mental commands and low-cost devices. To acquire the brain signals, a Mindwave Mobile 2 sensor is used, which provides the attention, the measurement and the user's blinking; these data allow operating a prosthesis built with six servomotors.

2 Methods and Materials

2.1 Design of the Proposal

The proposal is design to control an arm-hand prosthesis using inexpensive devices. Figure 1 shows the elements use for the control of the prosthesis through attention. It is observe that a computer connects the inputs with the outputs, and performs the processing of the information.

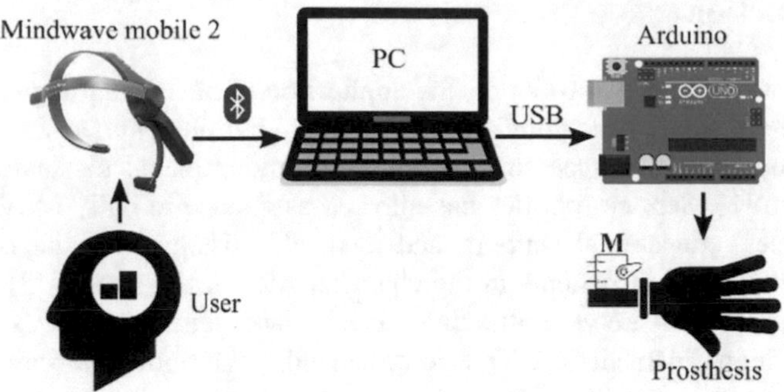

Fig. 1. Proposed system

The user must concentrate to send the orders of movement by means of the attention, the sensor of electrical signals of the brain (Mindwave Mobile 2) acquires this information and these data are send to the computer by means of Bluetooth. The computer is responsible for processing the information and coordinates the functionality of the movements. The orders are send to an electronic board (arduino) that activates the movements of the servomotors located in the prosthesis. It is propose to use Matlab as a programming language.

2.2 Development of the Proposal

The electrical connections of the system are realize using all the electronic components; Fig. 2 illustrates the electronic system and all its components. For the movement, low cost servomotors (ES08A II) strategically placed in the prosthesis are used, one servomotor realizes the movement of elevation of the arm and five servomotors are in charge of the movement of the fingers. The actuators are power by a 6-v battery and controlled with the PWM pins of the arduino.

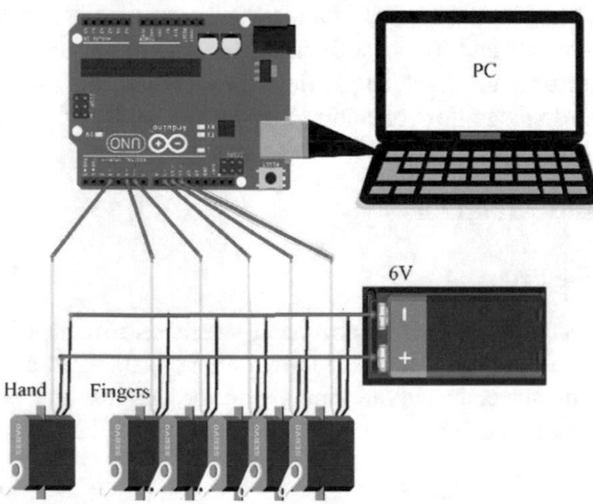

Fig. 2. Electric connections

The programming is develop in Matlab with the functions presented in Fig. 3, the orders of movements are based on the user's attention, for which a threshold is established that must be overcome to start the movements in the prosthesis, while the user does not exceed this threshold the actuators stay static.

Blinking allows changing the operating mode, initially controlling the elevation of the arm, and can reverse the direction by blinking, another additional blinking activates the control of the fingers, to open or close the hand as required, and these cases are repeat to control the prosthesis continuously.

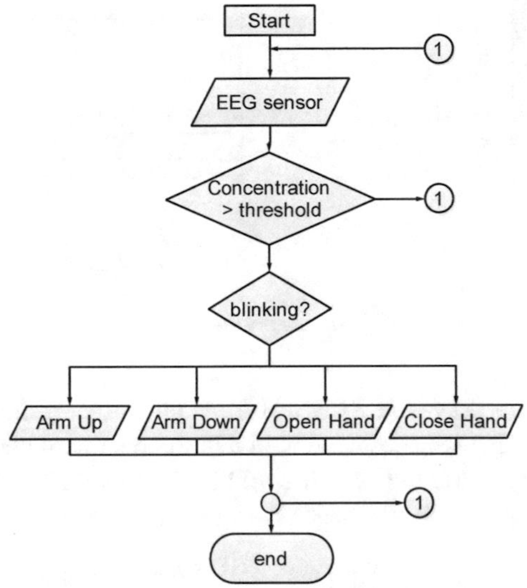

Fig. 3. Operation of the program

The programming has been implemented using the Thinkgear library, which allows to obtain the attention and meditation data, to determine the blinking, changes in the unprocessed information are detected, applying the Eq. (1) to data frames, a threshold is also established to distinguish a blinking (f > threshold).

$$f = \frac{\sum_{1}^{n} RAW_i}{n} \tag{1}$$

3 Results

The proposal is test in a patient with lack of hand and part of the forearm. Figure 4 presents the attention graphs, and the attention threshold established. In addition, the control changes in the prosthesis are observe as a new blink occurs. The results show changes in the angular position of the prosthesis servomotors only when the attention threshold is exceed, on the other hand, the blinking perform the selector function correctly. The Fig. 4 also shows the patient performing the experiment.

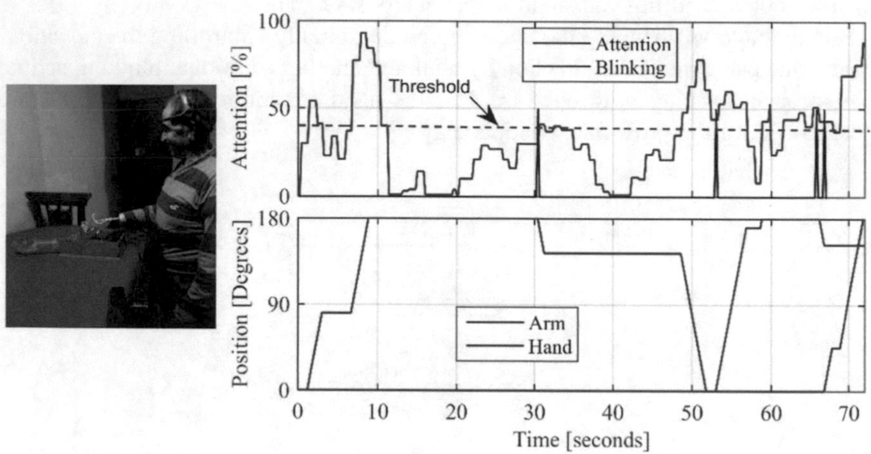

Fig. 4. Experimentation

In a complementary way, the SUS test is apply to determine the usability of the proposal, Table 1 details the scores obtained and the results, with a final score of 72.5, an acceptable usability value but which could be improved.

Table 1. Test SUS results

Question	Score	Operation
I think I would like to use this system frequently	4	3
I find this system unnecessarily complex	**1**	**4**
I think the system is easy to use	3	2
I think you would need technical support to make use of the system	**2**	**3**
I find the various functions of the system quite well integrated	3	2
I have found too much inconsistency in this system	**1**	**4**
I think most people would learn to make use of the system quickly	3	2
I found the system quite uncomfortable to use	**1**	**4**
I have felt very safe using the system	3	2
I would need to learn a lot of things before I can manage the system	**2**	**3**
Total		29
Punctuation		72.5

4 Conclusions

This work presents the control of a prosthesis through mental orders; the change from the control arm to the control hand, and vice versa, is made with the user's blinking; in addition, low-cost devices use to implement the proposal and the data is process using Matlab. The prosthesis consists of six servomotors which allow mobility and mental instructions are obtain of Mindwave Mobile 2 headband. In addition, the programming in Matlab coordinates the actions according to a threshold of attention and blinking for the change of command. The results show a good performance of the system, the attention signals are coordinate by the user and the blinking are detect correctly to change the operation mode, finally, the user indicates satisfaction with the characteristics of the proposal, as verified in the test SUS.

References

1. Iijima, T., Sato-Shimokawara, E., Yamaguchi, T.: Domestic robot system considering generalize. In: Proceedings of SICE Annual Conference 2010. pp. 390–391 (2010)
2. Hernández García, D., Esteban, P.G., Lee, H.R., Romeo, M., Senft, E., Billing, E.: Social robots in therapy and care. In: 14th ACM/IEEE International Conference on Human-Robot Interaction (HRI), pp. 369–370 (2019)
3. Cao, H., Chen, R., Gu, Y., Xu, H.: Cloud-assisted tracking medical mobile robot for indoor elderly. In: IEEE 3rd Information Technology and Mechatronics Engineering Conference (ITOEC), pp. 927–930 (2017)
4. Cipriani, C., Antfolk, C., Controzzi, M., Lundborg, G., Rosen, B., Carrozza, M.C., Sebelius, F.: Online myoelectric control of a dexterous hand prosthesis by transradial amputees. IEEE Trans. Neural Syst. Rehabil. Eng. **19**, 260–270 (2011)
5. Nisal, K., Ruhunge, I., Subodha, J., Perera, C.J., Lalitharatne, T.D.: Design, implementation and performance validation of UOMPro artificial hand: towards affordable hand prostheses. In: 39th Annual International Conference of the IEEE Engineering in Medicine and Biology Society (EMBC), pp. 909–912 (2017)
6. Zhang, X., Li, R., Li, Y.: Research on brain control prosthetic hand. In: 11th International Conference on Ubiquitous Robots and Ambient Intelligence (URAI), pp. 554–557 (2014)
7. Nguyen, B., Nguyen, D., Ma, W., Tran, D.: Investigating the possibility of applying EEG lossy compression to EEG-based user authentication. In: International Joint Conference on Neural Networks (IJCNN), pp. 79–85 (2017)
8. Holewa, K., Nawrocka, A.: Emotiv EPOC neuroheadset in brain - computer interface. In: Proceedings of the 2014 15th International Carpathian Control Conference (ICCC), pp. 149–152 (2014)
9. Oza, V., Mehta, P.: Arduino robotic hand: survey paper. In: International Conference on Smart City and Emerging Technology (ICSCET), pp. 1–5 (2018)
10. Lancheros-Cuesta, D.J., Ramirez Arias, J.L., Forero, Y.Y., Duran, A.C.: Evaluation of e-learning activities with NeuroSky MindWave EEG. In: 13th Iberian Conference on Information Systems and Technologies (CISTI), pp. 1–6 (2018)

Mechanical Design of a Spatial Mechanism for the Robot Head Configuration in Social Robotics

Jorge Alvarez[1(✉)], Mireya Zapata[2], and Dennys Paillacho[3]

[1] Center for Knowledge and Technology Transfer, Universidad Indoamérica, Machala y Sabanilla, 172103 Quito, Ecuador
jorgealvarez@uti.edu.ec

[2] Research Center of Mechatronics and Interactive Systems, Universidad Indoamérica, Machala y Sabanilla, 172103 Quito, Ecuador
mireyazapata@uti.edu.ec

[3] ESPOL Polytechnic University, Escuela Superior Politécnica del Litoral, Campus Gustavo Galindo Km. 30.5 Vía Perimetral, P.O. Box 09-01-5863, Guayaquil, Ecuador
dpailla@fiec.espol.edu.ec

Abstract. The manuscript presents the mechanical design of the head configuration in the Human Robot Interaction (HRI) used for the message transmission of emotions through nonverbal communications styles. The evolution of this structure results on a natural movement reproduction for the implementation of non-verbal communication strategies in a normal behavior, achieve the main patterns to evaluate the social interaction with the robotic platform. The mechanical design result from a biomechanical evaluation of the Pitch, Roll, and Yaw trajectories of the human head and neck. The spatial mechanisms, according to the Grübber formula for Spatial Robots, allows 4 degrees of freedom. The spatial chain has universal, prismatic, spiral and revolute joins of the mechanical model-ling. This CAD model permit the 3D print of cardan elements to performance the structure of the mechanisms. The appearance is friendly and the interface reach similar capabilities than a human would have for communication. Finally, human interaction through the head movement gives the opportunity in the future for the evaluation of more parameters of the social robotic interaction between robots-humans and robots-robots.

Keywords: Additive manufacturing · Spatial mechanism · Biomechanics · Social robotics · Mechanical design · Structural design

1 Introduction

Social robots are used in different places and daily there are habitats for the Human-Robot Interaction (HRI) [1]. The human head is a complex dynamic mechanism to mimic in order to reproduce or communicate emotions. When people talks, they often nod or tilt their heads to reinforce verbal messages and complement social communication. In this sense, Human-Robot Interaction (HRI) seeks to implement head

T. Ahram et al. (Eds.): IHSED 2019, AISC 1026, pp. 160–165, 2020.
https://doi.org/10.1007/978-3-030-27928-8_25

movement in Social Robotic which require to interact with humans, by identifying patterns in people on uncontrolled situations like answer, questions, interactive attention for elders and so on [2]. The implemented approach can perform orientation of the head in the trajectories around in the X, Y and Z coordinates axis (movements of Roll, Pitch and Yaw) through servomotors that allows the generation of 4 Degrees of Freedom. As well, a fish-eye camera is housed on the top of the head with an asynchronous motor in order to position and rotate the camera lens controlled through a remote operator. This design is a mobile platform with graphic and text interfaces according the level of thorax or head, consider similar human morphological. The interlocutor through this interfaces is able to see expressions displayed in the screen and read messages from the robot [3].

2 The Social Robot Human Interaction

The configuration of the robot structure for the gestural communication is implemented with stimulations of the human robot interaction. This stimulation is based in mimics for the interpretation of affirmative, negative or doubt answers. Through the camera housing in the top side of the head, the observation is registered by a camera to achieve feedback of the social interaction. The message depends of the gestural interpretation by the interlocutor [4] (Fig. 1).

Fig. 1. MASHI Robot: (a) Robot in SCEWC 2016, (b) Robot selfie in SCEWC 2016

3 Methods and Materials

The Method for the conception of the social robot is evolutive according the social expectation. The robot has 3 states until the final head configuration.

The structure evolution is implemented as first with a round base with 2 wheels for the translation. There is a joined bar for the body and subjection of features for the communication and display (Fig. 2).

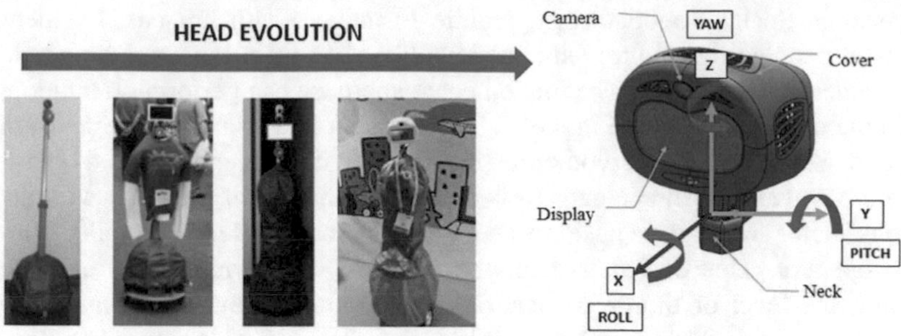

Fig. 2. Structure evolution and final head configuration

4 The Design Concept

4.1 Head Biomechanics

For the head mechanic configuration, the analyses of the head biomechanics begin with the cervical vertebrae [5]. The position of the head is positioned by the 7 cervical vertebrae, linked by cartilages, the movement of the head is flexible and natural with 3 degree of freedom. The rotation of the head has the limits in the gestural position for the answer interpretation: affirmation, negative or possibility doubt answer (Fig. 3).

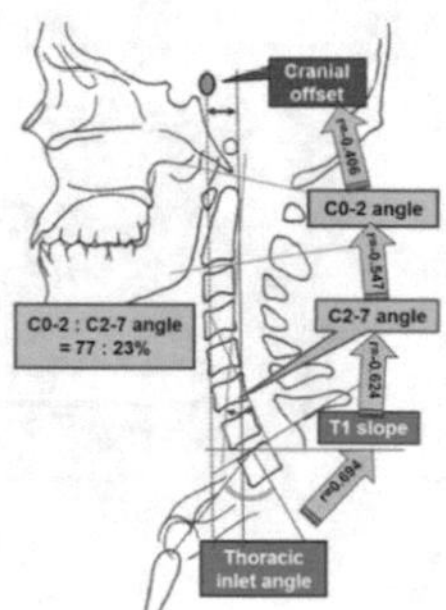

Fig. 3. Cervical vertebrae parts. Image modified from [6]

4.2 The Neck-Head Mechanism Design

The mechanism of the head and neck is using on the reference of the Cranial Spinal vertebral rotation of the base of the neck and the inertial force that the center of gravity of the head makes [7] (Fig. 4).

The spatial chain mechanism is implementing with 3 servomotors, there is one four linkages in the rare side and two four linkages. All linkages are joined with two lateral slider-crank mechanisms, the element connection in the rare side are the cardan configuration in both sides. The spatial chain has 10 links, 2 virtual links that permit the turn of the display assembled in the front side of the head. Each position allows the interlocution with different muster of interlocution [8].

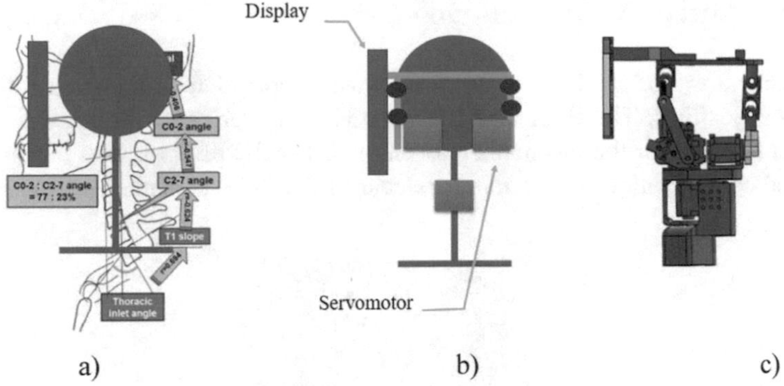

Fig. 4. Design of the mechanical mechanism: (a) Geometry definition, (b) structure concept of the links, nodes, servomotors and kinematics spatial chain, and (c) lateral view of the mechanisms implemented with servomotors (without cover).

4.3 The Spatial Mechanism

The position of the head around each axis is realized by the servomotors to bring in movement the connectors and links. The limits are the angles of the natural movement of the human neck between 15° in the Roll and Pitch, as well as 50° Yaw angles. The spatial positioning begin with the chain at the Z0′ axis (Fig. 5).

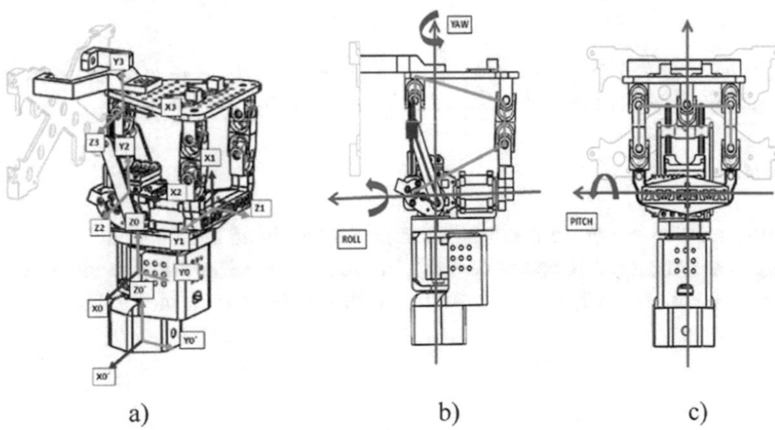

Fig. 5. 3D Model: (a) Coordinate axis and chain, (b) & (c) Slider-crank mechanism and spatial chain with the angles roll-pitch-yaw.

5 The Additive Manufacturing

The manufacturing of mechanical elements must be considered by the external forces o pressures exposition. The PLA or ABS material characterization define the volume of the part or element for the mechanism assembled. The 3D print of joins for revolute has not a good configuration in the filaments near of the Spheris elements (Figs. 6 and 7).

Fig. 6. ABS/PLA 3D printed parts as elements for the material characterization

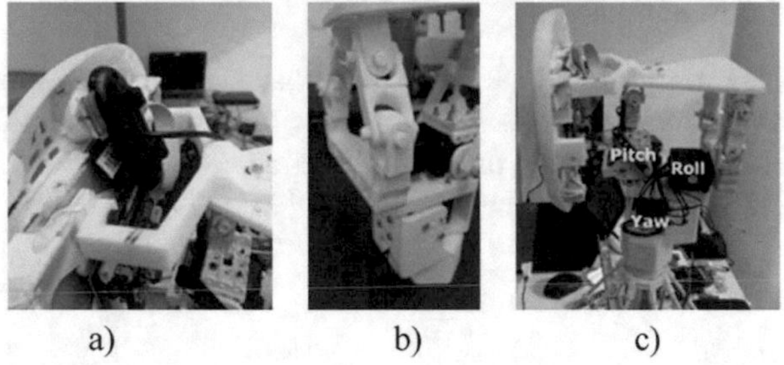

Fig. 7. 3D printing: (a) Camera assembly (without cover) and servomotor assembly with the spatial chain, (b) Joins of the spatial chain, (c) Base of the head structure [9].

6 Results

The results for the structure configuration about the head mechanical design has the following table of joins, connected with servomotors, programed with a set of the servomotors code. The steps corresponding to the angle of each motor as follow (Fig. 8 and Table 1):

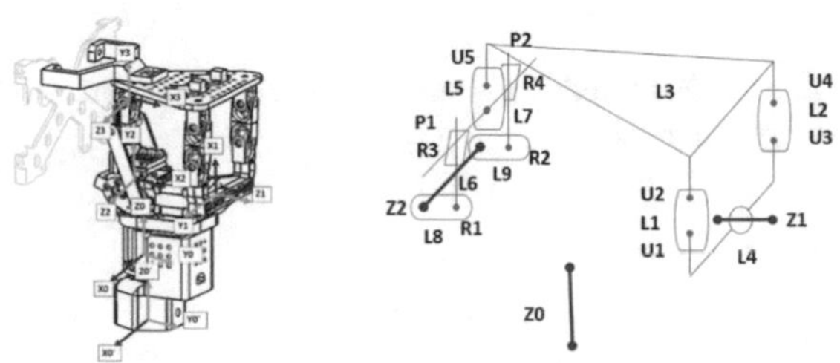

Fig. 8. 3D spatial chain (left), joins and nodes of the spatial chain (right)

Table 1. Join and nodes with 10 links of the robot head

Join	Name	Node
U1…U5	Universal	5
P1…P2	Prismatic	2
S	Spherical	0
R1…R4	Revolute	4

7 Conclusions

The social robotic is optimized in the mimic of the head movement trough the kinematic chain. For the configuration of the elements, the geometry and material has an important role. When the 4-bar and slider-crank mechanisms are joined, the spatial chain can be configured. In the implementation of the spatial chain, the mechanism allows the servo-motors to reproduce the position of the human head characterization which implemented in a not controlled ambient, where is permitted as well the communication in the human-robot interaction.

Acknowledgments. The authors would like to express their very great appreciation to L'Hospitalet City Hall and the BarcelonaTech in Barcelona, Spain for providing the research facilities used in this study. We are particularly grateful for the encouraging support of Mr. Ricardo Castro at La Bóbila Cultural Center. As well very thanks at the Universidad Tecnológica Indoamérica and the Escuela Superior Politécnica del Litoral.

References

1. Paillacho, D.: Designing a robot to evaluate group formations, Doctoral Thesis, Universitat Politécnica de Catalunya (2019)
2. Nordin, A.I., Hudson, M., Denisova, A., Beeston, J.: Perceptions of telepresence robot form, vol. 4 (2016)
3. Nuñez, V., et al.: Modelo vrml interactivo de un robot humanoide Bioloid, México, 1 Congreso interdisciplinario de Cuerpos Académicos 2013 (2013)
4. Nourbakhsh, I.R.: Robots and Education in the classroom and in the museum: on the study of robots, and robots for study (2000)
5. Edirisinghe, E.A.N.S., et al.: Design and simulation of a human-like robot neck mechanism. In: 2015 Electrical Engineering Conference [EECon], vol. 1893 (2015)
6. Scheer, J.K., et al.: Cervical spine alignment, sagittal deformity, and clinical implications. J. Neurosurg. Spine **19**, 141–159 (2013)
7. Ölçücüoğlu, O.: Human-like robot head design a thesis submitted to the graduate school of natural and applied sciences of Middle East Technical University (2007)
8. Danev, L., Hamann, M., Fricke, N., Hollarek, T., Paillacho, D.: Development of animated facial expressions to express emotions in a robot: RobotIcon. In: 2017 IEEE 2nd Ecuador Technical Chapters Meeting, ETCM 2017, pp. 1–6 (2018)
9. Hernández, X.R.: Rediscovering the experimental robotic platform MASHI. Thesis, p. 49, January 2017

Transportation Design and Autonomous Driving

Reclined Posture for Enabling Autonomous Driving

Dominique Bohrmann[1,2(✉)] and Klaus Bengler[2(✉)]

[1] Mercedes-Benz Technology Center, Daimler AG, Benzstr. Tor 16, 71063 Sindelfingen, Germany
Dominique.Bohrmann@daimler.com

[2] Chair of Ergonomics, Technical University of Munich, Boltzmannstr. 15, 85747 Garching, Germany
{Dominique.Bohrmann,Bengler}@tum.de

Abstract. As the future of mobility develops, automated vehicles (AV) will change road transportation and promise an improved quality of life. Within this development, however, the primary weakness, is the human per se. Due to physiological thresholds, many occupants react by developing symptoms of motion sickness (MS) when performing non-driving related tasks (NDRTs). This work describes approaches essential to mitigating MS with respect to interior design. Therefore, a real test-driving experiment with 25 volunteers was carried out at a test track in Sindelfingen. The effects of backrest angle and sitting direction were observed in consideration of predetermined NDRTs. The analysis showed that a reclined backrest angle leads to a significant ($p < 0.0001$) decrease in MS. Furthermore, the effect of seat direction appears likely to be less significant than the effect of backrest rotation. A second experiment was conducted on the Mercedes-Benz Ride Simulator in order to identify the acceptance of innovative sitting positions.

Keywords: Human factors · Autonomous driving · Motion sickness · Interior requirements · Non-driving related tasks · Sitting posture

1 Introduction

Autonomous driving (AD) has become known as one of the most challenging developments within the automotive industry ever. It has the ability to fundamentally change our understanding of road transportation. A number of advantages, e.g. improvement of traffic flow or occupational safety, are linked to increasing the level of automation (SAE level). Hence, it is necessary to understand the feasible benefits and associated challenges of AD while considering user expectations as well as their physiological restrictions. Many investigations have shown that drivers are likely to spend more time on NDRTs as the automation level rises [1]. In particular, when drivers become passengers and do not need to constantly monitor or control the path of the vehicle, activities increase in which their gaze deviates from the road.

In this situation, many occupants react with symptoms of pallor, dizziness, headache, sweating or vomiting while performing NDRTs. It is predominantly passengers

T. Ahram et al. (Eds.): IHSED 2019, AISC 1026, pp. 169–175, 2020.
https://doi.org/10.1007/978-3-030-27928-8_26

sitting in the rear of the vehicle who are already familiar with this phenomenon, which is called kinetosis or MS. A variety of factors contribute to the likelihood of experiencing MS symptoms. First, intra- and inter-individual susceptibility are main aspects concerning MS occurrence. Furthermore, the duration and type of perception stimuli also affect the nature and severity of symptoms [2]. Many researchers emphasize the importance of MS due to increased automation [3]. In order to prevent MS in AV, it is necessary to understand how various vehicle characteristics affect the behavior and well-being of physically active passengers. Therefore, the aim of the research activity described in this paper focused on fundamental approaches to mitigating MS as it relates to interior design. Several sitting positions were examined in detail under real test-driving conditions as well as under simulator conditions at Mercedes-Benz.

2 Method

2.1 Outline of the Real Test-Driving Investigation

A reasonable set of data were collected in a mixed design study of 50 driving events. The following metrics were applied in order to examine the influence of sitting direction and backrest angle on the severity of MS (Table 1).

Table 1. Design of the MS investigation

Seat direction (between)	**Backrest angle (within)** Upright (23°)	Reclined (38°)
Forward	N=13	N=13
Rearward	N=12	N=12

During a 20-min ride on a defined route, the subjects were asked to accomplish predetermined NDRTs while sitting in the rear of a Mercedes-Benz V-class vehicle. In this position, the occupants were unable to see the environment due to curtains in the cabin of the rear. This was necessary in order to standardize visual impact and, mainly, to observe the effect of head rotation and stabilization on vestibular changes. Every journey was divided into four equal segments of 5 min each. Between the four segments, short breaks were used for checking the individuals' well-being according to a single-item Likert scale between 0 and 20 (Fast Motion Sickness scale – FMS) and testing current mental acuity by performing the bdpq test [4, 5]. In this case, the letters "b", "d", "p" and "q" were presented in random order for 1 min. The goal was to mark as many correct pattern of "b" followed by a "q" as possible. In every segment, the subjects performed different NRDTs like reading e-books, watching movies, or playing games or quizzes on an Apple iPad (9.7 in.). The order of NDRTs was randomized among the participants according to the Latin square method. In order to compare the within-subject design measurements, the order of NDRTs was kept constant for every subject over the two exposures.

To ensure standardized conditions, e.g. head position, tablet orientation was provided by a prototyped fixture at the front seat and the c-pillar of the vehicle structure. During the testing, the subjects were monitored for indications of a pattern of physiological behavior, such as yawning, blinking, etc. In addition, respiration rate, heart rate and core temperature were measured before and during the test. Immediately after the driving event, MS occurrences were queried according to the Motion Sickness Assessment Questionnaire (MSAQ), a multidimensional questionnaire with 16 items [6].

2.2 Participants and Analysis in Real Test-Driving Condition

Since the focus of the research concerns the prevention of MS, subjects with high resistance to kinetogenic stimuli were excluded. A similar selection was made in previous studies to eliminate the influence of irrelevant factors that are not attributable to the subject of MS [7]. The total sample included subjects aged 21–56 years ($M = 42.46$, $SD = 11.96$), of whom 21 were male. The difference in experimental conditions was statistically significant neither in age, $F = 1.19$, $p = 0.39$, nor in the sex distribution, $\chi^2(1) = 0.40$, $p = 0.52$. Furthermore, it was ascertained that none of the participants had vestibular or gastrointestinal diseases. The mean value of the Motion Sickness Susceptibility Questionnaire (MSSQ) differed only 0.3 points between groups, averaging at $M = 12.80$, $SD = 8.21$. The MSSQ short consisted of 2 subscales with 9 items each and recorded prior experience with MS before the age of 12 years as well as in the past 10 years in a variety of passive transport systems [8]. Each subject experienced both the first and second trial dates, so all records are usable. Depending on the choice of appointment, there were from 24 h to 2 weeks between the two measurement times so that a complete regeneration of the initial test could be guaranteed. Due to the complex experimental design, linear mixed effects (LME) models were used as the main analysis method. All calculations were performed using R Studio and MATLAB 2018. Only the results of sitting conditions are presented in this paper. Further results of performance testing, physiological measurements etc. are published in the 27th Aachen Colloquium and the 34th VDI Conference of Driver Assistance and Automated Driving [9, 10].

3 Results and Discussion of Real Test-Driving Condition

The evaluation of the model parameters revealed backrest angle as a significant main effect according various measurements, e.g. the MSAQ $b = 18.95$, $t\ (24) = 5.72$, $p < 0.0001$, $r = 0.76$ (Fig. 1). Compared to the reclined position, the MSAQ value in the upright seated version increased by almost 19 points, based on a constant rate of susceptibility (MSSQ). The probability of MS termination predicted for the reclined position is 12%, whereas an 80% termination is likely when sitting in an upright position. Based on further analyses, the influence of seat orientation was considered for both levels of the measurement. In general, the influence of sitting direction seems to be slightly less than that of backrest rotation. In any case, in the group of upright sitting subjects (without termination), the sitting orientation had a significant effect, b = 4.27, t (19) = 3.45, p < 0.01, r = 0.62, whereas the orientation of the seat did not show any influence on the prone group.

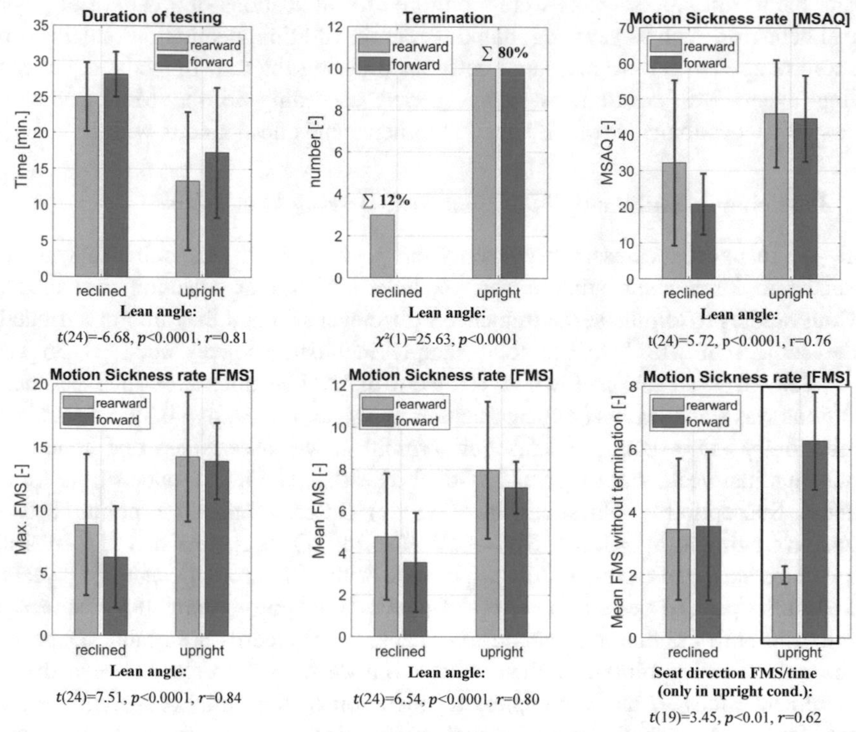

Fig. 1. Results of MS occurrence during the real test ride

The influence of backrest angle on MS remain similar between vans and sedans, although these have different vehicle dynamics. Short investigations have been conducted to confirm the same. Here Fig. 2 illustrates the movements of the Mercedes-Benz V-class measured at the seat structure of the subject. The road is characterized by sequential stimuli of longitudinal, lateral and vertical accelerations in a range of a naturalistic approach.

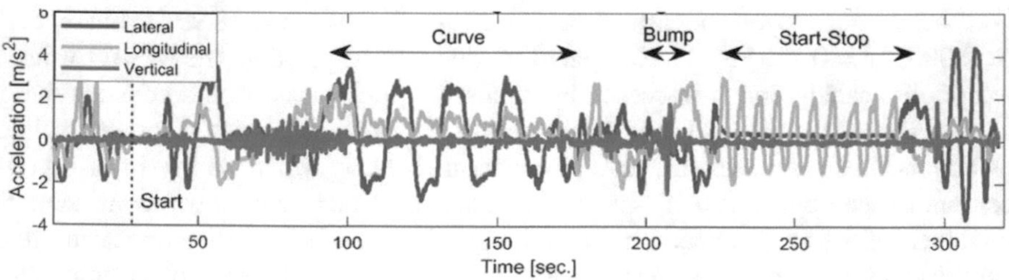

Fig. 2. Vehicle-acceleration – one segment

The evaluation of the chosen driving maneuvers according to MS severity is further shown in the following figure. The participants were asked in detail which of the experienced driving maneuvers provoked MS the most. Multiple answers were possible in this case (Fig. 3).

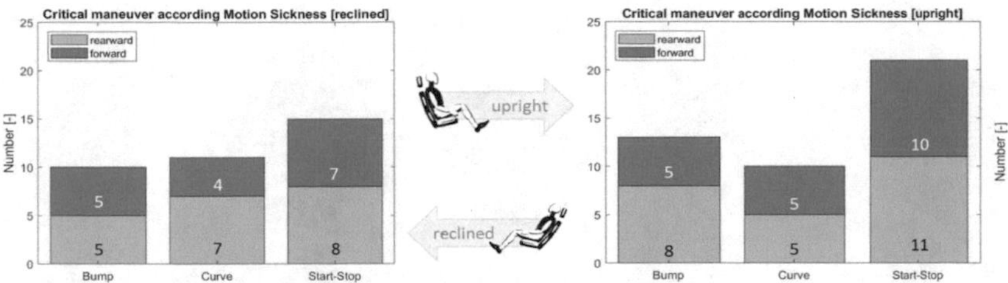

Fig. 3. Critical maneuver according MS

The effect of backrest angle only became significant when evaluating the start-stop-maneuver χ^2 (1) = 3.66, p = 0.06. No other condition showed a significant impact of seat direction or backrest angle.

4 Ride Simulator Study

16 healthy subjects (2 female, age: M = 36.6, SD = 11.8) with an average body height of 1792 mm (SD = 33.6 mm) and no history of vestibular or spine disorders participated in a within-subject study design at Mercedes-Benz Ride Simulator. The goal of this study in one respect was to assess under reproducible conditions the head-neck movements and, in another respect, the individual comfort feeling according to several sitting positions. Based on the real test-driving results, it is important to know whether reclined postures are also suitable for increasing comfort as well as reducing MS symptoms. Therefore, the participants were seating on a W206 (C-class) driver seat and were instructed to maintain their head in a comfortable and stable position while watching movies on an Apple iPad during the tests. Each test had a duration of 23 min within multisine waves, noise, as well as micro- and meso-shaking stimuli. The standardized accelerations were applied in 3 different sitting positions in a randomized order as shown in the following Fig. 4.

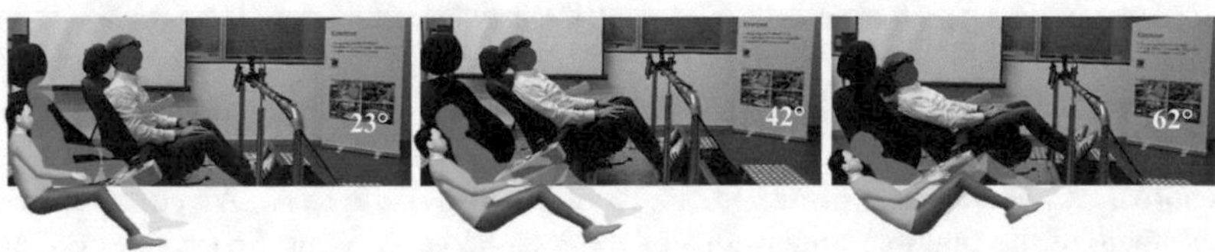

Fig. 4. Sitting conditions on ride simulator (within-subject design)

After all exposures, the participants were asked to evaluate the three postures under dynamic conditions regarding general comfort. The results, seen in Fig. 5, strengthen the potential of flat backrest angles with respect to the "watching movie task" as well as the various NDRTs expected in future vehicles, e.g. telephoning, relaxing or reading books. Observing, however, is preferred in upright sitting positions.

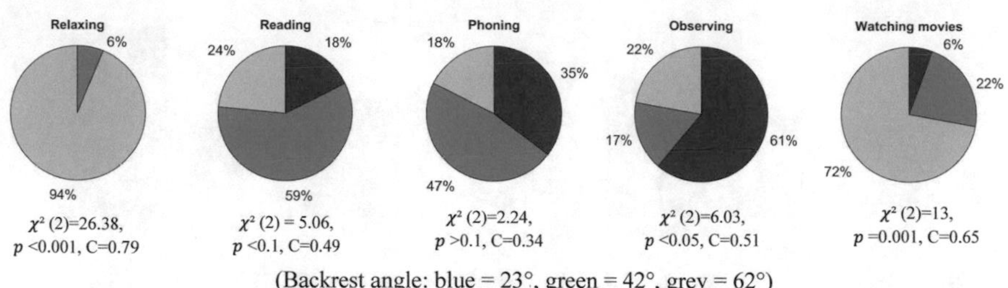

Fig. 5. Preference on sitting positions cross-referenced on NDRTs

5 Conclusion

The potential of reclined postures in relation to MS prevention and increased comfort has been proven within our research. Recent studies have demonstrated a significant effect of sitting direction on MS [11]. This effect is mainly observed in upright positions. This conclusion cannot be confirmed by the real test-driving events at Mercedes-Benz. Under visual standardization, our result is comprehensive due to the lack of anticipation. Therefore, it is recommended that the effect of sitting direction in a reclined posture have to be examined while providing the opportunity of visual cuing.

References

1. Kyriakidis, M., Happee, R., de Winter, J.C.F.: Public opinion on automated driving: results of an international questionnaire among 5000 respondents. In: Transportation Research. Part F: Traffic Psychology and Behaviour, vol. 32, pp. 127–140 (2015)
2. Mc Cauley, M.E., Sharkey, T.J.: Cybersickness: perception of self-motion in virtual environments. Presence **1**(3), 311–3185 (1992)
3. Diels, C., Bos, J.: Self driving carsickness. Appl. Ergon. **53**, 374–382 (2016)
4. Brickenkamp, R., Schmidt-Atzert, L., Liepmann, D.: d2-R: test d2-revision: Aufmerksamkeits- und Konzentrationstest. National Center for Biotechn. Information, Göttingen (2010)
5. Keshavarz, B., Hecht, H.: Validating an efficient method to quantify motion sickness. Hum. Factors **53**(4), 415–426 (2011)
6. Gianaros, P.J., Muth, E.R., Mordkoff, T., Levine, M.E., Stern, R.M.: A questionnaire for the assessment of the multiple dimensions of motion sickness. Aviat. Space Environ. Med. **72** (2), 115–119 (2001)
7. Bos, J.E.: Less sickness with more motion and/or mental distraction. J. Vestib. Res.: Equilibr. Orient. **25**(1), 23–33 (2015)

8. Lamb, S., Kwok, K.C.S.: MSSQ-short norms may underestimate highly susceptible individuals: updating the MSSQ-short norms. Hum. Factors **57**(4), 622–633 (2015)
9. Bohrmann, D., Lehnert, K., Scholly, U., Bengler, K.: Der Mensch als bestimmender Faktor zukünftiger Mobilitätskonzepte. In: 34. VDI/VW-Gemeinschaftstagung Fahrerassistenzsysteme und automatisiertes Fahren, vol. 2335, pp. 345–359. VDI Verlag, Düsseldorf (2018)
10. Bohrmann, D., Lehnert, K., Scholly, U., Bengler, K.: Kinetosis as a challenge of future mobility concepts and highly automated vehicles. In: 27th Aachen Colloquium Automobile and Engine Technology, Aachen, pp. 1309–1335 (2018)
11. Salter, S., Diels, C., Herriotts, P., Kanarachosa, S., Thakea, D.: Motion sickness in automated vehicles with forward and rearward facing seating orientations. Appl. Ergon. **78**, 54–61 (2019)

Implicit Communication of Automated Vehicles in Urban Scenarios: Effects of Pitch and Deceleration on Pedestrian Crossing Behavior

André Dietrich[(✉)], Philipp Maruhn, Lasse Schwarze, and Klaus Bengler

Chair of Ergonomics, Department of Mechanical Engineering, Technical University of Munich, Boltzmannstr. 15, 85748 Garching, Germany
{Andre.Dietrich, Philipp.Maruhn, Lasse.Schwarze, Bengler}@tum.de

Abstract. This Study analyzed the effects of AV's approach trajectories and the role of the vehicle's pitch on pedestrian's crossing behavior. 30 participants experienced an urban traffic scenario in the virtual reality simulator with vehicle convoys driving at 30 km/h. The decelerating vehicle approached the waiting pedestrian using three different kinematic trajectories, which were accompanied by four pitch conditions. The effect of an early or stronger vehicle pitch on the pedestrian crossing behavior was stronger when coupled with a defensive deceleration strategy. Overall, hard initial braking reduces the time, pedestrians need to understand an approaching vehicle's yielding intention. Active pitching might increase this effect, but requires further evaluation, as pedestrians link the vehicle's pitch to the perceived kinematics.

Keywords: Vulnerable road users · Automated vehicles · Implicit communication · Pedestrian simulator · Interaction

1 Introduction

Increasing automation capabilities in vehicles is a mutual goal within the automotive industry. Automated driving on highways is a well-researched topic and the first commercially available vehicles with higher levels of automation, in which the driver is no more in the loop, will be available soon for highway scenarios. Introducing automated driving onto urban environments creates new challenges – as the driver is not available, the automated vehicle (AV) needs to interact with other traffic participants. Especially vulnerable road users, such as pedestrians, will have to understand the intention of an approaching vehicle to ensure safe decision-making in street crossing scenarios. While some research suggests novel external Human Machine Interfaces to enable communication of AVs to pedestrians [1, 2], currently, such encounters are mostly resolved implicitly in urban traffic, i.e. through the kinematic motion of an approaching vehicle. As automated driving in urban environments is a novel topic, only little research has been conducted regarding the effects of implicit communication on

T. Ahram et al. (Eds.): IHSED 2019, AISC 1026, pp. 176–181, 2020.
https://doi.org/10.1007/978-3-030-27928-8_27

pedestrian crossing behavior. Studies have shown that early vehicle deceleration leads to earlier pedestrian crossings at crossroads [3] and that pedestrians are able to interact with driverless vehicles as long as the vehicle behaves in a predictable way [4]. Therefore, a yielding AV's kinematic approach needs to match the expectation of pedestrians. This study aims to gather insights regarding the potential of different braking strategies and artificial vehicle dynamics, namely the vehicle's pitch, to communicate the vehicle's intentions.

2 Methodology

The experiment was conducted in a virtual reality pedestrian simulator [2] to ensure a safe and controlled experimental setting. The virtual environment was hosted on a desktop PC (Intel Core i7-8700K, 32 GB Ram DDR4, Ge-Force GTX 1080 Ti 11 GB) and rendered on a HTC Vive head-mounted display (HMD) with 1080 × 1200 pixels per eye, 110° field of view, SteamVR Tracking and Vive Deluxe Audio Strap headphones. The virtual environment, created in Unity3D 2018, was modelled after a typical street of Munich's city center featuring a two-lane street (5,75 m) surrounded by houses and decorations (trees, parked cars, etc.). To increase immersion, virtual traffic spawned behind the buildings and entered the street via a traffic intersection at a 75 m distance.

An approaching vehicle on a straight road that intends to yield for a crossing pedestrian is likely to keep its heading while decelerating. Thus, the dynamics of a decelerating vehicle on a straight road can be reduced to its position in longitudinal direction, the vehicle's pitch and the corresponding time derivatives. To enable a comparison of different deceleration strategies an AV might use when yielding its right of way, the distance, at which the vehicle initiated its braking, was fixed at 21.5 m.

Yielding vehicles decelerated by inducing a constant jerk until a predefined maximum deceleration was reached, which was finally linearly reduced to zero at the time, the vehicle came to a full stop (see Table 1).

To evaluate the effects of deceleration strategies on pedestrian behavior two VR pre-studies with overall 10 participants were conducted to identify distinguishable, realistic maneuvers. Pre-study participants were shown various approaching vehicles with different deceleration strategies and had to identify the number of different trajectories as well as categorize the observed vehicle approaching behavior. 42 out of 60 presented maneuvers (70%) were identified correctly when three different deceleration strategies were shown, compared to 25 out of 75 (33%) with five strategies. This lead to

Table 1. Parameters describing the three braking maneuvers. Due to necessary thresholds, the time experienced in the simulation differs from the calculated time.

Name	Jerk induction	Deceleration maximum	Final jerk	Time calculated	Time simulated
Baseline	-2 m/s^3	-2 m/s^2	2 m/s^3	5164 ms	4770 ms
Defensive	-4 m/s^3	-2 m/s^2	0.4 m/s^3	6915 ms	6230 ms
Offensive	-1.17 m/s^3	-2 m/s^2	4 m/s^3	4433 ms	4140 ms

the conclusion that participants are able to distinguish three types of deceleration strategies. Thus, participants in the main study were presented with three deceleration strategies as seen in Table 1.

As the simulation recalculates the vehicles position, velocity and acceleration each frame, a threshold of v = 0.1 m/s was defined, at which the current velocity was equated to the set speed of v = 0 m/s for the yielding vehicle. This led to lower maneuver times observed in VR in comparison to algebraic values, especially for the defensive deceleration strategy.

To evaluate the effect of the vehicle pitch on pedestrians crossing behavior, four pitch strategies were included:

- No pitch – the decelerating vehicle slowed down without rotating
- Normal pitch – the deceleration lead to a vehicle pitch comparable to real driving behavior
- Boosted pitch – the pitch of the deceleration was amplified by four times
- Premature pitch – the vehicle pitched to −4° within 1.5 s before initiating the deceleration

Study participants stood at the curb of the street with convoys of vehicles approaching from the left. After the first vehicle had passed, participants were asked to cross the street whenever they felt safe to do so. As all decelerating vehicles came to a full stop, four further conditions were added, in which all cars drove at a constant speed with passable inter-vehicle gaps. This ensured that participants did not always wait for a decelerating vehicle. Each condition was repeated three times leading to a 3 (deceleration strategy) × 4 (pitch) design with overall (3 * 4 + 4) *3 = 48 crossings per participant. 30 participants took part in the VR experiment. As mostly students from the Department of Mechanical Engineering at the Technical University of Munich were recruited, the sample consists of rather young participants (M = 23.9y, SD = 3.17y) with a higher proportion of males (70% male, 30% female). The overall duration of the study was about 60 min and participants were compensated with 15 €.

3 Results

Crossing Initiation Time to Vehicle Stop (CIT_{VS}) served as dependent variable as the time difference in seconds from a full vehicle stop to the moment a pedestrian enters the encroachment zone (i.e. the area where (future) paths of encountering traffic participants overlap).

A 3 (trajectories) x 4 (pitch conditions) repeated measures ANOVA with a Greenhouse-Geisser correction determined that the mean CIT_{VS} showed a statistically significant effect of trajectory ($F(1.35, 39.27) = 995.56$, $p < .001$, $\eta_p^2 = .97$), pitch ($F(2.21, 63.96) = 9.87$, $p < .001$, $\eta_p^2 = .25$) as well as the interaction between these effects ($F(4.41, 0.1) = 2.7$, $p < .05$, $\eta_p^2 = .09$).

Bonferroni corrected post-hoc tests revealed that the defensive deceleration strategy leads to significantly earlier CIT_{VS} ($M = -1.2$ s, $SD = 0.87$ s) when compared to the baseline ($M = -0.25$ s, $SD = 0.85$ s, $p < .001$) or offensive strategy ($M = 0.54$ s,

$SD = 0.59$ s, $p < .001$). The difference between baseline and aggressive strategy was also significant ($p < .001$). Figure 1 illustrates the main effect of braking trajectories on CIT_{VS}.

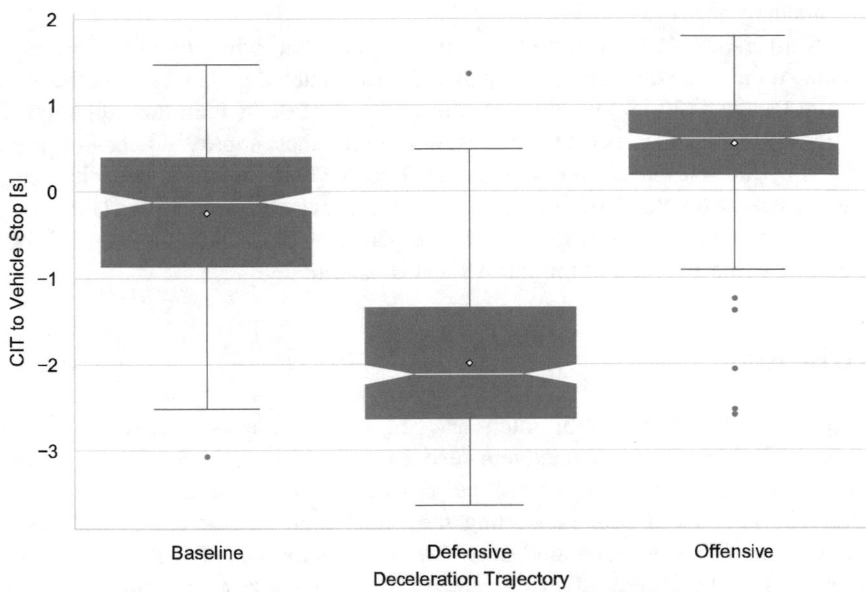

Fig. 1. Boxplots of Crossing Initiation Time to Vehicle Stop (CIT_{VS}) as a function of vehicle deceleration trajectories: baseline, defensive and offensive. White diamonds represent means, white lines medians and notches 95% confidence intervals.

Figure 2 shows a hybrid interaction of the two main factors pitch and trajectory, inhibiting a global interpretation of factor pitch. Where the pre-pitch increased CIT_{VS} for defensive and offensive braking compared to the boost pitch, a decrease was observed for the baseline condition. Simple main effects analysis of pitch revealed significant effects of pitch for trajectories baseline ($p < .001$) and defensive ($p < .001$) but no differences for trajectory offensive ($p = .28$).

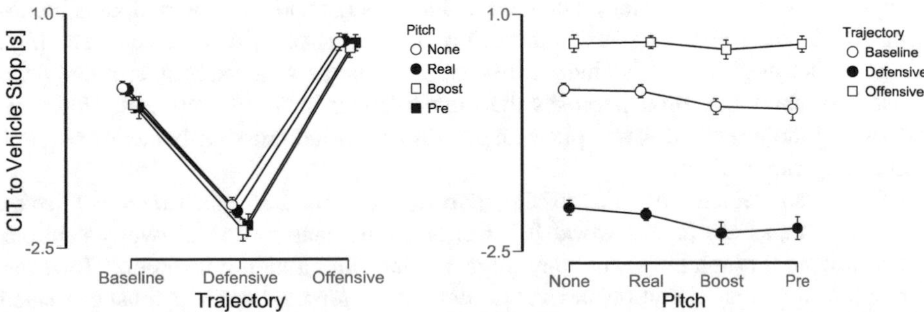

Fig. 2. Interaction diagrams of main factors pitch and trajectories. Error bars represent 95% confidence interval.

Post experiment interviews revealed that four participants did not consciously perceive the different pitch angles in the main test. A further four perceived differences but attributed them to the quality of the simulation and did not perceive them as part of a communication strategy. The 22 remaining participants were able to report the different nodding behavior.

17 test persons (56.7%) find the pre pitch unpleasant, while five (16.7%) rated it pleasant. At the same time, ten participants (33.4%) stated that they crossed the road at the earliest, eleven (36.7%) would cross the road at the latest within this condition. The realistic pitch with 12 votes (40%) corresponds to the most approved behavior, closely followed by the reinforced condition with 10 votes (33.4%). These two pitch angles also each receive 8 votes (26.7%) for the question of which condition leads to an early crossing maneuver. Only 3 people find the maneuver without pitch pleasant, 12 (40%) admitted crossing the road at the latest for this condition.

4 Discussion

The results indicate that both pitch and trajectory influence pedestrian crossing behavior. Thus, defensive maneuvers lead to pedestrians initiating their crossings sooner. However, the defensive maneuver costs more time than the aggressive one as the velocity is reduced sooner. Adding the actual time cost of the maneuvers (see Table 1) to the CIT_{VS} yields a total time cost of 4.24 s for the defensive, 4.52 s for the baseline and 4.69 s for the offensive maneuver, showing a similar trend as the CIT_{VS} but with diminished differences.

The CIT_{VS} is a theoretical value representing how long a car is fully stopped until the pedestrian starts to cross. In reality, drivers try to avoid full stops to maintain traffic flow. Decelerating early to indicate a yielding intention would lead to an earlier pedestrian crossing, enabling the driver to accelerate again before coming to a full stop at the encroachment zone. Therefore, the defensive maneuver could save even more time for a driver, compared to the aggressive one, which is also concluded in the theoretical modelling of Markkula et al. [5]. However, drivers might perceive the aggressive and defensive maneuvers as uncomfortable, as the absolute jerk values are higher. Thus, identifying an optimal approach strategy for future AVs relies on a variety of variables and should not be solely based on traffic flow.

The vehicle pitch significantly influences the CIT_{VS} but the effect of the trajectory is superior. Also, all artificial pitches were rated poorly and some participants misinterpreted the high pitch angles as emergency maneuvers or vehicle defects. The high artificial pitch angles may also induce discomfort in the passengers. If an artificial pitch is used for communication purposes than lower angles should be utilized. However, identifying the effects of lower pitch angles on pedestrian crossing behavior requires further experimentation.

The mean velocity of the crossing participants was 1.34 m/s (*SD* = .137 m/s), which is comparable to real world [6]. Hence, Participants were not overly cautious when moving through the virtual environment indicating a high immersion. However, due to the laboratory condition and VR limitations the reported findings must be treated carefully. We expect the tendencies to be replicable in real world experiments, however with different values for CIT_{VS}.

5 Conclusion

In the presented study implicit AV communication in the form of deceleration and vehicle pitch were evaluated in a VR pedestrian simulator based on the unity games engine. Three different deceleration trajectories were coupled with four different vehicle pitching behaviors to analyze their effect on the crossing behavior of pedestrians. Defensive braking reduces the crossing initiation times of pedestrians, even though the maneuver is slower than an aggressive trajectory. While artificial pitching of the vehicle also leads to faster crossings when coupled with a defensive trajectory, its effect is rather low. Furthermore, participants disliked high pitch angles and expected a change in the vehicle kinematics according to the pitching.

Overall, a defensive deceleration strategy is a good way to increase traffic efficiency in pedestrian crossing situations by conveying a yielding intention without using explicit means of communication. However, the comfort of AV passengers needs to be considered when designing approach trajectories.

Acknowledgments. This work is a part of the interACT project. interACT has received funding from the European Union's Horizon 2020 research & innovation programme under grant agreement no 723395. Content reflects only the authors' view and European Commission is not responsible for any use that may be made of the information it contains.

References

1. Fridman, L., Mehler, B., Xia, L., Yang, Y., Facusse, L., Reimer, B.: To walk or not to walk: crowdsourced assessment of external vehicle-to-pedestrian displays. ArXiv:1707.02698 (2017)
2. Dietrich, A., Willrodt, J.-H., Wagner, K., Bengler, K.: Projection-based external human machine interfaces – enabling interaction between automated vehicles and pedestrians. In: Proceedings of the DSC 2018 Europe VR, Driving Simulation and Virtual Reality Conference and Exhibition, pp. 43–50 (2018)
3. Schneemann, F., Gohl, I.: Analyzing driver-pedestrian interaction at crosswalks: a contribution to autonomous driving in urban environments. In: 2016 IEEE Intelligent Vehicles Symposium (IV), Gothenburg, pp. 38–43 (2016). https://doi.org/10.1109/ivs.2016.7535361
4. Rothenbücher, D., Li, J., Sirkin, D., Mok, B., Ju, W.: Ghost driver: a field study investigating the interaction between pedestrians and driverless vehicles. In: 2016 25th IEEE International Symposium on Robot and Human Interactive Communication (RO-MAN), New York, NY, pp. 795–802 (2016). https://doi.org/10.1109/roman.2016.7745210
5. Markkula, G., Romano, R., Madigan, R., Fox, C.W., Giles, O.T., Merat, N.: Models of human decision-making as tools for estimating and optimizing impacts of vehicle automation. Transp. Res. Rec. **2672**(37), 153–163 (2018). https://doi.org/10.1177/0361198118792131
6. Daamen, W., Hoogendoorn, S.P.: Free speed distributions—based on empirical data in different traffic conditions. In: Waldau, N., Gattermann, P., Knoflacher, H., Schreckenberg, M. (eds) Pedestrian and Evacuation Dynamics. Springer, Berlin (2007). https://doi.org/10.1007/978-3-540-47064-9_2

Non-driving Related Activities in Automated Driving – An Online Survey Investigating User Needs

Tobias Hecht(✉), Emilia Darlagiannis, and Klaus Bengler

Chair of Ergonomics, Technical University of Munich,
Boltzmannstr. 15, 85748 Garching, Germany
{t.hecht, emilia.darlagiannis, bengler}@tum.de

Abstract. Automated driving allows the driver to deal with non-driving related activities (NDRA). Surveys, observations in other modes of transportation and driving simulator studies reveal a high variance in possible activities and activity durations. An online study was thus conducted to investigate factors influencing the choice of NDRA. Privacy, storage option, travel duration and purpose were found to have an impact on the attractiveness of certain activities. Furthermore, we investigated the changing need for information that comes with activity engagement. When performing an intense NDRA such as working, results of the online survey indicate that information about current and upcoming maneuvers, surrounding traffic and current speed become less important, while reliability, system status and especially remaining time in current automation mode remain important for people engaged in a NDRA.

Keywords: Non-driving related activities · Automated driving · Human factors

1 Introduction

In automated driving (SAE Level 3–5), the driver will be allowed to disengage from the driving task with no need to supervise the automated driving system. Therefore, the user may engage in other, so called non-driving related activities (NDRA). Consequently, NDRA are seen as one of the main advantages of automated driving [1]. Online surveys, observations in other modes of transportation and driving simulator studies have been conducted to gain knowledge on desired activities, helping researchers and developers to design future car interiors and HMIs. Surveys amongst future users show that similar to today's manual driving, people want to chat with other passengers, listen to music, drink, eat, write messages and look at the surroundings [2, 3]. Studies investigating train or bus trips, using different methods such as (online) surveys or observing passengers on different modes of transportation reveal a high variety of activities, such as listening to music, talking, looking at the surroundings, doing nothing, telephoning, reading, writing, sleeping and the use of electronic devices like laptops, tablets and phones [2, 4]. Furthermore, two driving simulator studies investigating real-life NDRA in conditionally automated driving have been identified, also highlighting a large variety of activities and activity durations, e.g. smartphone use, watching videos, looking at the surrounding

T. Ahram et al. (Eds.): IHSED 2019, AISC 1026, pp. 182–188, 2020.
https://doi.org/10.1007/978-3-030-27928-8_28

traffic or listening to music [5, 6]. Some of these studies already reveal factors influencing the take-up of activities: On business trips, working is more important than sleeping and reading [7], also travel duration [8] and purpose [7] seem to affect the choice of activities. Pfleging et al. [2] expect privacy to have an effect on the likelihood of calls, talks and applications with speech input. Comfort seems to have an effect, especially on laptop use and working in general [4], but also eating and drinking might become more likely if there is a tray [8]. Additionally, socio-demographic factors like income, age and gender might influence the chosen activities [4, 6]. Furthermore, engaging in NDRA might affect the need for information during automated driving. Thus, we investigated the changing need for information when driving in an automated vehicle and engaging in a NDRA, based on findings by Beggiato et al. [9] and Diels and Thomson [10], as well as factors influencing the choice of activities.

2 Method

Lacking experience with automated driving functions most likely affects questions on NDRA in automated driving [2]. Thus, we introduced a train setting for the online survey for all questions on the attractiveness of specific activities. From the studies presented above, the following parameters possibly influencing the selection of activities were derived: privacy, comfort, purpose of travel, trip duration, age and gender. The influence of these parameters on the popularity of several activities was examined and detailed with the help of an online survey. For this purpose, questions about the attractiveness of various activities were asked twice, varying the above mentioned influencing factors. The attractiveness was measured using a 5-point Likert scale from "absolutely not" to "definitely". The effects of age and gender were investigated using the questions on privacy. Trip duration was assessed by directly asking for the uninterrupted travel time needed to start a specific activity. For questions about information need, items based on Beggiato et al. [9] and Diels and Thomson [10] were rated on a 5-point Likert scale from "not necessary" to "necessary". The statistical analysis was performed using t-tests for dependent samples and repeated measures ANOVA, Cohen's d_z (for t-test) and η^2 (for ANOVA) are given for effect size. When Mauchly's test indicated a violation of the sphericity assumption, degrees of freedom were corrected using Greenhouse-Geisser.

3 Results

A total of 200 people took part in the survey, which was online for four weeks in late 2018 and in German language only. The average age of the participants was 33.83 years ($SD = 14.77$), ranging from 18 to 70 years. 58% of the participants were female and 49% were students. Privacy (private vs. fully occupied train compartment) has a strong influence on the tendency to make phone calls while travelling ($t(199) = 12.14$, $p < .001$, $d = 0.86$). There is also a significant difference in the attractiveness of voice messages ($t(199) = 10.53$, $p < .001$, $d = 0.74$) between a private and a non-private compartment and the tendency to sleep also differed significantly ($t(199) = 7.58$,

$p < .001$, $d = 0.54$). Other significant differences were found for eating/drinking ($t(199) = 6.78$, $p < .001$, $d = 0.48$), relaxing ($t(199) = 6.97$, $p < .001$, $d = 0.49$), working ($t(199) = 5.38$, $p < .001$, $d = 0.38$), laptop/tablet use ($t(199) = 2.83$, $p = .01$, $d = 0.20$), smartphone use ($t(199) = 2.94$, $p < .01$, $d = 0.21$), and sending text messages ($t(199) = 2.29$, $p = .02$, $d = 0.16$), but with no more than small effect sizes (Fig. 1).

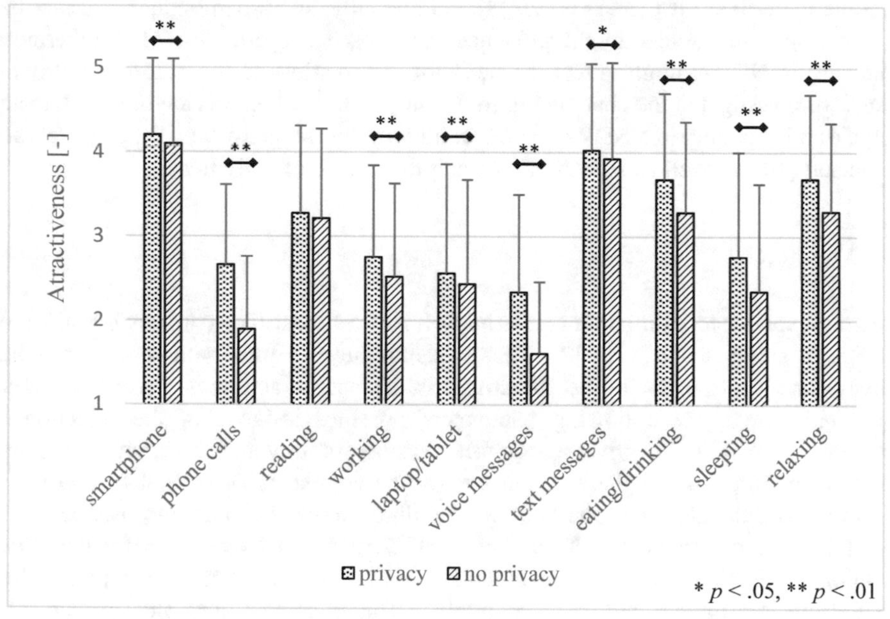

Fig. 1. Attractiveness (1: absolutely not; 5: definitely) of activities with and without privacy, error bars represent standard deviation

Comfort (table vs. no table) leads to higher use of laptops ($t(199) = 10.96$, $p < .001$, $d = 0.77$), higher willingness to work ($t(199) = 9.83$, $p < .001$, $d = 0.69$) and to eat or drink ($t(199) = 9.69$, $p < .001$, $d = 0.69$). Other differences show only small effect sizes. Furthermore, the purpose of the journey (work vs. private) has an influence: a laptop or tablet is used more frequently for work if the journey has a working background ($t(199) = 6.38$, $p < .001$, $d = 0.45$). The laptop is more likely to be used for leisure when it is a private trip ($t(199) = -6.70$, $p < .001$, $d = 0.47$). However, the use of smartphones for leisure is the most popular activity for both private and business trips. Reading ($t(199) = -10.32$, $p < .001$, $d = 0.73$) and relaxing ($t(199) = -9.25$, $p < .001$, $d = 0.65$) are more popular on private trips.

In order to assess the influence of the available time budget, participants were asked to state the minimum (uninterrupted) time needed to start a specific activity. The average time (in minutes) required for sleeping is the highest ($M = 76.08$, $SD = 83.89$). Followed by using the laptop for working ($M = 52.75$, $SD = 50.96$), using the laptop for leisure ($M = 45.06$, $SD = 46.96$) and reading ($M = 31.97$, $SD = 32.26$). Thinking/planning ($M = 11.95$, $SD = 12.28$), smartphone use for leisure ($M = 10.42$, $SD = 13.78$) and watching the surroundings ($M = 8.13$, $SD = 12.30$) require the least amount of time. In a repeated measures ANOVA, the chosen activity has a significant and large effect on the desired time span ($F(3.86, 767.58) = 62.06$, $p < .001$, $\eta^2 = 0.56$). However, time spans differ heavily among participants, especially for sleeping and using laptops/tablets. Half of the respondents need at least 60 min travel time to sleep. A further 25% need between 60 and 113 min. 50% of respondents would work on a laptop if the journey was at least 30 to 60 min. Smartphone use, one of the most popular activities, is particularly attractive for short driving times: 25% would already use it within one minute (Fig. 2).

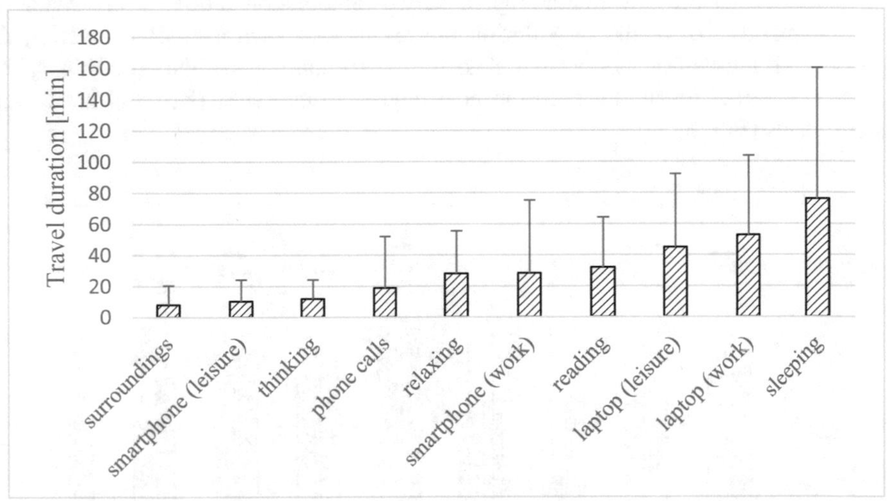

Fig. 2. Time spans to start a specific activity, error bars represent standard deviation

In addition, the survey examined the influence of the sociodemographic factors age and gender. Age effects were assessed using the first question of the survey (given privacy). For the following activities, a significant correlation with at least intermediate effects ($r = 0.24$) were found. The older people are, the more attractive reading is rated ($r = 0.35$, $p < .001$). Furthermore, age negatively affects the attractiveness of smartphone use ($r = -0.43$, $p < .001$) and of sending text messages ($r = -0.45$, $p < .001$). Taking into account both privacy conditions, significant effects for gender were found for working ($F(1, 198) = 5.44$, $p = .02$, $\eta^2 = 0.03$) and reading ($F(1, 198) = 7.47$, $p = .01$, $\eta^2 = 0.04$), both only reaching small effect sizes.

The survey also aimed at investigating the influence of an NDRA (reading to prepare for a professional appointment) on information needs during an automated car drive. For this purpose, an urban trip (from the stadium of FC Bayern Munich to the central railway station in the city center) with a duration of approx. 20 min was introduced. The automation system was presented as a level 3 system, capable of driving parts of the trip but with the need of a driver as fallback-ready user. Participants were asked about their preferences regarding necessary information, both with and without NDRA. The answers show a decreasing need for information on current and future maneuvers ($t(199) = 8.88$, $p < .001$, $d = 0.63$), as well as reasons for these maneuvers ($t(199) = 7.11$, $p < .001$, $d = 0.50$). Also information on the perception of surrounding traffic by car sensors ($t(199) = 8.38$, $p < .001$, $d = 0.59$) and on current speed ($t(199) = 9.22$, $p < .001$, $d = 0.65$) become less important. On the other hand, the need for information on navigation ($t(199) = 5.88$, $p < .001$, $d = 0.42$), the overview of the availability of automation on the entire route ($t(199) = 4.84$, $p < .001$, $d = 0.34$), the system status ($t(199) = 4.04$, $p < .001$, $d = 0.29$), and the reliability of the automation system ($t(199) = 3.91$, $p < .001$, $d = 0.28$) shows only small effects. Despite reaching significance, there is no effect ($t(199) = 2.45$, $p = .015$, $d = 0.17$) for information about the time remaining in current automation mode. This remains of constant importance for the user during the exercise of such a NDRA. Overall, the reliability of automation is considered to be most important in both conditions, followed by the time remaining until the next request to intervene (RtI) and the current system status (Fig. 3).

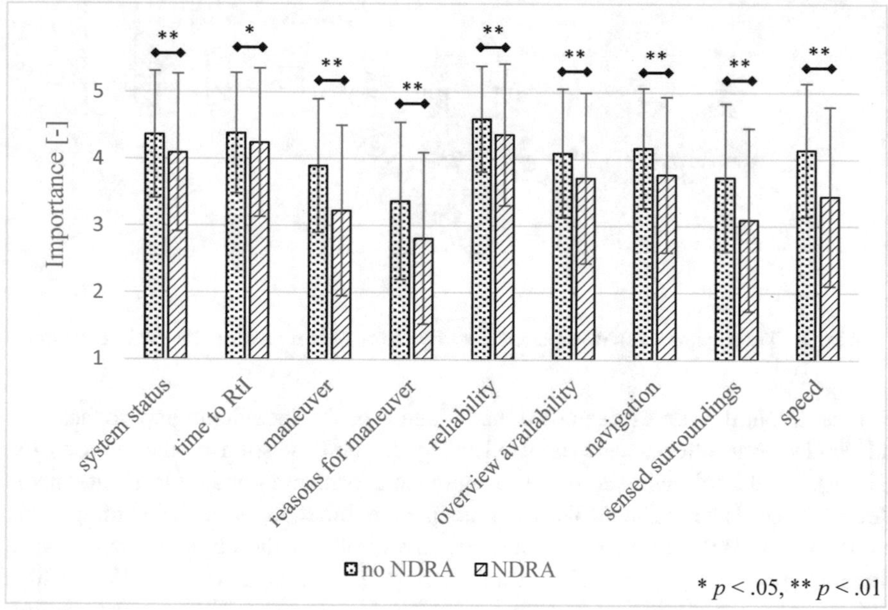

Fig. 3. Importance of information (1: not necessary, 5: necessary) with and without NDRA, error bars represent standard deviation

4 Discussion and Conclusion

The results of this online survey help understand the huge variety of possible NDRA in future automated driving cars. Privacy boosts the attractiveness of communicational activities like phone calls or voice mails. A table enhances activities like working, eating and drinking. For short trips, activities like watching the surroundings and smartphone use are especially popular, whilst relaxing, working and sleeping are more popular for longer trips. Furthermore, reading and relaxing are more attractive on private trips, but relaxing is also attractive for business trips. Overall, the activities phone use, relaxing and watching the surroundings are of highest attractiveness, no matter what the circumstances are.

The most important information needs for an urban automated drive are reliability of the automated system, remaining time in automated driving mode, and system status. When engaging in an important work-related activity while driving, most information loses importance compared to not engaging in a NDRA, especially information on maneuvers, surroundings and speed. Nonetheless, remaining time in current automation mode stays about equally important. However, with drivers mainly using external devices, such as phone, books or magazines, communication between driver and automation system becomes more difficult than with NDRA performed on internal devices. The need for the HMI to display the remaining time and the insight that available time spans are influencing the choice of NDRA highlight the importance of an HMI communicating the available time budget in automated driving mode in a way interfering as little as possible with the NDRA.

Acknowledgments. This report is based on parts of the research project @CITY-AF carried out at the request of the Federal Ministry of Economics and Energy, under research project No. 19A18003N. The author is solely responsible for the content.

References

1. König, M., Neumayr, L.: Users' resistance towards radical innovations. The case of the self-driving car. Transportation Research Part F: Traffic Psychology and Behaviour (2017). https://doi.org/10.1016/j.trf.2016.10.013
2. Pfleging, B., Rang, M., Broy, N.: Investigating user needs for non-driving-related activities during automated driving. In: Alt, F. (ed.) Proceedings of the 15th International Conference on Mobile and Ubiquitous Multimedia, MUM 2016, 12–15 December 2016, Rovaniemi, Finland, pp. 91–99. The Association for Computing Machinery, Inc, New York (2016). https://doi.org/10.1145/3012709.3012735
3. Sommer, K.: Continental Mobilitätsstudie 2013. Continental AG (2013)
4. Russell, M., Price, R., Signal, L., Stanley, J., Gerring, Z., Cumming, J.: What do passengers do during travel time? Structured observations on buses and trains. JPT **14**, 7 (2011). https://doi.org/10.5038/2375-0901.14.3.7
5. Large, D., Burnett, G., Morris, A., Muthumani, A., Matthias, R.: Design implications of drivers' engagement with secondary activities during highly-automated driving – a longitudinal simulator study. In: Road Safety and Simulation International Conference (RSS2017), The Hague, Netherlands, 17–19 October 2017 (2017)
6. Hecht, T., Feldhütter, A., Draeger, K., Bengler, K.: What do you do? An analysis of non-driving related activities during a 60 minutes conditionally automated highway drive. Article Submitted for Publication (2019)
7. Susilo, Y.O., Lyons, G., Jain, J., Atkins, S.: Rail passengers' time use and utility assessment. Transp. Res. Rec. (2013). https://doi.org/10.3141/2323-12
8. Ohmori, N., Harata, N.: How different are activities while commuting by train? A case in Tokyo. Tijdschrift voor economische en sociale geografie **99**, 547–561 (2008). https://doi.org/10.1111/j.1467-9663.2008.00491.x

9. Beggiato, M., Hartwich, F., Schleinitz, K., Krems, J., Othersen, I., Petermann-Stock, I.: What would drivers like to know during automated driving? Information needs at different levels of automation. In: 7. Tagung Fahrerassistenzsysteme (2015)
10. Diels, C., Thompson, S.: Information expectations in highly and fully automated vehicles. Advances in human aspects of transportation. In: Stanton, N.A. (ed.) Advances in Human Aspects of Transportation: Proceedings of the AHFE 2017 International Conference on Human Factors in Transportation. Springer, Cham (2018)

Yielding Light Signal Evaluation for Self-driving Vehicle and Pedestrian Interaction

Stefanie M. Faas[1,2(✉)] and Martin Baumann[2]

[1] Daimler AG, Leibnizstraße 2, 71032 Böblingen, Germany
Stefanie.Faas@daimler.com
[2] Department of Human Factors, Ulm University, Albert-Einstein-Allee 41, 89081 Ulm, Germany

Abstract. An external Human-Machine-Interface (eHMI) signaling the vehicle's intended movements facilitates pedestrians' encounters with self-driving vehicles (SDV). However, there is no standard for automated driving system (ADS) lamps today. This study compares the efficacy of a steady, a flashing and a sweeping light signal to communicate an SDV's intention to yield. The eHMI designs were evaluated at an unsignalized intersection with participants crossing in front of a yielding Wizard-of-Oz SDV. We analyzed crossing behavior and conducted questionnaires and structured interviews with $N = 30$ participants to identify eHMI design recommendations. Our research provides evidence that a steady and a flashing signal facilitate user experience, learnability and likeability more than a sweeping light. With a flashing signal, pedestrians tend to cross sooner compared to a sweeping signal, and thus improving traffic flow. Design adjustments to the present signals are proposed. This paper provides guidance in the development of a standardized yielding light signal.

Keywords: Automated vehicles · External Human-Machine-Interface · Automated driving system lamps · Pedestrian safety · Interface design

1 Introduction

Self-driving vehicle (SDV) technology is steadily developing. When the degree of autonomy increases to SAE Level 4 and 5, a human driver is no longer needed [1]. With manually steered vehicles, pedestrians are accustomed to interpreting vehicle movements and vehicle lights, but also communicating with the driver by eye contact, gesture and posture [2]. With the vehicle in control, pedestrians cannot rely on a present human driver anymore. There is a great body of studies showing that an external Human-Machine Interface (eHMI) supports the interaction between SDVs and pedestrians [3–7]. Previous research concluded that an eHMI communicating vehicle's intent is most effective to facilitate SDV and pedestrian encounters. Pedestrians wish to receive information about the SDV's next movements, e.g. whether the vehicle is stopping [8]. Displaying vehicle's intention to yield can help pedestrians to cross [3] and resolve deadlock situations [4]. It can be seen as a frontal brake light, which was shown to enable pedestrians to detect a braking maneuver more easily and to enhance road safety [9]. Furthermore, displaying vehicle's intention contributes to perceived safety [5], trust [4], a positive experience [2] and faster decision times [7].

T. Ahram et al. (Eds.): IHSED 2019, AISC 1026, pp. 189–194, 2020.
https://doi.org/10.1007/978-3-030-27928-8_29

The development of standardization guidelines for ADS lamps is in process. The UNECE set up the Taskforce “Autonomous Vehicle Signalling Requirements”, highlighting the use of light signals to communicate the vehicles driving mode and intention [10]. ISO emphasizes that communicating intention could improve societal acceptance and trust in SDVs [11]. SAE International [12] recently published a recommended practice, recommending the use of blue-green “Marker Lamps” to display the engagement of the ADS. The task force points out that further research is needed to evaluate the design of a light signal communicating the ADS yield message.

The current study compares the designs of a steady, a flashing and a sweeping light signal communicating an SDV’s intention to yield. This paper is a small contribution towards constructing research-based design principles for a standardized eHMI.

2 Method

2.1 Participants and Procedure

The final sample consisted of $N = 30$ participants (16 male, 14 female, $Mage = 42.97$ years, $SDage = 15.37$ years). The study was conducted as part of a larger research project at Daimler AG’s testing and technology center in Immendingen, Germany, from September 10th to September 24th, 2018. To create a Wizard-of-Oz setup of an automated driverless vehicle, we mounted four fake sensors on top of a Mercedes-Benz E-Class (Series W213) and a professional driver was maneuvering the vehicle wearing a seat costume [6]. The study was approved by Daimler AG’s Ethical Clearing Committee. Participants’ task was to cross an unsignalized four-way intersection when they feel safe to do so. In a controlled setting, the vehicle was approaching with a constant speed of 30 km/h, indicating the engagement of the ADS with a Marker Lamp as suggested by the SAE [12]. The vehicle yielded, allowing pedestrians to cross the road first. As soon as the vehicle started to brake, the yielding light signal was activated. In a randomized order, all participants encountered each yielding signal once (within-subject). After the study, participants were informed that the encountered vehicle had not been self-driving at any time.

2.2 Independent Variable: Yielding Light Signals

The vehicle indicated its intention to yield via a LED light stripe mounted on top of the windshield. B We chose to display the yielding light signals in blue-green (CIE chromaticity values: $x = .15$, $y = .38$) since this color is not yet associated with a specific meaning in traffic [13, 14]. The yielding light signals were connected to braking, hence they were activated as soon as the vehicle started to brake, analogous to rear brake lights. We explored the efficacy of three yielding light signals (Fig. 1): (1) *Steady light*: 30 LEDs in the middle of the stripe were turned on (adapted from [9]). (2) *Flashing light*: 30 LEDs in the middle of the stripe were flashing slowly with a frequency at 0.5 Hz following a sinusoidal waveform (adapted from [12]), (3) *Sweeping light*: Two 20 LEDs bars moved from side to side (adapted from [12, 15]).

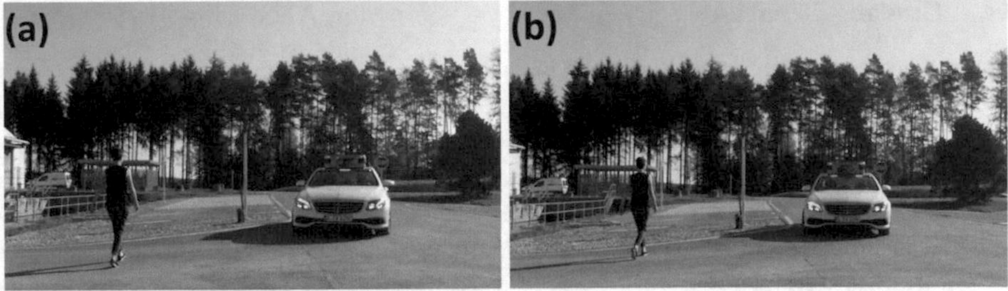

Fig. 1. Yielding light signals. 30 LEDS are steady or flashing (*a*) or two 20 LED bars are sweeping from side to side (*b*).

2.3 Dependent Variables

To gain insights into the crossing behavior, we analyzed video footage of the vehicle encounters. Two behavioral measures were computed: *crossing onset* represents the time in seconds pedestrians started to cross after the vehicle started to brake, indicated by the yielding light signals. *Crossing duration* is the time pedestrians take to cross.

After each vehicle encounter, participants rated the yielding light signals with the following 7-point scales: *User Experience* was assessed with the dimension "pragmatic quality" from the UEQ-S [16], Cronbach's $\alpha = .94$ to .97. *Learnability* was assessed with a single item: It is easy to learn that the light signal on the vehicle indicates "yielding" (strongly disagree - strongly agree). *Likeability* was measured with three semantic differentials (adapted from [17]): dislike – like, unfriendly – friendly, unpleasant – pleasant, Cronbach's $\alpha = .86$ to .95.

After all three vehicle encounters, participants were asked to rank the yielding signals on their preference and to explain the chosen order. The interviews were recorded on tape and transcribed. The total scope of the logs was 4355 words. Data analysis followed inductive category formation (based on [18]). The average inter-coder reliability was Cohen's $\kappa = .79$, reflecting an excellent level of clinical significance.

3 Results

Repeated measures ANOVAs were used to evaluate crossing behavior and subjective ratings of the yielding signals (see Table 1).

Table 1. Repeated measures ANOVA. * $p < .05$, ** $p < .01$, *** $p < .001$.

	Measurement	df1	df2	*F*-value	*p*-value	partial η2
Behavioral	Crossing onset	2	54	3.58	$p < .05$*	.12
	Crossing duration	2	56	2.56	$p = .086$	
Subjective	User experience	1.22	35.59	8.21	$p < .01$**	.22
	Learnability	1.27	36.75	9.18	$p < .01$**	.24
	Likeability	1.57	45.48	6.18	$p < .01$**	.18

3.1 Crossing Behavior

The results show that crossing onset was significantly affected by yielding signals, $p < .05$, partial η2 = .12, but not crossing duration, $p = .086$. Post-hoc tests with Bonferroni adjustment show that participants started crossing sooner with the flashing signal ($M = 1.43$, $SD = 1.77$) than the sweeping signal ($M = 1.87$, $SD = 1.52$), $p < .05$. There are no significant differences between the steady signal ($M = 1.62$, $SD = 1.59$) and the flashing signal, $p = .794$, or the sweeping signal, $p = .443$. Crossing duration does not differ among yielding signals (steady signal: $M = 7.12$, $SD = 1.25$; flashing signal: $M = 7.35$, $SD = 1.32$; sweeping signal: $M = 7.08$, $SD = 1.24$).

3.2 User Experience, Learnability and Likeability

All subjective measures were significantly affected by yielding signals, all *ps* < .01, partial η2 = .18 – .24. Post-hoc tests with Bonferroni adjustment show that participants rated User Experience higher with the flashing signal ($M = 1.73$, $SD = .20$) and the steady signal ($M = 2.00$, $SD = .21$) compared to the sweeping signal ($M = .83$, $SD = .34$), both *ps* < .05. There is no significant difference between the flashing signal and the steady signal, $p = .216$. Based on [16] the scores can be interpreted as excellent for the steady signal, good for the flashing signal, and below average for the sweeping signal. Analogously, participants rated learnability higher with the flashing signal ($M = 1.77$, $SD = .22$) and the steady signal ($M = 1.97$, $SD = .21$) compared to the sweeping signal ($M = .60$, $SD = .38$), both *ps* < .05. There is no significant difference between the flashing signal and the steady signal, $p = .791$. Participants rated likeability significantly higher with a flashing signal ($M = 1.88$, $SD = .19$) than with the sweeping signal ($M = .81$, $SD = .32$), $p < .01$. There is a trend that likeability is rated higher for a steady signal ($M = 1.71$, $SD = .19$) compared to a sweeping signal, $p < .10$. There is no significant difference between the flashing and the steady signal, $p = 1.00$.

3.3 Qualitative Content Analysis

Table 2 gives an overview of the derived category systems reflecting benefits and limitations that emerge from each yielding light signal.

Table 2. C = number of codings. Common stated categories (>5) are marked in bold.

Yielding light signal	Level	Benefits		Limitations	
		Category	C	Category	C
			73		71
Steady		Total	26	Total	16
	Affective	Pleasant	4		
	Cognitive	**Unambiguity**	**7**	**Inconspicuousness**	**7**
		Simplicity	4	**Illumination**	**5**
		Visibility	2	Ambiguity	4
	Conative	**Familiarity**	**9**		

(*continued*)

Table 2. (*continued*)

Yielding light signal	Level	Benefits		Limitations	
		Category	C	Category	C
			73		71
Flashing		Total	26	Total	13
	Affective	Pleasant	2	Annoying	2
	Cognitive	**Unambiguity**	**8**	**Non-permanence**	**5**
		Visibility	**7**	Ambiguity	4
		Alertness	4	Danger of confusion	2
	Conative	**Interaction**	**5**		
Sweeping		Total	21	Total	42
	Affective			Irritating	4
				Annoying	3
	Cognitive	**Visibility**	**9**	**Ambiguity**	**10**
		Alertness	**7**	**Gimmick**	**10**
		Unambiguity	3	**Overload**	**8**
				Danger of confusion	4
	Conative	Advice to cross	2	Vehicle priority	3

4 Discussion

It can be concluded that a steady or flashing light is more appropriate for an SDV to indicate its intention to yield to pedestrians than a sweeping light. They reflect a good to excellent user experience and are more learnable and likeable. A flashing signal even contributes to traffic flow, compared to the sweeping light. The earlier pedestrians start to cross, the less vehicles have to brake, and thus causing less delay to both pedestrians' and vehicles' flow.

Qualitative interviews allowed deeper insights into pedestrians' motivation. Though reflecting high visibility and alertness, a sweeping signal is ambiguous to pedestrians, implicating conflicting recommendations for action. Some participants rate the sweeping signal as advice to cross, whereas others interpret it contrarily as message of vehicle priority, taking right of way. Furthermore, a sweeping signal is perceived as an overloading gimmick, evoking negative emotions of feeling annoyed and irritated. The steady signal is appreciated for its simplicity and familiarity to the already existing rear brake lights. However, participants are concerned of its inconspicuousness and danger of confusion with illumination lights. To address these issues, we recommend using more than one lamp, analogous to the three rear brake lights. The flashing signal reflects high visibility, raising pedestrians' alertness, and is perceived to be suitable for pedestrian-vehicle interactions. However, pedestrians are concerned of overlooking the signal because of its non-permanence. Since the signal was flashing in a sinusoidal pattern, there were "dark" phases when the light signal could not be seen. We recommend adjusting the flashing signal in a way that the lamp is never turned off completely, but changes intensity in a way that the signal is always illuminated and detectable.

An eHMI will be added to the existing lamps of the vehicle, that are steady (e.g. braking lights, headlights) or flashing (e.g. turn indicator, hazard lights), too. The present study recommends staying in this design space, also having the claim of the ISO [11] in mind, stating that eHMI signals should be salient but not distractive.

References

1. SAE International: Taxonomy and definitions for terms related to driving automation systems for on-road motor vehicles (J3016) (2018)
2. Šucha, M., Dostal, D., Risser, R.: Pedestrian-driver communication and decision strategies at marked crossings. Accid. Anal. Prev. **102**, 41–50 (2017)
3. Mahadevan, K., Somanath, S., Sharlin, E.: Communicating awareness and intent in autonomous vehicle-pedestrian interaction. In: 2018 CHI Conference on Human Factors in Computing Systems, pp. 1–12 (2018)
4. Matthews, M., Chowdhary, G.V., Kieson, E.: Intent communication between autonomous vehicles and pedestrians (2017)
5. Lagström, T., Lundgren, V.M.: Automated vehicle's interaction with pedestrians (2015)
6. Rothenbücher, D., Li, J., Sirkin, D., Mok, B., Ju, W.: Ghost driver: a field study investigating the interaction between pedestrians and driverless vehicles. In: 25th IEEE International Symposium on Robot and Human Interactive Communication, pp. 795–802 (2016)
7. Stadler, S., Cornet, H., Theoto, T.N., Frenkler, F.: A tool, not a toy: using virtual reality to evaluate the communication between autonomous vehicles and pedestrians. In: tom Dieck, M.C., Jung, T. (eds.) Augmented Reality and Virtual Reality, pp. 203–216. Springer, Cham (2019)
8. Merat, N., Louw, T., Madigan, R., Wilbrink, M., Schieben, A.: What externally presented information do VRUs require when interacting with fully automated road transport systems in shared space? Accid. Anal. Prev. **118**, 244–252 (2018)
9. Petzoldt, T., Schleinitz, K., Banse, R.: The potential safety effects of a frontal brake light for motor vehicles. Intell. Transp. Syst. **12**, 449–453 (2018)
10. UNECE. https://wiki.unece.org/download/attachments/78742442/AVSR-04-05e.docx?api=v2
11. ISO: Road Vehicles: Ergonomic aspects of external visual communication from automated vehicles to other road users (ISO/TR 23049:2018) (2018)
12. SAE International: Automated Driving System (ADS) Marker Lamp (J3134_201905) (2019)
13. Dietrich, A., Willrodt, J.-H., Wagner, K., Bengler, K.: Projection-based external human machine interfaces: Enabling interaction between automated vehicles and pedestrians. In: 17th European VR, Driving Simulation and Virtual Reality Conference, pp. 43–50 (2018)
14. Werner, A.: New colors for autonomous driving: an evaluation of chromaticities for the external lighting equipment of autonomous vehicles. Colour Turn **1**, 1–15 (2018)
15. The Ford Motor Company. https://media.ford.com/content/fordmedia/fna/us/en/news/2017/13/ford-virginia-tech-autonomous-vehicle-human-testing.html
16. UEQ Data Analysis Tool. https://www.ueq-online.org/Material/Short_UEQ_Data_Analysis_Tool.xlsx
17. Bartneck, C., Kulic, D., Croft, E., Zoghbi, S.: Measurement instruments for the anthropomorphism, animacy, likeability, perceived intelligence, and perceived safety of robots. Int. J. Soc. Robot. **1**, 71–81 (2009)
18. Mayring, P.: Qualitative content analysis: theoretical foundation, basic procedures and software solution (2014)

How Should an Automated Vehicle Communicate Its Intention to a Pedestrian? – A Virtual Reality Study

Tanja Fuest[1(✉)], Anna Sophia Maier[1], Hanna Bellem[2], and Klaus Bengler[1]

[1] Chair of Ergonomics, Technical University of Munich, Boltzmannstr. 15, 85748 Garching, Germany
{tanja.fuest, sophia.maier, bengler}@tum.de
[2] BMW Group, New Technologies, 85748 Garching, Germany

Abstract. To analyze the influence of right of way and automated vehicle (AV) deceleration maneuvers on pedestrians' behavior, a virtual reality study was conducted. Participants were asked to press a button when they understood the intention of an approaching AV, and to rate the driving behavior after each trial. Results showed that the AV's driving behavior was able to communicate an intention. If the AV decelerates for pedestrians, it should decelerate early. When right of way is not defined, the AV should adapt to the expectations of pedestrians (e.g. not drive too fast). At a zebra crossing, participants expect the AV to communicate at an early stage that they are allowed to go first.

Keywords: (Automated) Vehicle-Pedestrian-Interaction · Virtual reality · Implicit communication · Mixed traffic

1 Introduction

According to BMW, highly automated vehicles (AV) should be available on the market by 2021 [1]. By then, AVs must be ready to interact safely with other human road users (HRU) such as pedestrians. In addition to sufficient sensors, AVs should also have well-designed trajectories. Driving behavior must not only be programmed with traffic safety in mind, but must be understood by HRUs in order to ensure traffic flow [2]. At best, this AV-to-HRU communication should be possible without adding additional communication channels, such as displays, and rely solely on an existing channel, i.e., driving behavior.

Therefore, we observed pedestrians crossing a street and asked them to explain what vehicle driving behavior shows that it is safe to cross the road. Based on these findings, we deduced driving maneuvers and implemented them in a virtual reality (VR) setup to explore the driving behavior in more detail and figure out whether it can be used to communicate an AV's intention. Pedestrians mentioned the braking maneuver in particular as an implicit communication method. This agrees with study results from [3–5], who mentioned that pedestrians are accustomed to interpreting a

T. Ahram et al. (Eds.): IHSED 2019, AISC 1026, pp. 195–201, 2020.
https://doi.org/10.1007/978-3-030-27928-8_30

vehicle's braking [4], even when it is smooth [5]. They also decide whether to cross the road depending on the velocity of the approaching vehicle [5].

Therefore, the first aim of the study was to evaluate whether different rights of way and, therefore, altered situations influence the intention recognition time (IRT) and the rating of the driving behavior. Second, we wanted to find out whether varied braking maneuvers affect the dependent variables.

2 Method

2.1 Preliminary Study

To evaluate the interaction between drivers and pedestrians, we asked the latter to explain what vehicle driving behavior shows that it is safe to cross the road. Therefore, we showed $N = 20$ participants ($M = 36.15$ years, $SD = 15.55$ years) four pictures, one with a normal vehicle and one with an AV, of two situations: a street with and without a zebra crossing. Moreover, we observed and interviewed $N = 20$ pedestrians ($M = 37.50$ years, $SD = 10.89$ years) crossing both types of streets. The results of the observation showed that all pedestrians needed to see a deceleration of the approaching vehicle and about 50% waited for a standstill, regardless of the street type (Table 1). The participants who saw the pictures indicated that 85% waited for a deceleration and 50% for a standstill in situations with a zebra crossing. Without a zebra crossing more pedestrians waited for a deceleration of the vehicle (95%), though only 40% waited for a complete standstill. However, the value for a standstill increased if no driver was sitting in the vehicle: 60% would not cross the zebra crossing and 90% would not cross a normal street if the vehicle was still driving.

As both preliminary studies showed, deceleration plays a central role in recognizing the intention of vehicles with and without driver. To explore the driving behavior in more detail, we implemented different AV behaviors in a VR setup.

Table 1. Identified driving behavior

		Driving Behavior		No Driver
		Deceleration	Standstill	Standstill
Observation	Both types of streets	100%	50%	–
Pictures	With zebra crossing	85%	50%	60%
	Normal Street	95%	40%	90%

2.2 Virtual Reality Pedestrian Simulator

The study took place at BMW in Munich (Germany). The pedestrian simulator consists of an HTC Vive Pro VR setup (head mounted display, two infrared trackers, and an HTC VIVE's remote control). The simulation software is based on Unity 3D and a BMW i3 is implemented. The investigator can manipulate the driving behavior by adding a trajectory path and maneuver points. The actual speed in km/h and the acceleration in m/s^2 are presented to the investigator on the computer.

2.3 Procedure

During the simulation, participants were immersed in an urban environment, standing at the side of a road with the intention of crossing the road seeing an AV approach from the left. They experienced nine driving profiles in three situations with different rights of way. The vehicle changed speed and made lateral maneuvers to communicate its intentions. However, in situations where the HRU had right of way, characterized by a zebra crossing, the AV decelerated in every trial to comply with road traffic regulations. Altogether, participants experienced 18 situations in an incomplete design in random order. In every situation, the participants were asked to press a button on the remote control held in their hand when they thought they recognized the intention of the vehicle. The simulation stopped simultaneously and the participants filled out a verbal questionnaire after each trial.

2.4 Independent Variables

We implemented two intentions for the AV. In the first, the AV had the intention to "Go first" and in the second, the AV was going to "Let the HRU go first". In all situations, the AV accelerated for the first 22 m to 49 km/h, while the HRU's position was 65 m from the AV's starting point.

For both intentions, the right of way was varied and was broken down into three conditions. In the first condition, we used a normal street, where the AV had right of way. A zebra crossing defined the second condition, so that the HRU had right of way. Due to a shared space situation, the right of way was not defined in the third condition.

For the intention "Go first", the AV maintained at a constant speed of 49 km/h. For the intention "Let the HRU go first", the AV decelerated in two different ways: an early slow and a late abrupt deceleration. In the early condition, the AV started to decelerate at 25 m from the pedestrian's position, with the brake pedal at a 20% incline. In contrast, the late deceleration started at 16 m distance from the HRU with the brake pedal at 40% incline. In both conditions, the AV came to a standstill 5 m from the HRU's position.

We also varied the lateral dynamics of the AV. However, the simulation generated unexpected swerves in driving behavior while approaching towards or moving away from the road edge. Therefore, we excluded this independent variable from the analysis. In addition, the intention behind the swerves was not clear to the participants: over 52% could not interpret the swerves.

2.5 Dependent Variables

IRT is the time elapsed from the moment the vehicle set off until the moment at which the pedestrians thought they recognized the vehicle's intention. The participants stopped the timing by pressing a button on the remote control.

After each trial, the participants filled out a questionnaire. One question inquired about the vehicle's assumed intention ("Let the HRU go first" or "Go first"). Then, we used a five-point Likert scale to let the pedestrians rate their certainty about the vehicle's intention (from very uncertain to very certain), and the vehicle's driving behavior (from very poor to very good). We also measured the perceived criticality of the situation using the criticality scale from [6]. However, for this question, participants had problems with the anchor points (e.g. "not controllable") from the pedestrian perspective and the results showed a large deviation in answers. Therefore, we excluded this variable.

2.6 Participants

In total, $N = 61$ pedestrians (41 male, 20 female) with a mean age of $M = 34.38$ years ($SD = 10.88$ years) were recruited. All participants worked for BMW. Each participant had a valid driver's license, although this was not a prerequisite, and participated as a pedestrian in traffic $M = 5.77$ h ($SD = 5.30$ h) per week.

2.7 Analysis

The data from one scenario of one participant were excluded from analysis due to technical issues. For the dependent variable "evaluation of driving behavior", data were collected only for 36 participants. If participants did not press the button after the vehicle had already passed by or forgot it completely, we adapted the IRTs. For the intention "Go first" the vehicle passed the pedestrian after 5.64 s, so we changed the IRT to this maximum time. For the intention "Let the HRU go first" with early deceleration, we adapted the IRT to a maximum of 6.73 s, and to 6.13 s for late deceleration because then the vehicle came to a standstill.

3 Results

3.1 Misinterpretations of Intentions

Each participant experienced every driving profile, resulting in a total of 61 valid data sets for each profile except for the driving profile "Go first, AV has right of way". As mentioned before, for the latter we have only 60 data sets due to technical issues. Altogether 487 trials were included in the evaluation.

Overall 88% of the pedestrians understood the intentions correctly. With regard to the intention "Go first", fewer than 1% misinterpreted the driving behavior. The rate for the intention "Let the HRU go first" with early deceleration showed similar results: only 3% misinterpreted intentions. However, late deceleration led to 28% of participants interpreting incorrectly (Table 2).

Only the answers of the correctly recognized intentions were used for further analyses.

Table 2. Misinterpretations for both intentions.

		Right of Way		
		AV	Undefined	HRU
Go first	No deceleration	0 (0.0%)	1 (1.6%)	–
Let the HRU go first	Early deceleration	1 (1.6%)	3 (4.9%)	2 (3.3%)
	Late deceleration	12 (19.7%)	18 (29.5%)	22 (36.1%)

3.2 Right of Way

For the different rights of way with the intention "Go first" we did not find significant differences in the IRT ($z = -1.86$, $p = 0.62$) and the subjective decision-making reliability ($z = -0.83$, $p = 0.41$). However, the participants rated the driving behavior better if the AV had right of way ($z = -3.90$, $p \leq 0.001$, $r = 0.64$, $Mdn_{AV} = 4.00$, $Mdn_{undefined} = 2.00$).

For the intention "Let the HRU go first" no significant differences were found between right of way in the IRT (early deceleration: $\chi^2(2) = 1.51$, $p = 0.47$, $n = 56$; late deceleration: $\chi^2(2) = 1.44$, $p = 0.49$, $n = 27$), the subjective decision-making reliability (early deceleration: $\chi^2(2) = 1.18$, $p = 0.55$, $n = 56$; late deceleration: $\chi^2(2) = 0.66$, $p = 0.72$, $n = 27$), and the evaluation of driving behavior (early deceleration: $\chi^2(2) = 0.41$, $p = 0.82$, $n = 36$; late deceleration: $\chi^2(2) = 1.68$, $p = 0.43$, $n = 16$).

3.3 Deceleration

To examine whether the different braking maneuvers led to different IRTs and evaluations of driving behavior, we used the mean of all rights of way because we found no significant differences between these conditions. Using the averaged values, we calculated a Wilcoxon test for every dependent variable.

Participants needed longer IRTs to understand the intentions of an AV if the deceleration started early ($z = -4.33$, $p \leq 0.001$, $r = 0.57$, $Mdn_{early} = 5.71$, $Mdn_{late} = 5.57$). However their subjective decision-making reliability ($z = -3.60$, $p \leq 0.001$, $r = 0.47$, $Mdn_{early} = 4.33$, $Mdn_{late} = 4.00$) and the evaluation of driving behavior ($z = -4.01$, $p \leq 0.001$, $r = 0.68$, $Mdn_{early} = 4.00$, $Mdn_{late} = 3.33$) were significantly better when the AV decelerated early.

4 Discussion

4.1 Driving Behavior

The results for the driving behavior showed that most pedestrians understood intentions correctly. Only one person misinterpreted the intention "Go first". It seems that maintaining speed at 49 km/h is a good way to communicate that the AV is going first. This agrees with the results from [3]. However, in situations where right of way is not defined, the AV should adapt its driving behavior to the expectations of pedestrians (e.g. not too fast) because participants rated driving behavior better when the AV had right of way and did not yield.

When the AV wants to "Let the HRU go first", pedestrians preferred an early, slow deceleration, regardless of right of way, even though they needed more time to understand the intention. By taking more time, pedestrians feel more confident in their decisions and evaluate driving behavior better. Additionally, the late, abrupt deceleration led to frequent misinterpretations at zebra crossings or when right of way was not defined. So, it seems that at zebra crossings, participants expect the AV to communicate at an early stage that the AV respects the right of way.

4.2 Virtual Reality Method

The VR setup was used so as not to endanger participants. However, we found major weaknesses in the programming of our VR setup: we were not able to program the driving behavior in detail. We could change the angle of the brake pedal in percentages but not control the results because of missing driving data. Due to the latter, driving behavior was only designed by watching the vehicle through the head mounted display. The deceleration was chosen in such a way that differences in braking were noticeable, but both looked like "normal" braking maneuvers. However, the early, slow deceleration had a value of $-6\ m/s^2$ and the late, abrupt deceleration of even $-11\ m/s^2$. As a comparison, a comfortable deceleration rate of $-3.4\ m/s^2$ from passengers' perspective is recommended [7]. The perception in VR seems to deviate from real values, which agrees with previous studies [8, 9]. The strong decelerations might have also led to the swerves in the driving behavior. As a result, we would highly recommend establishing basic requirements for VR studies. The output of driving data should be one of these requirements.

References

1. BMW Group: The path to autonomous driving. https://www.bmw.com/en/automotive-life/autonomous-driving.html
2. Fuest, T., Sorokin, L., Bellem, H., Bengler, K.: Taxonomy of traffic situations for the interaction between automated vehicles and human road users. In: Stanton, N.A. (ed.) Advances in Human Aspects of Transportation, vol. 597, pp. 708–719. Springer, Cham (2018). https://doi.org/10.1007/978-3-319-60441-1_68
3. Fuest, T., Michalowski, L., Traris, L., Bellem, H., Bengler, K.: Using the driving behavior of an automated vehicle to communicate intentions - a wizard of oz study. In: 2018 21st International Conference on Intelligent Transportation Systems (ITSC), pp. 3596–3601. IEEE (2018). https://doi.org/10.1109/ITSC.2018.8569486
4. Ackermann, C., Beggiato, M., Bluhm, L.-F., Löw, A., Krems, J.F.: Deceleration parameters and their applicability as informal communication signal between pedestrians and automated vehicles. Transp. Res. Part F Traffic Psychol. Behav. **62**, 757–768 (2019). https://doi.org/10.1016/j.trf.2019.03.006
5. Schneemann, F., Gohl, I.: Analyzing driver-pedestrian interaction at crosswalks: A contribution to autonomous driving in urban environments. In: 2016 IEEE Intelligent Vehicles Symposium (IV), pp. 38–43. IEEE (2016). https://doi.org/10.1109/IVS.2016.7535361

6. Neukum, A., Lübbeke, T., Krüger, H.-P., Mayser, C., Steinle, J.: ACC-Stop&Go: Fahrerverhalten an funktionalen Systemgrenzen. In: Maurer, M., Stiller, C. (eds.) 5. Workshop Fahrerassistenzsysteme, pp. 141–150 (2008)
7. AASHTO: A Policy on geometric design of highways and streets. Washington (2011)
8. Bhagavathula, R., Williams, B., Owens, J., Gibbons, R.: The reality of virtual reality: a comparison of pedestrian behavior in real and virtual environments. Proc. Hum. Factors Ergon. Soc. Annu. Meet. **62**, 2056–2060 (2018). https://doi.org/10.1177/1541931218621464
9. Hurwitz, D., Knodler, M., Dulaski, D.: Speed perception fidelity in a driving simulator environment. In: Conference: Driving Simulator Conference North America, pp. 343–352 (2005)

Digital Human Modelling, Occupant Packaging and Autonomous Vehicle Interior

Sibashis Parida[1,2(✉)], Sylvester Abanteriba[2], and Matthias Franz[1]

[1] BMW Group, Knorrstr. 147, 80788 Munich, Germany
{sibashis.parida,matthias.franz}@bmw.de
[2] RMIT University, 124 La Trobe Street, Melbourne 3000, Australia
sylvester.abanteriba@rmit.edu.au

Abstract. The biggest advantage of autonomous driving is the value added free time that the users would enjoy during travel. Research shows that the users would use this time to participate in different non-driving activities: which include resting, sleeping, using smartphone, reading for pleasure, working and the most common activity, i.e. looking outside the window and enjoying the landscape. The challenge for the automotive industry however is to develop vehicle interior and seating concepts to facilitate the different ergonomic seating postures, which would allow the users to pursue these activities for a longer period. The paper helps to understand why digital human modelling and occupant packaging would be of increasing importance in the interior development of autonomous driving vehicles. Discussed in the paper are proposed tools which when implemented would aid the interior development of vehicle interior in the virtual phase and would help to recognize vehicle integration and package problems earlier on in the vehicle development phase.

Keywords: Digital human modelling · Occupant packaging · Autonomous driving · Autonomous vehicles · Vehicle ergonomics

1 Introduction

Seats are the most important human-machine interface in a passenger vehicle. The objective of vehicle seating is to offer ergonomic optimal seating postures for a wide range of body shapes and sizes for the driving task, as well as for the passengers. Automotive industry is undergoing a major shift due to the increasing importance of autonomous (AV) and connected vehicles (CV). This creates new opportunities as well as challenges for automotive human machine interaction and vehicle seating [1, 2]. Research shows that value added free time, which the drivers would enjoy during level 4 and level 5 autonomous driving (AD) [3], is one of the biggest drivers of AD technology. Therefore, the AD vehicles would no longer be used as a mode of transportation, but would act as a living space for the users. According to surveys conducted in the recent past, the most common and preferred non-driving use-cases include; resting, sleeping, use of smartphone, reading for pleasure, working on a laptop, looking outside the window and enjoying the landscape [4–7]. The challenge for the automotive industry is to develop innovative vehicle interior and seating concepts,

T. Ahram et al. (Eds.): IHSED 2019, AISC 1026, pp. 202–208, 2020.
https://doi.org/10.1007/978-3-030-27928-8_31

which would facilitate these set of non-driving use-cases. The paper helps to understand how digital human modeling and occupant packaging would be of increasing importance and the need to implement interactive digital human design approach to develop the interior of AVs.

2 Background

With the increasing number of megacities around the globe, the travel time spent to and from work is drastically increasing [8–10]. AD with the value added free time offers to solve the transportation problem by allowing users to use the travel time productively [8]. The notion of wasted, unproductive time being turned into economically valuable time is a dominant argument in the debate on AD. The way users would spend their time while traveling in the future might not only affect the valuation of different transport modes but also the distances travelers are willing to overcome [11]. The automotive trend is also currently heading in the direction where automotive OEMs are presenting unconventional interior concepts and designs allowing the users to use the vehicle as a living space and perform the above stated non-driving activities [12, 13]. Discussed in this section are factors important in the interior development of AV.

2.1 Digital Human Modelling

Digital Human Modeling (DHM) is an emerging area that bridges computer-aided engineering design, human factors engineering and applied ergonomics [14]. As a technology, digital human modeling is a means to create, manipulate, and control human representations and human-machine system scenes on computers for interactive ergonomics and design problem solving. As a fundamental research area, digital human modeling refers to the development of mathematical models that can predict human behavior in response to minimal command input and allow real-time computer graphic visualization [15]. The implementation of digital human modeling technology allows easier and earlier identification of ergonomics problems, and lessens or sometimes even eliminates the need for physical mock-ups and real human subject testing [16–18]. While there is an additional implementation and training cost in the initial stage, the use of digital human modeling, digital prototyping, and virtual testing in a computer-aided design or ergonomics process can quickly lead to reduced cost and shorten time [15].

A number of organizations in a wide range of industries are facing the problem that human element is not considered early enough in the product development process. This has a negative effect on cost, time to market, quality and safety. Naumann and Rötting [14] suggest that implementing DHM and human factors engineering would bring competitive advantage and would realize benefits such as shorter design time, lower development costs, improved quality and enhanced productivity. Cappelli and Duffy [19] agree with the same and suggest that implementing DHM early in the development phase would reduce the need for production of real prototypes and would in return help in cost reduction and enhanced productivity.

2.2 Occupant Packaging

In the automotive industry, the term "packaging" is used to describe the integration of different components and systems in the vehicle architecture [20]. The main components include the vehicle powertrain system, including engine and gearbox; chassis; body in white; interior and exterior trim; the different electrical and electronic components; climate control and a whole lot more. In addition to these main components and systems, there are hundreds or even thousands of minor components to be included (such as individual switches, relays, fuses, etc.). Occupant packaging however focuses in the system integration of the occupants, that is, the vehicle driver and the passengers, with the emphasis on human anthropometry, biomechanics, psychology, statistics and so forth. Occupant packaging aims to ensure that there is a best possible fit between the vehicle, the driver and the passengers. Occupant packaging also ensures that a large range of occupants are comfortably accommodated in a vehicle [20]. Vehicle occupant packaging is the process of laying out the interior of a vehicle to achieve the desired levels of accommodation, comfort, and safety for the occupants [21].

Occupant packaging focuses primarily on the driver's workstation. Driver packaging not only includes the integration of steering wheel, seat, pedals, but also ensures the physical locations of controls and displays that the drivers interacts. Interior and exterior drivers vision, both direct and indirect using mirrors are also important aspects of driver packaging [21]. The passenger packaging is equally important in vehicle development. The passengers must have appropriate space inside the vehicle and must be able to sit in comfortable posture for a longer period. As the driver, the passengers also need to have adequate level of vision out of the vehicle [20].

The philosophy of occupant packaging is to fit the product to the user and not the other way round. Therefore, for AD vehicles, it is important to design vehicle interior that meets the user expectations and allows the passengers to make use of the benefits offered by autonomous driving. According to Gkikas [20] occupant packaging may be used as a vehicle or brand differentiator, especially in the premium sector. Thus, occupant packaging is extremely important in the development of autonomous driving vehicle interior.

3 Discussion

In a study conducted by Kent [22] even though the car was not necessarily perceived as faster than alternative transport, it was that the participants perceived time taken on trains, buses, or walking and cycling, as more of an investment, more frustrating, less comfortable and more disempowering than the time they spend in their car. It was important to the participants to spend their time being comfortable and in control than 'wasting' their time by, for example, waiting for bus connections or dealing with crowded public transport. The participants' expressed appreciation of the freedom, flexibility and reliability associated with the use of private car in comparison to using the public transportation or a shared vehicle [22]. While the advent of car sharing might be changing the ownership model of vehicles, for now, the car is often a secluded and personal context where we are shielded from the outer world [23]. In a study Parida et al. [24] considered the seating positions from their original environment; i.e. office chairs for the use of laptop, lounge seats/couch for the relax position and zero gravity position for the sleeping position. These seating positions were configured in a BMW 7-Series vehicle and the participants were asked to evaluate the positions. In contrast to expectation, the results were rather negative. This suggests that the seating positions for non-driving activities in a passenger vehicle is rather different to that of the original environment. Parida et al. [25] conducted another study with 51 participants to evaluate the seating positions of non-driving activities in a BMW 7-Series vehicle where the participants could select their seating positions according to their own preference. The participants were between the age of 23 and 60 and ranged from 158 cm to 203 cm. The activities included working on a laptop, using smartphone, reading for pleasure, relaxing/looking outside the window and sleeping. The study suggested, in order to perform these activities inside a vehicle the subjects preferred to be in seating positions other than the driving position. The most critical outcome of the study was the standard deviation (SD). The results suggested an extremely high SD. For the backrest angle the SD was more than 5°, for seat-pan angle, the calculated SD was more than 12° and for neck angle was 4° (Fig. 1 and Table 1). Therefore, there is no "one-fits-all" seating position for a given non-driving activity. This means when developing seats for AV, it is important to consider a wide range of seating positions. Hence, the challenge for the automotive industry is to develop the best possible occupant package for a wide range of use-cases and seating positions. The digital human models or manikins currently used in the automotive industry are static models, and are not efficient enough for an effective occupant packaging in AVs.

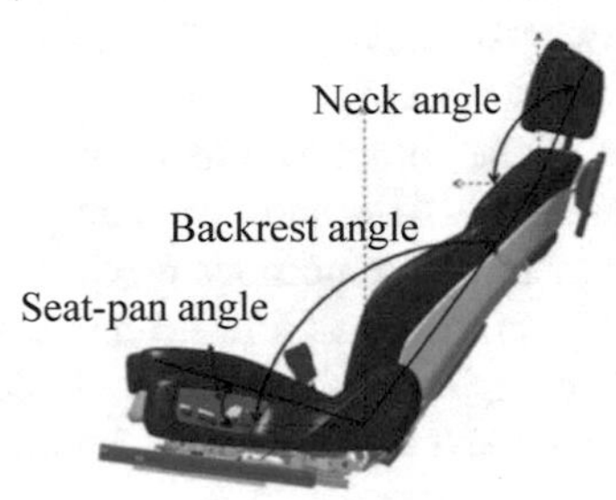

Fig. 1. Seating angles

Table 1. Seating positions for AD use-cases with SD values (Parida et al. [25])

AD use-case	Backrest angle		Seat-pan angle		Neck angle	
	Mean	SD	Mean	SD	Mean	SD
Use of Laptop	118°	6°	11°	15°	5°	4°
Smartphone	117°	5°	15°	12°	7°	4°
Reading	121°	7°	15°	13°	5°	4°
Window gazing	121°	5°	15°	14°	7°	4°
Sleeping	146°	8°	15°	12°	6°	4°

In order to develop an effective vehicle occupant packaging model for AVs, it is important to develop 3D dynamic human models which would not only have end positions as current DHM manikins, but would have real time kinematic movements in accordance to the different seating positions for the prospective use-case. Another important aspect would be to parametrize the different joints of the human models for a quick modification and visual representation, which again is extremely important for occupant packaging in the concept development phase. The important joints for an effective DHM include hip angle, knee angle and neck angle. These three angles have major influence on comfort [26–28].

4 Conclusion

Value added free time, which the drivers would enjoy during level 4 and level 5 AD, is one of the biggest driver of AD technology. Research shows that there is willingness among users to use this value added free time during automated driving to participate in non-driving secondary activities namely; resting, sleeping, using smartphone, reading for pleasure, working on laptop, relaxing and window gazing. The autonomous driving vehicle of the future would be more of a living space rather than just a mode of transportation. DHM and occupant packaging would play a very important role, in the development of autonomous driving vehicle interior. The authors propose a 3D kinematic human model as a DHM and occupant packaging tool to develop the interior of AVs. The implementation of the 3D DHM would bring benefits such as shorter design time, lower development costs, improved quality, and enhanced productivity.

References

1. Flemisch, F., Heesen, M., Hesse, T., Kelsch, J., Schieben, A., Beller, J.: Towards a dynamic balance between humans and automation: authority, ability, responsibility and control in shared and cooperative control situations. Cogn Tech Work (2012). https://doi.org/10.1007/s10111-011-0191-6
2. Kutila, M., Jokela, M., Markkula, G., Rue, M.R. (eds.): Driver distraction detection with a camera vision system. In: IEEE International Conference on Image Processing (2007)
3. SAE International: SAE International's levels of driving automation for on-road vehicles (2014)
4. Karlsson, I.M., Pettersson, I.: Setting the stage for autonomous cars. A pilot study of future autonomous driving experiences. IET Intell. Transp. Syst. (2015). https://doi.org/10.1049/iet-its.2014.0168
5. Llaneras, R., Salinger, J., Green, C.A.: Safety related misconceptions and self-reported behavioral adaptations associated with advanced in-vehicle system: lessons learned from early technology adopters. In: Proceedings of the Seventh International Driving Symposium on Human Factors in Driver Assessment, Training, and Vehicle Design
6. Naujoks, F., Purucker, C., Neukum, A.: Secondary task engagement and vehicle automation – Comparing the effects of different automation levels in an on-road experiment. Transp. Res. Part F: Traffic Psychol. Behav. (2016). https://doi.org/10.1016/j.trf.2016.01.011
7. Pfleging, B., Rang, M., Broy, N.: Investigating user needs for non-driving-related activities during automated driving. In: Häkkila, J., Ojala, T. (eds.) Proceedings of the 15th International Conference on Mobile and Ubiquitous Multimedia - MUM 2016. The 15th International Conference, Rovaniemi, Finland, 12 December 2016–15 December 2016, pp. 91–99. ACM Press, New York (2016). https://doi.org/10.1145/3012709.3012735
8. Frauenhofer Institute: The Value of Time (2016)

9. McKinsey Company: Automotive Revolution & Perspective Towards 2030. Auto Tech Rev (2016). https://doi.org/10.1365/s40112-016-1117-8
10. McKinsey & Company: Bloomberg New Energy Finance: An Integrated Perspective on the Future of Mobility. McKinsey & Company, Bloomberg New Energy Finance (2016)
11. Cyganski, R., Fraedrich, E., Lenz, B.: Travel time valuation for automated driving: a use-case driven study. In: TRB 94th Annual Meeting (2014)
12. Brown, A.: BMW's CES Concept Predicts the Self-Driving Car Interior of Tomorrow (2019). https://www.thedrive.com/tech/6810/bmws-ces-concept-predicts-the-self-driving-car-interior-of-tomorrow. Accessed 6 June 2019
13. Bloom, J.: Car Designers Show Off Futuristic Interiors for Tomorrow's Self-Driving Vehicles (2019). https://www.nbclosangeles.com/news/tech/Car-Designers-Show-Off-Futuristic-Interiors-for-Tomorrows-Self-Driving-Vehicles-504896031.html. Accessed 1 June 2019
14. Naumann, A., Rötting, M.: Digital Human Modeling for Design and Evaluation of Human-Machine Systems. MMI-Interaktiv (2007)
15. Zhang, X., Chaffin, D.B.: Digital human modeling for computer-aided ergonomics. University of Michigan (2005)
16. Badler, N.I., Phillips, C.B., Webber, B.L.: Humans: Computer Graphics, Animation, and Control (1993)
17. Morrissey, M.: Human-centric design. Mech. Eng. **120**, 60–62 (1998)
18. Zhang, X., Chaffin, D.B.: A three-dimensional dynamic posture prediction model for simulating in-vehicle seated reaching movements: development and validation. Ergonomics **43**, 1314–1330 (2000)
19. Cappelli, T.M., Duffy, V.G. (eds.): Motion Capture for Job Risk Classifications Incorporating Dynamic Aspects of Work. Digital Human Modeling for Design and Engineering Conference, Jul. 04, 2006. SAE International400 Commonwealth Drive, Warrendale, PA, United States (2006)
20. Gkikas, N.: Automotive Ergonomics. Driver-Vehicle Interaction. CRC Press, Hoboken (2012)
21. Parkinson, M.B., Reed, M.P. (eds.): Optimizing Vehicle Occupant Packaging. SAE 2006 World Congress & Exhibition, APR. 03, 2006. SAE International400 Commonwealth Drive, Warrendale, PA, United States (2006)
22. Kent, J.L.: Driving to save time or saving time to drive? The enduring appeal of the private car. Transportation Research Part A: Policy and Practice (2014). https://doi.org/10.1016/j.tra.2014.04.009
23. Pettersson, I., Ju, W.: Design techniques for exploring automotive interaction in the drive towards automation. In: Mival, O., Smyth, M., Dalsgaard, P. (eds.) Proceedings of the 2017 Conference on Designing Interactive Systems - DIS 2017. The 2017 Conference, Edinburgh, United Kingdom, 10 June 2017–14 June 2017, pp. 147–160. ACM Press, New York (2017). https://doi.org/10.1145/3064663.3064666
24. Parida, S., Mallavarapu, S., Abanteriba, S., Franz, M., Gruener, W.: User-centered-design approach to evaluate the user acceptance of seating postures for autonomous driving secondary activities in a passenger vehicle. In: Ahram, T., Karwowski, W., Taiar, R. (eds.) Human Systems Engineering and Design, vol. 876. Advances in Intelligent Systems and Computing, pp. 28–33. Springer, Cham (2019)
25. Parida, S., Mallavarapu, S., Abanteriba, S., Franz, M., Gruener, W. (eds.): Seating Postures for Autonomous Driving Secondary Activities. International Conference on Intelligent Interactive Multimedia Systems and Services, 17 June 2019–19 June 2019. Springer International Publishing (2019)
26. Babbs, F.W.: A Design Layout Method for Relating Seating to the Occupant and Vehicle. Ergonomics (1979). https://doi.org/10.1080/00140137908924606

27. Milivojevich, A., Stanciu, R., Russ, A., Blair, G.R., van Heumen, J.D.: Investigating psychometric and body pressure distribution responses to automotive seating comfort. In: SAE Technical Paper Series. SAE 2000 World Congress, MAR. 06, 2000. SAE International400 Commonwealth Drive, Warrendale, PA, United States (2000). https://doi.org/10.4271/2000-01-0626
28. Paddan, G.S., Mansfield, N.J., Arrowsmith, C.I., Rimell, A.N., King, S.K., Holmes, S.R.: The influence of seat backrest angle on perceived discomfort during exposure to vertical whole-body vibration. Ergonomics (2012). https://doi.org/10.1080/00140139.2012.684889

Evaluation of Display Concepts for the Instrument Cluster in Urban Automated Driving

Alexander Feierle(✉), Fabian Bücherl, Tobias Hecht, and Klaus Bengler

Chair of Ergonomics, Technical University of Munich, Boltzmannstraße 15, 85748 Garching, Germany
{alexander.feierle, fabian.buecherl, t.hecht, bengler}@tum.de

Abstract. Instrument clusters represent the primary human-machine-interface for displaying driving-related information. Due to the changing relevance of driving-related information in automated driving, a reconfiguration of the display should be investigated. A display concept for an instrument cluster for partially and highly automated driving was developed that adapts the display and positioning of information to the level of automation. The developed concept was compared with a conventional display concept. Thirty participants took part in an experiment performing an occlusion task and a choice reaction time task. The statistical analysis revealed no significant differences in the response accuracy of the occlusion task between the display concepts. The results of the choice reaction time task show a significantly faster reaction time and in contrast to the occlusion task, a significantly lower response accuracy for the adaptive display concept compared to the conventional display concept. The subjective analysis revealed a preference of the adaptive display concept.

Keywords: Human-Machine-Interface · Automated driving · Instrument cluster

1 Introduction

In automated driving, the relevance of driving-related information changes with the level of automation [1], which should be communicated to the driver. The instrument cluster (IC) is the primary driver-vehicle interface for presenting driving-related information in the vehicle. The information is to be communicated to the driver in an efficient and ergonomic manner to avoid a visual overload in the complex urban area with its high information density. Previous ICs are notably characterized by the circular instruments of the tachometer and speedometer, which have a high space requirement. The use of a freely programmable IC allows one to reconfigure the contents of the IC. Irrelevant information can be hidden and relevant information can be repositioned in order to bring important information more into the driver's focus. This situation-adapted presentation of information in the IC allows for the design of the display

T. Ahram et al. (Eds.): IHSED 2019, AISC 1026, pp. 209–215, 2020.
https://doi.org/10.1007/978-3-030-27928-8_32

contents according to the requirements of the current automation level. To investigate this potential, an adaptive IC (aIC) concept was developed for partially and highly automated driving [2] that adapts the display and positioning of information to the level of automation. The developed concept was compared with a conventional IC (cIC) concept, which showed no adaptation to the level of automation. We evaluated both display concepts in an occlusion task, choice-reaction-time task and by subjective ratings.

2 Method

2.1 Instrument Cluster Design

The cIC was a conventional concept developed for urban driving [3], which was expanded according to Götze [4] for the levels of automation. The aIC was developed taking into account several guidelines [5, 6]. Icons of the automation mode of the aIC were taken from Melcher et al. [7] and Jenke et al. [8]. The main differences in the concept of the aIC were the absence of a tachometer, the introduction of an automation scale [7], a digital speed display, the color coding of transitions and the hiding of the additional analog speedometer for highly automated driving (Fig. 1).

Scenario: Following a front vehicle in partial automation mode

Scenario: Lane change to the right in high automation mode

Fig. 1. Presentation of the cIC concept (left) and aIC concept (right) of specific scenarios in partially and highly automated driving mode

2.2 Experimental Design

The experiment took about 90 min per participant and was conducted in a modular driving simulator setup. However, participants did not drive. To evaluate the concepts, the experiment was divided into two parts. The experimental design of this study is based on Götze et al. [9].

In the first part, we used an occlusion task [10]. Eleven different stimuli were presented in the IC for 450 ms, 900 ms and 1300 ms. The presentation times of the stimuli result from the presentation times of an experiment evaluating head-up displays (200 ms, 250 ms, 300 ms) [9]. The time advantage of a head-up display over the IC is estimated at 0.25 s–1 s [11], which led to the minimum, mean and maximum presentation time for this experiment. Following each stimuli, the participants had to answer short questions for the specific stimuli. The participants perceived 11 different stimuli in both IC concepts. The answers were recorded verbally.

In the second part of the experiment, a Compensatory Tracking Task (CTT) in combination with a Choice-Reaction-Time Task (CRT) was used to simulate a continuous activity. In the CTT, the participants performed a two-dimensional follow-up task in which a moving point had to be positioned with a computer mouse (right hand) to a fixed position in the middle of the simulator's front screen. The software used was the Psychology Experiment Building Language (PEBL) [12]. When prompted by an acoustic signal after 3000 ms, 5000 ms or 7000 ms CTT time, the participants had to stop working on the CTT and carry out a CRT. The CRT simulates a secondary task during vehicle driving. In the CRT, the participant had to read information from the IC and enter an answer on a keypad (left hand). Data logging of the CRT was done in the driving simulation software SILAB. This results in a speed-accuracy tradeoff in which a decision must be made between a fast response time and a high degree of accuracy in the performance of the task [13]. We used seven different stimuli for the CRT.

The stimuli displayed information about the driving scenario, upcoming maneuver, automation level, automation transition, speed and navigation data in both tasks. The stimuli, presentation times in the occlusion task and CTT times were randomized.

Each trial was preceded by a training session and was followed by assessing the usability and subjective workload of the IC concepts by using the system usability scale (SUS) [14] and NASA-RTLX [15]. After all trials, participants filled out a follow-up questionnaire to complete the experiment.

3 Results

We conducted repeated-measures ANOVA to analyze differences in the response accuracy, reaction time and in the subjective ratings. When Mauchly's test indicated a violation of the sphericity assumption, the degrees of freedom were corrected using Greenhouse-Geisser. A significance level of 0.05 was applied for all statistical tests. We used JASP to carry out the statistical analysis.

3.1 Participants

Thirty participants (4 women, 26 men) with a mean age of 29.77 years (SD = 14.22 years, ranged from 19 to 63 years) took part in this experiment. Eight subjects (27%) had already experience in driving simulation studies on automated driving.

3.2 Occlusion Task

Statistical analysis of the response accuracy shows no significant effect of the IC concept ($F(1, 29) = 3.203$, $p = 0.084$; Fig. 2). We found significant differences for the presentation time ($F(1.363, 39.540) = 20.877, p < 0.001$, *partial* $\eta^2 = 0.419$). Bonferroni-corrected post-hoc tests show this effect between the presentation time of 450 ms compared to the duration of 900 ms ($p < 0.001$, $d = 0.866$) and 1300 ms ($p < 0.001$, $d = 0.893$). Furthermore, a significant interaction effect between presentation time and IC concept was observed ($F(1.428, 41.417) = 6.189$, $p = 0.009$, *partial* $\eta^2 = 0.176$; Fig. 2). The aIC resulted in a higher accuracy for the short presentation time of 450 ms compared to cIC.

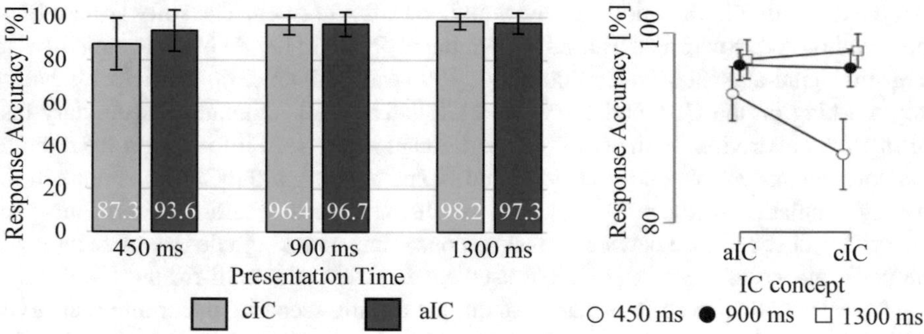

Fig. 2. Left: Mean and standard deviation of the response accuracy. Right: Interaction of IC concept and presentation time.

3.3 CTT and CRT

Within the CTT, no significant differences in the performance of the processing with respect to the display concepts could be observed ($F(1, 29) = 0.279$, $p = 0.601$) which excludes any influence on the evaluation of the display concepts. The statistical analysis of the response accuracy of the CRT showed a significantly higher accuracy of the cIC ($F(1, 29) = 6{,}158$, $p = 0{,}019$, *partial* $\eta^2 = 0{,}175$; Table 1). In contrast, the response time for the aIC was significantly shorter ($F(1, 29) = 5{,}608$, $p = 0.025$, *partial* $\eta^2 = 0.162$). No influence of CTT time on response accuracy ($F(2, 58) = 3.087$, $p = 0.053$) or response time ($F(2, 58) = 0.102$, $p = 0.904$) could be determined.

3.4 Subjective Ratings

The statistical analysis revealed no significant effects of the IC concepts on the SUS score ($F(1, 29) = 0.090$, $p = 0.767$; Table 2) or NASA-RTLX score ($F(1, 29) = 4.136$, $p = 0.051$; Table 2). However, in the follow-up questionnaire, 73% of the participants indicated that they would prefer the aIC, while 27% preferred the cIC.

Table 1. Mean and standard deviation of the response accuracy and reaction time in the CRT.

CTT time	IC concept	Response accuracy		Reaction time	
		M	*SD*	*M*	*SD*
3000 ms	aIC	91.4%	19.3%	1702 ms	545 ms
	cIC	93.8%	18.6%	1916 ms	588 ms
5000 ms	aIC	88.1%	19.9%	1720 ms	624 ms
	cIC	93.8%	17.9%	1868 ms	497 ms
7000 ms	aIC	89.5%	20.2%	1676 ms	599 ms
	cIC	89.5%	19.8%	1895 ms	612 ms

Table 2. Mean and standard deviation of the SUS and NASA-RTLX scores.

Score	IC concept	Occlusion task		CRT task	
		M	*SD*	*M*	*SD*
SUS	aIC	79.33	13.97	80.58	14.32
	cIC	74.17	15.82	74.25	16.95
NASA-RTLX	aIC	41.19	9.99	45.92	15.80
	cIC	40.92	12.59	46.83	14.26

4 Discussion

The higher response accuracy of the aIC for 450 ms presentation time in the occlusion task implies a higher readability and perceptibility for short viewing times on the IC. This result is contradicted by the lower response accuracy for the aIC in the CRT. However, within the CRT, the input accuracy is only limited suitable for the evaluation of the response accuracy, since there is also a speed-accuracy tradeoff of the CRT [13]. Here, a balance must be found between a fast reaction and a correct reaction. Consequently, the input accuracy cannot be directly attributed to the response accuracy of a piece of information, which limits the interpretability. The reaction time of the adaptive display concept was significantly reduced in the CRT. This underlines the potential of the aIC in particular for time-critical scenarios, such as those that can occur at the functional limits of partially automated driving. On average, the adaptive display concept resulted in a higher usability with a similar subjective workload, which implies that the participants felt more effective, efficient and satisfied with the presentation of the information by the aIC than supported by cIC. The final questionnaire showed a clear trend in the choice of the preferred display concept, with 73% participants preferring the adaptive display concept. This underlines the positive results for the aIC. In addition, it is important to emphasize the result to the effect that most participants are used to a display concept according to the cIC, which corresponds to an IC in most vehicles. The time needed to get accustomed to a vehicle system can be estimated at one to four weeks [16]. The participants were exposed to the aIC for the first time within the scope of this study, while it can be assumed that they were used to the cIC. This acclimatization asymmetry further enhances the positive results of the aIC.

Acknowledgment. This research was funded by German Federal Ministry of Economics and Energy within the project @CITY.

References

1. Beggiato, M., Hartwich, F., Schleinitz, K., Krems, J., Othersen, I., Petermann-Stock, I.: What would drivers like to know during automated driving? Information needs at different levels of automation. In: 7. Tagung Fahrerassistenzsysteme (2015)
2. SAE International: Taxonomy and Definitions for Terms Related to Driving Automation Systems for On-Road Motor Vehicles (J3016) (2018)
3. Rittger, L., Götze, M.: HMI strategy – recommended action. In: Bengler, K., Drüke, J., Hoffmann, S., Manstetten, D., Neukum, A. (eds.) UR:BAN Human Factors in Traffic, pp. 119–150. Springer Fachmedien Wiesbaden (2018)
4. Götze, M.: Entwicklung und Evaluation eines integrativen MMI Gesamtkonzeptes zur Handlungsunterstützung für den urbanen Verkehr. Dissertation, Technical University of Munich (2018)
5. Götze, M., Bißbort, F., Petermann-Stock, I., Bengler, K.: "A careful driver is one who looks in both directions when he passes a red light" – increased demands in Urban Traffic. In: Yamamoto, S. (ed.) Human Interface and the Management of Information. Information and Knowledge in Applications and Services, pp. 229–240. Springer, Cham (2014)
6. Naujoks, F., Wiedemann, K., Schömig, N., Hergeth, S., Keinath, A.: Towards guidelines and verification methods for automated vehicle HMIs. Transp. Res. Part F Traffic Psychol. Behav. **60**, 121–136 (2019)
7. Melcher, V., Rauh, S., Diederichs, F., Widlroither, H., Bauer, W.: Take-over requests for automated driving. Procedia Manufact. **3**, 2867–2873 (2015)
8. Jenke, M., Lassmann, P., Fischer, M., Reichelt, F., Othersen, I., Maier, T.: Fahrererlebnis beim Wechsel zwischen Automatisierungsstufen (SAE L2/L3) (2018). https://tangoversuch1.files.wordpress.com/2019/05/tango_poster_unis2018b.pdf
9. Götze, M., Schweiger, C., Eisner, J., Bengler, K.: Comparison of an old and a new head-up display design concept for urban driving. In: de Waard, D., Brookhuis, K.A., Toffetti, A., Stuiver, A., Weikert, C., Coelho, D., Manzey, D., Ünal, A.B., Röttger, S., Merat, N. (eds.) Proceedings of the Human Factors and Ergonomics Society Europe Chapter 2015 Annual Conference (2015)
10. ISO 16673: Road vehicles - Ergonomic aspects of transport information and control systems - Occlusion method to assess visual demand due to the use of in-vehicle systems (2017)
11. Gish, K.W., Staplin, L.: Human Factors Aspects of Using Head Up Displays in Automobiles. A Review of the Literature, National Highway Traffic Safety Administration (1995)
12. Mueller, S.T., Piper, B.J.: The Psychology Experiment Building Language (PEBL) and PEBL test battery. J. Neurosci. Methods **222**, 250–259 (2014)
13. Yellott, J.I.: Correction for fast guessing and the speed-accuracy tradeoff in choice reaction time. J. Math. Psychol. **8**, 159–199 (1971)

14. Brooke, J.: SUS - a quick and dirty usability scale. In: Jordan, P.W., Thomas, B., Weerdmeester, B.A., McClelland, A.L. (eds.) Usability Evaluation in Industry, pp. 189–194 (1996)
15. Hart, S.G., Staveland, L.E.: Development of NASA-TLX: results of empirical and theoretical research. Adv. Psychol. **52**, 139–183 (1988)
16. Weinberger, M., Winner, H., Bubb, H.: Adaptive cruise control field operational test—the learning phase. JSAE Rev. **22**, 487–494 (2001)

Providing Peripheral Trajectory Information to Avoid Motion Sickness During the In-car Reading Tasks

Yi-Ting Mu(✉), Wei-Chi Chien, and Fong-Gong Wu

Department of Industrial Design,
National Cheng Kung University, Tainan, Taiwan
yitingmu@gmail.com, chien.uxdesign@gmail.com,
fonggong@mail.ncku.edu.tw

Abstract. Carsickness is one of the most common types of motion sickness. It can be resulted by unanticipated body motion, which mismatches our anticipation of movement. However, related studies tend to increase user's visual information which instead affects the possibility of doing non-driving tasks. Therefore, matching sensory signals and engaging in the performance of non-driving tasks will be the key to improving user requirements. The experiment, which we choose to carry on a vehicle, analyses subjective ratings of motion sickness, in order that the user not only releases carsickness symptom, but also obtains a meaningful and rich traveling experience (in-vehicle reading). The result and experience shall be able to be applied in other related design research about physiological issues of future technology, such as motion sickness in autonomous driving or virtual reality.

Keywords: Motion sickness · Carsickness · Sensory conflict

1 Introduction

Long-distance journey or daily commute is common in our modern and urbanized daily lives. Traveling by vehicle is not only for transportation but also a way to gain experiences [1]. A survey in 2015 shows that 50 to 60% passengers prefer to engage in non-driving tasks while traveling by vehicle [2]. However, some preferred activities, which require intensive visual works, such as watching videos and reading, can easily cause dizzy and uncomfortable effect – the annoying carsickness. We are therefore motivated to design a supportive device that allows users to have a better reading experience in moving cars and the presented paper is the result of our first exploration. According to sensory conflict theory, the feeling of body movements and balance is an integrated perception through our visual, vestibular, and somatosensory systems. When our bodies are driven in an unexpected motion, these three systems are forced to process correspondent information that are asynchronous and conflicts to each other. As a result, we start to feel dizzy. When our brains keep processing such abnormal motion information for a while, we get motion sickness [3]. For instance, when sitting in a vehicle and looking down at a smart phone, our visual perception is fixed on

T. Ahram et al. (Eds.): IHSED 2019, AISC 1026, pp. 216–222, 2020.
https://doi.org/10.1007/978-3-030-27928-8_33

words. However, the vestibular and somatosensory systems consider, uncoordinatedly, our bodies as being moving. In a serious situation, according to poison theory [4], the brain recognizes the continuous dizzy feeling as being poisoned, and the central nervous system induces our bodies to remove toxins from the body by arousing nausea and vomit. Following this theory, synchronization of the three systems should reduce carsickness.

Our empirical evidences as well as researches show that drivers are less likely to get carsickness than passengers, because of their controllability of and predictability for the movement of cars [5]. Passengers, which cannot have direct control of the cars, have to adopt other strategies to increase their perception of vehicle movement and to avoid carsickness. They may gaze the landscape horizon or a distant fixed-point outside of the windows. Besides, passengers may see the direction of travel through the front windscreen to synchronize visual perception by predicting cars' motion trajectory [6]. As a strategy, keeping the outside world within peripheral area of passengers' sight of view is potential when designing interactive in-vehicle products [7]. In addition to the strategy of sensing the real horizon, another visual strategy uses artificial image (e.g. artificial horizon) to help reducing motion sickness [8, 9]. Another related and simple strategy enhancing motion prediction is to provide trajectory information in advance. For example, researches use light signals to indicate trajectory information. They argue for taking advantage of the human eyes' sensibility to a moving light spot [10, 11].

For sure, it is not passengers' goal, when using transportation means, to "fight" against motion sickness (unless they are fighter pilot). Passengers expect a comfortable journey without excessive mental workload and enjoy their non-driving tasks, such as reading, without trouble of carsickness. Strategies such as using moving light signals disadvantage in-vehicle experience by overloading visual information. Therefore, our approach is to provide the user with a real-time navigation information within the line of sight range by placing this navigation map system near the user's smartphone. Since the navigation system is dynamically providing signals corresponding to the happening movement, it is expected to improve not only the awareness of future motion trajectory but also the synchronization of three reception systems and this to reduce carsickness. Theoretical approaches presented above provide a wide scope for designing assistive in-vehicle reading devices using visual strategies. Still, the actual praxis of adopting transportation means is complex and often not driving-oriented. In our project, we focus on passengers' in-vehicle activities related to carsickness and design devices to reduce carsickness. So that we can move from theories and lab-based experiences to real field, we designed a simple solution against carsickness for in-car reading activities and conducted an experiment to study our design in the field.

2 Method

2.1 Device Design

The presented study aims to mitigate motion sickness when performing in-vehicle reading activities by providing navigation information aside the reading device. Passengers' awareness of motion trajectory aroused by the navigation system is expected to reduce their carsickness symptoms. The reading material was presented on a 5.2 in. screen (a Sony Xperia XZs) and the navigation information was presented on a 4 in. screen (an iPhone 5). The reading device was hold in the participants' hands and the distance between the user's eyes and the screen was kept around 40 cm with a field of view between 10° to 20°. The navigation system was designed as an in-car device and installed at the back side of the front seat. The navigation device was designed to be positioned above the passengers' reading device. This allows participants to read a text and simultaneously perceive the driving information (Fig. 1).

Fig. 1. Two conditions in the experiment, without navigation system (left) and with navigation system (right).

2.2 Experiment Design

We conducted a field experiment with 3 participants. They are three females (27, 31 and 59 years old) which had ever suffered from carsickness for the last five years. All participants included were free of vestibular disorders, and had not drink alcohol and not have meal in the last 12 h before the experiment. Participants were neither pregnant nor in their menstrual cycle. They were well informed about the purpose and procedure of the experiment. The content of the text was preselected by each participant before each ride to keep their motivation for reading. Participants were asked to perform an in-car reading task without our additional navigation system in the first ride (control ride) and then with navigation system in the second ride (test ride). Each ride took 10 min. The two rides were executed at least one day apart to avoid carry-over effect. Participants were instructed to maintain a relative intensive reading activity and keep their focus on the text. The riding experiments were done in a SUV (Ford Kuga) which was driven by the same driver on the same route (Fig. 2) in a suburb area around 8 am to 10 am. The participant sat behind the front passenger seat. The in-car temperature was conditioned to 25 °C during the experiment. Measure instrument, an 11-point misery scale (MISC, Table 1) [12], was used in the experiment. The scale contains symptoms representing slight to severe motion sickness, such as dizziness, nausea, retching and vomiting. The scale was explained to make sure participants fully comprehend the meaning of each item. Participants were asked to state the MISC before the experiment started. During each 10-min ride, participants were further asked to rate and state their MISC each minute.

Fig. 2. Itinerary of the experiment.

Table 1. 11-point MIsery SCale (MISC) [12].

Symptoms		MISC
No problems		0
Some discomfort, but no specific symptoms		1
Dizziness, cold/warm, headache, stomach/throat awareness, sweating, blurred vision, yawning, burping, tiredness, salivation, …but no nausea	Vague	2
	Little	3
	Rather	4
	Severe	5
Nausea	Little	6
	Rather	7
	Severe	8
	Retching	9
Vomiting		10

3 Results and Analysis

The self-reported MISC data from three participants in their two 10-min rides is presented above. The data at time 0 was measured before the ride started and those from time 1 to time 10 are measured each minute in-between the rides. Ride 1 is the control ride (reading in car without assistive device) and ride 2 the test ride (reading in car with navigation information in the peripheral visual area). Participant A and B encountered a middle level of carsickness (MISC 4–7) at the end of both rides (Figs. 3 and 4), although the use of our assistive device generally reduced the MISC level. Participant C (Fig. 5) seemed to be less sensitive and resistive to motion sickness. Still, she reported an improvement of MISC level. In the experiment of participant A (Fig. 3), although the navigation device improved the symptom of motion sickness, she reported deterioration in the last two minutes of the ride (from time 8 to time 10). In average, participants reported better MISC in their rides with the assistive navigation information (Fig. 6). This result shows a slightly positive effect against carsickness when reading in car through the presentation of navigation information in the peripheral area. It also shows a large range of individual difference in the development of carsickness. Participant A, for example, benefited from the provided navigation information at first;

however, it is possible that this information became annoying and made negative affect at the end of her ride. Participant B still gained carsickness when performing her reading task. The provided navigation information just slowed down the development of the symptom. Participant was rather resistive to motion sickness during her in-car reading task, and the navigation information did the best effect to reduce the happening of carsickness.

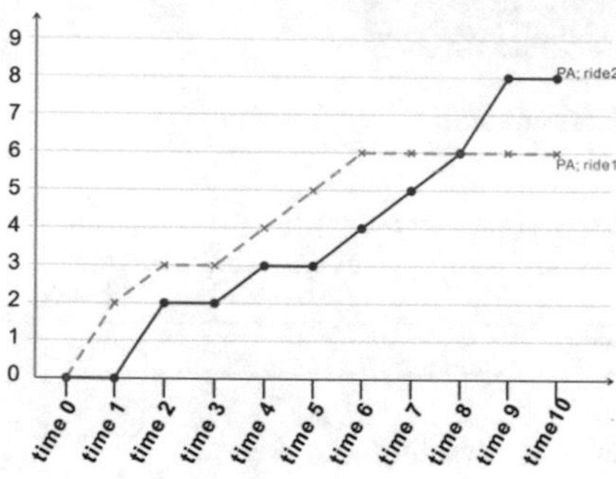

Fig. 3. (Left) data participant A.

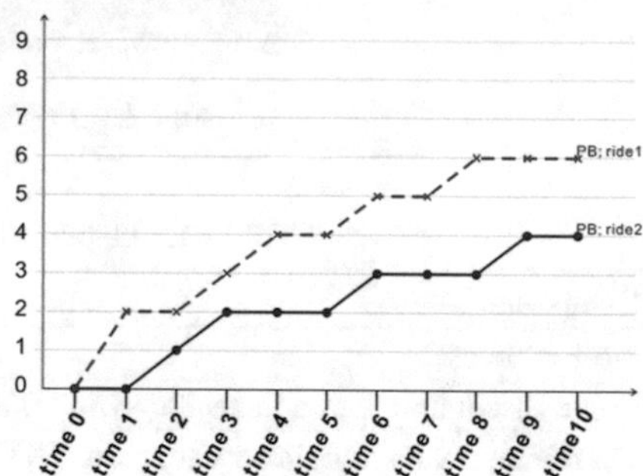

Fig. 4. (Right) data participant B.

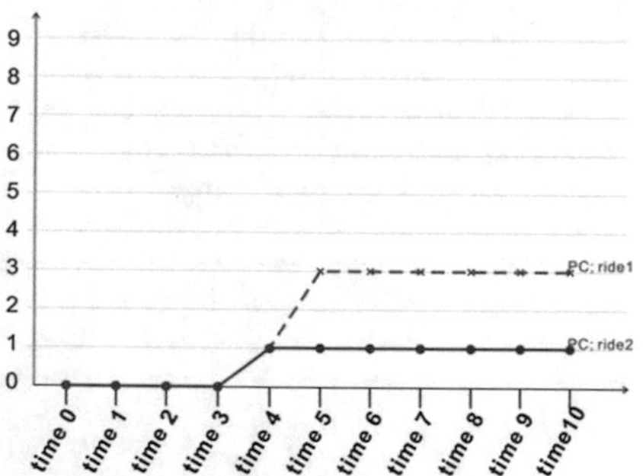

Fig. 5. (Left) data participant C.

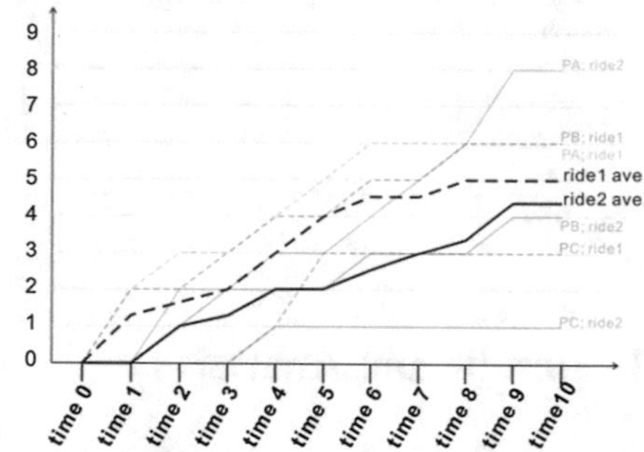

Fig. 6. (Right) average data.

4 Discussion

The presented study was done with a small sample size. We also only focused on a shorter-term in-car reading experience (10 min). These certainly limited the interpretation of the result. However, the contexts of the driving task, the reading material, as well as the environment in this small-scale experiment was qualitatively conditioned. In such experiment, we could analyze participants' task performances individually and we found interesting phenomena that should be studied in the future to create generalizable theories. They are: a. Awareness of car trajectory information through navigation system had the potential to decrease the symptom of carsickness. However, future studies should pay their attention on the different development patterns of the symptom. The effect could be less significant when the users are less resistive to motion sickness. b. To use navigation information to avoid motion sickness, the resulted information load through the device could play a crucial role. Extra information such as road names, arrival times and distances might become dominant and overwhelming to the users and worsen the symptom of motion sickness. Providing only necessary information that enhances the awareness of movement could be important to effectively reduce motion sickness. c. The effect of reducing carsickness through assistive information when reading in car could be limited only in a short time period. Existing studies of motion sickness usually focus on task performances of longer time. However, mini reading tasks, such as finding information on a smartphone, answering messages, or checking phone calls, are actually common in our daily lives. Preventing carsickness by these mini reading activities is with potential.

A more effective device to provide navigation information for in-car reading activities should be our future work. This includes studying different types of trajectory information, different presentation of the information, and the different types of reading tasks. More experiments in the field are also expected. We even expect our future design not only to prevent carsickness during the reading but also to improve reading performance and the related in-car experiences.

References

1. Meixner, G., Häcker, C., Decker, B., Gerlach, S., Hess, A., Holl, K.,…Orfgen, M.: Retrospective and future automotive infotainment systems—100 years of user interface evolution. In: Meixner, G., Müller, C. (eds.) Automotive User Interfaces: Creating Interactive Experiences in the Car. Human–Computer Interaction Series, pp. 3–54. Springer, Cham (2017)
2. Kyriakidis, M., Happee, R., de Winter, J.C.F.: Public opinion on automated driving: results of an international questionnaire among 5000 respondents. Transp. Res. Part F Traffic Psychol. Behav. **32**, 127–140 (2015)
3. Reason, J.T., Brand, J.J.: Motion Sickness. Academic Press, London (1975)
4. Treisman, M.: Motion sickness: an evolutionary hypothesis. Science **197**(4302), 493–495 (1977)
5. Rolnick, A., Lubow, R.: Why is the driver rarely motion sick? the role of controllability in motion sickness. Ergonomics **34**(7), 867–879 (1991)
6. Griffin, M.J., Newman, M.M.: Visual field effects on motion sickness in Cars. Aviat. Space Environ. Med. **75**(9), 739–748 (2004)
7. Kuiper, O.X., Bos, J.E., Diels, C.: Looking forward: in-vehicle auxiliary display positioning affects carsickness. Appl. Ergon. **68**, 169–175 (2018)
8. Krueger, W.W.: Controlling motion sickness and spatial disorientation and enhancing vestibular rehabilitation with a user-worn see-through display. Laryngoscope. **121**(2), 17–35 (2011)
9. Tal, D., Gonen, A., Wiener, G., Bar, R., Gil, A., Nachum, Z., Shupak, A.: Artificial horizon effects on motion sickness and performance. Otol. Neurotol. **33**(5), 878–885 (2012)
10. Karjanto, J., Md. Yusof, N., Wang, C., Terken, J., Delbressine, F., Rauterberg, M.: The effect of peripheral visual feedforward system in enhancing situation awareness and mitigating motion sickness in fully automated driving. Transp. Res. Part F: Traffic Psychol. Behav. **58**, 678–692 (2018)

11. van Veen, T., Karjanto, J., Terken, J.: Situation awareness in automated vehicles through proximal peripheral light signals. In: 9th International Conference on Automotive User Interfaces and Interactive Vehicular Applications, pp. 287–292. ACM Press, New York (2017)
12. Bos, J.E.: Less sickness with more motion and/or mental distraction. J. Vestib. Res. **25**(1), 23–33 (2015)

Influence of the Vehicle Exterior Design on the Individual Driving Style

Florian Reichelt(✉), Daniel Holder, and Thomas Maier

Institute for Engineering Design and Industrial Design (IKTD), Department of Industrial Design Engineering, University of Stuttgart, Pfaffenwaldring 9, 70569 Stuttgart, Germany
{Florian.Reichelt,Daniel.Holder, Thomas.Maier}@iktd.uni-stuttgart.de

Abstract. The individual driving style depends on different influences and is mainly affected by habitual manners, personal attributes and several external factors, such as the vehicle or traffic in general. The vehicle exterior, especially the design, is an essential part in both the vehicle in general and as one important customer requirement. Thus, it is reasonable to assume influences of the exterior vehicle design on the driving style as well. For investigating, whether such an effect occurs a study was conducted. Within different driving parameters, variations of the participants' driving style were measured, according to the different vehicle exterior designs they were shown. Overall, an influence of different vehicle exterior designs on the individual driving style was statistically proven.

Keywords: Automotive design · Driving style · Human-centred design

1 Introduction

In general, the driving style can be defined as "habitual way of driving, which is characteristic for a driver or a group of drivers." [1]. This habitual behaviour is dependent on both personality and environmental factors [2]. These factors include personal attitude, daily form and the current traffic situation. Additional effects of the vehicle on the driver's individual driving style are conceivable. Especially the vehicle exterior design which significantly determines external perception, is nowadays one important customer requirement [3]; [4]. Therefore, it is reasonable to assume that the vehicle exterior, as part of the vehicle design, also exerts an influence on the driving style, e.g. by conditioning the driver's expectations to the following ride.

This study investigates the question of whether and how interdependences exist between the influence of different vehicle exterior designs and the individual driving style. The results of this investigation can be used to draw conclusions about how the driving experience and driving style are influenced by different automotive exterior designs, thus making an important contribution to the analysis of the driving style. Especially in the context of increasing demand on driving conditions in automated driving, these results provide important insights.

T. Ahram et al. (Eds.): IHSED 2019, AISC 1026, pp. 223–228, 2020.
https://doi.org/10.1007/978-3-030-27928-8_34

2 Methods

Deduced from the previous considerations of an influence between driving style and vehicle exterior the hypothesis "The personal driving style varies between different vehicle exterior designs." can be formulated.

For investigating this hypothesis several aspects and requirements, such as the driving setting, the stimulus and the measurement must be considered for guaranteeing a successful scientific research.

Driving Setting. Despite the most common study settings of an analysis of driving style, this research used a driving simulator setup to ensure a reproducible and comparable study. A purely virtual car simulation combined with an exactly rebuilt vehicle interior characterizes the available driving simulator at the department of Industrial Design Engineering, University of Stuttgart [5]. Figure 1 shows the used simulator and gives an impression of how a high quality of the participant's immersion can be realized, due to its complete vehicle interior.

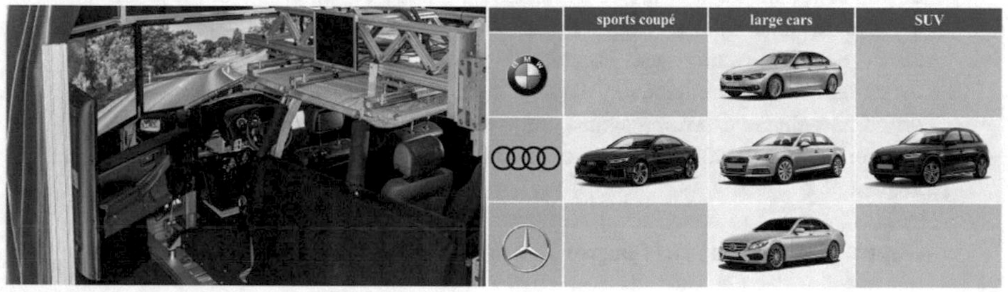

Fig. 1. Left: Exterior view of the used driving simulator; Right: Overview of the shown exterior design models

Stimulus. For achieving different perceptions and therefore different expectations and behaviours of the driver miscellaneous vehicle exterior design were necessary. Hence, the participants were presented different vehicle exterior designs on a display, placed outside of the simulator-cabin. Each shown perspective of the different cars was identical. As different stimuli, five real cars were chosen (Fig. 1): *Audi A4*, *Audi RS5*, *Audi Q5*, *BMW 3-series*, and *Mercedes-Benz C-Class*. Therefore, differentiations could be realized, e.g. between different brands (Audi, BMW and Mercedes) as well as between different vehicle classes (large cars, sport coupés and SUV). For generating a decided perception, the cars of different classes were chosen due to their divergent exterior design.

For the purpose of an objective, reproducible and comparative study these different vehicles were only used as visual stimuli, whereas the dynamic setting of the driving simulator was identical. In addition, a baseline drive was undertaken before the stimuli drives took place, for extrapolating the possible influence of the vehicle exterior on the driving style.

According to the five shown exterior models the participant had to drive through five comparable road maps. The course of each map had a length of 3.5 km and was identical. At last environmental elements, such as landscape, were different for each map. During the map setting both a city drive and a country road were included. Moreover, the country road was separated in different road sections with different traffic rules and settings, as to enable different driving patterns.

Measurement. The individual driving style can be measured both with objective and subjective methods [1]. A modified version of the MDSI by Taubman-Ben-Ari [6] was used to indicate the individual driving style in general. Besides this most common questionnaire [1], the following objective parameters were recorded as well:

- Speed
- Lateral and longitudinal acceleration
- Lateral position
- Acceleration pedal angle

Only the presentation of different vehicle exteriors was modified; thus, these various vehicle exteriors represent independent variables. Among the driving behaviour and the belonging driver-specific measurements as well as the responses of the questionnaire apply as dependent variables. These are influenced by the independent variables mentioned above [7].

3 Results

Participants. In total, 38 persons participated in the study. These were primarily students and academic staff. However, since the perception and significance of the vehicle exterior is not limited to these occupational groups, an exhaustive survey must be excluded; it is therefore a census study [7].
Due to motion sickness 7 participants had to quit the study and are not going to be analysed. Of the remaining 31 participants, 26% were female and 74% male. The majority of the participants were in an age group between 23–28 years of age, with a total of the continuous age range of 21–31 years.

As part of the questionnaire set the participants had to respond, whether they indicated a change of their individual driving style during the study: 81% of the participants noticed a change in their driving style depending on the different vehicle exteriors.

Measurements. Due to the characteristics of the objective parameters as metrical data and the primary scope of whether differences between various exterior designs exist, the data evaluation focused on an average analysis. In addition to a descriptive analysis, an inferential statistical analysis was undertaken.

As primary possibility to prove that an influence of the vehicle exterior could be detected, a so-called assessment sample can be used to show the regarded phenomena [7]. Figure 2 outlines the progression of driving speed for the participant No. 30. It can be outlined, that in dependence of the shown vehicle the needed time for the 3.5 km course varies; with the RS5 (red curve) No. 30 needed 150 s to accomplish the course. Whereas with the Q5 (yellow curve) he drove the slowest (240 s). Furthermore, there are clear differences between the maximum speeds driven by the participant. Based on these exemplary data a variation of the driving style can be confirmed. Moreover an interdependence between the shown vehicle exterior and the driving style can be detected.

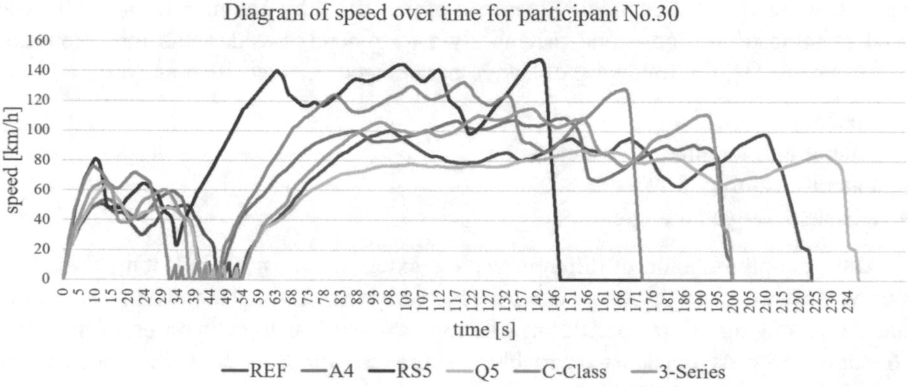

Fig. 2. Progression of the driving speed for the separate test drives (REF == Reference/ Baseline)

These results can be confirmed by the extended examination of the measurement data over the complete participant collective. Moreover, not only the measured data for the *speed* but also for the *lateral position*, the *lateral* and *longitudinal acceleration*. Nevertheless, there are different characteristics between the parameters, so that not all parameters provide the same clear differentiation between the stimuli. For instance, the data differences between the stimuli regarding the *lateral position* is more distinctive than for the *lateral acceleration*. In addition, the differentiations vary depending on the road's characteristic: on the country road the measurement data gives more clear results than in the city drive. All things considered, the results show a distinctive variation in driving style, depending on the perceived vehicle exterior.

In addition to this primary descriptive analysis a statistical analysis with inferential methods was undertaken as well. For analysing the metric data sets, the Friedman-Test was used to identify significance [8].

Within the Friedman-Test, it can be shown that not all determined differences show significances. But especially the previously observed clear differences within several parameters, such as *lateral position*, are significant. Table 1 shows an excerpt of the results gained with the Friedman-Test. In terms of the reference drive, without any perception-based influence on the driver's behaviour, the most significant differences of driving-style specific parameters can be found between each stimulus and this baseline. A significant differentiation between each stimulus can only be confirmed for the parameters *lateral position* and *acceleration pedal angle*. Overall the significance level of all parameters is close to significance.

Regarding the parameter *lateral position*, a pairwise comparison was undertaken to identify single significances between each stimulus. Concerning the baseline, a significance between each stimulus and the baseline drive can be verified. Within the strict comparison of each stimulus significances between the *Audi RS5* on the one hand and the *Audi Q5* and *Audi A4* can be marked as highly significant. For the pairs of *BMW 3-series* and *Audi RS5* there is a significant difference in the *lateral position* driven by the participants.

Ultimately, the inferential analysis undermines the results; the hypothesis can be confirmed for several parameters.

Table 1. Overview of the results gained with the Friedmann-Test

Driving style specific parameter	Road section	Significance (p)	
		With baseline	Without baseline
Speed	City	0.391	0.109
	Country road	0.045[a]	0.198
Lateral accel.	City	0.000[c]	0.435
	Country road	0.001[c]	0.231
Longitudinal accel.	City	0.524	0.431
	Country road	0.048[a]	0.639
Lateral position	City	0.008[b]	0.259
	Country road	0.000[c]	0.000[c]
Accel. pedal angle	City	0.092	0.644
	Country road	0.054	0.042[a]

[a]significant $p \leq 0.050$
[b]very significant $p \leq 0.010$
[c]highly significant $p \leq 0.001$

4 Conclusion and Outlook

Conclusion. As shown, the study setting of this research is appropriate to identify different driving styles and even constitute a variation of driving style, purely based on different visual perception. The results of the undertaken research to investigate whether the vehicle exterior conditions the driving style, show clearly that the appearance of a vehicle significantly influences the driving style. Therefore, the underlying hypothesis could be proven. An influence of the perception of the vehicle exterior on the individual driving style was confirmed.

Outlook. Due to the fact that not all recorded parameters for an objective driving style analysis show significant differentiations, it is assumable to both expand the numbers of exterior models and undertake a real drive study.

The special connection between the technical product car, the aesthetic design and the driving behaviour of people does not seem to have been in the focus of research projects, so far. An expansion of this research is conceivable, since this study regarded only a fraction of the national drivers and available vehicle designs and emotions. Further findings could be used, e.g., for marketing purposes. A control of customer emotions through purely optical changes, i.e. without expensive adjustments of vehicle dynamics, could make the car purchase of the future more efficient. The mentioned increase in the importance of design as a purchasing decision factor favours this development. The interaction between driving style and vehicle exterior is also relevant for automated vehicles. This raises the question of which driving style a customer expects suggested by the vehicle design. In addition to the currently highly focused research on kinetosis, the user's preference in terms of driving style is also an important factor for a user-friendly design of automated vehicles.

References

1. Sagberg, F., Selpi, G., Picinini, G.F.B., Engström, J.: A review of research on driving styles and road safety. Hum. Factors **57**(7), 1248–1275 (2015)
2. Flade, A.: Der rastlose Mensch. Springer Fachmedien, Wiesbaden (2013)
3. Achleitner, A., Antony, P., Burgers, C., Döllner, G., Ebner, N., Futschuk, H.D., Gruber, M., Kiesgen, G., Inderka, R.B., von Malottki, S., Mohrdieck, C.H., Noreikat, K.E., Urstöger, M., Schildhauer, C., Schulze, H., Wagner, M., Wolff, K., Wöhr, M.: Formen und konzepte. In: Pischinger, S., Seiffert, U. (eds.) Vieweg Handbuch Kraftfahrzeugtechnik, pp. 131–252. Springer Vieweg, Wiesbaden (2016)
4. Ranscombe, C., Hicks, B., Mullineux, G., Singh, B.: Visually decomposing vehicle images: exploring the influence of different aesthetic features on consumer perception of brand. Des. Stud. **33**(4), 319–341 (2012)
5. Mandel, R., Pomiersky, P., Maier, T.: Der vollvariable Fahrzeug-Ergonomieprüfstand: Absicherung des digitalen Auslegungsprozesses. In: Binz, H., Bertsche, B., Bauer, W., Roth, D. (eds.) Stuttgarter Symposium für Produktentwicklung 2015 (2015)
6. Taubman-Ben-Ari, O., Mikulincer, M., Gillath, O.: The multidimensional driving style inventory—scale construct and validation. Accid. Anal. Prev. **36**(3), 323–332 (2004)
7. Kosfeld, R., Eckey, H.-F., Türck, M.: Deskriptive Statistik: Grundlagen - Methoden - Beispiele - Aufgaben, 6th edn. Springer Gabler, Wiesbaden (2016)
8. Universität Zürich: Methodenberatung 18 Feb 2019. http://www.methodenberatung.uzh.ch/de/datenanalyse.html

Sensor Matrix Robustness for Monitoring the Interface Pressure Between Car Driver and Seat

Alberto Vergnano(✉), Alberto Muscio, and Francesco Leali

Department of Engineering Enzo Ferrari, University of Modena and Reggio Emilia, Via P. Vivarelli, 10, 41125 Modena, Italy
{alberto.vergnano, alberto.muscio, francesco.leali}@unimore.it

Abstract. An effective sensor system for monitoring the pressure distribution on a car seat would enable researches on Advanced Driver Assistance Systems (ADAS) and comfort of occupants. However, the irregularities of the seat shape or those of the occupant clothes challenge the robustness of such a sensor system. Moreover, the position identification of bodies of different percentiles by few pressure sensors is difficult. So, a higher resolution pressure pad has been developed. The number of sensors is significantly increased by means of a matrix scan strategy. Tests on the pressure pad with different occupants proves its robustness in scanning the contact area.

Keywords: Sensor matrix · Pressure · Driver monitoring · Car seat · Comfort · Advanced Driver Assistance Systems

1 Introduction

The body pressure distribution on the seat surface is a source of meaningful information in several lifecycle phases of a car. During design phases, experiments on seats for a prolonged time can evaluate the seat comfort [1, 2]. The relationship between pressure and temperature fields can also be investigated [3, 4]. The experimental equipment may also become an effective device to be embedded in the seats of production cars. In fact, the interface pressure monitoring may send information to safety systems committed to attention monitoring [5], up to adaptive systems for crash damage reduction [6]. Finally, the seat pressure can be processed by machine learning software for describing the driver patterns, in order to enhance the driving experience [7, 8].

Previous research proved the feasibility of this technology for driver monitoring [6]. A seat cover with pressure sensors is capable to monitor even the driver breathing when the car is stopped. When driving, a module for relative measurement monitors the driver, while an absolute one takes a reference on the car as a moving platform [9], whose accelerations significantly affect the pressure centre shift. Comparing the two

T. Ahram et al. (Eds.): IHSED 2019, AISC 1026, pp. 229–235, 2020.
https://doi.org/10.1007/978-3-030-27928-8_35

modules, this system can detect the Out of Position conditions for the driver or other passengers, to be communicated to an adaptive Airbag Control Unit in order to reduce the damage in case of crash. Two shortcomings were detected. First, the body size percentile significantly affects the pressure distribution. Then, a system with a small sensors area is too prone to the irregularities of the matching between the seat shape and the clothing of the occupant. As a result, the relative module must be calibrated for each driver and for each driving session too.

A higher resolution of the sensor matrix may improve these shortcomings. It would have multiple applications, from the aforementioned verification tool for comfort purposes, up to the integration in an Advanced Driver Assistance System (ADAS). So, a pressure pad has been developed in the present research. It has been tested on different occupants, proving its greater robustness for driver monitoring.

The paper is organized as follows. Section 2 presents the experiment materials and methods. The pressure monitoring experiments are reported and discussed in Sect. 3, while Sect. 4 draws the concluding remarks.

2 Pressure Pad

The pressure pad consists of a sensor matrix with a dedicated control system capable of scanning the load over its cells. The pressure pad is conceived as a flexible layer to cover the interface surface to be monitored.

Three superposed layers form the pressure pad assembly. Each outer layer consists of a number of parallel copper strips, the lower in row while the upper in column directions respectively, as shown in Fig. 1a and b. The middle layer is a Velostat sheet (Adafruit®, New York, NY, USA). The Velostat is flexible, conductive and pressure sensitive so that its squeezing reduces its electric resistance. The 30 × 25 strips matrix have 600 mm × 500 mm overall dimensions, and it is covered with 750 sensorized cells of 20 mm × 20 mm size. Cells scan and data communication to a laptop through

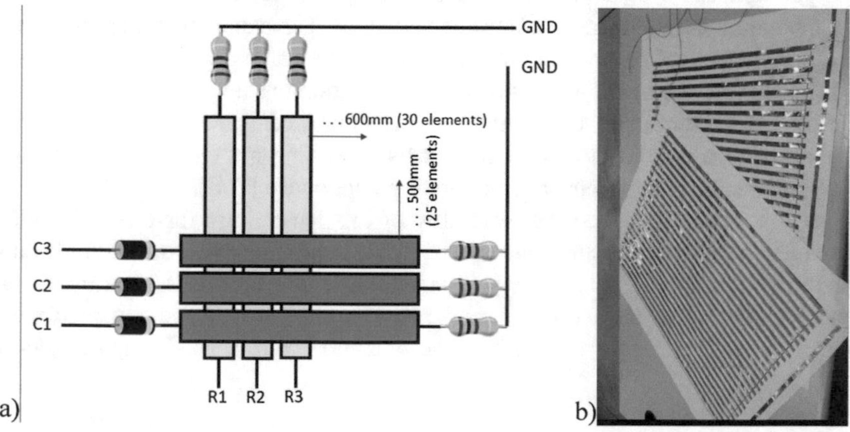

Fig. 1. Pressure pad: (a) sensor matrix concept and (b) copper strips on outer layers.

the serial port are controlled with an Arduino Mega 2560 controller (Arduino®, Turin, Italy). The matrix scan requires to power one by one the 25 Digital Outputs at 5 V and to read the 30 Analog Inputs at 0 ÷ 5 V for each powered DO. On the purpose, the DOs of the Arduino Mega 2560 are increased by a sequence of 4 74HC595 shift registers (Texas Instruments®, Dallas, TX, USA). The AIs are increased by 2 BOB-09056 multiplexers (SparkFun Electronics®, Niwot, CO, USA). In total, the cost for a pressure pad is comparable with that of the cover developed in [6]. For the experiments, two single pads are used for the seat bottom and the seat back, for a total of 1500 cells.

The controller powers one DO column at a time. The cells of each column are sampled by acquiring the voltage signal on all the AI rows. All the cell values are sent to a laptop through serial communication. The program structure is reported as follows:

```
libraries include();
variables declaration();
serial communication begin();
test loop() {
   ack();   //wait for the operator pushbutton to start
                 the test
   for (int j = 0; j < 25; j++) {   //scan the 25 columns
      column[j]=HIGH;
      for (int i = 0; i < 30; i ++) {  //scan the 30 rows
         cell[i,j] = analogread(row[i]);
         print(cell[i,j]); //through serial communication
      }
      column[j]=LOW;
   }
}
```

At present, this software enables to scan the pressure field once after the operator commands that with a pushbutton. That is, it is conceived as a design verification tool. The average scan time is 2,39 s. However, the system can also works as a stand alone device, as if it were an ADAS equipment. In this configuration the average scan time is 0,19 s. Thus, the serial communication is the most time consuming function for this controller.

3 Experiments and Discussion

In the experiments, the pressure on the driver seat of a 308 SW Peugeot® (Paris, France) car is monitored. Both the sensorized seat cover from [6], shown in Fig. 2, and the pressure pad, shown in Fig. 3, are tested with three different male drivers. Table 1 lists their body sizes. The testers are asked to take a normal driving position with hands on wheel and feet on pedals. After one minute for self adjustment in a comfortable position the pressure field is scanned.

Figures 4, 5 and 6 report the experiment results for the three drivers. The colour scale ranges from 0 (green) to 500 (red), while each sensor full scale would be 1024. The results reveal the qualitative pressure distribution, while for having a pressure result the cells should be calibrated. It can be clearly noticed the greater continuity of the pressure field as measured with the pad. The total 1500 cells are a very fine subdivision of the surface that results in continuous transitions between different loaded areas. For example, let us consider the different height of drivers and the consequent pressure on the seatback. The first driver is taller by 5 cm than the second one and by 20 cm than the third one. The different pressure distribution can be clearly identified with the pad, as in Figs. 4a, 5a and 6a. This difference would be much more difficult to identify by scanning the surface with the cover, as in Figs. 4b, 5b and 6b. Also, the experiment of Fig. 5b identifies an asymmetric position for the second driver. Also the lateral support is better monitored by the pad. In the seat bottom area the leg pressure is clearly identified. In fact, in Figs. 4a and 5a for the first and second driver with larger bodies, there are two lateral loaded bands, corresponding to the contact between the legs and the lateral supports. The pressure pad of (a) experiments accurately monitors the pressure field and the contact configuration with the seat. The sensorized seat cover of (b) experiments is not able to appreciate this contact area. On the other hand, the seat cover better identifies the different weight of drivers.

Fig. 2. Installed sensorized seat cover.

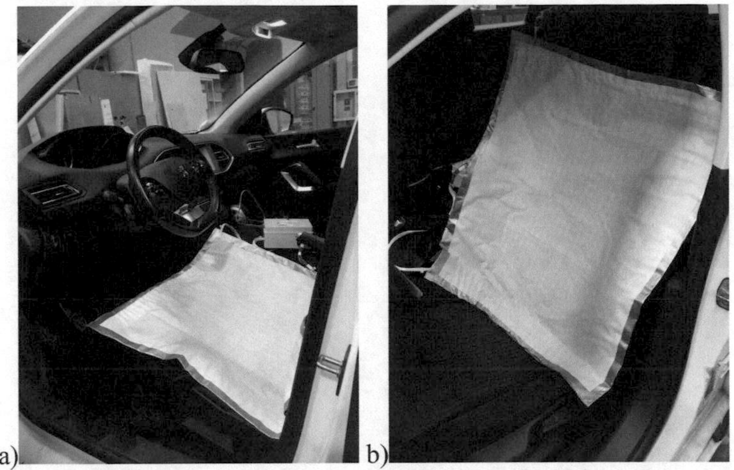

Fig. 3. Installed pressure pad on (a) the seat and (b) seatback.

Table 1. Body sizes of the test drivers.

Driver	Weight [kg]	Height [cm]	Age [y]
1	90	185	36
2	80	180	26
3	60	165	24

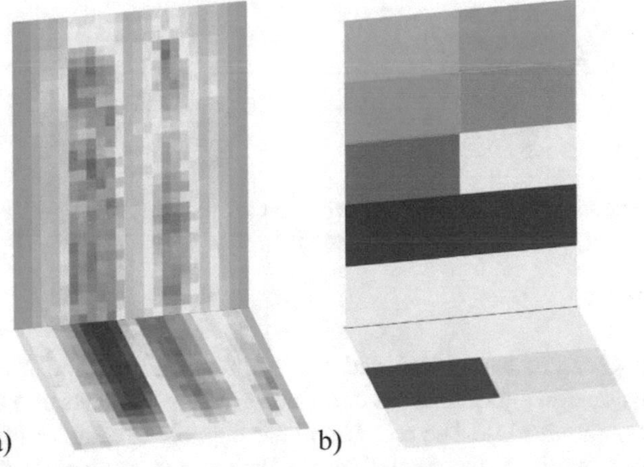

Fig. 4. Pressure fields for the first driver on (a) the pad and (b) the seat cover.

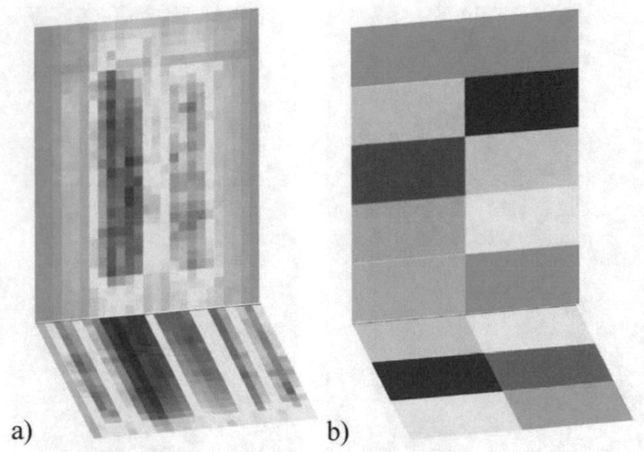

Fig. 5. Pressure fields for the second driver on (a) the pad and (b) the seat cover.

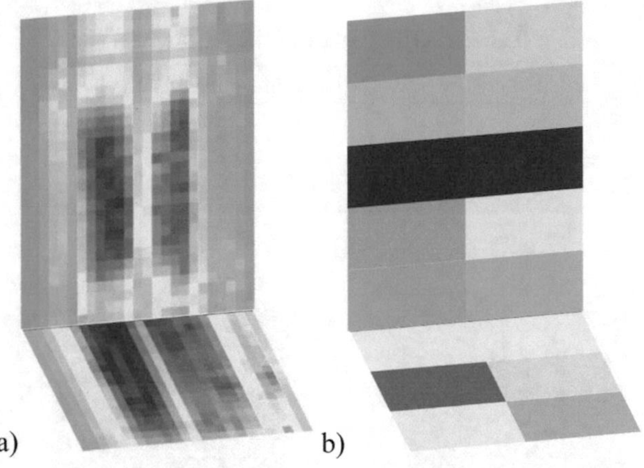

Fig. 6. Pressure fields for the third driver on (a) the pad and (b) the seat cover.

4 Conclusions

In the present research, a pressure pad has been developed and tested in order to provide a robust design equipment for comfort testing in car seats and a sensor system for car occupants monitoring. The pressure pad is capable to monitor the pressure of an occupant on his seat with a fine resolution. So, the pad is less prone to the irregularities of the seat shapes, clothes folds and accessories of the occupant. The present research proved the feasibility of a pad with a sensor matrix as a design verification tool and robust sensor tool to be integrated in an ADAS. On the other hand, the sensor robustness and calibration have to be improved. Moreover, a careful material selection would enable the pad permeability to the air.

Future work will merge the features of the pressure pad and the seat cover in order to synthetize both the advantages in an ADAS solution.

References

1. Na, S., Lim, S., Choi, H.S., Chung, M.K.: Evaluation of driver's discomfort and postural change using dynamic body pressure distribution. Int. J. Ind. Ergon. **35**(12), 1085–1096 (2005)
2. Kyung, G., Nussbaum, M.A.: Driver sitting comfort and discomfort (part II): relation-ships with and prediction from interface pressure. Int. J. Ind. Ergon. **38**(5–6), 526–538 (2008)
3. Nielsen, P.V., Jacobsen, T.S., Hansen, R., Mathiesen, E., Topp, C.: Measurement of thermal comfort and local discomfort by a thermal manikin. ASHRAE Trans. **108**, 1097 (2002)
4. Cengiz, T.G., Babalık, F.C.: An on-the-road experiment into the thermal comfort of car seats. Appl. Ergon. **38**(3), 337–347 (2007)
5. Vergnano, A., Leali, F.: Monitoring driver posture through sensorized seat. In: 1st International Conference on Human Systems Engineering and Design: Future Trends and Applications. Reims (2018)
6. Vergnano, A., Leali, F.: Out of position driver monitoring from seat pressure in dynamic maneuvers. In: 2nd International Conference on Intelligent Human Systems Integration: Integrating People and Intelligent Systems, San Diego (2019)
7. Zhang, Y., Lin, W.C., Chin, Y.K.S.: A pattern-recognition approach for driving skill characterization. IEEE Trans. Intell. Transp. Syst. **11**(4), 905–916 (2010)
8. Martínez, M.V., Del Campo, I., Echanobe, J., Basterretxea, K.: Driving behavior signals and machine learning: A personalized driver assistance system. In: IEEE 18th International Conference on Intelligent Transportation Systems. Las Palmas de Gran Canaria (2015)
9. Vergnano, A., Pegreffi, F., Leali, F.: Correlation of driver head posture and trapezius muscle activity as comfort assessment of car seat. In: 2nd International Conference on Intelligent Human Systems Integration: Integrating People and Intelligent Systems, San Diego (2019)

Feasibility Analysis and Investigation of the User Acceptance of a Preventive Information System to Increase the Road Safety of Cyclists

Oliver M. Winzer(✉), André Dietrich, Michael Tondera, Christoph Hera, Peter Eliseenkov, and Klaus Bengler

Chair of Ergonomics, Technical University of Munich, Boltzmannstr. 15, 85748 Garching b. München, Germany
{o.winzer, andre.dietrich, michael.tondera, christoph.hera, peter.eliseenkov, bengler}@tum.de

Abstract. This study aimed to investigate the user acceptance of a preventive Car-2-X communication warning system which could help drivers avoid potential collisions with cyclists. A User-Centered Design (UCD) process was used to develop a Human Machine Interface (HMI) as a warning system. Initially, an online survey (N = 153) was conducted to investigate user behavior and preference. To measure user acceptance, a 6-item survey on a 7-point Likert scale was conducted. User behavior was measured via 24 questions. Based on this, an HMI for the car driver was developed. The results of the acceptance score (M = 3.29) indicated rejection. However, most participants reported that they believe a system like this could improve road safety for cyclists. A functional prototype was developed and tested in a small field test. The results show that a 4G network and the GPS accuracy measured is sufficient for safety related applications.

Keywords: Road safety of cyclists · Car-2-X communication · HMI · Online survey · Prototyping · VRU

1 Introduction and State of the Art

Cyclist are part of the vulnerable road user (VRU) group. With their narrow silhouette and usually higher speed than pedestrians, they have a high risk of accident. Due to the increasing popularity of e-bikes over recent years, which have an even higher average speed than conventional bicycles [1], the accident risk is increased. Therefore, the potential risk of being injured as a cyclist in road traffic accidents is very high and will likely continue to increase in the years to come.

Analyzing statistics for bicycle accidents in Germany is challenging – the police is often not informed about minor accidents such as fender benders, so as to evade bureaucratic processes. Von Below [2] analyzed 2,768 accidents which involved cyclists. Those polled indicated that in just 12% of cases had the police issued a report

T. Ahram et al. (Eds.): IHSED 2019, AISC 1026, pp. 236–242, 2020.
https://doi.org/10.1007/978-3-030-27928-8_36

(for 30% of cases, this specification was missing). Therefore, the number of unreported cases is estimated at between 58% and 88% [2].

The preliminary official statistics from the German government in the year 2018 (between January and November) indicated that 13.6% more cyclists died in road accidents than in the previous year [3]. In combination with the rapidly increasing sales numbers for bicycles and e-bikes (increase of 36% in sales in 2018 [4]), it can be assumed that the amount of bicycle accidents will not decrease in the near future.

There are many possible ways to approach the challenge of reducing cyclist fatalities [5]. Post-crash solutions focus on reducing the seriousness of injuries, while pre-crash approaches attempt to lower the number of accidents. VRU airbags and cyclist airbag helmets are examples of post-crash solutions. These concepts minimize the damage inflicted upon the VRU, but they are of no help in avoiding crashes. One example of pre-crash solutions are improvements to the infrastructure, especially for cyclists. Infrastructure solutions are very expensive, localized, and have a long installation process. A very extensive overview is given in the project "XCYCLE" [6]. Jin et al. [7] recognized that a Vehicle-to-Vehicle communication solution via mobile devices could improve safety in road traffic. However, they concluded that the mobile network is too slow for safety-relevant functions. Prati et al. [8] also developed and tested an on-bike system (auditory warning and an LED blinking) to warn cyclists if a car appears. To locate the users, they used their own stationary system with local reference points. They did not use GPS because it lacked accuracy. Therefore, this system was limited to this specific location service.

This paper presents an approach to developing a preventive Car-2-X warning system without utilizing additional hardware. The behavior of cyclists and their acceptance of location sharing apps was identified via an online survey about developing a smartphone app that warns drivers whenever a collision is imminent.

2 Online Survey

An online survey was conducted to assess the acceptance of a collision-prevention app for cyclists, which requires location sharing in real time of all road users.

2.1 Participants

Of the 184 participants who took part in the online survey, 154 completed the survey. One participant was excluded from the survey due to them giving deliberately erroneous answers, resulting in an overall dropout rate of 16.85%. The remaining 153 participants have an age range between 17 to 83 years ($M = 30.01$; $SD = 10.72$). 90 indicated their gender as male and 60 as female (3 withheld their response). Most of the participants are in the process of studying for a university degree (52.94%).

2.2 Procedure and Measurements

The questionnaire "Steigerung der Sicherheit von Fahrradfahrern im Straßenverkehr durch Car-2-X Kommunikation" (English: Increasing the Safety of Cyclists through

Car-2-X Communication in Urban Areas) was created for native speakers of German. The translated questions are shown in Table 1. The survey was designed using the software LimeSurvey and was accessible from December 13, 2017 to January 15, 2018. It was distributed via e-mail and social networks. The data was collected anonymously, and the participants received no compensation.

The survey was separated into four parts, with 24 questions and 6 items on a 7-point Likert scale (1: strongly disagree; 7: strongly agree). In the first part, participants were asked about their user behavior as a cyclist. The second part focused on their general behavior around smartphone use – specifically, their attitude towards using location services and social media. Additionally, smartphone users were asked about their behavior while cycling. In the third part, participants were provided with a deeper introduction into the functionality of the application. A specific use case was described: a junction in a traffic-calmed area with parked cars obscuring the lateral view. After this introduction, the 6-item questionnaire to measure user acceptance was presented (see Table 1). The items correspond to dimensions of the Technology Acceptance Model (Intention to Use: Items 2 and 4; Perceived Ease of Use: Item 5; Perceived Usefulness: Item 6) [9]. Items 1 and 3 measured the users' feelings regarding data privacy. Finally, two qualitative questions and a demographic questionnaire were presented. All the data was analyzed using the software R Statistics Version 1.1423 and Microsoft Excel 2016. The qualitative data was analyzed manually.

Table 1. Items to measure user acceptance (English translation).

No.	Items	Scale
1	I have reservations about an app that shares my location with other road users	1…7
2	I would use the app planned here myself	1…7
3	If I can save insurance premium by using this app, I have nothing against my insurer receiving data about my cycling behavior	1…7
4	I would welcome it, if, in future, road users were legally obliged to use the Car-to-X communication in the form of a collision warning system	1…7
5	Using an app for cycling would be too much effort for me	1…7
6	I believe there would be an actual safety benefit from this idea	1…7

2.3 Results of the Questionnaire

Cyclists and Mobility. 40.5% participants reported that they use the public transport system to get to work. 31.4% use a car, 21.5% a bicycle, and 6.5% walk to work. 88.2% of the participants have their own bicycle and 7.8% are customers of a bike-sharing company. Approximately a quarter of the participants reported that they ride a bicycle to work (26.1%), 22.9% use it for sports activities, 19% for leisure, 9.8% for shopping, and 22.2% for "other" reasons. The frequency of bicycle use during the summer season was assessed using a 5-point scale (1: never; 5: very often). Nearly one third (31.37%) reported using bicycle often (4), 21.57% use a bicycle very often (5) and only 9.15% reported never using a bicycle (1). Only 4.6% reported that they have a smartphone holder on their bicycle, although 7.8% use a cyclometer. 5.9% of participants reported having had an accident as a cyclist within the last five years.

Smartphone Behavior. Nearly all participants own a smartphone (98%). The participants mostly use their smartphone for non-work purposes (84.3%). More than half of the participants reported that they do not use a smartphone app while cycling (54%). 21% listen to music and just 14% use it for navigation. 5% write text messages, 3% use it for phone calls, and 3% for "other" purposes. 59% use location services while an app is running, 21% always have a location service available, 13% never use location services, 5% don't know how to use them, and 2% "other".

Smartphone Behavior While Cycling. The results of the item analysis are presented in Table 2. The overall Cronbach's α of .804 represents, according to Field [10, p. 677], a good degree of reliability. The elimination of Item 3 (sending data to insurance companies) raised the alpha just slightly. The level of difficulty is good; none of the items are too easy (<.2) or too hard (>.8) [8]. Item 6 (belief in the usefulness) received the highest approval, with a mean of 4.21. The scree plot of the Principal Component Analysis shows that one factor alone can explain 51.33% of the variance (the second: 14.19%). The loadings of all items are higher than .56; Item 1 and Item 5 have negative loadings. According to Field [10, p. 639], the point of inflexion is at dimension two. Therefore, only the dimension to the left of the point of inflexion is selected. The mean acceptance score of all participants is 3.29 (SD = 1.29).

Table 2. Results of the item analysis (N = 153).

Item	Miss [%]	Mean	SD	Skew	Item difficulty	Item discrimination	α if deleted
1	0	3.10	1.82	0.60	0.44	0.565	0.773
2	0	3.10	1.83	0.17	0.49	0.792	0.718
3	0	2.70	1.80	0.72	0.39	0.416	0.806
4	0	2.86	1.77	0.49	0.41	0.547	0.777
5	0	3.48	1.92	0.29	0.50	0.517	0.785
6	0	4.21	1.77	−0.38	0.60	0.549	0.777
						Cronbach's α = 0.804	

Qualitative Data. In response to an open question, the participants were asked to give their own opinions regarding the claim that smartphones can be used to prevent accidents. In total 117 participants answered the question. The answers were categorized manually (see Table 3).

Table 3. Categorized results of the opinions of the participants regarding the smartphone system (N = 117; multiple entries permitted).

Category	Counts	Relative frequency
Enlarge safety/good idea	45	27.44%
Distraction for car drivers	28	17.07%
Concerns about data privacy	27	16.46%
Effort for using too high/Distribution is too limited	21	12.80%
Concerns about the technical realization	19	11.59%
Power consumption too high	14	8.54%
General misunderstanding of the idea	10	6.10%

3 Prototype Development

The HMI has been developed according to the UCD process described in ISO 9241-210:2010. According to the attention selection and multitasking model, auditory and visual stimuli broadcasted by the smartphone generate a reflex (automatic, bottom-up) action [11]. As this could have serious effects on the cycling performance of cyclists, the location-sharing app was designed to run in the background for cyclists. For drivers, an HMI was developed that incorporates the position of surrounding cyclists which the app. A driver receives a warning if cyclists ahead are on a collision course and can react, so that the cyclists indirectly benefit from sharing their positions.

3.1 User Interfaces

A user requirement analysis showed that an auditory warning of a potential collision is an important factor for drivers. Furthermore, the continuous tracking of cyclists on a map was identified as required information on the user interface.

The first digital low-fidelity prototype was developed using the Axure software. This software considered the recommendations from the ISO 15008:2017 as to the size of characters and letters, and the Google material design guidelines. Finally, the prototype (see Fig. 1) was effected using Android Studio 3.0.

Fig. 1. Human Machine Interfaces (left: Axure Prototype; right: Android Prototype).

3.2 Technical Implementation

To achieve the concept of a location-based collision-warning application, it is necessary to have a network of smartphones connected to a back-end server calculating trajectories and possible collisions. The mathematical algorithm in the back-end server compares the distance between cyclists and cars. If the distance between two participants closes to less than 200 m, their respective trajectories are used to calculate a point of intersection. At the next step, the time to arrival at the intersection point for both participants is computed. If the difference between the car and cyclist is less than or equal to three seconds, the car driver gets a warning message.

3.3 Proof of Concept

To evaluate the functionality of the prototype, a small field test with 19 tracks was conducted. The test area was a reduced speed zone in a residential area in Munich (Germany) with a T-junction, cars parked at the side of the road, with an obstructed view for drivers. Two LG Nexus 5Xs on a 4G network were used. The availability (M = 8.66) and accuracy (M = 3.71 m) of satellites was good. The average signal-transmitting time from the smartphones to the server and back was 459 ms. The field test showed that the system does indeed function.

4 Discussion and Outlook

The results of the online survey showed that user acceptance could be measured with the 6-item scale. The mean acceptance score (3.23) indicates that the participants are highly skeptical about this technology. However, they reported, in the open question that they support the idea of using smartphones to improve the safety of cyclists. In contrast to Jin et al. [7] we have come to the conclusion that the Car-2-X communication warning system is technically advanced enough for a safety-critical application. The prerequisite for this is a fast internet connection (4G). With the forthcoming expansion of the 5G mobile network, the connection times should become even faster.

Acknowledgments. This work is part of the "@city" project funded by the German Federal Ministry for Economic Affairs and Energy.

References

1. Schleinitz, K., Petzoldt, T., Franke-Bartholdt, L., Krems, J., Gehlert, T.: The German Naturalistic Cycling Study – Comparing cycling speed of riders of different e-bikes and conventional bicycles. Saf. Sci. **92**, 290–297 (2017)
2. von Below, A.: Verkehrssicherheit von Radfahrern – Analyse sicherheitsrelevanter Motive. Einstellungen und Verhaltensweisen. Bergisch Gladbach, Germany (2016)
3. Statistisches Bundesamt (Destatis): 2,7% mehr Verkehrstote im Jahr 2018 (2019)
4. Eisenberger, D.: Zahlen – Daten – Fakten zum Deutschen E-Bike-Markt 2018 (2019)

5. Silla, A., Leden, L., Rämä, P., Scholliers, J., Van Noort, M., Bell, D.: Can cyclist safety be improved with intelligent transport systems? Accid. Anal. Prev. **105**, 134–145 (2017)
6. XCYCLE: Present State of Affairs (2016). https://site.unibo.it/xcycle/en/deliverables
7. Jin, W., Kwan, C., Sun, Z., Yang, H., Gan, Q.: SPIVC: a SmartPhone-based inter-vehicle communication system. Transp. Res. Board 91st Annu. Meet. (2012). https://trid.trb.org/view/1130932
8. Prati, G., Puchades, V.M., De Angelis, M., Pietrantoni, L., Fraboni, F., Decarli, N., Guerra, A., Dardari, D.: Evaluation of user behavior and acceptance of an on-bike system. Transp. Res. Part F Traffic Psychol. Behav. **58**, 145–155 (2018)
9. Davis, F.D.: User acceptance of information technology: system characteristics, user perceptions and behavioral impacts. Int. J. Man Mach. Stud. **38**, 475–487 (1993)
10. Field, A.: Discovering Statistics Using SPSS (2009)
11. Engström, J., Victor, T., Markkula, G.: Attention selection and multitasking in everyday driving: a conceptual model. In: Regan, M.A., Victor, T.W., Lee, J.D. (eds.) Driver Distraction and Inattention: Advances in Research and Countermeasures, pp. 27–54. Ashgate Publishing (2013)

Interaction at the Bottleneck – A Traffic Observation

Michael Rettenmaier(✉), Camilo Requena Witzig, and Klaus Bengler

Chair of Ergonomics, Technical University of Munich,
Boltzmannstr. 15, 85748 Garching, Germany
michael.rettenmaier@tum.de

Abstract. At present, researchers and car manufacturers are focusing on the implementation of automated vehicles in road transport, which leads to mixed traffic containing automated vehicles and manual drivers. To realize this project, it is essential to understand the communicational processes and the interaction of contemporary road users. This paper focuses on investigating the interaction at bottlenecks caused by obstacles on both sides of the road, for instance due to double parking. The authors explore how drivers communicate to a simultaneously oncoming vehicle that they intend to pass through the narrow section first. As a method, a traffic observation was conducted. The results show that drivers almost exclusively use implicit means of communication. Moreover, four different interaction strategies could be found. Based on the underlying observation data, the authors provide recommendations on how automated vehicles might communicate implicitly by adopting the communication processes observed in the field study.

Keywords: Implicit communication · Traffic observation · Road bottleneck

1 Introduction

In the near future, automated vehicles (AV) will enter the transport system, which will initially lead to mixed traffic along with conventional road users. In order to reach the goals of automation, to enhance traffic safety and to increase time- and energy efficiency, communication between encountering AVs should be implemented in a comprehensible manner. Narrow situations in particular, in which negotiation between two oncoming vehicles is necessary, could cause serious problems for automated systems [1]. Therefore, the authors investigated a scenario in which two oncoming vehicles like to pass through a road bottleneck simultaneously, because this scenario is not regulated by German law [2], so both road users have to communicate.

One approach for implementing AVs in traffic is that they should communicate as present day manual drivers do by imitating traditional activities in traffic [3]. As a result, human road users will not have to learn new communication signals. For this reason, the authors analyze how manual drivers behave today and we performed an observation study in Garching near Munich to record both implicit and explicit communication at a road bottleneck scenario. In traffic manual drivers communicate explicitly for instance by using the turn indicator or implicitly by the trajectory of the car [4].

T. Ahram et al. (Eds.): IHSED 2019, AISC 1026, pp. 243–249, 2020.
https://doi.org/10.1007/978-3-030-27928-8_37

Imbsweiler *et al.* analyzed the strategies of drivers at road constrictions. The study discovered that human drivers use offensive and defensive strategies to cooperate in the bottleneck scenario. Moreover, the observed road users communicated mainly in an implicit way by changing the vehicle's velocity to signal their insistence on priority or to cede the right of way. [5]

This paper presents the interaction between simultaneously oncoming human drivers at road constrictions. The findings were used to derive communication strategies for automated vehicles, so that AVs will be able to express their intention to insist on priority and pass through the bottleneck first or to cede right of way and pass through it second.

2 Objectives

The paper researches various interaction strategies at road bottlenecks constricted by obstacles on both sides of the road. The authors analyze the amount and the specification of the various implicit and explicit means of communication. One research question the paper seeks to verify is the existence of offensive and defensive strategies used by human drivers to communicate ceding the right of way or insisting on it. Based on the underlying observation data, the authors give recommendations on how intuitive communication strategies might be implemented in AVs.

3 Method

3.1 Bottleneck-Scenario and Procedure

The scenario was a bottleneck caused by obstacles on both sides of the road due to double parking. Figure 1 shows an outline of the bottleneck scenario. The observation area measured 30 m in either direction from the observer's point of view, which was located in the center of the road constriction. Due to the fact that communication is important at lower speeds in particular, the bottleneck was located in a 30 km/h speed limit zone. A situation was recorded in the event that both oncoming vehicles (V1, V2) arrived at the observation area simultaneously. After observing the bottleneck for one day, we decided to collect data only from 8 a.m. until 11 a.m. and from 4 p.m. until 7 p.m. The traffic volume was more appropriate for observation during these periods.

The authors set the goal of enabling recording of the entire process of interaction between both oncoming vehicles, including the chronological sequence of the means of communication. Therefore, we decided to adapt the observation protocol Dietrich *et al.* [6] developed for logging the interaction process between different road users. The observation protocol was implemented in a HTML-based app performed on a tablet computer. The interrater reliability was calculated in a preliminary study in order to evaluate the method and the validity of the logged data. The main observation study includes the datasets of 50 situations when two oncoming human drivers entered the road constriction at the same time.

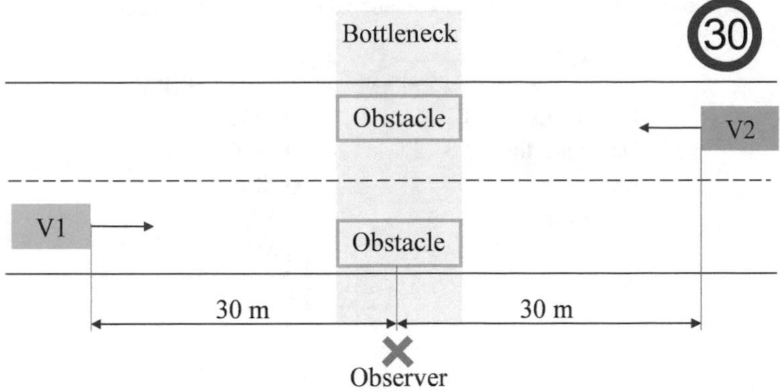

Fig. 1. Bottleneck scenario caused by obstacles on both sides of the road

3.2 Dependent Variables

The dependent variables are the means of communication the both oncoming vehicles used to negotiate which vehicle will pass through the bottleneck first. The authors differentiated between the implicit signals *accelerate*, *decelerate*, *maintain speed* and *stop* and the explicit signals *honk*, *flash headlight*, *use indicator*, *wave hand* and *raise hand*. The study also recorded which vehicle entered the scenario first and which one exited first.

3.3 Interrater Reliability

Before starting the observation study, the authors decided to train logging the interaction by using the observation tool. Two observers recorded 16 situations independent of one another to ensure data quality as well as the validity of the method. Table 1 contains the rater's agreement as well as the Cohen's kappa. Calculating Cohen's kappa is a suitable method of assessing the agreement of different raters. According to Döring and Bortz [7], the agreement of two raters is good if the kappa is from 0.60 to 0.75 and very good if it is from 0.75 to 1.00.

No explicit communication was observed during the preliminary study. Therefore, calculating Cohen's kappa was not possible, and the agreement of both raters was 100% for the explicit means. Thus, Table 1 only contains the values of the implicit communicational means. *Departure first*, *decelerate*, *maintain speed* and *stop* show a kappa higher than 0.75, which implies that the interrater reliability is very good. The values of *arrival first* and *accelerate* represent a good reliability between both raters. Consequently, we decided that it was appropriate to log the communication between two oncoming vehicles by using the observation tool.

Table 1. Interrater reliability of the communicational means using percentage of agreement and Cohen's kappa

	Agreement [%]	Cohen's kappa
Arrival first	81.25	0.6250
Departure first	93.75	0.8710
Accelerate	93.75	0.7168
Decelerate	87.5	0.7529
Maintain speed	96.88	0.9322
Stop	100	1

4 Results

4.1 Frequency and Differentiation of Communicational Means

Figure 2 shows the implicit and explicit means both oncoming vehicles used when interacting at the bottleneck scenario. It becomes obvious that, in total, the amount of implicit communication exceeds the amount of explicit means. Overall, 115 implicit means were recorded during all 50 encounters. Among these, the cues used most were *maintain speed* and *stop*, which were each logged 33 times. Thirty vehicles decelerated at the road constriction and 19 vehicles accelerated. The acceleration of the vehicle ceding right of way when leaving the situation second is not considered because the vehicle insisting on priority had already passed through the bottleneck and left the scenario. In contrast to the implicit cues, explicit means were used seven times in total, which were subdivided into *raise hand* (six times) to thank the other driver for giving right of way and *wave hand* (one time) to signal that the other driver should pass through the bottleneck first. The explicit means *honk, flash headlights* und *use indicator* were not used in any of the observed situations.

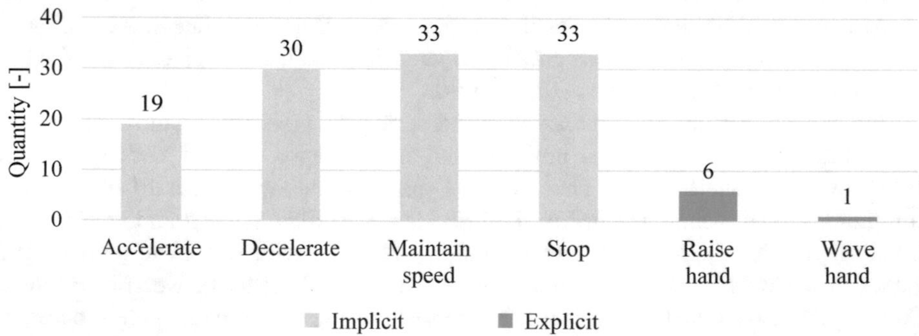

Fig. 2. Frequency and differentiation of the recorded communicational means

4.2 Cluster Analysis

In addition to the frequency of communicational means, we conducted a fuzzy clustering analysis to investigate whether the oncoming drivers have different interaction strategies. The results show four different strategies when passing through the road constriction. Two of them are of an offensive manner and two show defensive characteristics. Table 2 presents the centers of the four clusters. A value near zero states that the communicational mean is not distinctive in the cluster. A value near one means, that the signal is strongly represented in the cluster.

Vehicles communicating using the strategies of cluster 1 and cluster 2 have in common that they arrive second and depart second. Whereas a driver in cluster 1 decelerates for ceding the right of way, the drivers in cluster 2 come to a complete stop. The manual drivers using the offensive communication strategies are represented in cluster 3 und cluster 4. In both cases, the vehicles are entering and leaving the bottleneck scenario first. Whilst the drivers in cluster 3 are maintaining their speed, the drivers in cluster 4 are accelerating in order to insist on priority. Little or no explicit communication, which is shown by values near zero, is represented in any of the clusters.

In addition to the existence of the four different clusters, we analyzed how many drivers belong to each cluster. We defined a driver's membership in a cluster when the driver's communicational means showed the highest consistency with a particular cluster center. Nineteen drivers represented cluster 1, and there were 31 drivers in cluster 2. Regarding the offensive strategies, 33 vehicles maintained speed in cluster 3, and 17 drivers accelerated in cluster 4. Thus, 50 vehicles show defensive and 50 vehicles showed offensive strategies. Further analysis shows that each interaction has one defensive partner and one offensive partner.

Table 2. Fuzzy clustering analysis with four clusters

	Defensive strategies		Offensive strategies	
	Cluster 1	Cluster 2	Cluster 3	Cluster 4
Arrive first	**0.053**	**0.125**	**0.950**	**0.754**
Depart first	**0.029**	**0.020**	**0.989**	**0.954**
Accelerate	0.021	0.013	0.020	**0.930**
Decelerate	**0.968**	0.016	0.006	0.526
Maintain speed	0.010	0.009	**0.979**	0.051
Stop	0.029	**0.984**	0.016	0.063
Wave hand	0.003	0.010	0.001	0.003
Raise hand	0.011	0.036	0.027	0.032

5 Discussion and Recommendations

The observation study confirms the results of Imbsweiler *et al.* [5]. The communicational process at the bottleneck scenario consists almost exclusively of implicit communication. Both negotiating drivers solve the situation by adjusting their speed. The fact that a vehicle arrives the bottleneck first is decisive for maintaining speed or accelerating to pass through the narrow passage first. The vehicle arriving second decelerates or even stops so as to cede the right of way. One explanation for this behavior is that the both oncoming drivers do not arrive at the bottleneck at exactly the same time. The small differences are sufficient to regulate the bottleneck scenario so that one driver takes the offensive role and the other the defensive role. Thus, the offensive and defensive strategies do not need to use explicit communication, and no deadlock scenario arises. Another reason for a lack of explicit communication is that the distance between the interaction partners is too great, so explicit use of hand signals for communication does not make sense.

Researchers are presently investigating the potential of both implicit and explicit communication between AVs and other road users. If the approach will be to copy the traditional interaction between manually driven cars, we suggest that implicit communication will prevail in a bottleneck scenario. Nonetheless, Rettenmaier *et al.* [8] stated that AVs using explicit communication have the potential to support the manual driver while passing through the bottleneck. Using implicit communication, we propose using the observed means and to evaluate them in a driving simulator study. In case the AV wants to pass the narrow passage first, it should maintain its speed since accelerating will often not be possible due to the present speed limit. In case the AV cedes the right of way, it should decelerate to let the manual driver enter the bottleneck scenario first. If an AV enters the bottleneck second, it should cede the right of way in order to prevent accidents. According to the observed data, these strategies could help automated vehicles to avoid conflicts and to enhance traffic efficiency.

Acknowledgments. The German Federal Ministry of Economics and Energy funded this research within the project @City: Automated Cars and Intelligent Traffic in the City.

References

1. Metz, D.: Developing Policy for Urban Autonomous Vehicles: Impact on Congestion. Urban Sci. **2**(2), 33–43 (2018)
2. Straßenverkehrs-Ordnung: StVO (2013)
3. Juhlin, O.: Traffic behavior as social interaction - implications for the design of artificial drivers. Intell. Transp. Syst. (1999)
4. Fuest, T., Sorokin, L., Bellem, H., Bengler, K.: Taxonomy of Traffic Situations for the Interaction between Automated Vehicles and Human Road Users. In: Stanton, N. (ed.) Advances in Human Aspects of Transportation, pp. 708–719. Springer, Cham (2018)
5. Imbsweiler, J., Ruesch, M., Palyafári, R., Deml, B., Puente León, B.: Entwicklung einer Beobachtungsmethode von Verhaltensströmen in kooperativen Situationen im innerstädtischen Verkehr (2016)

6. Dietrich, A., et al.: interACT: designing cooperative interaction of automated vehicles with other road users in mixed traffic environments. interACT D.2.1 Preliminary description of psychological models on human-human interaction in traffic (2018)
7. Döring, N., Bortz, J.: Forschungsmethoden und Evaluation in den Sozial- und Humanwissenschaften, 5th edn. Springer, Heidelberg (2016)
8. Rettenmaier, M., Pietsch, M., Schmidtler, J., Bengler, K.: Passing through the Bottleneck - The Potential of External Human-Machine Interfaces. In: IEEE Intelligent Vehicles Symposium (IV) (2019, in Press)

Displaying Vehicle Driving Mode – Effects on Pedestrian Behavior and Perceived Safety

Philip Joisten(✉), Emanuel Alexandi, Robin Drews, Liane Klassen, Patrick Petersohn, Alexander Pick, Sarah Schwindt, and Bettina Abendroth

Institute of Ergonomics and Human Factors, Technische Universität Darmstadt, Otto-Berndt-Straße 2, 64287 Darmstadt, Germany
{p.joisten,abendroth}@iad.tu-darmstadt.de, {emanuel.alexandi,robin_christian.drews, liane.klassen,patrick.petersohn, alexander_klaus_peter.pick, sarah_selina.schwindt}@stud.tu-darmstadt.de

Abstract. The type and amount of information pedestrians should receive while interacting with an autonomous vehicle (AV) remains an unsolved challenge. The information about the vehicle driving mode could help pedestrians to develop the right expectations regarding further actions. The aim of this study is to investigate how the information about the vehicle driving mode affects pedestrian crossing behavior and perceived safety. A controlled field experiment using a Wizard-of-Oz approach to simulate a driverless vehicle was conducted. 28 participants experienced a driverless and a human-operated vehicle from the perspective of a pedestrian. The vehicle was equipped with an external human machine interface (eHMI) that displayed the driving mode of the vehicle (driverless vs. human-operated). The results show that the crossing behavior, measured by critical gap acceptance, and the subjective reporting of perceived safety did not differ statistically significantly between the driverless and the human-operated driving condition.

Keywords: Vehicle driving mode · Automation status · Pedestrian behavior · Perceived safety · Human machine interface · Human machine interaction

1 Introduction and Research Question

Today's road traffic system is constituted by a variety of different forms of implicit (e.g. trajectories) and explicit communication (e.g. gestures) between road users. In order to integrate in mixed traffic environments, autonomous vehicles (AVs) must have the capability to communicate different kinds of information to their environment. The vehicle driving mode is one information category that could help pedestrians to develop the right expectations regarding further actions of the AV [1, 2]. To provide this information external human machine interfaces (eHMIs) are currently under development and evaluation [3, 4]. This should contribute to safe, efficient and comfortable interactions between AVs and other road users. Unambiguous communication

T. Ahram et al. (Eds.): IHSED 2019, AISC 1026, pp. 250–256, 2020.
https://doi.org/10.1007/978-3-030-27928-8_38

between the vehicle and other road user is particularly necessary when the vehicle is driving without a driver (for instance driverless vehicles). Therefore, the aim of this research is to investigate which effects the display of the driving mode of a driverless vehicle have on the crossing decisions of pedestrians and compare those to ones made interacting with a human-operated vehicle.

Changes in the road transport system and their effects on safety have been studied under the theoretical framework of behavioral adaptation. Behavioral adaptation describes 'those behaviours which may occur following the introduction of changes to the road-vehicle-user system and which were not intended by the initiators of the change' [5, p. 23]. The theoretical framework of behavioral adaptation is promising to study the effects of introducing eHMIs on the behavior of pedestrians. Based on the research aim, following research question is formulated: Does displaying the vehicle driving mode of a driverless vehicle result in behavioral adaptation of pedestrians?

Theory suggests that the monitoring and attunement of risk plays a major role in the formation of behavioral adaptation [6]. If a vehicle with a given system (e.g. an eHMI) provides an improved feeling of control compared to a vehicle without the system, the assumed risk reduction might be compensated by a change in pedestrian behavior [7]. An eHMI can explicitly communicate the vehicle driving mode to pedestrians, therefore provide an improved feedback and feeling of control. Based on the research question and the brief insight into behavioral adaptation theory, two hypotheses are formulated:

- *H1*: An eHMI displaying the vehicle driving mode of a driverless vehicle increases the subjective feeling of perceived safety of pedestrians in comparison to a human-operated vehicle.
- *H2*: An eHMI displaying the vehicle driving mode of a driverless vehicle reduces the critical gap acceptance of pedestrians in comparison to a human-operated vehicle.

2 Method and Materials

2.1 Research Design, Procedure and Materials

To test the hypotheses a controlled field experiment was conducted on a test ground, a former airfield near Darmstadt, Germany. Participants were invited to experience a driverless and a human-operated vehicle from the perspective of a pedestrian. The scenario used in the experiment was an un-signalized crossing of a straight road with no obstructions of visibility and no other traffic participants present. Participants stood at the side of the road at the distance of 2.75 m from the middle of the traffic lane. The vehicle drove past the participants with a constant speed of 30 km/h in each trial.

The procedure of the within-subject design study was as follows: In the first trial the vehicle drove past the participants in the driverless and the human-operated driving condition (in permuted order). The participants stood with their back to the road and turned towards the vehicle when it was in a distance of 100 m. The task of the participants in the first trial was to observe the passing vehicle. After each passing of the vehicle the participants had to indicate which driving mode they experienced (manual, automated, driverless or other) and what caused them to make their decision. A written explanation of different terms related to automated driving was given to the participants before the experiment started.

In the second trial the participants experienced the driverless and the human-operated driving condition again in permuted order to avoid sequencing effects. Participants were informed about the different vehicle conditions and instructed to turn to the vehicle when it was 100 m away. Their task was to cross the road in front of the oncoming vehicle. For safety reasons the participants never actually crossed or stepped on the road in front of the vehicle. The critical gap acceptance and perceived safety were measured after each passing of the vehicle (see Sect. 2.3).

To simulate a driverless, fully automated vehicle (SAE Level 5), the Wizard-of-Oz approach "Ghostdriver" was used [8]. A seat-costume covers the driver so that outside road users cannot see him. In addition, a prototype eHMI was developed to explicitly communicate the vehicle driving mode to other road users (see Fig. 1). The eHMI (width: 420 mm, height: 300 mm) is positioned in front of the radiator grill. A translucent plate with an engraved symbol is placed behind a light which can be switched on and off. The color of the eHMI is turquoise.

Fig. 1. eHMI used to explicitly communicate the vehicle driving mode in the study. Left: activated eHMI with symbol representing the autonomous driving mode; Right: deactivated eHMI without symbol.

2.2 Independent Variable: Vehicle Driving Mode

As independent variable the vehicle driving mode was manipulated. The conditions were driverless (see Fig. 1 left) and human-operated vehicle (see Fig. 1 right). In the driverless condition the driver was covered by the seat-costume whereas he was well visible in the manual driving condition. The eHMI was visible in both conditions but only switched on in the driverless condition. In addition, a symbol representing an autonomous car was visible in the driverless condition. The symbol was based on a research on the labeling of autonomous systems.

2.3 Dependent Variable: Pedestrian Behavior and Perceived Safety

The dependent variable pedestrian behavior was measured by the critical gap acceptance which indicates the last moment the pedestrian is willing to cross the road in front of the vehicle [2]. Participants were given a stopwatch which they were instructed to start at the last moment they are willing to cross the road in front of the vehicle. The stopwatch was then handed over to a trained researcher who stopped the stopwatch when the vehicle reached a defined point on the road (25 m behind the participant).

To assess changes in perceived safety a questionnaire was immediately filled out by the participants after they experienced the driverless and the human-operated vehicle in trail two. Participants indicated their perceived safety on a 5-point scale (−2 "I feel unsafe", 0 – "Indifferent", 2 "I feel safe").

2.4 Participants

28 participants (21% female, 79% male, mean age = 25.2 years, SD = 2.86 years) took part in the study. 75% of the participants stated to participate in road traffic several times a day as a pedestrian. 18% stated to participate daily and 7% to participate approx. every second day in road traffic as a pedestrian.

3 Results

3.1 Identification of Vehicle Driving Mode

In the first trail participants experienced both (driverless and human-operated) driving conditions and indicated after each passing of the vehicle which driving mode they experienced. The answers of the participants are summarized in Table 1.

Table 1. Reported driving conditions by the participants in the first trail.

Answers	Driverless vehicle condition (N = 27)	Human-operated vehicle condition (N = 28)
Human-operated	–	20 (71%)
Autonomous	2 (7%)	8 (29%)
Driverless	25 (93%)	–

Participants were asked what caused them to make their decision regarding the vehicle driving mode. In the driverless vehicle condition the main reason reported was the empty driver seat (89% of the answers) followed by the eHMI (11%). In the human-operated driving condition, the main reasons reported was the visible driver.

3.2 Changes in Pedestrian Behavior

In the second trial, the behavioral change of the participants, measured by critical gap acceptance, was determined. In the driverless vehicle condition (Wizard-of-Oz seat-costume and activated eHMI) the mean critical gap acceptance was 5.3 s (SD = 1.6 s). The mean critical gap acceptance in the human-operated driving condition (visible driver and deactivated eHMI) was 5.1 s (SD = 1.4 s). The critical gap acceptance did not differ statistically significantly between the two conditions (dependent t-test, $|T| = 1.192$, $df = 27$, $p = .244$).

3.3 Changes in Perceived Safety

To assess changes in perceived safety a questionnaire was presented after the participants experienced the driverless and the human-operated vehicle in the second trial. Perceived safety was indicated on a 5-point scale (−2 - "I feel unsafe", 0 – "Indifferent", 2 – "I feel safe"). The results between the driverless vehicle condition (mean = 0.47, SD = 0.92) and human-operated vehicle condition (mean = 0.71, SD = 0.98) did not differ statistically significantly (dependent t-test, $|T| = 1.022$, $df = 27$, $p = .316$).

With regard to the eHMI participants were also asked how the eHMI influenced their feelings of safety. 57% of participants reported a positive effect of the eHMI on their perceived safety. 32% reported no effect and 11% a negative effect.

4 Discussion

The present study investigates the effects of communicating the vehicle driving mode through an eHMI on the crossing decision of pedestrians. The results show that neither the subjective ratings of perceived safety nor the measured critical gap acceptance differ statistically significantly between the driverless and the human-operated vehicle condition. In line with related research [see 2], hypotheses 1 and 2 are not supported.

Communication of vehicle driving mode could foster system transparency and thus help pedestrians to attune to the right expectations regarding their and the vehicle's further actions [1, 9]. The question of labelling the automation status might become increasingly important the more unclear it is who is driving the vehicle (human vs. automation). Learnt behavior plays a dominant role in the first interaction with AVs [2] which might suppress short-term effects of behavioral adaptation.

The identification of vehicle driving mode communicated via an eHMI must be unambiguous. The mere presence of the eHMI led to mode confusion among several participants. To prevent this, further development of eHMIs should explicitly consider possible effects of mode confusion caused by displaying (or not displaying) the automation status of the vehicle.

The influence of the design of the eHMI on the present study results must be discussed critically. Although the chosen symbol representing the automation status of the vehicle was seen as positive by participants it was not unambiguous for all. A possible solution for unambiguous communication could be standardized symbolism [3].

No changes in pedestrian behavior measured by critical gap acceptance were found in the present study. To draw the right conclusion, the method to obtain the measure of critical gap acceptance should be critically discussed. The method used via time-taking at defined points in the infrastructure is highly dependent on human influence (and human error). To minimize possible errors all researchers involved in the study completed a training. In future studies more reliable approaches to measure critical gap acceptance should be used.

Prior research shows that pedestrians perceive autonomous car traffic less risky than human-operated car traffic [10]. In the present study, the driverless vehicle was not perceived as more or less safe than the human-operated vehicle. Post-hoc interviews showed that no participant discovered the Wizard-of-Oz seat-costume. In contrast to survey studies [e.g. 10], participants experienced a real "driverless" vehicle.

Although no statistically significant effect of perceived safety between the driverless and the human-operated vehicle condition was found, the eHMI had a slightly positive effect on reported feelings of safety in post-hoc interviews. To distinguish the effect of the eHMI and driverless vehicle on perceived safety a comparison between a driverless vehicle with and without eHMI has to be carried out. This was beyond the scope of the present study.

5 Conclusion

This study gives first insights into the effects of displaying the vehicle automation status on pedestrian behavior and perceived safety. Overall, the eHMI displaying the vehicle automation status did not influence pedestrian behavior and perceived safety. Further evaluation of the eHMI design is necessary for an unambiguous interaction between AV and pedestrian.

The theoretical framework of behavioral adaptation seems promising for the investigation of the effects of eHMIs on pedestrian safety. Pedestrians' behavioral adaptation beyond the first interaction with an AV will be investigated in further studies.

Acknowledgments. The present study was supported by the project @CITY-AF which receives funding from the German Federal Ministry of Economy and Energy (BMWi).

References

1. Schieben, A., Wilbrink, M., Kettwich, C., Madigan, R., Louw, T., Merat, N.: Designing the interaction of automated vehicles with other traffic participants: design considerations based on human needs and expectations. Cogn. Technol. Work **21**, 69–85 (2019)
2. Rodríguez Palmeiro, A., van der Kint, S., Vissers, L., Farah, H., de Winter, J.C.F., Hagenzieker, M.: Interaction between pedestrians and automated vehicles: A Wizard of Oz experiment. Transp. Res. Part F Traffic Psychol. Behav. **58**, 1005–1020 (2018)
3. Ackermann, C., Beggiato, M., Schubert, S., Krems, J.F.: An experimental study to investigate design and assessment criteria: What is important for communication between pedestrians and automated vehicles? Appl. Ergon. **75**, 272–282 (2019)

4. de Clerq, K., Dietrich, A., Núñez Valesco, J.P., de Winter, J., Happee, R.: External human-machine interfaces on automated vehicles: effects on pedestrians crossing decisions. Hum. Factors (2019). https://doi.org/10.1177/0018720819836343
5. OECD: Behavioural adaptation to changes in the road transport system, Paris (1990)
6. Jiang, C., Underwood, G., Horwarth, C.I.: Towards a theoretical model for behavioural adaptations to changes in the road transport system. Transp. Rev. **12**, 253–263 (1992)
7. Vaa, T.: Psychology of Behavioural Adaptation. In: Rudin-Brown, C.M., Jamson, S.L. (eds.) Behavioural Adaptation and Road Safety, pp. 207–226. CRC Press, Boca Raton (2013)
8. Joisten, P., Müller, A., Walter, J., Abendroth, B., Bruder, R.: Neue Ansätze der Human Factors Forschung im Zeitalter des Hochautomatisierten Fahrens. In. Bruder, R., Winner, H. (eds.) Hands off, Human Factors off? Welche Rolle spielen Human Factors in der Fahrzeugautomation? 9. Darmstädter Kolloquium, pp. 69–88. Darmstadt (2019)
9. Selkowitz, A.R., Lakhmani, S.G., Chen, J.Y.C.: Using agent transparency to support situation awareness of the Autonomous Squad Member. Cogn. Syst. Res. **46**, 13–25 (2017)
10. Hulse, L.M., Xie, H., Galea, E.R.: Perceptions of autonomous vehicles: Relationships with road users, risks, gender and age. Saf. Sci. **102**, 1–13 (2018)

HUD Layout Adaptive Method for Fighter Based on Flight Mission

Xiaoyue Tian[1,2], Yafeng Niu[1,2(✉)], Chengqi Xue[1], Yi Xie[2], Bingzheng Shi[3], Bo Li[2], and Lingcun Qiu[3]

[1] School of Mechanical Engineering, Southeast University, Nanjing 211189, China
{tianxy,nyf,ipd_xcq}@seu.edu.cn

[2] Science and Technology on Electro-optic Control Laboratory, Luoyang 471023, China
eoei@vip.sina.com

[3] Shanghai Academy of Spaceflight Technology, Zhichun Road, Shanghai 201109, China
shibz87@126.com, qlcun@163.com

Abstract. In order to solve the problem of inadequate display of important information in the mission process, which leads to low cognitive of pilots, an HUD layout adaptive method based on flight tasks is proposed in this paper. Specific methods include the following steps: dividing HUD interface according to function; classifying the pilot's field of view; collecting the pilot's eye movement data in the task phase and calculating sub-region's interest value; gridding HUD interface; calculating fitness function; adjusting the layout according to rules, selecting the interface with the largest value as the final HUD output interface. This method can effectively improve pilots' attention allocation and provide important reference for HUD interface layout.

Keywords: HUD · Interest · Rasterization · Fixation point · Fitness function

1 Introduction

Head-up Display (HUD) was first used on military aircraft. Its role is to reduce the frequency of pilots' need to look down at instruments, thus avoiding attention interruption and loss of state awareness [1]. In the course of fighter piloting, the pilots pay different attention to HUD interface under different flight mission modes. Some of the key information needed cannot be fully displayed, resulting in lower cognitive performance of pilots.

Domestic and foreign scholars have conducted in-depth research on HUD interface layout optimization. Bai [2] conducted a quantitative research on Influencing Factors of pilot attention allocation for HUD based on SEEV model. Zhang et al. proposed a method to evaluate the readability of helicopter head-up display in complex visual environment based on the minimum discernible difference (PJND) theory [3]; Yang K also studied the behavior patterns [4] of pilots using head-up displays to perform different missions, and proposed a behavior recognition framework [5] that includes

T. Ahram et al. (Eds.): IHSED 2019, AISC 1026, pp. 257–263, 2020.
https://doi.org/10.1007/978-3-030-27928-8_39

various characteristics of pilot eye movement and hand movement. In the aspect of ergonomics evaluation, Lu discussed the differences [6] of eye movement patterns between experienced pilots and novice pilots in simulated directional flight tasks. Pei et al. also established a quantitative prediction model [7] for pilot human reliability based on the extended method of cognitive reliability [8] and error analysis (CREAM). In the part of flight simulation software, Peng et al. [9] studied the design of integrated avionics main flight display based on Flightgear, which provides some guidance for flight simulation.

Based on the attention distribution of the pilot, this paper proposes a HUD adaptive method to improve the cognitive performance of the pilot and reduce the layout research cost, through the interest degree calculation and the attention distribution intensity as the objective function.

2 HUD Layout Adaptive Method

This method mainly includes the following steps: dividing HUD interface area and pilot's visual field; collecting pilots' eye movement data and calculating interest value; rasterizing HUD interface and calculating fitness function; and adjusting layout iteratively according to interest degree.

2.1 Dividing HUD Interface Area and Pilot's Visual Field

Dividing the HUD interface into multiple sub-areas according to functions. g and h represent the length and width of the interface. The position and interest of the sub-regions [10] in the form of coordinates are expressed as follows: Functional sub-region set $\mathrm{R} = (r_1, r_2, \ldots, r_n)$, there are n sub-areas, Each sub-area is defined as $r_j = (x_j, y_j, \gamma_{r_j})$, (x_j, y_j) indicates the coordinates of the center point of the sub-area in the interface, γ_{r_j} indicates functional sub-regional interest (Fig. 1 shows an example):

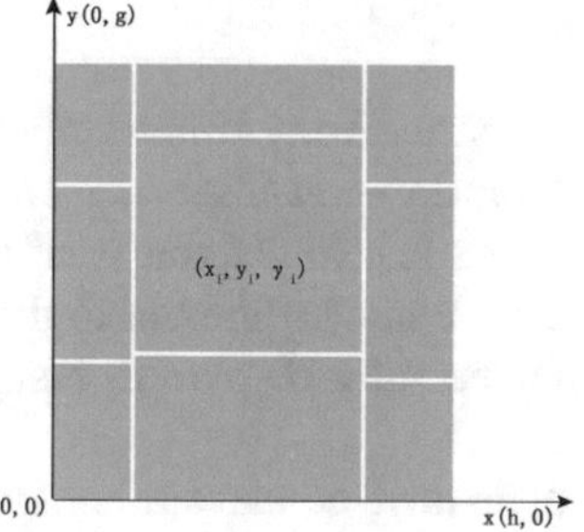

Fig. 1. HUD interface function partition map

Then, the visual field of 95-percentile human model is simulated and analyzed by using the "Open vision window" module of CATIA platform [11]. *A* region is the best visual field area, the horizontal range is about 30°, and the vertical range is about 35°;

B region is the effective area, the horizontal range is about 70°, and the vertical range is about 35°. Area C is the area except A and B (Fig. 2 shows an example).

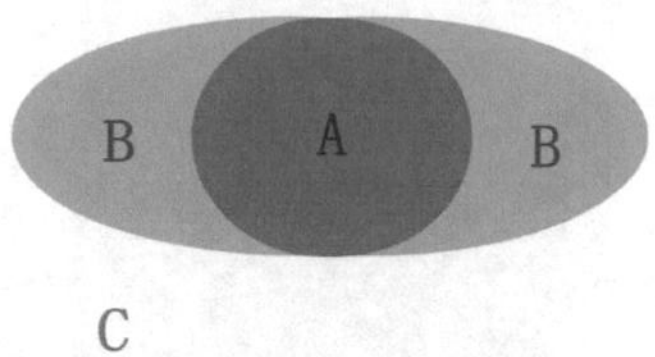

Fig. 2. Pilot vision orientation simulation

2.2 Collecting Pilots' Eye Movement Data and Calculate Interest Value

Based on Tobii eye tracker and Flightgear flight simulation software, subjects were recruited for different mission experiments, and recorded the number of fixation points, then calculated the interest value of each functional sub-area. The value is used to determine the relative importance of the target under different combat tasks and calculate the weight corresponding to the importance. The calculation formula of the interest value corresponding to each functional sub-area is:

$$\gamma_{r_j} = W_j \omega_{r_j} \tag{1}$$

Where 1 $\leq$ j$\leq$ n, W_j is the degree of importance of the sub-region:

$$W_j = \frac{t_j}{\sum_{j=1}^{n} t_j} \tag{2}$$

t_j indicates the number of gaze points in j functional subarea, $\sum_{j=1}^{n} t_j$ is expressed as the total number of gaze points in the HUD interface. The weight of the information amount in j functional subarea is expressed as:

$$\omega_{r_j} = \frac{1 - H_j}{n - \sum_{j=1}^{n} H_j} \tag{3}$$

H_j is information entropy:

$$H_j = -\sum_{j=1}^{L} p_j \log_2 p_j \tag{4}$$

Among them, H_j is the probability of occurrence of a gray scale in the entire image, L is the total number of gray level types.

2.3 Rasterization of HUD Interface and Calculation of Fitness Function

Centering on the visual center, the HUD interface is rasterized with the smallest element of the HUD interface information, and each functional area is numbered by a∼i (Fig. 3 shows an example).

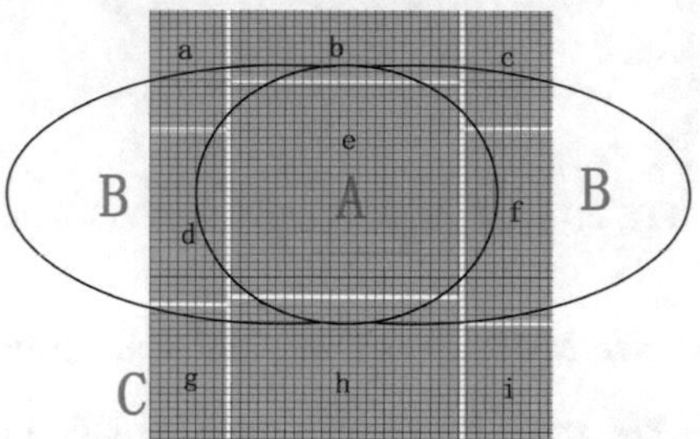

Fig. 3. Rasterized HUD interface diagram

Based on establishing the visual attention division model and the importance analysis of the area of HUD, the visual attention distribution of the interface layout is optimal as the objective function, and the interface fitness function is defined as follows:

$$Z = \sum_{j=1}^{n} \sum_{k=1}^{m} f_j e_{jk} S_{jk} \tag{5}$$

e_{jk} is the visual attention level of the area occupied by the functional sub-region j in the field of view k, S_{jk} is the area occupied by the functional sub-region j in the field of view k. $j = 1, 2, \ldots, n$, n is the total number of functional sub-areas, $k = 1, 2, \ldots, m$, m = 3, f_j is the weight of interest corresponding to the functional sub-region.

$$f_j = \frac{\gamma_j}{\sum_{j=1}^{n} \gamma_j} \tag{6}$$

2.4 Iterative Adjustment of HUD Interface Layout

Adjusting HUD interface layout according to the following rules:

(1) The size of the functional sub-area sequentially adjusted according to the order of interest degree, and the aspect ratio of the sub-area is maintained;
(2) After the adjustment, the attention distribution intensity value Z_i is calculated and compared with the previous Z_{i-1}; the interface with the large Z value is taken for the next adjustment.

(3) Each time, the area of the area is gradually increased according to the proportion of the smallest unit size when the interface is rasterized, and the sub-area layout with the largest Z value before reaching the size limit is finally adopted, and the minimum information unit size of the minimum functional sub-area size shall not be less than the national standard minimum size [12]: 0.7 mr.

3 Case Verification

Taking the HUD layout optimization design in the flight phase as an example, ten subjects with more than 30 h of simulated flight experience were invited to perform eye movement experiments based on the initial interface and the optimized interface. The following experimental hot spots and experimental data were obtained. Figure 4 is the hot spots of the two interfaces.

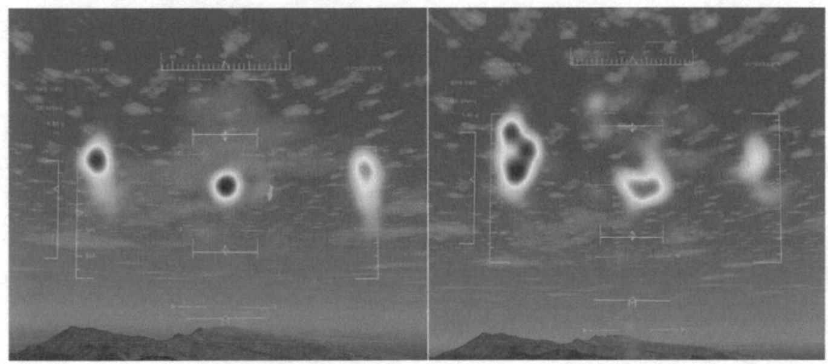

Fig. 4. Experimental hot spot map

According to the data analysis, the number of fixation points in the height zone and the speed zone increased by 38.4% and 22.7%, respectively, and the speed of completing the task was also increased by 12.5%. By analyzing the eye movement data and combining the subjective feedback of the test experiment: satisfaction, fatigue and easy to learn, verify the effectiveness of the adaptive method, and prove that the optimized HUD interface has good practicability.

4 Discussion

From the perspective of mission time, this study uses the simulated fighter HUD interface as the experimental material, carries out the take-off task, obtains the eye movement data, then adjusts the interface based on the algorithm, and performs the eye movement experiment to analyze and compare the eye movement data of the two groups of experiments. It is found that the number of gaze points in the height and speed zones is increased, and the task time is reduced. In addition, the algorithm can effectively combine the pilot's attention with the interface information, improve the attention distribution intensity, and improve the pilot performance to some extent. This verifies the effectiveness of the optimized interface and the accuracy of the fitness function. This study is not carried out in the real cockpit, the information is simulated, the data obtained is simulated, and the conclusions and methods can be used as a reference for layout design.

5 Conclusion

In order to highlight the important information of HUD interface, quantitative analysis HUD interface layout, this paper proposes a HUD layout adaptive method based on flight mission, and proves its authenticity and effectiveness with examples. However, during the research, the following problems were found: (1) Participants' background, this paper chose the subjects with certain flight simulation experience, and there is a certain gap in flight experience between the actual pilots and the follow-up. (2) This paper only uses the eye movement index to verify the adaptive method. In the future, we can increase research and further exploration in many aspects such as EEG and skin electrical technology to improve and optimize the method.

Acknowledgements. This work was supported jointly by National Natural Science Foundation of China (No. 71801037, 71871056, 71471037), Science and Technology on Electro-optic Control Laboratory and Aerospace Science Foundation of China (No. 20165169017), SAST Foundation of China (SAST No. 2016010), Equipment Pre research & Ministry of education of China Joint fund, Fundamental Research Funds for the Central Universities of China (No. 2242019k1G023).

References

1. French, G.A., Snow, M.P., Hopper, D.G.: Display requirements for synthetic vision in the military cockpit. Proc. SPIE. **4362**, 120–131 (2001)
2. Bai, J., et al.: Quantitative research on impact factors of pilot's attention allocation on HUD. J. CAUC **33**, 51–55 (2015)
3. Nazzal, A.A.: A new daylight glare evaluation method Introduction of the monitoring protocol and calculation method. Energ Buildings **3**, 257–265 (2001)
4. Hasse, C., Bruder, C.: Eye-tracking measurements and their link to a normative model of monitoring behavior. Ergonomics **58**(3), 355–367 (2015)
5. Rougier, C., Meunier, J., St-Arnaud, A., et al.: Robust video surveillance for fall detection based on human shape deformation. IEEE. Trans. Circ. Syst. Vid. **5**, 611–622 (2011)
6. Ohn-Bar, E., Tawari, A., Martin, S., et al.: Predicting driver maneuvers by learning holistic features. In: Intelligent Vehicles Symposium IEEE (2014)
7. Honjo, M., Numaga, J., Hara, T., et al.: The association between structure-function relationships and cognitive impairment in elderly glaucoma patients. Sci. Rep. **7**, 7095 (2017)
8. Bellenkes, A.H., Wickens, C.D., Kramer, A.F.: Visual scanning and pilot expertise: the role of attentional flexibility and mental model development. Aviat. Space Environ. Med. **68**, 569–579 (1997)
9. Peng, W.D., Liao, W.Y., Zhang, X., et al.: Design of integrated avionic primary flight display based on FlightGear. Aeronaut. CT. **48**, 87–90 (2018)

10. Stanton, N.A., Roberts, A.P., Plant, K.L., et al.: Head-up displays assist helicopter pilots landing in degraded visual environments. TIES **12**, 1–17 (2017)
11. Qin, P.Y.: Ergonomics research of Shipborne console based on CATIA. J. Mech. Design. **34**, 105–109 (2017)
12. Sudesh, K.K.: Development of HUD symbology for enhanced vision system. JAST **69**, 65–76 (2017)

Human-Centered Design for Healthcare

Visually Impaired Interaction with the Mobile Enhanced Travel Aid eBAT

David Abreu[1(✉)], Jonay Toledo[1], Benito Codina[2], and Arminda Suarez[2]

[1] Ingeniería Informática y de Sistemas, Universidad de La Laguna, La Laguna, Spain
alu0100047882@ull.edu.es
[2] Didáctica e Investigación Educativa, Universidad de La Laguna, La Laguna, Spain

Abstract. The visually impaired face problems in their daily life mobility. Despite the use of the white cane as the most popular aid, there is still no electronic aid to replace or complement its limitations. The eBAT (electronic Buzzer for Autonomous Travel) has been designed to provide information of the environment with a simple user interaction. Through a mobile phone, the blinds can perceive the distance to the obstacles with vibrations, using it as an haptic device. Results show that the device reduce the involuntary collisions at the cost of increasing the travel time.

Keywords: Visually impaired · Electronic aid · Mobile app · Haptic device

1 Introduction

According the World Health Organization [1], the number of visually impaired persons along the world were 36 million as in 2017. One of the main consequences of this disability is that the blind people have problems with their mobility in their daily life [2]. To improve their capabilities concerning the mobility, the most popular aid is the denominated white cane [3]. Despite its usefulness, it is limited to the length of the cane.

To increase the detection range of the cane or even to replace it, several electronic aids have been developed [4]. But no one of these devices have been adopted by the majority of the visually impaired. Trying to solve the lack of adoption, a low budget device has been developed centering the design in a simplified interaction with the user. Forgetting any use of the hearing (as it is the main sense for the blind), the tactile is the next easier available. The electronic Buzzer for Autonomous Travel (eBAT) uses the mobile phone as the Human Computer Interface (HCI) through vibrations.

2 Methods

2.1 Electronic Aid

The developed device is an Arduino nano microcontroller powered unit with two HC-SR04 distance sensors and a Bluetooth module. A picture of the device in Fig. 1. It is designed to be placed hanging from the neck of the user.

T. Ahram et al. (Eds.): IHSED 2019, AISC 1026, pp. 267–272, 2020.
https://doi.org/10.1007/978-3-030-27928-8_40

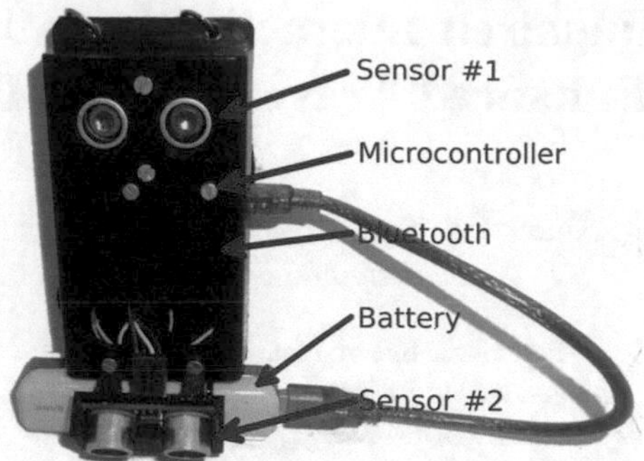

Fig. 1. Picture of the developed device prototype. Main parts labeled.

Each sensor calculates the distance to the obstacles using the registered time between the ultrasounds beam emission and the received echo, assuming stable sound speed. They can measure distance from few centimeters up to 4 m [5]. The upper sensor is intended to detect obstacles at head level, the lower one at floor level.

Once the distances are computed, the final data is then transmitted through Bluetooth to a mobile phone.

2.2 Mobile APP

In the mobile phone, a custom developed APP is executed. Connected to the eBAT device through Bluetooth. The design is as simple as possible, with only a button to activate/deactivate it.

Once the application is activated, the communication with the electronic aid is started. Distances calculated for the obstacles are then received for each sensor.

The interaction with the user is performed using vibration pulses of 100 ms duration. The frequency of the pulses is modulated proportional to the distance. Flow chart diagram of the interaction in Fig. 2.

For each sensor, the delay between vibration pulses varies from continuous vibration for obstacles below a predefined minimum distance to no vibration for a maximum. In the range between the minimum and the maximum, the delay is calculated with formula 1.

$$t_{delay} = \frac{d - d_{min}}{d_{max} - d_{min}} t_{max\ delay} \tag{1}$$

Where t_{delay} is the time between vibration pulses, d is the distance to the object, d_{min} the minimum distance before continuous vibration, d_{max} the maximum distance to provide vibrations and $t_{max\ delay}$ the maximum delay time between vibrations. The shortest delay calculated for each sensor is provided to the user as the final interaction.

Measured distance

Maximum distance

> Maximum distance

Yes

No vibration

No

Minimum distance

> Minimum distance

Yes

Vibration delay proportional to distance

No

Continuous vibration

Vibration pulse + delay

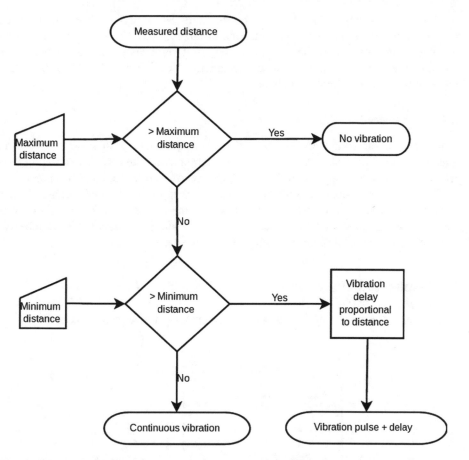

Fig. 2. Flow chart of the algorithm used to provide haptic interaction with the user. Processed for each of the two sensors.

2.3 Timing

The timing of the system is adjusted so the user receive continuous updates about the environment. The Fig. 3 shows the timing of the eBAT.

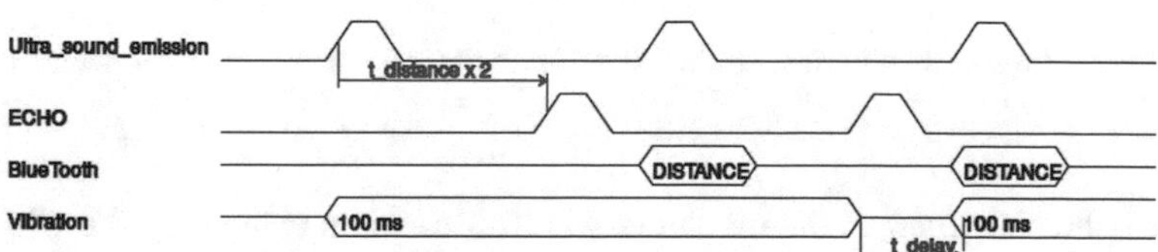

Fig. 3. Timing of the eBAT. After the ECHO is registered, the distance is transmitted through Bluetooth to the APP.

After the emission of the ultrasonic pulse, the ECHO is register. The time between the two signal corresponds to two times the distance traveled by the ultrasound. This distance is transmitted to the mobile phone. And right after the ECHO reception, another ultrasonic pulse is emitted. In the mobile, the APP calculates then the delay between the 100 ms vibration pulses.

2.4 Test Procedure

A set of 12 blind volunteers participated in the experiment. Half male, half female. Ages from 30 to 50 years. After a short explanation about the eBAT functioning they performed a course along a corridor with obstacles placed at floor and head level.

The course was done two times. A first one using only the white cane. Then a second time in reverse direction with the white cane and the eBAT. The phone was hold by the volunteers on the other hand (not handling the cane). The number of involuntary collisions and the time elapsed was measured.

3 Results and Discussion

In the Fig. 4, the results of the number of involuntary collisions are showed.

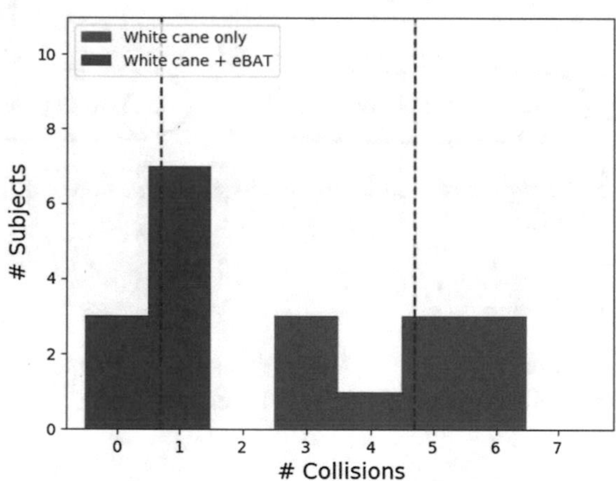

Fig. 4. Involuntary collisions histogram. Red for white cane only, blue for white cane and eBAT.

In red, the histogram of the involuntary collisions when using only the white cane. In blue, adding the eBAT. In vertical dashed lines, averages for each condition are plotted.

From an average of 4.7 collisions, they are reduced to 0.7 when the eBAT is used. The results of the time elapsed in the experiment are showed in Fig. 5.

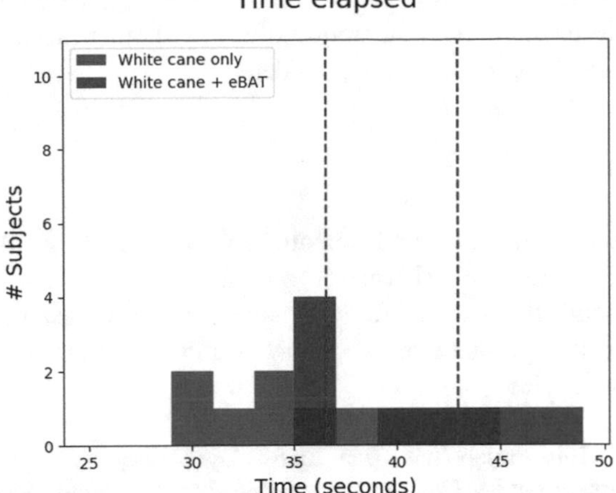

Fig. 5. Histogram of time elapsed in the course. Red for white cane only, blue for white cane and eBAT.

The histogram of the time (in seconds) elapsed for each volunteer is plotted in red (when using only the white cane) and in blue (using also the eBAT). Overlapped area has a violet (red + blue) color. Vertical dashed lines with average for each condition.

It is observed that for the case of the eBAT, the average time spent in the course is increased. From 36.5 to 43 s.

The eBAT obtains a good safety increasing as collisions are reduced by 85% (in average) but at the cost of increasing the time by 18%. Other devices studied by other authors [6], provide a lower decreasing in collisions (57% and 58%) but reducing also the time used (increasing the walking speed).

Voluntary contacts are also reduced when using the eBAT from an average of 1 to 0.1 contacts. Meaning that the larger time used for the second condition is not translated into more exploration of the surrounding but to process the information received by the haptic interaction.

Two hypothesis have been proposed to reduce the time used by the user of the eBAT when wearing it. The eBAT APP is too general and holding the phone confuse the user.

According to APP customization, As each user has their own preferences, trying to apply a configuration valid for all of them is not the best option. Should be possible to modify the minimum and maximum distance to be provided by the APP to the user as interaction. According to height and personal choices, can be configured before the use of the eBAT. Another option (not excluding the first one) is that the limits could be dynamically modified. Depending on the walking speed, that can be measured using mobile accelerators. The duration of the pulses and maximum delay is also another option to be customized. For some users the interaction of the APP can be too slow or too fast and can be adjusted using these parameters.

Apart from the customization of the APP, another source of confusion can be the place to receive the vibrations. As blind people is used to perform interactions with the environment using the free hand from the cane, having a phone there is maybe not the best choice. Should be study then the option of using another place for vibrations. The choice of a vibrating bracelet is being considered.

4 Conclusions

The eBAT has demonstrated a good potential to be a travel aid for the blind as the involuntary collisions are reduced when it is used.

On the other hand, the users are moving slower when wearing it. Two hypothesis has been stated for this problem and should be studied in future work to improve the performance of the device.

Acknowledgments. This project has been funded by Consejería de Economía, Industria, Comercio y Conocimiento to the ULL, partially financed in 85% by the Fondo Social Europeo.

References

1. World Health Organization: Blindness and Visual impairment (2017). http://wwww.who.int/news-room/fact-sheets/detail/blindness-and-visual-impairment
2. Lamoureux, E., Hassell, J., Keeffe, J.: The determinants of participation in activities of daily living in people with impaired vision. Am. J. Ophthalmol. (2004). https://doi.org/10.1016/j.ajo.2003.08.003
3. Carreiras, M., Codina, B.: Cognición espacial, orientación y movilidad: consideraciones sobre la ceguera. Integración **11**, 5–15 (1993)
4. Dakopoulos, D., Bourbakis, N.: Wearable obstacle avoidance electronic travel aids for blind: a survey. IEEE Trans. Syst. Man Cybern. Part C (Appl. Rev.) (2010). https://doi.org/10.1109/TSMCC.2009.2021255
5. Freaks: Ultrasonic ranging module (2016). http://elecfreaks.com/estore/download/EF03085-HC-SR04_Ultrasonic_Module_User_Guide.pdf
6. Roentgen, U., Gelderblom, J., de Witte, L.: User evaluation of two electronic mobility aids for persons who are visually impaired: a quasi-experimental study using a standardized mobility course. Assistive Technol. (2012). https://doi.org/10.1080/10400435.2012.659794

Palletising Support in Intralogistics: The Effect of a Passive Exoskeleton on Workload and Task Difficulty Considering Handling and Comfort

Semhar Kinne(✉), Veronika Kretschmer(✉), and Nicole Bednorz(✉)

Fraunhofer Institute for Material Flow and Logistics, Joseph-von-Fraunhofer-Str. 2-4, 44227 Dortmund, Germany
{Semhar.Kinne,Veronika.Kretschmer, Nicole.Bednorz}@iml.fraunhofer.de

Abstract. In logistical processes such as palletising and order picking, musculoskeletal disorders increase. As part of the INNOVATIONSLABOR research project, the latest model of a passive exoskeleton manufactured by LAEVO was investigated in a laboratory study. A final sample of $N = 37$ persons (73% men) from 20 to 64 years of age evaluated the exoskeleton regarding general comfort, local comfort in different body parts, handling characteristics such as adjustment possibilities, freedom of movement, efficiency, task support and task impairment with validated questionnaires. The analyses show that passive exoskeletons seem to have potential for static activities, but their wearing comfort and user handling should be further developed.

Keywords: Exoskeleton · Human factors · Ergonomics · Palletising · Musculoskeletal complaints · Comfort

1 Introduction

Musculoskeletal Disorders (MSD) are the most common cause of inability to work in Europe [1]. Therefore, preventive measures of occupational health to preserve long-term physical health and work ability of employees gain in importance. Due to required flexibility especially in the field of intralogistics, most work tasks are performed manually such as palletising. They involve a high risk of MSD in the region of the lower back due to repetitive or one-sided physical strain, forced postures or incorrect movement patterns when lifting and carrying heavy loads. The need for appropriate assistance systems to provide physical support has triggered the rapid development of exoskeleton technologies in recent years [2]. Exoskeletons are external supporting structures worn on the body that support or reinforce human movements.

The huge interest and the resulting research activities are confirmed in a review of Looze et al. (2016) regarding exoskeletons for industrial use [3]. The review examined 40 papers on the effects of different exoskeleton on physical relief. Passive exoskeletons were considered to have an overall positive effect to support the lower back during static holding activities and dynamic lifting and lowering. With active exoskeletons,

T. Ahram et al. (Eds.): IHSED 2019, AISC 1026, pp. 273–279, 2020.
https://doi.org/10.1007/978-3-030-27928-8_41

even greater effects on physical strain are possible for the lower body, trunk and upper body regions [3]. To support lower back, the passive exoskeleton LAEVO was developed, for which a positive effect on perceived stress reduction could be found in a field study in the automotive industry [4].

This study is based on a previous field study in cooperation with ADINA research project at a company for joining technology [5]. The results indicated that the full potential of the device can only be fully exploited in palletising with a difference in height between removal and placement, since an active flexion of the upper body takes place. The findings were incorporated into the experimental design of the present laboratory study under simulated, standardised working conditions at Fraunhofer IML.

2 Methods

2.1 Study Sample

The final study sample consisted of $N = 37$ persons (73% men) from 20 to 64 years ($M = 29.8$, $SD = 8.3$). Highest educational qualification of the sample is composed of participants with a higher education entrance qualification (Abitur: German equivalent of "A Levels") (71.7%) and persons with either an advanced technical college certificate or a secondary school leaving certificate (8.6%). Regarding measured technology competence [6, 7], participants had a high acceptance towards technologies ($M = 3.9$, $SD = .6$), high agency ($M = 4.6$, $SD = .5$) and control beliefs ($M = 3.9$, $SD = .6$) and a relatively high level of the need using technology ($M = 3.8$, $SD = .6$). Furthermore, the subjective potential threat of technologies is rather low on average ($M = 2.0$, $SD = .6$).

To ensure that the exoskeleton fits well and to minimize the time for adjustment, a body mass index in normal range (18.5 to 24.9 kg/m^2) was required for participation. Almost all participants met this target. The range of height was 1.63 to 1.91 m ($M = 1.79$ m), the range of body weight 52 to 93 kg ($M = 73$ kg).

2.2 Procedure

As part of the INNOVATIONSLABOR research project, the latest model of a passive exoskeleton manufactured by LAEVO (model series 2.5, sizes S to XL) was examined in a laboratory study. A total of 64 original boxes from the field study (app. 8 kg each) were placed on a europallet (Fig. 1).

Fig. 1. Experimental set-up of the study

Participants were introduced to pack all boxes from one pallet to another. The participants were instructed to work in the best possible way and with a normal speed to simulate real packaging processes during an 8-h workday.

There were two research conditions: palletising all boxes with the help of an exoskeleton and as a control condition packaging without using a mechanical device. Every participant was exposed to both conditions consecutively regarding a within-subjects design. To control position and thus learning effects, the groups were balanced between gender and age.

Before palletising with the exoskeleton, it was carefully adapted to anthropometric data (height and body weight). Furthermore, to get the participants used to the exoskeleton, they were asked to carry out special movement sequences (e.g. make squats, walk around the pallet).

2.3 Measurements

2.3.1 Objective Measurements

The individual settings of the exoskeleton and anthropometric measurements were documented. According to the manufacturer's recommendations, the correct size should be selected based on body size, whereby the use applies exclusively to men. However, weight and individual body shape are also important factors for an optimal fit.

Time for packaging the whole pallet was measured to compare both research conditions. For further evaluations with regard to individualisation of exoskeletons the participants chest and hip size were also documented.

2.3.2 Subjective Measurements

After each research condition, validated questionnaires were filled in via digitised tablet survey to evaluate the applied exoskeleton and the work process. The palletising task was examined regarding the subjective task difficulty and workload during packaging. Perceived task difficulty was queried using a visual analogue scale (VAS) from very easy (0) to very difficult (10) [8]. Overall workload was recorded by NASA Task Load Index (TLX) that consists of six subdimensions: mental demand, physical demand, temporal demand, performance, effort, and frustration level are rated from low (0) to high (100) [9]. Total workload (raw TLX score) was calculated by taking the mean value of all six subscales [10].

All participants evaluated the exoskeleton regarding general discomfort (0 = very comfortable, 10 = very uncomfortable) and local discomfort (0 = no discomfort, 100 = maximal discomfort) in different body parts on a VAS scale [8]. The examined body areas were chest, abdomen, upper back, lower back, upper legs ventral and upper legs dorsal side. The overall user impression regarding exoskeleton was assessed with the following subdimensions with the VAS scale: handling characteristics of the exoskeleton such as adjustment possibilities (donning and doffing, length adjustment, 0 = very easy, 10 = very difficult), range of movement (0 = not restricted, 10 = heavily restricted), efficacy (reduction of back loading: 0 = high reduction, 10 = no reduction; support of tasks: 0 = high support, 10 = no support; interference with tasks: 0 = no interference, 10 = high interference) [8]. In order to obtain an overall assessment of the exoskeleton, the participants assigned school notes.

Finally, sociodemographic data was queried that were in line with the BIBB/BAuA Employment Survey 2012 [11]. Overall technological competence of the sample was analysed with various 5-point Likert scales from 1 (strongly disagree) to 5 (strongly agree) due to technology commitment such as acceptance, control beliefs and agency beliefs according to new technologies in general [6] and the personal attitude towards new technologies like potential threat and the need for using technology [7].

3 Results

All examined scales were tested for normal distribution according to the Shapiro Wilk test. For detecting group differences between both research conditions nonparametric and therefore more powerful analyses with the Wilcoxon test were performed. Furthermore, nonparametric correlation analyses were used. Arithmetic mean values of the Raw TLX Score and the scores of the six subdimensions for both groups are shown in Fig. 2. The overall workload was rated significantly lower if the exoskeleton was worn in comparison to the control group without the mechanical device (*Raw TLX*: $Z = -3.1$, $p < .01$). In detail, analyses of the six subdimensions indicate that perceived physical demand ($Z = -2.9$, $p < .01$) and effort ($Z = -3.8$, $p < .001$) were reduced while wearing the exoskeleton than without one.

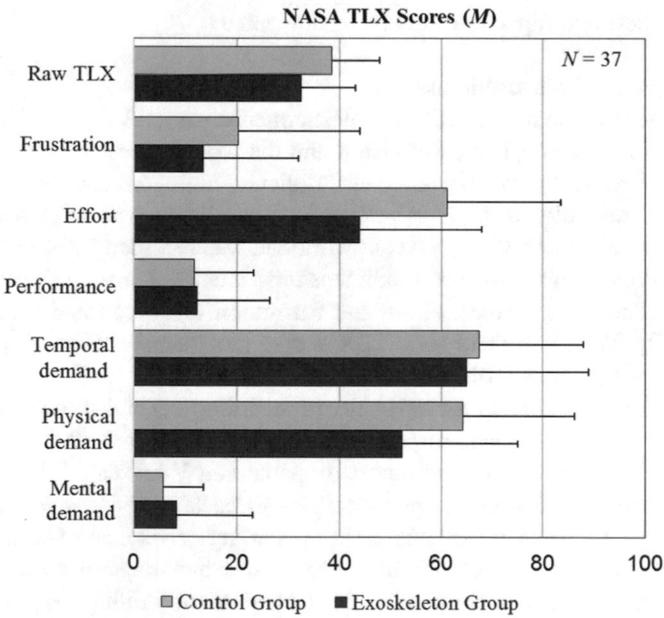

Fig. 2. Descriptive statistics of the workload of both research conditions

Furthermore, results indicate that the palletising task was significantly easier for the participants while they wore the exoskeleton as a support than without the help of a mechanical device ($Z = -2.0$, $p < .05$) (Table 1).

Table 1. Descriptive statistics of the perceived task difficulty of both research conditions.

	Min	Max	M	SD
Control group	0	100	47.2	24.2
Exoskeleton group	0	100	39.7	22.9

Note. Min = Minimum, Max = Maximum, M = Mean value, SD = Standard deviation.

Regarding correlation analyses we found out that the greater the overall discomfort was perceived, the greater the overall workload was rated (*Spearman's rho* = .35, $p < .05$). Likewise, task difficulty also rises with increasing discomfort (*Spearman's rho* = .38, $p < .05$). Workload was also significantly correlated to the feeling of local discomfort. Participants felt more stressed and impaired during palletising when the discomfort at the ventral thigh increased (Mental demand: *Spearman's rho* = .33, $p = .05$; Performance: *Spearman's rho* = .47, $p < .01$; Frustration: *Spearman's rho* = .33, $p < .05$; Raw TLX: *Spearman's rho* = .59, p < .001).

The physical demand increased significantly with a discomfort feeling in the chest area (*Spearman's rho* = .43, $p < .01$). The handling of the exoskeleton correlated significantly with the global workload (*Spearman's rho* = .34, $p < .05$) and perceived effort (*Spearman's rho* = .46, $p < .01$), although participants did not have to put on the exoskeleton themselves. The more the persons felt restricted by the device in their freedom of movement, the more they felt disturbed in accomplishing their tasks (*Spearman's rho* = .74, $p < .001$) and the greater the global workload (*Spearman's rho* = .70, $p < .001$), perceived effort (*Spearman's rho* = .54, $p < .01$) and frustration level (*Spearman's rho* = .39, $p < .05$) was. With increasing back relief, subjective physical demand decreased significantly (*Spearman's rho* = .41, $p < .05$).

The exoskeleton size had no significant influence on the dependent variables. An analysis of variance showed no significant size effect on the parameters queried.

However, significant sequence effects could be detected by using variance analyses. It became apparent that the task difficulty was rated higher during palletising without the use of an exoskeleton if participants started palletising without the mechanical device as if they started with the help of the mechanical device ($F = 4.5$, $p < .05$). The use of the exoskeleton was a kind of comparison anchor. The same effect tended to be observed for the perceived task difficulty during palletising with the help of an exoskeleton ($F = 3.0$, $p < .10$). The palletising task seemed more difficult if the participants had the direct comparison to the palletising task without the exoskeleton.

Time evaluation shows differences between the exoskeleton group and the control group. In the exoskeleton group, work was on average 51 s faster without device, with only one person taking longer in the second run. In contrast, the control group worked on average 8 s slower without exoskeleton. One third of the participants were slower in the second run despite learning effect.

4 Conclusion and Discussion

As part of the INNOVATIONSLABOR research project, the latest model of a passive exoskeleton was investigated in a laboratory study with 37 participants to analyse the effect on workload and task difficulty considering handling and discomfort. Based on a previous field study to investigate the use of exoskeletons during real palletising activities, we simulated the scenario as realistically as possible in the laboratory. The full potential of the device seemed to be fully exploited in palletising with a difference in height between removal and placement. Furthermore, we were interested whether and to what extent the exoskeleton can reduce physical demands and whether it can be seen as a personal protective equipment that is not perceived as disturbing. In order to validate these research questions, the laboratory study with a bigger final sample under controlled laboratory conditions was carried out at Fraunhofer IML.

In comparison to a control group, the palletising task itself as well as the overall workload and task difficulty was rated lower if the exoskeleton was worn. Passive exoskeletons seem to have potential for palletising tasks, but their wearing comfort and user handling should be further developed.

One third of the test persons in the control group without the device were slower in the second run with the exoskeleton. At the same time, the exercise of the task was rated easier. This suggests that despite the learning effect the movements with exoskeleton were performed more consciously, which indicates a positive effect on posture. These results should be considered in further studies. Furthermore, in future evaluations we will focus and describe gender differences and effects during palletising with and without the help of the exoskeleton.

References

1. European Agency for Safety and Health at Work: OSH in figures: work-related musculoskeletal disorders in the EU - Facts and figures. Office for Official Publications of the European Communities, Luxembourg (2010)
2. Young, A.J., Ferris, D.P.: State of the art and future directions for lower limb robotic exoskeletons. IEEE Eng. Med. Biol. Soc. **25**, 171–182 (2017)
3. de Looze, M.P., Bosch, T., Krause, F., Stadler, K.S., O'Sullivan, L.W.: Exoskeletons for industrial application and their potential effects on physical workload. Ergonomics **59**, 671–681 (2016)
4. Hensel, R., Keil, M., Mücke, B., Weiler, S.: Chancen und Risiken für den Einsatz von Exoskeletten in der betrieblichen Praxis. ASU **53**, 654–661 (2018)
5. Bednorz, N., Kinne, S., Kretschmer, V.: Ergonomieunterstützung in der Logistik – Industrieller Einsatz von Exoskeletten an Palettier- und Kommissionierarbeitsplätzen zur körperlichen Entlastung von Mitarbeitern. In: GfA-Frühjahrskongress, B.4.1 (2019)
6. Claßen, K.: Zur Psychologie von Technikakzeptanz im höheren Lebensalter: Die Rolle von Technikgenerationen. Dissertation, Heidelberg (2013)
7. Neyer, F.J.J., Felber, J., Gebhardt, C.: Kurzskala. Technikbereitschaft (TB) [Technology commitment]. In: ZIS - Zusammenstellung sozialwissenschaftlicher Items und Skalen (ed) (2016)

8. Baltrusch, S.J., van Dieën, J.H., van Bennekom, C.A.M., Houdijk, H.: The effect of a passive trunk exoskeleton on functional performance in healthy individuals. Appl. Ergon. **72**, 94–106 (2018)
9. Hart, S.G., Staveland, L.E.: Development of NASA-TLX (task load index): results of empirical and theoretical research. In: Hancock, P.A., Meshkati, N. (eds.) Human Mental Workload, 139-183. Elsevier, Amsterdam (1988)
10. Hart, S.G.: Nasa-task load index (Nasa-TLX) - 20 years later. In: Proceedings of the Human Factors and Ergonomics Society Annual Meeting, vol. 50 (2006)
11. Wittig, P., Nöllenheidt, C., Brenscheidt, S.: Grundauswertung der BIBB/BAuA-Erwerbstätigenbefragung 2012. BAuA, Dortmund (2013)

Understanding the Influence of Cognitive Biases in Production Planning and Control

Julia C. Bendul and Melanie Zahner(✉)

Chair for Management of Digitalization and Automation, RWTH Aachen University, Kackertstraße 7, 52072 Aachen, Germany
bendul@scm.rwth-aachen.de,
Melanie.zahner@rwth-aachen.de

Abstract. Production Planning and Control (PPC) requires human decision making in several process steps like production programme planning, production data management and performance measurement. Thereby, human decisions are often biased leading to an aggravation of logistic performance. Exemplary, the lead time syndrome (LTS) shows this connection. While production planners aim to improve due date reliability by updating planned lead times the result is even a decreasing due date reliability. In current research in the field of production logistics the impact of cognitive biases on the decision-making process in PPC remains at a silent place. We aim to close the research gap by combining a systematic literature review on behavioral operations management as well as cognitive biases and applying the Aachen PPC model. Based on a case study from the steel industry we show the influence of cognitive biases on human decision making in several phases of PPC.

Keywords: Cognitive bias · Human behavior · Production Planning and Control · PPS

1 Introduction

While in the area of psychology, anthropology and sociology human behaviour has been investigated in several research fields in the area of logistic and production planning only little research has been published [3]. Various models in the area of production logistic and operations management have been developed in order to support decision-making process within PPC to optimize logistic performance. Short lead times, high due date reliability, low WIP levels and high capacity utilization are thereby desirable key performance indicators measuring the performance of a manufacturing system [12]. The underlaying proposition of these models is often the theory of "homo economicus" - assuming a rational human behaviour in decision making process determined by maximizing their own possible outcome [7]. [20] challenged this assumption and show that human decisions are rather biased. Especially, when people are confronted with uncertainty and high complexity humans systematically take wrong decisions [1]. The LTS exemplarily shows how deviation from planned lead times, as a result of planners overreaction to decreasing due date reliability, even lead to an aggravation of

T. Ahram et al. (Eds.): IHSED 2019, AISC 1026, pp. 280–285, 2020.
https://doi.org/10.1007/978-3-030-27928-8_42

due date reliability [15]. In recent past human behavioral aspects became an integral part in economic research [19]. Nevertheless, especially in the area of a holistic process view of PPC human behavior remains an open research field.

In this research, we aim to improve the understanding of the role of cognitive biases in the field of PPC and its potential effects on logistics performance. Therefore, we combine a systematic literature review on behavioral operations management as well as cognitive biases by applying the integrated Aachener PPC model. Taking inspiration from a case study from the steel industry we show the influence of cognitive biases on human decision making in production planning and the potential impact on logistic performance. The remainder of this article is structured as follows: First we present cognitive biases in general in decision making process within the PPC process on the example of the Aachener PPC model. Afterwards, we show the influence of cognitive biases on logistic performance based on a case study from the steel industry and present several examples of the impact of cognitive biases on several tasks within the PPC process.

2 Cognitive Biases

[20] were the first who questioned the assumption of rational human behaviour and introduced the term of cognitive biases. They state that humans taking decisions especially in complex and uncertain environments systematically go wrong. In further experiments [8] deepen this research of the underlaying factors and describe the cognitive processes of intuition and reasoning. Based on the denomination of [16] of System I (intuition) and System II (reasoning) [8] further characterizes these cognitive routines. While System I acts automatic, fast, emotively, effortless and hardly controllable System II operates relatively slow, reflected and effortful [16]. System I creates spontaneous impressions and persuasions which forms the basis for further decisions and actions of System II. There are plenty of cognitive biases effecting decision making process. [1] categorized these biases in six relevant categories:

(1) Memory Biases: describe biases influencing the storage and the ability to remember information, (2) Statistical Biases: are the human tendency to over- or underestimate certain statistical parameter, (3) Confidence Biases: act to increase a person´s confidence in their prowess as decision maker, (4) Adjustment Biases: describe the human tendency to stick to the first available information or to a reference point when making decisions, (5) Presentation Biases influence humans in their decision making by the way how information is being displayed, (6) Situation Biases describe the way how a person responds to the general decision situation. [3] describe in their research how cognitive biases in PPC can lead to the LTS. Based on this research we want to extent these findings to all relevant process activities with the PPC process by applying the Aachener PPC model.

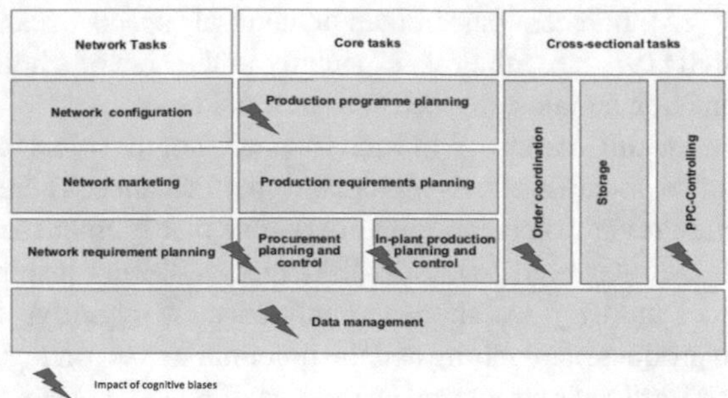

Fig. 1. Aachener PPC model with active cognitive biases

2.1 The Impact of Cognitive Biases on Decision Making Situations in PPC Based on the Aachener PPC Model

Within PPC several human decisions are required influencing logistic performance in positive or negative way [12]. Figure 1 shows the Aachner PPC model highlighting those phases of PPC where we identified active cognitive biases [3, 10].

The Aachener PPC model divides the PPC process in three involved subprocesses with relevant tasks. In this paper we focus on activities where we identified the influence of cognitive biases. During the production programme planning (PPP) the production sequence with the necessary material, capacity plans and due dates is set. Within procurement planning and control the goods receive in right quantity and quality on time is organized. Production Planner supervise the self produced parts within the process task "In plant PPC". Order coordination belongs to the cross-sectional task. During the process several disruptions like missing material or machine interruptions can occur which require several human decisions about updating orders. In every process step a lot of data is collected and requires a constantly data management like adjustment of master data, due dates, machine utilization times etc. Logistic and financial performance measurement is done in the task PPC Controlling [10].

3 Case Study Description

We take inspiration from a case study of the steel industry presented by [3]. The analyzed PPC process takes place within a R&D department. To compete in the global steel market a short time-to market is crucial. Especially in the R&D department production and analysis processes are hardly to plan and it is one of the major challenges of production planners to fulfill the customer requested delivery date. Samples of new alloys have to go through sequences of different tests before they can be launched in the market. In the analyzed R&D process first orders for different steel samples are placed through external and internal customers. After estimating the planned lead time for several manufacturing and analysis processes the orders get a due-date. To schedule the orders a supportive PPC IT system is used.

4 How Cognitive Biases Influence Decision Making in PPC

Based on the six presented categories of the cognitive biases presented by [1] we aim to map them to the PPC processes presented in the Aachener PPC model.

Memory Biases: The *availability heuristic* describes the tendency of people to overestimate the likelihood of events where they can easily restore the information [8]. As a result, people tend to overweight the outcome of the last decision as a basis for their decision making in their current situation. The *imaginability bias* describes the fact that people assume an event as more probable if it can be easily imagined by themselves [17]. We observe that planners tend to adjust planned lead times based on their intuition instead of entirely considering all influencing variables. This occurs mainly in the phase of PPP, order coordination, data management. The planners tend to connect their last updates to the planned lead time to any positive development of the logistics performance. In case of a negative development, the planners assume that external influences are responsible for the fact delayed due date reliability and they conclude that there is the need for another planned lead time adjustment.

Statistical Biases: [20] investigate that humans tend to overestimate the probability of two events occur together if this has already happened once in the past. This effect is described by the *correlation bias*. For example a change in material and an increase in throughput times within a machine can lead to the assumption that there is a correlation which actually does not exist. The *gambler's fallacy* describes the phenomenon of the assumption that future events are determined by the occurrence of past events [2]. This leads to an overestimation of possible events ignoring the really statistical possibility [7]. We observe statistical biases in the adaption of lead times within the phase of the order coordination. The planners, tend to assume that the coincident adaption of planned lead times and the positive development of due date reliability are correlated, although they are aware of the mathematical facts that a cause-and- effect are delayed by minimum four weeks.

Confidence Biases: The *illusion of control or overconfidence biases* describe the tendency of people to overestimate their ability to solve difficult problems [4]. The *conformation biases* lead people to seek for information which confirms their own estimation and hide information which are in contrary to their own perception [14]. Analyzing the case study, we found three examples of the *illusion of control*. (1) Planners tend to assume that their own procedures are better suited than the standard planning procedures. (2) When planning the lead times they behave as if the stable forecast of future incoming orders is really predictable and not only a prediction. (3) Planners increased the WIP level via the planned lead times in order to avoid that the production system could run into an idle state. We also found situations as an example for active *confirmation biases*. When the planners once had the intuition that updating the lead times would be the best option to increase due date reliability, they searched especially for information which confirms their first feeling. Obvious information which entails the result not to intervene in the planned lead times was ignored. We identified a strong impact of confidence biases in the task of PPP, in-plant PPC as well as in the order coordination.

Adjustment Biases: The *anchoring effect* explains the tendency to rely too heavily on an initially given information which influences further values [20]. [18] has shown that anchoring can create systematic errors in decision making situations. This is also what *conservatism bias* describes. Once taken estimations are not updated according to new information [11]. We found that lead times from previous years and from similar work systems seem to act as anchors. When setting planned lead times, the planner was justifying the extension of planned lead times with the numbers in the year 2014. Similarly, the planners tend to aim at the due date reliability that was given as the long-term goal of 95% and therefore seem to disproportionately extend the planned lead times.

Presentation Biases: The *ambiguity effect* describes the human tendency to favor simple looking options and avoid options which seems to be complicated [6]. According to the *primacy recency effect* information at the first and at last of a series can be restored at best whereas in the middle at worst [21]. We found that the planners were using the PPC system to study plenty of different analysis, such as the order forecasts for the next months, inventory levels etc. next to the data that was really important for decision making. The *primacy recency effect* became obvious when the planner was setting the planned lead time for a certain order to the double value of what was reasonable because he had just checked the current due date reliability and noticed that the value for the previous day was particularly low. In this, *presentation biases* influence decision-making in particular in the phases of data analysis within the In-plant PPC as well as the PPC controlling.

Situation Biases: The *complexity effect* describes that people are biased under time pressure or when information overload occurs [13]. The *ostrich effect* describes the habit of people to ignore an obvious negative information [9]. The *Bandwagon effect* describes the tendency to do things because many other do the same [5]. We found that the modern PPC systems provide a wide range of information. The amount of information was simply too much to consider for decision making. In particular, under time pressure the planners decided to extend lead times just to do anything. At the same time, they were ignoring the fact that this behavior of extending lead times influences due date reliability in a negative way. Moreover, we found that adjusting lead times was a common method to react on decreasing due date reliability. Planners who faced the situation of decreasing due date reliability therefore choose this method as it is a quite frequently used method of their colleagues.

5 Conclusion

We show the influence of several cognitive biases of different categories on decision making situation within PPC and the potential effect on logistic performance. The influences of these cognitive biases on specific key performance indicators in PPC can be a further research field. Due to the fact of a single case study approach these impacts of cognitive biases within PPC should be further investigated through behavioral experiments. Nevertheless, studies have shown that the awareness of cognitive biases does not eliminate them. This is also what the analyzed case study confirms. Even planners where aware of the LTS it occurs. Therefore, further studies should focus on debiasing techniques focusing on PPC process. Based on the fact that these human decisions are often supported through IT decision support systems, these findings should be taken into account when thinking about possible designs for these support systems [1].

References

1. Arnott, D.: Cognitive biases and decision support systems development: a design science approach. Inf. Syst. J. **16**, 55 (2006)
2. Barron, G., Leider, S.: The role of experience in the gambler‘s fallacy. J. Behav. Decis. Making **23**, 117–119 (2010)
3. Bendul, J., Knollmann, M.: The human factor in production planning and control: considering human needs in computer aided decision-support systems. Int. J. Manufact. Technol. Manage. **30**(5), 346–368 (2015)
4. Brenner, L.A., Koehler, D.J., Liberman, V., Tversky, A.: Overconfidence in probability and frequency judgements: a critical examination. Organisational Behav. Hum. Decis. Process. **65**, 212–219 (1996)
5. Carter, C., Kaufmann, L., Michel, A.: Behavioral supply management: a taxonomy of judgment and decision making biases. Int. J. Phys. Distrib. Logistics Manage. **37**(8), 631–669
6. Ellsberg, D.: Risk, ambiguity and the savage axioms. Q. J. Econ. **75**(4), 643–669 (1961)
7. Hogarth, R.M.: Judgment and Choice: The Psychology of Decision (1987)
8. Kahneman, D.: Maps of bounded rationality: a perspective on intuitive judgment and choice. Nobel Prize Lect. **8**, 449–489 (2002)
9. Karlsson, N., Loewenstein, G., Seppi, D.: The ostrich effect: Selective attention to information. J. Risk Uncertainty **38**, 95 (2009)
10. Luczak, H., Eversheim, W., Schotten, M.: Produktionsplanung und-steuerung Grundlagen, Gestaltung und Konzepte. Springer Verlag (1998)
11. Nelson, M.W.: Context and the inverse base rate effect. J. Behav. Decis. Making **9**, 23–40 (1996)
12. Nyhuis, P., Wiendahl, H.-P.: Fundamentals of Production Logistics: Theory, Tools and Applications. Springer, Heidelberg (2009)
13. Ordonez, L., Benson, L.: Decisions under time pressure: how time constraint affects risky decision making. Organ. Behav. Hum. Decis. Process. **71**, 121 (1997)
14. Russo, J.E., Medvec, V.H., Meloy, M.G.: The distortion of information during decisions. Organ. Behav. Hum. Decis. Process. **66**, 102–110 (1996)
15. Selçuk, B., Adan, I., de Kok, A., Fransoo, J.: An explicit analysis of the lead time syndrome: stability condition and performance evaluation. Int. J. Prod. Res. **47**(9), 2507–2529 (2009)
16. Stanovich, K.E., West, R.F.: Individual differences in reasoning: implications for the rationality debate? Behav. Brain Sci. **23**(5), 645–726 (2000)
17. Taylor, S.E., Thompson, S.C.: Stalking the elusive ‘vividness’ effect. Psychol. Rev. **89**, 155–181 (1982)
18. Teng, B.-S., Das, T.K.: Cognitive biases and strategic decision processes: an integrative Perspective. J. Manage. Stud. **36**, 757 (1999)
19. Tokar, T.: Behavioural research in logistics and supply chain management. Int. J. Logistics Manage. **21**(1), 89–103 (2010)
20. Tversky, A., Kahneman, D.: Judgment under uncertainty: heuristics and biases. Science **185**(4157), 1124–1131 (1974)
21. Yates, J.F., Curley, S.P.: Contingency judgement: primacy effects and attention decrement. Acta Psychol. **62**, 293–302 (1986)

Complete Block-Level Visual Debugger for Blockly

Anthony Savidis[1,2(✉)] and Crystalia Savaki[1,2]

[1] ICS, FORTH, Heraklion, Greece
as@ics.forth.gr
[2] CSD, University of Crete, Heraklion, Greece
kystalsav@csd.uoc.gr

Abstract. Blockly is a visual programming editor by Google, being open-source multi-platform and multi-language, while offering jigsaw-style program blocks. It is very popular and currently adopted by an increasing number of visual development solutions. However, as with similar earlier tools, it lacks a full-scale debugger. We present a complete visual debugger for Blockly, working over blocks, supporting the full range of debugging features as with typical source-level debuggers. To support all tracing functions we make no amendments to the underlying JavaScript engine, supporting all debugging operations through code instrumentation inserting invocations to a busy-wait debugger service loop. The latter affects only the source code that is output by Blockly.

Keywords: Visual programming · End-user development · Debuggers

1 Introduction

Visual programming languages and systems (VPLs) are amongst the most popular tools for end-user development (EUD), relying on interactive composition with graphic components that are mapped directly to programming code snippets. Such snippets may be very close to the actual programming constructs of one or more target languages, or may map to the main programming patterns of a target domain and an underlying application engine (such as games). Currently *jigsaws* are the most popular graphical style for VPLs, firstly appearing in the context of *Scratch* [1].

Blockly [3] (https://developers.google.com/blockly) is an extensible software library to build visual programming editors. It is a Google project, being open-source under the Apache 2.0 License, and is delivered with a number of premade visual blocks. It enables to introduce new types of blocks, ranging from primitives (leaves) to composite ones (containers) and to also define the way they target to text-based programming snippets. Currently, Google and the Scratch [4] team at MIT Media Lab, are collaborating on the development of a new generation of graphical programming blocks, called Scratch Blocks building on top of the Blockly technology.

In this context, the contribution of our work is twofold. Firstly, a full-scale block-level debugger offering all tracing (stepping) and inspection (watches) features in analogy to source-level debuggers. The latter not only compares with Blockly, but also with all popular visual programming systems either with blocks like Scratch and App

T. Ahram et al. (Eds.): IHSED 2019, AISC 1026, pp. 286–292, 2020.
https://doi.org/10.1007/978-3-030-27928-8_43

Inventor [5] or syntax-driven hierarchical composition like Touch Develop [6]. We show that block-based debugging allows finer tracing control compared to source-level debuggers. Secondly, we deliver all tracing functions through instrumentation of the source-code generated by the visual tool (in our case Blockly). The latter not only simplifies a lot the development of the debugger, but also makes it entirely portable.

2 Visual Debugger

2.1 Breakpoints

In source-level (text-based) debuggers, breakpoints are inserted per line, left to the editor area, usually at a special column reserved for custom icon annotations by the programming tools of the development environment. Usually, such annotations are inserted by the bookmarker, source manager, IntelliSense, and the debugger frontend.

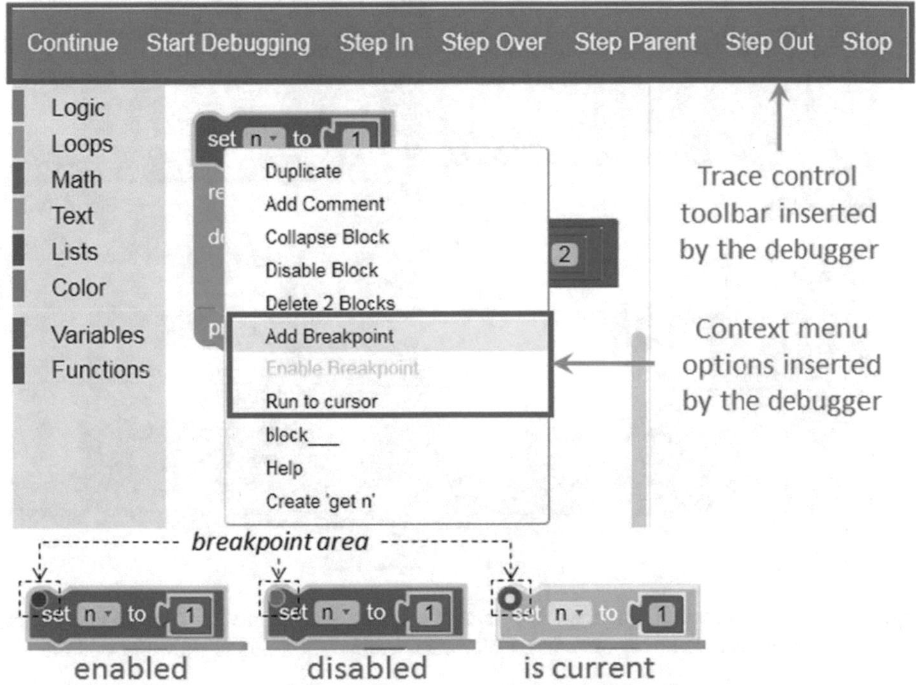

Fig. 1. Trace control toolbar and context menu with breakpoint management during debugging; breakpoints are associated to individual blocks and are shown at their top-left corner.

In the case of Blockly, we support breakpoints per individual block, with a typical breakpoint icon, located on the top-left of the associated block (see Fig. 1). Interestingly, this allows more fine-grained control for execution stop points compared to source-level debuggers. Breakpoint management is possible through a few extra options inserted as part of the original context menu of blocks. The state of breakpoints can be enabled or disabled, while once an enabled breakpoint is hit, it is highlighted. The association of breakpoints to individual blocks is implemented on top of the Blockly as follows: Internally, Blockly exposes the actual object reference of every single block. This is actually a well-documented and standard feature of Blockly library. We use it to directly associate, as part of the breakpoint manager, the block references to their breakpoint state. Then, as part of the code instrumentation, the code generated per block is decorated to post an event both to: (i) the Blockly library, with a request to highlight the block; and (ii) the breakpoint manager, to test if a breakpoint is hit – if the latter is true, meaning a stop point is met, execution will break and a trace command will be expected by the debugger User-Interface so as to proceed.

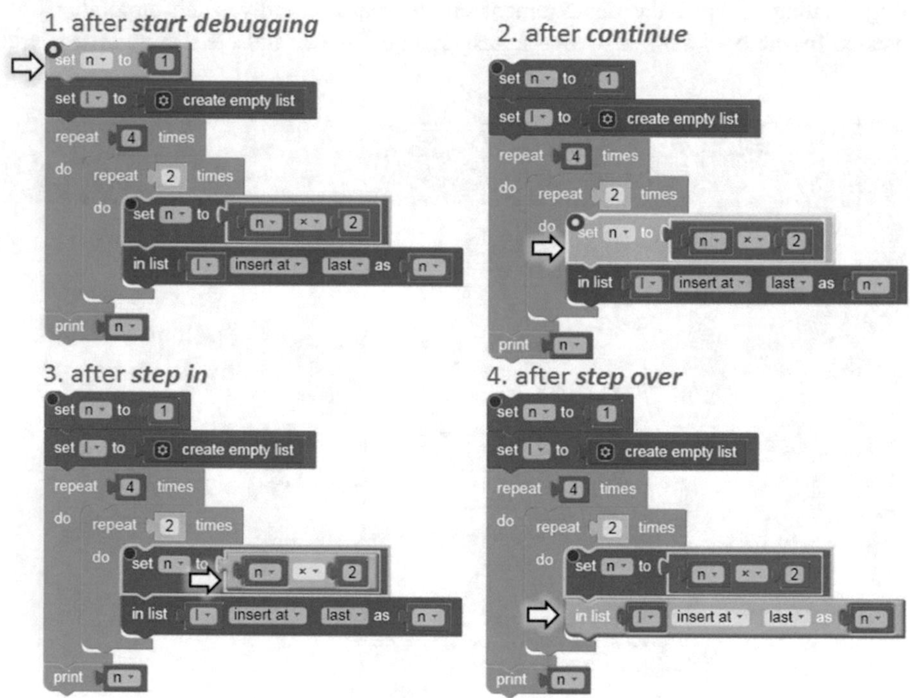

Fig. 2. Basic step functions (tracing) in a sample program execution with a single breakpoint; the visual appearance of breakpoints in enabled/disabled/current mode is shown at bottom.

2.2 Tracing

As already mentioned, block-level tracing is functionally similar to source-level tracing, however, with a few important differences. The first variation concerns the basic *Step In* and *Step Over* commands. These two operations, originating from source-level debuggers, control whether a function call expression is evaluated thoroughly (*Step Over*), or if the execution progresses by evaluating all actual arguments and then by stopping into the first instruction of the invoked function (*Step In*). In our case, besides this behavior regarding function invocations, these commands work as follows given a current block during debugging: *Step In* stops in the first inner (child) block, and *Step Over* enters the next sibling block. Otherwise, if no inner or sibling blocks exists, they stop in the next executing block, following the control flow. Interestingly, these variations are possible due to the hierarchical structure of code, enabling users skip entire blocks of visual code during tracing, something not possible when using typical source-level debuggers. In particular, to skip entire blocks of text code, programmers would have to either use the *Run To Cursor* command or, alternatively, place temporary breakpoints and then use the *Continue* command. However, for nested expressions this far from straightforward: positioning the cursor in a single line or setting a breakpoint is not precise enough to trace particular subexpressions, unless the source code is manually reworked to place one such subexpression per line.

In Fig. 2, the behavior of *Step In* and *Step Over* is shown once execution meets a breakpoint. In this example, the expression $n = n*2$ is actually split in two blocks: the outer assignment block and the inner multiplication block. The latter allows, as shown in Fig. 2, to separately evaluate $n*2$ with a *Step In* command, something not possible directly with a typical source-level debugger. The same mechanics apply to the *Run To* command as well, which works for the currently selected block and will cause execution to stop exactly before evaluating this block.

2.3 Watches

Inspecting program variables, commonly known as *watches*, is also possible in two ways. Via the *variables* pane, showing all variables at the current scope (sometimes designated as *autos* in various source-level debuggers), and the *watches* pane, in which inspected variables can be added or removed during debugging by the user.

Currently, all Blockly variables reside in the global scope, thus used throughout the entire visual program, meaning the presence of the watch pane is somehow redundant. However, it is still possible in Blockly to implement a custom block type for the declaration of a local variable, simile to the *let* specifier of JavaScript. In this case, *autos* will enumerate only the local variables at the current block scope, and *watches* will show the particular user-chosen variables. In our implementation, if no local variables exists, the *variables* window automatically displays all global program variables. In Fig. 3, the automatic display of program variables is shown in a debugging session, in a simple example program involving a single *n* variable.

Besides variable inspection, it is possible to manually evaluate entire blocks, something being more flexible and expressible than typical expression evaluation. For instance, in the example of Fig. 3, at *Step 3*, the manual reevaluation of the current block is chosen. This is actually an extra evaluation with respect to the normal program execution. As a result, the expression $n = n*2$ is executed once more, causing *n* to gain *4* value, meaning it is also allowed to change program variables via watches.

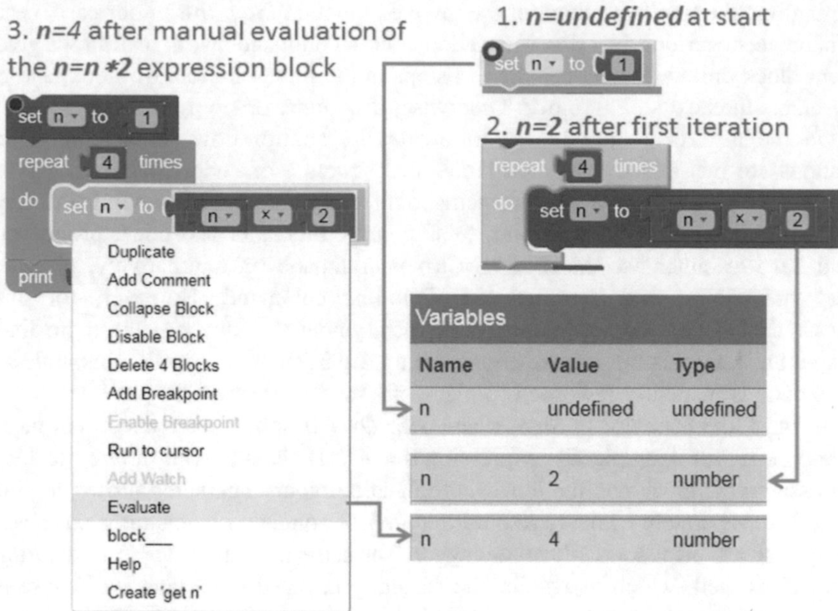

Fig. 3. Automatic variable inspection and the Evaluate operation which works for any kind of block, enabling to re-evaluate on-the-fly (during debugging) any code snippet.

3 Software Architecture

Code decoration (or instrumentation) is a technique that applies the insertion of extra special-purpose instructions, either at the source or at the binary level, with the intent of introducing additional mission-specific functionality, however, without altering the original observed behavior of the subject program.

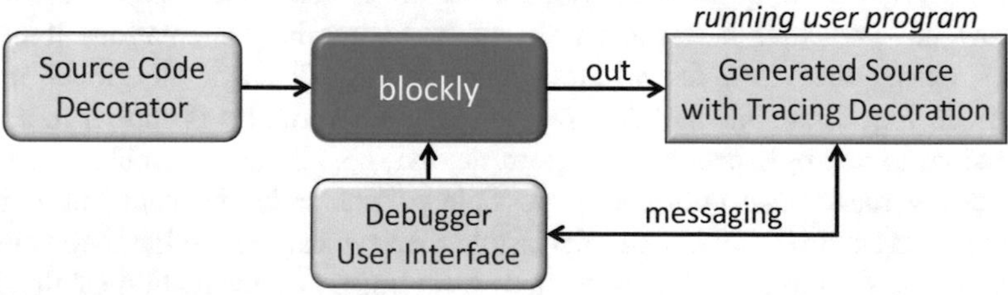

Fig. 4. Debugger architecture with two basic modules build on top of blockly, the code decorator and debugger User Interface that only communicates with the running user program.

Our technique (see Figs. 4 and 5) executes user code (the program) in a separate thread (*workers* in JavaScript) by instrumenting the target code of every block with a busy waiting loop that waits a trace message to proceed and execute the code that follows. In other words, such *wait()* invocations inserted before the code of blocks are essentially service loops for debugger trace commands. The unique block id is supplied to the instrumentation code, enabling post a request to highlight the block.

```
function Wait (nesting, block, editor) {
  let stopPoint = dbg.IsBreakpoint(block) || dbg.IsRunTo(block);
  if (!stopPoint) {
    if (dbg.InState("Continue")) { dbg.nesting = nesting; return; }
    if (dbg.nesting == -1) return;
  }
  if (dbg.InState("StepIn") || stopPoint || nesting <= dbg.nesting)){
     if (dbg.block == block && !stopPoint) return;
     if (dbg.InState("StepParent") && nesting == dbg.nesting &&
         && !dbg.InBreakPoint()) return;
     while(!dbg.proceedRequest)
         await get_next_message(); ← will set dbg.proceedRequest
     dbg.proceedRequest = false;
     dbg.block = block;
     if (dbg.InState("StepOut"))
       { dbg.nesting = -1; dbg.ResetState("StepOut"); }
     else
       dbg.nesting = nesting;
   }
}
```

Fig. 5. The key Wait() function, essentially the debugger service loop, being inserted by the instrumentation logic before every program instructions to support tracing.

4 Summary and Conclusions

Although block-based programming tools became popular as learning instruments, they increasingly gain attention as end-user programming tools. But even at such a relatively small programming scale, debugging is an essential feature of the programming process. We focused on Blockly, a popular general-purpose kit for block-based editors, and introduced a full-scale block-level debugger front-end offering the entire range of operations commonly met in source-level debuggers. We implemented the debugger backend by introducing source code decorators that insert invocations of the debugger service loop. This way, our implementation is made portable independently of the browser and its underlying language technology (virtual machine).

References

1. Resnick, M., Maloney, J., Monroy-Hernández, A., Rusk, N., Eastmond, E., Brennan, K., Millner, A., Rosenbaum, E., Silver, J., Silverman, B., Kafai, Y.: Scratch: programming for all. Commun. ACM **52**(11), 60–67 (2009)
2. Ousterhout, J.: Scripting: higher-level programming for the 21st century. IEEE Comput. **31** (3), 23–30 (1998)

3. Pasternak, E., Fenichel, R., Marshall, A.N.: Tips for creating a block language with blockly. In: IEEE Blocks and Beyond Workshop (B&B), Raleigh, NC, USA, pp. 21–24 (2017)
4. Maloney, J., Resnick, M., Rusk, N., Silverman, B., Eastmond, E.: The scratch programming language and environment. ACM Trans. Comput. Educ. **10**(4), 1–16 (2010). Article 16
5. Tissenbaum, M., Sheldon, J., Abelson, H.: From computational thinking to computational action. Commun. ACM **62**(3), 34–36 (2019)
6. Ball, T., Burckhardt, S., Halleux, J., Moskal, M., Protzenko, J., Tillmann, N.: Beyond open source: the touch develop cloud-based integrated development environment. In: Proceedings of MOBILESoft, 2nd ACM International Conference on Mobile Software Engineering and Systems, Florence, Italy, pp 83–93, 16–17 May 2015

Adequacy of Game Scenarios for an Object with Playware Technology to Promote Emotion Recognition in Children with Autism Spectrum Disorder

Vinicius Silva[1(✉)], Filomena Soares[1,2], João Sena Esteves[1,2], Ana Paula Pereira[3], and Demétrio Matos[4]

[1] Algoritmi Research Centre, University of Minho, Guimarães, Portugal
a65312@alunos.uminho.pt, {fsoares,sena}@dei.uminho.pt
[2] Department of Industrial Electronics, University of Minho, Guimarães, Portugal
[3] Research Centre on Education, Institute of Education, University of Minho, Braga, Portugal
appereira@ie.uminho.pt
[4] ID+ – IPCA-ESD, Campus do IPCA, Barcelos, Portugal
dmatos@ipca.pt

Abstract. Every human interaction is emphasized by the participants' emotional states and manifestations. Being unable to understand the emotional state of the others is a significant deficit for anyone. This is the situation children with Autism Spectrum Disorder (ASD) usually find themselves in. Thus, researchers are exploring different technologies, such as Objects based on Playware Technology and social robots, in order to improve the emotion recognition skills of individuals with ASD. Following this trend, an improved Object of Playware Technology to be used in a storytelling activity as an add-on to the human-robot interaction with children with ASD, is proposed in the present work. Additionally, an evaluation of the images developed for the game scenario was conducted through an online questionnaire. Overall, the storytelling scenarios were well interpreted by the participants, matching a mean accuracy of 90%.

Keywords: Human-robot interaction · Playware · Tangible interfaces

1 Introduction

Individuals with Autism Spectrum Disorder (ASD) are characterized as having repetitive patterns of behaviour, restricted activities or interests, and impairments in social communication. Therefore, they have difficulties to recognize and understand emotional states in themselves and in others. In order to mitigate these impairments, recently, researchers have been investigating different technological intervention tools such as social robots and Objects based on Playware Technology (OPT). These technologies have been used to promote emotion recognition with children with ASD.

T. Ahram et al. (Eds.): IHSED 2019, AISC 1026, pp. 293–298, 2020.
https://doi.org/10.1007/978-3-030-27928-8_44

Lund [1] proposed a new term “playware” that defines the combination of using intelligent hardware and software with the aim to produce new innovative play. These devices are expressed in different forms and shapes, such as tangible tiles [1], Lego-like building blocks [2], among others. Few works in the literature explores the use of objects with playware technology as an intervention tool with children with ASD. One example are tangible tiles [1], i.e., modular blocks that can be placed on the ground and on a wall where each contains multiple LEDs to display information to the user. A proposed game consisted in mixing the tiles in order to produce new colours. Another work [2], approaches the use of a Lego-like building blocks augmented with electronic modules with the intend goal to be used as a therapy tool to improve social and cognitive skills in children with ASD. In both works, studies were conducted with children with ASD. In general, the authors concluded that the OPT can be a playful tool for cognitive challenge children.

Regarding social robots, research has demonstrated that robots can improve social behaviours in children with ASD [3, 4]. More recently, researchers have been using social robots with a humanoid design, since it can offer a special occasion and potential for generalization, especially in tasks of imitation and emotion recognition [5].

Following this trend, in a previous work, the research group developed an initial prototype of an OPT (PlayCube) that was evaluated with typically developing children [6] and with children with ASD [7] in a school setting with an emotion recognition game. The present work proposes a second iteration of the OPT to be used in a storytelling activity in a triadic setup (child-robot-researcher). Additionally, the development and evaluation of the images for the game scenario (storytelling) to be displayed by the OPT are addressed in this paper. The activity consists in the robot telling a story and asking the participant to identify the robot emotional state in the story. Simultaneously, the OPT has a display where an image that illustrates the story scenario is shown.

In order to validate the game scenario (image and stories), an online questionnaire was conducted with a total of 138 participants (adults and typically developing children). In summary, this study allowed to conclude that, in general, the scenarios were well interpreted by the participants.

The present article is divided in four sections. The methods and methodology are presented in Sect. 2. Section 3 shows the results. Finally, Sect. 4 addresses the final remarks and future work.

2 Methods and Methodology

The following section starts by giving an overview of the proposed system, describing its different components – the humanoid robot and the OPT. Then, it is explained the storytelling activity. Finally, the evaluation methodology is addressed.

2.1 Proposed System

The proposed system is composed of a computer, the new OPT (PlayBrick), and the humanoid robot ZECA (Zeno Engaging Children with Autism), Fig. 1. The Zeno R50

RoboKind humanoid child-like robot ZECA (a Portuguese name that stands for Zeno Engaging Children with Autism) is a robotic platform that has 34 degrees of freedom: 4 are located in each arm, 6 in each leg, 11 in the head, and 1 in the waist. The robot is capable of expressing facial cues thanks to the servo motors mounted on its face and a special material, Frubber, which looks and feels like human skin, being a major feature that distinguishes Zeno R50 from other robots

Fig. 1. The experimental setup, starting from the left: the developed OPT PlayBrick, a computer, and the humanoid robot ZECA.

The developed OPT was designed to provide a tangible and adaptive experience, being easy and intuitive to manipulate through natural gestures (such as touch, tilt, rotation), with different sources of immediate feedback (haptic and visual). Therefore, the PlayBrick has a 5.0-inch touch screen, an Inertial Measurement Unit (IMU), a haptic driver with a Linear Resonant Actuator (LRA), and a LED RGB strip, Fig. 2.

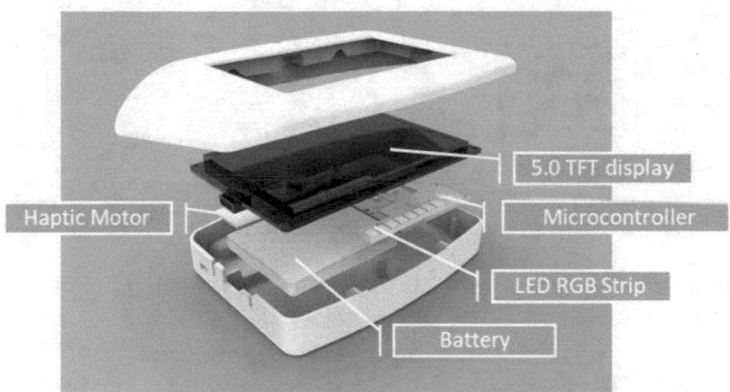

Fig. 2. The PlayBrick and its components.

2.2 Activity

The storytelling activity consists in ZECA randomly telling one of the fifteen available stories that are associated with an emotion and the child has to choose the correct facial expression matching the emotion. Simultaneously, as a visual cue, an image is shown representing the social context of the story. The child selects the answer by tilting the PlayBrick back or forward, scrolling through the facial expressions (common emoji) displayed by the OPT. In parallel, when the answer is selected, a positive or negative reinforcement is prompted by ZECA and the PlayBrick. The goal of this game scenario is to evaluate the affective state of a character at the end of a story.

2.3 Evaluation Methodology

Before conducting the tests with children with ASD, the experimental set-up and the proposed activity must be well-defined and tested. Included in this validation step, are the developed game scenarios of the story telling activity. So, in order to validate the game scenarios (images and stories), an online anonymized questionnaire was conducted with two target groups: typically developing children (7–11 years old) and adults (20–67 years old). The aim was to validate the new scenarios that were developed based on previous ones [5]. Figure 3 shows a visual comparison of a scenario (on the left the previous and on the right the new version) for the emotion surprise. The new version was developed taking into account the amount of stimuli, the colour palette, and the spatial location of the different components. For the target group, a simpler approach is more suitable. In the questionnaire, it was randomly presented the text and the image of each story from a total of 15 stories. The child had to read the story, observe the image, and select the emotion (anger, fear, happiness, sadness, and surprise) that matches the story. For example, Fig. 3 on the right shows the image of the following story: "Every day, I go to school. One day when I entered the classroom my colleagues screamed: Congratulations Zeca! They knew it was my birthday. I was so amazed." Prior to the study, and since the study involves working with children, it was established a protocol between the school and the research group.

Fig. 3. An example of the story game scenario Surprise. On the left: the previously used image. On the right: the new developed image (in Portuguese).

3 Results and Discussion

In the online questionnaire, it was tested a total of 15 scenarios, three for each emotion – happiness, sadness, anger, surprise, and afraid. Tables 1 and 2 addresses these results. A total of 138 participants – 69 typically developing children (56.5% female and 43.5% male) and 69 adults (69.6% female and 30.4% male) answered the questionnaire. In general, the average emotion matching accuracy was 91.8% for the adults and 89.9% for the typically developing children. Concerning the adult's results (Table 1), the scenario that had the lowest performance belongs to the sadness emotion with a matching accuracy of 81%. However, the matching accuracy for all stories stayed above 80%. Conversely, in the results obtained from the typically developing children (Table 2), two scenarios regarding the surprise emotion presented the lowest accuracy (60% and 69.3%).

Table 1. Adults' emotion matching accuracy in percentage (%) for the 15 stories for each emotion.

Story	Emotion				
	Anger	Fear	Happiness	Sadness	Surprise
Anger	**95.7**	0	0	4.3	0
	92.8	0	0	7.2	0
	98.6	0	1.4	0	0
Fear	0	**95.7**	0	1.4	2.9
	0	**97.1**	0	2.9	0
	0	**92.8**	0	4.3	2.9
Happiness	0	0	**100**	0	0
	5.8	0	**91.3**	2.9	0
	1.4	0	**97.1**	1.4	0
Sadness	1.4	15.9	0	**81.2**	1.4
	5.8	0	0	**94.2**	0
	10.1	7.2	0	**82.6**	0
Surprise	1.4	0	10.1	0	**88.4**
	0	0	14.5	1.4	**84.1**
	0	1.4	2.9	4.3	**91.3**

Table 2. Children's emotion matching accuracy in percentage (%) for the 15 stories for each emotion.

Story	Emotion				
	Anger	Fear	Happiness	Sadness	Surprise
Anger	**95.7**	0	0	4.3	0
	98.6	0	0	1.4	0
	98.6	0	1.4	0	1.4
Fear	0	**97.1**	0	2.9	0
	0	**97.1**	1.4	1.4	0
	0	**87**	0	13	0
Happiness	0	0	**97.1**	0	2.9
	2.9	0	**97.1**	0	0
	0	0	**100**	0	0
Sadness	1.4	4.3	0	**94.2**	0
	1.4	1.4	0	**97.1**	0
	13	2.9	0	**84.1**	0
Surprise	0	0	39.1	0	**60.9**
	0	0	33.3	0	**66.7**
	0	1.4	15.9	0	**82.6**

4 Final Remarks and Future Work

Individuals with ASD display social communication deficits such as difficulties in understanding facial expressions (and by association, emotional states) of others during interactions, leaving them at a disadvantage in social exchanges. In order to mitigate these difficulties, researchers are employing technological tools such as social robots and OPT. Therefore, the present work proposes an optimized OPT to be used in a storytelling activity in a triadic setup (child-robot-researcher) in a school setting. Additionally, the development and evaluation of the images for the game scenario (storytelling) to be displayed by the OPT were addressed in this paper.

By analysing the results, it is possible to perceive that in general, the scenarios were well interpreted by the participants, adults and typically developing children. Nevertheless, the two scenarios concerning the surprise emotion that presented the lowest accuracy (60% and 69.3%) in the children's opinion are going to be readjusted.

As future work, these scenarios are going to be tested with children with ASD in a school setting with the optimized OPT, PlayBrick. Moreover, a larger study will be conducted with the goal to evaluate how this hybrid approach (OPT and social robot) can be used as a valuable tool to promote emotion skills of children with ASD.

Acknowledgments. The authors would like to express their acknowledgments to COMPETE: POCI-01-0145-FEDER-007043 and FCT – Fundação para a Ciência e Tecnologia within the Project Scope: UID/CEC/00319/2019. Vinicius Silva also thanks FCT for the PhD scholarship SFRH/BD/SFRH/BD/133314/2017. The authors thank the teachers and students of the Elementary School of Gualtar (EB1/JI Gualtar) in Braga for the participation.

References

1. Lund, H.H., Dam Pedersen, M., Beck, R.: Modular robotic tiles: experiments for children with autism. Artif. Life Robot. **13**(2), 394–400 (2009)
2. Barajas, A.O., Al Osman, H., Shirmohammadi, S.: A serious game for children with autism spectrum disorder as a tool for play therapy. In: 2017 IEEE 5th International Conference on Serious Games and Applications for Health, SeGAH 2017 (2017)
3. Tapus, A., et al.: Children with autism social engagement in interaction with Nao, an imitative robot. Interact. Stud. **13**, 315–347 (2012)
4. Kim, E., Paul, R., Shic, F., Scassellati, B.: Bridging the research gap: making HRI useful to individuals with autism. J. Hum.-Robot Interact. **1**, 26–54 (2012)
5. Costa, S.: Affective Robotics for Socio-Emotional Skills Development in Children with Autism Spectrum Disorders. University of Minho (2014)
6. Silva, V., Soares, F., Esteves, J.S., Pereira, A.P.: PlayCube: designing a tangible playware module for human-robot interaction. In: Advances in Intelligent Systems and Computing (2019)
7. Silva, V., Soares, F., Esteves, J.S., Pereira, A.P.: Building a hybrid approach for a game scenario using a tangible interface in human robot interaction. In: Göbel, S. et al. (eds) Serious Games. JCSG 2018. LNCS. Springer, Cham (2018)

Priority Order of Single Gaze Gestures in Eye Control System

Yating Zhang[1], Yafeng Niu[1](✉), Chengqi Xue[1], Yi Xie[2], Bingzheng Shi[3], Bo Li[2], and Lingcun Qiu[3]

[1] School of Mechanical Engineering, Southeast University, Nanjing 211189, China
{zyt, nyf, ipd_xcq}@seu.edu.cn

[2] Science and Technology on Electro-Optic Control Laboratory, Luoyang 471023, China
eoei@vip.sing.com

[3] Shanghai Academy of Spaceflight Technology, Shanghai 201109, China
shibz87@126.com, qlcun@163.com

Abstract. The eye-control system uses eye movements to achieve human-computer dialogue. This paper designs an ergonomic experiment to solve the problems of high complexity and low interaction efficiency of Complex Gaze Gesture (CGG) in the eye-control system. The experiment concludes the order of eye Single Gaze Gesture (SGG) priority by exploring the ergonomic differences between the eye movements. According to the conclusion of the priority order of SGG, it provides a scientific theoretical basis for the design of CGG.

Keywords: Eye-control system · Eye movements · Single Gaze Gesture (SGG) · Complex Gaze Gesture (CGG) · Priority order

1 Introduction

In recent years, with the study of human-computer interaction has gradually been favorably received. For example, eye-movement interaction, speech recognition, gesture input, sensory feedback and so on. Interaction based on eye tracking has attracted much attention in the field of human-computer interaction [1]. Since vision is the gradual development and progress of human beings in the long-term natural evolution process, people can flexibly turn their eyes to gaze at different objects without special training [2]. Therefore, the gaze-based interaction will improve the naturalness and intuitiveness of human-machine dialogue. But so far, gaze-based interaction has not been widely used in our daily life. It is usually due to the expensive and low precision eye control equipment. The most typical one is the "Midas-touch" problem, that is, the eye control system cannot distinguish whether the user gazes the interface elements for interaction or for information. Some Scholars have proposed evasion methods for "Midas-touch" from confirmation mechanism, location calibration algorithm, selection probability model, multi-channel. But these methods require high hardware and algorithm, and bring about new problems such as visual fatigue, poor interaction experience and so on. Hyrskykari (2012) proposed a way of using saccadic behavior to

T. Ahram et al. (Eds.): IHSED 2019, AISC 1026, pp. 299–305, 2020.
https://doi.org/10.1007/978-3-030-27928-8_45

solve the low precision of the eye tracking device. Users are required to use saccadic to complete the interaction, which relies on the relative position relationship between the interface elements. It is defined as gaze gesture [3].

For gaze gesture, the number of saccades included in eye movements is defined as its length. According to the length, it can be divided into SGG and CGG [4] (Fig. 1 shows the SGG and CGG). SGG contains one saccade, involving two objects. CGG contains multiple saccades, involving more than two objects [5]. There are 12 SGG movements as shown in Fig. 2. CGG can be seen as the result of the combination of multiple SGG. Obviously, the CGG saccade process is more complex than the SGG, which will bring users more cognitive difficulties and physiological burden. Because it is difficult for users to remember the relationship between eye movements and input instructions, and it is also difficult to complete too many eye movements at one time. For SGG, although it is easy to learn and has low cognitive requirements for users, it can contain limited input instructions.

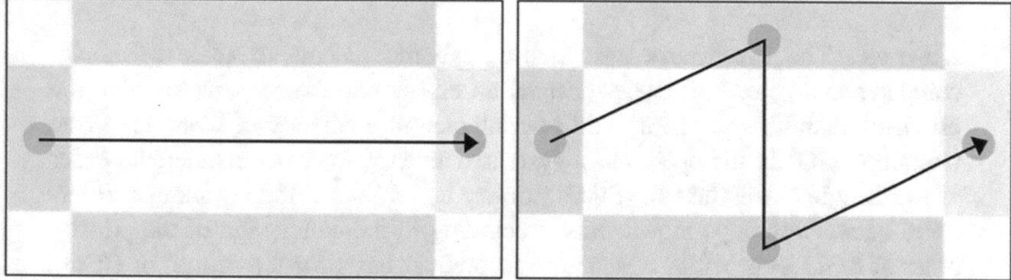

Fig. 1. Single Gaze Gesture and Complex Gaze Gesture

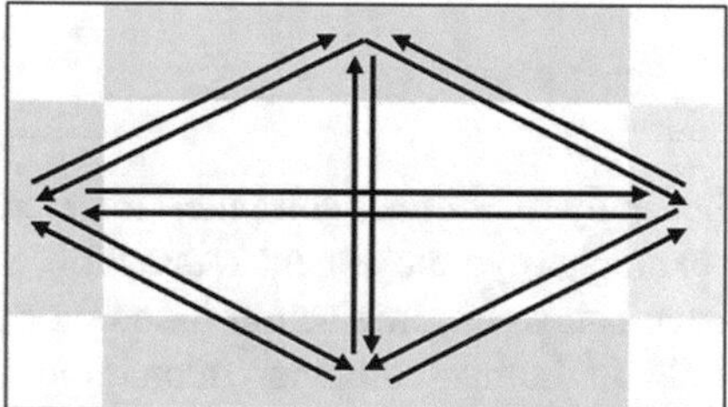

Fig. 2. SGG movements

Compared to gaze, the gaze gesture has many advantages. The duration of the saccade is significantly less than the gaze, so the interaction is faster than the gaze [6]; And the gaze can't achieve accurate positioning, not like the mouse. While the gaze gesture is based on the relative position relationship between two objects, which doesn't require high spatial accuracy. In addition, the gaze gesture has some drawbacks. Firstly, the gaze is a direct and simple way. The gaze gesture is not direct enough, which requires a lot of practice for users to master; secondly, the cognitive load of gaze interaction is low, and it is often used in a way of "what you see is what you get". The gaze gesture requires users to remember the relationship between eye movements and interaction instructions, which requires users to have a higher cognitive load.

Therefore, aiming at the difficult memory and high complexity of CGG, this paper intends to explore the ergonomic differences between different SGG movements, and obtains the priority order of SGG movements to determine the CGG. It provides a scientific theoretical basis for the design of CGG movements in eye control system, which can improve the interaction efficiency of eye control system

2 Paper Preparation

2.1 Purpose

In this paper, ergonomics experiments are designed to explore SGG. The purpose is to determine the priority order of SGG and provide a scientific theoretical basis for CGG design.

2.2 Participants

Twenty participants (12 males) with the age range of 22–28 years (M = 25.207; SD = 1.399) were recruited for the study. All participants had normal or corrected-to-normal visions, without astigmatism, or any other eye diseases. Meanwhile, all were free from any history of neurological or psychiatric disorders. Before the experiment, they all signed written consent forms voluntarily to participate in after fully informed the operations.

2.3 Apparatus and Definition of SGG Movement

The eye movement was recorded by a TobiiX2-30 eye- tracker system with an accuracy of 0.32° and a sampling rate of 30 Hz. The eye-tracker was composed of a recording laptop, a 24-in. LCD monitor (1920 × 1080 pixels, 60 Hz). A calibration procedure was performed to control the error of eye position within 0.5° at the begin of the experiment.

Eye movement is defined as eye movement 1 from left to right, eye movement 2 from left to top, eye movement 3 from left to bottom, eye movement 4 from right to left, eye movement 5 from right to top, eye movement 6 from right to bottom, eye movement 7 from top to left, eye movement 8 from top to right, eye movement 9 from bottom to left, eye movement 10 from bottom to right, and eye movement 10 from bottom to right. From bottom to right is eye movement 11, and from bottom to top is eye movement 12.

2.4 Procedure

Before the beginning of the experiment, in order to improve the efficiency and correctness of eye control interaction, it is particularly important to determine the appropriate experimental stimuli. According to the minimum size threshold (121 pixel)

[7] of eye-control interaction elements and the icon standard of Windows system, the size of eye-control interaction component was determined to be 256 pixel [8].

At the beginning of the experiment, the computer screen showed the experimental instructions. After reading, the subjects entered the experimental stage by pressing any key. There were 12 eye movements in the experiment. Each eye movement repeated five times, so there were 60 interactive tasks and 12 groups of experimental modules. In each module, the eye movement prompt interface first appears, which is used to display the eye movement interaction task. In the abstract material experiment, Azizizian et al. [9] set the presentation time of stimuli to 500 ms. In view of the need to recall the eye movements in the experiment, the presentation time of stimuli was set to 1000 ms and the stimulus interval was 500 ms to eliminate visual residues (Fig. 3 shows the experimental sequence). The whole experimental flow is about 20 min.

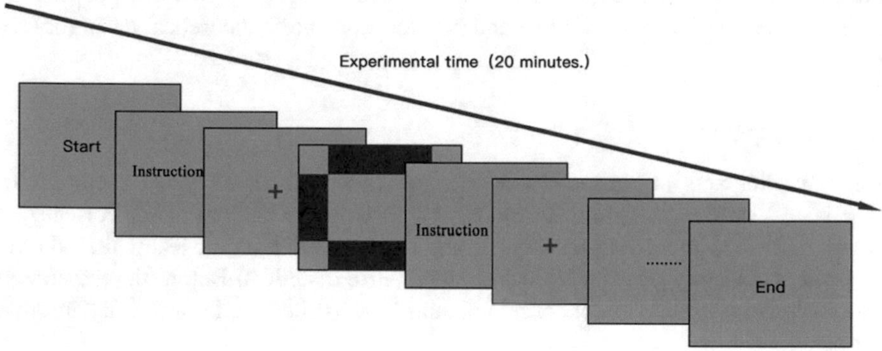

Fig. 3. Experiment sequence

3 Result

According to the statistics of experimental data, 1189 valid samples were collected. The data showed that the average saccadic speed of each SGG movement was quite different. The saccadic speed dispersion degree is small, and the accuracy rate is more than 99.5%. This shows that the subjects have maintained a high accuracy rate in 12 groups of experimental modules. Because the small difference in accuracy between the 12 groups of experiments, the analysis is no longer carried out.

The priority division of SGG movements is shown in Fig. 4. The above twelve eye movements can be divided into four priorities. The first priority saccadic speed is about 2000 pixel/s, including eye movement 1 and 4; the second priority saccadic speed is about 1100 pixel/s, including eye movement 7, 8, 10 and 11; the third priority saccadic speed is about 500 pixel/s, including eye movement 2, 5; and the fourth priority saccadic speed is about 400 pixel/s, including eye movement 3, 6, 9 and 12. The analysis of variance shows that eye movements have a significant effect on the average saccadic time, which indicates that different eye movements can be used as an effective index to classify the priority order.

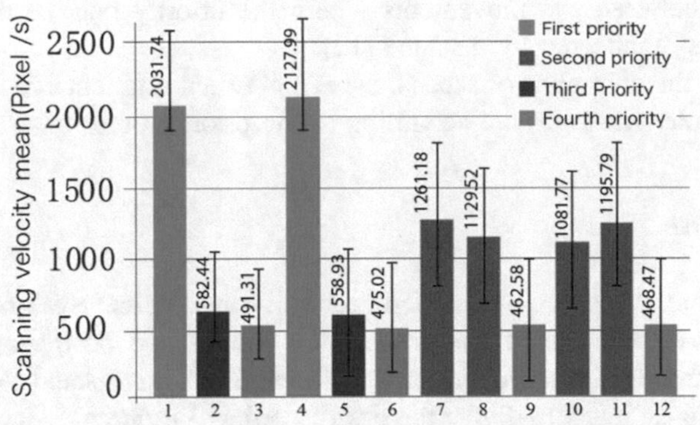

Fig. 4. SGG movements priority division

In order to further explore the differences of SGG movements within priority, variance analysis was carried out. The results showed that in the first priority, SGG movement had no effect on the average saccadic speed ($F = 1.114$, $P = 0.322 > 0.05$), In the second ($F = 2.743$, $P = 0.127 > 0.05$), the third ($F = 0.834$, $P = 0.494 > 0.05$), and the fourth ($F = 1.688$, $P = 0.230 > 0.05$), SGG movements had no significant effect on the average saccadic speed. For the analysis of variance between priorities, there are significant differences between 1–2 priorities, 2–3 priorities and 3–4 priorities ($p < 0.01$). Therefore, there is a significant difference between 1, 2, 3 and 4 priorities in the speed. Therefore, the above priority division of the twelve eye movements is reasonable.

4 Discussion

The purpose of this study is to prioritize SGG movements by comparing the differences of saccadic speed. The data analysis is used to compare the priority order of different SGG. When the SGG is horizontal, the saccadic speed is significantly better than the vertical direction. This finding is consistent with the findings of Mollenbach (2010). However, there are some rigorous places in this experiment. The experiment is carried out on a rectangular screen. In order to control the horizontal and vertical distances, the size of the starting area of the eye in these two directions is not uniform, and there is no good control variable [10]. In our paper, the experimental variables are strictly controlled to eliminate the serious difference in reaction time caused by the unequal distance. The saccadic speed (the saccadic time divided by the saccadic distance) is used as an effective index to divide the priority order of SGG. This method is very objective and accurate in evaluating the scientific nature of eye movement prioritization. In addition, this paper separates the experimental task interface and the description interface, thus ensuring the naturalness of gaze gesture interaction and following the principle of eye movement interaction design [11]. There are significant differences in the saccadic speed of different SGG movements, which indicates that the interaction efficiency of different SGG varies greatly. The reason why there is no difference in saccadic speed between eye movements within the priority may be that the time of saccade is almost unaffected by distance [12].

Considering the difference of saccadic speed between different eye movements, we can design complex gaze gesture according to the priority order of eye movements.

5 Conclusion

This paper explores the ergonomic differences between different SGG movements, and makes quantitative analysis of eye movements within and between priorities. The results showed that the saccadic speed of different SGG was related to their position. Saccadic speed as an index to divide the priority order of different SGG, can reflect the spatial position advantage of target recognition.

Of course, it is not enough for SGG to consider only speed and efficiency. In the next study, we will start from the functional semantics of interaction mode, and further explore the relationship between eye movement and interaction function.

Acknowledgments. This work was supported jointly by National Natural Science Foundation of China (No. 71801037, 71871056, 71471037), Science and Technology on Electro-optic Control Laboratory and Aerospace Science Foundation of China (No. 20165169017), SAST Foundation of China (SAST No. 2016010), Equipment Pre research & Ministry of education of China Joint fund, Fundamental Research Funds for the Central Universities of China (No. 2242019k1G023).

References

1. Chatzidaki, E., Xenos, M.: A case study on learning through natural ways of interaction. In: Global Engineering Education Conference. IEEE (2017)
2. Xuebai, Z., Xiaolong, L., Shyan-Ming, Y., et al.: Eye tracking based control system for natural human-computer interaction. Comput. Intel. Neurosci. **2017**, 1–9 (2017)
3. Hyrskykari, A., Istance, H., Vickers., S.: Gaze gestures or dwell-based interaction? In: ETRA, pp. 229–232 (2012)
4. Dybdal, M.L., Agustin, J.S., Hansen, J.P.: Gaze input for mobile devices by dwell and gestures. In: ETRA, pp. 225–228 (2012)
5. Drewes, H., Schmidt, A.: Interacting with the computer using gaze gestures. In: IFIP TC 13 International Conference on Human-Computer Interaction (2007)
6. Rozado, D., Agustin, J.S., Rodriguez, F.B., Varona, P.: Gliding and saccadic gaze gesture recognition in real time. TIIS **1**, 1–27 (2012)
7. Chengzhi, F., Mowei, S.: Eye tracking technology and its application in human-computer interaction. J. Zhejiang Univ. **29**, 225–232 (2002). (in chinese)
8. Windows Dev Center. https://docs.microsoft.com/en-us/windows/uwp/design/style/app-icons-and-logos
9. Azizian, A., Freitas, A.L., Watson, T.D., et al.: Electro-physiological correlates of categorization: P300 amplitude as index of target similarity. Biol. Psychol. **71**, 278–288 (2006)
10. Møllenbach, E., Lillholm, M., Gail, A.G., Hansen, J.P.: Single gaze gestures. In: ETRA, pp. 177–180 (2010)

11. Seffah, A., Taleb, M.: Tracing the evolution of HCI patterns as an interaction design tool. Innov. Syst. Softw. Eng. **8**(2), 93–109 (2012)
12. Heikkilä, H., Räihä, K.J.: Speed and accuracy of gaze gestures. J. Eye. Mov. Res. **3**, 1–14 (2010)

An Approach of Supporting Access to Educational Graphic of the Blind Students Using Sound and Speech

Dariusz Mikulowski(✉) and Andrzej Salamonczyk

Faculty of Science, Siedlce University of Natural Sciences and Humanities, 3 Maja. 54, 08-110 Siedlce, Poland
{dariusz.mikulowski, andrzej.salamonczyk}@ii.uph.edu.pl

Abstract. The access to graphical educational materials is very difficult issue particularly for the blind students. To solve this disadvantage, different methods such as braille printed drawings or swell paper are used. There are often expensive and time-consuming. As an alternative approach we offer method of audio and speech synthesis access to education graphics. In our solution appropriate sound signals and the synthetic speech descriptions are generated from graphic files saved in SVG format. The proposed method was implemented as a simple web application. It allows the teacher to prepare audio-labelled maps of countries and students to explore such maps. The student can listen to sound signals and descriptions representing object such as rivers, towns, when he/she slights his finger across the screen. The method was tested by few blind students. Tests have shown that the method makes the educational materials possible to prepare without using an expensive equipment.

Keywords: Sound access to graphic · SVG graphic · Sound and speech description · Teaching blind students

1 Introduction

The blind people in the similar way as other users are able to use devices with touch screens. They can do this using screen readers that are available on devices with the Windows, IOS and Android systems. These screen readers are programs working in background that reads all text information from the screen and describes small graphic symbols such as icons using text labels related to them. But the problem of easy access to graphical educational materials for the blind has not been sufficiently solved yet. To neutralize this disadvantage, different methods can be used i.e., braille printed drawings, convex graphics on swell paper [6] etc. There are, however, often expensive and time-consuming methods.

As an alternative approach to solve this problem we offer audio and speech synthesis access to education graphics. This method can be implemented as web application running on a laptop with an ordinary touch screen. In this method appropriate sound signals and the synthetic speech messages are generated from graphic files saved in universal SVG format. Such SVG file is the ordinary vector graphic that are

T. Ahram et al. (Eds.): IHSED 2019, AISC 1026, pp. 306–311, 2020.
https://doi.org/10.1007/978-3-030-27928-8_46

equipped with only small text descriptions and other additional elements. Such audio-labelled graphic can be interactively explored by blind student with his finger slighting on the touch screen. The choice of SVG follows from the fact that it is very popular format, it is part of the HTML5 standard. Moreover, each element of SVG graphic such as lines, triangles, borders etc. can be easily modified in order to adapt them to specific needs of the user. Due to this user can enhanced information about graphics for accessibility purposes by grouping elements and assign a specific description to them using desc or title tags. More over to promote accessibility on the Web, W3C provides Web Content Accessibility Guidelines, WCAG 2.0, [1]. According to this recommendation SVG can be also supported by WAI-ARIA tags for description of elements or groups and also for description of structural relationships. For these reasons we use these facilities to ensure the accessibility of SVG graphics.

The proposed method was implemented as a simple web application, which offers the student a feature of exploring maps of the countries as a sound schema and speech descriptions. This approach gives also the teacher a possibility of preparing such audio-labelled maps by adding descriptions and selecting proper classes such as river, town border etc. to lines and figures placed on the drawing. The method was tested by a few blind students. The tests have shown that the method makes it possible to more rapidly prepare the educational materials. The advantage of that method is that such graphic can be prepared and used by blind students without the presence of expensive equipment, such as Braille printers or machines for heating of swell paper. It means that our approach can be treated as an additional method of access to graphic for the blind students or as low-expensive, simpler and faster method of access to this kind of information for them. The proposed method is described in few following sections of this paper.

2 Related Works

There are many solutions that facilitate access to graphics for the blind. First of all there are exist traditional methods of printing graphics as a tactile drawings using so called Braillon [2] or swell heating paper [3] which offers haptic perception to the fingers of the user. Preparation of tactile graphic materials with Braillon technique requires sticking convex elements made of cardboard on a paper and covering this model with plastic film. Then this covered model is heated in vacuum what causes coping it's shape as a convex drawing on the plastic film. The swell paper method is little easier. It requires special kind of paper, on which a standard 2D printout is printed. Then this paper is heating with special machine what causes that the appropriate convexities are created on this paper.

Another way is to use Braille embossers, which are used to create tactile graphics based on 2D drawings. Methods of design of tactile graphics for these printers are described in [4, 5]. The main disadvantage of these solutions is the fact that they are expensive techniques (a braille embossers and paper hitting machines are required) and does not provide the opportunity to adding long descriptions of the context of the drawing due to the lack of sufficient space, for example, for the braille subtitles.

However, many educational materials are created using such techniques, although they require a lot of time devoted to developing these materials.

Using computers software in the preparation of graphic is a significant facilitation. A big breakthrough in the possibilities of learning graphics on computers by blind people was provided by touch screens. Among the file formats used to learn graphics on computer screens, the SVG format seems to be the most useful.

Using SVG graphics in learning geographic information it is not a new idea, [6], but it is not very popular and it is still developed and investigated [7, 8]. Some barriers are described in [9] and they concern, for example, the lack of a text description for the blind or improper contrast for the visually impaired. Broke et al. [10], use idea similar to our solution by replacing braille with simple audio-tactile interaction in geographic maps and significantly improved efficiency and user satisfaction.

In spite of existing approach, many tailor-made solutions are still missing, such as didactic materials for geography lessons, and it is worth taking care of this subject and developing accessible geographic graphics using the advantages of SVG format and screen-based computer screens with screen readers.

3 Exploring Graphic Using Sound

We propose an alternative method for access to educational graphic based on using of standard devices with touch screens. Our idea is that the user who slights his finger across the screen receives sound feedback from the computer about details of the graphics that is currently displayed. It works in such way that user touching a variety of graphical objects can listen to adequate sound signals generated by computer or played as adequate sounds from audio file. In addition, after performing a tapping gesture on the object, user can listen to a description of this object spoken by speech synthesizer. Unfortunately, it is not so simple, because user may easy lose his orientation and then do not know in what point of the graphic he is currently touching. Let us notice that blind student who explores the graphic using only one finger must build complete shape of the investigated object in his imagination what is an additional brain operation. Another question is proper recognizing of distance relations between objects that is also difficult with using only one finger.

A solution of this problem may be appropriate optimization of the graphics for this method of viewing it. It means that the items in the figure should not be too small, they cannot be too close each other, the lines should not be too thin, and there should not be too many of them in the picture. Moreover, the graphic can be equipped with additional items such as an orientation triangle, placed in the upper right corner or with borders allowing to recognize whether the user's finger has not left the picture. Another facility supporting the navigation on the graphic may be the using multi-fingered gestures. For example, we can imagine the situation when user holds a first finger at any point, and then explores it's the nearest surroundings using another finger. Due to this feature, user is able to capture the distance dependences between objects.

The format that seems to be the most suitable for manipulating of objects and providing these essential elements to graphic is Scalable Vector Graphic SVG. Applying SVG format for this purpose gives also a possibility to implemented aural access to graphic in the form of web application that can also be step in direction to versatility of the proposed method, so in the next section we will discuss the details of our method.

3.1 Access to Graphic Using Sound

As mentioned above, to ensure a good accessibility of the SVG graphics, it is necessary to prepare it in the right manner. Fortunately, the SVG standard allows to add additional elements to graphic file as well as tags that do not appear in the image preview, but carry with them the additional required information. This possibility is widely used in our approach. Thus, we reach SVG graphic with the following elements.

We divide drawing into layers (groups) in such way that each layer contains information of different type. For example, one layer can store text descriptions of the objects in the picture, other may store cartographic grid of the map, even additional items such as borders or orientation triangle placed in the upper right corner of the drawing. This feature gives the ability to easily manipulate whole sets of elements. For example, in a simple way we can show cartographic grid by assigning the visible property of its layer or change the font to braille of all text labels by assigning proper font family to its layer. Another facility could be to add text descriptions to objects by adding a <desc> element to its SVG node. This description can be then read with speech synthesizer, after user will tap in to designated object. We can also manipulate with classes and attributes of different objects. As an example, we will show a SVG <path> element which is a line that represents of a river. Such element saved in SVG file has the following code:

```
<path stroke="blue" stroke-width="2.5" style="fill: none;
stroke-opacity: 1; cursor: pointer;" id="path11351" d="m
690.81955,221.77965 c -2.16836,2.08195 -4.05611,4.44888
.... z" class="river" mapElementType="river"
mapElementDesc="River Vistula"></path>
```

We can notice that except standard attributes and sub elements this element is additionally marked with a class called “river” and the attribute “desc”. Using these items, an application can play the corresponding sound when the user will touch this line and read description from the value of desc attribute with speech synthesizer.

The method described above we have implemented as a web application that allows the blind student to explore maps of countries and teachers to make such audio-labelled maps. In the next section we will describe shortly how this application works.

3.2 How It Works

The solution of exploring graphic using sounds and speech descriptions was implemented as a sample web application. It can work in two modes: edit mode or view mode. The edit mode allows to load a usual map from SVG file and then edit its additional properties in the way to make this map as designed to the needs of the blind user. It means that each object on the map can be labelled as one of available class such

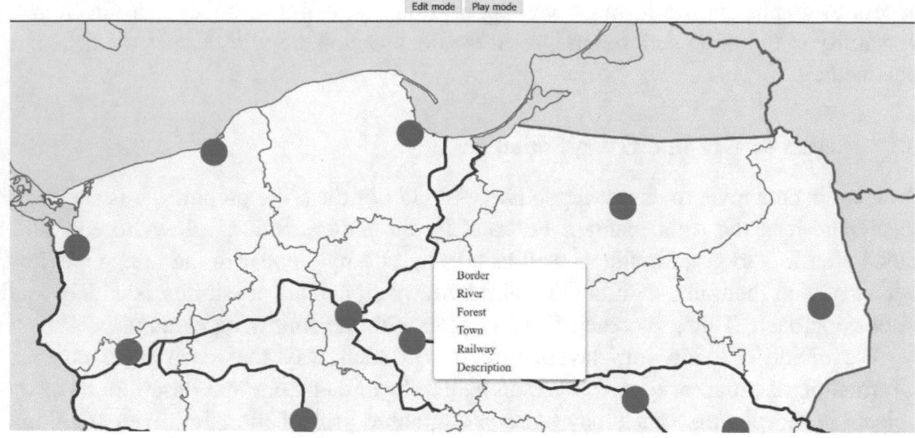

Fig. 1. Interactive map of Poland in edit mode

as: river, border, town, forest etc. Moreover, additional text description can be added to this object, i.e. object can be assigned as town and name of the town Warsaw can be added as a description. The application working in edit mode is presented on Fig. 1.

After labelling all objects map can be saved and application can be switched to view mode. A view mode is assigned for the blind student, who is able to "listen to" audio map during slighting his/her finger across the screen. It means that for example when user will touch the line representing the Vistula river, he listen to the sound of greasing water. In this moment, when he/she will tabs on screen, he will listen to text description related to selected objects.

4 Summary

In this paper we have presented an approach of making audio-described educational graphic for the blind students. As an example of this method the sample easy application allowing students to explore maps of countries was implemented.

Despite the fact that our method cannot completely replace the tactile graphics, but it can be treated as low-cost and easier alternative for expensive and time consuming widely used tactile graphic production methods. The approach is still developed, because i.e. not all objects on the drawing are recognized so quickly and sufficiently accurate by blind users as expected. For these reasons it is necessary to develop such rules regarding the size of objects, distances between them, and also add the use of multi-fingered gestures in order to ensure better navigation on the drawing. The method was tested and users have confirmed their desire to develop it in the future.

References

1. W3C: Web Content Accessibility Guidelines (WCAG) 2.0 (2008). https://www.w3.org/TR/WCAG20/. Accessed 05 Apr 2018
2. Gardner, J.A.: Tactile graphics: an overview and resource guide. Inf. Technol. Disabil. E-Journal **3**(4) (1996). http://itd.athenpro.org/volume3/number4/article2.html
3. Wijntjes, M.W.A., van Lienen, T., Verstijnen, I.M., Kappers, A.M.L.: The influence of picture size on recognition and exploratory behaviour in raised-line drawings. Perception **37** (4), 602–614 (2008)
4. The Blind Authority of North America. Guidelines and standards for tactile graphics (2011) https://www.prcvi.org/media/1125/guidelines-and-standards-for-tactile-graphics.pdf. Accessed May 2019
5. Gardiner, A., Perkins, C.: Best practice guidelines for the design, production and presentation of vacuum formed tactile maps. http://www.tactilebooks.org/tactileguidelines/page1.htm. Accessed May 2018
6. Brisaboa, N.R.: Improving accessibility of web-based GIS applications. In: Proceedings Sixteenth International Workshop on Database and Expert Systems Applications, pp. 490–494, August 2005
7. Poppinga, B.: Touchover Map: audio-tactile exploration of interactive maps, MobileHCI, Stockholm, Sweden, 30 August–2 September 2011
8. Bahram, S.: Multimodal eyes-free exploration of maps: TIKISI for maps. ACM SIGACCESS Accessibility Comput. **106**, 3–11 (2013)
9. Jiménez, T.C., Luján-Mora, S.: Web accessibility barriers in geographic maps. Int. J. Comput. Theory Eng. **8**(1), 80–87 (2016)
10. Brock, A.M., Truillet, P., Oriola, B., Picard, D., Jouffrais, C.: Interactivity improves usability of geographic maps for visually impaired people. Hum.-Comput. Interact. **30**(2), 156–194 (2015)

Inclusive Design for Recycling Facilities: Public Participation Equity for the Visually Impaired

Kin Wai Michael Siu(✉), Chi Hang Lo, and Yi Lin Wong

School of Design, The Hong Kong Polytechnic University, Hunghom, Kowloon, Hong Kong
{m.siu, sdpaullo, yi-lin.wong}@polyu.edu.hk

Abstract. Researchers and designers are increasingly advocating wider and fairer participation in recycling. However, many people are still excluded from this meaningful social participation. For example, visually impaired persons (VIPs) still face many barriers to participating in recycling independently. Since 2015, a series of studies has been conducted to explore the possibility of VIPs participating in recycling. Through a case study of designing recycling facilities for VIPs, this paper explores and discusses the difficulties and limitations that the visually impaired experience in participating in recycling, despite technological advancements that claim to help people with disabilities. It also identifies some directions for researchers and designers for future studies and professional practice.

Keywords: Equity · Fairness · Inclusive design · Public design · Recycling · Participation · Visually impaired persons (VIPs)

1 Introduction

Recycling has become a worldwide issue since human beings began to generate more waste than our planet can afford. According to Geyer, Jambeck and Law [5], in 2015 only around 9% of plastic waste ever manufactured was recycled, and 79% was disposed of in landfills or the natural environment. The recycling rate is low, but due to limited natural resources, it is imperative to promote recycling and waste reduction campaigns in communities. Wider participation in recycling is thus important, and approaches such as participatory action research have been developed rapidly to explore how it can be achieved [16]. Groups that are not included in recycling activities have drawn the attention of researchers and advocates. Children are one of these groups, and an increasing number of researchers are including children as subjects in research about sustainability activities and environmental issues (e.g., [1, 2, 8]). However, other groups such as visually impaired persons (VIPs) still face many barriers to participating in this meaningful social activity, and are excluded from active participation [13]. VIPs also face many difficulties in their daily life in accessing public events and facilities [12, 14–16]. Although different kinds of technology and city planning such as smart cities have been developed to help VIPs and other people with different levels of disability [10], no new inventions or devices have been introduced to

T. Ahram et al. (Eds.): IHSED 2019, AISC 1026, pp. 312–317, 2020.
https://doi.org/10.1007/978-3-030-27928-8_47

facilitate VIPs to participate in recycling activities as members of their communities. They are unable to contribute to the public environment by joining recycling activities.

Designers and governments in Western countries such as Canada, the U.K., and Australia have noticed this problem and have designed recycling bins that address the needs of VIPs. For instance, in the city of Calgary in Canada, a plaque with a raised letter "G" and its braille equivalent is attached to the lids of recycling bins [9]. The colours of the letter and the bin are in high contrast so that VIPs can easily recognize the bins. In the Derbyshire Dales in the U.K., designers found that the recycling bins were confusing for VIPs, and so they designed bins with braille on the top and two notches on the sides of the lid [7]. The local government of the Greater Geelong district in Australia provides tactile information on the lid of their recycling bins. A square tactile symbol is placed on bins for green waste, a triangle symbol on bins for recyclables, and a circle symbol on bins for garbage [3]. However, these bins are only provided on request. It is also unknown whether the bins really do help VIPs to distinguish between the different kinds of bins and encourage them to participate in recycling activities.

Since 2015, the Public Design Lab at the Hong Kong Polytechnic University has conducted a series of studies to explore the possibility of VIPs participating in recycling. It is found that involving VIPs in the design of recycling facilities encourages them to participate in recycling events and activities. VIPs are respected, and their opinions are valued in the process. This paper presents that design process, and discusses the importance and fairness of the participation of VIPs in recycling. The needs and preferences of the VIPs in recycling, and how new inclusive designs can fit their needs, are also addressed.

2 Method and Process

A case study of recycling facility design was conducted to investigate the difficulties and limitations of VIPs using recycling facilities. The key design research activity was participatory design workshops. Through the design process, several groups of VIPs introduced by non-governmental organizations (NGOs) were invited to provide design ideas, modify the designs, and test prototypes. A participatory design approach was adopted. According to Druin [4], participants in a participatory design process can play four roles: user, tester, informant, and design partner. The participating VIPs in this study took the roles of user and tester. Participatory design research is a recent design movement and research method in which designers design not for users but with users [11].

Researchers and designers who were specialists in designing inclusive facilities went through the design process with the VIPs and eventually designed a set of recycling bins. The set comprised a purple bin for general waste, a blue bin for paper recyclables, a yellow bin for aluminium recyclables, and a brown bin for plastic recyclables. The whole set was then brought to the VIPs to obtain their opinions and test the bins in use. Photographs were taken to record the test process. Semi-structured interviews were conducted after the tests to understand the usability of the bins and the VIPs' experiences of using the recycling facilities.

The findings of previous studies have shown that VIPs are often unable to identify different kinds of recycling bins and are unable to throw recyclables into the correct bins, despite the use of different colours. Therefore, the new recycling bins designed in this study had a new opening at the top of the lid, and a tactile pictogram and Braille were added to the edge of the lid so that VIPs were able to identify the bins by touching and did not need to open the lid to throw the recyclables inside.

After receiving comments and opinions from the VIPs after the tests and interviews, the researchers and designers were tasked with managing the data and optimizing the design. After three rounds of tests and interviews at three different stages of design using different test prototypes, the final set was produced. The test and interview data are organized and summarized in this paper.

3 Finding and Analysis

Figure 1 shows the testing of the size and shape of the bin openings. The recycling bins were designed to have openings of different shapes. The opening of the general waste bin (purple) was circular, that of the paper recyclables bin (blue) was a long oval, that of the aluminium recyclables bin (yellow) was square, and that of the plastic recyclables bin (brown) was hexagonal. The size of the openings was designed to match the maximum size of existing recyclables in each category. The VIPs were satisfied that the shape and size of the lids allowed them to distinguish the different bins by the shape of the opening.

Fig. 1. Testing prototype 1 with a VIP

Figure 2 shows the testing of the tactile information on the edge of the lid, and Fig. 3 shows the tactile pictograms and the Braille before modification. One of the VIPs commented that the tactile pictograms were too complicated and she could not understand the information; however, the Braille on the edge was clear. Figure 4 shows the final testing of the lid of the recycling bins. All of the VIPs who participated in the test were satisfied with the final design. The VIPs generally welcomed and were happy with the new design ideas.

The VIPs suggested that the main difficulty with using the recycling bins was that the bins were often covered in dust and dirt. They were reluctant to touch the lids and

Fig. 2. Testing prototype 2 with a VIP

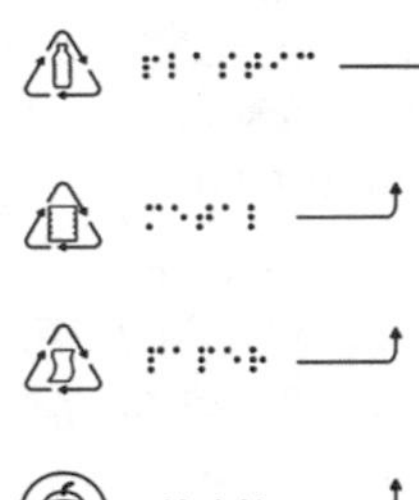

Fig. 3. Tactile pictograms and Braille on the lids of the recycling bins before modification

Fig. 4. Testing the final design with a VIP

open the bins because they were unable to see clearly. This discouraged them from using the facilities and thus hindered them from joining recycling activities. The poor hygienic condition of the facilities implied that other users of the facilities may not use them properly, and this was not adequately managed.

Some of the VIPs also expressed concerns that although the recycling bins were newly designed and made so that they could identify and recognize them easily, other facilities related to the recycling bins, such as a tactile guide path guiding VIPs to the recycling bins, were not well allocated or managed. Other pedestrian or street users may not understand the needs of VIPs, and thus may affect the usability and accessibility of recycling bins. These bins do not have fixed locations, and it is difficult for VIPs to find them if they are moved elsewhere. The VIPs also reported that Hong Kong people seldom offer help to them on the street. They have to be very independent, and engaging in recycling activities is just not an option for them when they have so many other issues to tackle in their daily lives.

The findings of the study show that the problem of deigning inclusive recycling facilities involves not only technical aspects, but also social, cultural and environmental aspects. It is important to address all of these aspects, because providing the proper facilities is not enough to encourage VIPs to engage in recycling. Other considerations need to be addressed so that they can participate in recycling independently and with dignity. Equity in public participation is the key to building a harmonious society.

Several suggestions to help VIPs be active in joining recycling activities in society are summarized below.

- In addition to a high contrast in colour and size, designs that require as little contact as possible with the facilities are favoured by VIPs.
- Standardization of design is essential for VIPs to become familiar with recycling facilities.
- The location of facilities should be fixed so that VIPs can locate them easily and independently.
- Researchers and designers should consider the entire system of accessing recycling facilities when designing them for VIPs.
- Information must be disseminated to VIPs about new recycling policies and facilities for them and their communities.
- NGOs that help VIPs and governments should work together to develop policies that address the needs of VIPs.
- More VIPs should be invited to join the design process for recycling facilities so that they are able to provide useful and constructive comments and opinions on the design.
- Education is required to inform other users of how to use recycling facilities properly and of the needs of the VIPs in this regard.

4 Conclusions

Through a case study of the design of a set of recycling bins for VIPs, this paper identifies several key limitations and difficulties that VIPs encounter when participating in recycling activities. The findings of the interviews with VIPs showed that apart from developing an inclusive design for recycling facilities, the corresponding social, cultural, and environmental coordination is needed to encourage VIPs to recycle. It is necessary to ease VIPs' worries about hygiene and management issues and educate other recycling users to be inclusive behaviourally and psychologically. This paper also identifies some design directions and approaches for researchers and designers that can be used as references for future studies and practical development. Designs that are generated through a participatory design process or other theory-based design process should be publicized and made visible and convenient for all users so that recycling rates increase [6]. Collaboration and coordination among different stakeholders as well as researchers and professionals from different disciplines (i.e., an interdisciplinary approach) are essential to cultivating an equitable and inclusive society that promotes public participation.

Acknowledgments. The authors thank The Hong Kong Polytechnic University (PolyU) for its research fund support for the study. The authors also acknowledge partial support from the Hong Kong Research Grants Council's Humanities and Social Sciences Prestigious Fellowship Scheme (RGC 35000316) during the data collection and the preparation of this paper. The

Eric C. Yim Endowed Professorship provided financial support for the data analysis. The authors also thank the visually impaired participants from the Hong Kong Blind Union, PolyU, and other NGOs and social service organizations for the visually impaired.

References

1. Borg, F., Winberg, M., Vinterek, M.: Children's learning for a sustainable society: influences from home and preschool. Educ. Inq. **8**(2), 151–172 (2017)
2. Casaló, L.V., Escario, J.J.: Intergenerational association of environmental concern: evidence of parents' and children's concern. J. Environ. Psychol. **48**, 65–74 (2016)
3. City of Greater Geelong (2017). http://www.geelongaustralia.com.au/bins/news/item/8d53d97920f52ef.aspx
4. Druin, A.: The role of children in the design of new technology. Behav. Inf. Technol. **21**(1), 1–25 (2002)
5. Geyer, R., Jambeck, J.R., Law, K.L.: Production, use, and fate of all plastics ever made. Sci. Adv. **3**(7), e1700782 (2017)
6. Largo-Wight, E., Johnston, D.D., Wight, J.: The efficacy of a theory-based, participatory recycling intervention on a college campus. J. Environ. Health **76**(4), 26–31 (2013)
7. Letsrecycle (2018). https://www.letsrecycle.com/news/latest-news/mgb-waste-systems-produces-bin-lid-for-the-blind/
8. Matthies, E., Selge, S., Klöckner, C.A.: The role of parental behaviour for the development of behaviour specific environmental norms – the example of recycling and re-use behaviour. J. Environ. Psychol. **32**, 277–284 (2012)
9. Peterson, K.: (2014). https://globalnews.ca/news/1689544/braille-to-help-visually-impaired-calgarians-with-recycling/
10. Ramirez, A.R., González-Carrasco, I., Jasper, G.H., Lopez, A.L., Lopez-Cuadrado, J.L., Garcia-Crespo, A.: Towards human smart cities: internet of things for sensory impaired individuals. Computing **99**(1), 107–126 (2017)
11. Sanders, E.B.-N: From user-centered to participatory design approaches. In: Frascara, J. (ed.) Design and the Social Science, pp. 18–25. CRC Press, London (2002)
12. Siu, K.W.M.: Accessible park environments and facilities for the visually impaired. Facilities **31**(13/14), 590–609 (2013)
13. Siu, K.W.M.: Innovation for diversity and fairness: inclusive design of recycling facilities for visually impaired people. Int. J. Community Divers. **12**(1), 51–65 (2013)
14. Siu, K.W.M.: Open space for the visually impaired: open or exclude? Int. J. Civic Polit. Community Stud. **10**(1), 7–21 (2013)
15. Siu, K.W.M., Wong, M.M.Y.: Promotion of a healthy public living environment: participatory design of public toilets with visually impaired persons. Public Health **127**(7), 629–636 (2013)
16. Siu, K.W.M., Xiao, J.X.: Public Facility Design for Sustainability: Participatory Action Research on Household Recycling in Hong Kong. Action Research, 30 March 2017. https://doi.org/10.1177/1476750317698027

Secure Visualization When Using Mobile Applications for Dementia Scenarios

Joana Muchagata[1(✉)], Pedro Vieira-Marques[1], Soraia Teles[2], Diogo Abrantes[1], and Ana Ferreira[1]

[1] CINTESIS - Center for Health Technology and Services Research, Faculty of Medicine, University of Porto, Porto, Portugal
{joanamuchagata, pmarques, djm, amlaf}@med.up.pt

[2] CINTESIS - Center for Health Technologies and Services Research, ICBAS, University of Porto, Porto, Portugal
stsousa@icbas.up.pt

Abstract. Wandering is a common behavior during all stages of Alzheimer's Disease (AD), but it can be distressing to both people with dementia and their caregivers. Ambient Assisted Living (AAL) solutions have been increasingly relevant to promote ageing in place, targeting both autonomous and dependent older adults. Despite the potentials of AAL solutions, their pervasiveness can raise several security and privacy challenges. With mobile mockups, this work presents a use-case with SoTRAACE - Socio-Technical Risk-Adaptable Access Control model, applied to the prevention of getting lost in AD. SoTRAACE functionalities are illustrated as embedded in an alarm navigation system aimed at monitoring the person's geolocation and alerting when a predefined safety perimeter is crossed. The model, applied to a healthcare scenario, assesses security and privacy risks at each moment of interaction and provides an innovative way for improved adaptable security visualization in AAL environments.

Keywords: Ambient Assisted Living · Alzheimer's Disease · Privacy · Security and visualization · Access Control · SoTRAACE · Adaptable risk assessment

1 Introduction

The world is experiencing demographic changes and the number of elderly people is growing. One effect related to this is the increasing number of people with dementia [1]. As this ailment affects memory, thinking, behaviour and the ability to perform everyday activities [2, 3], it tends to decrease the quality of life of both the individuals with dementia and their caregivers [3, 4]. Wandering and getting lost is common among persons with dementia, and may be problematic to both those who live with it, and their caregivers, potentially causing physical harm and emotional distress [5].

Advances in mobile technology applied to Ambient Assisted Living (AAL) solutions, Global Positioning Systems (GPS) and tracking devices, are opening up new means of tracking, thus, potentially increasing the level of safety by locating a person that can be lost [4]. Studies [4, 6], reveal that using GPS to locate persons with

T. Ahram et al. (Eds.): IHSED 2019, AISC 1026, pp. 318–324, 2020.
https://doi.org/10.1007/978-3-030-27928-8_48

dementia increases their safety while maintaining their autonomy, helping their family and caregivers to become less anxious about unsafe wandering.

Despite the potentials of AAL solutions, being able to geolocalize people raises several security and privacy challenges. SoTRAACE (Socio-Technical Risk-Adaptable Access Control) is a model proposed by Moura et al. (2017) [7], which can better adapt users' access control needs to each AAL security context and related challenges.

With mobile mockups, this work presents a use-case with SoTRAACE applied to a healthcare scenario, based on a persona described as an informal caregiver of someone living with Alzheimer's disease, in order to prevent the patient from becoming lost when wandering. Through an alarm navigation system, the persona (a caregiver and/or family member) can monitor the geolocation of the person with dementia, updated to the exact place where the patient's phone is located (with the navigation aid system installed), and launching an alert when a predefined safety perimeter is crossed.

2 Background

Solutions based on real-time sensor readings such as the GPS SmartSole [8] or Smart Watches and GPS trackers [9], can be useful to detect wandering trajectories or to prevent people from crossing predefined safe zones. Mobile applications are also beneficial tools to improve the quality of life for people living with dementia. iWander is an application [10] that collects data from the device's sensors such as GPS, time of day, weather, dementia stage and user's feedback. These data are evaluated and, depending on the probability that the person may be wandering, iWander automatically takes actions, such as notify caregivers and emergency services of the person's current location.

Despite the potentials of AAL solutions, human-computer interactions can raise several security and privacy challenges. SoTRAACE (Socio-Technical Risk-Adaptable Access Control Model) proposes to overcome those challenges [7] (Fig. 1). It is based on the standard RBAC (Role-Based Access Control) [11] and takes into account contextual (e.g. work, home, public places), technological (e.g. type of device, network connection) and user's interaction profiling, performing a quantitative and qualitative risk assessment analysis to support a smart decision-making on the most secure, private and usable way to access and display information. SoTRAACE integrates the Adaptable Visualization Module (AVM) module to improve availability, security and privacy of visualized data [12], Break the Glass (BTG) [13] for access in emergency or unanticipated situations, and Delegation for temporary accesses to an unauthorized user, performed on behalf of an authorized user.

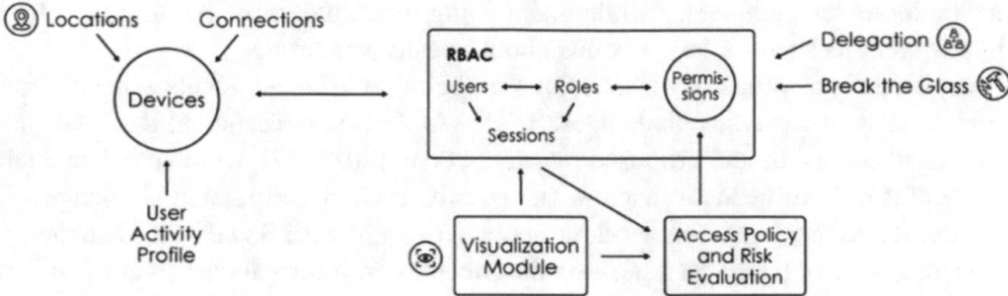

Fig. 1. SoTRAACE - Socio-Technical Risk-Adaptable Access Control Model [7].

3 Methods

In order to illustrate the application of the SoTRAACE model to an AAL solution, the process of defining working mockups starts with the definition of a persona (Sect. 4.1), who represents the main character of the created use-case (an informal caregiver of a person living with Alzheimer's Disease). SoTRAACE is then applied in two scenarios to test its adaptation to two different contexts and security risks (Sect. 4.2).

Scenario 1 focuses on the SoTRAACE application to a GPS locator device and mobile application, designed to address wandering behaviour. SoTRAACE entities/features illustrated in this scenario include 'Delegation'; 'Break the Glass' (BTG); contextual features analysis (type of connection, location and device); 'User Activity Profile' (UAP); privacy and security risk assessment (qualitative and quantitative) and 'Adaptable Visualization Module' (AVM).

Scenario 2 approaches the SoTRAACE application to a smart door lock and related mobile application, installed to provide a better awareness and control of who enters the care receiver's home. SoTRAACE entities/features illustrated in this scenario integrate the same as Scenario 1, but some these for completely different purposes such as on the most available and private way to display sensitive data regarding the door status (looked or unlocked), on the mobile screen.

4 Use-Case

4.1 Persona Description

Figure 2 describes a character who represents the target population (e.g., an informal caregiver), a persona that includes personality, demographic and other attributes that help better conceive what features and functionalities are required to provide usable, useful and quick assistance from a mobile application.

4.2 SoTRAACE Model Scenarios

4.2.1 Scenario 1: Navigational Aid Applied to Dementia

Mary receives an alert that her mother, named Rose, is outside the safety perimeter. Mary is at work, far from Rose's location. She accesses her application and uses

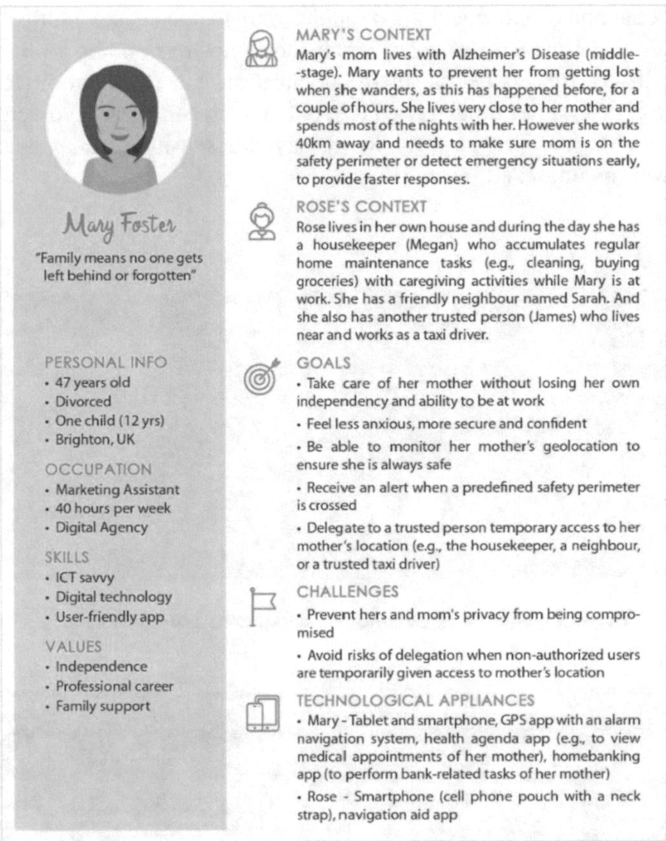

Fig. 2. Persona of the informal caregiver of a person living with AD.

SoTRAACE to delegate temporary access to Rose's location, to another person (e.g., the housekeeper, a neighbour, or a trusted taxi driver). Besides the monitorization of every access, SoTRAACE allows a Break the Glass (BTG) feature, which is a high-risk temporary access, that can be revoked once the emergency ends. Adding other contextual features such as Mary's type of connection, location and device, Mary and the delegatee's previous accesses and associated risks, as well as her mother's location, will allow for the calculation of the current privacy risk, which in turn, will help Mary decide on the best person to delegate that task, at that moment. Also, depending on the place Mary is accessing the application (e.g., work or restaurant), SoTRAACE can adapt visualization to improve both confidentiality (hiding unnecessary sensitive data) and availability (zooming mother's location) (Fig. 3).

4.2.2 Scenario 2: Smart Door Lock

Mary is trying to call her mother for more than three hours, but she is not responding her calls and the housekeeper is not working today.

Mary is concerned and checks her application to locate her mother, finding her inside the safety perimeter. However, since she is not responding to her calls, Mary accesses the list of trusted people in the application and calls Sarah (Rose's closest neighbour) to check on her mother. SoTRAACE assesses the risk involved to decide on the most available and private way to display the sensitive data regarding Rose's location as well as unlocked doors (Fig. 4).

Fig. 3. Scenario 1- Rose is outside the safety area and an alert is generated.

Fig. 4. Scenario 2 - Rose is inside the safety area but unreachable.

5 Discussion

One important aspect that hinders the full acceptance and use of AAL solutions is privacy and security concerns [14]. However, the authors could not find available solutions to assess risk for each interaction and adapt security decisions to context characteristics, as different situations represent different risk levels, as described in the presented use-case. SoTRAACE aims to provide an integrated access control decision solution with these features, adapt security visualization to each user's context and prevent sensitive data from being transmitted over unsecure channels, when this is not necessary. It also provides a variety of different features to adapt to the dynamics of each scenario requirements in terms of risk assessment, access control decision and secure visualization (e.g., BTG, delegation or other relationship, specific rules, etc.).

A limitation of the current work is that the presented mockups were not yet tested, but will be developed into functional ones, for this to happen quickly.

6 Conclusions

Presently, SoTRAACE model is an innovative proposal for an adaptable security visualization in AAL environments. Future work, includes the implementation and testing of the mockups (incorporating the main aspects and functionalities to cover the described persona and situations) with end users, to make them useful to the community.

Acknowledgments. This article was supported by FCT through the Project TagUBig - Taming Your Big Data (IF/00693/2015) from Researcher FCT Program funded by National Funds through FCT - Fundação para a Ciência e a Tecnologia. The author Soraia Teles holds a grant from the Portuguese Foundation for Science and Technology (FCT) (D/BD/135496/2018; PhD Program in Clinical and Health Services Research).

References

1. Alzheimer's Disease International: The Global Impact of Dementia: An analysis of prevalence, incidence, cost and trends. World Alzheimer Report 2015 (2015)
2. Alzheimer's Association: Alzheimer's & Dementia - What Is Dementia? (2018)
3. Alzheimer's Disease Education and Referral Center: Alzheimer's Disease Fact Sheet. NIH Publication (2016)
4. Oderud, T., Landmark, B., Eriksen, S., Fossberg, A.B., Aketun, S., Omland, M., Hem, K.G., Ostensen, E., Ausen, D.: Persons with dementia and their caregivers using GPS. Stud. Health Technol. Inform. **217**, 212–221 (2015)
5. Robinson, L., Hutchings, D., Dickinson, H., Corner, L., Beyer, F., Finch, T., Hughes, J., Vanoli, A., Ballard, C., Bond, J.: Effectiveness and acceptability of non-pharmacological interventions to reduce wandering in dementia: a systematic review. Int. J. Geriatr. Psychiatry: J. Psychiatry Late Life Allied Sci. **22**(1), 9–22 (2007)
6. Oderud, T., Landmark, B., Eriksen, S., Fossberg, A.B., Fossland Brørs, K., Mandal, T.B., Ausen, D.: Exploring the use of GPS for locating persons with dementia. Assistive Technol. Res. Ser. **33**, 776–783 (2013)
7. Moura, P., Fazendeiro, P., Vieira-Marques, P., Ferreira, A.: SoTRAACE - socio-technical risk-adaptable access control model. In: 2017 International Carnahan Conference on Security Technology (ICCST), Madrid (2017)
8. What Is GPS SmartSole? A GPS Tracker, Hidden In An Insole, Tracked. http://gpssmartsole.com/gpssmartsole
9. What Are the Top 10 Products to Best Track My Elderly Loved One's Location? https://www.caringvillage.com/product-review/top-10-products-track-elderly-location
10. Sposaro, F., Danielson, J., Tyson, G.: iWander: an Android application for dementia patients. In: 2010 Annual International Conference of the IEEE Engineering in Medicine and Biology, pp. 3875–3878 (2010)

11. Sandhu, R., Ferraiolo, D., Kuhn, D.: NIST model for role-based access control: towards a unified standard (2000)
12. Muchagata, J., Ferreira, A.: How can visualization affect security? In: ICEIS 2018 - 20th International Conference on Enterprise Information Systems. SCITEPRESS Digital Library, Poster Presentation in Funchal, Madeira - Portugal (2018)
13. Ferreira, A., Chadwick, D., Farinha, P., Correia, R., Zao, G., Chilro, R., Antunes, L.: How to securely break into RBAC: the BTG-RBAC model. In: 2009 Annual Computer Security Applications Conference, pp. 23–31 (2009)
14. Ferreira, A., Teles, S., Vieira-Marques, P.: SoTRAAce for smart security in ambient assisted living. J. Ambient Intell. Smart Environ. (2019)

Research on Color, Luminance and Line Width of HUD Symbols

Yitian Li[1], Haiyan Wang[1(✉)], Yafeng Niu[1,2], Yi Xie[1], Bingzheng Shi[3], and Ruoyu Hu[1]

[1] School of Mechanical Engineering, Southeast University, Nanjing 211189, China
742831345@qq.com, {whaiyan,nyf}@seu.edu.cn, eoei@vip.sina.com, 1848795747@qq.com
[2] Science and Technology on Electro-Optic Control Laboratory, Luoyang 471023, China
[3] Shanghai Academy of Spaceflight Technology, Zhichun Road, Shanghai 201109, China
shibz87@126.com

Abstract. Head-up displays (HUDs) can improve pilots' flight performance and efficiency. The pilot's cognitive performance is affected by the physical characteristics of HUD, such as character color, luminance, line width, and word height. Through experimental means, by taking contrast sensitivity as a measurement method, and the reaction time as the indicators. We made a comparison of the color and luminance of foreground characters in different sky background samples. The experiment showed that HUD has a better cognitive performance of the character color HSB value (120° 99% 33%) during the daytime, the modulation contrast between foreground luminance and background luminance should be greater than 0.5. The character height of HUD should be greater than 5.84 mrad, and the stroke line width should be greater than 0.73 mrad, which provides a scientific reference for the color design of the HUD interface.

Keywords: Head-up display · Modulation contrast · Contrast sensitivity · Luminance · Color design · Line width

1 Introduction

Due to the wide range of background luminance and the dynamic presentation of complex colors, effective express of the head-up interface information is becoming more difficult [1]. The physical properties of the display like color, salience, luminance, orientation, size, and clutter; the pilot's mission and expertise; and mission requirements determine the level of attention between the aircraft and external events [2]. Previous studies have only studied that the monochrome display characters of head-up are green, which meets the requirements of ergonomics for visual comfort [3]. Focusing on color and luminance, this paper studies the superior character line width and modulation contrast of the information in the superimposed field of view environment.

T. Ahram et al. (Eds.): IHSED 2019, AISC 1026, pp. 325–330, 2020.
https://doi.org/10.1007/978-3-030-27928-8_49

2 Basic Concept

In HSB color model, H (hues) is expressed as hue with a range of 0° to 360°; S (saturation) is expressed as saturation with a range of 0%–100%; B (brightness) is expressed as brightness with a range of 0%–100% [4].

Contrast sensitivity considers the luminance modulation contrast of spatial grating lines with the same width. When the color or luminance of the two lines is different, the modulation contrast of the two lines will change, which will affect the measurement results of contrast sensitivity [5]. For gratings, which are symmetrically higher and lower than the average brightness, the usual formula for calculating modulation contrast is defined as Formula (1). L_{max} and L_{min} represent the maximum and minimum luminance of the graph. The modulation contrast value is between 0 and 1, and the higher the modulation contrast, the higher the contrast sensitivity [6].

$$C = \frac{L_{max} - L_{min}}{L_{max} + L_{min}}. \tag{1}$$

3 Experiment

3.1 Experimental Variable

The object of this experiment is the luminance of HUD and background, and the purpose is to explore the appropriate color, line width, and word height of the HUD. According to the conclusion of the preliminary pre-experiment, the best green color with the HSB value (120° 99% 33%) is selected as the foreground color in the daytime [7]. The luminance value was 23.1 cd/m^2 measured by the CS-200 color luminance meter. Background, as a variable, needs to be selected according to luminance. Using software Photoshop to extract the most characteristic colors in the sky background image of the daytime as a pure color background [8]. The B (brightness) and S (saturation) are 33%, 66%, 99% respectively and the H (hue) are 0°, 30°, 180°, 210°, 240°, 270°. 54 background colors were obtained after combination (Fig. 1). Among the 54 colors measured by luminance meter, the value of HSB (240° 99% 33%) is the lowest, which is 3.1 cd/m^2, and the value of HSB (180° 33% 99%) is the highest, which is 333.7 cd/m^2. The luminance of the 54 colors is evenly distributed from high to low, ensuring the rationality of the selected color.

Therefore, the HSB value (120° 99% 33%) is chosen as the foreground green (Fig. 2), and the 54 colors are chosen as the background color representing the daytime environment. The independent variable is the modulation contrast between the foreground and the background luminance, and the dependent variable is the grating linewidth when the subjects responded, which could be calculated according to the reaction time of the subjects.

Fig. 1. 54 different color background images.

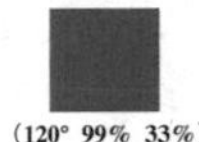

Fig. 2. The value of the selected green HSB is (120° 99% 33%).

3.2 Experimental Materials

The experiment is conducted to simulate the visual stimulus of pilots in the scene of flying missions, and to measure the performance of foreground characters by contrast sensitivity method [9]. Software Adobe After Effects (AE) is used in the experiment to make videos. The equal width grating lines of the two colors in the video are tapering (Fig. 3). The stripes become thinner, and in the extreme case, the two colors mix and become blurred. In order to avoid the data error caused by the repeated line width and image flicker, 54 videos are obtained by selecting 30° of grating lines.

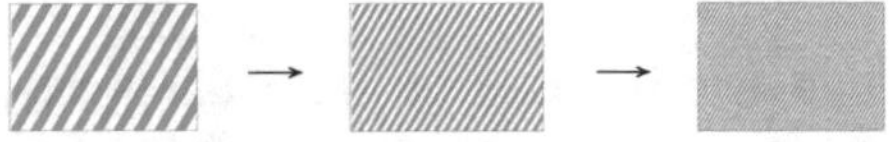

Fig. 3. Experimental video: The equal width grating lines of the two colors in the video are tapering.

3.3 Experimental Environment and Subjects

The experiment was programmed by E-prime2.0. The program automatically executed the random presentation of the experimental videos and recorded the experimental data. The experimental data were sorted out and summarized by Excel and were counted and analyzed by SPSS software.

The experiment was carried out on a Lenovo laptop with a resolution ratio of 1366*768. The subjects were 10 graduate students in a university, 5 female students and 5 male students, all of whom had not been exposed to similar experiments. They were between 22 and 27 years old, without color blindness and color weakness, and their corrected visual acuity was above 1.0.

3.4 Experimental Process

The reaction time of the subjects was measured according to the variables in the experiment. 54 videos were imported into E-Prime 2.0. The subjects needed to react to the dynamic raster video in the experiment. The grating line width was changed from wide to thin. When the lines of the two colors were indistinguishable, that is, when a uniform color was seen, any key of the keyboard was pressed to react. Distance between the subject and the screen was about 600 mm, the order of the video was random, and the reaction time was recorded by the computer program.

3.5 Experimental Result

The relationship between modulation contrast and the reaction time, background luminance and reaction time is obtained by the experimental data (Fig. 4).

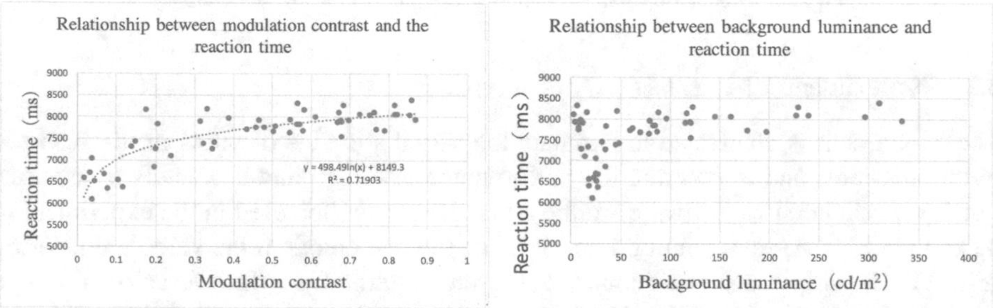

Fig. 4. Relationship between modulation contrast and the reaction time, background luminance and reaction time.

The longer the reaction time, the finer the line width of the grating. The higher the modulation contrast, and the higher the contrast sensitivity. As the modulation contrast increased, the contrast sensitivity increases first and then the upward trend becomes slower. As the background luminance increased, the reaction time of the subject decreases first, then increases, and the lowest background luminance is 21.4 cd/m^2. At this time, the modulation contrast is the lowest, corresponding to the background color (210° 33% 33%). The reaction time at this time is also the shortest, which is 6098 ms (Fig. 5).

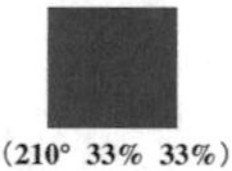

Fig. 5. HSB (210° 33% 33%).

The experimental data are processed according to the H value, that is, the hue (Fig. 6). The modulation contrast of the HUD display can be selected to be greater than 0.5, because it has high contrast sensitivity when the modulation contrast reaches about 0.5.

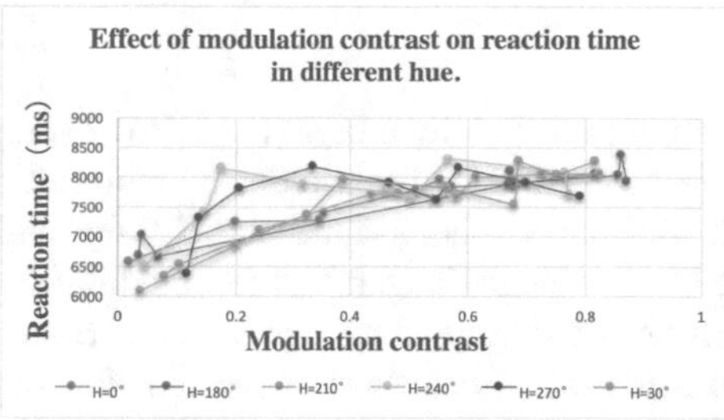

Fig. 6. Effect of modulation contrast on reaction time in different hue.

The result shows that the shortest reaction time is 6098 ms in the background of HSB (210° 33% 33%), and the longest reaction time is 8388.6 ms in the background of HSB (180° 66% 99%), the line width value is calculated according to the reaction time. The line width can be defined as Formula (2).

$$s = (10000 - t) \times \Delta l \times p. \tag{2}$$

The reaction time of the subjects is t, the unit is millisecond (ms); the line width is s, p is the width of each pixel, both units are millimeter (mm); Δl is the line width change speed, the unit is pixel/second (pix/s). In the experimental videos, the starting line width is 5 pix, and the video duration is 10 s then p is 0.2267857 mm. The line width can be defined as Formula (3).

$$s = (10000 - t) \times 0.0005 \times 0.2267857. \tag{3}$$

According to the formula, the minimum line width s_{min} under the 600 mm of sight is 0.18 mm, and the maximum line width s_{max} is 0.44 mm. The perspective can be defined as Formula (4).

$$\theta = \arctan\left(\frac{a}{L}\right). \tag{4}$$

Through calculation, θ_{min} is 0.3 mrad, θ_{max} is 0.73 mrad. When the grating line width is greater than 0.73 mrad, it is possible to distinguish the grating lines of the two colors corresponding to all the modulation contrasts. That is, the line width of 0.73 mrad is the minimum line width that can be adapted to all cases. According to the regulations, the line width of the character is one-eighth to one-sixteenth of the word height. It can be seen that when the line width is 0.73 mrad, the corresponding minimum word height is 5.84 mrad, and the maximum word height is 11.36 mrad.

This paper mainly studies the design of the color, luminance, line width and word height, which has a certain guiding effect on the interface design of HUD. The experiment was carried out in the laboratory, the choice of the background is pure color, not the real sky. At the same time, the luminance of foreground and background is not controlled separately. The actual application and inspection is not enough, which has certain limitations [10].

4 Conclusion

Head-up displays play an important role in flight [11]. However, proper character color, luminance and contrast, and suitable line width are required in different luminance environments [12]. In this paper, the luminance of foreground and background is measured according to the luminance meter, and the modulation contrast is calculated. The experimental result shows that contrast sensitivity is basically saturated after the modulation contrast exceeds 0.5. According to the reaction time of the subjects, the thinnest grating line width that can be discerned by the human eye is calculated to be 0.3 mrad, and in order to ensure that the line width is suitable for all cases, the display line width of the HUD should be greater than 0.73 mrad, HUD symbol word height should be greater than 5.84 mrad.

Acknowledgments. This work was supported jointly by Science and Technology on Electro-optic Control Laboratory and Aerospace Science Foundation of China (No. 20165169017), SAST Foundation of China (SAST No. 2016010), Equipment Pre research & Ministry of education of China Joint fund, Fundamental Research Funds for the Central Universities of China (No. 2242019k1G023).

References

1. Vinod, K., Yaduvir, S., Bajpai, P.P., Harry, G.: Study of attention capture aspects with respect to contrast ratio for wide background luminance range in head-up displays (2012)
2. Karar, V., Ghosh, S.: Soft computing based HUD brightness switching system for mitigating tunneling effect. J. Int. J. Electron. Commun. Comput. Eng. **3**(5), 1130–1136 (2012)
3. Kun, Y., Jing, D.: Analyses on the influences of head up display character color on head up display and head down display compatibility based on eye movement index. J. Sci. Tech. Eng. **1**(3), 179 (2018)
4. Stone, M., Szafir, D.A., Setlur, V.: An engineering model for color difference as a function of size (2014)
5. Mccormick, E.J., Sanders, M.S.: Human Factors in Engineering and Design. Tata McGraw-Hill, New York (1982)
6. Dowell, S.R., Foyle, D.C., Hooey, B.L.: The effect of visual location on cognitive tunneling with superimposed HUD symbology. In: Proceedings of the Human Factors and Ergonomics Society Annual Meeting (2002)
7. Hai-Yan, W., Jiang, S., Yan, G.: Color design of aiming interface for helmet-mounted display. J. Electr. Opt. Conf. (2016)
8. Stone, M.: In color perception, size matters. IEEE. Comput. Graphic Appl. **32**, 8–13 (2012)
9. Kang, H., Hong, H.K.: Measurement of minimum angle of resolution (mar) for the spatial grating consisting of lines of two colors. J. Displays **38**, 44–49 (2015)
10. Karar, V., Ghosh, S.: Estimation of tunneling effect caused by luminance non-uniformity in head-up displays. J. Chin. Opt. Lett. **12**, 73–78 (2014)
11. Tufano, D.R.: Automotive HUDs: the overlooked safety issues. J. Hum. Fac. Ergon. Soc. **39**, 303–311 (1997)
12. Jian, X., Jun, L.: The luminance and line-width measurement system for head up display. J. Sen. Tech. (2014)

Redesigning the Common NICU Incubator: An Approach Through the Emulation of Factors Resembling the Mother's Womb

Denisse Chavez-Maron, Alan Taylor-Arthur(✉), Sofía Olivares-Jimenez, Gabriela Durán-Aguilar, and Alberto Rossa-Sierra

Facultad de Ingeniería, Universidad Panamericana, Álvaro del Portillo 49, 45010 Zapopan, Jalisco, Mexico
{Denisse.chavez, Alan.anthur, Ana.olivares, Gaduran, Lurosa}@up.edu.mx

Abstract. Incubators are essential medical devices in Neonatal Intensive Care Units (NICU) with the purpose of regulating and maintaining the required humidity and temperature for neonates and preterm infants (PI): basic survival functions. Nonetheless, their optimal development must consider other fundamental factors such as positioning of the body, orientation in space, and other stimuli. The aim of this study was to compare the emotional visual impact variation caused by the common NICU incubator and a redesign proposal that takes into account all the factors previously mentioned. The emotional impact caused by these two devices was measured through a survey using PrEmo Tool©. The results showed that through the redesigned proposal, a higher positive response was achieved while decreasing its negative emotional impact. The proposal's aim for improvement was achieved through the emulation of the mother's womb both in form and function.

Keywords: NICU incubator · Preterm infants · Kangaroo position · Stimuli · Redesign · Emotional impact

1 Introduction

In Mexico, around two million children are born every year, of which 10% are premature [1]. According to the World Health Organization, prematurity is the leading cause of death during the first four weeks of life and the second leading cause of death among children under five years of age. This is an issue not only in Mexico but worldwide. Its relevance relies on the fact that the United Nations considers among its global goals to end all preventable deaths under five years of age [2].

Incubators are essential medical devices in the Neonatal Intensive Care Units (NICU) whose purpose is to regulate and maintain the required humidity and temperature for newborns and preterm infants (PI). It is said that an infant is premature when it is born before 37 weeks of gestation [3]. With the main user of incubators being the latter, it is understood that they need specialized medical care and must remain in incubators until their organs can function properly. However, NICU only

T. Ahram et al. (Eds.): IHSED 2019, AISC 1026, pp. 331–335, 2020.
https://doi.org/10.1007/978-3-030-27928-8_50

ensure their survival and not their optimal development. The abrupt change of environment suffered by PI after birth should be mitigated, having in mind that they still had a few weeks left in the womb until their gestation period was complete. This can be achieved by also contemplating other primordial factors such as the positioning of the body and stimuli of sound.

Preterm Infants Positioning. Due to immaturity, PI often lack adequate muscle tone and are at risk for developing abnormal movement patterns as well as skeleton deformation [4]. Proper position of premature infants may promote normal motor development while minimizing the development of abnormal movement patterns.

The positioning of PI includes supine, prone, side-lying, and head up tilted positions. Many studies have demonstrated a variety of outcomes, both positive and negative, affected by different body positioning of PI [5]. Newborns are born with a convex c-shaped spine so their thighs naturally pull up towards their chests. Laying them flat stretches out their natural position and can be stressful on their spines and hips [6].

However, the recent introduction of the Kangaroo Care© method has presented numerous benefits for the care and development of PI. Amongst these benefits, it was found that the kangaroo position, also called fetal position, is presumed to be the most optimal after its comparison with the other positions previously mentioned. The flexed position adopted during kangaroo position allows easier food digestion and breathing since less oxygen pressure and volume are required: as a result, the energy and calories of the PI are instead devoted to their own growth [6].

Sound Stimulation. Furthermore, it is well recognized that low-frequency maternal sounds, such as the mother's voice and heartbeat, are audible inside the womb early in gestation. By 25–26 weeks' gestation PI can already perceive and respond to sounds in their environment. When a premature birth occurs, the low-frequency maternal sounds in the amniotic environment are replaced by loud background noises [7]. Exposure to maternal sounds may, therefore, be crucial for healthy fetal development.

In the neonate, loud noise has been associated with hypoxemia and altered behavioral responses or cardiovascular symptoms. Noise can also have long term consequences for the newborns, by interfering with their psycho-biological balance resulting in the disruption of their normal growth and development. Additionally, it can cause sleep disorders and interfere with circadian rhythms [8].

Music is a non-pharmacological method used to reduce pain and stress in PI in the NICU [7]. The Original Sound© is a track created by a multidisciplinary group of neonatologists, pharmacologists, sound engineers, and artists. It includes a series of sounds that fetus experience during intrauterine life such as female and male voices, footsteps, electrical noises, traffic noises, TV and radio output, etc. Its use generates positive effects demonstrated by the regulation and reduction of heart rate, rise in oxygen saturation and improvement on behavioral state [7].

It is of high importance that physiological factors such as heart rate, body temperature, blood pressure, respiratory rate, and the circadian rhythm are stably preserved since the newborn's effort to self-regulate can increase the risk of tachycardia, bradycardia, increased intracranial pressure, and hypoxia. Therefore, it is essential that PI are relaxed and in a stress-free environment at all times to canalize energy expenditure for their growth and development.

NICU incubators are products that, to some degree, have an unpleasant emotional impact and this is generally not taken into account in the design of these products. Instead, they are designed on the basis of demands predominantly related to the user's survival contemplating only the basic needs previously mentioned. This proposal seeks to improve the visual and emotional impact, perceived by parents, through the emulation of the mother's womb both in form and function.

2 Materials and Method

53 middle and upper social class mothers from the city of Guadalajara, Mexico, ranging between 20 and 55 years of age were asked to answer an online survey. The survey used PrEmo Tool© to evaluate the emotional impact our design proposal generated. PrEmo© is a unique, scientifically validated tool to instantly get insight in consumer emotions. People can report their feelings towards a product whether they are positive or negative with the use of expressive cartoon animations instead of relying on the use of words [9].

Our survey consisted of 28 questions, each respondent answered it individually. First, it showed two pictures of a common incubator, the first one was the device by itself and the second one was the incubator in use. Then, the respondent had to answer some questions about how related they felt towards 14 different emotions represented by the PrEmo© cartoons. They had 3 options for each emotion: *I extremely relate to this emotion, to some extent I relate to this emotion*, and *I don't feel related to this emotion*. The cartoons were sequenced by types of emotion, showing 7 positive emotions (*admiration/respect, satisfaction/approval, attraction/desire, fascination/curiosity, hope/optimism, joy/happiness*, and *pride/self-esteem*) first and then 7 negative ones (*boredom/dullness, contempt/disrespect, disgust/aversion, dissatisfaction/anger, sadness/grief, shame/embarrassment,* and *fear/anxiety*).

After completing the first section of the survey, two pictures of our NICU incubator redesign proposal were shown. Again, the first one showed only the device and the second one showed how it would look when in use. The mothers had to answer another 14 questions in the same order as the first section.

3 Results

According to the results found by the survey, it is proven that in comparison, positive emotions obtained from the Incubator redesign were more prominent than those found in the Common Incubator. Emotions such as *attraction or desire* were found to be more relatable for users, and *fascination* or *curiosity* had a comparably important shift.

The results obtained from the comparison of positive emotions obtained from the visual impact of the common incubator and the redesign proposal show differences in variation in the *I don't feel related to this emotion* as great as 17% of difference, while on the *I extremely relate to this emotion* answer, a difference as great as 14% was obtained. Lastly, the *to some extent I relate to this emotion* answer, showed a difference of variation as great as 16.3%.

On the other hand, the results found from the stimuli on negative emotions, show that the respondents overall did not relate to strong negative emotions. However, emotions such as *sadness or grief* along with *fear or anxiety* were found to be the exception. According to the data obtained by the results, when shown the common incubator, these emotions previously mentioned were found to be strongly relatable to the respondents.

In addition, the data found representing the results from the visual impact on the proposed redesign model of the common incubator, stipulates that these two emotions above mentioned (*sadness or grief*, and *fear or anxiety*), had a significant variation compared to that in the common incubator.

Except from *sadness or grief*, and *fear or anxiety*, the remaining emotions did not show a comparable difference in variation and therefore, were not considered to be significant for further interpretation. the variation found on the *I don't feel related to this emotion* question was no greater than 5% of difference, while on the *I extremely relate to this emotion* question the difference was no greater than 2.5%. Finally, the *to some extent I relate to this emotion* question did not have a difference greater than 7.2%.

4 Discussion and Conclusions

A redesign proposal was made with the purpose of mitigating the abrupt environmental change that PI go through, while improving the non-optimal positions applied in the common NICU incubators as well as the unhealthy sound exposure. In addition, it aimed to enhance the positive emotional impact caused by its aesthetics since this is a factor that is usually not taken into consideration when designing this type of products.

The results obtained from the collected data demonstrated that it is possible to achieve a more positive perception of this devices while decreasing their negative visual impact through the bio mimesis or emulation of natural factors such as the mother's womb on the aesthetic appearance of the design. This is important since the parents' acceptance of the device is proven to come from the visual comfort that the NICU incubator should represent.

Furthermore, an important contribution from the proposed redesign includes functional aspects that would mitigate the abrupt change of the environment through the stimuli of sound following the literature proven by The Original Sound© and the recommendations following the literature of the Kangaroo Care© regarding the proper positions recommended for PI. However, further collaboration and approval from the authors of the former are yet to be obtained.

Moreover, the proposal did not aim to redesign any of the essential environmental factors needed for the proper functioning of the NICU incubator such as humidity, temperature, or oxygen. This factors are at the time, intended to be kept the same. It should be noted that the studies conducted on this project assessed the emotional impact of the common NICU incubator at one given moment. In the evaluation study of the redesign model of the NICU incubator, only one redesign proposal was tested.

A second limitation of both the first and the second study is that only the responses evoked by the appearance of the NICU incubators were measured. The experience of actually using the NICU incubator was not taken into account. In addition, the functional implementations of the proposed redesign were based on the literature found and are yet to be tested.

References

1. Gobierno de México: Secretaría de Salud. https://www.gob.mx/salud/prensa/en-mexico-nacen-cada-ano-200-mil-ninos-prematuros-inp
2. The Global Goals for Sustainable Development. https://www.globalgoals.org/3-good-health-and-well-being
3. World Health Organization. https://www.who.int/features/qa/preterm_babies/en/
4. Carayon, P., Kianfar, S., Li, Y., Xie, A., Alyousef, B., Wooldridge, A.: A systematic review of mixed methods research on human factors and ergonomics in health care. Appl. Ergon. **51**, 291–321 (2015)
5. National Center for Biotechnology Information: US National Library of Medicine. National Institutes of Health. https://www.ncbi.nlm.nih.gov/pubmed/27820087
6. Ludington-Hoe, S.: Kangaroo Care: The Best You Can Do to Help Your Preterm Infant. Bantam Books, New York (1993)
7. Tandoi, F., Francescato, G., Pagani, A., Buzzetti, G., Negri, E., Agosti, M.: "The Original Sound": a new non-pharmacological approach to the postnatal stress management of preterm infants. J. Matern.-Fetal Neonatal Med. **28**(16), 1934–1938 (2015)
8. Fernández Zacarías, F., Beira Jiménez, J.L., Bustillo Velázquez-Gaztelu, P.J., Hernández Molina, R., Lubián López, S.: Noise level in neonatal incubators: a comparative study of three models. Int. J. Pediatr. Otorhinolaryngol. **107**, 150–154 (2018)
9. Emotion Studio. https://www.premotool.com/

Experimental Study on Color Identifiable Area Threshold Based on Visual Perception

Yitong Pei, Haiyan Wang(✉), Chengqi Xue, and Xiaozhou Zhou

School of Mechanical Engineering, Southeast University, Nanjing 211189, China
poppyisz@163.com, {whaiyan, ipd_xcq, zxz}@seu.edu.cn

Abstract. Using color to encode data is a very common and important method in visual design. The rational use of color can improve the efficiency of information transmission. Therefore, we explored the effects of different levels of color on visual perception. From the perspective of minimum sensible difference and contrast sensitivity, the limit method is used to explore the minimum identifiable area of different levels of color, and the ability to recognize different levels of color is obtained. Based on the results, we made some suggestions: using the third and fourth level color coding key information, the seventh level coding the same level of information is not recommended. The results of this study can be used as a reference for interface color design to improve information readability.

Keywords: Human-machine interface color design · Attention capture · Color level · Color area

1 Introduction

Color is the most important and most complex of the elements that can mark data [1]. Since the color itself can carry quite a lot of information, the reasonable use of color for data encoding will make the visualization more accurate and expressive [2]. In addition, colors can represent concepts that other elements are not easily expressed, and help users understand and accept information [3]. Attention capture can be understood as a process in which attention is driven by stimulus-driven attention in visual selective attention. Attention to capture theory refers to the priority processing method that people's attention takes according to the salience of external objective objects or information, and is not affected by human subjective consciousness. According to Itti [4, 5], attention depends on the visual saliency of the object. Even when people have a clear search target, they will still be attracted by things with prominent colors [6]. Colors vary significantly with the size of the object being observed. The interaction between size and color is a key design factors. The ability of people to perceive color differences is significantly different among different marker types [7]. Stone [8] studied the effect of size on color perception in the case of discrete targets. The results show that the legibility of color changes with the change of target size [9]. Lee [10] pointed out that large color area suppression is small. Group, and predominate in image perception.

T. Ahram et al. (Eds.): IHSED 2019, AISC 1026, pp. 336–340, 2020.
https://doi.org/10.1007/978-3-030-27928-8_51

In order to avoid invalid color coding and improve the legibility of color application. We studied the minimum identifiable area of different levels of color. From the perspective of color saliency level, we explored the influence of different levels of color on attention capture, and proposed guidelines for interface color design, which provides a design basis for color to play a better visual guidance role in information interface.

2 Methodology

2.1 Materials

The division of the color level of this experiment is obtained by the gray-color mode conversion, and the color difference between the levels is 65. In the HSB model, the color difference values of the seven regions from top to bottom and white are: 0–65, 65–130, 130–195, 195–260, 260–325, 325–390, 390–441, marking the level from top to bottom as 1–7. To ensure that the sample covers the entire chromatogram, color picking takes place at important turns in each area. Each level contains 10 colors, and seven levels have 70 colors (see Fig. 1). The target color set is C1, C3, C5, C7, C9, and the control color is C1–C10.

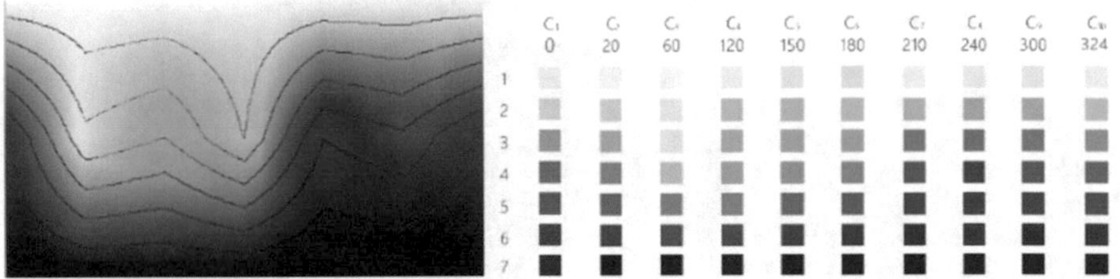

Fig. 1. 70 samples are selected in HSB space.

The device is MacBook 13.3 with a resolution of 2560 × 1600 px. The experimental software is E-prime 2.0, and the software resolution is consistent with the display. The operating area is the keyboard.

2.2 Subject Materials

Ten college students, including four male students and six female, whose ages range are 20–25 years old, are chosen as subjects. Among those subjects, corrected visual acuity are 1.0 or above. None of the subjects had previous experience in visual search, and the color vision was normal and there was no amblyopia.

2.3 Experimental Procedures

In the process of the formal experiment, a cross is first presented in the center of the screen for 1000 ms, which is used to remind the subject to pay attention, and then presents a white screen with a duration of 300 ms. Then the picture of the target color is displayed for 2000 ms, followed by the contrast color. In the video, the subject judges whether the color in the video screen matches the color of the target color just displayed. The subjects tried to operate the keyboard with both hands. If they are consistent, press "V" for the left hand, otherwise, press "N" for the right hand. When the subject responds, the video stops playing, showing a black screen, and then press the space to identify the next color.

Each group of experiments is cycled 50 times, having a total of 7 groups of 350 times. Each group is independent of each other and is provided with rest time to prevent the occurrence of visual fatigue and affect the accuracy of the experiment.

3 Results and Discussion

3.1 Results Analysis

The correct rates in the second, third, fourth, fifth, and six levels are not significantly different, which are higher than 90%, The correct rates of the first level and the seventh level are significantly lower than the previous five levels. Especially the seventh level, correctness rate of which is lower than 70% (see Fig. 2).

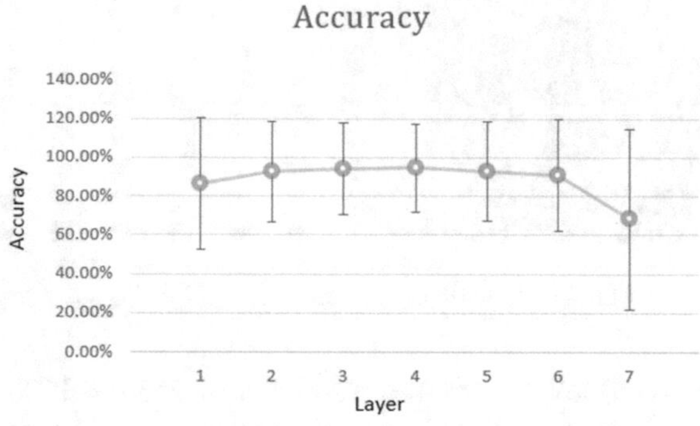

Fig. 2. Accuracy rate of color recognition

As shown in Fig. 3, the subjects respond longer to the first and seventh levels. The seventh level is the longest, duration of which is 36.57 ms. The second level and the sixth level are shorter than the first two levels. The third-level reaction is 12.85 ms, being the shortest reaction time of all.

The third, fourth and fifth levels have a small identifiable area, and the third level has the smallest identifiable area of 12.85 mm^2 (see Fig. 4). The minimum identifiable areas of the first and seventh levels are large, the minimum identifiable area of which 30.99 mm^2 and 36.57 mm^2, respectively.

The color level and area were analyzed by analysis of variance to obtain $P = 0.00055$. Since $P < 0.01$, the difference between groups was significant.

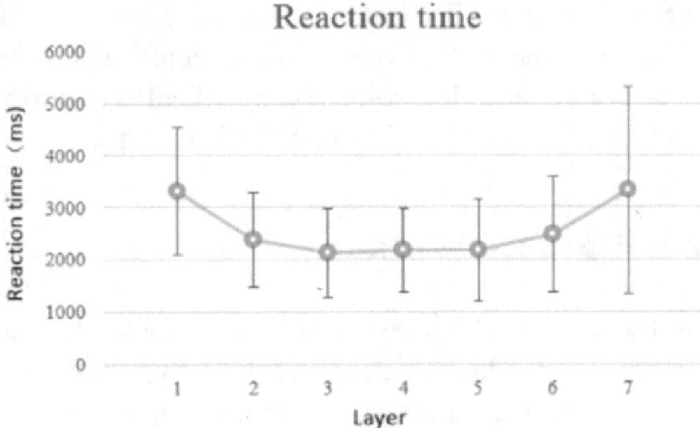

Fig. 3. Reaction time of color recognition

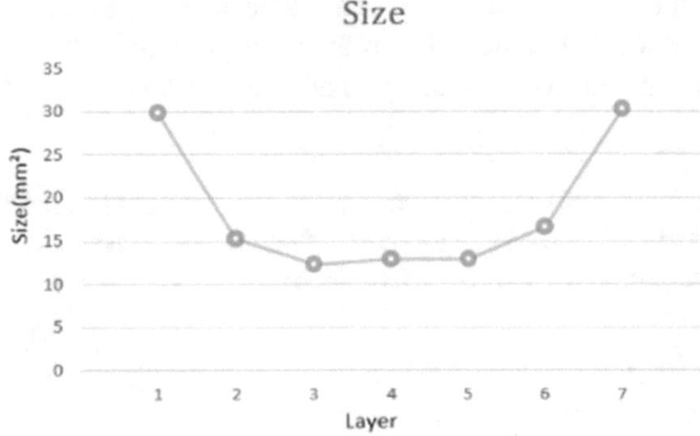

Fig. 4. Minimum identifiable area of color recognition

According to the minimum identifiable area and level of fitting analysis, the equations for the number of levels and the minimum identifiable area are:

$$S = -0.040l^3 + 2.545l^2 - 17.99l + 44.16 \quad (1)$$

3.2 Discussion

The variation of the minimum identifiable area of different levels is firstly reduced and then increased, which is related to the variation of color difference in the HSB color space. In the HSB model, the chromatic aberration in the first level to the seventh level is increased first and then decreased. The identifiable area decreases as the chromatic aberration increases. The first level of color saturation is low, and the color capture ability is weak, so it is not easy to recognize when the area is small, and can be used as a background to fill a large-sized area. The third and fourth levels of color have strong visual capture ability and are highly significant, so they can be used to emphasize key information. At the same time, the color difference between the colors of the third and fourth levels is large, and the difference between the colors is strong, so it can also be used to encode the classified data. The color gradation of the seventh level is low, and the color difference between the same levels is small, therefore, it is hard to distinguish. This level of color is not suitable for color coding of information in the same level, which is easy to cause information confusion and visual errors.

4 Conclusion and Future Work

The minimum identifiable area of different levels of color is obtained through behavioral experiments. From the results, the ability of the human eye to recognize different levels of color can be known. The data indicates that the third and fourth levels of high-intensity and high-saturation color are suitable as important information for accent color, and the second, fifth and sixth levels are suitable for encoding auxiliary information, and the first level of color is suitable as the background color. The seventh level of color is not representative and cannot be used as color coding for specific information. Future work includes exploring the impact of area on visual perception and providing reasonable color selection and area matching for different levels of information.

Acknowledgments. This work was supported jointly by National Natural Science Foundation of China (no. 71871056, 71471037), Science and Technology on Avionics Integration Laboratory and Aeronautical Science Fund (No. 20185569008).

References

1. Weiskopf, D.: On the role of color in the perception of motion in animated visualizations. In: Proceedings of the IEEE Conference on Visualization, pp. 305–312. IEEE Computer Society Press, Los Alamitos (2004)
2. Demiralp, C., Bernstein, M., Heer, J.: Learning perceptual kernels for visualization design. IEEE Trans. Visual. Comput. Graph. **20**, 1933–1942 (2014)
3. Tennekes, M., de Jonge, E.: Tree colors: color schemes for tree-structured data. IEEE Trans. Visual. Comput. Graph. **20**, 2072–2081 (2014)
4. Itti, L., Koch, C.: A saliency-based search mechanism for overt and covert shifts of visual attention. Vision Res. **40**, 1489–1506 (2000)
5. Itti, L.: Quantitative modeling of perceptual salience at human eye position. Vision Res. **14**, 959–984 (2006)
6. Christ, S.E., Abrams, R.A.: Abrupt onsets cannot be ignored. Psychon. Bull. Rev. **13**, 875–880 (2006)
7. Szafir, D.A.: Modeling color difference for visualization design. IEEE Trans. Vis. Comput. Graph. **24**, 392–401 (2018)
8. Stone, M.: In color perception, size matters. IEEE Comput. Graph. Appl. **32**, 8–13 (2011)
9. Stone, M., Szafir, D.A., Setlur, V.: An engineering model for color difference as a function of size. In: Proceedings of Color and Imaging Conference, pp. 253–258. Society for Imaging Science and Technology, Washington DC (2014)
10. Lee, S., Sips, M., Seidel, H.P.: Perceptually driven visibility optimization for categorical data visualization. IEEE Trans. Vis. Comput. Graph. **19**, 1746–1757 (2013)

Symbolic Similarity of Traffic Signals Based on Human Visual Perception

Yaping Huang, Haiyan Wang(✉), Chengqi Xue, Xiaozhou Zhou, and Yiming Shi

School of Mechanical Engineering, Southeast University, Nanjing 211189, China
2010hyp@sina.com, {whaiyan, ipd_xcq, zxz}@seu.edu.cn, 1198639063@qq.com

Abstract. Icons with identical features convey to users a semantic structure, but a strong visual grouping hint may hinder users from locating interface items. This study took traffic lights as an example to explore the degree to which similarities within a set of icons influence user decisions. The experiment simulated the situation of traffic lights with four types of symbols. Ten participants were asked to respond correctly to the green signal. The behavioral experiment showed that participants have the lowest reaction time to symbols exclusive elements that have no effect on semantics. The relationship between the reaction time and symbolic similarities of those four types of traffic lights was analyzed by the proposed method of computing item similarity. The results showed similarity values were highly consistent with the reaction time. The implications of the results are applicable to design a set of icons that have less effect on the user's quick recognition.

Keywords: Icon similarity · Decision-making · Traffic signals · Visual perception

1 Introduction

The similarity of graphics is of great significance to the design, recognition, processing and other fields of graphics. Visual attention theory emphases negative effects of nontargets resembling any possible target [1]. The similarity of objects can be expressed as a linear combination or contrast of their common features and salient features [2]. The key components of an icon contain border, background, element, symbol and text label [3]. The degree of icon similarity depends largely on these features. Similarity of two shapes can be obtained through the geometrical elements, the adjacency state and other topological structures, the geometrical shape and size constraints such as connection mode and symmetry are used as similar elements to calculate the similarity of these three aspects [4]. For the similarity calculation and search of similar pictures, Zauner [5] proposed a perceptual hash algorithm, in which hash value is not calculated in a strict way, but in a relative way to get the Hamming distance between two images.

T. Ahram et al. (Eds.): IHSED 2019, AISC 1026, pp. 341–346, 2020.
https://doi.org/10.1007/978-3-030-27928-8_52

The computer's pattern recognition and similarity analysis of natural images are investigated and developed maturely, which can achieve quite high precision. However, there are some problems in pattern recognition and similarity analysis of icons. Due to the special attribute of the semantics of icons and symbols, it is difficult to adopt the principle of image similarity calculation to match the degree of human perception similarity. Accordingly, icon similarity algorithm is rarely applied to icon design. Color similarity of applications can empower designers to develop new interactive modes on mobile devices that match the user's hedonistic and pragmatic needs [6]. If the computer can achieve high-precision pattern recognition and similarity analysis of icons are the same as natural images, it will bring great convenience to graphic icon design and evaluation. Designers can not only accurately grasp the graphical features based on the results given by the computer, but also find the icon visual identity and style in line with their design intention more quickly.

2 Icon Similarity Algorithm

The existing methods for computing graphic similarity are taking graphics as a whole, which cannot be applied to icons with multiple shape elements. Tversky proposed a ratio model of feature similarity representation theorem [2].

$$S(a,b) = f(A \cap B)/\{f(A \cap B) + \alpha f(A - B) + \beta f(B - A)\}, \alpha, \beta \; >= 0. \quad (1)$$

where similarity is normalized so that S lies between 0 and 1.

In line with his theory, no matter what features an icon consists of, as a type of image, icons could be evaluated, separated the contained shapes by edge detection. By calculating the similarity of each shape based on the differential hash algorithm or Hu Invariant Moment, integrated with the area proportion weighted, the similarity of the two icons can get the final value. The main innovation of this method lies in the idea of "deconstruction and integration" and weighted computing. By comparing the experimental results of the subjects, this method can get the similarity values which are in good agreement with human perception.

The limitation of the difference hash algorithm and the Hu invariant moment is due to the various shape composition and topological relationship of an icon. However, it may reduce the similarity precision if just taking all of elements in an icon as a whole to calculate. Therefore, if each element in an icon can be decomposed and extracted, the above methods could be used for similar elements of icons. The calculation accuracy can be greatly improved if the weighted integration is performed in an appropriate manner. In summary, the algorithm flow is described as follows.

Step 1. Specify two icon files to be compared.
Step 2. Execute the contour level acquisition module.
Step 3. Execute the area calculation module.
Step 4. Execute the shape segmentation module.
Step 5. Execute the peer comparison module.
Step 6. Execute the grading integration module and get the similarity value.

Step 7. If all the components need comparing of two icons that have been computed, then return the data and end the program. Otherwise, specify the next pair of components, then execute Step 2 to Step 6 again.

3 Methodology

Traffic lights are important indicators in the transportation system. In order to ensure traffic order and prevent accidents, traffic lights need to be conspicuous and clear. Luminance value of signals should be twice of the background. In addition, minimizing visual interference can help people detect the signal lights easily [7]. Different kinds of signal lights will have different effects on people. For example, people will be overwhelmed by unclear symbols, or make mistakes by too similar symbols. This experiment investigated the relationship of the reaction time that participants spent and the type of symbols in traffic lights.

3.1 Materials

The traffic signal light patterns prepared in this experiment are divided into four types. Each type has four directions: straight, right turn, left turn, and U-turn, with a total of 16 symbols, as shown in Table 1. Taking the straight symbol as an example, this symbol of type 1 is a low aspect ratio (value is 0.65) with thin lines, type 2 with high aspect ratio (value is 1) and thin lines, type 3 with a high aspect ratio and thick lines. Different from others, symbols of type 4 are based on type 3 but exclusive straight lines that have no effect on semantics.

Table 1. Traffic light symbols

Type	Straight	Right Turn	Left Turn	U-Turn
Type 1				
Type 2				
Type 3				
Type 4				

In addition to the 16 symbols in Table 1, a circular filled symbol is required as an experimental interference term. These symbols are available in red and green. In the experiment, these materials will appear as stimuli in the center of the screen in pairs of different combinations. Each trial contains only one type of symbols, in which the color combinations are red symbol and green one or red symbol and red one.

3.2 Subjects

Ten male undergraduate students, whose ages are all between 21–22 years old were chosen as subjects. All participants had a corrected visual acuity of 0.8 or rectified vision and normal color vision. In addition, in order to fully simulate the traffic lights in the driver's field of view, it is necessary to reflect the angle of view of the signal light into the experiment. Assuming that the distance between the driver and the traffic signal is 10 m and the diameter of the traffic signal is 40 cm, the angle of view α is calculated as below.

$$\alpha = \arctan\frac{40\ \text{cm}}{10\ \text{m}} = \arctan\frac{1}{25} = 2.29^\circ. \tag{2}$$

During the experiment, subjects were asked to keep their eyes 560–600 mm away from the screen was 60 cm and vertical perspective was controlled in 2.3°. According to the formula (2), the diameter of the traffic signal displayed on the screen was 24 mm. Reducing the size of the stimulus is not only to simulate the real scene, but to eliminate the effects of brightness.

3.3 Equipment and Procedures

The experiment was programmed by E-prime2.0, which collected behavioral data including reaction time. Experimental procedure is shown in Fig. 1. During the experiment, a white cross appeared in the center of the screen to remind the subjects, and continuously lasted for 1000 ms then disappeared. In the next moment, two traffic symbols appeared in the center of screen. The subjects must make a correct feedback according to the green traffic symbol. The key 'W' indicates straight. The key 'A' means. The key 'D' means right-turn. The key 'S' means U-turn. The filled circular symbol can be pressed with any key of the above. If subjects encountered two red symbols, they were asked to press the 'Space' key to enter the next trial. The interval time between two trials was 1200 ms. There were 130 different pairs of stimuli in the experiment. Each pair of stimuli would show up twice in a random sequence. The subjects were required to concentrate and make judgments quickly and accurately. The experiment recorded the time interval between the occurrence of the stimuli and the receipt of the correct response.

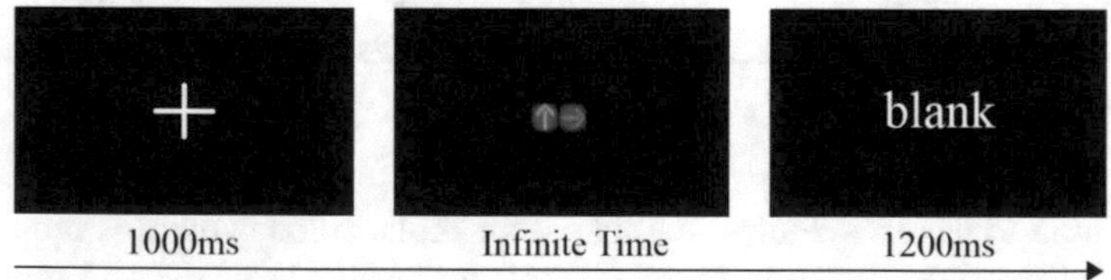

Fig. 1. Process of behavioral experiment

4 Results and Analysis

This study does not focus on the effect of the aspect ratio or the thickness of the line on the reaction time, but on the similarity, the degree of influence of these four types of signal indicators on the responses of the subjects. The experimental data and similarity calculation results are analyzed separately.

4.1 Behavioral Experimental Data Analysis

Because the participants are very familiar with the symbols of the traffic lights, the accuracy rate is very high. Here we mainly analyzed their reaction time to different types of signal lights. As is shown in Fig. 2, the mean reaction time over all subjects for four types of traffic symbols are: type1 (low aspect ratio and thin line) 527.41 ms > type 3 (high aspect ratio and thick line) 524.21 ms > type2 (high aspect ratio and thin line) 519.64 ms > type4 (similar to type 3 but exclusive straight lines) 492.52 ms. The mean RT of type 4 is significantly lower than the other types.

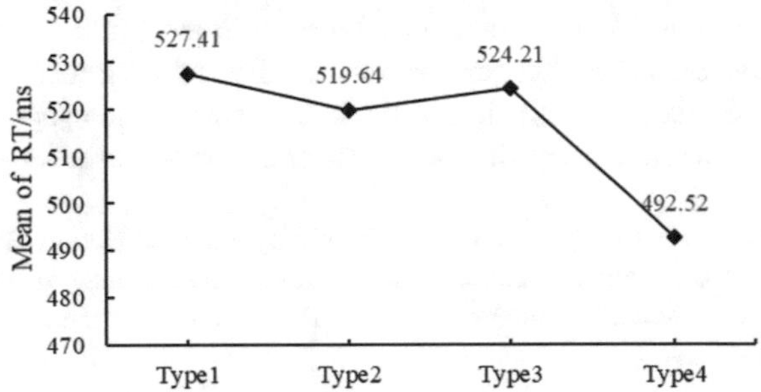

Fig. 2. Reaction time to four types of traffic signals

4.2 Analysis of Similarity Calculation Results

This study focused on the difference within symbols of each style. In the calculation phase, each type of symbols performed a pairwise calculation of their similarity values which would produce 6 different values. Combined the visual pattern recognition, similarities of traffic symbols in the same type were calculated separately by difference hash algorithm and by Hu moment invariants. First, all similarities of different symbols were calculated for the same type of traffic signals. After taking the average value, values were plotted in Fig. 3. Results of both methods showed that the similarity of type 4 is significantly lower than other types. Furthermore, the ANOVA results indicated the significant effect of similarity values gotten by Hu moment invariants on reaction time ($F = 11.83$, $p = 0.05$).

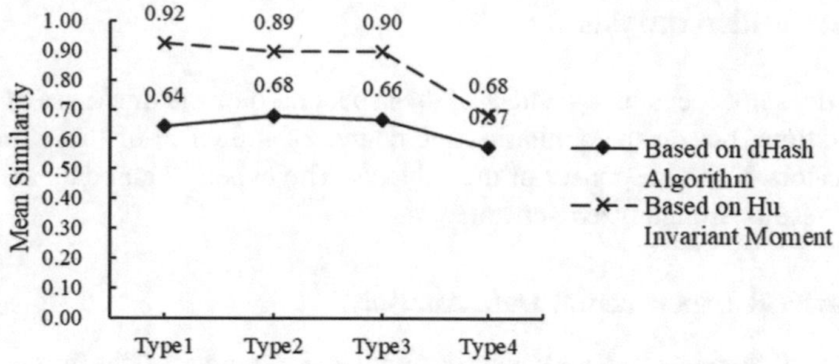

Fig. 3. Mean similarity of all the different pairwise symbols in the same type

5 Conclusions

This study found that different types of traffic symbols have different effects on reaction time, particularly exclusive elements that have no effect on semantics. The core of the icon similarity calculation method is "deconstruction and integration", which was briefly described in the paper. Compared with experimental data, similarity values of different types have the main effect on the reaction time of subjects. Our research was a trial to apply the similarity calculation to the icon design, which support the designers to create sets of icons uniformly that have less effect on the user's quick recognition.

Acknowledgments. This work was supported jointly by National Natural Science Foundation of China (no. 71871056), National Natural Science Foundation of China (no. 71471037), Science and Technology on Avionics Integration Laboratory and Aeronautical Science Fund (No. 20185569008).

References

1. Ducan, J., Humphreys, G.W.: Visual search and stimulus similarity. J. Psychol. Rev. **96**, 433–458 (1989)
2. Tversky, A.: Features of similarity. Psychol. Rev. **84**, 327–352 (1977)
3. Chi, C.-F., Dewi, R.S.: Matching performance of vehicle icons in graphical and textual formats. J. Appl. Ergon. **45**, 904–916 (2014)
4. Jianrong, T., Xiaoli, Y., Guodong, L.: Basic principles and methods of graphic similarity and their applications in structural pattern recognition. Chin. J. Comput. **25**, 959–967 (2002)
5. Zauner, C.: Implementation and benchmarking of perceptual image hash functions. Master's thesis, Upper Austria University of Applied Sciences, Linz (2010)
6. Trapp, A.K., Wienrich, C.: App icon similarity and its impact on visual search efficiency on mobile touch devices. J. Cogn. Res. Princ. Implic. **3**, 39–59 (2018)
7. Mccormick, E.J., Sanders, M.S.: Human Factors in Engineering and Design. Tata McGraw-Hill, New York (1982)

The Color Design of Driverless Bus Based on Kansei Engineering

Lulu Wu and Guodong Yin(✉)

School of Mechanical Engineering, Southeast University, Nanjing 211189, China
wl1950302@163.com, ygd@seu.edu.cn

Abstract. The color of the exterior of a vehicle is an important part of the overall aesthetic of the vehicle. From a perceptual point of view, body color is a way for vehicles to express their emotions and communicate with users. In order to improve the acceptance and satisfaction of the driverless bus for all people, this paper intends to take the color design research of driverless bus as an example to establish potential communication between the driverless bus and the users. Based on the theory of Kansei engineering (KE), perceptual terminologies were quantized by using semantic difference method. According to the result of perceptual experiment on Easymile-EZ10, a database of exterior color is established. Moreover, principal component analysis (PCA) is adopted to transfer multiple perceptual terminological variables into optimized integrated variables, and establishing the exterior color design priority of the driverless bus based on important terminologies.

Keywords: Exterior color · Kansei engineering · Driverless bus · Principal component analysis · Perceptual terminologies

1 Introduction

The exterior color of a vehicle is a vital element that drawing people's eyes. Vehicles establish emotional connections with users through their modeling, materials and body color [1]. The study intends to take the color design research of driverless bus as an example to establish potential communication between the driverless bus and the users. KE is a consumer-oriented product development technology. According to the prior studies, KE transforms the perceptual needs of target users into product design elements, and it has been widely used in the field of automotive design [2]. As for automotive color, one application was evaluating users' emotional response to the product, while another was establishing a database of emotional intent-design element mapping [3]. In 1990s, after Nissan, Mazda and Mitsubishi introduced the KE research method into the field of automobile development, relevant international companies have launched this research [4]. The application research on vehicle products is still in its infancy, however, researches in the design of exterior color design have less successful cases [5].

Principal component analysis of data uses mathematical dimensionality reduction methods to recombine multiple indicators into a few unrelated principal components,

T. Ahram et al. (Eds.): IHSED 2019, AISC 1026, pp. 347–353, 2020.
https://doi.org/10.1007/978-3-030-27928-8_53

achieving the purpose of simplifying data and revealing relationships between variables [6]. The grading weighting method determine the sample weights and perceptual indicators establishing a hierarchical weighting model of inductive variables and exterior color samples [7]. This research quantified perceptual terminologies by semantic difference method, and transferred multiple perceptual terminological variables into optimized integrated variables based on KE theory. Taking driverless bus - Easymile EZ10 as an example, the priority of exterior color design was concluded through perceptual experiment conducted with target users and the result of statistical analysis (Table 1).

2 Experiment

2.1 Color Samples and Perceptual Terminologies Screening

In short-distance traffic, low-speed driverless shuttles effectively assist humans in completing the last mile of traffic problems. In China, many campuses have introduced the French driverless bus - easymileEZ10 as the object of research or use.

The original vehicle was designed as a pure white body, and painted into various colors and patterns after being introduced by different countries. Due to the diversity of painting colors, a large number of color samples are provided for the study of exterior color design methods. Hue loop is an important way of color organization and color research method. Commonly, there are six kinds of hue rings, such as Oswalde, NCS, Munsell, PCCS, Eaton and H ribbon of PS color picker [8].

This study selected the Eaton hue loop that based on the three primary colors of pigment. Eaton hue loop consists of twelve colors, combined with driverless bus as a sample. (Fig. 1) A sample set of exterior color is created as Y, Y = $\{y_1, y_2, y_3, \ldots, y_{12}\}$.

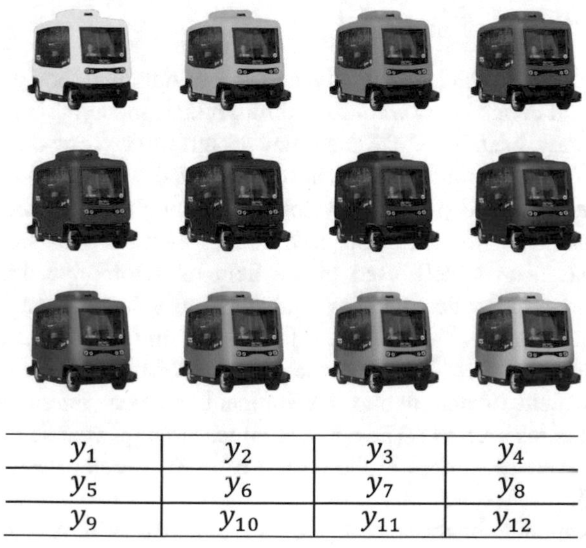

y_1 y_2 y_3 y_4
y_5 y_6 y_7 y_8
y_9 y_{10} y_{11} y_{12}

Fig. 1. Twelve samples of EZ10 exterior color

100 perceptual terminologies of exterior color were collected through public information and comments, and adjective that highly recognized and well-known to the design community and public. According to aesthetic attributes and time attributes, the perceptual terminologies can be classified into 11 categories, and filter out 21 pairs of terminologies [9]. After user acceptance questionnaire, six pairs of terminologies finally be selected, such as "a-a': safe-dangerous, b-b': brisk-depressed, c-c': kind-cold, d-d' simple-complex, e-e' business-casual, f-f': elegant-meretricious".

2.2 Consumers Perception on Exterior Color

The research conducted under school environment, firstly surveyed the basic information such as gender and age of the subjects to analyze the influence of factors such as age, gender and specialty on their selection of colors. Secondly, the semantic difference method is used to investigate the perceptual image of the twelve exterior color samples.

Take "safe-dangerous"as an example, pairs of terminologies describing the exterior color style are divided into five interval levels with scores of 1, 2, 3, 4 and 5, which correspond to very safe, safe, general, dangerous and very dangerous. The subjects' reaction intensity and perceptual knowledge are expressed by selecting different grades. The emotional image space is formed through selected color samples, the corresponding terminologies of the samples and the levels of each pair of terminologies. This experiment selected 12 color samples, each sample correspond to 6 pairs of terminologies, and each pair divided into 5 grades, a total of 360 imagery words constitute a perceptual image space.

3 Analysis

3.1 Trust Level Analysis

In order to test the validity of the questionnaire, this research used the SPSS to evaluate the consistency of the scores of each topic in the table. The larger the coefficient value of the data reliability analysis, the higher the reliability, and the more reasonable the questionnaire design is [10]. The output shows that there are 50 cases involved in the reliability analysis of the entire data file, and there are no missing values. The coefficient in the reliability statistic result of analysis is used to describe the reliability of the scale. The internal is well-consistent when the coefficient value is above 0.8. In this research, the reliability coefficient is above 0.8, which indicate that the questionnaire is highly reliable (Table 2).

3.2 Target Population Data Analysis

Through the preliminary analysis of 12 exterior color samples of driverless bus, the average of the corresponding data pairs between each sample and the perceptual terminologies is:

Table 1. The average of exterior color sample

Samples	Kansei index					
	a-a'	b-b'	c-c'	d-d'	e-e'	f-f'
y_1	2.92	2.88	2.58	3.14	2.90	2.40
y_2	2.88	3.83	2.70	3.44	3.18	2.20
y_3	3.24	2.84	3.10	2.78	2.78	2.82
y_4	3.52	2.68	3.12	2.48	3.24	2.74
y_5	3.54	3.10	3.54	3.02	3.60	3.50
y_6	3.14	3.70	3.16	3.24	2.82	2.84
y_7	3.28	3.32	3.34	3.18	2.92	2.96
y_8	2.68	3.48	3.38	3.62	2.80	3.18
y_9	2.74	3.34	2.96	2.82	2.82	2.96
y_{10}	2.58	2.88	2.72	3.02	3.22	2.84
y_{11}	2.38	3.32	3.20	3.02	2.64	2.68
y_{12}	2.60	3.02	2.84	3.34	2.54	2.84

$$\overline{x} = \begin{bmatrix} x_{1,1} & \cdots & x_{1,6} \\ \vdots & \ddots & \vdots \\ \overline{x_{12,1}} & \cdots & \overline{x_{12,6}} \end{bmatrix} \tag{1}$$

3.3 Hierarchical Weighting Model

A small number of perceptual terminologies can be extracted from multiple sets through principal component analysis method (Table 3).

Based on SPSS statistical analysis, the following conclusion are drawn:

Table 2. Principal component analysis results

Samples	Principal component load		
	I	II	III
y_1	0.347	−0.296	0.873
y_2	0.299	−0.444	0.818
y_3	−0.494	0.773	0.360
y_4	−0.854	0.275	0.338
y_5	−0.934	0.068	−0.213
y_6	0.743	0.514	0.248
y_7	0.381	0.830	0.385
y_8	0.920	0.066	−0.208
y_9	0.625	0.345	−0.449
y_{10}	0.229	−0.902	−0.041
y_{11}	0.841	0.246	−1.196
y_{12}	0.902	−0.065	0.039

The variance contribution rate is the weight of the different components. According to the calculation, the first three components are chosen to represent the data variable relationship for all samples. The eigenvalues corresponding to the first three components are 5.563, 2.918 and 2.215, and the variance is Var_1 = 46.357%, Var_2 = 24.317% and Var_3 = 18.461%. The table shows the number of principal component loads for the exterior color sample. The larger the absolute value of the load, the greater the correlation with the corresponding principal component. The principal component score coefficient, which is the weight of the three principal components variance contribution rates is calculated, and listed as a matrix:

$$\begin{bmatrix} a_{1,1} & \cdots & a_{1,3} \\ \vdots & \ddots & \vdots \\ a_{12,1} & \cdots & a_{12,3} \end{bmatrix} \tag{2}$$

A sample coefficient model is obtained by weighting the coefficients, and sample weighting is obtained after normalization:

$$v_{ai} = \frac{a_{i1} Var_1 + a_{i2} Var_2 + a_{i3} Var_3}{Var_1 + Var_2 + Var_3}, i = 1, 2, \ldots, 12 \tag{3}$$

$$w_{ai} = \frac{v_{ai}}{\sum_{i=1}^{12} v_{ai}}, i = 1, 2, \ldots, 12 \tag{4}$$

Table 3. The weight of the exterior color

Samples	Coefficient (v_{ai})	Weight (w_{ai})
y_1	0.151	0.075
y_2	0.109	0.054
y_3	0.065	0.032
y_4	0.097	0.048
y_5	0.225	0.112
y_6	0.280	0.140
y_7	0.270	0.135
y_8	0.184	0.092
y_9	0.130	0.065
y_{10}	0.099	0.049
y_{11}	0.197	0.098
y_{12}	0.194	0.097

The order of importance of different samples is: y_6, y_7, y_5, y_{11}, y_{12}, y_8, y_1, y_9, y_2, y_{10}, y_4, y_3.

Similarly, take y_6 as an example, the weighting model of kansei index of the exterior color sample is:

$$w_{(b_j|a_i)} = \frac{v_{(b_j|a_i)}}{\sum_{i=1}^{12} v_{(b_j|a_i)}}, i = 1, 2, \ldots, 12; j = 1, 2, \ldots, 6 \quad (5)$$

Visualization of hierarchy weighting model of the exterior color demonstrated as follow:

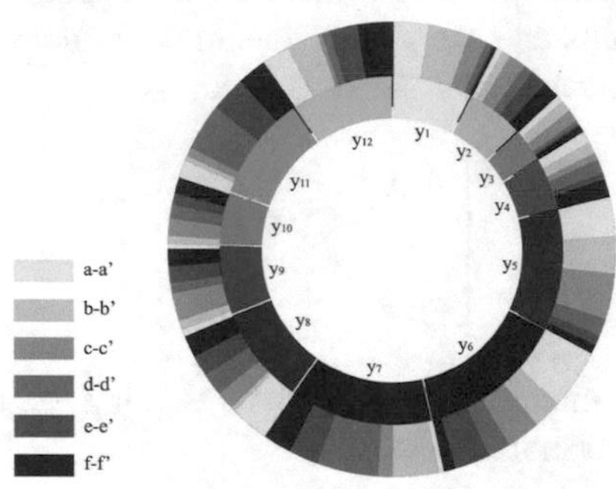

Fig. 2. Visualization of hierarchy weighting model

4 Conclusion

The study of user preferences through perceptual engineering and principal component analysis methods can objectively quantify subjective factors and provide a scientific basis for design. However, the differences in color perception between male and female may cause the deviations in experimental data, and it will be corrected in next stage (Fig. 2).

References

1. Tetsuya, O., Shinji, Y., Ken-Ichi, S., Yoshinori, O.: A CAD system for color design of a car, vol. 11, pp. 381–390 (1992)
2. Yao, X., Hu, H., Li, J.: Automotive body-side styling design based on kansei engineering. Packag. Eng. **35**(4), 40–43 (2014)
3. Zhang, Y.H., Wang, Z.Y., Jiang, M.M.: Research of the auxiliary design system of the product color design based on kansei engineering. Appl. Mech. Mater. **101–102**, 50–54 (2011)
4. Li, Y., Wang, Z., Li, D.: Research on the kansei engineering and application on the product development (2010)
5. Wang, Q., Shan, W., Li, Y.-Y., et al.: Discussion on exterior color in the vehicle research and development. Fashion Color, 113–117 (2014)
6. Zhao, H.-X., Wu, J.: Simply analyze the principal component analysis. Sci. Technol. Inf. (2009)
7. Hass, D., Bohm, G., Pfister, H.R.: Active red sports car and relaxed purple-blue van: affective qualities predict color appropriateness for car type. J. Consum. Behav. **39**(1), 88–89 (2014)

8. Hsiao, S.W.: Fuzzy set theory on car-color design. Color Res. Appl. **19**(3), 202–213 (2010)
9. Nagamachi, M.: Kansei engineering: the implication and applications to product development. In: IEEE International Conference on Systems (1999)
10. Ma, S., Yu, T., Wang, Z.: The improves method of the principal component analysis. Ship Sci. Technol. **34**(10), 21–23 (2012)

The Influence of the Threshold of the Size of the Graphic Element on the General Dynamic Gesture Behavior

Ming Hao, Zhou Xiaozhou(✉), Xue Chengqi, Xiao Weiye, and Jia Lesong

School of Mechanical Engineering, Southeast University, Nanjing 211189, China
630074044@qq.com, {zxz,ipd_xcq,230189776}@seu.edu.cn, 804855616@qq.com

Abstract. Nowadays, the image clarity and reality of augmented reality and virtual reality are constantly improving. However, The interaction in 3D space still relies on the handle or other mechanical objects to operate it. Therefore, how to interact with interfaces and objects in three-dimensional space in a natural way is a problem that current researchers will consider. The research direction of this paper is to explore the influence of the size of the graphic element in the interactive interface on the dynamic gesture behavior through the behavior experiment of the user. In the paper, the researcher observes the gestures when interacting with objects of the different size in the virtual space, and then, researchers analyzes the correlation between the size of object and the dynamic gesture. The correlation between the size of the graphic element presentation and the dynamic gesture behavior is obtained.

Keywords: Virtual reality · Natural gesture interaction · 3D interactive space · Graphic element rendering size · Dynamic gesture recognition · Leap motion · HTC vive

1 Introduction

With the advent of virtual reality technology and augmented reality technology, the concept of three-dimensional interactive space is gradually becoming familiar. Augmented Reality, a new technology that seamlessly integrates real-world information with virtual world information. Billinghurst [1] believed that virtual information should be applied to reality. The world was perceived by human senses to achieve a sensory experience that transcends reality. Ong [2] has discussed that AR technology can provided a natural method for modeling the actual working environment without the need to model the entire real world. Virtual reality technology is a computer simulation system that can create and experience virtual worlds. However, the interaction mode in the three-dimensional space will reduce the user experience and efficiency when using the user who needs to perform precise operations. This requires the intervention of natural human-computer interaction. The main core of human-computer interaction is that computers have the function of "humanoids". Swan [3] finds that according to the

T. Ahram et al. (Eds.): IHSED 2019, AISC 1026, pp. 354–359, 2020.
https://doi.org/10.1007/978-3-030-27928-8_54

concept of user experience, when an interface design conforms to the natural human-computer interaction, it should conform to ease of use, learnability, user task performance, Usability rules such as comfort and system performance. A piece of natural gesture interaction and human interaction. Chen [4] believe that reasonable of gestures can meet the various needs and laws of natural human-computer interaction. Mitra [5] believed that users can used simple gestures to control or interact with devices without touching them. Sun [6] considered that the gesture was static or dynamic.

2 Background

Previous researchers have done a series of research on the team's 3D interface and dynamic gestures. Yuan et al. [7] proposed a method in which a virtual object can be superimposed on any planar area formed by four points specified by the user. Sidharta [8] developed a new set of enhanced interfaces, including three new interface items, namely the cube interface, the physical interface and the desktop. Valentini [9] argued that in augmented reality, enhanced levels of interaction can be achieved with different levels of interaction. Gaffary et al. [10] found that the visual environment had a great impact on the perceived effect. James et al. [11] demonstrated the internal correlation between different displayed modes and user cognitive load. Chen et al. [12] used behavioral experiments to studied the distribution of operational performance of virtual interfaces for users' different ages. Robb et al. [13] showed that the use of mixed reality displayed to represent complex scientific problems can enhanced participation through post-event observations and interest. Feng et al. [14] established a secondary behavior model of basic gesture library and verified its usability on the interactive virtual assembly platform. Song et al. [15] used real-time image processing technology captured by a single network camera to improved the intuitiveness of the user's gesture operation in the AR system. Hou et al. [16] introduced the application of several representative deep learning models for speech feature extraction and acoustic modeling in speech recognition. Caggianese et al. [17] showed that the effectiveness of contactless interaction technology depended on the setting of task mode. Nickel and Stiefelhagen [18] expressed that most of the systems for gesture recognition worked on hidden Markov models.

3 Methodology

This paper needs to research the correlation between the elements of the 3D interface and the user's gestures. Therefore, the main purpose of the experiment is to observe the psychological consistency between the behavior of the subject and the dynamic vision. The experiment is designed by interacting with objects of different sizes of virtual objects, recording and observing the user's dynamic gestures by means of video recording, and then encoding the gestures to explore the correlation between gestures and graphic element sizes. The subjects selected for the experiment were 30 graduate students aged 21–25 years, including 21 male subjects and 9 female subjects. Each subject was individually placed in a virtual room of 300 cm * 300 cm * 300 cm for experimentation. In the experiment, Vive glasses were used as the hardware carrier for the experimental scene.

In the experimental scene, the experimenter selected the cube with the ratio of length to width of 1:1:0.3. In the experiment, 3 independent variables are defined: the size of the experimental body, which defines six horizontal dimensions; the spatial pose of the experimental body defines five levels of posture; the direction of motion of the experimental body defines four horizontal directions. There are five variables that need to be controlled in the experiment, which are the color, material, spatial shape, light and shadow effect, and animation effect of the experiment. The dependent variable in the experiment is the dynamic gesture behavior of the subject changing according to the change of the experimental body. During the experiment, the size, orientation and moving direction of the experimental body were observed by Bandicam recording software, and the user's dynamic gestures were captured by Bandicam and GoPro motion cameras, as shown in Fig. 1. By analyzing the motion segments of the experimental body, the motion trajectories and postures of the dynamic gestures are analyzed, and then the different dynamic gesture behaviors are classified by analysis. The collected gestures of the subjects are encoded. the coding of the dynamic gesture is from whether the user interacts with one hand or two hands, the gesture of the user gesture, the number of fingers touching the object, whether the palm touches the experimental object, whether the arm is needed for assistance, and the degree of opening of the finger. These aspects are recorded and coded.

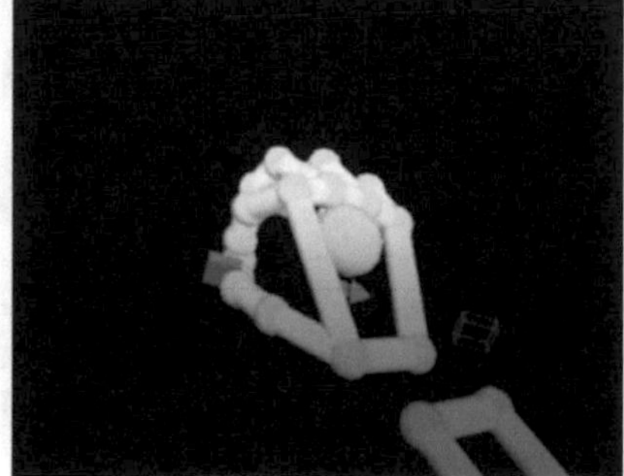

Fig. 1. Gesture collection determines the final dynamic gesture posture of the subject by observation in the real scene and virtual scene.

4 Results

The gesture of the subject was observed and recorded by playback of the behavior of subject, and then encoded the dynamic gestures which have been observed. After that, the researcher collected statistics about the collected behavior data, interface variables and subject information, and put the statistical test results into SPSS software for correlation analysis. In SPSS, the bivariate correlation was used for correlation analysis. The correlation coefficient used the Pearson coefficient. (In terms of Pearson correlation coefficient, |r| represents a correlation between 0.1 and 1.0. The larger the value, the higher the correlation), and the significance test was performed with two tails (the two-tailed test Sig is less than 0.05 is relevant). Variables such as gender, size of the experimental object, orientation of the object, and direction of movement of the object were considered as independent variables, and double/single hands (numOfHands), gesture pose (pose), number of fingers touching the object (numOfFinger), relationship between the palm and the finger (handFinger), whether using arm (arm) and finger opening degree (fingerOpen), etc. were defined as dependent variables, as shown in Table 1. By observing and analyzing the results of the experiments of the 30 graduate students, it was found that among the four independent variables, there was a significant correlation between gender, the size of the experimental object, and the moving direction of the object and the independent variable. This result includes:

1. There have the correlation between the variable of the experimental object size and all the six dependent variables.
2. Gender variables are related to the three variables of the user's single-hand selection, dynamic gesture gesture, and number of fingers touching the object.
3. The movement direction variable is significantly correlated with whether or not the arm is assisted.

Table 1. Correlation analysis of experimental results

		numOfHands	pose	numOfFinger	handFinger	arm	fingerOpen
Gender	Pearson	.218**	−.152**	−.070**	−0.024	−0.003	−0.036
	Sig.	0	0	0.001	0.271	0.897	0.092
	Number	2160	2160	2160	2160	2160	2160
Size	Pearson	.523**	.329**	.522**	.629**	.255**	.591**
	Sig.	0	0	0	0	0	0
	Number	2160	2160	2160	2160	2160	2160
Position	Pearson	0.022	−0.016	0.027	0.011	−0.02	0.015
	Sig.	0.313	0.459	0.214	0.625	0.355	0.49
	Number	2160	2160	2160	2160	2160	2160
moveDirection	Pearson	−0.028	−0.007	−0.008	0.002	−.069**	−0.008
	Sig.	0.194	0.76	0.721	0.937	0.001	0.698
	Number	2160	2160	2160	2160	2160	2160

5 Discussion and Conclusion

After observing the value of Pearson correlation coefficient through Table 1, it can be found that the difference of the moving direction and gender will have a certain influence on the dynamic gesture of the subject, but the effect of the influence is smaller than the effect of the size. The biggest impact on the dynamic gestures of the subjects is the size of the experimental body. As can be seen from Fig. 2, the 5 variables of the dynamic gesture are affected. The arm variable is the least affected (.255), and the order of influence from small to large is pose (.329); numOfFinger (.522); numOfHands (.523); fingerOpen (.591) and influence The most important affected is handFinger (.629).

Through the behavior experiment method, we conclude that the size change of the graphic element will affect the user's dynamic gestures. By sorting and observing the collected code content and gesture video of the tested gestures, you can find that When the subjects of 96 interacted with the objects presented by the 3D graphic element of different sizes, the trend of the dynamic gestures used was also the same, as shown in Fig. 2, so the conclusion can be drawn (the size is expressed in terms of the side length of the largest face, in cm):

1. When the size of the target object is less than 20, the dynamic gesture of the subject tends to interact with the object only with the finger; when the size is greater than 20, the subject tends to join the assistance of the palm.
2. When the target size is less than 20, the subjects tend to interact with one hand; when the target size is greater than 20, the user tends to interact with both hands.
3. When the target size is less than 20, due to the one-handed operation of the subject, the subject is more inclined to interact with 1–2 fingers. The gesture gesture is to clamp it with the finger; When the size is between 20 and 60, the subjects tend to use 3–4 fingers to interact with the object. The gestures used are also grabbed by hand; when the target size is greater than 60, the subjects tend to use The five fingers open the way to interact with the object, while the gesture gesture is the state in which the palm is open.

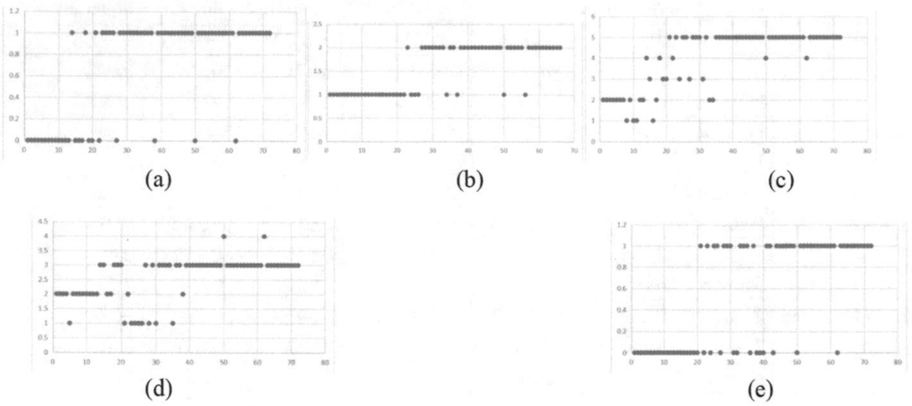

Fig. 2. Correlation between the size of the graphic element interface and the variables of the dynamic gesture. (a) is the variable of handFinger; (b) is the variable of numOfHands; (c) is the variable of numOfFingers; (d) is the variable of pose; (e) is the variable of fingerOpen.

Acknowledgments. The authors wish to thank Science and Technology on Avionics Integration Laboratory and Aeronautical Science Fund (No. 20185569008), supported by "the Fundamental Research Funds for the Central Universities", National Natural Science Foundation of China (no. 71871056), National Natural Science Foundation of China (no. 71471037).

References

1. Billinghurst, M., Clark, A., Lee, G.: A Survey of Augmented Reality. Now Publishers Inc., Delft (2015)
2. Ong, S.K., Yuan, M.L., Nee, A.Y.C.: Augmented reality applications in manufacturing: a survey. Int. J. Prod. Res. **46**, 2707–2742 (2008)
3. Swan II, J.E., Interrante, V.: Special section: best short papers from IEEE virtual reality 2015 guest editors' introduction. Presence **25**, iii (2016)
4. Chen, C.: Research on natural gesture interaction technology based on LeapMotion in industrial robot teaching. Doctoral dissertation
5. Mitra, S., Acharya, T.: Gesture recognition: a survey. IEEE Trans. Syst. Man Cybern. Part C **37**, 311–324 (2007)
6. Sun, H.M., Li, S.P.: The effect of user's perceived presence and promotion focus on usability for interacting in virtual environments. Appl. Ergon. **50**, 126–132 (2015)
7. Yuan, M.L., Ong, S.K., Nee, A.Y.C.: Registration using natural features for augmented reality systems. IEEE Trans. Visual. Comput. Graphics **12**, 560–580 (2006)
8. Sidharta, R.: Augmented reality tangible interfaces for CAD design review. Masters thesis, Iowa State University (2002)
9. Valentini, P.P.: Natural interface in augmented reality interactive simulations. Virtual Phys. Prototyp. **7**, 137–151 (2012)
10. Gaffary, Y., Gouis, B.L., Marchal, M.: AR feels "softer" than VR: haptic perception of stiffness in augmented versus virtual reality. IEEE Trans. Visual. Comput. Graphics **23**, 2372–2377 (2017)
11. James, B., Seung, Y.S.: Cognitive cost of using augmented reality displays. IEEE Trans. Visual. Comput. Graphics **23**, 2378–2388 (2017)
12. Chen, J.: Assessing the use of immersive virtual reality, mouse and touchscreen in pointing and dragging-and-dropping tasks among young, middle-aged and older adults. Appl. Ergon. **65**, 437–448 (2017)
13. Lindgren, R., Tscholl, M.: Enhancing learning and engagement through embodied interaction within a mixed reality simulation. Comput. Educ. **95**, 174–187 (2016)
14. Feng, Z., Yang, B., Xu, T.: Direct operation type 3D human-computer interaction paradigm based on natural gesture tracking. CJC **37**, 1309–1323 (2014)
15. Song, W., Cai, X., Xi, Y.: Real-time single camera natural user interface engine development. MTA **76**, 11159–11175 (2017)
16. Hou, Y., Zhou, H., Wang, Z.: Review of research progress in deep learning in speech recognition. JCA **34**, 2241–2246 (2017)
17. Caggianese, G., Gallo, L., Neroni, P.: Evaluation of spatial interaction techniques for virtual heritage applications: a case study of an interactive holographic projection. FGCS **81**, 516–527 (2018)
18. Nickel, K., Stiefelhagen, R.: Visual recognition of pointing gestures for human–robot interaction. IVC **25**, 1875–1884 (2007)

Fit and Comfort Perception on Hearing Aids: A Pilot Study

Fang Fu and Yan Luximon(✉)

School of Design, The Hong Kong Polytechnic University, Hung Hom, Kowloon, Hong Kong SAR
fang.fu@connect.polyu.hk, yan.luximon@polyu.edu.hk

Abstract. Fit and comfort issues in ergonomic designs contain both physical fit and users' comfort perception, which are important for products with increasing requirements based on user experience, such as hearing aids. In this study, 3D ear was modelled as a reference to determine the size and shape of in-the-ear hearing aids. Four models and twelves sizes of hearing aids were customized designed based on the anthropometric reference model. The prototypes were 3D printed with Acrylonitrile Butadiene Styrene (ABS), Polylactic Acid (PLA), and Nylon materials. The study revealed the differences of user experience with the changes of product parameters, including size, shape and material, which suggested scaling range and proportion for research in customized design of hearing aids. The study contributed to a better understanding of the relationship between anthropometry and hearing aid design for further research on generalizing comprehension of users' perception.

Keywords: Fit and comfort · User perception · Hearing aid design · 3D printing

1 Introduction

As science of design, ergonomics was highlighted when designing workplace, product, or system to fulfill human's demands on fit and comfort [1]. Researchers have investigated users' fit and comfort perception on different products, such as bra [2], footwear [3] and mask [4]. However, related studies on ear-related products mostly focused on audition performances [5, 6], while fit and comfort issues have not been sufficiently studies. Hence, there is a need to study users' fit and comfort perception on ear-related products, especially for long-wearing hearing aids.

Nowadays, ear-related products with various appearances and functionalities have been developed to meet customers' demands. With exploration of ear anthropometry, Lee et al. [7] applied the anthropometric data into ear-tip design with virtual fitting method, while Chiu et al. [8] investigated comfort perception on different models of Bluetooth earphones. Nevertheless, these findings cannot be directly utilized to guide hearing aids design considering the different types of hearing aids, including completely-in-the-canal (CIC), in-the-canal (ITC), in-the-ear (ITE), behind-the-ear (BTE), etc. As one of the main concerns in current hearing aids market [9], fit and comfort issues need to be addressed for related design use.

T. Ahram et al. (Eds.): IHSED 2019, AISC 1026, pp. 360–364, 2020.
https://doi.org/10.1007/978-3-030-27928-8_55

Modern techniques can assist related research efficiently and effectively. 3D scanning technology has been widely used to provide accurate and detailed information of ear anthropometry, and ear molding gives a solution to obtain the anthropometric data of the canal and some part of the concha. Ear anthropometry was conducted with a combination of ear molding and 3D scanning techniques [7, 10–12], as a theoretical reference for related product design. However, its application in product design needs to be explored with scientifically bridging the anthropometry with users' perception.

The main aim of the study is to test the influences of size and shape of hearing aids in fit and comfort perception, as a pilot study for further comprehensive investigation on users' preferences. In the study, 3D scanning was used to acquire the size and shape of external ear to seek for a proper physical fit, and prototypes in different sizes and shapes were 3D printed for user testing to understand users' fit and comfort perception.

2 Method

Prototypes of hearing aids were customized designed based on the ear anthropometric data of individuals. To acquire the anthropometric data related to hearing aid design, ear canal and concha were molded with ear impression silicone (®ABR), and the earmolds were scanned by a 3D scanner (®RangVision). The 3D model was then used as a reference to determine the original size and shape of the hearing aids.

Different types of hearing aids were designed including both full-shell and half-shell hearing aids, with and without the 1st bend of ear canal. Twelve sizes were generated with scaling the original size for each individual, with scaling proportions of 70%, 75%, 80%, 85%, 90%, 95%, 100%, 105%, 110%, 115%, 120% and 125%. Different materials,

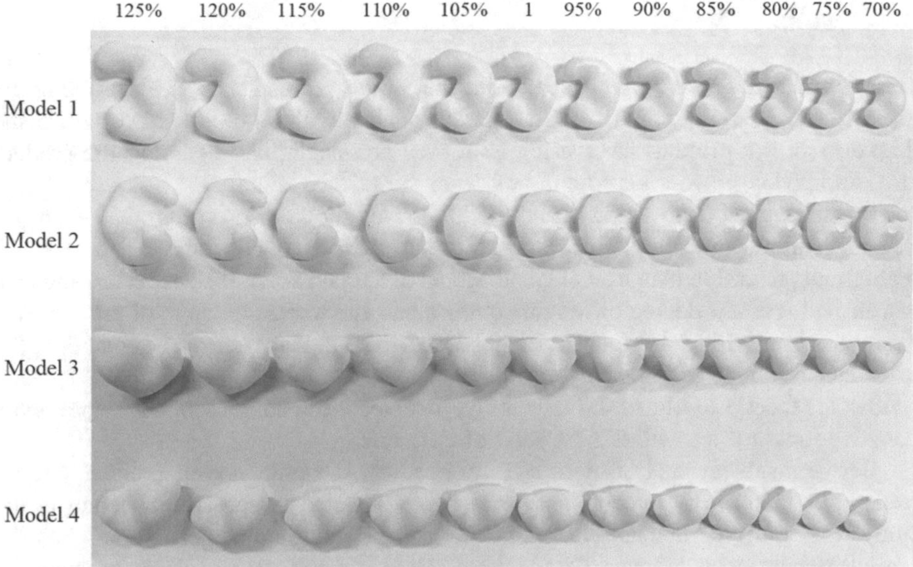

Fig. 1. Examples of prototypes: ***Model 1***: full-shell hearing aid without ear canal fitting; ***Model 2***: full-shell hearing aid with ear canal fitting; ***Model 3***: half-shell hearing aid with ear canal fitting; and ***Model 4***: half-shell hearing aid without ear canal fitting.

Polylactic Acid (PLA), Acrylonitrile Butadiene Styrene (ABS), and Nylon 12, were then 3D printed for further testing. Figure 1 presents the different shapes and sizes of hearing aid prototypes.

During the prototype test, the participant was asked to wear each prototype for 1 minute, and then fill a 7-point satisfaction Likert scale questionnaire on fit and comfort perception. There was a one-minute break between two tests. In the study, fit referred to the physical condition on the use of the prototype, for example, too lose or to tight was considered as dissatisfaction on fit, while comfort highlighted the cognitive preferences regarding comfort perception. Figure 2 shows examples of experiment setting.

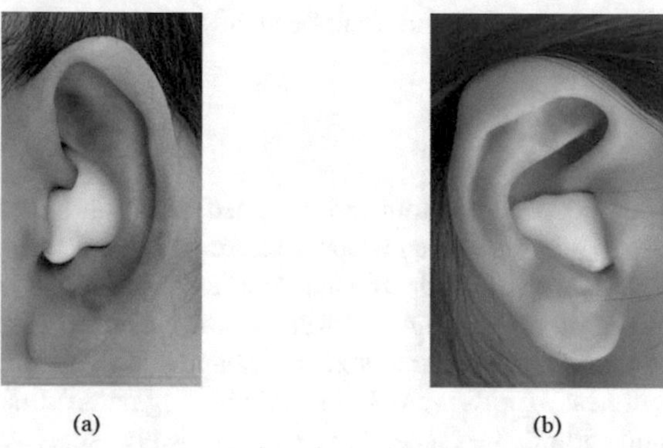

Fig. 2. Examples of experimental setting: (a) testing full-shell hearing aid; (b) testing half-shell hearing aid.

3 Results

Different materials were used to 3D print the prototypes of hearing aids. Among the three materials, ABS provided best printing quality with smooth surface, while the thickness of the printing layer using PLA was non-negligible regarding the product size, and Nylon powder kept falling off from the surface slightly.

Within-subject analysis was conducted to survey the influence of product shape on fit and comfort perception. According to the user experience, Model 1 and Model 4 were more preferable than Model 2 and Model 3. Certain sizes of selected hearing aids can provide certain degree of fit and comfort, and the tolerance range of product size decreased from Model 4, Model 1, Model 3, to Model 2. Under the same condition, the most satisfactory size varied upon different model, which decreased from Model 4, Model 1, Model 3 to Model 2. Particularly, participant felt anxious when the prototype contained part fitting with the 1st bend of ear canal.

Between-subject analysis was used to examine the influence of product size on users' fit and comfort perception. Model 1 was selected for prototype testing in the study. Generally, participants feel relatively bad fit and comfort perception when the scaling proportion is below 80% or above 110%. Among all the scaling proportions, 90% and 100% mostly meet the most satisfaction on fit and comfort perception.

4 Discussion

The study revealed the differences of user experience with the changes of product parameters, including material, shape and size of hearing aids. Application of these findings needs to be discussed with addressing the limitations and future application.

Differences among ABS, PLA, and Nylon to 3D print the hearing aid prototypes were intuitively observed. In the study, Fused Deposition Modelling (FDM) was used to print PLA, the accuracy of which was determined with the printing layer thickness (0.4 mm). Since the size of hearing aids is relatively small, the accuracy of PLA model obviously affected the prototype quality. Specifically, the prototype surface was rough which may influence uses' comfort perception, and even some surface cannot be printed fully. Using Selective Laser Sintering (SLS), Nylon material can provide higher accuracy (0.1–0.2 mm), but the principle of the technique brought same challenge of rough surface. ABS was 3D printed with Stereo Lithography Apparatus (SLA) technique with accuracy of 0.1 mm, which provided smooth surface of the complete hearing aids model. Therefore, ABS is more suitable for future hearing aid prototype test use.

Shape of hearing aids has noteworthy influence on users' fit and comfort perception. Model 2 and 3 provides fit on 1st bend of ear canal, while Model 1 and 4 deleted related part fitting with 1st bend of ear canal. Even though Model 2 and 3 can provide a better fit with the cavum concha and ear canal, Model 1 and 4 are more preferable than Model 2 and 3, which can be explained with the human sensitivity on ear canal. The tolerance size range of half-shell hearing aid is wider than the full-shell design, and the most satisfactory size of half-shell hearing aids is bigger than full-shell ones, which suggested different scales need to be explored for specific product shape.

Users' fit and comfort perception varied upon the product sizes. In the study, when the scaling proportion is below 80%, the prototypes were excessively loose to fall off from the external ear. With scaling proportion of 110% or above, the prototypes were extremely tight to congest the concha, and some of them even cannot be wear on the external ear. These findings suggested that proper fit can be further investigated within the scaling range of 80%–110% for full-shell hearing aids.

One of the limitations in the study was restricted material and equipment for 3D printing. With relatively less time-consuming and easy-to-use characteristics, ABS, PLA, and Nylon were selected in the study to create the hearing aids prototypes. The paper only discussed the application of 3D printing for hearing aids design use. Further study can be conducted on more types of hearing aids to generalize a more comprehensive understanding between anthropometry and user preferences.

5 Conclusion

The study provided a primary understanding of user's perception on hearing aids with influences of size and shape. With bridging the product parameters with fit and comfort perception in this study, future research can be conducted to generalize users' fit and comfort perception for designing ear-related products. In addition, these findings also can be considered as a validation of 3D anthropometry of external ear with 3D scanning techniques.

Acknowledgments. The research is fully supported by Hong Kong RGC/GRF project B-Q57F.

References

1. Pheasant, S.: Bodyspace: Anthropometry, Ergonomics and the Design of Work: Anthropometry, Ergonomics and the Design of Work. CRC Press (2014)
2. Lawson, L., Lorentzen, D.: Selected sports bras: comparisons of comfort and support. Cloth. Text. Res. J. **8**(4), 55–60 (1990)
3. Au, E.Y.L., Goonetilleke, R.S.: A qualitative study on the comfort and fit of ladies' dress shoes. Appl. Ergon. **38**(6), 687–696 (2007)
4. Lee, W., Kim, H., Jung, D., Park, S., You, H.: Ergonomic design and evaluation of a pilot oxygen mask. In: Proceedings of the Human Factors and Ergonomics Society Annual Meeting, vol. 57, no. 1, pp. 1673–1677. SAGE Publications, Los Angeles, CA (2013)
5. Hagerman, B., Gabrielsson, A.: Questionnaires on desirable properties of hearing aids. Scand. Audiol. **14**(2), 109–111 (1985)
6. Lunner, T.: Cognitive function in relation to hearing aid use. Int. J. Audiol. **42**, S49–S58 (2003)
7. Lee, W., Jung, H., Bok, I., Kim, C., Kwon, O., Choi, T., You, H.: Measurement and application of 3D ear images for earphone design. In: Proceedings of the Human Factors and Ergonomics Society Annual Meeting, vol. 60, no. 1, pp. 1053–1057. SAGE Publications, Los Angeles (2016)
8. Chiu, H.P., Chiang, H.Y., Liu, C.H., Wang, M.H., Chiou, W.K.: Surveying the comfort perception of the ergonomic design of bluetooth earphones. Work **49**(2), 235–243 (2014)
9. McCormack, A., Fortnum, H.: Why do people fitted with hearing aids not wear them? Int. J. Audiol. **52**(5), 360–368 (2013)
10. Chiou, W.K., Huang, D.H., Chen, B.H.: Anthropometric measurements of the external auditory canal for hearing protection earplug. In: Advances in Safety Management and Human Factors, pp. 163–171. Springer, Cham (2016)
11. Ji, X., Zhu, Z., Gao, Z., Bai, X., Hu, G.: Anthropometry and classification of auricular concha for the ergonomic design of earphones. Hum. Factors Ergon. Manuf. Serv. Ind. **28**(2), 90–99 (2018)
12. Unal, G., Nain, D., Slabaugh, G., Fang, T.: Customized design of hearing aids using statistical shape learning. In: International Conference on Medical Image Computing and Computer-Assisted Intervention, pp. 518–526. Springer, Heidelberg (2008)

Improvement in the Quality of Life of Patients with End-Stage Renal Failure Who Live Without Replacement Therapy in Mexico

Ana Paula Pelayo(✉), Montserrat Pelayo, and Gabriela Duran-Aguilar

Facultad de Ingeniería, Universidad Panamericana, Álvaro Del Portillo #49, Ciudad Granja, 45010 Zapopan, Jalisco, Mexico
0190632@up.edu.mx

Abstract. Chronic Renal Disease (CRD) is a failure of the kidney. CRD caused by diabetes mellitus is reported to be the third mortality cause in Mexico. The Public Health System in Mexico is deficient in covering all the Replacement Therapy (RT) sessions needed by the population, leaving more than half unattended in that matter, people who therefore, due to the high cost of these sessions and their inability to pay for them, experience a deteriorated Quality of Life (QOL) converting in death. A high relation between the social area and QOL has been found in cases studying palliative care with patients receiving RT. An aim in the emotional wellbeing, focusing on depression was then taken. A solution was created, a system with a data base using cross-reference designed by clinical experts as an attempt to provide Cognitive Behavioral Therapy in high proportions and a lower cost.

Keywords: Renal Insufficiency · Quality of Life · Emotional wellbeing · Lack of replacement therapy

1 Introduction

CRD is the loss of function in the kidneys [1]. Renal Insufficiency (RI) is the capability loss of blood waste filtration. It is progressive and irreversible, becoming deathly. Its main causes to 2014 in Mexico were diabetes mellitus 53,4%, hypertension 35,5%, chronic glomerulonephritis 4,2%, being the most affected age group, the one over 40 years old [2].

There are 5 stages in the disease, being stage number 5, the one were the disease is considered to be RI and it is then when RT is needed.

CKDs growth rate in Mexico is of 11% per year, without the diseased [3]. Between 2000 and 2012, overweight rates grew from 62% to 71% in the adult population. One out of three children have obesity; this is a significant cause associated to diabetes, a CKD main cause [10].

Available treatment could be divided in two; conservative treatment and replacement therapy (RT). Conservative treatment is such as diets, medication, electrolytic and hormonal correctors, and its objective is to prevent long term sequels and make the progression of the disease as slow as possible. There are three types of RT: peritoneal

T. Ahram et al. (Eds.): IHSED 2019, AISC 1026, pp. 365–369, 2020.
https://doi.org/10.1007/978-3-030-27928-8_56

dialysis (PT), hemodialysis and renal transplant. PD consists of the blood waste filtration through a catheter in the peritoneum, a fine membrane that surrounds the peritoneal or abdominal cavity. Hemodialysis consists of using a machine with a special filter in which blood flows through an access point, be it were it is found more convenient.

According to a case study by the UNAM [4] hemodialysis treatment represented a yearly expense of $159,000 Mexican pesos in the public sector and $168,000 Mexican pesos in the private sector. Peritoneal dialysis represents a lower expense than hemodialysis, but not all patients are candidate to peritoneal dialysis given the effectiveness of hemodialysis over PD. A renal transplant costs about $200,000 Mexican pesos in the public sector and doubles its price in the private sector.

In the Mexican Public Health System, because of the differentiation between health services, due to the different working status; not all the affected population has access to RT, and even those who do, do not always have access to all the sessions needed [7]. This leaves half of the population without RT and the other half with a low QOL [8], due to the lack of complete service from the Public Health System. The population left with no access to RT is therefore left to either pay their own RT in the private sector, sometimes with help of NGOs, or left to die.

According to a case study surveying the QOL in patients with PD and hemodialysis, PD was found to have a better QOL, being the physical sphere the most affected [5]. Even though the physical domain may have been recognized as the most affected by people in RT, the psychological domain is the main sphere related to QOL. It was also found depression to be a common suffering in people with RI [8], nevertheless, this area is not covered and receives little care besides medication (when given), even when this suffering is a meaningful cause in the detriment of QOL and doubles the probability of death or hospital treatment [9]. A lack of information regarding the QOL in people with no access to RT was strongly punctuated. Hence, is was speculated that these patients are not taken much into account.

As a result of the difficulties related with the illness, diagnosis and treatment; patients with CKD may experience significant emotional and psychological issues, without treatment, may have a negative impact in their health and wellbeing [6].

All taken into account, the aim of the study was in patients with RI with no access to RT given a lack of economic resources, be them their own or the government's that seek public medical attention in public health centers in Guadalajara, Jalisco, México. A solution to improve QOL in this patients was the main focus.

2 Methodology

Given the ambiguity of the general problematic, a design thinking approach was used while researching the final user, from an empathy perspective. This in order to obtain a deep approach towards the bad QOL causes and a correct definition of the specific problem.

An emotional design approach was looked for, given that emotions were thought to play a role in which decisions are made based on the necessity of the end-user, no matter how high the cost, the QOL of the end-user is thought to have a great value to the caregivers, thus a solution to improve their QOL is considered as desirable.

Total Design Methodology was used as a main structure for the project, in an attempt to make sure to cover all relevant aspects, with exception of the first and last phases, were design thinking and emotional design were also considered. Design thinking as an approach to the end-user, and emotional design in the last phases of definition of the product/service.

Interviews and observation of their surroundings were carried out with RI patients with access as well as patients with a lack of access to RT in the public health service in Jalisco, México. Interviews were also carried out with nephrologists and psychologists. It became known that the life expectancy of the people with no access to RT after diagnosis is of about 14–16 months. A strong emphasis was made towards the government to establish public policies for the palliative care of the patient, as for the psychological care, given that a need is expressed by nephrologists, as from patients and caregivers, in all patients to receive this type of attention, nevertheless, only extreme cases are referred to it due to a lack of economic possibility.

3 Results

Based on the research done, an aim towards patients with depression with no access to RT or any type of therapy was taken. The variables taken into account were the education level, obtained in the interviews, the constant lack of internet connection and their general condition, often as a terminal patient. We achieved a possible solution through a system. The aim is to provide the closest thing to a psychological therapy to patients without the need of a present therapist or internet connection. The system is composed of an app as service design and a set of products; a pillow and a lamp. The pillow (Fig. 1) was included thinking that terminal patients remain most of their time in bed. Its form was obtained from an already existing ergonomic pillow. It is made of 30 HD polyurethane foam in the core along with Memory Foam as a cover. The system in itself is based on a cross-reference program created among experts, which processes the entry given by the patient from the microphone in the pillow and with the help of a remote (Fig. 2) connected to the pillow, this entry is ran through the cross-reference system, included already in the remote, in order to obtain feedback, such as therapy. No immediate connection is required to access the data base, but in the case of not finding the right answer, the recording is saved so when connected to the app, it can be ran through a larger data base accessed through Wi-Fi, or as a last resource, sent to an expert. Each patient will have their profile, in which, after a first expert approach (via telephone or face to face); the smaller data base will be downloaded to their profile, to then be downloaded to the remote. The lamp (Fig. 3) has warm light, given that it was found to help soothe people [11]. This last consists of a LED lightbulb included in the package, and an outer casing made of the cardboard from the package itself.

Fig. 1. Pillow

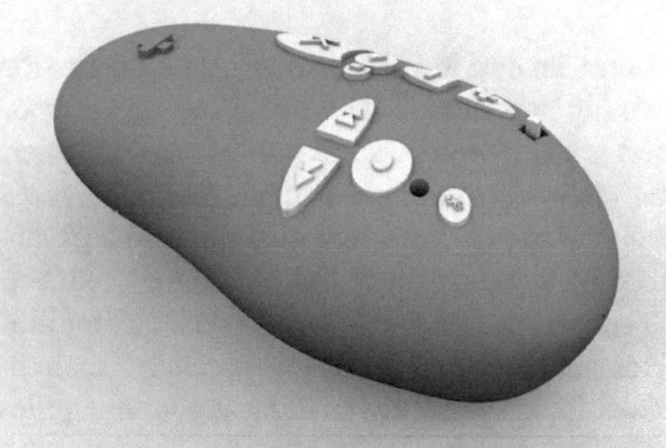

Fig. 2. Remote

Fig. 3. Lamp

4 Conclusion

During the development of the project, the need for a product that created accessibility for a service, otherwise unobtainable, was determined. After interviews and observations of the surroundings of the target user, an analysis was made, locating our product as a priority, but at a higher cost than desired. The function and objective of the product is still considered as necessary, given that according to the investigation, it is seen as an overlooked area, thus not covered yet in any way. The design will be rethought with the aim of a better and more accessible approach to the user, also given the existence of a limiting element, which is the lack of culture towards emotional care in Mexico, even though a need for it is shown.

References

1. National Institute of Diabetes and Digestive and Kidney Diseases. https://www.niddk.nih.gov/
2. Secretaría de Salud: Tratamiento sustitutivo de la función renal. Diálisis y Hemodiálisis en la insuficiencia renal crónica. Cenetec, Ciudad de México (2014)
3. Tamayo-y-Orozco, J.A., Lastiri Quiros, S.: La enfermedad renal crónica en México: Hacia una política nacional para enfrentarla. Intersistemas, Ciudad de México (2016)
4. Durán-Arenas, J.L., Ávila-Palomares, P.D., Zendejas-Villanueva, R., Vargas-Ruiz, M.M., Tirado-Gómez, L.L., López-Cervantes, M.: Direct cost analysis of hemodialysis units. Salud Pública de México **53**(4), 516–524 (2011)
5. Hernández Mariano, J.A.: Calidad de vida en pacientes con tratamiento sustitutivo renal: Diálisis Peritoneal Continua ambulatoria y hemodiálisis. Rev. Iberoamericana de Educación e Investigación en Enfermería, **4**, 67–74 (2014)
6. Taylor, F., Taylor, C., Baharani, J., Nicholas, J., Combes, G.: Integrating emotional and psychological support into the end-stage renal disease pathway: a protocol for mixed methods research to identify patients' lower-level support needs and how these can most effectively be addressed. J. BMC Nephrol. **17**, 111 (2016)
7. Tirado-Gómez, L.L., Durán-Arenas, J.L., Rojas-russell, M.E.: Las unidades de hemodiálisis en México: una evaluación de sus características, procesos y resultados. Salud Pública de México **53**(4), 491–498 (2011)
8. Leiva-Santos, J.P., et al.: Cuidados de soporte renal y cuidados paliativos renales: revisión y propuesta en terapia renal sustitutiva. R. Nefrología (Madrid) **32**(1), 20–27 (2012)
9. Hedayati, S.S., Yalamanchili, V., Finkelstein, F.O.: A practical approach to the treatment of depression in patients with chronic kidney disease and end-stage renal disease. Kidney Int. **81**(3), 247–255 (2012)
10. Domínguez, J., Gutiérrez, J.: Estudios de la OCDE sobre los Sistemas de Salud: México 2016. OECD Publishing, Paris (2016)
11. Kuijsters, A., Redi, J., de Ruyter, B., Heynderickx, I.: Lighting to make you feel better: improving the mood of elderly people with affective ambiences. J. Plos One **10**(7), e0132732 (2015)

The VITO (pn 20150100457, 2015): Novel Training Kit to Limit Down the Learning Curve of the Upper GI Endoscopy Operations

Constantinos S. Mammas(✉) and Adamantia S. Mamma

Fellowships of Excellence for Postgraduate Studies in Greece-Siemens Program 2014-16, State Scholarship Foundation, Vas.Sofia's Str. 92, 11528 Athens, Hellas
csmammas@med.uoa.gr

Abstract. The VITO (pn 20150100457, 2015) is a novel training kit which has been designed to be portable, light, mobile, and non-expensive, adjusted to the clinical needs of both beginners and advanced surgeons allowing a continuous and systematic self-centered and/or collaborative training in Surgical Endoscopy. The VITO assisted endoscopy training kit impact on the learning curve- in terms of the so called "machine learning curve" which refers to the assessment of the learning time limits computation- of specific upper GI endoscopy sessions by the part of a trainee in Endoscopy but specialist surgeon (considered as beginner) on a self-centered learning mode, is evaluated.

Keywords: Training · Education · Learning sciences

1 Introduction

Despite the great efforts it is established that the current training environment for the surgical endoscopy training has not changed dramatically even for the teaching hospitals in Europe. In fact there is training in endoscopy on the site and there is a single nominated training lead. The assessment team advises the unit to ensure that training lists are adjusted where appropriate to meet the needs of trainees [1, 2]. Also, the Health System Organizations are encouraged to identify some administrative resource to support the appropriate co-ordination of training lists. The feedback is obtained from trainees during training. Active use of the Endoscopy Training System (JETS) e-portfolio is strongly encouraged to capture trainee feedback [1, 3–5].

It is important that internet access is established in all endoscopy procedure rooms to support the use of JETS. The above mentioned are being currently addressed mainly to gastroenterologists while trainee surgeon do not share the same privileges except from some teaching hospitals units that traditionally adopted surgical endoscopy to their teaching program. On the other hand the training curve is even more steep for the so called Emergency or Acute Surgical Endoscopy. In this field trainees- even trainee Gastroenterologists- have limited exposure to elective and emergency endoscopy procedures at the Hospitals [2–5].

T. Ahram et al. (Eds.): IHSED 2019, AISC 1026, pp. 370–376, 2020.
https://doi.org/10.1007/978-3-030-27928-8_57

Taking into consideration the abovementioned standards in Surgical Endoscopy that are in power mostly for trainees and trainers Gastroenterologists and that according to the ASGE guidelines for optimizing quality in diagnostic and therapeutic Endoscopy the **frequency** of applied Endoscopic procedures is the most crucial indicator for **quality in endoscopy**, we invented the **Vito (pn:20150100457, 2015) training kit** and developed it clinically in the context of the "Program of Excellence 2014-16" –so as to give the opportunity to Surgical Endoscopy trainees and the specialists to learn the prerequisites and indications and above all to practice without limits their technical skills in a self-centered or mainly in a collaborative way facing the technical challenges of the clinical diagnostic and therapeutic Gastroscopy, ERCP and EUS on both elective and acute endoscopic operations, until the end of 2016 [3–14].

The first presentation and workshop on 15th of Mars 2018 focused on the description of the applied technology and the technological parts of VITO (pn 20150100457, 2015) training kit i.e. 3 training maps, a special device, and a user manual (Fig. 1. Depicts the 2D map and parts of VITO (pn 20150100457, 2015) training kit as demonstrated on 15th of Mars 2018).

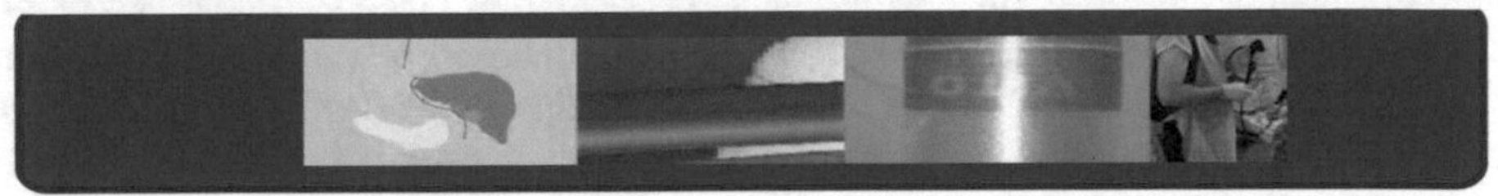

Fig. 1. The 2D map for ERCP VITO (pn 20150100457, 2015) training kit based ERCP training and the 3D printing parts of the training kit as partially is presented in the demo CD.

The VITO assisted endoscopy training kit impact on the learning curve- in terms of the so called ''machine learning curve'' which refers to the assessment of the learning time limits computation- of the initial maneuvers in endoscopy which are the initiation and the retrieval of the endoscope in Gastroscopy and in the ERCP respectively by the part of a trainee in Endoscopy- but a specialist surgeon- (considered as beginner) on a self-centered learning mode- is evaluated [1, 3, 4, 6].

2 Paper Content and Technical Requirements

Given that the machine learning curve is useful for many purposes including comparing different algorithms and choosing model parameters during design adjusting optimization to improve convergence, and determining the amount of data used for training the abovementioned assessment we standardized the amount of time needed for each operation separately to reach the "learning" limit in terms of time to perform the defined operation [3, 4]. Several series of a number of sessions of sets of operative trials with the VITO training kit on each training day is proposed and given that an amount of 8-10 series of ten sessions are necessary in average, beginners (these are surgeon that had never operated on any emergency operation [15–19] before or they have operated on a limited number of the abovementioned operations before) should spend the training time (two to three times per week) to train them-selves for each operation separately before going on to the next operation [9, 10].

As far as the advanced surgeons are concerned these are those who deserve perfection in both elective and emergency endoscopic surgery, and that is why is proposed that their training should include all types of the abovementioned therapeutic operations (therapeutic Gastroscopy, ERCP, EUS) on each training day by repeating a series as analyzed for two or better three training days per week in a period of a month [3–5, 20–22].

The operational criteria that have to be fulfilled in each operation by the part of the trainee are written in the manual of the kit and consists the diligent preparation of the prerequisites and the collection of instruments needed for each endoscopic operation and the reference of the safety criteria according to the third map of the kit before each operational repetition [6]. The applied technique is upon the necessary operational steps and the key maneuvers which are also described in the manual for each elective or emergency operation separately for both diagnostic and therapeutic Gastroscopy and ERCP. The VITO (pn 20150100457, 2015) training kit used in combination with the necessary set of devices for upper GI endoscopy i.e. Gastroscopy or ERCP (the central unit, the monitor and the flexible endoscope = Gastroscope or Duodenoscope according to the training mode) by a beginner trainee in flexible endoscopy, but specialist surgeon who had never applied any type of upper GI endoscopy in the past in a period of six months. During the experimental period the trainee assisted in upper GI clinical endoscopy sessions on human subjects without having the right to perform any by himself [6].

The metric that used for learning was the mean time for finishing the task in terms of the analysis of the mean time of each series spent in each training kit based set: i. Diagnostic gastroscopy session (to enter, to retrieve and the total) and ii. Diagnostic ERCP session (to enter, to retrieve and the total), to compute the sequential training session after which the mean time spent for training was no statistically different than the previous training sessions (ANOVA, SPSS 17.0).

2.1 Designing for Human Factors: VITO (Pn 20150100457, 2015) Based Gastroscopy Initial Machine Learning Curve and Clinical Gastroscopy Learning Curve (Initiation and Retrieval of the Endoscope)

Referring to VITO (pn 20150100457, 2015) based Gastroscopy (initiation and retrieval of the Endoscope) training, the trainee's trials consisted of fourteen series (N = 14) of ten (n1 = 10) kit based Gastroscopies sets of ten trials (n2 = 10) for each series in each session on several time points in a period of six months. Progressively, after the 8th series, which took place after a month of training no statistically significant difference was noticed among sessions (20,7 ± 1,8, p = 0.663) as depicted on Table 1. (Figure 2. depicts the measurements). The trainee managed to perform Gastroscopy on a patient after a month [3].

Table 1. Results of the statistical comparisons among measurements of the Total Time (sec) in the series of trials using the VITO (pn 20150100457, 2015) for Gastroscope initiation and retrieval in Gastroscopy training.

Total time	Measurement					
Measurement	1	2	3	4	8	9
1	–					
2	**<0,001**					
3	0,124	**0,010**				
4	**<0,001**	1,000	**<0,001**			
8	**<0,001**	**<0,001**	**<0,001**	**0,006**		
9	**<0,001**	**<0,001**	**<0,001**	**<0,001**	0,663	
13	**<0,001**	**<0,001**	**<0,001**	**<0,001**	1,000	1,000

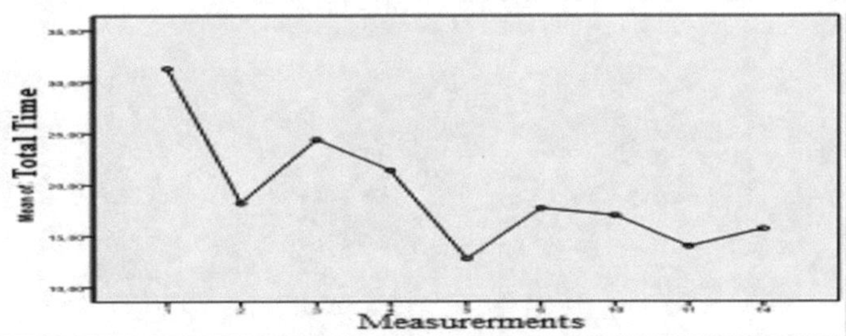

Fig. 2. After the 8th series of trials- in a series of 13 successive series of trials the statistical comparison of the mean Total Time of measurements- it was not observed any significant statistical difference among the following successive measurements (i.e. 9th, 10th, 11th, 12th, 13th) in a period of a month of training using the VITO training kit to learn the initiation and the retrieval of the Gastroscope in the diagnostic upper GI endoscopy.

2.2 Designing for Human Factors: VITO (Pn 20150100457, 2015) Based ERCP Initial Machine Learning Curve and Clinical ERCP Learning Curve (Initiation and Retrieval of the Duodenoscope)

Referring to VITO (pn 20150100457, 2015) training kit based ERCP trainee's trials using a Duodenoscope (initiation and retrieval of the Duodenoscope) in the same training session, consisted of eleven series (N = 11) of ten (n1 = 10) kit based ERCPs sets of ten trials (n2 = 10) for each series in each session, in the same period of time [4].

Progressively, after the 6th series, which took place also after a month of training no statistically significant difference was noticed among sessions (17,7 ± 1,5 s, p < 0,110) as given in Table 2. The trainee managed to perform a clinical diagnostic ERCP on a patient after six months (Fig. 3. depicts the measurements).

The VITO assisted endoscopy training kit impact on the learning curve of the initial or complex maneuvers in diagnostic and therapeutic endoscopy seems really promising not only for beginners but also for advanced surgeons and nurses under specific standards of its design and consequent ergonomics in the context of training [3–6].

Table 2. Results of the statistical comparisons among measurements of the Total Time (sec) in the series of trials of using the VITO (pn 20150100457, 2015) for Duodenoscope initiation and retrieval in the ERCP training.

Total time (sec)	Measurement							
Measurement	1	2	3	4	5	6	10	11
1	–							
2	**<0,001**							
3	**<0,001**	**<0,001**						
4	**<0,001**	0,356	0,553					
5	**<0,001**	**0,001**	**<0,001**	**<0,001**				
6	**<0,001**	1,000	**<0,001**	0,110	**0,004**			
10	**<0,001**	1,000	**<0,001**	**0,018**	**0,030**	1,000		
11	**<0,001**	**0,030**	**<0,001**	**<0,001**	1,000	0,110	0,553	
14	**<0,001**	1,000	**<0,001**	**<0,001**	0,683	1,000	1,000	1,000

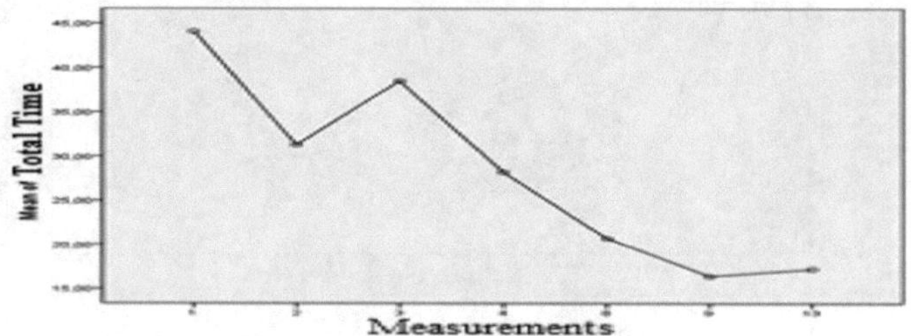

Fig. 3. After the 6th series of trials- in a series of 11 successive series of trials the statistical comparison of the mean Total Time of measurements- it was not observed any significant statistical difference among the following successive measurements (i.e. 7th, 8th, 9th, 10th, 11th, 12th, 13th, 14th) in a period of a month of training using the Vito training kit to learn the initiation and the retrieval of the duodenoscope for ERCP.

In fact, further endoscopic sessions are being developed and refer to the continuation of this particular innovative product engineering and improvement in the direction of advanced interventional Gastroscopy, therapeutic ERCP and diagnostic EUS in an hybrid type of self-centered and/or collaborative learning and training until the end of 2017 [3–6].

Currently and focused on the future training method and on the training standards of self- centered and/or collaborative training and learning using the VITO (pn 20150100457, 2015) training kit, is being tried in detail for standardization of: 1. The training technique, 2. The process of training and learning, 3. The ergonomics, and of 4. The results of its quality control integrated with the clinical performance of the trainees and/or the endoscopic teams in terms of clinical excellence in endoscopic performance with a priority to the technique, the time-point of the clinical performance from the part of the trainee and the therapeutic effect (clinical therapeutic effect and safety i.e. decrease in morbidity and mortality) in diagnostic and therapeutic endoscopy in Gastroscopy, in ERCP and in EUS of specific adult and pediatric elective and emergency endoscopic operations depicted on Table 3. [20–22].

Table 3. To further evaluate the VITO (pn 20150100457, 2015) training kit assisted learning curve in terms of "machine learning" for each elective and/or emergency endoscopy session on a self-centered learning mode and on a collaborative training mode the following elective and emergency endoscopic operations are being assessed:

I. ERCP	II. Upper-GI therapeutic endoscopy:	III. EUS-Diagnostic and therapeutic interventions
A. Sphincterotomy	A. Oesophageal stent placement	A. Pancreatic pseudocyst drainage
B. Biliary stone extraction	B. Management of ingested foreign bodies and food impactions	
C. Biliary stent drainage	C. Endoscopy in enteral feeding	
D. Pancreatic pseudocust drainage and pancreatic leak and fistulas management	D. Endoscopy in the management of acute non-variceal upper GI bleeding	
E. Papillectomy	E. Endoscopy in the management of variceal hemorrhage	
F. Pancreatic stent drainage and stone extraction	F. Endoscopy in gastroduodenal obstruction	

3 Conclusion

The VITO (pn 20150100457, 2015) training kit seems to be feasible and reliable for both beginners and advanced in diagnostic and therapeutic endoscopy surgeons in training to limit down the learning curve of initiation and retrieval of the endoscope in Gastroscopy and in ERCP. Its usefulness is based on the unique design to allow limitless practice so as to build a continuous self-centered and/or team endoscopy training and habituation pursuing excellence. Further experimental evaluation of its impact on the learning curve of specific diagnostic and therapeutic endoscopic interventions for advanced surgeons, combined with further industrialization and design's modification on a self-centered or collaborative learning mode in Gastroscopy, ERCP and EUS training, is being developed.

References

1. Defining and measuring quality in endoscopy Gastrointest Endosc, **81**, 1–2 (2015). http://dx.doi.org/10.1016/j.gie.2014.07.052
2. Quality indicators common to all GI endoscopic procedures Gastrointest Endosc, **81**, 3–16 (2015). http://dx.doi.org/10.1016/j.gie.2014.07.055

3. Quality indicators for EGD Gastrointest Endosc, **81**, 17–30 (2015). http://dx.doi.org/10.1016/j.gie.2014.07.057
4. Quality indicators for ERCP Gastrointest Endosc, **81**, 54–66 (2015). http://dx.doi.org/10.1016/j.gie.2014.07.056
5. Quality indicators for EUS Gastrointest Endosc, **81**, 67–80 (2015). http://dx.doi.org/10.1016/j.gie.2014.07.054
6. Steadman, R.H., Coates, W.C., Huang, Y.M., et al.: Simulation-based training is superior to problem-based learning for the acquisition of critical assessment and management skills. Crit. Care Med. **34**, 1, 151–157 (2006)
7. Ensuring competence in endoscopy prepared by the ASGE/ACG Taskforce on Quality in Endoscopy
8. Quality indicators for GI endoscopic procedures - complete set Gastrointest Endosc, **81**, 1–80 (2015)
9. Guidelines for safety in the gastrointestinal endoscopy unit Gastrointest Endosc, **79**, 363–372 (2014). http://dx.doi.org/10.1016/j.gie.2013.12.015
10. Quality indicators for gastrointestinal endoscopy units Video GIE, **2**(6), 119–140. http://dx.doi.org/10.1016/j.vgie.2017.02.007
11. Methods of Privileging for New Technology in Gastrointestinal Endoscopy Gastrointest Endosc, **50**, 899–900 (1999). http://dx.doi.org/10.1016/S0016-5107(99)70190-2
12. Guidelines for privileging, credentialing, and proctoring to perform GI endoscopy Gastrointest Endosc, **85**, 273–281 (2017). http://dx.doi.org/10.1016/j.gie.2016.10.036
13. Endoscopy by nonphysicians Gastrointest Endosc, **69**, 767–770 (2009). http://dx.doi.org/10.1016/j.gie.2008.11.006
14. Informed consent for GI endoscopy Gastrointest Endosc, **66**, 213-218 (2007). Reviewed and reapproved November 2014. http://dx.doi.org/10.1016/j.gie.2007.02.029
15. The role of endoscopy in enteral feeding Gastrointest Endosc, **74**, 7–12 (2011). Reviewed and reapproved May 2016. http://dx.doi.org/10.1016/j.gie.2010.10.021
16. The role of endoscopy in the management of variceal hemorrhage Gastrointest Endosc, **80**, 221–227 (2014). http://dx.doi.org/10.1016/j.gie.2013.07.023
17. The role of endoscopy in the management of acute non-variceal upper GI bleeding Gastrointest Endosc, **75**, 1132–1138 (2012). http://dx.doi.org/10.1016/j.gie.2012.02.033
18. The role of endoscopy in gastroduodenal obstruction and gastroparesis Gastrointest Endosc, **74**, 13–21 (2011). http://dx.doi.org/10.1016/j.gie.2010.12.003
19. Management of ingested foreign bodies and food impactions Gastrointest Endosc, **73**, 1085–1091 (2011). http://dx.doi.org/10.1016/j.gie.2010.11.010
20. Adverse events of upper GI endoscopy Gastrointest Endosc, **76**, 707–718 (2012). http://dx.doi.org/10.1016/j.gie.2012.03.252
21. Adverse events associated with ERCP Gastrointest Endosc, **85**, 32–57 (2017). http://dx.doi.org/10.1016/j.gie.2016.06.051
22. Adverse events associated with EUS and EUS with FNA Gastrointest Endosc, **77**, 839–843 (2013). http://dx.doi.org/10.1016/j.gie.2013.02.018

User Experience and Virtual Environments

Simulating Social Cycling Experience in Design Research

Nan Yang[1(✉)], Gerbrand van Hout[2], Loe Feijs[1], Wei Chen[3], and Jun Hu[1]

[1] Department of Industrial Design, Eindhoven University of Technology, 5600 MB Eindhoven, The Netherlands
{n.yang,l.m.g.feijs,j.hu}@tue.nl
[2] Obesity Centre, Catharina Hospital, 5623 EJ Eindhoven, The Netherlands
gerbrand.v.hout@catharinaziekenhuis.nl
[3] Department of Electronic Engineering, Fudan University, 200433 Shanghai, China
w_chen@fudan.edu.cn

Abstract. This paper shows an experimental simulation system that simulates the experience of social cycling in the process of design research. The system has been used in two iterative studies to evaluate different design concepts related to cycling and social interaction. We introduce the build and implementation of the simulation system and discuss the role of an experimental prototype in design research.

Keywords: Social interaction · Simulation · Design research · Prototyping · Experimental prototype

1 Introduction

The experimental simulation system presented in this paper is built to evaluate the design concepts of SocialBike—a digital augmented bicycle that aims to increase cyclists' motivation in physical activity by showing their quantified-self data to each other. The initial concept of SocialBike consists of a mobile application and a rear on-bike display. The mobile application can get data from various health tracking devices through Google Fit API (Fig. 1) [1, 2]. When using SocialBike, cyclists can select the type and form of their quantified-self data on the mobile application and send it to their on-bike display (Fig. 2).

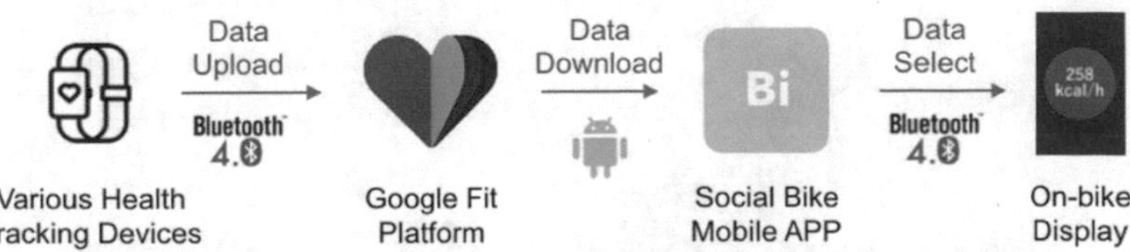

Fig. 1. Schematic of SocialBike's technical system.

T. Ahram et al. (Eds.): IHSED 2019, AISC 1026, pp. 379–384, 2020.
https://doi.org/10.1007/978-3-030-27928-8_58

Fig. 2. Prototype of SocialBike's on-bike display.

Although we have built a prototype of SocialBike, it is difficult to conduct rigorous control experiments in the real world, especially for design concepts related to social interaction and commuting [3]. Therefore, a simulation system was built to simulate the experience of social cycling and implement the concepts of SocialBike in a lab environment.

2 Build of the Experimental Simulation System

2.1 Simulating Social Cycling Experience

To maximise the simulation of actual riding experience. We connected a real bicycle to a software simulated scene built with Unity (Fig. 3) [4].

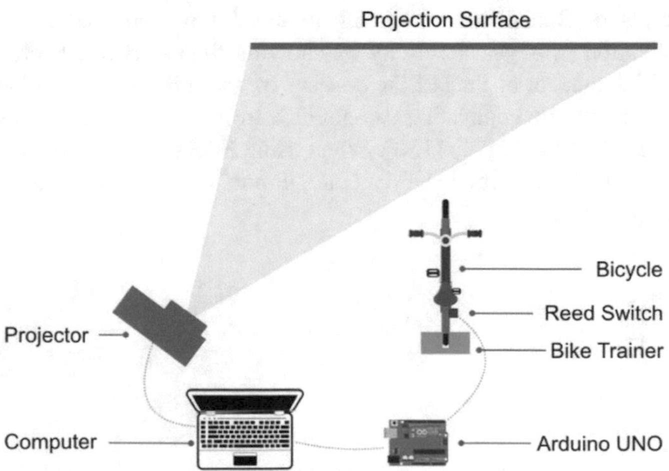

Fig. 3. Schematic of the experimental simulation system (Top View).

A reed switch was mounted on the bike frame near the rear wheel, and a magnet was clamped to the spokes of the rear wheel (Fig. 4). Each time the rear wheel turns 360 degrees, the reed switch sends a signal to the Arduino Uno board. Based on the interval between the received signals, we calculated the rotational speed of the rear wheels through an Arduino program. When the system is running, this speed value is continuously transmitted to the computer through serial port. Besides, A bike trainer is used to simulate the friction of the bicycle wheel rolling on the real ground (Fig. 4).

Fig. 4. The reed switch and magnet used to detect the rotational speed of bike wheel (left) and the bike trainer used to provide resistance (right).

The software part of the simulation system was built with Unity and running on a computer. The simulated scene in the program is projected on a large surface in front of the bicycle. When the computer receives the rational speed value from a serial port, the simulation program will convert it to line speed base on the diameter of the bicycle wheel. When the system is running, the user's view in the simulation program is changing according to the speed of the real bicycle (Fig. 5). During the ride, the user will also see other virtual cyclists riding at their own speed (Fig. 6).

Fig. 5. Setup of the simulation system (left) and the simulation system in experiment (right).

Since the studies were conducted in the Netherlands, we used the elements of the Dutch bicycle lane to make the scene as close as possible to the local appearance (Fig. 6). Other factors such as the inertia, friction, light and wind were also taken into account when building the simulation program.

Fig. 6. Screenshots of different scenes in the simulation program.

2.2 Implementing Design Concepts

Two iterative concepts of SocialBike have been evaluated through the simulation system. In the first iteration, we evaluated the influence of other cyclist's quantified-self data on participants' riding behaviour and psychological state. When riding on the simulation system, participants can see the virtual cyclists' rear on-bike display, showing the virtual cyclists' calorie consumption per hour at current speed (Fig. 7), 20 people participated in the first experiment. In the second iteration, a front on-bike display was added to the user's bicycle, shows the calorie consumption speed of participant and other cyclists in the same interface (Fig. 7). The positions of the circles in the interface represent the positional relationship between the cyclists on the road, 36 people participated in the second experiment. In both experiments, the speed of calorie consumption was estimated by cyclist's height, weight, age, gender, and riding speed [5, 6].

Fig. 7. Experimental setups of the first iteration (left) and the second iteration (right).

2.3 Recording Experimental Data

Since the simulation system is designed for experimentation, it has the function to record experimental data. At the end of each test, the system automatically generates a log file in ".csv" format. The log file contains all data that related to participant's riding behaviour in the experiment, such as position, speed and distance to the virtual cyclists. Depending on the research purpose, we can easily customise the type of data to be recorded. This feature makes it easier to organise and analyse experimental data.

3 Discussion

By building and implementing the experimental simulation system, we gained several insights about the experimental prototype and its role in design research.

Unlike engineering prototypes or conceptual models, prototypes built for experimental purposes have their unique properties [7, 8]. Researchers build experimental prototypes not only to present design concepts but also to create an environment in which participants can interact with the product or system [9, 10]. As we described in this paper, SocialBike's stand-alone prototype has presented its functionality and form. However, to evaluate its impact on the user's behaviour and psychological state, a simulation system was built to simulate the usage scenarios and social attributes of the design concepts.

The primary role of an experimental prototype is to serve design research and experiments, as long as the user experience is not affected, some experimental-related functions can be added to the prototypes. In SocialBike's simulation system, we used a logging function to record the user's riding behaviour. Since the data is exported directly from the system, it is more accurate than the results obtained through observation or video recording. In addition, most design research projects will be followed by further studies or iterations; experimental prototypes could have some flexibility to be adapted to similar concepts. The simulation system we presented in this paper has been used in two experiments with minor modifications and has the potential to be used in further studies related to cycling and social interaction.

Although the experimental simulation system allows us to conduct rigorous control experiments in the laboratory environment, it also has some limitations. No matter how many factors we consider in the simulation system, the simulated scenes are always different from the actual scenarios. Therefore, filed experiments can also be conducted to obtain some complementary results with the laboratory experiments.

References

1. Google Fit. https://developers.google.com/fit/
2. Yang, N., van Hout, G., Feijs, L., Chen, W., Hu, J. : Eliciting values through wearable expression in weight loss. In: Proceedings of the 19th International Conference on Human-Computer Interaction with Mobile Devices and Services, p. 86. ACM (2017)

3. Olsson, T., Jarusriboonchai, P., Woźniak, P., Paasovaara, S., Väänänen, K., Lucero, A.: Technologies for enhancing collocated social interaction: review of design solutions and approaches. Comput. Support. Coop. Work (CSCW), pp. 1–55 (2019)
4. Unity. https://unity3d.com/
5. Jette, M., Sidney, K., Blümchen, G.: Metabolic equivalents (METS) in exercise testing, exercise prescription, and evaluation of functional capacity. Clin. Cardiol. **13**(8), 555–565 (1990)
6. Maeder, M., et al.: Impact of the exercise mode on exercise capacity: bicycle testing revisited. Chest **128**(4), 2804–2811 (2005)
7. Zimmerman, J., Forlizzi, J., Evenson, S.: Research through design as a method for interaction design research in HCI. In: Proceedings of the SIGCHI Conference on Human Factors in Computing Systems, pp. 493–502. ACM (2007)
8. Lim, Y.K., Stolterman, E., Tenenberg, J.: The anatomy of prototypes: prototypes as filters, prototypes as manifestations of design ideas. ACM Trans. Comput.-Hum. Interact. (TOCHI) **15**(2), 7 (2008)
9. Buchenau, M., Suri, J.F.: Experience prototyping. In: Proceedings of the 3rd Conference on Designing Interactive Systems: Processes, Practices, Methods, and Techniques, pp. 424–433. ACM (2000)
10. Odom, W., Wakkary, R., Lim, Y.K., Desjardins, A., Hengeveld, B., Banks, R.: From research prototype to research product. In: Proceedings of the 2016 CHI Conference on Human Factors in Computing Systems, pp. 2549–2561. ACM (2016)

Customer eXperience: A Bridge Between Service Science and Human-Computer Interaction

Virginica Rusu[1], Cristian Rusu[2(✉)], Federico Botella[3], Daniela Quiñones[2], Camila Bascur[2], and Virginia Zaraza Rusu[2]

[1] Universidad de Playa Ancha, Av. Playa Ancha 850, 2340000 Valparaíso, Chile
virginica.rusu@upla.cl
[2] Pontificia Universidad Católica de Valparaíso, Av. Brasil 2241, 2340000 Valparaíso, Chile
{cristian.rusu,daniela.quinones}@pucv.cl, cbascurbarrera@gmail.com, rvzaraza90@hotmail.com
[3] Universidad Miguel Hernández de Elche, Avinguda de la Universitat d'Elx s/n, 03202 Elche, Spain
federico@umh.es

Abstract. User eXperience (UX) and Usability are key concepts in Human-Computer Interaction (HCI). Customer eXperience (CX) is traditionally related to Service Science, but lately becomes a relevant topic in HCI. It offers a more comprehensive approach to the user as customer of several products, systems or services that a company offers. We focused for years our research on tourist as user of dedicated software systems and digital products. We are now focusing on tourist as customer, involving not only interactive software systems, but also the whole range of products and services that she or he interacts with. The paper examines CX as an extension of UX, and as a bridge between HCI and Service Science, focusing on tourist as customer.

Keywords: Customer experience · User experience · Service science · Tourism · Online travel agencies

1 Introduction

Usability has been a key concept in Human-Computer Interaction (HCI), for decades. Most of HCI scholars and practitioners consider that User eXperience (UX) extends the usability concept. The ISO 9241-11 standard defines both usability and UX concepts [1]. Both concepts refer not only to (interactive software) systems, but also to products and services. Even if UX goes far beyond the classical usability dimensions (effectiveness, efficiency and satisfaction), it still focusses on user's interaction with a single system, product or service.

Customer eXperience (CX) centers on user's interaction with a range of systems/products/services. CX offers a holistic approach, dealing with customer (user) – company (or companies) interaction, through the brand's systems/products/services. Customer eXperience (CX) is a concept usually related to the Service Science field, but

T. Ahram et al. (Eds.): IHSED 2019, AISC 1026, pp. 385–390, 2020.
https://doi.org/10.1007/978-3-030-27928-8_59

there is an increasing interest on CX from the HCI point of view. The interest is natural, as CX involves a growing number of software systems and digital products.

We studied for years the usability and UX of tourism-related software systems, especially online travel agencies, virtual museums and national parks websites. We are using lately a more comprehensive approach, focusing on tourist as customer, involving not only interactive software systems, but also a whole range of products and services.

The paper examines CX as an extension of UX, and as a bridge between HCI and Service Science. We analyze the CX concept focusing on tourist as customer. The rest of the paper is organized as follows. Section 2 examines the CX concept as an evolution of UX. Section 3 describes the touch-points as basic elements that construct CX, examining a scenario that involves an online travel agency's customer. Section 4 examines the CX evaluation challenges and evaluates the possibility of developing CX heuristics. We present conclusions and future work in Sect. 5.

2 Customer eXperience: From User to Customer

The ISO 9241-11 standard defines UX as "user's perceptions and responses that result from the use and/or anticipated use of a system, product or service" [1]. There is no general agreed UX definition; however, most of the HCI scholars consider that UX extends the usability concept [2].

CX as concept has been highly discussed in recent years. However, there is no unique, clear, and common definition of what it is and what it represents. Joshi considers CX as the sum or set of experiences that a customer has with a brand during the time in which there is a service relationship [3]. CX is usually related to Service Science field, but it has a highly interdisciplinary nature. We definitely agree with Lewis, which identifies CX as a natural UX extension, and a possible bridge between HCI and Service Science [4].

UX focusses on user's perception of a single product/system/service. CX focusses on the cumulative experience of interactions between customers and companies, through all product/systems/services that a company offers [5]. The user of a product/system/service becomes company's customer, and CX offers a broader, holistic approach. Companies are increasingly aware that CX plays a key role in determining success. A good CX may increase customer's attraction and retention. Hence, CX research and CX evaluation are critical activities to understand how customers perceive the product, systems or services the companies offer.

3 Constructing the Experience: Tourist as Customer

Customers have experiences every time they "touch" any part of the product, service, or brand across multiple channels and at various points in time. CX is built through a process that consists of various "touch-points". A touch-point is any interaction (including encounters where there is no physical interaction) that could alter the way a customer feels about a product, brand, company, system or service [6]. Touch-points' nature is variable. Stein and Ramaseshan identified seven types of elements present at touch-points: atmospheric, technological, communicative, process, employee–customer interaction, customer–customer interaction, and product interaction [7].

Rusu et al. described a scenario where a tourist books a flight through online travel agency [5]. Bascur et al. identified a related scenario, using the Customer Journey Map [8]. Now we present a more complex scenario, detailing each touch-point (TP), the companies, and products, systems or services the traveler interact with, but also the nature of interactions at each touch-point (Table 1). A real-life situation may be much more complex. It includes touch-points of very diverse nature and interactions with different companies and many products/systems/services. Usability of almost all products, systems and services can be evaluated. UX with individual products, systems and services can also be evaluated. But assessing CX is much more challenging.

Table 1. Interactions traveler – companies when booking a flight and making a trip.

Touch-point	Company	Product, system, service	Nature of interaction
TP1: Exploring trip alternatives	Travel agencies, Airlines	Websites	Technological, Communicative, Customer-customer
TP2: Booking a ticket and a trip insurance	Travel agency, Airline, Insurance company	Websites	Technological
TP3: Flight changes, baggage policy	Travel agency, Airlines	Websites, Customer service	Technological, Customer-employee
TP4: Online check-in	Travel agency, Airlines	Websites	Technological
TP5: Airport check-in	Airport, Airline	Self-check-in machine, Airline counter	Technological, Customer-employee, Atmospheric
TP6: Departure	Airport, Airline	Customer service desks, Airline staff, Terminal transfer, Airport information systems	Customer-employee, Technological, Atmospheric
TP7: In-flight	Airline	Airline staff, In-flight entertainment/services	Customer-employee, Technological, Atmospheric, Product interaction
TP8: Connecting flight	Airport, Airline	Customer service desks, Airline staff, Airport information systems	Customer-employee, Technological, Atmospheric
TP9: Arrival	Airport, Airline	Airport information systems, Baggage claim, Customer service	Customer-employee, Technological, Atmospheric

(*continued*)

Table 1. (*continued*)

Touch-point	Company	Product, system, service	Nature of interaction
TP10: Incidents	Airline, Insurance company	Websites, Customer service, Email	Customer-employee, Technological
TP11: Post-travel	Travel agency, Airlines, Social networks	Websites	Customer-employee, Customer-customer

4 Assessing the Customer eXperience

CX is a complex concept, hard to synthetized in a single definition. Therefore, a common approach to CX is to identify its aspects. Several authors have acknowledged different CX aspects, factors, or dimensions. Vanharanta et al. identified three general CX features: CX involves subjective experiences; CX includes consumer's rational thoughts and emotions; and CX is a holistic concept [9]. Gentile et al. proposed six CX components: sensorial, emotional, cognitive, pragmatic, lifestyle, and relational [10].

Assessing CX is challenging for several reasons: CX is constructed through a sequence of touch-points; the nature of each touch-point is (very) different; the experience at one touch-point may (highly) influence experiences at other touch-point, across the entire customer journey; CX has several dimensions; and CX is highly personal. If we consider CX as an extension of UX, which is considered an extension of usability, UX and usability evaluation methods may also evaluate some CX aspects. Others CX aspects, which fall beyond UX, require specific methods. CX researchers and practitioners may benefit the lessons that we have learned as HCI scholars. UX researchers and practitioners are natural candidates to fill the need for CX research and practice. However, we should also learn lessons from Service Science field.

Thompson highlighted some key issues in CX research: undervaluing qualitative research and emphasizing only quantitative research inputs; undervaluing analysis and synthesis, not "making sense" of the collected CX data; and missing design skills and design training, missing the link between CX evaluation and design [11]. We definitely agree, and would argue (among others) for: CX assessment at each touch-point; the use of evaluation methods that address the specificity of each touch-point; a mixed qualitative and quantitative approached at each touch-point; attending all CX dimensions of interest; focusing on targeted customers; appropriate interpretation of collected data; and explicitly consider assessment outputs in the CX design process.

A common approach in CX evaluation, inherited from Service Science field is the use of scales to measure the quality of services. General scales as SERVQUAL [12], as well as particularized scales are available. Most of the CX metrics focus on measuring customer satisfaction [13]. Nevertheless, CX is much more than an overall satisfaction score. Assessing CX involves the emotions, perceptions, and behaviors of people with different tastes and thoughts at each touch-point. Palmer highlighted that, because customer experiences are very subjective and situation specific, a fully inclusive and absolute measuring scale is hard to achieve [14].

Developing new CX evaluation instruments is a necessity. Heuristic evaluation is a well-known usability evaluation method [15]. We think it could also be a powerful method to detect potential CX problems. However, specific heuristics would be necessary. Based on literature reviews, it is remarkable that there are no specific sets of heuristics to evaluate CX [16, 17]. As CX has unique and specific features, and usually involves interactions with several products/systems/services, through several touch-points, it is most likely that a single set of heuristics would not be sufficient to evaluate CX as a whole. For each touch-point, the set of heuristics should be able to capture their specific features and context. Creating sets of heuristics to evaluate CX is a challenge, since many variables and elements should be considered.

Many sets of usability heuristics were proposed, but the process of heuristics development is rarely documented. A methodology for establishing usability/UX heuristics was proposed by Quiñones et al. [18]. The methodology could be extended to allow creating CX heuristics. When creating CX heuristics we should consider at least: the CX specific features (aspects, factors, dimensions) that heuristics will attend; the touch-points that the customer journey may include; the specific products, systems or services that customer interacts with, at each touch-point; and the context of use of each product, system or service.

The goal of using heuristics as an evaluation instrument is to detect potential CX problems at touch-points. As in the case of usability, a CX heuristic evaluation would not cover all CX aspects, and it would not replace customers' opinion/perception. A CX heuristic evaluation should definitely be complemented with alternative assessment methods. However, it may offer interesting CX outputs, at a relatively low cost.

5 Conclusions and Future Work

UX refers to user's perceptions of a single product, system, or service. CX offers a holistic approach, examining the whole customer journey, and his/her cumulative experience of interactions with one or several companies, through several products, systems and/or services. Understanding, designing and evaluating CX is therefore more challenging.

The traditional approach to CX is the Service Science one, but there is an increasing interest from the HCI field. CX may build the bridge between HCI and Service Science. Some HCI lessons are valuable in CX; UX evaluation and design methods are also covering some CX aspects. In particular, we think CX specific heuristics could be developed.

As future work, we intend to particularize the methodology for developing usability/UX heuristics, and to develop and validate specific CX heuristics, focusing on tourist as customer.

Acknowledgments. Camila Bascur has been granted the "INF-PUCV" Graduate Scholarship.

References

1. ISO 9241-210: Ergonomics of human-system interaction — Part 11: Usability: Definitions and concepts, International Organization for Standardization, Geneva (2018)
2. Rusu, C., Rusu, V., Roncagliolo, S., González, C.: Usability and user experience: what should we care about? Int. J. Inf. Technol. Syst. Approach. **8**(2), 1–12 (2015)
3. Joshi, S.: Customer experience management: An exploratory study on the parameters affecting customer experience for cellular mobile services of a telecom company. Procedia – Soc. Behav. Sci. **133**, 392–399 (2014)
4. Lewis, J.R.: Usability: lessons learned… and yet to be learned. Int. J. Hum.-Comput. Interact. **30**(9), 663–684 (2014)
5. Rusu, V., Rusu, C., Botella, F., Quiñones, D.: Customer eXperience: is this the ultimate eXperience? In: Proceedings Interacción 2018. ACM (2018)
6. Interaction Design Foundation: Customer Touchpoints - The Point of Interaction Between Brands, Businesses, Products and Customers. http://www.interaction-design.org/literature/article/customer-touchpoints-the-point-of-interaction-between-brands-businesses-products-and-customers
7. Stein, A., Ramaseshan, B.: Towards the identification of customer experience touch point elements. J. Retail. Consum. Serv. **30**, 8–19 (2016)
8. Bascur, C., Rusu, C., Quiñones, D.: User as customer: touchpoints and journey map. In: Ahram, T., Karwowski, W., Taiar, R. (eds.) Human Systems Engineering and Design. IHSED 2018. AISC, vol. 876, pp. 117–122. Springer, Heidelberg (2019)
9. Vanharanta, H., Kantola, J., Seikola, S.: Customers' conscious experience in a coffee shop. Procedia Manuf. **3**, 618–625 (2015)
10. Gentile, C., Spiller, N., Noci, G.: How to sustain the customer experience: an overview of experience components that co-create value with the customer. Eur. Manag. J. **25**(5), 395–410 (2007)
11. Thompson, M.: The CX tower of Babel: what CX job descriptions tell us about corporate CX initiatives. Interactions **25**(3), 74–77 (2018)
12. Parasuraman, A., Zeithaml, V.A., Berry, L.L.: SERVQUAL: a multiple-item scale for measuring consumer perceptions of service quality. J. Retail. **64**(1), 12–37 (1998)
13. Grigoroudis, E., Siskos, Y.: Customer Satisfaction Evaluation: Methods for Measuring and Implementing Service Quality. Springer, Heidelberg (2010)
14. Palmer, A.: Customer experience management: a critical review of an emerging idea. J. Services Mark. **24**(3), 196–208 (2010)
15. Nielsen, J., Mack, R.L.: Usability inspection methods. Wiley, New York (1994)
16. Quiñones, D., Rusu, C.: How to develop usability heuristics: a systematic literature review. Comput. Stand. Inter. **53**, 89–122 (2017)
17. Hermawati, S., Lawson, G.: Establishing usability heuristics for heuristics evaluation in a specific domain: Is there a consensus? Appl. Ergon. **56**, 34–51 (2016)
18. Quiñones, D., Rusu, C., Rusu, V.: A methodology to develop usability/user experience heuristics. Comput. Stand. Inter. **59**, 109–129 (2018)

Forming Customer eXperience Professionals: A Comparative Study on Students' Perception

Cristian Rusu[1(✉)], Virginica Rusu[2], Federico Botella[3], Daniela Quiñones[1], Camila Bascur[1], Bogdan Alexandru Urs[4], Ilie Urs[5], Ion Mierlus Mazilu[6], Dorian Gorgan[7], and Stefan Oniga[7]

[1] Pontificia Universidad Católica de Valparaíso, Av. Brasil 2241, 2340000 Valparaíso, Chile
{cristian.rusu, daniela.quinones}@pucv.cl, cbascurbarrera@gmail.com

[2] Universidad de Playa Ancha, Av. Playa Ancha 850, 2340000 Valparaíso, Chile
virginica.rusu@upla.cl

[3] Universidad Miguel Hernández de Elche, Avinguda de la Universitat d'Elx s/n, 03202 Elche, Spain
federico@umh.es

[4] Babeş-Bolyai University, Universităţii 7-9, 400084 Cluj-Napoca, Romania
ursbogdan@yahoo.com

[5] Dimitrie Cantemir Christian University, Burebista 2, 400276 Cluj-Napoca, Romania
dr_ursilie@yahoo.com

[6] Technical University of Civil Engineering Bucharest, Bulevardul Lacul Tei 124, 020396 Bucharest, Romania
ion.mierlusmazilu@utcb.ro

[7] Technical University of Cluj-Napoca, Memorandumului 28, 400114 Cluj-Napoca, Romania
dorian.gorgan@cs.utcluj.ro, stefan.oniga@cunbm.utcluj.ro

Abstract. User eXperience (UX) and Usability are well-known concepts in Human-Computer Interaction (HCI). Traditionally related to Service Science, Customer eXperience also becomes a relevant HCI topic. It extends the UX concept, in a holistic approach. It focuses on customer's interaction with the whole range of products, systems and services that a company offers. Forming CX professionals may be challenging. The paper presents a comparative study on students' perception on CX. The survey includes students from Chile, Romania, and Spain. It compares the perception of students enrolled in different programs: Computer Science, Tourism, Medical Technology, Law, and Civil Engineering.

Keywords: Customer eXperience · User eXperience · Service Science · Curricula

T. Ahram et al. (Eds.): IHSED 2019, AISC 1026, pp. 391–396, 2020.
https://doi.org/10.1007/978-3-030-27928-8_60

1 Introduction

Customer eXperience (CX) is a concept usually related to marketing and Service Sciences. Lately there is an increasing interest on CX from the Human-Computer Interaction (HCI) community. There is no agreement on a unique CX definition; many definitions were proposed. For instance, Laming and Mason think that CX includes "the physical and emotional experiences occurring through the interactions with the product and/or service offering of a brand from point of first direct, conscious contact, through the total journey to the post-consumption stage" [1]. The ISO 9241-11 standard defines User eXperience (UX) as "a person's perceptions and responses that result from the use and/or anticipated use of a product, system or service" [2]. We may consider that CX extends the UX concept: it examines the whole customer journey and experiences with several systems, products or services that a company offers, instead of focusing on a single one.

CX is a highly interdisciplinary field. It may be the bridge between HCI and Service Science [3–5]. CX develops through a sequence of "touch-points" (interactions) between the customer and the company (or companies) that offer the product/system/service [6]. A good CX may increase customer attraction and retention, but a single touch-point may influence the whole customer's "journey", built from its (many) interactions with companies, through products/systems/services. There is an increasing demand for CX professionals [7, 8].

We are teaching HCI for years, at undergraduate and graduate level, mainly for Computer Science (CS) students. UX is one of the core topics in all our courses. We realized that including CX as topic in HCI courses is a necessity. We started teaching CX as optional course at graduate level in 2018. We will also teach the course at undergraduate level, in the second semester of 2019. We designed the course based mainly on our research and teaching experience, feedback from industry, and from our alumni.

The paper presents a comparative study on students' perception on CX, based on a survey that included students from Chile, Romania, and Spain. It compares the perception of students enrolled in different programs: CS, Tourism, Medical Technology, Law, and Civil Engineering. Section 2 describes the survey and the participants. Section 3 discusses the results. Finally, we present preliminary conclusions and future work in Sect. 4.

2 The Survey

We made an exploratory comparative study regarding students' perception on CX. Our goal was to assess mainly CS students' perception, but also to compare their opinion with the opinion of students from other fields of study: Tourism, Medical Technology, Law, and Civil Engineering.

We designed a specific questionnaire, iteratively. The questionnaire was validated trough a pilot study, with approximately 20 respondents (both academics and students). Its final version is described in Table 1.

Table 1. The questionnaire.

Section	Question	Type
Demographic	What is your field of study?	Selection
	Do you already get a university degree?	Yes/No
	If so, indicate it.	Open question
	Are you currently working?	Yes/No
	If so, indicate the field you are working in.	Open question
	Gender	Female/Male/Not revealed/identified
	What is your age?	20 y/o or less/21–25/26–30/over 30 y/o
	Country of residence	Chile/Romania/Spain
The perception of the CX concept and its relevance	Indicate the products/systems/services that you think you'll develop/offer as professional	Open Question
	How difficult it is to identify your customers? (P1)	Likert scale with 5 levels (1 - Very difficult, 5 - Very easy)
	How important do you think your customers' experience is? (P2)	Likert scale with 5 levels (1 - Very little important, 5 - Very important)
	How important it is an explicit approach to CX in the curricula? (P3)	Likert scale with 5 levels (1 - Very little important, 5 - Very important)
	Does the current curricula include CX related topics?	Yes/No/I don't know
	If so, indicate them.	Open question
Topics that a CX course should include	Products/systems/services that I'll develop/offer as professional (T1)	Likert scale with 5 levels (1 - Very little important, 5 - Very important)
	Costumers of products/systems/services (T2)	Likert scale with 5 levels (1 - Very little important, 5 - Very important)
	Customers' needs (T3)	Likert scale with 5 levels (1 - Very little important, 5 - Very important)
	CX design (T4)	Likert scale with 5 levels (1 - Very little important, 5 - Very important)
	CX evaluation (T5)	Likert scale with 5 levels (1 - Very little important, 5 - Very important)
	The company – customers relationship (T6)	Likert scale with 5 levels (1 - Very little important, 5 - Very important)
	CX importance for the company's success (T7)	Likert scale with 5 levels (1 - Very little important, 5 - Very important)
Comments	Indicate any CX related aspect that you would like to highlight.	Open question

3 Results and Discussion

We collected valid data from 202 students, from Chile (62.4%), Romania (25.2%), and Spain (12.4%). Most of them were in the final part of their studies, and 25.7% of them are already working. Most of the students were enrolled in CS programs (43.1%), followed by Tourism (20.8%), Medical Technology (17.3%), Law (12.9%), and Civil Engineering (5.9%). Gender distribution was relatively balanced (44.6% female, 53.5% male, and 1.9% did not revealed/identified their gender); however, males were predominant in CS and Civil Engineering. The predominant age group was 21 to 25 (59.9%), followed by up to 20 (23.3%), 26 to 30 (10.4%); 6.4% were over 35 y/o.

We did not follow a sampling procedure. Students' participation was voluntarily. The only field of study covered in all three countries was CS. None of the respondents has a CX course in their curricula. No CX definition was given to respondents. Results express students' unbiased perception. The number of participants was unbalanced, and the results cannot be generalized. Table 2 synthetizes the main quantitative results of the survey, the average scores of students' perceptions on CX relevance (P1, P2, P3), and on CX topics in a CX course (T1, T2, T3, T4, T5, T6, T7).

Table 2. Survey's quantitative results. Averages perceptions on CX, and on topics' relevance in a CX course.

Averages scores											
Country (No. of students)	Field of study (No. of students)	P1	P2	P3	T1	T2	T3	T4	T5	T6	T7
Chile (126)	CS (49)	3.35	4.20	4.00	4.35	4.33	4.59	4.51	4.47	4.27	4.39
	Tourism (42)	3.79	4.52	4.69	4.64	4.57	4.79	4.62	4.69	4.57	4.74
	Medical Technology (35)	4.11	4.29	4.69	4.49	4.46	4.83	4.23	4.51	4.46	4.66
Romania (51)	CS (13)	3.46	4.92	3.77	3.92	3.85	4.15	4.00	4.15	4.00	4.23
	Law (26)	3.19	4.62	4.62	4.31	4.27	4.69	4.15	4.23	4.65	4.58
	Civil Engineering (12)	3.08	4.42	4.42	4.25	4.25	4.50	4.00	3.83	4.50	4.25
Spain (25)	CS (25)	3.52	3.92	4.28	4.24	3.76	4.08	4.08	3.88	3.80	4.00
All (202)	All (202)	3.56	4.36	4.39	4.38	4.29	4.59	4.32	4.36	4.35	4.47

Identifying their customer seems remarkably easy for Medical Technology students, and rather difficult for Civil Engineering students (P1). It is somehow expected, as Medical Technology professionals have a direct contact with patients. However, even if for some students isn't easy to identify their customers, all of them think that customers' experience is crucial (P2 scores over 4.00, with only one exception). Almost all of them think that an explicit approach to CX in their curricula is important or very important (P3 scores 4.00 or more, with only one exception). All 7 proposed topics for a CX course have a high acceptance (T1 to T7 scores over 4.00, with few exceptions).

The Kolmogorov-Smirnov test showed that data does not have normal distribution. That is why we used non-parametric tests to analyze the survey results (Kruskal-Wallis, Mann-Whitney U and Spearman ρ). In all tests $\alpha = 0.05$ was used as significance level.

Comparison results are shown in Table 3. Kruskal-Wallis test shows significant differences among the opinion of students belonging to different field, in all items, excepting P2. Similar results were obtained were comparing groups of students by country. On the contrary, when comparing students by age group, significant differences occurs only for item P2. Mann-Whitney U test shows gender-related significant differences in all items, excepting P1. Mann-Whitney U test shows no significant differences between the students that are working and studying, and those that are only studying, in none of the 10 items.

Table 3. Comparisons between groups of students results.

p-values										
Groups/Test	P1	P2	P3	T1	T2	T3	T4	T5	T6	T7
By field of study/Kruskal-Wallis	.000	.058	.000	.008	.013	.000	.013	.000	.002	.001
By country/Kruskal-Wallis	.033	.000	.503	.018	.001	.000	.001	.000	.002	.002
By age group/Kruskal-Wallis	.233	.411	.000	.615	.414	.065	.811	.095	.257	.631
By gender/Mann-Whitney U	.147	.001	.000	.001	.000	.000	.002	.000	.000	.000
If working or not/Mann-Whitney U	.535	.551	.462	.572	.657	.188	.342	.385	.082	.936

Spearman ρ test shows significant correlations between almost all items. However, most correlations are weak to moderate. For space reasons, we are not presenting Spearman ρ test's results.

All students were able to identify products/systems/services that they think they will develop/offer as professionals. When asked if the current curricula include CX related topics, their answers are largely variable, even for students enrolled in the same program, and same courses. That means their perception of CX is rather blurry. When their answer is "yes", they usually indicate as topics/courses: HCI (CS students), Marketing (Tourism students), Public Health and Ethics (Medical Technology students), Civil Law (Law students), and Management (Civil Engineering students).

Very few students answered the last open question, regarding any CX related aspect that they would like to highlight. Some CS students referred to accessibility and social inclusion, innovation perceived by customers, customers' opinion, and CX knowledge as tool that could make a difference in their professional career. Some Tourism students indicated environmental awareness, carbon footprint, ethnic communities, and customer's trust. Some Medical Technology students mentioned service's impact on customer, and methods that could assess CX. Some Law students referred to customers' rights, the importance of being correctly informed as customer, and contractual and legal issues. Civil Engineering students did not make any comments. Students highlighted CX related issues based on their background/field of study. However, all suggested topics could be present (even if briefly) in a CX course.

4 Conclusions and Future Work

CX is a holistic, interdisciplinary concept, usually related to Service Science. Lately there is an increasing interest on CX from HCI scholars. As CX may build a bridge between HCI and Service Science, some HCI lessons are also valuable in CX. There is an increasing demand for CX professional, but forming them may be challenging.

Our exploratory study highlights that students are aware of the importance of CX and related topics. However, students enrolled in CS, Law, and Civil Engineering programs find rather difficult to identify the users/customers of the products/systems that they are developing, and their real needs. Identifying their "customers" seems to be easier for Medical Technology and Tourism students. Study's findings help us to better understand students' perception, and to better focus our CX course. All topics that we think a CX course should include are (very) important in students' opinion.

As future work, we intend to extend the survey, involving students from other fields, universities, and countries. We also intend to apply a survey to the students that will attend the CX course in 2019.

Acknowledgments. We would like to thank to all students involved in the study. Camila Bascur has been granted the "INF-PUCV" Graduate Scholarship. Authors from Pontificia Universidad Católica de Valparaíso and Universidad de Playa Ancha are participating in the HCI-Collab Project – The Collaborative Network to Support HCI Teaching and Learning Processes in IberoAmerica (http://hci-collab.com/).

References

1. Laming, C., Mason, K.: Customer experience – an analysis of the concept and its performance in airline brands. Res. Transp. Bus. Manag. **10**, 15–25 (2014)
2. ISO 9241–210: Ergonomics of human-system interaction — Part 11: Usability: Definitions and concepts. International Organization for Standardization, Geneva (2018)
3. Lewis, J.R.: Usability: lessons learned… and yet to be learned. Int. J. Hum. Comput. Interact. **30**(9), 663–684 (2014)
4. Rusu, V., Rusu, C., Botella, F., Quiñones, D.: Customer eXperience: is this the ultimate eXperience? In: Proceedings Interacción. ACM (2018)
5. Rusu, C., Rusu, V., Roncagliolo, S., González, C.: Usability and user experience: what should we care about? Int. J. Inf. Technol. Syst. Approach **8**(2), 1–12 (2015)
6. Interaction Design Foundation: Customer Touchpoints - The Point of Interaction Between Brands, Businesses, Products and Customers. http://www.interaction-design.org/literature/article/customer-touchpoints-the-point-of-interaction-between-brands-businesses-products-and-customers
7. CX Goes Mainstream: The Customer Experience Industry Report (2018). http://www.usertesting.com
8. Thompson, M.: The CX tower of Babel: what CX job descriptions tell us about corporate CX initiatives. Interactions **25**(3), 74–77 (2018)

Is a Virtual Ferrari as Good as the Real One? Children's Initial Reactions to Virtual Reality Experiences

Zbigniew Bohdanowicz(✉), Jarosław Kowalski, Katarzyna Abramczuk, Grzegorz Banerski, Daniel Cnotkowski, Agata Kopacz, Paweł Kobyliński, Aldona Zdrodowska, and Cezary Biele

National Information Processing Institute, al. Niepodleglosci 188 b, 00-608 Warsaw, Poland
{zbigniew.bohdanowicz, jaroslaw.kowalski, katarzyna.abramczuk, grzegorz.banerski, daniel.cnotkowski, agata.kopacz, pawel.kobylinski, aldona.zdrodowska}@opi.org.pl

Abstract. The Virtual Reality (VR) world is very suggestive as it intensely affects the senses of vision, hearing and—to a limited extent—touch. It can be expected that in the near future VR will be widely disseminated and used by people of all ages, including children. We decided to conduct a qualitative research project to assess children's (aged 7–12) first reactions to the use of VR. Children's opinions and reactions gathered during the interviews indicate that children highly appreciated the attractiveness of the virtual experiences, which were often assessed at a similar or higher level than their real-world counterparts. Our findings clearly suggest that children very easily adopt VR without any prior experience with that technology. We recommend that studies on children's behavior in VR are continued.

Keywords: Virtual Reality · Children · Adaptation · Behavior · Immersion · Interaction · User experience · Human-computer interaction

1 Introduction

The Virtual Reality (VR) technology has reached a greater degree of maturity in the last few years. The early dates of this technology date back to the 1960s, when the scientific study entitled 'The ultimate display' was published. It was assumed that this hardware, with corresponding software, would enable the creation of "the Wonderland into which Alice walked" [1]. From today's perspective, that equipment was very simple, even primitive, but the idea that guided its creators was identical to the goals pursued by the designers of modern VR devices: to create a virtual simulation that will be indistinguishable from the real world.

It can be expected that soon, along with other technological innovations, VR equipment will become more comfortable, easier to use and less costly. It is very likely that in the near future this technology will be widely disseminated and used by people

T. Ahram et al. (Eds.): IHSED 2019, AISC 1026, pp. 397–403, 2020.
https://doi.org/10.1007/978-3-030-27928-8_61

of all ages, including children. In view of these considerations, a decision was made to carry out a qualitative research project to provide a qualitative diagnosis of the reactions of children aged 7–12 to their experience in the VR environment. Reactions of children in this age group are interesting since it is difficult for the children to achieve dual representation which would allow them to understand the symbolic nature of media employing more salient or appealing symbols [1, 2]. Compared to adults, this may result in more intense experiences related to the "presence" in the virtual environment and "realness" of the virtual environment, conducive to positive assessment of such experience [3].

The diagnosis of children's reaction to VR was based on the Mütterlein [4] model (2018), which measures the quality of experience in VR in terms of satisfaction. According to that model, the level of satisfaction is determined by immersion, telepresence and interactivity. Immersion is defined as "feeling totally involved in and absorbed by the activities conducted in this or other place". Telepresence focuses on "a subjective feeling that one is in another place". Interactivity relates to the subjective "degree to which users of a medium can influence the form or content of the mediated environment".

2 Method

Subjects. Qualitative interviews with children aged 7–12 were conducted in April 2019. Each interview lasted approximately 60 min.

Materials. The virtual reality set used in the study was HTC Vive Pro with HTC Vive controllers, selected for its image quality and ease of use.

Stimuli. The following criteria for selecting the experience were used: the level of interactivity, the motion mode, the complexity and high quality of the application. Two experiences were created based on existing experiences available in WorldViz Vizard [5] environment, while the other three (two applications and a 360 video) are publicly available on Steam and YouTube. The experiences used were:

1. **A "Travel without leaving home" 360 video** [6]. A five-minute, stereoscopic 360° film that shows short shots from a journey to places with interesting nature. The aim of the simulation was to present the real world using VR technology.
2. **Dreams of Dali** [7]. An abstract world using motion based on predefined points with "gaze pointer" visual selection. The application presents unlimited possibilities of creating space in VR.
3. **Contemporary Loft Apartment.** A model of a flat in an apartment building with interactive elements (e.g. elements of interior design can be lifted), enabling movement and interaction with objects. The environment was developed in Vizard [5].
4. **Walk the Plank.** A simulation of a board suspended at a height that can be walked on. The environment was developed in Vizard [5]. Purpose of using this application was to test the reaction to a simulated situation of being at a significant altitude.

5. **Droid Repair Bay** [8]. A simulation of robot repair embedded in the Star Wars universe. The application offers advanced possibilities of interaction with the surroundings (control of devices, robots, manipulation of controllers).

Experimental Procedure. In the first part of the study, children were introduced to VR devices and given the opportunity to try out the VR simulations (in a random order). Then they were interviewed by a qualitative researcher. The qualitative interviews were selected as a method which is well suited for exploring areas that are under-researched. The data collected during the interviews were subjected to qualitative analysis for commonly recurring themes using thematic analysis [9].

3 Results

The collected insights, reactions, and remarks were analyzed in the context of aspects that assess the satisfaction with VR simulation, according to Mütterlein [4]: physical sensations, appeal of the virtual world, feeling of presence, interactivity, and locomotion.

3.1 Physical Sensations

The children quickly became accustomed to the new technology. After less than a minute, they were able to freely look around, move and capture objects in the virtual world. Children's first reactions were very positive, even enthusiastic. They liked the virtual world, and staying in it did not give rise to any negative psychophysical experience (such as dizziness, nausea, or problems with orientation).

3.2 Appeal of the Virtual World and Its Objects

Virtual experiences were often assessed at a similar or higher level than experiences from the real world. When the researcher asked in which reality the children would prefer to engage in their favorite activities, children mostly said that both realities would be equally good and often preferred the VR. The following verbatim quotations provide a clear illustration: *No one can foul me. Let's assume I have virtual players who are, kind of, playing with me. They can't touch me out there, so they won't do me any harm, I won't be fouled, I won't break my leg, nothing of the kind.*

One of the few examples of a notion that was more attractive in the real world was having a real pet. Children feel that the virtual world would not enable them to simulate real contact with a pet (*In real life, you can touch it, play with it. I know that now, in VR, you can throw a ball and the pet will run, for example. But you can't snuggle up with it. You want a hug, and suddenly you can't feel its body or touch*).

3.3 Feeling of Presence

For children, the virtual world was more than just an image displayed in the goggles. Children spontaneously said that this is the world they "move into" (*You can do everything and nothing will happen, you can make your dreams come true, like being in Hogwarts*). This rapid familiarization with the new digital reality may result from the fact that the children knew the imaginary world very well from the screens of phones, tablets, computers or TV sets. Children are familiar with the concept of the "world of pixels", and VR is just another level of development for this concept (*Out there, you could feel clearly that you were there. And with a console, you feel like you're sitting in a chair./It's like you're playing, but... not on a computer that has a square or rectangle, but there's a world around you*). The high appeal of VR also resulted from the fact that it is a new technology that children do not have free access to.

3.4 Interactivity

For children, one very important element is the interactivity: the possibility to grab objects, open doors, throw things, etc. The children actively searched for elements of the virtual world that could be manipulated (*I think what I liked the most was... repairing the robots. Because... I guess that's where you could do the most./Somehow I liked that room, you could walk around nicely, pick all kinds of things up*). Being in the virtual world, they did not have to be afraid of adults' reactions to their actions (*I could do various things that I couldn't do at home in everyday life*).

3.5 Locomotion

Movement in the applications used in the study was possible in three modes: walking around the simulated room, teleportation to indicated places, and moving a character using cursors on the controller.

Naturally, walking was the most intuitive way to move around, but when the situation required a different method, the children quickly adapted to the new mode and it was difficult to judge which one was preferred: using cursors to move or teleporting to the desired location. The only difficulty was to precisely point to predetermined locations for teleportation using gaze due to the relatively high weight of the goggles versus the child's body (only in the *Dreams of Dali* [7] application).

Notably, all the children, despite their engagement in the VR simulation, remembered the meaning of the red lines (representing the walls of a real room) and never crossed them.

3.6 Reactions to the Tested Applications in Detail

Application 1. The 360° Movie. This experience was rated positively, but without much excitement. During the projection, the children were sitting on a chair and, most likely, the fact of taking this position put them into passive viewing mode. The relatively low rating was due to the lack of interaction. Children reported that watching the world from a single point (the position of the camera) was a strange and rather unpleasant experience (*A strange feeling of having no body. I feel like somebody has taken me to another world and all I have left is my head. I was afraid I'd fall into the water*).

Application 2. Dreams of Dali. This application takes the user to an imaginary world inspired by the works of Salvador Dali. Elements and characters from Salvador Dali's paintings were placed in an endless desert, and the participants could move around in this world and look at them freely.

Children did not appreciate the abstraction and the unlimited possibilities of creating worlds in VR. They felt somewhat insecure in this strange, incomprehensible world. This experience was disturbing rather than inspiring for them (*I was afraid of that girl jumping on the skipping rope… There was some strange music and that terrible sight. / I went down the tower and there were those elephants there. I got a little scared, and I started laughing a bit*).

In this application, the participants were only able to see the virtual world, without the possibility to interact, which the children did not like: (…) *I couldn't lift anything, I couldn't fool around. All you could do was to teleport.*

Application 3. Contemporary Loft Apartment. This is one of the highest rated applications. From the very first moment, the children felt at ease in the virtual flat: the familiar space gave them a sense of security and, at the same time, it was a world where they could do things that are not allowed in a real flat, e.g. the children really liked making a mess (*It was cool to lift objects, you could turn them upside down, something my mom would forbid me to do at home*). Another attractive thing for the children was that they could pretend to be adults and feel independent: (*Someone else's home, I'm alone at home, my parents have gone away*).

The children intensively tested the technical limits of the application: they tried to carry objects through walls or even walk through the walls. At first, the experience of "being in a wall" aroused unpleasant feelings, but after a while the children who tried it freely looked for new barriers and tried to get outside the boundaries of the building.

Application 4. Walk the Plank. This application simulated the experience of being at a high altitude. The participants' task was to walk on a board suspended between two tall pillars.

Children reacted with excitement to altitude. All children bravely walked on a high-hanging plank and quickly got used to the open space. This experience was assessed as attractive, but the children wanted it to be more interactive.

Application 5. Droid Repair Bay. The application that children liked the most was a simulation of a robot repair station on a spaceship, reminiscent of the Star Wars movie. This simulation offered the greatest opportunities for interaction with the surroundings among all the simulations used in the study.

It should be stressed that most children did not need any instructions to operate freely in this simulated world. The children naturally entered the environment and moved freely around the room, scanning the surroundings while looking around. When they were asked to perform a task, i.e. repair a robot, the children could easily guess how to control the servicing station and what to do to replace the broken parts.

4 Summary and Discussion

In children's responses, the borderline between the real world and VR is very weak. For them, the value of virtual and material things is similar, and the attractiveness of experience in the two worlds is rated similarly. This confirms the observations by DeLoache and Troseth that children are unable to distinguish between symbolic and real concepts [1, 2].

The high level of satisfaction with the experiences, reflected in enthusiastic ratings, proves that the level of immersion in the VR environment was high. Based on the aspects identified by Mütterlein [4] we believe that the VR experience which is possible with modern equipment is characterized by both high telepresence (for children, VR is the world you move into) and interactivity (in the simulations that enabled it). This conclusion is also confirmed by the fact that children adapted to the VR environment very quickly. The rules of the new, virtual world were easily assimilated, and the children did not experience any unpleasant psychophysical sensations such as nausea, fatigue or dizziness while staying in that world.

Given the high immersion and satisfaction with the VR experience, the children did not focus on technical differences between the real and virtual worlds in their assessments because, from their perspective, the two worlds were comparable.

Our research has shown that children adopt VR technology uncritically, without being aware of the associated risks. It can be presumed that if a significant portion of a child's life experience is transferred to an illusionary world that can be adapted to the child's requirements and where it is easy to be under the illusion of having real skills, this can make it difficult for children to cope with challenges of the real world. One interesting direction for further research would be to assess to what extent VR can become an alternative world where people can escape to be who they would like to be, to have more skills than in reality and to have everything under control.

Another promising direction of research is to explore the potential of VR technology for practical applications such as education and communication. In the case of education, the unlimited possibilities for creating the world in VR can help to present educational content in an engaging and interactive way. Applied to communication, VR can create completely new, previously unknown methods of communication for people and redefine social interactions in the future.

References

1. DeLoache, J.S.: Dual representation and young children's use of scale models. Child Dev. **71**, 329–338 (2000)
2. Troseth, G.L., DeLoache, J.S.: The medium can obscure the message: young children's understanding of video. Child Dev. **69**, 950–965 (1998)
3. Sharar, S.R., Carrougher, G.J., Nakamura, D., Hoffman, H.G., Blough, D.K., Patterson, D.R.: Factors influencing the efficacy of virtual reality distraction analgesia during postburn physical therapy: preliminary results from 3 ongoing studies. Arch. Phys. Med. Rehabil. **88**, S43–S49 (2007)

4. Mütterlein, J.: The three pillars of virtual reality? Investigating the roles of immersion, presence, and interactivity. In: Proceedings of the 51st Hawaii (2018)
5. WorldViz: Vizard (2019)
6. BRIGHT SIDE: Travel Without Leaving Home|360 VR. Youtube. Accessed 8 Jul 2018
7. Dreams of Dalí. http://www.dreamsofdali.net/
8. Become a Resistance Mechanic in the New VR Experience Star Wars: Droid Repair Bay | StarWars.com. https://www.starwars.com/news/become-a-resistance-mechanic-in-the-new-vr-experience-star-wars-droid-repair-bay
9. Braun, V., Clarke, V.: Using thematic analysis in psychology. Qual. Res. Psychol. **3**, 77–101 (2006)

Designing Federated Architectures for Multimodal Interface Design and Human Computer Interaction in Virtual Environments

K. Elizabeth Thiry(✉), Arthur Wolloko, Caroline Kingsley, Adrian Flowers, Les Bird, and Michael P. Jenkins

Charles River Analytics, Cambridge, MA, USA
{ethiry,awollocko,ckingsley,
aflowers,lbird,mjenkins}@cra.com

Abstract. Advances in 3D immersive environments offer great potential for collaboration in circumstances that rely heavily on 3D visual models for collaboration, coordination, and data visualization (e.g., construction, medical, architecture). As existing platforms quickly evolve, become more widespread, and research aims to validate the benefits of AR/VR over more traditional platforms, user-centered approaches become more important, especially when supporting multiple users in 3D immersive environments. This article discusses our experience designing an adaptable, non-domain-specific, federated architecture for multimodal systems and virtual environments based on Human Factors and Human Computer Interaction methodologies.

Keywords: Human Factors · Multimodal interaction · Human centered design · Virtual and augmented reality · Cognitive Systems Engineering

1 Introduction

In recent years, the rapid development of state-of-the-art visualization devices have placed virtual reality (VR) and augmented reality (AR) at the forefront of information visualization and interaction, from remote social interaction and gaming, to interior design [1] and navigation. Advances in 3D immersive environments offer great potential for collaboration in domains that rely heavily on 3D models for data visualization, such as construction, medicine, architecture, gaming, and military command and control; new interaction types emerge with each new device (e.g., head-gaze, touchless gesture interactions). However, existing virtual environment systems are designed to support individual experiences, such as providing a single user with a personalized experience based on their preferences or context. As a result, these environments do not generalize well to multi-user applications. This limitation presents a fundamental need for perspectives from Human Factors (HF) and Human Computer Interaction (HCI) when designing for the future of multimodal virtual collaboration.

Ideally, users with different platforms and versions can collaborate together in the same 3D environment, but given the pace at which AR/VR technologies are evolving (and the often prohibitive costs), not all users have the platforms necessary to

T. Ahram et al. (Eds.): IHSED 2019, AISC 1026, pp. 404–410, 2020.
https://doi.org/10.1007/978-3-030-27928-8_62

participate. Therefore, a solution must have the flexibility to support virtual environment connectivity across platforms and rapidly changing devices. To ensure users can effectively connect and contribute, we can apply HF methodologies to the design of *federated system architectures*—architectures that define a group of systems and networks operating in a standard, and connected environment to support both synchronous and asynchronous collaboration, shared understanding, or interactions. Systems with these architectures can support virtual collaboration, even when individual users have constraints (e.g., unique hardware, users with glasses or those who experience vertigo, or have limited physical space available for movement). With HF and HCI, we can address these unique requirements and generate displays in 3D environments so users can use a variety of devices to interact in the same networked environment.

This paper explores how we assessed and defined effective human-centered design and interactions within spatiotemporal AR/VR display environments, facilitate integrated visualizations in those environments, and enable remote and co-located collaborations through multi-platform, networked viewing and interface elements. This exploration has been conducted under a number of R&D efforts that apply AR/VR technologies to promote enhanced shared situational understanding in complex domains, such as military command and control.

1.1 A Method for Holistic, Human-Centered Design

HF provides tools to identify and apply information about behavior, abilities, limitations and other characteristics to the design of machines, systems, tasks, environments, etc. for effective use [2]. An HF approach to HCI aligns the characteristics of the human user with those of the technology, enabling multimodal systems and interfaces to use the capabilities of each device while accommodating limitations [2].

Adding a 3D immersive environment to a multimodal system increases the complexity of the system. Based on our prior research integrating, developing, and evaluating AR/VR platforms, visualizations, and HCI techniques, we found that to successfully integrate novel displays such as AR and VR, we first had to clearly understand the individual elements that contribute to effective AR/VR experiences (e.g., natural interactions, human perception of AR/VR displays, visual perception in relation to information overload).

Cognitive Systems Engineering (CSE) techniques, such as Work Domain Analysis (WDA) and Abstraction Hierarchies, offer a principled and disciplined approach to identify relevant cognitive and perceptual tasks in operational workflows. CSE offers promising ways to support design requirements based on interrelationships across hardware, work domains, human perception, and cognition [3]. We found that methods derived from CSE can also provide structured representations of functional concepts and their relationships to determine the relevant information for a function and the best physical form to represent it. We also found that CSE methods can provide design guidelines for introducing novel modalities (AR, VR, Haptics) and efficiently tie them to tasks performed within work domains, eliminating the need to build expensive prototypes based on less formal design heuristics and then evaluate the prototypes to revise and refine design requirements.

Combining CSE, HF, and HCI methods yields a holistic perspective on rapid technology growth, environmental and situational context, multi-platform interface elements, and cognitive dimensions of users and domains. We can leverage this perspective to design new and effective multimodal interactions, immersive environments, and networked systems. This interdisciplinary and holistic approach characterizes diverse elements that reflect human use and implementation in both individual and multi-user settings, making it an ideal human-centered design approach.

1.2 Approach to Networked Collaborative Multimodal Virtual Environments

As existing platforms quickly evolve and become more widespread, research aims to validate the benefits of AR/VR vs. more traditional platforms [4]. Ideally, users with different platforms and versions could all collaborate in the same 3D environment. However, due to speed of AR/VR technology advancement, not all users may have the platforms required to participate. To support multi-user collaboration in 3D immersive environments for multiple contexts, we must consider factors that can affect usability. Such factors include device access, operational context, connectivity, physical space, and the effects of rapidly changing computing environments.

To address this issue, we developed a federated architecture that supports multi-platform delivery of visualization techniques and enables multiple users to simultaneously connect to a session across platforms (e.g., tablets, smartphones, PC, AR/VR HMDs) while providing integrated support for visualization using the IEEE standard (Distributed Interaction Simulation (DIS)) to enable live, virtual, entity visualization. However, this novel integration created another set of challenges: identifying the appropriate types and levels of interactions between users and AR/VR; the most effective interaction modalities between users and devices; best practices for conveying information between networked, augmented and virtual environments; human-centered design techniques for collaboration; and HF approaches to enable multimodal naturalistic interactions. Due to space limitations, we only briefly discuss aspects of multimodal systems and interactions, multi-user collaboration within federated architectures in virtual environments in Table 1.

2 Multimodal Virtual Collaboration

Multimodal virtual collaboration capabilities are well positioned to take advantage of the growing demand for novel content in both existing and emerging domains. However, the introduction of new technologies and increase in capabilities does not guarantee improved HCI and system performance [5].

With the rapid adoption of virtual collaboration solutions, the need to apply HCI becomes increasingly clear. Wrong applications and misuse of technology leads not just to poor or frustrating usability, but also to systems that are difficult to learn, increased workload for users, and errors that can have significant repercussions. Therefore, an effective approach must take full advantage of emerging developments in state-of-the-art technologies, support modalities and intuitive interactions, and support

Table 1. Multimodal integration challenges and considerations

Problem	Key Challenges	Considerations
Aspect: Supporting Multi-user Collaboration		
• Systems designed for a specific purpose do not generalize well for multi-user circumstances • Tendency to restrict systems to a particular device or software does not support fast changing technological environments (integration issues, software licenses, tool version updates, etc.)	• Cross-device configuration support • Systems designed using specific devices or computing environments • Configuration of device pairing, connecting, modality, and interaction combining • Support multi-user collaboration across different users and operational contexts • Flexibility to opt in or out of collaboration or multi-user setup • Identification of current devices and their strengths and limitations • Same environment across multiple platforms with different capabilities for representing content	• Effects of changing computing environments and technical advances • Human access to hardware necessary for collaboration • Feasibility and domain tracking from state of the art to off the shelf devices for particular applications and uses • Defined requirements for context of use, user actions and tasks, motivations • Representations across different types of users
Aspect: Multimodal Interaction: Input Modalities		
• Retrofitting and combining design elements that were developed for specific purposes is unlikely to result in a holistic HCI platform that provides multi-user and non-domain-specific support	• Integration between 3D immersive spaces and platform coexistence • Identifying multi-user and cross-device interactions and effective modalities [6] • Different consumption of content across platforms and changing devices • 2D and 3D information visualization elements, movement techniques, etc. • Determining what interaction modes are necessary or which is the most appropriate mode, etc.	• Content engagement and what techniques are aimed at interacting or exploring content, information, visualizations or interfaces across devices and platforms [6] • Input modality characteristics and multi-user representation • Appropriate modes of interaction across 3D space and real environments

(*continued*)

Table 1. (*continued*)

Problem	Key Challenges	Considerations
Aspect: Visualization		
• Content visualization across devices and environments as well as spatial considerations	• Designing for distributed interfaces and displays across dimensions • Display ecosystems and associated dynamic spaces, visualization across devices in both 2D and 3D and design for collaboration and personal use cases, etc. • Collaborative data visualization, modes of exploration, manipulation • 2D and 3D display change visualizations • Coordinated views, visualizations and graphics across spaces, environments and devices • Choice and relation of 2D and 3D visualization elements (e.g. icons) • Information overload	• Display, position and orientation, ecological interface design, etc. • Visualization and interface engagement based on 3D motion or device position • Collaboration requirements and best modalities • Displaying cues that correspond to virtual entities • Depth of real vs. virtual objects and conveying correctly to users across devices • Aligning content properly to reduce cognitive load to understand contents

individual and collaborative work across domains and applications. It must create a multimodal system that can support effective collaboration within virtual environments while successfully supporting devices to provide common ground in a contextually tailored and synchronized space.

3 Multimodal Interaction and Multi-user Collaboration

In the multimodal space, developers often retrofit and combine design elements developed for other, specific purposes; we have learned that this practice does not create a holistic HCI platform that can support a variety of operational situations, non-domain-specific applications, and multi-user capabilities. Instead, we aimed to design a non-domain-specific multimodal system. We first identified devices and their associated modalities, then defined requirements for effective collaboration in virtual environments. Next, we categorized and contextualized tasks used across domains and operational contexts and selected devices that could support collaboration in a synchronized space that can be contextually tailored [6].

Our experience has shown that to identify the devices and platforms that can support a domain, we must first recognize the scope of the task. Simply identifying hardware platforms and characterizing their display characteristics and capabilities is insufficient because developing effective AR/VR requires a thorough review of relevant factors such as display, visualization, and HCI techniques. We also found that our combined HF/CSE techniques provide a more efficient and human-centered alternative to the review process. For example, WDAs are usually applied to a specific domain, but are not limited to one; they can be applied more broadly to generate novel ways to approach diverse, non-domain-specific applications. A human-centered, holistic perspective on interactions between new displays/platforms and the cognitive dimensions of work environments integrates constraints from both of these aspects into a cohesive, human-centered design approach. By adapting and combining visualization and HCI techniques from various AR/VR supported domains with techniques from existing platforms (e.g., mobile, web, PC) we can create a hardware-agnostic solution.

4 Interaction Modalities

Each device has its own set of interaction modalities (Magic Leap: controller, HoloLens: gesture, tablet: touch); so for multi-user collaboration, we must create and maintain a synchronized virtual environment that supports the different modes of input, interfaces, and modalities that best suit each device's interactions. For example, editing text is difficult in AR, but well suited to a tablet. An environment that supports one person on a tablet who wants to input text is a straightforward design problem. However, if there is another person in the same environment using an AR headset who wants to complete the same task, how do we make sure that the interaction is natural? What is the best way to support a task across devices?

Our experiences within the virtual technology space for various domains indicate that the main benefits of AR/VR are centered on the display and interaction aspects of spatiotemporal information. However, when catering to a variety of domains and operational contexts, we must consider not only the physical dimensions of problem spaces, but also consider domains (e.g., military, architecture, construction, mechanical engineering) that are more complex and require more abstract reasoning. For example, one major consideration for more complex domains is ensuring all information is delivered to users without causing information overload. This means that the modes of information presentation should be intuitive and natural, and the amount of information should be enough to meet requirements while ensuring it does not overwhelm. Information and interactions must be compatible with existing modes of information display and work harmoniously with other tasks and users [7].

5 Conclusion

A non-domain-specific federated architecture for multimodal systems and virtual environments is difficult to design because it is infeasible to account for every possible operational context or domain. However, a combination of HF and HCI techniques can yield a stable architecture built on a foundation of user-centered design. Complex ecosystems, such as multimodal interactions, multi-device systems, or mixed-reality are challenging and costly to develop and test. A holistic, user-centered design approach elicits critical information regarding context of use, device configurations, and the meaning of interactions. It also supports iteration so that we can use both traditional and emerging HCI strategies to rapidly configure user information and identify visualization and interaction needs. This interdisciplinary approach reveals fundamental user perspectives over time, so we can create user-specified parameters and design elements. This approach is critical for emerging and evolving fields, such as AR/VR, where design and interaction standards are not yet established.

As technology and our understanding of how humans will collaborate using it continues to evolve, so must our methods and approaches. The aim of this paper is to provide a starting point for future discussion and evolution of an effective design approach to multimodal, collaborative work in a shared virtual environment.

References

1. Nasir, S., Zahid, M., Khan, T., Kadir, K., Khan, S.: Augmented reality application for architects and interior designers: Interno a cost effective solution. In: 2018 IEEE 5th International Conference on Smart Instrumentation, Measurement and Application (ICSIMA) (2018)
2. Lim, C., Pan, Y., Lee, J.: Human factors and design issues in multimodal (speech/gesture) interface. JDCTA **2**, 67–77 (2008)
3. Bisantz, A., Burns, C.: Applications of Cognitive Work Analysis. CRC Press, Boca Raton (2009)
4. Bonetti, F., Pantano, E., Warnaby, G., Quinn, L., Perry, P.: Augmented reality in real stores: empirical evidence from consumers' interaction with AR in a retail format. Augmented Reality Virtual Reality, 3–16 (2019)
5. Roth, E., Malin, J., Schreckenghost, D.: Paradigms for intelligent interface design. In: Handbook of Human-Computer Interaction, pp. 1177–1201 (1997)
6. Brudy, F., Holz, C., Rädle, R., Wu, C., Houben, S., Klokmose, C., Marquardt, N.: Cross-device taxonomy: survey, opportunities and challenges of interactions spanning across multiple devices. In: Proceedings of the 2019 CHI Conference on Human Factors in Computing Systems - CHI 2019 (2019)
7. Livingston, M., Azuma, R., Bimber, O., Saito, H.: Guest editors' introduction: special section on the international symposium on mixed and augmented reality (ISMAR). IEEE Trans. Vis. Comput. Graph. **16**, 353–354 (2010)

Editorial Design Based on User Experience Design

Carlos Borja-Galeas[2,7](✉), Cesar Guevara[3], José Varela-Aldás[1], David Castillo-Salazar[1,4], Hugo Arias-Flores[3], Washington Fierro-Saltos[4], Richard Rivera[5], Jairo Hidalgo-Guijarro[6], and Marco Yandún-Velasteguí[6]

[1] SISAu Research Group, Universidad Indoamérica, Ambato, Ecuador
{josevarela, davidcastillo}@uti.edu.ec
[2] Facultad de Arquitectura, Artes y Diseño, Universidad Indoamérica, Quito, Ecuador
carlosborja@uti.edu.ec
[3] Mechatronics and Interactive Systems - MIST Research Center, Universidad Indoamérica, Quito, Ecuador
{cesarguevara, hugoarias}@uti.edu.ec
[4] Facultad de Informática, Universidad Nacional de la Plata, Buenos Aires, Argentina
washington.fierros@info.unlp.edu.ar
[5] Escuela de Formación de Tecnólogos, Escuela Politécnica Nacional, Quito, Ecuador
richard.rivera01@epn.edu.ec
[6] Grupo de Investigación GISAT, Universidad Politécnica Estatal del Carchi, Tulcan, Ecuador
{jairo.hidalgo, marco.yandun}@upec.edu.ec
[7] Facultad de Diseño y Comunicación, Universidad de Palermo, Buenos Aires, Argentina

Abstract. This research deals with editorial design based on user experience design. The traditional editorial design has had to adapt to the new digital media composition, where multimedia audiovisual elements unthinkable a few years ago need to be integrated. The participation of the reader, as an external observer who only receives information through texts and images, now has new scenarios in which he can actively participate and decide what will come to his hands. In this study, a work methodology based on UxD User Experience Design is presented, in which will generate the editorial design of an educational book on environmental issues, which includes augmented reality for children from 6 to 8 years of age. The aim of this study is to know if an editorial product with augmented reality and developed from the user experience design can improve meaningful learning in a playful and active way. For its development, a composition model based on the Fibonacci sequence and the golden ratio will be used. Additionally, its graphic composition will be guided by the Massimo Vignelli canon and will be complemented by the reticular model of Beth Tondreau. The augmented reality markers position will also be based on the composition model previously mentioned, which will allow keeping the attention of the reader in the printed document and in the augmented reality animations. The user experience design will be applied with teachers, parents and

T. Ahram et al. (Eds.): IHSED 2019, AISC 1026, pp. 411–416, 2020.
https://doi.org/10.1007/978-3-030-27928-8_63

students from 4 schools in Quito and Ambato. Once the production is completed, the impact on teaching-learning process will be evaluated with a control and a test group, and the methodology with which they will work in the classroom with the educational material developed will be defined. At the end of the study, copies of the book will be delivered to the participating schools of this research for its implementation.

Keywords: Editorial design · Augmented reality · User experience design · Design Thinking

1 Introduction

In recent years, editorial design has had to adapt to the arrival of new electronic formats that, in turn, come with new multimedia technology tools that include animations in 2D, 3D, audio and digital video, augmented reality (AR), virtual reality or mixed reality, to name but a few of the existing ones that continue to be modified and improved. The schools have also begun to include these technologies, which is in the process of constant evaluation to obtain the expected learning results. A tool that allows to develop these educational products in line with the reality of the users is the user experience design (UxD). Involve the user in the development of educational material, take into account children and youth, teachers and parents, with their specific reality in certain contexts, uses of language, ways of seeing and telling the facts and how they want to be told the things to feel part of it.

The study by dos Santos and Tiradentes Souto [1], in which user experience design method was applied, demonstrates in its analysis that the design of graphic learning tools for primary schools had greater difficulty without the participation of children, that is, the contribution of children during the design process helped graphic designers to create more efficient learning tools: the appropiate level of complexity of the product for the audience, the suitability of the contents in relation to the audience knowledge and the use of creative design features were reinforced with this method of study. As conclusions, using user experience design both in print and product design is highly recommended by the authors, especially when the target audience is very specific.

In their research, Toledo, Garber and Madeira [2] explain the 5 stages of the Design Thinking methodology: 1. Empathise, that is to say, to deeply understand the needs of the users. 2. Define, that is, analyze the information received and filter the one that really adds value and leads us to find solutions to obtain innovative results. 3. Ideate, which means that every idea can be worth and contribute to generate innovative solutions. 4. Prototype, which leads to the construction of fast models that shape ideas or concepts. 5. Test or evaluate the prototypes that are connected to the Minimum Viable Product idea (MVP) of the Lean Startup methodology. On the other hand, Woloszyn and her group of researchers [3] present a study in which they affirm that the design of books, magazines and other editorial products starts from a concept that must take into account the target audience, their preferences, needs and limitations, and present a suitable design for the reader, the publishers and the market. Even before the emergence of new digital forms of publication - such as books and digital magazines for mobile devices - the editorial system expands and highlights the need for more careful studies of users, their habits and their relationship with technology. In this context, the public's understanding becomes essential for the development of editorial products.

In recent years, in Indian Schools have been embracing emerging technologies such as AR to enrich the quality of teaching and learning. These interactions, as Sarkar and Pillai tell in their study [4], require to be designed according to the expectations of the students, teachers and parents, in order to achieve a successful use experience. Understanding students perspectives towards technologies, as well as the expectations of the three user groups of an AR experience in Indian education was the main focus of this research. One of the most important conclusion of this work, is the dependance of the students to access these technologies.

The research of Michailidou [5] presents twelve methods that support designers, including UxD experts and non-experts, in these activities. These methods come from the disciplines of product development engineering, industrial design and interface design and have been applied and adapted in a research project with the collaboration of the industry. Regarding the current state of the art, the applied methods are reviewed and extended with recommendations of the researchers. These suggestions should help UxD professionals to choose the appropriate, usable and applicable methods that are most likely to result in a positive user experience.

Alhumaidan, Lo and Selby [6] in their research through co-design of Augmented Reality (AR) based teaching material, seek to enhance collaborative learning experience in primary school education. For this, an interactive AR book based on primary school textbook was introduced using tablets as the real time interface. The development of this AR book employs co-design methods to involve children, teachers, educators and Human Computer Interaction (HCI) experts from the early stages of the design process. As a conclusion, this study indicates that the subject of AR and its participation in education in Saudi Arabia have to be investigated more deeply.

The article of Sarkar, Pillai and Gupta [7] go in depth in the study of collaborative learning, which involves working in groups to solve a problem or perform a task. This learning was encouraged in rural schools to support the existing teaching methods. AR is one such technology that can provide a collaborative interactive experience. There is a lack of relevant research regarding parents' opinions on students' learning with emerging technology, such as AR. The work carried out by Cheng [8] is an initial attempt to explore parents' users experience (UX) of reading an AR book with their children from the perspectives of perceptions, expectations and intentions. 47 child-parent pairs were invited to participate in a shared AR book reading activity in this study. Some suggestions for the development and popularization of AR book systems were also proposed by the researchers.

The present article has a bibliographical analysis regarding the editorial design, user experience design, design thinking, augmented reality and the different methods that will be used for the development of the educational editorial product. The second section presents the methods used to develop the proposal. In the third section, the generated model with which the research will be carried out is presented. In the fourth section, the proposed model will be compared with those existing in the market, explaining the advantages and deficiencies at the moment of achieving a result. And finally, conclusions and future work related to the design of user experience and its relationship with the editorial design with AR are detailed.

2 Methods and Materials

The development of this project begin with the application of the Design Thinking methodology, developed by Tim Brown, which started as a software evaluation tool and currently is used in all areas, including the educational one that is our main focus. It began to develop theoretically at the Stanford University in California (USA) in the 70s, and its first application for profit as "Design Thinking" was carried out by IDEO design consultancy, being today its main precursor.

Additionally, Scrum will be used, a new teamwork strategy that divides its development in stages, and that until recently was used only for the creation of software, but is currently being implemented as a didactic tool in the educational field, making possible to create collaborative and group work environments, which will allow the development of high quality projects with students. Motivation, critical thinking, communication skills, innovation and aesthetics are enhanced with this strategy. In his article, Juan Lucas Onieva explains in detail a didactic proposal for its implementation in the classroom [9].

For the development of the printed product, the golden ratio will be kept in mind, which takes into account specific points of the design and composition field that have a major impact and attract readers' interest. Drawing the Fibonacci sequence in the work area, these golden points or the most visual impact ones will be found, which will be taken into account for the position of the augmented reality markers. The classification of Beth Tondreau's reticles will be used in order to generate the composition of the texts, visuals and AR markers [10]. The Vignelli canon [11] will be kept in mind to generate the design proposal.

3 Proposed Model

For the development of this research, the market segment for our educational material will be identified, in this case, children from 6 to 8 years old. The educational needs of this segment will be determined, for this the Design Thinking method will be used, which has 5 stages for its application: 1. Empathise, that is to say, to deeply understand the students' needs. The units of analysis will be 4 educational institutions: two public and two private. Two in Quito and two in Ambato. Each institution will receive an introduction to the work that will be carried out and the objectives that are expected to be achieved. Each institution will contribute with a room in which 12 students from 6 to 8 years old, 5 parents, 8 teachers and 8 students from the design career of the Technological University Indoamerica will participate. Each institution will be requested to provide a profile of each student, which will have information that will serve to develop this first stage of empathy. 2. Define, that is, the information received in the first meeting with students, teachers and parents will be analyzed and filtered, taking the one that really adds value and leads us to find solutions to generate innovative results. 3. Ideate, every idea that come up will be taken into account and analyzed as it may be important for the generation of innovative solutions. In the work groups the plot of the story will be created and models of possible sizes of the books will be presented, but the group of students, teachers and parents will be the ones who will define the physical structure of the story.

The designers will present some alternatives of composition and shape of the physical product, which will allow developing the most suitable model for its handling and use in the classroom. The golden ratio, the Fibonacci sequence, the division into grids, segments, columns, fonts, colors, materials for the book or if it should include any accessory will be exposed in a friendly language for its understanding. With the plot of the story, the scenes that the book will have per page will be defined and the graphic elements that they would like to see with augmented reality will be located. 4. The prototype is made, which will be in charge of the students of the digital design and multimedia career, who will work on the design and layout of the story, following the ideas of the children, teachers and parents. According to the work schedule, the design of the printed product, the animations that will be used for augmented reality and the development of the mobile application will be developed in four months. 5. Testing phase, that is to say, after making the most suitable prototypes, we will return to the educational institutions and will analyze if the product meet the requirements suggested in the ideate phase by children, teachers and parents, it will be collected all the information obtained, the prototype will be perfected according to the data received and the final product will be printed for each student (Fig. 1).

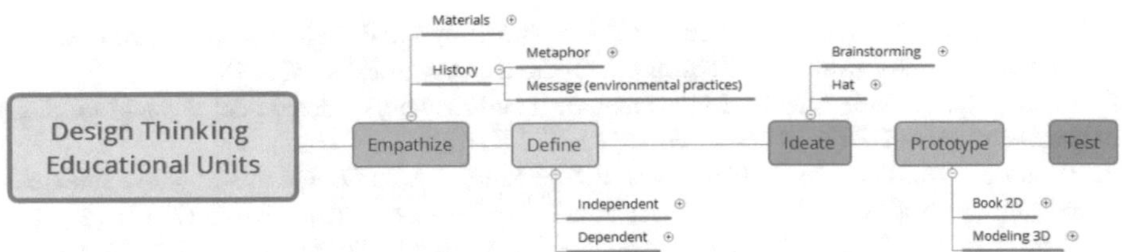

Fig. 1. Elaborated by Carlos Borja. Structure Design Thinking UE

4 Discussion

There are previous investigations that have worked in the area of editorial design with design thinking or user experience in schools, but combining them and making a larger sample will allow a greater quantity and quality of data. Including groups of students from public and private schools will also allow us to have a considerable number of variables to be analyzed. Many designers use the reticular model as the main design and layout option, giving less importance to the golden ratio and the Fibonacci sequence. This will also be a topic to be analyzed with the final product and the results it generates. The composition with reticles properly adapts to the design process and can be used in other studies as a starting point for the layout of printed products with AR and developed with user experience design.

5 Conclusions

Producing an editorial product using the user experience, combining design thinking and the scrum model for its development allows generating results user-centered. Designing a product, using the tools of composition and making it suitable for the suggestions of the users, will make easier their understanding, contact and appropriation. Including AR in the product design, makes the production phase longer, but generates greater empathy with the user, since a greater number of graphic elements for their interaction are combined, which can allow deeper learning results. Investigating the client and be clear about their ways of acting and learning, is fundamental for the application of the user experience model and the design thinking, since that information guides the way to proceed when developing the editorial product.

Acknowledgments. Special thanks to the institutions that opened their doors to apply this research: Ambato Teaching Units: Mario Coba Barona Public Education Unit and Suizo Private Education Unit of Ambato. Quito: Manuel Abad Public Education Unit and San Gabriel Private School, and all of those who participated in this project and who are motivated to generate new related research.

References

1. dos Santos, F.A., Tiradentes Souto, V.: Graphic design and user-centred design: designing learning tools for primary school. Int. J. Technol. Design Educ. (2018)
2. Toledo, L.A., Garber, M.F., Madeira, A.B.: Consideraciones acerca del design thinking y procesos. Rev. Gestão Tecnol. **17**(3), 312–332 (2017)
3. Woloszyn, M., Dick, M.E., Gonçalves, B.S., Fialho, F.A.P.: Design thinking no contexto do projeto editorial: contribuições instrumentais. DAPesquisa **13**(21), 59–75 (2018)
4. Sarkar, P., Pillai, J.S.: User Expectations of augmented reality experience in indian school education, pp. 745–755 (2019)
5. Michailidou, I., von Saucken, C., Kremer, S., Lindemann, U.: A user experience design toolkit, pp. 163–172 (2014)
6. Alhumaidan, H., Lo, K.P.Y., Selby, A.: Co-design of augmented reality book for collaborative learning experience in primary education. In: Proceedings of 2015 SAI Intelligent Systems Conference, IntelliSys (2015)
7. Sarkar, P., Pillai, J.S., Gupta, A.: ScholAR: a collaborative learning experience for rural schools using augmented reality application. In: 2018 IEEE Tenth International Conference on Technology for Education (T4E), pp. 8–15 (2018)
8. Cheng, K.-H.: Parents' user experiences of augmented reality book reading: perceptions, expectations, and intentions. Educ. Technol. Res. Dev. **67**(2), 303–315 (2019)
9. Onieva López, J.L.: Scrum como estrategia para el aprendizaje colaborativo a través de proyectos. propuesta didáctica para su implementación en el aula universitaria. Profesorado, Rev. Currículum y Form. del Profr., vol. 22, no. 2, June 2018
10. Ulled Nadal, T.: El gran espectáculo de la Realidad Aumentada. Fotogramas & DVD La Prim. Rev. cine, no. 2003, p. 8 (2010)
11. Vignelli, M.: The Vignelli canon. Design (2008)

A Cloud Based Augmented Reality Framework - Enabling User-Centered Interactive Systems Development

Anas Abdelrazeq(✉), Christian Kohlschein, and Frank Hees

Cybernetics Lab in Aachen, RWTH Aachen University, Dennewartst. 27, 52068 Aachen, Germany
{Anas.Abdelrazeq, Christian.Kohlschein, Frank.Hees}@ima-ifu.rwth-aachen.de

Abstract. Concepts for user interfaces are developed in order to enable humans exchanging information and having a better understanding of different processes. With the rise of technology, concepts of enriching our environments with computer-generated graphics appeared in the 1960s. The expansion of visual perception of the surrounding is the core of Augmented Reality (AR). Many AR applications have been appearing in different areas, opening the door for revolutionizing and changing interactivity models and the way people perform tasks. Even though AR systems' structures are well defined, applications are not yet widespread. One of the main reasons for this is that AR system developers face difficulties with developing their applications and adapting them to different hardware. Each hardware set (i.e. AR device) requires different software development kits and different technical requirements. That leaves AR concepts hard to be scaled and implemented in real life scenarios. One more additional challenge facing the implementing of AR is that AR mostly relies on complex vision-based algorithms. Based on that, this paper explores a lightweight, modular, and cross-platform solution. It presents a novel framework based on redesigning the AR pipeline, introducing a prototype of a solution where AR can operate within the cloud, converting all AR devices into thin clients with high performance capability. This also contributes directly to the Industry 4.0 aim of preparing the way for digitalized and optimized processes in a connected and integrated network.

Keywords: Augmented Reality · Cloud-Computing · Industry 4.0 · Intelligent interfaces · Mixed reality

1 Introduction

Augmented Reality (AR) technology opens the door for revolutionizing and changing the way people interact. In many contexts, AR enables its users to exchange information and have a better understanding of different processes [1]. In recent years, AR platforms and devices have experienced rapid development. Many applications of AR have started appearing in different areas of everyday life, particularly within workspaces and the way workers perform tasks.

T. Ahram et al. (Eds.): IHSED 2019, AISC 1026, pp. 417–422, 2020.
https://doi.org/10.1007/978-3-030-27928-8_64

AR is defined as a live view of the real-world environment enriched (augmented) with computer-generated three-dimensional (3D) models. It is common to refer to Azuma's definition outlining the technical requirements for AR. He defined three main properties for any AR system [2]: (1) combining real with virtual objects, (2) real-time interactivity, and (3) object registration in the 3D real-world.

Extending Azuma's definition into applying AR within the industrial field has led to the concept of Industrial Augmented Reality (IAR). The term IAR can be defined as the application of AR in order to support industrial processes [3]. Recently, many companies are seeking a shift towards Industry 4.0 and its principles where AR is seen as one of its technology pillars [4].

Even though AR systems' concepts are well defined, applications are not very widespread due to two main challenges. One of the main reasons is that AR system developers face difficulties with developing their application and adapting it to different hardware specifications [5]. Each AR hardware set (i.e. AR device) requires different development kits and has different technical requirements. This makes AR concepts difficult to be scaled and implemented in lifelike scenarios. Another challenge facing the implementation of AR that it mostly relies on complex vision-based algorithms. Considering this fact, available applications are only capable of providing basic algorithms to recognize simple markers. They are characterized with poor recognition and tracking capabilities for natural images and real objects that are needed for advanced scenarios [6]. This is equally true while using typical AR devices such as smartphones, tablets, and advanced devices. So far, all devices lack sufficient computational power along with weaknesses like limited battery life and limited storage capacity [7, 8].

Due to these challenges, applying AR is becoming restricted to dedicated, simple device-based and app-based solutions [6]. In order to cope with the rising demand for such applications, there is a need for a lightweight, modular, and cross-platform AR solution. The core idea of this paper is to solve the challenges by integrating cloud-computing technology for AR. This happens while keeping the requirements context directly connected to the Industry 4.0 goal of digitized and optimized processes in a connected and integrated network.

2 Framework Requirements

The AR cloud-based service specifications and demanded functionalities are analyzed based on two use cases from the industry. The use cases cover a broad range of applications that have been developed, including basic needs for visualization, interaction, and an extension for collaboration either between human and machine or human and human. From the use cases, a set of requirements for the demanded functionalities that need to be implemented by AR developers is defined.

The first use case uses AR to help employees by giving on the job instruction while working on producing units within their assembly cells [9]. In this scenario, complex tasks are required in order to produce the units that are not completely manually assembled, but in cooperation with a robot arm.

The second use case extends the first one by allowing collaborative work among workers, interacting with real-time information, and monitoring systems and processes [10]. As the technology is improving, AR appears to be taking a step towards emerging digital communications technology. AR is used to establish a space in which employees (i.e. users) are able to communicate about product designs and share their ideas. Instead of working on a real model due to its lack of availability, high cost, and travel distances, employees can work on virtual objects to establish their design together and simultaneously. AR technologies have the chance of bringing people together across long distances, masking the gap of physical distance by enabling face-to-face-like communication.

Our cloud-based AR framework should have the capability to support applications development for these two use cases. Modular functionalities should be able to run within the cloud, providing a real-time experience for the users in both cases.

3 Cloud-Computing for Augmented Reality

The concept of Cloud-Computing (CC) is formally defined by the National Institute for Standards and Technology (NIST). It is described as a model where a pool of configurable computing resources are accessible on demand via the network [11]. Our aim is to enable AR applications to operate within the cloud, converting all AR devices into thin-clients with low performance requirements. In order to achieve this, there is a need to redesign the AR pipeline, providing a new concept for AR software engineers (i.e. AR developers) for developing interactive and user-centered AR applications.

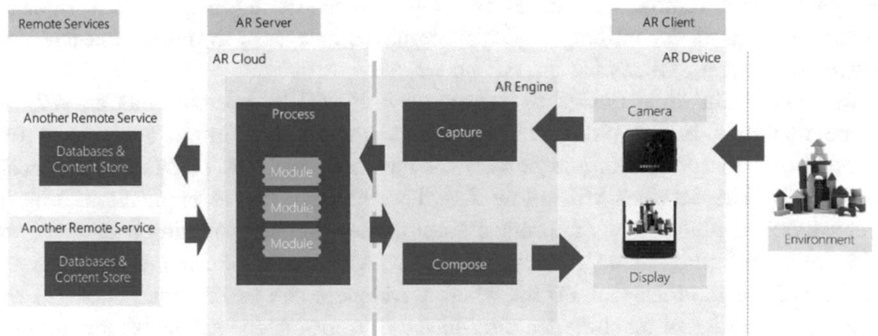

Fig. 1. Cloud-based augmented reality pipeline redesign.

"AR Engines" generate AR experience by applying a pipeline of steps on each received camera frame of the reality. Within the pipeline, received frames run into three main steps: capturing, processing, and composition. In the redesign, new architecture is proposed in order to implement the classical AR pipeline. Most of the middle-layered functionalities (i.e. processing) in the classical pipeline will be moved to run within a cloud service. The new AR cloud-based pipeline will form three main layers as shown in Fig. 1: AR Client, AR Server, and other remote services. The range of deployed functionalities over different layers is determined by the developers based on the targeted device's specifications.

In the runtime of AR application, the client's app is responsible for instantiating the communication with the AR service, registering the running application, and exchanging acquired data. It is possible that the client would do basic processing such as image resizing, color conversion, feature detection, and rendering.

The AR server is the host of different AR modules (i.e. functionalities). The server is accessible via a web service. The AR service runs over the suitable web protocols in order to be discovered and communicated. It can be viewed as an Application Programming Interface (API) where developers are able to create new functionalities, use them, and share them with others. Different functionalities reside on the service as templates to be cloned and reused. The service structure will allow developers to arrange different functionalities in different orders, creating a pipeline in dependency tree-like structures to run the AR processes through them. Functionalities are dynamic in their nature. They can include different third-party dependences or any needed internal or external database structure. This results in a highly modular structure for our AR cloud-based service.

By default, the AR server provides basic core AR modules to the developers that cover image processing, images matching, object recognition, 3D maps reconstruction, etc. It also manages the incoming requests, decides on the optimal pipeline deployment, and its running order. It also provides interfaces to upload new functionalities' source code to be complied and used. It manages users and authenticates them, monitors usage of different functionalities and their performance, and benchmarking.

As an extension, the AR cloud-based service allows a further connection to external third party online remote services. For example, some AR modules might need access to external datasets for training object recognition, or access to online text-to-speech services to generate voices for the AR scene, etc.

Eventually, the final design of the framework will be presented as an AR web service that runs in the cloud. This provides potentially "infinite" resources (e.g. computation, storage space, and power) for the connected AR clients meeting scalability needs. The services will allow AR developers to create applications that are inherently cross platform with no installation needed, hence providing a standard way for developing future AR applications. Primitive AR cloud modules are also presented as a basis for a modular framework where developers can extend their scenario with more functionalities of machine-learning-based concepts, such as objects recognition or multiplayer collaborative concepts.

4 Conclusion

The core contribution of this paper is to enable AR software engineers (AR developers) by means of a framework that makes it standardized for them to implement their desired AR functionalities. The framework extends AR developers' ability to build apps that overcome the challenges defined in the last section by being limited by the end device specifications. The proposed framework is a modular cloud-based AR platform that is exposed as a web service, which allows developers to utilize existing AR functionalities, code and upload new ones, and extend their computational scenarios based on their own implementations as well as shared code.

The proposed AR cloud-based framework thinks of its clients as lightweight terminals (i.e. Think-Clients) that transfer their heavy computation components and resource load to the cloud. The system will provide better performance in terms of storage, processing power, and computation time. In addition, scalable cross-platform applications will be easier to deploy with an easy way to update functionalities with no additional installations required. The cloud will allow content and context sharing, enabling collaboration between different connected clients. This is enabled by the fact that the content resources are centralized where data centers and model databases are located in the cloud. Developers are able to generate their own content and AR functionalities (i.e. code). The solution can enrich users' experiences via generating personalized content.

Acknowledgments. This research and development project is funded by the German Federal Ministry of Education and Research (BMBF) within the "Innovations for Tomorrow's Production, Services, and Work" Program (funding number 02L14Z000) and implemented by the Project Management Agency Karlsruhe (PTKA). The author is responsible for the contents of this publication.

References

1. Brooks, F.P.: The computer scientist as toolsmith II. Commun. ACM **39**(3), 61–68 (1996)
2. Azuma, R.T.: A survey of augmented reality. Presence: Teleoper. Virtual Environ. **6**(4), 355–385 (1997)
3. Fite-Georgel, P.: Is there a reality in Industrial Augmented Reality? In: 2011 10th IEEE International Symposium on Mixed and Augmented Reality (ISMAR 2011), Basel, Switzerland, 26–29 October 2011, pp. 201–210 (2011)
4. Fraga-Lamas, P., Fernandez-Carames, T.M., Blanco-Novoa, O., Vilar-Montesinos, M.A.: A review on industrial augmented reality systems for the industry 4.0 shipyard. IEEE Access **6**, 13358–13375 (2018)
5. Rajagopal, N., Miller, J., Reghu Kumar, K.K., Luong, A., Rowe, A.: Demo abstract: welcome to my world: demystifying multi-user AR with the cloud. In: 17th ACM/IEEE International Conference on Information Processing in Sensor Networks (IPSN), Porto, pp. 146–147 (2011)
6. Qiao, X., Ren, P., Dustdar, S., Chen, J.: A new era for web AR with mobile edge computing. IEEE Internet Comput. **22**(4), 46–55 (2018)
7. Luo, X.: From augmented reality to augmented computing: a look at cloud-mobile convergence. In: 2009 International Symposium on Ubiquitous Virtual Reality, pp. 29–32. GIST, Guangju (2009)
8. Huang, B.-R., Lin, C.H., Lee, C.-H.: Mobile augmented reality based on cloud computing. In: 2012 International Conference on Anti-Counterfeiting, Security and Identification, Taipei, Taiwan, pp. 1–5 (2012)

9. Daling, L., Abdelrazeq, A., Haberstroh, M., Hees, F.: Usability evaluation of augmented reality as instructional tool in collaborative assembly cells. In: The Twelfth International Conference on Advances in Computer-Human Interactions, ACHI 2019, pp. 199–205 (2019)
10. Schiffeler, N., Abdelrazeq, A., Stehling, V., Isenhardt, I., Richert, A.: How AR-e your Seminars?! Collaborative learning with augmented reality in engineering education. In: INTED2018 Proceedings, Valencia, Spain, pp. 8912–8920 (2018)
11. Mell, P., Grance, T., et al.: The NIST definition of cloud computing (2011)

Redesign of a Questionnaire to Assess the Usability of Websites

Freddy Paz[1(✉)] and Toni Granollers[2]

[1] Pontificia Universidad Católica del Perú, Lima 32, Peru
fpaz@pucp.pe

[2] Human-Computer Interaction and Data Integration Research Group (GRIHO), Polytechnic Institute of Research and Innovation in Sustainability (INSPIRES), University of Lleida, Lleida, Spain
antoni.granollers@udl.cat

Abstract. Usability is taking an essential role in the software development process. Nowadays, many teams incorporate methods to verify if the graphical user interfaces of a software product in development, meet the required degree of usability. However, most of these methods, instead of providing a numerical value about the accomplishment of this quality attribute, only give support to obtain a list of aspects that can be redesigned to increase the usability. Quantitatively measuring the ease of use is useful for companies which need to compare themselves with their main competitors, or when there are several choice design proposals for their website. In this study, we present the redesign of a questionnaire developed from the theoretical basis of two highly recognized sources, which define the concept of usability in a set of principles. The new assessment tool is a five-point version of the previous work with 60 items which attempt to cover all the spectrum of usability.

Keywords: Usability heuristics · Quantification method · Evaluation questionnaire · Assessment tool · Verification checklist

1 Introduction

Usability is defined as the degree to which a software product or system can be used by specified users to achieve specified goals with effectiveness, efficiency, and satisfaction in a specified context of use [1]. Nowadays, this quality attribute represents an important aspect to be considered during the software development process given that it can be determinant in a competitive market as the current one. In a context in which there are several alternatives of applications available on the Web, the extent of ease of use, as well as the capability of the system to be understood and attractive can be aspects that make the difference [2]. Because of this scenario, it is becoming more and more common to adopt assessment methods that help development teams to verify if the required degree of usability is reached.

Most of the techniques that are frequently used to evaluate the usability of graphical user interfaces allow specialists to obtain qualitative data [3]. The results that can be achieved through the use of these methods are usually design problems, that the

T. Ahram et al. (Eds.): IHSED 2019, AISC 1026, pp. 423–428, 2020.
https://doi.org/10.1007/978-3-030-27928-8_65

development team must fix afterward. However, although this process gives support to improve the usability of a specific system, there are scenarios in which it becomes necessary to determine the usability degree in numerical values. In some cases, for instance, when the technique "parallel design" is employed, several design proposals are generated as result, and at the end of the process, the development team must opt for the best option. It becomes difficult to perform a selection between different prototypes if there is not a clearly established quantitative criterion. In the same way, the companies are used to perform comparisons to identify how far they are regarding to their competitors. This situation was the main motivation to develop a questionnaire that provides specialists with a tool that quantitatively measures the usability degree of graphical software interfaces. In this study, the authors present the redesign of a previous proposal [4] which was developed from two recognized sources which widely describe the concept of usability.

2 Background

According to Fernandez [5], the questionnaire can be considered as a usability evaluation method. At present, there are some proposals [6] such as SUS, SUMI, and IBM approaches whose purpose is to evaluate the usability. Nevertheless, these recognized tools are more focused on the assessment of the final user's satisfaction and the quality in use than on the sub-characteristics of usability. This fact has led to the authors to design a new questionnaire based on a set of heuristics that widely cover the spectrum of the concept of usability [4]. Through an in-depth analysis of the guidelines proposed by Nielsen [7], and an examination of the principles of interaction design established by Tognazzini [8], an initial set of 60 YES-NO questions was proposed. However, the new approach that we are presenting in this study is a redesigned version that considers a Likert scale of five-point for each item of the questionnaire. This tool can be contemplated as an initiative based on the heuristic evaluation method, given that it is based on rules that usually are used by Human-Computer Interaction specialists to evaluate graphical user interfaces. Holding the same nature of the original technique, this questionnaire is oriented to be employed by expert professionals in the field of usability. The new assessment tool has been designed in a way in which evaluators must focus on specific aspects establishing a numerical value about the level of accomplishment of these identified characteristics. It is not the purpose of this study to provide with a questionnaire that can be used by final users. In this sense, it could contain aspects that would only be properly reviewed by specialists.

3 Redesign of a Questionnaire to Evaluate the Usability

The new redesign of the questionnaire provides 60 new items adapted to a five-point Likert scale ranging from 1 (strongly disagree) to 5 (strongly agree). The intention of this reformulation is to measure not only the features that are addressed in each item, but also the frequency of the accomplishment of the guideline in the entire system. All the questions are written in a positive way, establishing that the category of "strongly disagree" will be representing a negative degree of usability in the assessed item, and "strongly agree" a good level and perception of ease of use. The new questionnaire for the evaluation of usability of software applications is presented in Table 1.

Table 1. Proposed questionnaire for the evaluation of usability

ID	Question
Q01	*Does the application provide visible titles that clearly identify the different sections of the system?*
Q02	*Is it always possible to determine in which section of the system we are?*
Q03	*Is it possible to determine the actions the system is executing all the time?*
Q04	*Are the links provided by the system visible, clear and understandable?*
Q05	*Is it possible to quickly identify the actions provided by the system?*
Q06	*Is the information displayed in order and logical way in the system?*
Q07	*Is the design of the icons based on concepts extracted from the real context?*
Q08	*Is there a relation between the icons and the actions executed by the system?*
Q09	*Are the phrases and concepts used understandable and familiar?*
Q10	*Are there links that allow us to return to the home page from the different section of the system?*
Q11	*Does the system provide mechanisms to re-do and undo the changes?*
Q12	*Is it possible to easily return to a previous state of the application?*
Q13	*Is the name of the links related to the section to which they are redirecting?*
Q14	*Do the same actions in the system always lead to the same results?*
Q15	*Does the same icon have the same meaning throughout the system?*
Q16	*Is the information displayed consistently throughout the system?*
Q17	*Do the links have an appropriate color that distinguishes them from other elements in the interface?*
Q18	*Do the navigation elements such as buttons, checkboxes, among others have a standardized design?*
Q19	*Did you find it easy to use the first time you interacted with the system?*
Q20	*Is it easy to locate information or features that have already been sought before?*
Q21	*Is it possible to use the system without having to remember information from screens before the current one?*
Q22	*Is the information provided by the interface in the different parts of the process complete?*
Q23	*Is the information organized according to the logic of the "type" users?*
Q24	*Does the system have keyboard shortcuts to execute frequent actions?*
Q25	*Is it understandable how to use the keyboard shortcuts to perform actions in the system?*
Q26	*Is it easy to redo an action that has already taken place before?*
Q27	*Does the design adapt to the different sizes or resolutions of the screen?*
Q28	*Are visible the mechanisms that accelerate the different processes offered by the system?*
Q29	*Does the system prevent the user from incurring unnecessary waiting times?*

(continued)

Table 1. (*continued*)

ID	Question
Q30	*Does the system display appropriate confirmation messages when are requested actions in the application that are irreversible?*
Q31	*Does the system display messages in real time about the mistakes made?*
Q32	*Are the error messages understandable?*
Q33	*Does the system use any code to reference errors that appear on the screen?*
Q34	*Are appropriate confirmation messages displayed before performing relevant actions on the system?*
Q35	*Is the system explicit in indicating the information that must be entered in each field of a form?*
Q36	*Does the system respond appropriately to typographical and orthographic errors made in searches?*
Q37	*Is only relevant information displayed in the system?*
Q38	*Is the information displayed on the system short, concise and accurate?*
Q39	*Does each unit of information convey a unique and understandable message?*
Q40	*Is an organized text with short phrases and quick interpretation used in the interface?*
Q41	*Is there a clearly defined "help" section in the system?*
Q42	*Is the help section visible and easily accessible?*
Q43	*Does the help section offered by the system give support to solve problems?*
Q44	*Does the system have a clearly defined section of frequently asked questions?*
Q45	*Is the help documentation clear and provide examples?*
Q46	*Does the system allow the activities to be resumed at the point where they were left?*
Q47	*Does the system save the information in work automatically?*
Q48	*Does the system respond appropriately to external failures such as power cuts, Internet cutoff, among others?*
Q49	*Have the text sources an adequate size?*
Q50	*Does the contrast between the text and background colors allow an appropriate reading of the information?*
Q51	*Do the images or background colors make it easier to read the text?*
Q52	*Is the system interface appropriate for users with reduced vision?*
Q53	*Does always the system inform about the actions it is executing?*
Q54	*Does the system display updated information about what is executing?*
Q55	*Is it possible to customize the system interface according to personal preferences?*
Q56	*Does the system provide a mechanism that allows to reset the default values?*
Q57	*In case of returning to the default values, does the system clearly specify the consequences of performing this action?*
Q58	*Does the system use the term "default" to refer to default values?*
Q59	*Are the complex tasks executed by the system visible or notorious?*
Q60	*Is displayed the remaining time or some animation of the complex tasks that the system is executing?*

4 Conclusions and Future Works

The usability is a quality attribute that can represent the differentiating factor in a competitive market such as the current one. For this reason, there are several methods that allow specialists obtain qualitative and quantitative information about the degree of accomplishment of this aspect in the software products. However, the current techniques that quantify the usability such as questionnaires and metrics are related to quality in use. SUS, SUMI and TAM that are recognized and widely used proposals, are focused on determine the degree of the user's satisfaction more than on the evaluation of the concept of usability. Under this scenario, a new questionnaire was proposed considering two well-known sources that are commonly used in heuristic evaluations. The new approach presents 60 five-point Likert scale items that cover all the concept of usability. The assessment tool is intended to be used by specialists in Human-Computer Interaction who have a broad domain of the field and can answer in an accurate way from 1 (strongly disagree) to 5 (strongly agree) to each question. The usability degree of the application will be determined by a sum of the values obtained to each question. By providing a tool that allow specialists to obtain a numerical value about the usability of system, the software development teams will be able to evaluate and determine how far they are from an acceptable level of ease of use. In the same way, companies from the E-Commerce domain with this questionnaire will be able to perform comparisons and determine their strengths and weaknesses, with respect to the presented web applications by their main competitors.

Although the proposal has been defined, it is still necessary to perform some study cases to analyze the reliability of the questionnaire. In the same way, it is important to examine the validity of content, criterion and construct through statistical techniques. Given that the questionnaire has been developed to be used not only on web systems, but also on software products in general, it would be important to consider the execution of several evaluations in which system from different categories are assessed by specialists with different backgrounds.

Acknowledgments. This study was highly supported by the "Department of Engineering" of the Pontifical Catholic University of Peru – Peru and the "HCI, Design, User Experience, Accessibility & Innovation Technologies" Research Group (HCI-DUXAIT).

References

1. ISO: Systems and software engineering – Systems and software Quality Requirements and Evaluation (SQuaRE) – System and software quality models (ISO/IEC 225010:2011), Geneva, Switzerland (2011)
2. Paz, F., Paz, F.A., Pow-Sang, J.A.: Experimental case study of new usability heuristics, pp. 212–223. Springer International Publishing, Cham (2015)
3. Diaz, E., Arenas, J.J., Moquillaza, A., Paz, F.: A systematic literature review about quantitative metrics to evaluate the usability of E-commerce web sites, pp. 332–338. Springer International Publishing (2019)

4. Granollers, T.: Usability evaluation with heuristics. New proposal from integrating two trusted sources, pp. 396–405. Springer International Publishing (2018)
5. Fernandez, A., Insfran, E., Abrahão, S.: Usability evaluation methods for the web: a systematic mapping study. Inf. Softw. Technol. **53**, 789–817 (2011)
6. Sauro, J., Lewis, J.R.: Quantifying the User Experience: Practical Statistics for User Research. Morgan Kaufmann, Cambridge (2016)
7. Nielsen, J.: 10 usability heuristics for user interface design (1995). https://www.nngroup.com/articles/ten-usability-heuristics/. Accessed 15 June 2019
8. Tognazzini, B.: First Principles, HCI Design, Human Computer Interaction (HCI), Principles of HCI Design, Usability Testing (2014). http://www.asktog.com/basics/firstPrinciples.html. Accessed 15 June 2019

Edge Detection Method for the Graphic User Interface of Complex Information System

Yukun Song, Chengqi Xue(✉), Xinyue Wang, and Peiqi Zhang

School of Mechanical Engineering,
Southeast University, Nanjing 211189, China
466143950@qq.com, {ipd_xcq,230189401}@seu.edu.cn,
zhangpeiqi@fuji.waseda.jp

Abstract. There is such massive of redundancy remained in recent development of GUI about design, implementation, and evaluation that it is now urgent for researchers to make these tasks done automatically. Among the different stages of automatic GUI developing, the primary work is to get information of edges in GUI images in a suitable manner. It is crucial to obtain accurate information about the arrangement of the GUI component, which will be the key basis of automatic development of GUI. In this paper, optimizations have been proposed for GUI image after several experiments and evaluations. And it turns out that the optimizations work well.

Keywords: Graphic user interface · Edge detection · Image processing

1 Introduction

The constant change of information carried by the GUI and the continuous expansion of the application field have presented new challenges to researchers. The rise of deep learning methods and the popularity of artificial intelligence technology have provided us with new solutions [1]. Reducing duplication of labor and improving production efficiency are the main problems to be solved by developing science and technology. More and more researchers are also working on the frontier theory [2].

In this context, the issues related to the efficient and automated development of complex system GUIs still need to be improved: how to directly and automatically obtain corresponding results from design manuscripts or standardized descriptions, eliminating a lot of unnecessary work for developers, designers and development.

In the automatic generation of GUI programs, design documents and manuscripts are generally used as input. How to take appropriate measures to specifically quantify this input is the primary problem in the automatic generation of the entire GUI program [4]. In the field of image processing, the sensitivity and accuracy of edge detection directly affect the subsequent processing. The edge detection result of the GUI can be quantized into the target software interface, thus providing a clear task for the automatic design system [5].

The traditional evaluation methods mainly focus on two aspects, one is based on the actual test system [6], and the other is based on the statistical model test system. Edge detecting and obtaining advanced information such as GUI control type, position

T. Ahram et al. (Eds.): IHSED 2019, AISC 1026, pp. 429–434, 2020.
https://doi.org/10.1007/978-3-030-27928-8_66

and size based on the edge information is a key step based on automatic image detection and evaluation of GUI, which is of great significance for GUI automation development and evaluation.

2 Related Work

In the new era of information explosion, how to reduce the burden of workers and improve productivity is an important issue today.

In terms of implementation, researchers have made a lot of efforts and explorations on code generation based on deep learning methods. For example, Beltramelli et al. [7] trained end-to-end networks with deep learning methods, and generated different screenshots directly from GUI screenshots. The main problem with this approach is the need to design a domain-specific language autonomously, as well as a compiler between the language and the actual GUI programming engineering language.

Coincidentally, Wallner et al. proposed another end-to-end network [5], exploring the GUI code generation with the design draft as input, and directly generating the webpage code directly from the design template through the deep learning method. This method is more flexible than the pix2code [7]. However, the integrated code logic of this method is not clear enough to be modified or used multiple times.

In the above two methods of generating code, edge detection is an important part and the basis of the algorithm implementation.

Based on the traditional detection method, Zhao et al. [8] used the convolutional neural network to enhance the edge information of the target object and weaken the edge information of the surrounding environment in the problem of edge detection, thereby obtaining more practical edge images make the resulting edge information more valuable.

3 Methodology

This paper deals with screenshots of complex information system GUIs, aiming at obtaining high-precision image edge information based on analysis, and then obtaining GUI layout information that is useful for guiding GUI design, development and evaluation. Based on this, the research can be divided into the following steps:

First, to apply edge detection algorithms for GUI images. And to improve the algorithms for better performance.

Secondly, to combine and refine the algorithms to obtain layout information of the GUI controls of the complex information system.

3.1 Prewitt, Canny, and Sobel Edge Detection Performance in GUI Images

The Prewitt operator has not been widely used in natural pictures[9], but it has high performance in GUI pictures for complex information systems. Based on the analysis and experimentation of the experimental results, reasons are as follows. GUI images have no particularly obvious and complex changes at the pixel level compared to natural images.

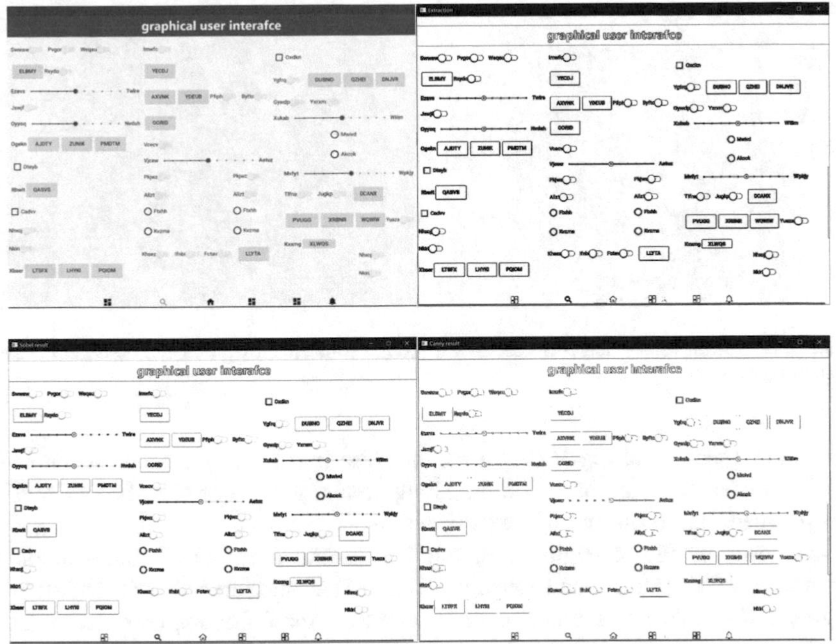

Fig. 1. The top-left: Complex information system interface model; The top-right: Prewitt operator output effect; The bottom-left: Sobel operator output effect; The bottom-right: Canny operator output effect

The Sobel operator is a widely used edge detection method in the field of image processing. Similar to other operators, the Sobel operator also integrates the local features of the image to obtain the gradient values of the image at the corresponding position. The Sobel operator uses the Gauss operator as the image preprocessing process. For the GUI images for complex information systems studied in this paper, the smoothness of image smoothing based on Gauss operator in Sobel operator is not unnecessary.

The Canny operator interpolation method in this subject uses linear interpolation, which assumes that the gradient value between different points changes linearly. The area between different gradient value points changes linearly, so as to calculate the assumed value of the intermediate region. In this paper, the upper and lower thresholds are set to 10 and 0 respectively, which maximizes the detection of edge information. As is shown in Fig. 1. Prewitt operator has the best output effect.

3.2 Layout Information Extraction Based on Image Edge Information

Elements Contour Detection

As mentioned above, GUI images for complex information systems have an unconventional shape, with most of the outlines being regular square or circular. Based on this feature, Hough transform is a very suitable detection method.

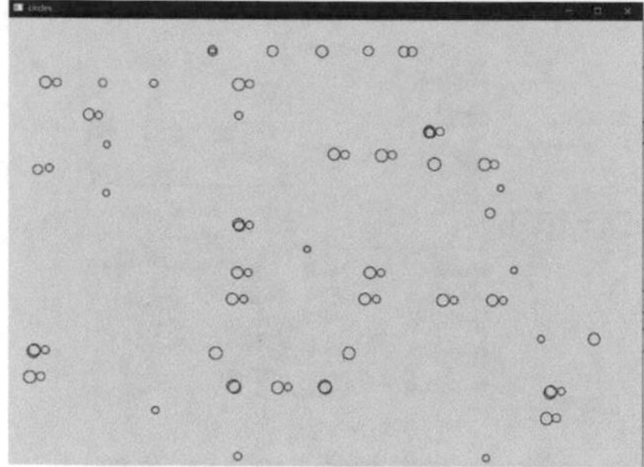

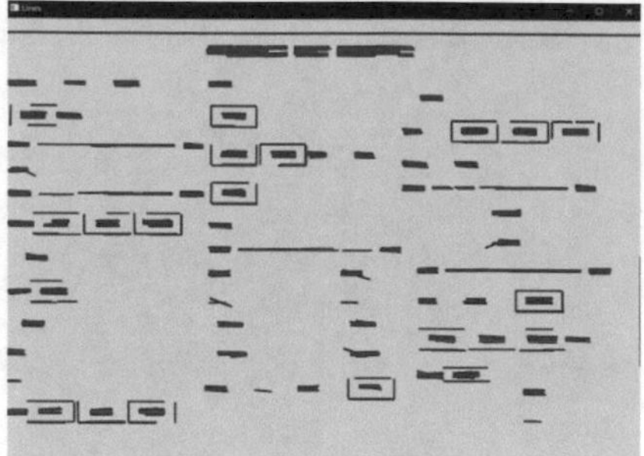

Fig. 2. left: Button circle detection based on Hough transform; right: Line detection based on Hough transform

As can be seen from Fig. 2 (left one), the noise interference is small. It can be seen that the circular detection based on the Hough transform is not disturbed by text after the adjustment based on multiple experiments.

As shown in Fig. 2 (right one), the Hough transform has a large amount of interference information when detecting lines. The interference information of the Hough transform detection line is mainly caused by the following reasons: on the one hand, the layout of the control is very complicated, and on the other hand, the interference from the text.

Satoshi et al. proposed a contour detection method based on a topological model of binarized images. Compared with the line detection based on Hough transform, the contour detection based on topology model has the following two advantages: first, the original wide-area-based detection is converted into local feature-based detection, which makes subsequent optimization and adjustment. Second, subject to the topological relationship, the contour detection based on the topology model does not produce false detection at the same position, and only internal or external false detection is generated. Such false detection can pass the parameter in the post-processing.

In addition, the binarization of the complex information system GUI is realized to provide the input binary image for the above algorithm, because in this example, there are many gray images, and the pixels on both sides of the contour are detected.

The contour detection based on the topology model has a good detection effect in this paper, and almost all rectangular elements can be detected. While the various controls on the GUI image contain different sizes, correct detections are accompanied by many false detections frequently.

Misdetection of Elements Closed Contour and Elimination of Retest

The Hough transform detects almost all of the circles in the graph, but the results of the edge detection are not perfect. The following two methods are mainly used to eliminate the false detection.

First, The re-detection of the same rectangle in a small range caused by edge detection often has a single-digit difference in the coordinate range. Therefore the round-based rounding operation can almost completely eliminate the re-detection caused by the edge information noise, and will not affect the accuracy of the original detection (Fig. 3).

Secondly, for the results of Hough transform and contour detection based on topological model. In this paper, a bubble sorting method based on the horizontal and vertical coordinates of the center point is realized, and different interface elements are integrated according to the coordinates. Then the weight matching based on the first step is eliminated, and a good effect is obtained (see Fig. 4).

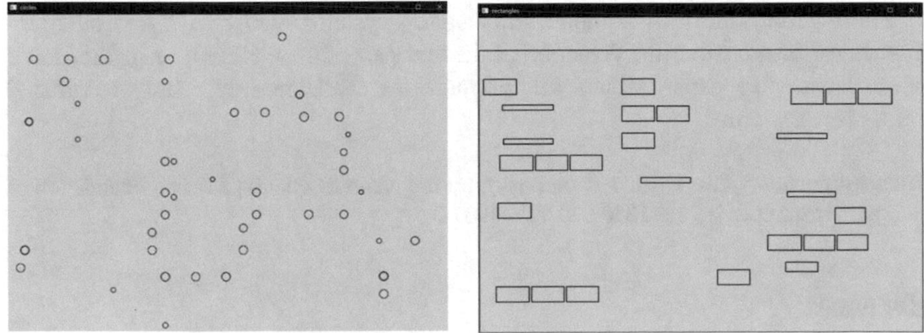

Fig. 3. GUI image after binarization

Fig. 4. GUI image after binarization

Extract Control Type Information From Image Outline Information

As can be seen from Figs. 3 and 4, the position and size of the circular elements and rectangular elements of the complex information system GUI have been detected more accurately, which has laid a good foundation for the final extraction of control type information.

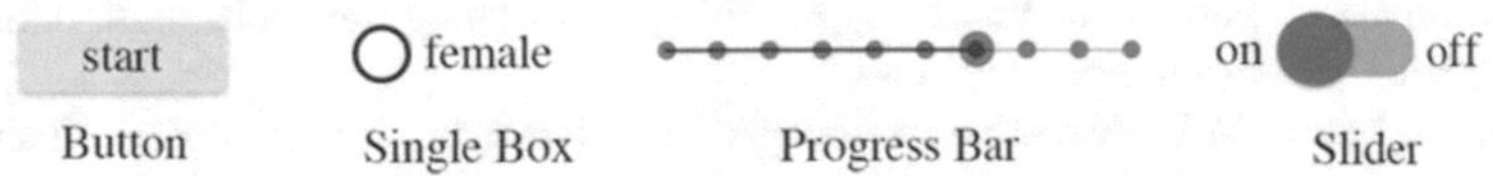

Fig. 5. Four types of controls

As shown in Fig. 5, based on the GUI model for complex information systems abstracted in this paper, there are four key interface control elements (sliding blocks, rectangular buttons, radio buttons and switches) in the screen.

Based on the rectangular geometric elements and circular geometric elements obtained above, the different combinations are analyzed and detected. Based on the reasonable setting of parameters, the final detection accuracy is over 90%.

Slider, rectangular button, radio button and switch buttons are all detected well. For sliding blocks and rectangular buttons, the algorithm can output its position, height (upper and lower maximum height difference of the sliding block) and length; for radio buttons and switches, the algorithm can output the position and radius of its central circle.

4 Conclusion

In the case of large-scale use of the GUI, the design, development and evaluation process still requires a lot of human labor. In the era of artificial intelligence, how to use intelligent and automated technology to replace repetitive labor in related processes is an urgent problem to be solved. Among them, the detection of the edge of the GUI image and the identification of the control are the key foundation for the realization of automation.

The methodology proposed above has strong robustness to the different layouts of GUI information control of complex information systems, which lays a good foundation. In the future work, the detection of the text part of the GUI will be included and combined with this paper, which will promote the realization of a fully automated development system.

Acknowledgments. This work was supported by the National Natural Science Foundation of China, project number is 71871056 and 71471037.

References

1. Yang, W., Prasad, M.R., Xie, T.: A grey-box approach for automated GUI-model generation of mobile applications. In: Cortellessa, V., Varró, D. (eds.) FASE 2013. LNCS, vol. 7793, pp. 250–265. Springer, Heidelberg (2013)
2. Toby, B.H.: EXPGUI, a graphical user interface for GSAS. J. Appl. Crystallogr. **34**, 210–213 (2001)
3. Hinton, G.E., Osindero, S., Teh, Y.W.: A fast learning algorithm for deep belief nets. Neural Comput. **18**, 1527–1554 (2006)
4. Chen, C., Su, T., Meng, G., Xing, Z., Liu, Y.: From UI design image to GUI skeleton: a neural machine translator to bootstrap mobile GUI implementation. In: 40th International Conference on Software Engineering, pp. 665–676. ACM, Gothenburg (2018)
5. Design Mockups Into Code With Deep Learning. https://blog.floydhub.com/turning-design-mockups-into-code-with-deep-learning/
6. Olivia, A., Mack, M.L., Shrestha, M., Peeper, A.: Identifying the perceptual dimensions of visual complexity of scenes. In: Proceeding of the Annual Meeting of the Cognitive Science Society, vol. 26 (2004)
7. Beltramelli, T.: pix2code: generating code from a graphical user interface screenshot. In: 10th ACM SIGCHI Symposium on Engineering Interactive Computing Systems, No. 3. ACM, Paris (2018)
8. Zhao, X.: Image edge detection method based on deep convolutional network. Modern Manuf. Eng. **2**, 144–149 (2018)

Identifying and Classifying Human-Centered Design Methods for Product Development

Gabriela Unger Unruh(✉) and Osiris Canciglieri Junior

Polytechnic School, Industrial and Systems Engineering Graduate Program (PPGEPS), Pontifical Catholic University of Paraná (PUCPR), Curitiba, Paraná, Brazil
gabriela.unruh@pucpr.br

Abstract. There are many human-centered design methods with different purposes and applications, but there is a lack of clarity that makes most people apply only traditional methods. The objective of this article is to map existing methods, identifying the appropriate moments and purposes for each application, in order to assist in choosing the appropriate technique. The survey of the methods was carried out through bibliographical research and the proposal is a table of methods categorized by tipe, with possible results indication, authors and product development process phase application.

Keywords: Human-Centered Design · User-Centered Design · Product Development · Methods

1 Introduction

Human-centered Product Development disciplines such as Ergonomics, Human Factors, User-Centered Design (UCD), Human-Centered Design (HCD), Usability and User Experience, have many methods and techniques to achieve product suitability to the human needs, preferences, safety, health and comfort. Many methods come from the discipline itself, but some from other fields, such as from psychology, anthropology and sociology, focused on research, analysis and evaluation. Therefore there are several references and authors that present these methods, in addition to international standards. Even so, in many of these references it is not clear how to decide which method is best to apply. Macedo [1] identified in his research that there is a lack of systematization of these methods, and this information was confirmed in a bibliographic review.

2 Methodology

In order to better understand this scenario, to identify existing methods and techniques, and to assist in the process of identifying them and choosing for application in product development, a bibliographical research was carried out.

Unruh [2] carried out a systematic bibliographic review in the capes journals portal (Brazilian portal with access to 532 databases), searching for articles with keywords related to the respective areas of product development and to the areas with a focus on

T. Ahram et al. (Eds.): IHSED 2019, AISC 1026, pp. 435–441, 2020.
https://doi.org/10.1007/978-3-030-27928-8_67

human being, seeking first articles on human-based product development processes, as well as existing related methods. It has identified a number of articles that present UCD methods and/or techniques, some of which are indications of when to apply, but most present only a limited number of different methods among them, and little detail that helps in the decision making of when to apply throughout a process.

Therefore, a complementary search in references of Unruh's research [2] and a specific search of the terms of UCD with the terms "methods" and "techniques" in the journals portal capes, being found only articles that speak on or carry out case studies of one or a few specific methods or techniques, or articles that indicate methods and tools by criteria that help decide which one to use. Then a few classic references were added and some papers were selected from the systematic review, taking into account the references that brought larger quantities of methods and greater detail of them, in application and/or indication of when to apply them. 9 references were selected, and its contributions are presented next.

3 HCD Methods

From the research described in item one, relevant bibliographical references and websites were identified. The websites identified have several well-detailed methods with information on how to apply them, most of them focused on Usability or UX area, and some also have a filter system that helps to indicate the most appropriate methods for specific situations.

Among the websites that present methods and their descriptions, the following were identified: (1) usability.gov, which presents only 19 methods, but very well detailed about its application and benefits; (2) design kit from Ideo (private organization), which presents methods for the entire product development process already stating the process step, however there is a mixture of methods, techniques and tips, because it gives the whole step by step project, which is very good, but leaves the application a bit confusing. A great contribution is that each of these items is presented in a very detailed way, including the execution stages. Among the methods available on this site, there are also 19 focused specifically on DCU.

Among these websites tree more were identified. The first is an automated tool still under construction, focused on software development: STRUM - Scheduling Tool for Recommendationing Usability Methods [3]. The second is the result of an academic research by Weevers [4], the UCDToolbox - ucdtoolbox.com, presenting 34 methods, which can be searched freely or through filters that help in selecting the most appropriate for each step and situation of a project. Besides the selection is excellent, all the methods have a detailed description, facilitating its understanding and application. The third was also developed by researchers in the area [5, 6] and is the "Usability Planner" - usabilityplanner.org, it is a platform for indicating methods for each stage of the project, so the site starts with a survey and then presents the list of recommended methods with a brief description and various auxiliary filters to reduce the number of indications. Of all the references found, this was the most complete and detailed, with 66 methods, it has a filter system that automatically indicates the ideal method for each project situation. In addition, there is an article with a very similar proposal to the previous ones, but as a quantitative method for UX techniques selection [7], that help in the decision process of which technique to use according to some parameters, even indicated in the ISO/TR 16892 [8], such as cost, time, objectives, type of product, user involvement, number of participants, among others.

Although the second and third websites cited above are ideal for applying HCD methods to product development processes, both are focus on digital products, and present a limited number of existed methods. So the fundamental basis for this analysis of methods categorization was made through bibliographic references. From the process described in the previous item, nine main authors were identified, who present a relevant quantity and quality of methods. The selected authors include classics as Nielsen [9], Jordan [10] and Hom [11] and others a little more recent [12–17] among which some provide the results of their survey of techniques on websites. These references were analyzed, and relevant data to clarify the understanding and application of methods were synthesized in Table 1, where the authors are identified by numbers:

1 - Nielsen [9];
2 - Jordan [10];
3 - Hom [11];
4 - Maguire [12];
5 - Bevan [13];
6 - Heinilä et al. [14];
7 - Cybis et al. [15];
8 - Bevan [16];
9 - Roto et al. [17].

The analysis also aimed to identify methods and techniques which could be applied to physical products, digital products and systems and/or services, and 103 methods were selected. As the objective is to generate clarity and help in the decision process of applying the methods, they were all listed, analyzed and classified in the following aspects: category of method and results obtained in the method, authors and stage of a product development process.

Analyzing the authors indications and the methods themselves, 10 categories were defined: (1) Inspection; (2) Diary; (3) Questionnaire; (4) Interview; (5) Observation; (6) Testing; (7) Synthesis; (8) Ideas Generation; (9) Others; (10) Prototyping.

Then the details of each method were analyzed to identify the results from its application, listed in the second column of the tables. Soon after in the tables the names of each method are listed and to the right the numbers of the respective authors that present them. Finally, the last five columns indicate the moment of the PDP (Product Development Process) in which each method can be applied, with each letter or number representing a step, as follows: (1) the initial stage of the project, where problems and opportunities that lead to project strategies and definitions are researched and identified; (2) product strategy formulation; (3) product development, where the product is developed and evaluated of the product during the PDP, usually after each stage of development (concept, initial design and detailed design); and (4) product in use (after market launch and sales).

Besides this analysis and classification that already helps in directing methods application, as verified in the references and in ISO/TR 16892 [8], there are some other variables that can help in the refinement of this targeting, they are: cost, objectives, type of product, time, user involvement, number of participants, among others.

Table 1. HCD methods.

Method	Author									PDP phase			
	1	2	3	4	5	6	7	8	9	1	2	3	4
1) Inspection (possible issues and opportunities)													
Heuristics/Expert evaluation/Properties checklist/Ergonomics inspection/Standards	X	X	X	X	X	X	X	X		X		X	
Diagnostic evaluat./Error prevention insp./Risk analysis					X		X	X		X		X	
Subjective/Analytical/UX eval.		X			X		X			X		X	
Functionality		X	X	X				X		X		X	
Hedonic utility scale (HED/UT)									X	X		X	
Perspective based inspection									X	X		X	
Task analysis/Functions mapping		X		X	X	X	X	X		X		X	
Cognitive walkthrough		X	X			X	X		X	X		X	
Macroergonomics structure analysis (MAS)/ Design (MEAD)					X					X		X	
Physical ergonomics								X		X		X	
Immerssion									X	X		X	
Valuation methods/Cost-benefit		X		X						X		X	
Requirements comparison								X		X		X	
Similar or competitors analysis	X			X	X		X	X		X		X	
2) Diary (situations, contexts, scenarios, behaviors, understandings)													
Structured									X	X		X	X
Self report										X		X	X
Private camera conversation		X							X	X		X	X
Audio narrative										X		X	X
Day reconstruction met. (DRM)									X	X		X	X
Context awareness									X	X		X	X
Affective diary									X	X		X	X
Incidents diary		X	X	X	X					X		X	X
3) Questionnaire (opinions, experiences, expectations and characteristics)													
Profile	X	X	X	X	X		X			X			X
Semantic differential										X		X	X
Product semantic analysis (PSA)									X	X		X	X
AttrakDiff									X	X		X	X
Intrinsic motivation inventory (IMI)									X	X		X	X
Experiment sampling met. (ESM)									X	X		X	X
Appreciation Question. of Geneva									X	X		X	X
QSA GQM/SUMI									X	X		X	X
Diferential of emotions (DES)									X	X		X	X
Perceived comfort evaluation									X	X		X	X
Cognitive workload/Mental effort				X				X	X	X		X	X
Completing the sentence									X	X		X	X
Aesthetic scale									X	X		X	X

(*continued*)

Table 1. (*continued*)

Method	Author									PDP phase			
	1	2	3	4	5	6	7	8	9	1	2	3	4
Affective grid										X		X	X
Product personality attribution									X	X		X	X
2DES									X	X		X	X
Emotion wheels of Geneva									X	X		X	X
Satisfaction	X	X	X	X	X		X			X		X	X
Emotions checklists/cards/Emocards/Emofaces/Emoscope									X	X		X	X
SE Expressing experiences/emotions									X	X		X	X
4) Interview (opinions, experiences, expectations and characteristics)													
Traditional	X	X	X	X	X	X	X		X	X		X	X
Contextual/Experience investigation/Exploratory test			X	X	X	X		X	X	X		X	X
Focus group	X	X	X	X	X	X	X	X		X		X	X
Post-experience/User feedback	X							X		X		X	X
UX Scale									X	X		X	X
This or that									X	X		X	X
Antecipated experience ev. (AXE)									X	X		X	X
Pluralistic walkthrough									X	X		X	
I.D.									X	X		X	X
5) Observation (situations, contexts, scenarios, behaviors, understandings)													
Controlled/Assisted assessment				X						X		X	
Field/Observation/Etnography	X	X	X	X	X	X	X	X	X	X		X	X
Living lab									X	X		X	X
6) Testing (situations, contexts, scenarios, behaviors, understandings)													
Traditional/Laboratory/Control.	X	X	X	X		X	X		X	X		X	
Co-discovery		X	X						X	X		X	
Field	X	X	X	X		X	X			X		X	X
Experience clip									X	X		X	X
7) Synthesis (research data synthesis)													
Storyboarding/Graphic narrative				X	X	X	X	X		X		X	
Mental mapping/Repertory grid									X	X		X	
Use contexts										X		X	
Use scenarios				X	X	X	X	X		X		X	
User profiles/Personas				X			X	X		X		X	
Stakeholders identification or consulting				X				X		X		X	
User journey										X		X	
Guidelines/Design/Style patterns				X	X	X		X		X		X	
8) Ideas generation (ideas)													
Braisntorming				X								X	
Card sorting			X	X	X	X	X					X	
Future workshop						X		X				X	

(*continued*)

Table 1. (*continued*)

Method	Author									PDP phase			
	1	2	3	4	5	6	7	8	9	1	2	3	4
Parallel design	X			X	X			X				X	
Affinity diagram				X	X		X		X			X	
RGT									X			X	
Participative workshops/evaluat.	X	X		X				X				X	
9) Others (experiences, reactions and emotions)													
Thinking aloud protocols										X		X	X
Performance metrics or model					X			X		X		X	X
Eyetracking									X	X		X	X
Physiological excitation via electrodermal activity									X	X		X	X
PAD									X	X		X	X
EMO2									X	X		X	X
Psychophysiological measurement									X	X		X	X
Emotions sample device (ESD)									X	X		X	X
Valencia method									X	X		X	X
Positive and negative range of affection (PANAS)									X	X		X	X
IScale/UX curve									X	X		X	X
Pregnancy scale									X	X		X	X
SAM									X	X		X	X
PrEMO									X	X		X	X
Sensual evaluation tool (of shape)									X	X		X	X
10) Prototyping/Simulation	X	X	X	X	X	X	X	X				X	

The cost is difficult to analyze in the literature and the types of product are physical, digital and systems, according to the selection made. So variables that are possible to analyze in the references are presented in Table 2.

Table 2. Resources for types of methods.

Type of method	Resources			
	Participants	Evaluators	Time	Environment
Inspection	0	2+	1–10 days	Office
Diary	5–10	2–3	3–24 weeks	Field
Questionnaire	50+	1–3	1–4 weeks	Office
Interview	10+	1–3	1–12 weeks	Field/Laboratory/Meeting room
Observation	5–10	2–3	1–24 weeks	Field/Laboratory
Testing	5–10	2–3	1–12 weeks	Field/Laboratory
Synthesis	0	1+	1–12 weeks	Office
Ideas Generation	0+	2+	1–24 weeks	Office

References

1. Macedo, V.D.: Métodos de avaliação da Experiência do Usuário (UX) com eletrodomésticos: um estudo exploratório. Dissertation, Federal University of Parná, Brazil (2014)
2. Unruh, G.U., Canciglieri Junior, O.: Human and user-centered design product development: a literature review and reflections. In: Transdisciplinary Engineering Methods for Social Innovation of Industry 4.0. IOS Press, Amsterdam (2018)
3. Cayola, L., Macías, J.A.: Systematic guidance on usability methods in user-centered software development. Inf. Softw. Technol. **97**, 163–175 (2018)
4. Weevers, T.: UCD Toolbox (2012). http://ucdtoolbox.com/browse-methods/
5. Ferre, X., Bevan, N., Escobar, T.A.: UCD method selection with usability planner. In: NordiCHI, Reykjavik, Iceland (2010)
6. Ferre, X., Bevan, N.: Usability planner: development of a tool to support the process of selecting usability methods. In: Proceedings of Interact (2011)
7. Melo, P., Jorge, L.: Quantitative support for UX methods identification: how can multiple criteria decision making help? Univ. Access Inf. Soc. **14**(2), 215–229 (2015)
8. Ergonomics of Human-System Interaction – Usability methods supporting human-centered design, ISO/TR 16892:2002 (2002)
9. Nielsen, J.: Usability Engineering. Morgan Kaufmann, San Diego (1993)
10. Jordan, P.W.: An Introduction to Usability. Taylor & Francis, London (1998)
11. Hom, J.: The Usability Methods Toolbox Handbook (1998). http://jthom.best.vwh.net/usability/usable.htm
12. Maguire, M.: Methods to support human-centred design. Int. J. Hum Comput Stud. **55**, 587–634 (2001)
13. Bevan, N.: UsabilityNet methods for user-centered design. Hum. Comput. Interact. Theor. Pract. **1**(1), 434–438 (2003)
14. Heinilä, J., Strömberg, H., Leikas, J., Ikonen, V., Iivari, N., Jokela, T., Aikio, K.P., Jounila, I., Hoonhout, J., Leurs, N.: User centered design: guidelines for methods and tools. The Nomadic Media consortium (2005)
15. Cybis, W., Betiol, A.H., Faust, R.: Ergonomia e Usabilidade: conhecimentos, métodos e aplicações. Novatec Editora, São Paulo (2007)
16. Bevan, N.: Criteria for selecting methods in user-centered design. I-USED (2009)
17. Roto, V., Lee, M., Pihakala, K., Castro, B., Vermeeren, A., Law, E., Väänänen-vainio-mattila, K., Hoonhout, J., Obrist, M.: All About UX (2012). http://www.allaboutux.org/

Systems Design and Human Diversity

Artificial Intelligence and Blockchain Technology Adaptation for Human Resources Democratic Ergonomization on Team Management

Evangelos Markopoulos[1(✉)], Ines Selma Kirane[1], Dea Balaj[1], and Hannu Vanharanta[2,3]

[1] HULT International Business School, Hult House East, 35 Commercial Rd, London E1 1LD, UK
evangelos.markopoulos@faculty.hult.edu, kinesselma01@gmail.com, db.deabalaj@gmail.com

[2] School of Technology and Innovations, University of Vaasa, Wolffintie 34, 65200 Vaasa, Finland
hannu@vanharanta.fi

[3] Faculty of Engineering Management, Poznan University of Technology, Poznan, Poland

Abstract. The integration of advanced technology in management and decision making is a continuous effort for organizational optimization. This paper focuses on developing a teaming requirements elicitation process using AI and Blockchain technologies under the democratic knowledge management perspective. AI will be primarily addressed via Machine Learning, Deep Learning, Expert Systems and Fuzzy Logic, but also through the integration of cognitive sciences that can contribute effectively on the selection of team members in team building. Furthermore, Blockchain maximizes the company's potential through personnel growth and diversity by supplementing AI on understanding deeper the capacity and competence teams have on responding to changes through continuous learning from each experience. This integration of AI and Blockchain in corporate teaming will be driven through the Company Democracy Model as the base framework for co-evolutionary processes development allowing an essential mutation from autocratic to liberal democratic leadership, and from skill-based to knowledge-based human capitalization.

Keywords: Team · Teaming · Artificial Intelligence · Blockchain technology · Company Democracy · Intellectual capital · Leadership · Management

1 Introduction

Advancements in Artificial Intelligence (AI) impact the way organizations utilize exponential data creation and data flow growth. This affects the management processes applied for the increase of operations optimization, performance and efficiency. In an ever-blending globalized world, companies' diversity, in both competencies and background, can be seen as a synonym of companies' potential. Consequently, reaching companies' optimum development can be best achieved through specific, measurable,

T. Ahram et al. (Eds.): IHSED 2019, AISC 1026, pp. 445–455, 2020.
https://doi.org/10.1007/978-3-030-27928-8_68

attainable, realistic and timely (SMART) implementation and adaptation of AI to understand opportunities, projects and initiatives. In the same sense, the use of AI can be extended on perceiving the capacity, potentiality, skills, ability, capability, competence, and maturity of the organizations' human resources to respond immediately and successfully to the opportunities given. Effective responses demonstrate the readiness and the optimization of the corporate human intellectual capital via effective teaming and team management per case and instance.

The dilemma between technology driven management and human driven managements does not really exist, and actually never existed. The evolution of the technology offers significant advantages but the human judgment and interference shall not be neglected either. It is a co-evolutionary relationship that human decision becomes better with the use of technology, and technology can deliver better results when properly used by humans. The co-existence of artificial and emotional intelligence with any form of automated or human analytics and judgment brings more precise results and decisions. The enhanced democratic teaming model with artificial intelligence and blockchain technology is a natural evolution of the X-management type of teaming process based on the employee's qualifications, to the Y-management type of teaming based on the project requirements regardless the employee's qualifications, seniority or rank.

Therefore, to preserve, support and safeguard the democratization element and Y-thinking approach on teaming and team management it is important to maximize its success, reduce its risk and eliminate its failure. The integration of AI and Blockchain in the democratic teaming model aims not only to optimize team efficiency but more than that to protect the democratic thinking in organizations and democracy itself.

2 The Comeback of AI in the Global Industry

By the beginning of 1950s, Alan Turing set on his book, entitled Computing Machinery and Intelligence, the question on whether or not machines can think, and if computers can be personified so realistically, as to interact intelligently as a human [1]. His interrogations were the premises of a long path of research and innovations in the field of technologies, rousing engineers around the world on forging intelligent machines and programs. The inauguration of the 'Machine Learning' era began by the 1980s, when Kunihiko Fukushima schemes the neo-cognitive machineries under the concept of "learning without a teacher", by presuming the capacity of machines to set stimulus of learning patterns to achieve conventional cognition [2]. Throughout this era, AI infiltrated numerous industries, such as military and academia. The win of Deep Blue, the chess-playing computer developed by IBM, against the world champion Garry Kasparov in 1996, brought constant debate on the capacity of machines to overthrown human cognitive abilities. With the enhancement of technologies of the 21st century, Deep Learning as an amplification of the human neural brain work, further inspired deep networking of programs, such as Siri and autonomous vehicles (shown in Fig. 1).

Owing to this complex history of AI's development, various misconceptions on the topic emerged, especially on the concept of Machine Learning and Business

Intelligence. Originally classified under the term "Machine Learning", the capacity of a systems to refine its analytical process as it received data, is nowadays identified as "Self-Programming".

Another misconceived notion is Business Intelligence, which is the process of translating historical data into "actionable intelligence" [3] that could help in predicting and strategizing future business operations.

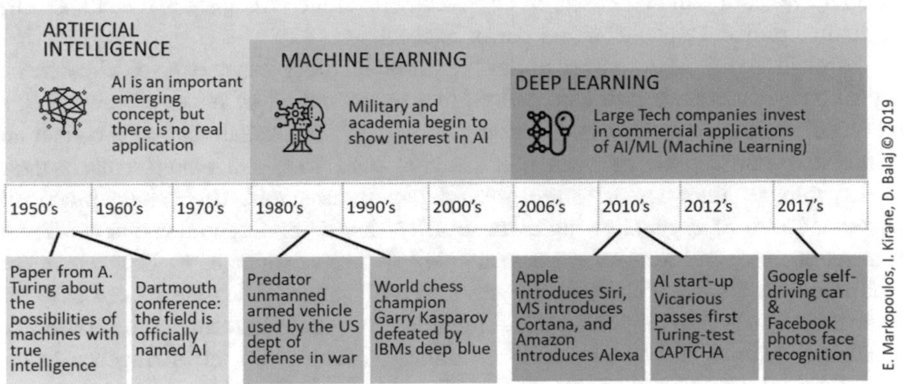

Fig. 1. Timeline of the Artificial Intelligence evolution eras

Those misconceptions have been facilitated by the dystopian popular culture, which merely portrayed Artificial intelligence as "evil" humanoid robots. Actually, emerging technologies in the business world is one of the most disruptive domains, due to the impact of AI in corporate management.

AI technologies are nowadays an essential part of modern management since the technology driven workplace management perspective (mid 1990s) is not only relying on Machine Learning, Deep Learning, Cognitive systems, Business Intelligence, but also on Expert Systems and Fuzzy Logic that impact tremendously managerial decision making and cooperate strategy formation. Fuzzy Logic Systems in particular, through their usage of semantic methods on knowledge processing, control effectively automatic response systems from the household appliances up to advanced robotics [4].

3 Blockchain Technology and AI in a Business Context

In 2018, Forbes Global 2000 has emphatically stated that at least 50 of the greatest public companies, such as ICBC, JP Morgan Chase & Co, Apple, and IBM, have explored and homogenized blockchain into their technologies, especially after being inspired by the Bitcoin movement [5]. Blockchain, being exploited after the establishment of cryptocurrencies, is defined as the system where data among computers is "linked in a peer-to-peer" network. Blockchain consists of data chain sets (blocks) that encompass multiple transactions, and thusly create "a complete ledger of the transaction history", which is further used for internal operations and analysis [6]. Due to a complex evolution and an intellectual effervescence around the subject, the blockchain technology has nowadays a myriad of applications on all kind of sectors. Governmental services, human contributions, contracts, and technology are the domains where corporations can find ideas in running their projects through SMART-strategies (Specific, Measurable, Attainable, Realistic, and Timely).

In this context, block chain can be combined with various emerging technologies in order to run human oriented tasks as well. Teaming for example, can use block chain for effective team building based on unbiased democratic principles that can maximize the team's project implementation performance.

Through Artificial Intelligence and its deep learning capacities, blockchain can improve its implementation, automation, and securitization of the data transacted, or analyzed. Further more with the 'Internet of Things', blockchain can also collect and secure data transactions between users, digital platforms, and amongst users themselves, through inter-corporate and intra-digital transactions. Blockchain practically acts as a Cloud Storage, allowing extra security, due to its decentralized network, low transaction cost, and unused space available [6]. The advantages of Blockchain (decentralized, verifiable, and durable), with the combination of flexible AI can empower companies to develop artificial general intelligence (AGI) platforms. The evolution of blockchain (shown in Fig. 2), and its combination with Artificial Intelligence creates new potential for advancements and applications in business management through smoother data management, document identification, and transaction verification among others.

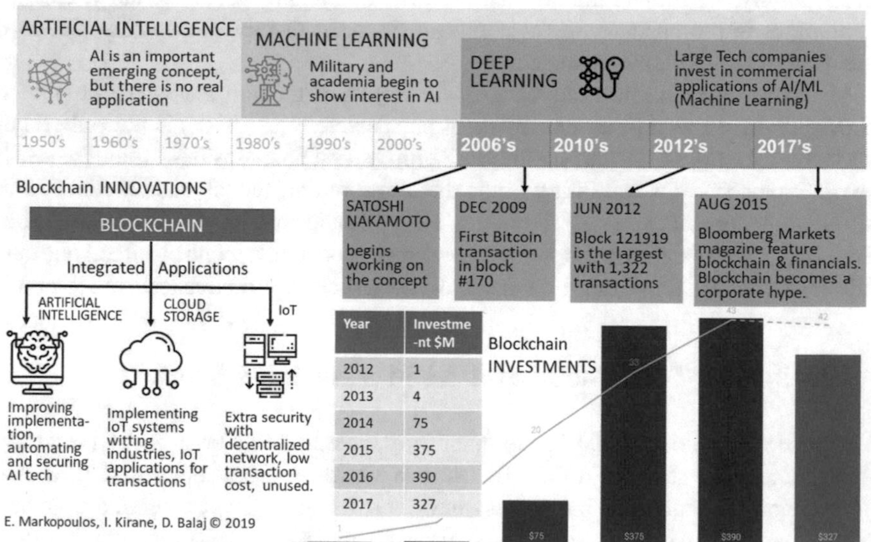

Fig. 2. Blockchain evolution timeline and integrated applications with AI.

4 The Challenge of Team Building and Team Management

A major challenge in the history of management has been the team building and the team management process. Teams are groups of individuals working together towards a common goal, but the degree of understanding the goal, the degree of commonality on the goal and the degree of expertise to provide valuable and meaningful contributions towards reaching the common goals is a barometer for project and individual success. Effective teaming is an essential democratic management and leadership practice, which helps managers shape the entity of their organization and generate valid knowledge among those who have it on behalf of those who need it. [7] Most well-established corporations are heavily relied on hierarchical structured teams, often set for a period of time. On the contrary startups, through their organic structure, rely more on their project as the framework space to compose their teams.

Thusly nowadays, teaming and team management has become binary, having the company structures either be formally driven or project driven. In either case understanding the project requirements is the initial step for team creation, regardless the size, the level of the company, or the teaming approach selected.

One way to address this challenge, is by using the Democratic Teaming Model, a project-driven evolutionary approach that allows managers to tailor agile team structures according to the project needs, for optimal performance and improved results.

The Democratic Teaming Model (DTM), is an alteration of the Company Democracy Model (CDM) [8, 9], which emphasizes on a democratic selection of the human resources to work effectively in a project and evolve though it in a co-evolutionalty, co-operative, and co-development way. The Company Democracy Model is based on the values and principles derive from the Delphic Maxims [10] and the ancient Hellenic knowledge to achieve a philosophy driven management and leadership [11]. Both CDM and DTM utilize organizational human intellectual capital through co-evolutionary knowledge-based democratic cultures [12]. Using ontologies and taxonomies for knowledge recording and analysis [13] the model continuously transforms organizational tacit knowledge into explicit [14].

Based on the CDM the Democratic teaming model maintains the six levels which in this case evolve the effectiveness of a team and its team members from the proper team selection at level 1, up to the maturity of the team to sustain the organization's competitive advantage globally at level 6 (Fig. 3).

The first level of the Democratic Teaming Model is the most significant one as the quality of the team selected to support a project is a catalyst in both the success of the project and the evolution of the team members in it. One of the innovative elements of the democratic team build is the role of the team builder and the way it operates on team formation and management (shown in Fig. 4). The teem builder is the controller of the team building process, an expert on understanding project needs and requirements but also one with deep knowledge on the organization's human resource's skills, abilities, weaknesses, limitations and drives.

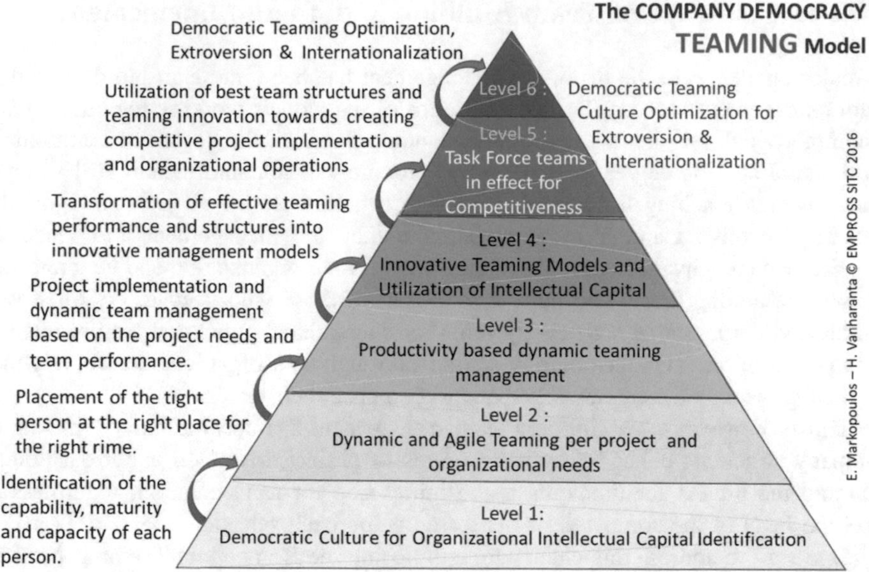

Fig. 3. The company democracy model for teaming.

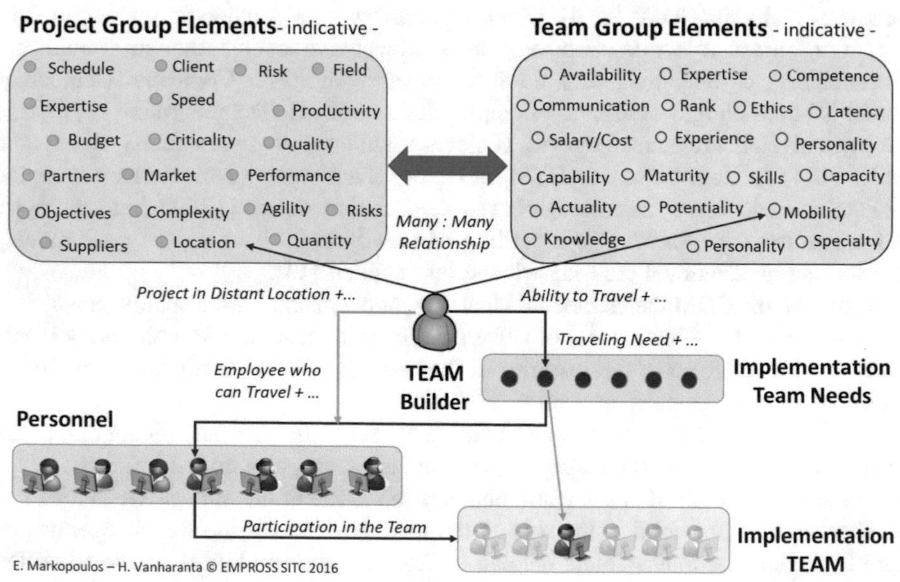

Fig. 4. The role of the team builder in the democratic project context.

The process begins after the team builder reads the project requirements and creates the project team positions to fulfil them. Each project need must be matched by at least one team member which derives from the total employees of the organization based on their skills and availability. The team builder takes into consideration not only the technical or formal qualifications of the available personnel to form a team for the specific project, but also creates various combinations for the best fit in order to reduce the size of the team without losing effectiveness and compliance to the project requirements.

The democratic element in this teaming model is based on the fact that the search for the most suitable team members is not restricted to the academic skills, seniority or popularity of an employee. The team builder uses the entire pool of employees to come up with the best matches unbiased from the profiles of each employee.

5 Approaching Teaming with AI and BC

The Y-theory management approach of the Democratic Teaming Model (DTM) transforms the teaming process through which companies perform more effective and innovative project implementation management. However, the model relies very much on the capability, maturity and capacity of the team builder as the sole decision maker in the team formation process. Lack of advanced processed information that could have been provided through the use of AI and blockchain restricts the teaming effectiveness on identifying the right person for the right place and for the right time.

This challenge can be addressed with the enhancement of the DTM and the integration of available advanced technologies to optimize teaming by reducing the associated risks on team member selection and project requirements understanding.

By integrating AI through expert systems, the organization can be receiving teaming recommendations deriving from the continuous monitoring of the data feed from the employee's activities, behavior, interests, experiences, etc. This can be supported by a blockchain system that secures the data feed and transactions to optimize its analytic output to the expert system. By providing a clear secured space for data storage outside the company's structure (ie. cloud), the artificial intelligence expert system can continuously and Socratically question its analysis and refine it as the project evolves within the implementation and management requirements (shown in Fig. 5).

In this case and in the enhanced version of the DTM, AI precedes the team formation and the team builder by performing the first matching process leading to several propositions of possible team compositions (team version propositions, v_1, v_2, … v_n etc.). Owing to its unique emotional intelligence capabilities, professional experience and expertise, the team builder takes control after this to select the final project team.

The use of AI expert systems is one of the AI dimensions that could optimize the Democratic Teeming Model effectiveness, but it is not the only one as other AI technologies such as machine learning, pattern recognition, case based reasoning and other, can form a wider AI infrastructure supported by Blockchain technology to act as a core decision making support mechanism for the team builder in the selection process.

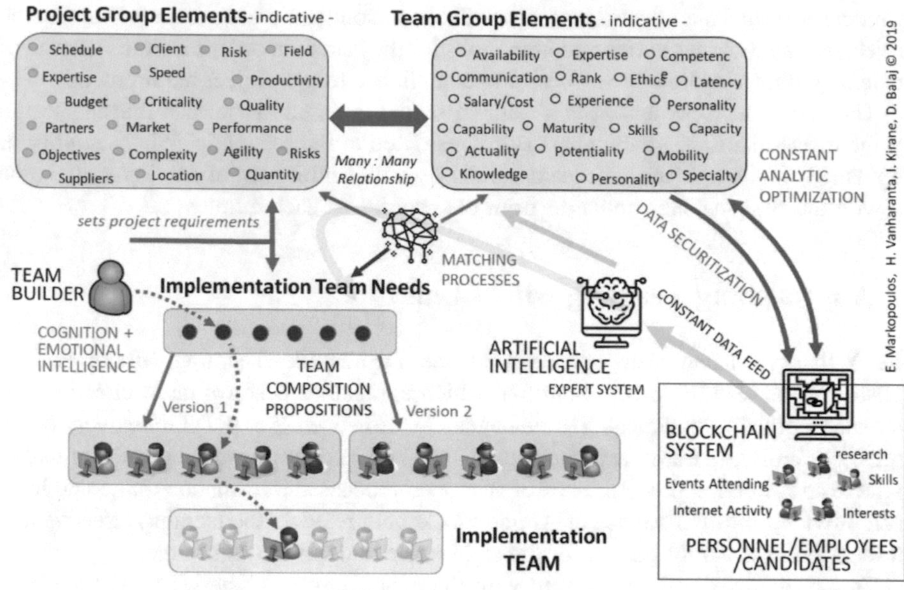

Fig. 5. AI & Blockchain in democratic project context

6 The Intelligent Dynamic Democratic Ergonomization Teaming Model

The AI-Blockchain integrated teaming process is a step toward optimal objective accomplishment and project performance. Indeed, managers should constantly question the adequacy of the team composed in terms of skills, knowledge, interpersonal relationships, time/space, organization, etc. The tech-based teaming ergonomization under the democratic philosophy is a Socratic strategy. The three team coordinators (team builder, artificial intelligence, and blockchain technology) are co-working on refining the team composition and management as the company's project changes. The company's project is symbolized by the pyramid and can be of various degree of complexity, from a new product launch in an existing line of business, to a new market penetration, and even the actualisation of a business itself. From Level 1 to Level 6, the team is initiating actions toward blue-ocean innovations, extroversion, and internationalization. This growth is relying on the collegial management of the coordinators (AI, BC, DTM) which are democratically adjusting the teams parameters on the evolving project needs and circumstances.

This triadic team building approach is characterized also by the degree of democracy, complexity and understanding the teaming process shall have. The degree of democracy is controlled by the team builder to make the final team composition providing the opportunity to all employees be part of a successful team. The degree of understanding is supported by AI Expert System which starts from a wide understanding of the project and ends up to the precise understanding. The same applies on the degree of criticality where in the early project levels are high risk, while at the lower levels the risk is eliminated due to the data gathered and analysis performed through the Blockchain technology.

In this relationship the lower levels of the pyramid indicate, the early stages of the project where democracy is used widely on team selection (many options to be considered) with a wide risk of uncertainty (unstable project requirements) from a wide criticality on properly understanding the project space (fuzzy project definition). As the levels evolve and the project becomes more complicated, the degree of risk, uncertainly and complexity are minimized due to the utilization of the knowledge gathered in the evolution of the project or from the company's projects databases, for a more precise project description and implementation management strategy. Figure 6 presents this tech-based Teaming Ergonomization under the Democratic philosophy.

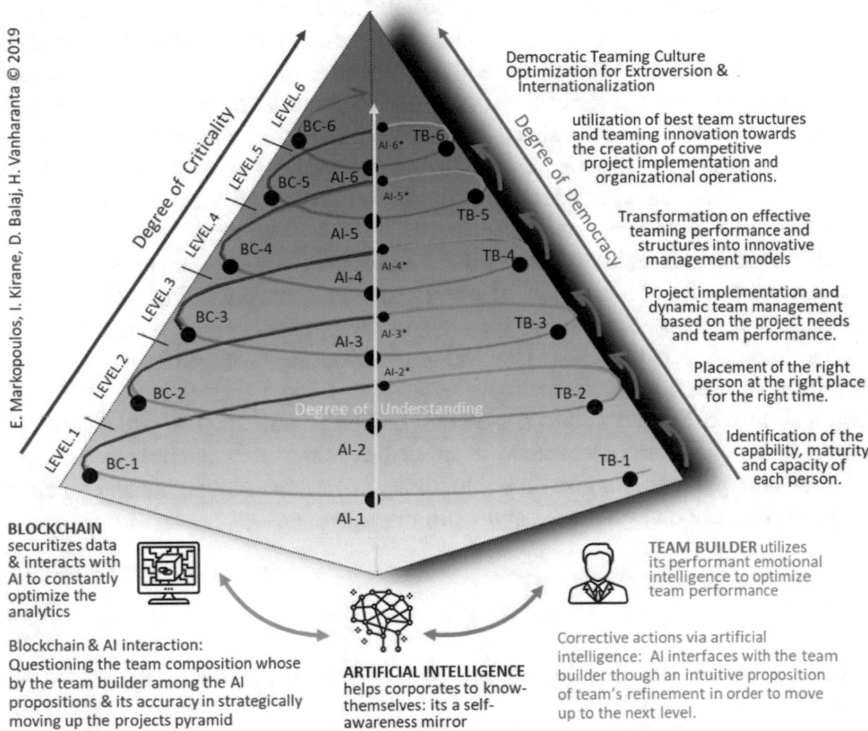

Fig. 6. Democratic co-evolutionary spiral process for intelligent team performance.

The proposed approach also resolves the DTM limitations on the selection of skills and competence according to standardized employee evaluations, based on experience, education, rank, availability, etc., that neglects personal characteristics and emotional intelligence values such as motivation, willingness, interest, loyalty, etc. Despite the proposed team selection by AI and the DTM' limitations, possible errors could happen when selecting team members with very similar skills and competences, or when the personnel does not perform as expected due to the uncertain level of the synthetic match by AI. Thusly, the DTM suggests changeable team structures and synthesis if teams indicate low performance.

In the process of evaluating the effectiveness of the team, it is crucial to have a break-off analysis on the value a team member, and especially an expert, can provide at a given time. This agile approach in team management composition can only be addressed via AI and blockchain that constantly monitors the performance of a team member on any instance, record deficiencies, and suggests corrective actions.

7 Conclusions

The proposed framework can be used to suggest alternative options to teaming challenges and to instruct and assists managers in justified decision making predictions on teaming with certainty or uncertainty factors based on a continuous data utilization from the personnel's past and current activities, behaviour and decisions. Furthermore, the combination of AI and Blockchain systems can contribute on creating, storing and transmitting such data to centralized knowledge bases that allows dynamic analytic optimization on the personnel's performance and effectiveness by composing the best possible teams that can satisfy long and short-term project teaming requirements.

This holistic approach, under the perspective of democratic management and leadership concepts, integrates AI to achieve organizational self-awareness with respect to metron-ariston for the creation of corporate ethos, by providing unbiased equal opportunities in companies' value creation to all employees and at all times. The proposed holistic approach on integrating democratic management and artificial intelligence can make significant improvements in critical sectors such as the pharmaceutical, medical, finance, and the defense/military industry. The paper introduced the integration of AI and Blockchain technologies within a democratic team requirement elicitation process, but also an evolutionary operations model that can potentially reshape the definitions of teaming and team management.

References

1. Harvard Business Review. https://hbr.org/2017/06/how-ai-is-streamlining-marketing-and-sales
2. Fukushima, K.: Neocognition: a self-organizing neural network model for a mechanism of pattern recognition unaffected by shift position. Biol. Cybern. **36**, 193–202 (1980)
3. CIO.com. https://www.cio.com/article/2439504/business-intelligence-definition-and-solutions.html
4. Lee, C.: Fuzzy logic in control systems: fuzzy logic controller, part II. IEEE Trans. Syst. Man Cybern. **20**(2), 419–435 (1990)
5. Forbes. https://www.forbes.com/sites/michaeldelcastillo/2018/07/03/big-blockchain-the-50-largest-public-companies-exploring-blockchain/
6. Nofer, M., Gomber, P., Hinz, O., Schiereck, D.: Bus. Inf. Syst. Eng. **59**(3), 183–187 (2017)

7. Markopoulos, E., Vanharanta, H.: Project teaming in a democratic company context. Theor. Issues Ergon. Sci. **19**(6), 673–691 (2018)
8. Markopoulos, E., Vanharanta, H.: Democratic culture paradigm for organizational management and leadership strategies - the Company Democracy Model. In: Proceedings of the 5th International Conference on Applied Human Factors and Ergonomics AHFE 2014 (2014)
9. Vanharanta, H., Markopoulos, E.: Creating a dynamic democratic company culture for leadership, innovation, and competitiveness. In: 3rd Hellenic-Russian Forum, 17 September 2013
10. Parke, H., Wormell, D.: The Delphic Oracle. Basil Blackwell **1**, 389 (1956)
11. Vanharanta, H., Markopoulos, E.: The applied philosophy concept for management and leadership objects through the Company Democracy Model. Theor. Issues Ergon. Sci. **20**(2), 178–195 (2019)
12. Kantola, J., Vanharanta, H., Karwowski, W.: The evolute system: a co-evolutionary human resource development methodology. In: Karwowski, W. (ed.) The International Encyclopedia of Ergonomics and Human Factors. CRC Press, Boca Raton (2006)
13. Paajanen, P., Piirto, A., Kantola, J., Vanharanta, H.: FOLIUM - ontology for organizational knowledge creation. In: 10th World Multi-Conference on Systemics, Cybernetics, and Informatics (2006)
14. Nonaka, I., Takeuchi, H.: The Knowledge-Creating Company: How Japanese Companies Create the Dynamics of Innovation. Oxford University Press, New York (1995)

Lean Application: The Design Process and Effectiveness

Tsung-Nan Wang(✉) and Yu-Hsiu Hung

Department of Industrial Design, National Cheng Kung University, No.1, University Road, East District, Tainan 70101, Taiwan
arne0619@hotmail.com

Abstract. Taiwan's trade mainly depended on exports, and the design of export products would affect the demand for clients to place orders. As the size of an enterprise and the number of the product categories it produced changed, the size of its design department was also affected. This study adopted the viewpoint of Toyota Production System (TPS) to re-explore the overall design process of the enterprise and plan a new design process. It aimed at proposing a framework with sound design efficiency. The results showed that after adopting TPS, the number of design cases undertaken by the design department increased, the time required to complete a case decreased, and the workload of each designer tended to be average. These results represented that the efficiency of the designers' execution of the design cases increased. Another finding of this study was that the working pressure of the designers reduced. In conclusion, TPS brought considerable benefits to the company.

Keywords: Toyota Productions System · Design process · Pull production

1 Introduction

In order to meet the needs of the market accurately, enterprises often introduce different management methods to change their internal organization. We can making some minor changes in the existing process, they can found the difference in efficiency before and after the introduction of new management methods. From the literature discussion and case study of enterprise management, it is known that if Toyota Productions System, pull production and the concept of Just in time are applied to improve the workflow of the Design Center, it may achieve the goal of producing high-efficiency design proposals.

The cooperative relationship between the units within the enterprise is moving towards the common goal. Related units can implement the tasks with clear goals just like the production lines, during which they can feel the dynamic sense of the product output. This study focuses on the management of design project undertaken by a Design Center in Thermaltake, a Taiwanese enterprise. The implementation methods and expectations are as follows: (1) Comparing the current situation in the process of undertaking design projects within enterprises with the difference after incorporating the Toyota Production System (TPS); (2) The target is to find out the key factors influencing the way in which each unit hands over its work and the process in which

T. Ahram et al. (Eds.): IHSED 2019, AISC 1026, pp. 456–461, 2020.
https://doi.org/10.1007/978-3-030-27928-8_69

the design unit undertakes the design plan; (3) The main purpose is to improve the designers' work efficiency, indirectly improve the design energy of designers, and bring more output value to enterprises.

2 Literature Review

The expert mentioned in his book Designing the Future: How Ford, Toyota, and other World-Class Organizations Use Lean Product Development to Drive Innovation and Transform Their Business that TPS was a system developed by TOYOTA, which introduced three concepts of real-time, automation and leveling into the production site, and subsequently developed various methods for gradual improvement of on-site production management and work and business processes with various departments to reduce waste of production system [1].

The scholar mentioned that the introduction of TPS should be based on 4P. With determined goals and consensus, TPS can be promoted with half the effort. The 4P proposed by TOYOTA are: (1) philosophy: focusing on long-term thinking; (2) process: focusing on process improvement and eliminating waste; (3) people/partner: cultivating employees and embracing leaders who can implement the philosophy of the company, maintaining mutual respect and teamwork; (4) perfection: building a learning organization with continuous improvement and a "genchi genbutsu" culture. In 4P, philosophy is the foundation of everything. Without a long-term philosophy, the company will not be able to implement the other three Ps (process, people/partners, philosophy), nor can it overcome the short-term benefits that must be sacrificed in the process of implementation and move towards a longer-term path to improvement [2].

Lean Production contains many ways to improve production processes. The authors of Lean Thinking, pointed out that enterprises that implement TPS thinking can meet customer demands with fewer resources (e.g. money, equipment, time and space). In addition, it also includes five basic principles: (1) Defining "value" for customers; (2) Draw and verify the Value Stream Mapping (VSM) of the product; (3) Strive for smooth value activities; (4) Adopt pull production system with production driven by market demand; (5) Pursue endless perfection [3].

From the above literature review, we know that lean production is a way for enterprises to continuously improve their production process. In order to successfully implement lean production management in enterprises, besides process improvement, it also requires the participation of all employees in the enterprise. Only when all the staff correct the old ideas and implement according to the lean production tool can the enterprise achieve maximum benefits with minimum cost in the production process and reach the state of absolute perfection.

3 Methods

This research takes Thermaltake as the research object. Thermaltake is a company specialized in selling computer peripheral products under three brands according to product categories: (1) Thermaltake, which mainly sells computer case, power supply, computer fan and so on. (2) Ttesports, which is a brand for selling e-sport products, including mouse, keyboards, headsets and e-sport chairs. (3) Luxa2, which mainly sells mobile phone peripheral products, such as portable power supply, headsets, Bluetooth stereo, etc. This study focuses on the approach of the Toyota Productions System and takes the Design Center of the case enterprise as the participant unit. Through the research, it proposes methods for improvement the efficiency of the Design Center in case undertaking upgrading design, and understands the difference between before and after improvement.

3.1 Status and Problem Description

3.1.1 Enterprise Workflow Status

At present, the work flow of an enterprise is: first, determine the task of product development after a meeting between CEO (Chief Executive Officer) and Sales department; then, the task is assigned to Product Manager, who shall issue a work order for the design project to the Design Center. When the Design Center receives a design project, it shall carry out a series of design and development processes, including R&D (Research and Development), MFR (manufacturer), MKT (market) etc. In this process, the Design Center, R&D department and MFR department will communicate with each other several times about the design in the development process and constantly revise their designs. The above workflow of the enterprise is drawn as shown in Fig. 1 below:

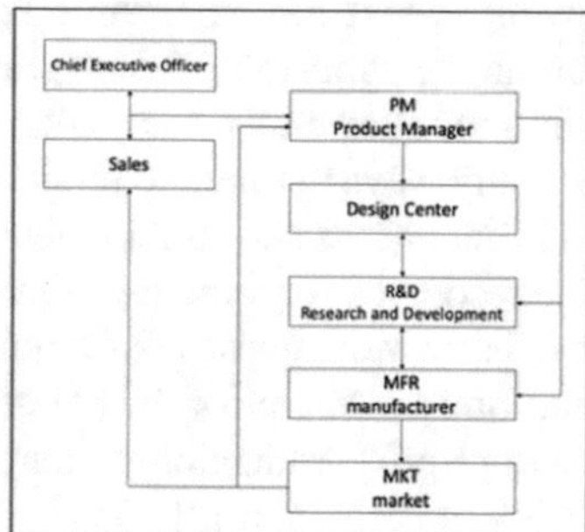

Fig. 1. Company workflow chart.

3.1.2 Enterprise Workflow Status of Design Center

At present, when the Product Manager gives the work order of the design project to the Design Center, the related personnel will cooperate directly according to the brand attribute or product category. The organization chart of the Design Center is drawn as shown in Fig. 2. However, the workflow diagram is shown in Fig. 3.

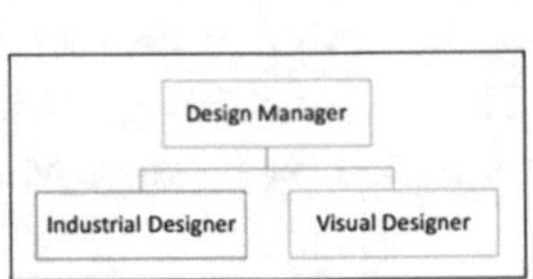

Fig. 2. Design Center member organization chart.

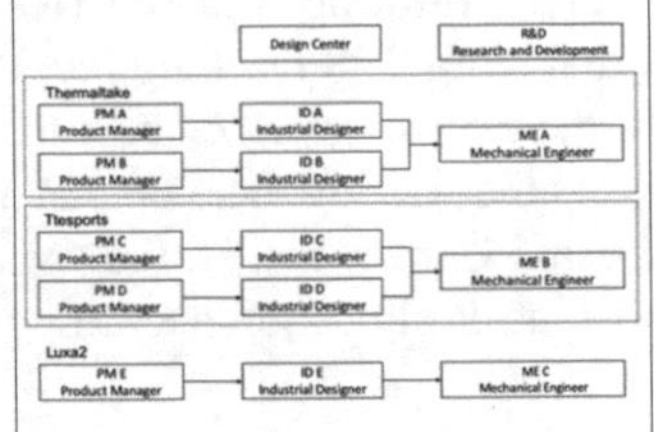

Fig. 3. The workflow status in Design Center.

This study records the work content and time required by Design Center for general types of projects in 2018. Table 1 is the work content and time required by Industrial Designers (ID) to undertake design development. As seen, the Industrial Designers are responsible for implementing the design project delivered by the Product Manager. In this work flow, the work assignment of Industrial Designers is uneven, which even cause the situation that when there is an urgent need for corresponding design plan in the Design Center, it will often be put on hold or delayed to submit the schedule, thereby affecting the work schedule of the next department and make the poor work inefficiency as a whole.

Table 1. The work content and time required by Industrial Designers to undertake design development.

Entry	Work projects	Operating hours/(day)
1	CMF (Color, Material& Finishing)	0.5
2	Product rendering	0.1
3	Product design	3
4	New product design	5
5	UI	1
6	UX	10
7	Photography	0.5

(Compiled by this study)

3.2 Integrate the Lean Production Principle

Based on the current workflow of the Design Center and the enterprise, integrate the principle of Lean Production to examine the way to adjust the workflow, amend the way in which personnel of Design Center acquire tasks, and carry out experiments.

3.2.1 Participants

The subjects of this study are the staff of the Design Center of Thermaltake. There are 8 members in this center, namely: 1 Design Manager, 5 Industrial Designers, 2 Visual Designers. The main participants in this study are the 5 IDs.

3.2.2 TPS Thinking into the Design Center

After introducing TPS thinking into the Design Center during its undertaking of design projects, there is not much change in its work flow. The main change is to make Design Manager as the window when the Product Manager gives the design project work order to the Design Center, that is, the "Manager" in Fig. 4. The Design Manager will make appropriate task allocation according to the work orders issued by Product Manager.

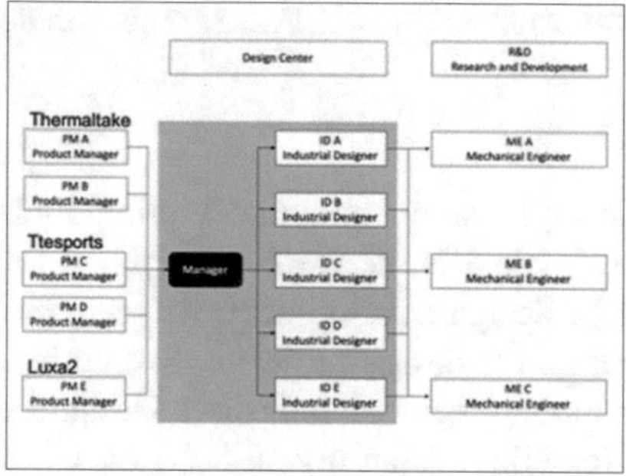

Fig. 4. The introduction of TPS in the workflow of the Design Center for undertaking of design projects chart.

4 Results

The impact of the work situation of the Design Center undertaking the design project on the overall work efficiency and that after introduction of TPS are described in detail as follows:

4.1 The Impact of the Workflow in Design Center for Undertaking of Design Projects

This study calculated the number of products developed by the Design Center in 2018. Based on the Table 2 data, the following factors are summarized:

1. Design requirements are delayed: Designers are busy with existing design work and have no time to handle other design projects delivered by the Product Manager that are more urgent.
2. The distribution of the workload is uneven: due to the inconsistency between the product development demand and quantity of each brand, the workload of designers is obviously different.

4.2 The Impact of the Introduction of TPS in the Workflow of the Design Center for Undertaking of Design Projects

From the Table 2 data, we can see that importing TPS into the workflow of the Design Center has really improved the work efficiency. It is found that:

1. The Product Manager delivers all the design requirements to the Design Manager, and the Design Manager uniformly assigns the work according to the working status of each ID designer in the Design Center.
2. In order to cope with the new management process and avoid unnecessary risks, the Design Manager may allocate the work according to the Industrial Designers' expertise in related products, so that the Industrial Designers can have opportunities to contact unfamiliar products and gain different design and development experience.

Table 2. The list of products developed in 2018.

2018						
Brand	New projects	Projects	Film maker	Total	Designer	Avg
Thermaltake	10	12	5	27	2	13.5
Ttesports	6	5	3	14	2	7
Luxa2	2	4	1	7	1	7
2019 January to June						
Thermaltake	6	5	2	13	5	5.6
Ttesports	4	4	3	11		
Luxa2	2	2	0	4		

(Compiled by this study)

5 Conclusions

The results of this study suggest that if an enterprise is willing to import TPS to its work flow, it should first implement in a small unit within the enterprise. In this way, employees can establish learning-based organization as the core concept for improvement. It is a better way to get employees familiar with the benefits of Toyota Productions System. Then it can continuously use PDCA (Plan Do Check Act) to proceed with the next processes. During the process, appropriate tools can be used to support continuous improvement and eliminate waste. If there is no improvement, it can find out the countermeasure to make correction. If the implementation is effective in the small unit, then it can be expanded to more organizations and finally form a brand new corporate culture.

References

1. James, M.M., Jeffrey, K.L.: Designing the Future: How Ford, Toyota, and Other World-Class Organizations Use Lean Product Development to Drive Innovation and Transform Their Business. McGraw-Hill Education, New York (2018)
2. Liker, J.K., Meier, D.: The Toyota Way Fieldbook. A Practical Guide for Implementing Toyota's 4Ps. McGraw-Hill, New York (2006)
3. Womack, J., Jones, D.: Lean thinking: banish waste and create wealth in your corporation. J. Oper. Res. Soc. **48**(11), 1148 (1997). https://doi.org/10.1057/palgrave.jors.2600967

The Anthropometry in Service of the School Furniture - Case Study Applied to the Portuguese Primary Schools

Maria Antónia Gonçalves(✉) and Marlene Brito

CIDEM - Research Center of Mechanical Engineering, School of Engineering of Porto (ISEP), Polytechnic of Porto, 4200-072 Porto, Portugal

Abstract. The focus of the current study was to set the benchmark for designing school furniture for Portuguese primary schools, taking into account the characterization and analysis of a previously selected sample of students. Considering the anthropometric data of the Portuguese children with ages between six and ten years old and its treatment, it was possible to set benchmarks for the design of school furniture for Portuguese primary schools, as well as to develop a methodological guide for adjusting the furniture according to the student's stature. The obtained results seem to be very relevant as they provide a scientific basis for the design and compatible with the anthropometric dimensions of the considered population of users.

Keywords: Anthropometry · School furniture · Ergonomic design

1 Introduction

Schools are subject to risk factors as much as any other workplace. But schools receive very young children who are unaware of the potential health and safety hazards that could result from the existing physical conditions in the classrooms, making them a particularly vulnerable group.

Some proposed activities in the classroom require a high concentration and the mechanisms for maintaining visual, auditory, motor and cognitive are constantly stimulated, which makes educational tasks complex [1]. These intellectual activities are held with students kept in the sitting position for extended periods of time [2–5]. It's at the age group of 6 to 10 years, which develops the habit of adopting incorrect postures, because at this point the children remain seated for long periods, without being able to move freely, being confined to inadequate furniture [6]. Considering that, the school furniture is a relevant element of the school organization, as it is a fundamental part of the school physical space.

A literature review on this subject seems to reveal that the mismatch between classroom furniture and student's body dimensions [4, 7–9] was caused, in the vast majority of cases, by the use of fixed-type furniture. This mismatch is strongly associated with back and neck pain reported by children during their school years [3, 10, 11], as well as to harmful effects at a cognitive level, such as hyperactivity, lack of

T. Ahram et al. (Eds.): IHSED 2019, AISC 1026, pp. 462–468, 2020.
https://doi.org/10.1007/978-3-030-27928-8_70

interest and consequent low learning performance [12, 13]. A systematic review of the literature, made by [14], shows that there is a clearly positive effect of school furniture dimensions on students' performance and physical responses.

In this study, we considered the adjustability of furniture based on anthropometric differences found both in several years of the first cycle and in the children of the same age who attend the same school year [15], thereby increasing comfort and decreasing the cases of complai8nts of body pain. This approach was used since it seemed unlikely that school furniture with fixed dimensions can be compatible with the vast majority of children [16], and the multiple adjustments seem to be the right solution [17].

This paper proposes a methodology for the design of school furniture for elementary schools, based in the relevant anthropometrics characteristics of the studied children, such as stature, popliteal height, buttock popliteal length, sitting elbow height, hip width and thigh thickness.

2 Methodology

An anthropometric study of the considered population was made, whose sample contemplated 432 children between 6 and 10 years (216 male, 216 female) of 9 schools, belonging to the first cycle of basic education, with ages between 6 and 10 years old (mean 8.5 ± 1.2 years). This sample represents a confidence level of 87,8%, with a maximum error of 5%.

For the sizing of school furniture, and taking into account other work done [11, 16, 18–20], the relevant anthropometric measures are listed in Table 1.

Table 1. Anthropometric measures and relationship with school furniture dimensions

Id	Anthropometric measure	Furniture dimensions
1	Popliteal Height (PH)	Seat Height (SH)
2	Hip Width (HW)	Seat Width (SW)
3	Buttock Popliteal Length (BPL)	Seat Depth (SD)
4	Sitting Elbow Height (SEH)	Desk Height (DH)
5	Thigh Thickness (TT)	Free space between the top of the thigh and the lower part of the desk (FSD)

The HW and TT represent relevant anthropometric measurements to space dimensions (min space or free space) and determine the min value for the SW and the free space between the seat and lower part of the desk, respectively. These are major limitations in one direction (one-way). Regardless of any other anthropometric measure, the SW and the free space between the top of the thigh and the lower part of the desk, should assume a min value corresponding to the value of a high percentile (e.g. 95th or 99th percentile).

For the remaining anthropometric measurements (PH and BPL), assumes a range, with a min and max value for having implications in posture. For this, it is necessary to consider the largest and the smallest children (limitation in both directions, two-way), depending on the size of the segments of the users' body. So, these are the dimensions which we pretend to be related.

The compatibility of school furniture, using anthropometry applied and according to ergonomic principles, the minimum (min) and maximum (max) limits between each dimension of the furniture is considered appropriate (two-way equations) or, for situations in which only required a min or max value (one-way equations). For this, were used the equations proposed by previous research (Table 2).

Table 2. School furniture and anthropometric measurements match equations

Id	Anthropometric measure	Furniture dimensions	Match equations		Authors
1	PH	SH	$88\%\,PH \leq SH \leq 95\%\,PH.$	(1)	[11]
2	HW	SW	$HW < SW.$	(2)	[21–23]
3	BPL	SD	$80\%\,BPL \leq SD \leq 95\%\,BPL.$	(3)	[11]
4	SEH	DH	$SEH \leq DH \leq SEH + 5.$	(4)	[20, 23, 24]
5	TT	FSD	$TT + 2 \leq FSD.$	(5)	[11]

For [25], some body segments can be expressed as a ratio of the stature (S). Following this, the study of the correlation between S and the relevant anthropometric measurements was carried out. The relationship between S and most anthropometric measurements revealed a statistically significant correlation coefficient ($p < 0.01$), thus being the main predictive factor for other anthropometric measurements (Table 3).

Table 3. Pearson Correlation for anthropometric measurements studied

Stature (S)	PH	BPL	SEH
Stature (S)	0,897**	0,840**	0,461**
	PH	0,724**	0,306**
		BPL	0,296**

**Significant correlation ($p < 0.01$)

Due to the high correlation observed between S and the other relevant anthropometric measures, a linear regression equation was established in order to predict the anthropometric measurements required from the S.

$$Y = a + b\,X + N\,(0,\ s). \tag{6}$$

Y = value of the anthropometric measure that is intended to be determined;
A = constant representing the intersection of the line with the vertical axis
b = constant representing the slope of the line
X = Independent variable (height)
N (0, s) = variable that includes all residual factors plus possible measurement errors

The result of applying this function to the sample data is translated by Eqs. 7 and 8.

$$\text{Popliteal height (PH)} = -5.89 + 0.302 * \text{Stature (S)}, \left(R^2 = 0.805\right). \quad (7)$$

$$\text{Buttock Popliteal Length (BPL)} = -2,919 + 0.313 * \text{Stature (S)}, \left(R^2 = 0.706\right). \quad (8)$$

The coefficient of determination, R^2, means that 80,5% and 70,6% of the PH and BPL, respectively, can be explained by S.

The S intervals were established after analysis of several combinations. In this analysis, it was sought that the amplitude of the PH values was lower than the height amplitude values. This resulted in five intervals for S (in cm): <120, [120–130[, [130–140[, [140–150[, ≥ 150. Smaller amplitude of the PH, makes more rigorous the dimensioning of the SH; On the other hand, the number of intervals found is reasonable to determine the adjustment levels, since many levels of adjustability can make the design of the seat impracticable and jeopardize the simplicity that is intended for the procedure to adopt for furniture adjust. When we analysed the S intervals, it was observed that the size of the sample in the upper extremity ($S \geq 150$ cm) has little expression (only 7 children, in a total of 432) and it was decided to encompass these children in the group of children with S greater or equal than 140 cm, resulting in the existence of only 4 levels.

3 Results and Discussion

To estimate the PH, according to S, Eq. 7 was applied. The estimated values were applied to the match criteria used by [11] (Eq. (1) and the resulting max and min for the SH were calculated. The values of the SH were established after testing the possible values, in the interval between the max and the min or, when there is only one of them, for values considered reasonable. The value with the highest match percentage was assumed as the sizing value of the SH. As result, with these levels of adjustability of the seat, it would have at least 59.6% of the children ergonomically accommodated, in a seat whose height is dimensioned according to the criteria defined by [11]. However, when using another criterion, that is, when analysing the percentage of children whose PH is lower than the SH (ergonomic criterion cited by a wide range of authors, such as [21, 26–28], the scenario is much more favourable, being the max value of mismatch 6.7%, registered in children of a lesser stature (<120 cm).

In relation to the SW, this must be able to accommodate individuals with the greatest HW and must be designed in line with the measure equivalent to the 95th percentile of this anthropometric measure [21–23]. The percentage of children accommodated comfortably is 94%. However, since the SW only represents incompatibility if don't accommodate children with wider hips, we can opt for a higher percentile value (99th percentile), increasing the number of children sitting comfortably (99,1%), and there is no need for adjustability.

The SD, has a mismatch with the BPL, according to the criteria of (mis)match used by [11], when this furniture measure is less than 80% or more than 95% of the considered anthropometric measure (Eq. 3). Thus, according to Eq. 8, we can reach the BPL measure through the S. To the obtained values, the criteria of (mis)match used by [11] were applied (Eq. 3) and the max and min SD was established for different S intervals. Following the same reasoning used in the sizing of the SH, the values of the seat depth were established after testing the possible values, in the interval between the max and the min or, when there is only one of them, for values considered reasonable. The value for which the highest match percentage was taken and was assumed to be the sizing value for the SD. The decisive dimension – stature – translates four levels for the SD. The dimensions provided for the four levels of SD, based in S, were tested according to the (mis)match criteria used by [11]. The results show that there is 8,9% max of mismatch, once more for the children of less than 120 cm stature. These results show that the dimensions established for the SD, based on distance BPL, have a large percentage of children properly accommodated. If we analyse the ergonomic criteria, used in several studies [21, 23, 29] that the SD should be lower than the BPL, the percentage of children who will be accommodated is greater (more than 98,1%).

The DH will always be in function of the SH, as this works as an integrated system. Thus, the DH is determined by the SH, plus the max TT, for each of the ranges of Ss. This value must be added by 2 cm (Eq. 5). It is also recommended that the DH (from the seat) is the 3 to 5 cm below the SEH. This last condition allows better support of the elbows, guaranteeing a more adequate trunk posture and a better efficiency of school tasks, such as reading, writing and drawing [23, 24]. As a result of several iterations with the values of the interval (between the min and max), were assumed values that presented greater compatibility (match), according to the criteria established in the Eq. 4. Thus, four levels of adjustability were established and determined the (mis)match to the values of the SHE. The match values were greater than 89% in the vast majority of children, whose S is between 120 and 140 cm. In children with a S equal to or greater than 140 cm, the compatibility decreases to 76%.

4 Conclusion

Stature proved to be a good predictor for the determination of other anthropometric measurements. Thus, using regression equations, with S as an independent variable, the values of the desired anthropometric dimensions (dependent variables) were determined. According to the principles of ergonomic suitability proposed by the literature, compatibility equations were used between the dimensions of the furniture and the anthropometric measurements, to reach dimensional values for tables and seats, for different levels of adjustability. When comparing the values obtained with the anthropometric data of the study sample, very interesting compatibility percentages are achieved.

Acknowledgments. We acknowledge the financial support of CIDEM, R&D unit funded by the FCT – Portuguese Foundation for the Development of Science and Technology, Ministry of Science, Technology and Higher Education, under the Project UID/EMS/0615/2016.

References

1. Marschall, M., Harrington, A.C., Steele, J.R.: Effect of work station design on sitting posture in young children. Ergonomics **38**(9), 1932–1940 (1995)
2. Geldhof, E., De Clercq, D., De Bourdeaudhuij, I., Cardon, G.: Classroom postures of 8–12 year old children. Ergonomics **50**(10), 1571–1581 (2007)
3. Murphy, S., Buckle, P., Stubbs, D.: A cross-sectional study of self-reported back and neck pain among English schoolchildren and associated physical and psychological risk fators. Appl. Ergon. **38**, 797–804 (2007)
4. Castellucci, H.I., Arezes, P.M., Viviani, C.A.: Mismatch between classroom furniture and anthropometric measures in Chilean schools. Appl. Ergon. **41**, 563–568 (2010)
5. Gonçalves, M.A., Arezes, P.M.: Postural assessment of school children: an input for the design of furniture. Work J. Prev. Assess. Rehabil. **41**(1), 876–880 (2012)
6. Viel, E., Esnault, M.: Lombalgias e Cervicalgias da posição sentada, 1st edn. Editora Manole, São Paulo (2000)
7. Castellucci, H.I., Arezes, P.M., Molenbroek, J.F.M.: Applying different equations to evaluate the level of mismatch between students and school furniture. Appl. Ergon. **45**(4), 1123–1132 (2014)
8. Dianat, I., Ali Karimib, M., Asl Hashemic, A., Bahrampour, S.: Classroom furniture and anthropometric measurements of iranian high school students: proposed dimensions based on anthropometric data. Appl. Ergon. **44**(1), 101–108 (2013)
9. Van Niekerk, S., Louw, Q.A., Grimmer-Somers, K., Harvey, J., Hendry, K.J.: The anthropometric match between high school learners of the Cape Metropole area, Western Cape, South Africa and their computer workstation at school. Appl. Ergon. **44**(3), 366–371 (2013)
10. Trevelyan, F., Legg, S.: The prevalence and characteristics of back pain among school children in New Zeeland. Ergonomics **53**(12), 1455–1460 (2010)
11. Parcells, C., Stommel, M., Hubbard, R.: Mismatch of classroom furniture and student body dimensions: empirical findings and health implications. J. Adolesc. Health **24**, 265–273 (1999)
12. Moro, A.R.P., Ávila, A.O., Nunes, F.P.: O design da carteira escolar e suas implicações na postura das crianças. Anais do VIII Congresso Brasileiro de Biomecânica, Sociedade Brasileira de Biomecânica: Florianópolis-SC, 125–130 (1999)
13. Mandal, A.C.: The prevention of back pain in school children. In: Lueder, R., Noro, K. (eds.) The Ergonomics of Seating, pp. 269–277. Taylor & Fancis, London (1994)
14. Castellucci, H.I., Arezes, P.M., Molenbroek, J.F.M., de Bruin, R., Viviani, C.: The influence of school furniture on students' performance and physical responses : results of a systematic review. Ergonomics (2016). https://doi.org/10.1080/00140139.2016.1170889
15. Gonçalves, M.A.: Análise das Condições Ergonómicas das Salas de Aula do Primeiro Ciclo do Ensino Básico PhD Thesis, Universidade do Minho, Guimarães (2012)
16. Panagiotopoulou, G., Christoulas, K., Papanckolaou, A., Mandroukas, K.: Classroom furniture dimensions and anthropometric measures in primary school. Appl. Ergon. **35**, 121–128 (2004)

17. Pereira, P.: Recomendações para o projecto de design de mobiliário escolar para o 1º ciclo do ensino básico em Portugal. Master Thesis, FMH, UTL, Lisboa (2010)
18. Garcia-Acosta, G., Lange-Morales, K.: Definition of sizes for the design of school furniture for Bogotá schools based on anthropometric criteria. Ergonomics **50**(10), 1626–1642 (2007)
19. Gouvali, M.K., Boudolos, K.: Match between school furniture dimensions and children's anthropometry. Appl. Ergon. **37**, 765–773 (2006)
20. Molenbroek, J.F.M., Kroon-Ramaekers, Y.M.T., Snijders, C.J.: Revision of the design of a standard for the dimensions of school furniture. Ergonomics **46**(7), 681–694 (2003)
21. Helander, M.: A Guide to the Ergonomics of Manufacturing. Taylor & Fancis, London (1997)
22. Mondelo, P., Gregori E., Barrau, P.: Ergonomía: Fundamentos, 3rd edn. Alfaomega Grupo Editor – UPC, México (2000)
23. Pheasant, S., Haslegrave, C.: Bodyspace: Anthropometry, Ergonomics and the Design of Work, 3rd edn. CRC Press (2006)
24. Poulakakis, G., Marmaras, N.: A model for the ergonomic design of office. In: Scott, P.A., Bridger, R.S., Charteris, J. (eds.) Proceedings of the Ergonomics Conference in Cape Town, Global Ergonomics. Elsevier Ltd., pp. 500–504 (1998)
25. Roebuck, J., Kroemer, K.H.E., Thomsonm, W.G.: Engineering Anthropometry Methods. Wiley, New York (1975)
26. Dul, J., Weerdmeester, B.: Ergonomics for Beginners – A Reference Guide. Taylor & Fancis, London (1998)
27. Molenbroek, J.F.M., Kroon-Ramaekers, Y.M.T.: Anthropometric design of a size system for school furniture, In: Robertson, S.A. (ed.) Proceedings of the Annual Conference of the Ergonomic Society: Contemporary Ergonomics, pp. 130–135. Taylor & Fancis, London (1996)
28. Oxford, H.W.: Anthropometric data for educational chairs. Ergonomics **12**(2), 140–161 (1969)
29. Milanese, S., Grimmer, K.: School furniture and the user population: an anthropometric perspective. Ergonomics **47**, 416–426 (2004)

Topological Properties of Inequality and Deprivation in an Educational System: Unveiling the Key-Drivers Through Complex Network Analysis

Harvey Sánchez-Restrepo(✉) and Jorge Louçã(✉)

Information Sciences, Technologies and Architecture Research Center (ISTA), ISCTE-IUL, 1649-026 Lisbon, Portugal
{Harvey_Restrepo, jorge.l}@iscte-iul.pt

Abstract. This research conceives an educational system as a complex network to incorporate a rich framework for analyzing topological and statistical properties of inequality and learning deprivation at different levels, as well as to simulate the structure, stability and fragility of the educational system. The model provides a natural way to represent educational phenomena, allowing to test public policies by computation before being implemented, bringing the opportunity of calibrating control parameters for assessing order parameters over time in multiple territorial scales. This approach provides a set of unique advantages over classical analysis tools because it allows the use of large-scale assessments and other evidences for combining the richness of qualitative analysis with quantitative inferences for measuring inequality gaps. An additional advantage, as shown in our results using real data from a Latin American country, is to provide a solution to concerns about the limitations of case studies or isolated statistical approaches.

Keywords: Complex network · Educational deprivation · Inequality · Large-scale assessments · Policy informatics

1 Introduction

Since the Sustainable Development Goals were launched [1], policymakers have been increasing the use of analytical and statistical models for improving their knowledge about complex dynamics of the educational systems, especially for those with high ethnic-linguistic diversity [2], facing the challenge of developing coherent multilevel theories for making decisions and designing public policies [3]. The wide number of nonlinear relationships exhibited by multiple agents in different hierarchical levels interacting in education, demands the development of new instrumental and thought tools for modelling the variety of fluxes of energy - financial, human, physical and social resources—as well as a better understanding and measurement of the parameters related with the emergence of self-organized patterns and groups in different scales [4].

Network theory has enormously developed in this decade and is a leading scientific field for descripting and analyzing complex social phenomena [5]. Educational systems

T. Ahram et al. (Eds.): IHSED 2019, AISC 1026, pp. 469–475, 2020.
https://doi.org/10.1007/978-3-030-27928-8_71

exhibit different properties in many scales coming from same dynamics, students stablish relationships with other students and teachers, which might provide some insights about social characteristics about preferences, choices and interest on learning [6]. Actually, many interventions show that interactions impact instructional quality and learning outcomes [7] and there are some evidences on how social network structure becomes an important intermediate variable in education and that cultural, social and economic variables are related with educational deprivation and learning outcomes [8].

2 Modelling Framework

In this research we incorporate a rich framework for analyzing topological and statistical properties of educational deprivation in a Latin American country, as well as its relationship with social determinants as Socioeconomic Status (SES), Rurality of area where the school is located (RA), Type of school (TS), and self-identify student's Ethnicity (ET), for unveiling the key factors driving inequality gaps in learning outcomes. The model is developed through a network with different levels for analyzing properties and nodes exhibiting centrality and non-equilibrium parameters than might help to better understand the structure of the system and phenomena behind them.

Data Sources. A multivariate dataset related with learning outcomes of every student who has completed the *k-12* education process, estimated by scoring based on a census-based large-scale assessment carried out in Ecuador for 39 219 students in 2017, through a standardized computer-based test with psychometric parameters estimated by Item Response Theory with 2P-Logistic model. Raw scores were re-scaled to a *Learning index* (LI), a monotonous transformation of ability's parameter θ^i, where higher levels of learning are more likely to have higher scores.

The model was developed in four phases, the first one was psychometrical analysis for estimating scores and identify deprived students (L_0)—those with a LI below of the cut point L_0—and the intensity of deprivation $\lambda(LI_i)$, given by the distance from LI_i to L_0. The next three phases are directly based on the level-network (LN). In this way, the 1-LN is for disaggregating L_0-group by SES, each student is represented by a node and edges are directed to one of the SES-decile nodes $\left\{\theta^i \rightarrow L_j^i \rightarrow \left(SES_k^i\right)\right\}\forall i$, weighted by λ. In 2-LN we extend 1-LN for including RA, TS and ET to analyze their effects through the sequence $\left\{\theta^i \rightarrow L_j^i \rightarrow \left(SES_k^i\right) \rightarrow \left(RA_k^i, TS_k^i, ET_k^i\right)\right\}\forall i$. The 3-LN amplifies and strengthens the network through more than one hundred educational and non-educational factors associated with learning achievements [9], trough the sequence $\left\{\theta^i \rightarrow L_j^i \rightarrow \left(SES_k^i\right) \rightarrow \left(RA_k^i, TS_k^i, ET_k^i\right) \rightarrow AF_{m,n}^i\right\}\forall i$, a multi-dimensional system exhibiting educational deprivation at different levels. Network analysis was carried out by Gephi 0.9.2 and statistical estimations with Orange 3.3.8.

3 Empirical Results

Estimates indicate that, from 39 219 students, 8 438 were deprived, an absolute prevalence of 21.5%. The L_0-group has a $LI = 6.32$ and intensity of deprivation $\lambda = 0.22$, i.e., in average, each L_0-student lacks 0.68 standard deviations (SD) to the minimum level of learning for not being deprived.

3.1 Socioeconomic Status and Learning Deprivation

In all cases, inequality means asymmetries, in conditions of total equity—where socioeconomic factors would not produce differences—we might expect equal distribution of L_0-edges over the network, but 1-LN specification detects SES effects in nodes grouping L_0-students by deciles, driving the system out of equilibrium with a negative correlation between LI and SES ($R = -0.58$ ($p < 0.001$)). A 2-LN model shown in Fig. 1 integrates the different self-identified ethnic groups and disaggregate them by Rural-Urban areas and Public-Private schools, to identify the magnitude with which the lower deciles dominate the interactions through the edges.

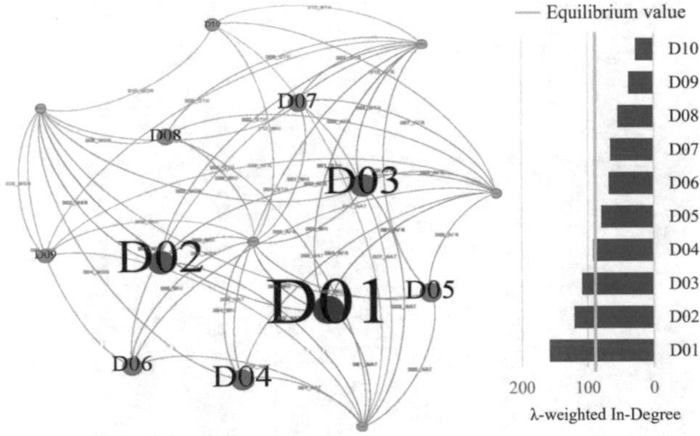

Fig. 1. Socioeconomic structure of learning deprivation (*left side*) and λ-weighted In-degree histogram (*right side*). Differential centrality of deciles shows a non-equilibrium L_0-system driven by ethnical groups dominated by poorest students.

As can be seen in Fig. 1, independently of RA, TS and ET, nodes corresponding to D01, D02 and D03—the poorest students—have stronger connections and dominate the network, being D01 the highest parameters in Prevalence rate ($DPR = 0.3860$), Hub ($H = 0.4887$), Weighted Degree ($WD = 158.9420$) and PageRank ($PR = 0.4289$). On the contrary, D10 is almost irrelevant for the network with parameters $DPR = 0.0800$, $H = 0.1133$, $WD = 27.222$ and $PR = 0.0289$.

3.2 Ethnicity and Type of School Financing

There is a negative correlation between SES and DPR ($R = -0.81$, $p < 0.001$) and, as can be seen in scatter-plot of Fig. 2, quintiles 1 and 2 have the highest lack of learning. The complementary Dendrogram was made by a hierarchical cluster analysis for $\left(SES_k^i, RA_k^i, TS_k^i, ET_k^i\right)$ and, in all cases, families are basically made of SES and TS where emerges an intra-class ranking ordered by ethnicity, meaning that closeness is a SES' function while distances are based on racial proximities, a structural inequality for the whole system. In this sense, scatter-plot shows that White-students (red circle) have the lowest deprivation rate among the poorest ($DPR = 0.402$), even lower than richest Afro-Ecuadorian students in blue circle ($DPR = 0.408$). For the whole network, Page Rank order of nodes and λ-intensity is based on students' ethnicity as follows: Afro-Ecuadorian (A), Montuvios (M), Indigenous (I), Other groups (O) and White (W). Furthermore, Private sector is dominated by richest students ($SES = 0.91$), while public schools serve to the poorest students ($SES = 0.33$) getting a $DPR = 0.257$, 2.12 times the rate of the private ones ($DPR = 0.121$), showing that SES is a key factor for educational deprivation due to the influence of cultural capital in learning outcomes.

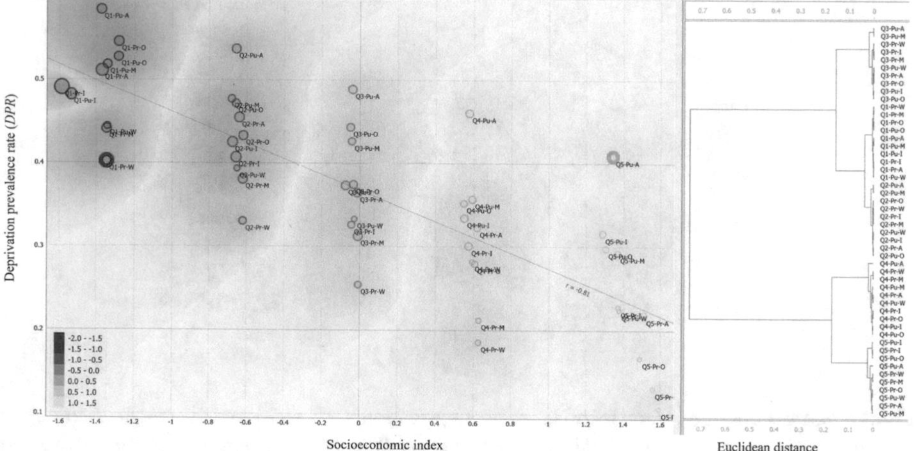

Fig. 2. Social determinants of learning deprivation. The scatter plot shows the strong negative correlation between SES and Deprivation prevalence rate (*solid line*). Highest deprivation rate in richest students (*blue circle*) is lower than the lowest rate of poorest students (*red circle*), pointing out a deep systemic racial discrimination.

3.3 Key-Factors for Public Policies

As 1-Ln and 2-LN are networks strongly connected, prevalence rate of L_0-students might be associated with Eigen-Centrality through measuring factor's influence for identifying how well connected a node is and how many links have its connections. In this way, building 3-NL for splitting L_0-students in communities becomes in a very valuable tool for developing group-oriented strategies to avoid implementing same actions for completely different populations and needs. One of the most useful strategies for improving learning and closing gaps is micro-planning, i.e., implementing different policies at local level, however, to select the most relevant needs is a great deal, mainly because they use to interact and 'ceteris paribus hypothesis' seems to be too naive; through network simulation, this can be solved in a very easy way varying *AF*-parameters for recognizing and ranking the most relevant nodes (factors) to be attended by policymakers.

Figure 3 shows the 3-LN after a Modularity process (with parameter 0.073 at resolution of 0.254) for splitting richest from poorest students to find key factors for educational deprivation in both groups. As shown, richest students' community (D10) has just a few factors (20) and most of them have very low connectivity. On the contrary, poorest students (D01) have a lot of different sources provoking deprivation (57), reflected in the Degree of authority of the D01 (0.98743) versus D10 (0.158058).

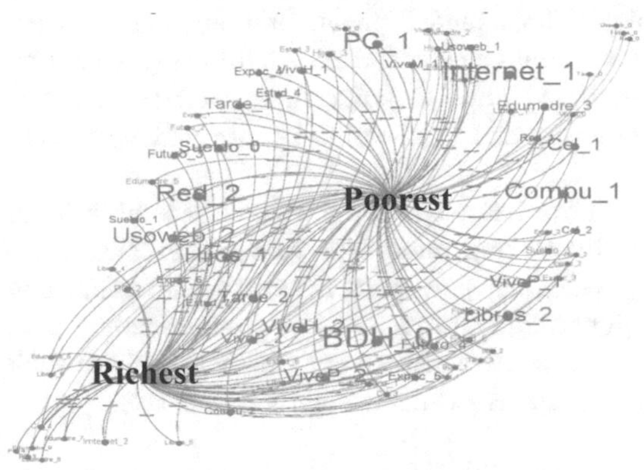

Fig. 3. Factor's network for richest and poorest students' communities. Centrality of nodes in each group points out the dissimilarities in factors provoking deprivation.

Thus, PageRank in 3-NL network helps to find and rank those factors dominating deprivation with the highest Degree of connectivity and centrality for each community ordering from the highest degree until the lowest for providing a very accurate and clear knowledge. In addition, it might be defined 'needs profiles' for focusing actions on those variables susceptible to be managed by policy, for example, a D01-profile might be: '*Members of a household who currently receives the Human Development Bond and needs to work for a wage. Their parents have a very low level of education, they do not have a desktop computer neither Internet connection, and also have no books or just a very few. In their school, teachers arrive late to class, are no committed with learning and have low expectations about student's future*'. As can be seen, this very detailed information is extremely helpful for developing based-evidence policies, to assign budget and have a successfully deployment.

4 Discussion

This approach provides a natural way to represent educational systems, bringing the opportunity of calibrating control-parameters for assessing order-parameters over time and in multiple territorial scales. Thus, the network analysis underpins current educational deprivation models [10] and provides a parametric way to estimate prevalence of inequality in learning outcomes for contexts with high levels of ethnic- diversity, a key aspect for understanding complexity in intercultural systems that also provides a solution to concerns about the limitations of case studies, the classical utility theory and isolated statistical analysis.

The findings offer evidence in the deep lack in equity that can help policymakers to identify those factors related with inequality in learning outcomes, as well as the magnitude of their relationship with deprivation. In this sense, running the modularity process for defining groups might help to identify those factors which are more relevant for one group and are not for others, avoiding statistical bias based on averages provoked by statistical multilevel modelling and bringing additional information about the order in which factors should be considered. Additionally, each LN can be stressed for calibrating boundaries and initial conditions, as well as to test policies at different levels, from-bottom-to-top and from-top-to-bottom and with real data, before being implemented.

These results confirm that network analysis is becoming fundamental for educational policy, specially linking microdata with other constructs and social macro-parameters. Finally, detecting how gaps in educational achievements are driven by students' context might highlight in a better way where policy can intervene properly, offering a series of unique advantages over classical analysis tools and allowing the intensive use of large-scale assessments and other datasets.

Acknowledgments. We would like to thank ISCTE's Programme in Complexity Sciences, the University of Lisbon and the Latin American Agency for Evaluation and Public Policy, for their support to the project as well as the authors of the articles mentioned in the references.

References

1. UN DESA. The Sustainable Development Goals Report 2018, UN, New York (2018)
2. Shee, N.K.: Karen Education Department's multilingual education for language maintenance, Kasetsart J. Soc. Sci. (2018)
3. Johnson, J.: Hypernetworks in the Science of Complex Systems. Imperial College Press, London (2015)
4. Sánchez, H.: Equity: the focal point of educational quality. Natl. Educ. Eval. Policy Gaz. **4**(10), 42–44 (2018)
5. Kivela, M., Arenas, A., et al.: Multilayer networks. J. Complex Netw. **2**(3), 203–271 (2014)
6. Mizuno, K., Tanaka, M., Fukuda, S., Imai-Matsumura, K., Watanabe, Y.: Relationship between cognitive functions and prevalence of fatigue in elementary and junior high school students. Brain Dev. **33**(6), 470–479 (2011)

7. Price, H.E., Moolenaar, N.M.: Principal-teacher relationships: foregrounding the international importance of principals' social relationships for school learning climates. J. Educ. Adm. **53**(1), 8–39 (2015)
8. Catalan Najera, H.E.: Multiple deprivation, severity and latent sub-groups: advantages of factor mixture modelling for analysing material deprivation. Soc. Ind. Res. **131**(2), 681–700 (2016)
9. OECD. Equity in Education: Breaking Down Barriers to Social Mobility, PISA, OECD Publishing, Paris (2018)
10. Gray, J., Kruse, S., Tarter, C.J.: Enabling school structures, collegial trust and academic emphasis: antecedents of professional learning communities. Educ. Manage. Adm. Leadersh. **44**(6), 875–891 (2016)

Public Opinion Divergence Based on Multi-agent Communication Topology Interconnection

Hui Zhao, Lidong Wang, and Xuebo Chen(✉)

School of Electronic and Information Engineering, University of Science and Technology Liaoning, Anshan 114051, Liaoning, China
xuebochen@126.com

Abstract. Nowadays, public opinion has become a hot topic in society. Correct knowledge and understanding of the evolution process of public opinion communication can guide mass emergencies accordingly. In this paper, a series of Matlab simulations are carried out according to the evolution process of group public opinion and the process of guiding group (multi-agent) guided by public opinion. First of all, based on the traversing simulation of the number of multi-agents from N = 15 to N = 50 and the number of communication topology interconnections from K = 1 to K = N − 1, the general conditions of bifurcation in the clustering process of multi-agents are summarized. Secondly, the bifurcation phenomenon is controlled by introducing single or multiple fixed individuals or changing the initial state of the cluster in the multi-agent cluster process simulation. Finally, in view of the actual social situation, the changing K value in the whole process is introduced, and the traversing simulation is carried out.

Keywords: Public opinion divergence · Multi-agent system · Cluster · Emergence · Bifurcation

1 Introduction

People often spontaneously express their personal views or opinions on a certain type of problem or phenomenon in the political, economic, cultural and social fields. When all kinds of views are disseminated and influenced each other through language, Internet, media and other means among the population, under the condition of eliminating the differences of individual opinions, Gradually in a certain social scope to form a reflection of social perception and collective consciousness, the majority of the common opinion [1].

In the natural world, cluster self-organizing movement, as the most common manifestation of group cooperation behavior, has become the carrier of information interaction, organization, cooperation and other behaviors between organisms [2]. Cluster self-organizing motion behavior is not only of great significance to the survival and reproduction of biological populations in nature, but also plays a very important role in the control of swarm robot systems and other research. In nature, from the

T. Ahram et al. (Eds.): IHSED 2019, AISC 1026, pp. 476–482, 2020.
https://doi.org/10.1007/978-3-030-27928-8_72

human population to the microbial community is usually based on group cooperation in order to more reflect the wisdom of the individual [3].

Multi-agent system (Multi-agent Systems, MASs) is an important branch of artificial intelligence, and it is the frontier subject of artificial intelligence in the world from the end of the 20th century to the beginning of the 21st century. With the rapid development of computer technology, artificial intelligence theory, control theory and the continuous exploration of modern science, multi-agent system has become one of the hot issues in different disciplines [4]. The bifurcation phenomenon of multi-agent communication topology interconnection cluster generally exists in the process of clustering, and is close to real life, which has far-reaching research significance. the divergence of public opinion is just like the bifurcation phenomenon mentioned above. At present, all kinds of practical problems and social contradictions in our country are relatively concentrated, especially with the rapid development and wide application of the network and information technology. Individual different views spread through a variety of network media, accelerate the formation of differences in opinions, easily lead to differences in opinions, affect social stability [5].

2 Establishment of Emergence Model

Definitions 1.1: D = ($V(D)$, $A(D)$ is a weighted digraph, $V(D)$ is $\{v_1, v_2, v_3., v_N\}$, $A(D)$ is one of the(v_i, v_j). (v_i, v_j) is treated as an edge set with the direction $i, j \in N = \{1, 2,.N\}$, $i \neq j$ from v_i to v_j. In this paper, the $N \times N$ correlation matrix $A = (a_{ij})$, is introduced, where the interconnection (v_i, v_j) of the edge set exists (v_i, v_j) $\in$ then the weight $a_{ij} > 0$ or $a_{ij} = 0$. Usually, $a_{ij} = a_{ji}$. A bidirectional graph D can be decomposed $\overleftrightarrow{D} = \overleftrightarrow{D}'$ into an undirected graph $\overleftrightarrow{D} = \left(V(D), A(\overleftrightarrow{D})\right)$. And a directed graph $\overleftrightarrow{D} = \left(V(D), A(\overleftrightarrow{D})\right)$, with $\overrightarrow{D} \cap \overrightarrow{D'} = \varphi$, $\overleftarrow{D}$ is the transpose of $\overrightarrow{D}$, so $D = \overleftrightarrow{D} \cup \overleftrightarrow{D}$. If the directed edge replacement a_{ij} is replaced by the basic graph obtained by the undirected edge e_{ij}, $G = \overleftrightarrow{D} \cup \vec{D} \cup \overleftarrow{D}$ can be deduced. In particular, $N_i^G = \{j \in v, j \neq i : (v_i, v_j) \in E(D)\}$ is defined as a collection of row elements for $E(D)$.

In order to describe the loose space about point $v_i = (x_i, y_i, z_i) \in R^3$ the distance relation matrix $P = (p_{ij}) = \left(\|v_i - v_j\|_2\right)$, where $i \neq j$ and $v_i \neq v_j$, give $v_i = (x_i, y_i, z_i) \in R^3$. for $\|\cdot\|_2$. Explore all the vertices and define a sequence $Q_i = (q_{ik})_{k=1}^{N-1}$ to describe the set of adjacent vertices which are correlated and become stronger from weak to strong on the i row vector Pi in P, so $q_{ik} \leq q_{ik+1}$ is obtained. the set of adjacent vertices on the i row vector Q_{ik} in P is defined as Q_{ik} to describe the set of adjacent vertices on the I row vector of P. The nearest neighbor set of the i-th fixed-point v_i, N_i, mimics two of the most common types of association rules in nature:

1. Distance relation model
2. Topological relationship interconnection

We only discuss the communication topology interconnection, so we do not delve into the distance relationship model for the time being. The topological relation model used here is $N_i^T = \{j \in N;\ q_{ik} = p_{ij}, K \leq n\}$, in which N is the number of groups and K is the nearest neighbor number of individuals. the model shows that the specific nearest neighbor numbers are interconnected with them. The interconnection rules of its critical matrix have asymmetric characteristics of $D = \overleftrightarrow{D} \cup \vec{D}$ communication topology interconnection.

In order to explore the emergence and control methods of emergence, the three-dimensional interconnection rules are selected as follows:

$$f_i = \sum_{j=N_i} f(\|v_j - v_i\|) \cdot e_{ij} \tag{1}$$

Among them, $e_{ij} = a_{ij} \cdot \frac{v_j - v_i}{v_j - v_i}$ shows that interconnection (v_i, v_j) has a unit vector of weight a_{ij}, while based on the bit vector, the force in this direction can be expressed by Lennard-Jones function. Let the simplest dynamic equation of cluster self-organizing system based on loose preference behavior rules be:

$$\dot{v} = f_i \tag{2}$$

Looking up the relevant literature, it is known that the premise of calling a multi-agent system a system is that the system must be connected. Therefore, in the model set up in this paper, all the interconnected weighted adjacency matrices can not be reduced. That is, the cluster system is interconnected and there is no solitary subgroup.

3 Simulation of Cluster Motion Under the Change of Individual Number and Nearest Neighbor Interconnection Number

3.1 Simplification of Related Models

In order to study the formation of public opinion and the formation of public opinion differences, the relevant models of the forces of close exclusion and long-distance attraction in this simulation are simplified as follows:

$$f(x)\begin{cases} -w_r \cdot (1/x - 1/r^*), r_l < r^* \\ w_a \cdot sin[(r_h - r^*)/(x - r^*)], r^* < x < r_h \\ 0, others \end{cases} \tag{3}$$

Here w_r and w_a are the parameters of the interconnection force, r^* is the expected distance between individuals (our value is 5.2 in this simulation) r_l is the minimum perceptual distance of the individual is the maximum perceptual distance of the individual r_h. In this simulation, we assume that it is carried out in an ideal state. usually, some emerging motion patterns in nature can be simulated by changing the parameters.

In order to explore the bifurcation phenomenon of multi-agent communication topology interconnection cluster, that is, the phenomenon of divergence of public opinion, the multi-agent cluster is placed in three-dimensional space for simulation by using Matlab simulation platform. Run 20000 steps per simulation.

Simulation 1: Cluster movement based on communication topology interconnection.

In order to explore the bifurcation condition of multi-agent communication topology interconnection cluster, N (the number of multi-agents in the system) is 15 to 50, K (communication topology interconnection number) is 1 to $N-1$, and the bifurcation law is found in the traversal of K and N. Because of the large number of pictures, we will no longer list them one by one. From the simulation results, we can see that it is possible to bifurcation at about $K = 1/5N$ to $2/5N$ [6]. So when we don't want to see bifurcation, we have to avoid this particular area.

4 Promote the Bifurcation of Cluster Motion by Adding Fixed Points

Simulation 2: Promote bifurcation by adding fixed points (randomly select the number of associations from K = 1/5N to 2/5N without bifurcation).

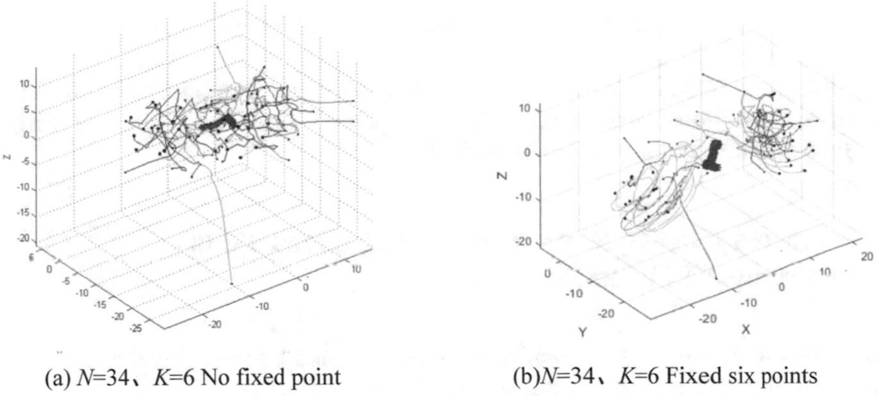

(a) N=34、K=6 No fixed point (b)N=34、K=6 Fixed six points

Fig. 1. Promoting cluster motion bifurcation by adding a fixed point

As shown in Fig. 1, the initial point on the left is in the normal distribution, and the right side is fixed at different points. It can be seen that by adding a fixed point, the cluster motion in the communication topology interconnection which would not otherwise produce a bifurcation can be forked, indicating that the bifurcation is controllable under certain conditions, By adding fixed points on both sides of the cluster, the movable points have two different targets, which diverge under the traction of the two targets.

5 Promote the Bifurcation of Cluster Motion by Changing the Initial Position

Simulation 3: Regulating bifurcation by changing the initial position of cluster motion under communication topology interconnection (promoting bifurcation) (Fig. 2).

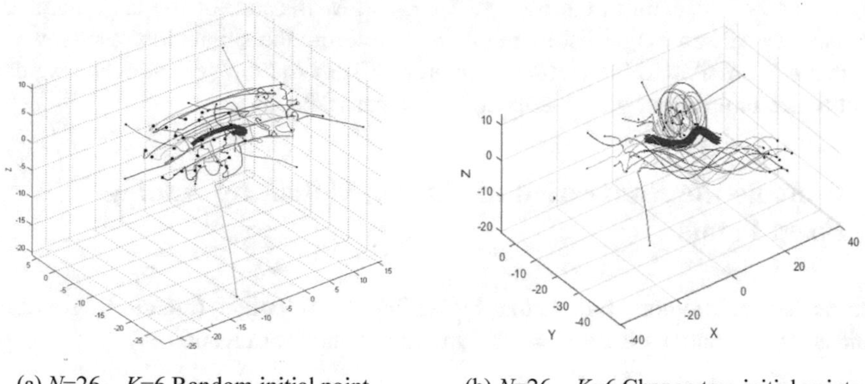

(a) N=26、K=6 Random initial point (b) N=26、K=6 Change two initial points

Fig. 2. Changing the initial position control bifurcation of cluster motion

From the Simulation 2, we can see that by changing the initial position of the cluster motion under the communication topology interconnection, the original unforked cluster motion can also be forked.

6 Suppress Cluster Motion by Adding Fixed Points to Produce Bifurcation

Simulation 4: Change cluster motion in communication topology interconnections by adding fixed points (suppress bifurcation) (Fig. 3).

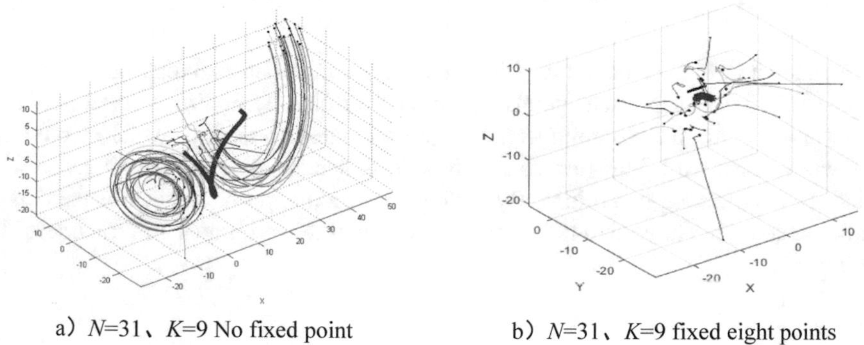

a) N=31、K=9 No fixed point b) N=31、K=9 fixed eight points

Fig. 3. Adding fixed points to suppress cluster motion bifurcation

From the Simulation 3, we can clearly see that by adding a number of appropriate fixed points, the original bifurcation of the population can be changed into an aggregation state.

7 Simulation of Cluster Motion with Individual Number Variation and Nearest Neighbor Interconnection Number Random

In view of the actual social situation, the number of communication topology interconnections in the process of clustering is not constant, that is, the number of each individual affected by other individuals is changing. So introduce the changing K values throughout the process, and iterate through the simulation results as shown in the following figure:

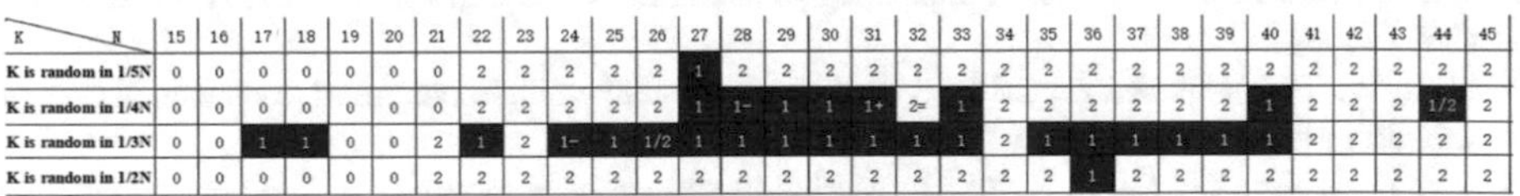

K \ N	15	16	17	18	19	20	21	22	23	24	25	26	27	28	29	30	31	32	33	34	35	36	37	38	39	40	41	42	43	44	45
K is random in 1/5N	0	0	0	0	0	0	0	2	2	2	2	2	1	2	2	2	2	2	2	2	2	2	2	2	2	2	2	2	2	2	2
K is random in 1/4N	0	0	0	0	0	0	0	2	2	2	2	2	1	1-	1	1	1+	2=	1	2	2	2	2	2	2	1	2	2	2	1/2	2
K is random in 1/3N	0	0	1	1	0	0	2	1	2	1-	1	1/2	1	1	1	1	1	1	1	2	1	1	1	1	1	1	2	2	2	2	2
K is random in 1/2N	0	0	0	0	0	0	2	2	2	2	2	2	2	2	2	2	2	2	2	2	2	1	2	2	2	2	2	2	2	2	2

Fig. 4. Bifurcation summary table

From Fig. 4, it can be seen that there is a divergence of public opinion between the random numbers in $1/5N$ and the random numbers in $1/2N$ in K.

8 Conclusion

In order to explore the conditions of public opinion differences and the control of public opinion differences, we have done a large number of simulations, from which we can draw the following conclusion:

1. Bifurcation can occur at approximately $K = 1/5N$ to $2/5N$ (divergence of opinion);
2. Differences of opinion can be promoted by adding fixed individuals (different fixed views) on both sides of the group (different views) in the process of multi-agent clustering. However, attention should be paid not to add fixed individuals at will, but to adjust the distance between fixed individuals and the original group or to increase or decrease the number of fixed individuals, that is to say, people's emotional will should be properly added.
3. By properly adding fixed individuals (unified views) in the process of multi-agent clustering, we can restrain the differences of opinion, but we should pay attention to the appropriate increase or decrease of their number, that is, the views should be properly close to people's emotional thinking.

4. By changing the initial position of the multi-agent, the disagreement of public opinion can also be suppressed.
5. Under the condition that K is completely random, there is a phenomenon of divergence of public opinion from the random number of 1/5N to the random number of 1/2N.

Acknowledgments. This research reported herein was supported by the NSFC of China under Grants No.71571091 and 71771112.

References

1. Bernays, E.L.: Crystallizing public opinion. Boni and Liverright, New York (1923)
2. Cluster self-organizing motion control of swarm robot system. Tianyun Huang. Dalian University of Technology (2015)
3. Cheng, D.Z., Chen, H.: From clustering to social behavior control. Sci. Technol. News (8), 4/7.4 (2004)
4. Ren, W., Beard, R.W.: Distributed consensus in multi-vehicle cooperative control. Springer, London (2008)
5. Zhao, Y., Kou, G., Peng, Y., Li, S.: Modeling and analysis of public opinion communication at the level of inconsistent trust in group emergencies. Theory Pract. Syst. Eng. **32**(05), 971–976 (2012)
6. Yuan, Y., Chen, X., Sun, Q., Huang, T.: Analysis of topological relationships and network properties in the interactions of human beings. PLoS ONE **12**(8), e0183686 (2017)

Exploring the Intersections of Web Science and Accessibility

Trevor Bostic(✉), Jeff Stanley, John Higgins, Daniel Chudnov, Rachael L. Bradley Montgomery, and Justin F. Brunelle

The MITRE Corporation, Bedford, USA
{tbostic, jstanley, jphiggins, dlchudnov, jbrunelle}@mitre.org, rachael@accessiblecommunity.org

Abstract. In this paper, we survey several ongoing research threads that can be applied to web accessibility solutions. We focus on the challenges with automatically evaluating the accessibility violations in websites that are built primarily with JavaScript. There are several research efforts that – in aggregate – provide insight into how users interact with websites; how to automate and simulate user interactions; how to record the results of user interactions; and how to analyze, evaluate, and map resulting website content to determine the relative accessibility. We close with a discussion on the convergence of these threads and the future of automated, web-based accessibility evaluation, and assurance.

Keywords: Accessibility · Web science · Web archiving

1 Introduction

The web is the prominent way information is exchanged in the 21st century. However, ensuring web-based information is accessible is complicated, particularly with web applications that rely on JavaScript and other technologies to deliver and build representations[1]. Static representations are becoming rarer [4] and assessing accessibility of web-based information is available to all users is increasingly difficult given the dynamic nature of representations.

In this work, we survey three ongoing research threads that can inform web accessibility solutions: assessing web accessibility, modeling web user activity, and web application crawling. Current web accessibility research is continually focused on increasing the percentage of automatically testable standards, but still relies heavily upon manual testing for complex interactive applications. Alongside web accessibility research, there are mechanisms developed by researchers that replicate user interactions with web pages based on usage patterns. Crawling web applications is a broad research domain; exposing content in web applications is difficult because of incompatibilities in web crawlers and the technologies used to create the applications. We describe research on crawling the deep web by exercising user forms. We close with a thought exercise

[1] A representation of a web page is what users see in the browser after visiting the page.

T. Ahram et al. (Eds.): IHSED 2019, AISC 1026, pp. 483–488, 2020.
https://doi.org/10.1007/978-3-030-27928-8_73

regarding the convergence of these three threads and the future of automated, web-based accessibility evaluation and assurance through a use case in web archiving. These research efforts provide insight into how users interact with websites, how to automate and simulate user interactions, how to record the results of user interactions, and how to analyze, evaluate, and map resulting website content to determine its relative accessibility.

2 Assessing Web Accessibility

As of 2010, about 15% of the world's population lives with some form of disability [33]. Ensuring these individuals can participate and contribute to society requires accessible technology. Section 508 dictates accessibility standards for government technology and electronic communications[2]. Additional legal requirements also prevent discrimination of individuals with disabilities through technology [9, 10]. As a reflection of accessibility's importance to production web applications, Garrison recommends accessibility testing be incorporated into the software development process alongside security evaluations [12]. Ensuring web content and applications are accessible, prevents discrimination against disabled users. It ensures information is available for users.

There are many efforts focused on web accessibility such as accessibility standards (e.g., W3C [32]) and training programs for testers and developers (i.e., Trusted Tester Program [8]). However, current automated tools are only capable of finding roughly 30% of accessibility issues and are often focused on static representations [31]. As a result, a government agency or business may spend hundreds of thousands of dollars a year attempting to conform to accessibility standards but still have a system that is inaccessible to a portion of the population.

3 Modeling Web User Interaction

In order to automatically evaluate the accessibility of a web application, a user's interactions may need to be simulated. The methods by which users and their interactions are studied and simulated are very important to automating accessibility assessments. Of particular interest are events on the client that are driven by JavaScript or Ajax since these are often hidden from traditional web-based automation tools.

Gorniak worked to predict future user actions by observing past interactions by other users [13]. This work modeled users as agents based on their interactions, and modeled clients based on potential state transitions. Each distinct configuration of content is considered a separate state, characterized by the underlying document object model (DOM) that is interpreted by the browser to render the webpage. The agents would make the next most probable state transition, indicating what action the user is expected to take to change the DOM, such as clicking a button. Leveraging user

[2] As cited by Kanta, government accessibility varies widely [18].

interaction patterns, Palmieri et al. generated agents to collect hidden web pages by identifying navigation elements from current selections [22]. They represented the navigation patterns from these agents as a directed graph and would train the agents to interact with navigation elements. This includes forms, which would be filled out based on repository attributes.

This body of work demonstrates that a broad domain of work exists to study the ways in which users interact with web pages. The application of these works include:

- Predicting user interactions or activities
- Automated web application testing
- Modeling information discovery behavior
- Methods by which events are constructed on a web application

4 Web Application Crawling

Web applications that rely on JavaScript and Ajax to construct dynamic web pages pose further challenges for both disabled users and automated evaluation of accessibility. These applications are increasing in frequency on the web [4]; meanwhile, websites are becoming less accessible over time [14]. Current automated tools are insufficient for mapping the states of dynamic web applications [3], leading to undiscoverable accessibility violations. Given the advancements and combination of web science and accessibility theory, it should be feasible to inform future ubiquitous, automated web accessibility tools through the cross-pollination of these domains.

Automatically crawling web applications (e.g., those driven by Ajax) for both testing and evaluation purposes is a rich area of research. These works can inform a model for mapping the states of a web application for accessibility evaluation; this may include the models for state equivalency [23–25].

From these efforts, one can observe a need for customized methods of automatically crawling JavaScript and Ajax-driven web applications. Traditional tools are unable to adequately execute JavaScript on a representation and – as a result – may miss important information on the page.

5 Crawling the Deep Web

Past works have focused on crawling and discovering deep web resources (e.g., content requiring a user to log in to a site or fill out a form). Crawling pages with input forms is particularly difficult given the near-infinite combinations of input texts that can be constructed. User interaction models will likely require interactions with – in part – forms and other methods of free-text inputs.

Kundu and Rohatgi used an approach to generate potential input queries to web forms to uncover hidden web representations [21] and a random forest classifier to construct the input text. Raghavan and Garcia-Molina have proposed frameworks for uncovering the deep web behind input forms using a variety of input generation techniques [28]. Lage et al. presented a set of user models that interact with web forms to uncover hidden content [22]. He et al. worked to sample and index the deep web [15]. Ntoulas et al. captured the textual hidden web with a guess-and-check method of supplying server- and client-side parameters [26].

There are various methods to simulate user input into a web form. However, they vary in their ability to be automated and the level to which entered information/data must be constructed.

6 A Use Case: Web Application Crawling for Archiving

Over time, web resources increasingly use JavaScript to load embedded resources [4]. Automatically testing, crawling, and evaluating representations of web resources that leverage JavaScript to load embedded resources and build representations often requires special tools beyond traditional crawlers. As such, the domain of web archiving is rich with examples of web application crawling approaches. For example, web archiving crawlers must leverage specialized tools to accurately archive a representation that use JavaScript (e.g., Heritrix [30] using specialized crawling techniques to "peek" into JavaScript to understand what it does [7, 17]) but often performs inadequately [2].

Browsertrix [19] and WebRecorder.io [20] are page-at-a-time archival tools for deferred representations and descendants, but they require human interaction and are not suitable for automated web archiving. Archive.is [1] handles deferred representations well but is a page-at-a-time archival tool and must strip out embedded JavaScript to perform properly, making the resulting representation non-interactive. Umbra is a tool used to crawl a human curated set of pre-defined URI-Rs [29] and Brozzler uses headful or headless crawling for archiving [16].

Our prior work showed that specialized crawling can expose the JavaScript-dependent events and deferred representations [6] using Heritrix and PhantomJS [5, 27]. This work was adapted from the research by Dincturk et al. [11] to construct a model for crawling applications by discovering all possible descendants and identifying the simplest possible state machine to represent the states. In our prior work, we define state transitions as user interactions (and that trigger JavaScript event listeners) and state equivalence is determined by the set of embedded resources required to build out the page [6]. This differs from other researchers' use of strict DOM or DOM tree equivalency to define a state. The web archiving domain and interpretations of the Dincturk et al. Hypercube model from the application testing domain inspire approaches that automatically crawl web pages in the web accessibility domain.

Based on the models being researched and implemented in the web archiving domain and the need for automated accessibility tools and approaches, a confluence of approaches and technologies must be employed. These research methods used to discover, crawl, map, and execute interactions on web applications can be adapted for accessibility. For each representation, the interactive components must be discovered, mapped, exercised, and evaluated with user interactivity models. The research being performed and integrated at the center of web archiving is a potential roadmap for performing the same work within automated web-based accessibility evaluations.

7 Conclusions

In this paper, we have provided a selected literature review of research threads (i.e., assessing web accessibility, modeling web user interactions, and crawling web applications) that can inform automated web-based accessibility evaluations. There is a need to reduce the manual and human-driven effort being dedicated to evaluating the accessibility of websites, if they are evaluated at all. Creating a service or tool that helps with these assessments given the challenges unique to web applications would be a major advancement toward assessing web accessibility.

We recommend further research be performed on the intersection of the research threads presented in this paper and demonstrate an applicable convergence of these research methods by discussing their impact on web archiving. Our goal is to stimulate the development of technology that improves accessibility of web information to disabled users.

Acknowledgments. We would like to thank Sanith Wijesinghe, the Innovation Area Lead funding this research effort as part of MITRE's internal research and development program (the MIP). We also thank the numerous collaborators that have assisted with the maturation of our research project.

References

1. Archive.is (2013). http://archive.is/
2. Berlin, J.: CNN.com has been unarchivable since November 1st, 2016 (2017). http://ws-dl.blogspot.com/2017/01/2017-01-20-cnncom-has-been-unarchivable.html
3. Brunelle, J.F.: Scripts in a frame: a framework for archiving deferred representations. Ph.D. thesis, Old Dominion University (2016)
4. Brunelle, J.F., Kelly, M., Weigle, M.C., Nelson, M.L.: The impact of JavaScript on archivability. Int. J. Digit. Libr. **17**(2), 95–117 (2015)
5. Brunelle, J.F., Weigle, M.C., Nelson, M.L.: Archiving deferred representations using a two-tiered crawling approach. In: Proceedings of iPRES 2015 (2015)
6. Brunelle, J.F., Weigle, M.C., Nelson, M.L.: Archival crawlers and JavaScript: discover more stuff but crawl more slowly. In: Proceedings of the 17th ACM/IEEE Joint Conference on Digital Libraries, pp. 1–10 (2017)
7. Coram, R.G.: Django-phantomjs (2014). https://github.com/ukwa/django-phantomjs
8. Department of Homeland Security. DHS Trusted Tester Program (2018). https://www.dhs.gov/trusted-tester
9. Department of Labor. Laws & Regulations (2018). https://www.dol.gov/general/topic/disability/laws
10. Department of Labor. Americans with Disabilities Act (2018). https://www.dol.gov/general/topic/disability/ada
11. Dincturk, M.E., Jourdan, G.-V., Bochmann, G.V., Onut, I.V.: A model-based approach for crawling rich internet applications. ACM Trans. Web **8**(3), 19:1–19:39 (2014)
12. Garrison, A.: Continuous accessibility inspection & testing. In: Proceedings of the 2018 ICT Accessibility Testing Symposium: Automated & Manual Testing, WCAG2.1, and Beyond, pp. 57–66 (2017)

13. Gorniak, P., Poole, D.: Predicting future user actions by observing unmodified applications. In: Proceedings of the Seventeenth National Conference on Artificial Intelligence and Twelfth Conference on Innovative Applications of Artificial Intelligence, pp. 217–222 (2000)
14. Hackett, S., Parmanto, B., Zeng, X.: Accessibility of Internet Websites through time. In: Proceedings of the 6th International ACM SIGACCESS Conference on Computers and Accessibility, pp. 32–39 (2003)
15. He, B., Patel, M., Zhang, Z., Chang, K.C.-C.: Accessing the deep Web. Commun. ACM **50** (5), 94–101 (2007)
16. Internet Archive. Brozzler (2017). https://github.com/internetarchive/brozzler
17. Jack, P.: Extractorhtml extract-javascript (2014). https://webarchive.jira.com/wiki/display/Heritrix/ExtractorHTML+extract-javascript
18. Kanta, S., Insights with PowerMapper and R: An exploratory data analysis of U.S. Government website accessibility scans. In: Proceedings of the 2018 ICT Accessibility Testing Symposium: Mobile Testing, 508 Revision, and Beyond, pp. 65–72 (2018)
19. Kreymer, I.: Browsertrix: browser-based on-demand web archiving automation (2015). https://github.com/ikreymer/browsertrix
20. Kreymer, I.: Webrecorder.io (2015). https://webrecorder.io/
21. Kundu, S., Rohatgi, S.: Generating queries to crawl hidden Web using keyword sampling and random forest classifier. Int. J. Adv. Res. Comput. Sci. **8**(9), 337–341 (2017)
22. Lage, P., da Silva, A.S., Golgher, P.B., Laender, A.H.: Automatic generation of agents for collecting hidden Web pages for data extraction. Data Knowl. Eng. **9**(2), 177–196 (2004)
23. Mesbah, A.: Analysis and testing of Ajax-based single-page Web applications. Ph.D. Dissertation, Delft University of Technology (2009)
24. Mesbah, A., Bozdag, E., van Deursen, A.: Crawling Ajax by inferring user interface state changes. In: Proceedings of the 8th International Conference on Web Engineering, pp. 122–134 (2008)
25. Mesbah, A., van Deursen, A., Lenselink, S.: Crawling Ajax-based Web applications through dynamic analysis of user interface state changes. ACM Trans. Web **6**(1), 3:1–3:30 (2012)
26. Ntoulas, A., Zerfos, P., Cho, J.: Downloading textual hidden Web content through keyword queries. In: Proceedings of the 5th ACM/IEEE-CS Joint Conference on Digital Libraries, pp. 100–109 (2005)
27. PhantomJS (2013). http://phantomjs.org/
28. Raghavan, S., Garcia-Molina, H.: Crawling the hidden Web. Technical report 2000-36, Stanford InfoLab (2000)
29. Reed, S.: Introduction to Umbra (2014). https://webarchive.jira.com/wiki/display/ARIH/Introduction+to+Umbra
30. Sigursson, K.: Incremental crawling with Heritrix. In: Proceedings of the 5th International Web Archiving Workshop, September 2005
31. Vigo, M., Brown, J., Conway, V.: Benchmarking Web accessibility evaluation tools: measuring the harm of sole reliance on automated tests. In: Proceedings of the 22nd International World Wide Web Conference (2013)
32. W3.org. Web accessibility initiative (2018). https://www.w3.org/WAI/
33. World Health Organization. World report on disability. Technical report, World Health Organization (2011)

Somatic Senses Required for the Emotional Design of Upper Limb Prosthesis

Luisa M. Arruda[1(✉)], Luís F. Silva[2], Helder Carvalho[1], Miguel A. F. Carvalho[1], and Fernando B. N. Ferreira[1]

[1] 2C2T – Centro de Ciência e Tecnologia Têxtil, University of Minho, Guimaraes, Portugal
luisamendesarruda@gmail.com
[2] MEtRICs – Mechanical Engineering and Resource Sustainability Center, University of Minho, Guimaraes, Portugal

Abstract. Despite the technological advances associated with prostheses, the total embodiment is still the great challenge in the segment of assistive technology. One of the main aspects is that the bionic member's sensibility is not responsive to the environment that surrounds it. The purpose of this study aims to identify the perceptual modalities of the somatic senses that are required for a more sensible prosthesis. The methodological strategy contemplates literary research and proposes an interrelationship between neuroscience with philosophical/cultural studies, which regards the different concepts of sensory experience. Such data were analyzed quantitatively and qualitatively. The main conclusion points out that it would be important to unite the nine physiological requirements identified in the state of the art, with the ontological image construction of the prosthesis, in order to create a sensory experience that, in addition to the perceptive organs, builds up by the visual areas of the brain.

Keywords: Emotional design · Upper limb prosthesis · Sensibility · Somatic sense

1 Introduction

The practice of limb amputation arose in the Neolithic period, 43000 B.C. [1] and the reasons for such an act were punitive, therapeutic, and ritualistic. Subsequent to the amputation procedure, the prostheses were developed to meet the requirements of functionality, aesthetic appearance and a psycho-spiritual sense of totality, around 1500 B.C. [2]. However, it is with the advent of the two great world wars that prostheses became modernized regarding their shape. In the 21st century, advances in additive manufacturing have allowed such devices to be more accessible, customizable and made from both flexible and rigid raw materials [3]; and the progress of nanotechnology, with the miniaturization of electronic devices - especially sensors and actuators - have added functionalities to the prostheses identified as degrees of freedom (DOF), which in turn have been allowing the amputee a greater diversity of activities of daily living (ADL), such as undoing buttons, simulating food cuts, moving an empty can, among others.

T. Ahram et al. (Eds.): IHSED 2019, AISC 1026, pp. 489–494, 2020.
https://doi.org/10.1007/978-3-030-27928-8_74

Although the mechanical and electrical advancements of prosthetic devices are irrefutable, upon asking a man with transhumeral amputation - who tested a technological prosthesis - what he considered to be improved on his artificial limb, he replied that he would like to hold a loved one's hand and really feel it, beyond the mechanical grasping movement [4]. In other words, the sensory character related to the functionality of prostheses as the crucial point to be developed by the assistive technology segment is evidenced.

The fact is that the act of touching a loved one's hand is a communicational action constructed in otherness, which means that such action does not generate fixed sensations; instead, they vary from enunciator to enunciator. In its turn, individual reactions go beyond the dimension of the somatic senses and express a combination of psychophysical mechanisms. Moreover, such combinations externalize codes that make possible the interpretation of affection by each enunciator of the communicational act (amputated or not).

In this sense, in order to enumerate assumptions for the design and development of a prosthesis that meets the diligence in question, this study proposes to investigate the sensibility concepts present in the current prostheses and interpose through an ontological approach of the sensory experience.

2 Methodology

The research strategy adopted in this paper is the exploratory study since one intends to enlighten the phenomenon in question - here referred to as the sensibility in prostheses - to conceive research through diverse techniques that involves bibliographical survey and exploration of patents. This methodology has the objective of defining the problem in question to then raise the somatosensory requirements required for the design and development of more emotional upper limb prosthesis.

Patent research was carried out using Derwent Innovation software (Clarivate Analytics). The keywords used were: upper limb prosthesis; bionic prosthetic and bionic limb. In turn, the sub-words searched were: tactile feedback; sensibility; sensing; sensation and affectivity.

Simultaneously, a bibliographical revision was made on the term "sensibility", aiming to define it according to a philosophical approach, specifically by the perspective of the Italian author Emanuele Coccia [5]. This author was evidenced as he deals with sensibility through the ontology of the images, dialoguing on an anatomical and prosthetic body and explores Fashion as a form of construction of the images and, therefore, of the sensibility in an autonomous sphere. Thus, Coccia contrasts essential concepts that allow us to elucidate about the prosthesis, beyond the medical devices, and closer to the emotional design.

3 Results

The patent search found 24 results, however, only 5 refer directly to sensibility. The first patent aimed at proposing sensory reconfiguration dates back to 1998. It distinguishes itself by being a method and device that creates tactile sensations through sensors and sends them to the individual. Such sensations relate to the texture and sliding of objects in contact with the prosthetic hand [6]. In addition, a system and method are cited for providing a sensation in a member with a prosthetic device or sensory deficiency: sensors positioned at the foot of the prosthesis contact the sock and produce a vibration stimulus in the residual limb, which is aided by an electrical unit to adjust and control the stimulus [7]; a mechanism and method for designing and constructing a mechanical interface between a wearable device and a segment of the human body [8]; and sensors that can maintain a comfortable temperature and humidity between the device and the residual limb of the individual [9].

In addition to these, another patent was also analyzed concerning a method and system to promote proprioceptive feedback, a characteristic present in the somatic system [10]. The others were discarded given that they do not fit the sensory system, or they refer to invasive procedures: 12 are directed to the identification and control of movements of the prosthetic limb; 1 refers to the detection of the error of human-machine interaction; 1 relates to a hospital control system; and 5 correspond to neuromuscular interference models.

4 Discussion

The somatosensory system allows the human being to obtain sensitive experiences with the environment that surrounds him, namely touch, pain, temperature and proprioception. Unlike other sensory systems present in the body, such as taste, smell, sight and hearing, the somatic system has receptors on distinct parts of the body.

Communication begins from contact with the outside world. The skin of the upper limb receives stimuli that are translated into neural signals. This information goes to the central nervous system and the brain realizes their perception. This means that the characteristics of the stimulus - such as intensity, duration, position and, sometimes, direction - are encoded by the sensory receptor. In turn, the central nervous system processes the activity of the receptors and uses this data to generate coherent perceptions. From here the brain uses sensory information to plan and coordinate movements [11].

As the research presented, 50% of the patents are directed to mimic the motor system of the human body. Concerning somatosensory awareness, only about 20% seek to achieve sensory mechanisms, including techniques such as sensory reeducation, tactile kinesthetic orientation, repetitive sensory practice or desensitization.

Concerning the sensory reeducation, the literature review pointed out that sensations are mimicked by detectors of temperature, texture, slip, vibration, pressure, torsion, impedance and position. Therefore, by the aspect of medicine, such survey pointed out that these nine characteristics are the desirable requirements for a sensory prosthesis.

However, human sensations have more than physiological components. For philosophy, for example, man survives through sensations, and the senses are what the individual trusts and depends on. Therefore, the interdependence of the experience of the sensible to the living human condition is noticed. However, unlike the medical approach, the sensitive ontology reports that the production of sensibility is not related to contact with the perceptual organ, on the contrary, it is emphasized that sensibility is beyond. This approach describes the sensible as being the image in the broad sense, and therefore, it is precisely through the ability to produce images that man is affected by the experience of sensibility. Moreover, the ontological idea of image is the representation of something out of its place, in an independent sphere. In other words, philosophy approaches the science of the sensitive and ratifies that the sensation is not caused by the perceptive organs but by the construction of images in our brain [5].

Not surprisingly, recent studies on the neurological representation of artificial limbs in amputees revealed that the more the individual is subjected to the daily use of prosthetics, the more visual areas of the brain in the lateral occipitotemporal cortex reorder and begin to respond to the prostheses' images [12]. This has been demonstrated by the increased connectivity between the visual and sensory-motor areas of the brain linked to the functionality of the hands. In addition, the same study revealed that the triad: activities performed by the bionic member; visual-motor connectivity; and frequent use of the prosthesis in daily activities can progressively cause the brain to reuse neural resources originally developed, from the visual representation, to reshape the motor system in order to control the prostheses. In other words, to live the experience of the sensitive, there is in both approaches (philosophical and neurological) the construction of images in the brain as a way of relating sensory stimuli to medical devices.

Philosophy describes that *phenomenon* such as dreams, fashion, and language are capable of producing images, and complements that when an object becomes a *phenomenon*, there is sensation. Therefore, if the prosthesis has not yet reached the degree of embodiment, it can be identified as an object external to the body. Then, the question arises: how can the prosthesis stop being an external object and, through the emotional design, approach the embodiment integrated into the anatomic body?

Through the discursive reasoning presented, it is worth thinking about new measures/materials/forms for the design and development of a prosthesis that approaches more of a wearable piece - integral to the corporeal visuality of the individual - than just a medical device. In addition to the related physiological requirements for the development of an "emotional" prosthesis, it is necessary for its design to avoid attempting to replace the missing limb as a replica of what it was, since such a point only evidences the loss. It is up to the designer to identify the spirit that singularly inhabits each individual in order to materialize a wearable piece/limb, which expresses such language. In addition, it is intended that such materialization will be at the same time the construction of an orthopedic image of all the limbs, through a skin constructed with unconventional materials and forms, beyond the design of the anatomical body.

5 Conclusions

For philosophy, the living one is, above all, the one who is able to give a sensible existence of that which dwells within him. This is because the body, if observed by the one who lives with it, is never a figure, a form, but a series of sensory states that interpenetrate themselves. In this sense, this paper proposed dialectic between cultural and physiological studies to highlight desirable assumptions for the design and development of an "emotional" prosthesis.

The state of art pointed out that there are many advances in the prosthetic segment, and has shown that the largest fraction is designated to motor functionalities. It confirmed that the crucial point to develop is the insertion of sensors that propose the somatosensory reconfiguration, and five patents addressed to this subject have been detected.

For the neurological aspect, nine desirable requirements for the development of a more sensory prosthesis were noted: compliance, temperature, texture, slip, vibration, pressure, torsion, impedance and position. By the philosophical approach, it is proposed to think of the prosthesis as a garment, with the use of nonconventional form and materials, that approach the object more of the *phenomenon* like Fashion, than of Medicine.

However, the relation of the living with the world is not purely ontological, not merely poetic and simply physiological. It is, therefore, necessary that the studied fields complement each other so that the individual can experience both the desired functionality of his or her prosthesis and their cognitive experiences.

Acknowledgments. This work is financed by Project "Deus ex Machina", NORTE-01-0145-FEDER-000026, funded by CCDRN, through Sistema de Apoio à Investigação Científica e Tecnológica (Projetos Estruturados I&D&I) of Programa Operacional Regional do Norte, from Portugal 2020 and by Project UID/CTM/00264/2019 of 2C2T – Centro de Ciência e Tecnologia Têxtil, funded by National Founds through FCT/MCTES.

References

1. Magee, R.: Amputation through the ages: the oldest major surgical operation. ANZ J. Surg. **68**, 675–678 (1998)
2. Padula, P.A., Friedmann, L.W.: Acquired amputation and prostheses before the sixteenth century. Angiology **38**, 133–141 (1987)
3. ten Kate, J., Smit, G., Breedveld, P.: 3D-printed upper limb prostheses: a review. Disabil. Rehabil. Assist. Technol. **12**, 300–314 (2017)
4. Tyler, D.J.: Restoring the human touch: prosthetics imbued with haptics give their wearers fine motor control and a sense of connection. IEEE Spectr. **53**, 28–33 (2016)
5. Coccia, E.: A vida sensível. Cultura e Barbárie Editora, Florianópolis (2010)
6. Chen, E., Eberman, B., Marcus, B.A.: Method and apparatus to create a complex tactile sensation (1998)
7. Sabolich, J.A., Ortega, G.M., Schwabe, B.G.: System and method for providing a sense of feel in a prosthetic or sensory impaired limb (2002)

8. Herr, H.M.: Mechanisms and methods for designing and manufacturing a mechanical interface between a wearable device and a segment of the human body (2016)
9. Altobelli, D.E., Perry, N., Christopher, E.: Arm prosthetic device system (2016)
10. Herr, H.M., Clites, T.R., Maimon, B., Zorzos, A., Carty, M.J., Duval, J.-F.: Method and system for providing proprioceptive feedback and functionality mitigating limb pathology (2019)
11. Bear, M.F., Connors, B.W., Paradiso, M.A.: Neurociências: Desvendando o Sistema Nervoso. ArtMed, Porto Alegre (2008)
12. Van Den Heiligenberg, F.M.Z., Orlov, T., MacDonald, S.N., Duff, E.P., Henderson Slater, D., Beckmann, C.F., Johansen-Berg, H., Culham, J.C., Makin, T.R.: Artificial limb representation in amputees. Brain **141**, 1422–1433 (2018)

Development of Behavior Profile of Users with Visual Impairment

Cesar Guevara[1,2], Hugo Arias-Flores[2,3(✉)], José Varela-Aldás[2,3], David Castillo-Salazar[2,3,5], Marcelo Borja[2,4,8], Washington Fierro-Saltos[2,5], Richard Rivera[2,6], Jairo Hidalgo-Guijarro[2,7], and Marco Yandún-Velasteguí[2,7]

[1] Mechatronics and Interactive Systems - MIST Research Center, Universidad Indoamérica, Quito, Ecuador
cesarguevara@uti.edu.ec

[2] Department of Informatics and Computer Science, Escuela Politécnica Nacional, Quito, Ecuador

[3] SISAu Research Group, Universidad Indoamérica, Ambato, Ecuador
{hugoarias, josevarela, davidcastillo}@uti.edu.ec

[4] Graphic Design Faculty, Universidad Indoamérica, Quito, Ecuador
carlosborja@uti.edu.ec

[5] Facultad de Informática, Universidad Nacional de La Plata, La Plata, Argentina
washington.fierros@info.unlp.edu.ar

[6] Escuela de Formación de Tecnólogos, Escuela Politécnica Nacional, Quito, Ecuador
richard.rivera01@epn.edu.ec

[7] Grupo de Investigación GISAT, Universidad Politécnica Estatal del Carchi, Tulcan, Ecuador
{jairo.hidalgo, marco.yandun}@upec.edu.ec

[8] Facultad de Diseño y Comunicación, Universidad de Palermo, Buenos Aires, Argentina

Abstract. The interaction of the user with visual impairment with assistive technologies, and in particular with screen readers, generates a group of actions and events during their navigation. These interactions are defined as behavioral patterns, which have a sequence that occurs at specific time slot. Understanding user behavior by analyzing their interaction with applications, in addition, details the characteristics, relationships, structures and functions of the sequence of actions in a specific application domain. The objective of this document is to find activity patterns from a set of commands used by the user, combining data mining and a Bayesian model. This model calculates the probability of the functions used with the screen reader and generates a behavior profile to improve the user experience. For this study, the screen reader JAWS version 2018, the Open Journal Systems platform version 3.0.1 and a computer with Windows 10 operating system were used. During the first phase, command history used by the user by interacting with the Open Journal Systems platform were collected. The result is that the accessibility of users with visual impairment to interact with the computer and its applications has been improved by applying this model.

T. Ahram et al. (Eds.): IHSED 2019, AISC 1026, pp. 495–500, 2020.
https://doi.org/10.1007/978-3-030-27928-8_75

Keywords: User · Visual impairment · Assistive technologies · Patterns

1 Introduction

Worldwide, it is estimated that 1300 million people live with some kind of visual impairment. Of these, 188.5 million have a moderate visual impairment, 217 million from moderate to severe, 36 million are blind [1] and 826 million people have near vision problems [2]. In Latin America and the Caribbean, it is estimated that between 1% and 4% of the population live with this condition, while in Ecuador 11.85% of people have visual impairment, that is 49,237 people that are part of this vulnerable group. These people require assistive technologies to maintain or improve their autonomy, promoting their welfare. Worldwide there are more than 1000 million people who need assistive products and only 10% have access to them, due to high costs and low availability.

To participate in education, the job market and social life, people with visual impairment must access information with assistive technologies. It is undeniable that, in order to access information, websites must be usable and accessible. The growing demand for learning materials, communication or digital media has been exponential. Likewise, its integration in the educational system has created the need to improve access. For many users with visual impairment, updated software versions or revised websites, apps and operating systems can create a situation where an interface that may have been accessible is no longer accessible, more difficult to use or more confusing [3].

Teachers and students can easily create and share knowledge through the use of interfaces, anywhere and with various devices. In this context, the research of [4], discuss the design of usable web interfaces for collaborative editing, focusing on how people with visual impairment interact with them. Based on the results, five guidelines are proposed for supporting collaborative editing, including "accessible awareness" for blind users, establishing that not considering accessibility in the design of interfaces, increases the difficulties for users of screen readers. Knowing the assistive technologies that people with visual impairment use while navigate the web, allow us to identify the need to rearrange web links, incorporate a search engine, a text version for all pages and more information about visuals [5].

On the other hand [6], explored the implementation of web accessibility in the Israeli higher education context, using an automated evaluation tool to measure the adherence of web pages to the accessibility standards. Results show that all examined web pages presented accessibility barriers and were non-compliant with the most basic standards. "Contrast" and "missing alternative text" errors were the most frequent problems identified in the evaluation. Likewise [7], show an integrated solution to simulate accessibility limitations. This tool allows for impairment simulations to be performed on Java, mobile and web applications. It also integrates two assistive technologies, a screen reader and a magnifier. As a result, user interface developers can experience how interaction would be affected from various impairments and also understand how their developments would be perceived by impaired users through assistive technologies, enabling the development of more accessible applications for users with disabilities [8].

Therefore, screen review software - assistive technology - allow visually impaired people to write and edit works, conduct research, gain access to information and develop job skills [9].

The article is structured as follows, in Sect. 2 the materials and methods used as a fundamental basis in the proposal are presented. In Sect. 3, the proposal in the conditional probability application in the command execution sequence on accessible platforms is presented in detail. Section 4 shows the results of the proposed model and a comparison with some related works. In Sect. 3, the conclusions obtained from the results and future lines of research are presented.

2 Materials and Methods

In this section, all the tools and theorems used for the development of the proposal of an intelligent model for user behavior profile in open access assisted reading tools are described in detail.

2.1 Tools

The tools used in this study were: screen reader JAWS version 18.0.2740, with the Open Journal Systems platform (version 3.0.1). The application uses the Windows 10 operating system. In the Open Journal System platform, a user with an editor role was used in order to carry out the different publication processes [10].

2.2 Materials

The sample consisted of 100 commands (Table 1), used during the process of initial review of a manuscript in the journal management platform. These data were collected during three months, addressing the processes of: submission review, articles download for reading and provide observations, adding discussion, provide answers to the discussions, adding participants to a discussion and assigning an editor for the manuscript.

Table 1. Processes collected from JAWS tool during a period of 3 months.

Process	Commands	Process	Commands
Submission review	25	Provide answers	17
Articles download	11	Adding participants	15
Adding discussion	23	Assign an editor	9

2.3 Methods

In this section, the application of the conditional probability and Bayes' theorem is detailed.

Bayes' Theorem

Thomas Bayes links the probability of an action A given an action B with the probability of B given A. Therefore, the following theorem for the calculation of conditional probabilities was developed:

Let $\{A_1, A_2, \ldots., A_i, \ldots, A_n\}$ be a set of mutually exclusive events whose union is the total, that is 1, and such the probability of each of them is different from zero [11]. Let B an event of which the conditional probabilities $P(B|A_i)$ are known. Then the probability $P(A_i|B)$, is given by the equation $P\frac{(A_i|B)}{P(B)} = \frac{P(B|A_i)P(A_i)}{\sum_{j=1}^{n} P(B|A_j)P(A_j)}$, where: $P(A_i)$ are the a priori probabilities. $P(B|A_i)$ is the probability of B in the hypothesis A_i. $P(A_i|B)$ are the a posteriori probabilities. This theorem, valid in all applications of probability theory, can serve to indicate how subjective probabilities should be modified when additional information is received from an experiment.

Development of Command Execution Profiles

In each of the tasks analyzed, different commands are applied (Fig. 1), the same ones that generate a sequence. The occurrence probability of each command will be described in Fig. 2.

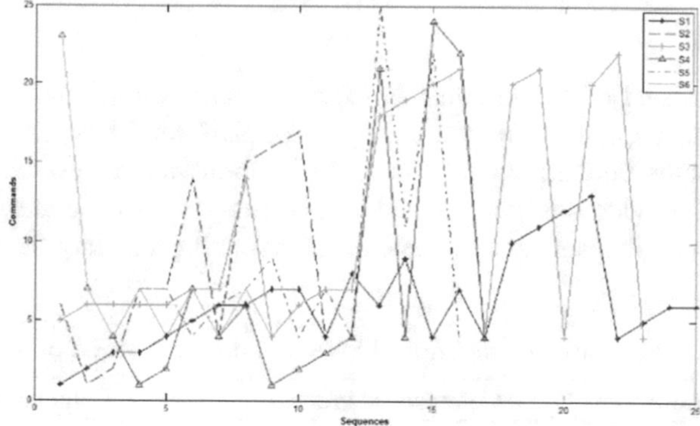

Fig. 1. Sequence of commands per task of JAWS application.

Based on this occurrence probabilities calculation, our study is based on identifying the probability of the sequences of commands C_1, C_2 more executed in the sequence's history obtained from JAWS, as presented in Table 2.

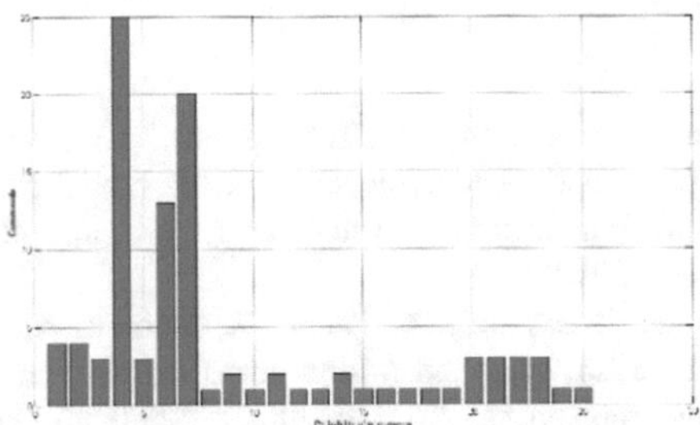

Fig. 2. Occurrence probability of JAWS commands.

Table 2. Conditional probability of the sequences of commands most used in the execution of JAWS commands.

Sequence of commands	Conditional probability
C_5, C_6	0.0723
C_6, C_7	0.0658
C_7, C_4	0.0471
C_{23}, C_7	0.0356
C_{22}, C_4	0.0307

Based on these sequences, a behavior profile of each of the users can be developed to determine a more accessible and efficient use for people with visual impairment.

3 Conclusions and Future Works

This work presents an application of mathematical models to identify the most executed tasks in an accessible platform for people with visual impairment. This model has obtained an optimal result in the detection of the most common commands and sequences in the command execution history. The application of conditional probability is essential to determine which sequences are going to be the ones that the user will execute during the use of the platform. With this, a work profile is determined, which facilitates the accessibility to the user with visual impairment, improving their time of performance and their learning.

As future research lines, it is expected to develop a predictive model of commands by applying polynomial functions and conditional probability to predict the tasks that the user wants to perform, to obtain a more dynamic and accessible result.

References

1. Bourne, R.R., Flaxman, S.R., Braithwaite, T., Cicinelli, M.V., Das, A., Jonas, J.B., Naidoo, K.: Magnitude, temporal trends, and projections of the global prevalence of blindness and distance and near vision impairment: a systematic review and meta-analysis. Lancet Glob. Health **5**(9), e888–e897 (2017)
2. Fricke, T.R., Tahhan, N., Resnikoff, S., Papas, E., Burnett, A., Ho, S.M., Naidoo, K.S.: Global prevalence of presbyopia and vision impairment from uncorrected presbyopia: systematic review, meta-analysis, and modelling. Ophthalmology **125**(10), 1492–1499 (2018)
3. Wentz, B., Lazar, J.: Exploring the impact of inaccessible redesign and updates. In: Designing Around People, pp. 3–12. Springer, Cham (2016)
4. Buzzi, M. C., Buzzi, M., Leporini, B., Mori, G., Penichet, V.M.: Collaborative editing: collaboration, awareness and accessibility issues for the blind. In: OTM Confederated International Conferences "On the Move to Meaningful Internet Systems", pp. 567–573. Springer, Heidelberg, October 2014
5. Menzi-Cetin, N., Alemdağ, E., Tüzün, H., Yıldız, M.: Evaluation of a university website's usability for visually impaired students. Univ. Access Inf. Soc. **16**(1), 151–160 (2017)
6. Nir, H.L., Rimmerman, A.: Evaluation of Web content accessibility in an Israeli institution of higher education. Univ. Access Inf. Soc. **17**(3), 663–673 (2018)
7. Giakoumis, D., Kaklanis, N., Votis, K., Tzovaras, D.: Enabling user interface developers to experience accessibility limitations through visual, hearing, physical and cognitive impairment simulation. Univ. Access Inf. Soc. **13**(2), 227–248 (2014)
8. Verma, P., Singh, R.: Developing speech-based web browsers for visually impaired users. In: Speech and Language Processing for Human-Machine Communications, pp. 107–118. Springer, Singapore (2018)
9. de Oliveira, S.T., Bozo, J.V., Okimoto, M.L.L.R.: Assistive technology for people with low vision: equipment for accessibility of visual information. In: Advances in Ergonomics in Design, pp. 701–710. Springer, Cham (2016)
10. Borchard, L., Biondo, M., Kutay, S., Morck, D., Weiss, A.P.: Making journals accessible front & back: examining open journal systems at CSU Northridge. OCLC Syst. Serv. Int. Digit. libr. Perspect. **31**(1), 35–50 (2015)
11. Swersey, A.J., Colberg, J., Evans, R., Kattan, M.W., Ledolter, J., Parker, R.: Decision models for distinguishing between clinically insignificant and significant tumors in prostate cancer biopsies: an application of Bayes' Theorem to reduce costs and improve outcomes. Health Care Manage. Sci. 1–15 (2019)

Study on Product Information Coding in the Context of Universal Design

Hongxiang Shan, Xingsong Wang(✉), Mengqian Tian, and Yuliang Mao

Department of Industrial Design, School of Mechanical Engineering, Southeast University, Nanjing 211189, China
18851668026@163.com, {xswang, tianmq, MaoYL}@seu.edu.cn

Abstract. This paper studies the design method of product information coding in the context of universal design, and puts forward an effective design method of information redundancy in the context of universal design. Determine the main body information and situation information of the product, and use redundant information to assist the design of the main body information and situation information, so that the product information can be fully conveyed. By determining the design method of information redundancy in product appearance modeling, this method is validated by taking massage chair design as an example, so as to determine the guiding significance of this method for product general design. Redundant information is used to assist product information coding, which helps to reduce the noise in the channel and to supplement the description of product functions; it is more inclusive and usable to help all users complete accurate solutions; ensures the consistency of information transmission; and reduces the interference of invalid information to users.

Keywords: Universal design · Information design · Information redundancy · Product design

1 Introduction

Universal design refers to a creative design activity for all users as far as possible considering product design and environment. Universal design mainly takes humanism as its design concept, and pays attention to the principles of reasonable function, flexible use, easy to understand information and tolerance of errors in the design process. Universal design concept involves many aspects of physiology and emotion in concrete operation, and is closely related to many disciplines such as ergonomics, psychology, behavioral science and so on. According to the universal design criteria, the product should have good cognitive function: through vision, through visual, through visual, and through visual. Sensory organs, such as touch and hearing, receive various information stimuli from objects to form the whole cognition, thus producing corresponding concepts. Cognitive psychology holds that the cognitive behavior of human brain consists of a series of continuous cognitive operation stages of information acquisition, coding, storage, extraction and use, and information processing

T. Ahram et al. (Eds.): IHSED 2019, AISC 1026, pp. 501–507, 2020.
https://doi.org/10.1007/978-3-030-27928-8_76

system is carried out according to a certain procedure [2]. Therefore, universal design and information theory have certain relevance.

Information theory is a subject which studies the transmission, storage and processing of information by using probability and statistics method. It has exerted a great influence on the development of modern science and technology. With the development of information technology revolution, human beings have made a breakthrough in information understanding, information processing and information dissemination. Information design is widely used in interface interaction design and mechanical design [3–5], but little research has been done on product modeling design. In product design, information in source and channel should be designed separately [6]. As shown in Fig. 1, in the product information transmission system, the designer codes the product information and passes it to the user through the product information carrier, the user is the receiver and decoder of the product information. Source is the functional information to be transmitted by the product. When designing the function, it is necessary to take the user's demand as the starting point of the main function design of the product, reduce the invalid redundant information, emphasize the main function of the product, and use the effective redundant information to distinguish the various functions.

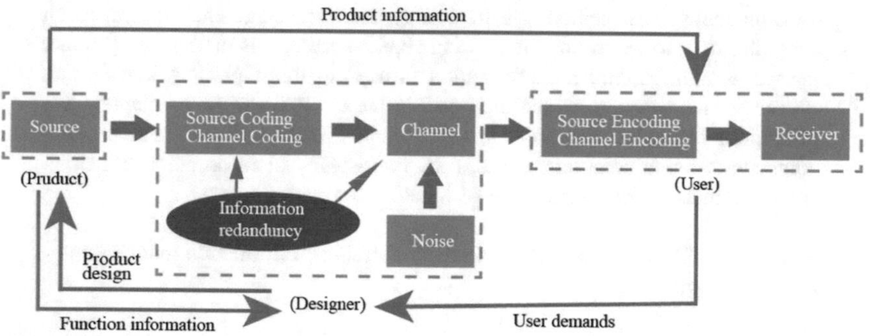

Fig. 1. Product information transfer system

2 Product Information Coding in the Context of Universal Design

2.1 Product Information Redundancy Design Method

In the formation of product information, it includes the main body information, situation information and redundant information of the product. The main body information refers to the modelling information that conveys the main functions of the product, including form, color and material [7]; the situation information refers to the environment information such as the space and state of the product used [8]; and the redundant information exists as auxiliary information to help the main body information transfer better. As shown in Fig. 2.

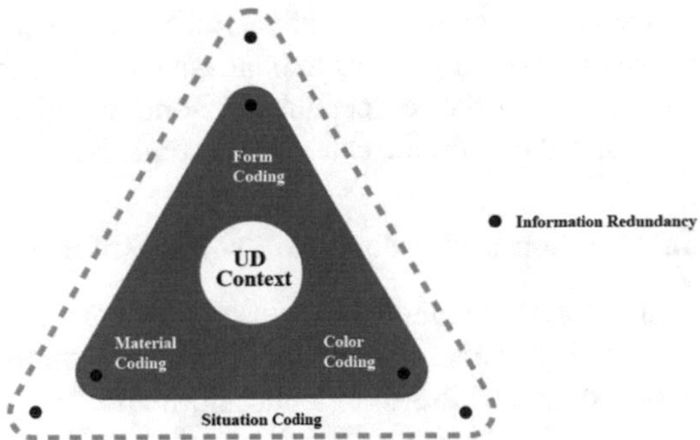

Fig. 2. Product information model in the context of UD

Redundant Information Design in Form Coding: The main function of the product determines the main form of the product. When designing several functional modules in the product, we should design around the main function. The design of redundant information in product form is mainly aimed at the possible obstacles in the use of vulnerable groups. For example, Panasonic inclined drum washing machine drum upward tilt about 30°, carefully designed easy to identify and operate buttons, whether adults, children, or disabled people, only need to reach out naturally, can be easily operated, and in the back of the washing machine fuselage, color and material design is based on the main body of the product information and situational information.

Redundant Information Design in Color Coding: Reasonable use of color can play a prominent role in highlighting product functions, while color can also bring physiological and psychological impact for people. When coding the color information, we need to take into account the functional characteristics of the product, so as to determine the main color of the product. For the redundant information in product color coding, its main function is to emphasize the function of the product and to distinguish different functional modules. Therefore, similar colors can be used to emphasize the function, and as the information part of differentiating function, it is necessary to adopt colors that contrast sharply with the main color. For example, the emergency stop buttons of machine tools are mostly red.

Redundant Information Design in Material Coding: Material needs to be selected according to the functional characteristics of the product, and redundant information can play a very good role in complementing the description. For the redundant information in product material coding, its main function is to supplement the product function, which is in line with the user's psychology, and is mainly reflected in the design of material texture. For example, bicycle handles are mostly made of rubber with texture; the bottom four corners of some important instruments are usually made of ground material.

Redundant Information Design in Situation Coding: Situation information is the external information of a product. Designers need to focus on the use of the product scenario, so that the product appearance and the surrounding environment are integrated. Redundant information in product context includes noise information which may be caused by temperature, sound, light and so on. Some special situations need to be taken into account when designing, so as to increase the safety and sustainability of the product. For example, when the temperature of some equipment is too high, the alarm light will flicker, and the operator can respond quickly.

2.2 The Role of Information Redundancy in Product Information Coding

In product design, information redundancy is used to better convey its functional information, which makes it possible for all users to receive information quickly and efficiently, complete decoding and use the product normally.

Reducing Noise and Supplementing Product Functions: Noise in the channel will cause loss of information, affect the accuracy of information transmission, affect the user's reception and decoding of complete functional information, resulting in the product cannot be used normally, and also does not conform to the universal design principle "perceptible information and tolerance for error". Redundant information can reduce noise, emphasize functional information of products, help users decode products quickly and use products successfully.

Inclusion and Availability to Help All User Complete Information Decoding: The design of information redundancy needs to be user-centered. Through the analysis of product function information, the user's needs, perceptual experience and the user's psychology and emotion when using the product are considered. The design of information redundancy fully reflects the "people-oriented" design idea, and conforms to the universal design principle "equitable and flexible use".

Ensuring Consistency in Information Delivery: The same product will have different users, because different users will have different knowledge composition. In order to effectively convey the functional information in product information design, redundant information will be designed in the design of the same functional information to ensure that the functional information can be transmitted to users in a complete and efficient manner, and ensure the transmission of functional information. Uniformity conforms to the simple and intuitive principle in universal design.

Reduce the Interference of Invalid Information to Users: In the process of user's use, it is necessary to distinguish the information of different functional modules and emphasize the main function information. The design of redundant information can help users distinguish functional information. Effective information redundancy can help users decode quickly and invalid redundant information, which seriously hinders the decoding efficiency of users and reflects the principle "tolerance for error" in universal design.

3 Application of Information Coding in Massage Chair Design

By coding the shape information of massage chair, the redundant information is applied to the product shape design to facilitate users to decode and use the product function information quickly. Massage chair can relax muscles, eliminate fatigue, and play an important role in ensuring health. Massage chair users are more extensive, there are some young who work under greater pressure and elderly groups, so it needs to be designed in accordance with the universal design concept. In the functional design of massage chair, the interference of invalid and redundant information on the main functional information should be reduced, and the product availability should be increased to meet the needs of all users. As shown in Fig. 3.

Coding Design of Product Form Information: In the design part of the massage chair, its main function is to massage some parts of the body. According to the structure, the massage chair can be divided into two parts: cabin and seat. The seat is ergonomically designed to ensure comfort. The cabin is used for mechanical structures such as engine core and air bag. In order to meet the aesthetic needs of most people, the cabin side body adopts streamlined design, which conveys the modern design concept. At the same time, the soft curved surface can avoid the harm of edges and sharp lines to users, reflecting comfort and security. Using redundant information, two cylindrical armrests are set on both sides of the seat. When the elderly and other vulnerable groups are in use, they can get up and sit down. According to ergonomics, a retractable pedal is designed under the seat. According to the length of different people's feet, a retractable pedal is set in front to increase comfort. The redundant information coding design of morphology embodies the universality of the design.

Coding Design of Product Color Information: In terms of color information coding, the streamlined cabin of massage chair uses white as the main color and red as the auxiliary color. White embodies the concept of simplified design, monochrome as the main color can also reduce visual pressure, in line with the aesthetic needs of most people. Adding red lines to the edge of the cabin can highlight the integrity and simplicity of the design. The redundant information is used to add red patterns to the armrest area of the massage chair cabin to emphasize the role of the armrest, which is easy to recognize. The redundant information coding of color reflects the concern of the elderly group and the inclusiveness of the design.

Coding Design of Product Material Information: In terms of material information coding, PVC and fabrics are used as the main materials, and the side body of the cabin is made of smooth and high hardness plastic, which is easy to clean. The seat surface is wrapped with fabric to increase comfort and air permeability. The seat is separated from the cabin body, which can be easily disassembled and replaced. Using redundant information coding, transparent black plastic material is used on the back of the fuselage, which increases the sense of science and technology and mystery of the massage chair. As you can see the fuselage interior, you can increase the new user's understanding of the function of the massage chair. The inner and outer handrails of the cabin are wrapped with uneven textures and rubber materials to increase the roughness and facilitate the grasp of the elderly. Therefore, the redundant information coding of materials fully embodies the universal design concept.

Coding Design of Product Situational Information: In product context design, two main wheels and one auxiliary wheel are used at the bottom of the massage chair to move and support. At present, massage chairs have been widely used in cinemas, waiting rooms, shopping malls and other places. These are the main situational information of massage chairs, in which most of the massage chairs are fixed and used by consumers. According to redundant information design method, massage chair can also be used in narrow space such as family and office. Because of the limitation of space, sometimes it needs to be placed in a corner after massage. Therefore, the mobile design increases the flexibility of the massage chair and can be used by more people. Thus, the research on redundant information coding for scenario information increases the flexibility and versatility of the design, which conforms to the universal design principles.

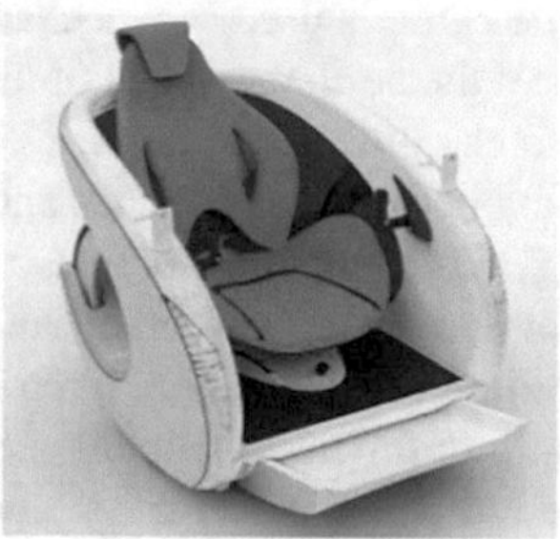

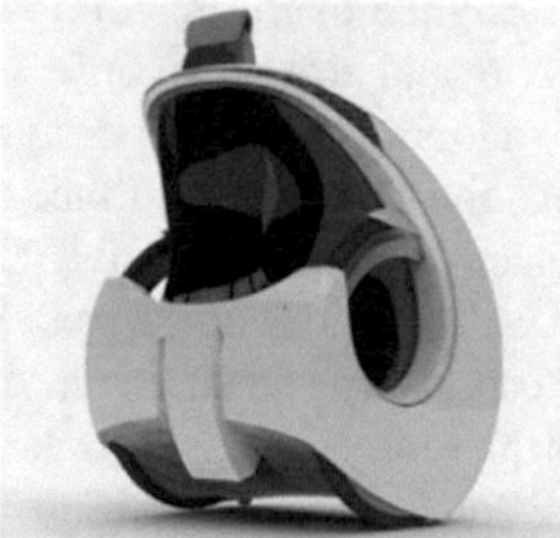

Fig. 3. Information coding of massage chair

4 Conclusion

In this paper, product information coding under the background of universal design is discussed, and the guiding significance of product information coding for universal design is proved theoretically and practically. By coding the redundant information of product shape, color, material and situation, and using the effective redundant information in product design can ensure the validity and reliability of product information transmission, and design more inclusive and humanized products. Users can decode products correctly, quickly grasp the functions and usage of products to convey, and establish a healthy and perfect information system between users and products.

References

1. Center for Accessible Housing: Accessible Environments: Toward Universal Design. North Carolina State University, Raleigh (1995)
2. Shao, J.: Research on the information encoding method of helmet mounted display system interface based on visual perception theory. Southeast University (2016)
3. Zhang, J., Xue, C.Q., Wang, J., Shen, Z.F., Zhou, L., Zhou, X.Z., Yun, L., Zhou, L.: Effects of cognitive redundancy on interface design and information visualization. Advances in Intelligent Systems and Computing, vol. 607, pp. 483–491. Springer (2018)
4. Braseth, A.O., Veland, Q., Welch, R.: Information rich display design. American Nuclear Society 4th International Topical Meeting on Nuclear Plant Instrumentation, pp. 1195–1206. American Nuclear Society (2004)

5. Li, J., Zou, X.J., Wang, H.J., Chen, Y.: A method for knowledge reasoning of mechanical product intelligent design using information entropy. Key Engineering Materials, vol. 522, pp. 313–318. Trans Tech Publications Ltd (2012)
6. Dai, X.L.: Information exchange of product. Hunan University (2002)
7. Tang, J.S.: The research on designing pleasurable products based on situation information. Hunan University (2003)
8. Tian, X.Y.: Information design research of mobile application based on “entropy” and “redundancy”. Design (2017)

Facilitating Storytelling and Preservation of Mementos for the Elderly Through Tangible Interface

Cun Li(✉), Jun Hu, Bart Hengeveld, and Caroline Hummels

Eindhoven University of Technology, Eindhoven, The Netherlands
{Cun.Li,J.Hu,B.J.hengeveld,C.c.m.Hummels}@tue.nl

Abstract. Mementos act as emotional companions that anchor stories. Current related research is not applicable for the elderly, as they focus more on digital mementos, while the elderly's mementos are normally physical. Additionally, digital devices supporting capture and recording are generally inaccessible for them. The above makes elderly's storytelling and preservation of mementos are still problematic. In response to this, a research prototype named Slots-Memento was designed and implemented. In the field study, six families were recruited. Semi-structured interviews and stories collected were transcribed, and thematic analysis was conducted, which form the foundation of insights on the research questions.

Keywords: Elderly · Memento · Storytelling · Tangible interface

1 Introduction

A memento is an object given or deliberately kept as a reminder of a person, place or event. It is directly meaningful to the owner's memories. The elderly generally have an abundant knowledge of family mementos, which could support family story reminiscence. Memento stories told by the elderly create meaning beyond the individual and facilitate positive identity [1]. Recalling memories of mementos also improves psychological well-being and helps older adults find meaning in their life. People hope they could be remembered. When death occurs, survivors are left with bundles of images, materials, objects, and wishes of the deceased [2]. However, memento sharing is problematic for the elderly. In response to this, we conduct a study, driven by *RQ1: What are the stories behind the elderly's mementos?* And *RQ2: How to promote the elderly to tell and these stories?* Our two contributions are: a tangible interface promoting elderly's memento storytelling, and related design considerations.

2 Research Prototype

Slots-Memento Prototype. We based the design of Slots-Memento on our previous work [3, 4]. Its design process included interview study, brainstorm, sketch and mock-up, and user consultation. It draws inspirations from the slots-machine, integrating

T. Ahram et al. (Eds.): IHSED 2019, AISC 1026, pp. 508–514, 2020.
https://doi.org/10.1007/978-3-030-27928-8_77

functions of memento photo displaying and story audio recording (Fig. 1). It includes of a tangible device and a USB flash disk, the former is operated by the elderly, and the letter is used by the young to copy memento photos into it. It is operated by one lever on its right side: when pulling the lever, a memento photo is displayed in a similar fashion as slots-machine. Raspberry Pi is the hardware platform.

Usage Scenario. To provide contextual richness and allow participants to understand the idea of Slots-Memento better, we further develop the following three usage scenarios: Share memento stories over a distance (Fig. 2①, ②, ③ and ④). Second, share memento stories face to face (Fig. 2⑤). Third, modify memento photos (Fig. 2⑥) (Fig. 3).

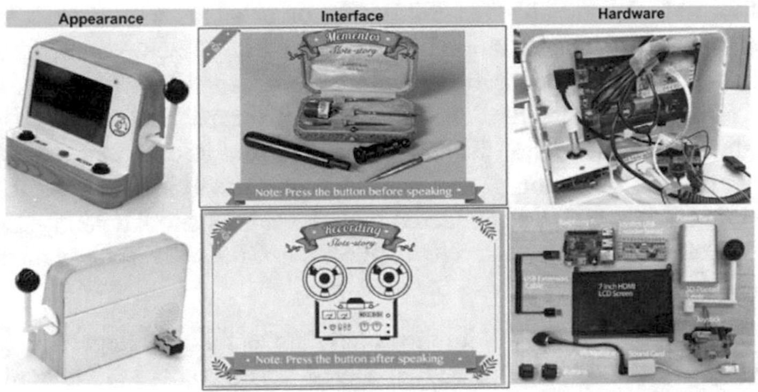

Fig. 1. Appearance of Slots-Memento, display interfaces, and hardware

Fig. 2. Usage scenario of Slots-Memento

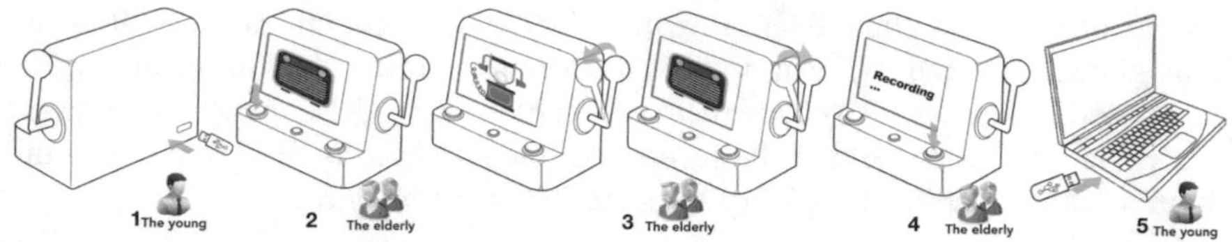

Fig. 3. Operating process of Slots-Memento

3 Field Study

Participants and Method. In field study, we recruited six pairs of participants (each consists of an old adult and his/her child). The older adults were from a local Dutch nursing home, and their ages ranged from 74 to 81. They were recruited through putting up posters in the nursing home, and recommended by caregivers. Firstly, we introduced our purpose to them, and they signed formal consent forms. Prototype together with a paper instruction for use was distributed to each pair, and we showed them the prototype's operation procedure. Each pair used the prototype for around a week. After that, interview was conducted with the elderly and the young. Handwritten notes were taken to aid analysis. We applied the Grounded Theory method [5] to analyze the data.

Findings. Findings of interview with the elderly are as follows: Regarding the validity of the prototype. The results indicated that Slots-Memento could promote the elderly's memento storytelling, and it was accepted by them. It also encouraged them to revisit and rediscovery their mementos. The reminiscence helped them to remember what they almost forgot. Additionally, audio recording lowers the cost of narrative for them, as they felt difficulty in writing. Regarding their preferences for the mementos, most preferred to talk about photos, especially which of family members and old friends: *"When I look at the old photos, I could see what I was like, it was different from today. The old photos bring me back to the time when I was young." "Some of the people on the photos were no longer alive. Every time when I see them, I could go back to that time."—P5, F* Second were artifacts received from others, such as inherited from parents, and from friends. While for the souvenirs gathered from travelling, the elderly just remembered where they bought, they already forgot the reason and situation of buying them: *"Maybe I had lots of reasons of buying them, but now I forgot, they were now just for decoration."* Regarding interaction with the prototype, metaphor of the slots-machine was accepted. They felt it was easy to understand and operate. They also felt recording stories was more labor-saving, compared with handwriting stories. Regarding comments for improvements, they suggested it should be able to keep running, instead of manual power on/off each time. Other suggestions included the display screen should be bigger, and adding functions of playing old videos, etc.

Findings of interview with their children are as follows: First, they learned new things of the elderly. Although they had known some of the family artifacts, they still felt that they learned new things from a different angle: *"I didn't know that pipe was from my grandfather until grandmother told me in the audio." "I think some of the photos, I have seen a long time ago. Now she tells the memories totally from her perspective."* Second, raising their awareness of preserving Mementos. Currently, they didn't have many chances to sit together with the elderly to listen to their stories. *"Occasionally my mother talked about her past stuff. This device makes me aware of the importance of preserving them. Otherwise, these memories will disappear forever."* Third, the recordings were treasure to pass on to next generation. They thought preservation of the stories was meaningful: *"The only way to know my grandmother's life is to ask my father. I will never have a chance to listen to my grandmother by herself because she has passed away. This device could solve this."* Finally, audio contains emotions and familiarity to them. The young found familiarity in the audios. Another advantage of using audio lies in conveying much more information than text, such as emotions, personalities, and feelings. The younger generation's preferences for memento stories were varied. Different young participant had different interests. The young also thought when they used prototype face-to-face with the elderly, it could prompt conversation topics. Regarding comments for Improvements. The flash disk transfer was not convenient, wireless network should be adopted.

Analysis of Mementos. The mementos were generalized into three main categories (**Artifact**, **Paper document**, and **Photo**) and 25 sub-categories. The proportion of each memento was calculated, helping us to understand their preference for mementos from an objective perspective. Mementos they talked the most were Photo (54.8%), among which "Family member" was the most, which was consistent with the result of interview with the elderly: old photos of family members and old friends could remind them of childhood memory. Followed by "Marriage", which was major life event. Next were friends, festival. 26.4% were regarding Artifact, among which the most was about "Gift from friends", which could remind the elderly of the relationship with friends. Followed by "Inherited from parents", which were normally very precious. Next was "Travelling souvenir", although they had quantity of travelling souvenirs, they didn't talk too much, the reason could be explained by the interviews with the elderly. 18.8% were related to Paper document, among which postcards and letters were the most. These kinds of mementos also represented relationships with the senders. Followed by certificates, including graduation and qualification certificates, which were normally the elderly were proud of.

Analysis of Stories. Transcription conventions and guidelines were based on Robert's method [6]. In the following section, we discuss different stories behind different mementos.

Artifact. The stories of artifacts normally start with a description of the artifact, such as what it is, what it is for, and where it is from. If it is from someone else, then the stories turn to describe the giver: such as who the giver is, stories happened between the elderly and the giver. In this case, the artifact reminds the elderly of more the relationship than the artifact itself. But for the souvenirs from travelling, stories behind were not as detailed as the above artifacts. As indicated in the interview, the elderly seemingly had no interests in telling them, despite the large quantity number of souvenirs they collected. There are also self-made artifacts. One participant is good at handicraft, and still makes handmade dolls. Stories of the self-made artifacts normally contain an introduction of how she made it. These kinds of skill instruction were also intangible treasure worth preserving. Second, **Paper documents**. The most typical paper documents in our field study are postcards, letters, drawing, map, certificate. They were normally also related to relationships with senders. Therefore, stories contain the memories of that person. Third, **Photo.** Photos were the elderly talked the most in our field study. For the photos of family members, stories contain introductions of people appearing on photo, time and location of the photo. In this case, date-stamps on photo are an important clue facilitating reminiscence. While for the photos of festivals, activities and major life events, stories normally start with a short description of that event, such as place and time of it, and the persons appeared in the photos. Then the stories continue with the scene at that time, and other memories the elderly could recall. For the photos of scenery, the photos could remind the elderly the situation of taking the photo, and an introduction of that scenery spots. Additionally, there are some photos capturing interesting moments (Table 1).

Table 1. Example stories behind the mementos

A. Inherited from parents
 *"This is an old bark, for smoking, from my father. In early days, the fabrics. they smoke, they put the fabric in it. this is from my father, very old. Uh hum ,but he didn't smoke, my father, he just bought it, and never used it so it is totally new." —P6, F*
B. Postcards
 *"This is postcard from India, you see the stamps, that's from a guy. Uh hum. He was here, he has leaved, he visited me for the haircut. once he said he went to Texas, so he was called "the Texas boy", ha-ha. and he said: before I leave, I visit you for a haircut, and you can do what you want. ha-ha. I said...then I was waiting, waiting, waiting, but he didn't apparat. and that's why he said: "sorry. I couldn't make it". –P3, M*
C. Scenery
 *"We went there by car, because it was far away. This is in Italy, the town was down in the water. Looking down it was so nice, so I took this photo. This town, I forgot its name, the town sank in the water, and people, in this town, they made up the town again. and they said people come out of town, then put water over there."—P5, F*

4 Discussion

Mementos of the elderly are normally in physical format, which are hard to preserve. Moreover, the mementos take up space, which leads to the elderly store them in hidden places, where out of the elderly's sights. One of the design guidelines for digital mementos is to integrate the digital content fully into the daily environment at home. Based on our findings, it also applies to the elderly's physical mementos. Our first strategy is integrating them into home environment by an unobtrusive tangible device. Therefore, the digitalization of physical mementos is necessary, contributing to memento preservation as well. Although the elderly have abundant knowledge of their mementos, they are lack of technology skills. Our second strategy is intergenerational collaboration: the young could play roles in digitalizing the mementos, while the elderly are the story producers. Intergenerational cooperation also brought communication. From the angle of design, our prototype is an interactive display employing tangible interface. The physicality of the tangible interface conveys advantages over conventional graphical interfaces in terms of its support for real-world skills, natural affordances, learning and memorization [7]. Slots-Memento aims to integrate existing operation that the elderly are familiar with, into a novel device, through employing the metaphor of slots-machine. The field study indicated that it was well understood and accepted by the participants. What's more, tangible device could also act as a physical reminder to encourage the elderly's storytelling. Adding external physical reminders to people's environments is an effective strategy to keep to their resolutions and remind them of their goals. Tangible materials could produce deeper engagement than the digital materials.

5 Conclusion and Limitation

In this paper, we report the implementation of a tangible storytelling system facilitating storytelling and preservation of mementos for the elderly. Insights on *RQ1* includes quantitative results, summarizing types of mementos the elderly kept, and their preferences for mementos. And qualitative results, summarizing different stories behind different mementos through specific story examples, Insights on *RQ2* could be concluded as: Integrating physical mementos into daily life through tangible device. Tangible interface employing metaphor makes technology accessible for the elderly, which also acts as a physical reminder to encourage the elderly to tell stories. Using audio recording could make the memento storytelling more labor-saving for the elderly, audio also contains familiarity and emotions of the elderly. Intergenerational cooperation promotes intergenerational communication, as well as contributes to the sustainability of memento story sharing. As pointed by the participants, organizing, digitalizing, and storytelling of mementos is a long-term job. Therefore, mementos and their stories collected in our field study were limited, the usage time of prototype was also limited.

References

1. Fivush, R.: Intergenerational narratives: how collective family stories Relate to adolescents' emotional well-being. Aurora. Revista de Arte, Mídia e Política 51 (2011). ISSN 1982-6672
2. Unruh, D.R.: Death and personal history: strategies of identity preservation. Soc. Probl. **30**, 340–351 (1983). https://doi.org/10.2307/800358

3. Li, C., Hu, J., Hengeveld, B., Hummels, C.: Slots-memento: facilitating intergenerational memento storytelling and preservation for the elderly. In: Proceedings of the Thirteenth International Conference on Tangible, Embedded, and Embodied Interaction. pp. 359–366. ACM, New York (2019). https://doi.org/10.1145/3294109.3300979
4. Li, C.: Designing a system to facilitate intergenerational story sharing and preservation for older adults. In: Proceedings of the Thirteenth International Conference on Tangible, Embedded, and Embodied Interaction, pp. 737–740. ACM (2019)
5. Corbin, J., Strauss, A., et al.: Basics of qualitative research: Techniques and procedures for developing grounded theory (2008)
6. Miller, R.L.: Researching Life Stories and Family Histories. Sage (1999)
7. Ishii, H.: Tangible bits: beyond pixels. In: Proceedings of the 2nd international conference on Tangible and embedded interaction, pp. xv–xxv. ACM (2008)

Safety Engineering and Systems Complexity

A Systemic Approach for Early Warning in Crisis Prevention and Management

Achim Kuwertz[1], Maximilian Moll[2], Jennifer Sander[1(✉)], and Stefan Pickl[2]

[1] Fraunhofer IOSB, Institute of Optronics, System Technologies and Image Exploitation, Karlsruhe, Germany
{Achim.Kuwertz,Jennifer.Sander}@iosb.fraunhofer.de

[2] Universität der Bundeswehr München, Werner-Heisenberg-Weg 39, 85579 Neubiberg, Germany
{Maximilian.Moll,Stefan.Pickl}@unibw.de

Abstract. Given the importance of early warning in crisis prevention this paper discusses both knowledge-based and data-driven approaches. Traditional knowledge-based methods are often of limited suitability for use in crisis prevention and management, since they typically use a model which has been designed in advance. Novel data-driven Artificial Intelligence (AI) methods such as Deep Learning demonstrate promising skills to learn implicitly from data alone, but require significant computing capacities and a large amount of annotated, high-quality training data. This paper addresses research results on concepts and methods that may serve as building blocks for realizing a decision support tool based on hybrid AI methods, which combine knowledge-based and data-driven methods in a dynamic way and provide an adaptable solution to mitigate the downsides of each individual approach.

Keywords: Early warning · Expert knowledge models · Deep Learning

1 Introduction

Early warning plays a vital role in effective crisis prevention as well as in crisis management, in cases where prevention is not (or only partially) possible. Of crucial importance is a timely detection of an emerging crisis in a reliable manner, where timeliness depends on the nature and context of the specific crisis. Both, timeliness and reliability in crisis detection can be increased by basing the implementation of early warning systems for crises on a methodic systemic approach.

In (crisis) early warning, data and information is collected, analyzed and interpreted with the aim of detecting and informing about potentially developing crises. To this end, it is most often necessary to apply quantitative as well as qualitative methods for data analysis and interpretation. In [1], a multi-step process for Crisis Early Detection (CED) is described, including the stages of data acquisition, analysis (filtering, trends etc.), interpretation (common factors, correlations etc.), prospection of future scenarios (foresighting), and, as an output, deriving indicators for observation and surveillance in early warning systems. Within this process, quantitative methods (e.g., mathematical

T. Ahram et al. (Eds.): IHSED 2019, AISC 1026, pp. 517–522, 2020.
https://doi.org/10.1007/978-3-030-27928-8_78

models) as well as qualitative methods (e.g., based on expert knowledge and human experience) can be employed in differing shades for the different stages.

The scenarios and indicators developed during this process constitute some form of explicated expert knowledge, which is the basis for developing formal scenario models for employment in knowledge-based early warning systems. These expert systems allow automating support to CED, e.g., processing of larger amounts of current information possibly containing evidence for crisis indicators.

However, it is often difficult to foresee all the relevant aspects of potential crisis situations. While similarities between past and current crises (and their emergence) may exist, it is unlikely that crises are ever identical. To improve the adequacy of models (semantic as well as mathematical) and to increase their generality (e.g., in the light of changing circumstances), the employment of current data and information to adapt these models might be beneficial. Furthermore, additional parts of the CED process could be automated using quantitative data-driven approaches and techniques from active learning.

Knowledge-based methods (Sect. 2) have the drawback of being static and requiring a sophisticated engineering effort for model creation. Data-driven approaches (e.g., based on Deep Learning; Sect. 3), require huge amounts of (annotated) data and may suffer from transparency issues regarding their decisions. As an answer to those drawbacks, hybrid AI methods (Sect. 4) may be considered, as they allow combining knowledge-based and data-driven methods in a dynamic way and provide an adaptable solution to mitigate the downsides of each individual approach.

2 Knowledge-Based Methods

Knowledge-based early warning systems can be employed in CED for recognizing and evaluating defined indicators of developing crisis situations. Here, the foreseeable different possibilities of how a crisis can evolve are each described by a scenario developed during the CED process. In this paper, it is assumed that these scenarios can at least partially be represented by formal models.

In combination with approaches for the extraction of relevant information from (various) data sources, such models can be employed in expert systems in order to support and partially automate CED based on quantitative methods. The information processing chain for such systems consists of accessing respective data sources for current data, extracting relevant information from these, correlating the information with the defined indicators for potential crisis scenarios, and, finally, assessing and evaluating the probability of existence for each scenario. At the heart of this approach for a continuous observation and early warning for crises are the formal models of crisis scenarios and indicators derived in the CED process. These models allow to correlate the extracted information in a controlled manner and, ideally, derive quantitative statements about probably unfolding or emerging crisis, together with statements of confidence and traceability information for such conclusions.

A formal model in this approach is based on elements such as concepts, relations, attributes and rules. Concepts prototypically represent real-world entities relevant to the crisis scenarios, such as objects, actors, areas, locations etc. They are detailed by attributes (describing features, properties etc. of a concept). The concepts of the model are instantiated within the processing chain from information suited as evidence for their existence of respective real-world entities. Attribute values for these instances are also derived from such information. Relations in the model can be defined either on concept-level (connecting two concepts for example by a subclass relation) or on instance-level, connecting two concept instances (e.g., actor A is associated with group G). The latter case of (the existence of) a relation is also derived from respective evidence information.

Concepts, relations and attributes form the basic level of the model for CED. On their basis, crisis indicators can be defined for the developed crisis scenarios as well as rules e.g. for drawing conclusions. Using such rules (defined based on model instances), crisis indicators can be connected with the developed scenarios. To a certain degree, it becomes possible to derive automatically probabilistic statements about the existence of a scenario given evidence information. An example of a knowledge-based system aimed at supporting the protection of critical infrastructure is described in [2]. This approach, based on a model of relevant elements in the environment of a critical infrastructure, uses rules in combination with probabilistic methods for performing a quantitative threat assessment.

Traditional knowledge-based systems often assume a closed world in their models. This is due to the fact that a model has to be created prior to operating a knowledge-based system, and thus, only those concepts, relations and attributes considered most relevant (e.g., during the CED process) will be included in the model. For CED, the relevance of model elements could for example be based on experience with previous crises. However, while similarities between past and current crises (and their emergence) may certainly exist, it has to be assumed that they are never identical. In addition, as models for CED are usually created by human experts in a time-consuming process (the CED process), they need to be focused on the (a priori) most relevant aspects. In consequence, there is always a possibility of such models being incomplete and insufficient for their intended application (e.g., after a while of system operation), especially when these applications consider dynamic domains. For knowledge-based early warning systems, thus, an approach for detection when the underlying models become insufficient and for performing or proposing model adaptation in such cases would provide a significant benefit.

Indications for the necessity of model adaptations can for example be given in the information fed currently into the system. A possibility for quantifying the suitability of the models employed by an early warning system with respect to the currently processed information is given by the approach of adaptive world modeling [3]. This approach rates the quality of an employed model by its ability to represent all the currently processed information (i.e., instantiate respective concepts contained in the model) and quantitatively detecting if there is any information (especially about entities) that can only be poorly represented by the model. Such information is accounted for by performing bookkeeping, and in the case when the occurrence rate of such information exceeds pre-defined thresholds (or, in other words, if the ability of the model for representing such information degrades too far), measures for model improvement are triggered. These measures include clustering the poorly represented information in a way such that it (i.e., the described entities) can be employed as training samples for finding new concept definitions. These new concept definitions then allow improving the model quality, again, when added to the model.

For CED, such newly acquired concepts can first be presented to human experts as proposed model extensions. After expert verification (and possibly adjustments), the verified concepts can be added to the model. Verification and model extensions are steps already contained in the CED process, as part of feedback loops for adjusting and improving on the initially developed crisis scenarios and their indicators. This step is intended at improving the predictive qualities for CED.

Besides adapting and extending the concepts contained in models for CED according to adaptive world modeling, it might also become necessary to adapt further models elements such as relations or rules to changing circumstances. Here, also data-driven approaches might be beneficial, e.g., for uncovering correlations contained in observed CED data.

3 Data-Driven Approaches

Data science in general can be mainly divided in descriptive, predictive and prescriptive analytics. Descriptive analytics tries to infer information about the past from collected data and hence tries to answer the question "what happened?". Predictive analytics, however, extends this and uses past data to obtain information about the future, in other words asks "what will happen next?". Prescriptive analytics, finally, takes a further step, by identifying actions required to reach a goal or situation of interest in the future. In the context of early warning, predictive algorithms are clearly of most interest. However, it can also be reasoned, that descriptive analytics plays an important role in understanding previous incidences.

One of the most prominent classes of data-driven algorithms in recent years has been Deep Learning, which has been used for early warning in various situations, in which the disaster is man-made, like stock market crisis [4], anthrax [5] or bomb [6] attacks, crowd disaster [7], voltage instability [8] or power outages [9]. However, a very important application is for crisis with natural causes due to the increased amount of uncertainty. One of the challenges here is often understanding data in various forms, like videos [10] or meteorological data [11] and draw fast and reliable conclusions for the near-term future. A particularly challenging area is earthquake early warning, since it has to work very rapidly and reliably to avoid huge losses [12]. Several different types of neural networks can be used here to distinguish between impulsive signals and noise [13] and to determine the distance to the source [14].

As can be seen, Deep Learning is most prominently being used to increase a systems capability to process and interpret a wide variety of input types. Furthermore, due to the online learning properties of neural networks, such systems can be continuously improved as new data becomes available. However, a significant amount of training data is required initially, which is often not available, depending on the exact type of crisis.

4 Opportunities and Challenges for a Hybrid Approach

As in every hybrid approach, each methodology should help to reduce the shortcomings of the other. In the context presented here, we want to focus on the fixed boundaries of the world model in expert systems and the lack of sufficient data for data-driven approaches. To this end, we suggest to change the typical model generation process: while it should still be initially generated by experts, it should be extended by data-driven methods. This will require a very close human-machine interaction.

To achieve this, we intend to take concepts from active learning. In this area of semi-supervised learning, algorithms usually start with unlabeled data for a supervised task and inquire labels for those data points that give the highest gain, thereby minimizing the amount of labeled data needed. In the context here, such methodology would be used to identify missing elements in the model and ask for expert opinions where necessary. This could alleviate the need for extensive historical data. At the same time, it focusses experts on the most important areas. Moreover, the system can also start to learn from those expert judgements and once enough confidence is gained, can shift the human attention to a new topic of importance.

Such an approach also aids in the interpretability of the resulting data driven models. While the field of explainable AI (XAI) has gained momentum in the last few years, it is still in its infancy, leaving most AI-based models in a black box. However, it is doubtful whether such models will be trusted in important applications like early warning. Starting from a human-build model and having a close machine-human interaction during its extension can make sure that enough interpretability is preserved.

While such a system provides many advantages, it comes with significant challenges. The more obvious ones relate to the human-machine interaction and communication, which needs to be made as intuitive and flexible as possible, since the experts cannot be expected to be necessarily experienced at using AI-based algorithms. More importantly, the general framework needs to be independent of the precise scenario being analyzed. However, considering that crises can range from natural disasters to attacks or political uprising, it is not immediately obvious whether and how a general process can be established. On the algorithmic side, care needs to be taken to make input and decisions as flexible as possible to be highly adjustable to each situation.

5 Conclusion

In this paper we contrasted qualitative and quantitative approaches to early warning in crisis situations. We discussed the benefits of a hybrid approach, which will have to draw on techniques from active learning as well as XAI.

To conclude this exposition, it needs to be stressed that an absolutely reliable prediction of crises will never be possible in general. Therefore, a realistically usable system for early warning has to take this fact into account. This implies that suitable means for user interaction have to be considered as well. Furthermore, extra care needs to be taken, that result presentation avoids over-confidence of decision makers and leads to adequate interpretation.

Acknowledgments. Inputs to the paper from Marian Sorin Nistor are gratefully acknowledged.

References

1. Roth, F., Herzog, M.: Strategische Krisenfrüherkennung – Instrumente, Möglichkeiten und Grenzen (Strategic Crisis Detection: Instruments, Possibilities and Limits). Zeitschrift für Außen- und Sicherheitspolitik **9**(2), 201–211 (2016)
2. Kuwertz, A., Mühlenberg, D., Sander, J., Müller, W.: Applying knowledge-based reasoning for information fusion in intelligence, surveillance, and reconnaissance. In: Multisensor Fusion and Integration in the Wake of Big Data, Deep Learning and Cyber Physical System, LNEE 501, pp. 119–139. Springer (2018)
3. Kuwertz, A., Beyerer, J.: Extending adaptive world modeling by identifying and handling insufficient knowledge models. J. Appl. Logic **19**(2), 102–127 (2016)
4. Chatzis, S.P., Siakoulis, V., Petropoulos, A., Stavroulakis, E., Vlachogiannakis, N.: Forecasting stock market crisis events using deep and statistical machine learning techniques. Expert Syst. Appl. **112**(1), 353–371 (2018)
5. Jo, Y., Park, S., Jung, J., Yoon, J., Joo, H., Kim, M.-H., Kang, S.-J., Choi, M.C., Lee, S.Y., Park, Y.: Holographic deep learning for rapid optical screening of anthrax spores. Sci. Adv. **3**(8), e1700606 (2017)
6. Zsifkovits, M., Moll, M., Pham, T.S., Pickl, S.W.: A visual approach to data fusion in sensor networks. In: Proceedings of the International Conference on Security Management (2017)
7. Nagananthini, C., Yogameena, B.: Crowd Disaster Avoidance System (CDAS) by deep learning using eXtended Center Symmetric Local Binary Pattern (XCS-LBP) texture features. In: International Conference on Computer Vision and Image Processing, pp. 487–498 (2017)
8. Zhang, W., Fu, S., Diao, Y., Sheng, W., Jia, D.: A situation awareness and early warning method for voltage instability risk. In: China International Conference on Electricity Distribution, pp. 1010–1014 (2018)
9. Khediri, A.: Deep-belief network based prediction model for power outage in smart grid. In: 4th International Conference of Computing for Engineering and Sciences (2018)
10. Lohumi, K., Roy, S.: Automatic detection of flood severity level from flood videos using deep learning. In: 5th International Conference on Information and Communication Technologies for Disaster Management (2018)
11. Huang, L., Xiang, L.-Y.: Method for meteorological early warning of precipitation-induced landslides based on deep neural network. Neural Process. Lett. **48**(2), 1243–1260 (2018)
12. Sihombing, F., Torbol, M.: Machine learning implementation for a rapid earthquake early warning system. In: 6th International Symposium on Life-Cycle Civil Engineering (2018)
13. Meier, M.A., Ross, Z.E., Ramachandran, A., Balakrishna, A., Nair, S., Kundzicz, P., Li, Z., Andrews, J., Hauksson, E., Yue, Y.: Reliable real-time seismic signal/noise discrimination with machine learning. J. Geophys. Res. Solid Earth **124**(1), 788–800 (2019)
14. Kuyuk, H.S., Susumu, O.: Real-time Classification of Earthquake Using Deep Learning. Complex Adaptive Systems Conference with Theme: Cyber Physical Systems and Deep Learning, pp. 298–305 (2018)

A Model-Driven Decision Support System for Aid in a Natural Disaster

Juan Sepulveda[1(✉)] and Jessica Bull[1,2]

[1] Universidad de Santiago de Chile, Santiago, Chile
juan.sepulveda@usach.cl
[2] Universidad Austral, Puerto Montt, Chile
jessicabull@uach.cl

Abstract. This article deals with the architecture of a support system for helping decision makers in optimizing aid distribution during natural disaster situations. In such type of events, one of the critical tasks is the transportation of staff, people, food and medicines. Given the complexity of the operations scheduling, the article proposes a model-driven decision support system with an embedded module for solving a vehicle routing problem.

Keywords: Natural disasters logistics · Decision support systems · Routing

1 Introduction

Natural disasters such as earthquakes, tsunamis, hurricanes, and volcanoes, are phenomena that significantly affect the world's population; these events can be very destructive and in many cases bring a large number of deaths [1]. Between 2004 and 2013, 830 disasters producing damages for 190 billion dollars were registered. In the period 2002–2011, the annual average was 107,000 casualties and 268 million were victims [2]. The trend of natural disasters occurrence shows in general an increase in frequency, although it can be observed a stabilization in the last years in the number of events, the magnitude of damages and the number of deaths. For instance, 342 disasters were recorded in 2016, somewhat below the 2006–2015 annual average of 376.4 [3]. Regarding the number of casualties caused by these events, in 2016 were recorded 8,733, the second lowest level since 2006, and largely under the 2006–2015 annual average (69,827). The economic damages in 2016 were estimated in US$ 153,93 billion, the fifth costliest since 2006. By different viewpoints, research and development in decision support for dealing with disaster management and prevention can be of great value. Each disaster occurrence is different and presents various cyclical stages; however, a majority of experts coincide on five phases: preparation, ocurrence, response, reconstruction of infrastructure, and mitigation [4]. The decision support system (DSS) addressed in this work concentrates on the preparation and response stages, and more specifically in the issue of aid delivery to communities affected by the emergency.

For the preparation and response stages, tools such as decision support systems play a key role in the design of resilient supply chains, in particular to evaluate response capacity of the logistic system either in a planning stage or during a disaster.

T. Ahram et al. (Eds.): IHSED 2019, AISC 1026, pp. 523–528, 2020.
https://doi.org/10.1007/978-3-030-27928-8_79

The role of a DSS is to tackle with the limitations of humans when dealing with many dynamic variables in extreme situations and help in the increase of their situational awareness in complex scenarios [5]. An example of this is assignment of tasks to the force in charge of the relief operations in very short periods of time. A feature of a DSS is to store descriptive knowledge and procedural knowledge referring the problem domain and its work with inference engines and communication facilities. A DSS also relies on database systems and knowledge bases, and finally it can take advantage of technologies of the Web 2.0 generation. Decision support systems can be organized into five types depending on the type of help they provide: (a) document-driven (search and retrieval of key documents such as maps or pieces of information), (b) communication-driven (use of network and communications technologies to facilitate collaboration and communication), (c) data-driven (time-series analysis of critical variables and prediction), (d) model-driven (analytics and optimization tools to advice actions), and, (e) knowledge-driven (specialized knowledge storage and elicitation in a given domain).

In this paper, the design of a model-driven decision support system for planning aid distribution during the early response stage is presented. The DSS architecture is based on three main sub-systems: (i) a mathematical model of the logistic system involved in the operation; (ii) a database of the logistics system with resource and geographical data; and, (iii) a GUI for users interaction and model management (objective selection and/or prioritization, constraints relaxation, parameters setting, among others).

2 Main Considerations for Aid Delivery Decisions

The main considerations for routing decision making for delivery of basic supplies in the response preparation stage require, first of all, a definition of response capacity. Experts agree on a definition based on 3 indicators: response time, coverage and relevance. That is, a rapid delivery is needed that covers the entire affected population and where delivery is made according to the needs of those affected. Moreover, there are three factors to be taken into account.

The first factor to evaluate a situation, is the variables associated with the aid delivery that must be identified during the planning stage:

(a) Risks associated with the planning area (identification of possible disasters)
(b) Areas of possible affectation
(c) Demand for basic needs: food, water, hygiene and shelter (quantity and type)
(d) Identification of collection centers and delivery areas
(e) Available capacity (transport, organizations involved and communications system)
(f) Demographic description of the zones with possible affectation

Second, a cadastre of air and land routes (including possible route cuts according to previous events). The third factor, is the characterization of available vehicles in case of an emergency (load capacity, performance, optimal load, among others). It is worth mentioning that different approaches are defined in the supply chain for transportation requirements, depending on requirements; for instance, delivery for basic needs, transfer of injured people, or evacuation of critical zones. Each of these approaches must be considered when defining transport needs; each, will have different interpretation of response capacity (response time, coverage and relevance).

3 Decision Support Systems for Natural Disasters Management

Decision support systems are an important aid for decision making in natural disaster management. Their proper design will allow the use of different types of models in an actual setting facilitating tasks for decision makers involved in planning and operating relief tasks. In the literature, DSSs have been used for almost three decades in many fields, such as natural resources protection [6] where the visual aspects and the functionality of the system plays a key role for their right use. Reference [7] proposes the integration of fuzzy logic into a medical assistant to make better decisions in diagnosis. In summary, the role of a DSS is "to relax the limits and constraints of the human decision-maker in making and taking a decision" [8]. DSS users are individuals or groups, such as the chief officer and the task force in charge of the relief operation in the emergency scenario.

Architecture. The system architecture is based mostly on three interconnected sub-systems, as shown in Fig. 1 (left-hand side). Three zones can be distinguished, each representing specialized sub-systems where the arrows represent the direction of the queries and responses. There is a database management system, the user interface, the computational solving engine (CPU) and the network connection. Through this scheme, the system receives historical data as well as recently occurred events to be considered during operations planning.

The user interface is the section where all the options and capabilities of the system are displayed, as shown in Fig. 1 (right-hand side).

4 Mathematical Model Embedded in the DSS

The system operates a deterministic model for a vehicle routing problem (VRP), which combines three extensions of the classic problem: a) HVRP (heterogeneous fleet), the SDVRP (split-delivery) and the MTVRP (multi-travel). Moreover, new features were included to adapt the formulation to a situation of humanitarian aid; for instance, the type of objective function to be optimized. The formulations implemented for the HVRP is based on [10], SDVRP on [11, 12] and MTVRP on [13]. The detailed model is presented in a previous work of one of the authors [14]. The model is of the multi-objective type, since it considers a first term with the total time for all vehicles and all of the tours (MTVRP) which allows to find the best routes to get the vehicles back to the depot, as quickly as possible, to be re-used; this also helps increase vehicle autonomy when fuel is scarce and encourages to use vehicles as long as there are units available at the depot, reducing the time in which each point of demand is attended. The second term considers the arrival times of each vehicle to the demand points which is also to be minimized. This translates into reducing the time in which the demand of

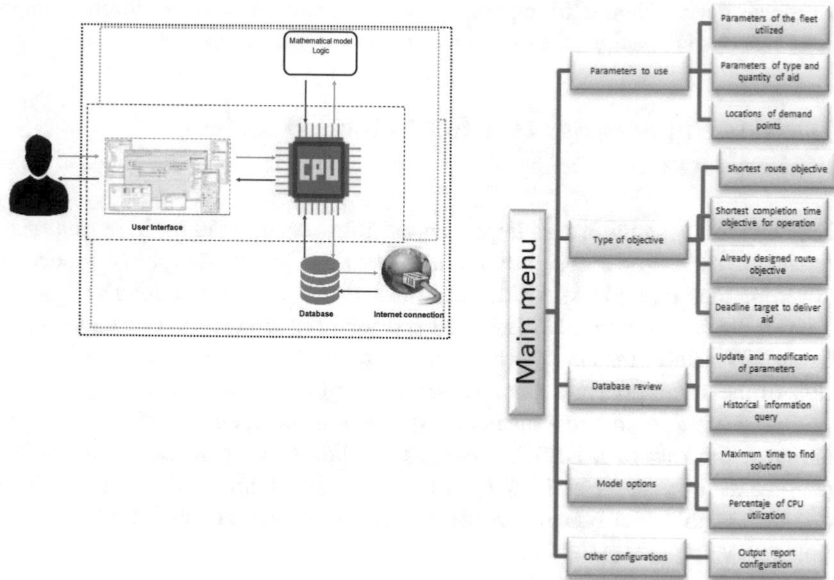

Fig. 1. System architecture and functionalities

each node is completed. Constants α and β are chosen to prioritize each of the objectives, such that $\alpha + \beta = 1$.

5 Case of the 27F Earthquake

On Saturday, February 27, 2010, at 3:35 AM, an earthquake measuring 8.8 on the Richter scale and subsequently a tsunami affected the coast of central Chile from the Valparaíso region to the Araucanía (aprox. 600 kms.). The event registered over 300 replicas over 4.9 on the Richter scale; 11,850 military personnel were mobilized to the affected areas; aircrafts were used for search and rescue, link, logistics transport and aeromedical evacuations (Evacam); aerial resources of the city of Concepcion´s Air Base included B-412 EP, Cougar, DHC-6, MD-530; and the aerial resources of the city of Talca's Air Base. During the activation of the air base in Concepcion, it had to be established in the 3rd Air Brigade, a logistics administrative center, to provide support to the troops and means deployed in the region of Bio-Bío and from the 2nd Air Brigade to support the base of Talca. There was a minimum of institutional ground support at the time of the force deployment as well as aerial resources, having to resort to the ground support from Lan Chile, main private airline. The facilities designed to support the lives of the forces, were structurally damaged by the earthquake and there was no potable for the staff. The commissioned forces, plus the support staff of other institutions, reached 240 men, who required approximately 18,000 L of water, for their basic needs (1 daily shower and use of bathrooms). Flights to the affected area began on February 27. The operations of aid to the population began on March 1. It was operated with a maximum rhythm in the development of operations. The work was carried out almost without rest motivated by the urgency of the situation.

Initially, there was the following equipment: 4 DHC-6, 4 Bell 412, 4 UH-1H, 1 Bell 206, 1 MD 530, 3 Cougar. Subsequently, three UH-60s were added, one Chilean and two Brazilian units. Upon the arrival of crews, the task force proceeded to execute tasks, such as: crew appointment, induction to the crews, briefing of procedures to the crews, first flights of foreigners with a Chilean pilot, total integration to operations and internal regime, recognition of landing zones in helicopters MD-530 and B-206, evaluation of security of landing sites for helicopters and planes, contact with authorities to organize the delivery of assistance, and marking of zones and population control during downloads.

The operations included day and night aerial patrols in the province of Concepcion in support of the internal security of the area, and day and night terrestrial patrols in the perimeter zone of the air bases. The results of the operations were: seven aeromedical evacuations, eight evacuations from critical zones, thirty three patients transported. The International Humanitarian Aid reception counted 76 flights with 1,285 tons. The general summary of the air bridge was: 1,980 air force men deployed, 73 aircrafts, 1,932.5 flight hours, 1,992.9 tons of cargo transported, 1,172 departures, and 12,335 people transported.

6 Computational Experiments

As seen in the situation described above, the number of transport operations may be sufficiently high to be simply resolved and this scenario calls for the decision support tool. Testing with the DSS has been made so far for small instances due to computing time, although ongoing research is addressing the problem with a heuristic solution. For instance, a scenario with 10 localities, 16,571 kgs of demand, six landing points close to the 10 localities, three types of aircrafts with six units in total and 13,255 kgs of load capacity, was tested. The model generated 1,285 continuous variables, 528 discrete variables and solved the routing in 7:16 h. Hence, the need for heuristics. Each aircraft made two trips and completed delivery in an average of 30.7 min. The total operations until the last locality was served lasted 58.6 min. The model was solved with GAMS on NEOS-SERVER and the DSS prototype coded in PHP.

7 Conclusions

In this paper, a decision support tool for an aid delivery problem in a natural disaster situation has been presented. The part corresponding the optimization problem presents acceptable resolution times for small instances (e.g., up to 5 nodes and 3 vehicles). Solution time becomes very high for more realistic instances, for example, up to seven hours for six nodes and ten vehicles, which is not helpful in emergency situations, and for this motive the effort is being made with a heuristic method. However, this technical issue, does not diminish the utility of the DSS in a more general sense.

Acknowledgments. The authors wish to thank the Industrial Engineering Department and DICYT of the U. of Santiago de Chile for their support. Special recognitions to former Col. Mr. Victor Drake of FACH for his cooperation during the scenario reconstruction.

References

1. Hallegatte, S., Przyluski, V.: The economics of natural disasters. CESifo Forum **11**(2), 14–24 (2010)
2. Manopiniwes, W., Irohara, T.: A review of relief supply chain optimization. Ind. Eng. Manag. Syst. **13**(1), 1–14 (2014)
3. Guha-Sapir, D., Hoyois Ph., Wallemacq P., Below. R.: Annual Disaster Statistical Review 2016: The Numbers and Trends. CRED, Brussels (2016)
4. Wan, M.: Public health emergencies. J. Pediatr. Child Health **2**(3) (2003)
5. Endsley, M.: Towards a theory of situational awareness in dynamic systems. Hum. Factors **37**(1), 32–64 (1995)
6. Costa Freitas, M., Xavier, A., Fragoso, R.: An integrated decision support system for the Mediterranean forests. Land Use Policy **80**, 298–308 (2019)
7. Malmir, B., Amini, M., Chang, S.: A medical decision support system for disease diagnosis under uncertainty. Expert Syst. Appl. **88**, 95–108 (2017)
8. Filip, F., Constantin-Bala, Z., Ciurea, C.: Computer-Supported Collaborative Decision-Making, pp 31–69. Springer (2017)
9. Power, D.: Supporting Decision-Makers: An Expanded Framework. Informing Science - Challenges to Informing Clients, A Transdisciplinary Approach, June 2001. http://www.dssresources.com
10. Bula, G., Gonzalez, F.A., Prodhon, C., Afsar, H., Velasco, N.: Mixed integer linear programming model for vehicle routing problem for hazardous materials transportation. IFAC-PapersOnLine **49**(12), 538–543 (2016)
11. Tavakkoli-Moghaddam, R., Safaei, N., Kah, M., Rabbani, M.: A new capacitated vehicle routing problem with split service for minimizing fleet cost by simulated annealing. J. Franklin Inst. **344**(5), 406–425 (2007)
12. Desrochers, M., Laporte, G.: Improvements and extensions to the Miller-Tucker-Zemlin subtour elimination constraints. Oper. Res. Lett. **10**(1), 27–36 (1991)
13. Cattaruzza, D., Absi N., Feillet D.: Vehicle routing problems with multiple trips. 4OR 3 (2016)
14. Sepúlveda, J.M., Arriagada, I.A., Derpich, I.S.: A decision support system for distribution of supplies in natural disaster situations. In: IEEE XPlore Digital Library, Proceedings of 7th International Conference on Computers, Communications and Control, Oradea, Romania (2018)

Maturity Analysis of Safety Performance Measurement

Aki Jääskeläinen(✉), Sari Tappura, and Julius Pirhonen

Tampere University, Management and Business,
Tampereen yliopisto, 33014 Tampere, Finland
{aki.jaaskelainen, sari.tappura,
julius.pirhonen}@tuni.fi

Abstract. Organizations have several indicators for their safety performance. However, the use of performance indicators often fails to create overall insights on the level of safety and the various factors affecting it. Performance indicators could be better utilized in safety-related decision-making. Maturity models have been presented in many different managerial fields, but no such models for safety performance measurement can be identified. Maturity analysis can provide information on why performance measurement utilization is flawed and how can it be improved. The aim of this paper is to design and test a maturity model for safety performance measurement. The study presents an approach for evaluating maturity which combines written descriptions of best practices, the overall satisfaction of employees in the evaluated aspects, and the experienced level of safety performance.

Keywords: Performance measurement · Safety performance · Maturity analysis

1 Introduction

Existing research classifies safety performance indicators into various dimensions, such as leading and lagging [1, 2]. Organizations have several indicators for their safety performance. However, it is often difficult to materialize the potential of performance measurement [3, 4]. The current use of performance indicators rarely creates overall insights on the level of safety and the various factors affecting it. Hence, performance indicators could be better utilized in safety-related decision-making [5].

In the recent decades, maturity models have been presented in many different managerial fields such as information management, strategy management, and performance management [6–8]. These models have been designed both for managerial and academic purposes. Maturity models typically define maturity levels which assess the completeness of the analyzed objects via different sets of multi-dimensional criteria, and describe essential attributes that would be expected to characterize an organization at a particular maturity level [9]. Maturity models can be used both as an assessment tool and as an improvement tool [9, 10]. There is indication in the literature that maturity models can improve organizational performance [11] by presenting desirable characteristics for operating.

T. Ahram et al. (Eds.): IHSED 2019, AISC 1026, pp. 529–535, 2020.
https://doi.org/10.1007/978-3-030-27928-8_80

Performance measurement literature has presented several models for the maturity analysis of organizational performance measurement [6–8]. In regard to safety management, maturity models for safety culture and risk management have been presented [12–14]. However, to the best of our knowledge, no one has applied maturity models for analyzing the status of safety performance measurement. The aim of this research is to design a maturity model for safety performance measurement based on literature review and analysis.

2 Materials and Methods

This study utilizes a design science approach in which the intention is to both develop scientific knowledge and solve practical problems [15]. Since the aim of this study is to design a maturity model for safety performance measurement and to give guidelines for its future implementation, design science is an obvious approach for this aim. The six main phases of design science process [16] can be found in Fig. 1. This study follows the first three steps while testing, deploying and maintaining are not in the scope.

Fig. 1. Main phases in constructing a maturity model [16].

In the first phase, the scope and target population of the model are defined. In this study, the scope was related to the safety performance measurement practices, commitment and culture supporting performance measurement and performance measurement usage. The model can be applied in different organizations without industry limitations.

The second phase consists of the definition of evaluation variables and execution plans. Evaluation variables can be identified both analytically (top-down) or by combining the existing literature (bottom-up). This study applied bottom-up approach and identified the evaluation variables based on appropriate literature. The analysis is designed to be implemented as a self-evaluation survey addressed to managers, supervisors and safety experts. The managerial perspective is deemed important in order to obtain a more reliable picture on the status safety performance measurement.

The main content of the model is defined in the third phase. A survey instrument was designed with maturity levels describing the alternative practices in each of the evaluated variables. In the fourth phase the model is tested. In our study, four persons of the intended population first evaluated the evaluation tool. Then it was further tested by two fellow scholars. Finally, a web-based survey tool was designed and tested with four persons of the intended population.

This study is mostly based on literature review and analysis. The performance measurement maturity model by Jääskeläinen and Roitto [6] was taken as a starting point and complemented by other existing performance measurement maturity models.

The adjustment of the model into safety management context was supported by the review of literature on safety management and safety culture maturity models and by the expertise of safety scholars.

3 Results

3.1 Maturity Model Framework

The model framework was divided into three main themes: safety performance measurement practices, commitment and culture to safety performance measurement and use of safety performance measurement. These reflect the three lifecycle perspectives of performance measurement including design, implementation and use [17]. In addition, the level of safety in an organization is measured.

In the following presentation, the number of evaluated items is presented alongside with the main themes of the model framework. Safety performance measurement practices represents performance measurement design and includes the most established content of the model. There are maturity several models concentrating the design of performance measurement. Performance measurement practices is further divided into two categories: performance measurement (10 items) and information systems (IS) (4 items). The importance of IS has been emphasized in parallel with the content of performance measurement [18].

Commitment and culture related to performance measurement is widely seen as an important success factor of performance measurement implementation [19, 20]. It is important that both managers (2 items) and employees (2 items) are committed to safety performance measurement. This aspect is closely related to safety culture [21].

There is no established definition on the content of performance measurement usage [22] which is also reliant on the field of management (e.g. safety management). In the actual usage of safety performance measurement, the first perspective of the presented model is communication of measurement results (2 items) which facilitate information flows and use of measurement information [19]. The extent of using performance measurement at different levels of the organization is also included with one item. The actual use of performance information in different managerial tasks is divided into use of information in planning (ex-ante perspective, 3 items) and management (ex-post perspective, 6 items). The management items were selected with a balanced approach related to the three perspectives: resource allocation (financial management), benchmarking and supply chain (processes) and competencies and rewarding (learning and growth).

Table 1 provides examples of items and related references in the three main dimensions of the model. Each of the three main perspectives is also evaluated in terms of a respondent's satisfaction towards the status of the perspective. By capturing satisfaction, the designed new model highlights purposeful objectives of developing safety performance measurement. It acknowledges that also more elementary measurement techniques may suffice if employees are satisfied. In this way, the model takes different contextual criteria for knowledge management practices into account.

Table 1. Exemplifying illustration of the maturity model framework.

Dimension	Example item	References
A. Performance measurement practices	Links between occupational safety performance measurement objects	[7, 12, 23–25]
B. Commitment and culture related to performance measurement	Employee commitment to occupational safety performance measurement	[12, 23, 26, 27]
C. Use of performance measurement	Defining action plans related to occupational safety	[3, 7, 12, 14, 23, 25, 26, 28]

3.2 Evaluation Instrument

The evaluation of the items in the model is carried out with four-step maturity levels representing the sophistication level in each item. This means written descriptions for the four evaluation levels (Table 2). The descriptions were based on the literature on best practices of performance measurement and management [e.g. 29, 30] and safety management [e.g. 12, 28]. In alignment with Maier et al. [10], the best and weakest practices were defined first. Authors's own expertise was needed in defining the levels 2 and 3.

Table 2. Example of written evaluation criteria in maturity levels.

Level	Item: links between safety performance measurement objects
Level 1	Linkages between measurement objects have not been considered
Level 2	Linkages between measurement objects are discussed
Level 3	Factors explaining the main measurement results are partially identified
Level 4	Linkages between measurement objects are analyzed and modeled (e.g. with a strategy map). There is a common understanding in the organization regarding the factors that should be improved in order to affect the main measurement results

Written evaluation criteria were chosen to differentiate the model from some earlier maturity surveys using Likert scales. The following benefits in written evaluation levels have been identified. First, written maturity levels provide clearer and more objective alternatives for the respondents in comparison to Likert scales [23]. Second, presentation of written maturity levels raises awareness of best practices, generates discussion and facilitates the identification of development areas already during the completion of the survey [31]. Third, written maturity levels decreases the need to use external consultants and knowledge on practices outside the own organization in the evaluation [32].

4 Discussion

The main contribution of this paper is a presentation of maturity model which can be utilized as a checklist in analyzing safety performance measurement. The model evaluates maturity by combining written descriptions of best practices, the overall satisfaction of employees in the evaluated aspects, and the experienced level of safety performance. Sophisticated performance measurement practices are useless if they are not beneficial for an organization. The perspective of employee satisfaction towards performance measurement acknowledges the need to fit the practices into contextual needs. The specific characteristics of the model is its balance between rigor (literature derived items) and relevance (written evaluation levels) reflecting the main idea of design science. To the best of our knowledge, the presented model is the first one specifically designed for the purposes of evaluation safety performance measurement.

The resulting model will benefit both the research and practice of safety management. Researchers may use the model in large-scale survey research (e.g. in identifying links between safety performance and the level of safety) and practitioners may utilize it in auditing performance management practices, for example, through group interviews or workshops. Based on the results, improvement means can be generated and prioritized in order to reach higher maturity levels [9]. The combination of maturity levels and satisfaction may be used in defining various profiles for the status of performance measurement in an organization, e.g. as follows: "Novice" (Low employee satisfaction and basic practices), "Facilitator" (High employee satisfaction and basic practices), "Experimenter" (Low employee satisfaction and advanced practices) and "Advanced exploiter" (High employee satisfaction and advanced practices). The profiling allows an easy way to position an organization in relation to other organizations in the three main perspectives of the model. Further research should test the presented model in practice and report the experiences of using the model in practice.

Acknowledgments. The authors acknowledge the research funding provided by the Finnish Work Environment Fund, participating companies, and Tampere University.

References

1. Podgórski, D.: Measuring operational performance of OSH management system – a demonstration of AHP-based selection of leading key performance indicators. Saf. Sci. **73**, 146–166 (2015)
2. Reiman, T., Pietikäinen, E.: Leading indicators of system safety – monitoring and driving the organizational safety potential. Saf. Sci. **50**, 1993–2000 (2012)
3. Bititci, U.S., Ackermann, F., Ates, A., Davies, J., Garengo, P., Gibb, S., MacBryde, J., Mackay, D., Maguire, C., van der Meer, R., Shafti, F., Bourne, M., Firat, S.U.: Managerial processes: business process that sustain performance. Int. J. Oper. Prod. Manag. **31**(8), 851–891 (2011)
4. Bourne, M., Franco-Santos, M., Kennerley, M., Martinez, V.: Reflections on the role, use and benefits of corporate performance measurement in the UK. Meas. Bus. Excell. **9**(3), 36–41 (2005)

5. Tappura, S., Sievänen, M., Jussila, A., Heikkilä, J., Nenonen, N.: A management accounting perspective on safety. Saf. Sci. **71, Part B**, 151–159 (2015)
6. Jääskeläinen, A., Roitto, J.M.: Designing a model for profiling organizational performance management. Int. J. Prod. Perform. Manag. **64**(1), 5–27 (2015)
7. Van Aken, E.M., Letens, G., Coleman, G.D., Farris, J., Van Goubergen, D.: Assessing maturity and effectiveness of enterprise performance measurement systems. Int. J. Prod. Perform. Manag. **54**(5/6), 400–418 (2005)
8. Wettstein, T., Kueng, P.A.: A maturity model for performance measure systems. In: Brebbia, C., Pascola, P. (eds.) Management Information Systems: GIS and Remote Sensing, pp. 113–122. WIT Press, Southampton (2002)
9. Goncalves Filho, A.P., Waterson, P.: Maturity models and safety culture: a critical review. Saf. Sci. **105**, 192–211 (2018)
10. Maier, A.M., Moultrie, J., Clarkson, P.: Assessing organizational capabilities: reviewing and guiding the development of maturity grids. IEEE Trans. Eng. Manag. **59**(1), 138–159 (2012)
11. Bititci, U.S., Garengo, P., Ates, A., Nudurupati, S.S.: Value of maturity models in performance measurement. Int. J. Prod. Res. **53**(10), 3062–3085 (2015)
12. Goncalves Filho, A.P., Andrade, J.C.S., Marinho, M.M.O.: A safety culture maturity model for petrochemical companies in Brazil. Saf. Sci. **48**, 615–624 (2010)
13. Kaassis, B., Badri, A.: Development of a preliminary model for evaluating occupational health and safety risk management maturity in small and medium-sized enterprises. Safety **4** (5) (2018)
14. Parker, D., Lawrie, M., Hudson, P.: A framework for understand the development of organizational safety culture. Saf. Sci. **44**, 551–562 (2006)
15. Van Aken, J.E.: Design science and organization development interventions aligning business and humanistic values. J. Appl. Behav. Sci. **43**(1), 67–88 (2007)
16. De Bruin, T., Rosemann, M., Freeze, R., Kulkarni, U.: Understanding the main phases of developing a maturity assessment model. In: Campbell, B., Underwood, J., Bunker, D. (eds.) Australasian Conference on Information Systems (ACIS), Sydney, 30 November–2 December (2005)
17. Bourne, M., Mills, J., Wilcox, M., Neely, A., Platts, K.: Designing, implementing and updating performance measurement systems. Int. J. Oper. Prod. Manag. **20**(7), 754–771 (2000)
18. Nudurupati, S.S., Bititci, U.S., Kumar, V., Chan, F.T.: State of the art literature review on performance measurement. Comput. Ind. Eng. **60**(2), 279–290 (2011)
19. Jääskeläinen, A., Sillanpää, V.: Overcoming challenges in the implementation of performance measurement: case studies in public welfare services. Int. J. Pub. Sect. Manag. **26**(6), 440–445 (2013)
20. Kennerley, M., Neely, A.: A framework of the factors affecting the evolution of performance measurement systems. Int. J. Oper. Prod. Manang. **22**(11), 1222–1245 (2002)
21. Fernández-Muñiz, B., Montes-Peón, J., Vázquez-Ordás, C.: Safety culture: analysis of the causal relationships between its key dimensions. J. Saf. Res. **38**, 627–641 (2007)
22. Tangen, S.: Demystifying productivity and performance. Int. J. Prod. Perform. Manag. **54** (1), 34–46 (2005)
23. Cocca, P., Alberti, M.: A framework to assess performance measurement systems in SMEs. Int. J. Prod. Perform. Manag. **59**(2), 186–200 (2010)
24. Das, A., Pagell, M., Behm, M., Veltri, A.: Toward a theory of the linkages between safety and quality. J. Oper. Manag. **26**(4), 521–535 (2008)
25. Marx, F., Wortmann, F., Mayer, J.H.: A maturity model for management control systems. Bus. Inf. Syst. Eng. **4**(4), 193–207 (2012)

26. Brondino, M., Silva, S.A., Pasini, M.: Multilevel approach to organizational and group safety climate and safety performance: co-workers as the missing link. Saf. Sci. **50**, 1847–1856 (2012)
27. Tung, A., Baird, K., Schoch, H.P.: Factors influencing the effectiveness of performance measurement systems. Int. J. Oper. Prod. Manag. **31**(12), 1287–1310 (2011)
28. Fernández-Muñiz, B., Montes-Peón, J., Vázquez-Ordás, C.: Relation between occupational safety management and firm performance. Saf. Sci. **47**, 980–991 (2009)
29. Neely, A., Mills, J., Platts, K., Richards, H., Gregory, M., Bourne, M., Kennerley, M.: Performance measurement system design: developing and testing a process-based approach. Int. J. Oper. Prod. Manag. **20**(10), 1119–1145 (2000)
30. Najmi, M., Rigas, J., Fan, I.: A framework to review performance measurement systems. Bus. Process Manag. J. **11**(2), 109–122 (2005)
31. Maier, A.M., Eckert, C.M., Clarkson, J.P.: Identifying requirements for communication support: a maturity grid-inspired approach. Exp. Syst. Appl. **31**(4), 663–672 (2006)
32. Garengo, P., Biazzo, S., Bititci, U.S.: Performance measurement systems in SMEs: a review for a research agenda. Int. J. Manag. Rev. **7**(1), 25–47 (2005)

Safety Evaluation of Steering Wheel LCD Screen Based on Ergonomic Principles and FEA

Zhi Cheng, Wenyu Wu(✉), Chengqi Xue, and Hongxiang Shan

School of Mechanical Engineering, Southeast University, Nanjing, China
1198117429@qq.com, wuwenyu1984@163.com,
18851668026@163.com, ipd-xcq@seu.edu.cn

Abstract. The safety of the steering wheel LCD screen is quite worrisome under normal situations or accidents. There is a worry that the screen breaks and hurt driver under the instant compressive stress of the airbag. This report integrates two analysis methods in both human factors and engineer factors, to find the visual influences and verify screen stability of the steering wheel LCD screen in normal situations and some accidents. The result helps to find out the safety of the steering wheel LCD screen. In the beginning, it comes with an ergonomic analysis to do the driver vision analysis to test the influences on the driver vision. Furthermore, the finite element analysis (FEA) about the steering wheel screen gives a preliminary conclusion about the stability of LCD screen in an accident to provide a foundation for future research and application.

Keywords: The steering wheel LCD screen · Ergonomic analysis · The finite element analysis

1 Introduction

With the development of artificial intelligence and advanced display technique, new functions have changed the vehicle industry and improved the operation experience profoundly. In the same way as the accident of the nuclear station, emerging technology often causes some unexpected consequences of a fault or terrible design. Furthermore, a study for accident probabilities showed that the accident probabilities respectively increase by 3 and four times when the drivers were chatting or typing on the phone during driving [1].

Attentional resources of a driver are limited, if the driver attempts to perform any secondary task, then the reallocation of the attentional sources may lead to deteriorated driving performance [2]. Furthermore, the influences come from both physiology and psychology. The study showed that different tests were relevant to the prediction of safe driving performance in various driving exercises. Contrary to previous research, logical reasoning showed significant effects [3].

Nowadays, it is an attempt to place the LCD screen on the steering wheel, which is a solution for innovative design and special applications like racing (Fig. 1 shows three examples). The screen provides a large amount of information and interaction. It provides

T. Ahram et al. (Eds.): IHSED 2019, AISC 1026, pp. 536–542, 2020.
https://doi.org/10.1007/978-3-030-27928-8_81

functions like navigation, instruments and buttons [5, 6], which requires the driver to gaze up to the screen and prompts messages sometimes. It is questionable whether a sudden message hint influences the driver's concentration [4] and decrease driving safety.

However, there are some doubts about whether the LCD screen collides with human or breaks in an accident to cause more damage. Furthermore, the airbag ejects in a car accident, and it puts instant pressure on the surrounding area where the LCD screen locates. The following finite element analysis is to assess the security of the steering wheel screen design, which includes both the LCD screen and airbag.

Hence, this report introduces two analyses, ergonomics and finite element analysis, to find out the security of the screen on the steering wheel. The first analysis adopts ergonomic methods. For instance, it uses the 50% dimensions of the human dimensions of Chinese adult in GB. The second analysis simulates a real-sized model by ANSYS, a finite element analysis software.

Fig. 1. Examples of the steering wheel with a screen or panel [5–7].

2 Research Methods

As shown in Fig. 2, there is a real-sized steering wheel with an LCD screen stimulated by Rhino (a parametric modeling software)

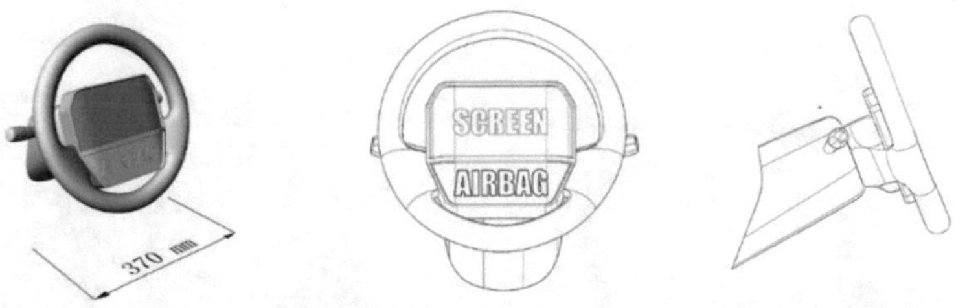

Fig. 2. A simulative steering wheel.

The analysis goes through 2 directions to estimate the safety of this LCD screen on the steering wheel. The experiment makes this process in a simulated condition by Rhino and ANSYS. The first step is the ergonomic visual analysis to evaluate the influence of the screen on vision. The next step is to assess the impact of the airbag on the LCD screen to check whether it cracks and hurt the driver.

A. Ergonomic vision analysis

The screen aims to represent the car's situation and provide shortcut keys, which include navigation, gear and windshield wiper [5–7]. However, the system unintentionally prompts messages that attract the driver's attention on the screen, which is dangerous in the busy traffic. It is a concern that the driver's sight is not rapid to transfer from the steering wheel screen to the front scene. Hence, this analysis is to verify that the steering wheel screen is in a good vision area that driver's sight moves rapidly.

Step 1 Prepare the data

First of all, it is necessary to standardize the size and position of human, screen and environment. The table below gives the Anthropometric data showing respectively of man and woman. Furthermore, the data is according to the 50% dimensions of the human dimensions of Chinese adult in GB [8], so that the result is equitable (Table 1).

Table 1. The Anthropometric data 50% dimensions of the human dimensions of Chinese adult in GB [8].

Dimension designation	GB number	GB value/cm
Shoulder height	3.4	59.8/55.6
Sitting body depth	3.9	45.7/43.3
Sitting height	3.1	90.8/85.5
Arm length	1.3 + 1.4	55.0/49.7
Shoulder height	3.4	59.8/55.6

Step 2 Simulation

Here are two demonstration graphs showing normal driving status. It employs an ordinary SUV interior, and the humans are respectively man and woman whose proportion comes from the 50% dimensions of the human dimensions of Chinese adult in GB [8] listed above.

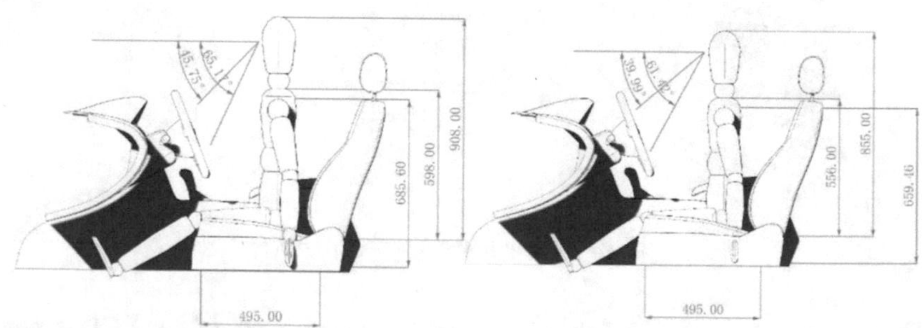

Fig. 3. The proportion of man (left) and woman (right) in the vehicle (unit: mm).

In the Fig. 3, the man and woman have different sizes. As a result, their eyes have different angles of view when looking at the screen on the steering wheel. The numbers of angle are listed below (Table 2).

Table 2. Table for the angle of view looking at the screen on the steering wheel

Gender	Minimum angle	Maximum angle
Man	45.75°	65.17°
Woman	39.99°	61.42°

When a person's head and eyes remain static, the vision he saw divides into the best vision area, good vision area and valid vision area. The angle of the best vision area is about 15° up and down of the horizontal line. The angle of the good vision area is about 30° up and down of the horizontal line. The angle of the valid vision area is about 25° up to the horizontal line and 35° down of the horizontal line [3, 4]. Compared to the table above, the minimum angle of the range of sight is larger than the valid vision area for both man and woman. Therefore, the driver is hard to transfer his eyesight to the front view when driving. It is improper and dangerous to put a message hint on the interface of the steering wheel screen.

B. Finite element analysis of airbag's compression stress

This part simulates the stress process between ejecting airbag and LCD screen, to test the safety of LCD screen under the instant stress from ejecting airbag. It runs on the ANSYS (a finite element analysis software).

In this test, the model is a steering wheel with a LED screen whose size and proportion refer to the standard size steering wheel in Fig. 2.

Step 1 Prepare the data

The ejecting airbag exerts a force to a point at the middle between the screen and airbag. The red point and its direction are signed in Fig. 4.

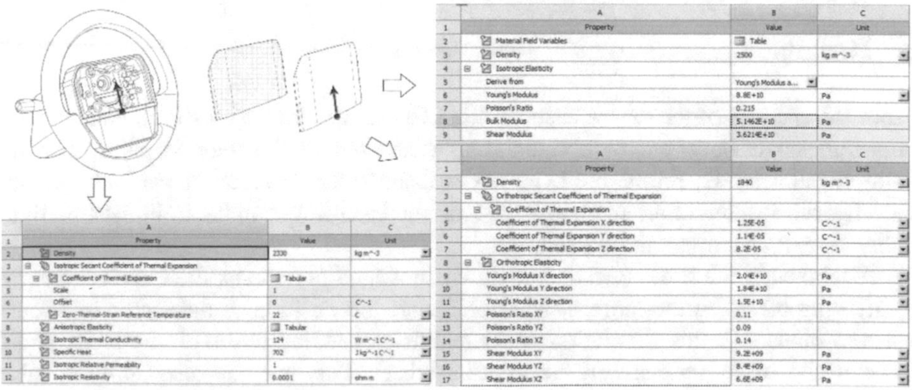

Fig. 4. The attributes of the plastic mold (left) and two layers of the screen (right).

Step 2 Analysis in ANSYS

The material of the steering wheel structure is the Acrylonitrile Butadiene Styrene plastic (ABS), which is universally in the design of the vehicle interior. Furthermore, the screen divides into two parts, resins and the protective shield. Glass is the primary material of protective shield. The attribute data of resins and ABS comes from the default option in the ANSYS. The attribute data of glasses comes from the study [9].

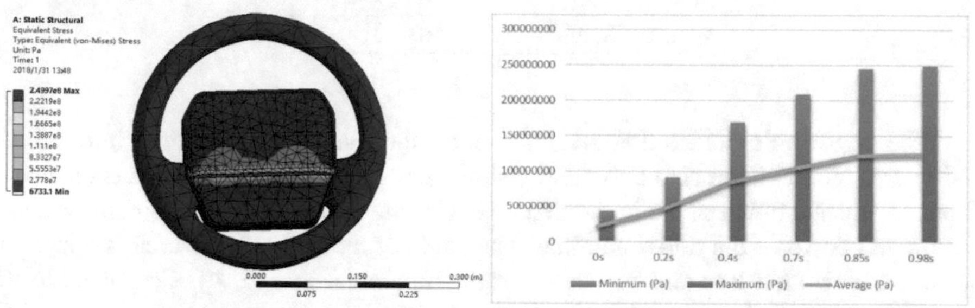

Fig. 5. The stress nephrogram of the steering wheel and the stress variation.

Figure 5 shows the analysis process and results. In 1 s, the compression stress increases to the highest value that ejecting airbag brings, to get the transform process and result of stress [9]. Since the elastic modulus of the screen is about (73.1 ± 4) GPa [10], which is far more significant than the statistics this experiment has got, the screen is stable under the pressure of airbag. From the stress picture, we can understand that the pressure focuses on the one-third area of the whole LCD screen. Even the maximum pressure value cannot achieve the allowable amount that causes the break of the screen. It shows that the pressure causes a slight influence on the LED, and it cannot cause any variation.

$$\sigma_{\max} = 2.4997 \times e^{8} < (73.1 \pm 4)\,\mathrm{GPa} \tag{1}$$

3 Results

From the analysis of the driver vision at the LCD screen, the result shows that even the minimum angle between the horizontal sight and the LCD screen is larger than the valid visual field. Therefore, the LCD screen disturbs the drive, and it can only provide hint by other interaction methods. The driver can only handle it in leisure time. Otherwise, it endangers the passengers.

As shown in the stress test, the pressure is not enough to break the screen and cause a safety problem. Furthermore, it will not change the position of the screen and cause harm to human. In this simulation, the screen is embedding in the plastic mold, and it has a series of components in the steering wheel. This test can also apply in other situations. For instance, there are many screens behind every seat on the airplane or train. They can set the airbag below the screen safely.

4 Discussion

According to the result, the test infers that the driver should not gaze at the steering wheel interface in driving, but the driver can still receive information from auditory or other display rather than visual. Hence, the auditory display is one of a right choice for the communication between the driver and vehicle system, and it is already in services in some situations. However, the LCD screen still can provide some unique hints and receive information input. For instance, the short-term or repeat operation is an excellent choice for the screen on the steering wheel. The high definition display provides vivid graphics, which helps the driver to be aware of the situation of the components of the vehicle. Furthermore, it is safe under the airbag's compression stress, and the screen itself can act like the cover of the airbag, which still requires the reality test.

5 Conclusion

The LCD screen module is a tendency of the steering wheel in the future with the development of the automotive drive and intelligent navigation, which provides helpful information and operation. As for the automotive drive, everyone is a passenger who doesn't need to observe the environment all the time. Thereupon, the screen on the steering wheel is a necessary experience. Furthermore, Driving is full of unpredictability and uncertainty [11]. Every detail in the car is remarkable. Therefore, the screen has to be stable when it comes across an accident. This test shows that the screen can defense the impact of the airbag in this condition. However, it is inappropriate to gaze at the steering wheel screen in driving, and it still requires more research to go a step further.

Acknowledgments. This work was supported jointly by National Natural Science Foundation of China (No. 71871056, 71471037).

References

1. Choudhary, P., Velaga, N.R.: Mobile phone use during driving: Effects on speed and effectiveness of driver compensatory behavior. Accid. Anal. Prev. (2017)
2. Parkes, A.M., Ashby, M.C., Fairclough, S.H.: The effects of different in-vehicle route information displays on driver behavior. In: Vehicle Navigation & Information Systems Conference IEEE (1991)
3. Sanders, M.S., McCormick, E.J.: Human factors in engineering and design. Ind. Robot Int. J. (1982)
4. Ding, Y.: Man-Machine Engineering, 5th edn. Institute of Technology Press, Beijing (2017). (in Chinese)
5. INTERIOR DESIGN. https://www.byton.com/m-byte-concept
6. Is this gesture-controlled steering wheel genius or madness? https://www.digitaltrends.com/cars/zf-touchscreen-steering-wheel/
7. Formula CSX. https://cubecontrols.com/product/formula-csx/

8. GB/T 12985-1991. General principles for the application of human body size percentiles in product design (1991)
9. Wu, K., Wang, H., Tan, W.: Measurement of the elastic modulus of glass by equal thickness interference. Phys. Exp. Coll. (1), 55 (2014). (in Chinese)
10. Zhang, X., et al.: Simulation and pressure analysis of CFD-based balloon deployment. Sci. Technol. Eng. 15, 15 (2015). (in Chinese)
11. Divakarla, K.P., Emadi, A., Razavi, S.: "Journey Mapping" - a new approach for defining automotive drive cycles. Department of Electrical and Computer Engineering, McMaster University, Hamilton, ON, Canada (2015)

Digital Human Modelling in Research and Development – A State of the Art Comparison of Software

David Pal Boros(✉) and Karoly Hercegfi

Department of Ergonomics and Psychology,
Budapest University of Technology and Economics,
Muegyetem rkp. 3, Budapest 1111, Hungary
{borosdavid, hercegfi}@erg.bme.hu

Abstract. This paper carries out a comparison of three Digital Human Modelling (DHM) software. They are compared to each other in the field of ergonomics with a larger focus on their ergonomic risk assessment capabilities. They were selected from a pool of ergonomic software. The selection was restricted to those programs which gave enough information to overview features and capabilities without purchasing the given software. The software included here are namely: Jack, Santos and ViveLab. Jack is a software by Siemens, Santos belongs to SantosHuman Inc. and ViveLab is newly developed Hungarian software by ViveLab Ergo Ltd. These software companies provide free material for their software overview and online tutorials free of charge to make the acquiring process easier for their customers.

Keywords: DHM · Digital Human Modeling · Santos · ViveLab · Jack · HFE · Risk analysis

1 Introduction

Human Factors/Ergonomics (HF/E) is increasingly implemented in many Industries' policy for the work environment, for risk factor analysis and for production performance [1–3]. It is an established fact, that in the case of large companies, workers' safety is beneficial to them, not just morally and ethically, but also financially [4]. Other than regulations, this fact gives an appealing incentive to apply HF/E. Keeping employees healthy, beneficial both short and long-term, there are many advantages of having long term employees [5]. Even if everything is done to satisfy the employee to stay at a company their health is still a factor that has to be maintained. It has been proven that there is a strong correlation between back disorders - such as Musculoskeletal Diseases (MSD) - and non-neutral trunk postures [6, 7]. MSD caused the highest number of sick leave in Germany in the 2000s, which was researched by Liebers, Brendler and Latza [8]. Work-related MSD also cost to the EU as high as 2% of the GDP [9]. With the help of DHM there is a developing field - Human Centered Automation (HCA) - to reform the established fully automated production line into a more machine-human integrated one [10].

T. Ahram et al. (Eds.): IHSED 2019, AISC 1026, pp. 543–548, 2020.
https://doi.org/10.1007/978-3-030-27928-8_82

There are several articles that give a state of the art review on Computer Aided Design (CAD) software, [11, 12] however, the authors did not find any from the recent years on DHM software in the context of this article. The most recent one on DHM software comparison is a report done by the Australian DOD in 2010 [13] and the in 2014 on the differences of Jack and Delmia has been looked at the University of West Bohemia [14]. The paper aims to fill some of the gaps between the widespread knowledge of CAD software and the much less known DHM solutions used in Research and Development (R&D) both in academics and in the industry. "Digital human modelling (DHM) is a term that designates a software tool that enables digital models of humans to interact with virtual workplaces or products in a digital CAD environment" [15].

2 Background and Related Works

DHM can be faster than pen-paper methods while used for the same tasks, - most researchers are using some kind of software these days even if it just a simple excel table - and more accurate for ergonomic risk assessment of factories, warehouses [16]. It can also be used for assessment of handling materials, tools and to make product assembly with more efficiency and less physical risk. One of the DHM products biggest advantage is speed and efficiency with ergonomic risk analysis methods. A few of the common methods are RULA, REBA, JSI, NASA and OWAS [17]. Although each software usually only has a few of these methods, once the models are set up it is very fast to run the different analyses on them and to generate reports on the results, which otherwise would have to be created manually. DHM has been used for decades as CAD has been, some of them like Car, Combiman&Crew Chief, Cyberman, Apolis and ADAPS lag behind and some like Santos [18], ViveLab, Jack, Ramsis [19] and Delmia are providing with new versions of their products with improved quality.

3 Methodology

The methodology of the review is an analysis of the materials received from the companies of these software and the information gained through email and phone interviews with their technical representatives. From the software Jack, the main information comes from their software which can be used with a trial license for 30 days or with a student license up to a year and in its help menu, one can find a 238 pages long Jack manual. From ViveLab, which is provided by ViveLab Ergo Ltd. most of the data can be gathered from their YouTube channel, – can be found by searching ViveLab on YouTube - which contains several presentations and tutorials for their product. The same can be said about Santos from SantosHuman Inc. The company provides visual presentation and tutorials on a wide variety of their product's capabilities.

4 Results

From the different source materials, the basic functions of these products can be described.

4.1 Jack

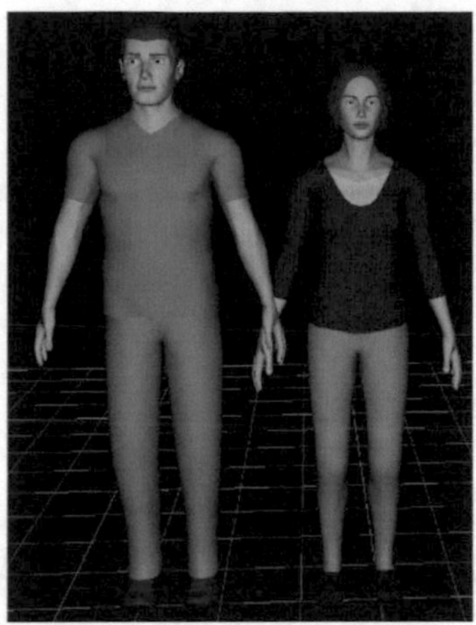

Fig. 1. The standard female and male human models in Jack 8.0.1

The Jack 8.0.1 version was tested during researching the software from Siemens. Jack provides two genders (Fig. 1) and in the trial version, they can be adjusted to percentiles, also adjusted by anthropometric databases, while in the student version their static anthropometrics are fixed. With the student version, only basic features can be used. Simple geometries can be created, and the two types of basic human and a reach analysis. Their joints can be moved and rotated separately or together, also there are preconfigured postures and arm positions in the software.

4.2 ViveLab

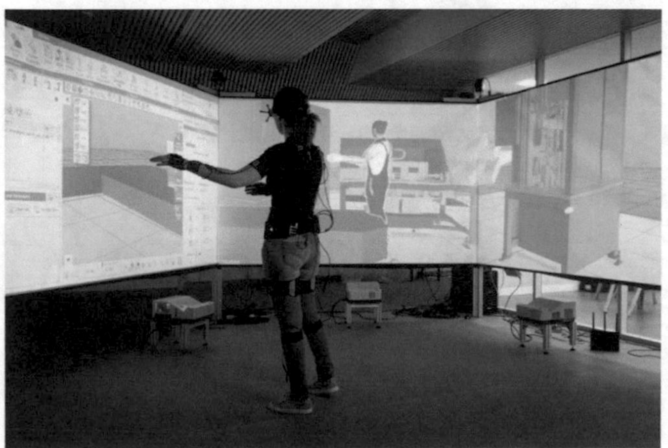

Fig. 2. Live motion capture with ViveLab [20]

ViveLab does not have a student. The research has been conducted through their tutorial videos and personal experience with the software through the companies help. It is similar to Jack, but it has a much more familiar user interface if one is accustomed to SolidWorks or similar CAD modelling products. Their anthropometric database can be adjusted by age, percentile, origin, weight and date of birth to count for the acceleration factor. Their product also has an excellent motion capture feature for which the hardware has to be bought separately from the software, but the active sensors the system use it allows live capture and analysis of any movement with high accuracy. As seen in Fig. 2. It can be used to simulate an augmented reality where the user can test out their surroundings before it is built that way, which can reduce cost and save time in designing assembly lines or workstations. The main ergonomic risk assessments in it are RULA, OWAS, NASA, ISO 11226 and EN 1005-4.

4.3 Santos

SantosHuman Inc. has a reputable history with the automotive industry and the military. Their product UI has a similar look to Jack's, but the software is more complex and the manipulation of the joints in it are easier than int the other two. Their YouTube page – SantosHuman – is very well presented, presenting most of the software features and it is frequently updated. In Santos one can tell the software to hold an object with a digital human while avoiding another one and it will anatomically correctly predict the posture, that one would follow to achieve that. For example, picking up a bucket with a chair in front of the figure as an obstacle (Table 1).

Table 1. Comparison of some features in the 3 software

Features	Jack (Trial)	ViveLab	Santos
Digital human models	Yes	Yes	Yes
Simple geometrics	Yes	Yes	Yes
Joint manipulations	Yes	Yes	Yes
Ergonomic risk analyses	Yes	Yes	Yes
Exporting analyses reports	Yes	Yes	Yes
Motion capture	With additional software (Full version only)	Yes	Yes
Live collaboration in the software	No	Yes	No
Anthropometric databases	No information	Yes	Yes
Predictive posture analyses	No	No	Yes

5 Conclusion

The paper does not aim to provide every single feature of these DHM products or to rank them into general usefulness but to give an insight into different solutions provided by these programs for many ergonomic problems. It can be concluded that Jack stands out in environmental ergonomics, Santos is prominent in the ergonomics of vehicle design and ViveLab is formidable in Augmented Reality and motion capture for ergonomic risk assessment. This can be a good guide to get to know the DHM type of software when someone decides to change from pen-paper methods or want to also implement DHM in their research.

Further study could be done in the future which can be more user experience testing as well as interviews with staff from the different companies who develop this software.

References

1. Bevilacqua, M., Ciarapica, F.: Human factor risk management in the process industry: a case study. Reliab. Eng. Syst. Saf. **169**, 149–159 (2018)
2. Fruggiero, F., Riemma, S., Ouazene, Y., Macchiaroli, R., Guglielmi, V.: Incorporating the human factor within manufacturing dynamics. IFAC-PapersOnLine **49**, 1691–1696 (2016)
3. Mearns, K.: Human Factors in the Chemical Process Industries. Methods in Chemical Process Safety, pp. 149–200 (2017)
4. Cagno, E., Micheli, G., Masi, D., Jacinto, C.: Economic evaluation of OSH and its way to SMEs: a constructive review. Saf. Sci. **53**, 134–152 (2013)
5. Fitz-enz, J.: The ROI of Human Capital. AMACOM, New York (2009)
6. Punnett, L., Wegman, D.: Work-related musculoskeletal disorders: the epidemiologic evidence and the debate. J. Electromyogr. Kinesiol. **14**, 13–23 (2004)
7. Punnett, L., Fine, L., Keyserling, W., Herrin, G., Chaffin, D.: Back disorders and nonneutral trunk postures of automobile assembly workers. Scand. J. Work Environ. Health **17**, 337–346 (1991)
8. Liebers, F., Brendler, C., Latza, U.: Alters- und berufsgruppenabhängige Unterschiede in der Arbeitsunfähigkeit durch häufige Muskel-Skelett-Erkrankungen. Bundesgesundheitsblatt - Gesundheitsforschung - Gesundheitsschutz **56**, 367–380 (2013)
9. Nguyen, T., Kleinsorge, M., Postawa, A., Wolf, K., Scheumann, R., Krüger, J., Seliger, G.: Human centric automation: using marker-less motion capturing for ergonomics analysis and work assistance in manufacturing processes. In: Proceedings of the 11th Global Conference on Sustainable Manufacturing (GCSM), pp. 586–592. Innovative Solutions, Berlin (2013)
10. Nguyen, T., Bloch, C., Krüger, J.: The working posture controller: automated adaptation of the work piece pose to enable a natural working posture. Procedia CIRP **44**, 14–19 (2016)
11. Goodacre, B., Goodacre, C., Baba, N., Kattadiyil, M.: Comparison of denture base adaptation between CAD-CAM and conventional fabrication techniques. J. Prosthet. Dent. **116**, 249–256 (2016)
12. Cuillière, J., François, V., Souaissa, K., Benamara, A., BelHadjSalah, H.: Automatic comparison and remeshing applied to CAD model modification. Comput.-Aided Des. **43**, 1545–1560 (2011)
13. Blanchonette, P.: Jack human modelling tool: a review. Sci. Technol. 1–37 (2010)
14. Polášek, P., Bureš, M., Šimon, M.: Comparison of digital tools for ergonomics in practice. Procedia Eng. **100**, 1277–1285 (2015)

15. Berlin, C., Adams, C.: Production Ergonomics. Ubiquity Press, London (2017)
16. Puthenveetil, S., Daphalapurkar, C., Zhu, W., Leu, M., Liu, X., Gilpin-Mcminn, J., Snodgrass, S.: Computer-automated ergonomic analysis based on motion capture and assembly simulation. Virtual Reality **19**, 119–128 (2015)
17. Stanton, N.: Handbook of Human Factors and Ergonomics Methods. CRC Press, Boca Raton (2006)
18. Standoli, C., Lenzi, S., Lopomo, N., Perego, P., Andreoni, G.: Using Digital Human Modeling to Evaluate Large Scale Retailers' Furniture: Two Case Studies. Advances in Intelligent Systems and Computing, pp. 512–521 (2018)
19. Premananth, S., Dharmar, G., Krishnan, H., Mohammed, R.: Development of Indian Digital Simulation Model for Vehicle Ergonomic Evaluations. SAE Technical Paper Series (2016)
20. ViveLab Ergo. https://gyartasergonomia.wordpress.com/2018/02/10/vivelab-ergo-3/

ErgoSMED: A Methodology to Reduce Setup Times and Improve Ergonomic Conditions

Marlene Brito(✉) and Maria Antónia Gonçalves

CIDEM - Research Center of Mechanical Engineering, School of Engineering of Porto (ISEP), Polytechnic of Porto, 4200-072 Porto, Portugal
mab@isep.ipp.pt

Abstract. The Lean manufacturing philosophy has been adopted by many different companies with the goal of responding to the economic recession which took place at the beginning of the twenty-first century. Two major motivators for this trend were reducing costs and improving customer satisfaction. Other than the known Lean methods, Human Factors and Ergonomics methods are also able to contribute to this emphasis. Thus, the creation of safe, effective workplaces emphasizing a human oriented approach, and the implementation of Ergonomics in business process management, represent two of the main conditions for the sustainable development of a company. Pinpointing the connections between Ergonomics and LPS enables us to address both at the same time. Besides, synergism between Lean and Ergonomics can be attained. The aim of this paper is to develop a tool in the form of a flowchart which will include the steps for integrating the single-minute exchange of die (SMED), a Lean tool, while considering Ergonomics. This flowchart (named "ErgoSMED") has been validated by several case studies. We were able to demonstrate that synergism between Lean and Ergonomics can be reached. The ErgoSMED flowchart developed in this study is innovative because it combines Ergonomics with a SMED tool and is ready to be used in any production area by professionals or researchers.

Keywords: Lean manufacturing · REBA · WMSD · SMED · Human Factors · Human-systems integration · Systems engineering

1 Introduction

These days, clients ask for products with high standards of quality, reasonable prices, and short response times. Thus, companies are forced to produce smaller lots. However, producing smaller batch sizes necessitates more changeovers. We can define both setup and changeover as the process of going from the production of one product or part number to a different one in a machine by changing parts, dies, molds, or fixtures [1].

Setup reduction is an essential program to expands production capacity use, and hence productivity, while increasing the level of flexibility of the plant regarding volume and variety of products [2]. Most initiatives for setup reduction time have been connected with Shingo's single-minute exchange of die (SMED) methodology. This methodology gives us a quick and efficient means of converting a manufacturing

T. Ahram et al. (Eds.): IHSED 2019, AISC 1026, pp. 549–554, 2020.
https://doi.org/10.1007/978-3-030-27928-8_83

process as soon as product changes. The SMED methodology renders it possible to execute equipment setup (changeover) operations in a time shorter than 10 min, i.e., a number of minutes expressed by a single digit [3]. We can see it as one of the tools of Lean manufacturing. It cuts waste and improves flexibility in manufacturing processes, making room for lot size reduction and improving flow. It reduces nonproductive time by streamlining and standardizing the operations for exchange tools, using basic techniques and easy applications. [4].

In the conventional SMED method, setup activities are essentially made through improvements upon machines; yet, not only machines but also operators participate in the setup process. Ulutas [5] noted issues related to worker safety and ergonomic principles during setups. Brito et al. [6] showed that it is possible to cut the setup time and improve ergonomic conditions concurrently.

The objective of this paper is to create a tool/methodology in the form of a flowchart which will include the steps for integrating the single-minute exchange of die (SMED), a Lean tool, while considering Ergonomic aspects. This flowchart (named "ErgoSMED") has been validated in several case studies, which showed that synergism between Lean and Ergonomics can be attained. In our limited literature research, we were not able to locate a flowchart or a tool demonstrating how to integrate SMED and Ergonomics.

2 Methodology

The flowchart proposed in this paper to with the goal of reducing setup times while improving ergonomic conditions was developed through several case studies. According to [7] a case study should be defined "…as a research strategy, an empirical inquiry that investigates a phenomenon within its real-life context." In accordance with this crucial idea, the case study, as a research methodology, makes it possible to understand, explore or describe a given system/problem in which a number of factors are involved at the simultaneously, in a real context.

In regard to ergonomic conditions, the team chose a postural analysis system - Rapid Entire Body Assessment (REBA) - to measure the level of MSDs risk. This was because that system provides a scoring system for muscle activity derived from static, dynamic, rapid changing or unstable postures [8], which fits well into setup activities. REBA gives us a quick and simple measure to evaluate a range of working postures for risk of work-related musculoskeletal disorders (WMSDs). It splits the body into areas to be coded independently, in accordance with movement planes; then, provides a scoring system for muscle activity across the entire body, be it stagnant, dynamic, fast changing or unsteady. REBA also provides an action level with a sign of importance and requires very little equipment, using the pen and paper method [8]. Table 1 depicts the REBA action levels.

Setup time is the time elapsed from the production of the last item of a product lot to the production of the first item of the following product lot. This definition was further developed by [9]. This new definition is represented in Fig. 1.

Table 1. Setup times and Reba Scores before and after improvements.

Action level	REBA score	Risk level	Action
0	1	Negligible	None necessary
1	2–3	Low	May be necessary
2	4–7	Medium	Necessary
3	8–10	High	Necessary soon
4	11–15	Very high	Necessary now

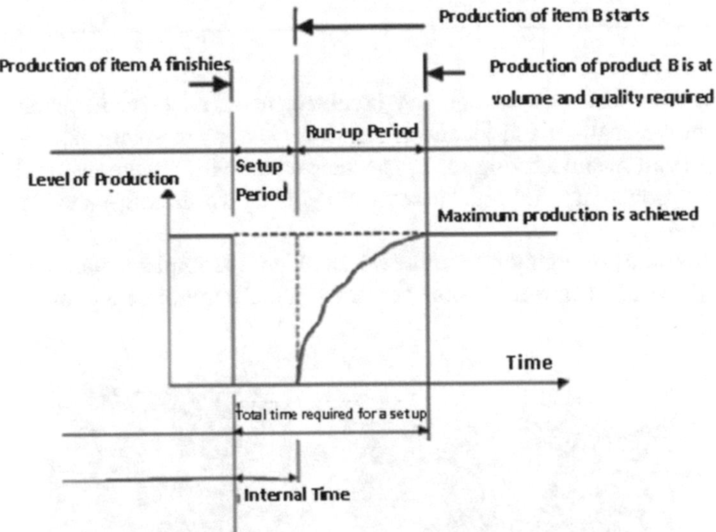

Fig. 1. Description of Setup time

In the SMED methodology, setup activities are divided into internal and external. External activities can be carried out during the normal operation of a machine while it is still running. For instance, one can prepare the equipment for the setup operation before the machine is shut down. It is only possible to carry out internal activities when the machine is off. For example, the dies are attached or removed. The internal and external setup tasks contain diverse operations which include preparation, after-process adjustment, checking of materials, mounting and removing tools, settings and calibrations, measurements, trial runs, and adjustments [10, 11].

3 Results

This study was conducted with the goal of presenting a flowchart with the methodology to be followed for an implementation of the SMED tool which considers ergonomic aspects. It took several cycles of the interactive process to get to the final presentation of the ErgoSMED flowchart. The tests were conducted in several productive areas of a metallurgical factory and wielded very positive results, both in reducing setup times and in reducing ergonomic risk (Table 2).

Table 2. Setup times and Reba Scores before and after improvements.

Production area	Setup time Initial situation	Setup time After improvements	REBA score Initial situation	REBA score After improvements
Tuning	105 min	57 min	9	4
Stamping	43 min	29 min	6	2
Foundry	67 min	43 min	7	4

Success was achieved through the involvement of all team members, including operators, in the analysis and identification of improvement solutions.

Several Lean manufacturing tools, together with SMED, were resorted to in order to improve productivity and reduce setup time, such as: the 5S, visual management, Kanban, TPM, etc.

Some solutions of ergonomic improvement were the implementation of setup cars (Fig. 2) and the replacement of tools for other, more ergonomic and automatic ones.

Fig. 2. Implementation of a car to improve ergonomic conditions.

Figure 3 depicts the ErgoSMED flowchart developed in this study.

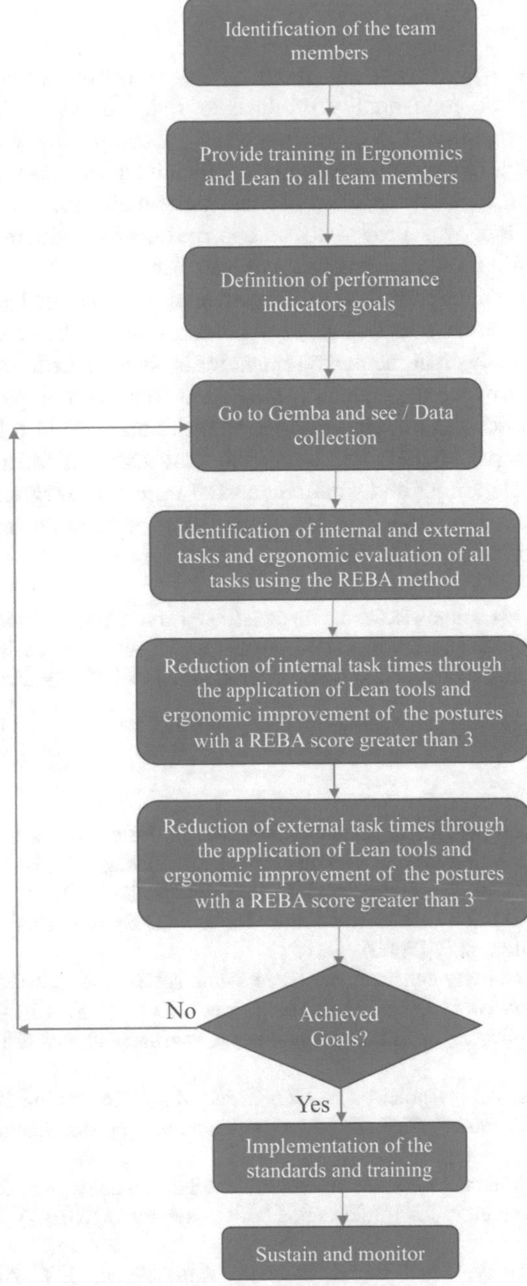

Fig. 3. ErgoSMED flowchart.

4 Conclusion

Companies struggle every day to survive in a very competitive market in which customers are looking for good quality products at reduced prices. Lean concepts and continuous improvement are the tools most used by companies which are focused in dealing with this situation in such a way that enables them to create products with more value for the customer, increase diversity and personalization, and shorten delivery times. The SMED tool was proved to be appropriate for reducing setup times and consequently cutting production batches and lead times.

Furthermore, companies deal with increases in occupational diseases related to musculoskeletal problems due to the poor postures of workers. Companies should include ergonomic analysis in the management tools, such as Lean manufacturing tools and continuous improvement projects, if they want to solve this problem.

This work proposes the integration of ergonomic analysis in a Lean tool used for the reduction of setups: SMED. Following the flowchart put forth in this work it is possible to reduce setup times and simultaneously improve the ergonomic conditions of workers. The authors have therefore concluded that ergonomic improvements often lead to waste reduction and vice versa.

Acknowledgments. We acknowledge the financial support of CIDEM, R&D unit funded by the FCT – Portuguese Foundation for the Development of Science and Technology, Ministry of Science, Technology and Higher Education, under the Project UID/EMS/0615/2016.

References

1. Benjamin, S.J., Murugaiah, U., Marathamuthu, M.S.: The use of SMED to eliminate small stops in a manufacturing firm. J. Manufact. Technol. Manage. **24**(5), 792–807 (2013)
2. Kester, J.: A lean look at ergonomics. Ind. Eng. **45**(3), 28 (2013)
3. Shingo S.: A Revolution in Manufacturing: The SMED System. Productivity Press, USA. Translated by Dillon, A.P. (2000)
4. Desai, M.S.: Productivity enhancement by reducing setup time – SMED: case study in the automobile factory. Glob. J. Res. Eng. Mech. Mech. Eng. **12**(5), 15–20 (2012)
5. Ulutas, B.: An application of SMED methodology. World Acad. Sci. Eng. Technol. **5**, 63–66 (2011)
6. Brito, M., Ramos, A.L., Carneiro, P., Gonçalves, M.A.: Combining SMED methodology and ergonomics for reduction of setup in a turning area. Procedia Manufact. **13**, 1112–1119 (2017)
7. Yin, R.: Applications of case study research. Sage Publications, Inc., California (2003)
8. Hignett, S., McAtamney, L.: Rapid entire body assessment (REBA). J. Appl. Ergon. **31**, 201–205 (2000)
9. McIntosh, R.I., Culley, S.J., Gest, G., Mileham, A.R., Owen, G.W.: An assessment of the role of design in the improvement of changeover performance. Int. J. Oper. Prod. Manage. **16**, 9 (1996)
10. Ferradás, P.G., Salonitis, K.: Improving changeover time: a tailored SMED approach for welding cells. Proceedia CIRP **7**, 598–603 (2013)
11. Boran, S., Ekincioğlu, C.: A novel integrated SMED approach for reducing setup time. Int. J. Adv. Manufact. Technol. **92**(9–12), 3941–3951 (2017)

Prediction of Failure Candidates of Technical Products in the Field Based on a Multivariate Usage Profile Using Machine Learning Algorithms Regarding Operating Data

Sebastian Sochacki(✉), Fabian Reinecke, and Stefan Bracke

Chair of Reliability Engineering and Risk Analytics, University of Wuppertal, Gaußstraße. 20, 42119 Wuppertal, Germany
{sochacki, freinecke, bracke}@uni-wuppertal.de

Abstract. In this paper an approach is presented, which gives the possibility to predict the probability for each specific product in a fleet to be affected by one or more failure modes over their lifetime. The approach considers the specific usage by the end consumer, which can be described by several usage variables. This gives the possibility to determine multivariate field load profiles within the product fleet. In addition, products, which are located in a usage area in the field, where several types of failure are likely to occur are assessed by corresponding probabilities of occurrence. Furthermore, it is possible to estimate the future risk probability for each product at the current field time as well as at upcoming field times. The approach is based on machine learning methods and is applied within a case study, which refers to a generated synthetic data set regarding the automotive industry.

Keywords: Product fleet · Failure candidates · Machine learning methods · Pattern recognition · Failure recognition · Field action strategy

1 Introduction

Nowadays, the number of innovations of technical products increases exponentially. Every new product generation is equipped with a variety of new functions, whereas they mostly maintain the same or smaller physical dimensions. While the development time of the products becomes shorter, at the same time, an ever-increasing product demand arises. This poses huge challenges for the product development of manufacturers and the component suppliers. These conditions, in association with a wide variety of different product usage by end customers, can lead to an increasing failure sensitivity and a high number of different types of failures, which can occur at each individual product depending on the kind of usage within the field time. In the case of planning field actions (e.g. recalls, maintenance actions), failure specific prognostics are difficult to perform for the manufacturers for each individual product in the field, especially if the specific failure types are described by more than one usage variable.

This paper presents an approach that gives the possibility to estimate the probability for each specific product in the field to be affected by one or more failure modes over

T. Ahram et al. (Eds.): IHSED 2019, AISC 1026, pp. 555–560, 2020.
https://doi.org/10.1007/978-3-030-27928-8_84

their lifetime. This can be done individually for each product based on their specific type of use by the end consumer, which can be expressed by several variables in a multivariate way. In addition, products, which are located in a usage area in the field, where several types of failure are likely to occur, can be prioritized by corresponding probabilities of occurrence. The presented approach is based on methods from the field of machine learning (ML). The data basis consist of warranty data, which contain different failure scenarios as well as cumulated data of usage variables for each element of a product fleet. The approach uses operating data as training data for a model building process. These data can be recorded for instance by condition monitoring or diagnosis systems using sensors and control units. Operating data of a product fleet are available for malfunctioned products and for those, which are not affected by any failure. The usage data are read out during recurring product maintenance actions (e.g. regular inspections). The approach is applied within a case study, presented in this paper, which refers to a generated synthetic data set regarding the automotive industry. The results obtained by the presented approach provide an estimation of possible failures occurrences of different failure modes of a product during field operation considering the individual product usage of the customer. The results can contribute to the improvement of products in the area of product optimization, component design and risk management. The approach is based on [1] but has the improvement of estimating probabilities for various failure modes as well as assessing fleet elements which are extrapolated regarding their field time.

2 Fundamentals

The presented approach in this paper uses the supervised machine learning technique of Gradient Tree Boosting [2] in order to learn a model out of labeled training data to classify new data instances. The technique uses Classification or Regression Trees (CART) [3]. A CART tree is grown by following a binary partitioning scheme, which means that the feature space S (training data) is splitted into two sub-regions. Then, the sub-regions can be further split in the same way until a stopping criterion (i.e. a maximum tree size) is reached. At each split, a specified split-point and a splitting variable is chosen. Regions R_j at terminal nodes of the tree indicate the results of the tree and can be used for classification (choosing the majority class in R_j) or regression (calculating the mean of the target variable in R_j) [4, 5]. Finding the best split at a tree node is done with the help of a splitting criterion (e.g. Gini index) [4].

The underlying algorithm for K-class classification uses a training dataset $\{(x_i, y_i)\}_{i=1}^{n}$, where x_i indicates an attribute vector and y_i a corresponding class and a differentiable loss function $L(y, F(x))$ as well as a specified number of iterations M as input. The idea behind gradient tree boosting is the formation of a predictive model by a sum of weak classifiers (e.g. CART trees), which is built following a stagewise procedure where at each step, optimal weak classifier coefficients are determined at each terminal leaf [4, 6]. The output are K different sets of coupled weak classifiers $f_{kM}(x), k = 1, 2, \ldots, K$ for each class.

The algorithm for K-class classification can be described as follows [4]: In step 1, an initializing model is defined for each class. Step 2a and 2b indicates the training stage and is repeated for M iterations for each class k. In step 2a. the probabilities $p_k(x)$ for each class k are calculated from the last generated class models $f_k(x), k = 1, 2, \ldots, K$ by using the normalized exponential [7], then step 2b is performed for each iteration m separately for each class k. In 2b, first of all, the pseudo residuals r_{ikm} are calculated by the negative gradient of the multinomial deviance. Subsequently, a decision tree is trained to the residuals r_{ikm} which results in the terminal nodes R_{jkm}. Then the optimal weak classifier coefficients γ_{jkm} are determined and the classification model is updated or expanded [4, 6]. A detailed explanation of the algorithm is given in [4].

3 Goals and Methodological Concept

The goal of the approach presented in this paper is to determine the risk probability of different failure modes for each individual fleet product based on the individual type of use, expressed by multiple variables. In order to determine the risky failure areas of the failure modes in the field, a ML model is trained based on data of products, which were affected by a specific failure mode in the past. By learning a decision tree model with Gradient Tree Boosting method out of the fleet data, failure probabilities of fleet products regarding different failure modes can be estimated. This gives the opportunity to detect changes in the failure behaviour of each product with the assumption of no changes concerning the usage profile. The results are product specific probability curves for each individually used fleet product, which contain the estimated dynamic failure probability evolution referring to the continuing field time. Based on the probability curve, products can be analysed regarding possible damage causality occurrences, which allows the derivation and planning of necessary field actions.

Figure 1 shows the general workflow of the approach in principle. In the first step, the warranty data and the operating data of fleet products are merged. In order to detect unnecessary variables that have no significance for the corresponding failure modes, the data is pre-filtered in a variable prioritization process. Subsequently, based on Sect. 2, a decision tree model is learned with the help of Gradient Tree Boosting. Once the model has been trained, using the variables of failed and non-failed products, the probabilities of the occurrence for each failure mode can be estimated for each product at the current field time as well as at upcoming field times. In the latter case, it is assumed that the operating data for future field times have been extrapolated linearly with respect to the field time.

Afterwards, risk analysis and assessment can be performed and changes of the reliability of each individual used product can be evaluated on the basis of the fleet knowledge respectively the training data. Finally, with the performed risk analysis, a product specific field action strategy can be carried out considering the product use.

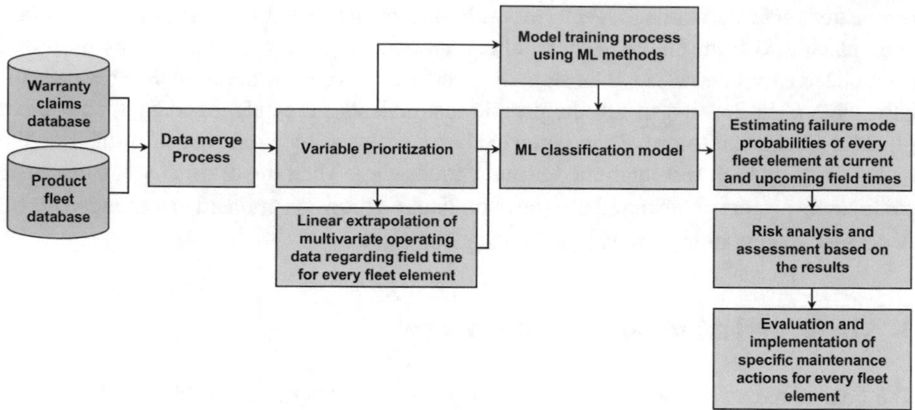

Fig. 1. General workflow of the methodological concept

4 Case Study – Base of Operations

The presented approach in this paper is based on a database, which consist of warranty data, merged to accumulated operating data of a technical product fleet. The operating data are recorded for each product in the fleet at various readout or field times. After the data merging process, a database exists which contains operating data for each product at a specific field time labelled to different failure modes including the faultless case. The database used in the performed case study of this paper was generated synthetically. It contains properties and data referring to the automotive industry and is inspired by a real data set. In total, the fleet database contains data of 4000 vehicles. For each vehicle 6 cumulated variables as well as the field time are read out, which describe the usage of different vehicle components during the use phase. If a vehicle is affected by a failure mode, the variables were read out at the field time of failure occurrence. Within the framework of this case study, a total of three different failure modes were generated with the help of manually predefined rules. Then, related operating data are labelled by the respective failure mode designations "FM1", "FM2" and "FM3". Vehicles without a failure are labelled with "FM0". Table 1 shows the general information of the database.

Table 1. Database information

Designation	Information
Production batch (automobile fleet)	$N_{PF} = 4000$
No. of failure modes	3
No. of training data	Failure mode 1: 300 Failure mode 2: 258 Failure mode 3: 335
Available variables for characterisation (accumulated)	Kilometrage, Engine operation time, Clutch cycles, Starter cycles, Start-Stop cycles, Door opening cycles

5 Case Study – Results

On the theoretical basis of the previous chapters, the methodology is applied to the described data set. The following Fig. 2 shows the result of the application of the method regarding a specific vehicle in form of probability curves of survival and failure probabilities over the field time.

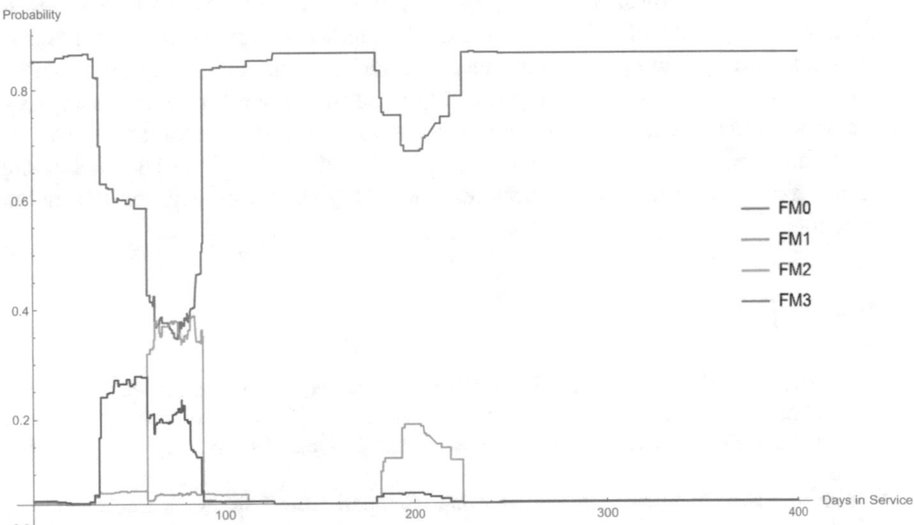

Fig. 2. Example of a vehicle failure mode profile based on the specific multivariate usage over the field time

The blue curve represents the general survival probability of the vehicle at the corresponding day in service. The yellow curve represents the probability over time of the occurrence of failure mode 1. The green and red curves indicate the probabilities of the occurrence of failure mode 2 and failure mode 3. It can be observed that there is a slight risk to be affected by a failure mode already at the beginning of the operating time of the vehicle. This usually indicates random failures. After 35 days of field runtime, the probability that the vehicle is affected by failure mode 3 increases significantly. The highest probability can be observed after 60 days in the field with 27.8% failure probability. After 61 days of field time, the probability that the vehicle is affected by failure mode 3 also increases. The highest probability to be affected by failure mode 3 is 38.2%. At this point, the vehicle shows high failure probabilities for FM2 as well as for FM3. It can be observed that FM2 and FM3 superimpose on each other with regard to the failure profiles. The lowest amount of the survival probability of the vehicle is 35.2%. If the vehicle is not affected by a failure in this field time interval, the failure probability decreases after 92 days for all failure modes below 2%. Then, the failure probability of FM1 increases between 180 and 210 days up to 21.2%. With this procedure, individual measures for each fleet element can be determined and implemented depending on the specific usage profile.

6 Outlook

This paper shows an approach for the individual evaluation of the failure probability of products with regard to specific failure modes on the basis of specific multivariate usage profiles. With the help of this procedure, the failure risk of a vehicle can be identified based on the type of use during field operation.

The presented variable prioritization of the data within the approach is not necessary at small data volumes, but it reduces the calculation time of the process significantly at high quantities of data. To optimize the prediction, several readout points of the products during field operation are recommended to obtain an adequate prognosis of future field usage: This gives the possibility to considering a further research work, the focus is on the extension of the approach regarding the application to Big Data as well as the application to real data sets of the automotive industry considering numerous failure modes, usage variables and a higher complexity of the failure behaviour.

References

1. Sochacki, S., Reinecke, F., Bracke, S.: Ansatz zur Anpassung von Wartungs- und Instandhaltungspaketen auf Basis maschineller Lernalgorithmen im Hinblick auf den zuverlässigen Betrieb technisch komplexer Produkte. In: Gesellschaft für Qualitätswissenschaft e.V. (2018, in press)
2. Friedman, J.H.: Stochastic gradient boosting. Comput. Stat. Data Anal. **38**(4), 367–378 (2002)
3. Breiman, L.: Classification and Regression Trees. Chapman & Hall, Boca Raton (1998)
4. Hastie, T., Tibshirani, R., Friedman, J.H.: The elements of statistical learning: data mining, inference, and prediction, 12th edn. Springer, New York (2017)
5. Awad, M., Khanna, R.: Efficient Learning Machines. Apress, Berkeley (2015)
6. Lin, L., Yue, W., Mao, Y.: Multiclass image classification based on fast stochastic gradient boosting. Informatica **3** (2014)
7. Bishop, C.M.: Pattern Recognition and Machine Learning, 8th edn. Springer, New York (2009)

A Systematic Review of Healthcare-Associated Infectious Organisms in Medical Radiation Science Departments: Preliminary Findings

D'arcy Picton-Barnes[1], Manikam Pillay[1,2], and David Lyall[1](✉)

[1] School of Health Sciences, The University of Newcastle, Callaghan, Australia
D.Picton-Barnes@uon.edu.au,
{manikam.pillay, david.lyall}@newcastle.edu.au

[2] Centre for Resources Health and Safety, School of Health Sciences, The University of Newcastle, Callaghan, Australia

Abstract. Healthcare-associated infections (HAIs) arise from exposure to infectious organisms within a healthcare setting, HAIs are a significant occupational risk for Medical Radiation Science (MRS) professionals which have a high degree of patient-contact. Methods: Systematic reviews and Meta-Analyses (PRISMA) guidelines were used in this research. The inclusion criteria were studies describing infectious organisms found directly in MRS departments or outlining a risk of infection to MRS professionals. Studies on infectious organisms on clothing or equipment used in these departments were also included. Results: Screening excluded 1655 articles that didn't meet the criteria, twenty-one articles were included in the final review. Preliminary findings, published between 1998 and 2018, were from eleven countries. Three studies were set in radiotherapy and thirteen in diagnostic radiography departments. No studies were published from nuclear medicine departments. Twenty different organisms were identified. Conclusion: Preliminary findings indicate a lack of related published studies within nuclear medicine departments.

Keywords: Hospital acquired infection · Nosocomial · Cross infection · Medical Radiation Science · Radiation therapy · Radiotherapy · Medical imaging · Radiography · Nuclear Medicine · PRISMA

1 Introduction

Healthcare-associated infections (HAIs) are infections that arise within a healthcare setting [1]. These include infectious organisms that patients are exposed to when receiving care, but also occupational risks of infection among healthcare workers [1]. HAIs are problematic as they result in greater rates of mortality and morbidity, increased antimicrobial resistance in microorganisms, and increased costs for healthcare systems and those receiving care [1]. Over the last 40 years, focus has shifted to hygiene and sanitation practices as the primary solution for preventing the spread of these infections [1, 2]. However, despite improvements in these practices, HAIs are still highly prevalent in healthcare facilities – affecting hundreds of millions of patients worldwide each year [1].

T. Ahram et al. (Eds.): IHSED 2019, AISC 1026, pp. 561–565, 2020.
https://doi.org/10.1007/978-3-030-27928-8_85

Appropriate infection control guidelines are imperative in reducing the number of HAIs, thereby reducing their burden and improving patient outcomes [1].

Medical Radiation Science (MRS) professionals have a high degree of contact with members of the community who are susceptible to infections; including children, the elderly, and those with already compromised immune systems. This level of interaction means that if an MRS worker develops an infection, there is a risk of the disease spreading to these susceptible patients. Reducing the risk of exposure to infectious organisms is the responsibility of both the healthcare institution and individual practitioners [1, 3]. Different organisms require different methods of infection prevention and control, such as sterilisation or disinfection. Accordingly, infection control methods should be specific to the healthcare setting in order to reduce the prevalence of HAIs [1]. Research into the nature of specific infectious organisms found within MRS departments may serve to guide infection control strategies.

Current research predominantly focuses on HAIs among patients only, and not on the infectious organisms that staff are exposed to. There is also a lack of research examining infectious organisms across MRS professions.

One study [4] examined the relationship between hospital staffing and HAIs; however, this study focussed on nursing and medical staff and not MRS. This is a significant gap. This systematic review therefore aims to collect and analyse the data from recent literature to determine the infectious organisms that MRS staff are exposed to within their departments; as well as examine the reservoirs and modes of transmission of these organisms.

2 Aim

The aim is to systematically review the literature to determine the common infectious organisms found within MRS departments, their reservoirs and modes of transmission.

3 Methods

A systematic review of the literature was performed to identify studies that explored the different infectious organisms found in MRS departments, and the risk of MRS professionals developing an HAI. Reporting of this systematic review followed the Preferred Reporting Items for Systematic reviews and Meta-Analyses (PRISMA) guidelines [5].

3.1 Protocol and Registration

The protocol for this systematic review is submitted for registration with international prospective register of systematic reviews, PROSPERO.

3.2 Search Strategy

Electronic databases Scopus, EMBASE, the Cochrane Library, Medical Literature Analysis and Retrieval System Online (MEDLINE), and the Cumulative Index to Nursing and Allied Health Literature (CINAHL) were searched for relevant studies published between 1984 and 2018. The search was confined to studies published after 1983 as a number of international infection control documents were published that year, including the Association for Professionals in Infection Control and Epidemiology Curriculum [6]. The search was performed on the 27th of January 2019, and was restricted to studies published in the English language.

The search strategy was developed using a combination of Medical Subject Headings (MeSH) and keywords associated with the topic, including "healthcare associated infection," "hospital pathogens," and "allied health". Additional studies were identified through a manual search of citations in retrieved studies.

3.3 Selection Criteria

The inclusion criteria were all studies examining infectious organisms found directly in MRS departments, or outlining a risk of infection to MRS staff. Studies on clothing or equipment commonly found in these departments were also included. Studies set in intensive care, critical care, surgical wards or maternity wards were excluded. Case studies were also excluded based on their inherently limited statistical validity (due to small sample sizes and no control groups to compare outcomes). Other exclusion criteria included non-peer reviewed literature, reviews, editorials, commentaries, articles on community-acquired infections, articles published prior to 1983, and articles not published in the English language.

Definitions

For the purpose of this systematic review, the following definitions were used:

- *Medical radiation scientists* were defined as healthcare professionals who perform diagnostic imaging studies or plan and deliver radiation treatments to patients. These professionals include diagnostic radiographers, nuclear medicine scientists and radiation therapists.
- *Healthcare-associated infections* comprise of infections acquired by persons in a healthcare setting. These include bloodstream infections and organism-specific infections (such as methicillin-resistant Staphylococcus aureus and Clostridium difficile infections) that were defined as being associated with healthcare in the literature.
- *Radiation department associated infections (RDAI)* refer to infections that MRS staff are at risk of contracting within their departments.

3.4 Study Selection

The titles and abstracts of all studies were assessed for relevance according to the systematic review aim, and articles that were not relevant were excluded. The full texts of remaining articles were acquired to further assess relevance based on the systematic review inclusion and exclusion and criteria. Articles with relevant study design and data were included. The study selection process was performed by researches using Covidence - an online systematic review management programme. Two researches reviewed the title and abstracts of studies, checking the level of agreement at several intervals. A third reviewer cross-checked included studies against the criteria at each stage of the study selection process. Any disagreements or discrepancies in the application of the selection criteria were resolved by discussion between all reviewers.

3.5 Data Extraction

Data were extracted using a Microsoft Excel spreadsheet. Data extracted from eligible studies included: title, author(s), year of publication, country, study population, healthcare department (setting), sample, type(s) of organisms, organism incidence and/or prevalence data, outcomes, and limitations. Extracted data were reviewed by a second researcher.

3.6 Data Analysis

This work is still in progress. Data from included studies will be synthesised and displayed in evidence tables. Summary tables will include studies examining the prevalence of infectious organisms, specific department and infectious organisms, and equipment and infectious organisms. Meta-analysis will be undertaken if sufficient homogeneity is found in the final set of articles.

4 Results and Discussion

The search across five electronic databases returned 2063 studies, reduced to 1676 once duplicates were excluded. Screening excluded 1655 articles that did not meet the inclusion criteria, resulting in twenty-one articles included in the final review. Initial findings suggest these articles, published between 1998 and 2018, were from eleven countries in Africa, Asia, North America and Europe. Three studies were set in radiotherapy and thirteen in diagnostic radiography departments. No studies were published from nuclear medicine departments. This is a significant gap five studies were non-specific but included information on equipment found in MRS (such as mobile phones or identification badges).

5 Conclusion

Identifying the infectious organisms MRS professionals are exposed to allows practitioners to be better informed when applying infection control practices. Initial findings of this review indicate a lack of related published studies within nuclear medicine departments. This is a significant gap and a subject of future research. Future studies will report on the different infectious microorganisms identified following physical sampling and the practitioner's perception of occupational risk to infectious microorganisms.

References

1. Report on the Burden of Endemic Health Care-Associated Infection Worldwide Geneva: World Health Organisation (2011)
2. Bolyard, E., Tablan, O., Williams, W., Pearson, M., Shapiro, C., Deitchman, S.: Guideline for infection control in health care personnel. Am. J. Infect. Control **26**(3), 289–327 (1998)
3. Hughes, R.G.: Patient Safety and Quality: An Evidence-Based Handbook for Nurses. MD Agency for Healthcare Research and Quality U.S. Department of Health and Human Services, Rockville (2008)
4. Mitchell, B.G., Gardner, A., Stone, P.W., Hall, L., Pogorzelska-Maziarz, M.: Hospital staffing and health care-associated infections: a systematic review of the literature. Joint Comm. J. Qual. Patient Saf. **44**(10), 613–622 (2018)
5. Moher, D., Liberati, A., Tetzlaff, J., Altman, D.: Preferred reporting items for systematic reviews and meta-analyses: the PRISMA statement. PLoS Med. 6(7) (2009)
6. Soule, B.: The APIC curriculum for infection control practice. Kendall-Hunt, Dubuque (1983)

Process Operator Students' Abilities to Assess OSH Risks

Noora Nenonen(✉), Sanna Nenonen, and Sari Tappura

Unit of Industrial Engineering and Management,
Tampere University, 33014 Tampere, Finland
{Noora.Nenonen, Sanna.Nenonen, Sari.Tappura}@tuni.fi

Abstract. Safety competence is an important process operator skill. Due to hazardous work assignment and environments, skills in assessing the risks related to occupational safety and health (OSH) are especially important. Carrying out risk assessments can be difficult, and several problems have been identified. The aim of this study was to discover how well process operator students are able to assess OSH-related risks. Risk assessment exercises with observations were carried out for students (n = 35) in three vocational education and training (VET) organizations. The results showed that all students were able to identify at least some hazards. The students identified the most probable, high-risk, and easily observable hazards. Those with previous training or experience in work and risk assessment were more capable of identifying a wide range of risks. We conclude that successful risk assessment requires related competence, which should be developed via theoretical and practical learning during VET.

Keywords: Hazard recognition · Risk identification · Risk assessment · Workplace safety · Manufacturing industry

1 Introduction

Safety competence is an important skill for process operators. In the safety-critical process industry, it is one of the key determinants of company performance. Although safety issues have increasingly gained attention, safety criticality is still emphasized in the process industry [1]. For example, the increasing complexity of processes makes the safety-focused aspects of the process operator role significant [2].

A basic understanding of safety and related practices is imparted to process operator students during their studies in VET organizations [3]. Companies expect that students have obtained this basic safety knowledge (e.g., risk assessment skills) before they proceed to training or employment [4]. Training for company-specific safety requirements, culture, and practices is provided in the workplace.

The European OSH legislation states that the employer should prevent occupational risks [5]. The practices of risk assessment and management are generally considered the foundation for OSH management, and they are widely used in workplaces [6]. In addition, in many workplaces, employees must carry out a short risk assessment before starting work. Process operators especially need skills in OSH-related risk assessment

T. Ahram et al. (Eds.): IHSED 2019, AISC 1026, pp. 566–572, 2020.
https://doi.org/10.1007/978-3-030-27928-8_86

because their work environment contains many hazards (e.g., dangerous materials, high temperatures and pressure), which can lead to major accidents [1]. In addition, young and inexperienced workers typically experience more injuries than others [7, 8].

However, carrying out risk assessments can be difficult, and several problems have been identified in previous research. For example, a considerable number of OSH-related hazards in the work environment seem to remain unidentified [9–11]. Moreover, deficiencies in risk assessment and management are often mentioned among the causes of occupational accidents [12, 13].

The aim of this study was to discover how well process operator students were able to identify and analyze OSH-related hazards and means to avoid or control related risks. The differences related to student age, stage of study, safety competence, and work experience are discussed.

2 Materials and Methods

In this study, risk assessment exercises (n = 15) with observations and short interviews were carried out for process operator students (n = 35) in three VET organizations in Finland. This study is part of a larger study [see e.g., 4] focusing on workplace learning carried out in cooperation with process industry VET organizations and companies. The VET organizations participating in this study are the organizations participating in before-mentioned the larger study. There were approximately 40–100 process operator students in the VET organizations.

The risk assessment exercises took place in the VET organizations' laboratories, and assessment targets were chosen from these laboratories. The assessment targets were chosen in collaboration with the teachers. The researchers wanted to ensure that the students participating in the study were familiar with the risk assessment targets. The students did the exercises one group at a time in groups of two or three persons. A pre-prepared checklist (see Appendix A) was used in the risk assessment. The list was compiled on the basis of checklists used in two process industry companies cooperating with the participating VET organizations. In addition, the Risk Assessment in Workplaces Workbook [14], which is a commonly applied tool for OSH-related risk assessment in Finland, was employed.

The exercise started with a short interview to gather students' background information. In addition, the researchers gave a short introduction on how to do the exercise and use the checklist. The students were asked to identify and describe hazards and their consequences, estimate the magnitudes of risks (on a scale from 1–3), and come up with actions to avoid or reduce the risks. However, actual training on risk assessment was not provided. The researchers observed the assessments but did not participate in them. The students were helped, and their questions were answered if they seemed to have problems carrying out the assessments. Once the group was ready with the risk assessment, the researchers went through the results and briefly discussed them with the students. Notes were taken throughout the entire exercise. For comparison, teachers in each organization did the exercises as well.

In one of the VET organizations, the risk assessment target was the sheet mold and press and the work carried out with them. In two organizations, the assessment target was work carried out in the crushing room. The risk assessments lasted between 10 and 52 min, with the average duration being 24 min. The background information of the students participating in this study is summarized in Table 1. In two VET organizations, the participating students were mainly younger first-year students who had not yet received workplace learning. There were, however, some students who had studied longer and had been on workplace learning period. For these students, the studies lasted three years. In one of the VET organizations, all of the participating students were adult students who had been through a short two-week workplace learning period. For these students, the studies lasted 15 months.

Table 1. Background information on the process operator students (n = 35).

Number of students	VET organization A (37%), B (26%), C (37%)
Age	Between 16 and 53 years, average: 23, median: 19
Gender	Men (89%), women (11%)
Years of study	First-year students (30), second-year (4), fourth-year (1)
Workplace learning	51% of the students had been in the workplace learning, 49% had not
Work experience	77% of the students had work experience, 14% did not (9% unknown)
Experience in risk assessment	Yes (40%), no (60%)
Safety training	All of the students had received some safety training
Familiarity with the assessment target	51% of the students had worked in the assessment target, 49% had not, but they were otherwise familiar with the target

3 Results

The checklist used in the risk assessment exercises contained a total of 28 different items in five categories. Both students and teachers identified hazards related to all categories. The students identified on average 13 (range: 5–28) different hazards for 12 (range: 5–20) different items. The teachers identified on average 24 (range: 18–30) hazards for 17 (range: 14–20) items. Table 2 summarizes the number of hazards identified in the two different assessment targets of this study.

Table 2. The number of hazards identified in risk assessments.

Target	Students	Teachers
Sheet mold and press	Hazards average 8 (range: 5–13)	18 hazards
	Items average 8 (range: 5–12)	14 items
Crushing room	Hazards average 17 (range: 8–28)	30 hazards
	Items average 15 (range: 8–22)	20 items

The Appendix A summarizes the results of the hazard identification by hazard category and item. With regard to the sheet mold and press, there was only one risk all student groups identified: the risk of fingers being crushed between the sheet press. Otherwise, the students identified mainly accident hazards, although some groups and teachers identified some other hazards as well.

In the risk assessments concerning the crushing room, all student groups identified the hazards related to small pieces of rocks being hurled from the crushing mill and noise from the devices used. There were also many other hazard items in the crushing room that most student groups identified (e.g., objects being dropped and unsafe activities).

The students identified some hazards that the teachers did not identify and vice versa. For example, the teachers did not recognize the hazard posed by the noise from a machine near the sheet mold and press, probably because the noise did not exceed limit value and there was an instruction to use hearing protectors while this machine was in use. Moreover, the students pointed out that safety orientation concerning emergencies (e.g., the location of the first aid equipment) in the laboratory was given to students only in the beginning of the school year. There were some students who had started their studies in the middle of the school year and therefore had not participated in the orientation. The teachers recognized, for example, the risk of falling from a safety ladder while opening the compressed air (a short person cannot reach the switch without a ladder).

Many student groups pointed out that there was no first aid equipment in the crushing room, whereas the teachers probably considered it sufficient that first aid equipment was available in the nearby laboratory. In addition, the students brought up the dust from rocks in connection with two items in the checklist: suffocation and dust and fiber. The teachers reported this in connection only with the item "dust and fiber." The students discussed the possibility of dust causing allergic reactions and therefore suffocation.

The groups with adult students (with previous work experience) identified more hazards than younger students. Moreover, students with more experience from work and study identified more risks than inexperienced first-year students. In many of the groups that identified many hazards, the students had some previous experience in risk assessment (e.g., school exercises). However, there were also groups in which students had risk assessment experience but did not recognize many risks, and there were groups in which the students had no previous risk assessment experience but still recognized many hazards.

The consequences of the risks were rarely described, although such descriptions were requested in the exercise. Moreover, some students mentioned that it was difficult to define actions to avoid or control risks. Often the students mentioned controls that were already in use (e.g., personal protective equipment). Some students mentioned that it would have been easier to do the assessment if they had been more familiar with the assessment target. In addition, the students often estimated a larger magnitude of risk when they considered the risk highly probable, although the consequences were also considered. The students also often estimated the risks as less severe than the teachers.

4 Discussion

The results showed that all the students were able to identify at least some hazards. The students identified the most probable, high-risk, and easily observable hazards. Among the students, risks with very small consequences, low probability of occurrence, or existing controls were not considered relevant risks by the students. The students discussed these issues in the assessments, but did not document them in the actual assessment tool. Students with previous training or experience in work and risk assessment were more capable of identifying a wide range of risks. Overall, the teachers seemed to identify more hazards and assess the magnitudes of the risks more highly than the students.

We conclude that successful risk assessment requires related competence, which should be developed via theoretical and practical learning during VET. Workplace learning supports process operator students' risk assessment and safety skills. However, safety training should begin in the early phases of the studies and continue systematically throughout the studies. Skills in OSH-related risk assessment are likely to support the assessment of risks related to process safety as well.

Acknowledgments. The authors gratefully acknowledge the Finnish Work Environment Fund for the funding of this study and the teachers and students of the cooperating VET organizations for their contributions.

Appendix A: Results of the Hazard Identification by Hazard Category

Categories and items in the checklist	Target 1[a]		Target 2[a]		Examples of identified hazards by category
Accident hazards	S[b]	T[c]	S[b]	T[c]	
Slipping, stumbling, falling (down/over)	4	Y	7	Y	Crushing fingers between the sheet press; Dropping the weights of sheet mold; Stumbling on the platform used on sheet mold; Pieces of rocks can hurtle from the mill; Suffocation because of dust and allergic reaction
Fall of a person/falling from height	4	Y	3	Y	
Electric shock or static electricity	3	Y	6	Y	
Reduction of oxygen, suffocation	0	N	5	N	
Goods transport and other traffic	1	N	4	Y	
Objects being dropped or falling over	5	Y	8	Y	
Objects/material being hurtled around or hit by a moving object	1	Y	9	Y	
Being crushed between objects or entangled in a moving object	6	Y	6	Y	
Being slashed, cut, or stabbed	0	N	6	Y	

(*continued*)

(*continued*)

Physical hazards and strain	S	T	S	T	
Noise	3	N	9	Y	Lifting heavy weights of the sheet mold in a circular motion repeatedly; The devices in the crushing room are noisy; Lifting the rock material in heavy buckets and using poor lifting postures; The crushing room can be hot
Hot and cold objects and surfaces	2	N	1	1	
General and local ventilation	0	N	3	Y	
Lightning	1	Y	1	N	
Vibration	1	N	6	Y	
Radiation	0	N	0	N	
Poor working postures, repeated movements, and lifting or carrying with hands	2	Y	7	Y	
Usability of tools, machinery, and devices	1	N	4	Y	
Organization and personnel activities	S	T	S	T	
Exceptional situations and disturbances (e.g., unexpected starting of machine)	2	Y	6	Y	Unexpected descent of sheet press when compressed air opened; Not obeying instructions
Unsafe activities	1	Y	8	Y	
Safety arrangements	S	T	S	T	
Personal protective equipment (PPE), safeguarding (condition and use)	1	Y	4	Y	Requiring more strictly the use of PPEs; Emergency exits are not clear (objects stored); Rehearsing emergency situations
Alarm and rescue equipment	2	Y	2	N	
Walkways and corridors and their safety and indicator lightning	2	Y	6	Y	
First aid arrangements and equipment	2	N	7	N	
Chemical and biological hazards	S	T	S	T	
List of chemicals	0	N	1	Y	Chemical register only available in teachers room in the laboratory; Dust from rock and ore material in the crushing room
Labeling of chemical packages	0	N	0	N	
Hazardous or harmful chemicals (allergenic, carcinogenic, flammable, explosive)	0	N	2	N	
Dust and fiber	1	N	7	Y	
Gases, vapor, fumes, and smoke	0	N	1	N	
Other hazard	0	Y	3	N	Cramped workspace

[a]Target 1: Sheet mold and press, Target 2: crushing room

[b]S: Number of student groups (in target 1 n = 6, in target 2 n = 9) that identified a hazard or hazards related to this item

[c]T: Did teachers identify a hazard or hazards related to this item (Y = yes, N = no)

References

1. Rodríguez, M., Díaz, I.: A systematic and integral hazards analysis technique applied to the process industry. J. Loss Prevent. Process Ind. **43**, 721–729 (2016)
2. Nazir, S., Sorensen, L.J., Øvergård, K.I., Manca, D.: How distributed situation awareness influences process safety. Chem. Eng. Trans. **36**, 409–414 (2014)
3. Tappura, S., Kivistö-Rahnasto, J.: Annual school safety activity calendar to promote safety in VET. In: Arezes, P.M., et al. (eds.) Occupational Safety and Hygiene VI: Proceedings of the 6th International Symposium on Occupation Safety and Hygiene (SHO 2018), pp. 131–135. CRC Press, London (2018)
4. Tappura, S., Nenonen, S, Nenonen, N.: Developing safety competence process for vocational students. In: Ahram T., Karwowski W., Taiar, R. (eds.) Human Systems Engineering and Design, IHSED 2018. Advances in Intelligent Systems and Computing, vol. 876, pp. 668–674. Springer, Cham (2018)
5. Council Directive 89/391/EEC on the introduction of measures to encourage improvements in the safety and health of workers at work (1989)
6. European Agency for Safety and Health at Work: Summary - Second European Survey of Enterprises on New and Emerging Risks (ESENER 2). Publications Office of the European Union, Luxembourg (2015)
7. Laberge, M., Ledoux, E.: Occupational health and safety issues affecting young workers: a literature review. Work **39**(3), 215–232 (2011)
8. Salminen, S.: Have young workers more injuries than older ones? An international literature review. J. Saf. Res. **35**(5), 513–521 (2004)
9. Albert, A., Hallowell, M.R., Skaggs, M., Kleiner, B.: Empirical measurement and improvement of hazard recognition skill. Saf. Sci. **93**, 1–8 (2017)
10. Bahn, S.: Workplace hazard identification and management: the case of an underground mining operation. Saf. Sci. **57**, 129–137 (2013)
11. Heikkilä, A.-M., Malmén, Y., Nissilä, M., Kortelainen, H.: Challenges in risk management in multi-company industrial parks. Saf. Sci. **48**, 430–435 (2010)
12. Carter, G., Smith, S.: Safety hazard identification on construction projects. J. Construct. Eng. Manage. **132**(2), 197–205 (2006)
13. Haslam, R.A., Hide, S.A., Gibb, A.G.F., Gyi, D.E., Pavitt, T., Atkinson, S., Duff, A.R.: Contributing factors in construction accidents. Appl. Ergon. **36**, 401–415 (2005)
14. Ministry of Social Affairs and Health: Riskien arviointi työpaikalla – työkirja. Ministry of Social Affairs and Health and The Centre for Occupational Safety (2015)

The Cost of Ensuring the Safety of Technical Systems and Their Service Life

Evgeny Kolbachev, Marina Perederiy, and Yulia Salnikova(✉)

Platov South-Russian State Polytechnic University (NPI),
Prosveshcheniya, 132, 346428 Novocherkassk, Russian Federation
kolbachev@yandex.ru, pmv__62@mail.ru,
yuliasalnikova@gmail.com

Abstract. The article proves that the condition for the creation of safe technical systems is the formation of their design appearance based on the cost parameters calculated using actuarial models. At the same time, the reasonable service life of the designed technical systems is to be determined by the cost of ensuring the safety of their operation. The article proposes a set of methods and relevant management tools for designing safe technical systems under the conditions of Industry 4.0 and re-industrialization in order to improve the quality of the technologies and increase human and social capital.

Keywords: Technical systems · Safety · Service life · Operating costs · Actuarial models

1 Introduction

The conditions of technical systems creation are determined by the relevant stage of the socio-economic development and the corresponding technological order. Currently, these conditions are associated with the processes of re-industrialization, the emergence of Industry 4.0, NBIC-convergence [1–3].

The authors of numerous studies draw attention to the fact that along with the new opportunities, development of Industry 4.0 and, especially, NBIC-convergence can cause potential threats to humans.

These risks and threats are mainly associated with the dangers of both physical and biological and psychological impact on a person [4–7].

It is obvious that the creation of NBIC technologies and production systems will require the formation of new management methods and tools in order to minimize risks and threats and to enhance their capabilities. In this connection, NBIC technologies and production systems should be considered as human-oriented.

Firstly, human-oriented technical system are to minimize risks and threats of system functioning, and secondly, to increase and cumulate the human and social capital [8]. In the context of our research, the attention should be focused on the first of these factors.

Speaking of Useful Life (which ensures the safety of a technical system as the main condition of its human-orientation), two aspects need to be considered: trouble-free operation during Useful Life (since any accident is more or less likely to harm human

T. Ahram et al. (Eds.): IHSED 2019, AISC 1026, pp. 573–578, 2020.
https://doi.org/10.1007/978-3-030-27928-8_87

health) and environmental aspect (as the absence of harm to human health during the stable operation period of the system).

It is obvious that the safe operation of the technical system should be carried out during the whole period of its useful use.

When determining the Useful Life of technical system, its cost parameters are equally important. This is important both for determining the regulatory Useful Life, which is calculated when designing a technical system, and for determining Remaining Useful Life, typical for current operation of a technical system when solving problems of its modernization and extending the service life.

2 Methods for Determining the Service Life of Technical System at the Stages of Technical Specifications and Technical Proposals

The most complete analysis of the known methods for determining the service life of technical systems is made by Mrugalska [8, 9]. These articles focus on Remaining Useful Life (RUL). It allows to estimate how long it will take the equipment until it reaches a failure threshold in the future operating conditions. Usually, this is done when deciding on the extension of the useful life of technical systems. However, the methods and approaches described in this article can be applied in other cases of calculating the service life. In particular, when designing new equipment and other technical systems.

In our previous works [10, 11], an approach was proposed to the design of technical systems based on their cost parameters. In our opinion, the number of these indicators should include Useful Life during which the safe operation of the technical system is ensured.

The functioning of the designed system (or its parts) can prevent various emergencies. Overcoming the consequences of such situations requires, over the lifetime of the system, the costs significantly higher than the costs of its operation. Such situations are quite numerous.

For example, the functioning of agricultural irrigation and drainage systems during their service life prevents situations when in dry years (which happen with a certain probability during the service life of the system) the cost of acquiring food exceed many times the cost of creating such systems [12]. It is obvious that the onset of dry years is probabilistic but very well described statistically.

Gorobets [13] considered this situation for the construction of electric locomotives. Unlike the previous example, the occurrence of man-made disasters on railways is also probabilistic in nature but there are no representative statistics on such disasters due to their comparatively small number.

The results obtained in the above-mentioned studies [12, 13] were used to calculate the service life of irrigation and drainage systems and locomotives.

The termination of the operation during the life of any system of the life support complex of the settlement will lead to a crisis, requiring costs for the implementation of measures for the alternative performance of relevant functions, etc.

When evaluating the performance of projected facilities or systems by the degree of risk reduction, the essence of the processes associated with their operation can be described using insurance models, and the corresponding cost parameters can be determined using actuarial calculations [14, 15].

Thus, as a working model, the situation is considered in which the economic result of the functioning of the designed object/system during the service life is assumed identical to the result of the creation of a special insurance fund, which allows to carry out measures to overcome emergency situations.

An analogue of the insurance rate (the rate of insurance premiums) here is the economically reasonable amount of the cost of operating the designed facility/system during its service life. It is assumed that annually an amount equal to insurance payments is required for its maintenance, which must be paid in order to compensate for the damage caused by the termination of the designed object.

In the process of justifying the cost of maintaining the designed object, the following indicators are calculated: the frequency of events, the risk-cumulating coefficient, the loss ratio, the average insurance amount, the risk severity, the insurance sum loss, the loss ratio, the frequency of damage, the damage.

Each of these insurance parameters corresponds to a constructive parameter affecting the service life of the technical system being built. The correspondence of these parameters is presented in the Table 1.

Table 1. Compliance of insurance and design parameters [11]

Parameter designation	For insurance conditions	For design conditions
P	The amount of insurance premiums collected	The total cost of the function "ensure security"
B	Insurance indemnity paid	The amount of costs associated with the elimination of the consequences of a single accident
C	Total insured amount of the insured objects	The amount of costs associated with the elimination of the consequences of accidents for all objects with a permissible probability for Bayes
Cm	Sum Insured per Policyholder	The sum of the costs associated with the elimination of the consequences of accidents on one object with a permissible probability for Bayes ($Cm = C/m$)
m	The number of affected objects as a result of the insured event	The number of accidents involving economic and legal consequences, the likelihood of which is excluded due to the cost of security.
L	The number of insurance events	The number of adverse factors that can lead to an accident
n	Number of insurance objects	The number of machines (objects) in the party (in operation)
Kt	The amount of insurance fund required to pay the insurance indemnity by the end of year t	The amount of capital investments in work to improve the safety of objects

In this case, the calculation is performed based on the fact that the value of P is identical to the value of operating costs for the maintenance of the designed system over its service life, and B - the value of the costs of overcoming crisis situations for the same period.

When predicting the hazard level of projected objects belonging to product groups for which there is no statistical information about emergencies occurring during the service life, the provisions of the theory of solutions can be applied, which is complemented by the Bayes formula [16].

In this case, the probability of the accident-free operation of the designed object during the lifetime of $P(H_1)$ in this case is determined by the formula:

$$P(H_1) = 1 - K$$

where H_1 is the hypothesis of the accident-free operation of the designed object during the entire service life; K - assessment of the probability of an accident, obtained by expert.

The probability of an emergency in this case is determined by the formula:

$$P(H_2) = K$$

where H_2 is an assumption of an emergency during the service life.

Regardless of whether actuarial calculations were performed on the basis of statistical information about the reliability and safety of technical systems like the one being designed, or the Bayes formula was used, the planned service life of the system can be defined as the quotient of P divided by K_t.

3 Experience of Using the Approach, Results Obtained and Directions for Further Research

The approaches described above were applied (fully and partially) in the design of electric locomotives and other railway equipment, mining equipment, equipment for land reclamation and water management.

Table 2 presents some of the parameters of products designed with a probabilistic method for determining the service life in the early stages of design.

It allows us to outline areas for further research and development for the development of the actuarial approach. For more reliable use of the Bayes method it is necessary to work out the algorithm for conducting an expert assessment.

In addition, the use of indicators of the parametric complexity [17] of structures to calculate their cost, conducted in parallel with the calculation of service lives, requires the formation of a regulatory framework. The solution of these problems will allow you to create an integrated system for the formation of digital twins of products used at all stages of their life cycle.

Table 2. The parameters of products designed with described methods for determining the service life

Parameters	Electric locomotive (for passenger service, AC)	Electric locomotive (for cargo transportation, AC, two sections)	Electric locomotive (for passenger service, dual standard, AC and DC)	Electropulse fish protector
Determined safety costs, thousand rubles	11 611	18 764	31 320	13,6
Product's estimated value on the level of draft proposal, thousand rubles	46 428	101 432	106 534	798
Estimated service life, years				
Variant 1	30,0	30,0	30,0	5,1
Variant 2	-	-	-	6,3
Variant 3	18,0	26,2	18,3	-

Variant 1 - calculated according to depreciation rates
Variant 2 - calculated on actuarial models based on statistical data
Variant 3 - calculated on actuarial models using the Bayes method

The solution of questions about the service life of equipment and other technical systems was significantly influenced by the emergence of M2M monitoring systems and NB-IoT systems [18], which allow to install sensors in hard-to-reach places of equipment, to receive information about the physical parameters of their work. In addition, it improves the quality of empirical information about the actual state of the technical system, which can be used in the design of such systems in the future.

4 Conclusion

The processes of re-industrialization and NBIC-convergence, the emergence of a new technological order influence and determine the conditions of modern technical systems creation. It is necessary to minimize their possible negative impact on a person. This allows creating human-oriented systems capable of ensuring the harmony of the technosphere and humanity. It is important that the design process focus on the interests and needs of humans and ensure the safety of operation of the designed system during its entire service life.

It was established that the calculation of the service life of a technical system, as well as the formation of its cost parameters at the conceive stage, should be carried out with a probabilistic approach. It lies in the fact that the probability of trouble-free operation of the designed technical system is estimated (by the actuarial method or on

Bayes formula) during its entire service life. Taking into account this probability, the costs of eliminating the consequences of possible accidents are calculated and on their basis (using the value engineering technique), the costs of operation and the price of the designed technical system and the period of its safe operation are determined.

References

1. Roco, M., Bainbridge, W.: Converging Technologies for Improving Human Performance: Nanotechnology, Biotechnology, Information Technology and Cognitive Science. Arlington (2004)
2. Roco, M.C.: Converging science and technology at the nanoscale: opportunities for education and training. Nat. Biotechnol. (2002)
3. Marsh, P.: The new industrial revolution: consumers, globalization and the end of mass production. Yale University Press, New Haven (2013)
4. Schmidt J.C.: NBIC-Interdisciplinary? A Framework for a Critical Reflection on Inter- and Transdisciplinary of NBIC-scenario. Georgia Institute of Technology. WorkingPaper No 26, April 2007
5. Nordmann, A.: Converging Technologies. Shaping the Future of European Societies. European Commission: Report (2004)
6. Simondon, G.: Du mode d'existence des objets techniques. Aubier, Paris (2001)
7. Nishimoto, Sh., et al.: Reconstructing neuron visual experience from brain activity evoked by natural movies. Curr. Biol. (2011)
8. Mrugalska, B.: Macroergonomics for manufacturing systems an evaluation approach foreword. Manage. Ind. Eng. VII–IX (2018)
9. Mrugalska, B.: Remaining useful life as prognostic approach: a review. In: Proceedings of the 1st International Conference on Human Systems Engineering and Design (IHSED2018), Future Trends and Applications, 25–27 October. CHU-Université de Reims Champagne-Ardenne, France (2018)
10. Kolbachev, E., Kolbacheva, T.: Biological and social factors that exert an impact on decision making during working-out of the convergent technologies. Adv. Intell. Syst. Comput. **722**, 255–260 (2018)
11. Kolbachev E., Salnikova Y.: Actuarial models in the design of human-oriented production systems and products. In: Report at Conference: Human Interaction and Emerging Technologies (IHIET 2019), Université Côte d'Azur, Nice, France, 22–24 August 2019
12. Gerard, D.: The law and economics of reclamation bonds. Resour. Policy **26**(4), 189–197 (2000)
13. Gorobets, D.G.: Economic features of working out dangerous or responsible products of machine building. Cost analysis and innovation of the enterprise. SRSPU (NPI), Novocherkassk, 23–24 (2000)
14. Trowbridge, C.-L.: Fundamental Concepts of Actuarial Science. AERF, Washington (1989)
15. Kaas, R., Goovaerts, H., Dhaeue, J., Denuit, M.: Modern Actuarial Risk Theory (2002). Springer. https://link.springer.com/book/10.1007/b109818
16. Kahneman, D., et al.: Judgment Under Uncertainty: Heuristics and Biases. University Press, Cambridge (2005)
17. Kolbachev, E.: Management of mechanical engineering design processes based on product cost estimates. SHS Web Conf. **35**, 01021 (2017)
18. Extended Coverage – GSM – Internet of Things (EC-GSM-IoT). https://www.gsma.com/iot/extended-coverage-gsm-internet-of-things-ec-gsm-iot/

Quantitative Nondestructive Assessment of *Paenibacillus larvae* in *Apis mellifera* Hives

David Lyall[1(✉)], Phil Hansbro[2], Jay Horvat[2], and Peter Stanwell[1]

[1] School of Health Sciences, The University of Newcastle, Callaghan, Australia
{David.lyall,Peter.stanwell}@newcastle.edu.au
[2] School of Biomedical Sciences and Pharmacy, The University of Newcastle, Callaghan, Australia
{Phil.hansbro,Jay.horvat}@newcastle.edu.au

Abstract. Paenibacillus larvae is a Gram-positive spore forming bacterium and is the causative agent of American Foulbrood (AFB), a highly contagious infectious disease of Apis Mellifera (the European honey bee), which results in hive death and significant economic losses within the apiary industry world-wide. AFB is typically diagnosed by visual identification which is labour intensive and highly operator dependent. In the absence of a low cost, rapid and reliable industry surveillance tool targeting AFB, the disease has spread around the world. Real-time polymerase chain reaction (rtPCR) is a potential AFB surveillance tool with a demonstrated high sensitivity for the detection and quantification of Paenibacillus larvae infections in Apis Mellifera hives. Current techniques result in the disturbance of the internal hive microclimate by opening and deconstructing the hive. Our research aims to develop a new, non-invasive, rtPCR technique for the detection of AFB in beehives without the need to open the hive.

Keywords: Paenibacillus larvae · American Foulbrood · AFB · Polymerase Chain Reaction · PCR · Non-destructive · Apis Mellifera · European honeybee · Bee

1 Introduction

Paenibacillus larvae are the causative agent of American Foulbrood (AFB), which is a highly contagious infectious disease of *Apis Mellifera* (the European honey bee), which results in hive death and massive economic losses within the apiary industry world-wide [1, 2].

Paenibacillus larvae is a Gram-positive rod-shaped, spore-forming bacterium. It is highly infectious in its only host, *Apis mellifera* larvae, with as few as 10 spores required to infect young bee larvae [3]. Each infected bee larvae have the potential to become a host that can produce millions of spores that are shed into the hive [4]. This has significant ramifications, not only for other larvae within the contaminated hive which rapidly become infected, but also for larvae in other hives given the propensity for the spores to be spread by adult bees as they drift to other hives. It has been reported that *Paenibacillus larvae* spores in the field remain infective for more than 35 years [5].

T. Ahram et al. (Eds.): IHSED 2019, AISC 1026, pp. 579–583, 2020.
https://doi.org/10.1007/978-3-030-27928-8_88

Paenibacillus larvae spores are found in the brood comb where new *Apis Mellifera* larvae hatch from eggs, in the honey fed to young *Apis mellifera* larvae and on hive parts *enclosing* the colony [2].

Whilst young larval stage bees are highly susceptible to *Paenibacillus larvae* within first 36 h after hatching, susceptibility rapidly declines after 72 h to the point where adult bees are resistant to the infection [6].

The pathogen develops in the bee larval gut and, dependent of the virulence of the four *Paenibacillus larvae* genotypes (ERIC I – IV) described, death occurs within 7–12 days with each death from the infection resulting in the production of millions of spores, which further infect the colony [7].

AFB is typically diagnosed by the appearance of darkened brood comb with a mottled appearance within the capped brood. Individual infected cell caps may appear darker, sunken and often have a hole. There is an associated sour smell. The "matchstick test" is often the definitive field test employed where a matchstick is pushed into the cell where the decomposing bee larvae has been transformed into a brown ropy mass, which forms a stringy substance between the cell and match as the match is withdrawn [4] (Figs. 1 and 2).

Fig. 1. Hive inspection [10]

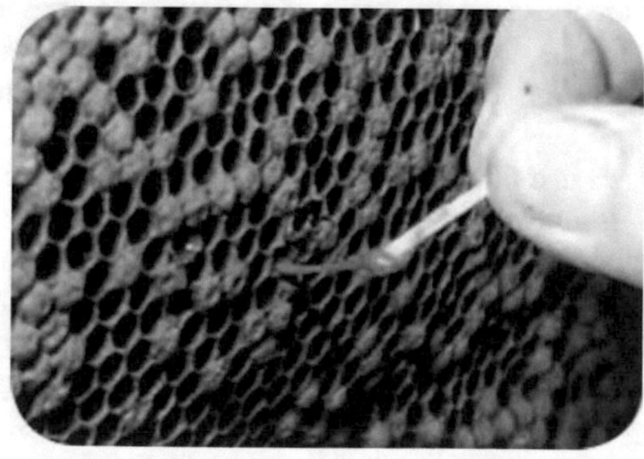

Fig. 2. Matchstick test [10]

These methods of identification and diagnosis are labour intensive and highly variable being dependent upon the apiarist (beekeeper) manually opening the hive and lifting every brood frame out of the hive and visually inspecting each cell on each frame. A single hive may have 8–10 brood frames, with 3,000–4,500 cells on each side of each frame. This method is highly operator dependent and typically requires a relatively high level of infection before detection occurs [8].

Apis Mellifera hives expend significant energy to maintain the desired internal microclimate of 34 °C at 50% humidity in a wide range of external environments, when the hive is opened for inspection this may significantly alter the internal hive microclimate, risking chilling the brood and accidently injuring or killing the Queen.

Microbiological techniques, therefore, play an important role in confirming AFB infection following initial discovery by visual inspection. Samples collected from honey, brood comb and other hive products are processed as stained samples for microscopic inspection and/or cultured to identify the infectious agent by description of the resulting colonies [2].

In the absence of a suitable industry surveillance tool targeting AFB, where the tool is highly sensitive and highly specific, has a low cost and able to generate results in rapid sequence, AFB has spread around the world to every country where bees are kept [7].

Martinez [5] and others have proposed real-time polymerase chain reaction (PCR) as a potential tool that is capable to be utilised as a screening tool for AFB. In a series of papers exploring the utility of real time PCR, results have shown that real-time PCR has a very high sensitivity for the detection and quantification of *Paenibacillus larvae* infections in *Apis Mellifera* hives and can effectively differentiate *Paenibacillus larvae* from closely related *Paenibacillus alvei* and other *Bacillus* microorganisms.

These studies have relied on samples collected from honey and/or brood comb, which still requires the disturbance of the internal hive microclimate and displacement of the bees by opening the hive and lifting frames and shaking bees from the frames for inspection and sampling [3].

This research will seek to develop and test a new, non-invasive, quantitative, real-time based PCR test for detecting the presence of *Paenibacillus larvae* spores in Apis Mellifera hives without the need to open the hive and remove bees or brood comb from the hive environment to visually inspect and/or collect of samples. The benefit of this technique is that it does not destroy the hive microclimate (34 °C, 50% humidity), avoiding harm to bee larvae.

In field PCR sample collection takes approximately 1 min, visual inspection takes approximately 30 min per hive. The PCR technique has the potential to save more than 90% of labour for field inspections. Over an apiary of 1,000 hives being inspected twice per year the labour saving has a significant economic benefit. The increased sensitivity and specificity of PCR will provide earlier identification of infection limiting the spread of AFB.

Datta [9], estimated that during an AFB outbreak incidence of new infections is exponential and earlier detection by as little as 1 month nearly doubles the chance of disease eradiation. Where burning of hives is the recommended response to an AFB outbreak [10] a delay in detection has a significant negative economic impact.

2 Hypothesis

A novel real-time quantitative PCR technique will have the sensitivity and specificity to detect and confirm an AFB infection caused by *Paenibacillus larvae* at an earlier stage than visual field inspection.

3 Expected Outcomes

The research will test the hypothesis, a novel real-time quantitative PCR technique will have the sensitivity and specificity to detect and confirm an AFB infection caused by *Paenibacillus larvae* at an earlier stage than current visual field inspection, microscopy and culturing. The technique has the potential to significantly reduce the labour associated with field visual inspection and potentially provide a verifiable hive health status for industry quality assurance programs before hives are involved in mass pollination events.

The significance of the proposed new sampling and quantitative PCR technique is it has the potential to be commercialised as an international standard and industry screening tool for the early detection and potential eradication of AFB from apiary industry globally. This is a critical issue when honeybee populations are declining globally, and pollinated agriculture is estimated to have an annual value of US$577 Billion [11].

References

1. Rossi, F., Amadoro, C., Ruberto, A., Ricchiuti, L.: Comparison of Quantitative PCR (qPCR) Paenibacillus Larvae Targeted Assays and Definition of Optimal Conditions for its Detection/Quantification in Honey & Hive Debris. Preprint (2018)
2. Govan, V., Allsopp, M., Davison, S.: A PCR detection method for rapid identification of paenibacillus larvae. J. Appl. Environ. Microbiol. **65**, 2243–2245 (1999)
3. Rusenova, N., Parvanov, P., Stanilova, S.: Development of multiplex PCR for fast detection of paenibaciilus larvae in putrid massess and in isolated bacterial colonies. J. Appl. Biochem. Microbiol. **49**, 88–93 (2013)
4. Bassi, S., Formato, G., Milito, M., Trevisiol, K., Salogni, C., Carra, E.: Phenotypic Characterization and ERIC-PCR based genotyping of Paenibacillus larvae isolates recovered from American foulbrood outbreaks in honey bees from Italy. J. Vet. Q. **35**, 27–32 (2015)
5. Martinez, J., Simon, V., Gonzalez, B., Conget, P.: A real time PCR based strategy for the detection of Paenibacillus larvae vegetative cells and spores to improve the diagnosis and screening of American foulbrood. Lett. Appl. Microbiol. **50**, 603–610 (2010)
6. Dainat, B., Grossar, D., Ecoffey, B., Haldemann, C.: Triplex real time PCR method for the qualitative detection of European and American foulbrood in honeybee. J. Microbiol. Methods **146**, 61–63 (2018)
7. Schafer, M., Genersch, E., Funfhaus, A., Poppinga, L., Formella, N., Bettin, B., Karger, A.: Rapid identification of differentially virulent genotypes of Paenibacillus larvae, the causative organism of American foulbrood of honey bees, by whole cell MALDI-TOF mass spectrometry. J. Vet. Microbiol. **170**, 291–297 (2014)

8. Knazovicka, V., Miluchova, M., Gabor, M., Kacaniova, M., Melich, M., Krocko, M., Trakovicka, A.: Using real time PCR for identification of paenibacillus larvae. J. Anim. Sci. Biotechnol. **44**, 424–431 (2011)
9. Datta, S., Bull, J., Budge, G., Keeling, M.: Modelling the spread of American foulbrood in honeybees. J. R. Soc. Interface **10**, 20130650 (2013)
10. Somerville, D.: American Foulbrood. NSW Department of Primary Industries, Prime Fact Sheet (2015)
11. Potts, S., Imperatriz-Fonseca, V., Ngo, H., Biesmeijer, J., Breeze, T., Dicks, L., Garibaldi, L., Hill, R., Settele, J., Vanbergen, A.: Editorial, Summary for policymakers of the thematic assessment of pollinators, pollination and food production. Biota. Neotrop. **16**, e20160101 (2016)

Green Light Optimum Speed Advisory (GLOSA) System with Signal Timing Variations - Traffic Simulator Study

Hironori Suzuki[1](✉) and Yoshitaka Marumo[2]

[1] Department of Robotics, Nippon Institute of Technology, 4-1 Gakuendai, Miyashiro, Saitama 3458501, Japan
viola@nit.ac.jp

[2] Department of Mechanical Engineering, Nihon University, 1-2-1 Izumicho, Narashino, Chiba 2758575, Japan
marumo.yoshitaka@nihon-u.ac.jp

Abstract. Green light optimal speed advisory (GLOSA) systems are vehicle-to-everything (V2X) communication applications that transfer signal information between vehicles and traffic lights in order to achieve higher time and energy efficiency together with safer traffic at signalized intersections. This paper focuses on the Red to Green Time (RGT) ratio of traffic lights and evaluates our GLOSA system performance levels and limitations using RGT ratio variations in traffic simulator experiments. Statistical analyses showed that when the RGT ratio is almost 1 (i.e. red time and green time are almost equal), the partial assistance (PA) mode, in which the GLOSA system only activates during red signals, is highly recommended in terms of traffic efficiency and environmental impact. In contrast, when the RGT ratio is less than 1, the full assistance (FA) mode, which assists the vehicle regardless of the signal phase, performed better than the PA mode provided that the travel time is not significantly increased.

Keywords: GLOSA · ADAS · Red to green time ratio · Travel time · Traffic simulation · Fuel consumption · CO_2 emissions

1 Introduction

Vehicle-to-everything (V2X) communication schemes have opened up new ways to not only enhance vehicle safety but also improve traffic efficiency especially in the vicinity of signalized intersections. The use of Green Light Optimal Speed Advisory (GLOSA) systems is an application of V2X communications that transfers signal information between vehicles and traffic signals in order to achieve higher time and energy efficiency together with safer traffic at intersections [1–7].

The GLOSA system concept was initially designed to advise drivers on their optimum speeds when approaching signalized intersections in order to avoid the need for complete stops. However, another GLOSA concept, which we proposed previously, calls for providing advice to drivers on the most appropriate *position* for their vehicles instead of the optimal *speed* [8, 9].

T. Ahram et al. (Eds.): IHSED 2019, AISC 1026, pp. 584–590, 2020.
https://doi.org/10.1007/978-3-030-27928-8_89

In our previous study [9], we tested our GLOSA system in traffic flows with variations of traffic demand and information distance, which we define as the distance between the assisted vehicle and the traffic light where the GLOSA system first becomes active. We found that 500 m of information distance, which is regarded as the upper limit of the V2X communication range, provides the best performance in terms of traffic efficiency. In addition to those parameters, the performance of the GLOSA system depends on the traffic light signal timing, especially the Red to Green Time (RGT) ratio. In this study, we prepared scenarios in which the RGT ratio was varied and then carried out traffic simulator experiments in order to investigate the performance and limitations of our GLOSA system in terms of RGT ratio variations.

2 Methodology

2.1 GLOSA System Outline

Our proposed GLOSA system assumes that connections exist between the vehicles and traffic lights via a vehicle-to-infrastructure (V2I) communication system and uses the system data in order to calculate the distance required to cross the intersection without making a complete stop. Please refer to [8] and [9] for additional details about our GLOSA system. As shown in Fig. 1, the GLOSA system consists of a combination of green and red rectangles placed adjacent to each other on the road surface.

The lengths of both rectangles are computed as:

$$d_e(t) = v(t) \times \text{TTR} \text{ and } d_d(t) = v(t) \times \text{TTG}. \quad (1)$$

Here, $d_e(t)$ and $d_e(t)$ are the lengths of the green and red rectangle shapes, respectively, while time to red (TTR) and time to green (TTG) are the remaining green and red times for the forthcoming signal. $v(t)$ is the current vehicle speed at time step t.

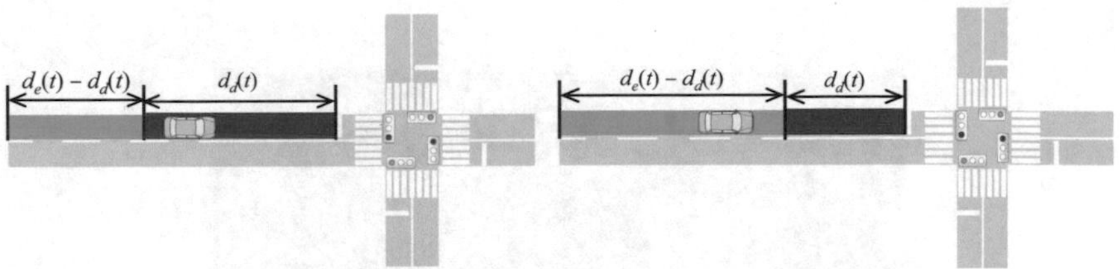

Fig. 1. Green and red rectangles placed adjacent to each other. The driver is advised to maintain his/her current speed when the vehicle is traveling on the green rectangle to pass through the signalized intersection without stopping (right). In contrast, if the vehicle is on the red portion, the driver is advised to decelerate until the vehicle enters the green portion (left).

The combination of green and red-colored rectangles are projected onto the road surface ahead of the vehicle. If the vehicle is on the green portion, the GLOSA system advices the driver to maintain his/her current speed in order to cross the intersection without making a complete stop. However, if the vehicle is on the red rectangle, the driver is advised to decelerate until the vehicle enters the green portion. Hence, our proposed GLOSA system provides the most appropriate position for drivers instead of the optimal speed. The front view image from the driving simulator (DS) cabin is shown in Fig. 2. In our DS experiment, we confirmed that our GLOSA system showed the good performance in terms of fuel efficiency, reduced driver distraction, and improved comfort in a limited environment with low traffic demand [10].

2.2 Traffic Simulation with Car-Following Model

A time-scanning traffic simulator (TS) was developed and validated by one of the authors of this paper, after which it was integrated into the GLOSA system. The vehicle acceleration is calculated by the car-following model given by Eq. (2):

$$\ddot{x}(t) = a\left(\frac{\Delta v}{d} \cdot v^{\alpha}\right) + b\left(\sqrt[\delta]{\beta \cdot v^{\gamma}} - \frac{v}{d}\right) + c\left(\frac{R}{1 + m\exp(n \cdot d)} - v\right) + \Phi. \quad (2)$$

where, $a, b, c, \Phi, \alpha, \beta, \gamma, \delta, R, m,$ and n are the parameters given in Table 1, d is the headway distance from the preceding car (m), and v is the velocity (m/s). Whenever a vehicle is generated, a set of parameters is assigned to the vehicle at random.

3 Numerical Simulation

3.1 Test Setup

Except for the information distance, the simulation test field and scenarios used in this study were identical to the ones that appeared in our previous study [9].

Geometry: A corridor consisting of four 500 m straight links was prepared. Traffic lights are installed at the end of each link. The speed limit was set to 16.7 m/s.

Fig. 2. Front view image from the DS cabin. The GLOSA system makes it possible for drivers to quickly recognize if they are traveling on either the red or green rectangles.

Table 1. Five parameter sets for a car-following model

ID	a	b	c	Φ	α	β	γ	δ	R	m	n
1	1.88	0.026	0.158	0.10	0.8	0.0027	1.55	3.2	14.1	35	–0.259
2	1.88	0.201	0.106	0.078	0.8	0.0015	1.70	3.0	15.0	55	–0.220
3	1.90	0.451	0.166	0.080	0.8	0.0040	1.65	3.0	15.0	25	–0.251
4	2.25	0.230	0.110	0.091	0.8	0.0020	1.80	3.0	18.1	48	–0.207
5	1.73	0.246	0.258	0.015	0.8	0.0030	1.55	3.3	14.5	25	–0.194

Simulation Parameters: The simulation duration was set at 2400 to 7200 s depending on the traffic demand and the signal phase. The time interval was set at 0.1 s. Vehicles were generated at the entrance of the first link but only injected into the road within the first 600 s.

Measurements: Traffic detectors for all four links were installed at locations 50 m in front of the traffic lights in order to measure the distributions of time headway, accelerations, spot speed, and link traffic count.

Information Distance: The information distance was set to 500 m because it was found that our GLOSA system performed better when the system was activated as early as possible [9].

3.2 Test Conditions

Signal Phase: To change the RGT ratio, the combinations of green (G), amber (A) and red (R) durations were set to G50-A3-R47 s, G62-A3-R35 s and G72-A3-R25 s. A pre-timed control was assumed for all signals.

Traffic Demand and Penetration Rate: As defined in our previous study [8, 9], traffic demands were assumed 100, 200, 400, 800 and 1200 vehicles per hour (VPH) and the penetration rate of the GLOSA-assisted vehicle was assumed to be 100%.

Assistance Mode and Control Condition: We prepared two assistance modes: full assistance (FA) and partial assistance (PA). In the FA mode, the vehicle is assisted by the GLOSA system, regardless of the signal phase whenever it is traveling inside the information distance. In contrast, the GLOSA system only activates during red signals when it is set in the PA mode, thus avoiding the extremely low driving speeds that sometimes result at green signals at times of heavy traffic [8]. The control condition is the scenario without GLOSA system assistance.

Measures of Effectiveness (MOE): The traffic flow characteristics with and without the GLOSA system were evaluated in terms of travel time (which includes the waiting time until the vehicle entered the first link), fuel consumption, and CO2 emissions.

3.3 Simulation Results

Travel Time Reduction: Figure 3(a) shows the travel time reductions for each RGT ratio, which is defined as the reduction rate from the control condition where none of the vehicles are assisted. If the green time is short (upper), the FA mode worsens the travel time even when the traffic demand is low (e.g., 100 VPH). As the green time is lengthens (middle to lower), the FA mode becomes even more capable of accommodating high volume traffic without a significant increase in travel time. When the RGT ratio is almost 1, the FA mode is not recommended, but the PA mode remains acceptable for all traffic demand levels. In contrast, when the green time is longer, both the FA and PA modes are acceptable because the travel time increase is limited to within a few percentage points.

Fuel Economy: Fuel economy improvements are shown in Fig. 3(b). Except for lower green time (upper), in which the FA mode did not perform well in higher level demand conditions, both FA and PA modes are acceptable for all traffic demand levels. However, as long as the travel time is not significantly increased, the FA mode performs better than the PA mode, as the green time is longer.

CO2 Emission Reduction: Figure 4 shows the CO_2 emission reductions. As the green time is extended (middle to lower), the FA mode reduced CO_2 emissions more than the PA mode. However, when the RGT ratio is almost 1 (upper), the PA mode is better and more stable than the FA mode.

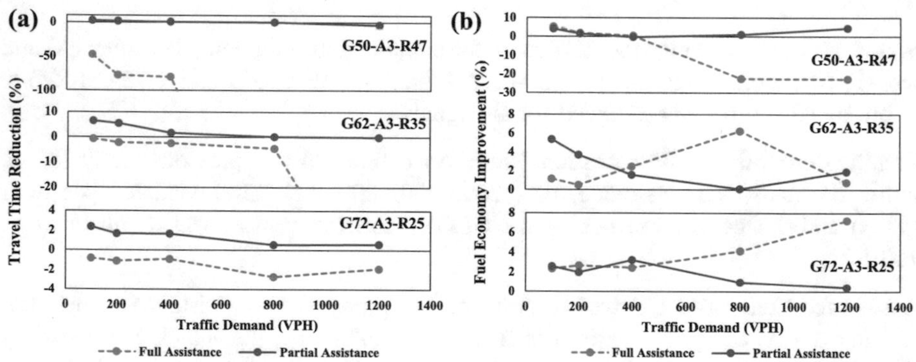

Fig. 3. Travel time reduction (a) and fuel economy improvement (b) from the control condition in which the GLOSA system assists no vehicle. When the green and red times are almost equal (upper), the PA mode had less effect on fuel efficiency. As the green signal time lengthens (middle and lower), the FA mode performed better unless travel time lengthens as well.

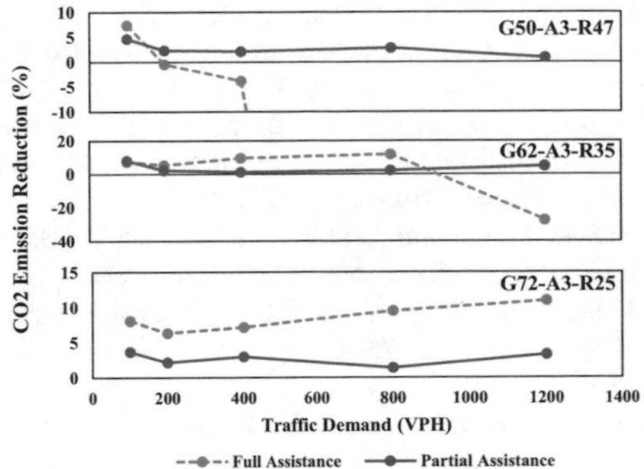

Fig. 4. CO2 emissions reduction from the control condition. Even when the RGT ratio is almost 1 (upper), the PA mode showed a few percentage points of reduction in terms of CO2 emissions. For the other RGT ratios (middle and lower), the FA mode performed better than the PA mode.

4 Conclusion

In this study, we showed that when the red and green times are almost equal, the PA mode, in which the GLOSA system activates during red signals only, is highly recommended in terms of traffic efficiency and environmental impact. In this case, the PA mode never increased the travel time but showed more than 2% improvement of fuel economy and reduced CO2 emissions. In contrast, when the green time is longer than the red time, the FA mode, in which the vehicle is assisted regardless of the signal phase, performed better than the PA mode unless there is a significant increase in travel time.

Since the efficiency by the GLOSA system depends significantly on the RGT ratio, it is clear that both the RGT ratio and traffic demand level should be considered before selecting either the FA or PA mode.

Acknowledgments. Research supported by the Japan Society for the Promotion of Science (JSPS) KAKENHI Grant-in-Aid for Scientific Research (B) JP17H02055.

References

1. Eckhoff, D., Halmos, B., German, R.: Potentials and limitations of green light optimal speed advisory systems. In: 2013 IEEE Vehicular Networking Conference, pp. 103–110 (2013)
2. Katsaros, K., Kernchen, R., Dianati, M., Rieck, D.: Performance study of a Green Light Optimized Speed Advisory (GLOSA) application using an integrated cooperative ITS simulation platform. In: 2011 7th International Wireless Communications and Mobile Computing Conference, pp. 918–923 (2011)
3. Nguyen, V., Kim, O.T.T., Dang, T.N., Moon, S.I., Hong, C.S.: An efficient and reliable green light optimal speed advisory system for autonomous cars. In: 2016 18th Asia-Pacific Network Operations and Management Symposium (APNOMS) (2016)

4. Seredynski, M., Dorronsoro, B., Khadraoui, D.: Comparison of green light optimal speed advisory approaches. In: Proceedings of the 16th International IEEE Annual Conference on Intelligent Transportation Systems (ITSC 2013), pp. 2187–2192 (2013)
5. Seredynski, M., Mazurczyk, W., Khadraoui, D.: Multi-segment green light optimal speed advisory. In: 2013 IEEE 27th International Symposium on Parallel & Distributed Processing Workshops and PhD Forum, pp. 459–465 (2013)
6. Bodenheimer, R., Brauer, A., Eckhoff, D., German, R.: Enabling GLOSA for adaptive traffic lights. In: 2014 IEEE Vehicular Networking Conference (VNC), pp. 167–174 (2014)
7. Bodenheimer, R., Eckhoff, D., German, R.: GLOSA for adaptive traffic lights: methods and evaluation. In: 2015 7th International Workshop on Reliable Networks Design and Modeling (RNDM), pp. 320–328 (2015)
8. Suzuki, H., Marumo, Y.: A new approach to Green Light Optimal Speed Advisory (GLOSA) systems for high-density traffic flow. In: Proceedings of the 21st IEEE International Conference on Intelligent Transportation Systems, pp. 362–367 (2018)
9. Suzuki, H., Marumo, Y.: Evaluating Green Light Optimum Speed Advisory (GLOSA) system in traffic flow with information distance variations. In: Proceedings of the 1st International Conference on Human Interaction & Emerging Technologies (2019, in printing)
10. Marumo, Y., Nakanishi, T., Yamazaki, K., Suzuki, H.: Driver assistance system to prevent unnecessary deceleration at signalized intersection by indicating deceleration required distance on road. Int. J. Automot. Eng. **10**(1), 100–105 (2019)

Detection of Student Behavior Profiles Applying Neural Networks and Decision Trees

Cesar Guevara[1(✉)], Sandra Sanchez-Gordon[2], Hugo Arias-Flores[3], José Varela-Aldás[3], David Castillo-Salazar[3,5], Marcelo Borja[4,8], Washington Fierro-Saltos[5], Richard Rivera[6], Jairo Hidalgo-Guijarro[7], and Marco Yandún-Velasteguí[7]

[1] Mechatronics and Interactive Systems - MIST Research Center, Universidad Indoamérica, Quito, Ecuador
cesarguevara@uti.edu.ec

[2] Department of Informatics and Computer Science, Escuela Politécnica Nacional, Quito, Ecuador
sandra.sanchez@epn.edu.ec

[3] SISAu Research Group, Universidad Indoamérica, Quito, Ecuador
{hugoarias, josevarela, davidcastillo}@uti.edu.ec

[4] Graphic Design Faculty, Universidad Indoamérica, Quito, Ecuador
carlosborja@uti.edu.ec

[5] Facultad de Informática, Universidad Nacional de la Plata, Buenos Aires, Argentina
washington.fierros@info.unlp.edu.ar

[6] Escuela de Formación de Tecnólogos, Escuela Politécnica Nacional, Quito, Ecuador
richard.rivera01@epn.edu.ec

[7] Grupo de Investigación GISAT, Universidad Politécnica Estatal del Carchi, Tulcán, Ecuador
{jairo.hidalgo, marco.yandun}@upec.edu.ec

[8] Facultad de Diseño y Comunicación, Universidad de Palermo, Buenos Aires, Argentina

Abstract. Education worldwide is a significant aspect for the development of the peoples and much more in developing countries such as those in Latin America, where less than 22% of its inhabitants have higher education. Research in this field is a matter of interest for each of the governments to improve education policies. Therefore, the analysis of data on the behavior of a student in an educational institution is of utmost importance, because multiple aspects of progress or student dropout rates during their professional training period can be identified. The most important variables to identify the student's behavior are the socio-economic ones, since the psychological state and the economic deficiencies that the student faces while is studying can be detected. This data provides grades, scholarships, attendance and information on student progress. During the first phase of the study, all the information is analyzed and it is determined which provides relevant data to develop a profile of a student behavior, as well as the pre-processing of the data obtained. In this phase, voracious algorithms are applied for the selection of attributes, such as greedy stepwise, Chi-squared test, Anova, RefiefF, Gain Radio, among others. In this

T. Ahram et al. (Eds.): IHSED 2019, AISC 1026, pp. 591–597, 2020.
https://doi.org/10.1007/978-3-030-27928-8_90

work, we apply the artificial intelligence techniques, the results obtained are compared to generate a normal and unusual behavior of each student according to their professional career. In addition, the most optimal model that has had a higher accuracy percentage, false positive rate, false negative rate and mean squared error in the tests results are determined.

Keywords: Education · Artificial intelligence · Student · Behavior profiles · Neural networks · Decision trees

1 Introduction

The data analysis in the educational field has become a very promising research area in recent years, where it has been proven that the academic information of each of the students provides highly relevant data on the evolution of learning. Currently, an area of learning analytics has been developed in the educational field, which has allowed identifying variables and problems in the student's evolution.

In the work carried out by [1], an analysis of data based on the interaction with information systems and educational technology is presented. This study details the understanding of the learning process and the interrelation with the student and the computer platforms, applying educational data mining. The study presents known cases in the literature, and also, critical issues that have a sustainable impact on the research and practice of learning and teaching. This article shows the importance of data mining in educational information and how it can be beneficial for problem identification and student evolution.

In the work published by [2], a learning analytics conceptual framework of learning designs is detailed. The study analyzes the information of interviews with teachers and students in technological platforms, and later, the learning activities and their relation with the pedagogical intent are evaluated. The analysis of variables consists of five dimensions: temporal analytics, tool-specific analytics, cohort dynamics, comparative analytics and contingency.

Another paper presented by [3], proposes to analyze the indicators based on the interaction between learning agents (student-student and active-passive). This information contributes to the individual development of the student and the relationship between the individual assessment grades and the interaction with the technological systems. This analysis allows to evaluate and monitor the students individual progress and prevents learning problems, putting corrective measures and decision making into practice. This model has obtained excellent results and has identified learning problems in students with a high rate of accuracy.

Our work focuses on the analysis of academic information, specifically the grades obtained in each subject of the architecture career of an Ecuadorian university. A detection model of educational issues by applying neural networks and decision trees is proposed.

The article is structured as follows: In Sect. 2 the methods and materials used in the creation of the classification model are presented. In section three, the proposed model and its application with neural networks and decision trees are described. Section 4 shows the results and a comparison with other previously published works. Finally, section five is presented, where the conclusions obtained in the study and future lines of research in the area are detailed.

2 Materials and Methods

In this section the materials, such as the database used for the study of student performance behavior, are presented.

2.1 Materials

The database used was compiled from the academic system of an Ecuadorian university, during the period between 2014 and 2018. This information provides the grades obtained from each of the students in their respective subject. These data allow showing the evolution of each student and their possible problems during the academic period. The database contains 42 attributes (headquarters, mode of study, degree, students ID, level, subjects, grades, etc.) and 11.494 records. An example is shown in Table 1. In addition, by regulations of the educational institution, grades are classified as good (7.0 points to 10 points) and low grades (0 points to 6.9 points).

Table 1. Example of the academic database with information of grades of the students of an educational institution.

Headquarters	Mode of study	Degree	Age	Marital status	Sex	Level	Subjects	GRADE1	GRADE2	FINAL_GRADE
1	1	1	21	1	1	4	3927	4,60	3,80	8,40
1	1	1	21	1	1	4	4134	4,70	4,00	8,70
1	1	1	21	1	1	4	6078	4,10	4,60	8,70
…	…	…	…	…	…	…	…	…	…	…
1	1	1	21	1	1	5	3875	3,90	4,30	8,20

The data distribution according to the final grade is shown in Fig. 1a, where the blue color represents the sample of the students who have a grade lower than 7 points. On the other hand, the color red represents the grade of the students with an outstanding grade higher than 7 points.

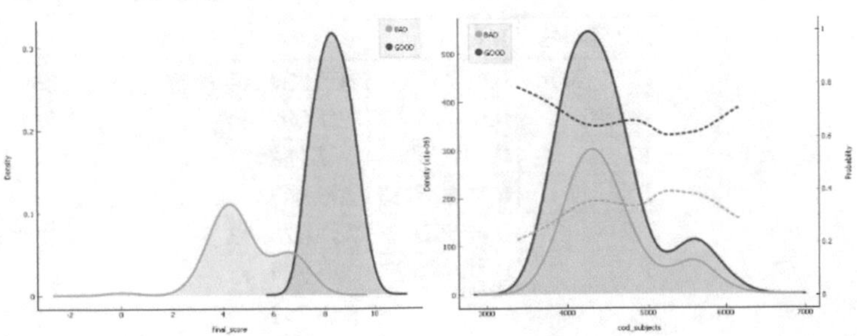

Fig. 1. Data frequency distribution (a) students grades (b) students grades by subjects.

On the other hand, the final grades for each subject are presented, as shown in Fig. 2a. These data shows that the sample is distributed among good students (red) and bad students (blue). In the same way in Fig. 2b, the final grades of the group of students for each level are shown and if they have good or bad grades.

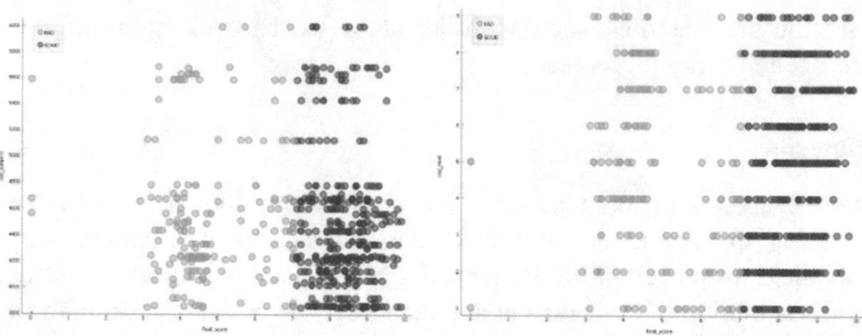

Fig. 2. (a) Chart of subjects versus final grade of the academic database (red color, good grades and blue color, low grades). (b) Chart of final grades versus the levels to which each of the students belongs.

With this data set, a pre-processing of data will be performed in order to identify the most relevant attributes for our study and in the same way apply filtering techniques for instances throughout the data set.

3 Model for Classification of Academic Information

In this section data preprocessing using variable selection and then the neural networks and decision trees for student learning classification is described.

Preprocessing

For variable selection, two methods were applied to the whole data set, Info Gain and Chi-Squared, and only 10 of the 42 attributes contained in the data set were selected. These techniques have obtained optimal results, as shown in Table 2.

Table 2. Result of the application of Info Gain [4, 5] and Chi-Squared [6] methods.

Attributes	Info. gain	χ^2
final_score	0.687442	303.838241
score2	0.64478468	291.498864
P7	0.51050945	237.168423
P6	0.50636735	266.867565
P9	0.47582343	263.625002
P10	0.46464855	220.204145

Likewise, Pearson's correlation method was applied, which has described the relationship between these 10 selected variables [7]. Table 3 shows the highest relationships among the variables.

Table 3. Correlation values of variables applying Pearson method.

	Value	Attributes
1	0.991	NOTA_SUP, SUPESC
2	0.991	NOTA_SUP, SUPPRA
3	0.965	SUPESC, SUPPRA
4	0.953	NOTA2, P6
5	0.95	NOTA2, final_score

In the next section, the application and configuration of classification techniques to identify students' academic performance are presented.

4 Classification Model

A neural network [8–10] has been applied to the 11 attributes (10 attributes of academic information and the identifying class), with the following network variable configuration: number of neurons in the hidden layers $NN_{hl} = 100$, logistic activation, regularization parameter $a = 0.001$ and number of interactions $i = 200$. Likewise, algorithm C4.5 [11, 12] was applied to the data set with the following configuration parameters: two instances in sheets, five instances in internal nodes, maximum depth division 100 and presence of binary trees. This model combines two classification techniques with optimal results to obtain a high final result with a higher accuracy rate. The results of each of the techniques are shown in Table 4.

Table 4. Results of the training in the application of neural networks and the algorithm C4.5 to the academic database.

Method	AUC	CA	F1	Precision	Recall
C4.5	0.983	0.955	0.955	0.956	0.955
Neural Network	0.99	0.967	0.967	0.969	0.967

To carry out the tests of this model, cross-validation method was applied with a k = 10, with the results shown in Table 5a and b. Table 5 presents confusion matrix of algorithm C4.5. Table 5b shows confusion matrix of the neural network.

Table 5. (a) Confusion matrix of application of algorithm C4.5 to the academic database with a $k = 10$, (b) Confusion matrix of application of neural networks to the academic database with a $k = 10$.

C4.5	BAD	GOOD	Σ
BAD	175	21	196
GOOD	5	379	384
Σ	180	400	580

Neural Network	BAD	GOOD	Σ
BAD	177	19	196
GOOD	0	384	384
Σ	177	403	580

The results obtained in this study have been quite optimal with an accuracy rate greater than 95% and a false positive and false negative rate of less than 4%. This proposed model is an approach to the classification of the students' academic performance based on their academic record. The conclusions obtained in the development of this work, as well as the future lines of research are presented in the following section.

5 Conclusions and Future Works

In the development of this work, parts of the key process for an optimal result of the proposal have been identified, one of these is the variable selection using the techniques of Chi-Squared and Info Gain. These techniques have allowed to identify the most important attributes for the study in a more effective way. On the other hand, the application of neural networks was a fundamental basis, because its use for the development of user profiles has been very suitable for the implementation of an efficient classification model. Finally, the C4.5 algorithm has efficiently reinforced the result of neural networks with a similar result and a fairly low false positive rate. As future research lines, the application of grades prediction using polynomial functions and the application of Deep Learning to identify the social, educational and financial profile of each student has been proposed.

References

1. Gašević, D., Dawson, S., Siemens, G.: Let's not forget: learning analytics are about learning. TechTrends **59**(1), 64–71 (2015)
2. Bakharia, A., et al.: A conceptual framework linking learning design with learning analytics. In: Proceedings of the Sixth International Conference on Learning Analytics and Knowledge – LAK 2016, pp. 329–338 (2016)
3. Fidalgo-Blanco, Á., Sein-Echaluce, M.L., García-Peñalvo, F.J., Conde, M.Á.: Using learning analytics to improve teamwork assessment. Comput. Hum. Behav. **47**, 149–156 (2015)

4. Azhagusundari, B., Thanamani, A.S.: Feature selection based on information gain. Int. J. Innov. Technol. Explor. Eng. **2**(2), 18 (2013)
5. Jovic, A., Brkic, K., Bogunovic, N.: A review of feature selection methods with applications. In: 2015 38th International Convention on Information and Communication Technology, Electronics and Microelectronics (MIPRO), pp. 1200–1205 (2015)
6. Thaseen, I.S., Kumar, C.A.: Intrusion detection model using fusion of Chi-square feature selection and multi class SVM. J. King Saud Univ. – Comput. Inf. Sci. **29**(4), 462–472 (2017)
7. Zhou, H., Deng, Z., Xia, Y., Fu, M.: A new sampling method in particle filter based on Pearson correlation coefficient. Neurocomputing **216**, 208–215 (2016)
8. Montavon, G., Lapuschkin, S., Binder, A., Samek, W., Müller, K.-R.: Explaining nonlinear classification decisions with deep Taylor decomposition. Pattern Recogn. **65**, 211–222 (2017)
9. Ashfaq, R.A.R., Wang, X.-Z., Huang, J.Z., Abbas, H., He, Y.-L.: Fuzziness based semi-supervised learning approach for intrusion detection system. Inf. Sci. (NY) **378**, 484–497 (2017)
10. Maggiori, E., Tarabalka, Y., Charpiat, G., Alliez, P.: Convolutional neural networks for large-scale remote-sensing image classification. IEEE Trans. Geosci. Remote Sens. **55**(2), 645–657 (2017)
11. Ngoc, P.V., Ngoc, C.V.T., Ngoc, T.V.T., Duy, D.N.: A C4.5 algorithm for English emotional classification. Evol. Syst., 1–27 (2017)
12. Mantas, C.J., Abellán, J., Castellano, J.G.: Analysis of Credal-C4.5 for classification in noisy domains. Expert Syst. Appl. **61**, 314–326 (2016)

Smart Sensor Technology and Infrastructure Safety

Mohamed Elhakeem[1,2](✉), A. N. Thanos Papanicolaou[2], and Walaa Gabr[3]

[1] Abu Dhabi University, Abu Dhabi, United Arab Emirates
mohamed.elhakeem@adu.ac.ae
[2] University of Tennessee, Knoxville, USA
[3] Benha University, Benha, Arab Republic of Egypt

Abstract. Sensor technology has become an attractive tool to facilitate remote and continues health monitoring of cities infrastructure. In this study, a passive Radio Frequency IDentification (RFID) system of low radio frequency band was configured for monitoring scour of bridge foundations. The performance of the RFID system was evaluated through a laboratory study to determine the maximum detection range of the system in terms of transponder orientation, housing material and maximum antenna-transponder detection distance. Analysis of the experimental results showed that the maximum antenna-particle detection distance was 0.73 m when the transponder is perpendicular to the antenna. Glass material provided better housing (cover) for the transponder when compared to material made of concrete-tungsten. It was also found that the maximum detection distance of the antenna did not change significantly for the buried particles when compared to the particles tested in the air. Thus, the low frequency RFID system is appropriate for monitoring bridge scour because its waves can penetrate water and sand bodies without significant loss of its signal strength. The pier model experiments showed that the RFID system was able to predict successfully the maximum scour depth when a single particle was used near a pier model where the scour-hole was expected.

Keywords: Sensor technology · Bridge health monitoring · Radio frequency identification · Bridge foundations scour

1 Introduction

Sensor technology contributed towards the development of several sustainable cities around the world. It has many applications, and becomes an attractive tool to facilitate remote and continues health monitoring of cities infrastructure. Monitoring of cities infrastructure can help in predicting structures deterioration to avoid sudden failure and ensure public safety (Fig. 1). In the United States alone, more than 3,000 bridges and other hydraulic structures are susceptible to scour of foundations, one of the main reasons for bridges failure [1, 2]. One challenging is the underwater monitoring of those structures foundations, which is addressed in this study by proposing the use of sensor technology, namely Radio Frequency IDentification (RFID) technology [3] to

T. Ahram et al. (Eds.): IHSED 2019, AISC 1026, pp. 598–604, 2020.
https://doi.org/10.1007/978-3-030-27928-8_91

monitor remotely scour around bridge foundations to limit the risk of life and property loss due to sudden failure. RFID is a wireless automated identification technology that utilizes waves at radio frequency (RF) to transfer information between a reader and a transponder via an antenna [4].

Fig. 1. Scour and failure of US-90 Biloxi Bay Bridge after Hurricane Katrina.

The main objective of this study is to evaluate the performance of a configured RFID system in monitoring scour of bridge foundations. The RFID system performance was evaluated by conducting laboratory experiments to estimate the maximum detection range of the system considering factors such as transponder orientation, transponder housing material and maximum antenna-transponder detection distance.

2 The RFID System

An off-the-shelf passive RFID system (Fig. 2) of low radio frequency band (134.2 kHz) developed by Texas Instruments is used in this study [5]. The reader of the system consisted of three modules, namely the frequency module, the control module, and the antenna tuning module. The frequency module (RI-RFM-008B-00) converts the radio frequency signals emitted by the transponder to electric signals that can be processed by the control module (Fig. 2A1). The control module (RI-CTL-MB2A-03) is the interface between the frequency module and the PC used to control the system (Fig. 2A2). Essentially the control module converts the RF signals received from the transponder to the serial number of the transponder and transmits this information over a serial line to the host PC. The antenna tuning module (RI-ACC-008B-00) tunes the antenna so that the outgoing RF signals from the frequency module are emitted at the correct frequency to ensure communication with the transponders (Fig. 2A3). Two antennas of different size were used in this study. The larger antenna (RI-ANT-G01E) has a rectangular loop shape with dimensions of 0.70 × 0.27 m, while the smaller antenna (RI-ANT-G02E) has a square loop shape with dimensions of 0.20 × 0.20 m (Fig. 2B). A 10 m long waterproof cable is used to connect the selected antenna to the reader. The entire RFID system is powered by a 12 V power supply.

The passive transponder low radio frequency (134.2 kHz) used herein was of the type RI-TRP-WEHP-30 with density of 2.22 g/cm^3 available from the economy line of Texas Instruments. The transponders are 23 mm long and 3.8 mm in diameter, protected by self-contained glass casings, thus allowing the use of the transponder without an additional casing (naked) if desired (Fig. 3). The transponder came with 80 bit Read/Write memory, which enabled the coding of a unique identification number on each transponder. The low acquisition cost of the transponder permitted the purchase of large number of transponders.

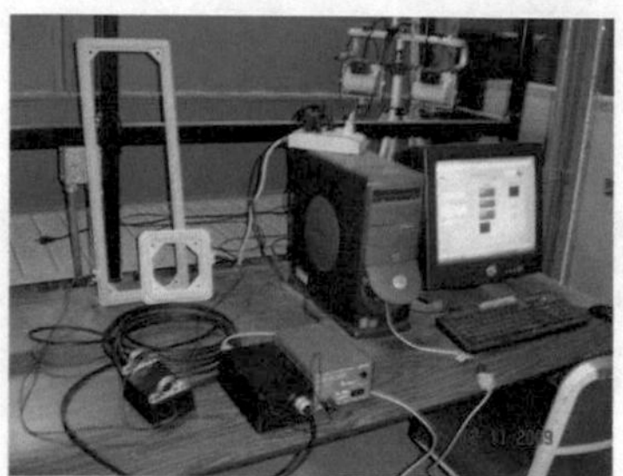

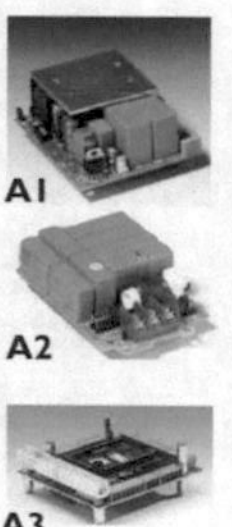

Fig. 2. General view of the RFID system and components of the TI 134.2 kHz RFID system used in this study: (A) RFID reader; (A1) reader frequency module; (A2) antenna tuning module; (A3) control module; and (B) antennas.

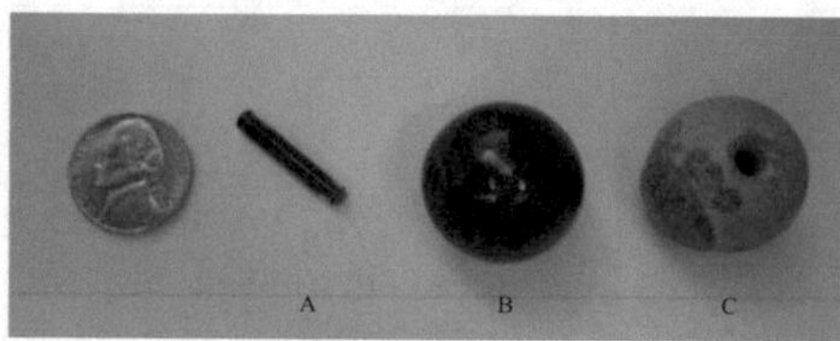

Fig. 3. Size comparison for the transponder and particles used. From left to right: (A) naked transponder, (B) glass particle and (C) concrete-tungsten particle.

3 Results

The performance of the configured RFID system was evaluated by conducting a set of laboratory experiments to determine the maximum detection range of the system. Factors that affect the detection range of an RFID system examined in this study are the transponder orientation, the transponder housing material (or particle type) and the transponder distance from the antenna.

Preliminary experiments were conducted in air using first a naked transponder to determine the optimal orientation of the transponder with respect to the antenna plane and the maximum distance between a transponder and the antenna for the transponder to be successfully detected. The maximum transponder detection distance of the RFID system was 0.68 m when the transponder long axis was perpendicular to the plane of the antenna and decreased by 37%, when the transponder was at its unfavorable orientation. The orientation experiments were repeated for the transponders by placing them in glass and concrete-tungsten spherical particles (Fig. 2) to examine the effect of these housing materials on the detection distance. The concrete-tungsten tagged particles had the smaller detection distance compared to the glass tagged particles. The glass particles had only 6% reduction in the detection distance compared to the naked transponders. Therefore, only glass housing was used for the next set of experiments with the transponders being buried in sand.

The next set of laboratory experiments were conducted in recirculating flumes to determine the permissible distance for a transponder to be successfully detected when it is buried in a sand-bed. Two sets of experiments were conducted in this stage: (1) to identify the maximum distance between a buried transponder and the antenna; and (2) to assess the applicability of the RFID system for monitoring bridge scour using a pier model.

3.1 Buried Depth Results

The first set of experiments was conducted in a recirculating flume 21.3 m long, 4 m wide, and 1.21 m deep with a maximum discharge of 0.211 m^3/s filled with a 0.61 m layer of natural sand (Fig. 4). The test section in the flume, where the antenna and the transponders were placed was 0.70 m long by 4.0 m wide, located approximately 12.0 m downstream of the flume entrance. Two steel cables, placed 0.64 m apart from one another traversed the flume test section. The RFID antenna was mounted on a wooden board attached to the cables via four metal hooks. The antenna could therefore sweep the entire test section of the flume, when towed along the cables. The water depth in the flume was kept 0.30 m for all experiments, and the bottom of the antenna was just above the water surface to keep it dry.

Eight (8) experimental runs were conducted to determine the maximum detection distance between a buried transponder and the antenna. In each run, sand was removed partially from the test section and 4 glass particles each containing a transponder were buried at a certain depth below the sand bed surface, 0.38 m apart from each other along the flume width. Then, the test section was leveled. During the placement of the particles, it was ensured that the transponders were placed vertically, i.e. at their favorable orientation in respect to the antenna plane as found from the preliminary experiments in air. The burial depth of the particles was increased in each run until no particles could be detected, which corresponded to the maximum detection distance between a buried particle and the antenna.

For each tested distance, the antenna was swept over the particles 16 times providing a total of 64 iterations to detect the particles. The percentage of the particles detected as a function of the antenna - particles distance, D_{ap}, is shown in Fig. 4. The figure showed that the particles can be successfully detected 95% of the time, when the antenna - particles distance is less than 0.71 m. This value was considered the maximum antenna-particle detection distance for the RFID system. An abrupt reduction in the percentage of particles detected rate took place when the antenna-particle distance was larger than 0.73 m. This abrupt reduction was attributed to the fact that the particles were outside the electromagnetic field of the antenna. It was also found that the maximum detection distance of the antenna did not change significantly for the buried particles compared to the particles tested in the air [5]. Thus, the low frequency RFID system is appropriate for monitoring bridge scour because its waves can penetrate water and sand bodies without significant loss of its signal strength.

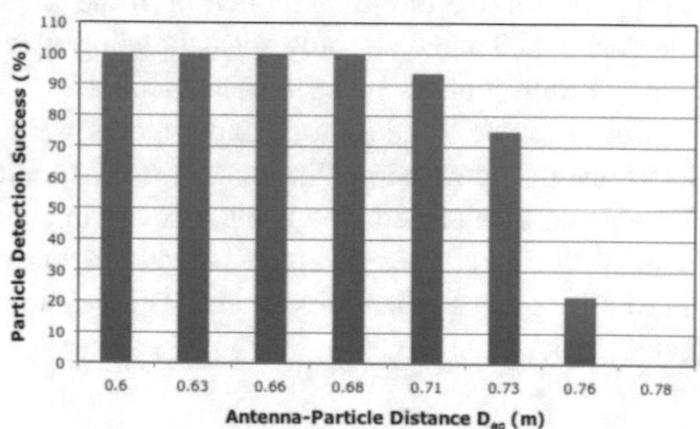

Fig. 4. Looking upstream of the recirculating flume. Particle detection success rate as a function of the antenna-particle distance, D_{ap}.

3.2 Pier Scour Results

The experiments described so far to evaluate the detection range of the RFID system were conducted without any structure in place. In the second set of experiments, a pier model was used to assess the ability of the RFID system in monitoring bridge scour. This set of experiments was based on the particle signal-loss principle and was conducted in the flume using a rectangular pier model 1.33 m long with a cross-section of 0.22 × 0.41 m. The pier was made of wood with an intermediate 0.15 m glass section to allow the placement of a video camera to monitor the scour occurring around the pier (Fig. 5). The pier was fixed into the bottom of the flume and buried in the sand up to 0.61 m of its height. A small antenna with a cross-section area of 0.20 × 0.20 m was fixed at the upstream face of the pier model as shown in Fig. 5. The sand bed surface was leveled around the pier and one glass particle was placed on the sand bed surface directly below the center of the antenna. The distance between the antenna and the particle was 0.20 m, which was 0.11 m less than the maximum detection distance of the antenna that was determined from previous experiments. It was anticipated that the antenna would lose contact with the particle, when the scour exceeded 0.11 m, thus providing an estimate of the maximum scour depth. The discharge and water depth during the test were 0.168 m^3/s and 0.18 m, respectively.

The experimental results showed that the particle remained stationary under this flow condition, while scour commenced underneath the particle. The particle in the scour-hole was detected via the antenna and was also monitored via a video camera from glass section of the pier. As the scour developed, the particle started to drop down into the scour-hole. Figure 5A–D shows a sample of images captured at different time

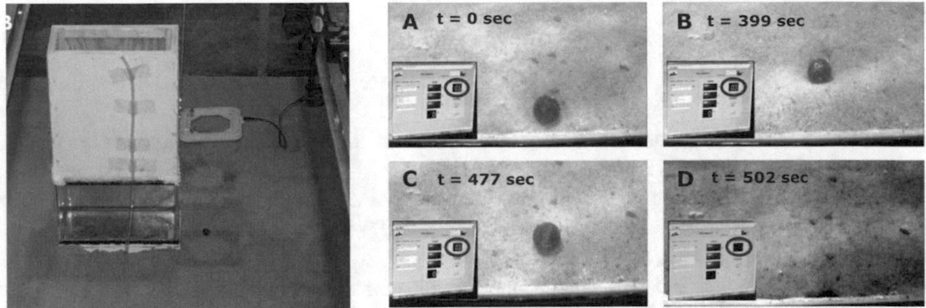

Fig. 5. Experimental setup of the rectangular pier with the antenna placed on the pier side and the RFID particle underneath. A–D: Snapshots of the movement of the transponder around the pier at characteristic time instances during the experiment.

instances during the experiment, while the particle was in communication with the antenna. When the distance between the particle and the antenna exceeded the maximum detection depth of the antenna (0.31 m), the contact between the particle and the antenna was lost (Fig. 5D). This indicated that the scour-hole had reached the anticipated maximum scour depth of 0.11 m. In Fig. 5D, the particle also went out of the camera field of view due to the scour and came closer to the pier model as the scour depth increased.

4 Conclusion

A passive Radio Frequency IDentification (RFID) system of low radio frequency band was configured for monitoring scour of bridge foundations. The performance of the RFID system was evaluated through a laboratory study to determine the maximum detection range of the system in terms of transponder orientation, housing material and maximum antenna-transponder detection distance. Analysis of the experimental results showed that the maximum antenna-particle detection distance was 0.73 m when the transponder is perpendicular to the antenna. Glass material provided better housing (cover) for the transponder when compared to material made of concrete-tungsten. It was also found that the maximum detection distance of the antenna did not change significantly for the particles buried in sand when compared to the particles tested in the air. Thus, the low frequency RFID system is appropriate for monitoring bridge scour because its waves can penetrate water and sand bodies without significant loss of its signal strength. The pier model experiments showed that the RFID system was able to predict successfully the maximum scour depth when a single particle was used near of a pier model where the scour-hole was expected.

References

1. Melville, B.W., Coleman, S.E.: Bridge Scour. Water Resources Publication, Highlands Ranch (2000)
2. Wardhana, K., Hadipriono, F.C.: Analysis of recent bridge failures in the United States. J. Perform. Constr. Facil. **17**(3), 144–150 (2003)
3. Tsakiris, A.G., Papanicolaou, A.N., Moustakidis, I.V., Abban, K.B.: Identification of the burial depth of Radio Frequency IDentification (RFID) transponders in riverine applications. J. Hydraul. Eng. **141**(6), 04015007 (2015)
4. Finkenzeller, K.: RFID Handbook and Applications in Contactless Smart Cards and Identification. Wiley, Hoboken (2010)
5. Papanicolaou, A.N., Elhakeem, M., Tsakiris, A.G.: Autonomous measurements of bridge pier and abutment scour using motion-sensing radio transmitters, Project-595, Iowa Highway Research Board Final Report (2010)

Considerations for the Strategic Design of the Humanitarian Supply Chain: Towards a Reference Model

Jessica Bull[1(✉)] and Juan Sepúlveda[2]

[1] Universidad Austral de Chile, Puerto Montt, Chile
jessicabull@uach.cl
[2] Universidad de Santiago de Chile, Santiago, Chile
juan.sepulveda@usach.cl

Abstract. The strategic design of supply chain is a key factor for organizations, we know that it is not possible to make correct operational decisions if they are not aligned with the strategy of the organization. This same rule applies to humanitarian supply chains, we will not have efficient answers if there is no prior planning. The objective of this research is to identify, on the basis of the Chilean experience and the literature review, the main considerations that should be taken into account when carrying out the strategic design of humanitarian supply chains in order to increase the capacity to respond to natural disasters. The study aims to provide the basic guidelines for the creation of a reference model for humanitarian supply chains, through the identification of the main challenges that humanitarian supply chain management faces today and the critical factors that increase the response capacity of the community.

Keywords: Humanitarian supply chain · Disaster risk management · Strategic design of humanitarian supply chain

1 Introduction

The earthquake and tsunami that occurred in Chile in 2010 marked the beginning of concern at the country level regarding risk management, although Chile is a country with recurrent natural hazards, the severity of the disaster that occurred opened the door to the question of how disasters in Chile were faced until that year. The United Nations Development Program [1] and the Pan American Health Organization [2] made evident the deficit in disaster management that the country had, the main statements of these organizations were related to socio-institutional weaknesses and lack of preparedness to respond in emergency and early recovery contexts, emphasizing also the need for a minimum organizational framework for all activities related to the provision of material assistance to people affected by disasters, as well as those supplies used by organizations in their assistance tasks.

The Chilean experience became a good starting point for the study of the management of the humanitarian supply chain since it has made it possible to identify the challenges that decision-makers must consider not only when facing a crisis caused by

T. Ahram et al. (Eds.): IHSED 2019, AISC 1026, pp. 605–610, 2020.
https://doi.org/10.1007/978-3-030-27928-8_92

a natural hazard, but also before the threat occurs. This, together with the international interest in risk management, understanding risk as the consequence of the interaction of the variables threat, vulnerability and capacity [3], provides the basis for the creation of a reference model for the strategic management of the supply chain. This study focuses on presenting the challenges obtained through the "Lessons Learned", the critical factors for the success of supply chain management and the considerations that must be taken into account when designing a strategic management model for the humanitarian supply chain in the event of a natural hazard.

2 Supply Chain and Disaster Risk Management

The international community's interest in reducing the effects of disasters caused by natural events has allowed its policy objectives to evolve. In 1989, when the decade of natural disaster reduction (1990–2000) was declared, its objectives were "to reduce, through concerted international action, especially in developing countries, the loss of life, material damage and social and economic disruption caused by natural disasters" [4] today after only three decades, the objectives are very different, the Sendai framework states as its objectives "to prevent new risks from occurring, reduce existing risk and strengthen resilience, as well as a set of guiding principles, including the primary responsibility of States to prevent and reduce disaster risk" [5]. The United Nations Office for Disaster Reduction (UNISDR) currently emphasizes: "disaster risk management rather than disaster management, hoping that these guidelines will prevent new risks, reduce existing risks and strengthen the resilience of communities" [5], understanding that disasters are not natural but the result of poor risk management.

Risk is conceived as the consequence of the interaction of the variables threat, vulnerability and response capacity (1) [3]. The threat cannot be managed and the scientific community has focused its studies on vulnerability and response capacity. Although the increase in interest of the scientific community has been noticeable, efforts are still concentrated on providing post-disaster tools considering response capacity as a variable to be worked on only at the aftermath. The risk management approach proposes management at all stages of the cycle (Prevention and mitigation, preparedness, response and recovery) [3], giving the guidelines to approach response capacity from a strategic perspective, incorporating it from prevention and preparedness, thus eliminating its emergency character. This leads to a shift in the focus or objective of future research from "seeking solutions for rapid response" to "reducing disaster risk by increasing the capacity of communities".

$$\text{Risk} = \text{threat} * (\text{Vulnerability}/\text{Response Capacity}). \quad (1)$$

One way to increase the capacities of communities is to better manage the humanitarian supply chain for natural hazards and thereby reduce risk.

Including risk management as the basis for the design of humanitarian supply chains for natural hazards requires consideration of temporariness in decisions, as there must be approaches and strategies at the prevention and mitigation stage; defining and using methods and tools in preparedness; and making use of tools in response and recovery, paralleling the commercial supply chain [6] and the risk management stages [3] (Table 1).

Table 1. Relationship between supply chain management and disaster risk management.

Management stages	Supply chain management activity	Stage of the risk management cycle
Strategic	Define supply chain focus and strategies	Prevention and mitigation
Tactical	Define and use supply chain methods and tools	Preparation
Operative	Make use of supply chain tools	Response and recovery

In this way we work under the same principle of commercial supply chains, where it is clear that without correct strategic planning, efficient operational decisions cannot be made, that is, we will not have resilient supply chains in the humanitarian response to a natural threat if there is no strategic design that allows us to plan in the preparation phase.

3 Lessons Learned - Experience in Chile

Critical analysis of the experiences of the different organizations involved in the response stage is essential to identify the challenges facing humanitarian logistics. There are currently two types of information sources in Chile that provide access to these experiences in a documented manner: international diagnoses [1, 2, 7–9] and documents on lessons learned conducted mainly by the Chilean armed forces [10–15]. International diagnoses describe the problem of risk management in Chile in a systemic manner, while lessons-learned documents from national organizations focus on the difficulties encountered in the response operation. By analyzing both sources of information, it is possible to identify nine challenges for the humanitarian supply chain in the event of a natural hazard: (a) Multiple actors or decision-makers, (b) diversity of objectives in conflict, (c) little or no coordination among the actors in charge of providing aid, (d) high uncertainty, (e) lack of logistical models incorporated in decision-support systems, (f) logistical systems affected by the environment, motivations and culture of the actors, (g) no strategic planning, (h) culture of incipient risk management and (i) strongly centralized decision-making.

4 Critical Success Factors for the Humanitarian Supply Chain

The concept of "critical success factors" (CSF) has been described by a large number of authors, agreeing that these are: "the conditions, characteristics or variables that when properly managed can have a significant impact on the success of an organization" [16]. In the case of critical success factors for the humanitarian supply chain (natural hazards), these will be defined as "conditions, characteristics or variables that, when efficiently managed, increase response capacity and therefore decrease disaster risk".

Despite the boom in research on risk management and the humanitarian supply chain caused by the large number of disasters associated with natural hazards in the last two decades, the scientific community has focused mainly on the study of post-disaster tools for the supply chain without incorporating the risk management paradigm. For this study, a bibliographic analysis was carried out through the web of science finding between 2000 and 2018: 59 articles in Disaster Logistics; 89 articles in Humanitarian Logistics; 47 articles in Relief Logistics; 42 articles in Disaster Supply Chain; 55 articles in Humanitarian Supply Chain and 23 articles in Relief Supply Chain. Within this analysis, 6 articles [16–21] that explicitly focus on success factors for the humanitarian supply chain or humanitarian logistics were chosen for in-depth study. These include the article "Identifying critical success factors in emergency management using a fuzzy DEMATEL method" [19] in which the authors identify 20 critical success variables that are grouped into the concepts of: Reasonable organizational structure, clear awareness of responsibilities, an effective emergency information system, the government's leadership unit for planning and coordinating as a whole, application of modern logistics technology and continuous improvement of the emergency management operating system. After the study of the 6 articles mentioned above, the critical success factors presented by the authors can be summarized in 12 factors: Coordination of actors, simple protocols, timely and distributed information, management of donations and volunteers, clear organizational structure, planning and preparation, education campaigns, use of technologies, international aid management, systemic vision, performance evaluation and common objective among the participating organizations.

Finally, contrasting the 12 factors found in the analysis of the literature with the main challenges obtained from the lessons learned and the characteristics of the supply chain according to the risk management paradigm, 9 constructs or critical success variables were identified that impact on the response capacity of the communities: Organizational structure, coordination, communication, quality of information, centralized decisions, available resources, education of the population, planning and preparation and management of donations. These constructs will be validated through a model of structural equations in a subsequent work.

5 Humanitarian Supply Chain Strategic Design Considerations

As mentioned above, research on the humanitarian supply chain and/or humanitarian logistics has been carried out with a clear emergency bias without incorporating the objectives set forth by the international community in the framework of Sendai [5], which bases its lines of action on the risk management paradigm. This makes evident the need to connect the recent research to the needs exposed by the community, making them be used by decision makers. Under this precept, the study focused on identifying supply chain challenges through the experience gained by Chile given the recurrence and high impact that natural hazards have in the country; and counteracting them with the success factors of the humanitarian supply chain exposed by researchers and with the characteristics of a supply chain design based on the incorporation of the stages of the disaster risk cycle in order to include this knowledge when designing a strategic humanitarian supply chain model in the event of natural disasters. In this way, when designing a strategic model of humanitarian supply chain management, classic supply chain management activities are considered within the stages of risk management, identifying their strategic, tactical and operational character as appropriate.

The strategic humanitarian supply chain management model should initially identify the supply chain focus to be designed; during the research carried out, three focuses were identified: supply of basic food and shelter products, evacuation of the population and transfer of critical wounded. The clarity of the focus will allow the final design objective to be identified, although for all strategic designs of humanitarian supply chains related to natural hazards the objective will be to increase the community's response capacity, its indicators - response time, coverage and relevance - may have different weighting factors depending on the supply chain focus. Once the approach has been defined, the supply chain strategies should be identified based on the critical success factors described in this work. This will allow the inclusion of management indicators, giving the possibility of quantifying the chain's performance, since each of the success factors is directly related to the final construct (response capacity). The next step will be to identify the models and tools to support the strategies defined above, currently it is possible to find in the literature different operational tools that will serve for this stage: tools for locating shelters, routing emergency vehicles, evacuation models, among others. The responsibility for the strategic design of the humanitarian supply chain should fall on the communities, strengthening the local capacity, the communities should be prepared based on their needs, culture and particular characteristics, which is why it is recommended that each community at the local level should know their hazards and vulnerability allowing them to better identify the approach that will have the strategic design; understand beforehand how the stages of risk management relate to the design of supply chain management, considering the temporality in the decisions, which will allow them to identify approaches and strategies in the stage of anticipation and mitigation; methods for the preparation and tools for the response stage; last but not least put special emphasis on the critical factors of success that impact on the response capacity considering these as future indicators of your management.

Designing a strategic humanitarian supply chain management model for natural hazards will enable the community to make informed decisions; to manage resources efficiently; to incorporate the role of the local authority by decentralizing decisions; to incorporate simple and effective protocols; to manage information with data integration; to allocate pre-position critical stocks that meet international standards; to know the capabilities of organizations; to manage volunteers and donations; to manage international aid; to establish organizational structure; to apply and use technology; to plan capacity; to manage resources; and to improve demand prediction.

Acknowledgments. The authors wish to thank: Department of Industrial Engineering and DICYT of the University of Santiago de Chile; and the School of Industrial Civil Engineering and DID of the Universidad Austral de Chile (UACh) for their support.

References

1. United Nations Office for Disaster Risk Reduction, Regional Office for the Americas: Diagnóstico de la situación de riesgo de desastres en Chile, pp. 96 (2010)
2. Organización Panamericana de la Salud: El Terremoto y tsunami del 27 de febrero en Chile, Crónica y lecciones aprendidas en el sector salud, pp. 23–73 (2010)
3. Municipalidad de Talcahuano: Estrategías territoriales para la reducción del Riesgo de desastre, pp. 10–16 (2016)
4. Oficina de naciones unidas para reduccion del riesgo de desastre (UNISDR). https://www.eird.org/americas/we/historia.html
5. Oficina de naciones unidas para reduccion del riesgo de desastre (UNISDR): Marco de Sendai para la reducción de riesgo de desastre 2015–2030, pp. 12–27 (2015)
6. Chavez, J., Torres-Rabello, R.: Supply Chain Management, Segunda edición, pp. 201–227 (2012)
7. Oficina de naciones unidas para reducción del riesgo de desastre (UNISDR): C Consideraciones para fortalecer una estrategia suramericana para la reducción del Riesgo de desastre, pp. 21–51 (2015)
8. Organización panamericana de la salud.: Manual para el manejo logístico de suministros humanitarios, pp. 11–119 (2000)
9. Granadilla, L.: Instituto Mexicano de transporte ISSN 0188–7297. Consideraciones para la gestión de la logística humanitaria postdesastre, pp. 3–27 (2015)
10. Ejercito de Chile: División Doctrina Lecciones aprendidas en una MOOTW Desastres naturales Terremoto (2009)
11. Ejercito de Chile: División Doctrina Lecciones aprendidas del incendio que afectó a la ciudad de Valparaíso el 12 de abril de 2014 (2014)
12. Ejercito de Chile: División Doctrina Lecciones Aprendidas del terremoto y tsunami que afectaron a las ciudades de Arica e Iquique el 1y 2 de abril de 2014 (2014)
13. Revista militar digital, foro las Américas: Dialogo con el capitán de navío Alberto Soto (2015). https://dialogo-americas.com/es/articles/las-lecciones-aprendidas-sobre-desastres-naturales-en-chile
14. Albornoz, C., Romero, H.: ¿Lecciones aprendidas?: Gestión de desastres y las erupciones volcánicas en Chile (2015). https://doi.org/10.13140/rg.2.1.3309.3603
15. Drake, V.: Fuerza Aérea de Chile Lecciones aprendidas producto de la catástrofe del 27 de febrero del 2010. Simposio: Gestión de respuesta a desastres naturales. Universidad de Santiago de Chile (2015)
16. Li, Y., Hu, Y., Zhang, X., Deng, Y., Mahadevan, S.: An evidential DEMATEL method to identify critical success factors in emergency management. Appl. Soft Comput. **22**, 504–510 (2014). ISSN 1568-4946
17. Oloruntoba, R.: An analysis of the Cyclone Larry emergency relief chain: some key success factors. Int. J. Prod. Econ. **126**(1), 85–101 (2010). ISSN 0925-5273
18. Yadav, D., Barve, A.: Analysis of critical success factors of humanitarian supply chain: an application of interpretive structural modeling. Int. J. Disaster Risk Reduct. **12**, 213–225 (2015). ISSN 2212-4209
19. Zhou, Q., Huang, W., Zhang, Y.: Identifying critical success factors in emergency management using a fuzzy DEMATEL method. Saf. Sci. **49**(2), 243–252 (2011). ISSN 0925-7535
20. Davidson, A.: Key Performance Indicators in Humanitarian Logistics, Massachusetts Institute of Technology, USA (2006)
21. Pettit, S., Beresford, A.: Critical success factors in the context of humanitarian aid supply chains. Int. J. Phys. Distrib. Logist. Manag. **39**(6), 450–468 (2009). https://doi.org/10.1108/09600030910985811

Prevalence and Risk Factors Associated with Upper Limb Disorders and Low Back Pain Among Informal Workers of Hand-Operated Rebar Benders

Sunisa Chaiklieng[1,2(✉)], Pornnapa Suggaravetsiri[3], Wiwat Sungkhabut[4], and Jenny Stewart[5]

[1] Department of Occupational Health and Safety, Faculty of Public Health, Khon Kaen University, Khon Kaen, Thailand
csunis@kku.ac.th

[2] Research Center in Back, Neck, Other Joint Pain and Human Performance (BNOJPH), Khon Kaen University, Khon Kaen, Thailand

[3] Department of Epidemiology and Biostatistics, Faculty of Public Health, Khon Kaen University, Khon Kaen, Thailand

[4] The Office of Disease Prevention and Control, 5th Nakhon Ratchasima, Nakhon Ratchasima, Thailand

[5] School of Rehabilitation and Occupation Studies, AUT University, Auckland, New Zealand

Abstract. Hand-operated rebar bender work involves physical exertion and repetitive movement of upper extremities that may cause upper limb disorders (ULDs) and low back pain (LBP). This cross-sectional analytic study aimed to investigate the prevalence and risk factors for ULDs and LBP among hand-operated rebar benders. Data were collected from 241 workers through questionnaires, lighting measurements, and physical fitness test. Risk factors were indicated by multiple logistic regression analysis using adjusted odds ratio (ORadj) at $p < 0.05$. During the 12 months up to data collection the highest prevalence of ULDs occurred at the wrists/hands (78.8%) and shoulders (46.9%). LBP prevalence was 68.9%. The risk factors significantly associated with wrists/hands pain were poor grip strength (ORadj = 2.69), smoking (ORadj = 4.44), work experience >5 years (ORadj = 2.34). The significant risk factors for LBP were age ≥ 50 years (ORadj = 1.88), work experience >5 years (ORadj = 1.89), and work hours ≥ 8 h/day (ORadj = 3.44). Risk factors for shoulders pain were workdays >5 days/week, work discomfort (SWI $\geq$ 2), work experience >5 years. The findings of risk factors are useful as a basis for surveillance to prevent occupational diseases among informal workers.

Keywords: Wrists pain · Low back pain · Informal worker · Subjective Workload Index

1 Introduction

Thailand is classified as an industrially developing country where the majority of the workforce is employed in small-scale home based work for informal workers. In 2017, the National Statistical Office of Thailand reported that 56% of a total employed

T. Ahram et al. (Eds.): IHSED 2019, AISC 1026, pp. 611–618, 2020.
https://doi.org/10.1007/978-3-030-27928-8_93

workforce (38.3 million) were informal workers [1]. Hand-operated rebar bender work is traditional home based work, making products for small and medium industrial enterprises in Thailand. This work involves physical exertion, awkward postures and repetitive movement of upper extremities and manual handling [2] which may cause musculoskeletal disorders (MSDs) [3]. A high incidence of upper limb disorders (ULDs) including hand/wrist, arm, and shoulder disorders and low back pain (LBP) among workers was previously reported in relation to work environmental factors in a company or an industrial workplace [4, 5]. Research has also shown that loss of working time due to injury [6] and cost of long term treatment have affected the income of workers and small business as well as gross national product [7].

An earlier study by the author found high prevalence of MSDs among this type of home-based worker [3]. This present study, therefore, aimed to indicate the prevalence and associated risk factors of ULDs and LBP among informal sector workers of hand-operated rebar bender in Thailand.

2 Materials and Methods

2.1 Recruitment of Subjects

A cross-sectional analytic study was conducted in four months among informal workers who used a hand-operated rebar bender. The study area was in the Non-Sung district of Nakhon Ratchasima province, Thailand. The sample size was based on an estimate of the population employed in this work as the exact number was unknown. A systematic random sampling method was used to recruit 241 subjects who met the study criteria. These were; Thai citizenship, a minimum of one year experience in this work, aged ≥ 18 years, no injury or disorders related to cervical, thoracic, or lumbar spine such as rheumatoid arthritis or degenerative disc disease, no birth related disability, not pregnant and agreeing to participate. This study obtained ethical approval from Khon Kaen University ethics committee, Thailand, No. HE542265. All subjects gave informed consent prior to entering the study.

2.2 Questionnaires and Measurements

Face-to-face interviews were conducted by a trained research assistant using the Standardized Nordic questionnaire (SNQ) for analysis of musculoskeletal symptoms [8] and work ergonomic factors questionnaire [9], which included demographic characteristics. These comprised age, gender, education level, marital status, work experience, health behavior (smoking and exercise), health status (chronic diseases) and body mass index (BMI). Psychosocial work factors were assessed by the Subjective Workload Index (SWI) [10]. Negative elements in this index were level of discomfort due to fatigue, perceived risks, mental concentration, task complexity, work rhythm and job responsibilities. Positive elements were interest in the job and autonomy in their work. The total SWI was calculated as follows: SWI = [(summation of load factors)−(summation of compensating factors)]/8. SWI $\geq$ 2 means discomfort at work.

A light meter (model LUTRON LX-105) was used to measure lighting intensity at workstations in home workplace. The physical fitness test was based on the International committee for the standardization of physical-fitness test (ICSPFT). Dynamometry was used to measure hand grip strength, back strength, and leg strength.

2.3 Data Processing and Analysis

Data were recorded by Epi-Info. for windows (Texas, USA, 2007) using a method of double data entry and analysed by STATA Version 10.1. Descriptive statistics were used to describe the prevalence of ULDs and LBP. Associations between ULDs and LBP and specified study factors were examined by univariate analysis. Factors with p-value less than 0.25 from univariate analysis were included in the multiple logistic regression analysis. Significant risk factors were screened in a backward stepwise manner using likelihood ration tests as selection criteria. Factors of age, gender and work experience were always included in the model as the control variables. Odds ratio (OR), adjusted odds ratio (OR_{adj}) and 95% CI were presented significantly set at $p < 0.05$.

3 Results

3.1 Personal Factors, Health Status and Work Conditions

Most informal sector workers were female (78.1%), aged 25–76 years (median = 49 years), married (84.2%), with primary school education (79.7%). Sixty-eight workers (28.2%) had underlying chronic disease. The most common diseases were hypertension diabetes, hyperlipidemia and dermal allergy (n = 11). Work experience ranged from one to 15 years (median = 5 years), with the largest group working between 6–10 years (n = 91, 37.8%). Forty-eight percent of subjects worked eight or more hours a day. The physical fitness test indicated that in 43.1% of workers had good grip strength (n = 105), while 52.7% had a fair level of back strength (n = 127) and 51.5% had a low level of leg strength (n = 124). Lighting intensities were lower than the standard requirement (300 lx) for 13 workstations (5.39%) when compared to Thailand's regulation on the standard of lighting intensity in the workplace [11].

3.2 Prevalence of Upper Limb Disorders and Low Back Pain

Data showed that the highest prevalence of ULDs during the 12 months up to data collection occurred at the wrists/hands (78.8%; 95% CI: 73.13–83.82) and shoulders (46.9% 95% CI: 40.45–53.40). Prevalence of low back pain was 68.9% (95% CI: 62.62–74.67). The highest prevalence of ULDs during the 7 days prior to data collection occurred at same sites as follows: wrists/hands 45.6% (95% CI: 39.24–52.16) and shoulders 27.4% (95% CI: 21.86–33.48). Prevalence of low back pain was 41.5% (95% CI: 35.20–47.99). Disorders at these three sites affected work and daily activities. Fifty-three workers reported pain in the wrists/hands and low back at some time during the twelve months prior to data collection and 31 workers reported pain at the same sites in the last seven-day period.

3.3 Risk Factors Associated with Wrists/Hands, Low Back and Shoulders Pain

Univariate analysis indicated that working for more than five years in this work was a significant risk factor ($p < 0.05$) for wrists/hands pain, shoulder pain, and low back pain. A further risk factor for wrists/hands pain was poor grip strength. The factor significantly associated with shoulder pain was psychosocial work factors with SWI 2 and age of 50 years and older was significantly associated with low back pain (Table 1).

Table 1. Correlation between personal factors, working characteristics and wrists/hands, lower back and shoulders pain by univariate analysis of (n = 241)

Risk factors of body parts	ULDs or LBP		OR (95% CI)	p-value
	No; n (%)	Yes; n (%)		
1. Wrists/hands				
Chronic diseases	42 (61.76)	26 (38.24)	1.61 (0.89−2.91)	0.114
Smoking	14 (58.3)	10 (41.7)	1.72 (0.72−4.04)	0.224
Poor grip strength	84 (61.7)	52 (38.4)	2.34 (1.30−4.19)	0.004*
Work experience > 5 years	66 (61.1)	42 (38.9)	2.01 (1.15−3.49)	0.014*
Workday > 5 days/week	111 (66.07)	57 (33.93)	1.69 (0.90−3.18)	0.102
2. Shoulders				
Workday > 5 days/week	127 (75.6)	41 (24.4)	1.82 (0.88−3.78)	0.109
Discomfort at work (SWI $\geq$ 2)	133 (75.1)	44 (24.9)	2.31 (1.02−5.23)	0.044*
Work experience > 5 years	76 (70.4)	32 (29.6)	2.38 (1.27−4.47)	0.007*
Workstation lower/above the elbow	50 (72.5)	19 (27.5)	1.60 (0.84−3.08)	0.156
Overweight/Obesity ($\geq$23 kg/m^2)	91 (75.2)	30 (24.8)	1.47 (0.79−2.73)	0.224
3. Low back				
Age $\geq$ 50 years	77 (68.1)	36 (31.9)	1.83 (1.04−3.29)	0.042*
Work $\geq$ 8 h/day	12 (60.0)	8 (40.0)	2.06 (0.08−5.31)	0.134
Work experience > 5 years	72 (66.7)	36 (33.3)	2.06 (1.14−3.69)	0.016*
Poor back strength	149 (72.33)	57 (27.67)	2.29 (0.85−6.21)	0.102

*The significant risk factors were set at $p < 0.05$.

Factors entered into the multiple logistic regression analysis were age, gender, work experience, body mass index (BMI), physical fitness parameters, chronic diseases, smoking, work hours per day, workdays per week, height of workstation and psychosocial work factors. Factors of age, gender and work experience were always entered into the model as the confounders. The significant risk factors for ULDs and LBP are shown in Table 2.

3.3.1 Wrists/Hands Pain

Informal sector workers who had poor grip strength had 2.69 times the risk of pain compared to those with good grip strength (95% CI:1.37–5.27; p = 0.004). Smokers were 4.44 times more likely to develop pain than non-smokers (95% CI: 1.18–16.69; p = 0.028) and those who had worked more than five years had 2.34 times the risk of pain compared to those working for less time (95% CI:1.26–4.35; p = 0.007).

3.3.2 Shoulders Pain

Workers employed more than five days a week were 2.25 times more likely to develop pain than those working five or less days (95% CI:1.03–4.95; p = 0.043) while those employed for more than five years had 3 times the risk of pain compared to their counterparts (95% CI:1.45–5.95; p = 0.003). Discomfort at work (SWI $\geq$ 2) increased the risk of pain by 2.23 times compared to no discomfort (95% CI:1.13–6.87; p = 0.026).

3.3.3 Low Back Pain

Workers aged 50 years and older were 1.88 times more likely to report LBP than their younger counterparts (95% CI:1.01–3.94; p = 0.043). Those who had worked more than five years had 1.89 times the risk of LBP (95% CI: 1.01–3.55; p = 0.045) and those working eight or more hours a day had 3.55 times higher risk of LBP compared to those working less time (95% CI: 1.08–10.97; p = 0.037).

Table 2. Factors significantly associated with wrists/hands, lower back and shoulders pain by multi variate analysis (n = 241).

Risk factors of body parts	ULDs or LBP		OR_{adj} (95% CI)	p-value
	No; n (%)	Yes; n (%)		
1. Wrists/hands				
- Smoking	14 (58.3)	10 (41.7)	4.44 (1.18−16.69)	0.028
- Poor strength	84 (61.7)	52 (38.4)	2.69 (1.37−5.27)	0.004
- Work experience > 5 years	66 (61.1)	42 (38.9)	2.34 (1.26−4.35)	0.007
2. Shoulders				
- Working > 5 days/week	127 (75.6)	41 (24.4)	2.25 (1.03−4.95)	0.043
- Work discomfort (SWI $\geq$ 2)	133 (75.1)	44 (24.9)	2.23 (1.13−6.87)	0.026
- Work experience > 5 years	76 (70.4)	32 (29.6)	3.00 (1.45−5.95)	0.003
3. Low back				
- Age $\geq$ 50 years	77 (68.1)	36 (31.9)	1.88 (1.01−3.94)	0.043
- Work $\geq$ 8 h/day	12 (60.0)	8 (40.0)	3.44 (1.08−10.97)	0.037
- Work experience > 5 years	72 (66.7)	36 (33.3)	1.89 (1.01−3.55)	0.045

Factors of age, gender and work experience were always entered into the model.

4 Discussion

This study demonstrates the most commonly experienced sites for musculoskeletal symptoms in workers of hand-operated rebar benders which was the area of the wrists/hands (78.8%), followed by the low back (68.9%). These findings indicate the adverse health effects for workers doing this physical work. The nature of this work requiring prolonged physical exertion, repetitive work, keeping unbalanced posture and/ or prolonged sitting might play an important role in disorders [2]. Consistent with studies in other countries, ULDs and LBP were a major problem found among different workers from various businesses, either in the industrial sector or informal economy [12, 13]. There were reports of repetitive movements, static-postural load and body bending causing repetitive strain injuries [14].

Although, gender was not a significant risk factor for ULDs and LBP in this study, longer work experience and aging were significant risk factors for MSDs. However, these factors are all considered variables for the surveillance of work-related MSDs. Working for a greater number of years, in parallel with aging, can increase the risk of ULDs and LBP in cases where the workplace has unsafe ergonomic conditions. As found in this study, the more workdays per week (>5 days) and the more work hours per day (≥ 8 h), the significantly higher the risks of the development of disorders. As shown in the previous study, workers exposed to a prolonged period of working hours had a high risk of the development of low back pain [13]. The exposure to this nature of hazards in the home based workplace with inappropriate workstations and inadequate conditions lower than the standard requirements for lighting intensity were identified at some workplaces of hand-operated rebar benders. Therefore, lighting intensity should be adjusted to meet the standard and workstations should be designed to suit the figure of each worker to avoid repetitively awkward posture and forceful exertion of the upper extremities which might later cause ULDs and LBP.

Postures including shoulder flexion, wrist extension and deviation, body twisting or repeated flexion, including the binding force of the work tools during rebar bending cause ergonomic risks [2] associated with the development of upper limb disorders and low back pain as supported by the previous study in the informal sector workers [12]. Physical factors associated with the occurrence of ULDs and LBP in this study were consistent with the previous study which mentioned that prolonged standing with frequent and repetitive movement of upper extremities may cause upper limb disorders and low back pain [14]. In addition to the matter of workers having a high prevalence of wrists/hands pain, poor grip strength was shown to be significantly related to wrists/hands pain among this kind of worker. It has been shown that good physical fitness has a protective effect against the development of musculoskeletal disorders. Therefore, it is suggested that workers may be able to avoid these risks by increased awareness of health promotion by regular exercise for good physical fitness as reported by the previous report in informal garment workers [15].

Interestingly, in terms of informal workers of hand-operated rebar benders, the factor of discomfort at work (SWI $\geq$ 2) significantly played an important role in shoulder pain. This might be explained by the nature of heavy workloads, work done at speed and monotonous work [2, 16], which might increase the risk of MSDs, accidents or injuries at work in small scale workplaces or the informal sector. Personal factors such as smoking significantly related to wrist/hands pain which might be an unexpected finding, even though Lei et al. [17] clarified that smoking for several weeks or longer than 12 months was associated with persistent of low back pain and the risk was increased by the duration of smoking through a dose-response relationship.

5 Conclusion

The high prevalence of pain in the areas of the wrists/hands, followed by the lower back among workers of hand-operated rebar benders had an important impact on the health. In particular, female and elderly workers who are a high proportion in rural areas of Thailand, had potential health risks for ULDs and LBP. Therefore, home workers in the informal economy of Thailand should be supported with occupational health services and the surveillance of occupational diseases. The findings of personal factors, physical factors and psychosocial factors associated with ULDs and LBP are very useful for setting guidelines on the prevention on musculoskeletal diseases and upgrading improvement of working conditions in order to create healthy workplaces.

References

1. National Statistical Office (NSO), Thailand: The report of informal employment survey 2017. Statistical forecasting division, Bangkok (2017)
2. Chaiklieng, S., Sangkhabut, W.: Applying BRIEF survey for ergonomic risk assessment among home workers of hand-operated rebar benders. J. Med. Technol. Phys. Ther. **26**, 57–66 (2014)
3. Sungkhabut, W., Chaiklieng, S.: Prevalence of musculoskeletal disorders among informal sector workers of hand-operated rebar bender in Non-sung district of Nakhon Ratchasima province. KKU J. Graduate Stud. **13**, 135–144 (2013)
4. Burton, A.K.: Work-relevant upper limb disorders: their characterization, causation and management. Occup. Health Work **5**, 13–18 (2008)
5. Shuval, K., Donchin, M.: Prevalence of upper extremity musculoskeletal symptoms and ergonomic risk factor at a Hi-Tech company in Israel. Int. J. Ind. Ergon. **35**, 569–581 (2005)
6. Department of Disease Control, Ministry of Public Health, Thailand: MSDs among out-patients and in-patients (2010). http://occ.ddc.moph.go.th/. Accessed 6 Nov 2011
7. Occupational Safety and Health Administration (OSHA): The need to reduce musculoskeletal disorders in America's work force (2000). http://osha.gov/pls/oshaweb/owadispshow_document?p_table=TESIMONIE&P_id=245. Accessed 28 Oct 2011
8. Kuorinka, I., Johnson Kilbom, B., Vinterberg, A., Biering, M., Sorenson, F., Anderson, G., et al.: Standardized Nordic questionnaire for the analysis of musculoskeletal symptoms. Appl. Ergon. **18**, 233–237 (1987)
9. Chaiklieng, S., Suggaravetsiri, P., Boonprakob, Y.: Work ergonomic hazards for musculoskeletal pain among university office workers. Walailak J. Sci. Technol. **7**, 169–176 (2010)
10. Vanwonterghem, K., Verboven, J., Op De Beeck, R.: Subjective assessment of workload. Tijdschrift voor Ergonomie **10**(3), 10–14 (1985)

11. Department of Labor Protection and Welfare, Ministry of Labour, Thailand: The standard of lighting intensity in the workplace, version (2018). http://www.labourgo.th/law/index.htm. Accessed 21 Feb 2018
12. Chaiklieng, S., Homsombat, T.: Ergonomic risk assessment by RULA among informal sector workers of Rom Suk broom weaving. Srinagarind Med. J. **26**, 9–14 (2011)
13. Guo, H.R.: Working hours spent on repeated activities and prevalence of back pain. Occup. Environ. Med. **59**, 680–688 (2002)
14. Chaiklieng, S., Suggaravetsiri, P.: Risk factors for repetitive strain injuries among school teachers in Thailand. Work **42**, 2510–2515 (2012)
15. Homsombat, T., Chaiklieng, S.: Physical fitness and muscular discomfort among informal garment female workers in Udon Thani province, Thailand. J. Med. Assoc. Thai. **100**, 230–238 (2017)
16. Chaiklieng, S., Juntratep, P., Suggaravetsiri, P., Puntumetakul, R.: Prevalence and ergonomic risk factors of low back pain among solid waste collectors of local administrative organizations in Nong Bua Lam Phu province. J Med. Technol. Phys. Ther. **24**, 97–109 (2012)
17. Lei, L., Dempsey, P.G., Xu, J.G., Ge, L.N., Liang, Y.X.: Risk factors for the prevalence of musculoskeletal disorders among Chinese foundry workers. Int. J. Ind. Ergon. **35**, 197–204 (2005)

Detection and Classification of Facial Features Through the Use of Convolutional Neural Networks (CNN) in Alzheimer Patients

David Castillo-Salazar[1,4(✉)], José Varela-Aldás[1], Marcelo Borja[2], Cesar Guevara[3], Hugo Arias-Flores[3], Washington Fierro-Saltos[4], Richard Rivera[5], Jairo Hidalgo-Guijarro[6], Marco Yandún-Velasteguí[6], Laura Lanzarini[7], and Héctor Gómez Alvarado[8]

[1] SISAu Research Group, Universidad Indoamérica, Ambato, Ecuador
{davidcastillo,josevarela}@uti.edu.ec

[2] Architecture, Art and Design Faculty, Universidad Indoamérica, Quito, Ecuador
carlosborja@uti.edu.ec

[3] Mechatronics and Interactive Systems - MIST Research Center, Universidad Indoamérica, Quito, Ecuador
{cesarguevara,hugoarias}@uti.edu.ec

[4] Facultad de Informática, Universidad Nacional de la Plata, La Plata, Argentina
washington.fierros@info.unlp.edu.ar

[5] Escuela de Formación de Tecnólogos, Escuela Politécnica Nacional, Quito, Ecuador
richard.rivera01@epn.edu.ec

[6] GISAT Research Group, Universidad Politécnica Estatal del Carchi, Tulcán, Ecuador
{jairo.hidalgo,marco.yandun}@upec.edu.ec

[7] Instituto de Investigación en Informática LIDI (CICPBA Center), Facultad de Informática, Universidad Nacional de la Plata, La Plata, Buenos Aires, Argentina
laural@lidi.info.unlp.edu.ar

[8] Universidad Técnica de Ambato, Ambato, Ecuador
hf.gomez@uta.edu.ec

Abstract. In recent years, the widespread use of artificial neural networks in the field of image processing has been of vital relevance to research. The main objective of this research work is to present an effective and efficient method for the detection of eyes, nose and lips in images that include faces of Alzheimer's patients. The methods to be used are based on the extraction of deep features from a well-designed convolutional neural network (CNN). The result focuses on the processing and detection of facial features of people with and without Alzheimer's disease.

Keywords: CNN · Alzheimer's · Algorithms · Images

T. Ahram et al. (Eds.): IHSED 2019, AISC 1026, pp. 619–625, 2020.
https://doi.org/10.1007/978-3-030-27928-8_94

1 Introduction

With the evolution of technology many paths have emerged in the scientific field such as the field of artificial intelligence (AI), expert systems and neural networks. This research details the development of a system that allows the recognition of images in Alzheimer's patients. Facial expressions in patients with neurological diseases reveal valuable information for the diagnosis and monitoring of neurological disorders [1].

The article written by [2] mentions the importance of the extraction of facial features related to the use of algorithms for face detection. CNN is one of the most representative network structures in deep learning in the field of image processing and recognition.

In the research developed by [3], a treatment is presented in patients with oral emotions in real time that is characterized by a face detector and a Key points as well as a CNN using DeepLearning to extract characteristics from the detected face.

In an article by [4], a new methodology for the detection of Alzheimer's using the TensorFlow Connective Neural Network (TF-CNN) is applied, which consists of three convolutional layers that allow for increased classification performance to accurately predict Alzheimer's disease on the basis of brain structural explorations.

In a study developed by [5], the importance of computer vision in the detection and recognition of some traditional characteristics such as posture, facial expression and illumination is presented, and CNN based on face recognition is proposed. The design of this network shows an accuracy of 87.0%.

2 Methods and Materials

The methods used are convolutional neural networks (CNN) seen as the architecture of artificial vision, important in image processing and categorization.

2.1 Methods

Methods based on deep learning are presented in [6]. CNN contain a series of convolution and grouping layers that are hidden, connected and specialized, this means that the first layers can detect lines, curves and specialize until deeper layers are reached.

The algorithm has a design consisting of several convolutional and operational maxpool layers that allow for reduced dimensionality, plus an exclusion layer to classify the extracted characteristics, as shown in Fig. 1.

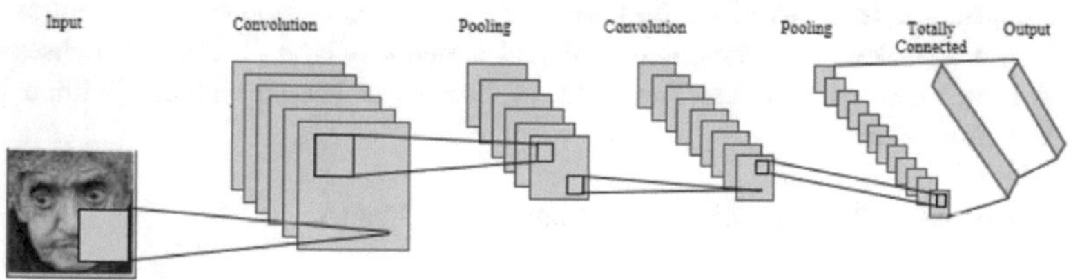

Fig. 1. CNN convolutional network

The convolutional network is a type of neural network that contains at least one layer using the convolution operation. This operation consists of making a multiplication of matrices between a matrix K called kernel that will be constant, and a window M taken from the image that will be of the same size as K. This operation is repeated for each pixel of the image where a window of the same size as K can be extracted, as shown in Fig. 2.

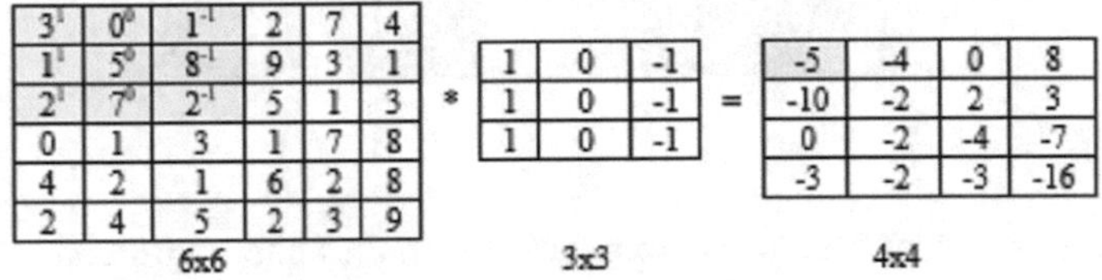

Fig. 2. Convolution operation

Its objective is to calculate the output size that an image will have after passing through a convolution. The following equation is required (1).

$$n_s = \left(\frac{n_e + 2 * p - k}{s}\right) + 1 \tag{1}$$

where n_s refers to the output size n_e in relation to the input size. p refers to the padding that is applied to the image, i.e. the number of pixels that are added to the image by counting only on one side. k is the size of the Kernel. s refers to the stride, the number of pixels that the window advances in each calculation; in the first image the stride was 1 as it performs the filtering in each of the pixels.

[7] shows the process of recognition of human facial expression. This process starts with input images that are assigned to the feature maps, then a grouping process is done, and finally the result of this process is registered as a vector. Table 1 exemplifies the transformations of each of the same cases that are recognized by the CNN classifier.

2.2 Materials

The images for this research have been taken from an existing database of a previous study that has a total of 55 people between men and women. These images are in grayscale at a size of 100 × 100 where the classification of the eyes, nose and lips is determined. [7] focuses on recognizing the smile as a meaningful facial expression by applying CNN.

Table 1. Set of processed images

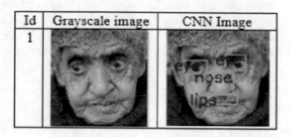

Id	Grayscale image	CNN Image
1		

The development procedure is complemented on a laptop with common features. Regarding the software, Python 3.16 was used with important libraries such as cv2, numpy, os.

3 Model Development

The proposed CNN model is a specific type of artificial neural network that is capable of learning to distinguish characteristics in a dataset through the calculation of convolutions. In [8], the characteristics for the recognition of emotion such as the forehead, eyebrows, eyes, cheeks and mouth areas are presented from a gray image as an entry, this process is identified in Fig. 3.

The stages of the model are indicated below:

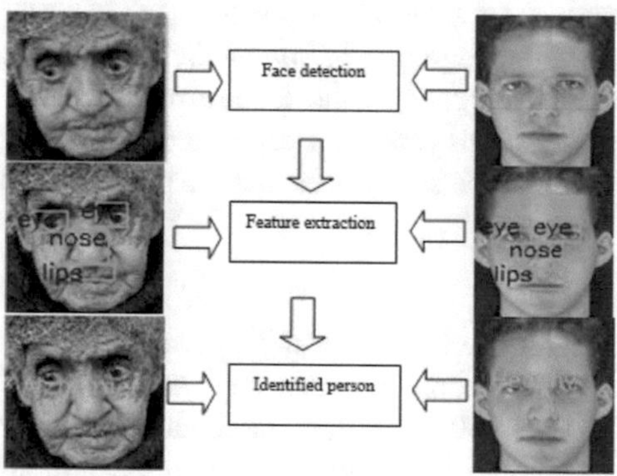

Fig. 3. Process of generated faces

Input layer, receives data from outside the CNN. In this case the data in images of the faces, the entries are three-dimensional, the dimensions are width, height and color channel in gray. [9] proposes a CNN which contains several layers of convolution, the maximum grouping, the residual block and the connectivity of the layers starting with the grayscale image.

Feature extraction layer, a series of layers that are responsible for obtaining relevant information to progressively build higher order features.

Convolutional layers, these aim to transform the input data through a set of neurons that are connected locally from the previous layer.

Pooling layers, which are placed between convolutional layers to reduce the dimensionality of the data passing through the different layers.

Classification layers, it is completely connected to its adjacent layer and its objective is to take the final characteristics calculated by the rest of the network and produce scores or probabilities corresponding to the trained classes.

The processing initially has an average loss of 13 ~ 15, the average is 0.05 once the training is finished. The average loss after training is 0.053 or less. In Table 2, a sample of the overall processing result is presented.

Table 2. General set of results obtained

Current network dataset	Average loss	Net present rate	Time	Image
5521:0.067291	0.053584	0.001000	3.321321	353344
5522:0.102180	0.058443	0.001000	3.587507	353408

4 Results

Fifty-five images of Alzheimer's patients with referential anxiety and depression features in the eyes, nose and lips were processed, and 40 images of normal people without any disease were processed for both men and women. The data obtained are presented in Table 3.

Table 3. Set of results obtained

Id	Grayscale image	CNN Image	Result
1	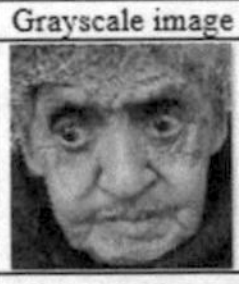	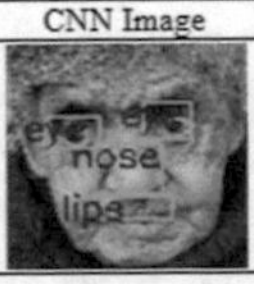 eye eye nose lips	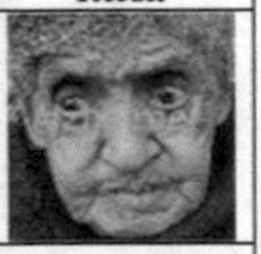
...	...	...	...
95		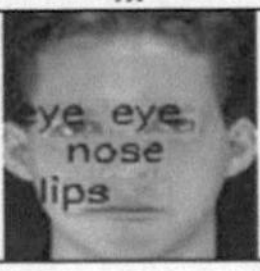 eye eye nose lips	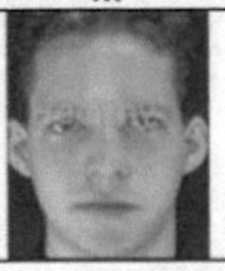

The procedure developed allows us to start from a grayscale image, process it with CNN-based characteristics to obtain a classification of the eyes, nose, lips with healthy and sick people determining the expected result according to the classification.

According to [10], facial expression recognition is easier in laboratory environments, facial expressions of 7 classes are shown with a proposed average rate of 71.6%, while the proposed research is developed with 6 classes and achieves an average of 53% in another type of context.

5 Conclusions

Relevant aspects of convolutional neural networks were applied in this research work, including the convolution process itself, the maximum grouping with completely connected layers, making its application interesting exclusively in Alzheimer's patients.

With the application of CNN in the trained data set, a correct tagging of eyes, nose and lips was obtained in 100 out of 100 images, both for Alzheimer's patients and for healthy people as opposed to other investigations that focus on the automatic detection of other components such as eyeglasses in facial detection.

It is important to reference the existing database of Alzheimer's patients from previous research, because it is not very common to find data with certain neurological characteristics.

References

1. Yolcu, G., et al.: Deep learning-based facial expression recognition for monitoring neurological disorders. In: Proceedings - 2017 IEEE International Conference on Bioinformatics and Biomedicine BIBM 2017, pp. 1652–1657, January 2017
2. Zhang, H., Qu, Z., Yuan, L., Li, G.: A face recognition method based on LBP feature for CNN. In: Proceedings 2017 IEEE 2nd Advanced Information Technology, Electronic and Automation Control Conference IAEAC 2017, pp. 544–547 (2017)
3. Gil, R.: Desarrollo de un sistema de reconocimiento de emociones faciales en tiempo real (2017). no. Cvc
4. Taqi, A.M., Awad, A., Al-Azzo, F., Milanova, M.: The impact of multi-optimizers and data augmentation on tensorflow convolutional neural network performance. In: Proceedings - IEEE 1st Conference on Multimedia Information Processing and Retrieval, MIPR 2018, pp. 140–145 (2018)
5. Yuan, L., Qu, Z., Zhao, Y., Zhang, H., Nian, Q.: A convolutional neural network based on tensorflow for face recognition. In: Proceedings 2017 IEEE 2nd Advanced Information Technology, Electronic and Automation Control Conference IAEAC 2017, pp. 525–529 (2017)
6. Murata, K., Mito, M., Eguchi, D., Mori, Y., Toyonaga, M.: A single filter CNN performance for basic shape classification. In: 2018 9th International Conference on Awareness Science and Technology iCAST 2018, pp. 139–143 (2018)
7. Qu, D., Huang, Z., Gao, Z., Zhao, Y., Zhao, X., Song, G.: An automatic system for smile recognition based on CNN and face detection. In: 2018 IEEE International Conference on Robotics and Biomimetics, ROBIO 2018, pp. 243–247 (2019)
8. Al-Azzawi, A., Hind, J., Cheng, J.: Localized deep-CNN structure for face recognition. In: Proceedings - International Conference on Developments in eSystems Engineering DeSE, September 2018, pp. 52–57 (2019)

9. Jain, D.K., Shamsolmoali, P., Sehdev, P.: Extended deep neural network for facial emotion recognition. Pattern Recogn. Lett. **120**, 69–74 (2019)
10. Wang, X., Peng, M., Hu, M., Jin, C., Ren, F.: Combination of valence-sensitive loss with restrictive center loss for facial expression recognition. In: Proceedings - 2018 10th International Conference on Advanced Computational Intelligence ICACI 2018, pp. 528–533 (2018)

Healthy Office by WELL Building Standard: Polish Examples

Anna Taczalska-Ryniak(✉)

Faculty of Architecture, Cracow University of Technology,
ul. Warszawska 24, 31-155 Cracow, Poland
ataczalska@pk.edu.pl

Abstract. The aim of the following article is a short presentation of rules of WELL Building Standard, which is a new building certification method. What differs it from the systems that are already in common use, is the fact, that it concerns mostly on human well-being and health. That is why the WELL Building Standard is designed to be a supplement of systems such as BREEAM or LEED and presumably may be the most popular in the future. Beside discussing general rules, author concentrate on presenting examples of building and interiors, that have been designed and built in Poland and already certified.

Keywords: Building certification · Office buildings rating · WELL Building Standard · Office work environment · Indoor Environmental Quality

1 Introduction

Since the beginning of the 21th Century green building rating systems have become so popular, that nowadays we can hardly imagine a building of public use, that is not assessed with one of the method. Systems such as BREEAM, LEED, DGNB or HQE have a particular application to office constructions as they allow to compare and judge potential lease areas objectively and according to many criteria. Certifications have almost replaced A, B, C – class rating that had been used before to define the quality of an office building. For example, in Poland more than 50% of all rated building area in 2018 was designed for offices. Furthermore, last year already 64,5% of new-built Polish offices have been rated [1].

All rating systems mentioned above rule similar principles. During certification process many different aspects are being evaluated. Checklists are divided into a few overriding categories. Among those groups the most valuable are: energy and water management and saving and materials selection according to natural resources respect. All of them are fundamentally dedicated to environmental protection.

Aspects of building users health and comfort are also being judged, but they are much less worth than those devoted to sustainable design. By LEED for Core and Shell (v2009) rating method, which is still the most popular for office building evaluation in Poland, an certificated object may not even collect a single point in *Indoor Environmental Quality* category to obtain the highest rate (Platinum) in certification process. In

T. Ahram et al. (Eds.): IHSED 2019, AISC 1026, pp. 626–630, 2020.
https://doi.org/10.1007/978-3-030-27928-8_95

this category there are only 12 points out of 110 (11%) to gather, while collecting 80 points in the whole certification process provide obtaining Platinum note.

For this reason, to increase the quality of indoor environment, as well as buildings' users health and comfort a new certification system have been compiled by American International Well Building Institute: WELL Building Standard. It have been in use since 2014 and is becoming more and more popular every year. Until today 2160 projects (covering all together 385 sqm of usable area) in 51 different countries have been already registered to be certified by this method [3].

WELL Building Standard rate criteria in seven categories dedicated to human well-being. Among them are: air, water, nourishment, light, fitness, comfort and mind. Due to narrowed scope of study it is often used as a supplement to other rating systems to certificate building or their interiors, that are not used by their owners but temporary occupants (workers or visitors). For this reason it is most successful by office buildings, that are judged with following schemes: *Core and Shell* and *New and Existing Interiors*. The third method of rating by WELL Building System is *New and Existing Buildings*, but it is not so popular by now [2].

2 Poland as a Leader in Building Certifications in Central and Eastern Europe

Among thirteen countries of Central and Eastern Europe (among them: Bulgaria, Croatia, Czech Republic, Estonia, Hungary, Latvia, Lithuania, Poland, Romania, Serbia, Slovakia, Slovenia, Ukraine) Poland is an undisputed leader of buildings' certification. At. the end of year 2018 there were 1268 certified objects in the whole region. 648 (51%) of them are situated in Poland (Fig. 1) [4].

Fig. 1. Complex of three office building pre-certified in WELL Building Standard: VARSO tower, designed by Foster and Partners, VARSO I, VARSO II, designed by Hermanowicz Rewski Architekci. All of the buildings are still under construction.

Exactly the same tendency appears in case of WELL Building Standard. At the beginning of the year there were 26 projects of Central and Eastern Europe registered in WELL system and 13 of them are Polish examples. At the end of May 7 Polish project were already certified with pre- or finale certificate. All of them are situated in Warsaw, which is the capital of the country and its central office market. The most of them (5) are pre-certificates for office buildings under construction, assessed with *Core and Shell* scheme. All of them have been rated GOLD (the second highest mark). The other two certificates have been given to office interiors designs, judged by *New & Existing Interiors* method. They both have already definitive status and are also marked GOLD [5].

3 Characteristic of WELL Building Standard Certification System on Polish Examples of Office Interior Design

The first definitive WELL certificate in Poland was given in 2018 to BuroHappold headquarter located in Spectrum Tower building in Warsaw. Interiors were designed by Kuryłowicz & Associates architectural studio together with the owner and user - BuroHappold Engineering. The participation of future user in the design process is obligatory by WELL Building Standard (Fig. 2).

Fig. 2. BUROHAPPOLD Engineering headquarter in Warsaw interiors, designed by Kuryłowicz & Associates in cooperation with BuroHappold Engineering, completed and certified in 2018.

Office area is divided into zones, that are designed for specific action (concentration, communication, chill) and separated one from another. Thanks to this it is really quietly in the office, although more than a hundred employees work there. Acoustic comfort is one the most important factor to improve people's concentration.

Well-being in particular zones is supported by intelligent ventilation system. In the whole office there are carbon dioxide sensors installed. Thanks to this, air streams to particular zones are automatically customized proportionally to the amount of users in a particular zone at the time.

To maximize people's comfort, the whole office area is also divided into several dozen of air-conditioning zones with an individual control.

Of course, the quality of air need to be carefully tested, the same as the quality of water. To provide it, equipment and finishing materials were carefully selected to minimize the presence of volatile organic compounds.

Air quality and people's comfort are succored by green walls, that are present in every office zone. Beside open-space offices, conference rooms, quite work rooms, architects provided a large kitchen combined with dining room, where employees can prepare and eat their own meals, as well as meet, talk and spend some time together.

The other important factor is light. Office plan is also designed in such a way, that all workstations have access to daylight. Thanks to external shutters and interiors blinders natural light can be adjusted to owns preferences. Also artificial light system have been designed with the utmost care. Its color and intensity change during the day, following the human daily cycle to keep employees awake and concentrated all the day long [6].

The other definitive WELL certificate in Poland was given in 2019 to SKANSKA headquarter located in Spark C office complex in Warsaw, designed by Kaim i Bursche Architekci. The whole building is still in the middle of certification process. SKANSKA declares, that all of their forthcoming office investment in Poland will be certified by WELL Building Standard in the future (Fig. 3).

Fig. 3. SKANSKA office in Spark C office building in Warsaw, designed by Kaim i Bursche Architekci, completed and certified in 2019.

SKANSKA's headquarter is an office full of greenery: plants and great areas of green walls. Beside precise selection of building materials, that are free from volatile organic compounds, it helps improving indoor air quality and people's mind comfort. For this purpose office space have been equipped with individually designed and prepared pieces of art.

The office is equipped with drinkers with fresh and clear (well controlled) water. Such solution is popular in Mediterranean countries, but still rather innovative in Poland, where climate is much colder. Drinkers helps here to reduce usage of plastic bottles and encourages people to live and eat healthier. A wide offer of fresh and healthy breakfasts accessible for employees together with a full nutritional value information and omnipresent posters also promote healthy habits.

SKANSKA's headquarter is probably also the first Polish office, that allows its users to control indoor environmental parameters (temperature, humidity of air, light intensity) by smartphones. Beside the Human Centric Lighting system, that automatically follows current conditions, people's activity and daylight cycle, it is a great facilitation in microclimate adjustment [7].

4 Summary

WELL Building Standard is the first certification system, that emphasize the significance of health and comfort of buildings' users. It focuses only on those criteria, that influence sense of well-being of people, who use the rated buildings.

It is beyond doubt, that it could really improve quality of office work environment. Among the most valuable aspects, that are rated by WELL system we can mention looking after the quality of light in the office: providing access to daylight to all workstations, managing the intensity and color of light according to natural human daily cycle as well as allowing users to adjust those parameters at their own directions. Increased control of air and water quality serve health of building users health. For those reason, we should be glad, that the WELL Building Standard is now becoming a popular supplement to green building rating systems such as BREEAM, LEED and DGNB. It proves, that more and employers understand, that a better designed and healthier office raise their employees efficiency.

Against, we should consider, if evaluating non-architectural criteria is needed. I suppose that equipping an office with vending machines, that serve only healthy food is still a part-architectural action, But planning methods of business trip organization is already a job for an manager rather than an architect.

References

1. Taczalska-Ryniak, A.: Healthy office according to WELL Building Standard. Polish Office Market 2019, pp. 180–181 (2019)
2. International Well Building Institute: The WELL Certification Guidebook (2018)
3. International WELL Building Institute. https://www.wellcertified.com/
4. Polish Green Building Council (PLGBC): Polish certified green buildings in numbers – 2019 Analysis
5. Polish certified buildings database by Polish Green Building Council (PLGBC). https://plgbc.org.pl/baza-budynkow-certyfikowanych/
6. Property Design: Warszawska siedziba BuroHappold z pierwszytm certyfikatem WELL w Polsce. http://www.propertydesign.pl/design/185/warszawska_siedziba_burohappold_z_pierwszym_certyfikatem_well_w_polsce,20544_1.html
7. Skanska: Spark C z certyfikatem WELL Interior. https://www.skanska.pl/o-skanska/media/informacje-prasowe/231285/Spark-C-z-certyfikatem-WELL-Interior-

Development Needs of the OSH-Related Risk Management Process: A Companies' Viewpoint

Noora Nenonen[1(✉)], Johanna Pulkkinen[1], Sanna Anttila[2], and Jouni Kivistö-Rahnasto[1]

[1] Unit of Industrial Engineering and Management, Tampere University, P.O. Box 541, 33101 Tampere, Finland
{noora.nenonen, johanna.pulkkinen, jouni.kivisto-rahnasto}@tuni.fi

[2] Environment and Health, HSEQ-Management, Ramboll Finland Oy, P.O. Box 718, 33101 Tampere, Finland
sanna.t.anttila@ramboll.fi

Abstract. Occupational safety and health (OSH) legislation, as well as OSH practitioners, generally consider the practices of risk management (RM) to be the foundation for OSH management in the workplace, and RM practices are commonly used in this setting. In practice, carrying out RM can be difficult. To ensure the effective improvement of RM practices, this paper examines the current development needs workplaces encounter in practice. Interviews (n = 46) were carried out in four large-scale Finnish enterprises operating in the manufacturing industry and facility services. The interviewees most often mentioned development needs related to the activity and competence of the persons involved in RM, and the process and tools used in RM. Companies would benefit from better descriptions and communication of the roles and responsibilities related to the RM process. It would be interesting to examine whether increased utilization of electronic systems can improve the success of OSH-related RM.

Keywords: Hazard recognition · Risk identification · Risk assessment · Workplace safety · Manufacturing industry

1 Introduction

Occupational safety and health (OSH) practitioners generally consider the practices of risk management (RM) to be the foundation for OSH management, and they are commonly used in the workplaces. European OSH legislation requires that employers take measures necessary for the safety and health protection of workers, including prevention of occupational risks [1]. Primarily, risk should be avoided, and if not possible, evaluated and treated at its source [1]. In Finland, the OSH legislation states that the employer shall, considering the nature of the work and activities, systematically and adequately investigate and identify the hazards and factors that can cause harm by the work, work premises, work environment, and working conditions [2]. If these

T. Ahram et al. (Eds.): IHSED 2019, AISC 1026, pp. 631–636, 2020.
https://doi.org/10.1007/978-3-030-27928-8_96

hazards and factors cannot be eliminated, their consequences for employees' safety and health should be assessed [2]. However, several statistics and surveys propose that occupational accident prevention and OSH promotion have not had the desired effects in Finland or Europe [3–6].

In practice, carrying out RM can be difficult, and several problems have been discussed in previous research. For example, a considerable number of OSH-related hazards in the work environment seem to remain unidentified [7–9]. Moreover, deficiencies in hazard identification are often mentioned among the causes of occupational accidents [10]. The problems related to defining the magnitude of risks are widely discussed in the literature [11–13], and those involved in OSH-related RM in the workplace consider it difficult [14]. Some studies have mentioned that the results of OSH-related risk assessments could be better utilized, and that RM efficiency is rarely estimated, or RM practices improved [14, 15]. In addition, methods and tools used in RM have been criticized as incomprehensive, subjective, static, and poorly applicable for different environments [16]. Limitations related to the competence of those involved in OSH-related RM in the workplace have also been mentioned [16].

Problems identified in RM hinder the successful realization of OSH-related RM in the workplace. To ensure effective improvement of RM practices and their support of OSH management, a better understanding of the practical development needs from the workplace viewpoint is required. This paper examines the current development needs of OSH-related RM that workplaces encounter in practice.

2 Materials and Methods

A qualitative approach [17] and a case study strategy [18] were employed in this research due to its descriptive nature. Congruent with the methodological approach, data for this study were collected from companies through interviews (see the summary in Table 1). Four large-scale companies from the manufacturing industry and facility services participated.

Table 1. Summary of the participating companies and data.

Companies (n = 4)	Industry	Product manufacturing (3 companies), facility services (1)
	Revenue *	Between €70 and 420 million
	Personnel *	Between 250 and 7,908 persons
Interviews (n = 46)	No. of interviews and interviewees	Company A (16, 19), Company B (10, 10), Company C (10, 16), Company D (10, 11)
	Role of the interviewees	Manager or supervisor (15), OHS representative (15), employee (12), HS or HR manager or specialist (9), occupational health service representative (5)

* Revenue and number of employees of the subject examined in the research project at the time of the research.

The companies were chosen based on their concrete need to develop, and interest in developing, their risk management practices. There were similarities and differences in the RM practices between the companies. All the companies were part of an international or Finnish larger corporation. However, the study focused on the Finnish operations, and on the manufacturing companies, in one factory or production facility.

The companies participating in this study used the Risk Assessment in Workplaces Workbook (the MSAH Workbook) [19], published by the Ministry of Social Affairs and Health, in their OSH-related RM. The MSAH Workbook provides basic guidelines in OSH-related RM for workplaces, and it is widely applied in Finland. The workbook was developed in a research project in the 1990s, and it has been updated to correspond to changes in the legislation and working life [19]. Practical implementation of the RM was described in more detail in Nenonen et al. [20].

The purpose of the interviews was to find out how risk assessment and management are carried out in practice, and what kind of good practices and development needs related to risk assessment have been identified by those who participate in risk assessment. The interviews were directed at personnel involved in RM, including OSH managers and experts, representatives of occupational health services, managers, and employees. The interviewees were selected with the companies' contact persons. The interviewees had varied risk assessment experience.

A total of 46 interviews were carried out, and 56 persons were interviewed. The interviews were conducted between March and May 2017. They were mainly individual face-to-face interviews. In addition, several group and phone or Skype interviews were carried out. The interviews were semi-structured, so that the themes and questions were considered in advance, but the order and form of the questions could vary, and additional questions were asked as needed. In addition, the questions varied according to the background of the interviewee. The interview questions were drawn up based on initial meetings with the companies, research questions, and the steps involved in the risk management process. All interviews were recorded and transcribed. The results of the interviews were categorized according to the themes that emerged.

3 Results

The interviewees most often mentioned development needs related to the competence of the persons involved in RM, the process and tools used in RM, the roles and activity of those involved in the RM, and the personnel's awareness of RM practices.

According to the results, OSH experts', supervisors', and employees' RM-related competence could be improved. There were notable differences, particularly in supervisors' competence and awareness related to RM practices. The interviewees mentioned that, for example, the terminology and definitions, and the purpose and significance, of RM could be better known. Employees' skills in identifying hazards, determining the magnitude of the risk, and being knowledgeable about conducting risk evaluations, were mentioned as development needs of employee competence. However, although RM-related competence was mentioned as a development need, interviewees did not wish more education or training. Safety issues should be better integrated into all education or training.

There seemed to be some contradiction between the OSH experts' and supervisors' expectations for the RM roles. Many interviewees mentioned that supervisors' resources were limited, and they did not have time to take care of safety issues or RM. In addition, lack of practical experience and the fact that analysis are done infrequently were seen as problematic. OSH experts called for supervisors to take a more active role in RM. However, supervisors wished for more support from OSH experts. The supervisors hoped the OSH experts would remind them about what should be done and when regarding RM. All roles and responsibilities in the RM process should be clearly specified and informed.

The interviewees hoped for a less complicated, more tailored process and tools for RM, especially for small, infrequent situations and frequent risk analysis processes. They mentioned that the processes related to RM overlapped. RM should be a continuous and systematic process. The risk analysis process is often performed by using paper forms, and then, the results are saved in electronic forms. Interviewees mentioned that this process was too complicated. There was a strong inclination toward the utilization possibilities of electronic systems in RM. For example, automatic reminders related to the implementation and recording of risk controls supported by electronic systems were seen as important.

In addition, development needs were mentioned in the use of the risk assessment results and up-to-datedness of the RM and related documents. Risks are often analyzed so rarely that the results are not up to date. Moreover, realization of controls and documenting them may be forgotten or is done with delay. The results of the risk analysis should be utilized much better than at present. All supervisors did not recognize their role as a utilizer of the RM results, or as an information distributor for employees. The need to discuss the results of risk analysis with employees came up in the interviews, especially the need to do it better.

4 Discussion

Most of the development needs that emerged in this study were similar to those mentioned in previous research. For example, similar to this study, development needs have been mentioned previously regarding competence and tools [16], and utilization of results and up-to-datedness [14, 15] of RM. However, several different and less discussed aspects were found.

For example, some studies have highlighted the significance of training and competence, and the characteristics of tools for successfully identifying hazards [16]. The same themes were emphasized in this research, although mainly for different reasons. The interviewees called for better understanding related to the theory, significance, and practices of risk assessment and management, in general. In addition, the interviewees wanted to develop the RM tool so that it would be easier to carry out RM. However, the need for skills to identify risks comprehensively was also mentioned, as emphasized in previous research.

Interestingly, one of the most important development needs in this study concerned a contradiction in the expectations related to the roles and initiative of those involved in RM. It was discovered that OSH experts wished supervisors had a more active role in

RM. However, the roles, responsibilities, and practices related to RM were not completely clear to supervisors and other personnel involved in the RM. Moreover, the supervisors seemed to lack resources, and there was a need for strong support on behalf of the OSH organization. Supervisors' roles in safety management and leadership have been widely discussed in previous research [21–23]. However, this issue seems to have received less attention in studies focusing on OSH-related RM.

Overall, it seems that the companies' representatives considered the development of OSH-related RM mainly from the viewpoint of the practical implementation of RM. The development needs the interviewees named emphasized easier realization of RM in practice. There was less concern about whether RM is successful, although there was some concern about, for example, the identification of good risk controls. This is an interesting detail, because the researchers of this study have previously pointed out the importance of better understanding of whether planned controls are effective and efficient [20].

It is noteworthy that the respondents were also content with some of the issues that were mentioned as requiring development. The respondents stated that they were satisfied, for example, with some aspects related to competence, cooperation, and tools in RM. At the same time, this may be interpreted as differences between companies' and respondents' needs, but as a real willingness to further develop OSH-related RM. For example, the respondents were happy that the cooperation and competence of employees of the risk assessment target and experts (e.g., occupational health care) were utilized when necessary. In addition, the RM tools enabled the basic process of the RM, although many further development needs were mentioned.

Based on the results, it seems companies would benefit in particular from better descriptions of, and communication regarding, the roles and responsibilities related to the RM process. Moreover, RM-related training and guidance directed at supervisors and employees would be beneficial. In the future, it would be interesting to see whether the utilization of electronic systems increases, and whether the systems can substantially improve the practical implementation and success of OSH-related RM in the workplace.

Acknowledgments. The authors would like to sincerely thank the Finnish Work Environment Fund for the funding for this study, as well as the companies that participated.

References

1. Council Directive 89/391/EEC on the introduction of measures to encourage improvements in the safety and health of workers at work (1989)
2. Finnish Occupational Safety and Health Act 2002/783 (2002)
3. Workers' Compensation Center: Työtapaturmien määrä ja tapaturmataajuus kääntyivät nousuun vuonna 2017. Workers' Compensation Center (2018)
4. Workers' Compensation Center: Työtapaturmatja tapaturmataajuus kääntyivät lievään nousuun vuonna 2015. Workers' Compensation Center (2016)
5. Workers' Compensation Center: Työtapaturmat – Tilastojulkaisu 2015. Workers' Compensation Center (2015)

6. Eurostat: Health and safety at work in Europe (1999–2007): a statistical portrait. Office for Official Publications of the European Communities, Luxembourg (2010)
7. Albert, A., Hallowell, M.R., Kleiner, B.: Enhancing construction hazard recognition and communication with energy-based cognitive mnemonics and safety meeting maturity model: multiple baseline study. J. Constr. Eng. Manag. ASCE **140**, 04013042 (2014)
8. Bahn, S.: Workplace hazard identification and management: the case of an underground mining operation. Saf. Sci. **57**, 129–137 (2013)
9. Carter, G., Smith, S.: Safety hazard identification on construction projects. J. Constr. Eng. Manag. ASCE **132**, 197–205 (2006)
10. Haslam, R.A., Hide, S.A., Gibb, A.G.F., Gyi, D.E., Pavitt, T., Atkinson, S., Duff, A.R.: Contributing factors in construction accidents. Appl. Ergon. **36**, 401–415 (2005)
11. Duijim, N.J.: Recommendations on the use and design of risk matrices. Saf. Sci. **76**, 21–31 (2015)
12. Johansen, I.L., Rausand, M.: Ambiquity in risk assessment. Saf. Sci. **80**, 243–251 (2015)
13. Goerlandt, F., Reniers, G.: On the assessment of uncertainty in risk diagrams. Saf. Sci. **84**, 67–77 (2016)
14. Lanne, M., Heikkilä, J.: Uutta riskienarviointiin! Tietopohjan merkitys ja uudistamisen keinot. VTT Technical Research Centre of Finland Ltd, Espoo (2016)
15. Niskanen, T., Kallio, H., Naumanen, P., Lehtelä, J., Liuhamo, M., Lappalainen, J., Sillanpää, J., Nykyri, E., Zitting, A., Hakkola, M.: Riskienarviointia koskevien työturvallisuus- ja työterveyssäännösten vaikuttavuus. Ministry of Social Affairs and Health, Helsinki (2009)
16. Albert, A., Hallowell, M.R., Skaggs, M., Kleiner, B.: Empirical measurement and improvement of hazard recognition skill. Saf. Sci. **93**, 1–8 (2017)
17. Denzin, N.K., Lincoln, Y.S.: Introduction: the discipline and practice of qualitative research. In: Denzin, N.K., Lincoln, Y.S. (eds.) The SAGE Handbook of Qualitative Research, pp. 1–19. SAGE Publications Inc., Thousand Oaks (2011)
18. Yin, R.K.: Case Study Research: Design and Methods. Sage Publications, Thousand Oaks (2003)
19. Ministry of Social Affairs and Health: Riskienarviointityöpaikalla – työkirja. Ministry of Social Affairs and Health and The Centre for Occupational Safety (2015)
20. Nenonen, N., Kivistö-Rahnasto, J., Anttila, S.: How workplaces actually carry out OSH-related risk assessment and management. In: Proceedings of the AHFE 2019 International Conference on the Applied Human Factors and Ergonomics, Washington D.C., USA, 24–28 July 2019
21. Conchie, S.M., Moon, S., Duncan, M.: Supervisors' engagement in safety leadership: factors that help and hinder. Saf. Sci. **51**, 109–117 (2013)
22. Hale, A.R., Guldenmund, R.F., van Loenhout, P.L.C.H., Oh, J.I.H.: Evaluating safety management and culture interventions to improve safety: effective intervention strategies. Saf. Sci. **48**(8), 1026–1035 (2010)
23. Tappura, S., Nenonen, N.: Categorization of effective safety leadership facets. In: Arezes, P., Carvalho, P. (eds.) Ergonomics and Human Factors in Safety Management, pp. 367–383. CRC Press, Boca Raton (2016)

A Review of the Risk Perception of Construction Workers in Construction Safety

Siu Shing Man(✉), Jacky Yu Ki Ng, and Alan Hoi Shou Chan

Department of Systems Engineering and Engineering Management, City University of Hong Kong, Kowloon Tong, Hong Kong
ssman6-c@my.cityu.edu.hk

Abstract. This research reviews the empirical studies on the risk perception of construction workers from January 2010 to April 2019. Results show that the risk perception of construction workers is an important construct in understanding their behavior at work. Previous studies have attempted to assess the cognitive risk perception of construction workers but have overlooked the affective risk perception of construction workers. These studies have also failed to report the content, convergent, discriminant, and criterion-related validity tests when they measured the construction workers' risk perception. These validity tests are vital for developing a valid scale because relevant literature lacks a psychometrically sound instrument, which measures the risk perception of construction workers. Further efforts are necessary to develop a reliable and valid scale for quantifying both the cognitive and affective risk perception of construction workers.

Keywords: Construction safety · Construction worker · Risk perception

1 Introduction

The construction industry has long been notorious for its high number of accidents and fatalities. Approximately 20% of the total (4,836) industrial fatal accidents were reported in the construction industry in 2015, showing an increase of 26% compared with that in 2011 in the U.S. [1]. Given the unsatisfactory performance of construction safety, previous studies have exerted substantial efforts to improve construction safety by proposing different methods, such as management-based, technology-based, and behavior-based approaches [2–4]. The behavior-based approach has obtained increasing attention from construction safety researchers because 80% of accidents result from the unsafe behavior of workers [5]. The risk perception of construction workers, which is the subjective judgment about risks, may play an important role in the behavior-based approach [4].

T. Ahram et al. (Eds.): IHSED 2019, AISC 1026, pp. 637–643, 2020.
https://doi.org/10.1007/978-3-030-27928-8_97

1.1 Risk

Risk is generally defined as involving uncertainty and a certain type of damage or loss possibly obtained in an event or situation [6] as well as a combination of uncertainties and events or consequences [7]. Risk is often denoted as the existence of potential circumstances, influencing the project's objectives, as an uncertain consequence. Risk is not constantly related to negative outcomes and involves opportunities as well. However, most risks often produce negative outcomes, leading people to focus exclusively on its negative side [8].

1.2 Risk Perception

Risk perception research originated from the studies on decision-making and judgment by Mosteller and Nogee [9] and Edwards [10]. Such study has continuous grounding in basic cognitive psychology [11]. In particular, risk perception is defined as the subjective judgment of people about particular risks [12].

The concept of risk perception has been applied to diverse research areas related to people's behavior. For example, Chen [13] studies the effects of perceived risk on the attitude and the intention of using mobile banking services and finds that perceived risk negatively affects such attitude and intention. Apart from consumer behavior, risk perception has been considered an important variable in driver-behavior research. Rhodes and Pivik [14] investigate the relationship between risk perception and risky driving behavior of teen and adult drivers in the U.S. and find that risky driving is negatively influenced by perceived risk. Risk perception has slowly gained considerable attention in construction safety by researchers. However, no review of this concept is available in construction safety, thus leaving a research gap. Therefore, the current study presents a review of the risk perception of construction workers in construction safety to provide a comprehensive and extensive knowledge about the risk perception of construction workers and to identify future research opportunities.

2 Method

The current research reviewed empirical studies about the risk perception of construction workers from January 2010 to April 2019. The selection criteria were set as follows: quantitative and qualitative studies were related to the risk perception of construction workers, the research subjects included construction workers, research had a clear description of research methodology and participant, and the research results were fully presented in English. Scopus, Social Science Citation Index, and Google Scholar were used as search tools. The abstracts or full texts of research papers were reviewed before their inclusion in the review based on selection criteria. A total of 19 studies were identified (Table 1).

Table 1. Empirical studies about the risk perception of construction workers.

Studies	Method	Subjects	Reference
Risk perception and risk-taking behavior of construction site dumper drivers	Quantitative questionnaire survey	40 construction site dumper drivers	[15]
Safety risk perception in construction companies in the Pacific Northwest of the USA	Quantitative questionnaire survey	51 construction workers	[16]
Perceptions of personal vulnerability to workplace hazards in the Australian construction industry	Quantitative questionnaire survey	176 construction workers	[17]
Alcohol consumption and risk perception in the Portuguese construction industry	Quantitative questionnaire survey	107 construction workers	[18]
A comparative study of the measurements of perceived risk among contractors in China	Quantitative questionnaire survey	76 construction project managers	[19]
Exploratory study to identify perceptions of safety and risk among residential Latino construction workers as distinct from commercial and heavy civil construction workers	Quantitative questionnaire survey	218 construction workers	[20]
Hazard recognition and risk perception in construction	Quantitative experiment and questionnaire survey	61 construction workers	[21]
Association between perceived risk and training in the construction industry	Quantitative questionnaire survey	177 construction workers	[22]
Risk perception of construction equipment operators on construction sites of Turkey	Quantitative questionnaire survey	198 heavy equipment operators	[23]
Quantitative analysis of the construction industry workers' perception of risk in municipalities surrounding Salvador	Quantitative questionnaire survey	160 construction workers	[24]
Work-health and safety-risk perceptions of construction-industry stakeholders using photograph-based Q methodology	Quantitative questionnaire survey	40 architects, engineers, construction managers, work health and safety professionals	[25]

(*continued*)

Table 1. (*continued*)

Studies	Method	Subjects	Reference
Stakeholder perceptions of risk in construction	Quantitative questionnaire survey	60 architects, engineers, construction contractors, and safety professionals	[26]
Influence of personality and risk propensity on risk perception of Chinese construction project managers	Quantitative questionnaire survey	152 project managers	[27]
Risk-taking behaviors of Hong Kong construction workers–a thematic study	Qualitative interview	40 construction workers	[4]
Do we see how they perceive risk? An integrated analysis of risk perception and its effect on workplace safety behavior	Quantitative questionnaire survey	120 construction workers	[28]
Effect of distraction on hazard recognition and safety risk perception	Quantitative experiment and questionnaire survey	70 construction workers	[29]
The risk-taking propensity of construction workers–an application of quasi-expert interview	Qualitative interview	16 safety professionals and six super-safe construction workers	[30]
Impact of safety climate on hazard recognition and safety risk perception	Quantitative questionnaire survey	287 construction workers	[31]
Construction worker risk-taking behavior model with individual and organizational factors	Quantitative questionnaire survey	188 construction workers	[32]

3 Results and Discussion

3.1 Influence of Risk Perception on the Behavior of Construction Workers

Risk perception has been identified as an important factor in the risk-taking behavior of construction workers in qualitative studies [4, 30]. Construction workers tend to take risks at work because they perceive the risk as low. When construction workers perceive that the risk is high, they tend not to take risks at work. This finding implies that the risk perception of construction workers might negatively influence their risk-taking behavior. Several quantitative studies, using a questionnaire survey, clearly proved that the risk perception of construction workers statistically and negatively influences their risk-taking behavior [15, 32]. Apart from the risk-taking behavior of construction workers, their risk perception has also been found to positively influence their safety behavior, such as safety compliance and safety participation [28].

Moreover, unsafe work habits, such as alcohol consumption, were found to be negatively related to the risk perception of construction workers. Specifically, construction workers with a lower level of risk perception tend to consume more alcohol [18]. Only the cognitive risk perception of construction workers, that is, "*probability*" and "*severity*" of experiencing an accident, has notably been considered in these quantitative studies. The affective risk perception of construction workers, that is, the feelings of "worry" and "unsafe" about the outcomes of risky scenarios, is also important in understanding their behavior [33]. Concerned safety researchers possibly overlook the importance of the affective risk perception in construction safety because they lack a comprehensive, reliable, and valid measurement to assess the affective risk perception of construction workers.

3.2 Quantification of Risk Perception of Construction Workers

The risk perception of construction workers (Table 1) and its measurement focused on the cognitive risk perception of construction workers without consideration of their affective risk perception in quantitative studies. Although the contents of the risk perception measurement in these studies were relevant to safety risks at work, such as being hit by falling materials, being electrocuted, and falling from a height, the development of such measurements failed to include validity assessment, which can be used to demonstrate the utility of developed measurements. Content, convergent, discriminant, and criterion-related validity of the measurement should be investigated in validity assessment. Previous studies on the risk perception of construction workers failed to conduct such validity assessment. For example, Xia et al. [28] use a newly developed measure to assess the cognitive risk perception of construction workers in investigating the relationship between risk perception and safety behavior of construction workers. Although Xia et al. [28] make significant contributions to the literature about the influence of the cognitive risk perception on the safety behavior of construction workers, these authors failed to consider the affective risk perception of construction workers in understanding their safety behavior and did not report the validity assessment for the risk perception measurement.

4 Conclusion

The role of risk perception of construction workers in understanding their behavior was deemed important. Construction workers with a high level of risk perception generally tend to work safely and engage in less risk-taking behavior. However, two limitations were identified concerning the empirical studies on the risk perception of construction workers in the last decade. First, the affective risk perception has not been considered in explaining construction workers' behavior. Second, construction safety researchers lack a comprehensive, reliable, and valid measurement to assess the cognitive and affective risk perception of construction workers. These limitations should be addressed in future studies to develop a reliable and valid scale for quantifying both the cognitive and affective risk perception of construction workers. Moreover, the extent of the cognitive and affective risk perception' influences on the construction workers' safety and risk-taking behaviors requires further investigation.

References

1. CPWR – The Center for Construction Research and Training. Construction Chart Book. https://www.cpwr.com/sites/default/files/publications/The_6th_Edition_Construction_eChart_Book.pdf
2. Guo, H., Yu, Y., Skitmore, M.: Visualization technology-based construction safety management: a review. Autom. Constr. **73**, 135–144 (2017)
3. Zhou, Z., Goh, Y.M., Li, Q.: Overview and analysis of safety management studies in the construction industry. Saf. Sci. **72**, 337–350 (2015)
4. Man, S.S., Chan, A.H.S., Wong, H.M.: Risk-taking behaviors of Hong Kong construction workers-a thematic study. Saf. Sci. **98**, 25–36 (2017)
5. Fleming, M., Lardner, R.: Strategies to Promote Safe Behaviour as Part of a Health and Safety Management System. HSE Books, Norwich (2002)
6. Kaplan, S., Garrick, B.J.: On the quantitative definition of risk. Risk Anal. **1**(1), 11–27 (1981)
7. Aven, T.: A unified framework for risk and vulnerability analysis covering both safety and security. Reliab. Eng. Syst. Saf. **92**(6), 745–754 (2007)
8. Baloi, D., Price, A.D.F.: Modelling global risk factors affecting construction cost performance. Int. J. Project Manage. **21**(4), 261–269 (2003)
9. Mosteller, F., Nogee, P.: An experimental measurement of utility. J. Polit. Econ. **59**(5), 371–404 (1951)
10. Edwards, W.: Probability-preferences in gambling. Am. J. Psychol. **66**(3), 349–364 (1953)
11. Slovic, P., Fischhoff, B., Lichtenstein, S.: Why study risk perception? Risk Anal. **2**(2), 83–93 (1982)
12. Slovic, P.: Perception of risk. Science **236**(4799), 280–285 (1987)
13. Chen, C.: Perceived risk, usage frequency of mobile banking services. Managing Serv. Qual. Int. J. **23**(5), 410–436 (2013)
14. Rhodes, N., Pivik, K.: Age and gender differences in risky driving: the roles of positive affect and risk perception. Accid. Anal. Prev. **43**(3), 923–931 (2011)
15. Bohm, J., Harris, D.: Risk perception and risk-taking behavior of construction site dumper drivers. Int. J. Occup. Saf. Ergon. **16**(1), 55–67 (2010)
16. Hallowell, M.: Safety risk perception in construction companies in the Pacific Northwest of the USA. Constr. Manag. Econ. **28**(4), 403–413 (2010)
17. Caponecchia, C., Sheils, I.: Perceptions of personal vulnerability to workplace hazards in the Australian construction industry. J. Saf. Res. **42**(4), 253–258 (2011)
18. Arezes, P.M., Bizarro, M.: Alcohol consumption and risk perception in the Portuguese construction industry. Open Occup. Health Saf. J. **3**, 10–17 (2011)
19. Lu, S., Yan, H.: A comparative study of the measurements of perceived risk among contractors in China. Int. J. Project Manage. **31**(2), 307–312 (2013)
20. Lopez del Puerto, C., Clevenger, C.M., Boremann, K., Gilkey, D.P.: Exploratory study to identify perceptions of safety and risk among residential Latino construction workers as distinct from commercial and heavy civil construction workers. J. Constr. Eng. Manag. **140**(2), 04013048 (2013)

21. Perlman, A., Sacks, R., Barak, R.: Hazard recognition and risk perception in construction. Saf. Sci. **64**, 22–31 (2014)
22. Rodríguez-Garzón, I., Lucas-Ruiz, V., Martínez-Fiestas, M., Delgado-Padial, A.: Association between perceived risk and training in the construction industry. J. Constr. Eng. Manag. **141**(5), 04014095 (2014)
23. Gürcanlı, G.E., Baradan, S., Uzun, M.: Risk perception of construction equipment operators on construction sites of Turkey. Int. J. Ind. Ergon. **46**, 59–68 (2015)
24. Carriço, A., Gomes, A.R.C., Gonçalves, A.P.: Quantitative analysis of the construction industry workers' perception of risk in municipalities surrounding Salvador. Procedia Manuf. **3**, 1846–1853 (2015)
25. Zhang, P., Lingard, H., Blismas, N., Wakefield, R., Kleiner, B.: Work-health and safety-risk perceptions of construction-industry stakeholders using photograph-based Q methodology. J. Constr. Eng. Manag. **141**(5), 04014093 (2014)
26. Zhao, D., McCoy, A.P., Kleiner, B.M., Mills, T.H., Lingard, H.: Stakeholder perceptions of risk in construction. Saf. Sci. **82**, 111–119 (2016)
27. Wang, C.M., Xu, B.B., Zhang, S.J., Chen, Y.Q.: Influence of personality and risk propensity on risk perception of Chinese construction project managers. Int. J. Project Manage. **34**(7), 1294–1304 (2016)
28. Xia, N., Wang, X., Griffin, M.A., Wu, C., Liu, B.: Do we see how they perceive risk? An integrated analysis of risk perception and its effect on workplace safety behavior. Accid. Anal. Prev. **106**, 234–242 (2017)
29. Namian, M., Albert, A., Feng, J.: Effect of distraction on hazard recognition and safety risk perception. J. Constr. Eng. Manag. **144**(4), 04018008 (2018)
30. Low, B.K.L., Man, S.S., Chan, A.H.S.: The risk-taking propensity of construction workers—an application of quasi-expert interview. Int. J. Environ. Res. Public Health **15**(10), 2250–2260 (2018)
31. Pandit, B., Albert, A., Patil, Y., Al-Bayati, A.J.: Impact of safety climate on hazard recognition and safety risk perception. Saf. Sci. **113**, 44–53 (2019)
32. Low, B.K.L., Man, S.S., Chan, A.H.S., Alabdulkarim, S.: Construction worker risk-taking behavior model with individual and organizational factors. Int. J. Environ. Res. Public Health **16**(8), 1335–1347 (2019)
33. Rundmo, T.: Safety climate, attitudes and risk perception in Norsk Hydro. Saf. Sci. **34**(1–3), 47–59 (2000)

Diffusing the Myth Around Environmental Sustainable Development Delivery in South African Construction Industry

Idebi Olawale Babatunde[1(✉)], Timothy Laseinde[2], and Ifetayo Oluwafemi[3]

[1] Department of Operations Management, University of Johannesburg, Johannesburg, South Africa
waleidebi@yahoo.com

[2] Department of Mechanical and Industrial Engineering Technology, University of Johannesburg, Johannesburg, South Africa
otlaseinde@uj.ac.za

[3] Postgraduate School of Engineering Management, University of Johannesburg, Johannesburg, South Africa
ijoluwafemi@uj.ac.za

Abstract. Environmental Sustainable Development (ESD) is the future of construction, following the necessity to preserve the environment in order to serve present needs and the future generation as well. This research investigates the reluctance of South African Construction Industry (SACI) designers in delivering ESD solution. The study will impact SACI to accelerate incorporation and compliance of ESD solution into its construction development delivery. A desktop research approach was employed by reviewing existing literatures to identify the wrong perceptions of SACI designers regarding ESD delivery. The major myth that surrounds reluctance to ESD as inferred is increase in building costs. This myth is due to misunderstanding of the ESD concept, lack of experience in ESD delivery, reluctance to construction technological change of ESD, lack of awareness of the benefits of ESD, and client's reluctance to engage in ESD. Diffusing the myth and it underlying factors will involve enacting ESD supporting policies, intensified by awareness of its benefits and importance.

Keywords: Environmental Sustainable Development · Construction industry · Green environment · ESD · Development myth

1 Introduction

The incessant human socioeconomics demands rely hugely on the continuous existence of the natural environment resources [1]. It is therefore imperative that the capability of the natural environment resources should be maintained.

The concept of environmental sustainability generally connotes the processes involved in ensuring that the natural environment maintains its ability to sustain human activities forever.

T. Ahram et al. (Eds.): IHSED 2019, AISC 1026, pp. 644–647, 2020.
https://doi.org/10.1007/978-3-030-27928-8_98

The singular effect of construction industry activities which include heavy consumption of; natural non-renewable resources such as coal mining for power generation, renewable resources such as timber for housing construction and sizable waste generation such as emission of CO2, among others, contribute largely to the causes of environmental degradation [2–4].

The concept of Environmental Sustainable Development (ESD), is, therefore, a response to address the dangers faced and to maintain the natural environment, to sustain human socioeconomic demands as a going concern.

ESD delivery focuses on three main action which according to Goodland are; "harvesting renewable natural resources at a rate which is maintainable or within regeneration, generate wastes and pollution within the assimilative capacity of the environment without impairing it and reduce the depletion rate of non-renewable natural resources equal to the rate at which renewable substitutes can be created.

From the above foregoing, the focus to employ ESD by the construction industry dominates the feasibility of accomplishing sustainable development.

2 Need for Environmental Sustainable Development

South Africa is a developing nation with a growing population and challenges like; unemployment sitting at over 25%, and about 40% living in informal or substandard housing according to UN standard and in dire need for stable electricity, clean water, among other issues, require infrastructure and economic development which depends hugely on natural environmental resources to meet these needs [1, 4].

According to Gibberd [3], South Africa is outside the environmental sustainability maximum ecological footprint of 1.8gha by 0.3gha and below the minimum human development index of 0.8% by 0.14%.

To bring South Africa into the allowable belt of the environmental indices, the South African construction industry has imbibed and incorporated the terms sustainable construction and/or Green Building as a means of ESD delivery. Having both terminologies' intended meaning and definition to connote; providing ESD focused construction infrastructure execution methodologies.

Despite the introduction of sustainable construction and/or Green Building, ESD delivery has not had much impact as it would have been expected in the South African construction industry as suggested by researchers [2].

3 Environmental Sustainable Development Implementation Challenges

The massive pressure on the South African government and developers to provide social infrastructure and amenities at all cost, on one hand, and the reluctance of construction professionals to proffers ESD solution, has and is still having its negative implication on the environment. This paper focusses on the reluctance of the construction professional in providing ESD solution.

The competencies of construction professionals in South Africa are not in doubt based on the institutionalized system of education and skill acquisition of the industry, but there are two major misconception in the industry that affect ESD delivery.

The first misconception stems from misinterpreting the true meaning of ESD to mean sustainable construction methods rather than environmentally focused construction solution. The second misconception by the industry practitioners is the general perception that ESD solutions are expensive. Other perceived reasons why construction professionals shy away from ESD delivery are lack of adequate knowledge of the benefits of ESD delivery, Lack of adequate skill to executed ESD solutions, the flow along with developers to stick with the traditional methods of construction.

4 Overcoming the Challenges

The construction professionals in South Africa, who have been legally entrusted with providing construction solutions, are well positioned to take advantage of the high socioeconomic demands for infrastructure development and steer the ship of the industry towards ESD solutions. The goodwill expected by politicians and the profit chased by developers, are there driving force in infrastructure development and hence are not challenged to explore other options than the traditional methods of construction which is heavily dependent on the depleting the natural environmental resources.

The knowledge base of construction professionals in South Africa needs to be broadened to include:

- A thorough knowledge of the meaning, importance, benefits, and need for environmental sustainability focus construction solutions.
- Having an overview of construction projects cost from life cycle analysis and not just the initial construction cost.
- The initial increased cost of the introduction of new methodologies will disappear when it becomes a norm.
- An understanding of the global and future value of environmental sustainability construction solution beyond cost and immediate gratification.

5 Conclusion

The traditional construction designs methods of infrastructure developments in South Africa is still high and it is negatively impacting on the environment. Without doubt, ESD is the future of construction industry in South Africa and will soon catch up, as it is currently trending in the developed world. The necessity to consider the comfort of the future generation is the singular driving force to employ ESD. The Government decision makers and developers might not be willing to consider a change to ESD due to the urgent pressure to meet socioeconomic demands and economic profits. Hence, the construction professionals in South Africa has an ethical responsibility to champion the drive for construction solutions to ESD focussed.

6 Recommendations

The South African construction industry professionals need a drastic change of orientation and focus, to ESD solution principles. This can be achieved by re-educating the current practicing professionals and incorporating into the curriculum of study of tertiary institutions that offer diplomas and degrees in construction courses in South Africa; the knowledge base of the concept of ESD solution principles. It is also recommended that there should be a legislative backing for construction professional to enforce ESD solution.

Acknowledgments. The authors wish to acknowledge University of Johannesburg as well as the National Research Foundation of South Africa (NRF) for funding the publication process through the Thuthuka funding mechanism award number TTK180805351820.

References

1. Nahman, A., Wise, R., Lange, W.D.: Environmental and resource economics in South Africa: status quo and lessons for developing countries. S. Afr. J. Sci. **105**(9–10), 350–355 (2009)
2. Aigbavboa, C., Ohiomah, I., Zwane, T.: Sustainable construction practices: "a lazy view" of construction professionals in the South Africa construction industry. Energy Procedia **105**, 3003–3010 (2017)
3. Gibberd, J.: Sustainable development criteria for built environment projects in South Africa (CSIR). Hum. Settl. Rev. **191**, 34 (2010)
4. du Plessis, C.: Action for sustainability: preparing an African plan for sustainable building and construction. Build. Res. Inf. **33**(5), 405–415 (2005)

RecogApp - Web and Mobile Application to Recognition Support

André Esteves[1(✉)], João Jesus[1(✉)], Ângela Oliveira[2(✉)], and Filipe Fidalgo[2(✉)]

[1] Polytechnic Institute of Castelo Branco, Castelo Branco, Portugal
{andre.esteves,pedro.jesus}@ipcbcampus.pt
[2] DiSAC – R&D Unit in Digital Services, Applications and Content, Castelo Branco, Portugal
{angelaoliveira,ffidalgo}@ipcb.pt

Abstract. Over time, with the advancement of technology and the number of devices increasing to be launched in the market, namely smartphones, tablets, hybrids and laptops, it is essential to have advances in digital security and in forms of authentication. These forms of authentication have become increasingly. This article presents a system consisting of a web platform and a mobile application for android, with the objective of supporting researchers in the development, improvement and testing of facial recognition and iris recognition algorithms. This system will store large quantities of images sent voluntarily, obeying the particularities of studies defined by researchers to which users adhere, these images are taken in environments with different characteristics, considering several accessories and in several places, in order to increase the number of cases of possible analyzes for development and/or use of recognition algorithms.

Keywords: People recognition · Image database · Images collection

1 Introduction

People recognition is a constant need these days [1–11], the same is done in several ways, being more traditional, such as ID card, passport, driving license, etc., and more recent ways such as digital pressure, face or iris recognition and even DNA.

Univocal identification becomes more and more important, hence failures tolerance becoming less and less acceptable, considering the implications underlying false identification.

Human characteristics can be used as identification form and/or access control [12]. Identifiers can be classified either as physiological characteristics, related to body shape - fingerprint, DNA, iris recognition, facial recognition, etc. or as behavioral characteristics - voice, marching, etc. [12].

Iris recognition begins with the discovery of the iris in an image, demarcating its internal and external borders on the pupil and sclera as well as detecting all borders and excluding eyelashes superimposed on the image or reflections of the cornea or glasses, for example [13]. It is currently considered one of the most accurate and potent non-invasive biometric recognition techniques since the amount of information that can

T. Ahram et al. (Eds.): IHSED 2019, AISC 1026, pp. 648–653, 2020.
https://doi.org/10.1007/978-3-030-27928-8_99

be measured through the iris is much larger than fingerprints, accuracy is greater than DNA and can be captured at a certain distance [12, 14, 15]. However, this process has to deal with the fact that the iris region of the eye is a relatively small, moist and constantly moving area due to involuntary eye movements and, in addition, the eyelids, eyelashes and reflexes are occlusions of the iris pattern that can cause errors in the segmentation process. These factors can therefore lead to errors in biometric recognitions and seriously reduce the final result [12].

Thus, automatic detection and recognition of the face and eyes as well as the monitoring of eye movement are currently important topics in the field of scientific research. This importance and relevance, which face and eye detection applications have today, and may continue to have in the near future, is highlighted by their ability to facilitate interaction with man-machine interfaces, improve the capabilities of surveillance and security systems or even helping people with motor difficulties.

However, the detection and monitoring of ocular movement are difficult to perform because of a variety of factors, such as face position or facial expression, which limit the efficiency of most sensing mechanisms.

Although efforts have been made to create efficient sensing methods, an increase in the need for accurate eye location makes this work relevant because it seeks to better understand the way of functioning and the main problems of the eye detection algorithms.

The various existing detection methods have advantages and disadvantages in relation to each other, but present similar problems. Thus, despite advances in detection methods, these still require further improvements. Thus, this work intends to contribute to the investigation in the area of the ocular detection algorithms, trying to fill one of the main limitations presented by these, the number of images available for tests, and, in order to have these images, there must be a system that allows the collection and that they can be used. The contribution of this work is intended to create studies and participation of users through the sending of photographs.

2 Web and Mobile Application

In order to improve recognition algorithms, it is fundamental to have a large number of photographs for facial recognition tests, so the project consists of two applications development: a web to manage and store the different types of studies and a mobile one to users can participate in the studies by sending photographs, the final objective being an database construction to support the recognition algorithm test. At the moment, both applications are in Portuguese language, but being prepared to move to the English language.

The web application was developed with the administrator's features: after login, it is possible to create several studies, studies that are identified by an name, and which may contain criteria/rules and filters; change studies, which have not yet been opened; open studies means the definition of an closure date, an opening date, and an date to extend the study if it does not reach the goals inserted when it was created; although it has the possibility to close studies since this is an automatic process but only in special cases will be closed manually, for example if the goal of the study is reached earlier than expected.

Studies include criteria to be met and filters, the first are the accessories, locations and facial expressions that users should have when they take photographs, the latter serve to select the people to whom the studies are directed, such as a certain color eyes, hair, type of eyes, skin tone and sex, as shown in Fig. 1.

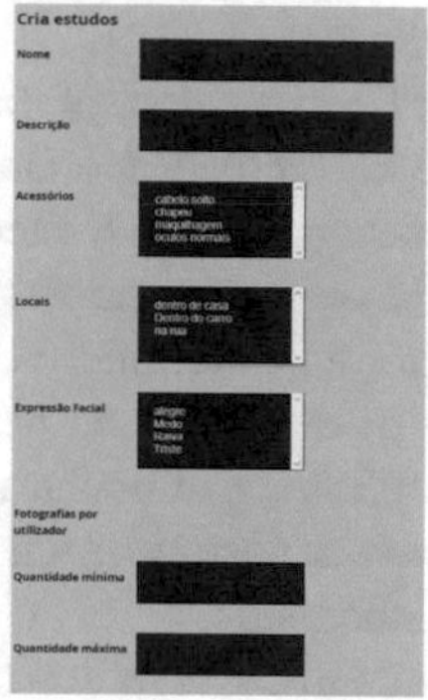

Fig. 1. RecogAPP - create studies

Another administrator feature was the management of these features, adding more or editing existing ones. And can also manage users, such as registering researchers, mobile users and administrators, also edit researchers, mobile users and administrator's information and enable or disable them. It is possible to obtain the images of the studies; the images are presented by study selecting the same, as shown in Fig. 2. The researcher has all the permissions of the administrator, with exception of the functions related to the management of users and only can access their studies, that is, while an administrator can access all the images of all the studies the researcher only has access to the images of the studies it promoted.

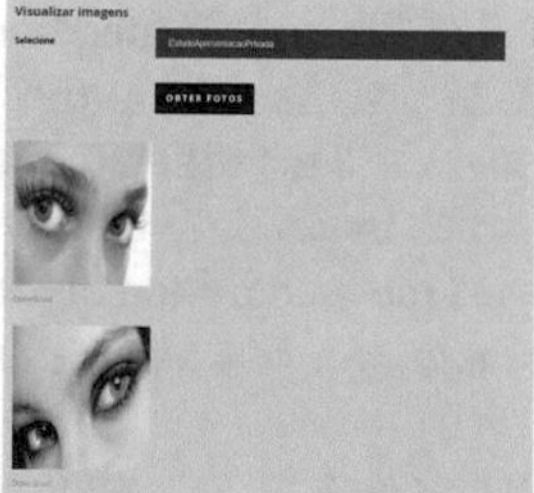

Fig. 2. RecogAPP – view images

In the case of unauthorized users, they will only have access to an initial page, where they can download the applications from the mobile application and access the shortcut to the Facebook group for the community that participates and is interested in knowing the concept, at the top of the page, as shown in Fig. 3, followed by an application explanation, how it works, and what gains with collaborating in this project. At the end of the page, in the part of contacts, the request for access to a researcher's account is allowed.

Fig. 3. RecogAPP – view images

The mobile application, shown in Fig. 4 was developed for android, is intended for users who wish to participate in available or suggested studies. In the beginning of the application, users find the possibility to read more about it or to carry out login. For new users, it is possible to register, in which the user must enter their usual data, such as their username, password and email, and specific data that will characterize the user to be invited to certain studies, such as age, gender, hair color, eye color, type of eyes and skin tone. All users need administrator validation. It also has a personal area with quick access to studies, profile and profile editing. In the subscribe studies and studies option, a list of the studies that have not been adhered to, is shown, as well those to users was already joined. By clicking on a study, they obtain information such as name, description, participation rules, closing date, notes regarding study particularities and the number of photos to send. If user have not yet joined to a study, an adhesion button will be displayed. If he has already joined, sending photos is allowed and displays the number of photos that the user has already sent. There is a ranking to highlight the most participative users, presenting the three that sent more photos, thus creating an incentive to participate in studies. Finally, there is the help center where it is how to use

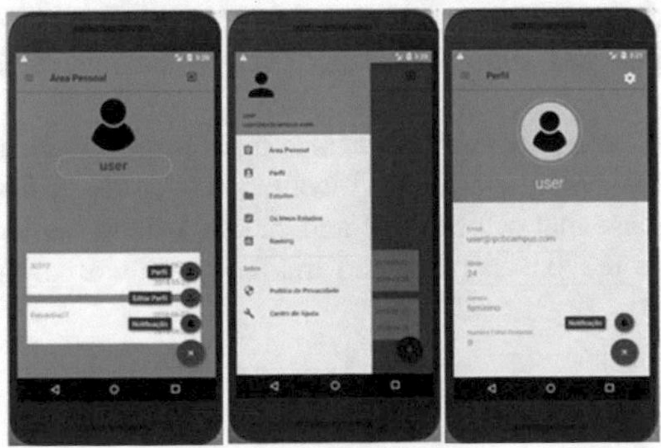

Fig. 4. Personal area, navigation menu and profile of mobile application

the application as well as the Policies and Privacy with the regulation. The application requires internet for its operation, however, to take the photos users use smartphone camera and later connect with the application and submit the photos.

3 Discussion

To determine the effectiveness of the web and mobile applications users' tests were performed, using two inquiries, one for each application, in which, those responses were obtained from users who did not know and never saw the applications. For web application, the inquire was made with possible investigators in the area and 12 responses were obtained so far. The assessment scale varies from one to five, considering the level of ease use, speed and intuition in performing each task and as well as an open question for suggestions.

The results of these answers already suggest some improvements to be made with appearance and the language selection, but denoting already, very positive aspects, considering the majority of users, That the accomplishment of tasks is quite easy and intuitive and the structure of the studies and the characteristics organization of information its pertinent and relevant.

For mobile application, inquires was conducted for the general public, to which 43 people responded and once again with issues ranging on a one to five scale, assessing the ease use, speed and intuition of each task. In general, for the tasks evaluation and as well the application, the results were quite satisfactory and there is some improvements in the presentation of the information, and how images were administered after submission, since users, in this version, would not have more access to them. They consider relevant to be able to exclude a photo after submission, also suggest the possibility of awards or points for exchange by products and not only ranking.

4 Conclusion

This article presents a solution whose goal is to help researchers in the area of facial recognition algorithms development, so that they can have a considerable set of images for development more efficient algorithms, using an open source platform.

Will be continuing to distribute the inquiries by the users and the opinions they provide, properly analyzed and included in the applications, if they are considered relevant.

The development of an additional platform that allows the inclusion of algorithms and respective tests is being analyzed. However, now, was in first place relevant to build an image database that can be used and shared around the world, by researchers and users to improve the existing face and iris recognition algorithms.

References

1. Abdel-Kader, R.F., Atta, R., El-Shakhabe, S.: An efficient eye detection and tracking system based on particle swarm optimization and adaptive block-matching search algorithm. Eng. Appl. Artif. Intell. **31**, 90–100 (2014)
2. Adwan, S., Arof, H.: Modified integral projection method for eye detection using dynamic time warping. Int. J. Innov. Comput. Inf. Control **8**(1A), 187–200 (2012)
3. Bhoi, N., Mohanty, M.N.: Template matching based eye detection in facial image. Int. J. Comput. Appl. **12**(5), 15–18 (2010)
4. Hassaballah, M., Murakami, K.: An automatic eye detection method for gray intensity facial images. Int. J. Comput. Sci. Issues (IJCSI) **8**(2), 272–282 (2011)
5. Jung, C., Sun, T., Jiao, L.: Eye detection under varying illumination using the retinex theory. Neurocomputing **113**, 130–137 (2013)
6. Liu, H., Liu, Q.: Robust real-time eye detection and tracking for rotated facial images under complex conditions. In: Proceedings of the 2010 6th International Conference on Natural Computation, ICNC 2010, vol. 4, no. ICNC, pp. 2028–2034 (2010)
7. Nanni, L., Lumini, A.: Combining face and eye detectors in a high-performance face-detection system. IEEE Multimedia **19**(4), 20–27 (2012)
8. Orman, Z., Battal, A., Kemer, E.: A study on face, eye detection and gaze estimation. Int. J. Comput. Sci. Eng. Surv. **2**(3), 29–46 (2011)
9. Park, C.W., Park, K.T., Moon, Y.S.: Eye detection using eye filter and minimisation of NMF-based reconstruction error in facial image. Electron. Lett. **46**(2), 130 (2010)
10. Soetedjo, A.: Eye detection based-on color and shape features. Int. J. Adv. Comput. Sci. Appl. **3**(5), 17–22 (2013)
11. Wu, Y., Ji, Q.: Learning the deep features for eye detection in uncontrolled conditions. In: Proceedings of the International Conference on Pattern Recognition, pp. 455–459 (2014)
12. Thomas, T., George, A., Devi, K.P.I.: Effective iris recognition system. Procedia Technol. **25**, 464–472 (2016)
13. Daugman, J.: New methods in IRIS recognition. IEEE Trans. Syst. Man Cybern. Part B (Cybern.) **37**(5), 1167–1175 (2007)
14. Wildes, R.P.: Iris recognition: an emerging biometric technology. Proc. IEEE **85**(9), 1348–1363 (1997)
15. Verma, S.J., Saxena, D., Gautam, R., Kaushal, L.: IRIS recognition system. Int. J. Eng. Adv. Technol. **2**(6), 239–244 (2017)

Lean and Ergonomics Competencies: Knowledge and Applications

Beata Mrugalska(✉)

Faculty of Engineering Management, Poznan University of Technology, ul. Strzelecka 11, 60-965 Poznan, Poland
beata.mrugalska@put.poznan.pl

Abstract. The paper refers to two components of competencies such as skills and knowledge. Its investigation is directed towards showing lean value and its relation to human. Particularly, it explores lean and ergonomics in reference to the required knowledge and practice which enhance these initiatives as they are highly inter-related. For this aim, an integrating model of lean and ergonomics competencies is shown. It presents these initiatives in different stages of their acquisition.

Keywords: Competences · Competencies · Ergonomics · Human factors · Lean

1 Introduction

The manufacturing industry has undergone extreme changes starting the era of global growth and innovation. Lean manufacturing has become a leading management philosophy. In spite of the fact that it originates from the Toyota Production System, and is utilized by the successful automotive manufacturer Toyota, many others have followed this philosophy. Moreover, it should be emphasized that lean management plays acrucial role in its operation and success. The willingness to adopt a lean leadership approach allows companies to fully benefit from the its implementation [1, 2]. Therefore, it is required that the lean managers are coaches, leaders and mentors rather than a 'boss'. They should be able to:

- give vision, focus & direct the organization,
- be active coach & problem solver,
- build good relationship,
- strive for development & improvement,
- be customer focus & value systems.

In order to achieve these requirements a focus should be paid on competences [3]. Competences create human potential, however, they require development and on the other hand, they are required to realize company's strategy and achieve established goals. It can be even stated that they include everything what is required for performance of tasks: knowledge, skills and attitude [4]. The knowledge can be obtained formally or informally, due to know-what, know-why, know-how and know-who.

T. Ahram et al. (Eds.): IHSED 2019, AISC 1026, pp. 654–660, 2020.
https://doi.org/10.1007/978-3-030-27928-8_100

The skills are determined by practical activity, including experience whereas attitude shows willingness and readiness to use the knowledge and skills in actions [5].

In the paper the author will refer to the first two components of competencies to show lean value and its relation to human. To gain a better understanding, this paper explores lean and ergonomics, in details, it presents the required knowledge and practice to enhance these initiatives as both approaches are highly inter-related. For this aim, an integrating model of competencies will be shown where these initiatives will be shown and discussed.

2 Competencies

The concept of competence and competency dominated the management strategy theory in the 1990s. The first notion generally refers to organization functional areas whereas the second one describes behavioural areas. However, the use of both notions is often inconsistent [6].

In reference to company level, competence is a key resource that couldbe used to gain competitive advantage [7, 8]. Itallows to identify the complex interaction of people, skills and technologies which "drive firm performance and address the importance of learning and path dependency in its evolution" [9]. Therefore, it includes the transference, synergy, coordination of technology but also human knowledge and skills to create and provide the value [10].

On the other hand, competencies encompass factors which are necessary for achieving important results in a specific job or work role in a particular organization. They can be considered from the following points of view, categories such as [11]:

- technical - job-related knowledge and skills,
- methodological - skills and abilities for general problem solving and decision making,
- social - skills and abilities of cooperating and communicating with others,
- personal - individual's social values, motivations, and attitudes.

They are defined as "instrumental in the delivery of desired results or outcomes" [12] dependable on knowledge, skills, abilities and personal characteristics referring to specific behaviors, which are shown performing these tasks in specific roles or positions [13]. Thus, it is possible to refer to five major components of competencies [13, 14]:

- knowledge- information and learning dependable on a person,
- skill-ability to perform a certain task,
- self-concepts and values- attitudes, values and self-image,
- traits - physical characteristics and responses to situations or information,
- motives - emotions, desires, physiological needs or similar impulses that prompt action.

Each competency consists of a number of levels within it. In a competency document for a particular role only the required levels necessary to demonstrate are described. However, it is possible to see the other levels and the competencies required for other roles in the company. Such information allows to build a career path assessing the areas a person has already gained and which have to be developed in the near future. Generally, competencies contain two levels: main and additional focus. The main level refers to the requirements needed to perform to work most of the time, whereas an additional level states a higher level of activity which is an important part of the role but required less often. It may also happen that more than one level of competencies is required what indicates that all levels are equally important to the particular role. As the competencies develop over time, it is acceptable that the same level of competency does not have to be required to someone new to a role might as someone who is more experienced [15].

3 Integration of Lean and Ergonomics Competencies

Many companies have implemented lean-related systems to control their productivity and quality, however, it has become clearly visible that a clear understanding of the potential of ergonomics is needed to contribute to the objectives of the lean initiative. It results from the fact that the core principles of lean focus on optimizing all aspects of the business, including working relationships between people, understanding and communicating the application of modern ergonomics methods in lean-related wastes [16, 17]. Therefore, the model representing integration of competencies in both approaches is shown in Fig. 1 whereas its deep analysis in reference to particular stages is presented in Table 1. It refers to lean competency system [18] and core competencies in ergonomics according to International Ergonomics Association [19].

Fig. 1. Model integrating lean and ergonomic competencies.

Table 1. Lean and ergonomics competencies in knowledge and practice areas: examples (Adapted from [18, 19]).

AWARENESS	
LEAN KNOWLEDGE	ERGONOMICS KNOWLEDGE
Origins and evolution of lean Related theoretical concepts and approaches Lean principles and frameworks Core elements of lean Awareness of human and strategic dimensions of lean thinking	Origins and evolution of ergonomics Related theoretical concepts and approaches Ergonomics principles and frameworks Core elements of ergonomics Awareness of human performance and quality of life

(*continued*)

Table 1. (*continued*)

AWARENESS	
LEAN PRACTICE	ERGONOMICS PRACTICE
Not applicable	Not applicable
ANALYSIS AND DIAGNOSIS	
LEAN KNOWLEDGE	ERGONOMICS KNOWLEDGE
Purpose, customer/stakeholder value identification and understanding Diagnostic and problem solving methods, techniques and approaches Basic data gathering/statistical techniques Planning and communication techniques Capacity and demand analysis and assessment	Purpose, demands for ergonomics design identification to ensure appropriate interaction between work, product and environment, and human capabilities and limitations Problem solving methods and techniques Basic data gathering, experimental design and statistical techniques Planning and communication techniques Validation techniques
LEAN PRACTICE	ERGONOMICS PRACTICE
Define problems and issues Undertake basic data gathering and statistical techniques to help diagnose and solve problems Analyze a process/value stream through the application of diagnostic methods, techniques and approaches Understand customer value and variation Understand how to improve processes and quality and plan them Develop 'current state' and 'future state' maps	Define problems and issues Evaluate products or work situations in relation to expectations for error-free performance Analyze current guidelines, standards and legislation, regarding the variables influencing the activity Document ergonomic findings Determine the compatibility of human capacity and planned or existing demands Develop a plan for ergonomic design or intervention
IMPROVEMENT AND IMPLEMENTATION	
LEAN KNOWLEDGE	ERGONOMICS KNOWLEDGE
Workplace organization and optimization techniques Standard operations Visual management and performance measures Management, scheduling and planning Enabling flow People, teams and sustainability	Nature of design improvement Workplace organization and optimization techniques Organizational management, scheduling and planning Needs of special groups
LEAN PRACTICE	ERGONOMICS PRACTICE
Apply workplace organization techniques Implement standard work principles and tools Use visual management and performance measures for effective communication and control	Develop simulations to optimize and validate recommendations Refer to user's body size, skills, cognitive abilities, age, sensory capacity, general health and experience

(*continued*)

Table 1. (*continued*)

IMPROVEMENT AND IMPLEMENTATION	
Implement capacity planning techniques and scheduling approaches Apply tools to enable flow Apply quality tools Apply techniques to understand the nature of demand and manage it effectively Use policy deployment techniques to plan, measure and communicate Use key people or 'soft' skills to effect change, lead, coach, participate and communicate	Apply design principles to design of products, job aids, controls, displays, instrumentation and other aspects of the workplace, work and activities Provide design specifications and guidelines for technological, organizational and ergonomic design or redesign of the work process, the activity and the environment Make decisions regarding criteria influencing a new designor a solution to a specified problem
LEADERSHIP	
LEAN KNOWLEDGE	ERGONOMICS KNOWLEDGE
Strategy formation and policy deployment techniques Design and deployment of effective and relevant performance measures Leadership skills for effective lean team management Supply chain management Advanced lean thinking knowledge and techniques, complementary approaches Sustainable change and continuous improvement Project management, implementation and control	Intervention strategy formation and legislation policy Design and deployment of effective and relevant performance measures Leadership skills for effective ergonomic team management Advanced ergonomic knowledge and techniques, complementary approaches Sustainable change and continuous improvement Project management, implementation and control
LEAN PRACTICE	ERGONOMICS PRACTICE
Play leadership role in workplace implementation Facilitate workplace change, guide and monitor improvement Apply approach required to meet the organizational improvement need or objective Plan and control effectively Use strategically aligned lean programs Deploy programs and communicating effectively Engage with people at all levels	Recommend staff choice and their training as part of a balanced sustainable solution to the defined problem Ethical practice and standards of performance, professional behavior Evaluate outcome of ergonomic recommendations Discuss with the client, users and management the design or intervention strategies available, their rationale, realistic expectations of outcome, limitations to achieving outcome, and the costs of the proposed ergonomics plan Establish effective relationships and collaborates effectively with professional colleagues in the development of ergonomic design solutions

4 Conclusions

In a knowledge-based economy, human resource determines mostly the success of organizations and itis needed to rely on competent employees. In order to achieved it, it is required to refer to the workforce competency, but also their evaluation and development on an ongoing basis to meet the global competition. Therefore, it is needed to define the knowledge and practice areas to be able to achieved it. In the paper it was done on the example on lean and ergonomics competences as they are highly inter-related. For this aim, an integrating model of lean and ergonomics competences is presented and these initiatives are shown in different stages of their acquisition.

References

1. Alefari, M., Salonitis, K., Xu, Y.: The role of leadership in implementing lean manufacturing. Procedia CIRP **63**, 756–761 (2017)
2. Wyrwicka, M.K., Mrugalska, B.: Mirages of lean manufacturing in practice. Procedia Eng. **182**, 780–785 (2017)
3. Elias, S.: The Five Essential Qualities of a Lean Manager. Lean Competency System (2016). https://www.leancompetency.org/lcs-articles/five-essential-qualities-lean-management. Accessed 5 Jun 2019 (2016)
4. Grzelczak, A., Kosacka, M., Werner-Lewandowska, K.: Employees competences for industry 4.0 in Poland - preliminary research results. In: DEStech Transactions on Engineering and Technology Research, pp. 139–144 (2017)
5. Garud, R.: On the distinction between know-how, know-what and know-why. In: Huff, A., Walsh, J. (eds.) Advances in Strategic Management, pp. 81–101. JAI Press, Greenwich (1997)
6. Le Deist, F.D., Winterton, J.: What is competence? Hum. Res. Dev. Int. **8**(1), 27–46 (2005)
7. Nadler, D.A., Tushman, M.: The Organisation of the future: strategic imperatives and core competencies for the 21st century. Organisational Dyn. **27**(1), 45–58 (1999)
8. Snyder, A., Ebeling, H.W.: Targeting a company's real core competencies. J. Bus. Strategy **13**(6), 26–32 (1992)
9. Scarborough, H.: Path(ological) dependency? core competencies from an organisational perspective. Br. J. Manage. **9**, 219–232 (1998)
10. Belasen, A.T.: Leading the Learning Organization: Communication and Competencies for Managing Change. State University of New York Press, New York (2000)
11. Hecklau, F., Galeitzke, M., Flachs, S., Kohl, H.: Holistic approach for human resource management in industry 4.0. Procedia CIRP **54**, 1–6 (2016)
12. Bartram, D., Robertson, I.T., Callinan, M.: Introduction: a framework for examining organizational effectiveness. In: Robertson, I.T., Callinan, M., Bartram, D. (eds.) Organizational Effectiveness: The Role of Psychology, pp. 1–10. Wiley, Chichester (2002)
13. Chouhan, V.S., SandeepSrivastava, S.: Understanding competencies and competency modeling- a literature survey. J. Bus. Manage. **16**(1), 14–22 (2014)
14. Tucker, S., Cofsky, K.: Competency-based pay on a banding platform. ACA J. **3**(1), 30–45 (1994)
15. Manchester Metropolitan University: Employer Liaison Officer: Competencies. https://www2.mmu.ac.uk/media/mmuacuk/content/documents/human-resources/a-z/competencies/employer_liaison_officer.pdf. Accessed 5 Jun 2019

16. Gibson, M., Mrugalska, B.: Lean thinking practices in ergonomics in industrial sector. In: Occupational Safety and Hygiene VI, pp. 529–534. CRC Press (2018)
17. Gibson, M., Mrugalska, B. Lean production and its impact on worker health: force and fatigue-based evaluation approaches. In: Realyvásquez, A., Maldonado-Macías, A.A., Karina C. (eds.) Advanced Macroergonomics and Sociotechnical Approaches for Optimal Organizational Performance, pp. 118–127. IGI Global (2019)
18. Cardiff University: Lean Competency System. https://www.lcbgroup.nl/wp-content/uploads/2019/02/LCS-level-descriptors-1609.1.pdf. Accessed 12 May 2019
19. IEA: Core Competencies in Ergonomics: Units, elements, and performance criteria. https://www.iea.cc/project/2_Full%20Version%20of%20Core%. Accessed 12 May 2019

Preprocessing Information from a Data Network for the Detection of User Behavior Patterns

Jairo Hidalgo-Guijarro[1(✉)], Marco Yandún-Velasteguí[1], Dennys Bolaños-Tobar[8], Carlos Borja-Galeas[3,7], Cesar Guevara[4], José Varela-Aldás[2], David Castillo-Salazar[2,5], Hugo Arias-Flores[4], Washington Fierro-Saltos[5,9], and Richard Rivera[6]

[1] Grupo de Investigación GISAT, Universidad Politécnica Estatal del Carchi, Tulcán, Ecuador
{jairo.hidalgo,marco.yandun}@upec.edu.ec

[2] SISAu Research Group, Universidad Indoamérica, Ambato, Ecuador
{josevarela,davidcastillo}@uti.edu.ec

[3] Facultad de Arquitectura, Artes y Diseño, Universidad Indoamérica, Quito, Ecuador
carlosborja@uti.edu.ec

[4] Mechatronics and Interactive Systems - MIST Research Center, Universidad Indoamérica, Quito, Ecuador
{cesarguevara,hugoarias}@uti.edu.ec

[5] Facultad de Informática, Universidad Nacional de la Plata, La Plata, Argentina
washington.fierros@info.unlp.edu.ar

[6] Escuela de Formación de Tecnólogos, Escuela Politécnica Nacional, Quito, Ecuador
richard.rivera01@epn.edu.ec

[7] Facultad de Diseño y Comunicación, Universidad de Palermo, Palermo, Argentina

[8] Grupo de Investigación GISS, Universidad Politécnica Estatal del Carchi, Tulcán, Ecuador
dennys.bolanios@upec.edu.ec

[9] Facultad de Ciencias de la Educación, Universidad Estatal de Bolívar, Guanujo, Ecuador

Abstract. This study focuses on the preprocessing of information for the selection of the most significant characteristics of a network traffic database, recovered from an Ecuadorian institution, using a method of classifying optimal entities and attributes, with the In order to achieve a complete understanding of its real composition to be able to generate patterns and identification of trends of behavior in the network, both of patterns that deviate from normal traffic behavior (intrusive), as well as normal, to detect with high precision possible attacks. Network management tools were used as a multifunctional security server software, as well as pre-processing of data tools for the selection of attributes, as well as the elimination of noise from the instances of the database, It allowed to identify which ins- tances and attributes are correct and contribute with effective information in the study. Among them we have: Greedy Stepwise

T. Ahram et al. (Eds.): IHSED 2019, AISC 1026, pp. 661–667, 2020.
https://doi.org/10.1007/978-3-030-27928-8_101

Algorithm (Algoritmo Voráz), K-Means Algorithm, Discrete Chi-square Attributes and the use of computational models as Evolutionary Neural Networks and Gene Algorithms.

Keywords: Intrusion detection · Server · GreedyStepwise · Algorithm · K-Means · Evolutionary neural networks · Genetic algorithms

1 Introduction

The data without doubt is the most important fixed asset for any organization, so the information that nowadays is generated worldwide and in particular in government entities and companies in Ecuador are very representative and extensive, so it is necessary to use artificial intelligence methods, tools and mathematical algorithms to classify and organize this information in an effective way and that allow to contribute in the decision making process through the generation of algorithmic models. The information in the convergent data networks come from several sources and in different ways, these networks group hierarchical designs, which must provide greater efficiency, greater speed and optimization of functions. Although the main defense of a computer system is its access control, in most cases we can not trust such controls. This study focuses on the application of a model for the selection of the most significant characteristics of the database of an Ecuadorian institution, using a method of classifying optimal entities and attributes [1], in order to achieve a complete understanding of its real composition to generate patterns and identification of trends in behavior in the network [2], both patterns that deviate from normal traffic behavior (intrusive), as well as normal, to detect possible attacks with high precision [3]. Network management tools were used as a multifunctional security server software, as well as pre-processing of data tools for both the selection of attributes and the elimination of noise from the instances of the database, it allowed to identify which instances and attributes are correct and contribute with effective information in the study. Among them we have: GreedyStepwise Algorithm (Algorithm Voraz), Algorithm K-Means, Atributos discretos Chi-cuadrado as well as the application of artificial intelligence techniques such as neural networks and genetic algorithms for detection, obtaining optimal results [4].

2 Materials and Methods

This section presents the database recovered from the multifunctional security server, which was used to generate the preprocessing of the information to generate detection of patterns of user behavior.

2.1 Materials

The data set for the analysis in this article is the database that was implemented in the Untangle server, which was used to manage and manage the traffic that was generated in the network. The database is composed of 111370 instances with 29 attributes, information that was recovered during a period of 3 months between November 2018 and January 2019. The network topology used provided coverage and wireless service to 15 users between mobile devices and laptops, through the use of an access point (AP) that was directly connected to the second network interface (eth0 int) of the server. The results obtained in this model have shown that the proposal is quite effective in the testing phase, with a high accuracy rate. It's shows that our model is effective and adaptable to the dynamic behavior of the significant characteristics of the information traveling in the data network through the server. The example of the proposed network scheme is presented in Fig. 1.

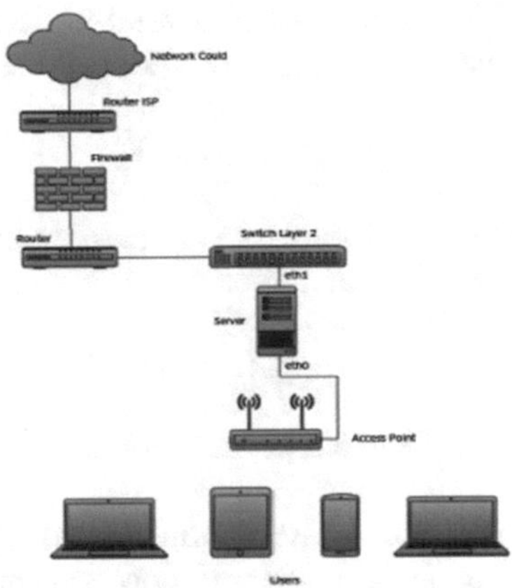

Fig. 1. Physical topology proposed network scheme Physical topology proposed network scheme

The table of the database used is identified with the name http_events, composed of 111370 instances and 29 attributes, the following is a part of the database in Table 1.

Table 1. Part of the records and attributes of the table http_events.

	time_stamp	c_server_addr	s_server_addr	request_id	session_id	c_client_addr	s_client_addr
1	1:23:20	52.169.82.131	52.169.82.131	101413646882457	101413646894482	192.168.2.189	172.20.4.151
2	1:23:19	13.107.4.52	13.107.4.52	101413646882456	101413646894477	192.168.2.189	172.20.4.151
3	1:23:26	13.89.187.212	13.89.187.212	101413646882458	101413646894490	192.168.2.189	172.20.4.151
.	.	.	.	.	.	.	.
.	.	.	.	.	.	.	.
111370	10:18:36	13.33.57.67	13.33.57.67	101453286876983	101453286996979	192.168.2.189	172.20.4.151

Lemmas, Propositions, and Theorems. The numbers accorded to lemmas, propositions, and theorems, etc. should appear in consecutive order, starting with Lemma 1, and not, for example, with Lemma 11.

2.2 Methods

In this section, we present the techniques and data mining algorithms that are used to perform the information preprocessing, as well as the algorithm of evolutionary neural networks and genetic algorithms that was used.

Information Preprocessing

Attribute Selection

The selection of attributes as a supervised technique allows selecting an attribute called class and seeks to determine if the attributes belong to a certain concept or not [5]. The function classifies a data element into one of several predefined classes. The object is described as a set of variables or attributes. (X, Eq. (1)),

$$X \rightarrow \{X_1, X_2, \ldots, X_n\} \quad (1)$$

Whose objective of the classification task is to classify the object within one of the categories of the class. (C, Eqs. (2) and (3)),

$$C = \{C_1, .., C_k\} \quad (2)$$

$$f : X_1 x X_2 x \ldots x X_n \rightarrow C \quad (3)$$

For the pre-processing of the database information, the c_server_addr field of the http_events table was used as the classifier attribute, obtaining as a result of 111370 instances 6 characteristics (no missing values) discrete class with 5429 values (no missing values) and 22 meta attributes 22.3% missing values.

The result of the GreedyStepwise voracious algorithm application, using the CfsSubsetEval evaluator attribute, it determines the attributes that provide the most information to the study by analyzing each attribute of the table against all others after incorporating an unsupervised classifier attribute (addcluster) [5] that adds one more attribute and groups according to similar characteristics. The result is presented in Table 2.

Table 2. Attributes that provide more information from the table http_events.

Number of folds		Attribute
10	(100%)	10 c_server_port
10	(100%)	11 s_server_port
10	(100%)	18 s2c_content_type

Preprocessing Using Evolutionary Neural Networks and Genetic Algorithms

Evolutionary neural networks use a model for the detection of behavior patterns and the tendency applying curve fitting, linear regression, quadratic functions [6], the adjustment of curves and linear regression applies the equation y = (ax + b) and the calculation of the values for a and b, which is expressed with equation [7]. (a, Eq. (4)).

$$a = \frac{n\sum xiyi - \sum xi \sum yi}{n\sum xi^2 - (\sum xi)^2} b = \frac{\sum yi}{n} - a\frac{\sum xi}{n} \quad (4)$$

Next the quadratic functions are applied. (y, Eq. (5)).

$$y = f(x) = ax^2 + bx + c \quad (5)$$

This method allows the detection of the patterns of behavior and even have a tendency to achieve it [8], it is necessary to have measurements or the extraction of the databases in periods or established time intervals and equal this is for xi in the function and of these measurements of the base in the same time interval, we obtain the values for yi that are the supplies to apply the adjustment of curves, the linear regression and the quadratic functions. With these results an extended matrix can be generated conformed by the totals of $xi, yi, xi^2, xi^3, xi^4, xiyi, xi^2y$ (Table 3).

Table 3. Attributes that provide more information from the table http_events.

	día	c_client_addr	time_stamp					
	xi		*yi*	x^2	x^3	x^4	xy	x^2y
1	1	192.168.2.189	0,05787037	1,000	1,000	1,000	0,05787037	0,05787037
2	2	192.168.2.189	0,05785880	4,000	8,000	16,000	0,11571759	0,23143519
3	3	192.168.2.189	0,05793981	9,000	27,000	81,000	0,17381944	0,52145833
4	4	192.168.2.189	0,05795139	16,000	64,000	256,000	0,23180556	0,92722222
5	5	192.168.2.189	0,05803241	25,000	125,000	625,000	0,29016204	1,45081019
	15	1	0,289652778	55	225	979	0,869375	3,188796296

Result of the application of the Gaussian method.

$$\left(\begin{array}{ccc|c} 1,79002E+15 & 0 & 0 & 0 \\ 0 & 791521500 & 0 & 36906,2594 \\ 0 & 0 & 157500 & 9113,48958 \end{array}\right)$$

Genetic Algorithms

A genetic algorithm is used for the resolution and determination of behavior patterns and even tendencies in the instances based on the selection ranking (ISR) [9], this algorithm is based on the algorithm of selection of dependent instances (WITS9, proposed by Morring and Martínez [10] (Table 4).

Table 4. Result matrix applying the genetic algorithm.

dia	c_client_addr	time_stamp1	time_stamp2	time_stamp3	time_stamp4
xi		*S11/s1*	*S21/s2*	*S31/s3*	*S41/s4*
1	192.168.2.189	0,05787037	0,057969	0,058011	0,058002
2	192.168.2.189	0,0578588	0,057924	0,057936	0,057919
3	192.168.2.189	0,05793981	0,057976	0,057893	0,057997
4	192.168.2.189	0,05795139	0,057988	0,057883	0,057924
5	192.168.2.189	0,05803241	0,057988	0,057991	0,058027

The obtained confusion matrix allows to calculate the sensitivity or positive precision (S, Eq. (6)), specificity or negative precision (E, Eq. (7)) and precision (P, Eq. (8)). The values of the matrix are: VP (true positives), VN (true negatives), FP (false positives), FN (false negatives).

$$S = \frac{\mathrm{VP}}{\mathrm{VP} + \mathrm{FN}} \tag{6}$$

$$E = \frac{\mathrm{VN}}{\mathrm{VN} + \mathrm{FP}} \tag{7}$$

$$S = \frac{\mathrm{VP} + \mathrm{VN}}{\mathrm{VP} + \mathrm{VN} + \mathrm{FP} + \mathrm{FN}} \tag{8}$$

Multilayer Neural Network MLP-CG-C

For the execution of the study, 20 records of the http_events table were used, obtaining the following results:

Set: training, Total percentage of classification by division k-fold 0.947%

Matrix of confusion. Set: test, Total percentage of classification 0.947%.

3 Conclusions and Future Jobs

When applying mathematical analysis based on Gaussian mutations and a model of neural networks that allowed to determine the trend of behavior patterns, in the case of the Gaussian application it takes as a supply the database. The database applied methods such as linear regression, application of quadratic functions and Gaussian application, obtaining a trend in the results that under normal conditions or patterns, its scatter plot and trend line in similar throughout the exercise, but when abnormal conditions are determined outside the patterns, it generates peaks in the scatter plot and trend line, making known the anomalies caused. As future lines of research, we intend to train and generate a Genetic Algorithm model that will allow us to incorporate methods to solve search and optimization problems.

References

1. Yin, C., Zhu, Y., Fei, J., He, X.: A deep learning approach for intrusion detection using recurrent neural networks. IEEE Access **5**, 21954–21961 (2017)
2. Reddy, R.R., Ramadevi, Y., Sunitha, K.V.N.: Effective discriminant function for intrusion detection using SVM. In: 2016 International Conference on Advances in Computing, Communications and Informatics (ICACCI), pp. 1148–1153 (2016)
3. Modi, C.N., Acha, K.: Virtualization layer security challenges and intrusion detection/prevention systems in cloud computing: a comprehensive review. J. Supercomput. **73**(3), 1192–1234 (2017)
4. Farnaaz, N., Jabbar, M.: Random Forest Modeling for Network Intrusion Detection System (2016)
5. Zarkami, R., Moradi, M., Pasvisheh, R.S., Bani, A., Abbasi, K.: Input variable selection with greedy stepwise search algorithm for analysing the probability of fish occurrence: a case study for Alburnoides mossulensis in the Gamasiab River, Iran. Ecol. Eng. **118**, 104–110 (2018)
6. Martínez-Estudillo, F.J., Hervás-Martínez, C., Gutiérrez, P.A., Martínez-Estudillo, A.C.: Evolutionary product-unit neural networks classifiers. Neurocomputing **72**(1–3), 548–561 (2008)
7. Yao, X.: Evolving artificial neural networks (1999)
8. Noguera, J., Portillo, N., Hernandez, L.: Redes Neuronales, Bioinspiración para el Desarrollo de la Ingeniería. Ingeniare **17**, 117 (2014)
9. García Vallejo, C.A.: Selección de Instancias y Atributos en Conjuntos de Datos mediante Algoritmos sobre Grafos (2012)
10. Morring, B.D., Martinez, T.R.: A nearest neighbor data reduction algorithm (2004)

Sports Design and Sports Medicine

Designing an e-Coach to Tailor Training Plans for Road Cyclists

Alessandro Silacci(✉), Omar Abou Khaled, Elena Mugellini, and Maurizio Caon

HES-SO, University of Applied Sciences and Arts Western Switzerland, Montreux, Switzerland
{alessandro.silacci, omar.aboukhaled, elena.mugellini, maurizio.caon}@hes-so.ch

Abstract. Athletes seek to constantly improve their performances pushing their limits and overwork is often the direct consequence of this behavior. This is not a common problem for professionals, who usually are followed by a coach, but it is a growing phenomenon in amateur sports, which drives people to get injured because of overtraining or incorrect movements. Meantime, advances in artificial intelligence enabled the creation of new tools increasingly capable of understanding the complexity of our world. We therefore propose a novel e-coaching system for road cycling athletes, able to automatically follow and tailor their training plans. This paper describes the design of the machine learning algorithm, its model based on reinforcement learning and the metrics that were adopted for the scoring system. Finally, we report our tests, which show that the virtual coach already can compete with human experts in making a proper personalized training plan.

Keywords: Road cycling · Artificial intelligence · Virtual coach · e-Coaching · Training plan

1 Introduction

In our century, customers are changing the way they consume goods, things have to be instantaneous and tailored to their needs. Therefore, the industry is implementing new services based on machine learning techniques in order to respond to this high and fast demand. The same goes for the sports domain, where people constantly want to perform better and want to do it faster than ever, unfortunately sports coaching services are human-dependent and expensive. When starting a new sport, no one really knows what she is capable of or which are her limits. Amateurs are likely to be early discouraged by not seeing any changes in their performance or by getting injured due to mistraining. Higher level athletes, such as professionals or semi-professionals, are also concerned about their performance and often consider paying for a personal coach. At the same time, the growing use of tracking sensors in the domain of sports is encouraging the application and creation of systems based on Artificial Intelligence (AI) capable of enhancing athletes' performance. Therefore, the creation of a virtual coach (or e-coach) to support athletes' training can focus on several different points [1]: (1) Enhancing

T. Ahram et al. (Eds.): IHSED 2019, AISC 1026, pp. 671–677, 2020.
https://doi.org/10.1007/978-3-030-27928-8_102

gear settings, (2) Strategy optimization, (3) Doping cases detection, (4) Training paths generation based on constraints, (5) Predicting results of athlete, (6) Detecting potential crisis of the athlete during a competition, (7) Simulation of opponents' behavior, (8) Injury Prevention.

Advances in artificial intelligence coupled with the high availability of tracking devices could be the solution to solve most of the issues in sports. Using new technologies to manage the data produced by human beings is a key aspect in order to create tailored services for any kind of athlete. Further developing such tools could help enhancing people's health, while increasing their performance on a daily basis.

This paper is composed of a first state-of-the-art review in the Related Work (Sect. 2), followed by a detailed description of our approach in the Methods (Sect. 3). Finally, results are demonstrated in the Results (Sect. 5) and argued in the Discussion (Sect. 6).

2 Related Work

Virtual coaching can be a high source of intrinsic and extrinsic motivation to encourage people into physical exertion [2]. Participants revealed a higher level of enjoyment while having a constant feedback on their current performance. E-Coaching is also a great tool to encourage immersion into systems based on virtual environments. Especially in the context where nowadays' market proposes a high range of home training devices [3]. E-Coaches do not just induce motivational gains but they can also be used for injuries prevention by analyzing and optimizing athletes' movements. Their capabilities are always increasing as ubiquitous computing technologies are growing fast and allow always deeper analysis of one's physiological status [4]. RunningCoach is a great example of how a virtual coach can reduce injury risk by optimizing one's running cadence [5].

Virtual Coaching has also been used to provide a training support system for athletes. Based on sensor-gathered data, systems can be created to increase one's performances. In that case, training plan creation is seen as an optimization problem. Thus, one can build an e-coaching system to try and find the optimal training sessions' arrangement [6]. Therefore, algorithms would have as a constraint to maintain the athlete's form and ensure his readiness for a set objective. Over the years, researchers have tried to propose exertion-measurement tools. Originally, the coaches and athletes often relied on the heartrate indications, but additional exertion indicators have revealed to be more efficient in defining the training load of a session [7]. Rating of perceived exertion is a scale rating tool to measure the training exertion experienced by the athlete, although it is more efficient than the heartrate to evaluate a session, it requires a lot of self-knowledge. Others, like Morton, use training impulses to detect and prevent overtraining [8].

In cycling, training is often relying on the heartrate. However, the power output reveals to be a great alternative for training load estimation. Allen and Coggan developed a full training management system using the power output of a cyclist [9].

Our literature review showed that many researchers focused on the problem of personalized training to improve performance avoiding overtraining. However, we could not find any work implementing machine learning to generate tailored training plans. For this reason, in this work, we wanted to develop an AI-based system capable of leveraging the large collections of data the sportsmen currently collect during their trainings in order to produce training plans that are made considering the athletes' conditions and performance measured during each training session.

3 Methods

Reinforcement learning (RL) in computer science is directly inspired from Pavlov's psychological experiment with dogs [11]. Skinner expressed a behaviorism theory where he stated that the consequence of a behavior change needs to be positive [12].

While coming from past psychological theories, RL is a growing sector in computer science. The core principle is to train a model to maximize a computed score. The model will learn what decisions can provide the best resulting reward. Thus, one can see an RL-based software as an optimization algorithm.

We based the system's core on a particular agent's architecture. In fact, the problem in a training management system is to maintain the athlete's form at its best. This implies that the model has to retain a part of the cyclist's past performance. In order to keep a contextual information, we decided to use the Deep Recurrent Q-Learning Network (DRQN) in a dueling mode [13].

In that case, the agent will optimize the q-values based on the answers from the environment. We added a first layer of Long-Short Term Memory (LSTM) cells to offer a proper handling of the past context to the agent. Figure 1 demonstrates our final agent's architecture.

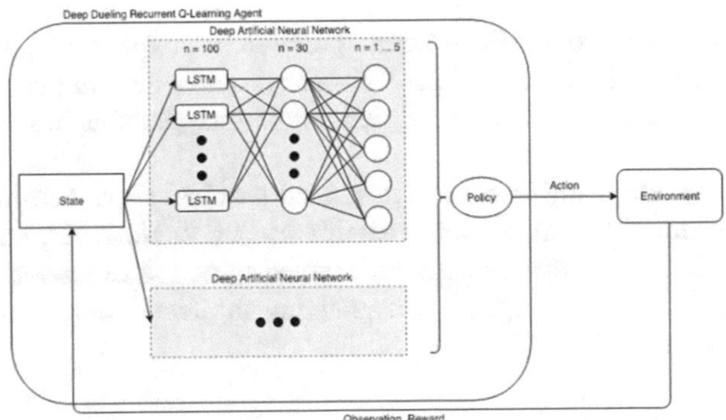

Fig. 1. The RL agent's architecture

The environment is where the decision made by the agent are evaluated and scored. To manage the form and the fatigue of an athlete, we based our computation on the power. Allen and Coggan introduced two mathematical functions based on the power output of a cyclist named Training Stress Score® (TSS) and Training Stress Balance® (TSB), respectively explaining the fatigue and the form of an athlete [9]. Both measures imply a certain number of recovery or rest days.

Obviously, the higher the intensity, the more time has to be given to the recovery of the cyclist. Coggan has provided a guide to estimate the number of recovery days induced by a certain amount of TSS generated with a training session.

Similarly, Friel proposed a training management based on the TSB where the TSB values are categorized in 5 different categories depending on their value [9],

We therefore combined these two theories to produce a weighted scoring system where TSS and TSB scores can be either equal or discriminated. This additional dimension has to be trained and further explored.

4 Experiments

To evaluate the output of our system, we designed an experiment involving professional sports coaches. We prepared AI-generated plans and used built plans from the reference platform TrainingPeaks. This platform is a well-known tool by the cyclists. It provides coach-made training plans or distant coaching services.

We used 3 different training plans from the platform and generated 3 others with our artificial intelligence using the same amount of days, respectively 6, 14 and 24 weeks. Experts needed to fill out a questionnaire based on a 5-point Likert scale (1 = very bad, 5 = very good).

Plans were provided in a random order for each training period. Thus, experts could not tell which of the two planning, for a given period, was generated or created by a professional coach.

5 Results

Results were collected from two different professional experts in training for road cycling. Interestingly, the results of the first question has always been 50% for each planning period. Both did not agree on defining if one planning was better than the other.

In terms of session distribution, the generated training plans are always one point behind the reference platform, as demonstrated in Fig. 2. However, one expert mentioned that the number of interval sessions seemed to be too excessive in the human-made planning. Note that the generated planning is always indicated with the "*" character.

Fig. 2. Results for the second experiment's question

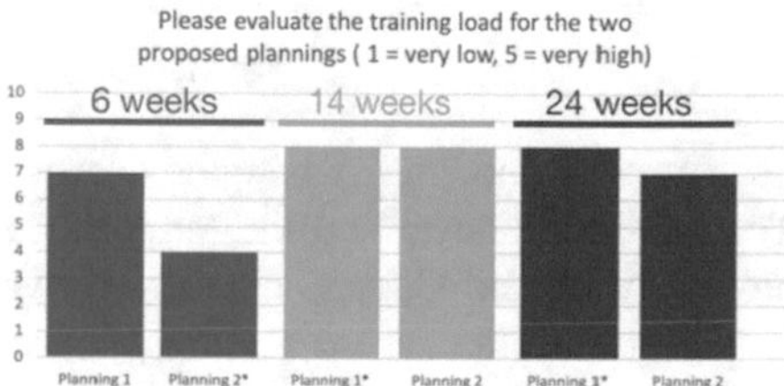

Fig. 3. Results for the third experiment's question

The AI-generated planning has been defined as more efficient on the session load aspect for the shortest period of time (6 weeks). But it revealed to propose a higher amount of intensive sessions on the longest training period (24 weeks). The system however has been qualified as equally intensive on the 14 weeks period. Figure 3 shows the different experts' scores gathered for the session load evaluation.

In Fig. 4, we observe that the resting time quantity has been evaluated as too high in the generated training plans compared to the ones coming from the reference platform. In longer training periods (14 and 24 weeks) both plans were providing too much recovery or resting days.

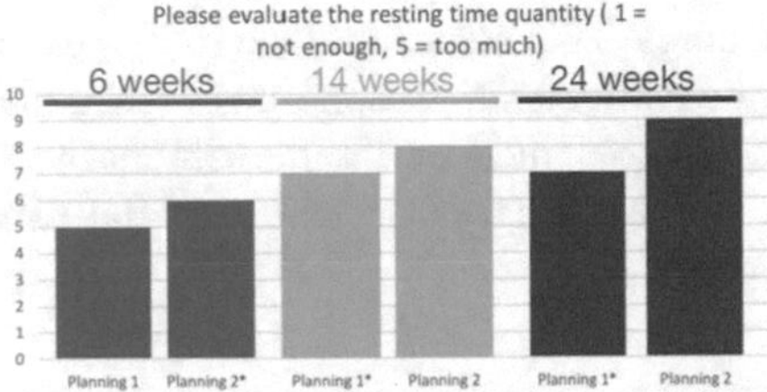

Fig. 4. Results for the fourth experiment's question

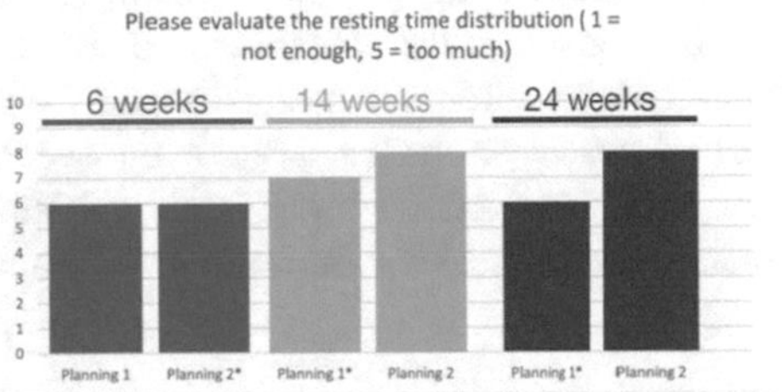

Fig. 5. Results for the fifth experiment's question

In counterpart, the system has been able to distribute in a more balanced manner the resting time. Figure 5 shows the results and experts have evaluated the generated planning being equally distributed in the 6 weeks planning, but more balanced for the two others.

Finally, the results in terms of efficiency are lower for the generated plans than the human coaches' ones. As demonstrated in Fig. 6, the scores are always 2 points lower for our system's training plan compared to the reference one.

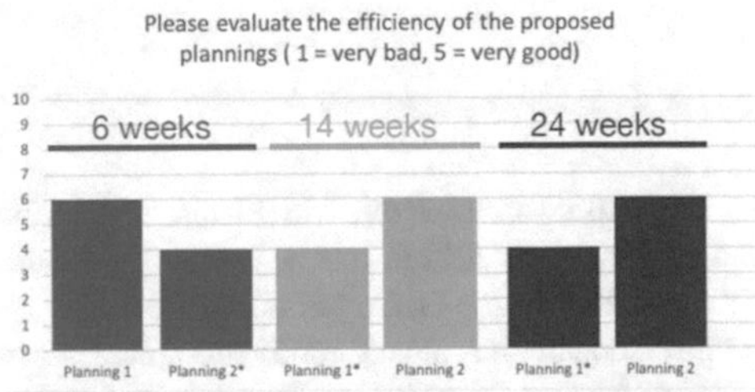

Fig. 6. Results for the sixth experiment's question

6 Discussion

The obtained results show that the virtual coach is not that far from beating human coaches in creating training plans with professional quality. Through this preliminary experiment on the prototype, we have identified problems in the training load of the plans proposed by the artificial intelligence. AI-generated plans where totally crafted upon the past data from the athletes. But, the fact that the selected user from the dataset had a high intensity session averaging 457 TSS made the algorithm use it several times during the different periods.

Results may be skewed as one of the participants was really frustrated about the training loads of the generated plannings. Thus, the AI has been given worse notes than the ones given to the professional coaches' plannings. This means that the results have usually been taken down drastically by one of the participants. Despite this aspect, the virtual coach was still near or even over the scores obtained by the human coaches' training plans. Interestingly, both participants could not agree on which of the two plans was the best for each proposed training period.

7 Conclusion

Through this paper introduces a novel solution for the training plan personalization and optimization dedicated to road cyclists. Based on their past trainings, any athletes can receive expert-level training advices adapted to their capabilities. The system is based on recent deep learning-based techniques designed to optimize the health of the user.

We carried out preliminary tests, which revealed that our AI system is able to produce training plans nearly as efficient as the ones made by human cycling coaches. These results will lead our next research and path of enhancements in order to outperform the human experts.

In the future, we plan to carry on the tests with cyclists and other sports experts in order to fine-tune our algorithm. We also intend to add more sports and training types in order to produce more efficient and diverse training plans.

References

1. Fister, I., Ljubič, K., Suganthan, P.N., Perc, M., Fister, I.: Computational intelligence in sports: challenges and opportunities within a new research domain. Appl. Math. Comput. **262**, 178–186 (2015). https://doi.org/10.1016/j.amc.2015.04.004
2. Eyck, A.: Effect of a virtual coach on athletes' motivation. In: IJsselsteijn, W.A., de Kort, Y.A.W., Midden, C., Eggen, B., van den Hoven, E. (eds.) Persuasive Technology, pp. 158–161. Springer, Heidelberg (2006)
3. Bonants, R., Kort, Y.D., Ijsselsteijn, W.: J. Westerink, M. de Jager, J. van Herk
4. Banos, O., Hermens, H., Nugent, C., Pomares, H.: Smart sensing technologies for personalised e-coaching. Sensors **18**, 1751 (2018). https://doi.org/10.3390/s18061751
5. Aranki, D., Peh, G.X., Kurillo, G., Bajcsy, R.: The feasibility and usability of runningcoach: a remote coaching system for long-distance runners. Sensors. **18** (2018). https://doi.org/10.3390/s18010175

6. Rauter, S.: New approach for planning the mountain bike training with virtual coach. Trends Sport Sci. 69–74 (2018). https://doi.org/10.23829/tss.2018.25.2-2
7. Rodríguez-Marroyo, J.A., Villa, G., García-López, J., Foster, C.: Comparison of heart rate and session rating of perceived exertion methods of defining exercise load in cyclists. J. Strength Cond. Res. **26**, 2249–2257 (2012). https://doi.org/10.1519/jsc.0b013e31823a4233
8. Pentland, A., Liu, A.: Modeling and prediction of human behavior. Neural Comput. **11**, 229–242 (1999). https://doi.org/10.1162/089976699300016890
9. Allen, H., Coggan, A.: Training and Racing with a Power Meter. VeloPress, Boulder (2010)
10. Strava | Run and Cycling Tracking on the Social Network for Athletes. https://www.strava.com/
11. Pavlov, I.P.: Lectures on conditioned reflexes. In: Conditioned Reflexes and Psychiatry, vol. II. International Publishers, New York (1941)
12. McLeod, S.: Skinner - Operant Conditioning, 12
13. Hausknecht, M., Stone, P.: Deep Recurrent Q-Learning for Partially Observable MDPs, 9

Half Scale Dress Forms from 3D Body Scans in Active Poses

Arzu Vuruskan[1(✉)] and Susan Ashdown[2]

[1] Department of Fashion and Textile Design, Izmir University of Economics, Izmir, Turkey
arzu.vuruskan@ieu.edu.tr

[2] Department of Fiber Science and Apparel Design, Cornell University, Ithaca, NY, USA
spa4@cornell.edu

Abstract. Sport is highly competitive, and there is a continual search for ways to improve performance. A good fit and comfort provided by activewear will be factors in performance, but providing good fit for the active position is challenging. This exploratory research is focused on creating half scale forms in active poses to facilitate activewear design, pattern making and fit testing. Digital data obtained from 3D body scanning in the active pose was reconstructed to create a valid body form as the basis for producing precise half scale body forms.

Keywords: Half scale body forms · 3D body scanning · Activewear · Apparel · Fit · Design

1 Introduction

Activewear currently represents an important segment of total apparel industry sales and is expected to continue to grow [1]. More use of casual dress in daily life and an increasing awareness of the benefits of a healthy lifestyle have been major reasons for the increasing popularity of activewear.

Due to strong associations between the garment and body image [2], aesthetics is a factor contributing to the impact of activewear as well as well-fitted and functioning garments [3]. Therefore, activewear design has been a trending topic, impacting athletes' performances both with functionality and expressiveness.

A major concern for user-centered design in functional clothing design, such as for protective clothing, workwear, and daily wear, is the understanding the three-dimensional (3D) body, and exploration of the relationship between clothing and body movements [4]. Likewise, the challenge in activewear design is to accommodate the changes to the body in motion by understanding the interaction of garments throughout this motion. One design strategy for creating effective activewear is to create clothing optimized for active positions rather than neutral standing position [5]. With this aim, in this research project, customized dress forms in active body poses are suggested as a new tool for understanding the 3D body, and allowing innovative/effective design development and fit evaluation of activewear. This paper provides an introduction to,

T. Ahram et al. (Eds.): IHSED 2019, AISC 1026, pp. 678–683, 2020.
https://doi.org/10.1007/978-3-030-27928-8_103

and overview of the development procedures of active half scale forms by using 3D body scanning techniques, and their uses. For this purpose, two activewear products in different sports categories are examined: bicycle shorts for the cycling position and golf pants for the golf swing pose.

2 Half Scale Body Forms for Apparel Industry and Education

Scaled body forms are recognized as a tool of design, manufacturing and display throughout the history of fashion. However, these traditional small-scale dress forms were general representations, rather than precise body forms; and therefore, were not adequate for the demands of technical design development or fit evaluation. 3D body scanning, as a recent technology, made it possible to create exact body forms, which can be used to create realistic dress mannequins. In a collaborative project between Alvanon and Cornell University in 2007, the first modern commercial half scale dress forms were developed [6]. As a follow-up project, development of customized half scale dress forms using in-house facilities were introduced by Ashdown, et al (2014) [7]. Timesaving prototyping techniques such as 3D printing and laser cutting were integrated for the production of half-scale dress forms. By reducing time, space and cost requirements, these tools made it possible to develop customized forms for all, including those with nonstandard body-shapes, such as plus size individuals and those with physical disabilities.

The benefits of half scale forms have been exploited in educational settings, and have possible uses in industry. In a contemporary design education context, “designers are seen as makers, and makers as designers”, and through the process of design, the design idea evolves in a process in which the designer continually alternates between two-dimensional (2D) and three-dimensional (3D) representations [8]. Within this context, half scale dress forms can make a contribution in the understanding of the 3D body for design and making, and for quick prototyping.

3 Half Scale Body Forms in the Active Pose

One main direction in our research, in parallel to the wider investigation of uses of half scale forms, has been the development of these forms in active body poses. In this paper, an overview of three main issues are introduced with the cases of cycling and golf clothing: (1) Production of half scale body forms in the active pose, (2) use of these forms for active wear fit testing, (3) and use for design development of active wear. The investigated sports categories, cycling and golf were selected as being different from each other from the perspective of apparel, thus increasing the range of knowledge gained. While cycling involves a symmetrical and continuous motion of legs, movements performed by golfers are asymmetrical. Since the motion in cycling occurs mainly in the lower body, tight fitting bicycle shorts were selected for investigation. Non-stretch woven golf pants were chosen for product development explorations for the golf swing pose.

3.1 Production of Half Scale Body Forms in the Active Pose

In order to produce half scale dress forms in the active pose, the first challenge was exploring 3D body scanning in various poses other than a standard standing pose. A digital copy of the human body can be obtained with 3D body scanning systems, however, these systems are designed to scan in a standing pose. Therefore, iterative methods were tested for the 3D body capturing. These investigations were conducted with a Human Solutions Vitus Smart XXL body scanner, by scanning athletes in different cycling and golf swing poses. The point cloud data acquired was incomplete due to data loss caused by occlusions resulting from the nonstandard standing pose in the scanner. After conducting various tests to optimize the scan, the missing areas could be reliably patched using 3D digital tools. Further explorations were conducted with a Structure Sensor, a hand held scanner, with promising results. However, the Human Solutions scanner provided better scanning accuracy in a shorter time, and was preferred for this research. Following the scanning procedures, the scan data were augmented using 3D software tools to obtain a valid body form [9, 10].

For the subsequent process using these digital body models, a variety of materials and methods were tested to create the optimal active half scale dress forms. The two main production techniques were 3D printing, and stacking of sliced foam materials with differing densities and thicknesses. In both methods, forms were covered with appropriate materials in order to maintain dress form requirements [9, 10].

Through these procedures, customized digital and physical body forms can be created in active poses for further investigations. Figure 1 illustrates a stacked-foam form produced from scanning a participant in the golf swing pose, and an example of a 3D printed form extracted from a cycling pose scan.

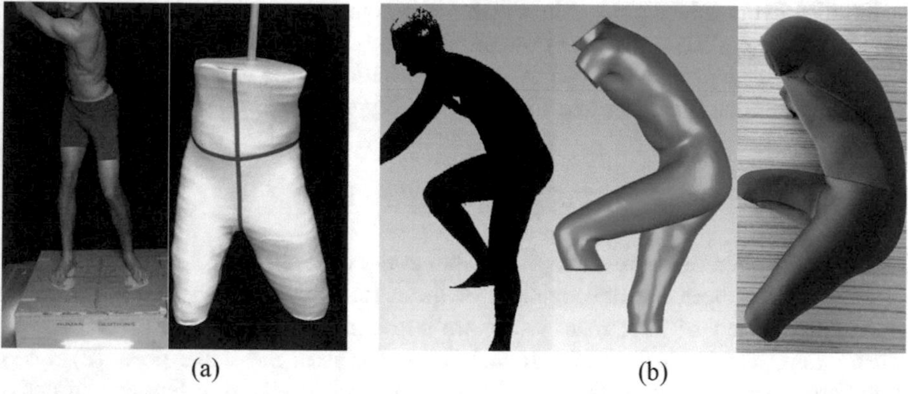

(a) (b)

Fig. 1. (a) Scanning a golfer and the stacked-foam half scale dress form (b) Point cloud data of a cyclist's scan, processed watertight digital model with 3D software, and 3D printed half scale body form covered with fabric

3.2 Half Scale Body Forms for Fit Evaluation of Activewear

The next stage was to test the half scale body forms in the active pose as fit evaluation tools. Basic cycling shorts and bibshorts were manufactured both in half and real scale, which allowed a comparison of fit with half scale dress forms and with the real participants whose bodies were scanned. Visual fit testing and the feedback obtained from the participants were the main sources of data (Fig. 2). The comparison of results revealed similarities in fit issues for the original athletes/participants and the half scale forms, and provided insights for pattern making in active body poses [9].

Air-pack compression sensors were used as a secondary testing method for fit evaluation, allowing the results from half scale garments on the forms to be compared with matching results from full-scale garments on the study participants (Fig. 2). Results were aligned within similar patterns in standing and cycling poses in half and full-scale tests [11].

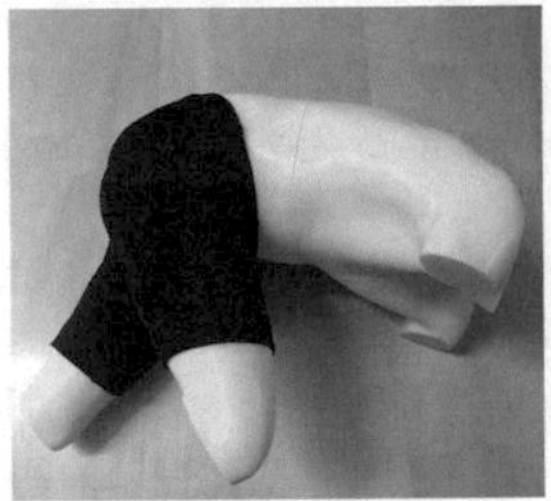
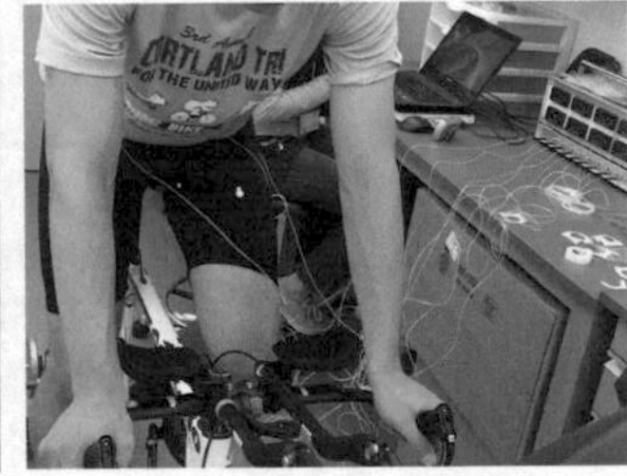

Fig. 2. Visual fit testing with half scale form and use of compression sensors on study participant

3.3 Half Scale Body Forms for Design Development of Activewear

Both 3D printed models and stacked-foam 3D models were tested/used for design development of cycling and golf clothing. Draping was the main pattern and design development technique, and 2D digital pattern making tools were used for refinement and scaling of patterns.

Design is referred to as problem-solving [8], and the particular problem in fashion is the complexity involved in understanding the 3D body. By using a 3D body form as a design development tool, these forms facilitate the fitting of the human body in the active pose, and matching the proportions of the active body. In designing for woven fabrics for the golf pants, nontraditional placement of seams provided good fit in the active position.

Our overall approach included design development on the active dress forms to improve fit. Half scale dress forms were produced in both standing and active poses for the athletes. This resulted in improved design, optimized in the active pose and acceptable in the standing pose. These tools lead to enhanced fit with a better understanding of the body's behavior in sport.

4 Conclusions and Future Work

An exploratory approach to the development and use of half scale dress forms in the active pose showed that custom forms can be produced economically using a variety of methods, with potential benefits for the activewear industry and education. The production and use of precisely half-scaled tools for garment testing was achieved and validated with the fit test results.

One innovative aspect of this research is creating active half scale forms using the digital data obtained from a conventional 3D body scanner. Athletes' 3D data will support product development by identifying the 3D shapes and proportions of the body in active poses. A more precise definition of the 3D body will allow design modifications based on active poses.

These convenient and affordable half scale body/dress forms will facilitate easier, faster and workable explorations and iterations for human centered design. A linear process of design is increasingly being replaced by circular flows for regenerating design ideas and the use of resources. Within this context, the methods and tools presented in this research have strong potential as rapid prototyping and designing tools for activewear, supporting the evolution of design in conjunction with pattern making, by improving understanding of the active 3D body.

One further direction for 3D body scanning the active body will be to explore the use of 4D scanners for activewear. Some recent scanning tools support dynamic assessment. Various high-speed scanning technologies, such as 3dMD, allow the possibility of 4D capture through simultaneous motion capturing; however, further research is needed into the usability of the obtained data from 4D capturing for apparel product development, and the extraction of useful measurements [3, 12].

Activewear, as a growing segment in apparel industry, is diversifying from purely functional activewear to a mix of both fashionable and functional garments [13]. Two activewear categories, golf wear and cycling wear in our project contributed to the overall purpose of improving activewear design and fit. Both these sports categories are conservative from the design perspective, without substantial opportunities for changes in design. However, new tools and methods, such as the ones suggested in this research project, may allow for cutting edge design alternatives in activewear markets. Nevertheless, it is important to consider the acceptability of new designs, and the assessment of their fit on the intended users, which is a further phase in this project.

References

1. The NPD Group: The Casualization of American Apparel: Will the Athleisure Trend Stay or Go? https://www.npd.com/wps/portal/npd/us/news/press-releases/2018/the-casualization-of-american-apparel-will-the-athleisure-trend-stay-or-go/ (2018)
2. Labat, K.L., Delong, M.R.: Body cathexis and satisfaction with fit of apparel. Clot. Tex. Res. J. **8**(2), 43–48 (1990)
3. Gill, S.: Body scanning and its influence on garment development. In: Hayes, S.G., Venkatraman, B. (eds.) Materials and Technology for Sportswear and Performance Apparel, pp. 311–325. CRC Press, New York (2016)

4. Watkins, S.M., Dunne, L.E.: Functional Clothing Design: From Sportswear to Spacesuits. Fairchild Books, Bloomsbury Publishing, New York (2015)
5. Ashdown, S.: Improving body movement comfort in apparel. In: Song, G. (ed.) Improving Comfort in Clothing. The Textile Institute, pp. 278–302. Woodhead Publishing (2011)
6. Ashdown, S., Vuruskan, A.: From 3D scans to haptic models: apparel design with half scale dress forms. In: 8th 3DBody.Tech Conference & Expo on 3D Body Scanning and Processing Technologies, pp. 31–41. Hometrica Consulting, Switzerland (2017)
7. Ashdown, S., Devine, C., Barker, T., Forker-Ruoff, J. Cegindir, N.: Development of half scale forms for the apparel industry. In: ITAA 2014 Annual Conference, Charlotte, NC, USA (2014)
8. Gully, R.: Cognition and process vs. design artifact in fashion design pedagogy. Cumulus Working Papers, Melbourne, pp. 40–45, Aalto University (2010)
9. Vuruskan, A., Ashdown, S.: Modeling of half scale human bodies in active body positions for apparel design and testing. Int. J. Clot. Sci. Tech. **29**(6), 807–821 (2017)
10. Vuruskan, A., Ashdown, S.: Reimagining design of golf clothing: addressing the asymmetrical pose. In: 8th 3DBody.Tech Conference & Expo on 3D Body Scanning and Processing Technologies, pp. 113–119. Hometrica Consulting, Switzerland (2017)
11. Vuruskan, A., Ashdown, S.: Fit analyses of bicycle clothing in active body poses. In: ITAA 2016 Annual Conference, Vancouver, British Columbia (2016)
12. Daanen, H., Psikuta, A.: 3D body scanning. In: Rajkishore, N., Padhye, E. (eds.) Automation in Garment Manufacturing. The Textile Institute, pp. 237–252. Woodhead Publishing, Cambridge (2018)
13. O'Sullivan, G., Hanlon, C., Spaaij, R., Westerbeek, H.: Women's activewear trends and drivers: a systematic review. J. Fash. Mark. Man. **21**(1), 2–15 (2017)

A Comparison of Heart Rate in Normal Physical Activity vs. Immersive Virtual Reality Exergames

José Varela-Aldás[1,2](✉), Esteban M. Fuentes[1], Guillermo Palacios-Navarro[2], and Iván García-Magariño[3]

[1] SISAu Research Group, Universidad Indoamérica, Ambato, Ecuador
josevarela@uti.edu.ec, tebanfuentes@gmail.com
[2] Department of Electronic Engineering and Communications, University of Zaragoza, Zaragoza, Spain
guillermo.palacios@unizar.es
[3] Department of Software Engineering and Artificial Intelligence, Complutense University of Madrid, Madrid, Spain
igarciam@ucm.es

Abstract. Regular physical exercise helps to maintain a good physical condition besides a healthy life, but the new working conditions and the needs of modern man makes hard to practice a sport, so a new tendency to practice sports periodically are the exergames, based in the immersive virtual reality, to allow new possibilities of games with different levels of physical activity. This work focuses on the comparison of the heart rate generated through normal physical activity compared with the obtained through by an immersive exergame, the physical activity employed was table tennis and the application was implement in the Gear VR with controller. The application was develop in Unity, using a mobile device compatible with virtual reality, and a Mi band 3 sensor, which acquires heart rate data. Finally, results indicated a decrease in heart rate in the case of the exergames, demonstrating that this technology does not substitute conventional physical activity in spite of the benefits, although the usability test was satisfactory.

Keywords: Exergame · Immersive virtual reality · Heart rate · Gear VR · Smartwatch

1 Introduction

Quality of life can be influenced by the physical and mental fatigue experienced day by day as a result of a speeded way of life and the adoption of sedentarism, appearing overweight and obesity troubles and even some mental affections such as depression or anxiety [1]. To avoid these bad health conditions, it is necessary to practice any sport activity, to help with the consumption or burning of calories ingested through the daily diet [2], besides to avoid illnesses of the cardiovascular system [3].

Different sports activities can be recommended according to the level of intensity required for a physical activity, however sometimes it is necessary to look for

T. Ahram et al. (Eds.): IHSED 2019, AISC 1026, pp. 684–689, 2020.
https://doi.org/10.1007/978-3-030-27928-8_104

alternatives that allow the realization of these activities with the help of technology, avoiding the use of physical spaces or specific sports material to practice a sport discipline. This is how the exergames appear, a mixture of videogame and physical activity, which is possible through virtual reality and immersive systems, where the players by means of the senses of sight and hearing are transported to scenarios in which you can explore, perform activities, among others, simulating real physical activity [4].

Nowadays, exergames have stopped to be purely a tool with playful purposes to become a very useful and applicable tool in various fields, as a tool to determine physical health [5], as a motivating agent for the development of physical activity in children [6], as well as for the development of the motor system in children [7]. In older adults who suffer from Parkinson's disease to improve their quality of life through different types of recreational activities [8], as well as training tool to avoid falls [9]. Finally the exergames are being used at the sports level, by means of training in virtual environments, studying the muscular development [10], improvement in the physical condition of the player [11]. There are studies which focuses on the relationship that exists between a player and an avatar within the player and the environment [12] and even to study changes or benefits that a player can have at a physiological level [13].

About the heart rate, it can be expressed as the number of contractions per unit of time (min), and depending on the age this can vary between 60 and 100 beats per minute while a person is in repose [14]. However not only physical activity can affect this vital sign, but it can also be affected by the perception of the senses, depending on the nervous system, that is why it is important to study the effects of immersive systems through Virtual reality over the vital signs.

So the main objective of this study was to compare heart rates obtained from the application of an immersive virtual reality system, through a head mounted display (HMD) device with a controller and ear buds and also through the classic way with the required equipment needed for the practice of table tennis. In both cases, the cardiac frequency data will be collected by a cardiac rhythm measurement band.

2 Methods and Materials

2.1 System Components

The structure of the proposed system uses a mobile device to visualize the virtual environment, the Gear VR as a HMD device, including the hand controller, and the Mi Band 3 wristband to acquire the user data. Figure 1 shows the structure of the system, where each component is presented.

The mobile device contains the virtual reality application developed in Unity and the application to acquire the user's heart rate data, whose data is stored in the Smartphone's memory. The Gear VR allow the immersion in the virtual environment and the hand controller simulates the table tennis racket. The player uses the Mi Band 3 wristband on the wrist while participating in the exergame; this device sends the information about the heart rate to the mobile device through Bluetooth conection.

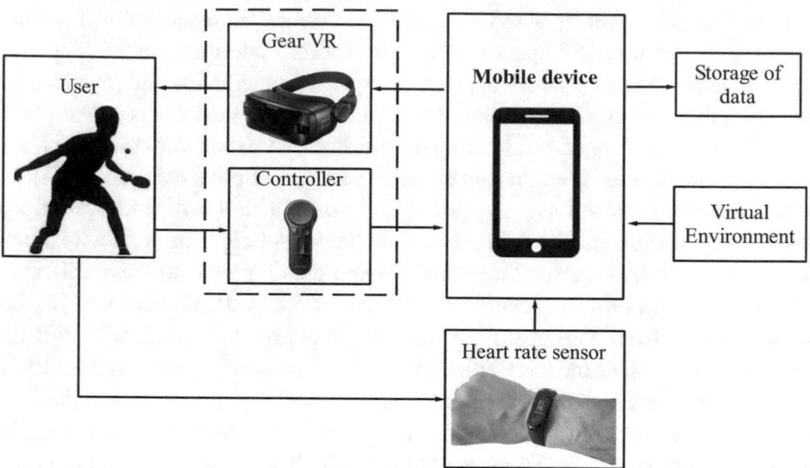

Fig. 1. System Components

2.2 Virtual Environment

The basic Unity tools were used to implement the application and the functions of the virtual elements, the exergame scene was implemented inserting different free-use prefabs found in the Asset Store, the main game objects are: the terrain, the racket, the tennis table, the ball and the opponent as shown in Fig. 2. At the scene of the exergame. In addition, the Oculus library was used to develop the game's control scripts, the camera allows the user to observe the first-person environment, the racket script imitates the movements of the controller (Gear VR), the colliders and rigid body produce the bounce of the ball and the response is controlled by the opponent's script.

Fig. 2. Scene of the exergame

3 Results

The results of the heart rate obtained through the application of the immersive virtual reality of the table tennis exergame were satisfactory. The results were evaluated by means of the data acquired by the heart rate sensor in 5-players and by applying a usability test of the application. For the operation of the Gear VR a Galaxy S8 + Smartphone was used and for the sound, the Z8 wireless headphones were employed. Figure 3 shows the user experimenting with the virtual reality system.

Fig. 3. Player using the exergame

The collected data from the tests of real table tennis vs. games in the virtual immersive environment are shown on the Fig. 4; observing a lower heart rate in the case of the exergame experiment, and greater pulsations when the individuals perform the actual physical activity.

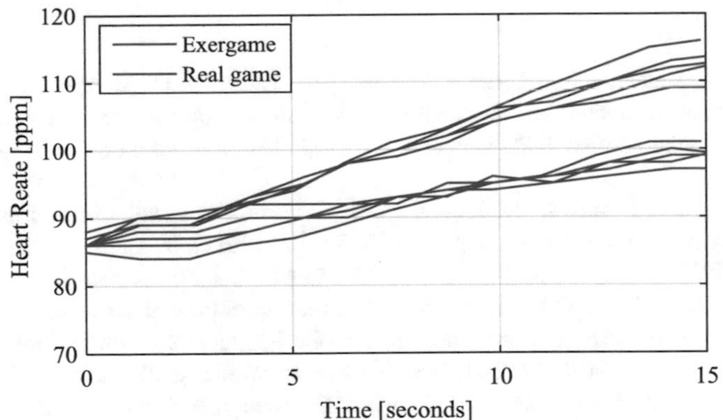

Fig. 4. Heart rate comparison

To evaluate the usability of the application, a SUS test was applied, using the procedures detailed by Quevedo, on his paper Market Study of Durable Consumer Products in Multi-user Virtual Environments [15]. Figure 5 shows the answers of the test, obtaining a final score of 77.5, a satisfactory value of usability for the application of virtual reality.

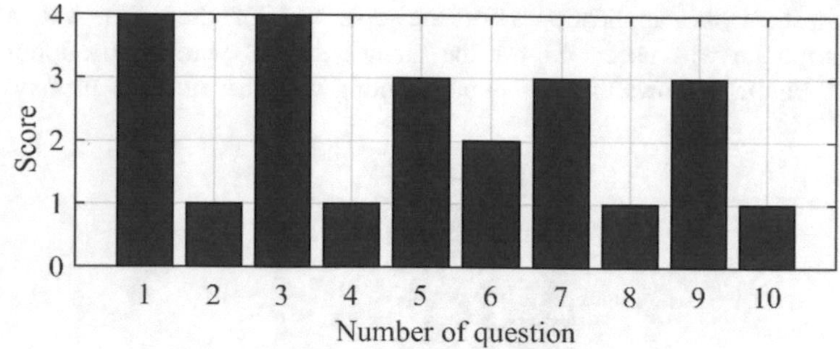

Fig. 5. Answers SUS

4 Conclusions

This work presents the comparison of the heart rate data obtained of a real table tennis game and an immersive exergame, using the Gear VR and a wearable heart rate sensor. The results showed very clear differences in the heart rate obtained on both activities; the virtual exergame produces a lower heart rate response compared with the heart rate obtained through real physical activity, this can be related with more movements and activities during the game. Making evident that the virtual reality application can be improved in software and hardware to achieve greater results with closer similarity to the real ones. In addition, the evaluation of the application through the SUS test demonstrated satisfaction from the user of the exergame.

References

1. Gross, A.C., Kaizer, A.M., Ryder, J.R., Fox, C.K., Rudser, K.D., Dengel, D.R., Kelly, A.S.: Relationships of anxiety and depression with cardiovascular health in youth with normal weight to severe obesity. J. Pediatr. **199**, 85–91 (2018). https://doi.org/10.1016/j.jpeds.2018.03.059
2. Schubert, M.M., Desbrow, B., Sabapathy, S., Leveritt, M.: Acute exercise and subsequent energy intake. A meta-analysis. Appetite **63**, 92–104 (2013). https://doi.org/10.1016/j.appet.2012.12.010
3. Xia, J.Y., Lloyd-Jones, D.M., Khan, S.S.: Association of body mass index with mortality in cardiovascular disease: new insights into the obesity paradox from multiple perspectives. Trends Cardiovasc. Med. **29**, 220–225 (2019). https://doi.org/10.1016/j.tcm.2018.08.006
4. Muñoz, J.E., Villada, J.F., Giraldo Trujillo, J.C.: Exergames: una herramienta tecnológica para la actividad física TT - exergames: a technological tool for the physical activity abstract. Rev. Médica Risaralda. **19**, 126–130 (2013)
5. Staiano, A.E., Calvert, S.L.: The promise of exergames as tools to measure physical health. Entertain. Comput. **2**, 17–21 (2011). https://doi.org/10.1016/j.entcom.2011.03.008
6. Rhodes, R.E., Beauchamp, M.R., Blanchard, C.M., Bredin, S.S.D., Warburton, D.E.R., Maddison, R.: Predictors of stationary cycling exergame use among inactive children in the family home. Psychol. Sport Exerc. **41**, 181–190 (2019). https://doi.org/10.1016/j.psychsport.2018.03.009

7. Gao, Z., Zeng, N., Pope, Z.C., Wang, R., Yu, F.: Effects of exergaming on motor skill competence, perceived competence, and physical activity in preschool children. J. Sport Heal. Sci. **8**, 106–113 (2019). https://doi.org/10.1016/j.jshs.2018.12.001
8. Ribas, C.G., Alves da Silva, L., Corrêa, M.R., Teive, H.G., Valderramas, S.: Effectiveness of exergaming in improving functional balance fatigue and quality of life in Parkinsons disease a pilot randomized controlled trial. Park. Relat. Disord. **38**, 13–18 (2017)
9. Choi, S.D., Guo, L., Kang, D., Xiong, S.: Exergame technology and interactive interventions for elderly fall prevention: a systematic literature review. Appl. Ergon. **65**, 570–581 (2017). https://doi.org/10.1016/j.apergo.2016.10.013
10. Soltani, P., Figueiredo, P., Fernandes, R.J., Vilas-Boas, J.P.: Muscle activation behavior in a swimming exergame: differences by experience and gaming velocity. Physiol. Behav. **181**, 23–28 (2017). https://doi.org/10.1016/j.physbeh.2017.09.001
11. Huang, H.C., Wong, M.K., Lu, J., Huang, W.F., Teng, C.I.: Can using exergames improve physical fitness? A 12-week randomized controlled trial. Comput. Human Behav. **70**, 310–316 (2017). https://doi.org/10.1016/j.chb.2016.12.086
12. Li, B.J., Lwin, M.O.: Player see, player do: testing an exergame motivation model based on the influence of the self avatar. Comput. Human Behav. **59**, 350–357 (2016). https://doi.org/10.1016/j.chb.2016.02.034
13. Mills, A., Rosenberg, M., Stratton, G., Carter, H.H., Spence, A.L., Pugh, C.J.A., Green, D.J., Naylor, L.H.: The effect of exergaming on vascular function in children. J. Pediatr. **163**, 806–810 (2013). https://doi.org/10.1016/j.jpeds.2013.03.076
14. Faust, O., Acharya, U.R., Ma, J., Choo, L., Tamura, T., Polytechnic, N.A.: Compressed sampling for heart rate monitoring. Comput. Methods Programs Biomed. **108**, 1191–1198 (2012). https://doi.org/10.1016/j.cmpb.2012.06.002
15. Quevedo, W.X., Benavides, O.J., Rocha, V.A., Gallardo, C.M., Acosta, A.G., Tapia, J.C., Andaluz, V.H.: Market study of durable consumer products in multi-user virtual environments. In: De Paolis, L., Bourdot, P. (eds.) Augment Reality, Virtual Reality, Computer Graphics. AVR 2018. Lecture Notes in Computer Science, pp. 86–100. Springer, Cham (2018)

The Impact of Ergonomic Design on Smart Garments

Rachel S. Boldt(✉), Luisa M. Arruda, Yao Yu, Helder Carvalho, Miguel A. F. Carvalho, and Fernando B. N. Ferreira

2C2T – Centro de Ciência E Tecnologia Têxtil, University of Minho, Guimaraes, Portugal
r.boldt@det.uminho.pt, rachelsagerb@gmail.com

Abstract. This paper reports the design process of a smart garment, which comprised 3-lead sEMG (Surface Electromyography) electrodes. The ergonomic design is central for a proper monitoring response because it is a related with the stability and very well contacted between the electrode and the user' body. For this, different body postures and the t-shirt behavior on the body was studied and simulated using a virtual prototype. This approach contributed to understanding ways to solving problems related to fit and the electrodes' stabilization. Furthermore, physical and electronic tests using a prototype on a human subject were conducted. The real prototype presented positive results on the EMG monitoring, showing the impact of ergonomic design on the smart garment. The EMG system was tested and presented good results, especially in regular movements. However, the system still needs to be improved in order to get a better signal when it comes to movements without pauses.

Keywords: EMG · Fit · Vital monitoring · Textile electrode

1 Introduction

Surface electromyography is a technique that records the electrical activity of muscles in a non-invasive way. The sEMG electrodes conduct biopotentials generated by the exchange of ions through the membranes of muscle fibers due to muscle contraction and relaxation [1, 2] to an appropriate signal conditioning system. This technology has been broadly utilized in muscle activity assessing of athletes and patients [3, 4]. In sports, EMG signal monitoring has potential benefits such as controlling repetitions, checking muscular fatigue, supporting the development of body awareness, tracking athlete's performance, among others [3, 5].

Although the possibility of remote transmission of EMG signals exists, the acquisition method is restrictive and uncomfortable to be conducted in out-of-lab settings. Thus, in recent years, many efforts have been developed by manufacturers and researchers aiming to incorporate this technology in clothes [6–8].

The present study reports the development of a smart garment, more specifically, a long sleeve t-shirt with embedded ECG (Electrocardiography) and EMG electrodes. Initially, two prototypes for cardiac and muscle activity monitoring were developed and

T. Ahram et al. (Eds.): IHSED 2019, AISC 1026, pp. 690–695, 2020.
https://doi.org/10.1007/978-3-030-27928-8_105

reported. However, the ECG and EMG systems were negatively impacted by the movement of the arms during use [9, 10].

Considering that a perfect fit is of paramount relevance in this kind of smart garment, contributing to the proper operation of the electronic system integrated into the clothing, new prototypes were developed considering the ergonomic aspects of the design. The fit is essential to control the contact, stability, and positioning of electrodes on the user's body and skin [11].

The development and validation of the t-shirt design was aided by tools such as a body scanner and a 3D CAD software. These tools improve the efficiency and effectiveness of the design process and are alternatives to traditional patterning methodologies, expanding the exploration of new project possibilities [12, 13].

The main objective of this article is to present the redesign and development process of an ergonomic sportswear garment – a long-sleeve t-shirt – that integrates textile sEMG (Surface Electromyography) electrodes for unobtrusive, real-time monitoring of a person's activity and physiological data.

2 Materials and Methods

In the development of the smart garment for EMG monitoring, design guidelines were focused on the fit and ergonomic-functional capacity of clothing, as well as on the improvement of contact and stabilization of the shirt to the body to create a better performance environment for electronic monitoring.

The design process was divided into ergonomic design conception, virtual try-on, and tests on a real prototype. For the design, the surface extraction of the 3D digital body obtained using 3D-2D flattening technology was used, through a 3D CAD software, Clo 3D version 4.1. After this, a dynamic virtual try-on analysis, exploring the different positions of the sensors, was performed to develop a virtual prototype.

Subsequently, a physical prototype was developed, using conventional and technical textiles. A Jersey fabric, using 95% polyamide and 5% elastane, was used for the t-shirt base, Fig. 1. A technical conductive knit *Shieldex Medtex P180+B* was uses as electrode and for conductive paths. For those, the knit was coated with a thermoplastic film (Bemis Exofliex 3900) and reinforced with TPU. The bonding conditions were: 130°, at 5.5 bar, for 20 s. The electrodes have a rectangular shape (5 cm × 3 cm), as shown in Fig. 2. The *BITalino Board kit*, and *Opensignals* software were used to carry out the experiments. A volunteer was selected to participate in the experiments.

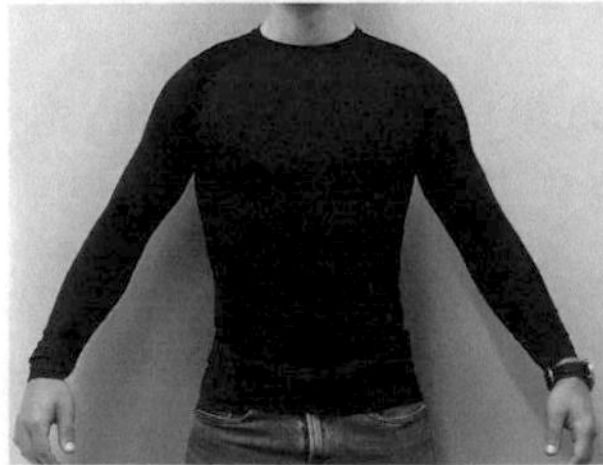

Fig. 1. T-shirt prototype at use and the electrode design positioned in the sleeve.

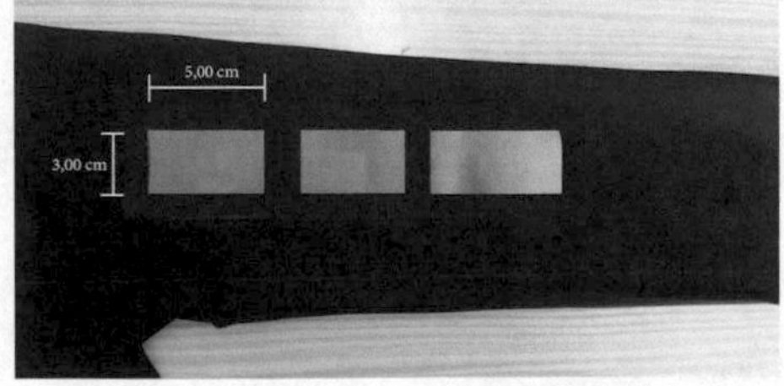

Fig. 2. T-shirt prototype at use and the electrode design positioned in the sleeve.

The acquisition procedure was delimited by the repetition of movements related to the muscles studied: biceps, triceps, and chest. The exercises performed were respectively, Dumbbell bicep curls, Kickback, and Push up, and each one was repeated five times by the volunteer. The tests were performed with the t-shirt worn on dry skin and no additional support system for the electrodes was used, ensuring that only the pressure of the t-shirt on the body was applied.

3 Results and Discussion

The first ergonomic pattern was designed using a flattening tool available in the 2D-3D Cad software *Clo 3D*. This tool creates "a flattenable mesh surface, which is a polygonal mesh surface that can be unfolded into a planar patch without stretching any polygon" [14]. The pattern flattening was done with the delimitation of lines directly over the body model. The tool helped the perception of the ease values necessary in different body postures. The results served as a reference for the design of the final pattern.

Furthermore, the 3D Cad software enabled the try-on prototype tests in dynamic postures. The available avatar was customized according to the volunteer's body measurements, and four body positions were pre-defined as follows: Position 01 - body rest (initial); Position 02 - bending elbows; Position 03 - up arms open, and Position 04 – body rest (final). Figure 3 presents the digital try-on test in the four dynamic postures. Additionally, the digital knit was parameterized using the weight, thickness, bend and stretch behavior data from the real knit.

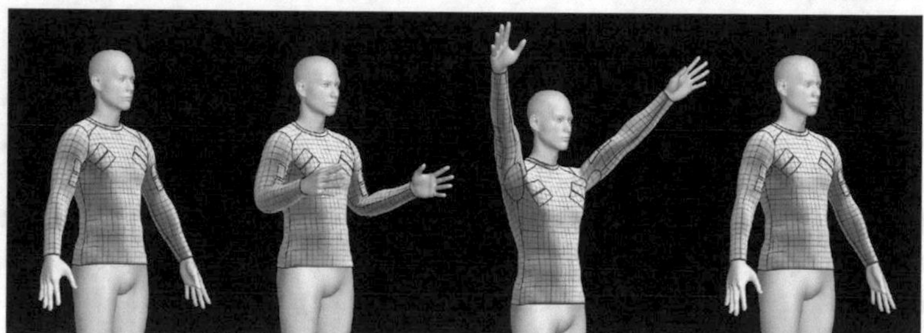

Fig. 3. Digital try-on test in dynamic postures

The results showed the most stable t-shirt zones, contributing to further defining the electrode locations. Figure 4 shows the overlapped images of the t-shirt in Position 1 and Position 4, after the body's movements.

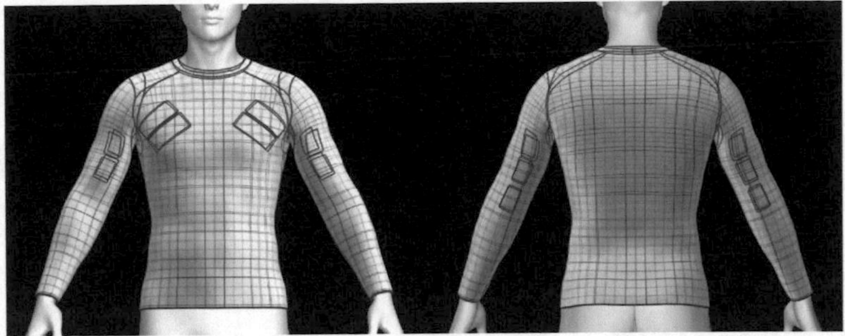

Fig. 4. T-shirt overlap in positions 1 and 4.

The virtual prototype presented good fit performance, despite a small variation of the t-shirt position on the body. But the instability impact was limited way on the electrode zone. The actual prototype proof presented a good fit, free body movement, and stability to the volunteer's body. However, in the area of the shoulders, chest, and upper back still some instability conditioned to the shoulder's movement. A complementary study of localized compression and ergonomic cut will be required for the overall stability of the t-shirt.

Figures 5 and 6 show the EMG signal obtained by Biceps, Triceps, and chest muscles.

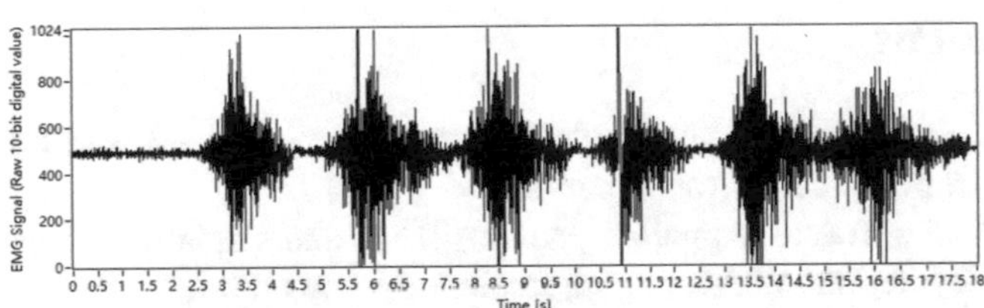

Fig. 5. EMG signal, Muscle: Biceps, Exercise: Dumbbell bicep curls; Repetitions: Six; Weight: 1.0 kg.

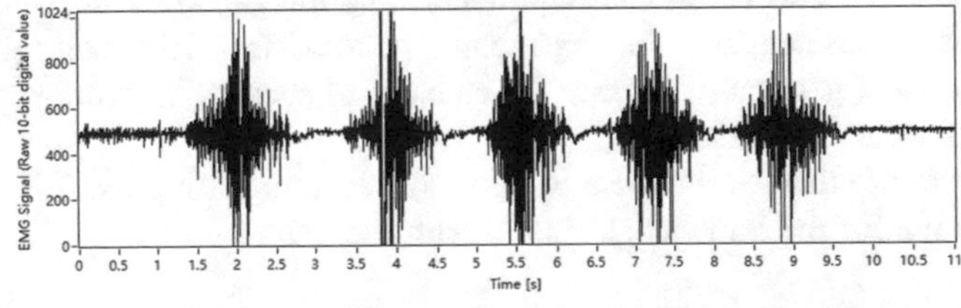

Fig. 6. EMG signal, Muscle: Triceps, Exercise: Kickback; Repetitions: five; Weight: 1.0 kg.

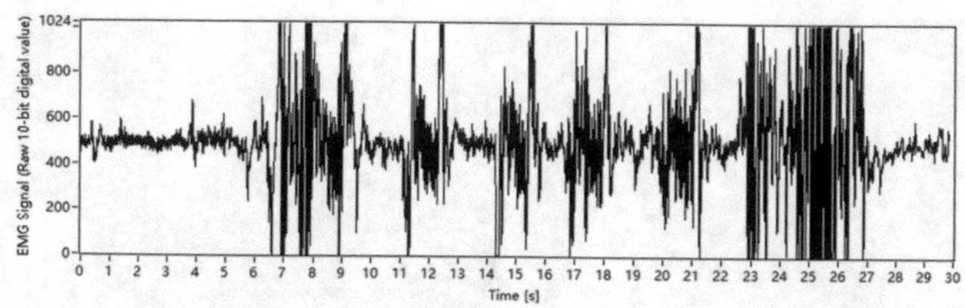

Fig. 7. EMG signal, Muscle: Chest, Exercise: Push up; Repetitions: six.

The electromyographic signals acquisition results can be considered very positive, especially the acquisitions related to biceps and triceps muscles. The waves captured were sharp and with low noise. The graphs showed clearly the muscle's activity and rest moments. These results are promising for future applications such as monitoring in sports and rehabilitation (Fig. 7).

It was also possible to capture the electric chest signals during pushups exercise. It was possible to identify the peaks of the exercise, but with a higher level of noise than other acquisitions. The noise may be related to the instability of the t-shirt, previously identified. The first and last spikes indicate the beginning and the end of the exercise that involved complementary body movements for the beginning and end of the Arm flexion exercise. The remainder of the exercise graph shows greater homogeneity in the signals, even with the presence of noise.

4 Conclusions

In this paper, the ergonomic design developed a smart garment, and EMG measurements result from the t-shirt was reported.

The 3D Cad software supported pattern design and virtual prototypes tests. This assisted the development of a fit and stable t-shirt, according to the smart garment demands. However, there are still instability points on the t-shirt. Future studies will be necessary to achieve solutions to render the electrode positioning less sensitive to body movements.

The *BITalino* device was used to carry out the experiments, but a device that follows the aesthetic and functional requirements of the garments shall be designed.

Finally, we consider that the ergonomic pattern, from the perspective of the demands of a smart garment, increases the chances of success in future applications of other types of wearable sensors.

We also want to thank colleagues Ricardo Moreira for testing the shirt on his body and André Paiva for the knowledge shared with the team.

Acknowledgment. This work is financed by Project "Deus ex Machina", NORTE-01-0145-FEDER-000026, funded by CCDRN, through Sistema de Apoio à Investigação Cientifica e Tecnológica (Projetos Estruturados I&D&I) of Programa Operacional Regional do Norte, from Portugal 2020 and by Project UID/CTM/00264/2019 of 2C2T – Centro de Ciência e Tecnologia Têxtil, funded by National Founds through FCT/MCTES".

We also want to thank colleagues Ricardo Moreira for testing the shirt on his body and André Paiva for the knowledge shared with the team.

References

1. De la Peña, S., Polo, A., Robles-Algarín, C.: Implementation of a portable electromyographic prototype for the detection of muscle fatigue. Electronics **619**, 2–15 (2019)
2. Jordanić, M., Rojas-Martínez, M., Mañanas, M., Alonso, J., Marateb, H.: A novel spatial feature for the identification of motor tasks using high-density electromyography. Sensors **17**(7), 1597 (2017)
3. MacLean, K.F.E., Dickerson, C.R.: Kinematic and EMG analysis of horizontal bimanual climbing in humans. J Biomech. (2019)
4. Kim, H., Lee, J., Kim, J.: Electromyography-signal-based muscle fatigue assessment for knee rehabilitation monitoring systems. Biomed. Eng. Lett. **8**(4), 345–353 (2018)
5. Trindade, T.B., de Medeiros, J.A., Dantas, P.M.S., de Oliveira Neto, L., Schwade, D., de Brito Vieira, W.H., et al.: A comparison of muscle electromyographic activity during different angles of the back and front squat. Isokinet Exerc. Sci. Pre-Press., 1–8 (2019)
6. Taelman, J., Adriaensen, T., van der Horst, C., Linz, T., Spaepen, A.: Textile Integrated contactless EMG sensing for stress analysis. In: 29th Annual International Conference of the IEEE Engineering in Medicine and Biology Society, pp. 3966–3969. IEEE Press, New York (2007)
7. Finni, T., Hu, M., Kettunen, P., Vilavuo, T., Cheng, S.: Measurement of EMG activity with textile electrodes embedded into clothing. Physiol. Meas. **28**(11), 1405 (2007)
8. Manero, R.B.R., Shafti, A., Michael, B., Grewal, J., Fernandez, J.L.R., Althoefer, K., et al.: Wearable embroidered muscle activity sensing device for the human upper leg. In: 38th Annual International Conference of the IEEE Engineering in Medicine and Biology Society (EMBC), pp. 6062–6065. IEEE Press, New York (2016)
9. Paiva, A., Catarino, A., Carvalho, H., Postolache, O., Postolache, G., Ferreira, F.: Design of a long sleeve t-shirt with ECG and EMG for athletes and rehabilitation patients. In: Machado, J., Soares, F., Veiga, G. (eds.) Innovation Engineering and Entrepreneurship. Lecture Notes in Electrical Engineering, vol. 505, pp. 244–250. Springer, Cham (2019)
10. Paiva, A., Ferreira, F., Catarino, A., Carvalho, M., Carvalho, H.: Design of smart garments for sports and rehabilitation. In: IOP Conference Series Materials Science Engineering, vol. 459, no. 1 (2019)
11. Harms, H., Amft, O., Troster, G.: Influence of a loose-fitting sensing garment on posture recognition in rehabilitation. In: Proceedings IEEE-BIOCAS Biomedical Circuits and Systems Conference, pp. 353–356. IEEE Press, New York (2008)
12. Hernández, N.: Does It Really fit? Improve, Find and Evaluate Garment fit, Stema Specialtryck, vol. 25. University of Borås, Sweden (2018)
13. Fan, J., Yu, W., Hunter, L.: Clothing Appearance and Fit: Science and Technology. Woodhead Publishing in Textiles, Cambridge (2004)
14. Zhang, Y., Wang, C.C.L., Ramani, K.: Optimal fitting of strain-controlled flattenable mesh surfaces. Int. J. Adv. Manuf. Technol., **87**(9–12), 2873–2887 (2016)

Biomechanics, Health Management and Rehabilitation

User Centered Design of a Pill Dispenser for the Elderly

Florian Reichelt[1(✉)], Peter Schmid[1], Thomas Maier[1], Nada Sahlab[2], Nasser Jazdi-Motlagh[2], Michael Weyrich[2], Gerd Meyer-Philippi[3], Günter Kalka[3], and Manfred Matschke[3]

[1] Institute for Engineering Design and Industrial Design (IKTD), Department of Industrial Design Engineering, University of Stuttgart, Pfaffenwaldring 9, 70569 Stuttgart, Germany
{Florian.Reichelt,Peter.Schmid, Thomas.Maier}@iktd.uni-stuttgart.de

[2] Institute of Industrial Automation and Software Engineering (IAS), University of Stuttgart, Pfaffenwaldring 47, 70569 Stuttgart, Germany
{Nada.Sahlab,Nasser.Jazdi, Michael.Weyrich}@ias.uni-stuttgart.de

[3] CompWare Medical GmbH, Robert-Bunsen-Straße 4, 64579 Gernsheim, Germany
{gmp,gk,mm}@compwaremedical.de

Abstract. The demographic change facing Germany and Europe leads to a growing importance of user-centered products for the elderly. Especially the increase of patient multimorbidity leads to a growing necessity of future healthcare solutions. In terms of medication management, a lack of reliable pill intake support for the elderly can be observed. This research project develops an approach for a fully automated pill management system. Special focus in this contribution lies on the user-centered design process for the elderly.

Keywords: Ambient Assisted Living · Healthcare · User-centered design · Design for the elderly · Pill dispenser

1 Object and Significance

Patient multimorbidity is a growing problem, especially with the demographic change facing Germany and Europe [1]. The largest population group in Germany in 2017 was the generation aged between 45 and 60. According to forecasts, this trend will shift until 2060 to the generation of the 60 to 70-year-olds. In relation to the total population, the proportion of the older population group will rise from 2017 to 2060 [2]. As a result of this shift in population groups, the everyday complaints and problems of the elderly are increasingly becoming the focus of political and economic attention.

An increasing elder population also leads to a further risk in the intake of medication by older people. According to BMBF estimates, about 2/3 of people over 60 years in Germany already take two or more different medications on average, and the trend is rising. Medication should enable patients to maintain a good quality of life,

T. Ahram et al. (Eds.): IHSED 2019, AISC 1026, pp. 699–704, 2020.
https://doi.org/10.1007/978-3-030-27928-8_106

however, the process of taking medications and the required corresponding medication management entails risks. According to studies, 20–50% of German patients treated for chronic diseases take prescribed pills incorrectly or not at all [3, 4].

Overall, the care of multimorbid elderly people represents a great economic and social challenge. One of the greatest wishes of elder people is to lead an independent life in their familiar home. A key aspect for independence in old age is the correct management of medications. The use of the original packaging and blister packs to manage the medication or the use of simple manual sorting boxes currently dominates the daily intake of pills. In order to prevent inadvertent false or non-intake and to increase the success of treatment, aids and measures are needed to ensure that the medication is provided and taken as required, especially for elderly patients [5].

Therefore, research projects have focused on Ambient Assisted Living (AAL), aiming at fulfilling this desire for independence in old age. This paper describes the approach of a user-centered design development of an age-appropriate pill dispenser for correct pill management for the elderly.

2 Methods

For developing an age-appropriate pill dispenser the user-centered design process is based on VDI 2221. To investigate central requirements on this special pill dispensing system a user survey was undertaken. After gaining these main requirements different functional concepts were designed. Using an expert evaluation, a decision for one general system design was made and several design concepts were subsequently shaped and evaluated.

User Experience Study. In a first step, currently available pill dispensers were analyzed in order to derive technical, physical and human-product requirements for the design of the future dispenser system. Current pill dispenser systems can be divided into three classes: tube blister systems, cup blister systems or the provision of pills in individual daily magazines. Within the scope of this analysis, the strengths and weaknesses of individual dispenser systems were identified. In a further study elder people were interviewed about their medication intake behavior and pill management. The questionnaire was completed online by the participants. The target group of this survey were persons between 48 and 80 years of age. The average age of the 38 participants (n = 38) was 59 years. The ratio male to female was 50/50. The questionnaire comprised 26 questions, including single-choice, multiple-choice and free text questions. Hereafter a user-centered survey on medication handling among older people was also conducted and evaluated. Using three standard dispensers with a reminder function, the usability was qualitatively analyzed in the form of a strengths and weaknesses analysis. The target group of the survey were people between 46 and 73 years of age. The average age of the 17 participants (n = 17) was 61 years.

Design Evaluation. Based on the fundamental results (Sect. 3) of the user research sixteen design variations were created. Within these variations, different design characteristics in layout, shape, color and graphics were chosen.

A primary evaluation of these different design concepts was undertaken in an expert survey. For the purpose of a consistent presentation and an identical level of detail the central perspective was selected as the basic view for evaluating the concepts. According to the recommendation of Reid et al. [6], the concepts are executed as renderings.

For extrapolating the different perception and valuation of the experts, a semantic differential was chosen. The semantic differential is characterized by bipolar scales, which are arranged as pairs of opposite describing items. These contrary items are arranged in such a way that the ends of the poles do not appear uniform in order to avoid an unreflected one-sided evaluation [7]. When creating the semantic differential for the tablet dispensing system, the following essential categories were considered: *performance*, *principle*, *purpose*, *handling*, and *age-appropriate design*. The scale for evaluating the opposite pairs is designed with three positive degrees in both directions (Fig. 1). Figure 1 shows an excerpt of the resulting valuation for one concept.

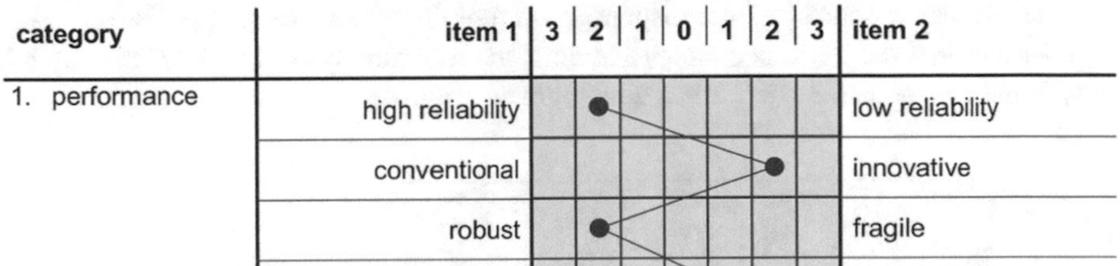

Fig. 1. Excerpt from the evaluation of the semantic differential for concept E.

In addition to the perception evaluation via semantical differential, the experts for medical and healthcare products also rated the different design concepts concerning the product's appeal. In total, six statements were shown to the experts and they had to decide whether they agree to them or not. Therefore, a seven-point Likert scale was used, using the maximum values fully agreed (1) or fully disagreed (7).

3 Results

This chapter summarizes all findings of both the user surveys and the expert review on the first design concepts.

User Experience Study. Within both user surveys it could be shown that several approaches are made to handle the underestimated risk of mishandling important medications. Indeed, any available solution was satisfying for the participants. Further various demands on an age-appropriate pill dispensing system could be recorded. The following requirements can be transferred:

- In Terms of *performance* aspects, the dispenser should be highly reliable, use innovative technology and must be robust.

- The functional *principle* should be characterized by a long-life functionality, a possibility for mobile usage; repair-, replace- and upgradeable.
- The *purpose* of the pill dispenser should be to automatically provide the right medication at the right time.
- Special requirements were made for the *handling* of the dispenser. The interaction should be intuitive. The system should not be complex to handle and consider physical weaknesses of elder people.
- Also, the *age-appropriate design* is a fundamental aspect. For guaranteeing an appropriate lifestyle without being recognized as a person with disease in public, a modern non-stigmatizing design is necessary.

Based on these requirements general functional solutions can be divided into three categories: a stationary pill dispenser with one mobile daily unit, a stationary pill dispenser with several mobile daily units and a modular system consisting of a stationary pill dispenser with two exchange units (several mobile daily units and a blister system which dispenses pill portions). All variants consider the connection of wearables as well as a control of the dispenser via mobile devices (e.g. app). Based on an expert review of the various concept variants, the modular system with several mobile daily units was selected for further development (Fig. 2).

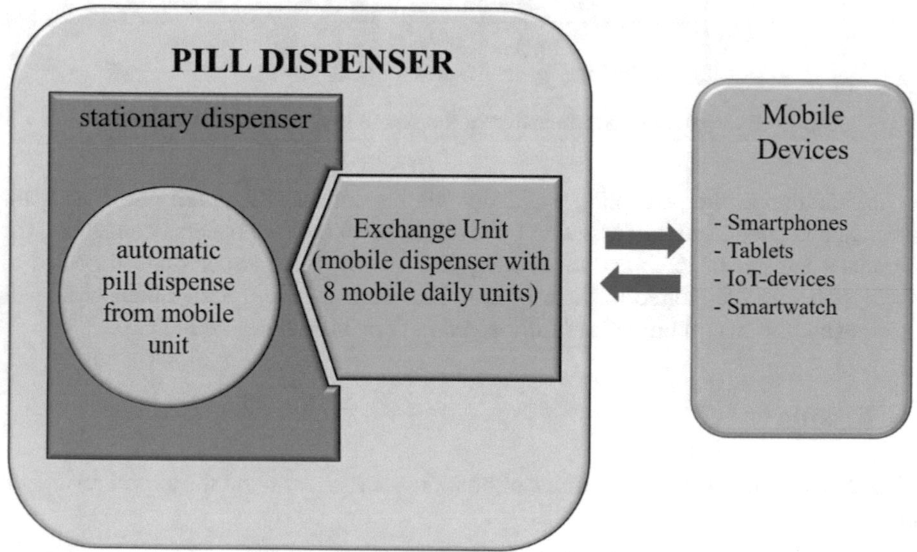

Fig. 2. Modular pill dispenser concept with several mobile daily units

Based on the chosen functional solution of the pill management system, various design drafts for the modular structure of a stationary pill dispenser with eight daily magazines were developed during the design phase. The design of the stationary dispenser can be described as a vertical, tower-like structure. A basic housing serves as a mounting for the modular insert, which contains the individual daily magazines.

In stationary operation, the dispenser automatically dispenses the pills at the time they are planned to be taken. This dispensing process starts after the patient has confirmed he is ready to take his medication. The individual daily magazines are stored on an inclined level in the stationary dispenser. At the time of dispensing, the corresponding compartment of the daily dispenser is controlled and opened by stepper motors. The pills fall out of the compartment via a tilted position of the magazines. The stationary dispenser can be operated either on the built-in 7" touch display or via the app. In addition to the requirements regarding icon sizes and color contrasts for elder users, a tactile feedback from the touch displays can be considered according to the user group. The user of the dispenser should be guided through the dispenser design to the dispensing opening. In addition, LED lights, placed in the dispenser housing, are also intended to provide optical user guidance. At the time of removal, the LED lights should draw attention to the removal opening via a dynamic flashing light.

Design Evaluation. After revealing the records of the expert evaluation, three of the sixteen shown design concepts reached high consent, based on the product's appeal (Fig. 3). Within the semantical differential the perception of these concepts can be characterized as *innovative*, *modern* and *non-stigmatizing*. At last many primary demands, which were extrapolated in the user-surveys, can be coped within these design concepts. Figure 3 shows the three best-rated concepts. It is evident that the handling of concept E is worse in comparison to the concepts I and J, due to the missing display angle. Regarding the ergonomic aspects this angle is necessary for an interaction with a touch-interface.

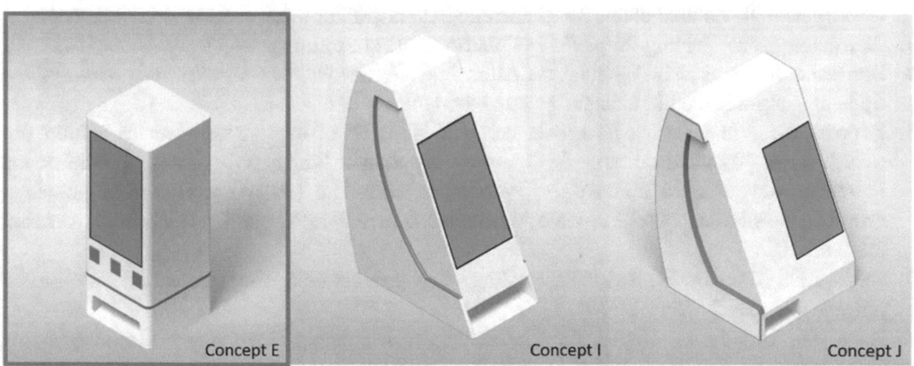

Fig. 3. Overview of the best-rated design concepts, with highlighted best-rated concept

4 Conclusion and Outlook

Conclusion. In the context of the determined survey analysis, it became apparent that the existing dispensers represent all gap-filling products for special user groups with corresponding advantages and disadvantages.

A holistic solution for a pill dispensing system could not be identified among state-of-the-art dispensers. Therefore, the shown functionality of the here planned system constitutes an innovative approach to support elder persons with their pill management and minimize mistaking or non-taking vitally important medicine.

Outlook. The primary evaluation of the design concepts by the experts shows how the approach can be realized in terms of the analyzed requirements on a pill dispensing system. Nevertheless, an evaluation with specific users must be undertaken to validate the concepts regarding the direct user input.

With this new set of user input the concepts are going to be rated and the most promising concept is going to be elaborated. In this phase the specific functional realization is going to be design, like the electronics and mechanics. At last the industrial design is going to be finished as well. The final concept is going to be built as prototype and demonstrator.

Acknowledgments. The presented research is funded by Zentrales Innovationsprogramm Mittelstand (ZIM). The goal of the 2 years research project is to develop an age-appropriate pill management system by combining hardware and software system components, which make it possible to considerably improve the medication management of multimorbid people.

References

1. Biermann, H., Weißmantel, H.: Regelkatalog SENSI -Geräte – Bedienungsfreundlich und barrierefrei durch das richtige Design. Technische Universität Darmstadt, Institut für Elektromechanische Konstruktionen, Regelkatalog (2003)
2. Schneider, N.F.: Zahlen und Fakten. https://www.demografie-portal.de/SharedDocs/Informieren/DE/ZahlenFakten/Bevoelkerung_Altersstruktur.html. Accessed 03 Jan 2019
3. Pharmazeutische Zeitung. https://www.pharmazeutische-zeitung.de/index.php?id=4148
4. Baumann, F., Funk, M.: Wohnen im Alter. https://www.wohnen-im-alter.de/zuhause/pflege-tipps/umgang-mit-medikamenten. Accessed 04 Jan 2019
5. Pöpperl, T.: Wie viele Medikamente nehmen Sie ein?: 60-Jährige nehmen im Schnitt fünf verschiedene Arzneimittel ein - das kann zu Problemen führen. Wie Ärzte, Apotheker und Patienten gegensteuern können. In: Apotheken Umschau (2016). https://www.apotheken-umschau.de/Medikamente/Wie-viele-Medikamente-nehmen-Sie-ein-526169.html. Accessed 22 Jan 2019
6. Reid, T.N., MacDonald, E.F., Du, P.: Impact of product design representation on customer judgment. J. Mech. Des. **135**(9), 1–12 (2013)
7. Raab, G., Unger, A., Unger, F.: Methoden der Marketing-Forschung - Grundlagen und Praxisbeispiele. Springer Gabler, Wiesbaden (2018)

The Importance of ICT and Wearable Devices in Monitoring the Health Status of Coronary Patients

Pedro Sobreiro[1(✉)] and Abílio Oliveira[2]

[1] Instituto Universitário de Lisboa (ISCTE-IUL), Lisbon, Portugal
pgsaf@iscte-iul.pt

[2] Instituto Universitário de Lisboa (ISCTE-IUL), ISTAR-IUL, Lisbon, Portugal
abilio.oliveira@iscte-iul.pt

Abstract. Cardiovascular diseases remain the leading cause of death in the world, despite its avoidable nature, as in the case of coronary artery disease. The healthcare technology market trends make access to information and communication technologies (ICT), including wearable devices, increasingly available to the general population. Understanding the perspectives of healthcare professionals in the use of these technologies in clinical settings for surveillance and promotion of the health status of patients with coronary diseases may help to bring technological advances closer to the expectations of clinical practice. After two sessions of focus groups, we present the results obtained from the textual analysis of the discussion between Nurses, Cardiac Physiologists, Physiotherapists, and Physicians. The findings of this study are important in the domain of adoption and acceptance of technology in health care and contribute (with a set of items) to the development of a questionnaire, to be used in a subsequent study.

Keywords: Health status · Coronary patients · Information and communication technologies (ICT) · Wearable devices · Adoption and acceptance of technology

1 Introduction

The World Health Organization identifies the cardiovascular disease group as the leading cause of death in the world [1]. In this context, there is a higher incidence of deaths due to coronary artery disease and cerebrovascular accidents [2]. Most of these deaths are preventable by lifestyle modification and by reducing risk factors such as smoking, diabetes, obesity, and physical inactivity, as well as promoting adherence to therapies involving patients in this process [3]. Considering the full adoption of preventive measures, it is estimated that about 80% of cardiovascular diseases can be avoided by eliminating health risk behaviors [4]. In this sense, exercise-based cardiac rehabilitation, along with secondary prevention programs involving education on risk factors, psychological support, and medication, prove to delay or even reverse the progression of cardiovascular disease [5]. However, frequent early withdrawal, highlights the needs to re-think innovative ways aligned with patients' preferences in order

T. Ahram et al. (Eds.): IHSED 2019, AISC 1026, pp. 705–711, 2020.
https://doi.org/10.1007/978-3-030-27928-8_107

to improve their acceptance and adherence to these programs [5]. The rapid growth of the market for health technologies and wearable devices, accessible both to patients and healthcare professionals, still raises several doubts about the real application of these technologies and devices in a clinical context, expected use by healthcare professionals, benefits for clinical practice, the security of data collected and privacy of users. Understanding healthcare professionals' perspectives on the use of ICT, including wearable devices, is therefore of particular importance, in order to bring the technological developments that have been achieved in this area to the needs, expectations and intentions of healthcare providers, to promote and monitor the health status of people with coronary artery disease. In this paper, we present the most relevant results obtained from the discussion, in the course of two focus groups, between nineteen healthcare professionals (Nurses, Physiotherapists, Physicians, and Cardiac Physiologists), according to the content textual analysis that were made, through Leximancer software, from their answers or quotations.

2 Literature Review

Digital health technologies represent a great potential for patient-centered action in promoting and changing lifestyles and in boosting adherence to therapies [3]. In health, ICT provides resources that help populations to obtain accurate information about cardiovascular diseases, enabling access to preventive services, as well as global benefits for the individual at the level of their self-care [4].

Features such as connectivity anywhere at any time allow health services to overcome geographical, temporal and organizational barriers, helping to solve emerging problems in health services, such as the growing number of chronic diseases, high costs of national health services and the need to empower patients and families to care for and promote their health [6]. However, obstacles to the integration of e-Health and telemedicine remain in daily clinical practice. In addition, there is an aggravating fact in the development of these technologies, still largely driven by a technical level, rather than by the needs and expectations of the users for whom the technology is intended [5]. This lack of incidence is clinically based on factors such as the absence of a clear structure for these new technologies and the reluctance of the professionals to adopt new forms of work, creating barriers or delays to their implementation [7].

Advances in mobile technology, have already increased the potential of its use in health [8]. In addition, mobile health applications in some cases show improvements in the efficacy of health services. However, abundance of different applications makes it difficult for users to choose, resulting a gap at the level of a comprehensive literature framework to help manage and evaluate mobile health applications [9].

Highlighting in the field of Internet of Things (IoT), wearables can have multiple applications, equipped with sensors and processors, allowing connectivity to Internet and connection between different operating systems [10]. In health, wearables become devices capable of monitoring physiological parameters, with the possibility of continuous measurements [11]. This ability to collect physiological data enables both the user and healthcare professionals to be better informed about the effects of the patients' own actions or treatments, the evolution of clinical status and decision support [12].

The research and development of health wearables can be characterized through technological antecedents, such as utility, functionality, compatibility, quality and cost, but also through research, using models of technological acceptance, with a view to the behavioral intention of consumers [10]. Though, the absence of clinical applications and the validation of measurements constitute barriers to the wide use in the clinical context of wearable technologies [12]. There's also increasing concerns about safety and health risks related to the use of these devices, privacy of the data collected and reliability of health professionals in the physiological estimates obtained [10].

3 Methodology

To evaluate the perceptions about the use of ICT and wearable devices in coronary patients, first, we explore the importance and uses given by healthcare professionals to ICT both in the context of their personal lives and their workplace.

We also intend to: verify the importance attributed by these professionals to the ICT in the provision of health care; determine their perceptions about the wearable devices, and apps, in the surveillance of the health status of coronary patients; verify what they consider to be the most relevant technological characteristics in a program to promote and monitor the health status of people with coronary disease.

A total of 19 participants (N = 19), nine women (47.4%) and ten men (57.6%) working daily with coronary patients, with eleven Nurses (57,9%), two Physiotherapists (10.5%), three Physicians (15.8%) and three Cardiac Physiologists (15.8%). These sessions involved the discussion of four major themes: (1) the importance and uses given by healthcare professionals to ICT both in the context of their personal lives and their workplace; (2) perspectives for the future use of ICT and coronary disease; (3) perceptions about the wearable devices, and apps, in the surveillance of the health status of coronary patients; (4) the most relevant technological characteristics in a program to promote and monitor the health status of people with coronary disease. Once the informed consent of each participant was obtained, all the interventions in the group discussions were recorded. Subsequently, the individual responses in each theme were analyzed through Leximancer, a text analysis software that provides a means to quantify and display participants' perceptions in a structured way through a conceptual map, representing the main themes and concepts (according to the most frequent words, sentences, sentiments and expressions, and the proximity between these terms found) and how they relate [13].

4 Results and Discussion

The first objective of this study was to explore the importance and uses given by healthcare professionals to ICT, both in the context of their personal lives and in their workplace. In the focus group sessions, the role of ICTs in their personal lives was consensual among all participants, assuming to be essential or of growing importance. Similar results were obtained when the relationship between ICT and daily practice was questioned, with a direct link between concepts as "important" and "essential". All participants reported daily use of ICT in their personal lives, with this contact being essentially attributed to the use of the smartphone that emerged as the main device for access to ICT.

The concept map generated (see Fig. 1) from the analysis of the text based on the answers obtained on the two focus group sessions. Related to the second objective of this study, this concept map reflects the future relationship and opportunities that healthcare professionals consider existing in the use of ICT as a means of monitoring and promoting the health status of coronary patients.

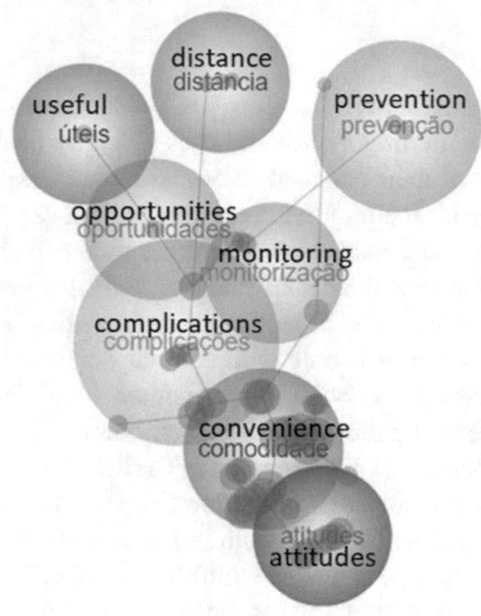

Fig. 1. Relationship between the possibility of future use of ICT and the coronary patient

We verify that the great majority of the generated terms are encompassed in the concepts of "convenience" and "attitudes". In fact, these two concepts are interconnected by the relationship identified in several responses that attribute to the use of ICT in the context of coronary disease, a potential capable not only of conferring on the patient greater convenience capable of changing attitudes in the management of their own disease, but also to healthcare professionals, who identify for themselves opportunities in the use of these technologies, with a potential that may bring more convenience to follow up these patients and consequently alter their own attitudes towards the use of these health resources.

Considering the connection between the concepts "convenience", "monitoring" and "prevention", we observe several references by the healthcare professionals to an opportunity for the ICTs to give greater convenience to themselves and to the coronary patients, giving greater knowledge about the health status of the person, through the use of devices capable of evaluating vital signs, enabling the monitoring of the health status of these patients, making the use of these ICTs a factor of prevention. The concept "prevention" gains special emphasis by noting the link in the border area between the concept's "complications" and "opportunities". From the analysis of the connection between these three concepts, we highlight that ICT has the potential to play an important role in the future in the prevention of complications, perceiving, through the point of intermediate connection between these three concepts, that to occur a real opportunity in the prevention of complications will be worth adding the monitoring of the health status of these patients, benefiting as previously said the use of devices capable of collecting vital signs.

Graphically, this can be confirmed by following the link between the point in the border zone that links "complications" and "opportunities" and the "useful" concept, thus demonstrating the usefulness perceived by healthcare professionals about the use of ICTs as an opportunity to prevent complications in the health status of the patients, using this for their monitoring, thereby ensuring greater comfort and contributing to adjust the patient's attitudes towards better management of their coronary disease. In the same sense, the concept of "distance" is graphically connected to the same point previously mentioned in the border area between "complications" and "opportunities", which allows, for example, to affirm the importance and potential of the use of ICTs in coronary disease, as a useful way of preventing complications related to distance between patient and healthcare providers with the use of monitoring technologies.

One of the objectives of this study was also to determine perceptions about the use of wearable devices and mobile applications in monitoring the health status of coronary patients. From the analysis of results, it was possible to understand that the majority of healthcare professionals rely on the evaluation's feed of vital signs collected by wearable devices or those based on manual insertion by the patient in a mobile application, nevertheless, possible problems were mentioned in the manual insertion of data. This tendency was evidenced by the clear preference of almost all the participants in the use of technologies that allow the automatic collection of health data. It was also verified that in parallel to the automatic collection it is important to allow in some way the involvement and interaction of the patient.

It was also objective for this study to verify the most relevant technological characteristics in a program to promote and monitor the health status of people with coronary disease. It was possible to verify that issues such as the security and privacy of data collected by mobile applications and wearables are important, raising many questions among healthcare professionals. Confidence in the security and privacy of the data collected was affirmed by half of the healthcare professionals. Among the others a clear denial of confidence or a neutral or unreliable feeling was perceived. An important finding of the technological characteristics for a coronary patient monitoring program was the evaluation of vital signs or assessments of information considered most important to perform: "heart rate", "blood pressure", "pain" and "heart stroke". Concepts associated with the monitoring of physical activity, such as calories, assessment of adherence to medication and the possibility of communicating symptoms by the patient were also mentioned.

5 Conclusions

ICTs are perceived as valuable very important for healthcare professionals, both in their personal life and work practice. ICTs are considered valuable vehicles for monitoring and promoting the health status of people with coronary disease. Healthcare professionals perceive the utility and opportunities in using these technologies not only to prevent complications in the health status of these patients, but also as way to offer comfort, and even help change attitudes towards better self-management of the disease. Wearable monitoring may be a very important aspect in a coronary patient surveillance program, given the preference of healthcare professionals for assessing vital signs automatically in deterioration of the user's manual insertion of values into mobile applications. Technological developments in this area should be very clear in terms of safety and privacy of health data collected, to raise no doubts in this domain.

A coronary patient health status surveillance and monitoring program should collect at least values for heart rate, blood pressure, cardiac rhythm, and perceived patient pain. So, a positive effort may be continuously made, by all health professionals, to conjugate the medical assistance with the technological facilities, for benefit of people, in this digital era.

References

1. Mendis, S., Weltgesundheitsorganisation, World Heart Federation (eds.): Global Atlas on Cardiovascular Disease Prevention and Control. World Health Organization, Geneva (2011)
2. OECD: Health at a Glance 2017. OECD Publishing, Paris (2017). https://doi.org/10.1787/health_glance-2017-en
3. Khan, N., Marvel, F.A., Wang, J., Martin, S.S.: Digital Health Technologies to Promote Lifestyle Change and Adherence. Current Treatment Options in Cardiovascular Medicine, vol. 19 (2017). https://doi.org/10.1007/s11936-017-0560-4
4. del Hoyo-Barbolla, E., Arredondo, M.T., Ortega-Portillo, M., Fernandez, N., Villalba-Mora, E.: A new approach to model the adoption of e-health. Presented at the (2006). https://doi.org/10.1109/MELCON.2006.1653319
5. Buys, R., Claes, J., Walsh, D., Cornelis, N., Moran, K., Budts, W., Woods, C., Cornelissen, V.A.: Cardiac patients show high interest in technology enabled cardiovascular rehabilitation. BMC Med. Inform. Decis. Making **16** (2016). https://doi.org/10.1186/s12911-016-0329-9
6. Silva, B.M.C., Rodrigues, J.J.P.C., de la Torre Díez, I., López-Coronado, M., Saleem, K.: Mobile-health: a review of current state in 2015. J. Biomed. Inform. **56**, 265–272 (2015). https://doi.org/10.1016/j.jbi.2015.06.003
7. Gund, A., Lindecrantz, K., Schaufelberger, M., Patel, H., Sjöqvist, B.A.: Attitudes among healthcare professionals towards ICT and home follow-up in chronic heart failure care. BMC Med. Inform. Decis. Making. **12** (2012). https://doi.org/10.1186/1472-6947-12-138
8. Olla, P., Shimskey, C.: mHealth taxonomy: a literature survey of mobile health applications. Health Technol. **4**, 299–308 (2015). https://doi.org/10.1007/s12553-014-0093-8
9. Sadegh, S.S., Khakshour Saadat, P., Sepehri, M.M., Assadi, V.: A framework for m-health service development and success evaluation. Int. J. Med. Inform. **112**, 123–130 (2018). https://doi.org/10.1016/j.ijmedinf.2018.01.003

10. Marakhimov, A., Joo, J.: Consumer adaptation and infusion of wearable devices for healthcare. Comput. Hum. Behav. **76**, 135–148 (2017). https://doi.org/10.1016/j.chb.2017.07.016
11. Gatzoulis, L., Iakovidis, I.: Wearable and portable eHealth systems. IEEE Eng. Med. Biol. Mag. **26**, 51–56 (2007). https://doi.org/10.1109/EMB.2007.901787
12. Pevnick, J.M., Birkeland, K., Zimmer, R., Elad, Y., Kedan, I.: Wearable technology for cardiology: an update and framework for the future. Trends Cardiovasc. Med. **28**, 144–150 (2018). https://doi.org/10.1016/j.tcm.2017.08.003
13. Leximancer Pty Ltd.: Leximancer User Guide. Release 4.5, p. 141 (2016)

Improvement of a Monitoring System for Preventing Elderly Fall Down from a Bed

Hironobu Satoh[1(✉)] and Kyoko Shibata[2]

[1] National Institute of Information and Communications Technology, Nukuikita 4-2-1, Koganei, Tokyo, Japan
satoh.hironobu@nict.go.jp

[2] Kochi University of Technology, Miyanokuchi 185, Kami, Kochi, Japan
shibata.kyoko@kochi-tech.ac.jp

Abstract. Elderly sometime falls down from the bed. And, elderly's thighbone is broken. This accident makes it that it is decline that the quality of life of elderly. Therefore, to solve this problem, we proposed monitoring system. The proposed monitoring system is not able to adapt individual differences. To solve this problem, we proposed a new learning method. From the results of the previous researches, the new learning method is adapted to the proposed monitoring system. And ability of the proposed monitoring system is increase. From the experimental result, when the initial learning is completed, detection rate of the dangerous behavior is 79.8% (399/500) and detection rate of the safe behavior is 82.4% (412/500). After proposed learning method is executed, detection rate of the dangerous behavior is 84.0% (420/500) and detection rate of the safe behavior is 91.0% (455/500) From the experimental results, it is concluded that the predicts rate increase.

Keywords: Deep learning · Human behavior · Fall down from a bed

1 Introduction

Elderly sometime falls down from the bed. And, elderly's thighbone is broken. This accident makes it that it is decline that the quality of life of elderly. Therefore, to solve this problem, we proposed monitoring system [1–7].

The purpose of this study is avoiding the accidents, that elderly falls down from the bed. We proposed monitoring system, which predicts that elderly will fall down from the bed. It is needed that the proposed monitoring system predicts dangerous actions of older people, and the proposed monitoring system monitors a target person while 24 h 365 days. Naturally, it is needed that the proposed monitoring system able to predicts the human behaviors in the dark rooms. The proposed system consists of Microsoft Kinect and personal computer. Kinect measures the target body. And, the measured data is taken into a computer using Kinect. Deep learning predicts that target person falls down from the bed.

However, the proposed monitoring system is not able to adapt individual differences. To solve this problem, we proposed a new learning method. From the results of

T. Ahram et al. (Eds.): IHSED 2019, AISC 1026, pp. 712–717, 2020.
https://doi.org/10.1007/978-3-030-27928-8_108

the previous researches, the new learning method is adapted to the proposed monitoring system. In this paper, the proposed learning method is verified.

2 Monitoring System

In this section, target person's behavior, hardware and software of the proposed monitoring system are indicated.

2.1 Human Behavior

(a) Pattern A (b) Pattern B

Fig. 1. Safe behavior

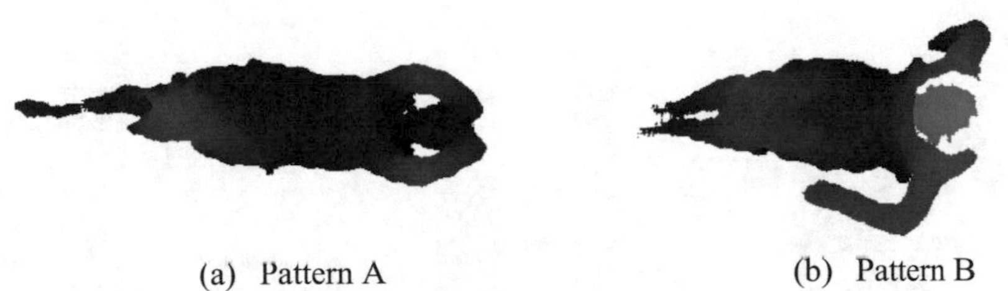

(a) Pattern A (b) Pattern B

Fig. 2. Dangerous behavior

Human behaviors are defined a two category. One is a safe behavior. Another is a dangerous behavior. Figure 1 shows examples of safe behaviors. Figure 2 shows examples of dangerous behavior.

The behavior leading to the fall down from a bed is defined as a dangerous behavior. The behavior that does not lead to the fall down from a bed is defined as a safe behavior. An example of the dangerous behavior is that the subject knees on the bed, sits on the bed etc. Those behaviors are told to be behavior leading to the fall down from the bed by nursing experts.

An example of the safe behavior is that the subject laying on a bed.

The proposed monitoring system predicts the dangerous behavior of the subject.

2.2 Hardware

In previous study, we proposed the monitoring system using Kinect. The monitoring system constructed of a PC (Personal Computer) and Kinect.

Kinect is developed by Microsoft for game. Kinect is able to measure a shape of human body while a day. It is needed that the monitoring system is able to predict human behavior. Kinect is able to measure the human body shape on the dark site.

The monitoring system has a PC. PC receives the measured data of subject's human body shape from Kinect. PC calculates predicting human behavior. When the calculation result is the dangerous behavior, PC warn care givers of that the subject is falling down from a bed.

2.3 Software

The monitoring system has a software. The software calculates prediction of dangerous behavior of the target person with DBN (Deep Belief Network) [8]. It is known that DBN has a high ability of the signal recognition. DBN is one of the Deep Learning algorithm.

The prediction algorithm is indicated as follows. First, the human body shape is measured by Kinect. The bed area is extracted of the measured data. The extracted position is established by the user. Next, the subject body is extracted of extracted data using the threshold which is height of the bed surface. The threshold is defined by user (Fig. 3).

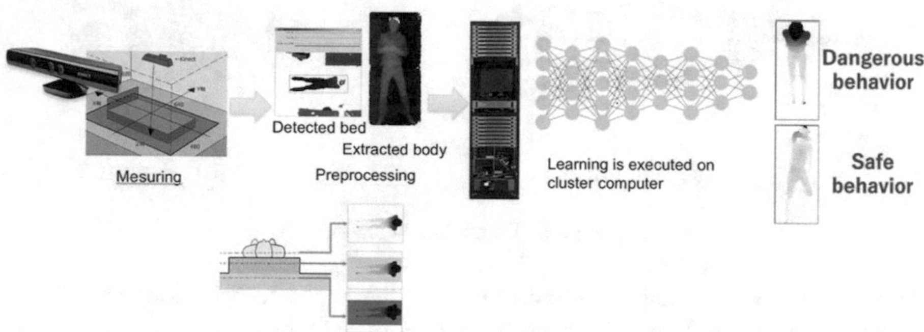

Fig. 3. Installation of Kinect and predicting flow

3 Proposed a Learning Method

The proposed system adapts predicting method for individual difference of behaviors. A new learning method is as follows. First, initial learning is executed in order to learn variation of physique and basic behaviors. Second, by Kinect target's distinctive behaviors are measured. And distinctive behaviors are collected by user. Therefore, the collected data include distinctive behaviors of user for additional learning of DBN.

Next, on first learning weight, a learning is executed based. A learning data include learned data of initial learning and a collected data by user. Initial learning data and collected data are mixed when the second time learning executed. The proposed system carries out a second time learning.

4 Experimental

On the proposed monitoring system, an experiment was conducted to verify the effectiveness of the proposed learning method. The prediction rate in the initial learning was compared with the prediction rate in the proposed learning method. 10 subjects behavior are measured by Kinect. There are 10 subjects on a bed. 10 subject's behavior are unique.

Table 1. The prediction rate of a previous method

Subject	Dangerous behavior [%]	Safe behavior [%]	Total [%]
a	84.0 (42/50)	90.0 (45/50)	87.0 (87/100)
b	88.0 (44/50)	92.0 (46/50)	90.0 (90/100)
c	78.0 (39/50)	80.0 (40/50)	79.0 (79/100)
d	86.0 (43/50)	86.0 (43/50)	86.0 (86/100)
e	90.0 (45/50)	88.0 (44/50)	89.0 (89/100)
f	80.0 (40/50)	86.0 (43/50)	83.0 (83/100)
g	76.0 (38/50)	80.0 (40/50)	78.0 (78/100)
h	36.0 (50/50)	78.0 (39/50)	75.0 (75/100)
i	40.0 (50/50)	76.0 (38/50)	78.0 (78/100)
j	32.0 (50/50)	68.0 (34/50)	66.0 (66/100)
total	79.8 (399/500)	82.4 (412/500)	81.1 (811/1000)

Table 2. The prediction rate of a proposed new method

Subject	Dangerous behavior [%]	Safe behavior [%]	Total [%]
a	88.0 (44/50)	94.0 (47/50)	91.0 (91/100)
b	86.0 (43/50)	90.0 (45/50)	88.0 (88/100)
c	84.0 (42/50)	92.0 (46/50)	88.0 (88/100)
d	86.0 (43/50)	86.0 (43/50)	86.0 (86/100)
e	92.0 (46/50)	96.0 (48/50)	94.0 (94/100)
f	84.0 (42/50)	94.0 (47/50)	89.0 (89/100)
g	78.0 (39/50)	88.0 (44/50)	83.0 (83/100)
h	80.0 (40/50)	90.0 (45/50)	85.0 (85/100)
i	82.0 (41/50)	92.0 (46/50)	84.0 (84/100)
j	80.0 (40/50)	88.0 (44/50)	87.0 (87/100)
total	84.0 (420/500)	91.0 (455/500)	87.5 (875/1000)

To verify the effectiveness of the proposed learning method, an experiment was carried out. The experimental results are shown as follows. When the initial learning is completed, detection rate of the dangerous behavior is 79.8% (399/500) and detection rate of the safe behavior is 82.4% (412/500). After proposed learning method is executed, detection rate of the dangerous behavior is 84.0% (420/500) and detection rate of the safe behavior is 91.0% (455/500) From the experimental results, it is concluded that the predicts rate increase (Tables 1 and 2).

5 Conclusion

In the home and the hospital, elderly fall down from a bed sometime. Feld elderly breaks the femur. This accidents cause declining the quality of life of an injury. Therefore, we develop a monitoring system. The proposed monitoring system avoids falling down from a bed utilize Deep Belief Network.

However, the proposed monitoring system has a problem. The previous system is not able to adapt individual differences behavior. To solve this problem, we proposed a new learning method to adapt the proposed system for individual behaviors.

In this paper, to verify the effectiveness of the proposed learning method, an experiment was conducted. From the experimental results, the proposed learning method has the ability of adapting the proposed system to the individual difference of a behavior.

In the future, apply proposed new learning method to the proposed monitoring system.¥

References

1. Yamanaka, N., Satoh, H., Shiraishi, Y., Matsubara, T., Takeda, F.: Proposal of the awakening detection system using neural network and it's verification. In: The 52nd Annual Conference of the institute of Systems, Control and information Engineers (2008)
2. Satoh, H., Takeda, F., Shiraishi, Y., Ikeda, R.: Development of a awaking behavior detection system using a neural network. IEEJ Trans. EIS **128**(11), 1649–1656 (2008)
3. Ikeda, R., Satoh, H., Takeda, F.: Development of awaking behavior detection system nursing inside the house. In: International Conference on Intelligent Technology 2006, pp. 65–70 (2006)
4. Matubara, T., Satoh, H., Takeda, F.: Proposal of an awaking detection system adopting neural network in hospital use. In: World Automation Congress 2008 (2008)
5. Satoh, H., Shibata, K.: Development of human behavior recognition for avoiding fall down from a bed by deep learning. In: International Conference on Brain Informatics & Health (2017)
6. Satoh, H., Takeda, F.: Verification of the effectiveness of the online tuning system for unknown person in the awaking behavior detection system. In: IWANN 2009, Proceedings of the 10th International Work-Conference on Artificial Neural Net- works: Part II: Distributed Computing, Artificial Intelligence, Bioinformatics, Soft Computing, and Ambient Assisted Living, pp. 272–279 (2009)

7. Satoh, H., Shibata, K., Masaki, S.: Development of an awaking behavior detection system with kinect. In: HCI International 2014 - Poster's Extended Abstracts, Proceedings, Part II, pp. 496–500, pp. 272–279 (2014)
8. Yoshua, B., Pascal, L., Dan, P., Hugo, L.: Greedy layer-wise training of deep networks. Adv. Neural Inf. Process. Syst. **19**, 153–160 (2006)

Ergonomic Design Process of the Shape of a Diagnostic Ultrasound Probe

Ramona De Luca[1(✉)], Leonardo Forzoni[1], Fabrizio Spezia[1], Fabio Rezzonico[2], Carlo Emilio Standoli[3], and Giuseppe Andreoni[3]

[1] Esaote SpA, Via di Caciolle 15, 50127 Florence, Italy
ramona.deluca@esaote.com
[2] Rezzonico Design, Via A. Diaz 14, 22100 Como, Italy
[3] Politecnico di Milano, Dipartimento di Design, Via Giovanni Durando 38/A, 20158 Milan, Italy

Abstract. This study presents a design process for the ergonomic design of a hockey stick ultrasound probe, based on clinical applications needs, competitors benchmarking and analysis of feedback about a commercially available probe. The design is performed in two phases: (1) design of the product and initial evaluation of subjective satisfaction of expert users; (2) the subsequent review of the project and the selection of the preferred probe grip design. This approach involves several successive stages allowing for progressively better results in terms of acceptability and comfort of the new product.

Keywords: Ultrasound probe · Design process · Subjective satisfaction

1 Introduction

Ultrasound (US) imaging is the most widely used diagnostic imaging modality worldwide as it is non-invasive, sufficiently harmless, portable, compact, low-cost and real time and it enables a variety of clinical applications. Due to the nature of US examinations, sonographers are prone to musculoskeletal disorders (MSDs): almost 90% of clinical sonographers experience painful disorders affecting muscles, nerves, ligaments and tendons [1]. The most frequent causes of MSDs in sonography include two core domains: the workspace environment and the equipment design. US systems consist of two main parts which are the console (including the monitor and the panel with controls, keyboard, and often a touch screen) and the probe [2]. Ergonomic designed probes aid to improve the use of proper force, sustained postures, ease of manipulation and contact pressure, which contribute to prevent MSDs of the hand, wrist and arm [3, 4]. The present paper aims to describe the study conducted for the ergonomic design and evaluation of an US probe to reduce the postural and muscular loads and improve subjective satisfaction of the operator [5, 6]. In designing a new handle of an US probe, we can identify several stakeholders (e.g., product marketing managers, engineers, sonographers, competition) that deal with different requirements which are strongly interdependent: aesthetics, function, and technology. The design path starts from the analysis of the needs (such as specific intended uses and diverse grasps of the probe) and the consequent definition of the product to provide a solution

T. Ahram et al. (Eds.): IHSED 2019, AISC 1026, pp. 718–723, 2020.
https://doi.org/10.1007/978-3-030-27928-8_109

appropriate to fit these needs [7]. In this preliminary stage, the competition overview and analysis and interviews of experienced sonographers and physicians regarding suggestions for improvements of the existing design, are crucial. Based on these data, the ergonomic design is developed as a trade-off between different grasps and users (due to diverse clinical applications) and technological specifications. The assessment of the new product is performed in two phases: (1) initial evaluation of handling during simulated US examinations with one mock-up and commercially available probes; (2) subsequent project review and second evaluation of subjective satisfaction. This process leads to progressively better results for obtaining a preferred probe design in terms of higher comfort and acceptance and therefore increased performance. As an example, we show the ergonomic design study of a hockey stick probe. It is a high frequency linear array, with an asymmetric thin handle and an angled transducer tip enabling to reach small acoustic windows and to be used in surgery. Based on the analysis of a commercially available probe and the intended uses of such a versatile probe, a first proposal was made and evaluated using a 3D-printed mock-up connected to a cable to mimic US examinations. The results of the evaluation by experienced subjects revealed that the newly developed handle was easier to be used, avoiding local contact pressure and with a good friction. Nevertheless, a project review was needed to maximize handle balance and good adherence. The evaluation of the new mock-ups had the outcome of the preferred probe grip design in terms of user acceptance and comfort.

2 From the Needs Roadmap to the Product Roadmap

The needs roadmap represents the understanding and the plan of the needs that the company is willing/enabled to solve/satisfy. It aids the product roadmap, that aims to define both a single product development and full portfolio over time. In this framework, the focus of the company lays on the analysis of the needs first and the definition of the product afterwards: the starting point is the request (i.e. the need) and not the proposal (i.e. the product). The definition of the needs roadmap and its continuous update are typically achieved with interviews to customers and non-customers, key opinion leaders, market and competition analysis, technical and clinical state of the art and tendencies analysis, as well as interviews to sales, marketing, application specialists. Specific additional topics/inputs of the needs roadmap are: inputs from technical (R&D) and clinical scenarios and inputs from application guidelines/standards/etc. (if any). The outcome is the needs, trends and tendencies roadmap, that is developed in 1-3-5 years according to priorities coming from relevancy and importance of inputs. In the specific case of a diagnostic US imaging company, it provides a worldwide scenario summary of the clinical situation and trends (for instance, the most commercially interesting applications) as well as a current picture of the company in different countries and a proposed path for the company to grow in terms of sales and branding per clinical application. The first step of the product roadmap is the analysis of the needs roadmap outputs, in order to understand the gaps to fulfill and define the project development to provide a product solution. Together with other stakeholders (such as R&D teams, quality and regulatory functions, service), resources, contents and timing schedule are planned [8, 9].

As an example, we show the case of an US probe for evaluating small superficial structures, that allows to improve maneuverability to reach reduced spaces acoustic windows (such as neonatal and pediatrics [10]) and to be used in surgery or difficult to be reached patient superficial areas (for instance, the supraclavicular approach of the subclavian vein [11]). This probe type consists of a very high frequency linear array, with a thin handle and an angled transducer tip (L-shape), namely a hockey stick probe. This probe covers multiple applications for diagnostic and interventional musculoskeletal, rheumatology and dermatology, that are areas in which US imaging has become more and more attractive in the last decade and is constantly used in the clinical routine.

3 Preliminary Design

Based on the request of such probe, a new probe handle was designed (Fig. 1) as a result of the benchmark analysis, the internal evaluation of the previous solutions of the company and interviews of professionals on the field. In order to provide ease of manipulation and avoid discomfort, the length of the handle and the thickness of the middle and frontal grips were properly designed to improve the grip fit. The length of the handle may generate peak pressure on the hand while the handle circumference may cause the lack of contact area, and therefore a high hand grip force supply for stable grip is needed. Furthermore, a short aperture 3 cm on average is preferred when dynamic analysis of the hand is performed. A mock-up of the newly-developed probe was printed and connected to a cable to simulate the realistic activity (Fig. 1) even though the mock-up was lighter than a complete transducer.

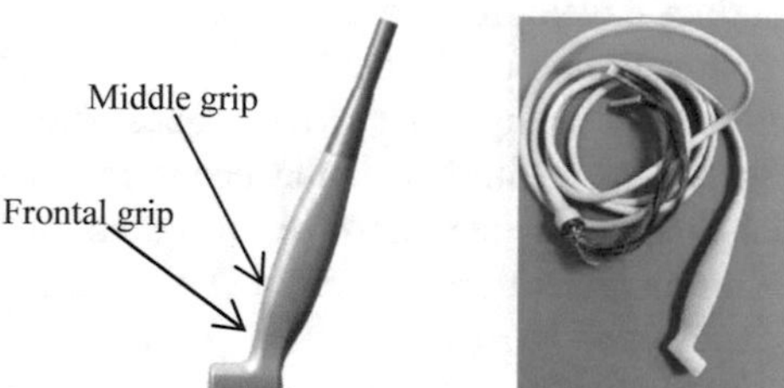

Fig. 1. Proposal 1 (left) and 3D-printed mock-up of the proposal, connected to the cable (right).

Eight subjects (u1, …, u8) were asked to simulate US examinations to compare and evaluate the newly-developed and existing handles. Since this probe type is dedicated to various clinical applications that require different grasps, sonographers/physicians with the following expertise were interviewed: 2 pediatric radiologists, 1 endocrinologist, 4 radiologists, 1 neonatal radiologist. Four subjects have no experience with this probe type, whereas three subjects constantly use the commercially available probe. A questionnaire was made containing questions relating to handle comfort and performance (Table 1) [12]. The subjects rated the handle comfort descriptors on a scale containing 5 discrete levels (from 1 = totally disagree to 5 = totally agree). The selected questions aimed to assess if the new design provides a high quality and easy to use handle. Question #1 "Is comfortable" referred to the overall subjective comfort. Question #2 "Fits the hand" aimed to understand if the shape and the size of the handle are suitable to the hand. Question #3 "Offers a nice grip feeling" referred to its overall grip comfort, whereas and question #4 "Needs low hand grip force supply for stable grip" aimed to check if the stability of the probe is due to the handle shape rather than the friction between the handle and the hand. In other words, subjects had to refer to the form of the handle and not to the material friction between the handle and the hand. Question #5 "Does not cause peak pressure on the hand" allowed to assess high contact pressure points.

Table 1. Subjective comfort rating questionnaire

		Totally disagree	Disagree somewhat	Neutral	Agree somewhat	Totally agree
1	Is comfortable	1	2	3	4	5
2	Fits the hand	1	2	3	4	5
3	Offers a nice grip feeling	1	2	3	4	5
4	Needs low hand grip force supply for stable grip	1	2	3	4	5
5	Does not cause peak pressure on the hand	1	2	3	4	5

Based on the obtained results from the subjective comfort ratings questionnaire, we can conclude that the newly developed handle has higher ratings for the overall handle subjective comfort indicators than the existing probe. It provides better fit, is easier in use, avoids local contact pressure and has a good friction due to the increase in the contact area. In particular, Fig. 2 shows that the new design exhibits comfort and acceptability compared to the old one, that has positive attributes only for users that usually employ it on the field and appreciate its image quality (Fig. 3).

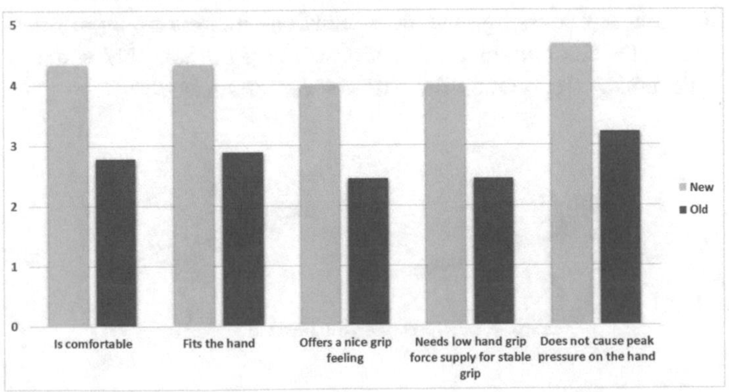

Fig. 2. Handle subjective comfort predictive ratings of the questionnaire shown in Table 1

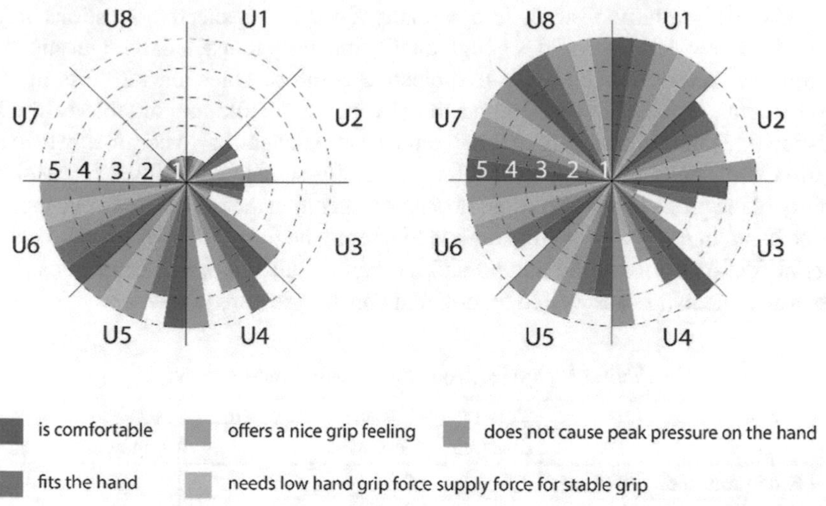

Fig. 3. Satisfaction test results: comparison of the existing (left) and newly-developed (right) handles

4 Redesign

To further improve the ergonomics of the handle, we developed other two proposals that present smooth features to maximize handle balance and good adherence. Internal and external evaluations of the proposals 2 and 3 (Fig. 4) revealed that these features are preferred to the radical shape of the proposal 1, mainly when gloves are worn. In particular, the most preferred probe grip design in terms of subjective satisfaction was the proposal 3. Nevertheless, a redesign of the handle 3 was needed to avoid sharp corners on the surface and to lower the "finger-sign" towards the frontal grip.

In addition to aesthetics (that may be related to the brand identity of the company) and function (that is related to usability and ergonomics), technology plays a big role in the design phase: small changes of the size of the probe head and the frontal grip were required to accommodate the transducer and the electrical connections.

Regarding the reprocessing, the new housing guarantees appropriate cleaning, disinfection and sterilization since there are no sharp corners, holes and protrusions, therefore liquid and gas can reach the whole external surface.

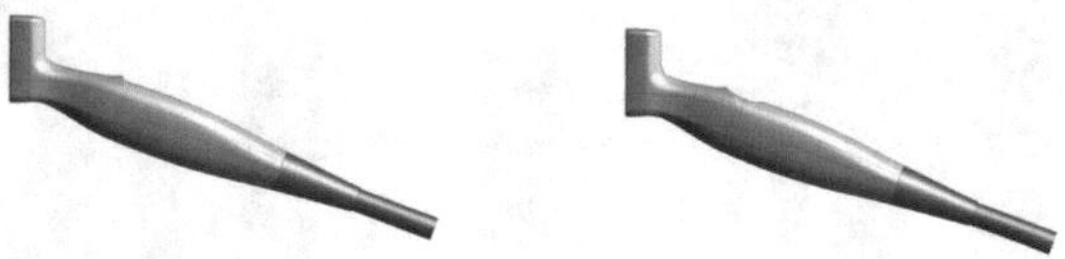

Fig. 4. Proposal 2 (left) and 3 (right)

5 Conclusion

The present study has shown a design process for the ergonomic design of a hockey stick US probe, based on clinical applications needs, competitors benchmarking and analysis of feedback about a commercially available probe. The design was performed in two phases: (1) design of the product and initial evaluation of simulated exams with one mock-up and the existing probe; (2) project review and the evaluation of new mock-ups. This approach allowed for subsequent redesigns of the probe for obtaining progressively better results. The main limit of the proposed design process is the lack of biomechanical measurements, such as motion analysis, eye tracking, superficial electromyography. However, the subjective satisfaction results effective to identify preferred features in detail.

References

1. Evans, K., Roll, S., Baker, J.: Work-related musculoskeletal disorders (WRMSD) among registered diagnostic medical sonographers and vascular technologists. J. Diagn. Med. Sonography **25**, 287–299 (2009)
2. Andreoni, G., Mazzola, M., Matteoli, S., D'Onofrio, S., Forzoni, L.: Ultrasound system typologies, user interfaces and probes design: a review. Procedia Manufact. **3**, 112–119 (2015)
3. Mazzola, M., Forzoni, L., D'Onofrio, S., Andreoni, G.: Use of digital human model for ultrasound system design: a case study to minimize the risks of musculoskeletal disorders. Int. J. Ind. Ergon. **60**, 35–46 (2017)
4. Paschoarelli, L.C., de Oliveira, A.B., Gil Coury, H.J.C.: Assessment of the ergonomic design of diagnostic ultrasound transducers through wrist movements and subjective evaluation. Int. J. Ind. Ergon. **38**, 999–1006 (2008)
5. Jung, H., Lee, N., Moon, S., Yang, X., Lee, S., Moon, J., Ha, K., Lim, J., You, H.: An ergonomic grip design process for vaginal ultra sound probe based on analyses of benchmarking, hand data, and grip posture. In: Proceedings of the 20th Congress of the International Ergonomics Association (IEA 2018), pp. 352–363 (2019)
6. Moon, S., Jung, H., Lee, S., Jeon, E., Moon, J., Lee, S., Ha, K., You, H.: An ergonomic evaluation of convex probe designs using objective and subjective measures. In: Proceedings of the 20th Congress of the International Ergonomics Association (IEA 2018), pp. 364–371 (2019)
7. UNI EN ISO 26800-2011, Ergonomics – General approach, principles and concepts, International Organization for Standardization (2011)
8. Forzoni, L., Buffagni, C., Guercini, S.: Understanding the customer needs in business network: an emerging model in the diagnostic imaging industry. In: 35th IMP Conference, IÉSEG School of Management, Paris (2019)
9. Forzoni, L., Nonino, F., Pompei, A., Buffagni, C.: Needs roadmap framework understanding and mapping needs in competitive markets. In: ISPIM Innovation Conference, Florence (2019)
10. Wakefield, R.J., D'Agostino, M.A.: Essential applications of musculoskeletal ultrasound in rheumatology. Saunders (2010)
11. Samra, T., Saini, V.: Ultrasound guided supraclavicular subclavian cannulation: a novel technique using "hockey stick" probe. J. Emergencies Trauma Shock **8**, 72 (2015)
12. Harih, G., Dolšak, B.: Comparison of subjective comfort ratings between anatomically shaped and cylindrical handles. Appl. Ergon. **45**, 943–954 (2014)

Discussion on the Effect of Bedding on Sleep Postures

Yu-Ting Lin(✉), Chien-Hsu Chen, and Fong-Gong Wu

Department of Industrial Design, National Cheng Kung University, No.1, University Road, East Dist., Tainan City 70101, Taiwan
yuting8312@gmail.com

Abstract. In all kinds of bedding, pillow is the most important tool of supporting the head during sleep, and it is also the key to get proper support for the cervical vertebra during this period. People change their sleep postures to avoid the discomfort caused by bad blood circulation derived from the prolonged compression of the nerves. It pointed out the discomfort caused by the prolonged compression of body, and pro-posed the appropriate relative positions where a pillow can properly support the body, thereby concluding the design factors of innovative bedding. This study was conducted in three directions, including user experience, expert interview, and market feasibility. This study aimed at proposing the principals of the design of pillows, so as to reduce the discom-fort caused by uneven pressure distribution in the body due to the changes of sleep postures. This result is intended to provide follow-up research on pil-low design as a reference.

Keywords: Bedding system design · Sleep postures · User experience · Expert interview · Market feasibility

1 Introduction

Most of the current studies on improving sleep quality and bedding designs were based on the measurement of back pressure. In addition to back and spinal pressure problems, pressure on cervical vertebra and sleep postures also had an impact on sleep, as well as the sleep quality. Therefore, whether the pressure on body can be evenly distributed during sleep was one of the key points that must be considered in bedding design. In consideration of nerves, blood vessels and other body parts affected by pressure during sleep, these parts should avoid continuous pressure. This is because prolonged compression at a single point can lead to paralysis and bad blood circulation, which may cause interruption of deep sleep.

In order to design innovative bedding that can improve sleep quality, this study aimed to explore whether the pressure on body parts can be evenly distributed during sleep to support shoulder and neck and to reduce the burden of them. Factors affecting sleep quality were summarized based on literature discussion and the interview. In the interviews, this study was conducted in three directions: user experience, expert interviews, and market feasibility. This study also observed and deeply understood the changes of sleep postures. Pillow was taken as the basis for the development of

T. Ahram et al. (Eds.): IHSED 2019, AISC 1026, pp. 724–729, 2020.
https://doi.org/10.1007/978-3-030-27928-8_110

bedding because pillow could support the head during sleep. Besides, this study found out the methods or strategies that can improve sleep quality. These results could be the design elements for bedding development in the future.

2 Literature Review

The main issue of this study was whether the pressure on body parts can be evenly distributed during sleep to support shoulder and neck and to reduce the burden of them. The following was the relevant literature for this issue.

2.1 Sleep Postures

In order to avoid prolonged compression of body parts, sleep was a dynamic process [1]. When people feel cold, they curl up their bodies; when they feel hot, they lift or kick the quilts. The change of sleep postures was to stretch the bodies, as well as to adapt to temperature variation [2]. These involuntary movements during sleep can be viewed as external sleep postures reflected by physiological states. People with normal sleep habit roll over 10–20 times a night, with too many or too few rolls reflecting sleep quality problems. But it varies from person to person. Sleep postures could be classified as major movement and minor movement. Major movement refers to so-called sleeping on one's back, sleeping on one's side, or sleeping on one's stomach [3]. The types of these movements were distinguished by the angle between the torso and the bed [4]. Because sleeping on one's stomach was a special sleep posture for very few people, it was not included in this study. Furthermore, the movement could also be divided into sleeping on the right side or sleeping on the left side according to the orientation of the torso contacting the bed surface. As shown in Fig. 1, the right side of the torso touched the bed surface represented sleeping on right side. On the other hand, there were many classifications for the types of minor movement in the literature, with different definitions for the postures of various body parts (especially limbs). Generally, it could be divided into bending, straightness or a state of relative positions of a certain body part, such as above the shoulder, parallel the shoulder, and below the shoulder. Positions of each part should be recorded in a table [5]. Through the observation of sleep postures, the study could discover the information and needs of physiology.

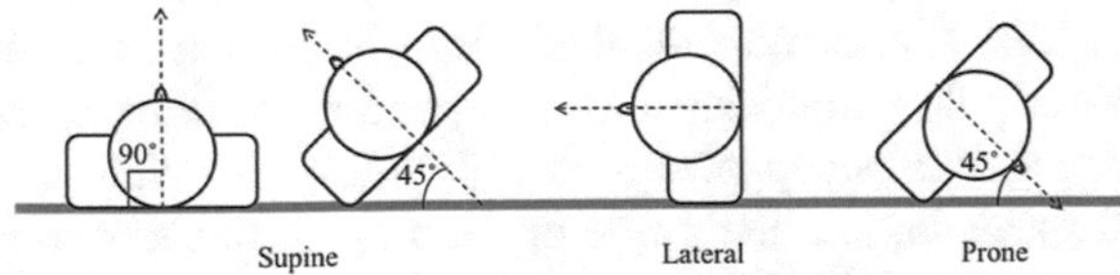

Fig. 1. Sleep postures.

2.2 Postures and Stress

Although changing sleep postures can reduce the compression time of the muscle tissue in the back and buttocks and avoid the formation of decubitus caused by blood clots in the microvessels, the deterioration of the body can cause symptoms such as adhesive capsulitis or arthritis in the shoulders. If a person with such symptoms sleeps on his/her side, he/she will feel painful on the affected part because of the pressure on the shoulder. People usually change their sleep postures several times in a sleep. That is, the pressure is not on a certain position for a long time, so that the muscle tissue oxygen and blood flow are sufficient. In contrast, if a person doesn't change the sleep posture for a long time, the risk of shoulder pain for him/her will be increased [6]. In addition, the shoulder is the joint between the human body and the arm. When the shoulder is pressed and deformed, the blood vessels, lymphatic vessels and nerves leading to the arm are also affected. They pointed out that when sleeping on one's side, the pressure on hips and shoulders was the greatest [7]. According to the research conducted [8], up to 80% people have had experiences with shoulder tense or painful shoulder and neck. Therefore, using appropriate bedding during sleep could make the cervical vertebra get proper support. Moderately changing the sleep postures with the support of suitable bedding could avoid bad blood circulation derived from the long-term muscle tension. These were the key to improving the quality of sleep in people.

2.3 Effects of the Sleep Bedding and Stress

In all kinds of bedding, pillow is the key to relieve the pressure on head, neck, upper back and shoulder. During sleep, the pressure points were mainly concentrated in head and neck [7, 9]. According to the experiment on pillow design, the height of the pillow would affect the pressure on head [10]. The difference in height of the pillow during sleep would cause different pressure area of head. So the design criteria for pillow should be increase the corresponding mechanisms of the changes of sleep postures to relieve the pressure on head [11]. The structural design for relieving pressure was created based on the purpose of making pressure distribution even and reducing pressure, with the consideration of pressure-reducing direction and groove of pillow and the function that increase ventilation. The design of pillow should not only consider the effect of pressure relief, but come up with the solution that would support the shoulder and neck and relieve the pressure on cervical vertebra for avoiding shoulder and neck pain under concentrated pressure [12]. However, the height of cervical vertebra was changed due to the movement of sleep, so that it is necessary to get the appropriate height for pillow that was able to respond to the different height of cervical vertebra in the changes of sleep positions [7]. When changing sleep postures, the position of head for sleeping on one's side or sleeping on one's back may be different, which vary in each person [13]. Pillow design pointed out that there is a height drop between sleeping on the side and back, which requires the flexible and corresponding structure to support the neck in the same area. As for the support design on the structure of bedding, the cervical vertebra is curved when sleeping on the back and the cervical vertebra is straight when sleeping on the side [14]. After receiving the support, the design improved tracheal opening and closing, so as to relieve sleep apnea [15].

3 Methods

This study was conducted in three directions, including user experience, expert interview, and market feasibility. The details of research process were shown as the following subsections:

3.1 User Experience

Through the observation and detailed analysis of the users, users' behavior patterns and habits of using bedding could be understood, and then a company could produce the products that meet the real needs of the users. This study used sleep monitoring to observe the users' sleep postures and record the changes of sleep postures. Users' sleep images were extracted and drawn into a 2D wireframe according to their major movement or minor movement during sleep. Subsequently, this study confirmed the postures with users at the interview and summarized the types of sleep postures.

3.2 Expert Interview

This study conducted the semi-structured interview with physiotherapists and experts with medical engineering background. The selection of physiotherapists was because they had actual clinical experiences for sleep postures, and the selection of experts with medical engineering background was for the overall consideration in biomechanics. The main content of the interview focused on the impact of human physiological mechanisms on normal sleep and the inducements triggered discomfort. The inducements were discussed based on the body compression caused by changes of sleep postures and the pressure distribution of cervical vertebra.

3.3 Market Feasibility

This study chose a bedding company that was popular among consumers as the research object to understand the market feasibility of pillows. The implementation in this part is to involve users in the process of product design. This company introduced the users' need into the design concept and invited the users to participate in the usage situation during the phase of design. Based on the market feasibility derived from business models, the company could design the products that meet the real needs of the users.

4 Results

Through the analysis of user experience, expert interview and market feasibility, this study integrated the results and concluded the design elements of the innovative bedding. The design elements concluded in the results of this study could be a reference for the follow-up product design.

The observation of the changes of sleep postures is to extract the users' sleep images and drew the users' major movement or minor movement into a 2D wireframe.

The wireframe was shown as Table 1. This study found out two phenomena: (1) the two most common sleep postures was lying on one's back and sleeping on one's side respectively; (2) the limbs were more able to be stretched when lying on the back, while the hands are mostly twisted around the body when sleeping on the side.

Table 1. Sleep postures

Sleep Postures	Sleep Postures of Picture
sleeping on one's back	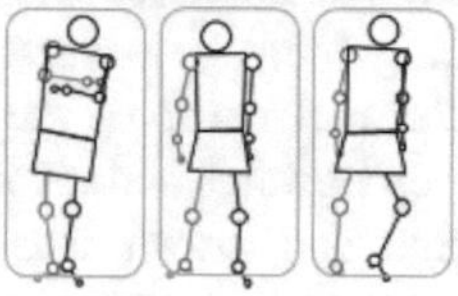
sleeping on one's side	

This study, through expert interview, concluded that the "sleep posture" did affect the quality of sleep. The related contents were as follows:

1. "The prolonged compression of body" indirectly affected "blood circulation" and "muscle compression". Therefore, the design of bedding must take into account the situation in which there are involuntary movements in sleep, and try to abate the discomfort caused by the prolonged compression of body in the changes of sleep postures.
2. Pillow is the most important tool of relieving the pressure on head, cervical vertebra and shoulder during sleep. By increasing the mechanism that can respond to physiological activity of the changes of sleep postures, the pressure on other parts of the body can be relieved. In particular, the neck and upper back muscles should be stretched easily in the changes of sleep postures to avoid the head and neck being pressed by the pillow for a long time. The design direction of pillow should interact with the users and related to the sleep posture.
3. The design of pillow must consider the pressure area and the degree of pressure of the head when lying on the back or sleeping on the side, as well as how to reduce the pressure on the neck and upper back.
4. The internal structure of the pillow could also be predicted from the pressure distribution on the surface of the pillow. The prominent part of the bone should be designed with a high internal structure to reduce the pressure on the surface and enhance the support for the head at the moment of changing sleep postures.

This study chose a bedding company that was popular among consumers as the research object to understand the market feasibility of pillows. The results of this part were concluded as follows:

1. Consumers no longer consider price as a priority for purchasing products. What they care about is whether the product is effective.
2. Consumers care about whether the product is designed to solve their problems of usage.

References

1. Borazio, M., Van Laerhoven, K.: Combining wearable and environmental sensing into an unobtrusive tool for long-term sleep studies. Paper presented at the Proceedings of the 2nd ACM SIGHIT International Health Informatics Symposium (2012)
2. Caldwell, J.A., Prazinko, B., Caldwell, J.L.: Body posture affects electroencephalographic activity and psychomotor vigilance task performance in sleep-deprived subjects. Clin. Neurophysiol. **114**(1), 23–31 (2003). https://doi.org/10.1016/S1388-2457(02)00283-3
3. Vaughn McCall, W., Boggs, N., Letton, A.: Changes in sleep and wake in response to different sleeping surfaces: a pilot study. Appl. Ergon. **43**(2), 386–391 (2012). https://doi.org/10.1016/j.apergo.2011.06.012
4. Defloor, T.: The effect of position and mattress on interface pressure. Appl. Nurs. Res. **13**(1), 2–11 (2000)
5. Desouzart, G., Filgueiras, E., Melo, F., Matos, R.: Human-bed interaction: a methodology and tool to measure postural behavior during sleep of the air force military. In: Marcus, A. (ed.) Design, User Experience, and Usability. User Experience Design for Everyday Life Applications and Services, vol. 8519, pp. 662–674. Springer (2014)
6. Zenian, J.: Sleep position and shoulder pain. Med. Hypotheses **74**(4), 639–643 (2010)
7. McCall, W.V., Boggs, N., Letton, A.: Changes in sleep and wake in response to different sleeping surfaces: a pilot study. Appl. Ergon. **43**(2), 386–391 (2012)
8. Bovim, G., Schrader, H., Sand, T.: Neck pain in the general population. Spine (Phila Pa 1976) **19**(12), 1307–1309 (1994)
9. Urden, L.D., Stacy, K.M., Lough, M.E.: Priorities in Critical Care Nursing. Elsevier Health Sciences (2015)
10. Di Lazzaro, V., Giambattistelli, F., Pravatà, E., Assenza, G.: Brachial palsy after deep sleep. J. Neurol. Neurosurg. Psychiatry (2014). https://doi.org/10.1136/jnnp-2013-306637
11. Gottesmann, C.: GABA mechanisms and sleep. Neuroscience **111**(2), 231–239 (2002)
12. Ni, H., Abdulrazak, B., Zhang, D., Wu, S., Yu, Z., Zhou, X., Wang, S.: Towards non-intrusive sleep pattern recognition in elder assistive environment. J. Ambient Intell. Humanized Comput. **3**(2), 167–175 (2012)
13. Tan, S.-H., Shen, T.-Y., Wu, F.-G.: Design of an innovative mattress to improve sleep thermal comfort based on sleep positions. Procedia Manufact. **3**, 5838–5844 (2015)
14. Brown, R.E., Basheer, R., McKenna, J.T., Strecker, R.E., McCarley, R.W.: Control of sleep and wakefulness. Physiol. Rev. **92**(3), 1087–1187 (2012)
15. Hallegraeff, J.M., van der Schans, C.P., de Ruiter, R., de Greef, M.H.: Stretching before sleep reduces the frequency and severity of nocturnal leg cramps in older adults: a randomised trial. J. Physiotherapy **58**(1), 17–22 (2012)

Design Culture Within the B2B Needs Roadmap

Leonardo Forzoni[1(✉)], Ramona De Luca[1], Maria Terraroli[2], Francesco Spelta[2], and Carlo Emilio Standoli[2]

[1] Esaote SpA, via di Caciolle 15, 50127 Florence, Italy
leonardo.forzoni@esaote.com
[2] Politecnico di Milano, Dipartimento di Design, Via Giovanni Durando 38/A, 20158 Milan, Italy

Abstract. The aim of the present paper is to support the reshaping and spreading of the role of the design culture within the B2B environment. The main target of the work is related to companies characterized by operating in fast-changing markets (in terms of needs, trends and tendencies change) while having to develop complex and long-time for completion products. Such characteristics force companies to a structured long-term forecasting approach where design culture adoption can fit mainly considering the purchase process multiple actors (user, customer, procurement office, technical-engineering department) involved within the product design activity since the early beginning. Design culture would, therefore, influence and surround several departments involved in product design and production, as well as concerning after sales product activities, becoming an innovative approach which can pass from being "at the end" of the process, to one of the major core company backbones.

Keywords: Design culture · Needs Roadmap · B2B market · Purchase process actors · Product design

1 Introduction

The Needs Roadmap concept has been recently developed as a framework with the major aim to catch, analyze, understand and translate into products the solution requisites, needs, trends and tendencies of specific B2B markets with the focus on healthcare capital equipment. The Needs Roadmap framework, which was conceptualized and initially implemented within the diagnostic imaging systems, is starting to show its possible practical applicability also within the industrial machinery market, where there are present similar purchase processes scenarios, as well as similar solution development time and market changes time circumstances [1–4].

T. Ahram et al. (Eds.): IHSED 2019, AISC 1026, pp. 730–736, 2020.
https://doi.org/10.1007/978-3-030-27928-8_111

2 Needs Roadmap, What Is It?

The Needs Roadmap is the identification of needs, trends and tendencies (usually and commonly indicated as "needs") over a certain time period. There is a key characteristic that shapes the roadmap: it is a living document that evolves and changes over time, within a 1-3-5-year coverage plan, consisting of the definition and prioritization of needs. The definition of the Needs Roadmap and its continuous update represent the main output of the strategic marketing function. This important strategic activity, aligned with the company vision and the related company strategy, aims at concretely elaborating the established goals. The Needs Roadmap is achieved (and perfected, on a revision-basis approach) with interviews to customers and non-customers (such as interviews to people on the field: sales, local marketing, marketing communication, etc.), key opinion leaders (KOLs; i.e. "influencers"), technical/clinical state of the art, and market, competition and tendencies analysis.

In terms of benefits, the Needs Roadmap framework has the aim to be useful for innovative companies that produce and sell complex products characterized by long time-to-market in technologically structured and competitive markets. The framework allows integrating complex client-supplier purchasing processes, where many stakeholders are involved, by creating a manageable and multidimensional space that outlines the current and future company market context. In fact, the actors of the purchasing process are involved since the very beginning of the procedure of needs understanding, solution definition and solution creation (and not left solely to the sales process interfacing at the end of the business chain). Keeping in mind the needs and tendencies, the designer takes the role of an interpreter: he/she is able in defining and understanding the company vision and strategy (and how it is practically sold on the market) and in translating the values and the insights of the Needs Roadmap in a more practical and tangible perspective. In this way it is possible to set up a specific plan, the product roadmap, to reach the desired goal, in terms of tangible outputs.

The "desired goal" should not be downsized or reduced because of the traditional expertise of the company. Innovation in an enterprise might affect its established structure and culture and might challenge the way in which people interact, the existing capabilities, and how things are accomplished in daily activities [5]. This should not stop the push towards innovation, but it should transform the company's mentality in a research of the balance between innovation and established ideas [5].

The Needs Roadmap is the place where the strategy is practically available [4]. The aim of the present paper is the one to understand the role of the designer within the B2B industrial market context (seen both as strategic designer and as product designer), therefore speaking at a more general level and under a "wider hat" of what done in the literature so far regarding the interaction between the product marketing-management/product roadmap and the strategic marketing/Needs Roadmap [1–4]. The major topic of the present work is the one to find the role of the designer (as written above, seen both as strategic designer and as product designer) within the Needs Roadmap framework and, in a more comprehensive approach, to hypothesize how and where the design culture can fit within an "average" industrial B2B company setup.

3 Design Culture in the B2B Business

The design culture is a unique system of knowledge and skills that keeps in mind multiple factors related to society, markets and technology, and enhances interaction between the world of production and the world of consumption.

In many companies, above all in the B2B and healthcare context, design culture is seen in its reductionist view as one of the skills required in the product development process to enhance the appeal of the products [5]. Such approach is requested usually in a quite late stage of the product design process, where the requisites and specifications are already set up, and most of the time just in order to enrich the aspects of user interface, ergonomics, aesthetics and look-and-feel characteristics. In a more up to date and design-oriented company, designers take care also of problem solving, new businesses and strategies definition, and optimization of product usability/pleasure in use, as a result of the application of design-oriented principles (e.g. design thinking methods and tools), in order to increase innovative and competitive qualities. In some other scenarios, the designer is completely absent and in such conditions his/her role is covered by the product marketing manager and/or the engineering roles within the R&D department (usually the ones covering the design activities of SW interfaces, system integration, mechanics and sometimes even production).

Even if the design's theoretical and practical dimensions, with respect to product (marketing) management, have been debated for a long time, and the discussion is far than being ended, the importance of design culture in a B2B environment concerning capital equipment (goods which are used to produce other goods or services) is experienced by the authors as constantly increasing, therefore gaining its own and personal attention in the whole process, also in non-consumer scenarios. For example, from some years now, the trend is to reduce the request of the certified ergonomist role in favor of ergonomically-oriented designers, who know how to take into consideration all the necessary aspects in the healthcare theme (in this case ergonomically-product designer oriented) [6]. Therefore, the general trend is to prefer roles with a good understanding of different areas and the ability to collaborate across different disciplines applying knowledge in areas of expertise rather than professionals whit a great depth of related skills and expertise in only a single field/topic.

The product design activity can be presented as linked to the product roadmap and, consequently, to the Needs Roadmap as shown in Fig. 1.

Being such definition and action space the multi-dimensional space where the product is conceived, designed and produced, starting from the overview to the outer and inner world which is defined by the Needs Roadmap framework, design culture should represent the framework where the needs and product roadmaps are conceived and where the related company departments more closely interfacing/addressing the product have to be "grouped", as in Fig. 2. This figure wants to graphically represent the unquestionable role as a bridge of the designer between the various company departments, due to his/her ability to "speak" many project languages (from marketing to production) [7].

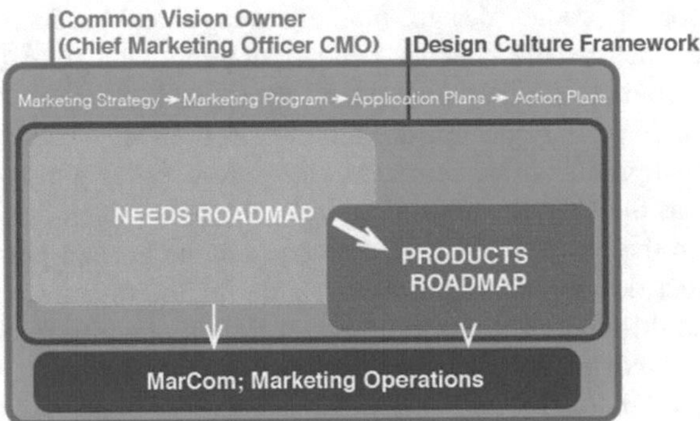

Fig. 1. *Needs Roadmap* relationships with respect to the vision owner framework (*Chief Marketing Officer – CMO*), new product introduction - NPI (*MarCom, Marketing Operations*) and *products roadmap*. The latter is completely "surrounded" by the *design culture framework* (rectangle in RED).

Fig. 2. The design culture framework (rectangle in RED) "surrounds" several departments involved in *product marketing* (MKTG), *design* and *production*, as well as concerning after *sales* product management. Such departments have different "temporal horizons" shaped by the market and trend and tendencies timing and which are indicated on the left side column, ranging from 1 month (m) to 10 years (y).

4 Purchase Process Actors in the Design Culture Framework

The specific approach presented in this paper investigates the role of design culture and processes within the B2B environment and organizations, considering the actors involved in the B2B product development process (users and customers, procurement office, technical-engineering department, and possible additional stakeholders), and the requested expertise (e.g. product design/management,…) within the organizations.

The approach implies the consideration of every actor since the early stages of the Needs Roadmap analysis, due to a participatory approach that defines the real values and needs of the actors. In this way, it is easier to understand and connect to which features to give specific "weight" in the product roadmap development.

From the early stages, the design approach must consider all these actors in a strategic process that includes reaching consensus and understand different priorities for the optimization of product design, market strategy, and product technical definition. As discussed before, the concept behind this approach is to follow the paths originated by the different stakeholders' values, needs, and ideas since the beginning of the Needs Roadmap, to the concept definition and product development, to finally converge to the final product to be delivered successfully in the purchase process. Such propelling effort has the outcome of driving design to be considered as an essential and leading role within the Needs Roadmap, delivering its inputs and priorities practically through the product development and as a vehicle for the organizational change and improvement. A multi-modal optimized design, weighting multiple factors for each one of the actors involved in the purchase process, can be presented with a multiple graphical biased model. Each one has a specific weight itself, depending on the variable importance that a certain actor has within a specific purchase process (Fig. 3). This study should start during the Needs Roadmap, in which practitioners are in direct contact with stakeholders (customers and non-customers, key opinion leaders, etc.) because for every new product introduction (NPI) they can spot and understand every actor's variable interest.

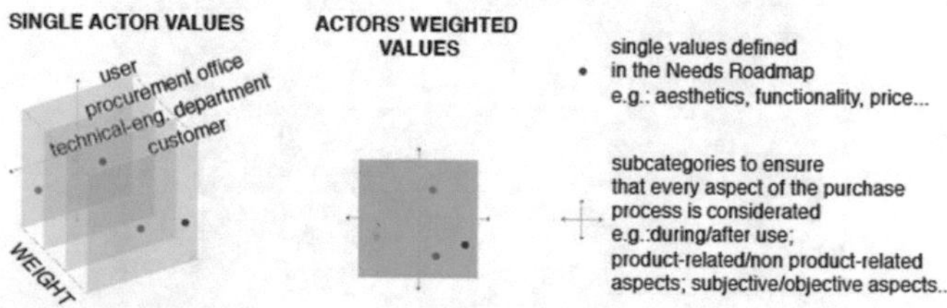

Fig. 3. Product design outcome represented as the optimization of a multi-dimensional space obtained by considering one weighted model for each actor involved in the purchase process

In a B2B multi-interface purchase process such as the one for healthcare capital equipment (within large institutions, clinics or hospitals), many actors are involved in the purchase process, [3] such as: User – the operator of the system; Customer – the final decision-maker; Technical/Engineering – analyzes and evaluates the system from a technical, installation, management and service perspective; Finance/Procurement Office – economical and feasibility of the purchase, capital equipment management, purchase, lease/loan/rent/pay-per-use options evaluation, etc.

An example of the features of the design weighted multiple factors intercepted during the process for defining the Needs Roadmap could be: User: aesthetics, functionality, usability, interaction, pleasure in use, technology features; Customer: price, productivity, usability/interaction; Procurement office: price, scalability, investment through time; Technical/engineering department: serviceability, reliability, technology.

The optimal product design outcome can be represented as the optimization of a multi-dimensional space obtained by considering one weighted model for each purchase process-involved actor who will deal with the product before, during and after use (and present within the purchase process). This is in opposition to the one related to a single point in a bi-dimensional environment, like the one to consider solely the user.

5 Conclusions

At the time of writing, the uniqueness of the presented approach is related to the different vision and involvement of the design culture within the Needs Roadmap, for which design should be involved in the company's strategy definition and not only relegated to the latest stages of the product development, merely committed to aesthetics and look-and-feel definition. This vision of the design culture within the Needs Roadmap framework is related to the consideration of the following points:

Purchase process multiple actors - The new conceptual and practical approach to design culture for B2B needs is to understand, analyze and implement all the designing inputs, as needs, trends and tendencies, for all the purchase process stakeholders. This approach recognizes not only the user or/and the customer, as traditionally proposed in the B2C market, but rather a complex system of multiple actors characterizing the considered B2B market, where all of them are considered since the early stages of the process. The concept behind this approach is to design natively for all these actors playing a role in the purchase process, modeling the solution around their differences in terms of needs, point of view and areas of expertise.

Short market needs changing time and long time development - The presented design culture-driven approach finds its "sweet spot" within complex markets and long product design processes, where product development can take roughly 3 to 5 years, while the market needs, trends, and tendencies can be expected to change much faster from 1 to 3 years, as a general indication. It is therefore intuitively understandable that an advanced level of forecasting is mandatory.

Entire product lifespan consideration - Thinking to the whole lifespan of the product/solution, while not designing solely with the NPI deadline in mind, represents a principal characterization of the presented approach. In such mood the forecasting of the possible/necessary upgrades or updates, that will be needed in the fast-changing markets in terms of reshaping or re-working are understood, planned and designed-for, since the early stages of the solution definition.

References

1. Forzoni, L., Buffagni, C., Nonino, F., Pompei, A.: Case study of a clinical marketing strategy within a global diagnostic imaging company. In: Proceedings of the 2018 International Conference on Education Technology Management - ICETM 2018, pp. 53–57. ACM Press, New York (2018)
2. Forzoni, L., Buffagni, C., Guercini, S.: Educational marketing strategic approach in a diagnostic imaging company: from global perspective to local implementation. In: Proceedings of the 2018 International Conference on Education Technology Management - ICETM 2018, pp. 28–32. ACM Press, New York (2018)
3. Forzoni, L., Buffagni, C., Guercini, S.: Understanding the customer needs in business network: an emerging model in the diagnostic imaging industry. In: 35th IMP Conference, IÉSEG School of Management, Paris (2019)

4. Forzoni, L., Nonino, F., Pompei, A., Buffagni, C.: Needs Roadmap Framework understanding and mapping needs in competitive markets. In: ISPIM Innovation Conference, Florence (2019)
5. Deserti, A., Rizzo, F.: Design and the Cultures of Enterprises. Des. Issues **30**, 36–56 (2014)
6. Costa, F.: Health Care Design. Territori Del Design, Poli.Design, Milan (2002)
7. Zurlo, F.: Le strategie del design. Disegnare il valore oltre il prodotto. Libraccio Editore, Milan (2012)

Masticatory Evaluation in Non-contact Measurement of Chewing Movement

Chika Sugimoto(✉)

Yokohama National University, 79-5 Tokiwadai,
Hodogaya, Yokohama, Kanagawa 240-8501, Japan
sugimoto-chika-zb@ynu.ac.jp

Abstract. Decline in masticatory force, which is caused by aging and disease, makes a negative effect on health conditions. As masticatory performance is greatly associated with eating habits, it is important to recognize the chewing states for checking the masticatory function. In this study, the method to classify the characteristics of ingested food, count the number of chewing per one bite, and verify the features of chewing states measuring the chewing movement with a non-contact sensor has been developed. The path and rhythm of chewing movement is evaluated by tracking the feature points on a face measured with an RGB-D camera. The importance of feature quantities extracted from chewing movement was analyzed using statistical approach and machine learning. The results suggested that the chewing states and masticatory function could be evaluated with some parameters such as cycle time, opening distance, and stability of path and rhythm measured with a simplified system.

Keywords: Mastication · Non-contact measurement · Healthcare · Features · Machine learning

1 Introduction

Mastication performs a crucial function in eating. As decline in masticatory force, which is caused by aging and disease, makes a negative effect on health conditions, it is important to maintain or recover the masticatory ability in the face of progressing in aging. Masticatory function is greatly associated with eating habits. Healthy eating habits are a key factor in maintaining good health. Thus, it is expected to check the masticatory function by monitoring the chewing state.

The characteristics of ingested food affect chewing patterns in the measurement of chewing states. The suitable number of chewing changes depending on the characteristics of the food such as hardness and size. It is required to analyze the chewing state taking account of the characteristics of ingested food. Therefore, we have been developing the method to classify the characteristics of ingested food, count the number of chewing per one bite, verify the features of chewing states, and evaluate the masticatory function measuring the chewing movement [1].

Conventional methods for measuring mandibular movement used by dentists are expensive and require a marker attached to the mandibular incisors. A simple system using a home camcorder was proposed to record the trajectories of an incisal marker

T. Ahram et al. (Eds.): IHSED 2019, AISC 1026, pp. 737–741, 2020.
https://doi.org/10.1007/978-3-030-27928-8_112

attached by solid wires with a precision equivalent to that of a conventional jaw tracking system such as a mandibular kinesiograph [2], which is considered to be one of the gold standard methods for jaw tracking. As a markerless method which requires neither an incisal marker nor headgear to record normal chewing, the validity of a markerless three-dimensional system using an RGB-D camera was also tested for tracking masticatory movement by comparing it with a conventional method using an incisal marker [3]. Other studies used an RGB video camera to track the markers attached on the chin for assessment of the jaw movement [4]. This study aims to measure chewing states and eating habits using a general-purpose sensor that can be easily installed in medical and welfare institutions and home. Therefore, the method to analyze the chewing movement and recognize the chewing states was developed and evaluated using an RGB-D sensor.

The paper proceeds with a description of the methods in Sect. 2. The methods are verified through experiments in Sect. 3. The paper concludes with some remarks in Sect. 4.

2 Methods

2.1 Measurement of Chewing Movement

The detailed information on a face is obtained using an RGB-D sensor which can measure the depth and RGB data of objects simultaneously. A Kinect V2 from Microsoft was employed as the sensor in this paper. The detection information of a face can be obtained using the library of HDFace. The facial points are determined based on the facial profile for the individuals and given as an array of vertices. 3D mesh of the face is constructed using the vertices as shown in Fig. 1. The vertex positions are mapped to the RGB frame following facial movements.

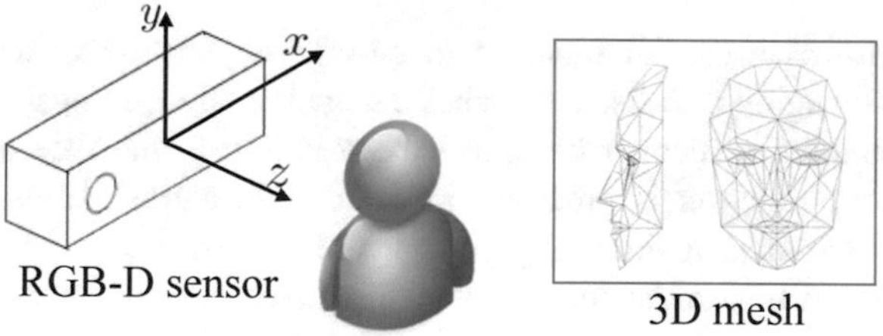

Fig. 1. Measurement system and 3D mesh of a face obtained

The chewing movement is measured using two facial points of the chin and nose tip based on the verification of the previous study [3]. The chin is defined as the most prominent point of the soft tissue region under the lip. The most prominent point in the mid face is defined as the nose tip to compensate for the movement of the head posture. 3D coordinate values of two facial points, $(x_{nose}, y_{nose}, z_{nose})$ and $(x_{chin}, y_{chin}, z_{chin})$, are calculated from the depth data. The distance between the nose tip and chin is calculated using the following Eq. (1). The horizontal displacement is calculated using the Eq. (2).

$$D = \sqrt{(x_{chin} - x_{nose})^2 + (y_{chin} - y_{nose})^2 + (z_{chin} - z_{nose})^2} \quad (1)$$

$$W = \sqrt{(x_{chin} - x_{nose})^2 + (y_{chin} - y_{nose})^2 + (z_{chin} - z_{nose})^2} - \sqrt{(y_{chin} - y_{nose})^2} \quad (2)$$

The system observes the distance 'D' and detects the peaks and local minimum values of the distance. 5-frame moving average of the distance is calculated for denoising before the peak detection. Food intake is judged by detecting the opening of mouth based on the distance between the upper lip and the bottom lip. Opening paths and closing paths are divided by their corresponding maximum and minimum vertical displacement. Occluding paths are observed following closing paths till the beginning of the opening paths. The number of the cycles of chewing movement is counted based on the displacement of the distance. Three parameters of opening distance, chewing width, and cycle time are calculated from the chewing movement path and the temporal information. Opening distance and chewing width are values shown in Fig. 2. Cycle time is the period from the beginning of an opening phase to the next opening phase. Each period (opening time, closing time, and occluding time) and movement velocity (including opening maximum velocity and closing maximum velocity) of an opening phase, a closing phase, and an occluding phase in a cycle time is also calculated. The stability of each period (stability of movement rhythm) is valuated to show the features of chewing states. Three indicators representing the stability of chewing movement path, the SD/OD (standard deviation divided by the opening distance) of the opening lateral, closing lateral, and vertical components, are calculated for examining the masticatory function following the previous study [3, 5].

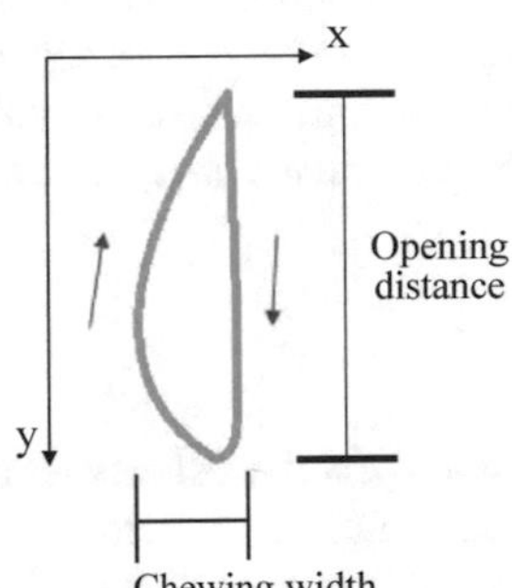

Fig. 2. Parameters of chewing movement path

2.2 Estimation of Food Characteristics

The influence on chewing patterns of ingested food is evaluated using machine learning. The characteristics of ingested food are learned and estimated by random forest (RF). The importance of feature quantities is also analyzed in order to find the impact on the way of chewing. The discriminator for judging hardness and size of ingested food on a two-level scale is built and tested. As the way of chewing is influenced by individual habits and oral functional properties, the accuracy is thought to be greatly affected by training data used for building the discriminator. Therefore, the difference in accuracy is verified comparing the case trained by using only the data collected from the individual with the case trained without using the individual data. 30 feature quantities which consist of three parameters in the first 10 cycles are used to classify the characteristics of hardness and size.

3 Evaluation

3.1 Experiments

This study was conducted in accordance with the Helsinki Declaration. The experimental procedures involving human subjects described in this paper were approved by the institutional ethical review board. Each participant gave informed consent before involvement in the experiments.

First, the experiment was performed to examine the method for estimating the characteristics of ingested food described in Sect. 2.2. The chewing movement was measured during each 10 times of 4 types of gummy jelly chewing in 4 healthy volunteers. Four types of gummy jelly have the characteristics of the combination of two levels of hardness and size (hard vs soft, large vs small). The parameters for RF were determined by grid search. The number of trees and the maximum number of features were set to 500 and 4, respectively. 5-fold cross validation was performed to evaluate the discriminator.

Next, the relations between the features of chewing states and the parameters obtained from chewing movement were evaluated. The chewing movement was measured during each 15 times of gummy jelly chewing under each condition of concentrating on chewing (doing nothing except for chewing), using a smartphone, and watching TV. 5 healthy volunteers participated in the experiments. Moreover, the chewing movement of one of the volunteers was measured during the repairing period of a tooth in the same way.

3.2 Results

The difference in recognition accuracy was verified comparing the case trained without using individual data with the case trained by using only the data collected from the individual. When only the individual data were used for learning, the average recognition accuracy of hardness and size was 75% and 91%, respectively. On the other hand, when the individual data were not used for learning, it was 64% and 74%, respectively. It is considered that the accuracy was not enough high due to the small amount of the training data. The way of chewing is influenced by individual habits as well as the characteristics of ingested food. The results show that it is required to use individual data for improving the accuracy if only the small amount of data are used. The feature quantities with high importance were cycle time (or opening time) and opening distance to discriminate the difference in hardness. On the other hand, that was chewing width to discriminate the difference in size.

The difference in the chewing states under the conditions of whether concentrating on chewing or not was represented in the stability of movement rhythm and the opening maximum velocity and closing maximum velocity. The result showed that the cycle time tended to increase under the conditions of not concentrating on chewing, especially using a smartphone, while the occluding time and the opening and closing maximum velocity as well as the stability decreased. The stability of movement path and rhythm, opening distance, and closing maximum velocity decreased when the masticatory function was abnormal due to the dental treatment.

4 Conclusion

The method to classify the characteristics of ingested food, count the number of chewing per one bite, verify the features of chewing states, and evaluate the masticatory function measuring the chewing movement with a non-contact simplified system was developed. The results of the experiments suggested that the chewing states and masticatory function could be evaluated with some parameters obtained from chewing movement. However, there are some persons who have little chin movement during eating. Therefore, the indicators representing the movement of the whole chin will be also examined in the future study.

Acknowledgments. This work was supported by Japan Society for Promotion of Science KAKENHI 17K00230.

References

1. Kushimiya, M., Sugimoto, C.: Characteristics identification of food ingested based on chewing movement path for non-contact mastication evaluation. IEICE-2017-HCGSYMPO, HCG2017-C-7-3, 4 pages (2017)
2. Kinuta, S., Wakabayashi, K., Sohmura, T., Kojima, T., Mizumori, T., Nakamura, T., Takahashi, J., Yatani, H.: Measurement of masticatory movement by a new jaw tracking system using a home digital camcorder. Dent. Mater. J. **24**(4), 661–666 (2005)
3. Tanaka, Y., Yamada, T., Maeda, Y., Ikebe, K.: Markerless three-dimensional tracking of masticatory movement. J. Biomech. **49**(3), 442–449 (2016)
4. Zafar, H., Eriksson, P.O., Nordh, E., Häggman-Henrikson, B.: Wireless optoelectronic recordings of mandibular and associated head–neck movements in man: a methodological study. J. Oral Rehabil. **27**, 227–238 (2000)
5. Uesugi, H., Shiga, H.: Relationship between masticatory performance using a gummy jelly and masticatory movement. J. Prosthodont. Res. **61**(4), 419–425 (2017)

Satisfaction of Aged Users with Mobility Assistive Devices: A Preliminary Study of Conventional Walkers

Josieli Aparecida Marques Boiani[1,2], Frode Eika Sandnes[2,3], Luis Carlos Paschoarelli[1], and Fausto Orsi Medola[1(✉)]

[1] Post-Graduation Programme in Design, Sao Paulo State University (UNESP), Av Eng Luiz Edmundo Carrijo Coube 14-01, Bauru 17033-360, Brazil
fausto.medola@unesp.br

[2] Department of Computer Science, Oslo Metropolitan University, Oslo, Norway

[3] Department of Technology, Kristiania University College, Oslo, Norway

Abstract. Elderly users' satisfaction with their walkers was evaluated. A sample of 13 institutionalized aged participants were interviewed and responded to the Quebec User Evaluation of Satisfaction with Assistive Technology (QUEST 2.0) questionnaire, with eight questions that address their level of satisfaction with different aspects of the device. In general, the users reported to be satisfied with their device, with the highest levels of satisfaction with durability and easiness of use. The lowest scores were associated with device weight, ease of adjusting and device stability and safety. The users' indicated the device safety, easiness of use and comfort as the most relevant aspects of the device. Elderly users' satisfaction with their mobility devices is likely to correlate with users' device needs and expectations.

Keywords: Walkers · Elderly · Mobility · Satisfaction · Assistive Technology

1 Introduction

The proportion of elderly is increasing globally due to the increase in life expectancy and lower birth and death rates. The aging process is associated with a decrease in physical function that may ultimately affect the ability to walk safety and independently. The study of Araujo et al. [1] found that impaired physical mobility among institutionalized elderly people was 100% of the sample, and that it was related to physical aspects such as muscular weakness, reduced strength and resistance, and cognitive impairments.

This context highlights the need for solutions that enable independent life [2, 3], and mobility function is a key part benefiting both social participation and quality of life. Assistive technology represents a potential solution to the maintenance and improvement of users' functionality and independence, contributing to a dignified life.

One of the devices most commonly used to assist the mobility of the elderly is the walker. The conventional walker design in Brazil is a folding frame made of aluminum

T. Ahram et al. (Eds.): IHSED 2019, AISC 1026, pp. 742–746, 2020.
https://doi.org/10.1007/978-3-030-27928-8_113

with four points of contact with the floor that increase the contact area, thus improving forward stability [4]. Although the main goal of conventional walkers is to improve stability and walking independence, some users do not satisfactorily recover mobility function. Despite the mobility benefits, walkers demand users to adapt an unnatural walking pattern, as the user needs to lift the walker, put it forward and then step forward, thus requiring the user to pay attention in locomotion [5].

Aspects of the walker design might be associated with this failure to provide means for independent mobility. Addressing the users' satisfaction with the mobility device may therefore shed some light onto the aspects of the walker design that is most influential to users' mobility. This study therefore evaluated elderly users' satisfaction with their walkers.

Table 1. Participants

Part.	Age (ys)	Gender	Weight (kg)	Height (m)	Time of use (months)	Device acquisition	Training of usage
1	96	F	65	1.56	84	Own resources	No
2	76	F	67	1.65	48	Own resources	No
3	68	M	63	1.7	9	Health professional	No
4	91	F	77	1.53	24	Family	No
5	66	M	50	1.65	2	Family	No
6	87	F	65	1.60	2	Family	No
7	87	F	48	1.55	24	Family	No
8	87	F	62.5	1.68	12	Health professional	Yes
9	75	M	47	1.65	18	Health professional	No
10	99	F	54	1.55	3	Health professional	No
11	85	F	48.5	1.52	60	Health professional	Yes
12	68	F	45	1.68	12	Health professional	Yes
13	76	F	65.5	1.60	12	Friend	No

F: female; M: male.

2 Method

The procedures involved interviews with elderly walker users from two institutions, namely Vila Vicentina in the cities of Bauru and Arealva. The procedures were approved by the Ethics Committee of the School of Architecture, Arts and Communication – FAAC-UNESP, Bauru (Process N. 1.835.531), and participants were informed about the study objectives and procedures and voluntarily signed an Informed Consent Form.

Thirteen persons over 65 years with a mean age of 81 years (±10.9) were recruited. They had used a conventional walker for a median time of 12 (min 2, max 84) months. None of the participants had cognitive impairment and were able to comprehend and respond the questionnaire. Table 1 presents the participants.

The evaluation was performed through interviews using the Quebec User Evaluation of Satisfaction with Assistive Technology (QUEST 2.0) questionnaire, translated and validated for Brazilian Portuguese by Carvalho et al. [6]. This instrument assesses user satisfaction with its AT device in two main domains: feature and services. We only applied the first eight questions referring to the satisfaction with the device in this study as most of the participants acquired the device by themselves or through their families. Data were analyzed descriptively by means of the frequency of answers of the participants' sample, using Microsoft Excel.

3 Results and Discussion

Overall users' satisfaction with their walkers was high. The device aspects with higher levels of satisfaction (very and quite satisfied) were durability (100%), ease of use (100%), comfort (92.3%) and effectiveness (92.3%). Accordingly, high levels of satisfaction with mobility assistive devices have been reported [7, 8]. On the other hand, the lowest levels of satisfaction (not satisfied at all and not very satisfied) were found with the aspects of adjustment (38.4%) and weight (15.4%). Table 2 presents the detailed analysis of the frequency of responses of satisfaction level for the eight aspects of the device.

Table 2. Item by item analysis of walker aspects affecting users' satisfaction

Items	Satisfaction				
	Not satisfied at all	Not very satisfied	More or less satisfied	Quite satisfied	Very satisfied
Dimensions	0 (0)	0 (0)	3 (23.1)	4 (30.8)	6 (46.1)
Weight	2 (15.4)	0 (0)	2 (15.4)	4 (30.8)	5 (38.4)
Adjustment	5 (38.4)	0 (0)	1 (7.7)	4 (30.8)	3 (23.1)
Safety	1 (7.7)	0 (0)	2 (15.4)	1 (7.7)	9 (69.2)
Durability	0 (0)	0 (0)	0 (0)	5 (38.4)	8 (61.6)
Ease of use	0 (0)	0 (0)	0 (0)	3 (23.1)	10 (76.9)
Comfort	0 (0)	1 (7.7)	0 (0)	7 (53.9)	5 (38.4)
Effectiveness	0 (0)	0 (0)	1 (7.7)	5 (38.4)	7 (53.9)

*n (%)

From the users' perspective, the three most important aspects of their walkers were safety (92.3%), ease of use (84.6%) and comfort (53.8%). This finding highlights the importance of providing users with a stable device that is simple and safe to operate (see Fig. 1). The study of Zhou et al. [9] found that impaired balance and the use of a walking aid are factors associate with a greater number of falls among older community dwellers.

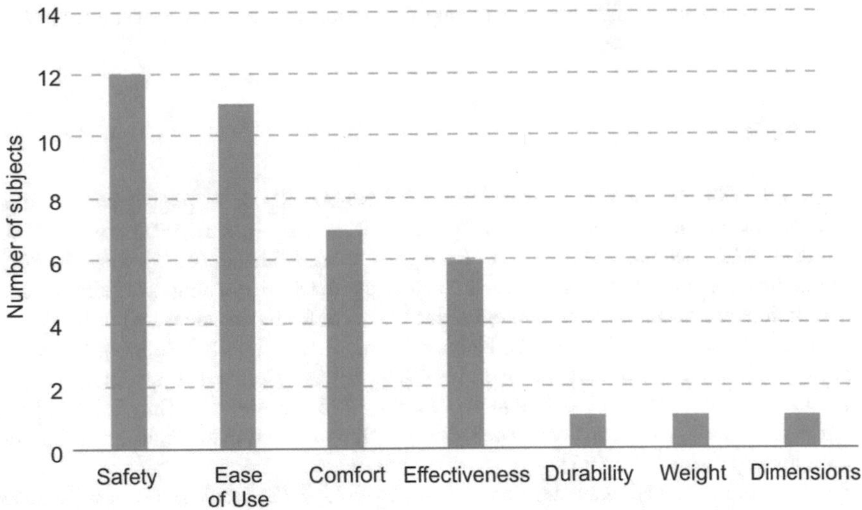

Fig. 1. Important assistive device characteristics from the users' perspective.

Several issues may have influenced the current findings. First, the small sample size may limit the extent to which the findings can be generalized to the whole population of walker users. Secondly, the responses may have been biased positively as some participants received the device from a family member or friend. Finally, this paper addressed satisfaction of aspects most associated with practical function. From a design perspective, it is interesting to explore products' aesthetical and symbolic functions, as it may play a role in technology acceptance. A recent study explore these aspects from the perspective of non users' perceptions about mobility devices [10].

4 Conclusion

This study explored elderly users' satisfaction with their walkers. We found durability and easiness of use were the device aspects associated with highest levels of satisfaction. Indeed, conventional walkers are simple, robust and very durable. Although the overall satisfaction with the device was high, important factors for the interface ergonomics such as weight, easiness of making adjustments and stability were associated with the lowest levels of satisfaction. Taking into account that walkers are most commonly used by aged users with decline in overall physical condition, providing a lightweight device that ensure stability and can be easily adjusted can positively influence one's ability to walk safely and independently. Additionally, the comprehension of the main aspects of the walker influencing users' satisfaction may contribute to the improvement of the design, prescription, provision and maintenance of mobility assistive devices.

Acknowledgments. The authors gratefully acknowledge the financial support from CNPq - National Council for Scientific and Technological Development (Process N. 427496/2016-0) and

Diku – the Norwegian Agency for International Cooperation and Quality Enhancement in Higher Education (UTF-2016-long-term/10053).

References

1. Araujo, L.A.O., Santana, R.F., Bachion, M.A.: Mobilidade física prejudicada em idosos: fatores relacionados e características definidoras. Rev. bras. enferm. **55**(1), 19–25 (2002)
2. Sandnes, F.E., Medola, F.O., Berg, A., Rodrigues, O.V., Mirtaheri, P., Gjøvaag, T.: Solving the grand challenges together: a Brazil-Norway approach to teaching collaborative design and prototyping of assistive technologies and products for independent living. In: Berg, A., Bohemia, E., Buck, L., Gulden, T., Kovacevic, A., Pavel, N. (eds.) Building Community: Design Education for a Sustainable Future, International Conference on Engineering and Product Design Education 7 & 8 September 2017, Oslo and Akershus University College of Applied Sciences, Norway. 1 edn. The Design Society; Institution of Engineering Designers, Glasgow, pp. 242–247 (2017)
3. Sandnes, F.E., Herstad, J., Stangeland, A.M., Medola, F.O.: UbiWheel: a simple context-aware universal control concept for smart home appliances that encourages active living. In: Proceedings of Smartworld 2017, pp. 446–451. IEEE (2017)
4. Medola, F.O., Bertolaccini, G.S., Boiani, J.A.M., Silva, S.R.M.: Mobility aids for the elderly: challenges and opportunities for the Brazilian market. Gerontechnology **15**(2), 65–97 (2016)
5. Stowe, S., Hopes, J., Mulley, G.: Gerotechnology series: 2. Walking aids. Eur. Geriatr. Med. **1**(2), 122–127 (2010)
6. Carvalho, K.E.C., Junior, M.B.G., Sá, K.N.: Tradução e validação do Quebec user evaluation of satisfaction with assistive technology (QUEST 2.0) para o idioma português do Brasil. Rev Bras Reumat **54**(4), 260–267 (2014)
7. Samuelsson, K., Wressle, E.: User satisfaction with mobility assistive devices: an important element in the rehabilitation process. Disabil. Rehabil. **30**(7), 551–558 (2008)
8. Wressle, E., Samuelsson, K.: User satisfaction with mobility assistive devices. Scand. J. Occup. Ther. **11**(3), 143–150 (2009)
9. Zhou, H., Peng, K., Tiedemann, A., Peng, J., Sherrington, C.: Risk factors for falls among older community dwellers in Shenzhen. China Inj. Prev. **25**(1), 31–35 (2019)
10. Boiani, J.A.M., Barili, S.R.M., Medola, F.O., Sandnes, F.E.: On the non-disabled perceptions of four common mobility devices in Norway: a comparative study based on semantic differentials. Technol. Disabil. **1**, 1–11 (2019). (in press)

Effect of Added Mass Location on Manual Wheelchair Propulsion Forces

Vitor Alcoléa[1], Fausto Orsi Medola[2(✉)], Guilherme da Silva Bertolaccini[2], and Frode Eika Sandnes[3,4]

[1] Department of Mechanical Engineering, Sao Paulo State University (UNESP), Bauru, Brazil
[2] Department of Design, Sao Paulo State University (UNESP), Bauru, Brazil
fausto.medola@unesp.br
[3] Department of Computer Science, Oslo Metropolitan University, Oslo, Norway
[4] Department of Technology, Kristiania University College, Oslo, Norway

Abstract. This study investigated the influence of mass distribution on the handrim forces during manual propulsion in four different mobility tasks: straightforward motion at self-selected speed; straightforward sprint; zero radius turn; and circular trajectory. A foldable-frame wheelchair was instrumented with a SmartWheel system placed on the right side. Three different positions of an additional mass of 7.8 kg were investigated: in the center of the rear wheels' axle; on the spokes of each of the two rear wheels; under the footrest. When mass is added in a centered position, there is little effect on the level of forces required to propel the chair, while when the additional mass is positioned distant to the wheelchair center, namely rear wheels' spokes and feet support, the effect on propulsion forces is increased. Optimizing wheelchair mobility efficiency requires an understanding on the effects of changes in equipment configuration on propulsion kinetics.

Keywords: Wheelchairs · Propulsion kinetics · Mass distribution · Mobility · Wheelchair configuration

1 Introduction

Manual wheelchairs are probably the most common means of mobility used by people with physical disabilities allowing the ability to walk independently and safely. The wheelchairs provide mobility and independence for the users, however the locomotion method used in manual wheelchairs is still mechanically inefficient and physiologically demanding [1]. Studies show that shoulder pain affects up to 73% of wheelchair users, and of these, around 84% tend to reduce participation in sports and leisure activities, and have difficulty performing other daily tasks [2].

Ergonomic approaches have been proposed to reduce the biomechanical loads during manual propulsion. Alternative systems such as lever propulsion [3], geared wheels [4] showed potential benefits in reducing biomechanical demands but requires the user to adapt to a new propulsion mode. Pushrim activated power assisted

T. Ahram et al. (Eds.): IHSED 2019, AISC 1026, pp. 747–753, 2020.
https://doi.org/10.1007/978-3-030-27928-8_114

wheelchairs also showed promising results in terms of reducing loads [5]. Additionally, innovative mechanism of providing complimentary torque by a single motor to both of the rear wheels has been proposed [6–8]. However, still the conventional manual wheelchair design remains the most globally used means of seated mobility. Therefore, investigating wheelchair design factors that are related to optimal mobility performance is necessary to improve efficiency and reduce the biomechanical risk related to prolonged use of manual wheelchairs.

Wheelchair professionals must be aware that there are several important aspects in the design and configuration of manual wheelchairs that may influence propulsion loads, stability and ultimately, mobility efficiency. The position of rear wheels' axle relative to the seat/user, wheels and tires types, frame design and accessories, casters' size and material are among the most relevant factors [9]. Mechanically, the configuration of these aspects in the wheelchair design determine the equipment mass, size and mass distribution. Mass distribution has been shown to influence resistive losses during manual wheelchair propulsion [10].

Altering weight distribution in daily use wheelchairs may result in changes in wheelchair configuration, use of accessories and the carriage of belongings. The influence of such changes in the mechanics of wheelchair design may affect required force to propel the chair in daily mobility. In order to explore the effects of wheelchair design on propulsion forces, this study was set out to investigate the influence of added mass location on handrim forces of the rear wheels.

2 Materials and Methods

2.1 Participants

A convenience sample of five male subjects without physical disabilities (average age of 31.8 ± 8.4 years, average height of 1.73 ± 0.06 m, and average weight of 77.5 ± 6.4 kg) was recruited at the São Paulo State University (UNESP, Bauru, Brazil). Participants met the following inclusion criteria: (a) minimum age of 18 years old; (b) had no upper limb pain, injuries or disorders that could influence the manual wheelchair propulsion. Prior to data collection, volunteers were informed about the study and signed an informed consent form that had been approved by Ethics Committee of the FAAC/UNESP (Process 800.500).

2.2 Equipment and Procedures

A manual rigid frame wheelchair (Starlite, ORTOBRAS, Brazil) (total mass 13 kg) and the instrumented wheel SmartWheel system (ThreeRivers CO, United States) were used. The SmartWheel system consists of a sensing system which enables the analysis of the forces and movements due to the manual forces applied on the rear wheels of a wheelchair. The system weights 4.3 kg, and since it was connected only to the right side of the chair (participants' dominant side), a compensation weight of 2.6 kg was added on the opposite wheel (1.7 kg) in order to avoid the asymmetry.

In order to analyze the impact of changes of the center of mass in the system wheelchair/user were used six similar weight plates (1.3 kg each) totaling 7.8 kg. In this sense the test had as a variable factor the center of mass of the total wheelchair-weight, represented by the weight distribution of the plates in three different situations. The wheelchair was configured with three weight plates distributed radially on each rear wheel (3.9 kg), as shown in Fig. 1(a). This configuration simulates a possible variation of weight in the wheels according to the different models and brands of rear wheels as well as tires and handrim. In the second configuration, all the additional weight (7.8 kg) was placed on the center of the rear axle of the chair (see Fig. 1(b)), in order to simulate different models and materials of the wheelchair frame. The third configuration comprised fixing the six weight plates on the footrest. This mass distribution simulated changes in the mass concentration in the anterior part of the chair as a result of different parts e.g., different caster models and/or different angles of the user's legs (see Fig. 1(c)).

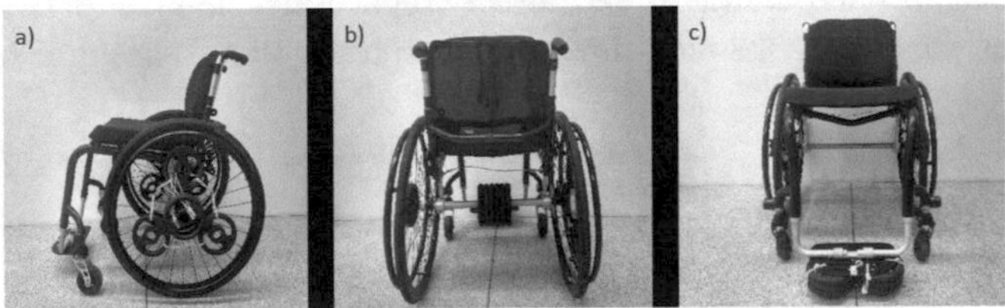

Fig. 1. Added mass on (a) the rear wheels' spokes; (b) on the center of rear wheels' axle; (c) under the foot support.

Participants familiarized themselves with the field test and the wheelchair before the data collection. The orders of the conditions were randomized to eliminate possible effects of learning or fatigue. The test comprised a sequence of four maneuvers performed in a flat surface: straightforward propulsion (five pushes) at self-selected speed; straightforward propulsion (five pushes) at maximum speed (sprint); 360° anti-clockwise rotation around the participants' own axis (zero radius turn) at a comfortable speed; one-meter radius anti-clockwise turn without a specific number of touches. Each of these maneuvers were repeated three times for each chair configuration. As the SmartWheel was placed on the right side, the turning maneuvers were performed in anti-clockwise direction so that the concentration of the forces occurred in the instrumented wheel.

All the trials of each participant were conducted during the same day, without the need for a rest between the maneuvers. In none of the collections were there complaints of the subjects due to the efforts employed in carrying out the maneuvers. The kinetic variables addressed in this study are: Peak total force (Peak Ftot [N]), Peak moment along the Z-axis (Peak Mz [N*m]), peak tangential force (Peak Ft [N]) Average total force (Ave Ftot [N]), average tangential force, peak/average force ratio.

3 Results and Discussion

Propulsion forces are important variables that directly affect the performance of manual wheelchair mobility. From the total force applied to the handrim, the component that most contribute to the wheelchair movement is the tangential force, although the axial and radial components are necessary to stabilize the hand-handrim coupling with enough friction for the push force actuation. From the perspective of handrim kinetics, manual propulsion efficiency refers to the ability of pushing the wheels with a greater tangential component. The analysis of the propulsion forces applied to the handrims during straight line displacement showed that the force in the first push was approximately 30% larger than the four consecutive pushes, which is due to the greater effort needed to put the chair in motion compared to the effort required to maintain the movement (see Fig. 2). Additionally, Fig. 2 shows that the handrim forces during the braking maneuver exhibited a similar pattern to the forward pushes, but in the opposite direction. From a clinical perspective, it is important to provide the user with lightweight wheelchairs and training of the correct propulsion technique for the first pushes, as high forces are factors that contribute with exposure to risk of injuries [11].

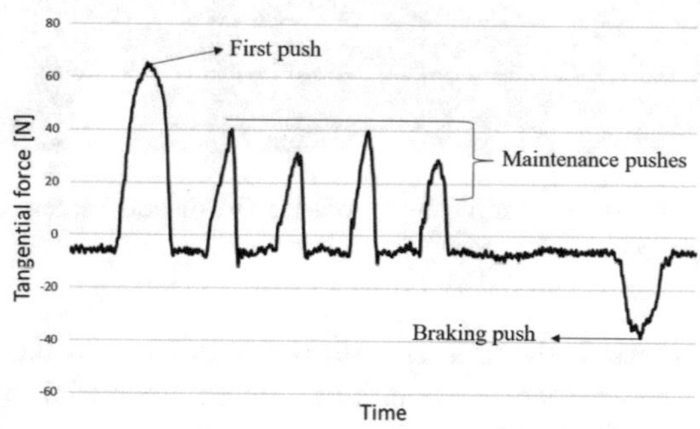

Fig. 2. Tangential force during the five consecutive pushes and a sixth action of breaking in straightforward propulsion.

Conversely, the addition of mass in the foot support led to an increase of 12.1% in the total peak force (Peak Ftot) in straightforward sprint, 2.2% in the straight line propulsion at self-selected speed, 14.3% for the zero-radius turn maneuver and 12.6% for the one-meter radius circle maneuver when compared to the configuration with the additional mass on the center of the rear wheels' axle. The analysis of the tangential component of the handrim force (Ft) showed an increase of 27.2% for the zero-radius turn and 17.2% for the 1 m-radius circle compared to the configuration with the additional mass on the center of the rear wheels' axle, demonstrating that the effect of the mass addition on the propulsion biomechanics depended on the location in the wheelchair geometry.

Table 1 demonstrates that, both peak and mean total and mean forces were lower when the wheelchair mobility tasks were performed with the wheelchair with the added mass located on the center of the rear wheels' axle in comparison to the added mass on the rear wheels' spokes and foot support. Furthermore, for most of the tasks, the condition with added mass under the foot support was associated with greater forces, possibly due to the weight concentration on the casters, which results in increased rolling resistance that, ultimately, demand stronger pushes from the user to maintain the expected velocity. Smaller wheel sizes are associated with increased rolling resistance [9]. Therefore, optimizing wheelchair configuration in a way that the wheelchair-user system's center of mass is positioned closer to the rear wheels can benefit propulsion efficiency by reducing rolling resistance.

Table 1. Average propulsive forces of the five participants, in the three conditions of added mass location and four different trajectories.

Configuration	Movement	Peak Ftot [N]	Peak Mz [N*m]	Peak Ft [N]	Ave Ftot [N]	Ave Ft [N]	Peak/Average force ratio
Weight on the rear wheels' spokes	Forward Self-selected speed	56.48	15.57	60.56	41.07	41.08	1.48
	Forward sprint	118.44	32.97	128.19	74.83	75.99	1.73
	Zero radius turn	44.17	10.95	42.57	33.49	29.62	1.42
	1 m radius circle	58.49	14.93	58.04	43.48	39.34	1.48
Weight on the rear wheel axle	Forward Self-selected speed	56.89	14.76	57.38	41.82	38.95	1.48
	Forward sprint	104.45	31.47	122.37	69.67	74.00	1.68
	Zero radius turn	40.58	9.56	37.18	31.88	26.58	1.37
	1 m radius circle	57.63	14.57	56.65	43.15	38.44	1.47
Weight under the footrest	Forward Self-selected speed	58.17	15.63	60.79	42.19	40.98	1.49
	Forward sprint	118.81	33.63	130.75	77.14	80.86	1.64
	Zero radius turn	47.37	13.13	51.07	34.48	31.76	1.64
	1 m radius circle	65.95	17.59	68.41	46.61	45.32	1.50

It is interesting to note the differences in the forces according to the mobility task. Considering the average total force, the less demanding situation was the zero-radius turn. When comparing straight line propulsion at self-selected speed and sprint, we found increases in the average total force of 66.6% with the added mass on the center of the rear wheels' axle, 75.7% with the mass under the foot support and 82.2% with the mass on the rear wheel spokes. A previous study demonstrated how trajectory influences kinetic energy of a wheelchair in motion [12].

Encouraging active wheelchair use can contribute to independence and social participation. A previous study demonstrated the benefits of practicing wheelchair sports for the users' quality of life [13]. In this context, a key factor is to provide wheelchairs with a design and configuration that allows optimal mobility performance.

This study has limitations that must be noted. The small sample size and the fact that participants had no previous experience with wheelchair usage limit the generalizability of the findings.

4 Conclusions

This study found that added mass location influences manual propulsion forces in different trajectories. Considering the same trajectory, lower handrim forces were found when the additional mass was positioned in a centered position on the rear wheels' axle, which is closer to the system center of mass. On the other hand, the condition that demanded greater forces was the one with the added mass under the foot support. As for the trajectory, forward sprint was associated with higher levels of force, followed by circular trajectory. The less demanding trajectory was the zero-radius turn. In the design, provision and adjustment of manual wheelchairs, one must consider the potential effects that changes in the mass and mass distribution as result of changes in wheelchair design, configuration and the use of accessories can have on the demand of propulsion forces. Optimizing wheelchair design to reduce handrim propulsion forces must be addressed in order to benefit mobility efficiency and minimize the biomechanical demand for the user.

Acknowledgments. The authors would like to thank FAPESP (Sao Paulo Research Foundation) for the financial support (Process. 16/05026-6 and Process).

References

1. Kabra, C., Jaiswal, R., Arnold, G., Abboud, R., Wang, W.: Analysis of hand pressures related to wheelchair rim sizes and upper-limb movement. Int. J. Ind. Ergon. **47**, 45–52 (2015)
2. Rossignoli, I., Fernández-Cuevas, I., Benito, P.J., Herrero, A.J.: Relationship between shoulder pain and skin temperature measured by infrared thermography in a wheelchair propulsion test. Infrared Phys. Technol. **76**, 251–258 (2016)
3. Lui, J., MacGillivray, M.K., Sheel, A.W., Jeyasurya, J., Sadeghi, M., Sawatzky, B.J.: Mechanical efficiency of two commercial lever-propulsion mechanisms for manual wheelchair locomotion. J. Rehabil. Res. Dev. **50**(10), 1363–1372 (2013)
4. Howarth, S.J., Polgar, J.M., Dickerson, C.R., Callaghan, J.P.: Trunk muscle activity during wheelchair ramp ascent and the influence of a geared wheel on the demands of postural control. Arch. Phys. Med. Rehabil. **91**(3), 436–442 (2010)
5. Lighthall-Haubert, L., Requejo, P.S., Mulroy, S.J., Newsam, C.J., Bontrager, E., Gronley, J.K., Perry, J.: Comparison of shoulder muscle electromyographic activity during standard manual wheelchair and push-rim activated power assisted wheelchair propulsion in persons with complete tetraplegia. Arch. Phys. Med. Rehabil. **90**(11), 1904–1915 (2009)
6. Medola, F.O., Purquerio, B.M., Elui, V.M., Fortulan, C.A.: Conceptual project of a servo-controlled power-assisted wheelchair. In: IEEE RAS & EMBS International Conference on Biomedical Robotics and Biomechatronics, pp. 450–454. IEEE (2014)
7. Lahr, G.J.G., Medola, F.O., Sandnes, F.E., Elui, V.M.C., Fortulan, C.A.: Servomotor assistance in the improvement of manual wheelchair mobility. Stud. Health Technol. Inf. **242**, 786–792 (2017)
8. Medola, F.O., Bertolaccini, G.S., Silva, S.R.M., Lahr, G.J.G., Elui, V.M.C., Fortulan, C.A.: Biomechanical and perceptual evaluation of the use of a servo controlled power-assistance system in manual wheelchair mobility. In: International Symposium on Medical Robotics (ISMR) (2018)

9. Medola, F.O., Elui, V.M.C., Santana, C.S., Fortulan, C.A.: Aspects of manual wheelchair configuration affecting mobility: a review. J. Phys. Ther. Sci. **26**(2), 313–318 (2014)
10. Sprigle, S., Huang, M.: Impact of mass and weight distribution on manual wheelchair propulsion torque. Assist. Technol. **27**(4), 226–235 (2015)
11. Morrow, M.M., Hurd, W.J., Kaufman, K.R., An, K.N.: Shoulder demands in manual wheelchair users across a spectrum of activities. J. Electromyogr. Kinesiol. **20**(1), 61–67 (2010)
12. Medola, F.O., Dao, P.V., Caspall, J.J., Sprigle, S.: Partitioning kinetic energy during freewheeling wheelchair maneuvers. IEEE Trans. Neural Syst. Rehabil. Eng. **22**(2), 326–333 (2014)
13. Medola, F.O., Busto, R.M., Marçal, A.M., Achour Junior, A., Dourado, A.C.: Sports on quality of life of individuals with spinal cord injury: a case series. Rev. Bras. Med. Esporte **17**(4), 254–256 (2011)

Exploration of TCM Health Service Mode in the Context of Aging Society

Hongwei Zhou[1(✉)], Ruifan Lin[1], Bin Wang[1], Ninan Zhang[1], and Qi Xie[2(✉)]

[1] Institute of Basic Research in Clinical Medicine, China Academy of Chinese Medical Sciences, Beijing 100700, China
zhouhw@ndctcm.cn
[2] China Academy of Chinese Medical Sciences, Beijing 100700, China
xieqixieqi@139.com

Abstract. With the development of medical conditions, aging of social population problem is becoming increasingly serious in China. For one thing, the elderly often at risk of functional decline. For another, because the distribution of medical resources is still unbalanced, contradiction between the problem of medical care and the aging population is increasingly apparent. How to effectively integrate old-age care and medical resources to improve the quality of life of the elderly has become a hot issue. Therefore, we should make full use of big data and other technologies to speed up the training of TCM rehabilitation nursing students. In addition, it is of great practical significance for the realization of "supporting the aged" and "treating the sick" through the development of TCM characteristic medical and nursing institutions.

Keywords: TCM · Health service mode · Aging society

1 Introduction

With the rapid development of China's economic and social undertakings, people attach more importance to the career of aging. Since the beginning of the 21st century, population of China has gradually presented a steady slowdown in the growth rate of aging and deepening the degree of fewer children. In order to layout the pension cause, China has issued important documents including Views on accelerating the development of old-age care service industry (No. 35 Document in 2013 of the State Council), Views on promoting the development of health services (No. 40 Document in 2013 of the State Council), Notice on accelerating the construction of health and old-age care service projects (No. 2091 Document in 2014 of development and reform investment) and Guidelines on promoting the integration of medical and health care services with elderly care. (No. 84 Document in 2015 of General Office of the State Council).

T. Ahram et al. (Eds.): IHSED 2019, AISC 1026, pp. 754–759, 2020.
https://doi.org/10.1007/978-3-030-27928-8_115

2 Present Situation of Population Aging in China

In 1956, the impact of aging and social economic from the United Nations proposed that when a country or region's population over 65 years old accounts for more than 7% of its total population, the country or region would belong to the aging society. In 1982, The first world congress on ageing adopted the Vienna International Plan of Action on Ageing. In 2005, the fifth national population census found that China's population aged 65 and above reached 88.21 million, accounting for 7.0% of the total population, China formally entered the aging society; It is predicted that by 2040, the proportion of 65-year-old people in China's total population will reach about 28% [1]. With the increase of the elderly, people who can't take care of themselves and empty-nesters, the problem of providing for the aged has become an important social problem to be solved.

Studies showed that with the increase of age, most of the elderly suffered a variety of chronic diseases such as respiratory, cardiovascular and endocrine diseases [2]. The main factors of the health status of the elderly include family situation, personal situation, health-related behavior and psychological cognitive factors [3]. Deng et al. [4] carried out a follow-up study on the disease status of 3,257 people in rural areas, urban suburbs and mountainous areas of Beijing from 1992 to 2004. This research unveiled that chronic diseases of the elderly mainly included hypertension, coronary heart disease, osteoarthropathy, cataract, chronic bronchitis, stroke and diabetes. Except for chronic bronchitis, the incidence of other diseases has been on the rise for more than ten years. Jiang et al. [5] conducted a questionnaire survey on 1,498 stroke patients over 60 years old living in changning district, Shanghai. The following results were obtained by studying the incidence of falls in the elderly in the past year. Among the 433 elderly patients who had at least one fall, 17 of them fell due to stroke, 36 of them fell due to environmental factors and 380 of them fell due to their own factors (including weak legs, dizziness, movement, lack of concentration and blurred vision). 273 pcoplc were injured by falls, 82 of them with fractures. Long-term insomnia can cause a variety of physical and psychological diseases. Xia et al. [6] conducted a cross-sectional survey on 312 elderly people in yangpu district of Shanghai by using Pittsburgh sleep quality index scale and Chinese medicine physique scale questionnaire and analyzed the data of the elderly who met the diagnostic criteria for insomnia. It was found that the insomnia rate of the elderly in yangpu district was 44.1%. The incidence of insomnia is higher in the elderly with Yang deficiency and qi deficiency and blood stasis. The incidence of insomnia in these populations is respectively about 23.6%, 22%, 15%.

3 Advantages of Traditional Chinese Medicine in Elderly Health Services

3.1 Fitness Exercises of Chinese Medicine

As one of the first national intangible cultural heritage projects in China, Tai Chi Chuan is characterized by the combination of Yin and Yang and the combination of hardness and softness. Wu [7] organized 60 elderly people to practice 24-style simple Tai Chi Chuan and 48-style Tai Chi Chuan for 50 min each time, 6 days a week. After a year, the mental and physiological measures of the elderly were compared before and after, and it was found that Tai Chi Chuan can significantly improve the elderly's ability of visual reaction and movement response, and improve the elderly's cardiopulmonary function. Huang et al. [8] has completed the systematic evaluation of the impact of Tai Chi Chuan on balance disorders. Through the analysis of four studies, it is shown that Tai Chi Chuan exercise has a positive impact on the dynamic posture stability of the practitioners in balance, which may improve the dynamic balance control and flexibility of patients with vestibular dysfunction, and is a useful vestibular rehabilitation treatment method. Apart from Tai Chi Chuan, there are also many different kinds of exercising ways about Baduanjin, six healing sounds (Liu-Zi-Jue), Yi-Jin-Jing and Rejuvenating health care exercises [9–11].

3.2 Traditional Chinese Medical Nursing

TCM nursing refers to the nursing work carried out under the guidance of the basic theory of TCM. Liu et al. [12] have achieved good results in nursing elderly patients with insomnia and improved the sleep quality of elderly patients suffered from this disease through the use of warm foot before going to bed, auricular acupoint, massage, medicinal diet conditioning and other nursing methods. At the same time, according to TCM syndrome differentiation, these old people are divided into Tanhuo Raoxin type, Tanre Neirao type, Yinxu Huowang type, Xinpi Liangxu type, and Xindan Qixu type. Li et al. [13] carried out a 24-week study on characteristic nursing of traditional Chinese medicine about 56 elderly patients. Before and after the nursing, 56 old people were surveyed by self - efficacy scale. Through the analysis of the questionnaire results, it is found that the characteristic nursing of traditional Chinese medicine (such as health care massage, traditional Chinese medicine fumigation, traditional Chinese medicine bath, foot scraping, cupping and moxibustion) has the advantages of cheap, convenient, experienced, and is deeply loved by the elderly. Jin et al. [14] attempted to analyze the safety and effectiveness of TCM nursing methods for patients with mild cognitive impairment from the perspective of evidence-based medicine. Through the study of 9 literature, it is found that although all literature prove the effectiveness of TCM nursing methods, most of them have problems such as short follow-up time and small sample size.

3.3 TCM Therapy

For a long time, TCM has accumulated extremely rich clinical experience in the field of prevention and treatment of senile diseases and formed a relatively complete theoretical system. Huang et al. [15] believed that the onset of diseases in the elderly is insidious, with atypical symptoms and poor self-perception. Once the disease occurs, it is urgent and critical. In clinical practice, it is necessary to restore the function of spleen, stomach and kidney. Fang et al. [16] made a retrospective analysis of traditional Chinese medicine treatment in the community of alzheimer's disease. It is suggested that in the community prevention and treatment strategy of alzheimer's disease, community physicians should be used as the leading dialectical treatment means, such as traditional Chinese medicine acupuncture and moxibustion. Patients can also carry out non-drug intervention activities such as exercise about diet therapy with the help of the guardianship staff.

4 The Integration of Pension and Healthcare to Innovate of TCM Health Service Model

The combination of medical and nursing care advocates the integration and promotion of medical resources and old-age care resources to meet the increasingly diversified health needs of China's aging population. Gong et al. [17] summarized the practice status quo of China's medical and old-age care integration mode, and proposed six integration modes of medical institutions and old-age care institutions. She believed that medical and nursing integration is an effective way to alleviate the contradiction between supply and demand of medical and nursing services. Xiao et al. [18] carried out a questionnaire survey on 60 nursing workers and 30 managers of 100 elderly people in 50 nursing homes in Beijing. The purpose was to seek their opinions on the status quo and the recognition of TCM nursing services in nursing homes. The results showed that only 11 of 50 nursing homes could provide TCM diagnosis and treatment services, and 17 nursing homes could provide TCM care services. This phenomenon is still in conflict with the old people's demand for TCM care services. In order to build the model of integrated medical and nursing service with traditional Chinese medicine, we need to start from three aspects about management mechanism construction, service personnel training and health service capacity improvement.

4.1 Establish and Complete the Mechanism for TCM About the Integration of Treatment and Care

We will encourage and support the establishment of traditional Chinese medicine clinics in elderly care institutions. Through the methods of promoting cooperation between old-age care institutions and hospitals of traditional Chinese medicine, the system of multi-point physician practice could be improved, the system of telemedicine services and referral green channel system could provide convenient service and high-quality service system for the elderly in pension institutions could be completed.

4.2 Strengthen the Training of TCM Elderly Care Personnel

Through short-term skills training, personnel could be sent to the hospital of traditional Chinese medicine for exchange learning to improve the ability of medical and nursing personnel in old-age care institutions. Government encourage institutions for higher learning in traditional Chinese medicine to develop personnel training programs for elderly care and guide doctors to provide long-term service in nursing institutions. By optimizing the performance management of pension service institutions, the salary of TCM pension service personnel will be improved.

4.3 Information Technology Will be Used to Improve TCM Health Services

Experts should make full use of big data, Internet of things, mobile communication and other technologies to build a health service platform for the elderly that is suitable for pension institutions [19], and provide health records or chronic disease management services for the elderly. At the same time, experts can provide personalized health consultation and TCM health service guidance based on the real-time health monitoring data of the elderly and the TCM knowledge base.

5 Prospect

TCM theory advocates prevention before disease onset, prevention disease form changing and prevention recurrence. This strategy is often used in the prevention and treatment of chronic diseases in the elderly (such as hypertension, diabetes, etc.). The methods involved in qigong, Daoyin, massage, foot bath and trophotherapy have the characteristics of simple, convenient, effective and inexpensive and easily accepted by the elderly [20, 21]. Traditional Chinese medicine health services provided in institutions for the elderly combining medical care with old-age care can significantly improve the quality of life of the elderly. The elderly in medical help to be able to get timely treatment, accomplish something and enjoy their old age.

Acknowledgments. This work was supported by the National Key Research and Development Program of China (Grant No. 2017YFB1002302), the National "Twelfth Five-Year" Plan for Science & Technology (Grant No. 2013BAH06F03) and the National Natural Science Funds of China (Grant No. 81403281).

References

1. Liu, W., Jiao, P.: Research on active aging from an international perspective. J. Sun Yat-Sen Univ. **55**(1), 167–180 (2015). (in Chinese)
2. Weng, D.: Advances and reflections on the health management of chronic diseases in the elderly. J. Tradit. Chin. Med. Manag. **27**(5), 202–203 (2019). (in Chinese)
3. Duan, W., Zhu, Y.: Health status and its influencing factors among elderly populations in China: a structural equation model analysis. Chin. J. Public Health **35**(7), 900–903 (2019). (in Chinese)
4. Deng, Y.M., Tang, Z., Wu, X.G.: Changes of disease spectrum of elderly patients in Beijing community. Chin. J. Gerontol. **29**(4), 867–870 (2009). (in Chinese)
5. Jiang, Y., Xia, Q.H., Hu, J.: Changes of disease spectrum of elderly patients in Beijing community. Chin. J. Gerontol. **29**(4), 867–870 (2009)
6. Xia, C., Zhu, Q.B., Hunag, F.: Traditional Chinese medicine constitution types in 127 elderly patients with insomnia: an investigation in communities of Yangpu District. Shanghai. J. Integr. Med. **10**(8), 866–873 (2012). (in Chinese)
7. Wu, Y.M.: Discussion of Taijiquan for health function in elderly. J. Changchun Univ. Chin. Med. **32**(1), 148–149 (2016). (in Chinese)

8. Huang, H.W., Nicholson, N., Thomas, S.: Impact of Tai Chi exercise on balance disorders: a systematic review. Am. J. Audiol. **2**, 1–14 (2019)
9. Zhou, H.W., Xie, Q., Liu, B.Y.: Progress on the impacts of Baduanjin exercise on physical and mental health of the elderly. Mod. Tradit. Chin. Med. Materia Med.-World Sci. Technol. **18**(04), 671–676 (2016). (in Chinese)
10. Tan, H.Q., Ai, Y.T., Li, Y.P.: Study on the effect of rejuvenating medical exercises on essential hypertension in the elderly. Lishizhen Med. Materia Med. Res. **25**(6), 1429–1430 (2014). (in Chinese)
11. Wu, W.W., Lan, X.Y., Kuang, H.R.: The effects of traditional exercises on sleep quality in older adults: a meta-analysis. Chin. J. Nurs. **57**(2), 216–224 (2016). (in Chinese)
12. Liu, S.M., Bai, R.C., Lu, X.: TCM nursing for elderly insomnia patients. Liaoning J. Tradit. Chin. Med. **42**(3), 617–618 (2015). (in Chinese)
13. Li, Z.Q., Wang, H.Y., Liu, Y.: Application research of TCM features care in elderly cadres. Lishizhen Med. Materia Med. Res. **25**(9), 2245–2246 (2014). (in Chinese)
14. Jin, Y.H., Shang, H.C., Xie, Y.L.: Tiao-She nursing of traditional chinese medicine for mild cognitive disorder: an assessment of clinical evidence. Chin. J. Evid.-Based Med. **15**(3), 346–352 (2015). (in Chinese)
15. Huang, F., Liu, T.H., Qiao, L.L.: Discussion on traditional chinese medicine treatment of geriatrics: based on the academic system of national geriatrics center of Beijing hospital. J. Tradit. Chin. Med. **60**(3), 265–267 (2019). (in Chinese)
16. Fang, R., Hu, J.Q., Ge, J.W.: Appropriate TCM intervention and its application strategies on Alzheimer disease patients in communities. World Chin. Med. **8**(6), 604–609 (2013). (in Chinese)
17. Gong, F.F., Sun, X.Z., Qiu, C.X.: Practical situation of the combination of medicine with pension in China. Mod. Hosp. Manag. **13**(2), 2–5 (2015). (in Chinese)
18. Xiao, J.L., Yang, Y.X., Zhong, Y.: Discussion on the nursing mode of integrated Chinese and western medicine in Beijing nursing homes. J. Qiqihar Med. Univ.—50 Nurs. Home Surv. **35**(22), 3351–3353 (2014). (in Chinese)
19. Zhou, H.W., Xie, Q., Liu, B.Y.: Design and application of Chinese medicine healthcare management system for senior housing. Mod. Tradit. Chin. Med. Materia Med.-World Sci. Technol. **18**(04), 688–691 (2016). (in Chinese)
20. Tong, Y.Y., Wang, Y.P., Su, Q.M.: Features, challenges and strategies of elderly TCM health services in China. Chin. J. Libr. Inf. Sci. Tradit. Chin. Med. **39**(5), 28–31 (2015). (in Chinese)
21. Yue, Q., Wang, Y.D.: Advantages and considerations of traditional Chinese medicine in community health care services for the aged. Adv. Considerations Tradit. Chin. Med. Commun. Health Care Serv. Aged **22**(2), 97–98 (2012). (in Chinese)

Standardized Research of Clinical Diagnosis and Treatment Data of Epilepsy

Ninan Zhang[1(✉)], Xinyu Cao[2], Liangliang Liu[3], Bin Wang[1], Huaxin Shi[4], Ruifan Lin[1], Yufeng Guo[5], Wenxiang Meng[6], Hongwei Zhou[1(✉)], and Qi Xie[7(✉)]

[1] Institute of Basic Research in Clinical Medicine, China Academy of Chinese Medical Sciences, Beijing 100700, China
znnzhangninan@163.com, zhouhw@ndctcm.cn

[2] Institute of Basic Standardization, Chinese Academy of Standardization, Beijing 100191, China

[3] School of Statistics and Information, Shanghai University of International Business and Economics, Shanghai 201620, China

[4] Hospital Management Office, China Academy of Chinese Medical Sciences, Beijing 100700, China

[5] Guang'an Men Hospital, China Academy of Chinese Medical Sciences, Beijing 100053, China

[6] Institute of Genetics and Developmental Biology, Chinese Academy of Sciences, Beijing 100101, China

[7] Department of Academic Management, China Academy of Chinese Medical Sciences, Beijing 100700, China
xieqixieqi@139.com

Abstract. Chinese medicine often uses natural language to describe disease symptoms, resulting in inconsistent symptom naming, affecting data mining analysis and clinical efficacy evaluation results. This study sorted out the terms of epilepsy symptoms by core symptom extraction, compound symptom resolution, semantic induction and semantic class method. The disease diagnosis terminology is compiled with reference to the International classification of Disease-Eleven Edition (ICD-11) Codes. Refer to the Pharmacopoeia of the people's republic of china to regulate the name of the Chinese medicine. The terminology for evaluation of efficacy is classified according to "healing", "improving" and "unhealing". The Standardized data can be used for data mining analysis.

Keywords: Epilepsy diagnosis and treatment data · Data standardization · Preferred term · Synonymous term

1 Introduction

Real-world study (RWS) refers to the research activities which using the information generated during the daily medical practice under the normal medical conditions [1]. Chinese medicine has a history of thousands of years, the most important factor for the inheritance is that the ancients kept their medical practice experiences in a record,

T. Ahram et al. (Eds.): IHSED 2019, AISC 1026, pp. 760–766, 2020.
https://doi.org/10.1007/978-3-030-27928-8_116

summarized in succession, inherited and innovated from generation to generation. It is because Chinese medicine characterized by attaches' importance to individual observation and accumulation that it is more in line with the value of human daily and real research. Objectively and truthfully sorting out the clinical information of Ying-Ao Yu–a famous prestigious Chinese physician, is an important part to inherit the academic thought and clinical experience of master Yu. It is the transformation from experience to evidence and valuable to improve the clinical efficacy of epilepsy. In this study, we use the diagnosis/treatment data of epilepsy which form the patient of master Yu, through direct extraction and indirect semantic extraction of symptom terms to dispose the data prepare for data mining. The main purpose of this study is to provide references to inherit the clinical experience of old-famous Chinese medicine.

2 Construction of Terminology Corpus for Clinical Diagnosis/Treatment Data of Epilepsy

First, we retrospectively obtained the medical records of the people with epilepsy from the capital famous clinic in Gulou Hospital (data from January 2014 to February 2018) and the Chinese medicine clinic in China Academy of Chinese Medical Sciences (data from January 2016 to February 2018). Then, we extract chief compliant, the current disease history, the past disease history, and prescription. Afterward, we take the patient ID number as the main index and use the comma as the separator to divide the 345 medical records into 3029 corpus data. Finally, the data were deposit into a worksheet and manual proofreading to and ensure accuracy and integrity.

3 Standardization of Terms of Epilepsy Symptoms

Symptom terminology is the carrier of clinical data. It is the main basis of disease differentiation and syndrome differentiation, especially the accuracy and standardization of the terminology. Although the real-world data is huge and derived from the real medical environment, its diversity of expression, natural language attributes, and unstructured features, which make it difficult to mine the data. Therefore, the standardization of the epileptic symptoms is particularly important.

3.1 Direct Acquisition of Symptoms

Core Symptom Extraction. Compared with the acquisition of symptom terms from standards, the symptom terms based on clinical medical records are more diverse and complicated, most including descriptions of symptom attributes, such as the nature, extent, duration of pain and aggravating or remission factors. Outpatient medical record data involves not only the main symptoms, complications, accompanying symptoms of epilepsy but also other diagnoses. These terms are more complex in conceptual and hierarchical than the standard documents. This study first extracted the 1674 core symptoms from the corpora database. For instance, sudden fainting at the time of the seizure, retaining core symptoms removal modifiers, and the term “sudden fainting” is obtained according to the core symptom extraction method.

Splitting the Compound Symptom. For the extracted 1674 core symptom terms, 675 symptom terms were obtained after deduplication, and 65 compound symptoms were selected. Mostly, the compound symptoms were composed of two symptoms and these symptoms were appear not simultaneously in clinical. After splitting it, the symptoms can be separate statistics, and it also could analysis after build the strong correlation between the two symptoms due to the term relationship. This is the reason why we split the compound symptom. For example, “dry mouth and thirst” is split into “dry mouth” and “thirst”.

3.2 Semantic Induction Indirect Acquisition of Core Symptoms

Semantic induction refers to summarizing the clinical symptoms according to the content described in natural language. It is a descriptive explanation of the symptoms and will certainly affect the clinical prescription. Based on the corpus of past electronic medical records, the corpus has narrative characteristics, such as “patient who came to see a doctor two years ago” does not contain symptom terms. Another example, although “the stool is every 3–4 days” is a narrative corpus and no core symptoms, but it can refine the core symptom from the semantic induction to “constipation”. The study obtained five symptom terms by inductive semantics.

3.3 Semantic Class

Semantic class is one of the types of semantic relations, which is classified according to the meaning of the word [2]. This study classifies semantically consistent or similar terms into a class, for example, classify forgetful, having a bad memory, memory loss, poor memory into a category. This study formed a symptom term of 204 by semantic classification. There are two types of semantic relations involved in semantic class. One is that the semantics are completely consistent, and the other is that the semantics are similar.

3.4 Preferred Terminology

Under the condition that the concept has both proper and alias name, according to the clinical terms habit or frequency of use, choose only one as the preferred term, and the rest are synonymous terms. The formulation of preferred terms should meet the following principles: (1) clear source; (2) conform to the Chinese medicine expression habits, modern expression is preferred when modern expression and ancient language expression co-occurrences; (3) semantic integrity; (4) semantic unique (5) self-made according to research needs; (6) when the source is unknown, the conditions (2) (3) (4) should be satisfied at the same time.

This study is based on the “Terminology working principles and methods” and “principles and methods for the examining and approving the Terms in Chinese Medicine”. Under the principles of singularity, conciseness and convention, we select the “WHO Western Pacific Regional Traditional Medicine Terminology International

Standards" [3], "Chinese Medicine terms" [4], "Chinese Medicine Common Terminology Dictionary" [5], "Chinese Medicine Dictionary" [6], "Chinese Medicine Symptom Diagnosis" [7], "Chinese medicine clinical common symptoms terminology" [8] as the reference to standardize the terminology of epileptic symptoms. The terms with clear origin are selected as standard terms in many similar expressions. There is also a type of terminology which is not included in the references and various knowledge platforms, such as Chinese Linguistics Research Center of Peking University, Terminology Online, WordNet, etc., one of which is selected as the preferred term according to the literature level, frequency of use, language habits, etc. A total of 163 terms for symptom selection were obtained in this study.

According to the principle of the preferred terms, this study removes the symptomatic attributes and retains the core symptoms as the preferred term when selecting the preferred term according to the research needs. Other terms with attributes are used as synonymous terms for the time being. There are 4 types of attribute types involved, including part attribute, degree attribute, feature attribute, and factor attribute. The preferred term is used for data mining analysis. The principle of preferred terms is selected according to the need of the research.

Table 1. Preferred term

Symptom acquisition method	Number	Symptom examples
Observation	35	**Head facial features:** Blink, Biting tongue, facial paralysis, etc. **Symptoms of consciousness:** coma, dull, slow in reacting, etc. **Limb symptom:** limb trembling, limb atony, foot shaking, etc. **Skin symptoms:** dry skin, furuncle, wheal, etc. **Secretory symptom:** salivate, excessive phlegm, Yellow sputum, etc.
Auscultation	18	**Sound symptoms:** Bowel sound, snoring, burp, etc. **Smell symptoms:** defecate smelly
Interrogation	109	**Cold and fad symptoms:** fever, cold, Hot body at night, etc. **Sweating symptoms:** Excessive sweat, No sweat, Night sweats, etc. **Pain symptoms:** Back pain, Shoulder pain, Rib pain, etc. **Head, chest, abdomen, limbs:** Brain sound, Chest tightness, belly swell, etc. **Ear and eye symptoms:** Dry eyes, Diplopia, Blurred vision, etc. **Sleep symptoms:** Sleepiness, Restless sleep, etc. **Dietary taste:** partial eclipse, Poor appetite, Like sweets, etc. **Defecation symptoms:** Dry stool, Urinary frequency, diarrhea, etc. **Emotional symptoms:** dementia, timid, Upset, etc. **Female symptoms:** Breast swelling, Dysmenorrhea, Irregular menstruation, etc. **Male symptoms:** Erectile dysfunction, etc.
Palpation	1	**Palpation:** Ankle edema, slippery pulse, taut pulse, etc.
Total	163	

3.5 Preferred Term Classification of Epilepsy TCM Symptoms

After the screening of the epilepsy corpus, we obtained 163 symptom terms, which were classified according to the classification framework of symptoms and signs [9], it is divided into 4 categories (Table 1).

4 Standardization of Diagnostic Terms

In the process of data sorting, the medical records of epilepsy patients which have other precise diagnostic results other than epilepsy were sorted. The diagnosis of Western medicine refers to the "International Classification of Disease-Eleven Edition (ICD-11) Codes". This study collated 23 diagnostic terms.

5 Standardization of Efficacy Evaluation Terms

The efficacy evaluation term is used to evaluate the overall condition of the patient, including a comprehensive response to the body and psychology, or to evaluate the relief or aggravation of a patient's independent symptoms and to respond a specific effect. Therefore, this study classifies the efficacy evaluation words into three categories according to their degree, which are "healing", "improving" and "un-healing". This study collated 21 efficacy evaluation terms.

6 Standardization of Chinese Medicine Names

If the Chinese herb has the only name in "The 2015 Edition Pharmacopoeia of the people's republic of china", the multiple aliases of this herb in the epilepsy medical record data are regulated by the standard. For example, the herb "Hei Chou" is standardized as "Qian Niu zi". Otherwise, "The 2015 Edition Pharmacopoeia of the people's republic of china" has multiple nomenclature, one of which is selected as the preferred term for Chinese herb. For example, "Yen Yang huo" "Xian Ling pi". We chose the "Xian Ling pi" as a preferred term.

7 Discussion

The clinical data is derived from the clinical outpatient medical records, and the terminology of the actual diagnosis and treatment is various. It is challenging to support data mining and analysis directly. Thus, the standardization of outpatient medical record data is essential to high-quality clinical research data. This study specifies 557 symptom terms, of which 163 were preferred, 394 were synonymous; 23 were diagnostic, and 21 were evaluated.

The actual clinical data of TCM is generated in the way of patient narrative, and the doctor collects the information through observation, auscultation, interrogation and palpation–four methods of diagnosis. In this process, because of the difference between the patient's idiom and the educational background, the doctor failed to convert it into medical terms in time, resulting in the diversity of the outpatient medical record data, the data is not standardized, so it is difficult to mining and analysis.

Li [8] collected a large number of literature and clinical medical records, compiled 2069 common symptom terms, and defined the terms; Zhang [10] and others take the ancient and modern masters' medical books and works to studied the symptom of TCM, and obtained 399 symptoms with the elementary connotation. And classify the symptom terms according to the clinical manifestation of the abnormalities of the five internal organs; Wang [11] and others obtained more than 100 symptom terms, more than 1500 kinds of Chinese and Western medicine diseases through the literature census method. Various scholars and clinical staff have different descriptions of this disease. There is no unified and accepted standard for TCM symptoms. Based on the previous exploration results, term research is still not satisfied. Therefore, from the perspective of data mining analysis, this paper sorts out the symptom terms, diagnostic terms, and names of herb in the medical record data, and the workload is mainly concentrated on the collation of symptom terms.

The preferred term for data mining analysis should be based on the mining needs to select the level of the term, such as insomnia, difficulty falling asleep, hard to sleep after wake up are the preferred terms. Insomnia is the superior term, sleep difficult and wake up after sleep is the inferior term of insomnia. During the data analysis, it needs to be based on the research needs. According to the characteristics of master Yu's clinical evidence, the study selects the superior terms in the preferred terms for data mining. There are 22 hyponymy relationships of terms involved in preferred term.

Acknowledgments. This work was supported by the National Key Research and Development Program of China (Grant No. 2017YFB1002302), the National "Twelfth Five-Year" Plan for Science & Technology (Grant No. 2013BAH06F03) and the National Natural Science Funds of China (Grant No. 81403281).

References

1. Xie, Q., Wen, Z., Liu, B., et al.: Technical General Guideline for Real-World Study of Chinese Medicine. China Association of China Medicine Group Standard, Peking (2017). (in Chinese)
2. Wang, H.: "Semantic class" "class Semantic" "Semantic field". J. Standard Sci. (11), 84–87 (2012). (in Chinese)
3. WHO Western Pacific Regional Traditional Medicine Terminology International Standards. Peking University Press, Peking (2009)
4. Chinese medicine terminology review committee. Chinese Medicine terms. Department of Science Press, Peking (2005). (in Chinese)
5. Li, Z.: Chinese Medicine Common Terminology Dictionary. China Press of Traditional Chinese Medicine, Peking (2001). (in Chinese)
6. Guangzhou University of Chinese Medicine, China Academy of Chinese Medical Sciences. Chinese Medicine Dictionary. People's Medical Publishing House, Peking (2005). (in Chinese)
7. Yao, N.: Chinese Medicine Symptom Diagnosis. People's Medical Publishing House, Peking (2004). (in Chinese)

8. Li, J., Ma, L.: Chinese medicine clinical common symptoms terminology. China Medical Science Press, Peking (2005). (in Chinese)
9. Dong, Y., Cui, M.: Discussion on classification of "symptoms and signs" in clinical terminology system of TCM. J. Chin. J. Med. Libr. Inf. Sci. **24**(10), 77–80 (2015). (in Chinese)
10. Zhang, Q.: TCM Symptom Research. TCM Publishing House, Peking (2013). (in Chinese)
11. Wang, Y., Wang, Z.: Standardized basis for pathological terms. People's Medical Publishing House, Peking (2015). (in Chinese)

Experience Design: A Tool to Improve a Child's Experience in the Use of Vesical Catheters

Natalia SantaCruz-González(✉), Mariana Uribe-Fernández, and Gabriela Duran-Aguilar

Facultad de Ingeniería, Universidad Panamericana, Álvaro del Portillo #49, 45010 Zapopan, Jalisco, Mexico
{natalia.cruz,mariana.uribe,gaduran}@up.edu.mx

Abstract. Urinary catheterization has been performed for more than three thousand years to drain urine. Over the years, its shape and materials have been changing to enhance the facility of usage and to reduce the possibility of infection. The perspective of the user experience has been taken little account in the procedure and the design of the device. This study is focused on the use of vesical catheters by children in Guadalajara, Jalisco at ages of 6 to 10 years old. The aim is to improve the user's experience during a vesical catheterization. This experience does not only include the insertion procedure, but the whole medical procedure. It is hoped that the proposed service and product will enhance the experience during the vesical catheterization. The results could help to be used as reference to improve the experience of diverse invasive or uncomfortable circumstances such as medical procedures.

Keywords: Vesical catheter · Medical service · Pediatric introduction · Medical psychology · Experience design

1 Introduction and Antecedents

Urethral catheterization is a manual procedure that involves inserting a catheter or probe through the urethra into the bladder to drain urine. It may be used for diagnostic purposes or therapeutically, to relieve urinary retention or incontinence, to monitor diuresis, or for hemodynamic control [1].

There are three types of urinary catheters depending on the time of use: intermittent catheter, as an in-and-out procedure for immediate drainage; short-term indwelling catheter, with a duration of less than 30 days; and long-term indwelling catheter, with a duration of 30 days or more [2].

Initially, the catheters were generally rigid and designed for intermittent catheterization in men. The earliest record of treatment of urinary retention dates of the year 1500 BC, with an ancient Egyptian papyrus (the Ebers papyrus) which shows the use of transurethral bronze tubes, reeds, straws and curled-up palm leaves. Since then, the materials and shape were changed to make it less uncomfortable and insecure [3].

T. Ahram et al. (Eds.): IHSED 2019, AISC 1026, pp. 767–772, 2020.
https://doi.org/10.1007/978-3-030-27928-8_117

In 1850, Auguste Nélaton developed a latex catheter, with a solid-tip and a single-eye. This catheter could be retained by adhesive tape or by a stitch. Then, in 1855, Jean François Reybard invented a self-retaining catheter, which had two channels, one for draining the urine and the other to inflate a balloon close to the tip to retain the catheter in the bladder. Dr. Frederic Foley later developed the "modern" balloon-based self-retaining catheter, which was placed on the market in 1933. This catheter made both short- and long-term catheterization feasible for both males and females. Its biggest problem was that it caused encrustation and infection with longer-term catheterization. Fortunately, in 2001, chemical impregnation and "antimicrobial" coating, particularly silver, was introduced aimed at inhibiting the formation of surface biofilms and encrustation [3].

Nowadays the indwelling Foley catheter is still one of the most used vesical catheters. This type of catheter is straight, has two or three ways/lumens and include a fixation balloon. The other type of catheter, the Nélaton catheter, is straight, semi-rigid and one-way, without fixation system. The materials currently used are latex, silicone, PVC and polyurethane [2].

Patients of all ages may require urethral catheterization. This is the reason for having different sizes. The calibers are given in Frenchs (Fr.), and they go from 3Fr. for newborns, to 18Fr. for adult men. The procedure is the same for children as for adults, though special attention must be paid on the delicacy needed [4].

The procedure for inserting the catheter is not very complex, but it should be performed very carefully. The privacy of the patient should be assured during the procedure. First, it is necessary to ensure the hygiene of the area and instruments to prevent infections. Then, the catheter is inserted into the urethra until the urine begins to flow and the balloon is slowly filled with sterile water to fixate it to the bladder if necessary. Finally, the catheter is fixated to the thigh to prevent movement [4].

2 State of Art

Over the years, people have made changes to improve urethral catheter design and its user's experience. The most common problems when using urinary catheters are infections and catheter blocks, therefore these are the most researched fields. For example, Dongsheng Dong created a urinary catheter capable of eliminating blocks using a dredging wire [5]. Also, Dong worked on a catheter with a flexible nipple-shaped segment that facilitates the placement of the urethral catheter in patients with physiological and pathological stenosis. [6]

According to a study made by O'Connell-Long et al., training and assessment is necessary to have a successful catheter placement. This training helps to avoid errors, which could lead to a bad experience [7]. Products, such as the one created by Meddings et al., aim to ease this process. This product is an assisting device for female catheter insertion, which includes a labia separator with flared wings having a curved profile and a posterior shield that covers the perineal tissues to prevent contamination of the sterile catheter with bacteria [8].

Using a urethral catheter is often met with significant ambivalence and even resistance by patients and their families. This has led to search for new options in which the patient can be explained the situation and accept it easily. One of these options is the Self-Cathing Experience Journal (SC-EJ), an online resource that collects stories and personal experiences from patients and their families about how it has been like to use clean intermittent catheterization. According to Holland et al., this journal appears to be safe, feasible, and useful to patients and families [9].

Also, a study made by the Radboud University Nijmegen Medical Centre showed how helpful it is to teach children and their families the catheterization process. They used a tell/show/do method within group interactions. This training includes sharing of mastery and difficulties with other children/parents, cognitive restructuring to enhance understanding and motivation, handling and trying out of devices, relaxation as a response to physical stress, and supporting parental guidance [10]. This process has proven to be a valuable additional option but it is not regulatorily used.

3 Problem Delimitation

This research was performed from February to May 2019. It is focused on children in Guadalajara, Jalisco at ages of 6 to 10 years old.

4 Justification

In Mexico, a country with a children population of 112'336,538 [11], only the 7.85% of the Health Research protocols are focused on pediatrics. From these pediatrics investigations, the 2.79% and 3.9% correspond to psychology-psychiatry and nephrology respectively [12]. The lack of investigation may be caused by the low expenditure in health issues. To illustrate, in 2016, only the 5.55% of the Gross Domestic Product (GDP) was expended in health questions [13].

It is estimated that 10% of the patients admitted to a hospital undergo bladder catheterization, and 10% of them will suffer a urinary infection [14]. In spite of the frequency of usage of a urinary catheterization, history demonstrates how the catheter has not changed drastically since the Foley and Nélaton's design. The improvements have been focused in the materials and its antibacterial properties. Little has been made to enhance the user experience.

Another important aspect is how the Mexican Norm NOM-052-SSA1-93 only specifies the catheter characteristics, but makes no mention of the procedure and how it should be performed or explained [15]. Also, the Secretary of Health in Mexico explains step by step the procedure, yet it does not specify how to approach the patient about the topic. It only requests to ensure that the patient, relative or caregiver has information about the cause of the catheterization, the revision plan and the possibility of removal [4]. The psychological approach is very significant, and especially in kids experiencing an invasive procedure such as a catheterization.

5 Statistical Methodology and Interviews

A hundred and forty-two people were surveyed, of whom 66.9% were female and 33.1% were male. The average age was of 38 years old, 45.1% were parents.

In regard to the subject of vesical catheters, 40.1% did not know what it is and 31% just had an idea about it. In terms of placement, the numbers decreased because just 17.6% knew how it is done. It is important to mention that 7.7% of the surveyed people had worn a vesical catheter and 52.1% of them knew someone who had worn it.

Focused on vesical catheters in children, on the assumption that the surveyed had a kid between 6 and 10 years old, 74.6% of the surveyed people will support him or her in everything they could. Also, 73.9% of them believes that the kid will react with fear, 32.4% with anxiety, and 23.2% with shyness.

Finally, 83.8% believe it is very important an adequate explanation when a kid is about to use a vesical catheter.

In addition, two interviews were performed to a urologist and a child psychologist to delve into the subject. The results showed the delicacy of the topic and the importance of an adequate approach to the patient and his or her family to embrace the situation. The need of an understandable explanation and a reduction of stress, embarrassment and/or discomfort during the whole procedure was highlighted.

In conclusion, urethral catheterization is not very known and understood. Also, it is perceived as a delicate subject, but it is important to do everything possible to improve it.

6 Service and Product Design

The proposed service is based in four principal aspects: privacy, hygiene, adequate explanation, and personalization. As mentioned before, privacy and hygiene are elemental for the catheterization process. The doors or curtains must be closed and the only people inside the cubicle must be the doctor/nurse, the patient, and one or two family members.

The personalization and adequate explanation were included with the aim of improving the experience for every user considering the age and personality differences. An adequate explanation refers to perform it before, during and after the catheterization process assuring understandability for all actors. In order to enhance this explanation a product will be added to the service.

The proposed product is a urinary drainage bag intended for 6 to 10-year-old users. The bag consists of two parts, the main part and the hanger. The main part comes with thematic graphics of the child's liking (Fig. 1 shows some examples). The hanger is also a toy of the same theme with the aim of distracting and relaxing the user in times of stress.

Fig. 1. Possible themes for proposed product.

At the beginning, the user chooses the theme he or she likes and the hanger/toy is given to him or her so that he or she becomes familiar with it. Depending on the chosen topic, when making the explanation to the patient, an analogy is used to relate the subject to the use of the urethral catheter. During the insertion and removal of the urethral catheter, the analogy is reinforced and the child is allowed to play with his or her hanger/toy so that he or she is distracted from unpleasant experiences.

During the use of the urethral catheter, the toy can remain as a hanger to place the bag where desired. In addition, the graphics of the main part of the bag are made with a special ink that is thermochromic, meaning that it changes color with a temperature change (Fig. 2). This way, filling and emptying the bag with urine becomes a more interactive and fun process for the child.

For example, if a kid chose the marine theme, an analogy about going fishing could be made. When the sterilization is done, it could be explained as when all the fishing equipment is being prepared. When the catheter is inserted, it could be explained as when the fishing net is thrown. When the drawings of the fish in the bag change color, it would mean that those fish were caught, and emptying the bag would be as taking them out of the net. Checking the urine would be as checking the captured fish. Finally, removing the catheter could be explained as if the fishing excursion was over. Various analogies could be done with different topics that children could relate to.

Fig. 2. Proposed product with thermochromic demonstration.

7 Conclusions

It was discovered that the vesical catheterization is a medical procedure that is usually associated with discomfort or seen as a delicate subject. Also, the majority of the latest improvements on this topic have been about preventing infections by changes in the materials, but the perspective of the user experience has been taken little account in the procedure and the design of the device. Additionally, the fact that children commonly experience fear to medical affairs makes the importance of taking into account the user experience more relevant.

The new service is focused on the improvements of the whole user experience by making sure that the patients and their family fully understand the process and avoid the fear to the unknown. Also, the product was designed with the aim of reducing the discomfort of this medical procedure by making it more understandable and familiar, and by providing a distraction to the uncomfortable situations and adding a positive, interactive and fun experience.

References

1. Romo, A., Remesz, O.: Infografía del procedimiento: Cateterismo Vesical.pdf - Google Drive. Barranquilla: Universidad del Norte (2018)
2. Jiménez Mayorga, I., Soto Sanchez, M., Vergara Carrasco, L., Cordero Morales, J., Hidalgo Rubio, L., Coll Carreño, R.: Protocolo de Sondaje Vesical Sondaje. Biblioteca Lascasas (2010)
3. Feneley, R., Hopley, I., Wells, P.: Urinary catheters: history, current status, adverse events and research agenda. J. Med. Eng. Technol. **39**, 459–470 (2015)
4. Secretaría de Salud de México: Procedimiento para el cuidado del paciente con sonda vesical durante la instalación, mantenimiento, retiro y detección de casos enfocado a la prevención de IAAS (2017)
5. Dong, D.: Urinary Catheter Capable of Eliminating Block, (2018)
6. Dong, D.: Stenosis Dilated Urinary Catheter (2018)
7. O'Connell-Long, B., Ray, R., Nathwani, J., Fiers, R., Pugh, C.: Errors in bladder catheterization: are residents ready for complex scenarios? J. Surg. Res. **206**, 27–31 (2016)
8. Meddings, J., Delancey, J.O., Ashton-Miller, J.A., Fenner, D.E., Saint, S.: Catheter placement assist device and method of use (2018)
9. Holland, J., DeMaso, D., Rosoklija, I., Johnson, K., Manning, D., Bellows, A., Bauer, S.: Self-cathing experience journal: enhancing the patient and family experience in clean intermittent catheterization. J. Pediatr. Urol. **11**(187), e1–187.e6 (2015)
10. Le Breton, F., Guinet, A., Verollet, D., Jousse, M., Amarenco, G.: Therapeutic education and intermittent self-catheterization: Recommendations for an educational program and a literature review. Ann. Phys. Rehabil. Med. **55**, 201–212 (2012)
11. INEGI: Población total por entidad federativa y grupo quinquenal de edad según sexo, 1990 a 2010. Instituto Nacional de Estadística y Geografía. INEGI (2010)
12. IMSS: Capítulo IX. Investigación en Salud (2017)
13. Secretaria de Salud de México: Gasto en Salud, 1990–2016 (2017)
14. Hospital General de México: INFORME DE AUTOEVALUACIÓN DEL DIRECTOR GENERAL DEL 1 DE ENERO AL 31 DE DICIEMBRE 2012 (2012)
15. Secretaría de Salud de México: Norma Oficial Mexicana NOM-052-SSA1-93., Ciudad de México (1994)

An Assistive Application for Developing the Functional Vision and Visuomotor Skills of Children with Cortical Visual Impairment

Rabia Jafri(✉)

Department of Information Technology,
King Saud University, Riyadh, Saudi Arabia
rabia.ksu@gmail.com

Abstract. Several characteristics of cortical visual impairment (CVI) can be resolved with careful assessment and planned, consistent intervention. Unfortunately, computer-based tools to support such intervention are few and provide limited activities, customization and data recording options. This paper, therefore, presents a novel mobile application that provides several activities to develop the functional vision and visuomotor skills of children with CVI. Several options are offered to customize the visual content and sensory stimuli (e.g., the visual complexity of the scene, the size, color, motion direction and speed of objects, and the audio/visual rewards) to each child's specific needs. The system also records data related to the child's performance and allows multiple teachers to keep track of their students' individual progress. The system is being developed in consultation with teachers for children with CVI to ensure that its design takes typical CVI characteristics into account appropriately and supports actual teaching practices for the targeted skills.

Keywords: Cerebral visual impairment · Visual attention · Visual tracking · Motor skills · Assistive technologies · Educational software · Children

1 Introduction

Cortical visual impairment (CVI) refers to all forms of visual impairment due to damage or malfunction of the visual pathways or visual processing centers in the brain, which interferes with communication between the brain and the eyes - the eyes may be structurally normal and be able to see, but the brain is unable to consistently interpret what is being seen [1–3]. CVI is currently the leading cause of pediatric visual impairment in developed countries and is becoming increasingly prevalent in developing nations [4].

The visual function of children with CVI can be affected in a wide variety of ways depending on the extent and severity of damage to the brain [3]. Children with CVI usually prefer specific colors, have delayed responses in looking at objects, exhibit atypical visual field preferences and visual reflexes, and have problems with visual complexity, distance viewing, visual novelty and viewing stationary objects; they also may have poor visuomotor skills finding it challenging to look at an object and reach for it at the same time [1, 3].

T. Ahram et al. (Eds.): IHSED 2019, AISC 1026, pp. 773–779, 2020.
https://doi.org/10.1007/978-3-030-27928-8_118

It has been shown that with careful assessment and planned, consistent intervention over an extended period of time, many of the CVI characteristics may resolve (i.e., diminish, have reduced effect or completely disappear) [3, 5]. The intervention requires a teacher/therapist to conduct several one-on-one sessions in which the child is carefully monitored to identify and record his visual preferences and characteristic behaviors and is then systematically provided with appropriate visual stimuli and environments tailored to his unique needs that encourage him to learn to use his vision in a sustained manner. Such sessions necessitate the provision of several customized visual materials, protracted observation, meticulous record keeping and long-term tracking of the child's progress by the teacher. Creating a computer-based application to support intervention sessions can alleviate some of these demands on the teacher. Such an application can conveniently and inexpensively provide a wide variety of visual stimuli and environments which can, furthermore, be adapted to the child's specific needs. It can also record data related to a child's responses either automatically (e.g., logging the time elapsed between a stimulus appearing on the screen and the child touching it) or through manual entry (i.e., the teacher can enter observation notes); it can further analyze and compile this data and present it in the form of summary statistics, graphs and tables which would aid the teacher in discerning long-term patterns in the child's progress. Moreover, to simplify session set up, the application can be configured to save and restore each child's preferences, thus freeing the teacher from the hassle of having to enter this information every time a new session is started.

However, currently only a few applications designed specifically to improve the skills of children with CVI exist [5] and those, too, are very limited both in terms of the variety of activities provided as well as options for customizing the activities to the child's age and specific needs; moreover, they do not offer any utilities for the teacher to record and track children's individual progress.

We have, therefore, designed a mobile application that provides several activities with a variety of visual content and cognitive challenges to develop the functional vision and visuomotor skills of children with CVI. The application is unique in that it offers several options to customize all aspects of the activities according to each child's specific needs and also records data related to the child's performance. Furthermore, it allows multiple teachers to create accounts for their students and keep track of their individual progress. Adopting a user-centered design (UCD) approach, the system has been developed in consultation with teachers at the Perkins School for the Blind, Watertown, MA, USA to ensure that the activities as well as the utilities for the teachers are compatible with and complement and support actual teaching practices for the targeted skills.

The rest of this paper is organized as follows: Sect. 2 provides an overview of existing applications, devices and research solutions for developing the functional vision and visuomotor skills of children with CVI. Section 3 outlines the requirements collection procedure and describes the initial design of the system. Section 4 concludes the paper and specifies some directions for future work.

2 Related Work

Though some computer-based products are available for developing the skills of children with ocular visual impairments [6–8], since CVI is brain-based rather than vision-based, it requires different teaching strategies which are not adequately and effectively supported by these tools [3]. Despite the increasing prevalence of pediatric CVI, only a handful of technological solutions exist that are specifically geared towards children with CVI. Tap-N-See Now [9] is a mobile application which teaches children with CVI color recognition, visual tracking and cause and effect; touching an object floating across a plain background enlarges it and plays a sound reward; the image, color, size and speed of the object, the background color and the sound can be changed using predefined settings. Big Bang Pictures [10] is a similar mobile application in which tapping an object triggers animation as well as sound effects; options to change the object's image, visual complexity and color, the background color and the length of the reward are provided. Vision 6 Lightbox [11] is an Android application which turns a tablet into a "light box"; options for selecting built-in and uploaded images and switching control between the child and the caregiver are provided; the light's intensity and color and the viewing time and number of displayed images can be adjusted. Several other simple mobile apps for developing the visual skills for young children have been recommended for children with CVI (e.g., [12, 13]) but these are not specifically designed for CVI characteristics. LightAide [14] is a device featuring a grid of 224 LEDs whose colors can be changed by the teacher to teach visual tracking, sequencing and organizational skills, literacy skills (such as alphabet and numbers) as well as social skills (such as turn-taking); up to four learners can simultaneously interact with it by pressing large switches. Our investigation into research efforts to develop engaging computer-based tools for children with CVI yielded only one study describing the initial design of a game-based system [5] which requires the child to find specified targets among arrays of colored shapes; clicking on a target triggers audio effects and occasionally visual effects.

Dedicated devices like LightAide [14] require the purchase of a new gadget and have fairly high prices which may place them beyond the financial reach of several of the intended users. Software applications developed for general-consumer mobile devices are more convenient both in terms of cost and accessibility but, as discussed above, only a few such applications exist and those, too, support just a couple of activities and offer very limited customization options; for instance, with the exception of Vision 6 Lightbox [11], none of the solutions reviewed above let the teachers/caregivers upload their own images or sounds. Moreover, none of these solutions save data related to the child's responses or offer any functions to support the teacher's ability to track the child's progress.

3 System Development

3.1 Requirements Collection

An unstructured interview was conducted with two teachers at the Perkins School for the Blind, Watertown, MA, USA to gain some insight into the technologies currently being utilized by them for developing the visual function and visuomotor skills of children with CVI, the challenges and limitations posed by these systems and desired features to include in such solutions. We found that the teachers currently use tools meant for children with low vision as well as some of the solutions (like Tap-N-See Now [9]) mentioned in Sect. 2. Their main frustration was that none of these systems allow them to upload their own images, sounds and animations; they asserted that such customization is crucial given the wide variability in the visual preferences and sensitivity to sensory stimuli among children with CVI. Another major concern was that these tools are not appropriate for children of all ages - the built-in images and sound/music are geared towards babies and pre-teens and thus, fail to evoke and sustain the interest of older children (14 to 16 year-olds); they suggested allowing teachers to define the visual and other sensory stimuli and providing more challenging tasks like puzzles, mazes and mystery solving games for resolving this issue. They also indicated that more touch options (tap, touch down, touch up, physical switch control) should be provided for visuomotor skills' development and agreed that recording and viewing the children's data would be beneficial.

An initial description of the application was then developed based on the teachers' recommendations and reviewed with one of the teachers; several aspects of the system were refined and modified based on her feedback. For instance, she recommended allowing the teacher to add notes after each activity session and letting her print out the child's data and to avoid sparkling effects (which we had suggested for one of the activities) since this would distract the child. When asked if audio or written instructions should be provided, she recommended adding this an option since it would give some independence to children who are able to understand such instructions. In response to our suggestion that the system automatically adjust the complexity of the visual stimuli within each activity based on the child's performance, she explained that the assessment of the child's progress is based mainly on the teacher's observations and is influenced by a number of factors, the combined effect of which may necessitate very different settings for the intervention tasks – hence, it would be preferable to give the teacher the control to customize all aspects of the activities according to her interpretation of the child's needs rather than have the system make that assessment.

3.2 System Design

The system is being developed for iPad devices since these are currently the most popular tablet computers by sales [15] and the school that we consulted with is already using these for its students with CVI to good effect. The software consists of two main components:

i. Activities for the children: Three activities are currently provided that require the child to do the following: (1) touch a brightly colored object floating on the screen, (2) select and move a star-shaped object so that it leaves behind a colored trail, (3) select one among two objects appearing on the screen and slide it towards the other one (e.g., give the ball to the cat). Data related to the child's progress such as the specific colors and objects he was able to perceive, touch and manipulate and the time taken to complete the task is also recorded. Audio and visual rewards (such as music, sound effects, confetti and fireworks) are offered for successfully completing a task in activities 1 and 2. Several customization options to tailor the objects (image, size, color, direction of motion, speed, and touch action), background color, background sound (audio content and volume) and the rewards (audio and visual) to the child's unique needs are provided; several pre-defined categories of objects (e.g., animals, flowers, etc.) of varying complexity as well as colors, sounds and visual effects are built-in; furthermore, the teachers can upload their own image and audio files for these attributes. Since visual complexity is the CVI characteristic with the strongest effect and is also the most protracted one [3], visually simple monochromatic object images with highly saturated bright colors and contrasting monochromatic plain uncluttered backgrounds have been provided as defaults for all the activities.

These activities target the following skills: visual tracking, visual discrimination (color recognition, object recognition), visuomotor skills (hand-eye coordination) and learning about cause and effect (touching an object or moving it to a specified target causes an audio/visual effect to occur). Furthermore, by observing the child while doing these activities, the teacher can also assess the child's color preferences, visual field preferences, visual latency, and need for movement. Note that there is a progression in the visual and motor challenges in the activities: since many children with CVI cannot handle visual crowding and are attracted to motion [3], the first activity provides a single moving object to help capture and sustain the child's attention; the second activity still features a single object – however, the object is stationary and the child has to perceive it, touch it and move it by keeping his finger or a pointer (if a switch is being used) on it; the third activity increases the number of objects to two, both of which are stationary – the child has to visually attend to both objects and maintain that attention while touching one object and moving it to the other one by keeping his finger or a pointer on it.

ii. An interface for the teachers: This allows a teacher to register and log in and to create, edit and delete accounts for her students. The teacher can also select a student and manage his activities: she can start an activity for him and give him the tablet, stop the activity when he is done and enter observation notes; she can also customize each activity according to the child's needs and preferences; these settings are saved and restored the next time the activity is started for that student. Furthermore, she can view and print data related to the child's progress; the data can be sorted and filtered based on various criteria such as date, time, activity type, etc.

4 Conclusion

We hope that our unique customizable assistive UCD-based software solution to support intervention for improving the visual function and visuomotor skills of children with CVI will practically benefit such children and their teachers and will inspire further research into developing technological solutions to enhance the skill sets of children with CVI and other neurological disorders. We plan to conduct usability testing of the system with students with CVI and their teachers at local schools for the visually impaired and enhance the activities based on their feedback. The system will be extended in the future to include more visually and cognitively challenging tasks with audio/textual instructions to accommodate older children and those with more advanced visual function. We also intend to further automate the monitoring and analysis of the children's visual and motor behavior – e.g., using an eye-tracking system and visual camera to record their hand and eye movements which can be analyzed using artificial intelligence-based techniques to gauge their visual alertness, visual attention, visual field preferences and hand-eye coordination.

Acknowledgements. We would like to extend our sincere thanks to Nathalie de Wit and Lindsey Lush from the Perkins School for the Blind, Watertown, MA, USA for their invaluable advice and suggestions for designing our system. We also want to sincerely thank Bedour Al-Rashed, Hadeel Al-Tammami, Raghad Al-Qassem and Sara Al-Fifi for their efforts to implement an initial prototype of the system.

References

1. Chokron, S., Dutton, G.N.: Impact of cerebral visual impairments on motor skills: implications for developmental coordination disorders. Front Psychol. **7**, 1471 (2016)
2. What is Cortical Visual Impairment (CVI)? http://littlebearsees.org/what-is-cvi/. Accessed 17 May 2019
3. Roman-Lantzy, C.: Cortical visual impairment: an approach to assessment and intervention. American Foundation for the Blind (2007)
4. Ospina, L.H.: Cortical visual impairment. Pediatr. Rev. **30**, e81–e90 (2009)
5. Linehan, C., Waddington, J., Hodgson, T.L., Hicks, K., Banks, R.: Designing games for the rehabilitation of functional vision for children with cerebral visual impairment. In: Proceedings of the extended abstracts of the 32nd Annual ACM Conference on Human Factors in Computing Systems, pp. 1207–1212. ACM, Toronto (2014)
6. Jafri, R.: Electronic braille blocks: a tangible interface-based application for teaching braille letter recognition to very young blind children. In: Miesenberger, K., Fels, D., Archambault, D., Peňáz, P., Zagler, W. (eds.) ICCHP 2014. LNCS, vol. 8548, pp. 551–558. Springer, Cham (2014)
7. Jafri, R., Aljuhani, A.M., Ali, S.A.: A tangible user interface-based application utilizing 3D-printed manipulatives for teaching tactual shape perception and spatial awareness sub-concepts to visually impaired children. Int. J. Child Comput. Interac. **11**, 3–11 (2017)

8. Alhussayen, A., Jafri, R., Benabid, A.: Requirements' elicitation for a tangible interface-based educational application for visually impaired children. In: Soares, M., Falcão, C., Ahram, Z.T. (eds.) International Conference on Applied Human Factors and Ergonomics (AHFE 2016). Advances in Ergonomics Modeling, Usability & Special Populations, pp. 583–596. Springer, Cham (2017)
9. Tap-N-See Now, Little Bear Sees. http://littlebearsees.org/cvi-ipad-app-tap-n-see-zoo/
10. Big Bang Pictures, Inclusive Technology, Ltd. http://www.inclusive.co.uk/big-bang-pictures-p2271
11. Vision 6 Light Box, Google Play. https://play.google.com/store/apps/details?id=air.Lightbox002a
12. Five Little Aliens, Inclusive Technology, Ltd. http://www.inclusive.co.uk/counting-songs-1-p4911
13. Peekaboo Barn, Night & Day Studios, Inc. http://www.peekaboobarn.com/
14. LightAide, Perkins School for the Blind Solutions, Philips. http://www.lightaide.com/
15. Kelly, M.: Global tablet sales decline, with only Apple and Huawei showing growth. The Verge (2018)

Structural Analysis of Spinal Column to Estimate Intervertebral Disk Load for a Mobile Posture Improvement Support System

Kyoko Shibata[1(✉)], Yu Suzuki[1], Hironobu Satoh[2], and Yoshio Inoue[1]

[1] Kochi University of Technology, Tosayamada, Kami, Kochi 782-8502, Japan
{shibata.kyoko, inoue.yoshio}@kochi-tech.ac.jp,
157sybaseball89@gmail.com

[2] National Institute of Information and Communications Technology,
Nukui-Kitamachi, Koganei, Tokyo 187-8795, Japan
satoh.hironobu@nict.go.jp

Abstract. In this study a support system to prevent by himself/herself before suffering from lumbago is developed. The load on the intervertebral disk is indirectly estimated using the measured shape of the lumbar part for that. In this paper, assuming that the lumbar vertebrae is a beam, the weight of the upper body and the rotation angle of the pelvis are considered as factors affecting its shape when the body bends forward. So structural analysis is performed. As a result, it was found that the upper moment of the spinal column has an effect of inducing a bad posture, the lower moment has an effect of maintaining the curvature of the spinal column to keep a good posture.

Keywords: Spinal column curvature · Beam · Deflection curve · Lumbar intervertebral disk load · Prevention of lumbago

1 Introduction

In recent years, patients complaining of lumbago have increased irrespective men and women of all ages. As one of the methods to prevent lumbago, this study is developing a system to support posture improvement. With system feedback, the user notices a bad posture and corrects. By continuing this, the user can maintain a good posture. Therefore, it is necessary to judge whether the current posture is good or bad. So we consider the load on the lumbar intervertebral disk as one of the indicators.

So far, Nachemson [1] had proposed the method to measure the disk load directly. This conventional method is highly accurate. However, because this method requires surgical operations, it is not easy and not always safe. Therefore, we have proposed a method to estimate disk load [2, 3]. The load is decided indirectly from the measured body surface shape at lumbar part. The results in this indirect method are qualitatively the same as those in the direct method [1].

On the other hand, in fact, the human spinal column is composed of the pelvis becomed the base, the sacrum where five are fused, the five lumbar vertebrae, the

T. Ahram et al. (Eds.): IHSED 2019, AISC 1026, pp. 780–786, 2020.
https://doi.org/10.1007/978-3-030-27928-8_119

twelve thoracic vertebrae, and the seven cervical vertebrae. The spinal column in sagittal plane depicts a loose S-shaped curve to receive the heavy head. Therefore, the rotation of the pelvis in the lower part and the weight of the upper body in the upper part are considered to affect the lumbar vertebrae so in the proposed estimation method, it is assumed that these effects are reflected in the shape of the lumbar part, then the lumbar load ratio is estimated from the measured body surface shape of lumbar part. Regarding the effect from the lower part that is the angle of the pelvis tilt, in previous report [4], multiple regression analysis using a simplified model revealed that the estimation accuracy is improved by considering it. In this report, we consider whether the proposed method is suitable. Focusing on the standing posture, first, the shape of the spinal column when the body is tilted forward is measured. Based on them, the effect on the shape of the lumbar part is examined by assuming that the spinal column is a beam and separating into the lower and the upper elements, where the lower element is the bending moment by the forced displacement of the pelvis, and the upper element is the bending moment due to the weight of the upper body.

2 Proposal of the Estimation Method

The overview of the proposed method to estimate disk load is described. Details are stated in the previous report [3]. As we change our posture, the vertebrae move and each intervertebral disk deforms, which change the gap between the vertebrae individually. We consider that the only factor that changes the gap is disk deformation. Thus, by using the gap change, the relative the intervertebral disk load can be derived.

We assume the following three things. First, the vertebra is a rigid body. Because, the Young's modulus of the vertebra is much larger than that of the intervertebral disk. Second, the facet joint of each vertebra is a pin joints. If we don't have something heavy, this axial force is negligibly small. Therefore, elastic deformation can be ignored. Third, the end of the vertebrae exists near the back of the body surface. By this, the motion of the vertebra can be guessed from the body surface. The motion of the vertebral ends on the body surface is read as position coordinates, and the lumbar part is curve-fitted to a quadratic function y(x). From y(x), the radius of curvature ρ is calculated.

$$\frac{1}{\rho} = \frac{\mathrm{d}^2 y(x)}{\mathrm{d}x^2} \tag{1}$$

An increase in radius of curvature is equal to a decrease in curvature of the lumbar part. This is equivalent to narrowing the gap between vertebrae. For example, tilting the body forward increases the disk load. At this time, the width between the vertebrae becomes smaller. Hence, the disk load can be estimated by using the radius of curvature, that is, by deriving the quadratic function of the lumbar part.

3 Measurement and Derived Approximated Cubic Function

First, we confirm the shape of the spinal column when the upper body is tilted forward at standing posture. There are many methods of measuring the lumbar vertebrae of the body surface. In this time, we use motion capture (MAC3D, Motion Analysis Corp.).

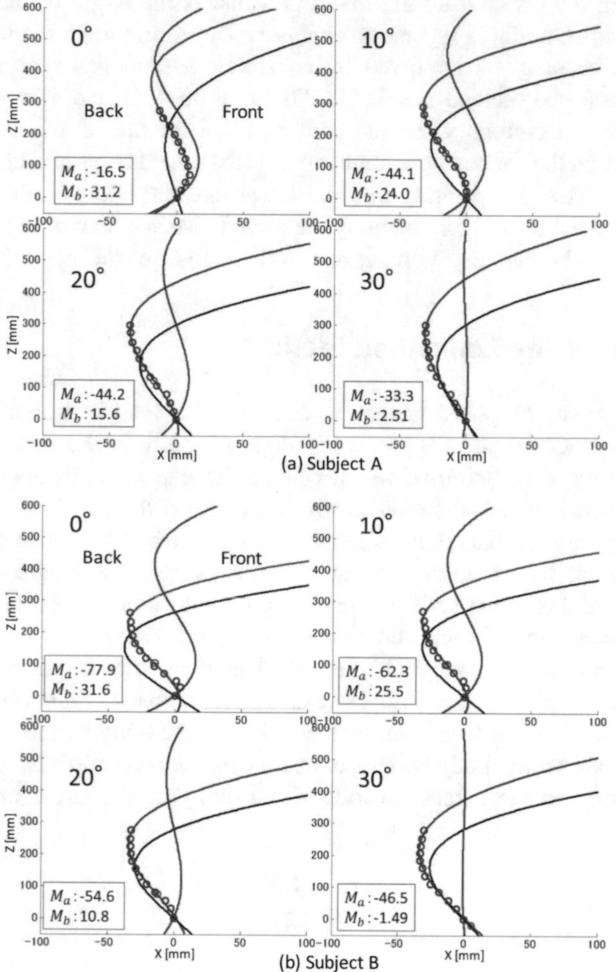

Fig. 1. Shape of spinal column and divided deflection curves for each posture. The red circles indicate 12 markers, the blue line indicates the approximated cubic curve, the blue circle indicates the inflection point of the blue line, the black line indicates δ_a due to M_a, and the red line indicates δ_b due to M_b.

Two healthy male volunteers in 20's participated. In this experiment we used twelve 6[mm] markers. These are pasted at positions on the body surface presumed to be vertebral ends in 6 lower parts of thoracic vertebrae, 5 lumbar vertebrae and one sacrum. Then the position coordinates of markers are obtained. We measured four standing postures which forward-bend angle is 0, 10, 20, or 30°. The sampling frequency in this experiment is 100 Hz. Each posture was measured for 3 s. In each case, it was measured twice at standing posture. The position coordinates of the obtained marker are averaged. Next, 12 markers averaged positions were approximated by a cubic function, and the shape of the spinal column was derived. The subject experiment was approved by the Kochi University of Technology Ethics Review Committee.

The red circles, blue lines and blue circles in Fig. 1 show the results for one trial of subject. In Fig. 1, the lowerest marker was taken as the origin, and the vertical line with this and the sixth cervical vertebra as the new Z-axis. The red circles in figure indicate the positions of 12 markers, the blue line indicates the approximated cubic curve, and the blue circle indicates the inflection point of the cubic curve. Since the position of this inflection point is the point at which the direction of the curvature of the spinal column changes, the inflection point may go down towards the pelvis when the spinal column becomes a "slouching" (in other words "stoop"). It can be seen from Fig. 1 that as the upper body is bended forward from 0 to 30°, the inflection point drops toward the pelvis in both subjects.

4 Effects from Upper and Lower on the Shape of Lumbar Part

The spinal column is modeled with both ends simply supported beam as shown in Fig. 2, and it is assumed that moments are applied from both ends. Figure 2 depicts the spinal column sideways, showing the upper part on the right and the lower part on the left. Then, the cubic function derived in Sect. 3 is separated into displacement waveforms caused by the moments at both ends. The lower effect means acting the bending moment by the forced displacement due to the tilt angle of the pelvis, and the upper effect means acting the bending moment by its own weight. Each of these two moments is back-calculated to determine how it affected the formation of the cubic curve.

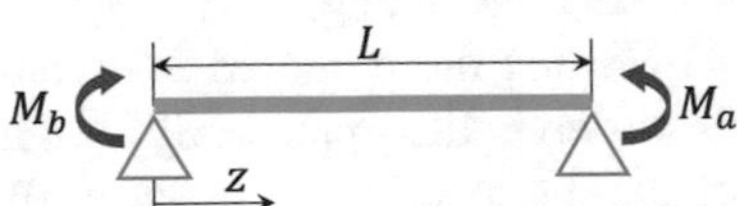

Fig. 2. The model of spinal column with beam.

In Fig. 2, the coordinate axis Z takes S1 (first sacrum vertebra) as the origin, and takes the spinal column upward direction to the right. M_a is the moment from upper, and M_b is the moment from lower. Assuming that the cubic function derived in Sect. 3 is generated by the two external force moments M_a and M_b, the cubic function can be considered as the deflection δ. Moreover, δ is represented by a linear combination of the deflection δ_a by M_a and the deflection δ_b by M_b. Here, if Young's modulus is E, moment of inertia of area is I, and length of the spinal column (T7 ~ S1) is L, then two deflections are stated by newly defined parameters a and b as

$$\delta_a = \frac{M_a L^2}{3EI}\left(\frac{1}{2L}z - \frac{1}{2L^3}z^3\right) = a\left(\frac{1}{2L}z - \frac{1}{2L^3}z^3\right) \tag{2}$$

$$\delta_b = \frac{M_b L^2}{3EI}\left(\frac{1}{L}z - \frac{3}{2L^2}z^2 + \frac{1}{2L^3}z^3\right) = b\left(\frac{1}{L}z - \frac{3}{2L^2}z^2 + \frac{1}{2L^3}z^3\right). \tag{3}$$

From Eqs. (2), (3)

$$\delta = \delta_a + \delta_b = \frac{(a+2b)}{2L}z - \frac{3b}{2L^2}z^2 + \frac{-a+b}{2L^3}z^3. \tag{4}$$

In other words, it became possible to separate the shape of the spine into upper and lower parts from the cubic curve derived from the measurement. Therefore, a and b were determined using the method of least squares so that each coefficient of Eq. (4) would be a coefficient of cubic function, then M_a and M_b were derived. Here, the Young's modulus and the moment of inertia of area were set to $E = 1.7 \times 10^8$[Pa] and $I = 1.45 \times 10^{-5}$[m^4] in any vertebra according to Refs. [2, 5]. In addition, the length L of the spinal column was the actual measured value from T7 to S1 for each subject.

The functions of Eqs. (2), (3) and (4) are shown again in Fig. 1. In figures, the blue line is a cubic function, that is, Eq. (4), the black line is the deflection δ_a due to the external torque M_a, i.e. Eq. (2), and the red line is the deflection δ_b due to the external torque M_b, i.e. Eq. (3).

The same tendency was found for both of subjects. First, focusing on the red line, an S-shaped curve is formed at a standing position of 0°, but as the body is tilted forward, the S-shaped curve gradually breaks and the curvature decreases. Focusing on the black line, as the body leans forward, the curvature gradually increases and the posture gradually changes slouching. Comparing the moment values, as the upper body bends forward, the lower moment M_b decreases and the upper moment M_a increases. Therefore, from Fig. 1, it can be considered that the lower element has the effect of making an S-shaped curve to keep a good posture, and the upper element has the effect of breaking into a bad posture like slouching.

From the above, it is shown that the proposed intervertebral disk load estimation algorithm reflects the effects of pelvic tilt, upper body weight, etc., and it is concluded that it is appropriate to estimate the load by the shape of the lumbar part.

5 Consideration

From the observation of the lower moment in Sect. 4, it was shown that the load on the lumbar vertebrae is increased because the force of maintaining a good posture is weakened. The greater the forward-bend angle, the greater the load on the lumbar vertebrae, which is clear both clinically and from its own sense. Here, we will examine the results of Sect. 4 further by comparing with the conception in this study presented in Sect. 2.

Each disk load is determined based on the method described in Sect. 2. In Sect. 2, one radius of curvature was determined from a quadratic function that was curve-fitted to 5 lumbar vertebrae, but in this report each 12 points including 5 lumbar vertebrae is used to approximate a cubic function, so a radius of curvature on between each point, that is each disk is determined respectively. These are used to derive the respective disk load ratios.

Figure 3 shows the results for two subjects. As for the results of both subjects, the load ratio increases as the forward-bend angle increases, and in addition, this tendency is more pronounced toward the lower disk. From Sect. 4, as the lower moment M_b weaken, the curvature of the spinal column decreases. On the other hand, as described in Sect. 2, when the curvature of the lumbar vertebrae decreases, the radius of curvature becomes large and the gap between vertebrae narrows, that is, the thickness of the disk decreases, so the load ratio increases. Consequently, both results were in agreement.

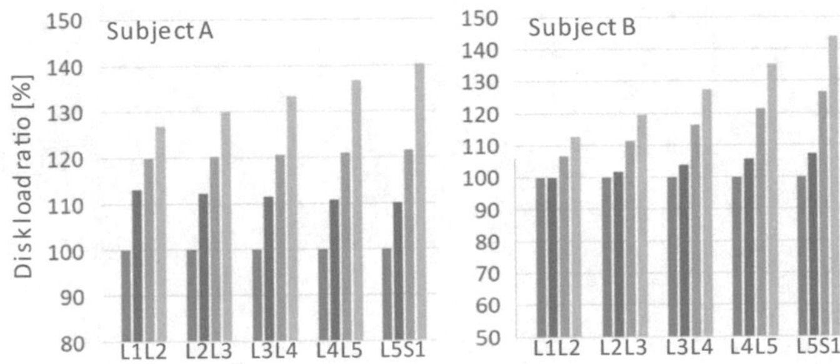

Fig. 3. Intervertebral disk load ratio.

6 Conclusion

In this report, structural analysis of the spinal column was performed to examine the validity of the load estimation algorithm proposed so far. As a result, it was found that the pelvic tilt works to keep a good posture, and the weight of the upper body works to guide a bad posture, and it is possible to estimate the load by the shape of the lumbar part. Based on this result, in order to prevent lumbago, the estimation algorithm is further enhanced and integrated into the proposed posture support system.

Acknowledgments. This work was supported by JSPS KAKENHI Grant Number JP18K11106.

References

1. Andersson, B.J.G., Örtengren, R., Nachemson, A., Elfström, G.: Lumbar disc pressure and myoelectric back muscle activity during sitting, I. Studies on an experimental chair. Scand. J. Rehab. Med. **6**, 104–114 (1974)
2. Shibata, K., Inoue, Y., Iwata, Y., Katagawa, J., Fujii, R.: Study on noninvasive estimate method for intervertebral disk load at lumbar vertebrae. Jpn. Soc. Mech. Eng. **78**(791), 2483–2495 (2012). (in Japanese)

3. Suzuki, Y., Shibata, K., Sonobe, M., Inoue, Y., Satoh, H.: Noninvasive estimation of lumber disk load during motion to improve the posture. In: AHFE 2017, pp. 578–588 (2017)
4. Shibata, K., Tsuyoshi, Y., Inoue, Y., Satoh, H., Sonobe, M.: Noninvasive estimation of lumbar intervertebral disk load using multiple regression analysis to consider the pelvic tilt. In: IHSED 2018, pp. 933–938 (2018)
5. Kudo, N., Yamada, Y., Ito, D., Honjou, A.: A basic examination of lumbar burden evaluation based on elastic beam theory. In: Proceedings of the 2016 JSME Conference on Robotics and Mechatronics, 16–2, 1A2-13b4 (2016). (in Japanese)

Human Cyber Physical Systems Interactions

Automated Decision Modeling with DMN and BPMN: A Model Ensemble Approach

Srđan Daniel Simić(✉), Nikola Tanković, and Darko Etinger

Faculty of Informatics, Juraj Dobrila University of Pula, Pula, Croatia
{ssimic,ntankov,detinger}@unipu.hr
https://fipu.unipu.hr/fipu/en

Abstract. Plethora of available heterogeneous transactional data and recent advancements in machine learning are the key forces that enable the development of complex algorithms that can reach human-level performance on an increasing number of tasks. Given the non-linear structure composed of many layers of computation, these highly accurate models are usually applied in a black-box manner: without a deeper understanding of their inner mechanisms. This hinders the transparency of the decision-making process and can often incorporate hidden decision biases which are potentially present in the data. We propose a framework for generating decision-making models conforming to Decision Model & Notation standard based on complexity-reducing techniques. An ensemble of decision-tree classifiers in a layered architecture is proposed to control the bias-variance trade-off. We have evaluated the performance of the proposed method on several publicly available data-sets tightly related to socially sensitive decision-making.

Keywords: Machine learning · Automated decision making · White-box models

1 Introduction

Understanding the behavior of intelligent agent algorithms is the key step in understanding what effects their outputs or decisions will have on our business processes. Generally, there is a need to reaping their benefits without accessing their harmful parts. Thus, a whole new field of science on intelligent machines as actors with behavior and their whole surrounding ecology was recently proposed [1]. The ecology part of machine learning algorithms is mostly considered to be their evolution which is affected through interaction with their surroundings and exchanging new information among themselves and their business operators.

In this paper, we are interested in the decision-making automation by applying Decision Model & Notation (DMN) [2] coupled with well-known BPMN process automation tools [3]. The primary outcome of this research is a framework for modeling decision-making processes based on the previous decision outcome data by applying machine learning techniques in a white-box manner. Understanding the model outputs, and the reasoning behind that output is the key requirement in avoiding unintentional bias.

T. Ahram et al. (Eds.): IHSED 2019, AISC 1026, pp. 789–794, 2020.
https://doi.org/10.1007/978-3-030-27928-8_120

Our contributions are the following: (1) a machine-learning ensemble method which is semantically compatible with the DMN specification, and thus human understandable, (2) a framework for applying this novel method on the existing decision logs - transactions of decision outcomes, (3) an algorithm for the automatic transformation of our proposed model into DMN artifacts.

The rest of the paper is organized as follows: Sect. 2 brings the definitions and motivation behind our contributions, Sect. 3 disseminates our contributions. Section 4 evaluates the proposed methods on three decision-making data-sets, and Sect. 5 concludes the paper with suggestions for future research.

2 Background and Motivation

The implementation of algorithmic operational decisions, an important driving force for a modern organization, should not be left solely to the software engineering teams. Automated decision-making should be carefully implemented, evaluated and continuously updated with the organizational business goals. Hence the implementation of optimal business decisions should be handled on the management level since documenting business requirements and offloading the implementation to software engineers is less efficient and error-prone due to misunderstandings. There is a level of model-driven automation present, but still it has not reach wider adoption [4].

The transition from a classical to a digital economy has brought new challenges: the three V's of Big Data: volume, velocity, and variety make the process of decision-making a challenging one and in a need for a continuous adaptation. Machine learning models in the role of predictive analytical models are the cornerstone in seeking insight into internal and external organizational factors.

Replacing multiple decision points embedded in business processes by pro- viding a method to derive DMN decision tables from the corresponding machine learning model was implemented in previous works [5]. But an important challenge still remains: how to extract knowledge from increasingly complex machine learning models that try to optimize the results based on historical transactions data in which unavoidable past decision biases are deeply concealed? Failure in efficiently implementing machine learning models can lead to undesired con- sequences [6]. The goal of Business Decision Management should be to gather insight from predictive methods, evaluate the biases and find the best trade-off in order to confidently deal with the risk of model drifts, regulatory compliance, and decision outcome performance [7].

3 Decision Framework Proposal

The overall framework steps can be observed in Fig. 1. The prerequisites for the proposed framework is the implementation of the ETL process (extract-transform-load) of data from the relational database into the Decision Tool modeling environment.

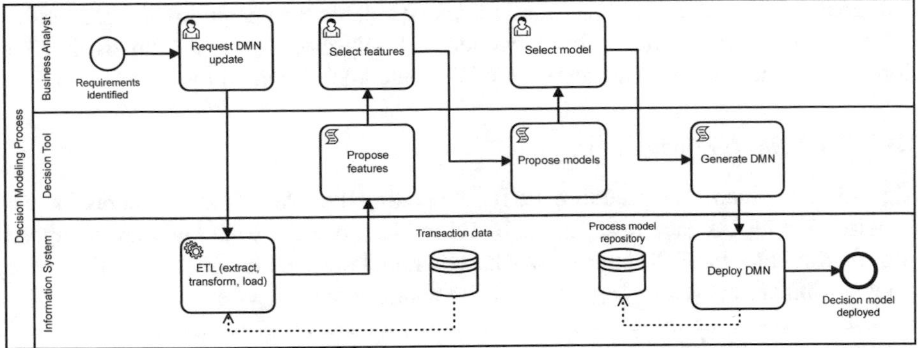

Fig. 1. The workflow of proposed decision framework

The framework is composed of three automated steps: (1) Feature proposition - based on the data provided, important features are provided to the business analysts (2) DMN model proposition - a layered machine-learning model is recommended based on the chosen features, (3) DMN table generation - automated generation of DMN tables from the proposed layered model.

3.1 Feature Proposition Task

The importance of individual features $f_i \in \Phi$, is determined based on training a Gradient Boosting Machine (GBM) classifier [8] on the data resulting from the ETL process. GBM consists of a desired number (n) of weak decision tree learners $wl_i,\ i \in [1, n]$ where feature importance corresponds to the number of splits throughout wl_i decision trees, or on the average gain of the feature when it was used in the model. The score of each feature $S(f_i)$ is reported in the interval [0, 1]. Decision tool then reports the proposed features to the user in the descending order of importance $S(f_i)$ and automatically suggests an optimal number of features. Optimal number of features is generating in a following way: $\bar{\Phi} = \{\mathrm{f}_i | \mathrm{S}(\mathrm{f}_i) > \psi * \bar{\mathrm{S}}\}$ where $\bar{S}$ is the mean feature importance and ψ is the inclusion parameter, which we recommend to set in the range [0.1, 0.5] based on the case studies we conducted. In this step special care should be given to exclude the features that should not contribute to the model according to regulations in the given domain (e.g. gender, ethnicity).

3.2 Model Proposition Task

Based on the agreed set of features with the business analyst, the decision framework tool is able to propose an ensemble of N_L decision-tree models: $\mathbb{E} = \{tm_i | i \in [1, N_L]\}$. The parameter N_L is configurable and reflects the overall complexity of the desired ensemble model. The models are organized in a layered style so that model tm_i suggests a decision $D(tm_{i,}\ \mathbf{x_i})$ for an input vector $\mathbf{x_i}$ only if the purity of the decision leaf $\mathrm{l}(tm_{i,}\ \mathbf{x_i})$ is higher than the configurable threshold π_i for that level, that is $purity(l(tm_{i,}\ \mathbf{x_i})) >= \pi_i$. Otherwise, the decision is offloaded to the next model $tm(i + i)$. In case that $\mathrm{i} + 1 > N_L$ the final decision is left to the human operator. Such a layered composition achieves two goals: (1) It a achieves a human-understandable composition of decision-tree models that can automatically be converted to DMN models, (2) It offloads the decisions where there is lower uncertainty of the outcome to human operators.

3.3 DMN Model Generation

Algorithm 1 contains the pseudo-code for generating the rules of DMN models for each decision-tree tm_i on each level. Final DMN models, one per each layer are composed into the final runtime BPMN process where the final decision in the case of uncertainty is left to human operators. Figure 2 depicts a case where $N_L = 2$.

Algorithm 1: Create DMN table rows

```
procedure generateTableRow(mlDictionary, annotation, isLastTable)
Input: mlDictionary is dictionary containing decision tree features, thresholds and logic
    for className, value in mlDictionary do
        newRule ← create new sub element of decision table
        if isLastTable == True then
            newAnnotation ← create annotation for rule with value annotation
        foreach featureName in value do
            thresholdSignList ← list of thresholds for each featureName
            if len(thresholdSignList) == 0 then
                ruleCell // empty in this case

            else if len(thresholdSignList) == 1 then
                ruleCell ← contains threshold and threshold sign
                // if threshold sign is "is" or "not" threshold becomes string value
            else
                if thresholdSignList[0] is "<=" then
                    if threshold[0] > threshold[1] then
                        ruleCell ← create range cell
                if thresholdSignList[0] is ">" then
                    if threshold[0] < threshold[1] then
                        ruleCell ← create range cell
                if thresholdSignList[0] is "not" or "is" then
                    ruleCell ← if positive signs in list, write them first, negation last
        end
    end
    newOutputCell ← rule sub element, contains className
```

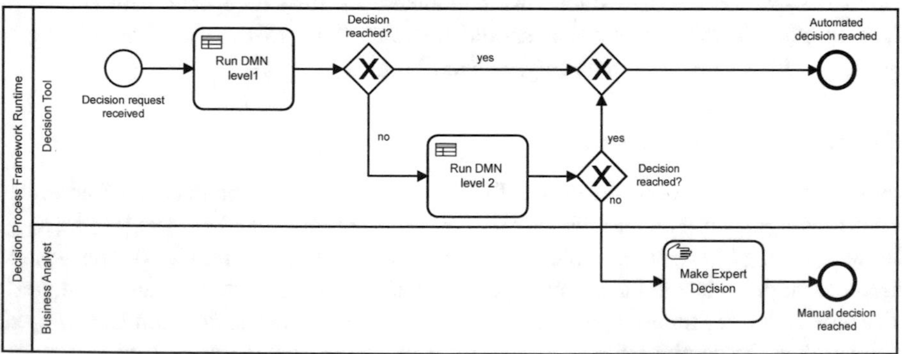

Fig. 2. Decision process framework runtime

4 Evaluation

We have evaluated our layered ensemble model on three distinct data-sets: (1) Credit risk assessment dataset[1], (2) Medical insurance claims dataset[2] and (3) Risk of prisoner recidivism dataset[3] and compared the results of the layered model against the simple decision-tree classifier and the black-box GBM model. The results are expressed through a *F1* score [9] and presented in Table 1. Results are obtained through a 10-fold leave-one-out cross-validation technique. We can observe that from our preliminary results, our layered model with $N_L = 5$ obtains more accurate results than simple decision-tree models. Such results are closer to more complex, but incomprehensible GBM models with 300 trees in ensemble. A much more comprehensive evaluation is left for future studies. It is our goal not to compare only the accuracy and the recall of the model (*F1* score) but also the complexity and present the trade-offs when tuning the parameters of our layered model. Each decision generated by the GBM model in these examples is calculated from the whole set of 300 decision trees, where our approach can map a decision only to a single tree (from a specific layer). We argue that this is a significant step forward in model comprehensibility.

Table 1. Preliminary evaluation results based on the F1 score.

	Decision-tree	Layered model	GBM
Credit-risk	0.700 ± 0.023	0.804 ± 0.028	0.830 ± 0.030
Insurance claim	0.942 ± 0.021	0.952 ± 0.018	0.987 ± 0.011
Recidivism	0.644 ± 0.036	0.650 ± 0.019	0.654 ± 0.022

5 Conclusion

In this paper we have presented a layered decision-tree model that is semantically compatible with the DMN specification. We have proposed a Decision-making framework and a tool for automatically building layered models and translating them into the within BPMN processes. In a brief preliminary evaluation, we have also showed that our approach reaches higher accuracy than the simple decision tree models, support-vector machines and linear models. In the future research we plan to conduct a more thorough evaluation of our proposed methods, as well as provide an optimization framework in tuning our model parameters.

[1] Available at https://www.kaggle.com/uciml/german-credit.

[2] Available at https://www.kaggle.com/easonlai/sample-insurance-claim-prediction-dataset.

[3] Available at https://github.com/propublica/compas-analysis.

References

1. Rahwan, I., Cebrian, M., Obradovich, N., Bongard, J., Bonnefon, J.-F., Breazeal, C., Crandall, J.W., Christakis, N.A., Couzin, I.D., Jackson, M.O., Jennings, N.R., Kamar, E., Kloumann, I.M., Larochelle, H., Lazer, D., McElreath, R., Mislove, A., Parkes, D.C., Pentland, A., Roberts, M.E., Shariff, A., Tenenbaum, J.B., Wellman, M.: Machine behaviour. Nature **568**(7753), 477–486 (2019)
2. Decision Modelling Notation (DMN) 1.2 Specification, Version 1.2 (2019)
3. Business Process Model and Notation (BPMN) 2.2 Specification, Version 2.2 (2014)
4. Tanković, N., Vukotić, D., Žagar, M.: Rethinking model driven development: analysis and opportunities. In: Proceedings of the ITI 2012 34th International Conference on Information Technology Interfaces, pp. 505–510. IEEE (2012)
5. Etinger, D., Simić, S.D., Buljubašić, L.: Automated decision-making with DMN: from decision trees to decision tables. In: Proceedings of the 42nd Inter- national Convention on Information and Communication Technology, Electronics and Microelectronics. Croatian Society for Information and Communication Technology, Electronics and Microelectronics (2019)
6. Dressel, J., Farid, H.: The accuracy fairness, and limits of predicting recidivism. Sci. Adv. **4** (1), eaao5580 (2018)
7. Sinur, J., Geneva, R., Debevoise, T., Taylor, J.: The MicroGuide to Process and Decision Modeling in BPMN/DMN: Building More Effective Processes by Integrating Process Modeling with Decision Modeling. CreateSpace Independent Publishing Platform, October 2014
8. Ayyadevara, K.V.: Gradient boosting machine. In: Proceedings of the Machine Learning Algorithms, pp. 117–134. Apress (2018)
9. Goutte, C., Gaussier, E.: A probabilistic interpretation of precision, recall and f-score, with implication for evaluation. In: Proceedings of the 27th European Conference on Advances in Information Retrieval Research, ECIR 2005, pp. 345– 359. Springer, Heidelberg (2005)

Potential of Industrial Image Processing in Manual Assembly

Alexander Nikolenko(✉) and Sven Hinrichsen

Ostwestfalen-Lippe University of Applied Sciences and Arts, Campusallee 12, 32657 Lemgo, Germany
{Alexander.Nikolenko,Sven.Hinrichsen}@th-owl.de

Abstract. As customers' options for configuring products to match their requirements increase, the number of assembly variants grows. Due to this large number of variants, assembly processes often cannot be automated in an economical way, and manual assembly remains highly important. Additional support options must be implemented to continue completing manual assembly processes reliably in the future. Image processing systems are one promising approach. The purpose of this paper is to establish the potential offered by industrial image processing in manual assembly, building on fundamental concepts, as well as to identify requirements and provide recommendations for selecting and arranging system components.

Keywords: Industrial image processing · Manual assembly · Assistance systems · Machine vision

1 Introduction

Manual assembly is becoming more and more complex. Drivers of this complexity primarily include an increasing number of product variants and components to be mounted, as well as an increase in mechatronic products [1, 2]. In light of this development, fully automated assembly processes often cannot be implemented economically. Manual assembly will remain highly important into the future. Humans, with their cognitive and motor skills, are able to complete complex assembly tasks. However, studies show that error frequencies increase as tasks become more complex [3]. Therefore, additional systems should be used to support humans in reliably carrying out manual assembly processes. Image processing systems are one promising approach. They support employees with continuous objective inspections, supplementing human visual and cognitive abilities through "machine vision". In manual assembly, however, the use of image processing systems is a unique challenge, since the lighting situation can change throughout the day, for instance, a large number of different testing features must be taken into consideration, or a camera zoom function must be used.

Against this background, the goal of this paper is to illustrate the potential of industrial image processing for manual assembly. To do so, requirements for the use of image processing systems in manual assembly are identified based on a case study, and recommendations for selecting and arranging system components are provided. To

T. Ahram et al. (Eds.): IHSED 2019, AISC 1026, pp. 795–800, 2020.
https://doi.org/10.1007/978-3-030-27928-8_121

achieve this objective, the process is divided into four steps. The theoretical foundations of industrial image processing are described in the first step. In the second step, key requirements for image processing systems in manual assembly are illustrated using a case study, and potential uses and limits for industrial image processing in manual assembly are discussed. In the third step, recommendations for selecting and arranging the individual system components are formulated. Finally, a conclusion is drawn in the fourth step.

2 Image Processing Systems

In its simplest form, an image processing system consists of the components camera, lens, lighting, image processing program, and computer [4]. The camera is the most technically complex component. It creates photographs of the real assembly scene to be inspected, which are analyzed by the computer [5]. The sensor is the main component of a camera. It is responsible for converting light (photons) into electrical signals (electrons). The sensors used in industrial image processing are based on CCD (Charge Coupled Device) and CMOS (Complementary Metal-Oxide Semiconductor) technology, whereby CMOS sensors are increasingly replacing CCD sensors [6]. Especially in light of its noise properties, the freely configurable pixel access, lower production costs and avoidance of blooming and smearing effects, CMOS technology has seen extensive development [7]. Blooming is the term for a light overexposure in the image. Smearing is an effect where a vertical line emanates from a very bright area of the photograph, stretching across the entire height of the photo [8].

In addition to the sensor technology, the sensor size and format play an important role. The same number of pixels can be displayed larger on big sensors. If pixels are the same size, more of them can be displayed. The advantage, for instance, of a 2/3" sensor in comparison to a 1/1.8" sensor is that larger pixel surfaces are available due to the larger surface of the sensor. More light can be absorbed over the larger surface. Consequently, sufficiently bright photographs can be taken with a shorter exposure time despite poor lighting conditions [9].

The selection and arrangement of lighting are key to the success of an image processing application [10]. Without sufficient lighting, even high-quality camera systems will deliver only a mediocre image, or one that cannot be analyzed. Object features that are poorly lit typically cannot be made visible later on, even using software [11]. The work station must be lit sufficiently in a stable manner over time and homogeneously throughout the space. However, glare at the work station is unpleasant and should be avoided. Light sources should be shielded and not installed in the worker's field of view [12]. The trend in industrial image processing is clearly moving towards light emitting diodes (LEDs) [6, 13]. This is due to the following properties of LEDs: Lower power consumption, long service life, high shock and vibration resistance and broad spectrum of colors. In addition, each LED can be activated individually, making it possible to illuminate an object from different directions successively [13].

3 Requirements Analysis

Testing was completed at a manual assembly work station using a complex product. The product was an assembly from a milling aggregate. It consisted of 34 individual parts, with different material properties and shapes. Assembly was monitored using a monochrome 2 MegaPixel area scan camera – equipped with a 1/1.8" CMOS sensor – aligned vertically to the assembly device. The lens used had a fixed focal length of 12 mm and an f1.6 aperture. The lighting consisted of two fluorescent tubes shining directly on the work station without a diffusor.

One important requirement for image processing systems in manual assembly is to select and arrange all of the components of the system so as not to create any additional strain for employees, and allow them to achieve their best possible work performance. In addition, taking photographs in line with requirements is a unique challenge, since assemblies typically include a large number of components, resulting in many different product features that need to be taken into consideration. In addition, the shapes, properties or sizes of the individual objects to be inspected typically vary. For example, it makes a difference whether a matte, glossy, even or uneven surface needs to be illuminated when selecting lighting technology. In addition, lighting circumstances are often not constant, since daylight influences them. An overview of the specific requirements of manual assembly is provided in Table 1.

Table 1. Requirements for image processing systems in manual assembly

No.	Requirement
1	Large number of parts, each with different features in terms of shape, properties or size
2	Significant change in distance between the camera and product during assembly (typically, a product is assembled from bottom to top)
3	Different product variants, each with different inspection processes
4	Changing lighting circumstances, in particular due to the influence of daylight
5	Occupational safety requirements (such as no use of UV light)
6	Employees with different skills and different requirements for support from an image processing system
7	No disruption to human movement processes due to the arrangement of system components/compatibility between image processing system and work process

4 Work Station Design and Component Selection

In order for the inspection to be successful, the image must be taken in a manner appropriate for the specific application. The camera must be installed so that it does not interfere with workers during assembly, but can still record each assembly step. Positioning the camera above the work station, aligned vertically to the inspected object, offers the best potential for manual assembly. Ensure that the camera is firmly fixed, since a deviation from its original position can result in problems when carrying out inspection tasks. As an alternative to a fixed camera arrangement – like the one used in the test – multiple cameras can be installed, especially with large and complex assemblies, so that the assembled object can be inspected from multiple sides without having to handle it.

Lenses in industrial image processing typically have a fixed focal length for a fixed object size. If the lens is attached to a specific position which does not change, this may result in small objects not being displayed large enough for the analysis. Zoom lenses, in contrast, have a focal length that can be changed. These are suitable for use in dynamic environments (such as different object sizes, changing distances between the assembled object and the lens). Accordingly, due to its high flexibility a zoom lens is the best choice for fulfilling the requirements of manual assembly.

Depth of field is another important variable. To achieve the best possible depth of field, the lens's aperture must be mostly closed. Closing the aperture results in less light hitting the camera sensor. The image is darker. Accordingly, the aperture must be selected and set, based on the product, so that the depth of field covers the entire height of the manually assembled product. For taller products, the aperture must be mostly closed, resulting in less light hitting the sensor and requiring a longer exposure time. The size of the sensor is key for an image processing system in manual assembly in this respect, since the worker must be taken into consideration when designing the lighting, in contrast to automated production. The assembly work station cannot be lit extremely brightly, since this could affect the worker in their assembly work, for instance by creating glare. Accordingly, the image could appear too dark if a small sensor is used or if the image has a large number of pixels.

The lighting must illuminate the assembly work station in a stable manner over time and homogeneously with sufficient brightness. Furthermore, a sufficiently bright image with enough contrast must be generated for the image processing system. Components look different when they are illuminated from different directions. Likewise, light is reflected differently by a matte surface than by a glossy one. The arrangement of the lighting must allow the worker to work effectively and without disruption, while ensuring a bright and homogeneous image for the image processing system. It can be helpful to attach a light source in an oblique angle to the left and right over the work area. The light sources will then point obliquely on the assembly device or the product on it. The assembly device must therefore be fixed in the center between the light sources.

LEDs are an appropriate light source. They can emit visible, ultraviolet (UV) or infrared light (IR). UV light is not an option for use in manual assembly, since it can damage human skin and eyes. IR lighting can, however, be used in addition to visible light. The visible light can be blocked using a daylight blocking filter, and only IR light can be used for the inspection. In this way, ambient light will not influence the image. If LED lighting in an IR wavelength range of 200 nm to 3000 nm is used, standard DIN EN 62471 [14], photobiological safety of lamps and lamp systems, must be observed.

Key information on using image processing systems in manual assembly is provided in a morphological box (see Fig. 1). Recommendations on selecting components are highlighted in white.

<table>
<tr><th>Components</th><th colspan="6">Specification</th></tr>
<tr><td rowspan="2">Camera type</td><td colspan="2">Area scan camera</td><td colspan="2">Line scan camera</td><td colspan="2">TOF camera</td></tr>
<tr><td colspan="3">Monochrome camera</td><td colspan="3">Color camera</td></tr>
<tr><td>Arrangement of the camera</td><td colspan="6">Above the work station, vertically aligned to the inspected object</td></tr>
<tr><td>Sensor type</td><td colspan="3">CCD</td><td colspan="3">CMOS</td></tr>
<tr><td>Lens mount</td><td colspan="3">CS mount</td><td colspan="3">C mount</td></tr>
<tr><td>Lens type</td><td colspan="2">Macro</td><td colspan="2">Fixed focal length</td><td colspan="2">Zoom</td></tr>
<tr><td>Lamp</td><td colspan="2">White LED</td><td colspan="2">Fluorescent light</td><td colspan="2">Halogen lamp</td></tr>
<tr><td>Wavelength range</td><td colspan="2">Ultraviolet</td><td colspan="2">Visible light</td><td colspan="2">Infrared</td></tr>
<tr><td>Light propagation</td><td colspan="2">Direct</td><td colspan="2">Diffuse</td><td colspan="2">Structured</td></tr>
<tr><td rowspan="2">Arrangement of lighting</td><td colspan="3">Incident light</td><td colspan="3">Transmitted light</td></tr>
<tr><td colspan="6">Left and right, oblique to the inspected object over the work area</td></tr>
</table>

Fig. 1. Recommendations for selecting and arranging components

5 Conclusion

Image processing systems offer great potential to support workers in manual assembly and achieve a high level of process capability in principle. Image processing systems can be used to detect progress, monitor the completion of individual assembly steps, and determine the current level of completion. Continuously reviewing the assembly process can, in particular, improve process capabilities. This can also help reduce cognitive strain for employees. Changing lighting circumstances can be a challenge in implementing such a solution. In addition, implementation is associated with a large amount of expense, since there are relevant testing features to consider in each assembly step. Therefore, especially in assembly processes with a large number of variants, it is a good idea to complete a targeted selection of tests based on a Failure Mode and Effects Analysis (FMEA).

Acknowledgment. The authors acknowledge the financial support by the Federal Ministry of Education and Research of Germany (BMBF) and the European Social Fund (ESF) in the project Montexas4.0 (FKZ Grant no. 02L15A260).

References

1. Bächler, A., Bächler, L., Kurtz, P., Krüll, T., Heidenreich, T., Autenrieth, S.: Assistenzsysteme für manuelle Industrieprozesse. In: Weidner, R., Redlich, T., Wulfsberg, J.P. (eds.) Technische Unterstützungssysteme, pp. 206–207. Springer, Heidelberg (2015)

2. Bornewasser, M., Bläsing, D., Hinrichsen, S.: Informatorische Assistenzsysteme in der manuellen Montage: Ein nützliches Werkzeug zur Reduktion mentaler Beanspruchung? Zeitschrift für Arbeitswissenschaft **72**, 264–275 (2018)
3. Falck, A.-C., Örtengren, R., Rosenqvist, M.: Assembly failures and action cost in relation to complexity level and assembly ergonomics in manual assembly (part 2). Int. J. Ind. Ergon. **44**(3), 455–459 (2014)
4. Bretschneider, J., Haider, H., Keinath, M.: Kameratechnik und digitale Bildverarbeitung. In: Neumann, H., Schröder, G., Löffler-Mang, M. (eds.) Handbuch Bauelemente der Optik, pp. 553–592. Hanser, München (2014)
5. Berndt, D.: Montage-Arbeitsplatz – Visuelle Assistenz und optische Prüfung. In: Schenk, M. (ed.) Produktion und Logistik mit Zukunft, pp. 84–108. Springer, Heidelberg (2015)
6. Eichin, D.: Modellbasiertes Konzept zur vollautomatisierten Montageprüfung von asynchron angetriebenen Getriebemotoren im lastlosen Zustand. In: Schriftenreihe des Instituts für Angewandte Informatik, Automatisierungstechnik, Karlsruher Institut für Technologie, Band 51. KIT Scientific Publishing, Karlsruhe (2015)
7. Beyerer, J., Puente León, F., Frese, C.: Automatische Sichtprüfung. Springer, Heidelberg (2016)
8. Kuroda, T.: Essential Principles of Image Sensors. CRC Press, Boca Raton (2016)
9. Bauer, M.: Kamera Kaufberatung: Was man vor dem Kauf unbedingt wissen sollte. Books on Demand, Norderstedt (2016)
10. Erhardt, A.: Einführung in die Digitale Bildverarbeitung. Vieweg und Teubner, Wiesbaden (2008)
11. Holst, K.: Beleuchtung für die industrielle Bildverarbeitung. In: Weissler, G.A. (ed.) Einführung in die industrielle Bildverarbeitung, pp. 23–66. Franzis, Poing (2007)
12. Görner, B.: Beleuchtung von Arbeitsstätten – Stand der Regelsetzung. Bundesanstalt für Arbeitsschutz und Arbeitsmedizin, Dortmund, Berlin, Dresden (2008)
13. Demant, C., Streicher-Abel, B., Springhoff, A.: Industrielle Bildverarbeitung: Wie optische Qualitätskontrolle wirklich funktioniert. Springer, Heidelberg (2011)
14. DIN, EN. 62471: Photobiologische Sicherheit von Lampen und Lampensystemen. Deutsches Institut für Normung und Beuth, Berlin (2009)

Relationship Between Facebook Fan Page and Trust of Fans

Yu-Hsiu Hung[1], Chia-Hui Feng[1,2(✉)], and Chung-Jen Chen[3]

[1] Department of Industrial Design, National Cheng Kung University, No. 1, University Road, Tainan City, Taiwan R.O.C.
p38041075@ncku.edu.tw

[2] Department of Creative Product Design, Southern Taiwan University of Science and Technology, No. 1, Nan-Tai Street, Yung Kang Dist. Tainan City, Taiwan R.O.C.

[3] Department of Visual Communication Design, Southern Taiwan University of Science and Technology, No. 1, Nan-Tai Street, Yung Kang Dist. Tainan City, Taiwan R.O.C.

Abstract. This study explored the influence of fan pages on the fans' trust, and the changes in users' trust in fan pages through online interaction, communication and sharing. This study used online questionnaire and collected 231 valid samples. Among those, 158 respondents joined 1 to 50 fan pages (68.4%); 192 respondents joined these fan pages for more than one year (83.1%). Young and middle-aged users have higher degree of acceptance of the online community, most of the fan pages are design to attract young people. The most common type of fan page is about brands or commodities (122 people, 50.8%), followed by restaurants, stores or sightseeing spots (106 people, 44.2%). This study used stepwise regression to analyze the questionnaires. The results showed that community communication and sharing have a positive relationship with the trust of fan pages, and the influence of community interaction on the trust of fan pages is insignificant.

Keywords: Social media · Facebook · Facebook fan page · Trust level

1 Introduction

Socialbakers [1], a social media analytics company, conducted a survey and found that in July 2017, the numbers of monthly global active users on Facebook were 190 million and the daily active users were 120 million, of which 92% were using mobile devices. Fans can choose to follow specific celebrities or brands' fan pages, and then the posts on these pages can be instantly shown on the fans' timelines.

Since 2016, some Internet celebrities have started to live-stream on the fan page and embed advertising on it. The fan page administrators and paid advertisement on fan pages allow the marketed products to be penetrated into consumer behaviors. Furthermore, fan page administrators can instantly browse and reply fans' messages to increase interaction. Social media has completely changed the way people interact and the way they consume. Facebook has become one of the important platforms for the fan pages of celebrities, brands and commodities.

T. Ahram et al. (Eds.): IHSED 2019, AISC 1026, pp. 801–807, 2020.
https://doi.org/10.1007/978-3-030-27928-8_122

Fan pages can attract online users, the status of active management or outsourcing of fan pages have been increasing. Understanding the communication of fan pages is important. This study aimed to explore whether the community interaction, sharing and communication between fan page and fans can increase fans' trust on fan pages.

Fan page administrators also can view the traffic statistics of fan pages and members' information from the administrator system to update their contents of sharing. The purpose of fan pages is for enterprises to develop public relations with their target and potential customers through open communication.

1.1 Trust Theory

Trust is a degree that a group believes the reliability and integrity of a trading partner. It is a psychological state. Once both parties have faith in each other's confidence and integrity, trust exists mutually. After the trust is generated, there is a close interaction between the two parties. One party believes that it is important to maintain the relationship with the other party, so it will try its best to maintain this relationship. That is, the party who made a commitment believes that the relationship between them is worth maintaining, and it guarantees to indefinitely continue this relationship [2]. Many scholars and experts have proposed theories, models and structures related to commitment. Morgan and Hunt proposed the Commitment-Trust Theory, also known as the Key Mediating Variable Model [2].

1.2 Community Communication

According to Socialbackers' statistics on Facebook users in May 2017, the largest number of fans on the fan pages in the United States is Walmart, with 33 million fans; the fastest-growing online community in Taiwan was Shopee, with 2.42 million fans.

In an online community, people can share their own or others' positive or negative experiences concerning products and services, show their current status, upload photos of food they enjoy, and check in a location. People can also repost good articles and share their own or others' experiences. The messages recommended by fans are considered to be a more credible or important source of information, as compared to the messages and conversations that were not from fan pages.

Hendriks [3] suggested that sharing information is a process of communication. When a person wants to learn new knowledge from others, users must first have basic information to acquire it. It is a kind of reconstruction behavior. Mogavero and Shane [4] suggested that information sharing is a feedback from the constant interaction between the message provider and the receiver.

1.3 Trust

Trust is a degree that a group believes the reliability and integrity of a trading partner. It is a psychological state. Once both parties have faith in each other's confidence and integrity, trust exists in both parties. After the trust is generated, there is a close interaction between the two parties. Chiou [5] found that fans' trust in the service provider of a website directly influences the fans' intention to use the service provided.

Ridings, Gefen and Arinze [6] argued that the trust between people is mutual. For the users, the intention to exchange information could help them to achieve interpersonal trust.

1.4 Research Hypotheses

According the previous literature, the experiences from fans estimated community interaction, community sharing and community communication may enhance the relationship with the community, and then gain the trust towards the community. Hence, this study proposed the research hypotheses.

H1: Community interaction has a positive relationship with the trust of fan pages.
H2: Community sharing has a positive relationship with the trust of fan pages.
H3: Community communication has a positive relationship with the trust of fan pages.

This study aims to find the mediating effect of fan pages to community interaction, sharing and communication.

2 Method

This study focused on the relationship associated with the trust of fan pages. The research method is divided into three phases as follows:

The first phase was literature review. Based on the result, the questionnaire was designed and the variables used in the study were determined.

The second phase was the questionnaire survey. In the part of basic personal information, the questions included gender, age and experience of usage. I the scale of trust level, the dimensions of community interaction, community sharing and community communication are integrated to measure user's trust in fan pages.

The third phase was to analyze the questionnaire results. This study used descriptive statistics, reliability analysis and stepwise regression analysis to analyze data, and test the relationship of community interaction, community communication and community sharing with users' trust in fan pages.

After conducted reliability test by SPSS, this study used Cronbach's Alpha, the most common method of Likert scale, to test the reliability of the scale.

3 Results

This study analyzed the questionnaire results with statistical methods to verify the proposed hypotheses. The specific statistics were implemented as follows:

1. Conduct descriptive statistical analysis on users and Facebook fan pages.
2. Conduct reliability analysis on the four variables of the samples.
3. Test the hypotheses of the four variables.

3.1 Description of Basic Information of Samples

A total of 240 samples were collected in this study. As 9 respondents never joined any of Facebook fan page, they were eliminated from the analysis, thus there were 231 valid samples, with a valid return rate of 96.2%. Among the 231 valid samples, 132 are female (57.1%), and 99 are male (42.9%). In terms of age, 113 respondents are 20-24 years old (48.9%). As for educational background, 161 have undergraduate degrees (69.7%). In the sample structure, most respondents are younger than 24 years old, 37.7% of whom are students, followed by 28 respondents aged 25 to 29 (12.1%), and 27 respondents who are over 45 years old (11.7%).

3.2 Facebook Fan Page Usage

Among the 231 valid samples, 158 respondents joined 1 to 50 fan pages (68.4%), followed by 51 to 100 fan pages (33 people, 14.3%), and 500 fan pages (6 people, 2.6%). Moreover, 192 respondents joined these fan pages for more than one year (83.1%). The number of respondents followed fan pages has continuously grown. As the young and middle-aged people have higher degree of acceptance of the online community, they are the largest proportion to join and be involved in fan pages.

About the types of the fan pages, the most common are those related to brands or products (122 people, 50.8%), followed by restaurants, stores or sightseeing spots (106 people, 44.2%), entertainers, music bands or public figures (102 people, 42.5%), and entertainment (80 people, 33.3%). In addition, 197 respondents have joined the Facebook fan pages more than 12 months (82.1%), 64 respondents spend 3–5 h to surf the Internet every day (26.7%), and 73 respondents browse the fan pages for less than 30 min (30.4%). Due to the popularity of smartphones, many people use smartphones to surf the Internet or go on Facebook (207 people, 86.3%).

3.3 Reliability Analysis

After conducting reliability test by SPSS, this study used Cronbach's Alpha, the most common method of Likert scale, to test the reliability of the scale. If the Alpha value is greater than 0.7, it means that this scale has high reliability. There were four dimensions in this study, including community interaction, community sharing, community communication and trust of community. The results showed that the Alpha values of these four variables were all greater than 0.7, indicating that the questionnaire of this study has a good reliability. The Alpha value of each variable was shown as Table 1.

Table 1. Reliability of each variable.

Variable	Number of questions	Cronbach's alpha
Community interaction	5	0.846
Community sharing	5	0.857
Community communication	5	0.889
Trust of community	5	0.942

3.4 Hypothesis Testing

The hypothesis of this study is to explore the relationship between community interaction, community sharing and community communication and the trust of community. Therefore, the stepwise regression analysis was used to verify the hypotheses that whether community interaction, community sharing and community communication can impact the trust of community, and whether the sense of identification of fan pages can influence the trust of fans. In addition, this study also identified the independent variable that has a significant impact on the dependent variable. The analysis used the adjusted R-squared and Beta coefficient to verify the hypotheses.

3.5 Test the Relationship Between Variables

This study used stepwise regression to analyze the questionnaire results. According to the research structure, this study conducted stepwise regression analysis between the independent variables and the dependent variable. The results listed the degree of recognition of the trust of community and the impact of the independent variables on the dependent variable, as shown in Tables 2 and 3.

Table 2. Stepwise regression analysis of variables.

Model	Unstandardized coefficients		Std. coeff.	t	Sig.	95% confidence interval of β	
	β	Std. error	Bata distribution			Lower bound	Upper bound
1 (constant) Community communication	1.064 .823	.194 .046	.764	5.482 17.919	.000 .000	.682 .733	1.446 .914
2 (constant) Community communication Community sharing	.583 .562 .353	.199 .062 .060	.522 .338	2.933 9.134 5.909	.004 .000 .000	.191 .441 .235	.974 .684 .471

Note: a. dependent variable: trust of community

Table 3. Stepwise regression analysis of variables

Model	R	R^2	Adjusted R^2	Estimated standard error
1	.764a	.582	.582	.92792
2	.799b	.636	.636	.86600

Note: a. predicted variable: (constant), communication.
b. predicted variables: (constant), communication, sharing.
c. dependent variable: trust of community

According to Tables 2 and 3, community communication and community sharing in Mode 1 are positively correlated with users' trust in fan pages and reached the level of significance (Beta: 0.522, 0.338; Adjusted R^2: 0.582; P:0.00), and there is no collinearity between variables (VIF:2.041). In Mode 2, the identification of community has a positive relationship with users' trust in fan pages and reached the level of significance (Beta: 0.717; Adjusted R^2:0.683:P:0.00). Community communication and sharing are positively correlated with users' trust in fan pages, thus H2 and H3 are valid. However, the relationship between community interaction and the trust of community is insignificant, so H1 is not supported.

4 Conclusion and Discussion

This study explored the relationship of community interaction, sharing and communication of Facebook fan pages with users' trust in fan pages. The results showed that community communication and sharing are positively correlated with users' trust in fan pages.

1. In terms of community sharing, people may post and share messages on fan pages. In the process of sharing, they would generate the recognition of the fan pages. As long as fans agree with the content shared by the fan pages, their trust in the fan pages would enhance.
2. Community communication is positively correlated with users' trust in fan pages. It means that people can obtain immediate feedback through communication and inspire the solutions of problems. They also generate trust in the information provided on the fan pages, so as to have greater loyalty towards the fan pages. Many people join the fan pages, especially the professional communities, because they want to solve problems at work [7].
3. The relationship between community interaction and users' trust in fan pages is not significant. The reason is that most fans just browse the posts, or occasionally share or like the comments. Fans usually do not actively reply messages, so it is difficult for fans to maintain a close interaction.

This study still had some limitation. It is because this study only analyzed the users of general fan pages and did not classify the differences between different types of fan pages. It is suggested that future research can take the Internet celebrities, brands or commodity companies on different types of fan pages as subjects. Moreover, this study only used one factor to test the mediating effect. In the future, researchers can conduct the analysis with multiple factors.

References

1. Social Media Regional Reports. https://www.socialbakers.com/resources/reports/
2. Morgan, R., Hunt, S.D.: The commitment-trust theory of relationship marketing. J. Mark. **58**, 20–38 (1994)
3. Hendriks, P.: Why Share Knowledge? The Influence of ICT on the motivation for knowledge sharing. Knowl. Process Manag. **6**, 91–100 (1999)
4. Mogavero, L.N., Shane, R.S.: Technology transfer and innovation. Marcel Dekler Inc, New York (1982)

5. Chiou, J.: The antecedents of consumers' loyalty toward internet service providers. Inf. Manag. **41**, 685–695 (2004)
6. Ridings, C., Gefen, D., Arinze, B.: Some antecedents and effects of trust in virtual communities. J. Strateg. Inf. Syst. **11**, 271–295 (2003)
7. Chiu, C.-M., Hsu, M.-H., Wang, E.T.G.: Understanding knowledge sharing in virtual communities: An integration of social capital and social cognitive theories. Decis. Support Syst. **42**, 1872–1888 (2006)

Effects of the Use of Smart Glasses on Eyesight

Natasa Vujica Herzog[1] and Amer Beharic[2](✉)

[1] Faculty of Mechanical Engineering, University of Maribor, Maribor, Slovenia
[2] Healthcare Center Dr. Adolfa Drolca Maribor, Maribor, Slovenia
Amer.Beharic@gmx.net

Abstract. Smart glasses, also named as data glasses, are an example of Head Mounted Display (HMD) with a great potential in different manufacturing environments. They collect data from a wireless network and project information directly into the user's eye. It is evidenced from the literature that HMD can cause headaches, pressure in the eyes, problems with focusing and difficulties with text reading. Therefore, we tried to research whether use of smart glasses could be harmful for the human eye or not. To study the addressed problems, a research was performed in testing warehouse environment at the Faculty of Mechanical Engineering. Users comfort was tested through questionnaire and several eyes investigations before and after use of smart glasses in Healthcare Center Maribor. The results of the performed ophthalmologic tests show that there are some statistically significant differences between tests results performed before and after use of smart glasses that cannot be overlooked.

Keywords: Human-systems integration · Smart glasses · Industry 4.0 · Warehouse

1 Introduction

Smart glasses are wearable devices that we wear like regular glasses or in combination with regular glasses. They have the ability to merge the physical environment with virtual information within the view field of the user [1]. Usually they are equipped with various sensors that gather information about the user's situational context, a WiFi-antenna to receive and send online information, a small memory, a processing unit and a small screen located in front of one eye. The processing unit allows the smart glasses to operate various recognition technologies to give the user context relevant information on his/her social and spatial surrounding [2, 3].

It is evidenced from the literature that use of smart glasses can have several benefits for the worker (hands-free access to computer-generated info, routing to storage locations, eliminating the need to carry handheld scanners or written documents), but there are also possible problems evidenced [4–7]. HMD can cause headaches, pressure in the eyes, problems with focusing and difficulties with text reading [8–10]. In our research, we tried to answer the question whether use of smart glasses could be harmful for the human eye or not. To study the addressed problems, a research was performed as the result of cooperation between two faculties from University of Maribor, Faculty

T. Ahram et al. (Eds.): IHSED 2019, AISC 1026, pp. 808–812, 2020.
https://doi.org/10.1007/978-3-030-27928-8_123

of Mechanical Engineering and Faculty of Logistics with Ophthalmologists from Healthcare Center dr. Adolfa Drolca Maribor, Maribor.

Most order-picking systems used in practice are manual and more than 80% of all orders processed by warehouses are picked manually [11]. Since this process is the most laborious and the most costly activity in a typical warehouse, we decided to test use of smart glasses as supporting tool in testing warehouse environment.

The research was funded by The Public Scholarship, Development, Disability and Maintenance Fund of the Republic of Slovenia and was later selected for ESF success story - European Commission - Creative Path to knowledge, among 186 projects performed in 2018.

In the paper, results of descriptive statistics will be present in detail although other statistical tests (tests of Normality, Paired samples t test and Wilcoxon Paired samples test) were performed, too.

2 Methodology

Effects of using Vuzix M300 smart glasses on users` comfort during order picking activities were researched in a testing warehouse environment at the Faculty of Mechanical Engineering, Maribor. The protocol of performed research is summarized in Fig. 1.

14 persons tested selected smart glasses, owned by the company Špica International. The testing period lasted four hours. Before and after use of smart glasses several ophthalmologic tests (visual acuity, contrast sensitivity, visual field testing and color test) were performed, therefore we got 28 measures altogether.

3 Descriptive Statistics

Results of statistical analysis for all the performed analyses are summarized in Table 1, explained by:

- the mean value of each variable, and
- the standard deviation.

Results of visual acuity for the right eye show that the mean value before work with smart glasses was 0.82, and after 0.76. Visual acuity was lower after using smart glasses for $\Delta = 0.062$ (7.6%). It means that tested persons` sight was weaker after use of smart glasses than before.

For the left eye, the results of visual acuity are similar as for the right eye, but there are slight differences. The mean value before work with smart glasses was 0.84, and after 0.76. Visual acuity after using smart glasses was also lower by $\Delta = 0.056$ (6.7%), meaning that tested persons' sight was weaker after use of smart glasses.

Comparison between right and left eye show that visual acuity of the right eye reduced by 7.6%, and 6.7% for the left eye. All tested persons had the visual display of the smart glasses in front of their right eye, therefore the weaker sight on the right eye could be caused by using smart glasses.

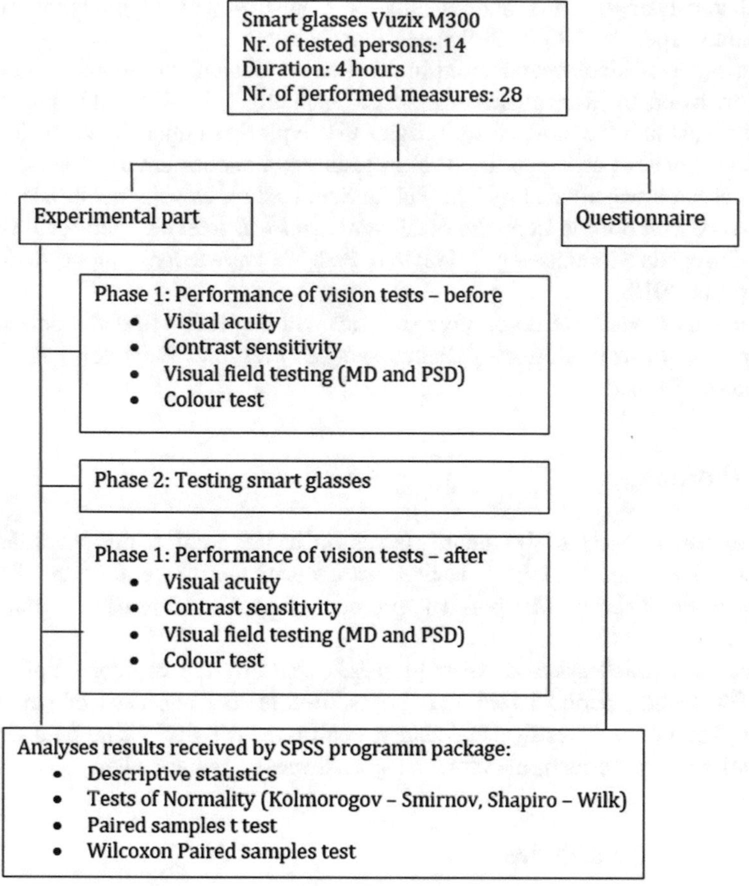

Fig. 1. Experimental protocol of testing Vuzix M300 smart glasses

Table 1. Results of descriptive statistics before and after use of smart glasses

	Before			After		
	Mean	SD	St. Err. mean	Mean	SD	St. Err. mean
VAR	0.820	0.242	0.064	0.757	0.245	0.065
VAL	0.842	0.210	0.056	0.786	0.234	0.062
CSR	1.628	0.054	0.014	1.596	0.095	0.025
CSL	1.617	0.063	0.017	1.564	0.096	0.025
CS	1.778	0.080	0.021	1.746	0.111	0.029
CT	13.143	4.737	1.266	13.14	4.737	1.266
MD	−2.542	1.950	0.521	−2.67	1.171	0.313
PSD	2.596	0.675	0.180	2.622	0.409	0.109
ST	0	0	0	0.428	0.513	0.137

VAR - Visual Acuity, right eye, VAL - Visual Acuity, left eye, CSR - Contrast sensitivity, right eye, CSL - Contrast sensitivity, left eye, CS - Contrast sensitivity of both eyes, CT - Colour test, MD - Mean Deviation, PSD - Pattern Standard Deviation, ST – Scotoma Threshold 30-2.

The mean value of contrast sensitivity for the right eye before work with smart glasses was 1.63, and after 1.60. Contrast sensitivity was lower after using smart glasses by $\Delta = 0.03$ (2%). It means that the tested persons' contrast sensitivity was weaker after use of smart glasses than before.

A similar situation is evidenced for the left eye, where the mean value of contrast sensitivity changed from 1.62 to 1.56, $\Delta = 0.05$ (3%), meaning that the tested persons' contrast sensitivity was weaker after the use of smart glasses than before.

The mean value of contrast sensitivity for both eyes changed from 1.78 to 1.75, which means that contrast sensitivity was lower after using smart glasses by $\Delta = 0.03$ (1.8%). Tested persons perceived lower contrast sensitivity after using smart glasses, but since the measured value was not lower than 1.5, use of smart glasses is not harmful for eye contrast sensitivity [12]. The mean value of perceiving contrast sensitivity was binocularly better than monocular, which is also verified with other studies [13].

The mean value of the Mean Deviation test (MD test) of visual field is −2.54 before using smart glasses and −2.67 after. The value after 4-h use of smart glasses was lower by $\Delta = 0.13$ (5%).

Our results were lower than the normal range before testing, which can be influenced by the patient's psychological state, concentration and cooperation during the test. We must be aware of the state when the tested individual performs the first test of the visual field.

The mean value of the Pattern Standard Deviation test (PSD test) before work with smart glasses was 2.6, and after 2.62. The tested value after 4-h use of smart glasses was slightly higher by $\Delta = 0.03$, or less than 1.

High values of PSD represent the scotomas [14]. The Treshold 30-2 Programme confirmed the increased level of scotoma in the right eye, but the Driver's license test gave us a lower level of scotomas. Even the second test could not confirm the results of Treshold 30-2 totally, so it can be concluded that the presence of scotoma in the right eye can be the result of load caused by using smart glasses.

4 Conclusions

The results of the performed ophthalmologic tests (visual acuity, contrast sensitivity, visual field testing and color test), show that there are some statistically significant differences between tests results performed before and after use of smart glasses that cannot be overlooked. The difference between the results of visual acuity for the left and right eyes is small, but, in both cases, the visual acuity is lower after use of smart glasses. The contrast sensitivity and color test did not show any statistically significant differences, but the additional test of the visual field did.

From the results of the visual field test, we found out that none of the tested individuals had the scotoma (dysfunction in central and peripheral vision) in the right eye (inferior quadrant where the projection of smart glasses was performed) before using smart glasses. After the test, scotomas were present in the same quadrant in 43% of cases. This might indicate that use of smart glasses for four hours during work can cause scotomas and, thus, impairment in the visual field and vision.

Our pilot test was performed with limited resources on 14 persons and 28 measures, therefore, rigorous, empirically based research, performed on a greater number of tested persons, would be of benefit before implementing it in warehouses as a part of everyday equipment for workers.

Acknowledgement. This work was supported by the Slovenian Research Agency in the framework of Grant P2-0190. The research was funded by The Public Scholarship, Development, Disability and Maintenance Fund of the Republic of Slovenia.

References

1. Hein, D.W.E., Jodoin, J.L., Rauschnabel, P.A., Ivens, B.S.: Are Wearables Good or Bad for Society? An Exploration of Societal Benefits, Risks, and Consequences of Augmented Reality Smart Glasses, Chapter 1 in Mobile Technologies and Augmented Reality in Open Education, IGI Global, USA (2017). https://doi.org/10.4018/978-1-5225-2110-5.ch001
2. Rauschnabel, P.A., Brem, A., Ivens, B.S.: Who will buy smart glasses? Comput. Hum. Behav. **49**, 635–647 (2015). https://doi.org/10.1016/j.chb.2015.03.003
3. Rauschnabel, P.A., Ro, Y.K.: Augmented reality smart glasses: an investigation of technology acceptance drivers (2016)
4. Peli, E.: Visual and optometric issues with head-mounted displays. In: Proceedings of the IS&T Optics & Imaging in the Information Age, The Society for Imaging Science and Technology, pp. 364–369 (1996)
5. Klein-Theyer, A., Horwath-Winter, J., Rabensteiner, D.F., Schwantzer, G., Wultsch, G:. The impact of visual guided order picking on ocular comfort, ocular surface and tear function. PLoS One **11**(6) (2016). https://doi.org/10.1371/journal.pone.0157564
6. Josefsson, P., Lingegard, S.: Potential of Smart Glasses in a Spare Parts Distribution Center. Chalmers University of Technology, Gothenburg, Sweden (2017)
7. Peli, E.: Visual issue in the use of a head mounted monocular display. Opt. Eng. **29**(8), 883–892 (1990). https://doi.org/10.1117/12.55674
8. Schuff, D., Corral, K., Turetken, O.: Comparing the understandability of alternative data warehouse schemas: an empirical study. Decis. Support Syst. **52**(1), 9–20 (2011). https://doi.org/10.1016/j.dss.2011.04.003
9. Jungmin, H., Seon Hee, B., Hyeon-Jeong, S.: Comparison of visual discomfort and visual fatigue between head-mounted display and smartphone. In: Proceedings of the IS&T International Symposium on Electronic Imaging, Human Vision and Electronic Imaging, Society for Imaging Science and Technology, pp. 212–217 (2017). https://doi.org/10.2352/issn.2470-1173.2017.14.hvei-146
10. Rosenfield, M.: Computer vision syndrome: a review of ocular causes and potential treatments. Ophthalmic Physiol. Opt. **31**, 502–515 (2011)
11. De Koster, R., Le-Duc, T., Roodbergen, K.J.: Design and control of warehouse order picking: a literature review. Eur. J. Oper. Res. **182**(2), 481–501 (2007)
12. Parede, T.R., et al.: Quality of vision in refractive and cataract surgery, indirect measurers: review article. Arq Bras Oftalmol. **72**(6), 386–390 (2013)
13. Karatepe, A.S., Köse, S., Eğrilmez, S.: Factors affecting contrast sensitivity in healthy individuals: a pilot study. Turk. J. Ophthalmol. **47**(2), 80–84 (2017)
14. Ramulu, P., Salim, S.: Standard Automated Perimetry. http://eyewiki.aao.org/Standard_Automated_Perimetry

Distributed Data and Information Management for Crisis Forecasting and Management

Barbara Essendorfer[1], Jennifer Sander[1](✉), Marian Sorin Nistor[2], Almuth Hoffmann[1], and Stefan Pickl[2]

[1] Fraunhofer Institute of Optronics, System Technologies and Image Exploitation IOSB, Karlsruhe, Germany
{Barbara.Essendorfer,Jennifer.Sander, Almuth.Hoffmann}@iosb.fraunhofer.de

[2] Universität der Bundeswehr München, Werner-Heisenberg-Weg 39, 85579 Neubiberg, Germany
{Sorin.Nistor,Stefan.Pickl}@unibw.de

Abstract. Crises forecasting and management require different kinds of information (management) processes. In this publication, we give an introduction into Coalition Shared Data (CSD), which is a concept of standardized information distribution, and describe means for its integration into a more comprehensive high level system architecture. Our publication intends to stimulate the potential use of the respective findings in the field of crisis prediction and management. To this aim, we put also particular emphasis on visual analytics approaches for integrating automated data analyses and interactive data visualizations to offer more insights into complex data and information.

Keywords: Distributed data and information management · Coalition Shared Data (CSD) · Interoperability · Crisis forecasting and management · System architectures · Visual analytics · Content awareness · Situation awareness

1 Introduction

Crises forecasting and management require different kinds of information (management) processes to support analysts and decision makers [1–3]. Of fundamental importance is the ability to provide the right information at the right time to the persons needing it and enabling these persons to draw the right conclusions. To this aim, relevant aspects in distributed data and information inventories have to be identified, placed in the overall context, prepared and analyzed appropriately, as well as made available as needed.

Nowadays, technology allows to disseminate data and information in near real-time. Technologies in the areas of sensors and platforms as well as network technology and storage capability have evolved to a level where mass data can also be easily shared and disseminated. In order to exploit these emerging opportunities, there is a need for information systems/system networks that can interact with each other in a

T. Ahram et al. (Eds.): IHSED 2019, AISC 1026, pp. 813–819, 2020.
https://doi.org/10.1007/978-3-030-27928-8_124

well-defined interoperable [4] way. The basis for the creation of such systems/system networks are coordinated services, interfaces and formats for data and information exchange and the ability to correctly interpret and correlate data and information.

Data and information that can be exploited to forecast and manage crises are usually of heterogeneous nature and gained by different departments/organizations. At data and information exchange, cross-sectional aspects such as network restrictions and, in particular, information security and data protection requirements must be taken into account. There may also be the need for the original owners of the data and information to be able to retain their sovereignty.

Cooperative information systems fulfilling the given technical and operational requirements can be used for data and information management and dissemination in the field of crisis forecasting and management. Methods and technologies to support operators with their tasks to identify relevant patterns in huge amounts of data and to give indications of developing crises need to be integrated with the systems.

In this publication, we give an introduction into CSD, which is a concept of standardized information distribution and describe how it supports higher-level information management processes by integrating it into a more comprehensive high level system architecture. The aim is to support tailored information provision and to increase content awareness. In addition, we put particular emphasis on visual analytics approaches with the aim to integrate automated data analyses and interactive data visualizations to offer more insights into complex data and information in the field of crisis forecasting and management.

2 Coalition Shared Data (CSD)

CSD [5, 6] has its origin in multinational military-oriented Intelligence, Surveillance and Reconnaissance (ISR) projects. However, although the community, workflows and specific data formats differ between the civilian (security) and the military environment, the underlying principles apply to both. In particular, the CSD concept offers a promising approach to enable the information flow between the military domain and the civilian sector in terms of information distribution and management processes, which may be a prerequisite for the use of cooperative information systems in the field of crisis forecasting and management.

CSD aims at sharing data and information within a coalition, i.e., among a set of participants that are interacting with each other. Conceptually, it is based on having a network of physically distributed sites (network nodes) which are connected using the available network infrastructure. Among the sites, each node shares information about the persisted content using metadata entries. By distributing the metadata across all instances of participating sites using adequate synchronization protocols, a global awareness of available data and information is achieved. Combining this awareness with the ability to retrieve data and information products on demand and link it to other products and tasks ensures ubiquitous access capabilities for data and information stored on any of the sites in the entire network with reduced network traffic. CSD clients can retrieve data and information products even from indirectly connected nodes in the network in an efficient way.

The CSD concept enables sharing of static, finished data and information such as (finished) documents, reports, images, and video clips. It also includes the ability to share dynamic, mutable data – such as we can encounter in collaborative business processes where multiple parties modify a common piece of data – and constantly changing data such as video streams. As further described in [7], technical solutions being based on the CSD concept can support different security aspects.

Fraunhofer IOSB has developed concepts, systems and services for CSD on multiple levels. It also has added aspects as role-based user management with support of different access roles and security settings for each user. Our focus of research in the field of CSD is now on the concept's improvement as well as on dedicated aspects of data and information management and cross-domain transfer. To enable the usage of the CSD concept in applications that connect different security domains and to achieve multilevel security in an operational setting, as it may be also the case at crisis forecasting and management, additional interfaces and new services are developed.

CSD can be adapted to include data protection concerns and other security standards as well as to add other COI (Community of Interest) specific services [8]. Its transfer into the area of crisis forecasting and management may be promising. The data types and the information sources themselves are often similar and there are thematically related standards. Additional sources of information (e.g., specific OSINT (Open Source Intelligence)), as well as specific types of information products (e.g., specific documents or report formats), can be flexibly integrated into the existing concept, e.g. via CSD data model extensions [7] or by the integration into a more comprehensive high level system architecture as described in the next section.

3 High Level System Architecture

To support tailored information provision and to increase content awareness additionally, a high level system architecture has been worked out by Fraunhofer IOSB [9]. It makes use of different sources, one of them being CSD, and has integrated information models from the CSD environment. As illustrated in Fig. 1, the architecture consists of four horizontal layers, separating the concerns of data and information acquisition, warehousing, analysis, and result presentation/access to information elements and functionalities of the architecture. Two additional cross-cutting layers address overarching concerns in terms of information quality management and general management tasks, including tailoring.

The characteristics of the field of crisis forecasting and management (see e.g. [2, 10]) require modular, scalable methods and processes for (higher level) data and information fusion and analysis. They have to be applied in environments where information being of heterogeneous nature and possessing different quality has to be shared and processed. The proposed high level system architecture may get adjusted to these as well as further constraints and demands the field of crisis forecasting and management. In particular, it can support the creation of an expressive knowledge model for the considered application domain in combination with different methods of reasoning used for information integration and conclusion drawing [11]. In addition to the incorporation of different techniques for data and information extraction, fusion, and knowledge-based reasoning, the incorporation of techniques from the field of visual analysis as described in the next section of this publication is considered to be promising. These can be realized in different kinds of (semi-)automated assistance functions that can improve the abilities of human operators for analyzing data and information and for reaching a high level of content awareness as well as situation awareness. Implementing visual analytics functionalities requires a close interaction between a respective visualization module in the Presentation & Access Layer and the respective underlying layer and/or, depending on the task, with the Information Quality layer of the proposed architecture.

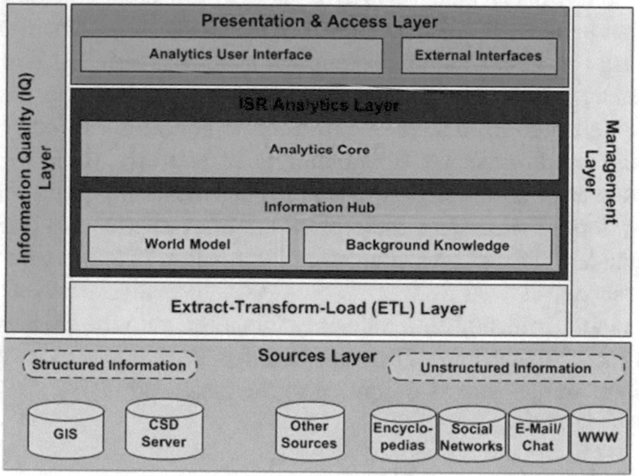

Fig. 1. Layer-based structure of the proposed high level system architecture.

4 (Semi-)Automated Assistance by Visual Analytics

The data collected from different sources in heterogeneous forms is only usable after processing and turning it into information. It becomes valuable only after it is judged by decision-makers according to pre-established sets of criteria. Hence, any (semi-) automated assistance functions developed with visual analytics techniques [12] and tools enhances the decision-making process.

Information visualization is the root field with a wider focus on abstract data visualization. Its recent branch, visual analytics, supports automated analysis algorithms combined with interactive visualizations to help the decision process in reaching valuable information in the shortest time possible [12]. In other words, it answers the *Why?* and not just the *What?* questions addressed to the data. A key aspect in the field of crisis forecasting and management is to understand the decision-assistance functions to avoid any gray boxes in the process.

The types of data handled in this process vary from space and time (e.g., geospatial, temporal, spatio-temporal), multivariate (e.g., projection-based methods), text (e.g., topic-based methods), to graph and network data. Each type uses different algorithms and layouting techniques tailored to the type of assistance required. Here, visual analytics for network data is addressed due to the capability of modelling complex systems with heterogeneous information collections using modular and scalable methods [13].

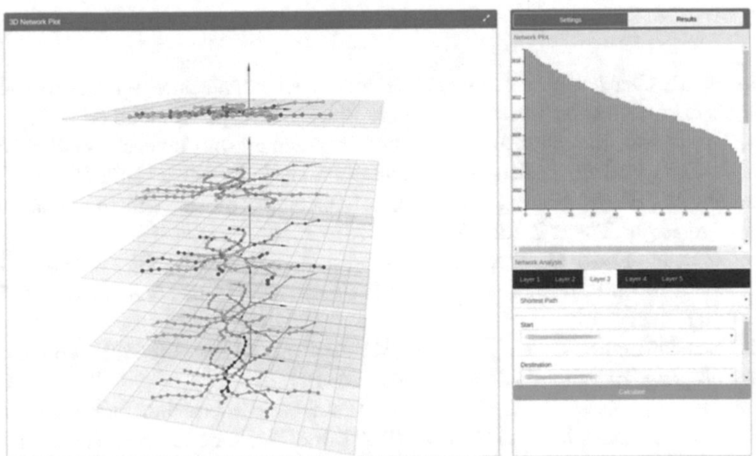

Fig. 2. Visual analytics dashboard with a multiplex visualization.

Suitable techniques and tools are researched and developed by Nistor [14], which focus on multilayer and multiplex visualizations. A multilayer visualization exploration of data enables the decision-maker to combine different (e.g., sensor) networks in a comprehensive view, while a multiplex visualization is focusing on one network at a time. Such exploration is usually chosen to analyze different aspects, or stages, of the network on multiple layers in the same view. A detailed description of capabilities and application limitations can be found in [15, 16].

A substantial advantage of this type of visualization is seen by performing different analyses at the same time. The decision-makers receive additional insights on the output by visual comparison as well. The visualizations can be integrated into an analytics dashboard (e.g., see Fig. 2) to aid in the network exploration as an intermediary step. Each layer of the network representing a different analysis enables the visual comparison to have multiple steps and better insights before the dissemination of the final results.

5 Conclusion

In this publication, an overview on CSD has been given and we described how it can get integrated into a more comprehensive system architecture. Our publication aims at stimulating the potential use of the respective findings in the field of crisis prediction and management. To this aim, we put also particular emphasis on visual analytics approaches to integrate automated data analyses and interactive data visualizations which give additional insights into complex data and information.

Acknowledgment. The CSD concept is part of multinational projects that were funded by the BMVg (Federal Ministry of Defence).

The authors would like to thank their colleagues, who have contributed to parts of the work presented in this paper, especially Dr. Maximilian Moll and Achim Kuwertz.

References

1. Usländer, T., Denzer, R.: Requirements and open architecture for environmental risk management information systems. Inf. Syst. Emerg. Manag., 344–368 (2009)
2. Roth, F., Herzog, M.: Strategische Krisenfrüherkennung-Instrumente, Möglichkeiten und Grenzen. Zeitschrift für Außen-und Sicherheitspolitik **9**(2), 201–211 (2016)
3. Zsifkovits, M., Meyer-Nieberg, S., Pickl, S.: Operations research for risk management in strategic foresight. Planet@ Risk **3**(2), 281–288 (2015)
4. Essendorfer, B., Kerth, C., Zaschke, C.: Adaptation of Interoperability Standards for Cross Domain Usage. In: Proceedings of the SPIE 10207, Next-Generation Analyst V, 102070E, SPIE (2017)
5. NATO Standardization Office (NSO): STANAG 4559, Edition 4 (2018). http://nso.nato.int/nso/nsdd/stanagdetails.html?idCover=8838&LA=EN
6. Essendorfer, B., Kuwertz, A., Sander, J.: Distributed information management through Coalition Shared Data. In: NATO STO-MP-IST-160, Big Data and Artificial Intelligence for Military Decision Making (2018)
7. Haferkorn, D., Klotz, P., Rodenbeck, R.: Application of a military data dissemination standard in a civil context. In: Proceedings of the SPIE 11015, Open Architecture/Open Business Model Net-Centric Systems and Defense Transformation 2018, 1101505, SPIE (2019)
8. Zaschke, C., Essendorfer, B., Kerth, C.: Interoperability of heterogeneous distributed systems. In: Proceedings of the SPIE 9825, Sensors, and Command, Control, Communications, and Intelligence (C3I) Technologies for Homeland Security, Defense, and Law Enforcement Applications XV, 98250Q, SPIE (2016)
9. Essendorfer, B., Hoffmann, A., Sander, J., Kuwertz, A.: Integrating Coalition Shared Data in a system architecture for high level information management. In: Proceedings of the SPIE 10802, Counterterrorism, Crime Fighting, Forensics, and Surveillance Technologies II, 108020F, SPIE (2018)
10. Rogova, G.L., Scott, P.D.: Fusion Methodologies in Crisis Management. Springer, Chambridge (2016)
11. Kuwertz, A., Mühlenberg, D., Sander, J., Müller, W.: Applying knowledge-based reasoning for information fusion in Intelligence, Surveillance, and Reconnaissance. In: Lee, S., Ko, H., Oh, S. (eds.) Multisensor Fusion and Integration in the Wake of Big Data, Deep Learning and Cyber Physical System, MFI 2017. LNCS, vol. 501, pp. 119–139. Springer, Heidelberg (2018)
12. Keim, D.A., Mansmann, F., Oelke, D., Ziegler, H.: Visual analytics: combining automated discovery with interactive visualizations. In: Jean-Fran, J.-F., Berthold, M.R., Horváth, T. (eds.) Discovery Science: 11th International Conference, DS 2008, Budapest, Hungary, 13–16 October 2008. Proceedings, pp. 2–14. Springer, Heidelberg (2008)
13. Nistor, M.S., Pickl, S.W., Zsifkovits, M.: Visual analytics of complex networks: a review from the computational perspective. In: The 2015 International Conference on Modeling, Simulation and Visualization Methods, Las Vegas, NV, USA, pp. 10–15. CSREA Press, San Diego (2015)

14. Nistor, M.S.: Proof of concept of a visual analytics dashboard for transportation network analysis. In: 51st Hawaii International Conference on System Sciences (HICSS-51) (2018)
15. Kivelä, M., Arenas, A., Barthelemy, M., Gleeson, J.P., Moreno, Y., Porter, M.A.: Multilayer networks. J. Complex Netw. **2**(3), 203–271 (2014)
16. Boccaletti, S., Bianconi, G., Criado, R., Del Genio, C.I., Gómez-Gardenes, J., Romance, M., Sendina-Nadal, I., Wang, Z., Zanin, M.: The structure and dynamics of multilayer networks. Phys. Rep. **544**(1), 1–122 (2014)

Visual Representation Strategy of Flow Line in Flow Maps Visualization

Linzheng Shang[1], Chengqi Xue[1(✉)], Yun Lin[1], and Jiang Shao[2]

[1] School of Mechanical Engineering, Southeast University, Nanjing 211189, China
1299842041@qq.com, {ipd-xcq,yunlin}@seu.edu.cn

[2] School of Architecture and Design, China University of Mining and Technology, Xuzhou 221116, China
shaojiang@cumt.edu.cn

Abstract. Flow maps are usually used to express the flow relationship of space and are commonly used in population migration, transportation networks, trade links, etc. The flow line is an important medium that helps people to understand the node relationship and the flow level. Its visual representation affects the cognitive effect of data information. This paper firstly discusses the characteristics of flow line symbols, then studies the design method of flow line symbols from the perspective of visual channels, and puts forward the visual mapping model of flow lines. Finally, this paper presents two design schemes and uses real data sets to generate visualized flow maps. The visualized flow maps are evaluated employing a questionnaire survey. The results show that the overall cognitive effect of the flow line symbols is good.

Keywords: Flow line · Visual channels · Flow maps · Data visualization

1 Introduction

With the advent of the era of big data, various forms of data visualization have emerged. Flow maps, as one of the forms of data visualization, are often used to express the flow relationship of natural objects in space and help people understand the spatial flow patterns of objects [1]. In flow maps, the nodes generally represent the geographical location and are connected by flow line symbols. The flow maps are widely used in population migration, transportation networks, trade links, etc.

Most of the research on flow maps focus on algorithm research and optimization of graph rendering, such as flow graph layout via spiral trees [2]. Some researchers also focus on the effectiveness of visual variables in flow maps, such as evaluating the effectiveness of expression of line width, color and other factors [3, 4]. Although some studies focus on the visual perception of flow maps, there is a general lack of theoretical research on flow line design methods. Now, the diversity of image rendering methods also provides more choices for the visual representation of flow maps. From the perspective of visual variables, this paper systematically discusses the visual design

T. Ahram et al. (Eds.): IHSED 2019, AISC 1026, pp. 820–826, 2020.
https://doi.org/10.1007/978-3-030-27928-8_125

methods of flow line symbols and establishes the mapping relationship between attribute feature and geometric feature of the flow line. Finally, two novel visual expression methods are proposed.

2 Characteristic Analysis of Flow Line Symbols

The flow line symbol is a typical motion symbol [5], which is generally used to express natural things with spatial displacement properties and show the direction, trajectory, quantitative characteristics, etc. The specific characteristics of the flow line symbols are as follows.

(1) Expressing directionality
The flow maps need to reveal the flow pattern from the origin to the destination, so the directional representation method needs to be considered.
(2) Expressing magnitude
Flow graphs need to describe the number of objects.
(3) Expressing classification

Flow maps may involve different subject attributes which are often categorized by different colors or shapes.

3 Application of Visual Channels in Flow Line Design

3.1 Concept Description

Visual coding consists of two parts: graphical elements and visual channels that control the graphical elements. Graphic elements are usually geometric figures such as points, lines, planes, etc. Visual variables are the visual elements that make up graphical symbols. In the design of map symbols, it is through the changes of these variables to express the differences between geographical phenomena. After the data attributes are visualized by graphic elements, the visual channels are used to control the presentation of the graphic elements. Different variables represent different visual channels. Visual variables include position, size, shape, orientation, hue, saturation, brightness, etc. [6]. Table 1 summarizes the visual variables commonly used in the flow line.

Table 1. Visual variables in flow line

Line type	
line width	
Shape	
Solid and dotted	
hue	
transparency	
texture	
gradual change	
blur	

3.2 Application Principle of Visual Channels in the Flow Line

The visual channels can be divided into three types. The first type is qualitative or classified, that is, describing what or where the object is; the second type is a quantitative visual channel that describes the magnitude of objects. For example, the shape is a typical qualitative visual channel. People usually distinguish shapes into circles, triangles, or rectangles, rather than describing them as size or length. In turn, the area is a typical quantitative visual channel. The user intuitively uses different areas or lengths to describe different values; the third type is grouping, and the basic grouping channel is proximity. The sensing system can automatically assign objects that are close to each other to the same class.

The quantitative data in the flow maps is generally expressed by the line width and the large width means that the flow level is high. Weihua Dong determined the validity of the linewidth which expresses flow level [3]. Color gradients can also represent different levels of magnitude. Gradient variations in color can increase the readability of dense flow maps. Colors with high saturation or transparency can be used to represent larger magnitudes, and colors with lower purity or transparency represent smaller magnitudes. Although the texture density can also be expressed in magnitude, its effectiveness has not been verified. The qualitative information in the flow graph is mainly direction information and category information. The direction is generally indicated by an arrow, and the shape of the arrow can be changed by controlling the shape channel, such as changing the curvature of the arrow [7], using a full arrow or a half arrow. Besides, according to the symbol design theory, the map symbols use different geometric elements to express the semantic information [8]. Therefore, the arrow as a basic geometric element can also be replaced by the symbol form, which provides a variety of choices for the flow line. The classification information in the flow maps mainly expresses different themes through different colors, shapes or textures.

When expressing some complicated spatial relationships, the organization of the flow line also directly affects the visual performance of the flow maps. The inappropriate organization will increase the cognitive load of users. The flow line needs to minimize the number of intersections and overlaps [9]. Generally speaking, the curve can bring less overlap than the straight line and the curve should be as smooth as possible. The angle and curvature of the curve also should be taken into account. Figure 1 shows some suitable layouts of flow lines.

Based on the theoretical study, this part selects the appropriate visual channel for the flow line and constructs the mapping model from the attribute space to the geometric space. This model presents visual channels to express different information types and gives some correct organization forms of flow lines thus providing guidance for the real flow line design. The mapping model is shown in Fig. 1.

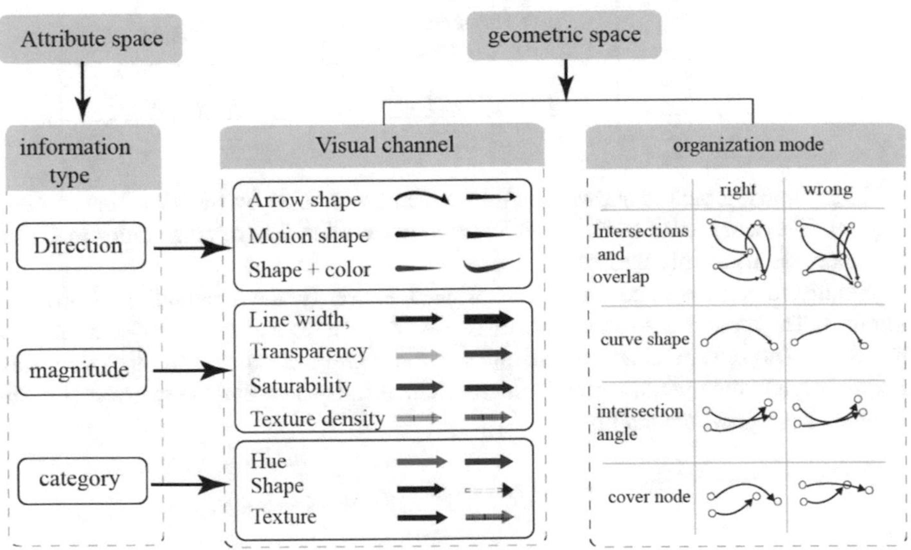

Fig. 1. Visual mapping model

4 Design Practice and Verification of Flow Line Symbols

Based on the previous theoretical analysis, this section gives two visualization schemes of the flow line. The first scheme uses a motion symbol instead of the arrow symbol. The scheme has a higher curvature at the starting point and then the curvature is continuously reduced. The tail of a flow line disappears when contacting the destination. The scheme uses a constantly changing curvature to represent the direction and the gradient color acts as a redundant variable representing the direction. The color of the flow line near the starting point is red and is blue near the endpoint so that the aggregation mode of the starting point and the destination can be observed more macroscopically. The head and tail can effectively avoid overlap. This scheme expresses the magnitude by the line width. In the second scheme, the direction is indicated by the shape of half arrow and the flow magnitude is expressed by the color gradient. Table 2 shows specific design schemes.

Table 2. Design schemes of the flow line.

Scheme	Express direction	Express magnitude
	Varying curvature; Color gradients act as redundant variables	Line width
	Half arrow shape	Transparency

The following picture shows visual flow maps generated by two flow line symbols (Fig. 2). The data comes from a certain commodity trade data of some cities in China. This example uses only part of the data.

Finally, a simple experiment is designed to verify the readability of the two schemes. The readers are asked to browse two flow maps and then answer a series of questions. The questions are set mainly for the identification of flow direction, flow magnitude, and the overall aesthetic degree. The results show that the two visualization schemes have good readability.

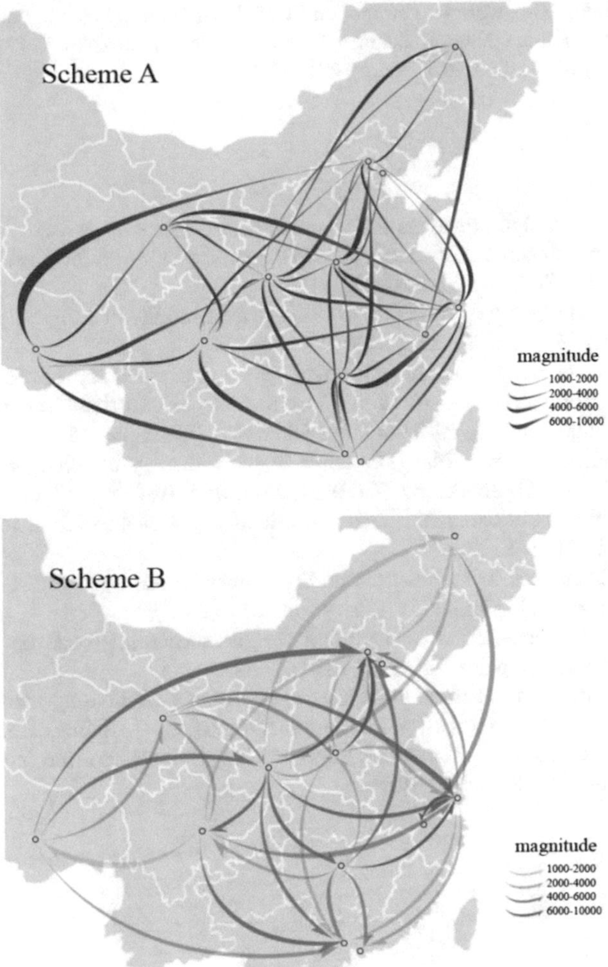

Fig. 2. Commodity flow maps

5 Conclusion

The flow line is important form expressing spatial relations. Its visual design affects people's cognition. Based on the theory of visual channel, this paper analyzes the visual variables that affect the visual representation of the flow line and constructs the mapping model from attribute space to geometric space. At the end of this paper, two visualization schemes of flow line symbols are presented. The validity of the scheme is verified by generating real visual flow maps. In the future, more effective experimental methods can be used to verify the effectiveness of the flow line. Besides, the layout of the flow line can be studied to increase the readability.

Acknowledgments. This work was supported by National Natural Science Foundation of China (No. 71871056), National Natural Science Foundation of China (No. 71471037), and Basic Research Program of Xuzhou, 2017 (No. KC17071).

References

1. Jenny, B., et al.: Design principles for origin-destination flow maps. Car. Geog. In. S (2016)
2. Buchin, K., Speckmann, B., Verbeek, K.: Flow map layout via spiral trees. IEEE Trans. Vis. Comput. Graph. **17**(12), 2536–2544 (2011)
3. Weihua, D., et al.: Using eye tracking to evaluate the usability of animated maps. Sci. China (Earth Sciences) **57**(3), 512–522 (2014)
4. Hornsby, K.S., et al.: Evaluating the effectiveness and efficiency of visual variables for geographic information visualization. In: Spatial Information Theory. LNCS, vol. 5756 (2009). https://doi.org/10.1007/978-3-642-03832-7. Chapter 12
5. Ware, C., Kelley, J.G.W., Pilar, D.: Improving the display of wind patterns and ocean currents. Bull. Am. Meteorol. Soc. **95**(10), 1573–1581 (2014)
6. Kubícek, P., et al.: Cartographic design and usability of visual variables for linear features. Cartographic J. **54**, 1–12 (2016)
7. Kurata, Y., Egenhofer, M.J.: Structure and semantics of arrow diagrams. In: International Conference on Spatial Information Theory. Springer (2005)
8. Wei, L.I., et al.: Personalized map symbol design mechanism based on linguistics. Acta Geodaetica Et Cartographica Sinica (2015)
9. Huang, W., Huang, M.: Exploring the relative importance of crossing number and crossing angle. In: ACM Press the 3rd International Symposium - Beijing, China (2010.09.28-2010.09.29) Proceedings of the 3rd International Symposium on Visual Information Communication - VINCE 2010 (2010)

Business Analytics, Design and Technology

Democratizing New Product Development Through an Industry-Society Entrepreneurial Partnership

Evangelos Markopoulos[1(✉)], Emma Luisa Gann[1], and Hannu Vanharanta[2,3]

[1] HULT International Business School, Hult House East, 35 Commercial Rd, London E1 1LD, UK
evangelos.markopoulos@faculty.hult.edu, emma.gann@hotmail.de

[2] School of Technology and Innovations, University of Vaasa, Wolffintie 34, 65200 Vaasa, Finland
hannu@vanharanta.fi

[3] Faculty of Engineering Management, Poznan University of Technology, Poznan, Poland

Abstract. New products derive from the industry and the society as well. They can be created in the mind of anyone who sees a need that people can buy, as an outcome of knowledge creation and sharing. This paper presents the structure, concepts, methods and operations of a proposed framework that addresses a new approach on product knowledge registration, evolution and utilization. It explains the co-evolution of a democratic industry-society relationship between the large or small organizations and individuals and provides the process model to apply it in practice. The proposed framework adapts and uses the Company Democracy Model as the base for the development of a practical approach through the evolution of the new product development process over its six escalated stages. Starting from the first level of the idea creation within a knowledge sharing culture, until sixth level of the models which deals with globalization and internationalization strategies.

Keywords: New Product Development · Innovation · Society · Consumer · Company Democracy · Shared value · Entrepreneurship · Management · Leadership

1 Introduction

New Product Development (NPD) is a significant concept that drives innovation and growth with direct impact on the society, the markets, and the economy. There are no limits on the size or type of organizations that develop new products. It seems that NDP is a privilege of the organizations who actively invest in R&D for new products and/or services. However, start-ups tend to drive this concept as their existence depends heavily on insights, ideas, innovations, and uniqueness their products shall have. Thus, large corporations are challenged from the pressure to continuously come up with new products to sustain their market shares, while start-ups are challenge from capital needs to fund their creativity, transform it into new products and place them in the markets.

T. Ahram et al. (Eds.): IHSED 2019, AISC 1026, pp. 829–839, 2020.
https://doi.org/10.1007/978-3-030-27928-8_126

This dual challenge on NPD can be approached with strategic collaborations, partnerships, mergers, and acquisitions between large-scale organizations and start-ups, but these cannot be considered the most productive, profitable, efficient and effective approaches.

The new product requirements that can make an impact and succeed in both the industry and society arise primarily from the consumers themselves. Market analysis, consumer preferences and behaviors, surveys, promotion events and other date marketing activities do not always assure the desired outcome on the data, information, and knowledge needed to come up with new products. However, understanding customers' added value needs and creating new rewarding channels of communication and cooperation can ignite profound circles of collaborative innovation activities.

The creation of a generic framework to provide a democratic innovation culture and operations through which individuals will be able to promote their new product ideas to the industry could be a potential solution in the found problem. In this new suggested approach data, information, knowledge, desires, needs, and wishes could be recorded in an independent organization, that can link the companies search for new product ideas directly with the consumers. Such a framework can benefit large organizations to reduce R&D costs, avoid mergers and acquisitions, and commit to long term investments by providing entrepreneurial opportunities to individuals who desire to share their ideas in practice. Furthermore, financial and entrepreneurial incentives can be provided to individuals in order to cooperate and co-develop, to the extent of their contribution, new useful and innovative products. This win-win relationship can be executed through a democratic approach, which provides equal opportunity for anyone to turn any idea into an actual product.

2 NPD in Management Science

NPD sustains long term organizational growth, increases profits and financial stability due to the continuous presence of the organization in the market. The need to remain visible and present in the eyes of the consumers is what primarily drives NPD. This continuous relationship with the customers, sustains and increases the market share. NPD can be mostly considered as a management strategy rather than a product management process. It incorporates product improvements, adaptations, extensions, and innovation on new or existing products, while at the same time incorporates the marketing, the communication, the finance and the entrepreneurial discipline as well, in a well-balanced effort to maintain organizational market lead and impact.

Today more than 25% of the industry total profits are generated through new product launches [1]. New products contribute on increasing brand loyalty of existing customers via its evolution and alignment on the market needs and trends, but also on the creation of completely new markets. Furthermore, NPD can operate as a risk diversification strategy through which selective products test the market needs and experiment with ideas that can change the route of a product or an industry. Fast changing market demands require agility on the creation or evolution of the products [2]. This agility reduces the risk of products being outdated with controlled evolution in minor or major features, or even with the use of effective communication strategies when the change is not on the product itself, but on its packaging, appearance or content volume [3].

Based on the organizational strategy, NPD can be used on a global scale, to drive socio-economic and environmental change, technological advancements and improve efficiency by responding to prevailing challenges [4]. New products can, and have, been designed to support new ideas, concepts, lifestyles and needs that reflect to the fast evolving environmental, political, social and cultural changes.

Even that NPD is not necessarily related with innovation, it is that the most innovative organizations can lead NPD in either originality on their products or meaningful innovative variations on them. Bloomberg gives the global innovation lead to Korea with a score of 87.3, followed by Germany with 87.3 and Finland 85.5 [5]. Figure 1 indicates that innovation is not quite related to the country size or its position in the global economy. Innovative countries do not necessarily derive from the leading economies but they can be considered the ones with the most global impact from the innovations they generate regardless when and by whom these innovations will be applied into new products or services. It is the quality of innovation and not the quality. Qualitative innovations can generate multiple products or applications. Canada for example is 8th on patents activity but 20th in the innovation index, while S. Korea is 1st on innovation and 20th on patents.

2019 Rank	2018 Rank	YoY Change	Economy	Total Score	R&D Intensity	Manufacturing Value-added	Productivity	High-tech Density	Tertiary Efficiency	Researcher Concentration	Patent Activity
1	1	0	S. Korea	87.38	2	2	18	4	7	7	20
2	4	+2	Germany	87.30	7	3	24	3	14	11	7
3	7	+4	Finland	85.57	9	16	5	13	9	8	5
4	5	+1	Switzerland	85.49	3	4	7	8	13	3	27
5	10	+5	Israel	84.78	1	33	8	5	36	2	4
6	3	-3	Singapore	84.49	13	5	11	17	1	13	14
7	2	-5	Sweden	84.15	4	15	9	6	20	5	25
8	11	+3	U.S.	83.21	10	25	6	1	43	28	1
9	6	-3	Japan	81.96	5	7	22	10	39	18	10
10	9	-1	France	81.67	12	41	13	2	11	20	15
11	8	-3	Denmark	81.66	8	21	15	12	19	1	28
12	12	0	Austria	80.98	6	11	12	24	8	9	18
13	14	+1	Belgium	80.43	11	26	10	9	41	16	9
14	13	-1	Ireland	80.08	32	1	1	16	15	14	38
15	16	+1	Netherlands	79.54	16	29	21	7	42	12	12
16	19	+3	China	78.35	14	13	47	11	6	39	2
17	15	-2	Norway	77.79	17	49	23	15	17	10	11
18	17	-1	U.K.	75.87	20	45	26	14	5	21	19
19	18	-1	Australia	75.38	19	56	17	20	18	15	6
20	22	+2	Canada	73.65	22	39	27	22	31	19	8
21	20	-1	Italy	72.85	24	22	20	19	29	29	26
22	21	-1	Poland	69.10	36	20	40	18	16	38	37

Fig. 1. Bloomberg 2019 innovation index.

3 Major NPD Challenges

Maintaining a successful flow on new products is a very challenging goal that can lead to great financial and reputation disasters if not performed effectively. The effort on inventing new product features or communication practices is highly expensive and risky due to the diversity of the customers that cannot be captured through ordinary data collections activities. Major challenges on NDP can be categorized into lack of human intellectual capital, lack of innovation, and lack of collaborative thinking.

Specifically, human intellectual capital challenges can be considered the organizational struggle on finding human resources that have the knowledge as well as the skills to contribute in the development of new products that fulfil the organization's objectives and simultaneously be relevant to the market.

Furthermore, the lack of innovation is a challenge faced primarily by mature organizations that have difficulties to change the winning recipe that drive their operations around a successful business model based on existing products or services [6]. Such organizations learned to expect certain profit from their operations, therefore new products development that might cause disturbance to their cashflow and planning, is systematically avoided.

Lastly, lack of innovation has been attempted to be tackled with strategic collaboration initiatives, but this brings up the challenge of collaborative thinking which requires both parties to be aligned on the same point of view, mentality, goals and visions for such partnerships to work successfully. Collaborative thinking is a multi-dimensional challenge as it is related to the organizational needs on exchanging knowledge and combining practices towards generating new products under the company itself or through joint product lines. Successful products that derive from collaborative thinking have been found primarily in the design industries with collaborations between various organizations such as Adidas, Porsche, Goodyear, H&M, Karl Lagerfeld and others. In these cases, new products were developed by merging only ideas and not organizational entities of any type. On the other hand, new products have also been developed after significant corporate mergers and acquisitions mostly on high tech or high-cost product development industries such as the telecommunications (Sony-Ericsson, Fujitsu-Siemens), the automobile industry (Mercedes-Benz), or other critical industries (Shell-Royal Dutch Petroleum) [7]. NPD approaches through collaborative products can be highly expensive on either profits distribution or merging investments. The return on investments has to be quite significant for such cases but this does not always seem to work. Failures, such as the Daimler –Chrysler or the America On Line (AOL)-Time Warner merger, indicate that collaborative NPD can be of high cost and risk [8].

4 Democratizing the NDP Process

The NPD barriers can possibly be avoided if a close relationship of the industry with the society can be achieved, since most NDP challenges are innovation and collaboration oriented. As of today, practices such as questionnaires, surveys, focus groups, and product sample distribution, among others are employed by the majority of the organizations globally. Even that there is no specific research indicating the effectiveness of these practices in NPD, it can be considered that they mostly contribute to the dissemination and evolution of existing products instead of creating new ones.

There is tremendous cost on original knowledge generation with direct impact on the society, as this requires well-structured communications units, R&D product units, corporate innovation culture and strategic partnerships.

What organizations seem to neglect in this effort is the society itself as a source of ideas and knowledge generation engine for NPD. The society is the people and the people have needs that can use used to inspire NPD. The elicitation of these needs can be turned into valuable knowledge that can be transformed into new products. However the challenge in this case in the incentivized engagement of the people to provide valid, true and valuable knowledge instead of information.

This challenge can be addressed with the democratization of the NPD process through an integrated framework that can link any person with any idea to any organization seeking to use such an idea for new and valuable products. This framework needs to be based on a number of incentives for both parties.

The initial reward for those who have such ideas is that they can benefit and improve their lives with the new products and services that will derive from their implementation. In addition, citizen must be assured that their knowledge is registered under their name and therefore any type of utilization from any organization will return the related benefits according to their involvement. This concept empowers people to actively create conceptual solutions for the daily life problems. Organizations can support this engagement by providing the relevant resources and infrastructure to the ones having the ideas in order to develop their desired products.

5 The Democratic New Product Development Model (DeNPD)

The democratization of the New Product Development process can be achieved through the Company Democracy Model as the base framework for the adaptation of a society-industry democratic and co-evolutionary relationship.

The Company Democracy Model (CDM) [9, 10], is an evolutionary process framework based on the ancient Hellenic wisdom through which the interpretation Delphic maxims such as 'Know Thyself', 'Metron Ariston' and 'Miden Agan' [11] are applied in a business context. The model is executed in a spiral formation and is aligned with the Evolute methodology for the generation of knowledge and its transformation into innovation [12]. Using ontologies and taxonomies the knowledge is classified and evolved based on the capacity, capability, competence, and maturity of the human resources [13], turning organizational tacit knowledge into explicit [14]. The adaptation of the CDM into a Democratic New Product Development framework follows the six levels pyramid structure in which the organization and the knowledge contributor co-evolve in the transformation on an idea into a global innovative new product (Fig. 2).

At level 1, the model creates the democratic culture and the process framework needed for ideas for new products to be generated, recorded and shared. It is the most critical stage of the model where the participation of the society impacts its overall execution and success. Level 2 is based on the utilization of the human capital with the identification of the best knowledge contributors and the invitation to them to join a team with corporate experts in order to develop their ideas furthermore. In this stage, the idea holder has the option to participate actively by taking the lead in the maturity of the idea and the design of the new product or stay less active by following the process by providing feedback. Once the idea of the new product has been designed and approved the level 3 takes place where the actual new product is being initially

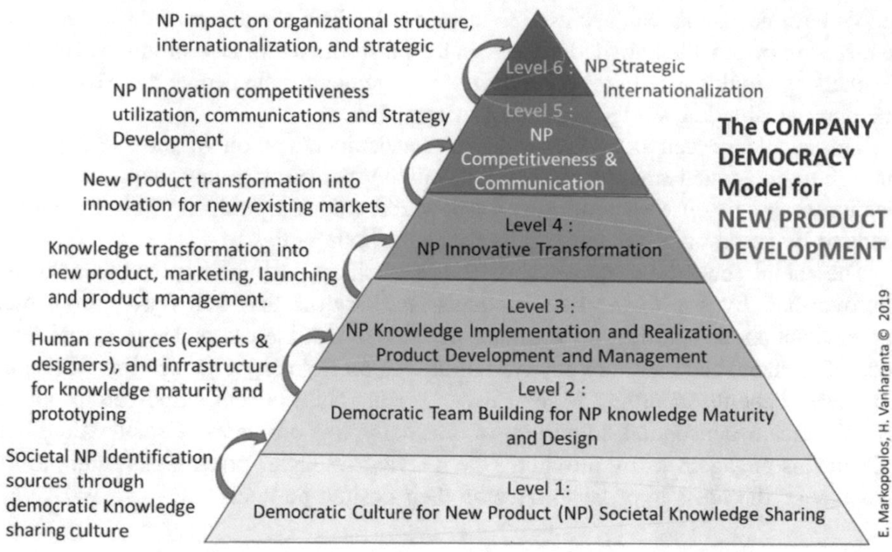

Fig. 2. The Democratic New Product Development Model (DeNPD)

developed as a prototype and then as a fully developed product, ready to be launched in the market. This is the product engineering level and the level that in most cases closes the new product development cycle unless the product behaves exceptionally well in the market, indicates innovative characteristics, and wide consumer acceptance.

In this case, process moves to level 4 where the innovative characteristics as supported furthermore making the product robust and solid on what leads its innovative impact on the market and the industry. With a highly innovative product, the process enters level 5 where the unique competitive elements of the product are emphasized in the communication, marketing and dissemination strategies impacting the global marketing strategy, sales, and operations. The execution of a competitive based strategy leads to level 6 where the product makes an impact on the organization itself due to its success and becomes part of the organizational structure as a new product line, business unit, spin-off, part of a strategic alliance via mergers and acquisitions or through other forms of international and global success utilization.

The evolution of an idea from the citizen up to the creation of a world-class innovative product is achieved over a period of time that is defined by the idea's potential. In this journey, the relationship of the citizen with the company can also go through a number of progressive stages related to the success of the product. The relationship of the product success and the commitment of the citizen are aligned with the benefits the citizen receives from this process. The evolution of the benefits offered to a participating citizen in the DeNPD can be represented by an inverse pyramid where the benefits are limited in the start but increase over the success of the product (shown in Fig. 3).

In this inverse, benefits, pyramid, level one provides the minimum benefits as the idea is submitted in draft format, the documentation is limited, and the effort is

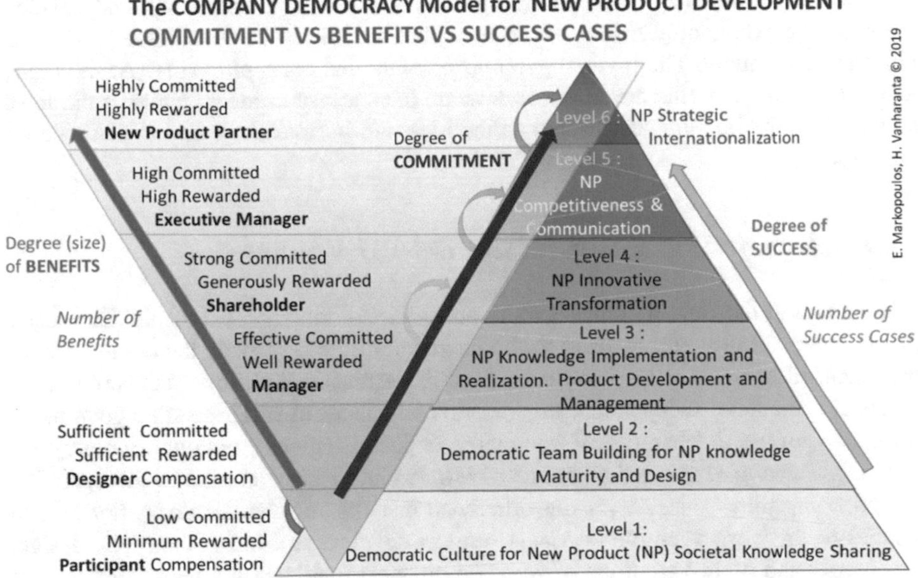

Fig. 3. Degree of commitment, success and benefits in the DeNPD

restricted to the completion of such documentation. Compensation can be granted in the case once an organization requests further documentation. In this sense, the commitment of the citizen is limited as well. The work submitted is registered under the citizen which is an alternative form of intellectual, instead of monetary, compensation.

At level two the new product idea has been selected by and organization and the citizen is required to participate in the evolution of the idea to a mature product design. In this case, the commitment of the citizen, if the offer is accepted, increases with the participation in the design team receiving the compensation for the work executed. However, the citizen can withdraw from the process. In this case the company continues with their own resources.

At level three the citizen takes the role of the project manager, based on the commitment and contribution placed in level two. The compensation can be relevant to the time and effort placed in the development of the idea into the actual product. Until this point, no profit shares are provided to the citizen as the success of the new product in uncertain. Level four is activated once the product achieves significant success and can justify its transformation effort into an innovative product. In this case the citizen receives part of the intellectual property shares and also the option to be part of the innovation transformation process.

Level five increases the organization's commitment on the product, but also participation of the citizen in the strategic management and communication on the utilization of the new innovative product. Being part of the journey from the product's initial idea until its global utilization, the citizen can receive an executive role on the management of the product. Level six is after the product international success where citizen can receive partnership on the organizational schema formed to support the product.

The model indicates that existing opportunities for personal growth and development can be exponential from level to level, but only a few ideas become world-class innovative products. The inverse pyramid presents this concept clearly. At the higher levels (4–6) the benefits are many as few are the successful ideas, while at the lower levels (1–3) the benefits are limited as the ideas are in draft stage and their success is uncertain.

6 Implementation Forms of the DeNPD Framework

The success of the DeNPD relies heavily on a society-industry relationship that must be driven by credibility and security to the citizen in order to provide the confidence and trust needed to participate. Such a goal can be achieved with the establishment of a regional or national independent organization which could either be state governed or privately governed (chamber of commerce or local, regional, national, industry federation). The organization shall act as a New Product House, similar to the operations of the companies houses with the difference that the registered information will be related to the work submitted on new product development ideas, concepts, designs, prototypes and other variations of work on products that have not been commercially launched yet.

This New Product House (NPH) house shall act responsibly for the credibility of the overall process. The purpose of this organization is to ensure the matching of the right new product idea with the right organization request. The NPH operates at the first level of the DeNPD, setting up a democratic culture, environment and infrastructures for all those who wish to contribute. It must be noted that the knowledge, idea or work contributors do not always seek entrepreneurial or corporate careers, but mostly expect to see in the market a product or service they feel a need to have. Figure 4 presents a high-level process of this democratization concept on the New Product Development, with the NPH to orchestrate this industry society relationship.

Access to the NPH can be granted to both the citizens and the organizations. Companies can request feedback on new product needs or access ideas, designs or even prototypes. These requests can be visible to all, and the collaboration between the industry and the society can be established through the NeNDP framework. This win-win relationship provides the opportunity to all citizens to share their new product ideas, and to participate in corporate projects initiatives and activities on new products development. It also provides access to all organizations to search or request such knowledge by returning the relevant rewards, incentives and opportunities to utilize it.

The Democratic New Product Development concept combines the three main stakeholders (Society, New Product House, and Industry) and creates a knowledge-sharing environment between the society and the industry under the coordination, security and knowledge safeguarding of an independent institution formed to support this goal through the democratization of new product development challenge.

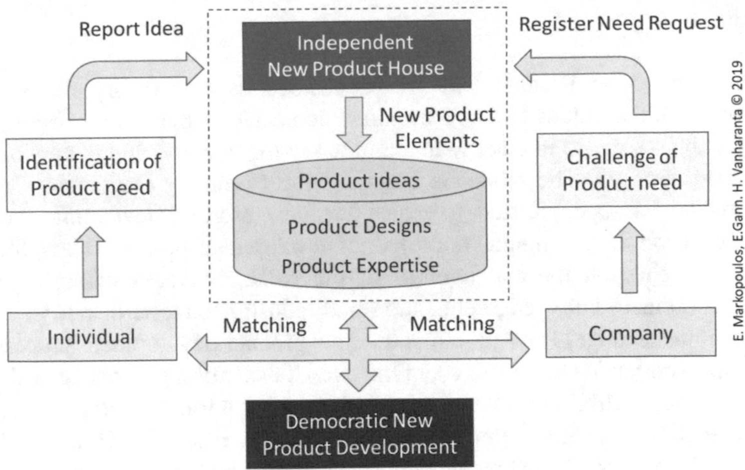

Fig. 4. The new product house operations.

7 Industry-Society Partnership. Social and Shared Value Innovation

The DeNPD framework extends its scope into a more social dimension by empowering individuals to contribute to the economy to the, society but also to their personal development and growth. The framework is aligned with the shared value innovation and development principles in a purely democratic way which assures credibility, recognition and rewards to all who wish and can contribute.

Share value innovation describes the process that aims mutual gains for the industry and the society, with organizations to generate economic returns by addressing social issues and adding value to the society [15].

The social product development model today extends the open innovation and moves beyond the traditional customer-involvement models towards the social engagement of individuals [16, 17]. Under democratic procedures the people, as fully involved actors in the development of new products, participate is all stages, from the ideas generation to the actual product realization.

The framework inspires both the people and the organizations to collaborate in a co-evolutionally relationship forming a society industry partnership. This ethical dimension creates the added value on this approach and redefines the contribution of the companies to the society not only as product providers and therefore profit makers, but also as opportunity providers and therefore profit givers. The profit-making-giving circular approach is what democratic social shared value new product development is all about.

This re-definition of the shared value innovation under the democratic concept and towards new product development can impact tremendously the way new products are conceived, developed and even used. The world strives for the application and adaptation of the democratic values that extend the freedom of speech to the freedom of act. All who can and wish to contribute towards making this world better need and must have the opportunity of a fair share on this effort.

8 Conclusions

The Democratic New Product Development model is a Y-theory liberal thinking model, based on the values and opportunities democracy offers to all. It provides the channel for the society to interact with the industry for shared value in new or existing products and services. The model is based on the Company Democracy Model for innovation and has been altered to match the new product development challenge through a co-evolutionary human and product development process. The effectiveness of the model relies on the effectiveness of the NPH, the intermediate organization needed to coordinate knowledge elicitation and industry requests or needs. The organization sets the standards and monitors the new product development process. It also validates and verifies the knowledge and the overall execution process needed to create the security and credibility that will attract the participation and commitment of the society. The DeNPD is at a theoretical stage and further research will be conducted to identify the legal, social, technical, cultural and financial elements for the process optimization that will lead it to a pilot adaptation in the industry.

References

1. McKinsey. https://www.mckinsey.com/business-functions/marketing-and-sales/our-insights/how-to-make-sure-your-next-product-or-service-launch-drives-growth
2. Leite, M., Braz, V.: Agile manufacturing practices for new product development: industrial case studies. J. Manufact. Technol. Manage. **27**(4), 560–576 (2016)
3. Lee, Y., O'Connor, G.C.: The impact of communication strategy on launching new products: the moderating role of product innovativeness. J. Product Innov. Manage. **20**, 4–21 (2003)
4. Global Entrepreneurship Index. https://thegedi.org/wp-content/uploads/dlm_uploads/2017/11/GEI-2018-1.pdf
5. Bloomberg. https://www.bloomberg.com/news/articles/2019-01-22/germany-nearly-catches-korea-as-innovation-champ-u-s-rebounds
6. Forbes. https://www.forbes.com/sites/tendayiviki/2018/11/04/why-large-companies-continue-to-struggle-with-innovation/#47a503bc67b4
7. Daimler. https://www.daimler.com/company/tradition/mercedes-benz/history.html
8. Köhler, H.D.: From the marriage in heaven to the divorce on earth: the daimlerchrysler trajectory since the merger. In: Freyssenet, M. (ed.) The Second Automobile Revolution. Palgrave Macmillan, London (2009)
9. Markopoulos, E., Vanharanta, H.: Democratic culture paradigm for organizational management and leadership strategies - the company democracy model. In: Proceedings of the 5th International Conference on Applied Human Factors and Ergonomics AHFE 2014 (2014)
10. Vanharanta, H., Markopoulos, E.: Creating a Dynamic Democratic Company Culture for Leadership, Innovation, and Competitiveness. 3rd Hellenic-Russian Forum, 17 September 2013
11. Parke, H., Wormell, D.: The Delphic Oracle, vol. 1, p. 389. Basil Blackwell, London (1956)
12. Kantola, J., Vanharanta, H., Karwowski, W.: The evolute system: a co-evolutionary human resource development methodology. In: Karwowski, W. (ed) The International Encyclopedia of Ergonomics and Human Factors. CRC Press, Boca Raton (2006)

13. Paajanen, P., Piirto, A., Kantola, J., Vanharanta, H.: FOLIUM - ontology for organizational knowledge creation. In: 10th World Multi-Conference on Systemics, Cybernetics, and Informatics (2006)
14. Nonaka, I., Takeuchi, H.: The Knowledge-Creating Company: How Japanese Companies Create the Dynamics of Innovation. Oxford University Press, New York (1995)
15. Markopoulos, E., Vanharanta, H.: The company democracy model for the development of intellectual human capitalism for shared value. Procedia Manufact. **3**, 603–610 (2015)
16. Forbes, H., Schaefer, D.: Social product development: the democratization of design, manufacture and innovation. Procedia CIRP **60**, 404–409 (2017)
17. Abhari, K., Davidson, E.J., Xiao, B.: Measure the perceived functional affordance of collaborative innovation networks in social product development. In: 49th Hawaii International Conference on System Sciences, pp. 929–938 (2016)

Development of a Concept for the Use of Humanoid Robot Systems with the Example of a Logistic Support Process

Tim Straßmann[1(✉)], Daniel Schilberg[1], and Anna-Lena Wurm[2]

[1] Institute of Robotics and Mechatronics, University of Applied Science Bochum, Lennershofstr. 140, 44801 Bochum, Germany
{tim.strassmann, daniel.schilberg}@hs-bochum.de

[2] Daimler AG, Mercedes-Benz Werk Düsseldorf, Rather Strasse 51, 40476 Duesseldorf, Germany
anna-lena.wurm@daimler.com

Abstract. In this paper the advantages and disadvantages of a humanoid robot system at this time are pointed out. Therefore the requirements to the humanoid robot system are first derived from the use case. This enables us to analyze the state of the art of humanoid robot systems. Finally we present a theoretical concept for the use of Humanoid robot system in a convenient large-scale production application. This leads us to an evaluation of the feasibility.

Keywords: Humanoid robotic · Humanoid · Human-systems integration · Systems engineering

1 Introduction

Humanoid robots are mostly developed for large assemblies like the aviation industry [1] or for emergency measures [2]. But are they also useful for large-scale productions and the related task? This paper tries to answer this question based on the use case.

The aim to use new and innovative technologies with the benefits of saving personnel costs and gaining more efficiency is existing ever since especially in the leading industry of Germany the automotive corporations. In order to do that you need a suitable application and a suitable product with the new technology. We found a use case at Daimler in Düsseldorf that satisfies the first condition by having the advantages of simplicity and being an unergonomic task which therefore would be good to substitute with an automatic system. Now it must be clarified if the feasibility for the implementation with a currently available humanoid robot system is given. The main achievement in this paper will be a general list of requirements to the humanoid robot system for applications similar to the use case.

Despite there might be better options this paper only considers bipedal humanoid robots and their specifications and abilities for the concept. It is also a condition that there are no adjustments to the production system or the facilities. Most humanoids are radio-controlled and it will be a novum and difficulty of this application to run the system autonomous as a usual industrial robot system does.

T. Ahram et al. (Eds.): IHSED 2019, AISC 1026, pp. 840–846, 2020.
https://doi.org/10.1007/978-3-030-27928-8_127

2 The Use Case: A Logistic Support Process

The use case is a logistic support process in the paint shop of the Sprinter production in Düsseldorf. It is about the needs based supply of pre-commissioned material boxes onto the floor of the body. The cases contain damping pads which will be installed in a later step by additional workers. There are four different kinds of boxes because of the many options and features of the final product the customer can choose from. The boxes need to be carried over approx. 110 m from a commission place to the production line. A ramp and stairs are obstacles on this route. The humanoid robot needs to lift and drop the boxes with a maximum weight of 12 kg in heights from ground level up to 1.50 m. About every third body needs the damping pads of the boxes which leads to an average cycle time of approx. 4.5 min and the daily work time is 21 h [3].

3 Requirements to the Humanoid Robot System

The requirements to the humanoid robot system can be divided into four categories which collect a group of coherent abilities or specifications. Some points may appear multiple times because they are important in different ways.

The first category is *safety and protection* of the system itself and its surrounding. These requirements arise from general standards of the corporation as well as from international and national standards. This category contains e.g. an emergency stop for dangerous situations that immediately interrupts the movement of the robot and prohibits further motion of the humanoid. This will turn out to be a difficulty because of the complex stabilization of these robots that preempt to just do an easy and safety shutdown of the drives. Another stop to prevent danger to the robot, humans or the environment is the safety stop and is essential for industrial robot systems. This stop should be activated when something enters the safety zone – which also needs to be defined - and interrupts the robots motion for the time until the safety zone is free again. Further the design and the construction of the torso and all other parts of the humanoid robot should not have a shape that can cause injury to the operator. In addition limitations for the drives are required so that the power cannot constantly increase. A collaboration mode need to guarantee stricter constraints when the humanoid interacts with humans. Additionally a suitable communication between the robot system and the human co-workers or general surroundings contributes to successful prevention of dangerous situations. The last point of the safety and protection is the stabilization of the humanoid robot itself. Humanoids entail the high and dangerous risk of falling while they are in motion or just stand. The control needs to ensure the lowest probability of the robot falling at any point.

The next category summarizes all required *abilities* of the humanoid robot. The abilities arise from the use case as the load capacity of the humanoids arms for example must be at least 12 kg. The robot also must reach the working space of 0 m to 1.5 m with his arms and ensure enough power at all positions. Additionally the humanoid requires image processing to distinguish between the different types of boxes. The theoretical velocity of the robot must at least be 3.5 kph which is a roughly calculation from the use case. Furthermore import for an autonomous working robot system is the recognition of the environment in collaboration with the navigation which has to direct the robot through the production facilities. This requires corresponding sensors and control. The operating time of the humanoid must be at least 21 h a day.

The third category deals with the *design and construction* of the humanoid robot system. These specifications contribute more to the effectiveness than the general feasibility of the humanoid robot at the application. Among others the mobility of e.g. the arms need to ensure that the humanoid reaches all the places of the boxes and can grab them too. The dimensions of the humanoid also contribute to its mobility and also flexibility and walking speed. The gripper must be designed so that they can grab the cases. The design of the head should not only contributes to the technical specifications needed but also enable a good physical cooperation and communication with its surrounding. The selection of the drive system should provide an easy control and a support of the abilities. It should also contributes to the durability and prevent maintenance.

The most complex category of requirements is *control and software*. It is not easy to describe which kind of algorithms or control the robot requires because there are different solutions with similar results. The important aspects to ensure are safety, stability while standing and in motion as well as effectiveness and efficiency. How they are implemented will distinguish from humanoid to humanoid and considers the design of the robot. The humanoid robot system needs to be connected to the production system to get the production information needed and communicate to the system about its status. An user friendly graphic user interface can help operators to correct issues by themselves which is an important advantage in the daily industrial live to prevent long downtimes. Generally the control of the robot need to be well-engineered and reliable.

The table below lists the previous points to give an overview for the further consideration (Table 1).

Table 1. Requirements to the humanoid robot system

Category	Specification/Ability	Comment
Safety and protection	Emergency stop	Immediate “Shutdown”
	Safety stop	Immediate stop
	Safe construction	No danger spots
	Power limitations	Of the drive system
	Stability	No fall down
Abilities	Power and mobility	Of drive system and construction
	Image processing	For cases identification
	Velocity	When walking with or without load
	Endurance	21 h workday
Design and construction	Mobility	Of limbs
	Dimensions	Height etc.
	Head design	Sensors and design
	Drive system	Reliability and power
Control and software	Control	General
	GUI	Maintenance and modifications

4 State of the Art

The state of the art in humanoid robotic is defined by the research and development of specialized institutes. The economic use of humanoid robots does not occur yet but they are tested in the field and are constantly upgraded. This chapter will mention some humanoid robots with their special technologies and also some promising research fields.

The DARPA Robotics Challenge [2] was a turning point in the recent history of the humanoid robotic. DRC-HUBO+ [4] was the winner of this challenge and convinced with his two locomotion modes. He can walk when he needs to reach more highly situated objects or has to climb stairs and drive on wheels attached to his knees if he needs to get to lower objects and gains more stability through the lower center of gravity. Air-cooling and cinematic redundancy increases the power and mobility of his limbs to the best weight/power ratio of the participants. A hybrid position/force control enabled him to be the only one who solved all tasks while he was radio controlled by an external operator. Most noticeable was the deficit of the long time all robots needed to do the challenges [4]. Atlas was also used in the DARPA Robotics Challenge and got recently updated an now has impressive athletic skills. A demonstration video of the hydraulic humanoid shows what could be physically possible but Atlas is unsuitable for a convenient use yet [5].

Two of the latest humanoid robots are WALK-MAN [6] form the IIT and HRP-5P [7] which is a prototype of an upcoming humanoid from the AIST. They both have an outstanding power density and are cinematically efficient designed. Especially HRP-5P shows impressive skills as he managed to manipulate a gypsum boards at construction site [7]. His physical structure seems to be the strongest at that time but the main issue is still the software.

An example for an industrial useful applicable humanoid is ARMAR-6 [8] from the KIT in Germany. His proposed usage is as an assistance to maintenance technicians. It has an artificial intelligence that enables it to understand the needs of the worker and its task-space impedance controller brings the advantage of safe collaboration with humans and a simplification of grasping actions. [8].

This leads to the general question and research topics of the control of humanoid robots. The latest rudiments in the field of humanoid robotic deal with artificial intelligence [9] and biological inspired structures [10]. The overall goal is to improve the stability and reliability of humanoid robots which is often done by implementing imitations of human or general natural mechanisms. Further computation time has to be reduced and AI and deep learning seem to be key technologies at this point but progress is naturally made in little steps. Now it is the time to find applications where humanoids can really be usefully implemented with the capabilities the state of the art offers.

4.1 Evaluation: Requirements and State of the Art

The requirements have been evaluated as they are fulfilled by the state of the art or not. Through the study of over 50 publications, videos and articles the aspects could be assigned by points from 1 to 4. 1 Point is defined as fulfilled and ready for the application whereas 4 Points stand for an specification or ability which was not mentioned or will not be mature in the next few generations of humanoids yet (Table 2).

Table 2. Evaluation: status of requirements

Category	Specification/Ability	Points	Comment
Safety and protection	Emergency stop	3	No safe power-of-position
	Safety stop	2	Needs to be specified
	Safe construction	1	E.g. 3D printed covers allows it
	Power limitations	1	Mechanical or programmable
	Stability	2	Constantly improved (continuous process)
Abilities	Strength a. Mobility	2	High but not in every posture
	Image processing	2	Theoretically possible
	Velocity	4	Too slow and difficult to improve
	Endurance	2	Not possible through running times but alternatively
	Mobility	1	Possible
Design and construction	Dimensions	1	Human-size humanoids
	Head design	1	Fully equipped
	Drive system	1	High power and small design
Control and software	Control	3	No similar application before
	GUI	4	Not mentioned yet

The data in the table show that some requirements are already viable but the more specific they get concerning the application the more detailed and even new solutions have to be developed.

5 The Concept

The humanoid used in the concept will be HRP-5P or one with similar specification since the important features are strength and utility and the control will be most likely developed especially for the application. HRP-5P proved his abilities impressively at construction site in an experiment of the AIST. A set of rechargeable batteries which can be changed easily without danger for the worker are required and will be implemented into the humanoid. A kneeling posture can be used for getting access to the humanoid's torso as well as when the drives are disabled at safety stop. The abilities in locomotion are dissatisfying yet and so an auxiliary (a hand-wagon) will be used. It will shorten the cycle time to 27 min which leads to an achievable walking velocity for the robot. Additionally he can use it as a support and has less energy consumption thus it can carry the boxes. A redundant safety zone monitoring is applied and will operate from outside and on the robot to ensure no humans or obstacles are overlooked. The safety zone will be in the falling radius of the robot which is about 2 m to every side. At places where the humanoid and humans collaborate the zone as to be dynamically and the robot should lower is overall activity. The process will be divided in sequences which occur depending on the production status – when damping pads are needed for the body. Boxes will be provided with QR codes to simplify the identification process. The robot will need to use the production environment as support for his tasks as he is not able to walk up stairs with boxes in his hands. A rack near the production line will be used as a buffer to maintain the material feed.

6 Conclusion

First in this paper the use case is described and the requirements like emergency stops are derived. After the state of the art is thematized it leads to an evaluation of the feasibility of individual abilities/specifications with a table of rated attributes as the result. A concept arises from the deliberations made and mentions adjustments that have to be made. The breakdown shows that an application of a humanoid robot in a logistic support process is theoretically possible when auxiliaries are used like it is done in the concept. The effectiveness and more over the efficiency must be doubt. Humanoids do not seem to reach human-like capabilities in this kind of application soon. Unfortunately the availability of humanoid robot systems is limited and often only research and development institutes build and use them. But to make progress applications like this must be implemented and experience has to be made to also make a point about reliability and need of maintenance. However the benefits of a humanoid working in this sector are huge and important in the changing process from traditional modern factories to the smart factory of the future because they are flexibly usable and almost no adjustments to facilities or the existing systems have to be made.

References

1. Aero Telegraph. https://www.aerotelegraph.com/humanoider-roboter-soll-airbus-maschinen-bauen
2. Spenko, M., Iagnemma, K., Buerger, S.: The DARPA Robotics Challenge Finals: Humanoid Robots to the Rescue. Springer, USA (2018)
3. Daimler, A.G.: Standard operating procedure. Düsseldorf (2019)
4. Lim, J., Bae, H., Oh, J., Lee, I., Shim, I., Jung, H., Joe, H.M., Sim, O., Jung, T., Shin, S., Joo, K., Kim, M., Lee, K., Bok, Y., Choi, D.-G., Cho, B., Kim, S., Heo, J., Kim, I., Lee, J., Kwon, I.S., Oh, J.-H.: Robot System of DRC-HUBO+ and Control Strategy of Team KAIST in DARPA Robotics Challenge Finals. Springer, South Korea (2018)
5. Boston Dynamics. https://www.bostondynamics.com
6. Ackerman, E., Guizzo, E.: A lighter and more efficient redesign of IIT's disaster robot can fight industrial fires. Spectrum IEEE (2018). https://spectrum.ieee.org/
7. Kaneko, K., Kaminaga, H., Sakaguchi, T., Kajita, S., Morisawa, M., Kumagai, I., Kanehiro, F.: Humanoid robot HRP-5P: an electrically actuated humanoid robot with high power and wide range joints. IEEE Robot. Autom. Lett. **4**(2), 1431–1438 (2019)

8. Asfour, T., Kaul, L., Wächter, M., Ottenhaus, S., Weiner, P., Rader, S., Grimm, R., Zhou, Y., Grotz, M., Paus, F., Shingarey, D., Haubert, H.: ARMAR-6: a collaborative humanoid robot for industrial environments. In: 2018 IEEE-RAS 18th International Conference on Humanoid Robots, pp. 447–454. IEEE, Beijing (2018)
9. Tran, D.-H., Hamker, F., Nassour, J.: A humanoid robot learns to recover perturbation during swinging motion. In: IEEE Transaction on Systems, Man, and Cybernetics: Systems. IEEE (2018)
10. Nassour, J., Hoa, T.-D., Atoofi, P., Hamker, F.: Concrete action representation model: from neuroscience to robotics. In: Artificial Intelligence, Computer Science, Chemnitz University of Technologie. IEEE, Chemnitz

Production Management Model Based on Lean Manufacturing Focused on the Human Factor to Improve Productivity of Small Businesses in the Metalworking Sector

Jonathan Huamán[1(✉)], José Llontop[1(✉)], Carlos Raymundo[2(✉)], and Francisco Dominguez[3(✉)]

[1] Ingeniería Industrial, Universidad Peruana de Ciencias Aplicadas (UPC), Lima, Peru
{u201314536,pcapjllo}@upc.edu.pe

[2] Dirección de Investigación, Universidad Peruana de Ciencias Aplicadas (UPC), Lima, Peru
carlos.raymundo@upc.edu.pe

[3] Escuela Superior de Ingeniera Informática, Universidad Rey Juan Carlos, Madrid, Spain
francisco.dominguez@urjc.es

Abstract. Micro and small businesses in the metalworking sector present a high percentage of informality, and the majority of their production processes are not documented nor standardized. The production approach of Toyota, the enterprise-resource-planning system, and the lean manufacturing system were created for large organizations with sufficient financial resources to successfully implement these models. However, the human factor is a critical component in implementing these models. Therefore, a production management model has been proposed, which is based on lean manufacturing and focuses on the human factor, to increase productivity in the production area of a metalworking company, a small business. The result was that the company's mentality changed, reaching a productivity of 70% through the use of lean manufacturing tools in the production area.

Keywords: Productivity · Lean manufacturing · Production management model · Human factor

1 Introduction

The majority of Peruvian manufacturing companies consists of micro and small businesses, which make up approximately 99.6% of the total number of businesses. The metalworking sector ranks third among manufacturing companies and represents approximately 15% of operating companies, according to the Socio Economic Labor Observatory [1]. The main problem of this sector is that informality is very high, and its production processes are neither documented nor standardized, which affects their productivity. The metalworking sector is very important to the country's economy, not

T. Ahram et al. (Eds.): IHSED 2019, AISC 1026, pp. 847–853, 2020.
https://doi.org/10.1007/978-3-030-27928-8_128

only due to its contribution to added value and technological development but also because it is a sector that supplies to other key industrial sectors, acting as a link in the productive framework of different economic sectors [2]. Currently, the use of production models such as the just-in-time manufacturing of the Toyota Production System, materials requirements planning, or lean manufacturing need to be adapted on the basis of the real situations of growing businesses [3]. According to Takami, these models are based on the period, market need, country, level of research and innovation, economy, policies, and cultural aspects different from that of the Peruvian manufacturing reality.

The proposed production management model uses lean manufacturing; it focuses on the human factor and, at the same time, employs additional elements to improve the productivity of the production area of small businesses in the metalworking sector. This study contributes to proposing a production model adapted to the Peruvian reality, which is focused on the human factor and utilizes the lean manufacturing model as a reference to increase small business productivity.

2 State of the Art

2.1 The Human Factor in Business

The effectiveness and efficiency of applying the lean model has been proven by many successful cases around the world, such as Toyota and Nissan. However, according to Budyunek [4], successful lessons from developed countries help to provide insight into the fact that it is not possible to uniformly apply any business management approach. Based on its situation, each company must find the proper management approaches for its business and be flexible in their application. In his article, Ngyuyen [5] mentions the fact that in Vietnam, a sustainable lean manufacturing model for Vietnamese companies had to be adjusted to their reality and a model for his country's manufacturing companies was proposed. Valencia indicates that lean tools are known, and there are people who are knowledgeable about their implementation, but they have not worked out as expected [6]. The authors mention that the human factor is important to successfully implement different production models. A sustainable lean manufacturing model is proposed for business development, which would recognize the importance of the role of human resources on the impact of implementing the lean model in businesses.

2.2 Production Implementation Models

In their article, "A Conceptual Model of Lean Manufacturing in its Dimensions," Abdul, Mukhtar, and Sulaiman [7] state that there are seven main dimensions for lean models, including the manufacturing process and equipment, manufacturing planning and scheduling, visual information systems, supplier relationships, customer relationships, workforce, and product development and technology. These dimensions are important in contextualizing the model to a reality closer to that of small businesses. Ajallim [8] also explains the relationships of these dimensions with their residuals and the critical factors for their implementation. These authors explain that the majority of companies that have put the lean model into practice and did not achieve its implementation, failed in large part due to internal issues.

2.3 Implementation Strategies for the Lean Manufacturing Model

There are different production models of the lean manufacturing philosophy. According to Mohammad [9] in his article "Lean Application Frameworks, Challenges for Small Businesses," there are a number of frameworks and road maps to implement the lean model [10]. However, many companies have difficulties putting the model into practice, given that using these frameworks does not guarantee correct implementation. Shingo mentions that for the successful implementation of the lean production model, there are critical factors within the organization that must also be considered, such as the culture of the organization and ownership [11]. As such, the application of a model for small businesses can be implemented by analyzing an appropriate road map and adapting it to the reality of the business.

3 Contribution

3.1 Proposed Model

The proposed production management model for small businesses is based on lean philosophy. To this effect, the human factor adapted to the organizational culture of the proposed model is crucial. In general, small companies understand the concept, its nature, and its implementation methods. However, they have problems normalizing and sustaining the model over time. Therefore, the novelty of the model is in that it focuses on creating added value for companies by using inputs from the employees to continuously improve a sustainable production model. Figure 1 shows the diagram of the proposed model. According to Abdul [11], the human factor is important; the participation of every employee of an organization is important to the implementation of the production model. It reinforces the concept of production management on the basis of the lean manufacturing mentality of zero waste and the realization of process standardization in small businesses to collaborators, leaders, external and internal model specialists, and the manager being important to the success of the model.

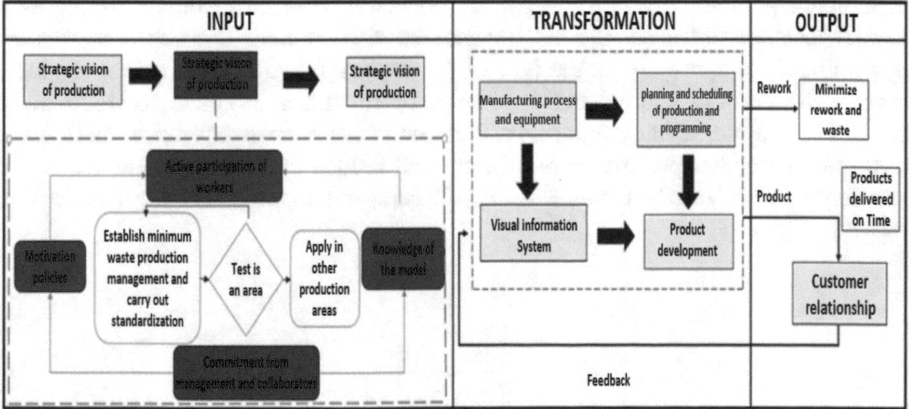

Fig. 1. Production management model diagram overview

4 Validation

4.1 Company Description

Grupo Obregón SAC is a small metalworking business that is dedicated to the manufacturing and reconstruction of metal pieces for the mining, textile, metallurgy, and paper sectors. It is located on Calle San Andrés Mz G Lote 1, Urbanización Santa Martha in the district of Ate- Lima, Perú. Current productivity of the production areas is 60%. Similarly, the workers' active participation and knowledge of the model through training sessions and motivation policies were mediocre, scoring 50% for each.

4.2 Model Implementation

4.2.1 Human Factor

Each person who plays a role in the organization is important for the implementation of the proposed production model. The implementation of the change to the production model is shown below.

- Knowledge of the model: The expert on the proposed production model explains the importance of everybody involved to learn about the production management model that is to be implemented. Education and training by the managers, area leaders, and collaborators on the concepts of minimizing waste and realizing standardization is proposed.
- Motivation policies: The human resources department and the production manager hold meetings to develop motivation policies so that workers implement the model and contribute their own ideas or improvements in their work areas. The objective is that all collaborators feel that they benefit from the implementation of the model.

4.2.2 Manufacturing Process and Equipment

According to Asgharizadesh [8], to ensure that a company's many standards and production methods are complied with, it is necessary to choose the most suitable support tools to support production, in order to have a continuous flow, diminish manufacturing times, and manage precise operation. Nguyen recommends the most popular tools for SMEs such as 5S, kaizen, and visualization. The aim is to seek the most suitable support tools for the type of production. It is important for the model to promote the use of technology and innovation by small firms, to increase productivity, and to become increasingly competent. Currently, there are many machines that can make the production process more efficient and reduce future investment costs. The company standardized the manufacturing of inserts and applied 5S at work stations, as shown in Fig. 2.

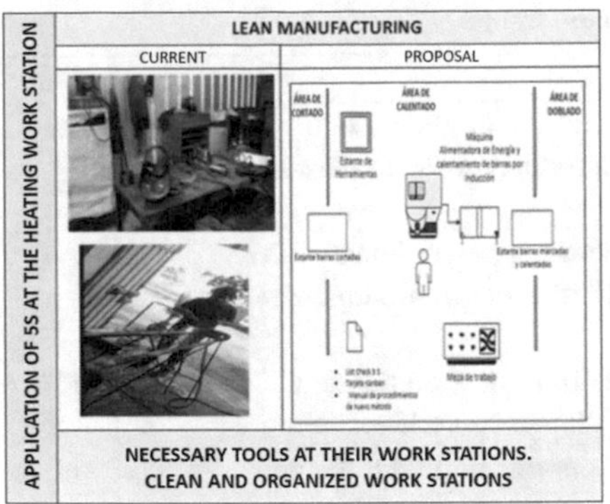

Fig. 2. Application of 5S at the heating area workstation

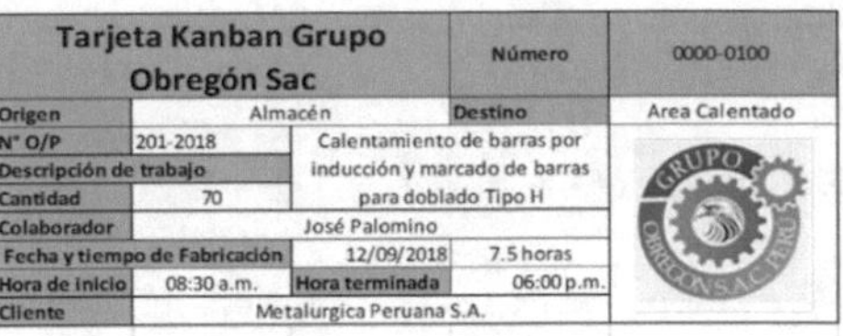

Tarjeta Kanban Grupo Obregón Sac			Número	0000-0100
Origen	Almacén		Destino	Area Calentado
N° O/P	201-2018	Calentamiento de barras por inducción y marcado de barras para doblado Tipo H		
Descripción de trabajo				
Cantidad	70			
Colaborador	José Palomino			
Fecha y tiempo de Fabricación		12/09/2018	7.5 horas	
Hora de inicio	08:30 a.m.	Hora terminada	06:00 p.m.	
Cliente	Metalurgica Peruana S.A.			

Fig. 3. Kanban production card

4.2.3 Visual Information System

An information system enables relevant decision-making for rapid feedback and to carry out corrective action. The objective is to treat and administer data and information for its subsequent use. To this end, the information system works on the basis of information, people, and resources. Shared information is important to monitor the proposed objectives and make decisions on the same. Likewise, visual information in the production area makes it possible to visualize the current production state of the work area and take immediate corrective action, as per Mohamed [6]. In this case, an adapted Kanban card was used to display information for the manufacturing staff, such as materials to be used, production time, and work method, among others. This information was supplied by other areas and delivered to the production area (Fig. 3).

The information conveyed to the production area is important. Therefore, the implementation of the visual control tool was necessary for production. The quantity to be produced, the collaborator executing the work, the description of the work, quantity to be made, and production time are detailed on the Kanban card.

4.2.4 Indicator Results

The results of the indicators of the lean manufacturing production model implementation focused on the human factor are shown below. Productivity increased to 70% after the implementation of the proposed model. Worker participation to reduce waste increased to 80%, and the knowledge of the model was imparted to all employees in the production area reaching 100%. Likewise, employee satisfaction with the improvements in motivation policies reached 80% due to their good quality work being recognized (Table 1).

Table 1. Indicator results

Human factor	Indicator	Result
Commitment from management and collaborators	Production productivity with the new model/Production productivity with the former model	70%
Active participation	Num. of collaborators that contributed with suggestions or new methods/total num. of collaborators	80%
Knowledge of the model	Num. of collaborators trained on the new model/Total num. of trained collaborators	100%
Motivation policies	Num. of collaborators satisfied with the new policies/Total num. of collaborators in the company	80%

5 Conclusions

In this case study, the proposed production model based on lean manufacturing focused on the human factor for small companies, managed to increase productivity to 70% in the manufacturing of inserts through improved staff management. Support from management and recognizing, educating, and providing training on the proposed model and teaching a lean mentality to company employees was successfully implemented, thereby achieving the benefits of the production system based on lean manufacturing.

Despite the positive result shown for the application of the proposed model, it is possible to reduce rework and waste within companies using the human factor for the correct operation of other lean manufacturing tools such as kaizen and just-in-time manufacturing.

Given the results obtained and the sound performance achieved in the manufacturing of inserts, replicating the implemented production model in the manufacturing of other company products is recommended.

References

1. Vilela, V.: Colección Investigación y Desarrollo. Estudio N° 03-2014 OSEL Lima Norte (2014). http://repositorio.ucss.edu.pe/bitstream/handle/UCSS/299/Victor_Vilela_OSEL_estudio_N3_2014.pdf?sequence=1&isAllowed=y. Recuperado el 20 de mayo de 2018 de
2. Montañez, J.: Dinámica de la cadena priorizada del Sector metalmecánico del departamento de Santander (2016). http://revistas.ustabuca.edu.co/index.php/LEBRET/article/view/1693. Recuperado el 10 de Setiembre de 2018
3. Takami, T.: Production Engineering Strategies and Metalworking at Toyota Motor Corporation. Revista Elsevier Ltd. (2014). https://www.sciencedirect.com/science/article/pii/S1877705814012247. Recuperado el 12 de Setiembre de 2018
4. Budyunek, M., Celinska, E., Dybikowska, A., Kozak, M., Ratajczak, J., Urbano, J.: Strategies of production control as tools of efficient management of production enterprises. Universität Potsdam, Germany (2016). https://www.degruyter.com/downloadpdf/j/mspe.2016.21.issue-1/mspe-03-01-2016/mspe-03-01-2016.pdf. Recuperado el 10 de abril de 2018 de

5. Niguyen, D., Niguyen, T.: "Made in Vietnam" Lean Management Model for Sustainable Development of Vietnamese Enterprises. Revista Elsevier Ltd. (2016). https://www.sciencedirect.com/science/article/pii/S2212827116001566. Recuperado el 20 de Setiembre de 2018
6. Calderón, G., Valencia, J.: Perfil cultural de las empresas innovadoras. Un Estudio de caso en empresas metalmecánicas. Red de Revistas Científicas de América Latina, el Caribe, España y Portugal (2013). http://www.redalyc.org/articulo.oa?id=20503408. Recuperado el 02 de mayo de 2018
7. Abdul, A., Mukhtar, M., Sulaiman, R.: A Conceptual Model of Lean Manufacturing Dimensions (2013). https://www.sciencedirect.com/science/article/pii/S2212017313004817. Consultado el 15 de Octubre de 2018
8. Asgharizadeh, E., Ajalli, M.: Identification and Ranking the Key Dimensions of lean manufacturing using New Approach in Gas Industry (2016). https://socrd.org/wpcontent/uploads/2016/08/7BKK141Identification-and-Ranking-the-Key-Dimensions-of-Lean-Manufacturing-using-NEW-Approach-in-Gas-Industry.pdf. Consultado el 19 de Octubre de 2018
9. Mohamed, A., Konstantinos, S., Yuchum, X.: Lean Implementation frameworks: the challenges for SMEs. Revista Elsevier Ltd. (2017). https://www.sciencedirect.com/science/article/pii/S2212827117303529. Recuperado el 28 de Setiembre de 2018
10. Guillen, K., Umasi, K., Quispe, G., Raymundo, C.: Lean model for optimizing plastic bag production in small and medium sized companies in the plastics sector. Int. J. Eng. Res. Technol. **11**(11), 1713–1734 (2018)
11. Anvari, A., Zulkifli, N., Mohd, R., Mohammad, S., Yusof, I.: A proposed dynamic model for a lean roadmap. Afr. J. Bus. Manage. (2011). https://pdfs.semanticscholar.org/4655/806e350849bacc9386a1804bffbf084bc0d3.pdf. Consultado el 17 de Octubre de 2018

Intelligent and Innovative Solutions in Supply Chains

Sylwia Konecka(✉) and Anna Maryniak

Faculty of Management, Department of Logistics and Transport, Poznań University of Economics and Business, al. Niepodległości 10, 61-875 Poznań, Poland
{Sylwia.Konecka, Anna.Maryniak}@ue.poznan.pl

Abstract. The article outlines an innovative and intelligent supply chains with respect to synonymous terms. The main aim of the research was to identify the technological solutions used in supply chain management based on literature review, surveys and on the example of two selected supply chains. The research pointed to a low level of implementation of technologies which might result from a poor knowledge of these technologies. In the context of the presented topic, the focus was on factors related to technological aspects that, from the point of view of organizations, are important for obtaining competitive advantage.

Keywords: Innovations in manufacturing and logistics · Supply chains · Intelligent supply chains · Information technologies

1 Introduction

The purpose of the research included identification an innovative and intelligent solutions in supply chains based on selected examples. Both the type of modern technologies implemented and the ways of data storage were studied. Among other things, it was established that the nature of the supply chain (links with suppliers and customers, the type of goods moved etc.) significantly influences the needs for creating chains with innovative and intelligent features. The study pointed to future research directions, including the need to systematize concepts related to information technologies such as: *Internet of Things* (IoT), big data, *Artificial Intelligence* (AI), blockchain, business intelligence, traceability, machine learning, data science e.t.c. and (in the future research) development of a set of indicators allowing to set thresholds for individual levels of development of innovative and intelligent supply chains, to finally create a map of the level of advancement in the use of information technologies in the supply chain management process with particular emphasis on warehouse and transport areas. Finally assessment of individual technologies in terms of the need for implementation thereof on the example of selected companies.

T. Ahram et al. (Eds.): IHSED 2019, AISC 1026, pp. 854–859, 2020.
https://doi.org/10.1007/978-3-030-27928-8_129

2 Innovative and Intelligent Solutions in the Context of Using them in Supply Chains Management

Evidence for innovation in supply chains can be found in any logistics-related process or service from the basic to the complex. Innovation is seen as new and advantageous to a particular focal audience [1] and it may also be fundamental for some companies, e.g., in the transportation industry [2]. Abdelkafi and Pero [3] distinguish two types of innovation supply chain concept: innovation and technological innovation.

Practitioner research also argues that, in the future, supply chains will be autonomous and will have predictive capabilities [4–6]. Using IoT sensors, quintillion bytes of data will be generated across supply chain operations. AI will be deployed to analyse information in real time, monitor operations across the globe, predict the future with minimum error rate, and take actions to adjust to rapidly changing environments [7].

In spite of the promising benefits of the self-thinking supply chain found in practitioner literature, academic research on this and related topics is scarce. Calatayud, Mangan and Christopher [8] in their systematic literature review found no articles exploring the self-thinking supply chain and only 28 articles referring to related concepts such as “autonomous”, “predictive”, “smart” or “intelligent” supply chain. These articles were spread across different fields, including SCM, computer science, engineering and economics. The analysis of the selected articles gave insights into in particular two new digital technologies that are associated with autonomous, predictive, smart or intelligent supply chains, namely, IoT and AI.

According to Wu et al. [7] smart supply chains collectively possess not only intelligent and innovative characteristics but also the following distinctive characteristics: instrumented: information in the next generation supply chain is overwhelmingly being machine-generated, for example, by sensors, RFID tags, meters, and many others; interconnected: the entire supply chain, including business entities, and assets, IT systems, products, and other smart objects are all connected in a smart supply chain; automated: smart supply chains must automate much of its process flows by using machines to replace other low-efficiency resources including labor; integrated: supply chain process integration involves collaboration across supply chain stages, joint decision making, common systems and information sharing.

Among all critical resources, information systems continue to play a critical role in SCM as supply chain performance is often characterized and facilitated by the real-time collaboration and sophisticated integration. SCM would not even be possible without the advances in information systems and technology.

Literature review shows are that intelligent solutions that apply to supply chains in the literature of subject are discussed above all at the level of: various types of algorithms - genetic algorithms or fuzzy-logic systems [9], intelligent systems in intermodal transport [10], in urban and extra-urban transport [11], regarding the use of intelligent algorithms for flow management in the field of reverse logistics [12], evolution of the ITS system [13] and safety of their use [14], intelligent warehouse management system [15], *Cloud of Things* [16], RFID [17], Kiva robots [18], *Voice picking* and *Pick-by-Light* technologies [19], autonomous vehicles [20].

As a result of the own research, it was established that in the future companies will compete by means of technology mainly in four basic areas: technology resulting from the use of innovative and intelligent solutions, information systems, data analysis and storage, goods identification and tracking systems. Such systematization has been established for the needs of further own research.

3 Research Methodology

Given the wide variety of real needs in terms of the level of technology integration within supply chains, it was considered a priority to characterize the supply chain as a prerequisite for proper evaluation of research. Therefore, case study was considered to be the best form of research. It fits into various ontological and epistemological approaches. Therefore, it will be possible to embed the reflections in fundamental research within a modernist, interpretative or realistic paradigm.

The in-depth interview and structured questionnaire were used as a research tool. The research was of a pilot character. It was conducted among Polish manufacturing companies operating in the food and non-food sector. In total, 36 interviews were conducted with the employees of IT, logistics and research and development departments. In the first stage of the research, basic information on the supply chain (input and output links, industry competitiveness, flow objects, reverse integration, forward integration, etc.) was obtained. Then the respondents were asked to identify the technologies that were promising in terms of improving the competitiveness of the supply chain in a given industry. Subsequently the Likert scale was used to assess the implementation of these technologies, where: 1 meant – we are using them, 2 – we are going to implement them, 3 – we do not know them well, so it is hard to tell, 4 – we are not using them and we probably will not implement them and 5 meant – we are not using them and we will certainly not implement them.

4 Research Results

In this study, eight most popular items were distinguished in each thematic group.

In terms of the first area – technology resulting from the use of innovative and intelligent solutions, these were: devices for viewing analyses/reports in 3D; eye-glasses that allow 3D viewing of the content of the warehouse; smartphones and tablets/PDAs for viewing the warehouse in 3D; 3D printers – 3D technology; robotics (e.g. use of advanced robots, machines with programmed functionalities); autonomous vehicles; cyber-physical systems (CPS – combining mechatronic, electronic and communication systems and software); temperature, air, humidity, etc. sensors placed on goods.

The surveyed companies were not familiar with technological innovations, e.g. the ways to facilitate the management of their transport fleet or scanning of entire warehouses.

Within the second aspect (information systems), the following items were identified: material demand planning – production plans based on the availability of goods in the warehouse (MRP); planning production resources (MRPII) – including materials, finances, people); enterprise resource planning (ERP) – MRP II + marketing, sales, extensive accounting and HR; use of the Supply Chain Management System (SCM); customer contact support (CRM); automatic planning of routes (TMS); tracking, analysis and control of warehouse (WMS); the use of electronic data interchange (EDI). The less frequently mentioned functionalities included tracking and supervising production and material flow (MES), supporting contacts with suppliers (SRM), inventory management (IMS) or providing information and knowledge for supporting the process of decision making (DSS).

In terms of the way of analyzing and storing data, the following solutions were most frequently mentioned: data analysis available at any time, based on historical data from a single system of records without a data base (grouping data from different systems); data analysis available at any time, based on historical data (data bases grouping data from different systems); e.g. to explain why something went wrong – BI; analysis of large data sets available at any time, the data might often be disorganized and obtained from various sources in real time; forecasting the results of actions and presenting development scenarios; big data – software enabling remote data storage and processing; use of SMAC IT platform integrating social media, mobile, analytical and cloud solutions; using the Internet of Things – i.e. a system in which objects, equipped with special sensors, communicate and exchange data with computers and other devices, e.g. between the factory, transport, customer; machine learning (automatic process of data acquisition and analysis, discovery of hardly predictable interdependencies); cloud computing that enables online access to resources, e.g. networks, servers, software; enabling remote data storage and processing. Moreover, the companies listed their own solutions developed through many years of experience or those copied from their subsidiaries.

The tools supporting the tracking of goods and the assessment of their environmental impact throughout their life cycle included: barcodes; GPS satellite navigation system; data transmission technologies GSM/GPRS, 3 G or 4 G; forward tracking (web-based tracking) identification of goods from sourcing to their end sale; tracking back (online tracking) to retrieve information on the history of the shipment and to identify the source of the quality problems; radio frequency identification of goods – RFID; EPC automated identification of an item in the supply chain as a combination of the Internet and RFID; GS1 goods tracking system.

In addition, some companies mentioned the solutions that can only be applied at the warehousing stage. In order to illustrate the technologies used in different market conditions, two examples of supply chains from different industries are presented.

Supply chain A is established in the beer industry. Craft and regional breweries are expected to gain in importance in the future. The specificity of the industry means that supply chain management is supported by ISO 22000, HCCP and OHS standards, and the production is monitored in detail. The initial stage of beer making requires malt whose supplier, for the sake of logistics, is located in the vicinity of the brewery. Over 80% of beer is purchased via the off-trade channel, i.e. in hypermarkets, discount stores, specialized shops, online shops. The remaining product is distributed through the on-trade channel, i.e. through discos, pubs or clubs.

Supply chain B specializes in manufacturing, sale and rental of work clothing. It is a short chain, because the supply comes from domestic wholesalers, while the distribution is made to a few large, regular customer and to over a dozen of smaller clients. The chain is very specific because it is the so-called closed loop chain. Firstly, the manufacturer is integrated into a company that sells clothing and rents clothing to institutional customers. Then, when the clothing becomes worn out, these companies return it to the previous supply chain either to refresh it or return it to the manufacturer (the supplier's supplier) who recycles the clothing and reintroduces the raw material into their production. Each item of clothing is assigned a code identifying the user so that it eventually returns to the same user.

The A supply chain is transparent because it uses not only barcodes but also systems to track goods from the bottom up the supply chain and from the top down the chain. The supply chain uses an the ERP system that combines the functionalities of MRP and MRPII. However, no specialist GS1 codes are known to decision makers that could significantly improve the identification of the flow of goods. Supply chain management is supported by data analysis technologies, yet the respondents are familiar neither with the most up-to-date solutions nor with any utility technologies.

The looping of the B chain increases its complexity, but the specificity of the product and the small circle of customers mean that technologies supporting supply chain management are hardly used. Advanced IT systems are not used in the supply chain, and data collection and analysis methods are not known to the parties involved in the movement of goods. Utility technologies and tracking systems are not used in the chain either. The only technological solution is a special code that identifies the user. It should be noted that for this type of supply chain there is no justification for introducing modern technologies. Only a significant increase in the scale of operations and a longer supply chain would justify such investments.

5 Conclusion

In future, it would be useful to determine the impact of individual technologies on the competitiveness of supply chains and to identify the directions of influence of individual technologies. In addition, it is reasonable to determine which of the moderators has a significant impact on the relevance of implementing modern technologies. The obtained results should be additionally correlated with the moderators, whose proposals are included in the research scheme. The research is a pilot study, therefore it will be necessary to conduct a study on a representative research group divided into particular industries.

Nowadays, with the development of industry 4.0 and the extension of its ideas along the supply chain, the trend for the implementation of modern technologies is developing. Based on the research, four key technological areas can be identified whose competitiveness can be improved by supply chains. Among the identified factors, the most significant are those related to IT systems. Therefore, one can say that the level of technological advancement of the studied chains is low. The comparison of two different types of chain clearly shows that the scope of potential needs is determined by their character.

Acknowledgments. The Poznań University of Economics and Business statutory fund.

References

1. Flint, D.J., Larsson, E., Gammelgaard, B., Mentzer, J.T.: Logistics innovation: a customer value-oriented social process. J. Bus. Logist. **26**(1), 113–147 (2005)
2. Wagner, S.: Innovation management in the German transportation industry. J. Bus. Logist. **29**(2), 215–231 (2008)
3. Abdelkafi, N., Pero, M.: Supply chain innovation-driven business models: exploratory analysis and implications for management. Bus. Process Manage. J. **24**(2), 589–608 (2018)
4. IBM: The smarter supply chain of the future, USEN.PDF (2015). http://www-03.ibm.com/innovation/us/smarterplanet/assets/smarterBusiness/supply_chain/GBE03215
5. DHL: Logistics trend radar. http://www.dhl.com/content/dam/downloads/g0/about_us/logistics_insights/dhl_logistics_trend_radar_2
6. World Economic Forum (WEF): Impact of the fourth industrial revolution on supply chains (2018). http://www3weforum.org/docs/WEF_Impact_of_the_Fourth_Industrial_Revolution_on_Supply_Chains_.pdf
7. Wu, L., Yue, X., Jin, A., Yen, D.C.: Smart supply chain management: a review and implications for future research. IJLM **27**(2), 395–417 (2016)
8. Calatayud, A., Mangan, J., Christopher, M.: The self-thinking supply chain. Supply Chain Manage. **24**(1), 22–38 (2019)
9. Ngai, E.W.T., Peng, S., Karen, P.A., Moo, K.L.: Decision support and intelligent systems in the textile and apparel supply chain: an academic review of research articles. Expert Syst. Appl. **41**(1), 81–91 (2014)
10. Mondragon, C., et al.: Intelligent transport systems in multimodal logistics: a case of role and contribution through wireless vehicular networks in a sea port location. Int. J. Prod. Econ. **137**(1), 165–175 (2012)
11. Ehlers, U.C., et al.: Assessing the safety effects of cooperative intelligent transport systems: a bowtie analysis approach. Accid. Anal. Prev. **99**, 125–141 (2017)
12. Yan, Z., et al.: Intelligent optimization algorithms: a stochastic closed-loop supply chain network problem involving oligopolistic competition for multi products and their product flow routings. Math. Probl. Eng. **2015**, 1–22 (2015)
13. Wang, X.: Developmental pattern and international cooperation on intelligent transport system in China. Case Stud. Transp. Policy **5**(1), 38–44 (2017)
14. Erdogan, G.: Towards transparent real-time privacy risk assessment of intelligent transport systems. In: Großmann, J., Felderer, M., Seehusen, F. (eds) Risk Assessment and Risk-Driven Quality Assurance. RISK 2016, vol. 10224, pp. 11–18. Springer, Cham
15. Mao, J., Xing, H., Zhang, X.: Design of intelligent warehouse management system. Wireless Pers. Commun. **102**(2), 1355–1367 (2018)
16. Yan, J., Xin, S., Liu, Q., Xu, W., Yang, L., Fan, L., Chen, B., Wang, Q.: Intelligent supply chain integration and management based on cloud of things. Int. J. Distrib. Sens. Netw. **10**, 624839 (2014)
17. Zhao, Z., Fang, J., Huang, G.Q., Zhang, M.: Location management of cloud forklifts in finished product warehouse. Int. J. Intell. Syst. **32**(4), 342–370 (2017)
18. Bogue, R.: Growth in e-commerce boosts innovation in the warehouse robot market. Ind. Robot. **43**(6), 583–587 (2016)
19. Fager, P., et al.: Kit preparation for mixed model assembly – efficiency impact of the picking information system. CAIE **129**, 169–178 (2019)
20. Graham, L, Troni, E., Hall, R.: The changing face of distribution. The shape of things to come, Cushman@Wakefield (2019)

Plant Layout Model for Improving Footwear Process Times in Micro and Small Enterprises

Nataly Gutierrez[1(✉)], Wendy Jaimes[1(✉)], Fernando Sotelo[1(✉)], Carlos Raymundo[2(✉)], and Francisco Dominguez[3(✉)]

[1] Ingeniería Industrial, Universidad Peruana de Ciencias Aplicadas (UPC), Lima, Peru
{u201312983,u201314271,fernando.sotelo}@upc.edu.pe

[2] Dirección de Investigación, Universidad Peruana de Ciencias Aplicadas (UPC), Lima, Peru
carlos.raymundo@upc.edu.pe

[3] Escuela Superior de Ingeniera Informática, Universidad Rey Juan Carlos, Madrid, Spain
francisco.dominguez@urjc.es

Abstract. The manufacturing specifications used by small leather and footwear businesses are neither up to date nor based on innovative production and new technologies. Consequently, their production times are higher, rendering them unable to sufficiently compete against other developed countries. Thus, this study seeks an innovative way to reduce times that do not add value to the product, such as reducing travel distances between stations, and proposes improving procedures for a better monitoring of the resources required. As a whole, this proposal focuses on designing a model to minimize the displacement of different processes. This practice is currently being implemented in the manufacturing sector, but not in the footwear sector. The proposal provides broad results, saving around 33% in transfer times and 85% in distance traveled in footwear production.

Keywords: Footwear company · Design · 5s method · Value chain mapping

1 Introduction

The problem in the leather and footwear industry arises from the fact that the total amount of footwear sold in the local market is imported from China and Brazil. Moreover, according to the Institute of Statistical and Social Studies (IEES, 2018), the footwear industry does not employ a trained workforce or have innovative production of new technologies, which hinders it from competing against the specialized industry in other countries.

According to Ulutas and Islier (2015), another problem is the poor layout in leather and footwear production plants, which translates into long travel distances that cause production downtimes and delays. Similarly, the placement of machines is not optimal, and materials are inefficiently handled. The authors emphasize that machine placement issues are critical at facilities since material handling costs are directly related to the

T. Ahram et al. (Eds.): IHSED 2019, AISC 1026, pp. 860–866, 2020.
https://doi.org/10.1007/978-3-030-27928-8_130

location of workstations and machines. Therefore, this study uses industrial engineering tools to reduce the times that do not add value to the product and fulfill the demand within the timeframe established by customers.

2 State of the Art

2.1 Systematic Layout Planning (SLP)

Recent research related to plant layout concludes that poorly designed company work areas generate a greater transfer time from one workstation to another. This results in more production time and, in particular, time that does not add value. Consequently, the use of the SLP method is proposed since it is the most efficient method for small businesses [1, 2].

2.2 Lean Tools

Other supporting studies are related to Lean tools, since there are a variety of Lean tools, including Kanban, 5s Philosophy, Takt Time, and Just-in-Time. Such tools help reduce delivery times and high inventories; however, their implementation becomes complex since Lean is a philosophy. This means that workers are required to change their mentality from what their traditional methods, causing an adverse response to change, lack of trust, limited agility, and excessive costs [3–5].

3 Contribution

A comparison chart of plant layout models has been developed, as can be observed in Table 1. Here, we assessed criteria from other study cases found in the State of the Art, such as the following: the applicability to micro and small enterprises (MSEs), whether the studies were developed in the footwear sector, how they could be combined with Lean tools, travel and transfer time reduction, economic viability, and ease of implementation.

Based on this comparison, the model that met all the criteria required for beneficial results was selected.

The contribution of this study consists in the application of methodologies, techniques, and processes that may be adapted in essence to a footwear company. Thus, we added a component that may be adapted to the particularities of the company to facilitate production.

3.1 Systematic Layout Planning (SLP)

The most widely used plant layout model in small and medium companies is the SLP model. This is because the SLP model is a less complex model that allows for greater accessibility than an algorithm that generates high investment costs and long implementation and testing times. Therefore, the contribution of this study will be a SLP plant layout with 5s methodologies. In addition, a rail will be added to facilitate the footwear assembly sequence (Fig. 1).

Table 1. Comparison of plant layout models

Dimensions	Models			
	Mobile	Systematic Layout Planning (SLP)	Systematic Layout Planning Modified (SLP)	Algorithmic
Applicable to MSEs	✓	✓	✓	
Developed in the footwear sector		✓		✓
Combines with lean tools	✓	✓		
Reduces travel or transfer times	✓	✓	✓	✓
Economic viability	✓	✓	✓	
Ease of implementation	✓	✓	✓	

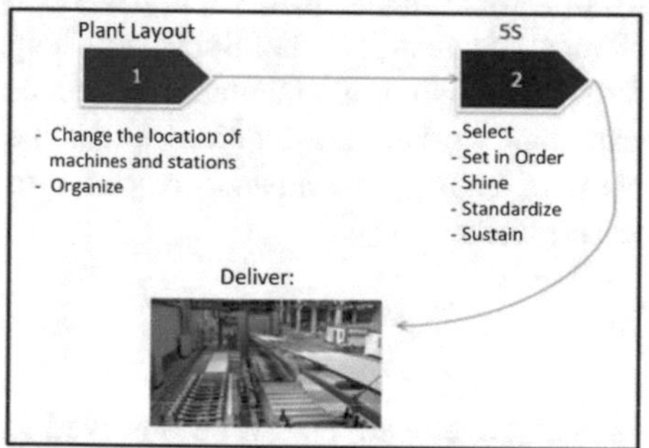

Fig. 1. Proposal design

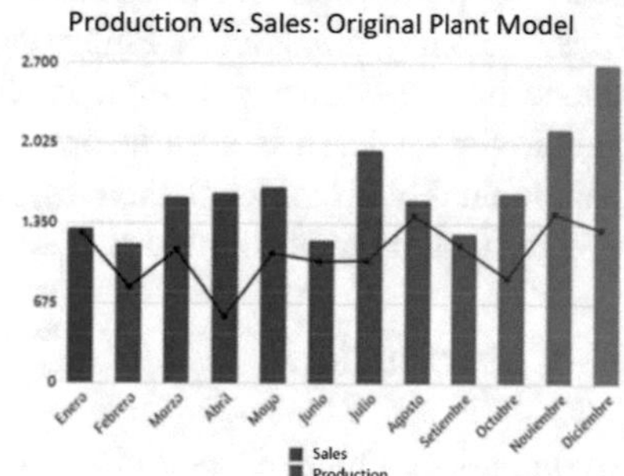

Fig. 2. Production vs. sales: original plant model

4 Validation

The following Project will be implemented in "Creaciones Laguna EIRL," a company under the MSE regime owing to its annual sales that ranges from 150 UIT and 1700 UIT. This company produces women's leather shoes in the city of Lima, Peru. It should be noted that most of its products are distributed to Peruvian provinces, such as Huancayo, Tarma, Trujillo, and Arequipa, in addition to being sold in their own store in downtown Lima.

4.1 Data Collection

The data collected included production quantities, the demand, sales, and cost per footwear style. It should be noted that there are seven footwear categories, with the most popular style being the Original Plant (PO) model. Therefore, this investigation will focus on the PO model.

4.2 Problem Identification

First, the data collected were analyzed to identify the problem: Orders were not being fulfilled since they were not being delivered on time. Further, their delivery was being postponed to the following month, which at the end of the year represented an opportunity cost since the orders were not being fully fulfilled, as shown in Fig. 2.

The Value Stream Mapping (VSM) reported high transfer times between processes, representing a total of 0.75 days for each batch of 100 pairs. Moreover, during the cutting process and in the warehouse area, large variations were reported in the times taken due to delays when searching for materials, as everything was cluttered and disorganized (Fig. 4).

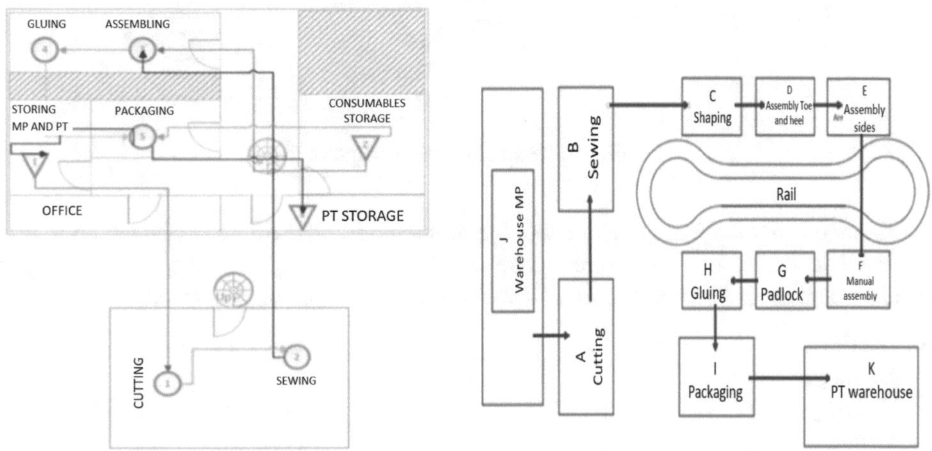

Fig. 3. Path diagram

Fig. 4. New plant layout

In addition, in order to conduct a more exhaustive analysis, a path diagram was developed, where many crosses may be observed in the workstations, which prevent the production process from becoming more fluid (see Fig. 3).

4.3 Identification of the Solution

Based on the previous analysis, which identified high travel times and a large amount of disorder in the cutting and storage areas as the main problems, we decided to apply a SLP plant layout with the 5s methodology (Fig. 5).

First, the plant layout was applied through the following steps:

- Data Collection
- Material Stream
- Assessment of Optimal Stations
- New Layout Proposal
- Evaluation of Alternatives
- Selection of Best Alternative

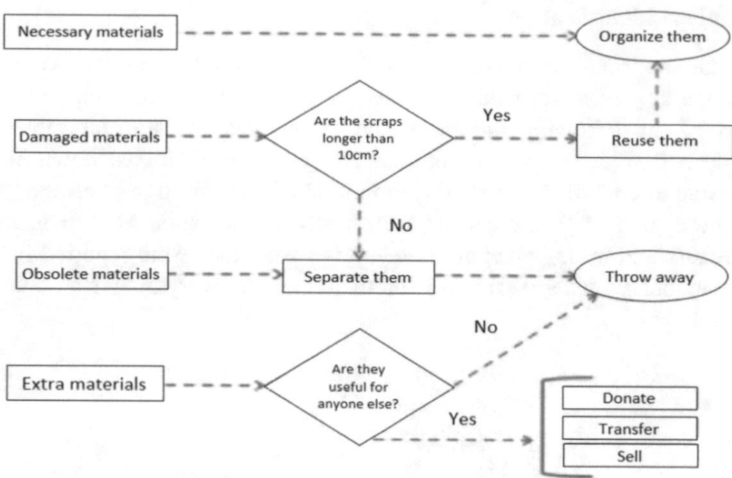

Fig. 5. Material stream diagram

Once the plant layout is completed, the 5S methodology will be implemented in the cutting and warehouse area, since these are the most critical areas in terms of organization and disorder. For this, the 5s methodology will be implemented:

- Sorting
- Setting in Order
- Shining
- Standardizing
- Self-discipline

4.4 Proposal Implementation

A new rail, measuring 9 m long and 1.50 m wide, was ordered as the first step for changing the plant layout. Since the rail is critical for footwear assembly, these measurements complied with the area that the assembly workstations would occupy.

Subsequently, the 5s methodology was implemented in accordance with the manual, which will be available at the company's administrative area. In addition, to generate greater commitment, a copy of this manual will be provided as required reading for all company employees.

4.5 Results Validation

After completing implementation of the new plant layout and the 5s methodology, a new VSM was produced for result visualization.

The new VSM showed that production times were significantly reduced from 0.75 days to 0.5 days for each batch of 100 pairs following an optimal redistribution of workstations. Hence, a 33% improvement was recorded in production times.

In addition, the distances between stations were also reduced. The previous travel distance of 174 m was reduced to a mere 29 m of travel time for the whole process. This is reflected in a production increase of 10,000 additional pairs of shoes per year. Therefore, after the improvements were implemented, the previous production of 20,000 pairs per year increased to 30,000 pairs per year, thus generating more income. This can be seen in Fig. 6, where it may be observed that all incomplete orders are now fulfilled; in addition, a new sale opportunity for 2,517 pairs is produced.

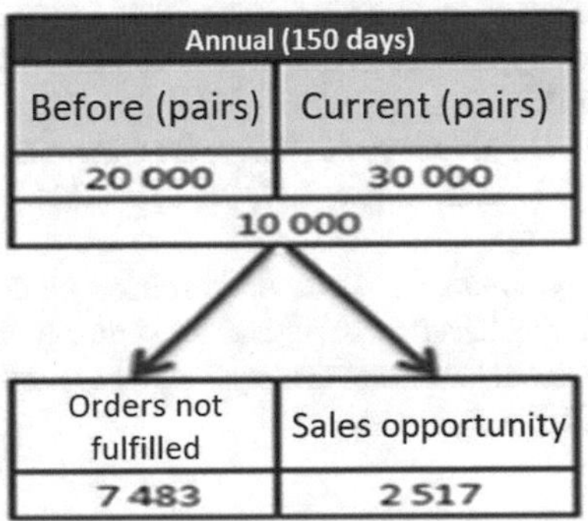

Fig. 6. Annual production in Pairs

5 Conclusions

With the new workstation layout, transfer downtimes are eliminated, and travel distances are reduced from 174 m to merely 29 m in total. Similarly, production process times were reduced from 0.75 days to 0.5 days, which is reflected in a 33% reduction in production time. This increased production with 10,000 additional pairs per year, which enabled the company to complete all previously unfulfilled orders and generate a sales opportunity.

- It is important to keep in mind that there are several expenses related to the implementation of an improvement project. For this case, training costs included the monthly salaries of the workers, since all training hours were paid. This implied a total cost of S/. 911.50 for all training sessions. Furthermore, for the implementation of the new plant layout, the expenses incurred included training, electrical installations, rail installation, and layout activities, which represented a total cost of S/. 6762.60. Overall, a total investment amount of S/. 7674.10 was reported.
- Subsequently, additional training sessions were scheduled for the following month. Here, the production manager will be responsible for ensuring these are training sessions are held frequently so that employees may take the initiative and continue with the discipline of maintaining the implementation of the 5s methodology.

References

1. Ali Naqvi, S.A., Fahad, M., Atir, M., Zubair, M., Shehzad, M.M.: Productivity improvement of a manufacturing facility using systematic layout planning. Cogent Eng. **3**(1) (2016). https://doi.org/10.1080/23311916.2016.1207296
2. Yang, T., Su, C., Hsu, Y.: Systematic layout planning: a study on semiconductor wafer fabrication facilities. Int. J. Oper. Prod. Manage. **20**(11), 1359–1371 (2000). https://doi.org/10.1108/01443570010348299
3. Alvarez, K., Aldas, D., Reyes, J.: Towards lean manufacturing from theory of constraints: a case study in footwear industry. In: 2017 International Conference on Industrial Engineering, Management Science and Application (ICIMSA), pp. 1–8. IEEE (2017). https://doi.org/10.1109/ICIMSA.2017.7985615
4. Ahmad, M.O., Dennehy, D., Conboy, K., Oivo, M.: Kanban in software engineering: a systematic mapping study. J. Syst. Softw. **137**, 96–113 (2018). https://doi.org/10.1016/j.jss.2017.11.045
5. Rojas, C., Quispe, G., Raymundo, C.: Lean optimization model for managing the yield of pima cotton (gossypium barbadense) in small-and medium-sized farms in the peruvian coast. In: 2018 Congreso Internacional de Innovacion y Tendencias en Ingenieria, CONIITI 2018 - Proceedings 8587062 (2018)

Public Sector Transformation via Democratic Governmental Entrepreneurship and Intrapreneurship

Evangelos Markopoulos[1(✉)] and Hannu Vanharanta[2,3]

[1] HULT International Business School, Hult House East, 35 Commercial Rd, London E1 1LD, UK
evangelos.markopoulos@faculty.hult.edu

[2] School of Technology and Innovations, University of Vaasa, Wolffintie 34, 65200 Vaasa, Finland
hannu@vanharanta.fi

[3] Faculty of Engineering Management, Poznan University of Technology, Poznan, Poland

Abstract. Human capital utilization fails under the integration of the knowledge management and leadership disciplines. Public sector organizations traditionally lack effective human capital utilization due to bureaucratic operations and structures that restrict knowledge sharing incentives and initiatives. This, however, can be achieved with knowledge democratization methods that should be related to the obligation public servants have to share knowledge and experiences for the effectiveness and sustainability of their organization. The co-evolutionary organizational culture of the Company Democracy Model can be used to implement such an approach. This paper evolves the Company Democracy Model (CDM) into the Democratic Governmental Intrapreneurship Model (DeGIM) and extends it to the Democratic Governmental Entrepreneurship Model (DeGEM). Furthermore, it proposes an organizational structure through which DeGIM and DeGEM can be applied at the local or national level through a centralized authority that can empower the contribution of the public sector to the national economy.

Keywords: Government · Management · Leadership · Public sector · Transformation · Entrepreneurship · Intrapreneurship · Innovation · Knowledge · Democracy · Applied philosophy

1 Introduction

The global workforce is categorized primarily into the employees of the public sector, the private sector and the self-employed individuals. The development and prosperity of a nation is based on the efficiency and productivity of both the public and private sector, through their contribution to the GDP.

The private sector and the self-employed workforce operate in a highly competitive environment, are self-sustained, and do not burden the national budget. On the contrary they contribute progressively to the society and the development of the whole nation, qualitatively and quantitatively, with their innovations, automations, profitability, insurances and tax payments. On the other hand, the contribution of the public sector,

T. Ahram et al. (Eds.): IHSED 2019, AISC 1026, pp. 867–877, 2020.
https://doi.org/10.1007/978-3-030-27928-8_131

remains questionable without direct productivity metrics that can be transformed into tangible results for the economy and the society.

The average of the global workforce employed in the public sector reaches the an impressive 30%, making the government to be, by far, the largest employer with more than 1.2 billion employees globally [1]. Therefore, the efficiency and productivity of the public sector employees can be a critical factor in the development of the government, the nation's financial stability, and the national fiscal management. Citi indicates that nearly 50% of the Chinese workforce is somehow in the government sector, with the Japan to be at 3%, the US at 17%, Greece at 28% and Germany at 31% [2].

There are many factors that affect the size of the public sector in each country. Such factors are based on their cultures, progressiveness, political system, temperament, and others. There are countries with a small public sector that obtain similar or even higher development rates from countries with large public sector and vice versa.

The challenge is not to justify the size of the public sector but first to acknowledge the fact that the effectiveness of the public sector contributes significantly to the economy and the stability of a nation. Furthermore, it is important to emphasize the ways the public sector can become more competitive by operating within a similar framework to the private sector.

2 The Public Sector – Economy/Society, Relationship

In most countries employees entering the public sector obtain permanent job protection and/or professional asylum. Such conditions and privileges eliminate personal and organizational motivation, ambition, innovation and creativity in the workplace and disincentive hard work for career development. This results in low or negative productivity which transforms the public sector into a national cost center with significant negative impact on the national fiscal management and economy and the national development [3].

Increasing the competitiveness and productivity of a nation requires the collective utilization of all productive recourses, including the public sector whose size crucially affects the nation's ability to generate operational added value. Effective public sector authorities can contribute to the quality of administrative services such as on issuing licenses, approvals, assessments, etc. which in turn can speed up investments, employment, extroversion, production and much more. This efficiency also returns to the society not only from the quality of services the citizens receive but also through the job opportunities that are created in a prosperous and progressive nation.

This overall and holistic role of the public sector in a national economy is evolved around the citizen as a double accelerator for the public sector's effectiveness in producing both social shared value [4] and economic wealth (Fig. 1).

3 Public Sector Human Capital Utilization Preconditions

The knowledge which exists in the public sector can be categorized on the impact it has in on the society and the economy. Emphasis shall be given to the organizational knowledge that impacts directly the quality of services provided primarily to the society, which is both the citizens and the organizations of the private sector, that belong to the citizens. However, public sector organizations that already operate as private sector enterprises selling their products or services to the clients or other private organizations are expected to run as private sector organizations in any case, with established knowledge and human capital utilization methods that effectively and successfully work.

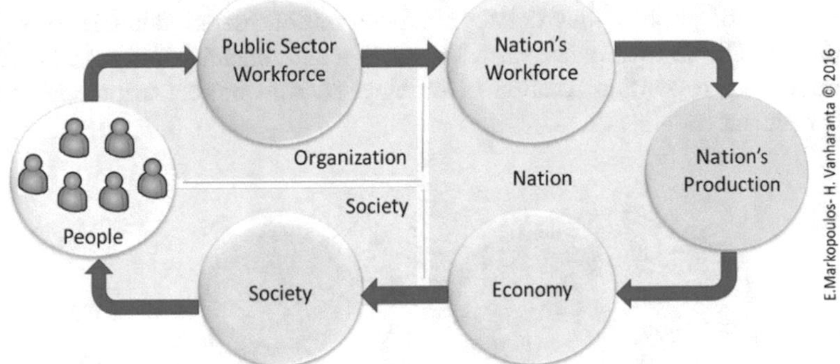

Fig. 1. Cyclical public sector contribution to the national economy and society.

Organizational knowledge is generated over the daily operations and activities of each employee. People can think and act upon their thoughts and learn from their acts. This basic learning process must be supported by a collaborative and co-evolutionary culture which provides the space needed for knowledge elicitation and sharing that will collectively be utilized on each organizational challenge [5]. This culture can easily be established as the collective development and growth ideology, especially in the public sector where the employee solidarity is at a much higher degree than in the private sector. Therefore, the desired human intellectual capital utilization environment that can be effectively applied in public sector organizations shall be based on six staged preconditions which address all the issues from the knowledge creation to the knowledge utilization (Fig. 2).

The first precondition is related to the development of the organizational culture in which all employees can be guided on obtaining active roles based on their professional, and not political skills, ideas, views, and capabilities. Through this culture freedom of speech and act shall be promoted above all. Free speech however must be supported with the practical reasoning for the benefit of the organization and the society. Likewise, freedom to act must be supported with the ability one has to realize, and implement the speech. Freedom of speech and act are fundamental democratic values, a privilege for those willing and wishing to progress, but also a punishment for those who directly, indirectly or creatively stay in the shadows, harming and not helping the organization.

The second precondition is related on applying quantitative and qualitative metrics to the ideas, thoughts, innovations, and initiatives proposed by each employee within a given timeframe. These metrics can identify the overall employee knowledge acquisition indexes to determine promotions, staffing needs, educational needs, or even penalties for those who think and/or act destructively. This approach gives each employee career development control and the opportunity to reach top management positions based on their work and contribution, and not only on academic attainments or political and social networks.

The third precondition is related to the promotion of a co-evolutionary framework where the right teams, composed from the right people, for the right project, and for the right time, can work together. Employees can be temporarily or partially extracted from their roles and responsibilities and assigned to manage, lead, or participate in the implementation of projects they created, initiated or considered valuable to be part of. This approach offers experiences, work satisfaction and financial compensation through a continuous and rewarding rotation of employees from project to project as project leaders or project members.

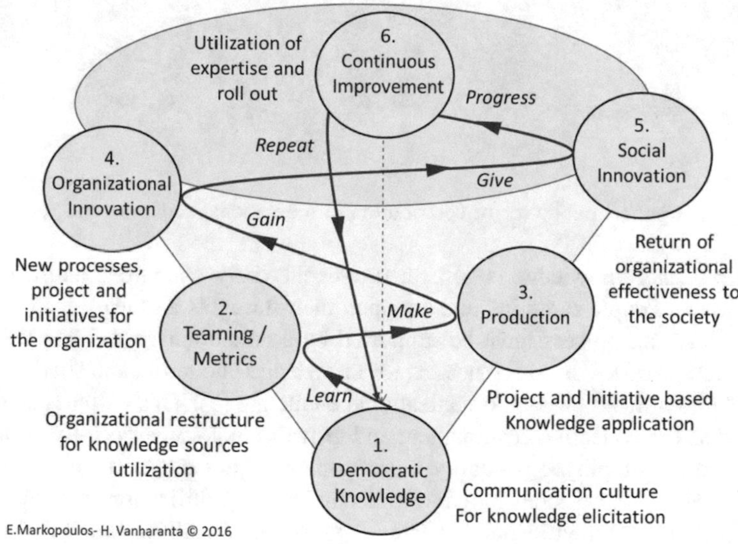

Fig. 2. Public sector knowledge utilization preconditions.

The fourth precondition is based on sharing the idea that the quality of the work produced in an organization is completely related to the lifespan of the organization. This concept can assure qualitative services to be delivered to the nation, the people and the society. Organizations with poor results have no reason to exist and must be shut down or restructured with new employees, releasing the existing ones to the open market of the private sector. It is certain that capable employees will not face challenges re-entering the workforce in the public or private sector.

The fifth precondition is related directly to the society, as public sector organizations are being formed to serve it. Ideas and innovations, through acts and projects, must always have a social impact to the citizens who actually pay and fund the existence of the public sector organizations in the first place. Public sector employees also called ‘public servants’, since their first and foremost duty is to serve the public. Such shared value effectiveness can be the reduction of bureaucracy, speeding up various processes, reduction of paperwork, corruption restriction and anything that can improve the lives of the citizens and increase the efficiency of the public sector.

The sixth and final precondition is based on a continuous integration of the knowledge generated in the organization back to the into the public sector operations, products and services. This integration ignites the iteration of the five preconditions towards a continuous optimization of the overall process under a democratic and effective knowledge management, business management and strategic leadership.

4 Knowledge Democratization in the Public Sector

Effective management and leadership rely significantly on knowledge management and sharing [6]. Knowledge sharing however seems to be a massive challenge in the public sector. The permanent positions provided by the governments to their public servants, reduces the motivation needed for their personal development and growth, negatively impacts the organization and the country but also withholds the application and adaptation of any management/leadership theory, model or framework that could be beneficial. Lack of motives however does not necessarily mean and lack of knowledge which probably exist in many public sector employees. The public sector can be, and should be, lead with an emphasis on the utilization of its human assets for high efficiency and effectiveness. These targets can be achieved through the six knowledge management preconditions whose adaptation can democratize the knowledge that exists in the public sector and provide opportunities to the public servants to control their career path, compensation, benefits and opportunities.

The quantity and quality of the public sector in productivity and service execution is a barometer for the development of a country. It is what attracts investments or ignites de-investments of any kind, not necessarily financial. Human capital investments can be more profitable than financial capital investments as brain-drain can be more disastrous than money-drain. However, to achieve such strategic goals the government and the public sector in general shall embrace the values of a democratic co-evolutionary economy through governmental intrapreneurship and entrepreneurship.

One way to approach such a challenge is through the application of the Company Democracy Model (CDM) [7, 8], a framework based on the wisdom of the ancient Hellenic Delphic maxims of ‘Know Thyself’, ‘Metron Ariston’and ‘Miden Agan’ [9]. The model is aligned with the Co-Evolute methodology that generates intellectual capital by empowering knowledge-based democratic cultures [10]. This co-evolutionary spiral method uses ontologies to identify and capture the capacity, capability, competence, and maturity of the human resources needed to turn such data, information, and knowledge into rewarding innovations [11]. Through a spiral process the model identifies and defines the degree of democracy needed in organizations for their successful evolution through a constant transformation of their tacit knowledge into explicit knowledge [12].

Adjustments on the CDM operations levels can develop a new framework that can empower a Democratic Governmental Intrapreneurship Model (DeGIM) and further extend it into the Democratic Governmental Entrepreneurship Model (DeGEM). At the first stage, the DeGIM attempts to promote the utilization of the human capital to achieves internal operations optimization of the public sector organization with direct or indirect impact on its target groups or the economy the organization serves. The DeGEM, follows as an attempt to further utilize the DeGIM human capital and to support the development of entrepreneurial initiatives by turning the existing knowledge into public sector spin-off organizations or enterprises that deliver new products and services to the existing target groups or to new markets.

Both models treat the public organizations and the governmental authorities as industry service enterprises that can generate profitability through effectiveness or innovation and serve the private sector, citizens, investors, and other target groups. DeGIM and DeGEM attempt to set new standards and a paradigm on the way countries can manage their public sector through the utilization of their usually underutilized workforce which turns out to be the larger running cost on the national budget.

5 Democratic Governmental Intrapreneurship Model - DeGIM

The structure of the Democratic Governmental Intrapreneurship Model (DeGIM), maintains the six levels of the Company Democracy model adjusted to the public sector knowledge utilization strategy preconditions (Fig. 3).

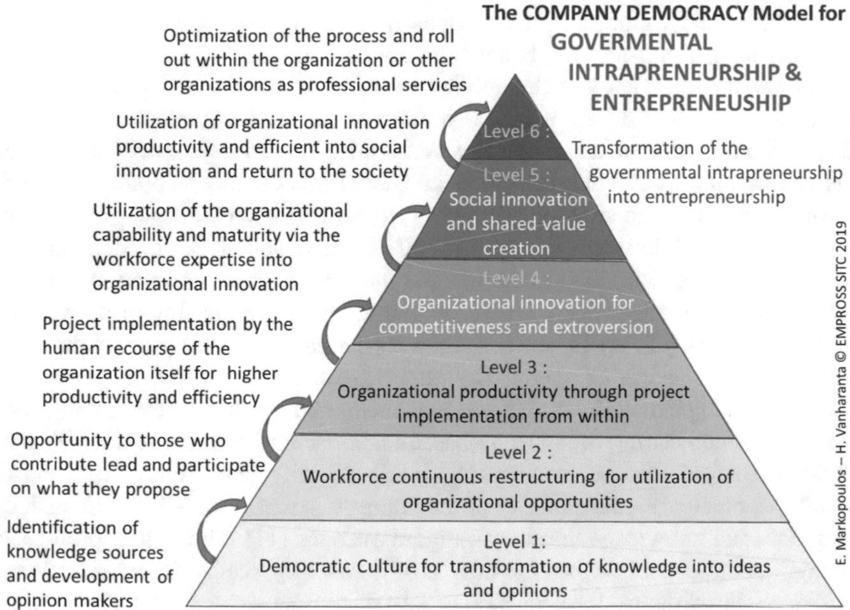

Fig. 3. The democratic governmental intrapreneurship model.

The first level of the model aims the creation and maintenance of an organizational culture under the principles of democracy and equality. As all employees, public or private, are capable to think, they should be capable then to generate knowledge from their acts. The public sector organizations cannot be an exception to this, as knowledge does exist in them but what seems to be missing is the organizational culture and process to transform this knowledge into valid opinions, actions, ideas, projects and strategies. This first level establishes the co-evolutionary culture as a foundation for organizational knowledge elicitations and utilization.

The second level promotes the co-development and co-operation concept. Since public sector employees have high syndicate bonds, the teaming concept is the best way to promote effective collaborations via co-opetition and not competition. This can be achieved by team formations based on the knowledge identified in the first level of the model. The teams composed to utilize this internal knowledge are lead from the ones the knowledge derived. It is common in public organizations ideas and visions generated by the employees themselves to be managed and implemented, with significant cost, by external organizations of the private sector. Level two promotes the concept that the public sector employees are capable to implement their ideas and proposals, receive credit for their contributions, and benefit from their successes.

The third level addresses the actual implementation of the ideas, insights, initiatives, innovations, and projects, derived from the public servants themselves at level one, and staffed with their colleagues at level two. This level allows public servants and organizations to demonstrate their skills, ability, capability, and competence to actually implement effectively and successfully their own projects, and bring significant benefits to the organization, the economy, and society. If public sector organizations can implement most, if not all, of their projects, then the funds allocated for outsourcing project implementation and management can be either fully saved or reduced to a minimum. Such savings impact directly the organization's budget, and the nation's expenditures. Furthermore part of these savings can either be returned to the employees who participated in the implementation of each project as productivity bonus, or used to hire term-based personnel to support project implementation. This approach reduces unemployment and increases the reputation of the organizations and the government to the society.

The fourth level emphasizes on the value returned to the organization in terms of competitiveness, innovation, and quality of services. It is the benefits harvesting period and the recognition on the efforts placed by the right people on the right project for the right time. It is also the level in which the organization can design and execute the best possible communication strategy to promote the effectiveness of the democratic governmental entrepreneurship/intrapreneurship programme. Furthermore, it is the level where the organization can build on the results of level three to optimize the achieved results and turn them into innovative processes, products and/or services.

The fifth level extends to the transformation of the achievements accomplished at level four into shared value innovations to maximize the return to the society by integrating feedback and knowledge from the society itself. Such shared value can be the design and implementation of services, and products as well, with social impact or for social purposes where the benefit for the organization and the society can be mutual and multiple. Social innovation can derive from shared value innovations which can derives from product/service innovation.

Lastly the sixth level of the model is related to the continuous improvement and the extroversion of the results achieved by the public sector organization through the Democratic Governmental Intrapreneurship Model. At this level the model turns from a Democratic Governmental Intrapreneurship (DeGIM) to a Democratic Governmental Entrepreneurship Model (DeGEM). This is achieved with the extroversion of the successful intrapreneurial activities and their transformation into organizational spin-offs, start-ups, business units, product lines or services lines. This is also the level where the effectiveness of the organization can also be exported internationally, to foreign governments, as new expertise, know-how, products or services.

The operations framework for the DeGEM have actually the same processes found in level six of the Company Democracy Models for Innovation Competitiveness and Extroversion. Once the public sector organization enters the DeGEM spiral, the team, the business unit or the department that runs a project, product or service is transformed into an autonomous organizational unit that shall function now under the private sector standards in management, compensation and operations in order to achieve high returns on investments and operations self-sustainability.

6 DeGIM Implementation Challenges

The implementation of the overall framework of democratic governmental intrapreneurship and entrepreneurship is a Y theory model based on strategic management and leadership skills. However, implementing such models in the government require a number of legislative and legal support activities to resolve challenges that might arise on setting up the execution infrastructure.

One of the key challenges the model must overcome is the power an organization needs to move employees into positions related to their knowledge contribution, abilities and expertise. The model rewards those who have the knowledge and want to apply it by offering career opportunities within the organization's hierarchy outside the traditional career progression path. Furthermore, it is also important for the organization to be able to properly reward the employees who share organizational knowledge, document best practices, participate in the governmental entrepreneurship with ideas, make suggestions and work hard to implement them. Flat hierarchy models with legislative restrictions on the promotion processes and compensations schemas, restrict the implementation of the model.

The identification of the department or division in the organization that can first apply the DeGIM is one more challenge. It is wise to apply the model first on the most mature department of the organization which can be used as a success story or example of progressiveness. Ideal departments that can be used as pilots are those that heavily interact with the citizens and the society or produce a product or service. In these cases qualified and quantified success indexes can be applied. In the first case the department can come up with new services that benefit the citizens, while the in the second case the effectives of the model can be related with the revenue increase and/or the profitability of the organization that derived from its clients. Both types of departments can produce different results for different goals, but with the same success factors and impact.

7 Democratic Governmental Entrepreneurship Authority - DeGEA

Once the DeGIM and DeGEM models get rolled out to more public organizations a critical challenge can be on their credibility and application consistency between the various organizations.

This can be managed with the creation of the Democratic Governmental Entrepreneurship Authority (DeGEA), an independent authority that can standardize the DeGIM and DeGEM models as structured methodologies on which public sector organizations can be assessed and score on process effectiveness and standards compliance. DeGEA can obtain ownership of the models, turn them into national standards, document its execution processes in detail, formalize an assessment process and create a pool of auditors/inspectors to inspect each organization over specified time periods. These audits shall be delivered by qualified inspectors on the effective and proper execution of each model where an official compliance score will be given. According to the organization's performance, level-based certificates can be issued each time an organization fulfills the requirements of a level.

This DeGEA shall not only assign quality management inspectors to assess the organizations, but must also provide consulting, coaching, support and training to the organizations committed on the models. Furthermore DeGEA can be responsible for handling the legislation challenges that could be faced by the public organization on the implementation of the model (Fig. 4).

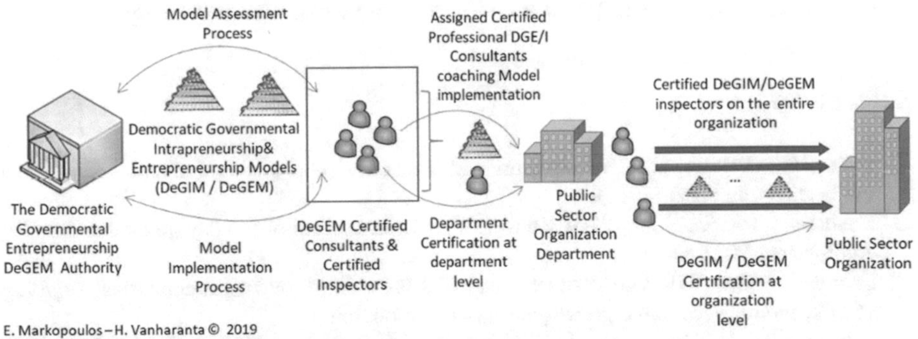

Fig. 4. Proposed operations of DeGEA.

Besides the certificates offered to the organization for successfully completing a DeGIM level, the certification can be extended to the employees as professional skills certificates based on their model participation scorecard. This can verify the degree of their commitment to successfully contribute and implement the model within the organization. These certificates can be associated with salary increases and other professional and career development benefits and privileges.

The design for the implementation of the Governmental Entrepreneurship/ Intrapreneurship Model in central government and its support to the public sector organization has low-financial cost strategic initiative. However, it is a high-commitment cost initiative for the political leaders that will be called to promote such transparent public sector management models, which can return them high reputational benefits.

8 Conclusions

The DeGEM, the DeGIM and the DeGEA can be considered as strategic transformational initiatives on the management and operations of the public sector. The value of the human intellectual capital that resides on the public sectors can be national wealth and valuable fuel for the growth and the development of a country [13]. It is also a leadership challenge to tackle issues that have not been managed effectively for centuries [14]. The job security offered by the public sector organizations and the lack of effective performance evaluation of not only the personnel, but at the organizations themselves is what primarily blocks any improvement transformation regardless how effective and absolutely necessary can be for the private sector. It is therefore a matter of strategic and authentic leadership that can move such initiatives from theory to practice, with pilot projects initially and with national programs after.

The transformation of public sector operations management towards private sector standards, under governmental leadership is neither a wise nor an unnecessary concept. It is an absolute necessity and a demand, to a great extent, of the people, the citizens, who invest on the public sector performance and existence. It is a necessity for all private sector organizations to generate multiple types of shared value and establish a culture that can lead a society and a nation to progression and prosperity.

References

1. James, P., O'Brien, R.: Globalization and Economy: Globalizing Labour, vol. 4. Sage Publications, London (2016)
2. Business Insider. http://www.businessinsider.com/chart-of-the-day-government-sector-employment-2011-11
3. Economics Help. UK Office of National Statistics. https://www.economicshelp.org/blog/7617/economics/economic-growth-during-great-moderation/
4. Markopoulos, E., Vanharanta, H.: The company democracy model for the development of intellectual human capitalism for shared value. Procedia Manufact. **3**, 603–610 (2015)
5. Markopoulos, E., Vanharanta, H.: Space for company democracy. In: Kantola, J., Barath, T., Nazir, S., Andre, T. (eds.) Advances in Human Factors, Business Management, Training and Education. Advances in Intelligent Systems and Computing, vol. 498. Springer, Cham (2017)
6. Asrar-ul-Haq, M., Anwar, S.: A systematic review of knowledge management and knowledge sharing: trends, issues, and challenges. Cogent Bus. Manage. **3**(1), 1127744 (2016)
7. Markopoulos, E., Vanharanta, H.: Democratic culture paradigm for organizational management and leadership strategies - the company democracy model. In: Proceedings of the 5th International Conference on Applied Human Factors and Ergonomics AHFE 2014 (2014)

8. Vanharanta, H., Markopoulos, E.: Creating a Dynamic Democratic Company Culture for Leadership, Innovation, and Competitiveness. 3rd Hellenic-Russian Forum, 17 September 2013
9. Parke, H., Wormell, D.: The Delphic Oracle, vol. 1, p. 389. Basil Blackwell, Oxford (1956)
10. Kantola, J., Vanharanta, H., Karwowski, W.: The evolute system: a co-evolutionary human resource development methodology. In: Karwowski, W. (ed.) The International Encyclopedia of Ergonomics and Human Factors. CRC Press, Boca Raton (2006)
11. Paajanen, P., Piirto, A., Kantola, J., Vanharanta, H.: FOLIUM - ontology for organizational knowledge creation. In: 10th World Multi-Conference on Systemics, Cybernetics, and Informatics (2006)
12. Nonaka, I., Takeuchi, H.: The Knowledge-Creating Company: How Japanese Companies Create the Dynamics of Innovation. Oxford University Press, New York (1995)
13. Bontis, N., Fitz-enz, J.: Intellectual capital ROI: a causal map of human capital antecedents and consequents. J. Intellect. Capital **3**(3), 223–247 (2002)
14. Kelly, C.: Bureaucracy and government. In: Lenski, N. (ed.) The Cambridge Companion to the Age of Constantine. Cambridge Companions to the Ancient World, pp. 183–204. Cambridge University Press, Cambridge (2005)

Three Dimensional Visualization and Interactive Representation of Carbon Structures and Compounds to Illustrate Learning Content

Tihomir Dovramadjiev(✉)

MTF, Department Industrial Design, Technical University of Varna, str. Studentska N1, 9010 Varna, Bulgaria
tihomir.dovramadjiev@tu-varna.bg

Abstract. To illustrate the learning content of models invisible to the human eye, digital and real patterns of carbon structures and compounds are developed. They occupy an important place in modern science. Of particular interest are the structures of diamond, fullerene, graphene, nanotube, and the molecular compound of methane. An approach to conventional and parametric computer design has been developed through the Blender open source system and specialized application Add-ons. The technical parameters and geometric characteristics of the models are conformed. 3D digital variants of polygonal-mesh and solid models are designed. There are presenting animations and 3D printed (real) models created. For the purpose of promotion, an interactive approach to three-dimensional models is built in Facebook environment.

Keywords: Design · 3D · Blender · Carbon · Nanostructures · Facebook · glTF

1 Introduction

The present times are characterized by the rapid development of technology. This covers many areas of our lives and activities. In parallel with the direct application and the specificity provided by the technological innovations, new opportunities are opened up to the educational process, where the study of science is a priority, this is successfully applied in the "Advanced Technologies in Design" engineering discipline [1]. Different approaches and methods of sharing and transferring information are proposed, as well as ways to optimize the visualization of the teaching material taught in schools and universities [2]. Undoubtedly, modern social networks occupy an important place both privately and socially [3]. On this basis, finding an approach for associating this resource in the educational process creates conditions for faster and

The Advanced Technologies in Design course, covering a number of modern technological processes, has been present in the course of the Industrial Design specialty, MTF, Technical University of Varna for several years.

T. Ahram et al. (Eds.): IHSED 2019, AISC 1026, pp. 878–884, 2020.
https://doi.org/10.1007/978-3-030-27928-8_132

more effective teaching of information. By addressing the exact sciences where computer programming, modeling, two-dimensional and three-dimensional visualization are combined, it is possible to predict the application and integration of the developed models into virtual social networking environments and to engage directly a wide range of learners [4–6]. It is also a good opportunity for rapid and qualitative prototyping to be provided by modern three-dimensional printers where, by obtaining digital data on a three dimensional geometry, reference models are reproduced, which can be used in the educational process to illustrate particular models by reference [7].

One of the main goals of this report is to illustrate models of carbon structures and compounds that are invisible to the human eye and to optimize the information taught in the educational process by finding innovative, interactive approaches. Figure 1 shows a process of illustrating three-dimensional models invisible to the human eye using modern technological means.

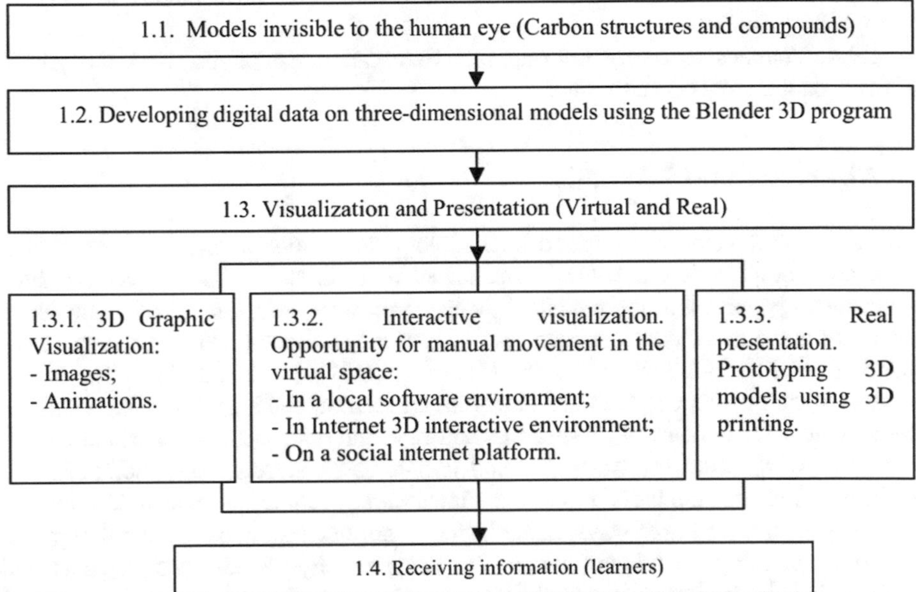

Fig. 1. Optimized methodology for visualizing and presenting virtual and real 3D models invisible to the human eye

The process includes the following steps in sequence:

1.1. Models invisible to the human eye. In this case, specific carbon structures and compounds are the subject of this study. These are models of diamond, fullerene, graphene, nanotubes, and the molecular compound of methane;

1.2. Development of digital data of three-dimensional models of carbon structures and compounds using the Blender 3D program. Various methods, approaches and techniques of design work are applied here, as well as the use of additional Add-ons applications to aid the generation of certain geometric volumes through parametric and other generating functions to help automate the design process;

1.3. Visualization and Presentation. This stage contains the following aspects: virtual (3D graphic and interactive) and real visualization.

1.3.1. Virtual visualization can be presented in a number of different formats, as the leading ones are 3D graphics representation of three-dimensional models in the form of conventional images (stationary still images, data carriers and presented in proper form) and animations that are loaded into the respective video players and sites that support the required formats;

1.3.2. as well as interactive presentation in virtual environments where manual movement, moving, rotation, scaling, etc. are possible in real time, in a local software environment (these are specialized viewers supporting certain file formats), internet interactive 3D environment (either on a social Internet platform - for example, Facebook);

1.3.3. Real Presentation. This refers to three-dimensional models prototyped through 3D printers;

1.3.4. Trainees receiving information. They can range pupils, students, private learners on the subject, and others.

2 Materials and Methods

Following the developed optimized methodology the studied models of carbon structures and compounds are made in sequence. In order to obtain the correct three dimensional geometry of the models, it is necessary to make a literature review where details of the state of carbon structures of Diamond, Fullerene C60, Graphene, Nanotube and carbon-hydrogen compound Methane (CH4) [8] to be provided. The design process of the developed 3D models includes different methods of operation mainly in Blender 3D environment [9]. Using this software, the main benefits of which are the optimized work interface, accessible and easy-to-use features, tools, modifiers, and additional Add-ons, the basis of the hree-dimensional geometry is created. The models are designed in a variety of ways, including conventional design, parametric design and hybrid engineering techniques, where the 3D geometry blends, complements and completely is built. Depending on the assignments, three-dimensional data is exported to specialized programs that support 3D prototyping for rapid prototyping or directly to devices if the 3D model is in Solid state (i.e., a developed density that is attached to the main three-dimensional surfaces). Good available software in this regard is Autodesk MeshMixer [10]. The other main application directions of the 3D data received in Blender software are the import of digital models into specialized graphics and video visualization programs. A modern way of virtual presentation of three-dimensional models is to visualize 3D models in desktop programs - viewers, similar internet viewers, or directly on social media platforms such as Facebook that support glTF file formats, allowing interactive manipulation of three-dimensional models such as moving, scaling, rotating and more [11–13].

The creation of the listed patterns of carbon structures and compounds vary in complexity with the process and solutions involved. It is relatively easy and fast to create three-dimensional graphene and nanotube models using the specialized parametric application CNT Add-on (License: GPL 2.0) [14, 15]. The values and parameters of the default carbon compound model at CNT Add-on startup are: *Wrap factor: 0.00; Index of Graphene/CNT cell): m: 5 and n: 5; Count of the x-array and the y-array modifier: Nx: 1, Ny: 1; C-C bond length: 0.25; C-C bond radius: 0.01; C atom radius: 0.04*. Depending on the requirements, precise settings are defined, with a value of *Ny-array modifier: 10* is given for the 3D graphene models and for the nanotube *Ny-array modifier: 10* and *Wrap factor: 1.00*, Table 1.

Table 1. Values and parameters of 3D Graphene and Nanotube models received through CNT Add-on in Blender software environment.

CNT Add-on. Values and Parameters	3D model of Graphene	3D model of Nanotube
Wrap factor:	0.00	1.00
m:	5	5
n:	5	5
Nx:	1	1
Ny:	10	10
C-C Bond:	0.25	0.25
C-C Bond Radius:	0.1	0.1
C Atom Radius:	0.4	0.4

Visually, the 3D graphene and nanotube models obtained with the CNT Add-on are shown in Fig. 2.

(a) (b)

Fig. 2. Digitally developed in Blender software and CNT Add-on 3D Carbon Pattern Models: *(a) Graphene adn (b) Nanotube*

Using the parametric capabilities of the Geodesic Domes Add-on (License: GPL) [16] and conventional design, the three-dimensional Fullerene C60 model is built. The necessary settings for making the original 3D primitive are: *Object: "Geodesic"/Types; Class: Class 1; Hedron:Icosahedron; Point: PointUp; Shape: hex; Round: Spherical; Object Parameters: Frequency: 1; Radius: 1.00; Eccentricity: 1.00; Squish: 1.00; Square x/y: 2.00; Rotate x/y: 0.00; Square z: 2.00; Rotate z: 0.00; Dual; superformula u (x/y); superformula v (z): Non- active.* This three-dimensional model is raw and for the making of Fullerene C60 3D model it is necessary to further manually process the resulting three-dimensional geometry by adding 60 carbon atoms to the vertices, as well as to build a skeletal structure with the required thickness. Creating the cubic carbon structure of a diamond is done by conventional design, while meeting the necessary requirements that are: *Fd* $\bar{3}$ *m*- diamond's cubic structure/space group); *8* carbon atoms at the corner, creating a cube; *6* carbon atoms in the faces create an octahedron; *4* internal carbon atoms at ¼ of the distance along body diagonal forming a tetrahedran; *a* (edge) $\frac{a}{4}(\hat{x}+\hat{y}+\hat{z})$, [17]. In similar way (conventional), the carbon structure of methane (CH4) is produced, keeping the angles between the compounds that are *109.4712°* [18]. Figure 3 shows the finished 3D models of Fullerene C60, Diamond and Methane (CH4).

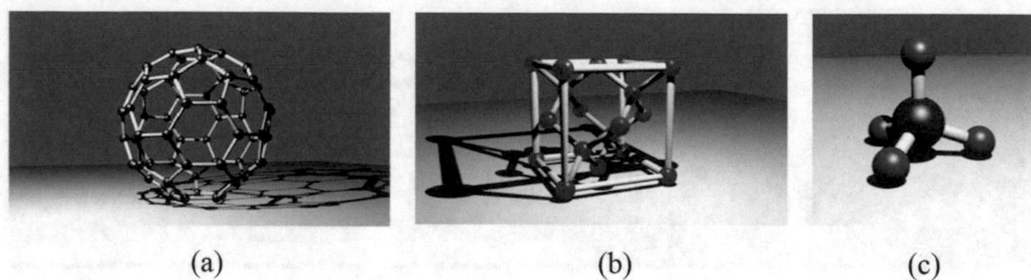

Fig. 3. Digitally developed in Blender software environment 3D models of carbon structures: *(a) Fullerene C60 (with Geodesic Domes Add-on support), (b) Diamond Cubic and (c) Methane (CH4)*

The resulting 3D geometry of the digital models using Blender software is used as intended (Fig. 4): it is materialized by 3D printing (a solid structure is added), used for interactive visualization and animations, and popularized on Facebook.

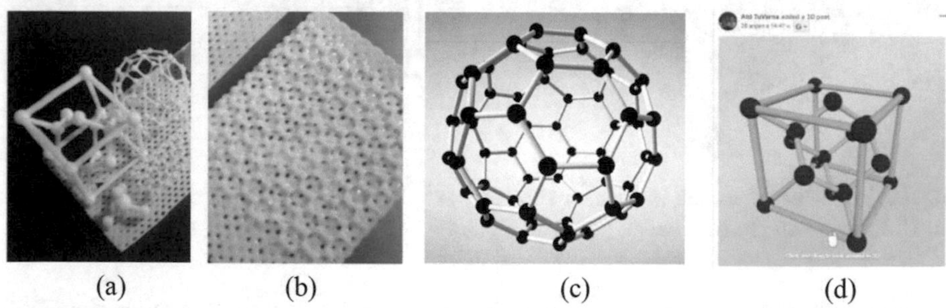

Fig. 4. Real and interactive presentation and popularization of the developed 3D models of carbon structures and compounds: *(a) and (b) 3D printed models of Diamond, Fullerene C60, Methane (CH4), Nanotube and Graphene, (c) interactive 3D model of Fullerene C60 in glTF Viewer environment and (d) interactive 3D model of Diamond Cubic in Facebook environment*

3 Conclusion

This study aims to create a correct three-dimensional geometry of popular carbon structures and compounds that in an appropriate contemporary fashion are presented and used by trainees and authorized persons working and engaged in the subject. The patterns are chosen according to their purpose and importance in contemporary science and direct application in practice.

Acknowledgments. This paper (result) is (partially) supported by the National Scientific Program "Information and Communication Technologies for a Single Digital Market in Science, Education and Security (ICTinSES)" (grant agreement DO1-205/23.11.18), financed by the Ministry of Education and Science.

References

1. Dovramadjiev, T.: Advanced Technologies in Design. Technical University of Varna, Bulgaria, p. 228 (2017). ISBN: 978-954-20-0771-5
2. Starichenko, B., Antipova, E., Slepukhin, A., Semenova, I.: Design of teaching methods using virtual educational environment. In: SHS Web of Conferences CILDIAH, vol. 50, p. 01176 (2018). https://doi.org/10.1051/shsconf/20185001176
3. Alexandre, C., Porto, C., Santos, E.: Facebook and Education (post, like & share). EDUEPB, Campina Grande – PB (2016). ISBN EBOOK – 978-85-7879-313-5
4. Saridakis, R.K.M., Dentsoras, A.J.: Soft computing in engineering design – a review. Adv. Eng. Inform. **22**, 202–221 (2008). https://doi.org/10.1016/j.aei.2007.10.001
5. Stoeva, M., Bozhikova, V.: A shape based approach for biometrical analyses. In: Proceedings of International Conference on Computer Systems and Technologies-CompSysTech 2009, vol. 433, pp. V11–V15 (2009). ISSN 1313-8936
6. Gardiyawasam, T., Dias, G.K.A.: An online virtual educational center. Int. J. Res. 22–23 (2017)
7. Kaminsky, W., Snyder, T., Moeck, P.: 3D printing of crystallographic models and open access databases (2014). https://doi.org/10.1107/s205327331408721x. MS84.P08, Acta-Cryst USA
8. Slepička, P., Hubáček, T., Kolská, Z., Trostová, S., Kasálková, N., Bačáková, L., Švorčík, V.: The Properties and Application of Carbon Nanostructures, INTECH, pp. 175–202 (2013). http://dx.doi.org/10.5772/51062
9. Blender software. Blender Foundation. GNU General Public License (GPL). https://www.blender.org/
10. MeshMixer.Autodesk. http://www.meshmixer.com/
11. glTF Viewer. https://gltf-viewer.donmccurdy.com/
12. p3d.in. Global Scanning Denmark A/S TM + . https://p3d.in/
13. Facebook. https://www.facebook.com/
14. GNU General Public License. https://www.gnu.org/copyleft/gpl.html
15. CNT Add-on. Object Create CNT Add-on; License: GPL 2.0. https://github.com/bcorso/blender-cnt

16. Houston, A., Gragert, P.: Geodesic Domes Add-on. Mesh: Geodesic Domes Add-on: GPL. https://archive.blender.org/wiki/index.php/Extensions:2.6/Py/Scripts/Modeling/Geodesic_Domes/
17. Virendra, K.V.: Crystallography: Diamond Structure. Madanapalle Institute of Technology and Science (MITS) (2014). https://www.slideshare.net/virendrave/diamond-structure
18. Robert, W.G.: Encyclopedia Polyhedra Tetrahedron (2007). http://www.rwgrayprojects.com/rbfnotes/polyhed/PolyhedraData/Tetrahedron/Tetrahedron.pdf

Agile Start-up Business Planning and Lean Implementation Management on Democratic Innovation and Creativity

Evangelos Markopoulos[1(✉)], Onur Umar[1], and Hannu Vanharanta[2,3]

[1] HULT International Business School, Hult House East, 35 Commercial Rd, E1 1LD London, UK
evangelos.markopoulos@faculty.hult.edu, onurumar@hotmail.com

[2] School of Technology and Innovations, University of Vaasa, Wolffintie 34, 65200 Vaasa, Finland
hannu@vanharanta.fi

[3] Faculty of Engineering Management, Poznan University of Technology, Poznan, Poland

Abstract. Startups are the entities developed by innovative minds aiming to change the world and disrupt the industry. However, those minds are often trapped in what they want to do and on what they have to do. Planning a business based on disruptive visions can contradict the logic that all business must be fast profitable and sustainable. One way to resolve this challenge is by democratizing the business planning process for the knowledge elicitation and commitment confirmation. This paper provides to start-ups, young entrepreneurs and ambitious innovators a practical methodological framework for democratic and agile business plan development, supported with templates and metrics for collective thinking and lean management. Under the applied philosophy for management and leadership concepts and driven by the Delphic maxis of miden-aga, metron-ariston and know-thyself, the framework balances first the nurturing of brilliant minds and then the management and leadership of their brilliant ideas, insights and innovations.

Keywords: Business plan · Company democracy · Agile · Lean · Management · Applied philosophy · Entrepreneurship · Startup · Methodology

1 Introduction

Globalization redefined the way how business established and operates. The last ten years the entrepreneurial roadmap has been changed massively and impressively due to the start-up phenomenon. The opportunities offered through the .com dream, especially to younger generations to reach success with an application or change the world with a smart idea creates a global competition of dreams accomplishment than actual business proposition and development. The gap creates between the desired and the feasible outcome, can be bridged with realistic business planning which combines the innovative creativity needed for the futures state but also the conditions, and restrictions needed to be considered at the present stage. Emphasizing and focusing on what can be

T. Ahram et al. (Eds.): IHSED 2019, AISC 1026, pp. 885–895, 2020.
https://doi.org/10.1007/978-3-030-27928-8_133

deliver under specific time, budget and recourses is what differentiates the ones determined to succeed who see seek a shoot to the dream.

However, what determines a well thought business plan remains a fuzzy issue. The completeness of a business plan based on what investors expect or require does not assure the correctness of the information used, projected and forecasted over an operations period. The number or assumptions in a business plan is high and quite accepted from the investors as a plan is a plan, but the degree of realism cannot quite measurable, and this is where the challenge begins.

Today there are many plan methods, templates, tool and practices guiding young entrepreneur, and not only, towards planning their business property. Nearly all of them expect a holistic planning approach with deep understanding of the target market, operations management, marketing, production, sales, financial planning and much more. Such expectations require planning expertise rarely in startups and SME's who seek more practical approached to balance what they want to do today with how they will be in the future or when they can achieve high returns on investment. This would not have been a major issue if the SMEs were not the backbone of every economy [1].

To tackles this issue and support the entrepreneurship movement, business planning shall be approached with methodological frameworks that can adjust the planning process by balancing the creativity and operations, One way to achieve such differentiation is the introduction of democratic, agile and lean business planning frameworks and not actual methodologies. The element of democratization provides the space needed for knowledge to mature in a co-evolutionary way with all involved the project, idea or initiative. Collective thinking can generate effective knowledge which can be transformed into realistic busies plan via agile planning process and later on via lean project implementation management.

2 The Startup Entrepreneurial Revolution

The evolution of technology, the computing power, the rise of the internet, the computing mobility, accessibility, user friendliness, the global connectivity, the digital services and many other factors which got aligned at the same time with the same pace created the circumstances for a new business revolution. The startup revolution is led primarily by the millennials, or the liberal Y-generation, which embraces the concept as an opportunity to crave more meaning in their lives through their vision to change the world or live their myth. They find it easier than ever to start a business and an opportunity to take control over their lives, as they do not trust their leaders. They feel that they do not have anything better to do and nothing to lose. They are not afraid to fail knowing that most of the startups fail. Today, 99% of the startups who request funding are rejected, and nearly 75% of those who receive venture capital funding never return their investment [2]. Therefore, they can safely go after their dream without counting a failure as a failure.

The entrepreneurial movement which also grows tremendously in the universities worldwide has supported this trend. New academic programs with emphasis on entrepreneurship of any kind teach millions of students each year how to utilize their skills through the opportunities startup offer. Today nearly 133 million startups are created annually [3]. This is 4 startups per second, with only 0.05% of them to reach basic funding. More than half of the funded ones, nearly 50.000 are in the US, and about 1.500 of them reach VC funding which is 1 out of 88.667 or 0.00112%.

The United States leads this startup revolution with tremendous amounts of investments each year. During the period of 2011–2014 nearly $140 billion were invested by venture capitalists on nearly 12 thousand deals. This is an average of $35 billion per year and an average of $12 million per deal [4]. The technology sector (Internet, mobile, software, electronics & hardware) absorbing in 2014, 75% of all VC deals, making an increase of 73% from 2013. However it is not all the US that spends such amounts on startups. The top 5 states with the most deals in 2014 were California (1.631), New York (422), Massachusetts (346), Texas (143) and Washington (97), while 23 states, nearly half the country, had under ten deals each.

3 Business Plan Methods, Model and Frameworks

The first step on the startup entrepreneurial journey is the creation of a business plan through which the overall idea is mapped on it, to be evaluated for its technical, financial and investment feasibility. To cope with this revolution, Business Plan templates have been developed from various organizations, institutions and private corporations, with different business planning approaches and similar steps, differentiated in the order of execution and on the depth of the requested documentation. The Business Model Canvas, the Lean Canvas, the McKinsey Template, the HSBC Template and the IBM Watson Build Business Plan Template are briefly presented and analyzed.

The Business Model Canvas (BMC) is a chart-based template that visually documents nine key areas of new or existing business. The canvas collects information for each area and provides an organizational blueprint based on which further planning and action can take place. The BMC areas are the Key Partners, Key Activities, Key Recourses Value Proposition, Channels, Customer Segment, Cost Structure, and Revenue Streams [5, 6].

The Lean Canvas is a simpler version of the Business Model Canvas with emphasis mostly on the idea than the business operations. The lean canvas maintains the nine areas of the Business Model Canvas but replaces the ‘Key Partners’ with the ‘Problem’, the ‘Key Activities’ with the ‘Solution’, the ‘Key Recourses’ with the ‘Key Metrics’ and the ‘Customer Relationship’ with the ‘Unfair Advantage’. The Problem-Solution-Metrics-Advantage structure reduces the operations complexity of the BMC and makes business planning more practical and simplified [7].

The McKinsey Business Plan structure is related with the McKinsey three horizons of innovation. The planning model has specific elements that get completed in depth and gradually over the maturity of the innovation. This completion takes place in three stages where in the first stage key elements and individual topics are addressed were at the last stage the integration of all the elements form the total business plan. The elements are the Product Idea, Competencies and Profiles of the management team, Product Marketing possibilities, Company Operations, Detailed Schedule of the company realization, Financial Planning and the Risks Involved [8].

The HSBC Business Plan template has a more financial approach. The template includes a Brief Business Description, a Business Overview, Markets and Competition, Sales and Marketing, Management, Operations, Forecasts and Financing. As the model derives from the financial sector it emphasizes more on the tangible and objectives financial elements than the subjective ones related to the innovation, market disruption and potential growth [9].

The IBM Watson Build Business Plan Template is a tool under the IBM's Watson Build global initiative that has been designed to accelerate the adoption of Watson by the IBM Business Partners' and encourage the development of AI solutions on the IBM Cloud. The main elements of the business plan are the Big Idea, Solution, Target Market, Market Need, Competition, Team, Budget and Sales Goals, Sales and Marketing, Risk Assessment, and Milestones [10].

4 The Dilemma Between Planning the Business and Business Planning

Startups and entrepreneurs are usually in the dilemma to plan their business based on the vision or on the profit. Planning a business to disrupt the market and attract visionary investors, does not necessarily has to be profitable in the early periods of its operations. However, planning a business that can create fast and quick profit, which can attract finance-oriented investors, need a solid financial plan regardless the degree of innovation or disruptiveness. Therefore, the dilemma entrepreneurs have is either to plan their business on do business planning. The first one is based on the vision and disruptiveness while the second is based on the financial sustainability and profitability.

Shining startups are the ones who can see the innovation, the vision, the disruption. There are the ones who create blue oceans and generate new eras in the economy, the markets and the society impacting the end users, the buyers, and the consumers.

Prior emphasizing on how fast a startup will generate profit, return on investment, and wealth, it is more important to emphasize on innovation itself [11]. It is the 'wow' that shall proceed the 'how'. There have been numerous success stories of startups that made global impact with very poor initial financial plans [12], and others that operated in loss for many years, prior reaching profitability and success [13].

Business plans and financial plans above all are pure estimations based on research under hypothesis rarely able to be proved sufficiently. This, however, makes sense as startups being in early stages cannot provide solid business plans, with actual financial data from business operations that can be used to develop future estimations through which investments will be attracted.

Startups have the charisma to think out of the box and of course out of the numbers. The pressure to justify every business operation with financial data restricts innovative thinking, vision and ideas. Start-uppers have the dream, and this is where innovators shall emphasize in the initial stage. Today there is surplus of funds to be invested but there is a scarcity of ideas with meaningful impact to attract investors who first want to be change makers and then to obtain profits from their success. According to Forbes, the eight factors investors are seeking to see in a startup are all related with the vision and the opportunity [14]. Specifically, they look for the person as an entrepreneur and the overall team, the mission, the size of the opportunity, the quality of the presentation, presentation, the value investors can add, the timing of the market, the research conducted and if anyone else is investing.

An alternative evaluation of a startup idea valuation has to do with the degree of uncertainly the idea has. This degree can be calculated by various economic, political, social, technological, legal, and other factors. The degree of acceptance an idea has, can affect heavily investment decisions. Also, the maturity (readiness) of the idea has to be safely and widely adopted in the market, in the legal framework that can support such operations or in the political environment that can create the stability in the target country the startup intends to operate initially, impacts as well the investment decision-making process [15]. The degree of such considerations can be calculated with probabilities, but again this remains a subjective approach to accurately identify the uncertainty of the business plan and reject it as insufficient. High uncertainty does not necessarily mean high investment risk. On the contrary, high uncertainly can mean low investment risk as the profit loss from not investing on something that not quite clear and miss the opportunity to disrupt the industry, can be tremendous.

In both approaches, the vision-based oriented one, and the uncertainty-based one, the financial analysis pays the least significant role. According to data, cases and experiences, a strong vision and market, makes the financial estimations less important [14]. Therefore, the dilemma remains between planning the business (vision driven) and business planning (finance driven). The entrepreneur shall decide vision discounts to be more affordable and fast profitable or insist on the vision hoping to find mind-like investors.

5 Democratizing the Business Planning Process

One approach to overcome the planning the business or business planning dilemma is to democratize the innovative thinking in the sense that a co-evolutionary business planning can be achieved with the contributions of all, starting from the vision all the way to the well thought financial estimations or operations. This collaborative thinking is a continuous process or knowledge sharing and experimentation towards the optimizations of a business idea.

Such a democratization can be achieved through the Company Democracy Model (CDM). The Company Democracy Model [16, 17] uses the wisdom of the ancient Hellenic Delphic maxims [18], with emphasis on the 'Gnothi Seauton', 'Metron Ariston' and 'Miden Agan'. The model operates in a spiral evolutionary process through which a knowledge-based democratic culture aligns personal and organizational goals. This co-evolutionary approach identifies the achievements, the capacity, capability, competence, and maturity of the human resources, and turns such data, information, and knowledge into innovations. The model is aligned with the Evolute methodology which optimizes knowledge management for continuous innovation development [19].

The CDM can be adjusted to contribute on resolving this startup dilemma. Figure 1 presents the six (6) levels of the adjusted model and the activities that take place in each. The six levels of the startup CDM are divided into two groups of activities that represent the vision planning and the financial planning approach.

Levels 1–3 emphasize on the creative and innovative part of the business plan. This is achieved initially on Level 1, with the optimization of the feasibility of the idea in terms of innovation and disruptiveness through continuous democratic knowledge elicitation from all who can contribute. Level 2 emphasizes on team building and management for maximum performance optimization and the development of working prototypes. Level 3 aims on extending the development of the working prototypes into operational prototypes, as well as product/service validation and verification that will clarify the realization of the innovation and contribute to its capitalization.

Levels 4–6 can contribute of the business operations dimension. Level 4 is based on the innovation effectiveness in order to identify the markets to operate best and based on that to analyze the operations costs and expected revenues. Having identified and tested both the innovation and the market, level 5 creates the competitive advantages and the communication strategy that will be disseminated by marketing initiatives to attach clients and partners. Lastly, level 6 identifies the breakeven point for further investors, global operations, IPO, strategic partnerships, etc., or the exit or de-investment strategy if that is more suitable per case.

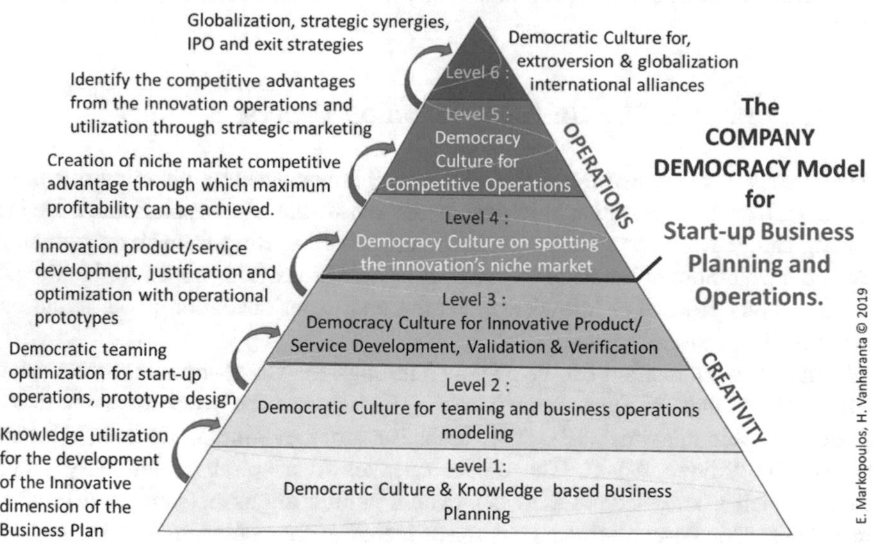

Fig. 1. The adjusted company democracy model for the democratization of startup business planning and operations

6 The Democratic Agile Start-up Business Plan (DASBP)

The Democratic and Agile Business Plan (DASBP) can be a new business-planning model to address the startup dilemma. The model approaches the planning process first for the creativity dimension and then from the financial. The creativity approach is aligned with the idea that the startups must emphasize first on the innovation that will support the business. Lack of innovation in products or services cannot assure long and or fast success and sustainability. Business plans with weak innovative products or services can be mostly considered theoretical or ambitious. Having defined, developed and tested the innovation, business planning becomes more realistic and predictable.

This balance between the innovative and operations dimension of a business plan is characterized by the agility offered to the innovator/entrepreneur to develop first the idea and then define the business based on the degree of innovation. This impacts the degree of competitiveness and helps safe operations predictions to be made. The two dimensions can be view as two layers of the business planning process. The first layer of the DASBP has six key elements that must be completed prior moving to the second layer. They are: Value, Growth, Transparency, People, Market and Agility (Fig. 2).

The value of the startup indicates the uniqueness or the idea, the degree of the innovation, and the benefit to the clients and the market. Growth indicates potential growth or the sectors in terms of innovation disruptiveness. The market indicates the existing total available market segmentation. The team indicates capability, maturity and capacity of the start-up human recourses and teaming structure. The transparency is related with the clarity on the innovation in terms that innovation is well defined and communicated. The agility is related with the degree of flexibility the innovation has when it needs to be changed or adjusted to various circumstances.

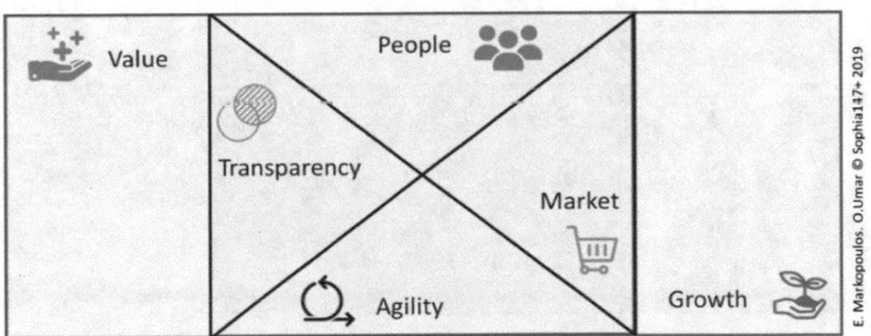

Fig. 2. The first, innovative, layer of DASBP

The second layer of the DASBP has also six key elements, which are the partners, revenue, suppliers, price, client relation, and operations cost (Fig. 3). Partners indicate the strategic partnerships made or need to be made for the start-up success. Revenue presents the revenue expected streams. Suppliers indicate the material and operations support needed. Price indicates the pricing scheme. Client indicates the relations communication, dissemination, marketing and client support. Operational costs indicate the fixed, contingency and risk management costs.

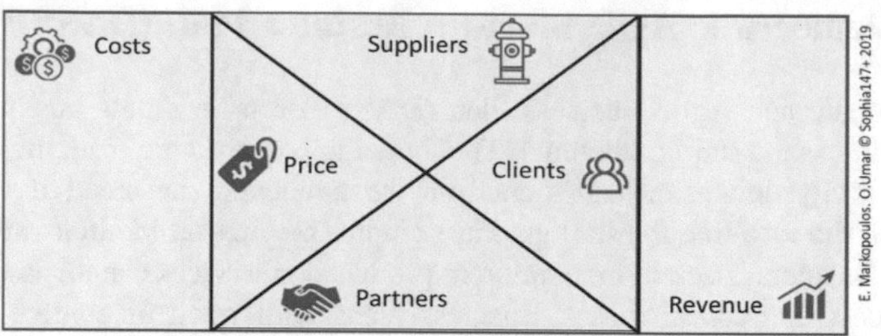

Fig. 3. The second, operations, layer of DASBP

The integration of the two layers form the holistic version of the DASBP model. The model is read from bottom to top going through the Startup CDM version evolution of knowledge maturity to innovation with the proper team structure and a series of prototypes (levels 1–3) that verify innovation existence and acceptance, prior the identification of the market segments, the competitiveness and the global sales strategy (Fig. 4). The distance between layers one and layer two is related with the time needed for the maturity, development and validation of the innovation.

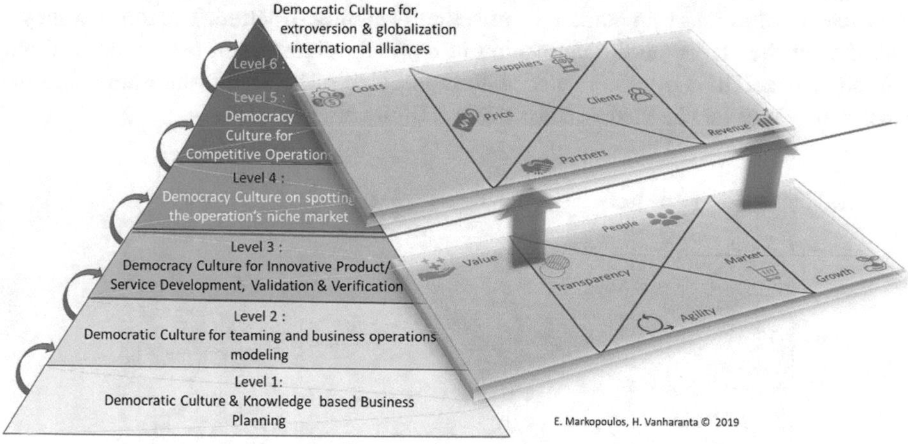

Fig. 4. The DASBM models in the CDM framework.

7 DASBP Application Metrics and Indexes

The effectiveness of a business plan can either be illusionary or have a subjective value based on the one who creates it or judges it. The Business Model Canvas, which is today a widely used framework does not record any data and does not provide any value indicators that could be used to score the effectiveness or the overall value of the business plan. No attempt has been indicated so far to quantify the effectiveness of a business plan, however the agility and staged development offered through the DASBP minimizes the business plan elements, groups them, prioritizes them, and makes the development of metrics feasible.

Based on this, a three dimensional DASBP metric model can be developed to deal with the importance of the elements of each layer, the correctness and the degree of completion. This scoring can be achieved by initially providing a weight to each element. This weight can be related with the potential of the industry which can be found in the Bloomberg, S&P500, and other ratings. The total weight of the elements based on the industry value can provide the overall weight, in terms of the business plan value.

Furthermore, the degree of complements can be calculated by counting the number of requirements each layer's element requires to be completed. As of today, no business plan provides a minimum set of required elements in each planning category. This creates a great degree of fuzziness in the correct execution of the business planning process. The last metric can be associated with the correctness of the completion. This can be justified with documentation provided for the fulfillments of each requirement of each element of a DASBP layer. The documentation needed can also be guided by a set of minimum references that justify the qualitative completion of an element requirement (Fig. 5).

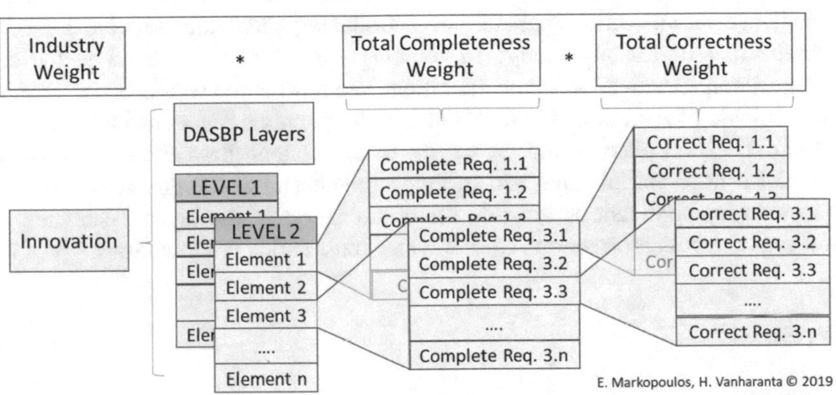

Fig. 5. Computation elements structure in the DASBP weighing system.

It must be noted that incomplete element requirements impact the business plan weight. In the same sense, completion requirements, or corrections requirements not completed, impact as well the business plan score. Equation 1 presents the weight calculation process (1). The complete total business plan weight has all its elements weighted with the maximum values. However, and as this an agile and staged approach the weight can be divided in the two layers and investments can be made based on the planning progress through the completion of each DASBP layer.

$$\text{Business Plan Weight} = \text{Industry Value} * \text{Total Elements Requirements Weight} * (\text{Total Completeness} * \text{Total Correctness}). \tag{1}$$

The DASBP weight approach is a theoretical start to business plans validation and a contribution towards standardizing the startup business planning process which distributes today tremendous investments through subjective or bureaucratic judgement.

8 Conclusions

The challenge of business planning or planning the business has been and will remain a major concern on investment management and market evaluation. Regardless the stage of an organization, in terms of operations, the business plan either it is for a startup or for the realization of global strategies in an established organization, contains to a great degree a number of estimations that impact significantly the outcome of its execution. Staged, agile and weighted business planning can reduce such risks, to an extent, due to the distribution of the planning commitments in stages and the investment commitments on the correctness of each stage. The DALBP model attempts to contribute on this challenge as an initial approach by introducing three innovations. First is the democratization of the knowledge in thinking and planning, second is the staged planning from the innovation to the operations and third is the weighting calculation to size the volume of the effort. DALBP links of the previous two contributions with the investment decision making and the degree of investment that can be made per case. The model is at an initial stage and further research and development will follow, as well as industry applications towards the optimization of the layer's elements, completion requirements, correction requirements, and metrics computations.

References

1. OECD Observer. http://www.oecd.org/cfe/leed/1918307.pdf
2. Forbes. https://www.forbes.com/sites/85broads/2013/11/18/the-millennial-startup-revolution/#536ecfa9622d
3. Global Entrepreneurship Monitor. https://www.gemconsortium.org/report
4. The 2014 US Venture Capital Report. Year in Review. Financial and Exit analysis. CP Insights (2015)
5. Oyedele, A.: Emerging market global business model innovation. J. Res. Mark. Entrepreneurship **18**(1), 53–62 (2016)
6. Osterwalder, A., Pigneur, Y.: Business Model Generation: A Handbook for Visionaries, Game Changers, and Challengers Paperback, NJ, USA (2010)
7. Arwa, A., Jamaludin, I.: Business model canvas, the lean canvas and the strategy sketch: comparison. Int. J. Sci. Eng. Res. **9**(1), 871 (2018)
8. Kubr, T., Marchesi, H., Ilar, D.: Starting-Up. Achieving Success with Professional Business Planning. McKinsey & Company, Netherlands (1998)
9. HSBC. https://www.knowledge.hsbc.co.uk/business_plan/embed

10. IBM. https://www-356.ibm.com/partnerworld/wps/servlet/ContentHandler/watson-build
11. Harvard Business Review. https://hbr.org/2012/05/great-businesses-dont-start-wi
12. Forbes. https://www.forbes.com/sites/kevinkruse/2018/07/30/the-no-plan-business-plan-for-entrepreneurs/
13. Entrepreneur Europe. https://www.entrepreneur.com/article/272710
14. Forbes. https://www.forbes.com/sites/alejandrocremades/2019/01/09/how-investors-decide-to-invest-in-startups/
15. Jean-Charles Pomerol, J.-C.: Business uncertainty, corporate decision and startups. J. Decis. Syst. **27**(sup1), 32–37 (2018)
16. Markopoulos, E., Vanharanta, H.: Democratic culture paradigm for organizational management and leadership strategies - the company democracy model. In: Proceedings of the 5th International Conference on Applied Human Factors and Ergonomics (2014)
17. Vanharanta, H., Markopoulos, E.: Creating a Dynamic Democratic Company Culture for Leadership, Innovation, and Competitiveness. 3rd Hellenic-Russian Forum, 17 September 2013
18. Parke, H., Wormell, D.: The Delphic Oracle, vol. 1, p. 389. Basil Blackwell, Oxford (1956)
19. Kantola, J., Vanharanta, H., Karwowski, W.: The evolute system: a co-evolutionary human resource development methodology. In: Karwowski, W. (ed.) The International Encyclopedia of Ergonomics and Human Factors. CRC Press, Boca Raton (2006)

Model for Improving Post-sales Processes Applying Lean Thinking to Reduce Vehicle Delivery Times at an Automotive Company

Osben Vizcarra[1(✉)], Fernando Sotelo[1(✉)], Carlos Raymundo[2(✉)], and Francisco Dominguez[3(✉)]

[1] Ingeniería Industrial, Universidad Peruana de Ciencias Aplicadas (UPC), Lima, Peru
{u201315434, fernando.sotelo}@upc.edu.pe

[2] Dirección de Investigación, Universidad Peruana de Ciencias Aplicadas (UPC), Lima, Peru
carlos.raymundo@upc.edu.pe

[3] Escuela Superior de Ingeniera Informática, Universidad Rey Juan Carlos, Madrid, Spain
francisco.dominguez@urjc.es

Abstract. This paper proposes the application and implementation of the lean thinking methodology at a company dedicated to selling vehicles, spare parts, and post-sales services. For these purposes, tools such as process maps, flowcharts, indicators, system layout planning tools, and 5S implementation are used, supported by training sessions for process participants. As a result, customer complaints were reduced by 0.13% per month, and the number of work orders handled by the site increased by 0.38% per month. Moreover, a 0.32% increase in man/hour productivity was reported, leading to a 44.32% decrease in Duster vehicle delivery times.

Keywords: Lean services · 5S Methodology · Operational performance · Value stream mapping · PDCA cycle

1 Introduction

Change seems to be commonplace for the worldwide automotive industry, which has driven progress toward globalization processes, which has resulted in the emergence of a single coherent industry. However, a clearer understanding of the degree and pace of the changes taking place in the industry, and whether convergence and integration has reached a global level, requires more than mere globalization. For example, simply the structural volatility of the sector provides enough evidence that simple consolidation into a smaller group of multi-brand and multi-market companies is unlikely to arise in the near future, either at the vehicle manufacturer or vendor level. Furthermore, complicated government interventions at multiple levels also erode any suggestion of the industry being shaped solely by "pure" economic forces [1].

T. Ahram et al. (Eds.): IHSED 2019, AISC 1026, pp. 896–902, 2020.
https://doi.org/10.1007/978-3-030-27928-8_134

Due to the large number of markets and brands offering similar goods or services within the hemisphere and due to technological advances easily transcending borders, companies require mechanisms to differentiate themselves from their competitors.

Regarding contribution, this study focuses on the application of the lean thinking methodology in an attempt to eliminate and reduce waste not only in the flow of materials but also in information and documentation.

However, most companies, national governments, and universities do not put much energy into service research, innovation, or education. Bitner calls this the "service imperative," where companies and nations that embrace the service imperative will prosper and benefit, as will individuals who do the same [2].

2 Prior Art

2.1 Implementation of Lean Service for Improving Quality at an Automotive Center

Larraín [3] proposes a diagnosis based on the analysis of the current situation. First, the study identifies customer expectations and perceptions to determine the current issues the company faces, and their causes, through a cause–effect analysis. Then, the author provides proposals for improvement, indicators for measuring these improvements, and a monitoring system to oversee, control, and guarantee service quality.

The main problems detected therein were customers not receiving enough information, poor promise fulfillment, untrained workers, unstandardized practices, little customer awareness of the new post-sales system, and spare part availability issues. Hence, the improvement proposals focused on conducting training to strengthen worker engagement and provide a more customer-centric approach, creating activity checklists, defining roles in the sales area, providing better planning of post-sales spare part availability, and implementing a monitoring system [4, 5].

2.2 Implementation of 5S Techniques in the Mechanical Area

Author Cabrera [6] proposes improving time management at both the senior management and the operations staff levels. This proposal includes determining the days and hours of low service demand, defining clear objectives, clearly sorting priorities, providing adequate motivation for the activities established, and preventing interruptions and unscheduled meetings to foster a successful execution of the research process. From this research process, we expect to learn efficient observation, analysis, reflection, demonstration, and verification methods, which include the use of industrial engineering tools, resources, information sources, simulators, databases, time management, field research, and better understanding of the behavior and dynamism of the services sector.

3 Contribution

3.1 Overview

Figure 1 denotes the proposal phases as well as all stages with short-term deliverables.

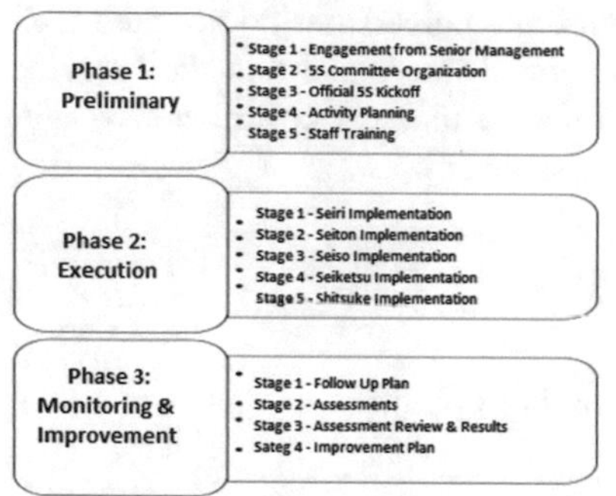

Fig. 1. Overview of the implementation plan

Fig. 2. 10.000 km vehicle maintenance delivery times for DUSTER models.

3.2 Specific View

As per the implementation plan from the previous section, each of these phases will be described below. The Preliminary phase basically establishes an energetic and affirmative determination from senior management, since without its full support this process may be interrupted and subsequently cancelled, which would be unfavorable for the company. Based on its organizational structure, senior management must form a work team tasked with leading the 5S implementation process within the company. The official launch of the 5S strategy is the point of departure for the implementation process within the company, where senior management must share decisions regarding this issue and their expectations with the rest of the staff. Before starting the 5S implementation process, a timetable or work plan must be defined. This plan must describe the different activities to be executed as well as their timeframe, location, and owner to guarantee that their development is fully effective.

In the Execution phase, the Seiri (Sorting) stage seeks to group the following activities: developing photographic records, sorting and assessing items, issuing waste notices, and preparing waste notice reports.

The Seiton (Set in Order) stage was focused on plant layout to properly manage work bays and unused spaces.

During the Seiso (Shine) stage, the following actions were implemented to improve physical appearance while preventing losses and accidents at work caused by dirt: determining the application scope and performing the corresponding cleaning activities.

The following steps, Seiketsu and Shitsuke, require maintenance of post-application visual controls as well as an internalization of the methodology from service bay operators within the workshop.

Finally, in the Monitoring and Improvement phase, monitoring and assessment plans are established for the solution of future problems or improving other processes.

3.3 Indicators

The proposal is supported by the following indicators:

- Vehicle Delivery Time Efficiency index
- Non-conformity index
- Claims per Workshop index

4 Validation

4.1 Company Description

- Reference Name: Service Center, Republic of Panama
- Location: Av. San Lorenzo 1001–Surquillo
- Company: MAQGAMA S.A.C.
- Brand: Renault

4.2 Current Company Situation

The biggest issue for the Republic of Panama service center was the delay in the delivery times for Duster vehicles. This delay progressively expanded by 15.46% with each passing month. At the same time, these delays generated confusion and unease among users of the Renault brand (See Fig. 2).

4.3 Description of Implementation Stages

4.3.1 Stage 1: Previous Assessment

The greatest issues identified were poor work area distribution and poor process management due to large variations regarding Duster vehicle delivery times.

4.3.2 Stage 2: Selection of Implementation Strategy

This stage takes the requirements identified in the previous stage as a reference for developing an implementation plan. This regime aims to be easy to understand for shift operators and consultants to streamline processes and facilitate order within the workshop.

4.3.3 Stage 3: Selection of Implementation Tools

Before the implementation, we assessed the engineering tools available for this specific case. Based on this assessment, the following tools will be used:

4.3.3.1 System Layout Planning. Systematic layout planning consists of a framework of phases through which each layout project passes, a pattern of procedures for systematic planning, and a set of conventions for identifying, visualizing, and rating the various activities, relationships, and alternatives involved in any layout project [7]. It will be implemented as per the following aesthetics: Phase I - Location, Phase II - General Design, Phase III - Detailed Design Plans, and Phase IV - Installation.

4.3.3.2 5S Methodology. Developed as a tool for continuous lean management process improvement, its task is to create a highly efficient, clean, and ergonomic work environment. It is a collection of five simple rules for visually controlling the workplace. The 5S tool is composed of five basic elements: Seiri (Sort), Seiton (Set in Order), Seiso (Shine), Seiketsu (Standardize), and Shitsuke (Self-Discipline).

4.3.4 Stage 4: Development of Proposal

In this stage, the selected tools will be applied. Support from managers, shift consultants, and bay operators will be available to obtain the best possible results.

First, new layouts were assessed and new processes were implemented to keep Duster vehicle services in line. For the development of the scenarios, the Rockwell Arena V1.0 simulation program was used since it provides highly realistic simulation rooms and excellent control variables (Fig. 3).

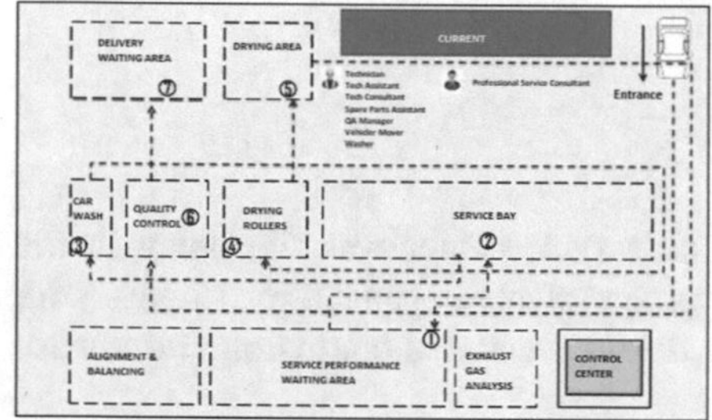

Fig. 3. Initial flow chart for the 10,000 km DUSTER vehicle maintenance

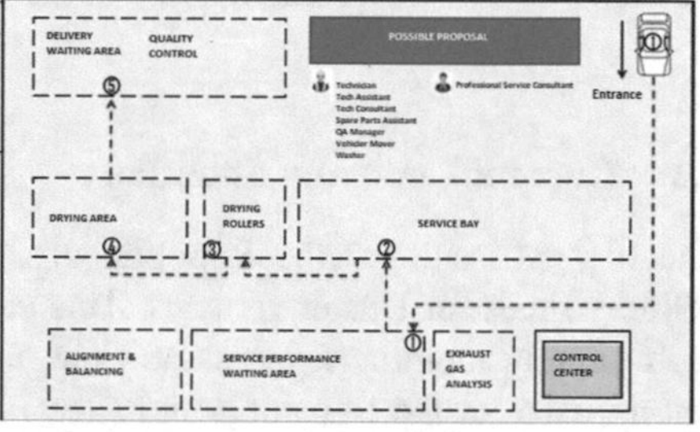

Fig. 4. Final flow chart for the 10,000 km DUSTER vehicle maintenance

After this procedure, initial sample sizes and the corresponding pre-implementation layouts were calculated (Fig. 5).

YEAR	PROCESS	DISTRIBUTION
2017	TRANSFER TO FINISHED VEHICLE AREA	NORM(0.208512 , 0.0827882)
2017	SERVICE EXECUTION	NORM(1.72985 , 0.931583)
2017	CAR WASH ROLLERS	NORM(0.685372 , 0.517467)
2017	CAR WASH	LOGN(1.20464 , 0.628398)
2017	DRYING	NORM(0.778644 , 0.31921)
2017	QUALITY CONTROL	NORM(0.524569 , 0.231422)
2017	TRANSFER TO DELIVERY AREA	LOGN(0.207492 , 0.132136)

Fig. 5. Initial process layouts with 85.00% NC

YEAR	PROCESS	DISTRIBUTION
2018	TRANSFER TO FINISHED VEHICLE AREA	LOGN(0.0915444 , 0.0489694)
2018	SERVICE EXECUTION	LOGN(1.01296 , 0.293233)
2018	CAR WASH ROLLERS	LOGN(0.208875 , 0.0287941)
2018	CAR WASH	NORM(0.710168 , 0.171515)
2018	DRYING	LOGN(0.859539 , 0.230319)
2018	QUALITY CONTROL	LOGN(0.290217 , 0.125286)
2018	TRANSFER TO DELIVERY AREA	LOGN(0.125542 , 0.0708457)

Fig. 6. Final process layouts with 85.00% NC

4.3.5 Stage 5: Results

In this stage, we proceeded to recalculate the final data based on the new layout proposed for the premises (see Fig. 4). Moreover, since times had improved for this scenario (see Fig. 6), a new process layout was also proposed, which in turn led to a new scenario.

5 Results Analysis

5.1 Number of Vehicles in Process

After the final simulation run under the new standardized and improved processes, only four vehicles were included in the process, which generated losses of only S/. 3,600.00. Therefore, the implementation of the proposal produced savings of 63.00% with respect to the initial scenario at the service center. As shown below, the number of claims decreased by a monthly trend of 0.13, which leads to savings of S/. 125.00 per month. Moreover, proper service center flows increased the number of work orders being fulfilled by 0.38%, which translates into S/. 3,500.00. Finally, although the man-hour efficiency improved considerably for the process, constant training on recording process of times, or the implementation of timing sensors, is highly recommended (Fig. 7).

PROCESS	INCOMING VEHICLES	OUTGOING VEHICLES	WIP	WAITING TIME IN QUEUE
WAITING AREA FOR SERVICE EXECUTION	1168.00	1167.00	1.00	0.00
SERVICE BAY	1167.00	1166.00	1.00	0.25
EXHAUST GAS ANALYSIS	573.00	573.00	0.00	0.00
CAR WASH ROLLERS	1166.00	1166.00	0.00	0.00
MANUAL CAR WASH	1166.00	1165.00	1.00	0.03
DRYING AREA	1165.00	1165.00	0.00	0.12
QUALITY CONTROL	1164.00	1164.00	0.00	0.00
ALIGNMENT & BALANCING	229.00	229.00	0.00	0.00

NUMBER IN	1168.00
NUMBER OUT	1164.00
WIP	4.00

Fig. 7. Final simulation results with 85.00% NC

6 Conclusions

The main results from this study were extremely encouraging since average delivery times show a 46.32% reduction against pre-study indicators, and a trend for even greater reduction. This means that the implementation of the chosen methodologies had a significant impact on the entire project, besides reducing the number of claims per month due to vehicle delivery delays and streamlining and standardizing the times within service bays.

Conversely, it must be noted that there were difficulties regarding method implementation related to the creation of design alternatives for the proposed layout. In fact, existing limitations restricted design options due to the enormous structural changes, such as the car wash rollover or tire alignment equipment. These two installations, for example, cannot be moved to other places within the facilities due to high costs that would generate negative values in the balance sheet of the proposal.

References

1. Bitner, M., Brown, S.: The service imperative, Business Horizons (2008)
2. Zima, L.: Service quality in the automotive industry, North University of Baia Mare (2014)
3. Larraín, A.: Diseño de una propuesta de mejoramiento de la calidad de servicio en una empresa del rubro automotriz, Universidad de Chile (2012)
4. Cabrera, H.: Propuesta de mejora de la calidad mediante la implementación de técnicas Lean Service en el área de servicio de mecánico de una empresa automotriz, Universidad Peruana de Ciencias Aplicadas (2016)
5. Guillen, K., Umasi, K., Quispe, G., Raymundo, C.: Lean model for optimizing plastic bag production in small and medium sized companies in the plastics sector. Int. J. Eng. Res. Technol. **11**(11), 1713–1734 (2018)
6. Steinmann, H., Schreyögg, G., Koch, J.: Grundlagen der Unternehmensführung Konzepte - Funktionen - Fallstudien, Management (2013)
7. Falkowski, P., Kitowski, P.: The 5S methodology as a tool for improving organization of production, Wydawnictwo Informacji Zawodowej WEKA (2015)

S-FES: A Structure-Driven Modeling Strategy for Product Innovation Design

Jinyu Lin, Wenyu Wu(✉), and Chengqi Xue

School of Mechanical Engineering,
Southeast University, Nanjing 211189, China
stonergreen@126.com, wuwenyu1984@163.com,
ipd_xcq@seu.edu.cn

Abstract. In this paper, based on the evolution of the functional solution model, a number of different (function-behavior-structure) FBS functional design mapping models are presented. It has found many limitations of the existing function solving model. The reuse of knowledge and mapping of post-redesign are separated. The order of mapping is difficult to judge in the image, cannot be integrated into the innovation redesign stage. The order of mapping is blurry. And the changing environment is not expressed in the image. Therefore, this paper optimizes the design of the FBS mapping model for knowledge reuse, as well as images. An innovative design model S-FES including the above structure or behavior is proposed, which overcomes the above shortcomings and improves the efficiency of the FBS model. The practicality of the new model is illustrated by taking a shoe washing machine as an example.

Keywords: Functional solution model · FBS · Mapping model · Innovation design

1 Introduction

Conceptual design plays an extremely important role in the development of new products. The conceptual design can determine the cost and performance of 70% of the product, while the cost only accounts for 10% of the development [1].

The purpose of conceptual design is to meet the requirements under certain constraints by finding effective functional structures. It is also a conceptual design approach by referencing existing suitable variables to the new design. 75% of design activities include reusing existing knowledge, which explains why designers spend 30% of their work time on knowledge acquisition during product development [2]. Such behavior is known as knowledge reuse, especially in the early stages of design, and it is part of the product life cycle (PLM). And the discussion improves the current reuse of knowledge in product design [3].

T. Ahram et al. (Eds.): IHSED 2019, AISC 1026, pp. 903–908, 2020.
https://doi.org/10.1007/978-3-030-27928-8_135

2 Related Research

2.1 Evolution and Development of Functional Solution Models

Firstly, the realization of conceptual design requires the establishment of a process model. There are currently three most influential design studies, namely the system design method, the FBS model [7, 8], and the axiomatic design method [2–5].

Zhang and Li summarized the hierarchical functional solution framework. In this framework, there are four design areas, including functionality, working principles, behavior, and structure. The reasoning process of product function design includes four mapping modes, namely, solving mapping mode, reconstructing mapping mode, decomposing mapping mode and derivative mapping mode. And basic mapping relationships is total of 15 [6].

In the design process, the designer will continue to deepen the design process because of the designer's own experience, the various processes, structural principles, details of the project, or the changes required [8]. We can divide the environment into three categories: the outside, the interactive and the expected. Outside world: the world in which products are ultimately designed to achieve their functions. Interactive world: the sensory world that been established by designers themselves, mainly from the design input or information provided by the Party A known by us and impact of environment on designers' designing process. It is the world that the outside world projects in the designers' mind. Expected world: the world in which designers' imaginary products are ultimately to achieve their function or to explain and describe the existing external world which is usually geared towards near-future scenarios [7].

In addition, mapping design based on positively extracting design knowledge in the repository. In some cases, the variable targets need to be introduced through existing designs. In the current design, analog-based designs gain new ideas for possible structures and related operations from existing designs with similar functions, similar behaviors or structures. Therefore, there is a need for a design knowledge organization that can help with functional and behavioral indexing when variables are not known at the outset [4].

Designers must not only meet the initial design inputs, but also discover new requirements that arise during the design process. In combination with the above point, there are two ways to invent a design problem: one way is to retrieve explicit knowledge or past cases, and to generate questions or requirements. The other way is to through the current situation, spontaneously construct an objective function or reason to invent design problems and requirements [9].

When a designer analyzes an existing product, it may be borrowed directly from the structure. It may be borrowed from behavior, or it may be learned by analogy or causality through functions [4].

2.2 Product Function Modeling Software

Current commercial CAD applications can only capture the geometric features of the product, not the product features that dominate the upstream design activities [10]. One of the roles of this paper is to help develop computer-aided design programs that helping upstream design [9].

2.3 Limitations of Existing Theories

The existing various models discuss various concrete models for finding effective structures in conceptual design. But most of them only involve the problem-driven innovative design model created based on the release requirements before the design behavior begins [11].

Previous studies have considered the targets triggered from similar cases. But the above methods are separated from the FBS concept design mapping process. It is necessary to complete the reuse then using FBS mapping for subsequent design which is inconveniently.

The images of the existing models are vague for the order in which the mapping occurs. The form of environmental factors in the image cannot express the characteristics of the environment evolving constantly. On the image, the order of the existing model mapping is more complicated.

Therefore, this paper provides clear product innovation design steps, which helps: (1) to help to standardize the product innovation design education method, (2) to capture the design process, and thus improve the efficiency of concept design from the conceptual design process, (3) to emphasize the importance of reusing existing designs, especially existing design structures, and provide standardized reuse knowledge to the complete process of generating new designs.

3 New Theory

The S-FES model summarizes the function-behavior-structure (Function-Behavior-Structure, FBS1) [4] model and the Gero and Udo Kannengiesser‘s function–behavior–structure framework for dynamic world model by the expansion of the structural framework. In particular, impact of recursive relationships between environments on the various elements of the three worlds [8].

Two new categories of variable need N and requirements R [11] are added during the identification and definition phase of the requirements. That make the requirements and needs more specific.

This model places the FBS three domains in the same dimension in each mapped domain, meaning that the destination domain of each mapping may be any one of the FBSs. While the existing model sets each domain to a dimension on the image. At the same time, in order to improve the difficulty of judging the order of the existing model images in the reciprocating mapping, the image of this paper is carried out in the form of upper and lower layers, and then the original reciprocating image is expressed as a one-way from top to bottom, which is convenient for Judging the mapping order. S-FES takes the start and end of each mapping as a dimension. Because the environment is changing constantly. Different environments are represented by different colors. So that each mapping or information flow is clearly visible in the three model. The efficiency is improved compared to other mappings.

The S-FES structure driving model can be roughly divided into five solution modes: reuse mapping mode (UMP), solution mapping mode (SMP), reconstruction mapping mode (RMP), decomposition mapping mode (DecMP) and derivative mapping mode (DerMP) [12]. These five maps are mapped in the design process space over time. UMP is used to find design goals or solutions from existing designs: Sx-Nx-Fx-N-Fi/Ni/Si (where *e* stands for the outside world, *i* stands for the world) or RN-Sx-Nx-Fx -Fi/Ni/Si where Sx, Nx, Fx refer to existing structures, behaviors, and functions. SMP is used to achieve one-pass and by-pass conversion. SMP: R-N-F-B-S, where F, B, and S may undergo several decompositions, and may experience bonding in S. In order to optimize behavior or structure, their functions are integrated to obtain the most compact design to prevent structural explosion. Refactoring mapping pattern (RMP) is used when the actual behavior is inconsistent with the expected behavior or cannot be adapted to meet the design requirements, including: Se-Si, Se-Bi, Se-Fi. Decomposition mapping pattern (DecMP), which is used to decompose complex elements into simpler elements. DecMP includes: Bi-SBi, Fi-SFi (S indicates the next layer). When obtaining results must rely on additional element implementations, we consider using the derivative mapping pattern DerMP, which includes: Bi-SFi and Si-SFi [6]. It should be noted that the environment variables will change after each mapping. As shown in Fig. 1 is a modeling image.

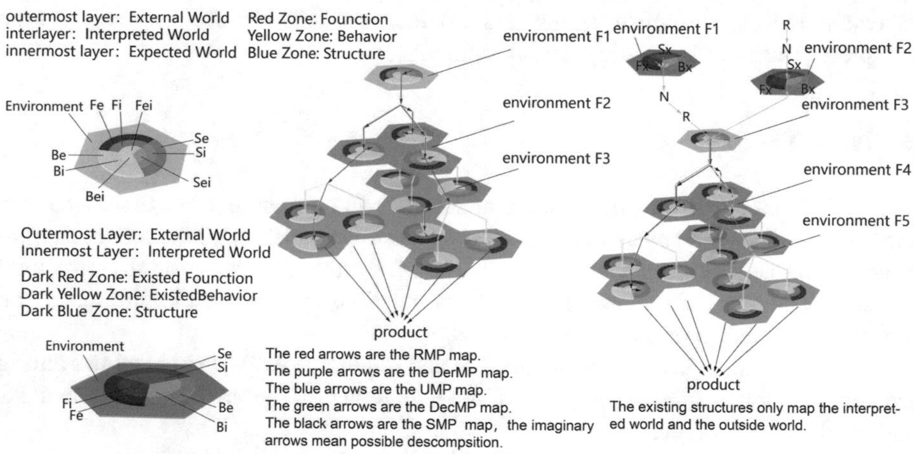

Fig. 1. The S-FES modeling strategy mapping pattern

4 Case Study

The methods discussed in this paper can only roughly illustrate the innovative approach. For this type of device, our focus is not on the novel solution, but on how to develop the design solution through the mapping proposed in this paper [6].

Taking the conceptual design of the shoe washing machine as an example, we can see how the S-FES works. Where *x* means the dishwasher and the existing design. The *b* means the shoe washing machine. The *ei* means the mapping of the expected world. And *i* means the mapping to explain the world. The *e* means the mapping of the external world. In the function of the dishwasher Fxe1, the dirt on the bowl would be washed away by the kinetic energy of the water through the Nxe1. It is easy to think of using Nxe1 to clean other items which means the Nbei1. Where UMP mapping is used to discover new requirements from Fxe1 and to derive new need (Nbei1), namely: UMP: Fx-Nx-Nei-Fei. Thus, Fbei1 has the function of washing shoes. And UMP reuses various structures and behaviors in the dishwasher, for example, the dishwasher supports the bowl and the tray support frame (Sxe1) for the purpose to support the (Fxe2) bowls. In the design of the shoe washing machine, it needs to be improved into the size and shape of the Fbei2 support shoes: Sxe-Sbei-Nbei2-Fbei2. In this way, the function of the shoe washing machine is completed. As shown in Fig. 2, that is the case description.

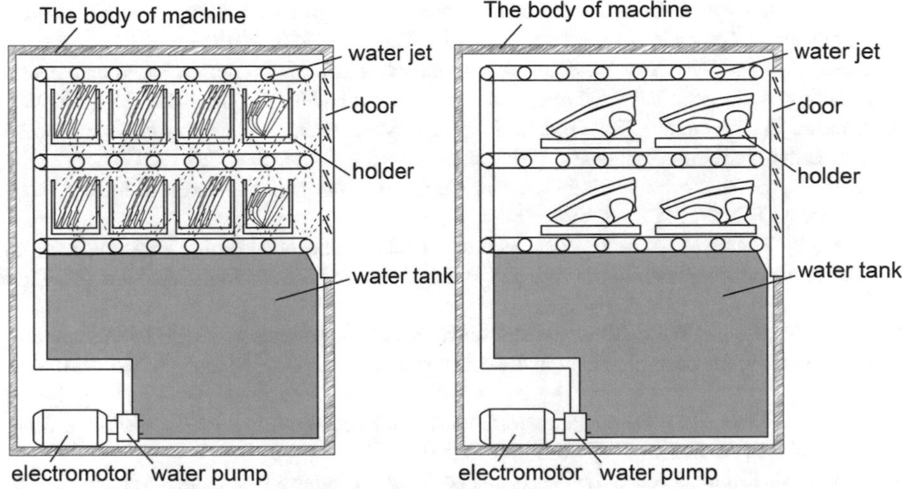

Fig. 2. The structure of dishwasher and shoes washing machine

5 Conclusion

This paper summarizes a variety of FBS innovation mappings by predecessors. By analyzing the advantages and disadvantages of the various methods, it is found that the existing mappings are separated about knowledge and FBS mapping in the innovation process, cannot be integrated into the post-innovation redesign stage. The order of mapping is unclear on the image. The changing environment is not expressed in the images. Therefore, the structure-driven S-FES product innovation mapping model is proposed. The knowledge reuse part is added to the mapping to emphasize the influence of the existing structure on the new design. Each mapping is set one dimension to emphasize the order of mapping and the changing environment. The three domains F, B, and S are designed as rings to emphasize the uncertainty of the mapping direction. The practicality of the new model S-FES is illustrated by taking a shoe washing machine as an example.

There are still some shortcomings and deficiencies in this model. Firstly, this does not fully explain the process of design innovation. Secondly, there are still many design cases in the model that cannot be fitted. Thirdly, there is no specific description for its operation on the product function modeling software, which requires more works to further optimize it.

Acknowledgments. This work was supported by the National Natural Science Foundation of China, project number is 71871056 and 71471037.

References

1. Weisbrod, G., Kroll, E.: Idea-configuration-evaluation (ICE): development and demonstration of a new prescriptive model of the conceptual engineering design process based on parameter analysis and C–K theory. Res. Eng. Des. **29**, 1–23 (2017)
2. Chen, Y., Zhao, M., Xie, Y., et al.: A new model of conceptual design based on scientific ontology and intentionality theory. part II: the process model. Des. Stud. **38**, 139–160 (2015)
3. Aubriot, O., Fernandez, S., Trottier, J., et al.: Water technology, knowledge and power. addressing them simultaneously. Wiley Interdisc. Rev. Water **5**, e1261 (2017)
4. Qian, L., Gero, J.S.: Function–behavior–structure paths and their role in analogy-based design. AI EDAM **10**, 289(1996)
5. Chen, Y., Zhang, Z., Xie, Y., et al.: A new model of conceptual design based on Scientific Ontology and intentionality theory. part I: the conceptual foundation. Des. Stud. **37**, 12–36 (2015)
6. Zhang, M., Li, G., et al.: A hierarchical functional solving framework with hybrid mappings for supporting the design process in the conceptual phase. J. Eng. Manufact. **226**, 1401–1415 (2012)
7. Gero, J.S., Fujii, H.: A computational framework for concept formation for a situated design agent. Knowl.-Based Syst. **13**, 361–368 (2000)
8. Gero, J.S., Kannengiesser, U.: The situated function–behavior–structure framework. Des. Stud. **25**, 373–391 (2004)
9. Suwa, M., Gero, J., Purcell, T.: Unexpected discoveries and S-invention of design requirements: important vehicles for a design process. Des. Stud. **21**, 539–567 (2000)
10. Zhang, W.Y., Lin, L.F., et al.: B-FES/o: an ontology-based scheme for functional modeling in design for semantic web applications. In: Proceedings of the Ninth International Conference on Computer Supported Cooperative Work in Design 2005, vol. 2, pp. 24–26. IEEE Press, Coventry (2005)
11. Cascini, G., Fantoni, G., Montagna, F.: Situating needs and requirements in the FBS framework. Des. Stud. **34**, 636–662 (2013)
12. Zhang, W.Y., Tor, S.B., Britton, G.A.: A graph and matrix representation scheme for functional design of mechanical products. Int. J. Adv. Manufact. Technol. **25**, 221–232 (2005)

Proposal for Process StandarDization for Continuous Improvement in a Peruvian Textile Sector Company

Miguel Arévalo[1(✉)], José Montenegro[1(✉)], Gino Viacava[1(✉)], Carlos Raymundo[2(✉)], and Francisco Dominguez[3(✉)]

[1] Ingeniería Industrial, Universidad Peruana de Ciencias Aplicadas (UPC), Lima, Peru
{U201310238,U200710537,pcingvia}@upc.edu.pe

[2] Dirección de Investigación, Universidad Peruana de Ciencias Aplicadas (UPC), Lima, Peru
carlos.raymundo@upc.edu.pe

[3] Escuela Superior de Ingeniera Informática, Universidad Rey Juan Carlos, Madrid, Spain
francisco.dominguez@urjc.es

Abstract. This article explores the application of continuous improvement tools in company M4C1 to improve processes and production procedure for a production line of polo shirts and the distribution of workstations to provide a more efficient flow, thereby reducing non-productive times of materials and people transportation. The steps to determine the correct approach of the procedures to control excess in production times and reprocesses for its correct implementation, flow charts, and work indicators to measure the production processes of polo shirts will be described. The Arena Simulation tool will then be used to simulate the operation of the production process of polo shirts and to compare the results before and after implementation. Finally, the conclusions of the project will be presented.

Keywords: Systematic layout planning (SLP) · Textile · Value chain · Cotton · Lean introduction

1 Introduction

Company M4C1 is dedicated to manufacturing clothing. It is located in the district of San Juan de Lurigancho. Its main clients are Saga Falabella, Ripley, and Oeschle, and it has 30 sewing and cutting machines for the patterns of polo shirts. Moreover, it has 39 workers, and 32 of them are assigned to the production process. At the end of 2017, the company had an annual turnover of USD 583,486. However, the company has additional expenses for labor, machine use, consumption of raw materials, and penalties for failure to comply with orders within the date agreed with the client. These orders represent 54% of the total orders delivered during 2017. Therefore, the aim of the improvement project is to propose a model to enhance the company's production process and mitigate the problems caused by non-compliance with the orders placed.

T. Ahram et al. (Eds.): IHSED 2019, AISC 1026, pp. 909–915, 2020.
https://doi.org/10.1007/978-3-030-27928-8_136

2 Prior Art

Reference [1] defines that the lean concept can be summarized as the systemic and regular application of a set of manufacturing techniques that seeks to improve production processes through the reduction of all types of waste, focusing on the creation of flows to deliver maximum value for customers. Moreover, [2] explains that lean manufacturing is a well-established set of principles for waste reduction. It aims at reducing waste, optimizing workflows, reducing costs, and improving quality. According to reference [3], lean classifies waste by the following criteria: waiting time, inventory, transport, movement, defect, process, and overproduction [4].

The 5S methodology is considered one of the operational practices that show the best results in world-class manufacturing studies due to its contribution to the improvement of processes focused on productivity and quality and work environment, with fast results and low implementation costs [5].

This methodology is feasibly included in management practices and contributes to maximizing efficiency and effectiveness and improving profitability. Furthermore, it can be developed regardless of the size of the company and it helps discover problems that usually go unnoticed by company employees [6].

The systematic layout planning (SLP) methodology consists of establishing a more flexible distribution where the areas that are more closely related are placed near each other and with easy access to each other [7].

Upon completion of the analysis and determination of the new standard production time, the line balance will be developed to ensure the continuous flow of the products and the balance of the workload of the operators. For this reason, the line balance is defined as the critical factor for the productivity of a company to determine a distribution of the right capacity to ensure the continuous and uniform flow of products. Besides finding ways to make work times on all stations equal, make the most of labor and equipment utilization, thus reducing or eliminating idle time [8].

3 Methods

To identify the problem within the company, a problem tree was created, which also identified the causes generating the problem within the production area. Afterwards, a fishbone analysis was conducted to break down the causes by each factor that composes it. Finally, using Pareto's method, it was determined which of the causes had the greater influence in the problem of orders delivery beyond the established deadline, which represents 54% of the total orders placed during 2017.

Next, a solution is proposed based on a time study and the SLP method. This proposal was simulated in the Arena software. The previous results were compared with the results obtained in the simulation and indicators were established to better visualize the results obtained with the improvement implementation.

4 Results

4.1 Innovative Technical Proposal

The development model of the research project proposal is described in stages as shown in Fig. 1.

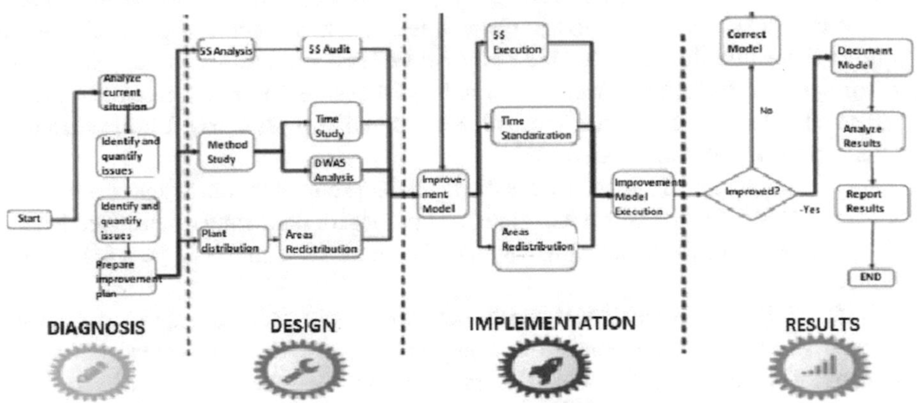

Fig. 1. Proposal model. Adapted from [9]

To establish the theoretical production time of polo shirts adjusted to the monthly demand and the total hours per month, the takt time is established, which helps to identify whether the company can cover the monthly demand. Moreover, under this takt time analysis we can evaluate the performance of each of the workstations in the production area, defining a problem that requires observation and studying what is taking up so much time.

Table 1 Analysis of the monthly production time expressed in minutes for the calculation of the monthly demand to obtain the production takt time.

Table 1. Calculo Takt-time

Concepto	Cantidad
Demanda (und/mes)	4684
Turnos laborales x mes	23
Line rate (und/turno)	203.65
Tiempo total por turno (hrs)	10.6
Tiempo total por turno (min)	600
Numero de turnos x día	1
Número de días x mes	23
Total turnos x mes	23
Tiempo disponible(min/mes)	13800
Número de Líneas (corte)	3
Demanda(und/mes)	4684
Takt-Time (min)	8.84

Table 1 shows the disposition in hours and minutes of the company's tailoring area and its monthly demand. This data will help obtain the takt time. It will also make it possible to analyze the production time of each one of the stations of the production area and determine which areas do not comply with the standard production time.

Currently, company M4C1 takes approximately 9.88 min to manufacture one polo shirt, and the takt time established for the analysis of the improvement is approximately 8.84 min. Therefore, the manufacturing area is identified as the area where the process improvement research should focus, through the new distribution of the work area to improve times and the unproductive movement of raw materials, and the planning of the correct manufacturing procedures to reduce losses, reprocessing, and overtime.

The takt time was calculated to establish the maximum parameter for the operations in the companies to work correctly.

As a solution to mitigate the causes of excessive production times, reprocessing and losses, a new distribution of work stations was proposed, consisting of three production lines. This will speed up the flow of materials and personnel and increase current production.

Additionally, a relational table and its corresponding relational diagram were made to better visualize the relationship and importance of each of the areas within the company: dispatch, administration, design, warehouse, cutting, sewing, and finishing areas, among others. Then, a total redistribution of the work areas was made so that the areas that are closely related and have a high flow of operators and raw material between them are placed next to each other (SLP).

Once the implementation of the SLP methodology was developed within the company, it was deduced that, under a standard production time of 10.45 min per polo shirt and 11 effective working hours, between the three production lines implemented, 4,206 units could be produced. This means that 478 polo shirts would still have to be produced to cover the monthly demand by incurring 29 overtime hours in the entire production plant with an overtime cost of operators of 6,609 USD per month.

Following the design of the established proposal, a time study was developed of the different workstations of the company in the production process of greatest demand: polo shirts. It focuses on analyzing the cycle times of the workstations in the garment area where the greatest operational flow is located.

Time was considered from each workstation and activities within the area of cutting, dressmaking, and finishing, with an average of 100 observations for the tailoring area, where the most issues due to time loss occur (see Fig. 2). The production time of the sewing area for manufacturing a polo shirt is 9.88 min and the total, between the three production areas, is of 10.45 min.

SEWING AREA	
OPERATION	TS (min)
Join shoulders	0.62
Close collar	0.54
Sew collar	0.78
Shoulder to shoulder	2.61
Sleeves hem	1.38
Sew sleeves	1.38
Close sides	0.88
Tail hem	1.01
Sew tag	0.40
Total	9.88

Fig. 2. Production time

SEWING AREA	
OPERATION	TS (min)
Join shoulders	0.62
Close collar	0.54
Prepare collar	0.30
Sew collar	0.78
Shoulder to shoulder	1.13
Sleeves hem	-----
Sew sleeves	1.18
Close sides	1.02
Tail hem	1.01
Sew tag	0.40
Total	6.92

Fig. 3. Enhanced production time

To enhance the production time of the sewing area, the *tapetero* method, which consists of preparing manuals to conduct this activity in an efficient manner, was established. Moreover, unnecessary inspections were eliminated in the activities of "shoulder to shoulder" and "sewing sleeve," along with the activity of "sewing sleeve" as it is considered that this process can be conducted in another activity such as "closing sides" (see Fig. 3).

By analyzing the determination of the new standard time, it was found that the time with an efficiency of 89% for the sewing area is 6.92 min and the total time for the production of a polo shirt is 8.36 min. This new standard time is 20% less than the production time before the implementation of the improvement and less than the takt time of 8.84 min that determines the ability of the company to produce a polo shirt and be able to meet the demand of 4,684 units.

4.2 Results After the Implementation

The data collection took place in the production area of the polo shirts from the moment they entered the process, from the cutting area followed by the sewing area and ending in the finishing area. For the functional validation, the data obtained through the Arenas [10] software was analyzed to show the accuracy of the diagnosed data and the production time of each of the workstations. This will demonstrate the viability of the improvement proposal in company M4C1.

By means of a sensitivity analysis, it was determined that under 99% sensitivity, the optimum number of replications to execute the program must be 34. This number of replications will help simulate our project and the results obtained from the simulator will be viable.

5 Results Analysis

The analysis shown in the following table proves the optimal situation of the result and the analysis between the difference of the current situation and the proposal. With the implementation of the improvement, company M4C1 will not only mitigate the problem of orders delivered beyond the deadline but also increase production by 16%. Conversely, the efficiency of each of the areas where the production time of a polo shirt is 8.36 min (i.e., 20% lower than the previous one) was increased (Table 2).

Table 2. Indicators

Indicators		Current	Proposed	Value	Variation
Orders compliance		46%	116%	+16%	Increase
Production time reduction		10.45 min	8.36 min	–20%	Reduction
Manufacturing time reduction		9.88 min	6.92 min	–30%	Reduction
Penalties cost reduction		16,684 Dolars	0	–100%	Reduction
Overtime cost reduction		43,156 Dolars	0	–100%	Reduction
Number of units produced		4684 units	5445 units	+16.3%	Increase
Distance made	Cut	73 m	39 m	–46.5%	Reduction
	Sewing	147 m	102 m	–30.6%	Reduction
	Finishing	30 m	15 m	–50%	Reduction

The functional validation of the proposal can be proven because in an optimal scenario, the company's problems due to non-compliance with orders are solved.

6 Conclusions

- With the implementation of the improvement tools, 5,445 polo shirts can be manufactured, which is an increase of 16% of the current production.
- The cost of implementing the project is equivalent to 4% of the sales revenue of the polo shirts in the year of implementation.
- Delays in orders were eliminated, which used to represent 54% of the total orders, and included penalty costs, overtime, and machine use, which represented 27% of the total polo shirts sale costs.
- The new standard production time found through time study and plant redistribution using the SLP method reduced by 20% the previous production time, from 10.45 min to 8.36 min.
- With this result, the net current value of the project amounts to 17,200 USD with an internal return rate of 61% in the period of one year.

References

1. Sarria Yépez, M.P., Fonseca Villamarín, G.A., Bocanegra, C.C.: Modelo metodológico de implementación de lean manufacturing. Rev. EAN **83**, 51–71 (2017)
2. Srinivasan, S., Ikuma, L.H., Shakouri, M., Nahmens, I., Harvey, C.: 5S impact on safety climate of manufacturing workers. J. Manufact. Technol. Manage. **27**(3), 364–378 (2016)
3. Vargas-hernández, J., Muratallabautista, G., Jiménez-castillo, M.: Lean manufacturing ¿una herramienta de mejora de un sistema de producción? steadiness approach and change approach in perspective of industrial engineer. Exploratory Study Decisional Propensity **17**, 153–174 (1856)
4. Bellido, Y., Rosa, A.L., Torres, C., Quispe, G., Raymundo, C.: Waste optimization model based on Lean Manufacturing to increase productivity in micro- and small-medium enterprises of the textile sector. In: CICIC 2018 - Octava Conferencia Iberoamericana de Complejidad, Informatica y Cibernetica, Memorias, vol. 1, pp. 148–153 (2018)
5. Lamprea, E.J.H., Carreño, Z.M.C., Sánchez, P.M.T.M.: Impact of 5S on productivity, quality, organizational climate and industrial safety in Caucho Metal Ltda. Ingeniare. Rev. Chil. Ing. **23**(1), 107–117 (2015)
6. Omogbai, O., Salonitis, K.: The implementation of 5S lean tool using system dynamics approach. Procedia CIRP **60**, 380–385 (2017)
7. Alarcón-Grisales, D.R., PeñaOrozco, D.L., Rivera-Rozo, F.J.: Análisis dinámico de la capacidad de respuesta de una cadena de suministros de productos tecnológicos. Caso Samsung*. Entramado **12**(2), 254–275 (2016)
8. López, M.D.R., Grajales, M.H., Corrales, M.E.V.: Lean construction – LC bajo pensamiento lean. Rev. Ing. Univ. Medellín **16**(30), 115–128 (2017)
9. Sundar, R., Balaji, A.N., Satheesh Kumar, R.M.: A review on lean manufacturing implementation techniques. Procedia Eng. **97**, 1875–1885 (2014)
10. Nyemba, W.R., Mbohwa, C.: Modelling, simulation and optimization of the materials flow of a multi-product assembling plant. Procedia Manufact. **8**, 59–66 (2017)

Technology Roadmap for Business Strategy and Innovation

Kazuo Hatakeyama(✉)

Enterprise Consulting Office, Av. Miguel Navarro y Cañizares, 31, apto 701, Salvador, BA, Brazil
khatakeyama875@gmail.com

Abstract. Companies to meet the needs of a highly competitive global market are utilizing product families and platform-based product development to increase the diverse variety of demand, shorten lead-times, and reduce costs. Current research in the area of product family design focuses on a market-driven approach aiming to cost-savings to search for the margin of profit and market share. The technology roadmap approach to sustain the modern business process is in the stake. The overall findings show that in the global marketplace scenario, the normative-driven approach plays a key role to support sustainable business practices along with market-driven and technology-driven approaches.

Keywords: Market-driven · Technology-driven · Normative-driven · Business innovation

1 Introduction

Nowadays, companies to meet the needs of a highly competitive global market are utilizing product families and platform-based product development to increase the diverse variety of demand, shorten lead-times, and reduce costs. Current research in the area of product family design focuses on a market-driven approach aiming to cost-savings to search for the margin of profit and market share. As the few existing design methods integrate market considerations with product development efforts in their formulation, the inclusion of a technology-driven approach considered an important factor. The crucial point to integrate the market considerations with traditional product family concerns is the scope of the design problem that should consider the normative values to be included in the line of a products' positioning problem.

2 Characterization

The market-driven approach introduced as the factor to systematically examine the impact of increasing the variety in the product offerings across different market segments and explore the cost-savings associated with commonality decisions [1].

Technology-driven approach based on the belief that information technology offers a powerful source of competitive advantage. Popularized by, and based on the work of [2], there is a corresponding emphasis on planning and the use of generic strategies.

T. Ahram et al. (Eds.): IHSED 2019, AISC 1026, pp. 916–921, 2020.
https://doi.org/10.1007/978-3-030-27928-8_137

The competency driven approach, on the other hand, holds that each company possesses unique sources of advantage, which it must exploit to derive maximum benefits [3, 4] (Fig. 1).

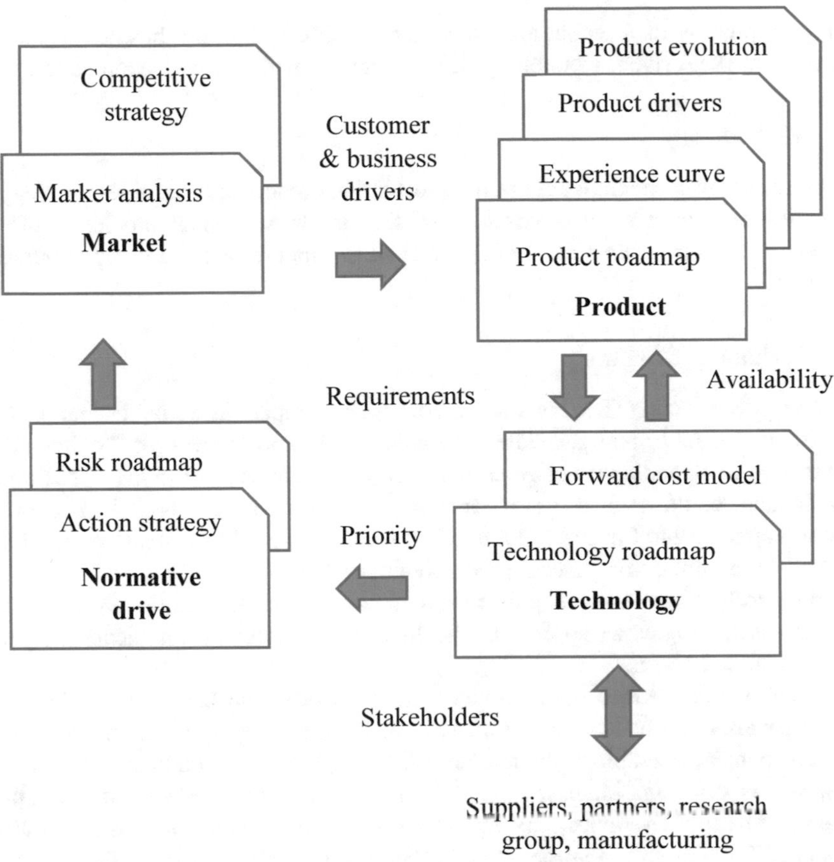

Fig. 1. Roadmapping process (adapted from [5])

Normative-driven is a value-based approach to building communities, based on the assumption that all people have a need to belong, want to have a sense of purpose, and want to experience success. Every member of a normative community carries equal importance in developing a set of norms for living for the community, and in taking responsibility for living those norms and holding others accountable for doing so. Using the group process as a means of accountability for behaviors and productivity of individuals allows for vicarious learning. Maintaining a focus on the shared mission statement gives a common purpose to all community activities, and depersonalizes feedback, creating an environment that is safe, both physically and emotionally. Sometimes, the government institutions could act as guarantor to keep the normative features of successful and sustainable business endeavor. Thus, such actions are prone to risk roadmap verification based on known successful record or sometimes on subjective approach with evidence of minimum setback.

2.1 Objective

Study the process of a technology roadmap to differentiate the business innovation regarding market-driven, technology-driven, and normative-driven approaches.

2.2 Methodology

Literature survey on references regarding technology roadmap for business strategy and innovation on products and services. Books, journals, web sites, articles in the proceedings of Congress and conferences searched to support the most likely approach on the theme.

2.3 Technology Roadmap

Technology Roadmap (TRM) is a broad concept encompassing activities that primarily focused on product or process related technologies. Several variants of TRM could find applications in operational activities depending on reference objectives. A threefold classification could appear as recommendable to consider the TRM approach. (a) roadmap for central-pace setting and key technologies; (b) roadmap for application systems; (c) roadmap for general productive output [6].

Some technology acts as pull trust concerning the application in products of emerging technologies as applied to the fuel cell, internet communications, genetic engineering, and so on.

As far as a concern to the roadmap for central-pace and key technologies can be seen as paramount issues for to company's manager to keep up as forerunners of new products to make available to the market. Likewise, the use of roadmap for application systems is experiencing increase year by year such as for tomorrow's office, intelligent buildings, driverless vehicles, city mobility equipment, internal professional training, and so on. The study in the productive output of companies is another focus of TRM, not limited only to products but also to services.

2.4 Purpose of the Technology Roadmap

The TRM accounts for the diversity of objectives attributed to specific group-related needs or for the general interest. The TRM can exert the supervision of the intra-corporate R&D units to follow up on the performance of new products or services as the business strategy for competitiveness. The TRM could enable the coordination of inter or extra corporate R&D activities aiming for extensive cooperation with external procurements. In the globalized market scenario, the individual companies supplying products worldwide have the option to join forces in devising the technology roadmap to get the support for the common interest to better the business performance. Such options happened in the semiconductors suppliers with their industries spread globally to attend every growing demand due to the surge of emerging design of products using the key items of different types to fulfill the performance of the embedded goods. The TRM is not the ultimate approach as it can experience several limitations between prognosticating technical developments and their interactions. Technology can happen by accident in the try and error chase to get breakthrough results. Upon learning about successful results applying the technology approach, managers make use of TRM as a step forward to the present stage of the state of the art. The documented existence of fullness could now be followed by the search for options of advancement and commercialization that might serve as the theoretical framework. The TRM is not new in so far, practiced by the experts without reverting to any graphical or flowchart illustrations, presently is advancing to consolidate as of a successful approach for the corporate executives. Current discussion among the experts on the roadmap can be marked by [7]:

the extraction from the mind by documenting and communicating of TRM;
the generation of roadmap across and beyond the boundaries of the department;
the use of intelligible tools.

The first aspect characterized by transforming the implicit knowledge generated to the explicit documentation for communication purposes. The second feature is concerned to institutionalize the closed shop procedure as options to include suppliers, stakeholders, partners, and key clients. The third feature underlines the use of instrumental linking functions combining with emerging tools and approaches such as balanced scorecard, time-to market-management strategy, and other strategic approaches in the business field, mainly the procedures of computer-based analysis and evaluation [7].

The TRM seems to continue strong fulfilling the demand for theoretical orientation for technology-based products for business competitiveness.

2.5 Market–Driven Roadmap

The issue on the market-driven roadmap is a complex issue as it involves two different objectives: the forecasting the future developments and planning of own approaches and actions [8]. The strategic decisions depend on prospective outlooks based on present-time knowledge. Requires the concrete information on established market and customer requirements upon technology innovation should apply relating to the medium and long-term perspective of the enterprise.

An enterprise plan is the integrated combination of all the functional plans representing that organization's strategy [9]. Functions within the business own responsibility for gathering, evaluating, and prioritizing certain elements of strategy on an ongoing basis. Marketing might own the responsibility of sensing trends, evaluating customer needs, and prioritizing opportunities in a specific market segment. Marketing should also own the strategic elements containing that information and would update, validate, and maintain those elements as part of their basic business function [9]. This ownership includes the creation and maintenance of a roadmap containing those strategy elements on a time baseline. Ownership also implies a responsibility for creating and maintaining the necessary relationship linkages between strategy elements from other functions. These linkages are created in response to meeting specific business objectives; for example, linking all the product and technology development elements required to satisfy a particular key business opportunity "owned" by marketing.

2.6 Normative-Driven

Concerning the normative-driven issues on TRM for new products sometimes is prone to government intervention due to the hazard to the human use and environmental damages, regardless of the size of the company. Regarding small and medium-size enterprises (SME) led to the government introducing numerous policies to facilitate the formation of new firms and to offer support to aid their survival and foster improved rates of growth [10]. In Japan, for instance, TRM exercises typically aimed at government especially to increase public sector R&D funding for a given field [11]. Highlighting that contributions of the government can be beneficial if involved early in TRM in that it provides data and analysis, garners support and participation from other departments, agencies, etc.

The normative-driven roadmap presents the concerns and benefits of TRM and helps industry bring in the requisite skills, act as a meeting facilitator or roadmap manager, liaise with other government departments or agencies that influence and monitor progress and disseminate the results knowledge [12].

2.7 Discussion

The government has been increasingly using TRM, particularly in the context of energy policy and sustainable energy [13]. There has been an evident shift from industry to public policy; the nature of TRM activities has been ongoing partly as a process of setting directions for the social goals.

TRM has already gathered interest from government, who have been interested in promoting roadmaps to facilitate the development of competitive industries and to push science and technology forward, with a growing interest in the development of new and emerging technologies [11]. TRM tools have been widely used to study and plan the development of industries in which the government has been involved, e.g. foresight vehicle TRM in the UK, Industry in Canada [12, 13]. An example of governmental TRM activities and its use are relatively small but the Ministry of Economy, Trade, and Industry in Japan have actively involved itself since 2003 [14]. In the aeronautical industry, for instance, TRM with continuous innovation strategy is mandatory to keep up the business competitiveness. Since the start of the new design to the final product by means of integration of know-how of producing aircraft, in complex partnerships between parts suppliers and assemblers, TRM, the market-driven and the normative-driven approaches plays the key role for the success of the enterprise. Several other sectors of the industry and services rely on TRM, such as the producers of drugs to control the spread of several epidemic diseases [15] demanding continuous agile technology roadmap [16].

2.8 Results

The overall findings show that the change of technology is presently occurring at a speed never experienced before. Implementing new technology on already existing is the challenge as the market scenario changes constantly requiring the business innovation posture through a systematic roadmap of the main approaches. In the global marketplace scenario, the normative-driven approach plays a key role to support sustainable business practices along with market-driven and technology-driven approaches.

References

1. Jaworski, B., Kohll, A.K., Sahay, A.: Market-driven versus driving market. J. Acad. Market. Sci. **28**(1), 45–54 (2000)
2. Porter, M.: Competitive Advantage: Creating and Sustaining Superior Performance. Free Press, New York (1985)
3. Rothwell, W.J., Lindholm, J.E.: Competency identification, modeling and assessment in the USA. Int. J. Training Dev. **3**(2), 90–105 (1999)
4. Schmidt, A., Kunzmann, C.: Sustainable competency-oriented human resource development with ontology-based competency catalog. In: Cunningham, C. (eds.) eChallenge 2007, Den Haag (2007)
5. Albright, R.E., Kappel, T.A.: Technology roadmapping: roadmapping the corporation. Res. Technol. Manage. **46**(2), 31–40 (2003)
6. Phaal, R., Muller, G.: An architectural framework for roadmapping: towards visual strategy. Technol. Forecast. Soc. Chang. **76**(1), 39–49 (2009)
7. Moehrle, M.G., Isenmann, R., Phaal, R.: Basics of technology roadmapping. In: Technology Roadmapping for Strategy and Innovation: Charting the Route for Success. Springer, Heidelberg (2013)
8. Kappel, T.A.: Perspectives on roadmaps: how organizations talk about the future. J. Prod. Innov. Manage **18**(1), 39–50 (2001)
9. Whalen, P.J.: Strategic and technology planning on a roadmapping foundation. Res. Technol. Manage. **50**, 40–51 (2007)
10. Probert, D., Radnor, M.: Technology roadmapping: frontier experiences from industry-academia consortia. Res. Technol. Manage. **46**(2), 27–59 (2003)
11. Laat, B.D., McKibbin, S.: The effectiveness of technology roadmapping - Building a strategic vision. Dutch Ministry of Economic Affairs, Holland (2003)
12. Kaplan, G.: New roadmap flags electronics industry roadblocks. Res. Technol. Manage. **44**(3), 4–5 (2001)
13. Amer, M., Daim, T.U.: Application of technology roadmaps for renewable energy sector. Technol. Forecast. Soc. Chang. **77**(8), 1355–1370 (2010)
14. Yasunaga, Y., Watanabe, M., Korenaga, M.: Application of technology roadmaps to governmental innovation policy for promoting technology convergence. Technol. Forecast. Soc. Chang. **76**(1), 61–79 (2009)
15. Mallet, L., et al.: MAN.08 - the role of technological roadmaps in shaping the future strategy for Bio-Manguinhos. In: VI International Symposium on Immunobiologicals, VII Annual Scientific and Technology Seminar (2018)
16. Carlos, R., Amaral, D.C.: Framework for continuous agile technology roadmap updating. Innov. Manage. Rev. **15**(3), 321–336 (2018). ISSN: 2515-8961

An Order Fulfillment Model Based on Lean Supply Chain: Coffee's Case Study in Cusco, Peru

Jorginho Gomez[1(✉)], Gino Alburqueque[1(✉)], Edgar Ramos[1], and Carlos Raymundo[2(✉)]

[1] Ingeniería Industrial,
Universidad Peruana de Ciencias Aplicadas (UPC), Lima, Peru
{u201213220,u201214314,pcineram}@upc.edu.pe
[2] Dirección de Investigación,
Universidad Peruana de Ciencias Aplicadas (UPC), Lima, Peru
carlos.raymundo@upc.edu.pe

Abstract. Coffee is one of the most important cash crops in Peru, a significant source of employment and income, and a great demander of inputs, goods, and services. We synthesized and analyzed the findings to propose improvements and foster the long-term growth of small- and medium-sized cooperatives. Our improvement proposal is based on the supply chain models implemented in the industry. The research also relied on information sources from researchers with experience in the assessment and analysis of the perishable food supply chain in different scenarios worldwide.

Keywords: Lean supply chain management · Agribusiness · Order fulfillment · Supply chain management · Coffee

1 Introduction

According to estimates from the International Coffee Organization, the world demand for coffee will increase from 155 to 205 million bags between 2015 and 2030. Coffee production at the national level is expected to increase from 4.0 to 7.5 million during that period, and thus, Peru will maintain its position among the world's top 10 coffee producing countries. Retailers deal with an important operational challenge in fulfilling online orders while managing their traditional distribution processes [1].

This research has addressed the diagnosis and proposed improvement in demand management with the aim of improving the fulfillment of coffee orders in Quillabamba, Cusco.

Development and continuous improvement is essential; therefore, companies in many developing countries are implementing the Lean methodology. Lean manufacturing [2] is a process improvement methodology used to maximize customer value while eliminating waste and non-value-added (NVA) activities [3], to improve value stream processes using a holistic approach and minimize activities that do not add value to the process [4].

T. Ahram et al. (Eds.): IHSED 2019, AISC 1026, pp. 922–928, 2020.
https://doi.org/10.1007/978-3-030-27928-8_138

Supply chain management (SCM) is deemed to be the integration of upstream and downstream processes [5].

2 Literature Review

2.1 Supply Chain Management

According to the Council of Supply Chain Management Professionals (CSCMP), [6] SCM can be defined as the planning and management of all activities related to sourcing and procurement, conversion, and logistics management. It also encompasses coordination and collaboration with channel partners to meet customer requirements [7]. Furthermore, SCM is a type of integrated management, i.e., a process of coordination and control over material flow, information flow, and capital flow between suppliers and demanders [8].

Experts claim [8, 9] that supply chains have become increasingly dynamic and consequently advise companies to redesign their supply chain more frequently, particularly because the design affects the types of relationships between supply chain partners, their performance measurement systems, and the overall supply chain vulnerability.

2.2 Lean Supply Chain Management

Experts [10, 11] provide an empirically validated instrument to assess lean supply chain management (LSCM) bundles [12] and their impact on key supply chain performance indicators with no parallel in existing literature.

The first objective of their paper was to provide a framework for defining the exact practices and bundles to be considered related to LSCM [12].

Secondly, they proposed to test empirically the positive association between LSCM bundles and supply chain performance [11].

2.3 Demand Management

Demand management is defined [13] as a process that is assigned limited capital and human resources necessary for the overall benefit of the business, and allows the relation between IT and business to be improved. In the model, the author indicated six key demand management mechanisms: strategic planning, portfolio management, delegated authority, financial planning, prioritization, and value management [13].

3 Case Study

3.1 Current Situation

Although San Martin is the region with the highest market participation in coffee production and commercialization, we chose Cusco as the center of study because most of the small- and medium-sized cooperatives are found in this region and we aim to propose a model for their growth and development (Fig. 1).

3.2 Cooperative XYZ, Quillabamba

In the Cusco region, there are about 1600 coffee farmers grouped into small and medium cooperatives. The cooperative XYZ, a medium-sized cooperative with approximately 130 farmers, is located in the La Convención Province.

3.3 Problem Identification

Currently, the main problems experienced by Peruvian coffee producers are low production levels, limited technological innovation, insufficient production infrastructure, limited human resource training, limited business management support, and poor SCM planning.

These problems generate low competitive levels and poor management. The cooperative does have a system for documenting the processes affected and also faces a lack of process standardization. Following meetings with the cooperative XYZ's employees, the main problem we identified was failure to comply with purchase orders over recent years.

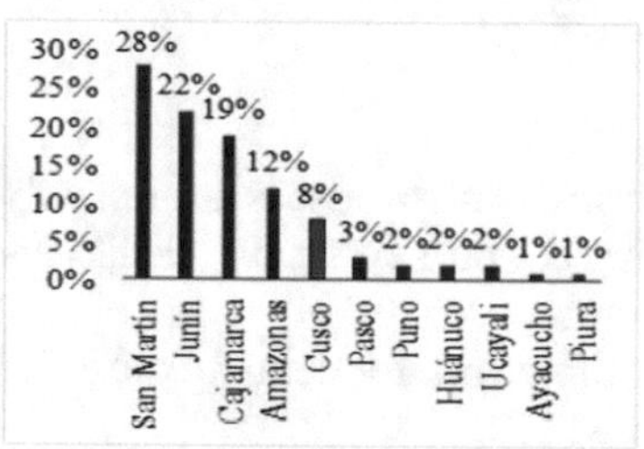

Fig. 1. Participation by region of national coffee production

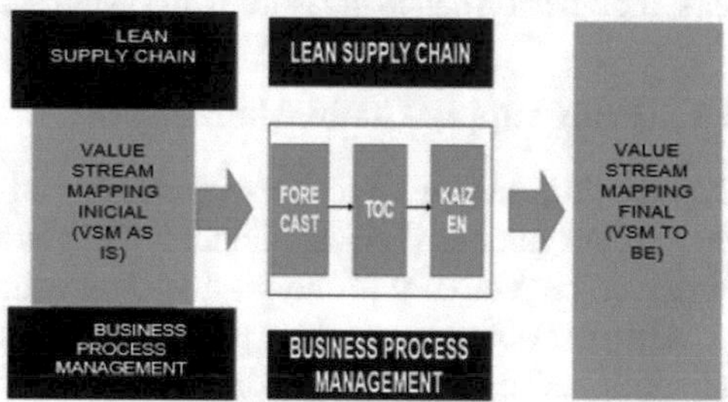

Fig. 2. Proposed model

4 Proposed Model

The proposed model relies on lean supply chain methodology and has been adopted based on solutions implemented in other countries. The main cooperative's goal is to achieve the capacity to compete in international markets, which requires an increase in its service level to improve internal processes (Fig. 2).

LSCM encompasses all links in the supply chain between the customer and the company with the aim of achieving continuous improvement [12] analyzing Procurement, Production, Warehouse Management, and Order Fulfillment areas.

The first step comprises drawing a current state Value Stream Mapping (VSM) to identify value-added (VA) processes and their downtime.

Based on the current state VSM, we identified the processes and times that add value to the processes (Fig. 3).

ITEM	PROCESO	TIEMPO DE OPERACIÓN			% NVA
		TIEMPO VA	TIEMPO NVA	LEAD TIME	
1	LAVADO	60	44	104	42%
2	SECADO	164	91	255	36%
3	INSPECCION	0.1	0.9	1	90%
4	CLASIFICADO	43	22	65	34%
5	TRILLADO	7.8	4.2	12	35%
6	EMPAQUETADO	5.1	0.9	6	15%
7	CARGA	3	1	4	25%
TOTAL		283	164	447	

Fig. 3. VSM times

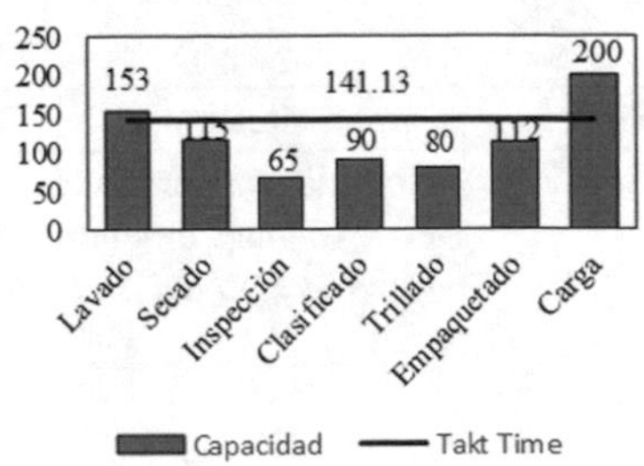

Fig. 4. Takt time

After analyzing the VA and NVA times, we developed a Takt Time plan to establish the pace of production to meet customer demand and eliminate process bottlenecks [14] (Fig. 4).

Using Takt Time, we managed to identify business process bottlenecks to implement the Kaizen tool to reduce downtime and eliminate NVA activities [15] (Fig. 5).

ITEM	PROCESO	LEAD TIME ANTIGUO	LEAD TIME NUEVO
1	LAVADO	104	65
2	SECADO	255	171
3	INSPECCION	1	1
4	CLASIFICADO	65	45
5	TRILLADO	12	8
6	EMPAQUETADO	6	5
7	CARGA	4	3
TOTAL		447	298

Fig. 5. Time reduction

DESCRIPTION	AMOUNT (USD)
Net Income	106.260
Operating Expenses	65.366
Other Operating Expenses	25.300
Other Non-Operating Expenses	3.425
Inventory Valuation	1.518
Accounts Receivable	2.000

Fig. 6. Input data

4.1 Comparison: Current Situation vs Future Situation

Herein, we compared the proposed model with the current model.

Table 1. Current situation vs Future situation

Process	Current situation	Future situation
Plan	-Lack of a strategic plan -Poor communication and interaction among farmers -Very static policies and strategies over time	-Develop a production strategy -Promote farmers' integration -Develop policies for each low and high season
Source	-Irregular delivery times -Empirical production method	-Standardize delivery time -Provide production manuals
Make	-Lack of activity standardization -Lack of training for farmers -Lack of activity documentation	-Standardize the production process -Develop a training program for farmers -Documentation of procedures

(continued)

Table 1. (*continued*)

Process	Current situation	Future situation
Enable	-No indicators are used -Poor training program -Defective products are not recorded	-Develop KPIs -Develop a weekly activity schedule -Monitoring and control of KPIs and farmers

5 Validation

We will validate the proposal using the DuPont analysis to determine the proposal's effectiveness.

5.1 Validation Methodology

The DuPont analysis determines how operations affect the company from an economic standpoint. The DuPont model provides us with four different scenarios: the worst-case scenario, the current scenario, a favorable scenario, and a very favorable scenario, facilitating decision-making and goal setting.

5.2 Methodology Implementation

To implement the DuPont model, we needed various input data such as that shown below (Fig. 6).

Using the data shown in Fig. 7, we calculated the return on investment (ROI), return on assets (ROA), and return on equity (ROE) considering the current 70% service level.

Figure 7 shows the cooperative's current DuPont analysis, i.e., at a 70% service level. As mentioned above, we will consider four different scenarios to compare the indicators.

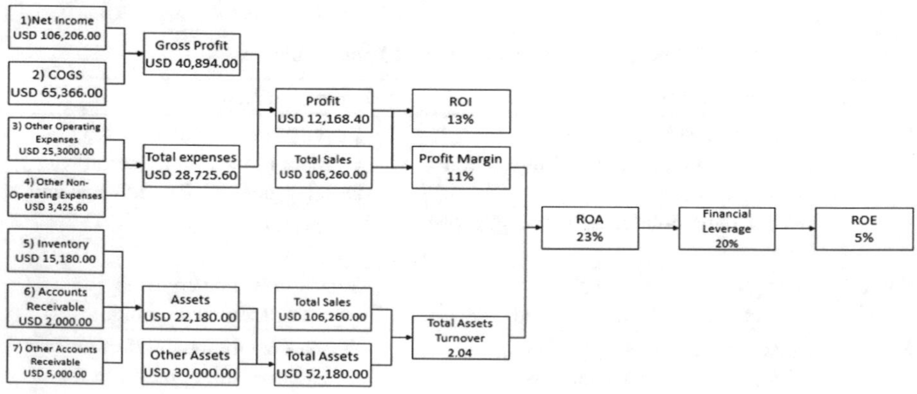

Fig. 7. ROI analysis at a 70% service level

We simulated the DuPont model using 11 different service levels, considering 100% as the most favorable scenario (Fig. 8 and Table 2).

Table 2. Results

Scenario	Service Level	Profit	ROI
1	100%	21%	27%
2	97%	20%	26%
3	93%	19%	24%
4	90%	19%	23%
5	85%	17%	21%
6	83%	17%	20%
7	80%	15%	18%
8	76%	14%	16%
9	73%	13%	15%
10	70%	11%	13%
11	65%	9%	19%

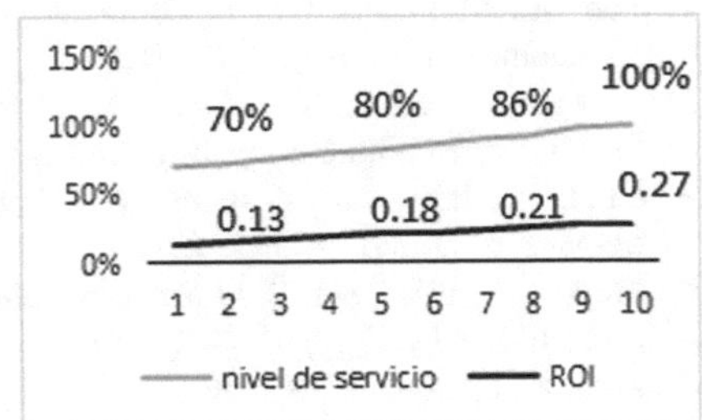

Fig. 8. DuPont analysis at a 70% service level

As shown in Table 1, the increase in the cooperative's profit, as well as in the ROI, ROA, and ROE profitability ratios is directly proportional to the increase in service level. Currently, the cooperative has a service level of 70% with a profit of 11% and aims at a favorable scenario of 85%, which will entail a 6% increase in its profits.

6 Conclusions

Proper SCM allows companies, enterprises, cooperatives, and entities to increase competitiveness and sustainability over time. Furthermore, companies that count on an integrated supply chain will increase their flexibility, entailing business process optimization, reduced operating expenses, and increased service level through responsiveness.

Thc current situation of the coffee supply chain is deficient compared to international competition. The main inefficiencies identified in this supply chain are poor or non-existent planning processes, non-existent process integration, and a lack of process standardization.

First of all, the proposed LSCM model allowed us to show the supply chain processes using the VSM tool. The main issues found were lack of standardization and controls. Then, we performed an analysis of NVA times.

In terms of results, the production cycle time was reduced by 33%, which generates an increase in the company's service level by 10%, since a shorter cycle time allows the company to establish an optimal delivery time to fulfill orders on time. Furthermore, the reduction of the cycle time generates an increase of 15% and 18% in the cooperative's profit and ROI, respectively.

References

1. Tan, A., Mitra, S.: Order fulfilment process for a large online retail in Singapore. Int. J. Autom. Logist. **1**(4), 400 (2015)
2. Ahmad, A., Lee, C., Ramlan, R., Ahmad, F., Husin, N., Abdul, M.: The Hybrid Lean System to Improve Manufacturing Environment (2017)
3. Gupta, S., Jain, S.K.: A literature review of lean manufacturing. Int. J. Manage. Sci. Eng. Manage. **8**(4), 241–249 (2013)
4. Majava, J., Ojanperä, T.: Lean production development in SMEs: a case study. Manage. Prod. Eng. Rev. **8**(2), 41–48 (2017)
5. Zare, R., Chavez, P., Raymundo, C., Rojas, J.: Collaborative culture management model to improve the performance in the inventory management of a supply chain. In: 2018 Congreso Internacional de Innovacion y Tendencias en Ingenieria, CONIITI (2018)
6. Ellram, M., Cooper, C.: Supply chain management: it's all about the journey, not the destination. J. Supply Chain Manage. **50**(1), 8–20 (2014)
7. Autry, W., Rose, J., Bell, E.: An exploratory assessment of the global supply chain framework. Working Paper, University of Tennessee (2013)
8. Wang, L., Chen, C., Zhang, Z.: Research on the hybrid supply chain of lean production and agile manufacturing. Appl. Mech. Mater. **220**, 49–52 (2012)
9. Vanichchinchai, A.: Supply chain management, supply performance and total quality management: an organizational characteristic analysis. Int. J. Organ. Anal. **22**(2), 126–148 (2014)
10. Kennedy, I., Plunkett, A., Haider, J.: Implementation of lean principles in a food manufacturing company. In: Advances in Sustainable and Competitive Manufacturing Systems, pp. 1579–1590 (2013)
11. Anand, G., Kodali, R.: Development of a framework for implementation of lean manufacturing systems. Int. J. Manage. Pract. **4**(1), 95–116 (2010)
12. Tortorella, L., Miorando, R., Marodin, G.: Lean supply chain management: empirical research on practices, contexts and performance. Int. J. Prod. Econ. **193**, 98–112 (2017)
13. Aguilar, I., Carrillo, J., Tovar, E.: Description of the framework's structure of the process of IT demand management. Int. J. Inf. Manage. **37**(1), 1461–1473 (2017)
14. Brioso, X., Murguia, D., Urbina, A.: Teaching takt-time, flowline, and point-to-point precedence relations: a peruvian case study. Procedia Eng. **196**, 666–673 (2017)
15. Santosa, A., Sugarindra, M.: Implementation of lean manufacturing to reduce waste in production line with value stream mapping approach and Kaizen in division sanding upright piano, case study. In: PT. X. MATEC Web of Conferences, vol. 154, p. 01095 (2018)

Strategic Sourcing Toward a Sustainable Organic Coffee Supply Chain: A Research Applied in Cuzco

Elizabeth Carbajal[1(✉)], Jordy Rivera[1(✉)], Edgar Ramos[1(✉)], and Carlos Raymundo[2(✉)]

[1] Ingeniería Industrial, Universidad Peruana de Ciencias Aplicadas (UPC), Lima, Peru
{u201412003,u201411602,pcineram}@upc.edu.pe
[2] Dirección de Investigación, Universidad Peruana de Ciencias Aplicadas (UPC), Lima, Peru
carlos.raymundo@upc.edu.pe

Abstract. This paper is a research on recent studies on the importance of global supply chains from the perspective of farmers and suppliers. For the specific case of coffee in Peru, a survey model was used for assessing integration levels at the cooperatives in Quillabamba, Cusco, which revealed that the lack of alliances between first-level suppliers and farmers/cooperatives significantly affects sustainability in supply chain management. Therefore, this study proposes a strategic sourcing model where an intermediary integrates or strengthens the relationships between smallholders and improves interrelationships within the supply chain, thereby increasing productivity, yielding higher income, and improving product quality.

Keywords: Coffee · Integration · Sustainability · Strategic sourcing · Supply chain

1 Introduction

Among the most important commodities in 2015, coffee is only second after oil, with large volumes of production in Brazil (42%), Vietnam (19%), and Colombia (9%), as a result of the growing demand and competitiveness among producers [1].

One of the largest problems affecting the commercialization of agricultural products occurs at the start of the supply chain (SC) for raw materials, where economic losses are generated by the little integration between cooperatives and farmers and the lack of benefits for the farming community [2]. In addition, farmers' dependence on lenders and large buyers hinders their possibilities for independence, as well as the generation of interdependent collective actions [3].

During this research, we noted the disparity between opportunities and growth for small farmers and their influence on production yields, since smallholdings are critical for ensuring success throughout the SC. Thus, to achieve a sustainable flow of products, services, information, and capital for the generation of value to all stakeholders,

T. Ahram et al. (Eds.): IHSED 2019, AISC 1026, pp. 929–935, 2020.
https://doi.org/10.1007/978-3-030-27928-8_139

an adaptation of the model known as Sustainable Supply Chain Management Integration (SSCMI) [4] is proposed. It is supplemented by another model for making strategic sourcing decision with multiple suppliers based on their association value and using the Kraljic Matrix for selecting potential suppliers, identifying behavioral factors for decision-making, and assessing effectiveness during the development of a value co-creation system [5].

2 Literature Review

2.1 Supply Chain Management

An SC determines the suppliers, the warehouse, and the transfer system required to meet customer demands [6]. Within the SC, the buyer–supplier relationship is important because of the effect of the buyer's investments on the supplier [7] and how this affects the product development routine. Once this relationship is understood, uncertainty costs are reduced [8]. In addition, supplier costs may be reduced through simultaneous sourcing [9].

2.2 Coffee Supply Chain

As per previous studies on Peruvian coffee associations, alliances encourage the application of market access standards, thereby fostering the growth of farming human capital to facilitate certification in addition to improving price premiums for producers [10]. Without external support, barriers such as the lack of financing and asset strengthening exist.

3 Problem Description

In terms of equipment, tools, labor, and fertilizers, farmers report unmanaged relationships with all their suppliers. On the other hand, suppliers report controlled relationships with their customers, mainly at the initiative of the latter, such as cooperatives, associations, or private companies. These companies assign greater value to their customer relationships and seek proper management from them. However, cooperatives and associations only conduct control actions on their customers, either for local or international markets.

3.1 Supply Chain Mapping

Due to the current status of the SC, farmers required up to one month to obtain the financing and seeds necessary to start planting. Further, farmers required up to one week for securing trucks, fertilizers, and hired labor because the process of purchasing supplies or services is not managed. From the implementation of the supply strategy, the cooperative—which assumes a new intermediary role—is expected to manage the farmer's purchase orders so that the new supplier network may source the farmer at a fixed time and cost before planting.

3.2 Indicators and Metrics

Based on a survey used to assess the level of sourcing collaboration at the beginning of the SC, the below mentioned six dimensions were defined. These dimensions were then measured with support from farmers and managers. Logistics: assessed whether the information exchanged was complete, timely, and relevant, as well as the inventory currently available and increases on customer demands. Price Information: assessed private price information, reasons behind price changes, and current information on the prices of raw materials. Exception Management: assessed whether solutions are found to order exceptions and the joint establishment of contractual clauses. General Administration: assessed the scope of the synchronizing decisions between cooperatives and farmers. In addition, this indicator assessed authorizations to provide contract amendment suggestions. Shared Risk: assessed the possibility of granting subsidies if the price of coffee declines, as well as financial assistance in case of crop rejections. Technical Support: assessed training programs and technical assistance. After understanding the study focus for each dimension, the following results were obtained.

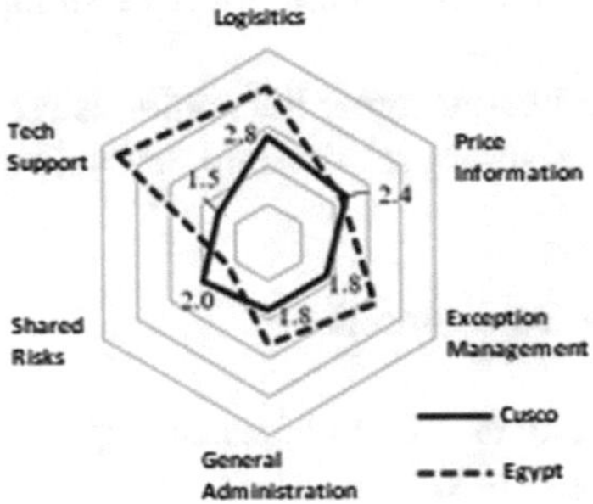

Fig. 1. Collaboration level radar between Cusco (coffee) and Egypt (cotton)

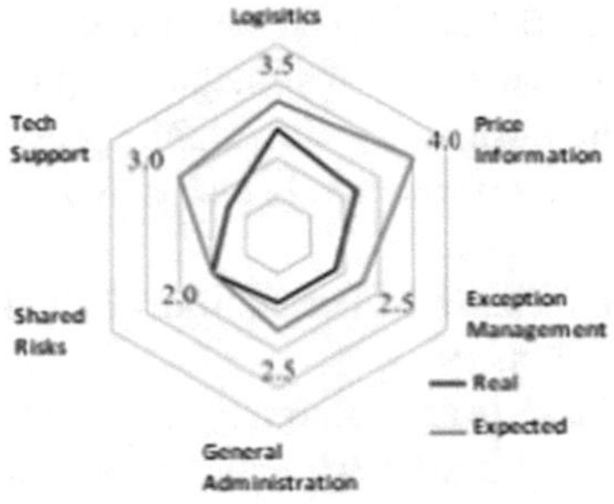

Fig. 2. Expected collaboration level radar in Cusco

As the figures show, the Price Information dimension is low in Cusco because stock exchange prices are taken as a reference. In the same way, the Technical Support dimension value reveals that there are farmers with little agricultural knowledge and inefficient technological knowledge. On the other hand, Egypt's agricultural system is better and has better-trained personnel (Fig. 1).

4 Proposed Model

4.1 Design

The model comprises four phases: identification of expectations from farmers and cooperatives, selection of strategic suppliers, mediation between suppliers and farmers, and measurement of co-creation activities through indicators, in the order presented [11] (see Fig. 3).

4.2 Implementation Phases

4.2.1 Phase 1: Identification of Expectations from Farmers and Cooperatives

Both—the cooperatives and farmers—seek opportunities for innovation and alignment of cost reductions to their income flows [12]. A lead team is defined, and a strategic supply program is implemented to improve coffee crop sourcing efficiencies. Management tools are used to define the most important organizational interests, job profiles, required competencies, and process cards to define the indicators and establishments for the activities at the start of the chain.

4.2.2 Phase 2: Selection of Strategic Suppliers

A supplier analysis is performed on the basis of the [13] purchase portfolio model. In this model, purchase costs and potential growth rates are taken into consideration. The lead team must understand and physically develop a matrix operation using quadrants containing leverage items, non-critical items, strategic items, and bottleneck items, influenced by the risk of supply and its impact on results. Then, suppliers are located in a quadrant, and strategies are defined and placed in the matrix, followed by an establishment of the factors that affect sourcing in terms of the number of suppliers, prices, market, logistics costs, etc.

The selection of strategic suppliers for the supplier–customer integration is critical. On the customer's side, building and maintaining a strong relationship that will result in high performance is also considered important.

4.2.3 Phase 3: Mediation of the Expectations from Suppliers and Farmers

The lead team communicates with the representatives of the selected suppliers to explain the strategy and invite them to the program kickoff. Next, the lead team facilitates a workshop for supplier representatives who agree to participate in the program. The workshop should provide a frank overview of its objectives of commitment to a cultural change from customer command and control to the development of an open collaboration culture with strategic suppliers through a set of behaviors observed by all members.

4.2.4 Phase 4: Measurement of Value Co-creation Activities

A new set of agreements and improvement plans linked to the balanced scorecard is established. Strategic suppliers embark on a series of initiatives, and quarterly progress review workshops are held.

For the monitoring of model execution results, three key indicators—Return Ratio, Quantity Discount, and On Time In Full (OTIF)—are proposed, which assess three different sourcing areas: Quality, Price, and Order Fulfillment [14].

Return Ratio assesses the number of supplies returned to the supplier due to noncompliance with the requirements; Quantity Discount compares purchase prices under normal conditions against the discount received from the established association; and OTIF seeks to assess the number of purchase orders fulfilled on time and in full among the total number of orders [14].

The Strategic Supply model (Fig. 3) seeks the sustainability of the SC in the production of organic coffee. In this case, the model generates positive environmental, social, and economic impacts by reducing defective products, increasing the level of collaboration, and reducing operating costs, respectively.

Fig. 3. Strategic agricultural supply model based on Nudurupati et al. (2015)

5 Results

The results were evaluated using the ProModel Student simulation software version 9.3.1.2081. As the case includes several factors affecting real life, the simulation simplifies these factors to the input and output variables with the greatest impact. On the one hand, the input variable is the productive capacity of the coffee plant with adequate and timely fertilization. On the other hand, the output variables are the defective products generated from the harvesting of the coffee cherries to the piling of green coffee and the quantity of green coffee produced. The simulation will be limited to one plantation hectare with 2,500 plants (aged 4–14 years) during the harvest season—from April to August. Per the simulation results, the production of green coffee increased by 27% and that of defective products reduced by 22%. The production increase is mainly due to the increase in the production capacity of coffee plants and the reduction of defective products is due to better crop fertilization control that prevents waste from fertilizer excess or shortages (Table 1).

Table 1. Simulation results from the ProModel student software

Product	Current (kg/ha)	Proposal (kg/ha)	Delta
Defective Green	1006	1280	+27%
Coffee	642	500	−22%

Table 2. Comparison of the Economic Return on Investment (ROI)

Metrics	Current ($/ha)	Proposal ($/ha)
Direct costs	1070	963
Indirect costs	294	334
Income	1761	2240
Margin	397	943
ROI	29%	73%

On the basis of the results, we develop a comparative table (Table 2) for the Return on Investment (ROI) generated, considering that for the purpose of the proposal, we may reduce direct costs by 10% as confirmed by the Productive Chains department of the Ministry of Agriculture and Irrigation (Minagri) at La Convención. For this, the price offered to farmers is defined at $1.75/kg, according to the La Convención Agricultural Agency (2017) and all the green coffee produced is considered sold. In addition, there is a $40/ha increase in the indirect costs related to the intermediary who manages the strategic supply. With this, the ROI increases from 29% to 73% (2.5 times more) due to higher income and lower total costs (Table 2).

Besides the economic (ROI) and environmental (Waste) indicators assessed, to achieve sustainability, the positive social impact is expected to increase with improvement in the collaboration level, reaching a total rating of 2.9 when raising all dimensions, with the exception of the Risk Shared whose level may be maintained with the applied model (Fig. 2).

6 Conclusions

As a contribution, this research structures a proposal [15] for the reorganization of value chains. In addition, it reinforces the importance of research on sourcing, a level above the producers of raw materials, to achieve a true sustainable SC [16].

By creating dependent relationships among suppliers, the product delivery reliability levels will improve in terms of adequate time, quality, and quantity. The sourcing lead time will be reduced to a fixed and specific period.

At the collaboration-among-suppliers level, suppliers will share and face risks together for better results. They will also be able to access complementary resources, increase productivity, and improve learning and innovation capacity levels.

Cooperative purchases aim to reduce costs in dynamic and competitive environments through price reduction, low management costs, low logistics costs, low transaction costs and greater flexibility in inventories.

References

1. Florêncio De Almeida, L., Zylbersztajn, D.:Int. J. Food Sys. Dyn. **8**(1), 45–53 (2017). www.fooddynamics.org. https://doi.org/10.18461/ijfsd.v8i1.814
2. Zare, R., Chavez, P., Raymundo, C., Rojas, J.: Collaborative Culture management model to improve the performance in the inventory management of a supply chain, In: 2018 Congreso Internacional de Innovacion y Tendencias en Ingenieria, CONIITI 2018
3. Emery, S.B.: Independence and individualism: conflated values in farmer cooperation? Agric. Hum. Values (2014). https://doi.org/10.1007/s10460-014-9520-8
4. Wolf, J.: Sustainable supply chain management integration: a qualitative analysis of the German manufacturing industry. J. Bus. Ethics (2011). https://doi.org/10.1007/s10551-011-0806-0
5. Nudurupati, S.S., Bhattacharya, A., Lascelles, D., Caton, N.: Strategic sourcing with multi-stakeholders through value co-creation: an evidence from global health care company. Int. J. Prod. Econ. **166**, 248–257 (2015)

6. Coskun, S., Ozgur, L., Polat, O., Gungor, A.: A model proposal for green supply chain network design based on consumer segmentation. J. Clean. Prod. (2016). https://doi.org/10.1016/j.jclepro.2015.02.063
7. Fan, Y., Stevenson, M.: Reading on and between the lines: risk identification in collaborative and adversarial buyer–supplier relationships. Supply Chain Manag. **23**(4), 351–376 (2018). https://doi.org/10.1108/SCM-04-2017-0144
8. Agrawal, A., Van Wassenhove, L.N., De Meyer, A.: The sourcing hub and upstream supplier networks. Manufact. Serv. Oper. Manage. (2014). https://doi.org/10.1287/msom.2013.0461
9. Mols, N.P.: Concurrent sourcing and supplier opportunism. Int. J. Procurement Manage. **10**(1), 89 (2017). https://doi.org/10.1504/IJPM.2017.080917
10. Bitzer, V., Glasbergen, P., Arts, B.: Exploring the potential of intersectoral partnerships to improve the position of farmers in global agrifood chains: findings from the coffee sector in Peru. Agric. Hum. Values (2013). https://doi.org/10.1007/s10460-012-9372-z
11. Nudurupati, S.S., Bhattacharya, A., Lascelles, D., Caton, N.: Strategic sourcing with multi-stakeholders through value co-creation: an evidence from global health care company. Int. J. Prod. Econ. **166**, 248–257 (2015). https://doi.org/10.1016/j.ijpe.2015.01.008
12. Dicecca, R., Pascucci, S., Contò, F.: Understanding reconfiguration pathways of agri-food value chains for smallholder farmers. Br. Food J. **118**(8), 1857–1882 (2016). https://doi.org/10.1108/BFJ-05-2016-0194
13. Kraljic, P.: Purchasing Must Become Supply Management (2018). https://hbr.org/1983/09/purchasing-must-become-supply-management. Accessed 19 Nov 2018
14. Chen, Y.-J.: Structured methodology for supplier selection and evaluation in a supply chain. Inform. Sci. **181**(9), 1651–1670 (2011). https://doi.org/10.1016/j.ins.2010.07.026
15. Fayet, L., Vermeulen, W.J.V.: Supporting smallholders to access sustainable supply chains: lessons from the Indian cotton supply chain. Sustain. Dev. **22**(5), 289–310 (2014). https://doi.org/10.1002/sd.1540
16. Pagell, M., Shevchenko, A.: Why research in sustainable supply chain management should have no future. J. Supply Chain Manage. **50**, 44–55 (2014)

Narrative Perception in the Exhibition Space-Studying of Multimedia Technical Device Design

Ming Zhong(✉), Ren-Ke He, and Dan-Hua Zhao

School of Design, Hunan University, Changsha 410082, China
41645397@qq.com

Abstract. Concepts of mixed reality and virtual environments and their supporting technology are brought together and developed further. Continuous multi-sensory interaction and digital description put audiences into a narrative environment and unifies the emotion and the perception of informations. This paper proposes design possibilities that multi-sensory and multi-time-and-space interactions of information lead audiences into the narrative process. The perceptual environment will produce various information feedback based on audiences' behaviors, ultimately achieve unity in physiological emotions and perceptual behaviors and creating a new multi-dimensional perceptual experience space. This perceptual environment will get the audience more involved in the narration with the application of multi-media devices, and audiences will be more relaxed and free to participate in the spatial narrative.

Keywords: Multimedia technology · Spatial narrative · Information perception · Exhibition space design

1 Introduction

In the era of multi-media, Continuous multi-sensory interaction and digital description put audiences into a narrative environment, which provides a interactive connections between audiences and informations. People and their perceptions are the center of the exhibition. In this paper, we will discuss how to utilize the multi-media technology in the exhibition space design to transmit obscure perceptual informations to audiences in spatial narrative space. Multi-media devices can transform fragmented informations into chains of emotional cognition by filtering, organizing and perceiving. The time and the space in narrative make it interlinked with the time and the space of the buildings. Therefore, audiences are vested with more deeper abilities of perception.

2 Multimedia Technologies and the Exhibition Space Design

The exhibition space design is a method based on spatial narrative. During the exhibition, the space involves enormous informations which will attract audiences' attention for a long time. In the era of multi-media, Exhibition space design are strongly

T. Ahram et al. (Eds.): IHSED 2019, AISC 1026, pp. 936–941, 2020.
https://doi.org/10.1007/978-3-030-27928-8_140

supported by multi-media technologies, which has transformed the passive display into a narrative display, and reduce the cognitive resistance by models, space, color, light, sound, experience. Audiences get multi-sensory display informations, extend their senses and thinking, which is the trend of human-centered design.

3 Narrative Features of Multi-technology in the Exhibition Space Design

In the process of narrative exhibition space design, the most difficult part is the inadequacy of 5 sense perceptual experience. Therefore, in the narrative exhibition space design, it is essential to concentrate on the human-machine interface of the multi-media technology and the perceptual experience.

3.1 Multi-sensory Spatial Narrative

Multi-media technologies are key technical support for the multi-level spatial narrative. Traditional space gives the priority to the vision and the original state of exhibition information. Weak interatctions between human and space make audience receive passively. Multi-sensory narrative supported by multi-media technology makes it easier for audiences to communicate with the informations by the body, languages and motions, which is the most natural, acceptable and the simpliest method of interaction. Therefore, multi-media technology will provide the audience with multi-sensory spatial narrative (As shown in Fig. 1).

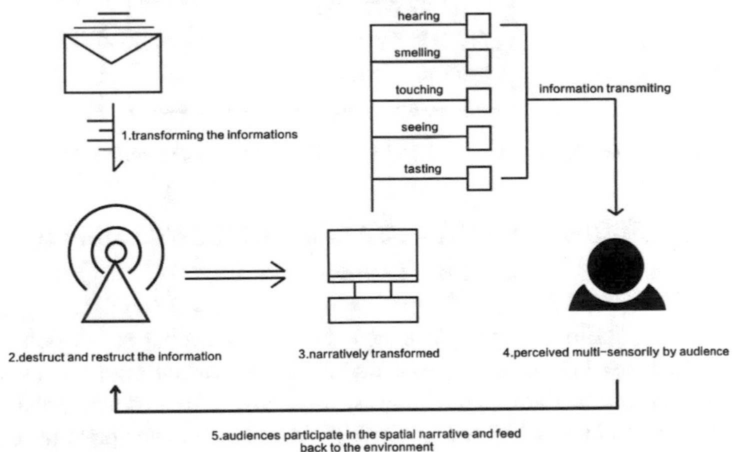

Fig. 1. Multi-sensory multi-channel narrative cognitive flow chart for multimedia devices (All Drawn by Author)

3.2 Perceptual Association

The perception is divided into sensation and perception. Sensory organs, as receivers of external information, receive external information and produce sensation. Because of internal differences and objective conditions, different experiences will be produced. Perceptual experience will be upgraded to level of thinking and experiencing in the spatial narrative, which is defined as perceptual association.

3.3 Information Interaction of Multi Time and Space

The single form of perception leads to the unnatural experiences and the cognitive impairment. However, using multi-media software and hardware effectively, display spaces will become multi time and space and multi-level narrative possibilities unable audiences to perceive the original cognitions correctly and enhance people's desire for the perceptual experience (As shown in Fig. 2). Multi time and space generates cognitive experiences in the spatial narrative continually and extensively.

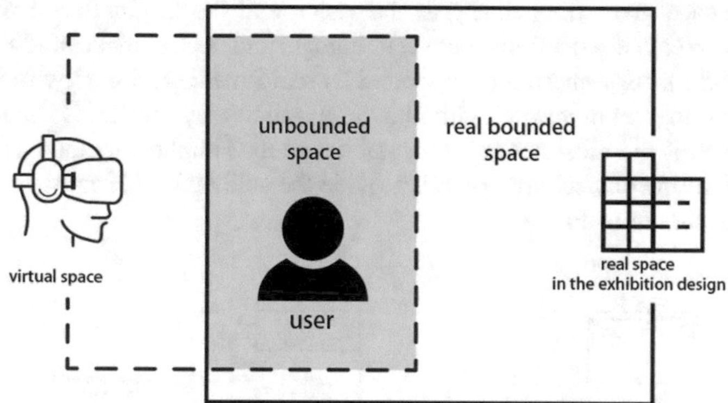

Fig. 2. The overlap of virtual and realistic space (All Drawn by Author)

4 Narrative Information Transmission of the Situational Perception and Perceptual Interaction

Variance between 2-dimensional plane and the 3-dimensional space will cause the cognitive differences [7]. In the spatial narrative, the virtual environments transfer physical informations to the organ of senses; five sense superposition generates sense of identity which are inspired by the narrative plots and experience patterns in audience subconscious; eventually, rational cognitions confirm the common between specific virtual and corresponding real environment, then physically feedback to the virtual environment (As shown in Fig. 3); the information interaction cycle emerges, multiply and derive into an enormous information net to make the situational perceptual interaction happen.

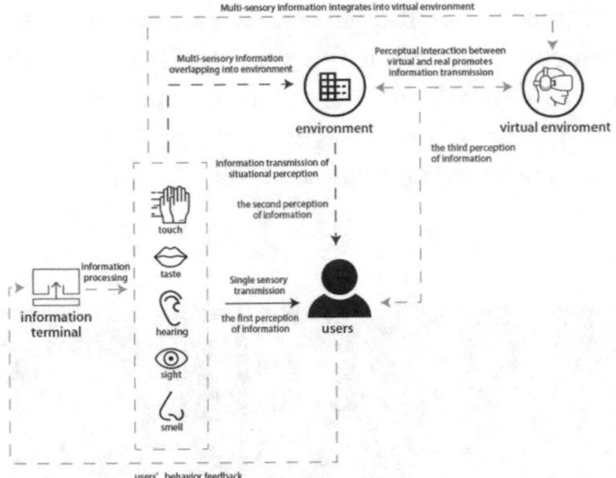

Fig. 3. Flowchart of three information transmissions (All Drawn by Author)

5 Practical Case - the Design Project of Exhibition Hall of CITC Electronic Information Headquarters

In September 2014, at the invitation of the client and the Saudi government, our group began to design the electronic information headquarters exhibition hall and the interactive multi-media of the Saudi Telecommunications and Information Technology Commission (the CITC). This project took multi-media technologies as the core. We concentrated more on audiences' perception and perceptual interactions trying to build a narrative space.

5.1 Perceptual Experience of "Community of Life"

We utilized the body language to interact perceptually with multi-media devices, and then destructed and recombined the spatial narration in accordance with the optical joint sensing. Multi-sensory interactive experiences will be triggered by human/body perceptual informations through infrared somatosensory capture equipment. Individual behavior will be connected with the perception of spatial narrative by dynamic capture (As shown in Fig. 4).

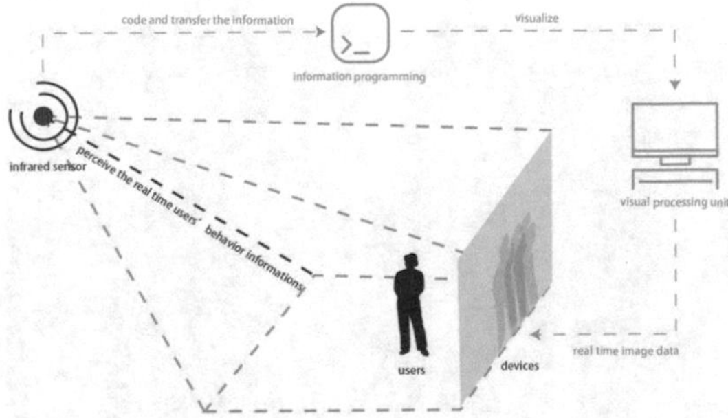

Fig. 4. Schematic diagram of the theoretical architecture of the time channel device (All Drawn by Author)

In this project, the space take the audiences' behaviors as the receiving organ of multi-sensory communication between the space and individuals, and then feedback perceptual informations of individual behaviors to the screen. With the multi-media technology, the human figure outline image can be produced, which achieves the real interactions between the audience and the space as a "community of life" (As shown in Fig. 5).

Fig. 5. Time channel composed of two layers of infrared somatosensory capture devices (From: Gallery in CITC Electronics Headquarter)

5.2 The Experience Space Mixed with the Real and the Virtual

Steve Mann, Canadian internationally recognized pioneer of the wearable computer technology, suggests:the feature of wearable devices is that it belongs to the individual, controlled by the wearer and has the continuity of the operation and interaction. Therefore, [10] this equipment will produce maximum various perceptual information feedback to satisfy multi-level experience demands. In the CITC project, the time machine of the future zone was built for the audience to perceive the touchable future of the telecommunications industry with multi-sences, the audience's real-time information feedback to the environment and change the environment with multi-media devices like VR equipments (As shown in Fig. 6). In the meantime, the motion datas of audiences will be transformed into a real-time visual interface.

Fig. 6. Time machine (From: Gallery in CITC Electronics Headquarter)

6 Conclusion

Multi-media technologies diversify methods of spatial narrative and make the interactive perceptual informations transmit more efficiently. Display informations are multi-dimension and multi-sensory information interaction. The multi-media technology inspire audiences to participate in spatial narrative. Multi-media devices increase the channels of information perceptions and promotes the audiences' perceptual cognition. In the meanwhile expression of the of the narrative informations and exhibition space design is more complete, and unifies the human emotions and the informations perception.

References

1. Li, Y., Wu, X.: Spatial narrative study of Jiangnan sericulture museum in Wuxi. **2019**(03), 123–125 (2019). Dazhongwenyi
2. Zhou, P.-C., Ye, P.: Deep experience of digital interactive display design based on the general narrative concept. Packag. Eng. **39**(04), 01–06 (2010)
3. Jin, X.-Y.: Study of contemporary museum narrative. Zhejiang University, Hangzhou (2013)
4. Ascott, R.: Art and transformation technology. CANS **2002**(10), 01–03 (2002)
5. Li, Q.-R.: The narrative construction of memory formulation in urban public space. Jiangnan University, Wuxi (2009)
6. Zhang, L.: Analysis on structure of virtual experience design. **2009**(11), 84–85 (2009). Zhuangshi
7. Tang, S.-M., Gu, H.-Z.: Multimedia virtual reality—sensory experience design of online brand stores. Art Des. **2008**(09), 17–19 (2008)
8. An, L.-L.: Graphics Language and Digital Media Studies. Guangxi Normal University, Guilin (2010)
9. Tan, Y., Fan, X.: Research on the characteristics of interactive display space based on perceptual experience. **2019**(02), 73–74 (2019). Dazhongwenyi
10. Zeng, L.-X., Jiang, X., Dai, C.-Q.: Design principle of gesture interaction in the wearable device. Packag. Eng. **36**(20), 135—138 + 155 (2015)

Debunking Limitations Hindering Continuing Professional Development Imperatives in South African Construction Industry

Idebi Olawale Babatunde[1(✉)], Timothy Laseinde[2], and Ifetayo Oluwafemi[3]

[1] Department of Operations Management, University of Johannesburg, Johannesburg, South Africa
waleidebi@yahoo.com

[2] Department of Mechanical and Industrial Engineering Technology, University of Johannesburg, Johannesburg, South Africa
otlaseinde@uj.ac.za

[3] Postgraduate School of Engineering Management, University of Johannesburg, Johannesburg, South Africa
ijoluwafemi@uj.ac.za

Abstract. Continuous Professional Development (CPD) is a universally accepted mechanism that bridges the gap between formal higher education knowledge transfer and work-place competencies. CPD reinforces existing knowledge and facilitates acquisition of new skills, and intricate competencies required for innovation. The study investigates the challenging factors South African Construction Industry (SACI) practitioners encounter which limits participation in CPD rated events and formal trainings. The study significantly contributes to the identification of the barriers to the subject matter and the recommendations that emerged may be explored by SACI practitioners towards becoming globally competitive. The research involved extensive review of existing literature to identify CPD participation challenges, and construction industry practitioner's competency challenges. It further analyzed secondary data obtained from industry actors as means of benchmarking industry expectations for SACI. The research identified financial cost of CPD, workload of personnel, and lack of suitable CPD activities as prevalent challenging factors in SACI. The study recommends that organizations need to support staff by financing CPD training costs, and consciously creating work-recess for staff to attend CPD trainings.

Keywords: Continuous professional development · CPD · Competencies · Participation · South African Construction Industry · Work place competency

1 Introduction

Continuous professional development (CPD) can be referred to as a process of on-the-job continuous learning, which entails both formal and informal forms of training [1–3]. CPD has become a widely used tool adopted by business organizations globally to consistently equip their personnel with improved skills, knowledge and expertise required to be competitive [4].

T. Ahram et al. (Eds.): IHSED 2019, AISC 1026, pp. 942–946, 2020.
https://doi.org/10.1007/978-3-030-27928-8_141

The involvement of CPD in an organization assists in; bridging the gap between institutional learning and workplace competence, focus learning, as well as dealing with new and complex methodologies that ensure competent delivery of an organization's goods and services at an utmost standard that safeguard business success as a going concern [4, 5].

CPD is key in an ever-evolving industry like the construction industry, to assist participants to upgrade, broaden and deepen their knowledge and skills constantly to match new technologies and methodologies throughout their career life [4, 5].

It has been observed that despite the various organized CPD programmes by the various professional bodies and accredited CPD providers in the South African construction industry (SACI), the level of participation by the industry professionals are still low. This paper aims to identify the challenges attributed to the low participation in CPD in SACI and suggestions of how to make it attractive for high patronage.

2 Literature Review

Act No. 43 of 2000 that establishes the Council for the Built Environment (CBE) In South Africa, also empowers the council to coordinate the administration of the six statutory professional bodies namely:

- The South African council for the Quantity Surveying Profession (SASQSP)
- The South African Council for the Architectural Profession (SACAP)
- The South African Council for the Landscape Architectural Profession (SACLAP)
- The Engineering Council of South Africa (ECSA)
- The South African Council for the Property Valuers Profession (SACPVP)
- The South African Council for the Project Management Professions (SACPCMP)

CBE in conjunction with theses professional organisation under the Built Environment (BE) have the mandate through the Act to develop conditions relating to professional competence by re-education and training through CPD [6]. Since, through the Act, the professionals are given a legal and exclusive rights to practice the profession in South Africa, CPD is a responsibility on the part of the professional bodies to organise competency focused programmes to protect public interest against incompetency of professionals. In response to the above responsibility CBE developed and published in 2007 a CPD policy document for guidance for development of CPD policies by the professional bodies for the purpose of development of professional competence in order to deliver quality service in the BE and as well contribute to South Africa national development goals. The professional bodies have put together CPD policies and have accredited providers and indeed have organised CPD programmes. The common CPD techniques that have been employed and encouraged by the BE professional bodies and providers include; workshops, seminars, conferences, colloquiums, educational designated short courses, mentoring, further studies, teaching, facilitation of training programmes and research and publications among others [7].

CPD programmes have been found to have positive impacts such as: deepening and broadening the knowledge base of the participant, aligning participants knowledge to practice, improves participants performance, creates participants job confidence and invariably participants overall competence [1]. It must however, be noted that CPD structured towards participants field and job skills achieve such results.

Despite several CPD programmes that are organised by the professional bodies and accredited service providers, tailored towards gaining new competencies to match evolving technologies and methodologies in the industry, there are yet an increased in cry of dissatisfaction of SACI's performance [4]. This therefore, questions the positive impact of enhancing improved competencies as expected by the CPD activities.

Research has identified low participation in CPD programmes as one of the major shortcomings of CPD effectiveness. Further to this, researchers have attributed the non-compulsion of CPD for all participants in the industry except for the registered persons, non-specific framework, lack of personal interest, lack of funds and lack of time among other issues, as attributes contributing to the CPD low participation [1, 4, 8].

The resonance from literatures reviewed, suggested a relook at CPD implementation strategies by the BE professional bodies and business organisations within the industry in order to attract high participation in CPD programmes. The following were the suggested construct, which could form the basses of CPD implementation strategies: creating ways of attracting personal interest in CPD programmes, making CPD programmes affordable, making CPD content gap and competency driven, CPD to be mandatory for all employees in the industry, business organisation to make out time for their employees to attend CPD, CPD programmes to be tailored towards targeted audience within an accessible location [1, 4, 8].

3 Research Methodologies

Qualitative research approach was used in this study, contents of existing literatures were reviewed to identify the importance SACI legislation placed on CPD within the industry including policies and acceptable techniques of CPD administration. The review also considered the major challenge causing ineffectiveness of CPD in the SACI and the underlying factors triggering the challenge. The study concluded with recommended possible intervention that could increase CPD participation in SACI. The literatures reviewed, were studies conducted within the South African construction industry in particular and other South Africa industries in general.

4 Research Findings

Findings from the research identified low participation in CPD as a major cause of CPD ineffectiveness in SACI. While the importance of CPD is not downplayed within the industry, the underlying reasons for low participation as deduced form literature are:

- Non-compulsion of CPD for non-professional registered persons: the number of non-registered professionals working the SACI outnumber the professional registered personnel and they felt no obligation to attend CPD programmes hence affecting low participation from the industry personnel

- Non-specific Framework: the existing CPD framework are generic and are not coconsciously structured to target audience, hence unattractive to some
- Lack of personal Interest: Naturally, with people that do repeated job activities, they feel comfortable with their knowledge and skills and have no urge to learn new ways, thereby lacks interest in CPD
- Lack of Funds: There are general concern within SACI that the cost of CPD programmes are too expensive and most importantly for the low-income earners and new entrants into the industry. Further to this most business organisations do not support their employees financially to attend CPD programmes
- Lack of Time: Employees in SACI have the challenges of getting time off their busy schedule to attend CPD programmes because their employers are not willing to give them time off work. This particularly have grave impact on CPD participation in SACI

The recommended interventions to overcome the observed challenges low CPD participation in SACI are: -

- Creating Personal Interest: CBE professional bodies must create continuous awareness of the benefits of CPD and also create incentives for participation.
- Making CPD Affordable: for increased participation CPD programmes must be affordable to individuals. One of the suggestions for affordability is to subsidise the cost and sourcing governmental and organisation funding for CPD courses.
- Creating Appropriate Time: Organisations must be made aware and encouraged to prioritise CPD attendance for their employees.
- CPD for All: All employees within the SACI should be mandated to participate in CPD programmes, though duration and content may vary for different categories.
- CPD Programme Content: CPD programme must not just be generic but target driven based on consistent competency needs analysis.
- CPD Framework: CPD methods to be engaged must consider the affordability of the target audience and be organised at a close and accessible location.

5 Conclusion

CPD is an important component of re-education apart from institutional learning to acquire work place knowledge skills and competencies, particularly in an ever-evolving industry like the construction industry. The importance of CPD transcends just equipping individuals with right skills and competencies, but also strongly link to organisations business performance and success.

Despite the key role CPD places in the equipping individuals' competencies and organisation success, low participation from SACI was observed.

The responsibility to make CPD attractive rests majorly on CBE and the professional bodies in SACI saddled with the mandate of organising CPD policy and programmes. One critical issue that stand out from the research that will require further probing, is the part played by SACI employers in discouraging CPD participation, by their attitude of not supporting their employees with funding and time availability.

6 Recommendations

CBE and professional bodies within SACI saddled with the responsibility of formulating CPD policies and implementation strategies, must with urgency integrate the recommended interventions from the study to attract high patronage.

Acknowledgement. The authors wish to acknowledge University of Johannesburg as well as the National Research Foundation of South Africa (NRF) for funding the publication process through the Thuthuka funding mechanism award number TTK180805351820.

References

1. Lombard, A.: Research findings and implementation challenges of the continuing professional development (CPD) policy for the social work profession in South Africa. Soc. Work/Maatskaplike Werk **46**(2), 121–143 (2010)
2. Collin, K., Van der Heijden, B., Lewis, P.: Continuing professional development. Int. J. Train. Dev. **16**(3), 155–163 (2012)
3. Goodall, J., Day, C., Lindsay, G., Muijs, D., Harris, A.: Evaluating the Impact of Continuing Professional Development (CPD). Department for Education and Skills London (2005)
4. Kwofie, T.E., Mpambela, J.S.: Challenges facing construction professionals' compliance with continuing professional development (CPD) in South Africa (2017)
5. Frank, L.D., Greenwald, M.J., Winkelman, S., Chapman, J., Kavage, S.: Carbonless footprints: promoting health and climate stabilization through active transportation. Prev. Med. **50**, S99–S105 (2010)
6. SACPCMP: Continuing professional development (CPD) policy framework (2017)
7. SACPLAN: Continuous professional development (CPD) policy and procedure (2018)
8. Mummenthey, C., Du Preez, R.: Implementing efficient and effective learnerships in the construction industry. SA J. Ind. Psychol. **36**(1), 1–11 (2010)

Designing a Procurement Management Model to Reduce Project Delays in a Hydraulic and Automation Systems Company

Melanie Vereau[1(✉)], Jose Rojas[1(✉)], Daniel Aderhold[2(✉)], Carlos Raymundo[2(✉)], and Francisco Dominguez[3(✉)]

[1] Ingeniería de Gestión Empresarial, Universidad Peruana de Ciencias Aplicadas (UPC), Lima, Peru
{u201311427, jose.rojas}@upc.edu.pe

[2] Dirección de Investigación, Universidad Peruana de Ciencias Aplicadas (UPC), Lima, Peru
{daniel.aderhold, carlos.raymundo}@upc.edu.pe

[3] Escuela Superior de Ingeniería Informática, Universidad Rey Juan Carlos, Madrid, Spain
francisco.dominguez@urjc.es

Abstract. This study focuses on the factors that cause project delays in an industrial hydraulic company. More specifically, the article also addresses the development of on-demand engineered products and the impact of the procurement process on delays. As part of the investigation, the authors assessed the historical data of the projects completed in 2017 before determining the critical path for each project in an attempt to identify which project stages may be improved. After further assessing the resulting data, the planning and procurement procedures were deemed as requiring improvement. Next, the article proposes a training plan and new procedures based on the Project Management Body of Knowledge guidelines for Engineer-to-Order projects. In conclusion, the results showed that delays were effectively reduced after implementing these new procedures in both Project Management stages.

Keywords: Delays · Project Management · PMBOK · ETO · Hydraulics

1 Introduction

In Peru, the Hydraulic and Automation sector approximately moves US$80 million, with mining being one of the top areas consuming products and services (Rumbo Minero, 2018), closely followed by industrial companies, such as steel mills and/or the food industry.

In this light, delivery times and delays become critical. Moreover, the projected growth rate for this sector is quite significant as, according to the Ministry of Energy and Mines, the country currently holds a portfolio of 48 projects, representing investment amounts of US$51,102 million. However, the latest BCR (Banco Central de Reserva del Perú) Inflation Report for the month of December reports that there

T. Ahram et al. (Eds.): IHSED 2019, AISC 1026, pp. 947–952, 2020.
https://doi.org/10.1007/978-3-030-27928-8_142

were only three important mining extensions, as well as just three new mining projects, in the entire 2018–2019 period.

Therefore, project performance must be effectively measured to reduce delays, thus increasing efficiency and productivity while preventing losses. Very few studies, such as the study proposed by Mello, Strandhagen, and Alfnes [1], focus on Peruvian companies operating in the Peruvian sector. These studies are usually single projects with different specifications for each client, known as Engineer-to-Order (ETO), which had been developed as case studies for the shipping industry. Hence, the research aims to develop a Procurement Management model to reduce delays in ETO projects.

2 Theoretical Framework

According to Ghiasi, Kaivan, Arzjani, and Arzjani [2], most of the development projects carried out worldwide face a 50% delay. Likewise, as project development is expected to generate economic growth for companies, communities, or countries, deadline or schedule violations in large-scale projects may cause irreparable damage. These damages may translate into higher costs, loss of project relevance, customer dissatisfaction, economic losses, and opportunity costs, among others. Consequently, a research study conducted in Saudi Arabia not only identified delay causes for two case studies but also proposed linear programming models. The methods used by Ghiasi [2] were Failure Mode and Effects Analysis (FMEA) and Fuzzy Risk Priority Number (FRPN). The FMEA method served to determine root causes while also being used as a prevention technique for the identification of potential failure factors, whereas, the FRPN technique prioritized the causes identified, assigning each cause an impact value or weight in the project. In this case study, the authors interviewed construction professionals and specialists with extensive expertise on understanding the context and factors of the field to identify delay causes. Further, Pham and Hadikusumo [3] introduce the Engineering, Procurement, and Construction business model, a contract in which the contractor becomes accountable for designing and building the project, as well as for procuring the materials required. The main advantage of this system is that it reduces project schedules as a single party undertakes the main phases of the project. However, despite this theoretical benefit, delay continued to arise in petrochemical projects in Vietnam. In an attempt to find and assess delay causes, the authors relied on two methodologies and collected information through Data Triangulation, which involves time, space, and persons. After the data were properly collected and assessed, the Cross-Case analysis was used to compare several cases against each other. Another project business model is the ETO. Mello, Strandhagen, and Alfnes [4] claim that the ETO method is mostly used by companies that produce capital goods, such as equipment, production lines, and buildings. These companies do not stock finished products as all products are developed as per client specifications; therefore, all materials and components will be exclusively procured for each particular project. In other words, the company does not keep any materials and components in stock either. Finally, Mello [4] assess the project delays reported by ETO companies to subsequently evaluate and understand the role of coordination in their corresponding mitigation.

3 Contribution

The Improvement proposal has been designed based on the following components, as collected from the articles reviewed:

Project Management Body of Knowledge (PMBOK) Procurement Management Guidelines: As per the quintessential Project Management guidelines, the PMBOK's Knowledge Management area will be used as a frame of reference.

ETO Coordination Mechanism: Mechanism used by projects aimed at producing capital goods, such as equipment, production lines, and buildings. No finished products are stocked as all products are developed as per client specifications. Therefore, delays strongly impact both the supply chain and the order fulfillment.

Critical Path Method: An important Project Management tool for effectively scheduling activities, monitoring their development as per the project's timetables, and reducing delays to a minimum.

FMEA: An analytical prevention technique used for (A) identifying and prioritizing potential failure modes in a system, product, or process; (B) defining and implementing measures for the elimination or mitigation of potential failures; and (C) collecting assessment results for future reference (Fig. 1).

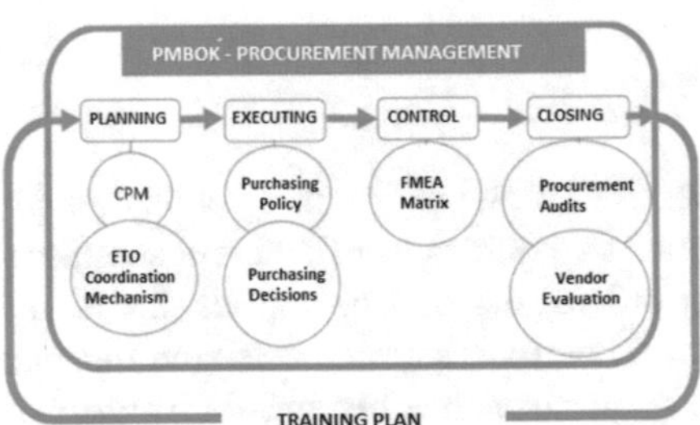

Fig. 1. Improvement proposal model

	Month		
USD 145,665	0	1	2
Sales		USD 70,298	USD 75,368
Subtotal		**USD 70,298**	**USD 75,368**
OPEX			
Training	-USD 2,000		
Mechandise Costs		-USD 35,207	-USD 4,103
Labor			-USD 7,789
General Costs		-USD 325	-USD 23,378
Subtotal - Costs	-USD 2,000	-USD 35,531	-USD 35,269
Gross Profit	-USD 2,000	USD 34,766	USD 40,098

Fig. 2. Results from improvement proposal

This Procurement Management model, which uses PMBOK's Procurement Management as a general framework, is divided into the following five phases:

3.1 Phase 0: Training

This first phase comprises a training plan addressing the main topics engaged by the model. The covered topics are Project Management as per PMBOK guidelines, the ETO project model, IT Tools and Systems, and Procedures and Policies. In this phase, a training schedule is established and total training costs are determined.

3.2 Phase 1: Planning

To solve the problems previously identified, the Critical Path Method project management technique and the ETO Supply Chain coordination mechanism are used [5]. At this stage, a Gantt chart is developed for the project. Additionally, the project's critical path is identified, as well as the key factors for each project stage for the implementation of the coordination matrix. Herein, the coordination mechanism is used following the model proposed by Mello [1] (Table 1).

Table 1. Coordination mechanism according to effort level

Coordination effort	Mechanism	Pre-project	Engineering	Production
−	Standard	Compatibility	Rules of design, technical knowledge	Production flexibility
↓	Plans	Development schedule as per scope	Approvals	Exception plan
	Mutual adaptation	Coordination committee	Production design reviews	Engineering changes
+	Teams	Joint development	Joint teams	Transition teams

3.3 Phase 2: Execution

In this stage, all purchases are performed as per the following company policy guidelines: Purchase decisions must be based on at least two quotes from different suppliers. As per company policy, payment terms must be at least net 30. For imports, Incoterms and means of transportation must be established. Therefore, supplier selection must be as per the following factors: production capacity, performance history, delivery reliability, on-time delivery, continuity, flexibility, quality, technical capability, quick response time, economic and financial conditions, and price/method of payment.

3.4 Phase 3: Control

Control will be FMEA-based, considering scenarios, which may arise throughout the procurement process to subsequently assess the severity, occurrence, and detection of each potential failure and their effect on project deadlines. Moreover, response actions are set forth.

Some possible potential problems may be as follows: errors when creating the bill of materials, failure to submit technical specifications to suppliers, factory delays, oversized merchandise, flight delays/rescheduling, customs delays, packages or merchandise damaged or in poor condition, wrong materials and/or products, and costing and/or administrative process delays.

Then, these problems are rated based on the following criteria: severity, occurrence, and detection.

With these criteria, each potential problem is scored for the calculation of the Risk Priority Number (RPN).

$$RPN = Severity \times Occurrence \times Detection \tag{1}$$

Since all ratings are on a 1–4 scale, the RPN provides the priority assigned to each problem from 1 to 64. If the RPN exceeds 16, preventive and corrective actions must be implemented as quickly as possible for this particular problem.

3.5 Phase 4: Closing

Finally, this phase includes procurement audits to determine the compliance level with the guidelines previously established in the Execution stage. Further, this stage includes an evaluation of the suppliers engaged in the project, which will be rated as per the following criteria: contract compliance, goods, work or service specifications quality meeting the project's quality requirements, supplier responses consistent with supervisor requests, requirement response times suited to our needs, Just-in-Time delivery (suppliers of goods), and deadline compliance (service providers).

4 Validation

The proposed model has been implemented as a case study in a company that develops hydraulic and automation systems. This company mainly produces Power Supply Units and Piping Systems for hydraulic drives.

The main problem reported by this company, which for the purposes herein shall be known as Company X, was delays in 100% of its projects, with an average delay of 64 days per project. Likewise, of all the projects completed, only 17% managed to reach or exceed a 40% profit margin, which is the minimum margin expected by the Company. The improvement proposal was applied to Project No. 0446, which focused on the development of a hydraulic piping system for a Peruvian mine. The entire project was developed on-site at client's work site.

The main project details are as follows:

- Sales Value: USD 145,665
- Scheduled Term: 55 days
- Expected Minimum Margin: 40%

Figure 2 shows the results obtained at the end of the project. The project required 56 days to complete and yielded a profit of US$72,864, which translates into a 50% profit margin.

5 Conclusions and Recommendations

- The coordination mechanism matrix identifies the stakeholders in the main project stages, as well as the appropriate coordination mechanisms required for the project's key factors.
- Any topics and areas that may require reinforcement must be properly identified according to the roles defined for each profile for the development of an effective training plan.
- The tools proposed were able to reduce delays from an average of 64 days to 1 day, as per the plan developed for Project 0446.
- At the end, the project yielded a 50% profit margin, which exceeds company expectations in 10%.
- Therefore, the proposed matrices and tools are able to provide value to the organization in terms of process improvement.

References

1. Mello, M.H., Strandhagen, J.O., Alfnes, E.: The role of coordination in avoiding project delays in an engineer-to-order supply chain. J. Manuf. Technol. Manag. **26**(3), 429–454 (2015). https://doi.org/10.1108/JMTM-03-2013-0021
2. Ghiasi, V., Kaivan, E., Arzjani, N., Arzjani, D.: Analyzing the causes of delay in development projects by fuzzy analysis. Int. J. Qual. Reliab. Manag. **34**, 1412–1430 (2017). https://doi.org/10.1108/IJQRM-08-2016-0134
3. Pham, L.H., Hadikusumo, H.: Schedule delays in engineering, procurement, and construction petrochemical projects in Vietnam. Int. J. Energy Sect. Manage. **8**(1), 3–26 (2014)
4. Mello, M.H., Strandhagen, J.O., Alfnes, E.: Analyzing the factors affecting coordination in engineer-to-order supply chain. Int. J. Oper. Prod. Manag. **35**(7), 1005–1031 (2015)
5. Zare, R., Chavez, P., Raymundo, C., Rojas, J.: Collaborative culture management model to improve the performance in the inventory management of a supply chain. In: 2018 Congreso Internacional de Innovacion y Tendencias en Ingenieria, CONIITI 2018 - Proceedings (2018). 8587073

On-Demand Warehousing Model for Open Space Event Development Services: A Case Study in Lima, Peru

Christian Balcazar[1(✉)], Christian Chavez[1(✉)], Gino Viacava[1(✉)], Edgar Ramos[1(✉)], and Carlos Raymundo[2(✉)]

[1] Ingenieria Industrial, Universidad Peruana de Ciencias Aplicadas (UPC), Lima, Peru
{u201413475, u201410532, pcingvia, pcineram}@upc.edu.pe
[2] Dirección de Investigación, Universidad Peruana de Ciencias Aplicadas (UPC), Lima, Peru
carlos.raymundo@upc.edu.pe

Abstract. This study focuses on the idle space within a warehouse which arises due to an inadequate identification of optimum materials, hindering their storage and reuse. Herein, an on-demand warehousing model is developed based on knowledge management, ideal design of warehouse facilities, and continuous monitoring of warehouse processes and activities for achieving an adequate material flow, cost minimization, high customer service levels, and better working conditions. Results show that the developed model reduced the idle warehouse space and operating costs by 72.14% and 58.55%, respectively.

Keywords: On-demand warehousing · Knowledge management · Systematic layout planning methodology · 5S philosophy · Good storage practices

1 Introduction

Warehouses are a point in the supply chain where the product stops, even if briefly, and can be accessed physically. Herein, materials in the warehouse are in constant movement and interaction owing to the use of on-demand business model [1]. Therefore, idle space becomes a problem owing to the improper storage management and high number of square meters used to store waste [2]. In warehouse management, making the right decisions is paramount since several performance aspects, such as material handling costs, space leasing costs, and warehouse storage capacity, may get affected [3].

Since the idle space increases the operating costs and decreases profitability, this study aims to improve the warehouse management for reducing warehouse operating costs. Considering that the warehouse is leased per square meter under an annual lease renewal agreement, which can be amended to the required number of square meters for generating significant savings, significant attention should be paid toward decreasing the leasing costs.

T. Ahram et al. (Eds.): IHSED 2019, AISC 1026, pp. 953–959, 2020.
https://doi.org/10.1007/978-3-030-27928-8_143

This study further attempts to develop knowledge management for facilitating the integration of the selected solution techniques, develop a solid methodological foundation, and achieve sustainability in long-term warehousing [4]. Herein, the aim is to implement an on-demand warehousing model by blending knowledge management, systematic layout planning (SLP) methodology, 5S philosophy, and good storage practices (GSPs).

2 State of the Art

2.1 On-Demand Warehousing

The on-demand warehousing or warehouse management under stochastic demand, wherein behavior is nondeterministic, has been addressed in previous studies [5, 6], the demand varies over time and on the basis of the customer portfolio of an organization. These studies have solved warehousing issues, such as on-demand warehouse configuration design and resource allocation streamlining for warehouse management [5], to optimize resource location areas, reduce operating costs, and increase process agility.

2.2 Inventory Management

Inventories have been defined as company assets that gain advantages for facilitating the continuous offering of the company's products or services [7]. To streamline the inventory management, several authors have proposed different solutions for securing the inventory numbers needed at the required time. Several inventory management methods determining the exact number of materials available in the warehouse to minimize total costs have been proposed. This technique can be applied in scenarios where the demand is predictable and stochastic demand environments, which are the focal point of this study [8].

2.3 Optimum Warehouse Layout Design

Another typology is associated with the SLP methodology. The implementation of this methodology has been described for different company areas [2]. However, it is specially implemented in warehouses to solve layout design issues due to cost increasing idle spaces. Other problems identified are reducing the amount of work in process and waste from material flows. Finally, in all these studies, warehouse layout designs were streamlined, which revealed great potential for improving resource performance, such as reducing non value adding spaces, eliminating wastes, and exerting continuous control of materials flow. A reduction of 23.3% in total order preparation times at a warehouse has been achieved.

3 Research Methodology

The methodology used herein, as shown in Fig. 1, not only seeks a new ideal warehouse layout design but also implements an on-demand warehousing model aimed at developing and sustaining an orderly, clean, organized, and safe workplace environment that allows workers to properly perform their duties over time.

Fig. 1. On-demand warehousing model structure

3.1 Change Management Phase

Senior management must define proper strategies for knowledge management within the interactive warehouse, which must be understood for generating new improvement knowledge, disseminate it among all employees, and transform it into innovative ideas.

3.2 Warehouse Layout Phase

This component seeks an ideal layout design for the interactive warehouse areas, wherein only the materials required for event development may be found.

3.2.1 Sort (Seiri)

The first stage of the 5S philosophy (sorting) removes unnecessary items from required items through the red tagging strategy.

3.2.2 Optimum Layout (SLP)

The SLP methodology identifies, assesses, and observes all elements engaged in the development of the ideal warehouse layout.

Herein, the warehouse has been divided into seven areas. The structures area reports 80% of the movement at the warehouse since medium and large metal structures are used for any type of event. Based on this information, we implemented different SLP methodology tools, mainly the activity tables and relationship diagram, which provide a better view of which areas should be closer due to their proximity; therefore, decreasing the number of crosses between the areas that exhibit higher relation intensity.

Another useful instrument is the space calculation method (m^2), since it provides the number of square meters required in each warehouse area for the proper development of work activities. In this way, the ideal warehouse layout was designed, as shown in Fig. 2.

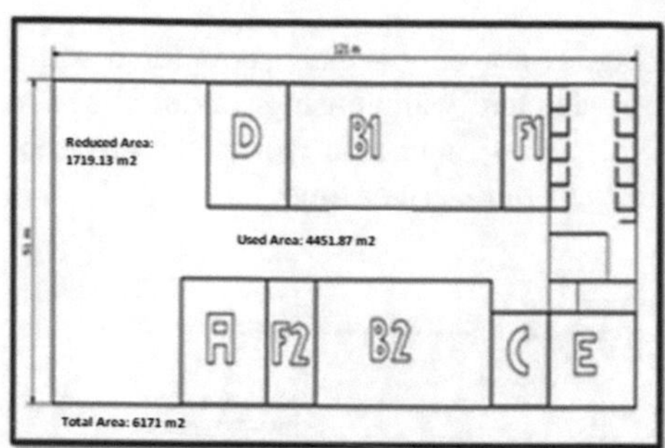

Fig. 2. Ideal materials warehouse layout

Fig. 3. Proper material storage structure

3.3 Warehousing Activity Standardization Phase

The 5S philosophy seeks to generate radical changes in work spaces by improving the environment and implementing a proper use of resources based on a culture of discipline [9]. In this component, the remaining stages of the 5S philosophy, i.e., Seiton, Seiso, Seiketsu, and Shitsuke, will be deployed. (See Fig. 3).

3.3.1 Set in Order (Seiton)

This step basically simplifies work activities. In the first place, sorting will be performed every time a new material enters the warehouse. This process uses stickers to identify the serial number of the material and ensure it is properly managed by using records.

3.3.2 Shine (Seiso)

This step identifies the sources of dirt within each warehouse area to take the corresponding corrective actions. To supplement this initiative, color-coded recyclable bins will be placed for adequate waste disposal.

3.3.3 Standardize (Seiketsu)

This step focuses on the standardization of the previous activities and steps. For these purposes, different signs or posters will be posted to prioritize the fulfillment of these steps, fostering the culture of cleanliness and accountability within the organization by establishing specific daily cleaning schedules.

3.3.4 Sustain Improvements (Shitsuke)

This step validates whether the results obtained actually fulfill the objectives defined at the onset of the project, documenting conclusions, and recommendations for future endeavors. Finally, area heads will hold a quarterly meeting to assess different continuous improvement proposals.

3.4 Procedure and Record Control Phase for Establishing a Safe Workplace Environment

GSP sets forth a series of standards which guarantee that warehouse operations do not represent risks in terms of quality, efficiency, safety, and functionality [10].

3.4.1 Creation of Procedures and/or Records

A new material usage procedure must establish all the measures to be implemented from the moment the materials arrive to an event to their exit and possible reuse.

3.4.2 Implementation of Safety Measures

This factor is taken into account due to the constant risk to which every warehouse worker is exposed. On the contrary, high number of accidents reported deems this problem as important within the organization. Therefore, different solutions will be proposed to improve the current situation of this resource.

3.4.3 Hazard Identification

Hazards will be identified based on the different activities performed by the workers. Currently, this warehouse employs operators in charge of fabrics and metals.

3.4.4 Risk Assessment

According to the hazards identified for each area, risks are assessed based on the ratings from four categories, generating a probability score. This score is then added to the severity score, resulting in the risk level used to determine its significance.

3.4.5 Risk Mitigation Controls

Based on the risks assessed, different controls, such as administrative, engineering, or EPPs, will be implemented to replace or eliminate risks and reduce occupational accidents and incidents.

3.4.6 Contingency Plans

After the hazard identification, risk assessment, and control implementation stages, contingency plans will be prepared. These plans will serve as a guide for efficient occupational health and safety management within the warehouse under study.

4 Validation

Our research methodology was validated through a case study conducted at a materials warehouse in the city of Lima [11]. This organization stages events in open spaces, which means that the warehouse is interactive due to the constant material inputs and outputs. In addition to all the waste and/or trash generated, the warehouse stores large quantities of materials from past events, i.e., their company policy is to store all materials used in previous events.

4.1 Methodology Application

The implementation of the on-demand warehousing model in the case study started with knowledge management, which is the contribution of the research methodology. This stage comprises establishing a point of departure for continuous warehouse management and control through a series of training sessions on the 5S philosophy and GSP. Then, all intervening factors are considered for the second stage of the methodology, corresponding to the ideal warehouse layout design by means of the SLP method. Then, during the third stage, all the documentation required, such as procedures and records, were prepared to organize, manage, and control the warehouse activities and foster continuous improvement. Finally, the fourth stage deployed GSP through safety procedures and measures in all warehouse areas to safeguard the physical integrity of all workers, as well as demanding compliance with process standardization documentation.

4.2 Metrics

The performance of these management indicators will be assessed based on the metrics or reference levels from other studies to determine whether the results obtained are optimum or not (Table 1).

Table 1. Metrics

Indicators	Before	Expected	After
Idle space reduction index	22.87%	69.79%	72.14%
Waiting orders index	43.24%	10.43%	13.51%
Incomplete orders index	34.73%	5.07%	9.10%
Monthly profit growth index	0.00%	2.45%	5.54%
Training efficiency index	1.20%	65.50%	52.22%

Table 2. Warehousing cost reduction

Improvement aspects	Decrease (%)
Idle spaces	27.86%
Nonproductive travel	4.62%
Occupational accidents	5.00%
Waste	6.33%
Work efficiency	14.74%
Total	**58.55%**

The idle space reduction index exhibited the greatest impact on the improvement since the total warehouse area was reduced by 72.14%, which translates into monthly savings of approximately S/11,500 in leasing costs. In addition, the waiting orders index had a considerable decrease, i.e., 13.51%, even when it did not reach the proposed target metric [12]. Therefore, further follow-up would be required to effectively meet the proposed objective. It should be noted that, as long as better results are obtained in these indicators, we are achieving sustainability, which means that the research methodology is proven effective. Hence, operating costs have been successfully reduced, particularly in terms of warehousing costs, which are divided into different aspects of improvement, as listed in Table 2. Based on this cost reductions, the monthly profit growth index was assessed, which show a 5.54% increase in net profits for the month of September.

5 Conclusions

Herein, an on-demand warehousing model that acquires sustainability and competitiveness over time has been developed. This methodology is fundamentally based on knowledge management, optimum layout design of warehousing facilities, and continuous monitoring of warehousing processes and activities. As a result, a 58.55% reduction in operating costs was reported, as well as a 5.54% net profit gain against the previous month.

Sustainability was achieved in warehousing activities and for the entire study case dedicated to staging events in open spaces, as validated by the performance indicators proposed by the research methodology returning optimum values. Hence, sustainability is achieved through a critical balance in its three pillars, namely economic, environmental, and social. In comparing with the previous methods, the proposed ideal warehouse layout, as developed by the SLP methodology, removed 72.14% of the idle space existing in the materials warehouse. In addition, this research study decreased the number of stations in each work area to minimize travel time between facilities.

References

1. Kembro, J.H., Norrman, A., Eriksson, E.: Adapting warehouse operations and design to omni-channel logistics: a literature review and research agenda. Int. J. Phys. Distrib. Logist. Manag. **48**, 890–912 (2018)
2. Roodbergen, K.J., Vis, I.F.A., Taylor, G.D.: Simultaneous determination of warehouse layout and control policies. Int. J. Prod. Res. **53**(11), 3306–3326 (2015)
3. Mirabelli, G., Pizzuti, T., Macchione, C., Laganà, D.: Warehouse layout optimization: a case study based on the adaptation of the multi-layer allocation problem. In: Proceedings of the Summer School Francesco Turco Industrial, pp. 49–58 (2015)
4. De Paula, N., Melhado, S.: Sustainability in management processes: case studies in architectural design firms. J. Archit. Eng. **24**(4), 05018005 (2018)
5. Mao, J., Xing, H., Zhang, X.: Design of intelligent warehouse management system. Wirel. Pers. Commun. **102**(2), 1355–1367 (2018)
6. Hilmola, O.-P., Tolli, A.: Warehouse layout implications on picking distance: case of human factor. World Rev. Intermodal Transp. Res. **6**(1), 43–58 (2016)
7. Bravo, C., Ortiz, S., Raymundo, C., Torres, C., Quispe, G.: Maturity model for the strategic management of the corporate scaling of family businesses in the services sector. In: CICIC 2018 - Octava Conferencia Iberoamericana de Complejidad, Informatica y Cibernetica, Memorias (2018)
8. Maddah, B., Noueihed, N.: EOQ holds under stochastic demand, a technical note. Appl. Math. Model. **45**, 205–208 (2017)
9. Randhawa, J.S., Ahuja, I.S.: 5S – a quality improvement tool for sustainable performance: literature review and directions. Int. J. Qual. Reliab. Manag. **34**(3), 334–361 (2017)
10. de Vries, J., de Koster, R., Stam, D.: Safety does not happen by accident: antecedents to a safer warehouse. Prod. Oper. Manag. **25**(8), 1377–1390 (2016)
11. Oey, E., Nofrimurti, M.: Lean implementation in traditional distributor warehouse - a case study in an FMCG company in Indonesia. Int. J. Process Manag. Benchmarking **8**(1), 1–15 (2018)
12. Dukic, G., Opetuk, T., Lerher, T.: An integrated model of storage and order-picking area layout design. In: 13th International Conference on Industrial Logistics, ICIL 2016 - Conference Proceedings, pp. 45–55 (2016)

A Descriptive Review of Carbon Footprint

Omoniyi Durojaye[1(✉)], Timothy Laseinde[1], and Ifetayo Oluwafemi[2]

[1] Mechanical and Industrial Engineering Technology Department, University of Johannesburg, Johannesburg, RSA
durojayerapheal@gmail.com, otlaseinde@uj.ac.za

[2] Postgraduate School of Engineering Management, University of Johannesburg, Johannesburg, RSA
ijoluwafemi@uj.ac.za

Abstract. In the last couple of years, the expression 'carbon footprint' has been in use among meteorologist, especially in the United Kingdom. The usage of the expression is not unconnected with the fact that the challenge of climate change has been one of the topmost issues on political as well as corporate agenda. Therefore, calculations of carbon footprint have been in high demand. Scholars have proposed so many approaches to providing estimates. The methods runs from uncomplicated online calculations to other ones like complex life-cycle examination otherwise called input-output-based technique and tools. In spite of its ever-present use, it is clear that the term 'carbon footprint' does not have a generally acceptable academic definition. Thus, there is a dearth of clarifications of what 'carbon footprint' is in the scientific literature despite the fact that there are several studies or researches on energy as well as ecological economics which should have measured what precisely 'carbon footprint' is. Therefore, this particular paper is an attempt to explore the obvious inconsistency in the use of some terminology in both public and in the academic contexts. This has led to a suggestion of academic definitions based on generally accepted principles of accounting and modeling approaches. The paper treats questions of methodology like comprehensiveness, completeness, system boundaries, as well as units, also, the durability of most indicator.

Keywords: Direct footprint · Direct carbon footprint · Environmental accounting · Ecological Footprint · Indirect carbon emissions

1 Introduction

The expression 'carbon footprint' has been widely used not only among meteorologists but also by people generally in public debates where the responsibility, as well as the abatement action in combating the threats of global warming and climate change, are being discussed. For some years now or even decades, the use of the expression has been on the increase because of its importance and even urgency. These days, the expression is not only used across different media but also in the business and government domains.

However, the question of what precisely 'carbon footprint' means has come to the forefront. In spite of its ever-present use, it is clear that the term 'carbon footprint' does

T. Ahram et al. (Eds.): IHSED 2019, AISC 1026, pp. 960–968, 2020.
https://doi.org/10.1007/978-3-030-27928-8_144

not have a generally acceptable academic definition. In addition, there exists some levels of confusion and possible misunderstanding regarding what the term means, what it evaluates as well as the precise unit to be adopted. Although the expression is entrenched in the Ecological Foot printing language, yet the generally accepted view is that the carbon footprint represents a specific quantity of gaseous emissions that have been considered significant to climate change and also associated with the production as well as consumption activities of human beings. However, this is just about the only area of convergence because there is no agreement regarding the measurement or quantification of carbon footprint. Thus, one of the definitions stems out of the direct CO2 emissions to full life-cycle greenhouse gas emissions but the actual units of measurement are not really clear [3].

Some of the questions we must ask are as follows: Should the carbon footprint consist of only carbon dioxide emissions, and probably any related greenhouse gas (GHG) emissions like methane be included? Should carbon footprint be limited to only carbon-based gases, perhaps, other substances like N_2O that doesn't have any carbon in their molecules should be inclusive? Should carbon footprint remain limited to elements having much greenhouse warming potential? Some gaseous emissions such as carbon monoxide exist, which are not only built on carbon alone, but are also pertinent to both human health as well as environmental wellbeing. Also, it is possible to convert carbon monoxide into CO2 with the use of chemical processes available in the atmosphere. Another thing that must be ascertained is whether the measure should include the various sources of emissions without excluding the ones like CO2 emissions from soils, which originate from fossil fuels.

One other important question is to ascertain if the carbon footprint requires only the direct emissions personified in the upstream production procedures, perhaps, it also requires the indirect one.

It should also be asked if carbon footprint replicate most of the life-cycle influences associated with both goods as well as services exploited. If that is the case, there should be a determination of where the boundary should be drawn as well as the impacts which will be quantified.

On a final note, the word 'footprint' appear to imply a form of measurement in area-based units. By and large, the 'Ecological Footprint', which is a related word can be measured or expressed in hectares, i.e. 'global hectares'. From this perspective, it can be asserted that the questions raised above have very serious implications on the overall assessment of the entire process because it results to the initial resolution which proposes that carbon footprint ought to be seen as an ordinary 'pressure' indicator signifying only the number of carbon emissions. Several of the questions raised have been presented in ecological economics as a discipline. Consequently, a number of answers are readily available. This implies that scholars and researcher have not applied the term carbon footprint, which strongly suggests that there is no apparent definition at the moment.

This paper, therefore, focuses attention on the answers to the questions asked above so the paper attempts to clarify some important issues. Thus, this paper presents a literature review, proposes a definition that suits the word 'carbon footprint' before discussing the methodological implications of the concept.

2 Literature Review

The concept of carbon emission has been researched into by so many researchers and scholars. Most the research works on the topic focus on the questions seeking to know the extent at which CO2 emissions may be ascribed to a particular product. However, no researchers or study has been able to give a generally acceptable definition of the expression carbon footprint. Be that as it may, this paper will attempt to define the expression carbon footprint. To start with, we may define carbon footprint as the quantity of CO2 emitted due to human daily activities. Such activities range from driving vehicles to using heavy construction machinery, from doing laundry with washing machines to microwaving food items at fast food joints and many more.

Patel [4], in his study, stressed that calculating the carbon footprint can be done when we measure the CO_2 corresponding emissions released from its environment, waste generation and disposal, construction sites and many more. Defining carbon footprint, Pandey [5] has referred to it as a methodology employed in estimating the entirety or full amount of the emission of GHG from a specific product running through the aforementioned life-cycle, right from initial unprocessed material stage employed in the production process to the actual production of the finished good. Additionally, Pandey submitted that it is a method used to identify and measure each GHG emissions activity found in a chain process. The expression carbon footprints also covers the framework to attribute any of these activity to each output. Furthermore, Energetic report [6], referred to carbon footprints as the full extent of the direct as well as the indirect CO2 emissions resulting from various business activities. Similarly, Kumar [7], perceives Carbon Footprint as the gauge of the effects of human activities on natural environment from the perspective of the quantity of greenhouse gases produced and measured in tonnes of carbon dioxide.

Grub *et al.* [9] itemized carbon footprint as the gauge of the quantity of carbon dioxide emitted in the process of combustion of fossil fuels. When compared to a business organization, carbon footprint can be referred to as the total or quantity of CO2 emitted by the organization on daily basis either directly or indirectly. Apart from the foregoing, carbon footprint has also been defined by the Parliamentary Office of Science and Technology, Allen [10], as the total amount of CO2 as well as other greenhouse gases that are emitted over the full life cycle of a product or even a process. Carbon footprint is usually expressed as grams of CO2 equivalent per kilowatt-hour of generation (gCO2 eq/kWh). From this perspective, carbon footprint accounts for the diverse global warming effects caused by some other greenhouse gases. Therefore, 'carbon footprint' can be expressed as common substitute for the emission of CO_2 or any other GHG expressed as carbon dioxide equivalents.

As mentioned by Haven [11], the carbon footprint analysis of a particular piece of office furniture like an office table possesses a life-cycle assessment that may be considered in the lines of being a material item, being manufactured at some point, transported to another location for use, used at the location and eventually disposal at the last stage of development. Considering carbon footprint from this perspective, it can be referred to as a more inclusive approach, not often described in other articles. Nevertheless, the term neither has a specific definition nor a particular methodological description. Thus, an examination carried out on carbon footprint formed during business process, does not focus only on calculating energy consumption rather it focuses on increasing each scrap of the data from all available aspects of the business practices [12].

It has been realized that the academia has mostly neglected the issue of defining carbon footprint. Instead, there has been a focus on ideas like businesses, consultancies, etc. for which the government has provided its own definitions and other supporting ideas. In the general literature, there are several descriptions and definitions. Some of those definitions have been illustrated previously in this paper. For instance, in the United Kingdom, it is the aim of the Carbon Trust to develop a more general perceptive of what carbon footprint of a particular product deals with. Besides this, the organization has also been circulating a draft methodology for consultation [13]. Therefore, the advocacy is that only three items should be included. The items are those that are directly connected with the product and the items are as follows, input, unit processes as well as output. There are many documents that contain the life-cycle thinking, and this has been developed into one characteristic of carbon footprint estimates. Thus, standardization procedure has been commenced by the Carbon Trust and Defra.

An organization saddled with the responsibility of compiling 'National Footprint Accounts' annually, The Global Footprint Network [14] has seen the carbon footprint as an inseparable part of the general or overall Ecological Footprint. Therefore, carbon footprint has been seen as a synonymous expression for similar expressions like the 'fossil fuel footprint' or the demand on 'CO2 area' or 'CO2 land'. One of the terms can be defined as the demand on biocapacity required to sequester with the use of photosynthesis or the carbon dioxide (CO2) emissions from fossil fuel combustion. This, it should be noted, consists of the biocapacity, characteristically that of unharvested forests required to absorb that portion of fossil CO2 which has not been absorbed by the ocean. On the other hand, some individual documents have employed a particular land-based definition, which is the Scottish Climate Change Strategy [15]. However, this has refused to change the general perception of the carbon footprint as a gauge of CO_2 emissions in the literature.

3 Meaning of Carbon Footprint

For the purpose of this paper, our proposed and the employed description of the expression 'carbon footprint' is as follows: For the purpose of this paper, our proposed and the employed description of the expression 'carbon footprint' is as follows:

> *"Carbon footprint can be defined as a measure of the whole CO_2 emissions which are either caused directly or indirectly via accumulated activity of a product."*

This also covers activities performed by individuals, groups, organizations, companies, governments, and production industries among others. In this sense, the term product covers both goods and services produced and rendered by different companies. As it is, both the direct as well as indirect must be considered. While the direct emission refers to the on-site or internal emission, the indirect emission is the off-site or external emission.

Our working definition for this paper provides answers to some of the questions raised at the beginning of the paper. Thus, just CO2 has been included in our analysis. However, it should be noted that there are other material items that have the same greenhouse warming effects or potentials. On the other hand, many of the materials are not based on carbon while some are more difficult to quantify due to the availability of data. It is easier to include methane but whatever information that is added from a partly accumulated indicator may include only two of the several relevant GHGs. An all-inclusive greenhouse gas indicator is expected to comprise all these gases. The gases may be called 'climate footprint'. Therefore, we have adopted only CO2 as the practical or authentic solution to the greenhouse effects when the 'carbon footprint' is concerned.

The definition given so far has refrained from mentioning the carbon footprint as a particular or specific area-restricted indicator. From the earlier stated definition, 'whole CO2' can be substantially measured in mass units. Based on this, there is no conversion available to an area unit (ha, m^2, km^2, etc) that can take place. The conversion into a land area must, expectedly, be based on a diversity of assumptions coupled with the level increases in uncertainties as well as errors associated with a footprint estimate [16]. Therefore, accountants normally make several attempts to avoid unnecessary conversions. They also try to voice out any observable fact in the most appropriate amount unit [17]. With this rationale at the back of one's mind, a land-based measure cannot be considered appropriate so these researchers have a preference for the more precise illustration based on tons of carbon dioxide.

Although the idea of 'carbon footprint' may be perceived as all-encompassing, including every possible cause that usually gives rise to carbon emissions yet it is similarly of high importance to expatiate on what this entails. The accurate dimension of carbon footprints can be said to have gained more importance as well as precariousness when considered from the perspective of carbon offsetting. Therefore, it is very clear that an apparent definition of scope and boundaries is very important when researches aimed at reducing emissions are proposed and sponsored. In a bid to account for indirect emissions, there is the need to adopt or apply methodologies that can be used to avoid under-counting together with double-counting of emissions. As a result, the lexeme 'exclusive' is included in the description. In addition, full life-cycle examination of products implies that every stage of the aforementioned life cycle must be appraised appropriately without disrupting the meaning of the original expression. Methodological implications of the aforementioned requirements are the next things to be considered in the next subsection.

4 Methodological Challenges

The duty of calculating or determining carbon footprints may be undertaken methodologically from two distinct approaches. The first is a bottom-up approach, and this is based on Process Analysis with the acronym PA. The second is top-down approach with its focus on Environmental Input/output with the acronym EIO. whichever approach is adopted, the challenges highlighted above must be captured in its entirety. This means that a full Life Cycle Analysis/Assessment popularly shortened as LCA must be conducted. In this instance, a very brief impression of some of their major advantages and shortcomings will be provided.

One of the approaches that employ the bottom-up method is the process analysis often shortened as PA. PA method is developed to gain the knowledge of environmental impacts of different products in their life cycle, i.e. from cradle to the grave of the product. When products are approached from the bottom-up angle, such products often suffer from what is known as a system boundary problem. This because just on-site, first-order, or some second-order impacts will be considered [18]. To derive carbon footprint estimates using PA-LCAs, a thorough emphasis has to be given to the recognition of suitable system boundaries so that the truncation error minimized. PA-based LCAs usually encounter a lot of problems when carbon footprints for larger entities like the industrial sector, general households or governments are to be established in the process. Despite the fact that different estimates may be obtained by extracting the information found in life-cycle databases, the outcomes often get ever patchier because the procedures normally need the assumption that an integral part of individual products is used to represent a larger product groups. Therefore, the available information for use from diverse databases will become generally inconsistent [19].

The EIO or environmental input-output investigation is capable of providing an substitute for the top-down method used instead of carbon foot printing [20]. Both input and output tables are economic accounts that give the picture of every economic activity at the meso or sector level. The input and output tables may be used to set up carbon footprint estimates in a complete and vigorous manner. They are also used to set up all higher order impacts including setting the whole economic system as a boundary. However, it should be noted that this completeness does not come without its expense of details. The appropriateness of input-output analysis in the environment to assess microsystems like products or processes is often limited because it does not only assume sameness in prices but also in outputs and their carbon emissions at the sector level. Even though some sectors may be disaggregated in order to be subjected to further analysis and bring it closer to a microsystem, it has been observed that this possibility is limited, especially when considered from a larger scale of activities. Despite the aforementioned, it has been noted that the input-output based approaches have their advantage and the advantage is that the approach requires much smaller amount of time as well as manpower the moment the model has been put in place.

In order to have a detailed and comprehensive analysis, the most appropriate option is to combine the strength of both methods, and this can only be done by using a hybrid approach [21–23], such that the PA and input-output methodologies will be integrated. Using this approach will give opportunity to preserve the detail as well as the accuracy of the bottom-up approaches in lower order stages. At the same time, higher-order requirements will be covered by the input-output element of the model. Such a **Hybrid-EIO-LCA** method, embedding procedure systems within the input-output tables, is the current state-of-the-art in ecological economics. Since the literature on this subject is just emerging, very few practitioners have been able to acquire the requisite skills to carry out that kind of hybrid appraisal or evaluation. However, speedy progress and better-improved methods will be expected in the nearest future.

The method to be chosen will be greatly dependent on the rationale of the inquiry as well as the availability of both resources and data. Therefore, it will be assumed that environmental input-output analysis is of higher importance in setting up carbon footprints in macro and mesosystems. From this perspective, a carbon footprint of individual businesses or larger production groups, industrial sectors or general households, government or the average citizen or an average member of a socio-economic group can easily be performed by input-output analysis [24]. Process analysis possesses obvious benefits for looking at microsystems: a process, an individual product or a relatively small group of individual products.

5 Illustrative Examples

It should be noted that there has been an establishment of carbon footprints for countries. Therefore, sub-national regions and institutions like schools [25], organizations, businesses, products, investment funds, and even services have been covered by researches into carbon footprints.

This subsection is used to present dual practical instances of a carbon footprint analysis which have been observed to align with the definitions suggested so far. The analyses presented here were those carried out by researchers of the **Stockholm Environment Institute at the University of York**. The researchers adopted the input-output based approach for their study.

According to the 'UK Schools Carbon Footprint Scoping Study', [25] projected that the totality of every school in the United Kingdom has been discovered to have a carbon footprint of 9.2 million tonnes of carbon dioxide as far back as 2001. This number is an equivalent of 1.3% of the entire UK emissions. Just about an average of 26% of this complete carbon footprint has been linked to onsite emissions from the heating of premises, while the remaining three quarters come from indirect sources of emission like electricity which has 22%, school transport which is estimated to be 14%, as well as other transport that has 6%, chemicals with about 5%, furniture with about 5%, paper with 4%, and other manufactured items having 14%. Finally, mining and quarrying, as well as other products and services, have about 2% and 3% respectively.

The second instance deals with calculating the carbon footprint of households in the United Kingdom, with a focus on both the direct and indirect emissions taking place on the UK territory as a result of both production and consumption activities. It also covers the indirect emissions, for instance, those emissions that are released from imports into the UK territory. The results, shown in the 'Counting Consumption' report [26], imply that an average household in the UK has the carbon footprint of about 20.7 tonnes of CO2 as far back as 2001. On the one hand, heating at different locations and car use across the territory account for direct emission. On the other hand, those emissions that occur during the generation of electricity or during the production of goods and services account for indirect emissions.

6 Conclusion

Judging by the available scientific literature, other publications, and general statements from both public and private sectors and a consideration of general media, it is certain that the term 'carbon footprint' has really penetrated the public domain despite the fact that it has not been plainly defined in the scientific literature. Therefore, this paper has suggested a definition of the expression 'carbon footprint'. With the submission from this paper, the writers believe that different academic debates will be stimulated concerning the concept as well as its process of carbon footprint examination or assessments.

Our argument is that it is imperative for a 'carbon footprint' to consist both direct and indirect CO2 emissions. In addition, it is expected that a mass unit of measurement is supposed to be employed, instead of the inclusion of other greenhouse gases The suitability of two main methodologies or process analysis has been considered in this paper. Thus, it has been asserted that input-output analysis can provide both comprehensive and healthy carbon footprint assessments of not only the production activities but also consumption activities, especially at the meso level. The paper, therefore, suggests a Hybrid-EIO-LCA approach. Through this approach, different life-cycle assessments will be combined with input-output analysis to produce the expected result. This is an approach where on-site, first- and second-order process data on environmental impacts will be collected for the product or even the service system being understudied, at the same time as higher-order requirements will be covered by the analysis of input-output procedure.

Whatsoever method we adopt to calculate carbon footprints will become important in order to avoid double-counting along the chains of supply or life cycles of the products as well as services. This is as a result of the fact that there are important repercussions originating from the practices of carbon trading as well as carbon offsetting.

References

1. Heidari, N., Pearce, J.M.: A review of greenhouse gas emission liabilities as the value of renewable energy for mitigating lawsuits for climate change-related damages. Renew. Sustain. Energy Rev. **55**, 899–908 (2016)
2. Wackernagel, M., Rees, W.: Our Ecological Footprint: Reducing Human Impact on the Earth. New Society Publishers, Gabriola Island (1989)
3. Rolfe, A., Huang, Y., Haaf, M., Pita, A., Rezvani, S., Dave, A., Hewitt, N.: Technical and environmental study of calcium carbonate looping versus oxy-fuel options for low CO2 emission cement plants. Int. J. Greenhouse Gas Control **75**, 85–97 (2018)
4. Patel, J.: Green sky thinking. Environ. Business **122**, 32 (2006)
5. Pandey, D., Agrawal, M.: Carbon footprint estimation in the agriculture sector. In: Assessment of Carbon Footprint in Different Industrial Sectors, vol. 1, pp. 25–47. Anonymous Springer (2014)
6. Energetics Report: The reality of carbon neutrality, London (2007)
7. Kumar, M., Sharm, L., Vashista, P.: Study on carbon footprint. Int. J. Emerg. Technol. Adv. Eng. **4**(1), 345–355 (2014)
8. Global Footprint Network: Ecological footprint glossary, Oakland, CA, USA (2007)

9. Grubb, E.: Meeting the carbon challenge: the role of commercial real estate owners, users & managers, Report, Chicago, USA (2007)
10. Allen, S., Pentland, C.: Carbon Footprint of Electricity Generation: POSTnote 383 (2011)
11. Haven, J.: Environ. Bus. **129**, 27 (2007)
12. Von Weizsacker, E.U., Hargroves, C., Smith, M.H., Desha, C., Stasinopoulos, P.: Factor Five: Transforming the Global Economy Through 80% Improvements in Resource Productivity. Routledge, London (2009)
13. Carbon Trust: Carbon footprint measurement methodology, version 1.1, 27 February 2007, London, UK (2007)
14. Wackernagel, M., Monfreda, C., Moran, D., Werner, P., Goldfinger, S., Deumling, D., Murray, M.: National footprint and biocapacity accounts 2005: the underlying calculation method (2005)
15. Scottish Executive: Changing our ways: Scotland's climate change program (2006)
16. Lenzen, M.: Uncertainty in impact and externality assessments-implications for decision-making (13 pp). Int. J. Life Cycle Assess. **11**(3), 189–199 (2006)
17. Stahmer, C.: The magic triangle of input-output tables. In: 13th International Conference on Input-Output Techniques, p. 25 (2000)
18. Lenzen, M.: Errors in conventional and Input-Output—based Life—Cycle inventories. J. Ind. Ecol. **4**(4), 127–148 (2000)
19. Tucker, A., Jansen, B.: Environmental impacts of products: a detailed review of studies. J. Ind. Ecol. **10**(3), 159–182 (2006)
20. Wiedmann, T., Minx, J., Barrett, J., Wackernagel, M.: Allocating ecological footprints to final consumption categories with input-output analysis. Ecol. Econ. **56**(1), 28–48 (2006)
21. Suh, S., Lenzen, M., Treloar, G.J., Hondo, H., Horvath, A., Huppes, G., Jolliet, O., Klann, U., Krewitt, W., Moriguchi, Y.: System boundary selection in life-cycle inventories using hybrid approaches. Environ. Sci. Technol. **38**(3), 657–664 (2004)
22. Heijungs, R., De Koning, A., Su, S., Huppes, G.: Toward an information tool for integrated product policy: requirements for data and computation. J. Ind. Ecol. **10**(3), 147–158 (2006)
23. Bullard, C.W., Penner, P.S., Pilati, D.A.: Net energy analysis: handbook for combining process and input-output analysis. Resour. Energy **1**(3), 267–313 (1978)
24. Foran, B., Lenzen, M., Dey, C.: Balancing act: a triple bottom line analysis of the Australian economy (2005)
25. Sydney GAP: UK Schools Carbon Footprint Scoping Study (2006)
26. SEI and WWF: Counting Consumption-CO2 emissions, material flows and Ecological Footprint of the UK by region and devolved country (2006)

Results-Based Process Management Model Applied to NGOs to Promote Sustainability and Reliability in Social Projects

Joel Heredia[1(✉)], Luis Quispe[1(✉)], Fernando Sotelo[1(✉)], Carlos Raymundo[2(✉)], and Francisco Dominguez[3(✉)]

[1] Ingeniería Industrial, Universidad Peruana de Ciencias Aplicadas (UPC), Lima, Peru
{U201120772,U201221903,fernando.sotelo}@upc.edu.pe
[2] Dirección de Investigación, Universidad Peruana de Ciencias Aplicadas (UPC), Lima, Peru
carlos.raymundo@upc.edu.pe
[3] Escuela Superior de Ingeniera Informática, Universidad Rey Juan Carlos, Madrid, Spain
francisco.dominguez@urjc.es

Abstract. This article describes the application of an information collection, monitoring, handling, and dissemination model which may provide objectivity to the results obtained by social organizations. Through Process Management, a model has been designed to reduce these identified gaps, which lead to the de-financing of sector's operations. By applying Digital Marketing Strategies, 74% more editable evidence has been collected as per the mission of the organization, in addition to contacting more volunteers, begin negotiations with three financial entities and increase the level of maturity of the Movimiento Peruanos sin Agua NGO.

Keywords: Third sector management · NGO sustainability · Social marketing · Project reliability

1 Introduction

A key International Cooperation objective is the efficient, transparent, and sustainable management of the scarce resources acquired through an institutional model founded in 2002 as the Peruvian Agency for International Cooperation-APCI. This information denotes that the resources from international cooperation, including resources raised by NGOs, are being poorly invested in Peru since the more poorer regions tend to receive fewer resources proportionally. This is the case of Movimiento Peruanos sin Agua (MPSA, a Non-Governmental Organization (NGO) dedicated to the implementation of projects in vulnerable communities with limited access to water services. Its projects mainly include the installation of fog collection systems, devices that capture water in foggy areas through particle condensation in Raschel nets. It has been identified that the project implementation times in communities exceed 4 months due to inadequate fundraising resources for their implementation. In light of this problem, the causes

T. Ahram et al. (Eds.): IHSED 2019, AISC 1026, pp. 969–974, 2020.
https://doi.org/10.1007/978-3-030-27928-8_145

influencing this phenomenon were assessed. It was observed that, in the last quarter of 2017, the organization stopped covering its fixed costs because of low project implementation activities during this period. However, the number of contacts made with private organizations is not sufficient to gain the funds required. To date, the NGO has stopped receiving more than 130 thousand soles since the projects planned for this year were not completed.

2 State of the Art

2.1 Reengineering Marketing Processes in Service Companies

The study by [1] develops and applies methodology related to process reengineering in service companies. The model shown integrates partial reengineering prospects through delimiting modifications and the system's global particularities, i.e., it partially evaluates the process that requires improvement through the system's general characteristics as a whole. The objective of the study is to describe the relevance of partial reengineering in organizations with limited participation in strategic areas as well as the relevance of the marketing area as a key and strategic process in service companies. Finally, recommendations are given on applying partial reengineering in customer focused companies in a way that guarantees the creation of value that may benefit the stakeholders.

2.2 Process Alignment Applied to NGO Services Management

In the study carried out by [2], the improvement process for health program aid in Ethiopia-Africa is explained through process restructuring. In the same vein, [3] assesses the impact of process orientation at organizations and suggests that a research gap exists that proves a direct correlation between process orientation and improvement in the company's performance. Non-governmental organizations, as in any context laden with shortages unrecognized by the State, maintain an active relationship with their beneficiaries, which is why they are critical actors in the development of these vulnerable communities. These organizations possess knowledge about their needs and requirements for integrating their beneficiaries into society. The MEDCAP civil medical program expects to be sustainable over time with approval from the US embassy and coordination from USAID. For the development of this work, unexpected delays in the formulation of aid projects have been identified. For this reason, a process restructure was proposed so that a more optimal system may become available. For example, the coordination and deployment of requirements translated into long waiting times for the military doctors to begin activities in the area. In addition, including these specialists in project development also meant that Schedules had to be delayed.

2.3 Tool for Measuring the Maturity Level Applied to the Social Economy

The research study conducted by [4] presents a Process Management Maturity model for third sector organizations. First, a theoretical diagnostic tool was developed aimed at assessing the level of management maturity within the organization, with the authors using exploratory-qualitative research. Then, as a second stage, the validation instrument is applied to a total of 13 social organizations in Brazil. According to the theoretical model proposed, social organizations evidence five management dimensions that must be considered. The tool designed by the authors is a 17 question questionnaire, covering the 5 dimensions mentioned. In contrast, the study further refers to the aggregate nature of the maturity level; therefore, a score system must be defined for each established dimension for its identification [5].

3 Contribution

3.1 Foundation

The method applied in the improvement proposal presents the application of four dimensions: Schedule Management, People Management, Process Management, and Marketing Strategies. It is by harmonizing the interaction between these dimensions that we seek to increase the maturity level of the organization studied.

3.2 General View

In the diagram developed in Fig. 1, the specific dimensions for each of the implementation phases of the model are observed (Fig. 2).

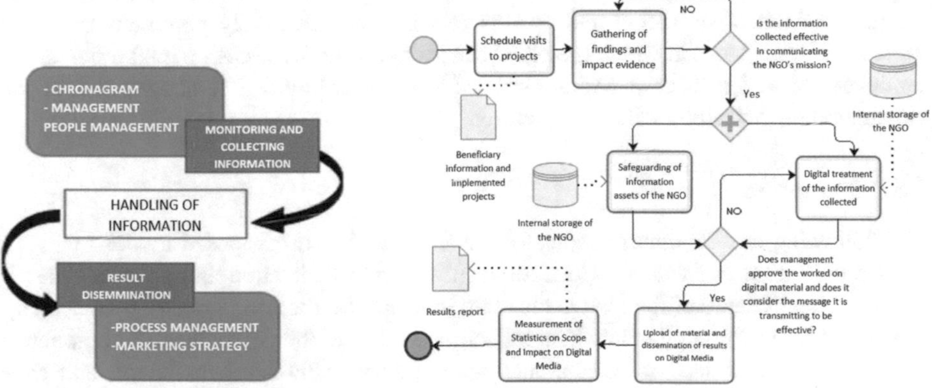

Fig. 1. General view of proposal

Fig. 2. Application flow of the Proposed Model

3.3 Specific View

Each dimension will be defined and detailed based on the model developed. Regarding Schedule Management, the organization must commit to the development of a project visit or implemented initiatives schedule to comply with the monitoring of the activities in which the NGO is involved. Conversely, in order to carry out this task, the required human capital and resources must be allocated to ensure proper evidence collection within the framework of the organization's activities. Next, a transition phase has been proposed as a supplementary part of the model, whose main activity is managing information that has been collected, as evidence of the achievements or impacts attained by the organization. Further, through the application of Process Management, specific work distribution guidelines need to be established, in addition to the allocation of responsibilities according to staff profile. Moreover, as part of the process control method, indicators must be defined for each of the stages of the established system.

Finally, the Digital Marketing Strategy will directly influence the impact that the results and collected evidence can convey to the external stakeholders of the organization. This dimension will be assessed through the traffic generated in digital media, i.e., the clicks/actions recorded after the dissemination of the evidence, while, in turn, assessing results based on the scope, number of visits or number of users logged in to the "fanpages" of the NGO.

3.4 Process View

As a first phase, the schedule of visits to the organization's projects is provisioned (regardless of the mission or turn of the NGO). Then, relevant findings must be gathered about the impact that the organization's actions are having on the direct beneficiaries.

Finally, as a final application phase, content must be uploaded to the selected Digital Media, applying a Digital Marketing Strategy established by the Senior Management of the organization and always focusing on effectively communicating the benefits of the system in which the NGO carries out its activities. As a final process, the achievement and impact statistics obtained from stakeholder interactions within the organization's activities will be monitored.

3.5 Indicators

The following are the control indicators established for the proposed model:

Effectiveness in Evidence Collection: Information Collection and Control Phase.

The effectiveness of the visit is the main indicator in a scoring system influenced by the percentage of compliance when collecting evidence, the percentage of compliance when interviewing leaders, compliance when interviewing the beneficiary, and generating a report that serves as a pertinent reference to the project status. Finally, the physical status of the project assets will be inspected. Compliance with these 5 activities is only evaluated within the framework of conducting inspections and collecting evidence.

3.5.1 Digital Media Impact and Scope Statistics: Dissemination of Results

To assess impact, we used digital platform indicators provided by the channel where news about the organization's activities are published, with their delivery method depending on the path of each platform.

The scope indicator is defined as the number of people who have viewed any publication related to NGO activities on their screen. In contrast, clicks/actions on publications will also be monitored, counting the number of clicks on the page's contact information and on the Call to Action button.

4 Validation

4.1 Case Study

4.1.1 Company Description

MPSA-"Movimiento Peruanos sin Agua" is an organization operating in the third sector of the social economy, which is dedicated to the implementation of projects in vulnerable communities with limited access to water services.

4.2 Description of Implementation Stages

4.2.1 PHASE 1: Information Collection and Control

During the first phase, active projects were inspected. Later, during the research phase, each of these visits was coordinated and planned through the leader of the beneficiary community, generating a visit schedule. The projects visited were:

- Google Project:
- APROVIRFA project
- Children's Garden Project
- Cerro Verde Project

PROGRAMME OF VISTS AND EVIDENCE TAKEN – PILOT PLAN																				
PROJECT	ACTIVITY	SEPTEMBER								OCTOBER								PROGRESS		
		1		2		3		4		1		2		3		4				
		P	E	P	E	P	E	P	E	P	E	P	E	P	E	P	E	PLANNED	EXECUTED	%FULFILLED
ASOCIACIÓN APROVIRFA	VISIT CARRIED OUT																	2	2	100%
	COLLECTING OF EVIDENCE EXECUTED																			
CERRO VERDE	VISIT CARRIED OUT																	2	2	100%
	COLLECTING OF EVIDENCE EXECUTED																			
INFANCIA PACHACAMAC	VISIT CARRIED OUT																	2	2	100%
	COLLECTING OF EVIDENCE EXECUTED																			
PROYECTO GOOGLE	VISIT CARRIED OUT																	2	2	100%
	COLLECTING OF EVIDENCE EXECUTED																			
TOTAL OF ACTIVITES PLANNED AND EXECUTED																		PLANNED	EXECUTED	CURRENTLY FULFILLED
% FULFILLED																		8	8	100%

Fig. 3. Projects visited in Phase 1 of the pilot plan

As a final measure, the multimedia information and evidence found were included in a summary report for each visit, which was disseminated through the internal storage medium used by the NGO (Fig. 3).

4.2.2 TRANSITION PHASE: Treatment of Information

From the visits and inspections, a total of 151 photos and 55 videos were obtained and used as part of the results dissemination. Next, design tools were used to improve the image quality of the material. These measures were taken in order to effectively communicate the mission of the organization: To promote the development of peri-urban communities through innovative fog collection project implementation.

4.2.3 PHASE 2: Result Dissemination

Two means of dissemination were chosen in coordination with Senior Management: Facebook and Instagram. Conversely, through Facebook's automatic post scheduling tool, Fanpage updates were scheduled from early October to December 14 of this year. Additionally, the administrative assistant was tasked with posting on the NGO's Instagram feed.

5 Conclusions

From the research study, it may be concluded that the proposed hypothesis was validated by attacking the root causes and designing and validating the proposed processes for the collection, processing and dissemination of information, in an attempt to mitigate the lack of interest from companies. To date, 15 volunteers and 3 companies have contacted the NGO to express their interest in participating in these projects. At the process level, the level of maturity of the company for the processes executed has improved to level 2 since they are already properly executed and monitored.

References

1. Kohlbacher, M., Reijers, H.: The effects of process-oriented organizational design on firm performance. Bus. Process Manag. J. **19**(2), 245–262 (2013)
2. Portella Tondolo, R., Tondolo, V.A.G., Rösing Agostini, M., Bessa Sarquis, A., Teixeira de Mello, S.P.: Modelo de madurez en gestión para organizaciones del tercer sector. In: Centro Latinoamericano de Administración para el Desarrollo (CLAD), pp. 195–224 (2016)
3. Navarro, A., Dinorah, C., Martinez, E.: Diseño de un modelo de reingeniería parcial en marketing basado en empresas centradas en servicios. Manag. Rev. **2** (2017)
4. Miles, C.S., Malone, C.J.: Perspectives from Ethiopia regarding US Military Humanitarian Assistance: how to build a Better Medical Civil Action Project (MEDCAP). Military Med. **178**(12), 1349–1352 (2014)
5. Bravo, C., Ortiz, S., Raymundo, C., Torres, C., Quispe, G.: Maturity model for the strategic management of the corporate scaling of family businesses in the services sector. In: Octava Conferencia Iberoamericana de Complejidad, Informatica y Cibernetica, Memorias 1, CICIC 2018, pp. 96–101 (2018)

Telecommunications Tower Kits Manufacturing Model Based on Ikea's Approach to Minimize the Return Due to Missing Parts in a Metalworking Enterprise Kit

Katia Lavado[1(✉)], Williams Ramos[1(✉)], Edgard Carvallo[1(✉)], Carlos Raymundo[2(✉)], and Francisco Dominguez[3(✉)]

[1] Ingeniería Industrial,
Universidad Peruana de Ciencias Aplicadas (UPC), Lima, Peru
{U201310226,U201213568,pcinecar}@upc.edu.pe

[2] Dirección de Investigación,
Universidad Peruana de Ciencias Aplicadas (UPC), Lima, Peru
carlos.raymundo@upc.edu.pe

[3] Escuela Superior de Ingeniera Informática,
Universidad Rey Juan Carlos, Madrid, Spain
francisco.dominguez@urjc.es

Abstract. This article proposes a work method reform in the production line of self-supporting telecommunication towers of a Peruvian metalworking enterprise. The problem raised considers the cost overruns caused by the delivery and dispatch of incomplete tower kits. The main process evaluated is the assembly of tower kits, which presented multiple deficiencies that we analyzed with multiple engineering tools using the eight steps of practical process improvement. We evaluated by a large scale definition the work table in the assembly and stacking of parts, defining by specific zones each part type and profile, and optimizing the total process times so as not to affect the scheduled dispatches. Thus, it is possible to minimize the number of customer returns.

Keywords: Manufacturing model · LEGO · IKEA · Telecommunications · Metalworking

1 Introduction

Over the centuries, numerous structures have been built worldwide, including the construction of metal towers, which are at present very important for global development. At the beginning, these structures were made of wood but with the emergence of the steel industry, their quality improved greatly. The demand for this type of structures has increased due to the requirements and needs of their use in the telecommunications sector, mainly in rural areas. Telecommunication towers are vulnerable structures due to certain factors, such as the high number of errors observed in their structure compared to other structures of similar economic and social importance.

T. Ahram et al. (Eds.): IHSED 2019, AISC 1026, pp. 975–980, 2020.
https://doi.org/10.1007/978-3-030-27928-8_146

The high failure rate is caused by poor design, which results in unsafe structures that can fully collapse [1] due to strong winds and because they are located in highly exposed places.

In addition, there is a great number of final customer complaints related to the manufacturing and subsequent stacking processes of parts produced and packaged in a kit for subsequent dispatch, since the required quantity of parts is not found at the time of site installation.

Thus, a study and analysis of the manufacturing, dispatch, and delivery processes of self-supporting towers is of paramount importance. The research focuses on the self-supporting towers, used as a structure for the installation of telecommunication equipment, such as radio frequency and microwave antennas. These towers have different heights and models, depending on customer's preferences and the installation site. The percentage of towers that arrived in incomplete condition to the local warehouses during the study period is approximately 92% of the total dispatches. Although these products can be completed or their missing parts can be sent to the logistics nodes, this generates additional costs for the enterprise that comprise expenditures, such as transportation, raw materials, services, labor.

In view of the problems and the low effectiveness of the solutions proposed in similar industries of the same size, we propose the eight steps of practical process improvement focused on the corporate philosophy of the IKEA case which establishes an order and strategy according to the parts to be assembled, considering their size and weight.

2 State of the Art

This study focuses on the manufacturing of self-supporting towers, very rigid free-standing structures, with a proper foundation to resist the forces to which they are exposed. The geometry of these towers depends on the tower's height, location, and manufacturer. Prior to their implementation, these towers are assembled in a kit of parts for their subsequent relocation and final implementation. Since a complete kit is required, we propose some methodologies applied to research.

The complete kit concept consists of components, tools, and information. Starting a job with an incomplete kit entails more labor time, low throughput, poor quality, increased storage area, and failure to meet the kit delivery date. As past experiences in manufacturing and subsequent assembly of parts into an incomplete kit show, 40% more labor time is spent if you work with an incomplete kit. Therefore, we focused on the philosophy of the eight steps of practical process improvement in order to improve processes, plant design, reduce hu-man effort, and reduce the amount of raw materials used [2].

We apply the methodology developed in the IKEA case study since it does not present packaging or logistics problems related to missing parts, given its control over most of the supply chain, which allows to study the impact of product and packaging design decisions in the supply chain [3]. An IKEA structure can be mounted in a structured environment due to the appropriate process of visual and tactile localization, movement planning, control force, and bimanual part coordination, and significant progress in robotic research, which encompasses the vision, planning, control, integration, and implementation of robotics software [4]. By considering "the whole package", i.e., the product, the packaging, and the supply chain characteristics in the product development stage, large savings can be achieved thanks to this methodology.

3 Contribution

Kit assembly is a very complex process in companies that ship to a customer who does not have a face-to-face meeting with the supplier, as the final customer is not aware of the status of their orders until receiving the product and checking whether the order is complete or not. In such cases, many companies pursue excellence in the process of assembling their packages, to ensure the quality of their shipment and reduce costs regarding reprocessing or reshipment of products that have already been delivered to the final customer.

3.1 The IKEA Method

IKEA is the world's best-known home furnishing retailer, since it is present in several markets with either physical or online stores. IKEA's production method consists of and is based on the design of each of its products, i.e., each new product design is accompanied by a specific production, selection, packaging, and distribution line. Therefore, each package is different and adapts to the types of parts of the final product. In that respect, we should note that IKEA's success mainly consists of the design of its products from the beginning, as it adapts the plant, which has sufficient manufacturing flexibility, to the different product designs defined.

Based on these kit assembly methods, which have proven to be very successful worldwide, we present the comparison matrix, which analyzes the dimensions taken into account for each method and the one related to the proposal.

Table 1. Proposed model

	Design-focused	Process-focused	Automated	Requires a management software	Requires more techniques within the reach of SMEs	Applies to any similar industry
LEGO		x	x	x		
IKEA	x		x	x		x
Proposed	x	x			x	x

As shown in the Table 1, the proposed method to be detailed in the development section mainly stands out due to the adaptation to small companies and management through more simplified engineering methods that do not generate the extra costs of a sophisticated software.

3.2 Overview and Components

Below is the main diagram of the proposed process, explaining each of its components. See Fig. 1.

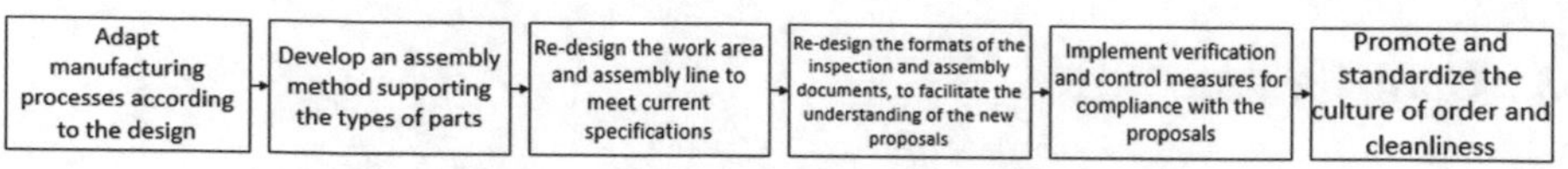

Fig. 1. Main proposal diagram

3.3 Process View

Below is the flowchart with the main processes highlighted. See Fig. 2.

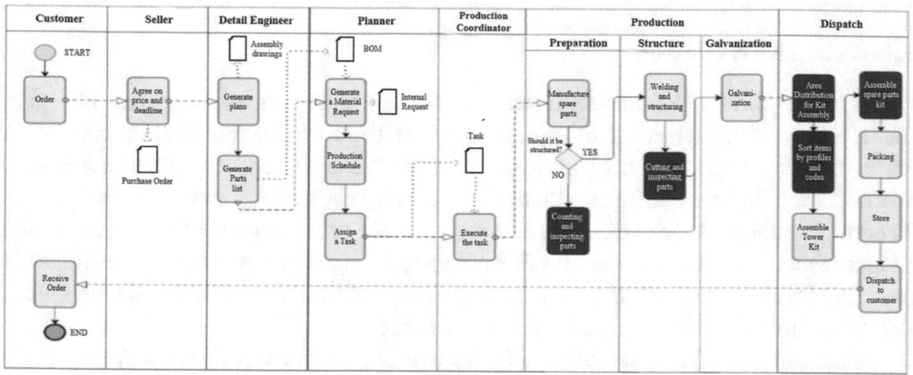

Fig. 2. Implementation design flowchart

3.4 Result Measurement Indicators

To validate the results, we will use the following indicators: % of properly assembled towers and % of poorly manufactured parts.

Using these indicators, we will analyze the number of towers properly assembled on the assigned site, to check that all the kits arrived in complete condition.

3.5 Development

We developed each of the proposals for improvement together with the time and motion study methodology, given that the root cause is a deficient working method.

3.6 Work Area Redistribution Using SLP

Since the parts do not currently have an established order, they are scattered in random places, thus operators do not know exactly in which places to look for and place the parts at the time of tower kits assembly. Then, we segmented the parts by type and specified the stacking area.

After characterizing the types of parts, we created the part relationship table to identify the positions in which each part should be placed (Fig. 3).

After that, we developed the part relationship diagram so as to properly locate the parts according to the most necessity relationships between them.

Finally, we adapted the parts according to the part relationship diagram (see Fig. 4).

Considering everything that was calculated with the SLP method, the new Packing List format aims to help the operator understand and place the right parts based on the 3 main order vectors that facilitate proper stacking.

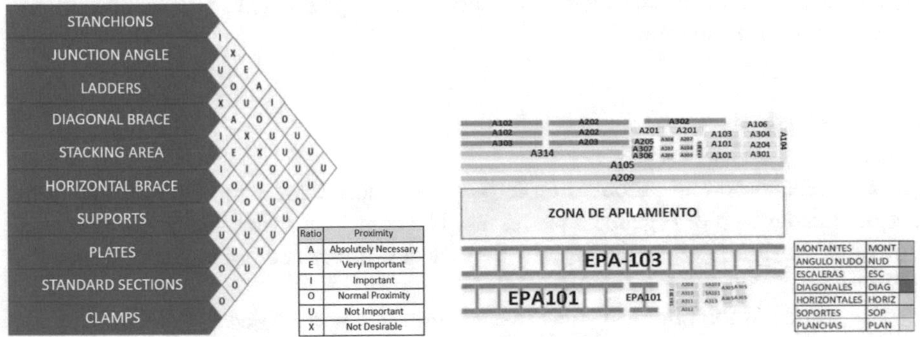

Fig. 3. Part relationship table for work area distribution

Fig. 4. Proposed work area distribution for 15-T1 tower

4 Validation

So as to validate the solution proposal presented in the previous chapter, we executed a pilot plan consisting of the implementation of our improvement method in the assembly area of telecommunication towers kits.

For this purpose, we drew up a schedule with the main activities to be carried out. First, in the planning, we established that the activities would be developed in 10 days, but in the execution process, the schedule was readjusted. The activities were as follows: 1 Proposal review, 2 Meeting with management for proposal presentation, 3 Planning of H 15 type 1 tower area, 4 Training and talks on the new work area distribution, 5 Training and talks on the new packing list design, 6 Training and talks on the new design of parts kit assembly and stacking, 7 Tower assembly with new work area distribution, 8 Pilot test, 9 Analysis of pilot results, 10 Proposal control, 11 Follow-up and comments, 12 Proposal adjustment as required, 13 Impact review, 14 Analysis of economic impact, 15 Conclusions.

We experimented with 15 towers to which we applied the proposed distribution and the redesign of their master lists, in order to implement the improvement.

As can be seen, after applying the proposal, we achieved a benefit-cost ratio of 2.42, which translates into a return of 2.42 Soles for each Sol invested (Table 2).

Table 2. Economic validation of the proposal

Current		
Non-quality costs	S/. -42,500.00	Monthly
Proposed		
Implementation cost	S/. -12,397.00	To be implemented
Profit	S/. 42,500.00	Monthly
Net profit	S/. 30,103.00	
Benefit-cost ratio	2.42	

Therefore, the proposal is valid and should be implemented in all associated processes to improve the return.

5 Conclusions

We have diagnosed and proposed an improvement in the kit assembly area due to the lack of standardized or organized processes. The contribution applied to the case study showed that the problem could be solved by means of engineering tools and a time and motion study.

The results of the pilot test show that the improvement project entailed a reduction in the number of missing parts of telecommunication tower kits, resulting in 100% effectiveness of the proposed improved method.

References

1. Travanca, R., Varum, H., Vila Real, P.: Los últimos 20 años de estructuras de telecomunicaciones en Portugal. Engineering Structures, 48, 472-48 (2013)
2. Ronen, B.: The complete kit concept. Int. J. Prod. Res. **30**(10), 2457–2466 (1992)
3. Zare, R., Chavez, P., Raymundo, C., Rojas, J.: Collaborative culture management model to improve the performance in the inventory management of a supply Chain. In: 2018 Congreso Internacional de Innovacion y Tendencias en Ingenieria, CONIITI 2018 - Proceedings 8587073 (2018)
4. Suárez-Ruiz, F., Zhou, X., Pham, P.-C.: Can robots assemble an ikea, School of Mechanical and Aerospace Engineering, Nanyang Technological University, Singapore (2018)

An Analysis of Critical Success Factors of Implementation of Green Supply Chain Management in Indian Tube Manufacturing Industries

Abhyuday Singh Thakur[1(✉)], Sagarkumar Patel[2(✉)], Aditi Chopra[1(✉)], and Vinay Vakharia[1]

[1] Padit Deendayal Petroleum University, Gandhinagar, Gujarat, India
{abhyuday.tmc14, aditi.cch14, vinay.vakharia}@sot.pdpu.ac.in

[2] G. H. Patel College of Engineering and Technology, Anand, Gujarat, India
sagarchandravadan@gmail.in

Abstract. Green Supply Chain Management (GSCM) has emerged as an organizational philosophy to strive for sustainability excellence through the integration of ecological concepts with Traditional Supply Chain Management (TSCM) principles. Due to increasing environmental, economic and regulatory issues there is an urgent need to implement GSCM but Indian tube manufacturing SMEs are struggling to do it. This paper aims to identify various critical success factors to implement GSCM in Indian tube manufacturing SMEs by consultation with the industry experts and academicians. Interpretive Structural Modeling (ISM) technique is adopted to propose a structural model that not only helps us to understand the interrelationships between the different CSFs but also explains their interdependence in implementing GSCM. MICMAC analysis is performed to determine the driving and dependence power of the CSFs and classify them according to their significance. This approach is applied to 10 tube-manufacturing industries in the western region of India.

Keywords: Green Supply Chain Management · Interpretive Structural Modelling · Critical success factors · Sustainability · MICMAC analysis

1 Introduction

There is no denying the fact that environmental issues like pollution, climate change, resource depletion, and degradation have pervaded today's business world. With the increase in environmental awareness, some industries especially in the manufacturing sector have started taking steps towards cleaner production. Implementing Green Supply Chain Management (GSCM) is one such innovative methodology that helps to reduce the environmental impact of the supply chain. Green Supply Chain Management is a combination of environment-sensitive thinking and supply chain management, encompassing product design, selection & sourcing of material, manufacturing

T. Ahram et al. (Eds.): IHSED 2019, AISC 1026, pp. 981–993, 2020.
https://doi.org/10.1007/978-3-030-27928-8_147

processes, final product delivery to the consumer and end-of-life management of the product [1]. Herren and Hadley [2] have mentioned the fact that small and medium enterprises (SMEs) face a number of barriers while initiating and implementing environmentally friendly activities in their supply chain, so there is an indisputable need for the industries to analyse these barriers/critical success factors to improve the sustainability of the organization. Research from recent years has indicated that sustainable supply chain creation is a difficult and challenging task [3–11]. The manufacturing SMEs have a lot of potential to convert their traditional supply chain into a better and greener supply chain but they are not quite able to achieve the intended result. The primary objectives of this study are:

1. Identification of CSFs to implement GSCM in Indian tube manufacturing SMEs.
2. Analysing the contextual relationship between the identified CSFs and determining the dominant CSFs by developing a structured ISM technique based model.
3. Classification of CSFs into four categories based on driving and dependence power by using MICMAC analysis.

2 Literature Review

2.1 Background of GSCM in Indian Context

During the past decade, there has been a steady increase in environmental concerns which has promoted a consensus that environmental pollution issues due to industrial development should be tackled in conjunction with supply chain management, thereby being conducive to green supply chain management [12].

Mathiyazhagan et al. [13] analysed the barriers to implement GSCM in Indian auto component manufacturing industries located in the state of Tamilnadu, India. Mudgal et al. [14, 15] pointed out in their research that due to the presence of different barriers and CSFs, implementation of green practices is difficult. Muduli and Barve [16, 17] explored the green issues in the Indian mining industry and differentiated between challenges faced by mining industries to green their supply chain. Luthra et al. [18] suggested 26 barriers to the adoption of GSCM practices in Indian automobile industries and also proposed an ISM technique based model. Mohanty and Prakash [19] empirically examined the adoption of GSCM practices in Micro, Small and Medium Enterprises in India and proposed that Indian MSMEs are being pressured by external stakeholders significantly to adopt GSM practices.

There are different critical success factors carrying varying impacts on the successful implementation of GSCM/SSCM practices. Therefore, a compelling need arises to develop a hierarchical model that differentiates between more dominant and less dominant CSFs. There has not been any attempt made to develop such a model for Indian tube manufacturing industries, mainly MSEs, in the state of Gujarat (western region of India), specifically.

2.2 Interpretive Structural Modeling Approach

Interpretive Structural Modeling is a decision-making tool that employs a combination of knowledge and experience of participants to identify and summarize contextual relationships between considered elements associated with the problem to be analyzed, and build a comprehensive structured model [61].

There have been many studies to explore the relationships between barriers, drivers and critical success factors using the ISM technique. Ravi and Shankar [62] identified 11 barriers to adoption of reverse logistics in automobile industries and used ISM technique to analyse the inter-relationships among the barriers. Kannan and Haq [63] analysed the relationships between criteria and sub-criteria influencing supplier selection for built-in supply chain using ISM. Singh et al. [64] used ISM to analyse elements in knowledge management in manufacturing industries. Diabat and Govindan [65] adopted ISM methodology to identify and analyse barriers to green supply chain management.

3 Problem Description

TSCM has economic profitability as its sole objective whereas GSCM has a more balanced approach and takes into account both ecological and economic considerations. The Indian tube manufacturing market is growing day by day. Due to globalization, a major chunk of manufacturing capacity is catering towards the export market also. It is a very profitable business but the competition is rising exponentially. Auto-mobile manufacturers, heat exchanger industries, refineries and many MNCs in the western region of India tend to buy their pipes and tubes from local manufacturing plants to increase their profit margin. However, these bigger companies are very dedicated to the cause of sustainability and hence, want their supplier cooperation in this mission. They want their suppliers to be environmentally conscious and adopt GSCM in their practices also. Thus, the Indian tube manufacturing industries are under a lot of pressure to green their supply chain. The reason to analyse the CSFs is due to the fact that the tube manufacturing industries are struggling to implement GSCM. The resource utilization is ineffective coupled with increasing industrial pollution, their transformation to a sustainable supply chain is very difficult.

4 Solution Methodology

4.1 Interpretive Structural Modeling

The ISM methodology is decision-making technique that clearly structures unclear models of systems. It establishes a hierarchy of importance of elements/factors and presents it in a well-defined form. Though ISM methodology has a number of benefits but its usage also has disadvantages like the bias of the person who is judging the variables in the system. The judgement relies on the past experience of the person, familiarity of the operation and the type of industry which can lead to deviation in the final model.

4.2 Questionnaire Development

In order to further analyze the identified CSFs, 15 Indian tube manufacturing industries mainly located in the state of Gujarat, India, were approached by e-mail, phone calls, and direct visits. Through this initial contact, industry experts were explained about the concepts of green supply chain management and objectives of the study. After frequent discussions with the experts, only 10 agreed to participate and contribute to the research. All the industry experts selected had more than 10 years of experience in the supply chain domain. In the same way, we contacted, 6 academic experts out of which 4 expressed interest to be a part of this research. This led to a formation of a decision team of 14 members constituting 6 supply chain general managers, 4 environmental managers and 4 university professors specialized in the field of sustainable development. Initially, 20 CSFs were selected from the literature review. Then a questionnaire was developed with a list of identified CSFs and their descriptions, asking whether these CSFs were significant in tube manufacturing industries. This questionnaire was sent through the mail to the decision team of 14 members. The panelists that agreed to participate came to the conclusion that 16 CSFs mentioned in the list were appropriate and can be used for further investigation. Table 1 shows the list of 16 CSFs and studies from which they were identified.

Table 1. List of 16 critical success factors

CSF #	Critical success factors	References
1	Initiation and commitment showed by top management	[20–24]
2	Government regulations and legislation	[24–28]
3	Supply chain department's awareness	[29–31]
4	Company's mission towards environment	[32–34]
5	Proper human resource management	[34–36]
6	IT enablement	[37–39]
7	Environmental collaboration with suppliers & vendors	[40–48]
8	Enhancement of brand image	[49–52]
9	Customer support and encouragement	[50–53]
10	Lack of natural resources	[49–59]
11	Societal pressures/issues	[53, 54]
12	High energy consumption rates	[55, 56]
13	Reusing and recycling materials and packaging	[55–59]
14	High cost for disposal of hazardous materials/products	[55, 56]
15	Economic advantages	[57–59]
16	Company's competitiveness	[57–60]

4.3 Data Collection

The ISM method uses expert opinions and judgements to identify the contextual relationships between the CSFs. Thus, in this research, to establish these relationships several meetings and discussions were done with the decision-making team members where they were asked to analyse and establish the contextual relationships of 'lead to' type which means one variable (i.e. barrier) leads to another. In accordance with the principle, the contextual relationships between the CSFs were developed.

4.4 SSIM Development

After the formation of contextual relationships of 'leads to' type, the next step in the ISM method consisted of developing a self-interaction matrix (SSIM) which indicates the pairwise relationship among CSFs. In order to do this, the following four symbols were used to denote the direction between the barriers (i and j) (Table 2):

Table 2. Notations used in SSIM

V	Barrier/CSF i will help or support to achieve barrier/CSF j
A	Barrier/CSF j will help or support to achieve barrier/CSF i
X	Barrier/CSF I and barrier/CSF j will help or support to achieve barrier/CSF j
O	Barrier/CSF i and barrier/CSF j are unrelated

The SSIM for CSFs to implementation of GSCM is summarized in Table 3.

4.5 Reachability Matrix Formulation

In this step, we derive the initial reachability matrix from the SSIM developed in the previous step by replacing V, A, X and O by 1 and 0. This conversion was done according to the following rules:

- If the (i, j) entry in the SSIM is V, the (i, j) entry in the reachability matrix is set to 1 and the (j, i) entry is set to 0.
- If the (i, j) entry in the SSIM is A, the (i, j) entry in the reachability matrix is set to 0 and the (j, i) entry is set to 1.
- If the (i, j) entry in the SSIM is X, the (i, j) entry in the reachability matrix is set to 1 and the (j, i) entry is set to 1.
- If the (i, j) entry in the SSIM is O, the (i, j) entry in the reachability matrix is set to 0 and the (j, i) entry is set to 0.

After developing the initial reachability matrix, the following step involved the development of final reachability matrix by application of transitivity rule [65]. The final reachability matrix is shown in Table 4.

Table 3. Self-interaction matrix

# CSFs	16	15	14	13	12	11	10	9	8	7	6	5	4	3	2	1
1	V	X	A	V	A	A	A	V	V	V	V	V	V	V	A	.
2	V	V	O	V	A	A	A	V	V	V	V	V	V	V	.	
3	V	V	A	V	O	A	A	V	V	V	V	V	A	.		
4	V	V	A	X	A	A	A	V	V	V	V	V	.			
5	V	V	O	V	O	A	A	V	V	V	A	.				
6	V	V	A	V	O	A	A	V	V	V	–					
7	V	V	A	V	A	A	A	V	V	–						
8	X	X	A	A	O	A	A	A	–							
9	V	V	A	A	O	A	A	–								
10	V	V	V	V	V	V	–									
11	V	V	V	V	V	–										
12	V	V	O	X	–											
13	V	V	A	–												
14	V	V	–													
15	X	–														
16	–															

Table 4. Final reachability matrix

CSFs No.	1	2	3	4	5	6	7	8	9	10	11	12	13	14	15	16	Driver power
1	1	0	1	1	1	1	1	1	1	0	0	1	1	0	1	1	13
2	1	1	1	1	1	1	1	1	1	0	0	1	1	0	1	1	13
3	1	0	1	1	1	1	1	1	1	0	0	1	1	0	1	1	12
4	1	0	1	1	1	1	1	1	1	0	0	1	1	0	1	1	12
5	1	0	1	1	1	1	1	1	1	0	0	1	1	0	1	1	12
6	1	0	1	1	1	1	1	1	1	0	0	1	1	0	1	1	12
7	1	0	1	1	1	1	1	1	1	0	0	0	1	0	1	1	11
8	1	0	1	1	1	1	1	1	1	0	0	0	1	0	1	1	11
9	1	0	1	1	1	1	1	1	1	0	0	1	1	0	1	1	12
10	1	1	1	1	1	1	1	1	1	1	1	1	1	1	1	1	16
11	1	1	1	1	1	1	1	1	1	0	1	1	1	1	1	1	15
12	1	1	1	1	1	1	1	1	1	0	0	1	1	0	1	1	13
13	1	0	1	1	1	1	1	1	1	0	0	1	1	0	1	1	12
14	1	0	1	1	1	1	1	1	1	0	0	1	1	1	1	1	13
15	1	0	1	1	1	1	1	1	1	0	0	1	1	0	1	1	11
16	1	0	1	1	1	1	1	1	1	0	0	1	1	0	1	1	11
Dependence power	16	4	16	16	16	16	16	16	16	1	2	12	16	3	16	16	

4.6 Level Partitioning

From the final reachability matrix, we obtained the reachability set and antecedent set for every CSF [61]. The reachability set for an individual CSF includes the CSF itself and other CSFs, which it may help to achieve. The antecedent set for an individual CSF includes the CSF itself and other CSFs, which may help in achieving it. From the reachability and antecedent set, we derive the intersection set. The CSF for which the reachability set and antecedent set are the same is given the top level position and is considered level I in ISM hierarchy [63]. This level of partitioning concludes the first iteration. After the completion of the first iteration, the CSFs forming the level I are discarded and the same procedure is performed with the remaining CSFs. The last iteration is completed when levels of all the CSF are reached, see Table 5. The partition levels help in building the digraph and finally the interpretative structural model.

Table 5. Level partitioning result of CSFs

CSFs #	Reachability Set	Antecedent Set	Intersection Set	Level
1	1,3,4,5,6,7,8,9,12,13,15,16	1,2,3,4,5,6,7,8,9,10,11,12,13,14,15,16	1,3,4,5,6,7,8,9,12,13,15,16	I
2	1,2,3,4,5,6,7,8,9, 12,13,15,16	2,10,11,12	2,12	II
3	1,3,4,5,6,7,8,9,12,13,15,16	1,2,3,4,5,6,7,8,9,10,11,12,13,14,15,16	1,3,4,5,6,7,8,9,12,13,15,16	I
4	1,3,4,5,6,7,8,9,12,13,15,16	1,2,3,4,5,6,7,8,9,10,11,12,13,14,15,16	1,3,4,5,6,7,8,9,12,13,15,16	I
5	1,3,4,5,6,7,8,9,12,13,15,16	1,2,3,4,5,6,7,8,9,10,11,12,13,14,15,16	1,3,4,5,6,7,8,9,12,13,15,16	I
6	1,3,4,5,6,7,8,9,12,13,15,16	1,2,3,4,5,6,7,8,9,10,11,12,13,14,15,16	1,3,4,5,6,7,8,9,12,13,15,16	I
7	1,3,4,5,6,7,8,9,13,15,16	1,2,3,4,5,6,7,8,9,10,11,12,13,14,15,16	1,3,4,5,6,7,8,9,13,15,16	I
8	1,3,4,5,6,7,8,9,13,15,16	1,2,3,4,5,6,7,8,9,10,11,12,13,14,15,16	1,3,4,5,6,7,8,9,13,15,16	I
9	1,3,4,5,6,7,8,9,12,13,15,16	1,2,3,4,5,6,7,8,9,10,11,12,13,14,15,16	1,3,4,5,6,7,8,9,12,13,15,16	I
10	1,2,3,4,5,6,7,8,9, 10,11,12,13,14,15,16	10	10	IV
11	1,2,3,4,5,6,7,8,9, 11,12,13,14,15,16	10,11	10,11	III
12	1,2,3,4,5,6,7,8,9, 12,13,15,16	1,2,3,4,5,6,9,10,11,12,13,14	1,2,3,4,5,6,9,12,13	II
13	1,3,4,5,6,7,8,9,12,13,15,16	1,2,3,4,5,6,7,8,9,10,11,12,13,14,15,16	1,3,4,5,6,7,8,9,12,13,15,16	I
14	1,3,4,5,6,7,8,9,12,13,14,15,16	10,11,14	14	II
15	1,3,4,5,6,7,8,9,13,15,16	1,2,3,4,5,6,7,8,9,10,11,12,13,14,15,16	1,3,4,5,6,7,8,9,13,15,16	I
16	1,3,4,5,6,7,8,9,13,15,16	1,2,3,4,5,6,7,8,9,10,11,12,13,14,15,16	1,3,4,5,6,7,8,9,13,15,16	I

4.7 Formation of ISM Model

With the help of the level partitioning show in Table 5, the ISM Structural model of the various critical success factors important in implementing green supply chain management for the manufacturing industries under study is produced and shown in Fig. 1.

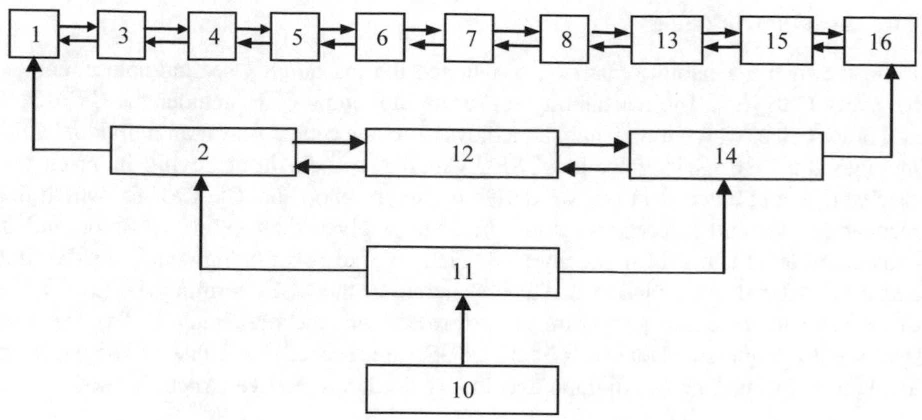

Fig. 1. Interpretive structural modelling for CSFs

4.8 MICMAC Analysis

In MICMAC analysis, the objective is to investigate the driving and dependence power of the CSFs under consideration (Ravi and Shankar 2005) by developing a digraph. According to Kumar et al. (2016), the summation of entries of possibilities of interaction in the rows gives us the driving power of the individual barrier/CSF whereas summation of entries of possibilities of interaction in the columns gives us the dependence power of the individual barrier/CSF. The driving and dependence powers calculated for each CSF is shown in Table 6. On the basis of the above analysis, CSFs are classified into four sectors given in Fig. 2.

Table 6. Classification of the four sectors in MICMAC analysis

Sector	Type	Driving power	Dependence power
I	Autonomous	Weak	Weak
II	Dependent	Weak	Strong
III	Linkage	Strong	Strong
IV	Independent	Strong	Weak

5 Results and Discussions

The objectives of this research study were to gain and provide knowledge in relation to the critical success factors that exert influence on the successful implementation of GSCM practices in Indian tube manufacturing industries. By application of ISM technique were successfully able to identify 16 CSFs, developed contextual relationships between them, and produced a hierarchical model for the CSFs to implement GSCM towards sustainability.

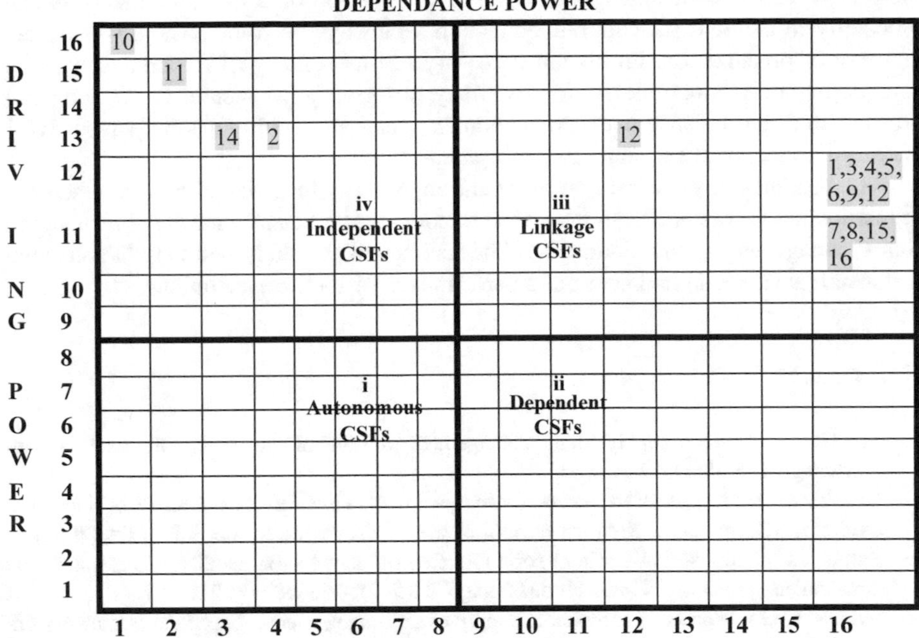

Fig. 2. MICMAC graph of CSFs to implement GSCM towards sustainability

As evident from Table 4, "Lack of natural resources" CSF has the highest driving power and the lowest dependence power. Thus, it is at the bottom-most level in the structural model. "Initiation and commitment shown by top management", "Supply chain department's awareness", "Company's mission towards environment", "Proper human resource management", "IT enablement", "Customer support and encouragement", "Reusing and recycling materials and packaging", "Environmental collaboration with suppliers & vendors", "Enhancement of brand image", "Economic Advantages" and "Company's Competitiveness" CSFs have the highest dependence power and hence, will be present at the highest level of the structural model. In the level partitioning of CSFs, we observed that the 16 CSFs were partitioned into four different levels as shown in Table 5. CSFs with highest driving power was placed at the lowest level and CSFs with the highest dependence power were placed at the uppermost level. MICMAC analysis has also been performed on the CSFs. According to the MICMAC graph, it has been observed that majority of the critical success factors lie in the linkage category. Thus, the majority of the critical success factors have strong dependence and strong driving power both.

In our research study, lack of natural resources are found to be the most important CSF and has the highest driving power. Scarcity of natural and non-renewable resources coupled with the increase in the prices of these resources will promote the replacement of these resources with recyclable and renewable resources. Industries will realize the strategic advantage and profitability that they can gain through waste management. The manufacturing sector in India has shown an increase in general awareness about sustainable practices but still, a lot of awareness issues remain especially in the SMEs. Economic benefits have always been the basic driver for any industry or organization. Profits are necessary for the running of the organization but considering only economic perspective may not give good results in the long run. Striking the right balance between economic, social and environmental perspective is the key to successful sustainable development.

The present study provides an excellent opportunity to Indian Tube Manufacturing Industries in the western region of India, to adopt and efficiently manage green supply chain management in their industries. The findings of the study will help in enhancing the sustainable, social and economic performance of their organizations.

References

1. Srivastva, S.: Green supply chain management: a state of the art literature review. Int. J. Manage. Rev. **9**(1), 53–80 (2007)
2. Herren, A., Hadley, J.: Barriers to environmental sustainability facing small businesses in Durham, NC, pp. 1–44. Nicholas School of the Environment, Master's Projects (2010)
3. Sarkis, J.: Manufacturing's role in corporate environmental sustainability—concerns for the new millennium. Int. J. Oper. Prod. Manage. **21**(5–6), 666–686 (2001)
4. Svensson, G.: Aspects of sustainable supply chain management: conceptual framework and empirical example. Supply Chain Manage. Int. J. **12**(4), 262–266 (2007)
5. Carter, C.R., Rogers, D.S.: A framework of sustainable supply chain management: moving toward new theory. Int. J. Phys. Distrib. Logist. Manage. **38**(5), 360–387 (2008)
6. Abdallah, T., Diabat, A., Rigter, J.: Investigating the option of installing small scale PVs on facility rooftops in a green supply chain. Int. J. Prod. Econ. (2013). https://doi.org/10.1016/j.ijpe.2013.03.016
7. Diabat, A., Kannan, D., Kaliyan, M., Svetinovic, D.: An optimization model for product returns using genetic algorithms and artificial immune system. Resour. Conserv. Recycl. (2013). https://doi.org/10.1016/j.resconrec.2012.12.010
8. Kannan, D., Khodaverdi, R., Olfat, L., Jafarian, A., Diabat, A.: Integrated fuzzy multi criteria decision making method and multi objective programming approach for supplier selection and order allocation in a green supply chain. J. Cleaner Prod. (2013). https://doi.org/10.1016/j.jclepro.2013.02.010
9. Preuss, L.: Addressing sustainable development through public procurement: the case of local government. Int. J. Supply Chain Manage. **14**(3), 213–223 (2009)
10. Faisal, M.N.: Sustainable supply chains: a study of interaction among the enablers. Bus. Process Manage. J.l **16**(3), 508–529 (2010)
11. Abdallah, T., Farhat, A., Diabat, A., Kennedy, S.: Green supply chains with carbon trading and environmental sourcing: formulation and life cycle assessment. Appl. Math. Model. **36**(9), 4271–4285 (2012)
12. Sheu, J.B., Chou, Y.H., Hou, C.C.: An integrated logistics operational model for green-supply chain management. Transp. Res. Part E Logist. Transp. Rev. **41**(4), 287–313 (2005)
13. Mathiyazhagan, K., Govindan, K., Noorul Haq, A., Yong, G.: An ISM approach for the barrier analysis in implementing green supply chain management. J. Clean. Prod. **47**, 283–297 (2013)
14. Mudgal, R.K., Shankar, R., Talib, P., Raj, T.: Greening the supply chain practices: an Indian perspective of enabler's relationship. Int. J. Adv. Oper. Manage. **1**(2–3), 151–176 (2009)

15. Mudgal, R.K., Shankar, R., Talib, P., Raj, T.: Modeling the barriers of green supply chain practices: an Indian perspective. Int. J. Logist. Syst. Manage. **7**(1), 81–107 (2010)
16. Muduli, K., Barve, A.: Developing a framework for study of GSCM criteria in Indian mining industries. In: Proceedings of 4th International Conference on Environmental Science and Development, vol. 5, pp. 22–26 (2011)
17. Muduli, K., Barve, A.: Sustainable development practices in mining sector: a GSCM approach. Int. J. Environ. Sustain. Dev. **12**(3), 222–243 (2013)
18. Luthra, S., et al.: An analysis of interactions among critical success factors to implement green supply chain management towards sustainability: an Indian perspective. Resour. Policy (2015). http://dx.doi.org/10.1016/j.resourpol.2014.12.006i
19. Mohanty, R.P., Prakash, A.: Green supply chain management practices in India: an empirical study. Prod. Plan. Control, 1–16 (2013). https://dx.doi.org/10.1080/09537287.2013.832822
20. Govindan, K., Kannan, D., Mathiyazhagan, K., Jabbour, A.B.L.D.S., Jabbour, C.J.C.: Analyzing green supply chain management practices in Brazil's electrical/electronics industry using interpretive structural modeling. Int. J. Environ. Stud. **70**, 1–17 (2013)
21. Min, H., Galle, W.P.: Green purchasing practices of US firms. Int. J. Oper. Prod. Manage. **21**(9), 1222–1238 (2001)
22. Zhu, Q., Sarkis, J., Lai, K.H.: Initiatives and outcomes of green supply chain management implementation by Chinese manufacturers. J. Environ. Manag. **85**(1), 179–189 (2007)
23. Zhu, Q., Sarkis, J., Lai, K.H.: Green supply management: pressures, practices and performance within the Chinese automobile industry. J. Clean. Prod. **15**(11–12), 1041–1052 (2007)
24. Zhu, Q., Sarkis, J., Lai, K.H.: Institutional-based antecedents and performance outcomes of internal and external green supply chain management practices. J. Purch. Supply Manag. **19**(2), 106–117 (2013)
25. Zhu, Q., Sarkis, J.: An inter-sectoral comparison of green supply chain management in China: drivers and practices. J. Clean. Prod. **14**(5), 71–74 (2006)
26. Zhu, Q., Cordeiro, J., Sarkis, J.: International and domestic pressures and responses of Chinese firms to greening. Ecol. Econ. **83**(1), 144–153 (2012)
27. Luthra, S., Kumar, V., Kumar, S., Haleem, A.: Green supply chain management: a literature review approach. J. Inf. Knowl. Res. Mech. Eng. **1**(1), 12–20 (2010)
28. Mitra, S., Datta, P.P.: Adoption of green supply chain management practices and their impact on performance: an exploratory study of Indian manufacturing firms. Int. J. Prod. Res. **52**(7), 2085–2107 (2014)
29. Lazuraz, L., Ketikidis, P.H., Bofinger, A.B.: Promoting green supply chain management: the role of the human factor. In: 15th Panhellenic Logistics Conference and 1st Southeast European Congress on Supply Chain Management, Greek Association of Supply Chain Management (EEL of Northern Greece), pp. 1–13 (2011)
30. Rao, P., Holt, D.: Do green supply chains lead to competitiveness and economic performance? Int. J. Oper. Prod. Manag. **25**(9), 898–916 (2005)
31. Kumar, S., Luthra, S., Haleem, A.: Customer involvement in greening the supply chain: an interpretive structural modelling methodology. J. Ind. Eng. Int. **9**(1), 1–13 (2013)
32. Tseng, M.L., Tan, R.R., Siriban-Manalang, A.B.: Sustainable consumption and production for Asia: sustainability through green design and practice. J. Clean. Prod. **40**(1), 1–5 (2012)
33. Giunipero, L.C., Hooker, R., Denslow, D.: Purchasing and supply management sustainability: drivers and barriers. J. Purch. Supply Manage. **18**(4), 258–269 (2012)
34. Muduli, K., Govindan, K., Barve, A., Kannan, D., Geng, Y.: Role of behavioural factors in green supply chain management implementation in Indian mining industries. Resour. Conserv. Recycl. **76**, 50–60 (2013)

35. Lee, S.: Drivers for the participation of small and medium-sized suppliers in green supply chain initiatives. Supply Chain Manag. Int. J. **13**(3), 185–198 (2008)
36. Luthra, S., Garg, D., Haleem, A.: Identifying and ranking of strategies to implement green supply chain management in Indian manufacturing industry using analytical hierarchy process. J. Ind. Eng. Manag. **6**(4), 930–962 (2013)
37. Tseng, M.L., Wu, K.J., Nguyen, T.T.: Information technology in supply chain management: a case study. Procedia Soc. Behav. Sci. **25**(1), 257–272 (2011)
38. Chan, H.K., He, H.W., Wang, W.Y.C.: Green marketing and its impact on supply chain management in industrial markets. Ind. Mark. Manag. **41**(4), 557–562 (2012)
39. Chen, C.C., Shih, S.H., Shyur, H.J., Wu, K.S.: A business strategy selection of green supply chain management via an analytic network process. Comput. Math. Appl. **64**(8), 2544–2557 (2012)
40. Yuang, A., Kielkiewicz-Yuang, A.: Sustainable supply network management. Corp. Environ. Manage. **8**(3), 260–268 (2001)
41. Zhu, Q., Sarkis, J., Lai, K.H.: Initiatives and outcomes of green supply chain management in China. J. Environ. Manage. **85**(1), 179–189 (2007)
42. Zhu, Q., Sarkis, J.: Relationships between operational practices and performance among followers of green supply chain management practices in Chinese manufacturing enterprises. J. Oper. Manage. **22**, 265–289 (2004)
43. Zhu, Q., Sarkis, J.: An inter-sectoral comparison of green supply chain management in China: drivers and practices. J. Clean. Prod. **14**(5), 472–486 (2006)
44. Zhu, Q., Sarkis, J., Lai, K.H.: Green supply chain management implications for closing the loop. Transp. Res. Part E Logist. Transp. Rev. **44**(1), 1–18 (2008)
45. Zhu, Q., Sarkis, J., Cordeiro, J.J., Lai, K.H.: Firm-level correlates of emergent green supply chain management practices in the Chinese context. Omega **36**, 577 (2008)
46. Zhu, Q., Sarkis, J., Geng, Y.: Green supply chain management in China: pressures, practices and performance. Int. J. Oper. Prod. Manage. **25**(5/6), 449–469 (2005)
47. Zhu, Q., Sarkis, J., Lai, K.: Green supply chain management: pressures, practices and performance within the Chinese automobile industry. J. Clean. Prod. **15**(11), 1041–1052 (2007)
48. Zhu, Q., Sarkis, J., Lai, K.: Confirmation of a measurement model for green supply chain management practices implementation. Int. J. Prod. Econ. **111**(2), 261–273 (2008)
49. Poksinska, B., Dahlgaard, J.J., Eklund, J.: Implementing ISO 14000 in Sweden: motives, benefits and comparisons with ISO 9000. Int. J. Qual. Reliab. Manage. **20**(5), 585–606 (2003)
50. Testa, F., Iraldo, F.: Shadows and lights of GSCM (green supply chain management): determinants and effects of these practices based on a multinational study. J. Clean. Prod. **18**(10), 953–962 (2010)
51. Faisal, M.N.: Sustainable supply chains: a study of interaction among the enablers. Bus. Process Manag. J. **16**(3), 508–529 (2010)
52. Gandhi, N.M.D., Selladurai, V., Santhi, P.: Green productivity indexing: a practical step towards integrating environmental protection in to corporate performance. Int. J. Product. Perform. Manage. **55**(7), 594–606 (2006)
53. Maloni, M.J., Brown, M.E.: Corporate social responsibility in the supply chain: an application in the food industry. J. Bus. Ethics **68**(1), 35–52 (2006)
54. Luthra, S., Garg, D., Haleem, A.: Identifying and ranking of strategies to implement green supply chain management in Indian manufacturing industry using analytical hierarchy process. J. Ind. Eng. Manage. **6**(4), 930–962 (2013)

55. Govindan, K., Kaliyan, M., Kannan, D., Haq, A.N.: Barriers analysis for green supply chain management implementation in Indian industries using analytic hierarchy process. Int. J. Prod. Econ. **147**, 555–568 (2014)
56. Gunasekaran, A., Spalanzani, A.: Sustainability of manufacturing and services: investigations for research and applications. Int. J. Prod. Econ. **140**(1), 35–47 (2012)
57. Ageron, B., Gunasekaran, A., Spalanzani, A.: Sustainable supply management: an empirical study. Int. J. Prod. Econ. **140**(1), 168–182 (2012)
58. Gopalakrishnan, K., Yusuf, Y.Y., Musa, A., Abubakar, T., Ambursa, H.M.: Sustainable supply chain management: a case study of British Aerospace (BAe) Systems. Int. J. Prod. Econ. **140**(1), 193–203 (2012)
59. Nimawat, D., Namdev, V.: An overview of green supply chain management in India. Res. J. Recent Sci. **1**(6), 77–82 (2012)
60. Chen, Y.S., Lai, S.B., Wen, C.T.: The influence of green innovation performance on corporate advantage in Taiwan. J. Bus. Ethics **67**(4), 331–339 (2006)
61. Warfield, J.N.: Developing subsystem matrices in structural modeling. IEEE Trans. Syst. Man Cybern. **SMC-4**, 74–80 (1974). https://doi.org/10.1109/tsmc.1974.5408523
62. Ravi, V., Shankar, R.: Analysis of interactions among the barriers of reverse logistics. Int. J. Technol. Forecast. Soc. Change **72**(8), 1011–1029 (2005)
63. Govindan, K., Haq, A.: Analysis of interactions of criteria and sub-criteria for the selection of supplier in the built-in-order supply chain environment. Int. J. Prod. Res. **45**, 3831–3852 (2007). https://doi.org/10.1080/00207540600676676
64. Singh, M.D., Shankar, R., Narain, R., Agarwal, A.: An interpretive structural modeling of knowledge management in engineering industries. J. Adv. Manage. Res. **1**(1), 28–40 (2003)
65. Diabat, A., Govindan, K.: An analysis of the drivers affecting the implementation of green supply chain management. Resour. Conserv. Recycl. **55**, 659–667 (2011). https://doi.org/10.1016/j.resconrec.2010.12.002

Construction of a Simple Management Method in Production Using a Digital Twin Model

Masahiro Shibuya(✉)

Tokyo Metropolitan University, Tokyo, Japan
mshibuya@tmu.ac.jp

Abstract. In Industry 4.0, advocated by the German government, all the processes related to the manufacturing industry are being optimized using cyber physical systems (CPSs). However, a large initial investment is required to ensure that a factory is Industry 4.0 compliant. Small and medium-sized enterprises cannot expect a return on their investment. Many of their manufacturing methods are make-to-order production and production of numerous models in small quantities, which means that the optimization of processes is difficult. Therefore, in this research, I examine a digital twin model that enables work improvement support by creating analog-type elements, such as visualizing known integrals, and by reconciling a real factory with its virtual one instead of aiming for higher productivity using excessive digitization and automation, such as that in Industry 4.0. I have developed an architecture that makes it possible to find solutions or answers by projecting effective internet of things (IoT) and digital manufacturing (DM) uses at production sites and how to evaluate manufacturing processes on a digital twin model, using production activity based on the plan-do-check-action (PDCA) cycle widely used in the Japanese manufacturing industry. In this paper, the outline and problems of our proposed architecture are explained.

Keywords: Digital twin · Kaizen · Digital manufacturing · PDCA

1 Introduction

In Japan, the manufacturing industry has been developing rapidly, and approximately 97% of the Japanese manufacturers are small and medium-sized enterprises (SMEs). Because meeting MONOZUKURI (or manufacturing) demand is essential, their manufacturing method is the make-to-order production type, and individually, the production of numerous models is achieved in small quantities. What is manufactured is a small number of items, ones with short delivery times, those having short lifecycles, those requiring many design or delivery alterations, and other customizations. Therefore, it is difficult to optimize processes to eliminate waste. To achieve wasteless production, it is necessary to gather production-site information and alter the production plan as early in the process as possible, but the present production management system cannot handle such change easily.

For example, visualization of the operation conditions of production machines by introducing internet of things (IoT) requires a structure to output the conditions directly

T. Ahram et al. (Eds.): IHSED 2019, AISC 1026, pp. 994–999, 2020.
https://doi.org/10.1007/978-3-030-27928-8_148

through sensors, etc. There is a strong desire to introduce IoT, but the ability to implement IoT has not been accumulated, and SMEs have not been able to adopt IoT.

However, in the Japanese manufacturing industry, the method in which workers independently pursue their goals is by using various kinds of management tables of their own and repeating the plan-do-check-action (PDCA) cycle. As a result, the manufacturing department (shop floor) has managed to handle difficult demands from the planning department (head office). At production sites, analog-type management using management boards is still in use. However, there are no production management systems that enable effective support of the PDCA cycle using digital techniques. Further, very few SMEs have introduced production management systems by themselves.

Digital Twin, which is a concept to reproduce the manufacturing physical world as it is in real time by digitization to ensure easier control and management of real factories, is thus being developed extensively. In the German Industry 4.0 scheme, optimization of all the processes related to the manufacturing industry by cyber physical systems (CPSs) is undertaken. Some large enterprises (e.g., Siemens, Dassault, and PTC) use Digital Twin to support their services for customers with state-of-the-art technology (e.g., aerospace for aircraft maintenance).

F. TAO classifies the factory evolution process to achieve coalescence of physical and virtual spaces into four stages. Digital Twin is the 4th stage, and the present is the 3rd, in which physical and virtual spaces interact with each other. What is imperative in the 3rd stage is production technique support using 3D CAD, called digital manufacturing (DM). This method was developed to support the shortening of development periods by front-loading the activities from product development by concurrent engineering to production preparation and process mounting. The IEC 62832 (digital factory: DF) standard, whose aim is the integrated management of virtual reality and actual world by digitizing the entire factory, is advocated. There are no case studies reported for this standard, but this method is expected to be utilized by feeding the actual operation and production information to the virtual reality production model to re-evaluate and connect it to production factories in operation.

I aim to develop a platform to support smaller enterprises' continuous improvement activities using information and communication technologies (ICT). It is necessary to create a structure that enables the easy use of a management table as the interface of Digital Twin in analogue-type work sites. It is also necessary to be able to interact at each phase of the PDCA cycle, so that the manager can alter the production plan considering the visualized processes. Moreover, I am developing tools that are usable at each phase. In this research, by using existing ICT, I developed a system to convert the production plan made in the "planning operation" stage to various kinds of standardized work tables in a certain format using digital technique, and then, based on the converted tables, created a DM model. Furthermore, I examined the possibility of creation of an autonomous control system by re-evaluating the modified model using the data obtained at the manufacturing site. In this paper, the outline of our system its usage are explained.

2 Digital Twin Model

2.1 Shop-Floor Management Model

First, the characteristics of SMEs are made clear, and then a shop-floor model is constructed. The smaller manufacturers and factories that I investigated were always forced to take "multiproduct variable quantity production" with short delivery times to actualize large enterprises' stern demands. In factories with frequent changeovers, they have no time to practice the PDCA cycle because they proceed to the manufacture of the next item soon after the first item is completed.

Figure 1 shows a production management model of smaller manufacturers, based on our investigation results. In some-scale manufacturers, the production management (office work) and manufacturing management (factories) are separate. Suppliers called Tier 3 or Tier 4 do not need the upper part of the model, and conduct their production management as in the lower part, because they conduct their manufacturing management according to order instructions.

For example, on the shop-floor, the factory manager makes a process capacity table, and he is also supposed to grasp the varying production capacity of each process. Based on the production plan (PLAN00) made at the office side, the manager modifies the plan for practical production by experience, makes three kinds of standardized work tables, and decides takt time. The shop-floor is operated on the basis of these standardized work tables. The output is separately recorded and utilized for evaluation and improvement.

Furthermore, the manufacturing contents of many SMEs, in most cases, do not end with simple assembly, but are attended by more complicated processes such as material processing. Further, till the moment an order is received, the product specification is not confirmed in some cases, and not much preparation information is available in advance. This means materials requirement planning is not usable, as shown in the upper part of Fig. 1.

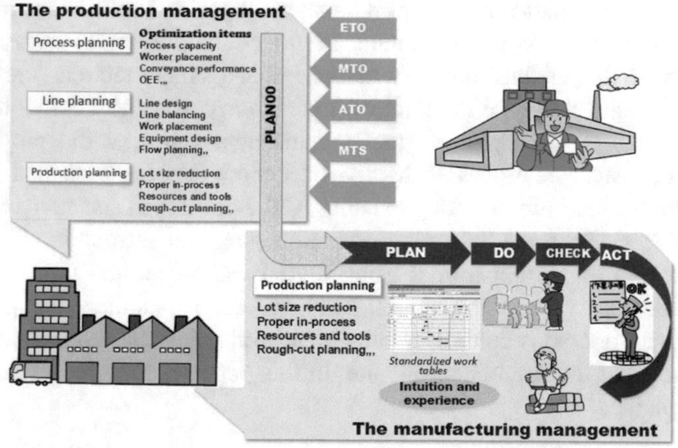

Fig. 1. A production management model of SMEs.

2.2 Physical Model

Figure 2(right) shows a generalized model of the manufacturing department in Fig. 1. This model's characteristic is that three kinds of standardized work tables are used as interfaces between the manager and his workers. One weak point here is that there is conversion work from the production plan made at the office to standardized work tables before the start of production.

What smaller enterprises wanted is

- "Standardized work combination tables" usable as a production site improvement tool
- A tool that enables the easy verification of graphical production results by inputting necessary items at the site

The figure looks good at first, but the making of manufacturing plans is a burden because the factory manager is overtasked with his routine work. To overcome this barrier at the production sites with the production of variable models in variable quantities, Genbaryoku, that is, the capability to find and solve problems in the field in which the manufacturing is good, is the key, which means that there is no choice but to expand Genbaryoku from manufacturing to production design and increase the added value.

2.3 Virtual Model

Figure 2(left) shows virtual model constructed from physical model using ICT. Here, necessary items are picked out from the production plan made at the office side. Then another production plan is made by use of a Gantt chart. DM is used for evaluation of this production plan. When a new factory is built or when a new line is introduced, DM is utilized for construction of a virtual line and its examination on the computer. It's an excellent tool. However, it has some weak points. DM is premised on use in "planning operation," and is not utilized in "production operation". DM is hard to utilize properly, and it is now experts' tool. As shown in the right, it is necessary to conduct processing work correlating DM with standardized work tables for the use of DM in the Phase "D".

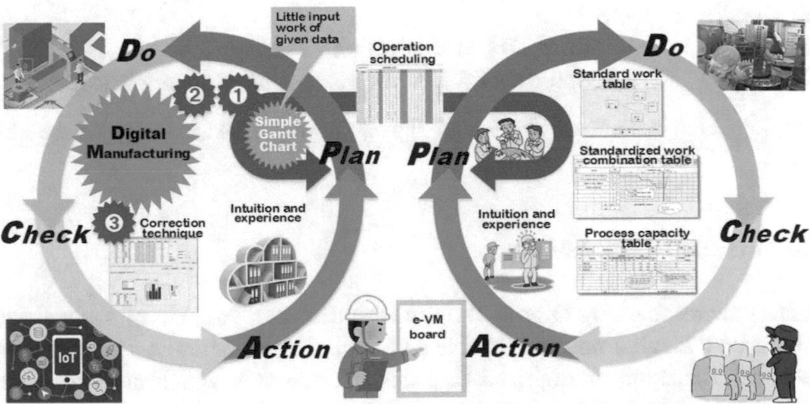

Fig. 2. Digital Twin model. Interaction exists between virtual space (left) and physical space (right).

3 Design of a Virtual Space Model

For the virtual space model, DM is utilized, but it has some problems. For example, let's take a look at the manufacturing department. The CAD data of the design department turn to bill of materials (BOM), and are linked with the manufacturing processes. The devices and jigs used in the manufacturing processes are linked as resources. DM can be utilized well only when every element has been linked.

Therefore, it is necessary to define each department accurately. At factories with the production of numerous models in small quantities, they cannot take time for this. Further, at factories with frequent production plan changes from external causes, the situation is the same. Such enterprises need to create a digital twin model, taking into consideration the fact that there is a barrier between the virtual world and the real world.

This model aims to support production management operation at the shop floor at each phase of the PDCA cycle. The requirements to actualize the model are as follows:

- Production planning with little input for the given data (items, machines, work in progress, delivery requests, etc.) and adjustment parameters
- Support for easy making of minor adjustments for production preparation, improvement, and so forth, on the basis of information obtained through various devices as well as observation, by using three kinds of standardized work tables as a user interface
- Effective use of IoT
- Use of appropriate software at the production site

I examined the structure in Phase "P" and Phase "C" so that the manager unaccustomed to ICT could easily use it. By re-designing a production plan made at the office for a more suitable one for the production site, making a standardized work combination table on the basis of the results and constructing a virtual line on DM by use of three kinds of standardized work tables, I conducted our research to develop a structure that enabled the evaluation of production plans using various tools. The following shows the procedure of function development.

(1) Investigate a production planning support method using three kinds of standardized work tables
(2) Investigate how to utilize DM in production planning
(3) Investigate the development of a tool for visualizing production work

The numbered wheels on the left of Fig. 2 show the relationships of the above 3 functions.

4 Results and Discussion

The factory represented by Digital Twin is recorded with the help of various sensors, and integrated with all kinds of available data from the production network IT systems. To deal with this kind of digitized factories, the research which aims to create the "Digital Shadow", an accurate depiction of reality is the focus. However, SMEs cannot afford to introduce IoT to all their machines because of the cost. I developed a check function which enabled the display of the difference between the real output values and the target values in Phase "Do" in mid-course or in Phase "Check" after production without use of IoT.

If IoT is introduced to the work site, the output of mid-production of a processing machine is obtainable from the sensors at intervals set beforehand. However, if not, there is no choice but to use data in the daily production report. In order to emulate the output of a manufacturing line model without IoT, I developed an adjustment function to bring the values of the acceleration speeds and other criteria of machines, conveyors and relevant equipment close to the real values for Quest. By comparing the results of a production plan simulated as it is with its revised results by use of this adjustment function, i examined whether the system was usable in Phase D or Phase C.

Using CAD on Quest, i created devices such as machines and conveyors, arranged them on the floor, and developed a function to send the report data to the devices and re-calculate. For the examination i used a daily work report for 1 item at an ice-cream factory. The manufacturing of ice cream is comprised of approximately 12 processes, and i investigated 11 processes excluding the process "Storage." The manufacturing processes from "Material blending" to "Filling process" to put frozen ice cream in ice-cream containers that were in a liquid state. I had to tax our ingenuity in "counting" for DM. Ice-cream varies in shape from powder to liquid, finally to solid matter in the course of processing. IoT devices for measuring fluid except for solid matter are too expensive for SMEs.

I reproduced a filling machine, set the speed of the conveyor for ice-cream-filled containers, and created a factory model so that the work of 3 workers conducting packing work could be truly reproduced. The number of the production estimate obtained by simulation based on scheduling results was very different from that of finished products. In the factory the ice cream in containers was dumped instead of being sent to the quick freezer until the ice cream reached the quality standards. The difference was caused by the fact that the amount of dumped ice-cream was unstable and the yield varied.

I found it necessary to conduct simulator re-calculation using information about the stirring of materials, inferior-production information and so on so as to bring in-operation values close to the model. After re-calculation, the graphed display of the gap between the simulation model and the emulation model using actual values made it possible to grasp the difference.

5 Conclusion

I created a digital twin model of the smaller manufacturers that i investigated and proposed a model using the standardized work combination table as the means of interaction between a virtual space and physical space. I created a prototype and evaluated it using factory data. From the results, I can conclude that it is possible to automatically make a DM model from three kinds of standardized work tables and examine the execution results using our visualization tool.

Privacy Concern in Mobile Payment: A Diary Study on Users' Perception of Information Disclosure

Jiaxin Zhang and Yan Luximon(✉)

School of Design, The Hong Kong Polytechnic University, Hung Hom, Kowloon, Hong Kong SAR
jx.zhang@connect.polyu.hk, yan.luximon@polyu.edu.hk

Abstract. Providing transactions with a mobile device in various use contexts, mobile payment transactions have prompted privacy concerns. However, there is still little knowledge about how users' perceptions match with the actual information disclosure in different payment scenarios in China. This study investigated users' perceived information disclosure in different use contexts of mobile payment transactions using a diary method. The results revealed that participants had serious misperception of information disclosure in the offline transactions. Findings from this study provide directions to reduce the misconception about perceived information disclosure for better adoption of mobile payment services.

Keywords: Perceived information disclosure · The actual information disclosure · Mobile payment · Quantitative diary study

1 Introduction

Mobile payments, which allow convenient and effective commerce transactions in various contexts, have raised increasing privacy concerns. In 2018, China consumers association reported that mobile applications had excessively collected users' personal information, including phone numbers, names, portraits, address lists, location data, identities, bank accounts and others [1]. Compared to other m-services, mobile payment services involve more sensitive information, such as identities, payment accounts, consumption information and asset information, and research has revealed that privacy concerns was a factor affecting mobile payment adoption in recent years [2, 3]. Despite both academic and public concerns about privacy, users' privacy awareness in mobile payment services is unclear yet. Previous research has focused on user intention to information disclosure and actual disclosure behaviors [4, 5], but very few studies have been done to understand users' awareness of information disclosure in mobile payment services.

On the other hand, context is an important factor for privacy concerns and privacy awareness [6], since mobile services can be accessed in anywhere anytime. Mobile payment services allow transactions in both offline and online scenarios, and personal

T. Ahram et al. (Eds.): IHSED 2019, AISC 1026, pp. 1000–1006, 2020.
https://doi.org/10.1007/978-3-030-27928-8_149

data could be collected through different ways. Previous research has investigated the influence of privacy risks on online services [7], while the offline services could involve more privacy risks because of the variety of use contexts.

This paper aims to investigate users' perceived information disclosure in different payment contexts by answering these research questions: Does users' perceived information disclosure match with the actual disclosure? How does the misperception of information disclosure vary from the payment contexts? To address these questions, the quantitative diary study collected users' perception of information disclosure in real life and get insight into the influence of payment contexts to users' misperception of information disclosure. This paper contributes the insights of users' perceived information disclosure in different payment contexts and investigates the gap between perceived information disclosure and actual information disclosure in mobile payment services. The findings can contribute knowledge for privacy risk education and context awareness tool design, which help users with decision-making in information disclosure.

2 Method

In order to investigate users' perceived information disclosure across different payment contexts, an online quantitative diary study was conducted. With quantitative diary method, participants could record the payment events and their perception with a pre-coded questionnaire [8]. Participants were required to record the types of information which perceived as being disclosed and the payment context in a pre-coded questionnaire. To be specific, ten types of information involved in mobile payment services were determined in this study, namely payment accounts, passwords, identities, phone numbers, address lists, social network accounts, consumption information, asset information, location data and cookies. The payment contexts were generated based on current popular payment methods in China [9, 10]. They are QR code pay, QuickPay, M-payment platform pay and In-app pay.

Participants were recruited through the social media. Each participant would firstly report their demographic information and evaluated the importance of each type of information, and then recorded payments event during 5 to 15 days. Each participant was required to report at least 10 payment events during their recoding period. Gift money was offered to compensate participants.

2.1 The Information Disclosure in Reality

The privacy policy of mobile payment platforms such as Alipay and WeChat Pay [11, 12], were investigated to allow details of information collected by the third parties in mobile payment transactions. According to the policy, only current consumption information is known by the third party when transactions are settled in offline payment scenarios (QR code pay and QuickPay), while the information disclosure in online payment scenarios (M-payment platform pay and In-app pay) is more complicated and context-based. This is because users might give away their information when they do online shopping in mobile phones (Table 1).

Table 1. The actual disclosure of information in different payment contexts.

Payment context	Types of information									
	Payment account	Password	Identity	Phone number	Address list	Social network account	Consumption information	Asset information	Location data	Cookie
QR code pay (Offline)	N	N	N	N	N	N	Y	N	N	N
QuickPay (Offline)	N	N	N	N	N	N	Y	N	N	N
M-payment platform pay (Online)	U	N	U	U	U	U	U	U	U	U
In-app pay (Online)	N	N	U	U	U	U	U	U	U	U

*Y: the information is disclosed; N: the information is not disclosed; U: the information disclosure is unclear. Whether the information is disclosed or not depends on the situation.

3 Result

Descriptive statistics analysis was applied to investigate the importance of information in mobile payment services. Then, we calculated and compared the proportion of misperception in disclosure of each type of information in different payment contexts.

3.1 Participants and Payment Events

There are 67 participants with 1094 payment events used for this paper. Twenty-seven of them were male and forty of them were female participants. The majority of participants were aged from 18 to 30 (82.1%), part of them were aged from 31 to 40 (11.9%) and 6% of them were above 40 years old.

Users also evaluated the importance of each type of information with 5 point-Likert scale. It revealed that users were concerned about all types of information, since all of information were evaluated above 4 points of the mean of importance. However, it was obvious that users considered that password, identity and asset information were the most important (the means were above 4.5). Social network account, consume information and location data were regarded as less important information (the mean was below 4.1).

3.2 Comparison of Perceived Information Disclosure and Actual Disclosure

Figure 1 shows the proportion of misperception and correct recognition of information disclosure in four payment contexts. According to the Fig. 1, users had the greatest misperception of information disclosure in payment accounts and consumption information in the offline payment contexts of QR code pay. Users mistakenly believed that payment accounts were disclosed in around 51.9% of payment events, and mistakenly thought the current consumption information was not disclosed in 53.2% of payment events in QR code pay. Users also had high proportion of misperception in location data disclosure in QR code pay, with 33.1% of events inconsistent with reality. Users correctly recognized that asset information and address list were not disclosed in most of the payment events (more than 90%) in QR code pay. In comparison, the misperception of disclosure of location data, payment accounts and consumption information were the most serious in QuickPay. Users correctly recognized that their consumption information were given out in about 53.6% of payment events in QuickPay, but they wrongly thought that payment accounts and location data were disclosed in around 50% of payment events in QuickPay. There was a big proportion of misperception in cookie disclosure in QuickPay, with 34.3% of payment events inconsistent with reality. Similar to QR code pay, users' perceptions of disclosure in asset information and address list were matched with reality in around 90% of payment events in the payment context of QuickPay. Regarding all payment events recorded in offline payment scenarios, only 13 payment events of QR code pay and 8 payment events of QuickPay were correctly recorded with the information disclosure of payment accounts and consumption information.

On the other hand, passwords would not be disclosed in both online payment scenarios of M-payment platform pay and In-app pay. Users' payment accounts are disclosed to payees in some payment contexts in M-payment platform pay, but would not be disclosed in the payment context of In-app pay according to the privacy policy of Alipay and WeChat Pay. However, users mistakenly believed that their payment accounts were disclosed in 67.7% of payment events in In-app pay. Users also mistakenly thought that their passwords were disclosed in 15.4% of payment events in M-payment platform pay and 17.2% of payment events in In-app pay. Because the disclosure of identity, phone number, address list, social network accounts, consumption information, asset information, location data and cookies were highly context-dependent, it was difficult to identify whether users' perceptions were consistent with reality or not in this study.

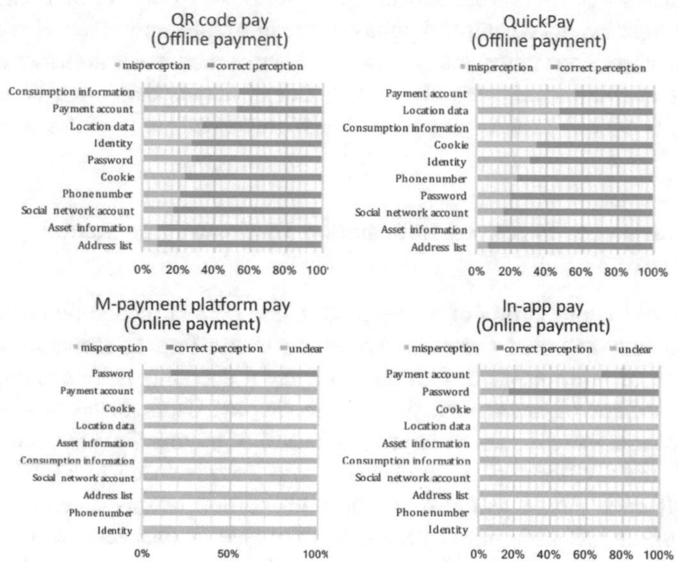

Fig. 1. Users' perception of information disclosure in four payment contexts.

4 Discussion

This study compared perceived information disclosure and actual information disclosure to understand users' privacy awareness in mobile payment services. The findings of this study revealed that users did not clearly realize what types of information were disclosed when using mobile payment services in most of the time. In the offline payment situation, users mostly believed that many types of information were disclosed, but it was not true. Many users mistakenly believed that payment accounts would be disclosed and current consumption information would not be disclosed, and location data would be disclosed in offline payment. Luckily, these three types of information were rated as less important types of information. However, the proportions of misperception in passwords and identities disclosure were remarkable, and between 18.9% and 30% of events were mistakenly believed to disclose these two types of information in offline payment scenarios. Since passwords and identities were rated as the most important information, users' misperception could increase perceived risk, which could lead to low trust to the service [13]. The results reflected that users were lacked of understating of information disclosure in offline mobile payment services. It suggested the necessity of improving users' awareness of information disclosure in mobile payment services, because perceived privacy risks could influence on the continuance adoption of mobile payment services [3]. Interface design could be a possible solution to help users better evaluate the information disclosure in mobile payment services. Notification and feedbacks about the information disclosure could be provided during the mobile payment transactions. In addition, mobile payment platforms should provide more privacy settings, and encourage users to customize the settings and control the privacy level.

This paper did not study the information disclosure in online payment contexts. The reason is that users' information disclosure behaviors in online payment contexts are highly context-dependent. Different online merchants may request users to authorize different privacy information to access the services. Since users believed that the context of In-app had involved the most information disclosure, further research about the actual information disclosure in this context would help to address users' privacy concerns better.

5 Conclusion

This paper examined users' perceived information disclosure in mobile payment services with the quantitative diary study. Comparing users perceived information disclosure and actual disclosure in offline payment contexts, results suggested that users had serious misperception in information disclosure in offline payment scenarios, and they were more likely to mistakenly perceived information disclosure in the payment context of QuickPay. This study revealed that users have not clearly recognized what kinds of information were disclosed in mobile payment services, which proposed related stakeholders an opportunity to educate or help mobile payment users with better perception of information disclosure.

Acknowledgments. The authors would like to thank the financial support provided by the UGC Funding Scheme from the Hong Kong Polytechnic University.

References

1. China consumer association. http://www.cca.org.cn/xxgz/detail/27939.html. Accessed 01 Dec 2018
2. Johnson, V., Kiser, A., Washington, R., Torres, R.: Limitations to the rapid adoption of M-payment services: understanding the impact of privacy risk on M-Payment services. Comput. Hum. Behav. **79**, 111–122 (2018)

3. Gao, L., Waechter, K., Bai, X.: Understanding consumers' continuance intention towards mobile purchase: a theoretical framework and empirical study - a case of China. Comput. Hum. Behav. **53**, 249–262 (2015)
4. Barth, S., de Jong, M.: The privacy paradox - investigating discrepancies between expressed privacy concerns and actual online behavior - a systematic literature review. Telematics Inform. **34**(7), 1038–1058 (2017)
5. Zimmer, J., Arsal, R., Al-Marzouq, M., Grover, V.: Investigating online information disclosure: effects of information relevance, trust and risk. Inf. Manag. **47**(2), 115–123 (2010)
6. Martin, K., Shilton, K.: Putting mobile application privacy in context: an empirical study of user privacy expectations for mobile devices. Inf. Soc. **32**(3), 200–216 (2016)
7. Casaló, L., Flavián, C., Guinalíu, M.: The role of security, privacy, usability and reputation in the development of online banking. Online Inf. Rev. **31**(5), 583–603 (2007)
8. Ciere, Y., Jaarsma, D., Visser, A., Sanderman, R., Snippe, E., Fleer, J.: Studying learning in the healthcare setting: the potential of quantitative diary methods. Perspect. Med. Educ. **4**(4), 203–207 (2015)
9. Payment Products of Alipay. https://intl.alipay.com/open/product.htm. Accessed 01 Aug 2018
10. Multiple Payment Methods of WeChat Pay. https://pay.weixin.qq.com/wechatpay_guide/intro_method.shtml. Accessed 01 Aug 2018
11. Alipay privacy policy. https://cshall.alipay.com/lab/help_detail.htm?help_id=201602172661. Accessed 01 Dec 2018
12. WeChat privacy policy. https://www.tenpay.com/v3/helpcenter/low/privacy.shtml. Accessed 01 Dec 2018
13. Olivero, N., Lunt, P.: Privacy versus willingness to disclose in e-commerce exchanges: the effect of risk awareness on the relative role of trust and control. J. Econ. Psychol. **25**(2), 243–262 (2004)

Democratization of Intrapreneurship and Corporate Entrepreneurship Within the McKinsey's Three Horizons Innovation Space

Evangelos Markopoulos[1(✉)], Vasu Aggarwal[1], and Hannu Vanharanta[2,3]

[1] HULT International Business School, Hult House East, 35 Commercial Rd, London E1 1LD, UK
evangelos.markopoulos@faculty.hult.edu, vaggarwal2016@student.hult.edu

[2] School of Technology and Innovations, University of Vaasa, Wolffintie 34, 65200 Vaasa, Finland
hannu@vanharanta.fi

[3] Faculty of Engineering Management, Poznan University of Technology, Poznan, Poland

Abstract. Knowledge democratization is essential to innovation strategy formulation and execution. It is the culture in which organizational strategies are embraced to create shared added-value. This paper provides a framework through which companies can develop democratic corporate entrepreneurship and intrapreneurship operations and strategies. The Company Democracy Model is used as the method based on which knowledge democratization is built by providing a structured path to satisfy the pre-conditions, post conditions and evolution of such initiatives. In this attempt, the Company Democracy Model integrates the McKinsey's 3 horizon model for organizational growth. The integration of the two models creates a knowledge based corporate entrepreneurship and intrapreneurship transformation strategy, supported by phases, stages, and goals. Furthermore, this integration is projected in a 3-dimensional space where the horizontal business development, of the McKinney 3 Horizons, affects the vertical organizational maturity, of the Company Democracy Model, through innovation development inside or outside the organization.

Keywords: Company Democracy · Corporate entrepreneurship · Intrapreneurship · Horizons · Knowledge · Innovation · Strategy · Management

1 Introduction

In the world of uncertainty, maximizing innovation and staying competitive is crucial for organizational sustainability and operations. The existing management systems empower employees of high educational, social and rank level to explore innovation opportunities on establishing corporate entrepreneurship and intrapreneurship projects and initiatives within the business. This, however, is not the most effective approach as the non-privileged employees are usually unheard and their opinions are often discarded.

T. Ahram et al. (Eds.): IHSED 2019, AISC 1026, pp. 1007–1017, 2020.
https://doi.org/10.1007/978-3-030-27928-8_150

Democratization within corporations is essential to lay a strong foundation for future innovation explorations that will prevail in the coming years. Corporations cannot fully rely on selective employees and management systems that limit the organization's potential by ignoring the intellectual capacity on their human resources. Companies must embrace a democratic process to consolidate knowledge and transform it into meaningful and applicable innovations through corporate entrepreneurship and intrapreneurship initiatives for all. However, to apply such democratization strategies it is important first to identify the human recourses willing and able to innovate, to tackle lack of confidence on those with insecurities, to overcome personal interests, and to manage the ego of the senior managers primarily.

2 Utilization of Human Intellectual Capital via Corporate Entrepreneurship and Intrapreneurship

Corporate entrepreneurship and intrapreneurship is premised upon the utilization of people's, goals, drives and intellectual capital towards their transforming it into corporate entrepreneurial assets. To identity, develop and generate such assets, it is imperative to identify and measure the degree of intellectual capital that resides in an organization and the degree of the organizational infrastructure to support it. Intellectual capital is a relatively new term in the field of business management comprised from the intangible assets of knowledge, skills and information [1]. Intellectual Capital or Human Intellectual Capital (HIC) can de defined with various terminologies, but the essence primarily revolves around the knowledge, expertise, brain power and other aspects within an organization, which is quite complex to size and measure [2]. Over the last ten years there has been tremendous effort to utilize human intellectual capital and its impact in the creation of new corporate business activities.

Organizations today divert efforts and resources towards the transformation of 'employee experience' to 'human experience', by understanding and acknowledging the aspirations of employees and by embracing the democratic workforce concept. Deloitte highlights three domains of change towards the human experience in a 3×3 matrix with rows the futures of workforce, the organization and the HR, and with columns the refresh, rewire and recode effort needed to change the status quo in the organizational world (Fig. 1) [3].

	Refresh	Rewire	Recode
Future of the workforce	• Leadership	• Alternative workforce	• Superjobs
Future of the organization	• Human experience • Rewards • Teams		
Future of HR		• Talent access • HR cloud	• Talent mobility • Learning

Fig. 1. Three domains for reinvention with approaches to change [3].

The elements that compose the matrix contribute towards improving the human intellectual capital concept for a shared, organizational and the employee's, development benefit.

3 Shared Value Innovation and Operations Optimization

Shared value innovation is an innovational dimension which integrates the shared benefits organizations and society can be obtained by working under a co-evolutionary philosophy. Effective intellectual capital utilization can lead to shared value innovation and open up new markets [4]. As the employees are part the society, their knowledge on improving the effectiveness of the company in both operations' management and new product/service development, returns back to the society in which they belong. The effective communication of such concept enables the transformation of the organization's workforce into corporate entrepreneurs where everyone has the ability, option and opportunity to make a difference for the organization, themselves and the society above all. The world is improving when the people improve it, and since not all people have the financial and organizational power to do it, this can be achieved with the proper utilization of their knowledge.

Figure 2 presents this integration of the company, market, society and the employees through intellectual capital on delivering shared value innovation under a co-evolutionary and co-developmental philosophy. The circular flow of knowledge is originated by the employees themselves. Through corporate entrepreneurial and intrapreneurial support they generate new intellectual capital-based products and services for the market to reach the society. The circle continues as the society inspires them with new ideas that ignite this repetition.

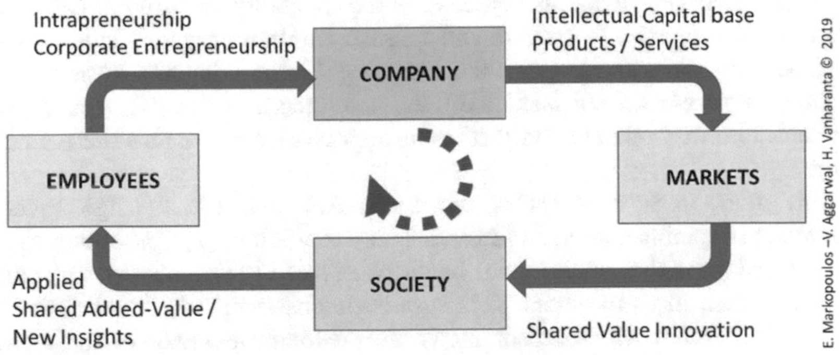

Fig. 2. Creation of shared added-value innovation for the society.

The effective support on such a circular shared value innovation approach impacts the success of the human intellectual utilization a company has. It is very difficult to measure the degree of human intellectual capital in an organization without the infrastructure needed for the employees to present their knowledge and apply it as well.

The human intellectual value of an organization is related with the success employees record on transforming their knowledge into organizational benefits either with new products and services or with organizational optimization procedures and activities.

4 Democratizing Corporate Entre- and Intrapreneurship

Industry data (shown in Fig. 3), indicates significant organizational benefits that can derive from highly engaged employees [5]. Engaged business units increase the profitability by 21% [6]. However, to achieve such results it is important to provide the freedom and the space where employees can contribute to such a goal.

Fig. 3. Employee engagement organizational benefits [5].

Practical, and actual, freedom of speech is essential for the transformation of the tacit knowledge into explicit knowledge. However, there are major significant barriers that need to be managed first, mostly dealing with effective leadership. Authentic leaders motivate and support employees to actively and effectively commit, share and communicate their knowledge. Irrespective of the capability, maturity or capacity, it is imperative for companies to operate and flourish within a corporate culture that recognizes, acknowledges and respects the employees' knowledge and ideas once properly stated. As people do not lack creativity, intelligence or passion, it is mainly the corporate culture they operate into that limits their charismatic nature and their critical thinking.

In this effort, freedom of speech can be an Aristotelian knowledge driver from observation, to experimentation and from learning to wisdom [7]. The ability of people to share their knowledge makes them better observers in their attempt to justify and communicate their ideas to others. This systematic observation is the inception to the generation of wisdom where reasoning is the predominant factor of the associated thinking. Employees who believe their voice is heard are five times more likely to feel empowered and perform at the highest potential [8]. Y-Theory believes employees are not passive and companies should provide ample opportunities and room for employee's development to harness their intellect and channelize it towards meeting strategic organizational goals [9]. Other scholars propose the Z-Theory which addresses human needs like belongingness, affiliation and trust [10]. Both theories set their based on the co-exitance of employees and organizations through the freedom to transform their knowledge into action. Such approaches on employee engagements have as common denominator the democratization of knowledge. Freedom of speech is a classic democratic characteristic in societies which can also be applied in corporations.

Over the last decade, there has been a knowledge democratization movement as corporations begin to realize the value and impact of intellectual capital that resides within them. Several management models, theories and frameworks have been developed to manage knowledge elicitation for innovation management and operations optimization, while others directly deal with the democratization concept. Such a model is the Company Democracy Model (CDM) designed on the applied philosophical principles. The model integrates several democratic values that contribute to ethical and effective utilization of the human intellectual capital [11].

CDM operates through an evolutionary process framework [12], which is based on the ancient Hellenic wisdom of the Delhpic Maxims. The model applies several maxims but mostly the 'Know Thyself', 'Metron Ariston'and 'Miden Agan' to control self-awareness, ego and exaggeration [13]. CDM uses ontologies and taxonomies to classify organizational knowledge [14]. The model works in a spiral flow which continuously turns tacit knowledge into explicit [15]. Another model which also promotes democratic knowledge creation and utilization and is aligned with the CDM is the co-Evolute methodology [16]. The base idea of the model evolves on the concept that organizations can support their employees' personal growth, vision and development in order to improve the core competencies and both sides [17].

5 The Applied DeCEIM Structure

Based on the principles of the Company Democracy Model (CDM), the Democratic Corporate Entrepreneurial and Intrapreneurial Model (DeCEIM) can be generated by altering the CDM towards a more entrepreneurial dimension for the creation of new corporate ventures. The DeCEIM sustains the levels and philosophy of the CDM which start from the organizational knowledge elicitation to its total transformation into corporate business units, startups or spin-offs (shown in Fig. 4). The first three levels of the DeCEIM follow the concepts of the relevant levels of the CDM and emphasize on the identification of the knowledge and the maturity of the employee. The other three emphasise on the entrepreneurial dimension and the knowledge evolution.

Specifically, the first level creates a culture where knowledge is selected continuously and unbiased by all employees. Engagement motives are given for all to participate in this organizational culture. A very brief business planning is required mostly on the benefits of the idea/knowledge and less on the financial or technical dimension. On level two, the ones whose knowledge contributions have been selected to evolve are supported with entrepreneurial training and the proper experts to develop their ideas furthermore into full and well thought business plans. The third level implements the idea into new organizational product or service and launches it in the markets to judge it. The degree of success will determine if the next levels will be followed.

Upon the desired success, the organization transforms the new product or service into an innovative new business unit or product line. The innovation remains within the organization but operates as a different organizational internal product or service line. The success of level four moves the product/service to level 5 where an organizational spin off is created. It is the stage where the product/service is considered mature and strong enough to get detached from the organization and be managed as a separate organization but within a distance from the base. The effectives of level five brings the product or service to level 6 where the spin-off is fully detached from the organization and becomes an independent company within the group of companies of the organization. It is the stage where the product or service expanded its functionality, client base, partnerships, etc. and can totally stand in the market on its own.

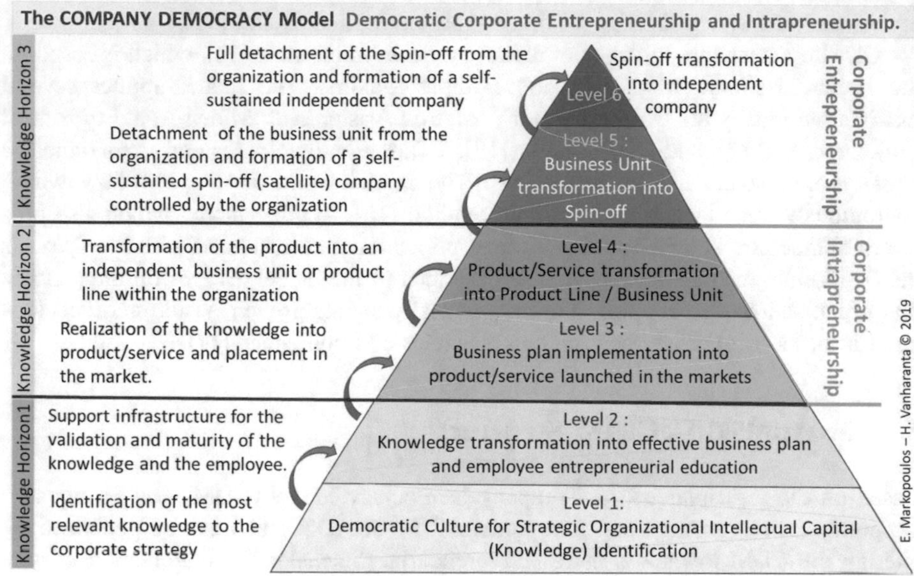

Fig. 4. The applied DeCEIM structure with the detailed activities

The organizational commitment on the DeCEIM on the knowledge utilizations is stronger in every dimension. Financial, organizational and administrative supports is fully provided, however the evaluation of the knowledge that enters DeCEIM is more demanding and rigorous. It must be noted that utilization and transformation of the organizational knowledge does not stop at level 3 where the prototypes are developed and successfully operate. Upon such success, organizations must target reaching each higher level where new corporate business units, product lines or spinoffs are created.

DeCEIM practically divides the evolution of the knowledge from an idea to an actual independent organization. However, there is another point of view in this evolution which can be extended into three knowledge maturity stages, or horizons. In the first stage the knowledge is transformed into a complete business plan (levels 1, 2), in the next the idea enters the market and obtains sustainability via corporate intrapreneurship (levels 3, 4), and the third stage detaches the operations of the idea from the organization giving it business independence, via corporate entrepreneurship (levels 5, 6). It is a journey from the stage of today to the stages of tomorrow.

6 Alignment of the DeCIEM with the McKinsey Horizons of Innovation

This triadic staged evolution of the DeCEIM horizons can be related with the McKinsey Horizons Model for organizational competitiveness.

In DeCEIM, the screening of ideas and their implementation over a time period must be related with the strategic vision of the organization. What is possible might not be feasible. Feasibility can be related to the expected success of a business initiative in specific time periods. McKinsey's 3 Horizon Model is premised upon addressing growth and innovation by assessing potential opportunities for growth in future without neglecting the performance of the present [18]. The model divides the timeframe into three horizons ranging from the core business (short-term) to the future business opportunity (long-term), with an intermediate horizon that acts a transition phase for the implementation of the innovation (shown in Fig. 5).

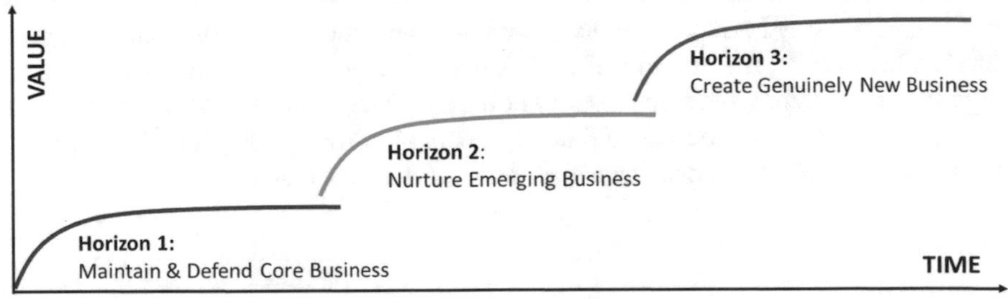

Fig. 5. McKinsey's 3 horizon model representing the three horizons of growth.

Horizon 1 refers to the core activities, assets and business of an organization. This horizon is responsible for activities that provide the current best profits. Horizon 2 refers to the emerging businesses, entrepreneurial ventures and opportunities that could result into future profits via investments. Horizon 3 refers to the creation of new businesses and opportunities that do not currently exist but can be achieved, with the activities of horizon 2, to create uncontested market spaces, or Blue oceans [19].

It is imperative to acknowledge and undertake a gap analysis between horizon 1 and horizon 3 to identify horizon 2. There are conditions that must be satisfied in every transition process. Firstly, it is important to gather a deep understanding of the core business (horizon 1) and key assets that drive revenue. As the business environment constantly changes with a dynamic nature, an organization can either imagine how to compete in horizon 1 or plan the desired future stage at horizon 3 through horizon 2, the bridge stage from the planning to the achieving.

A common denominator on the horizon's evolution equation is the quality of knowledge on understanding the current state, the future state and consequently what needs to be done in the intermediate stage. This corporate knowledge cannot be obtained effectively and accurately without the contribution of the organisation's human resources. It is the employees who are the driving force for the effective execution of business activities that take place in each horizon. Therefore, a collective understanding of the current and future state through the employees is imperative to propel the company towards growth and innovation by capitalizing on the knowledge the organization generates in the 3 horizons.

7 Implementing of the DeCEIM Within the McKinsey Horizons

The integration of the DeCEIM with the McKinsey Horizons can effectively be achieved with the utilization of the organization's intellectual capital that resides in the DeCEIM knowledge horizon 1. The democratic culture in DeCEIM horizon 1 identifies and engages the employees that can participate towards the realization of the McKinsey horizon 3. This implies that the vision will be divided into separate strategic goals and be achieved by competent and ambitious employees who are agile enough to drive this visionary change through their innovations and entrepreneurial journey within the DeCEIM. McKinsey horizon 2 involves the implementation of the strategic knowledge gained in horizon 1. The success of each initiative selected in horizon 1 at horizon 2 brings the organization one step closer to horizon 3. The degree of success of horizon 2 determines the degree of success in time and goals of horizon 3. Figure 6 highlights the role of the DeCEIM Horizons (levels) in each McKinsey Horizon.

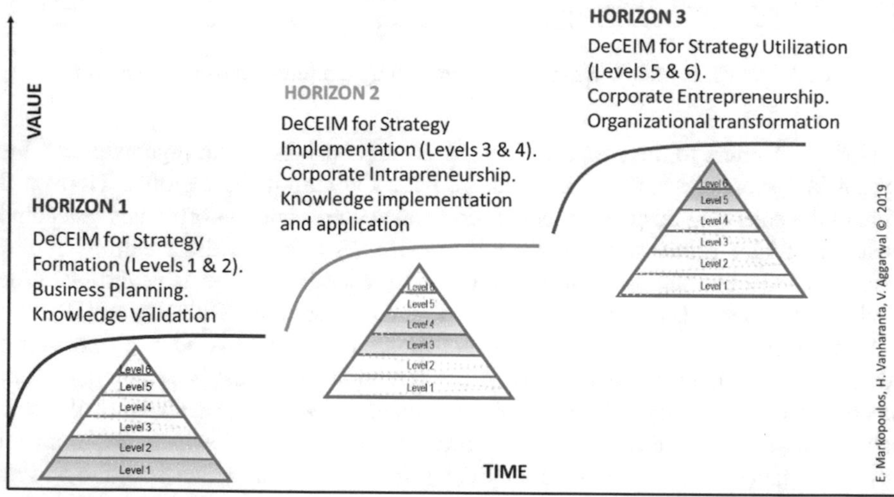

Fig. 6. A transition flow for companies under McKinsey's 3 horizon model by using the DeCEIM.

The horizon's transitions are best achieved once the organizations know the desired stage (McKinsey horizon 3). Knowing what shall be archived helps direct the activities of DeCEIM horizon 1 towards the identification of the intellectual capital that will be selected and matured in horizon 1, implemented in horizon 2 and integrated in horizon 3. A second condition to achieve McKinsey horizon 3 more effectively and minimize risk of corporate failure, is to decompose the vision of level three into quite small segments where the human recourses are able to direct their knowledge and skills through DeCEIM. Gatherings small knowledge contributions helps employees come up with more focused and valid knowledge and work (Fig. 7).

It must be noted that in the proposed horizon transition process there can be knowledge initiatives and projects that even if they could be well selected and matured in the DeCEIM levels 1 and 2, might not succeed in levels 3 and 4. This means that not all activities of McKinsey horizon 1 can move into horizon 2. This might either leave the goal on horizon 3 incomplete or extend the completion time until all the elements are in place.

Another effect this approach has is the impact of the McKinsey horizon 2 to horizon 3. As the elements of horizon 2 assimilate and integrated when reaching horizon 3, the new corporate vision, strategy might be different from the one designed in horizon 1.

This difference can be positive or negative based on the quality, the time, the innovation and the market's response indicated on each activity of horizon 2. However, it is up to the organization to accept such deviations or insist on the original goal over time. This transition flow within the McKinsey's 3 horizon, allow companies to learn from the activities of horizon 2 that could not sustain value, but also leverage upon such activities in the emerging business landscape that might be crucial for future strategy formulation and implementation.

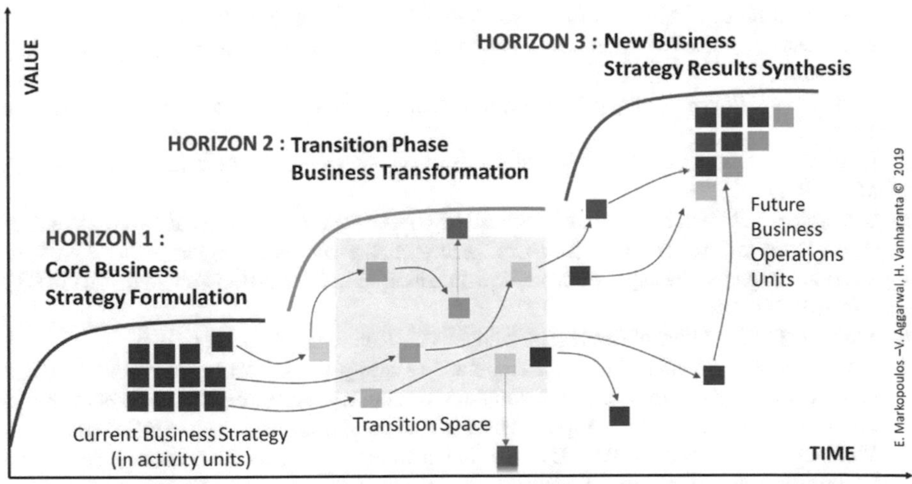

Fig. 7. A transition flow for companies under McKinsey's 3 horizon model by using the DeCEIM.

8 Conclusions

The integration of the Company Democracy Model's DeCEIM Horizons and the McKinsey's 3 Horizon model can generate a democratic business transformation strategy based on the practical utilization of the organization's human intellectual capital. Pre-conditions for such a strategy include the establishment of an ethical knowledge share culture for the two models to co-operate. Based on that, the Company Democracy Model, through DeCEIM, is evolved over the McKinsey's 3 horizon model, allowing the model to move from the current activities (horizon 1), to business transformation via corporate intrapreneurship (horizon 2) and to new and future market development via corporate entrepreneurship (horizon 3). Post conditions is the post implementation analysis of each maturity journey towards repeating the process for higher achievements and continuous development.

This knowledge driven democratic framework enables organizations to succeed, create and sustain competitive advantage by becoming confident in horizon 1, competitive upon horizon 2 and innovative upon horizon 3, within a knowledge based democratic corporate environment.

References

1. Stewart, T.A., Loose, S.: Your company's most valuable asset: intellectual capital. Fortune **130**, 68–74 (1994)
2. Abdolmohammadi, M.J.: Intellectual capital disclosure and market capitalization. J. Intellect. Cap. **6**(3), 397–416 (2005)
3. Deloitte. https://www2.deloitte.com/insights/us/en/focus/human-capital-trends.html
4. Markopoulos, E., Vanharanta, H.: The company democracy model for the development of intellectual human capitalism for shared value. Proc. Manuf. **3**, 603–610 (2015)
5. Raconteur. https://www.raconteur.net/hr/empower-and-inspire-the-right-people-and-they-will-excel
6. Gallup. https://www.gallup.com/workplace/236366/right-culture-not-employee-satisfaction.aspx
7. Oravec, C.: Observation' in aristotle's theory of epideictic. Philos. Rhetor. JSTOR **9**(3), 162–174 (1976)
8. Salesforce Research. https://c1.sfdcstatic.com/content/dam/web/en_us/www/assets/pdf/datasheets/salesforce-research-2017-workplace-equality-and-values-report.pdf
9. Versita. https://www.degruyter.com/downloadpdf/j/cris.2013.2013.issue-2/cris-2013-0012/cris-2013-0012.pdf
10. Morden, T.: Principles of Strategic Management. Routledge, London (2017)
11. Markopoulos, E., Vanharanta, H.: Democratic culture paradigm for organizational management and leadership strategies - the company democracy model. In: Proceedings of the 5th International Conference on Applied Human Factors and Ergonomics, AHFE (2014)
12. Vanharanta, H., Markopoulos, E.: Creating a dynamic democratic company culture for leadership, innovation, and competitiveness. In: 3rd Hellenic-Russian Forum, 17 September 2013
13. Parke, H., Wormell, D.: The Delphic Oracle, vol. 1, p. 389. Basil Blackwell, Oxford (1956)

14. Paajanen, P., Piirto, A., Kantola, J., Vanharanta, H.: FOLIUM - ontology for organizational knowledge creation. In: 10th World Multi-Conference on Systemics, Cybernetics, and Informatics (2006)
15. Nonaka, I., Takeuchi, H.: The Knowledge-Creating Company: How Japanese Companies Create the Dynamics of Innovation. Oxford University Press, New York (1995)
16. Kantola, J., Vanharanta, H., Karwowski, W.: The evolute system: a co-evolutionary human resource development methodology. In: Karwowski, W. (ed.) The International Encyclopedia of Ergonomics and Human Factors. CRC Press, Boca Raton (2006)
17. Kantola, J., Imran, F.: Evolute system approach and identification of talent. In: Proceedings of 86 th IASTEM International Conference, Rawalpindi, Pakistan, 29–30 October 2017, pp. 11–16 (2017)
18. McKinsey. https://www.mckinsey.com/business-functions/strategy-and-corporate-finance/our-insights/enduring-ideas-the-three-horizons-of-growth
19. Kim, W.C., Mauborgne, R.: Blue ocean strategy. Mass. Harvard Business School Press, Boston (2005)

Research on Enterprise Monopoly Based on Lotka-Volterra Model

Honghao Liu, Jian He, and Xuebo Chen(✉)

School of Electronic and Information Engineering, University of Science and Technology, Liaoning, Anshan 114051, Liaoning, China
xuebochen@126.com

Abstract. Monopoly can cause enormous market problems. The common characteristics of an enterprise evolution are much like those of a biodiversity evolution. In this paper, using the complex system theory and the ecological population evolution theory, we study and discover that an enterprise scale evolution trend accords with a Logistic model. After establishing a Lotka-Volterra evolutionary model of competition between two enterprises, we establish a Lotka-Volterra model of competition among multiple enterprises in a market. Through the analysis of these models, we can understand the essence of enterprise monopoly. By the analyses above, the government can forecast enterprise monopoly for a market in order to avoid corresponding risks early.

Keywords: Monopoly · Lotka-Volterra (LV) model · Logistic model

1 Introduction

Monopoly brings losses to the state and consumers, and it also breeds corruption. At the same time, monopoly will also hinder the development of a country into an innovative power. In this paper, the essence of monopoly is understood by modelling the competitive relationship of enterprises in the market. The earliest record of monopoly appeared in the third and fourth centuries B.C. In 1838, Cournot introduced the formal monopoly theory into economics [1]. Kaman and Schwartz argue that monopoly and competition together promote technological innovation [2]. The Marshall Conflict points out that a social function is facing the choice of either one or the other, either abandoning the scale economy and taking the road of free competition, or losing the free competition and taking the road of scale economy [3]. Based on the complex system theory, this paper introduces Lotka-Volterra (LV) model to simulate the interaction between N enterprises (providing the same products) in a commercial market. The complex interaction of enterprise competition can be used to understand the monopoly problem.

T. Ahram et al. (Eds.): IHSED 2019, AISC 1026, pp. 1018–1022, 2020.
https://doi.org/10.1007/978-3-030-27928-8_151

2 Enterprise Evolution and Ecological Population Evolution

Population is one of the most important concepts in population ecology. (1) Population characteristics: When a single population is in a limited resource environment, it presents an "S" curve. (2) Population relationship: (a) Population competition relationship. (b) Predatory relations. (c) Mutual beneficial symbiosis. (d) Parasitic relationship. (3) Population and environment: The adaptability of population to environment is the result of its evolution. The ecological pressure acting on the population determines the change of its evolutionary direction and adaptability [4]. The growth of population and enterprise scale shows an "S" curve.

3 Logistics Model of Single Population Evolution

The whole evolutionary process presents an S-shaped curve, which is consistent with the growth and evolution law of the biological population [5]. When the user size (cumulative user) is X_t at time t, the rule of X_t changing with time can be described by the following differential equations [6]:

$$\frac{dX_t}{dt} = rX_t\left(1 - \frac{X_t}{N}\right) \tag{1}$$

Assuming that the state value X_0 of the initial stage of enterprise evolution at $t = 0$ in formula (1) is known, the integral solution (1) can be obtained.

$$X_t = \frac{N_t}{1 + Ae^{-rt}} \tag{2}$$

In formula (2), R represents the growth rate of an enterprise; A is a constant; N represents the maximum value of the enterprise users $(\alpha > 0, N \gg 0)$; $\frac{X_t}{N}$ is the ratio of the enterprise user size to the maximum user size that the enterprise can achieve, while $\left(1 - \frac{X_t}{N}\right)$ represents the ratio of the enterprise user size to the enterprise maximum user size that has not yet been realized [7].

4 Lotka-Volterra Model for Evolution of Two Species

The same level of enterprise is a competitive relationship. In the process of enterprise evolution, it embodies the mutually beneficial symbiotic relationship. Weaker companies are slowly declining, predatory relationships [8].

The number of customers is limited. The market share acquired by Enterprise 1 will adversely affect the growth of Enterprise 2. $\left(1 - \frac{X_t}{N}\right)$ to express the promotion effect of enterprise 1 and enterprise 2 on each other's scale growth. To sum up, the Lotka-Volterra evolution model of two species growth is established [12]:

$$\begin{cases} \frac{dx_1(t)}{d(t)} = r_1 x\left(1 - \frac{x_1}{N_1} + \alpha \frac{x_2}{N_2}\right) \\ \frac{dx_2(t)}{d(t)} = r_2 y\left(1 - \frac{x_2}{N_2} + \beta \frac{x_1}{N_1}\right) \end{cases} \tag{3}$$

Among them, α and β are the influence coefficients of enterprise evolution.

This paper studies the evolution results of enterprise 1 and enterprise 2, i.e. $t \to +\infty$. The stable equilibrium point can represent the competition result of two enterprises [9]. Let $\frac{dx_1(t)}{d(t)} = 0$, $\frac{dx_2(t)}{d(t)} = 0$. The four equilibrium points are obtained respectively: $E_1(N_1,\ 0)$, $E_2(0,\ N_2)$, $E_3\left(\frac{N_1(1+\alpha)}{1-\alpha\beta}, \frac{N_2(1+\beta)}{1-\alpha\beta}\right)$, $E_4(0,\ 0)$. E_1 indicates that Enterprise 1 gains all market share and makes Enterprise 2 withdraw from the market. E_2 means that Enterprise 2 gains the whole market share and Enterprise 1 withdraws from the market. E_4 means that the scale of the enterprise becomes zero. This means that Enterprise 1 and Enterprise 2 have reached a balance in the evolutionary interaction process. The state of E_1, E_2 and E_4 is not conducive to the healthy development of enterprise system.

According to the criterion of stability of equilibrium point, Establish the model of Formula (4):

$$\begin{cases} \frac{dx_1(t)}{d(t)} = r_1 x_1\left(1 - \frac{x_1}{N_1} + \lambda_1 \frac{x_2}{N_2}\right) \\ \frac{dx_2(t)}{d(t)} = r_2 x_2\left(1 - \frac{x_2}{N_2} + \lambda_2 \frac{x_1}{N_1}\right) \end{cases} \tag{4}$$

Research is on enterprise evolution. $x_1(t)$ and $x_2(t)$ are the users of enterprise 1 and enterprise 2 at time t, respectively.N_1 and N_2 are the maximum users of Enterprise 1 and Enterprise 2 respectively. The growth rates of enterprise 1 and enterprise 2 are r_1 and r_2 respectively. λ_1 is the evolutionary influence coefficient of enterprise 2 scale change on enterprise 1. λ_2 is the evolutionary influence coefficient of enterprise 1 scale development on enterprise 2.

5 Multi-population Evolution Model

The Lotka-Volterra ecological model of enterprise competition in the same market is constructed, which is shown in the following formula (5):

$$\frac{dx_i(t)}{dt} = x_i(t)\left(B_i + A_{ii}x_i - \sum_{j=1, j\neq i}^{n} A_{ij}x_j\right),\ A_{ij} > 0, A_{ii} > 0 \tag{5}$$

Among them, $\frac{dx_i(t)}{dt}$ represents the rate of change of the number of users i at time t. $x_i(t)$ denotes the number of user population i at time t, $i = 1, 2, \cdots$, n. B_i represents the current growth rate of user population i. A_{ij} denotes the interaction index between user population i and user population j. $j = 1, 2, \cdots, n$; A_{ii} denotes the impact index of user population i within the same enterprise. Formula (5) $\frac{dx_i(t)}{dt} = 0$ should be satisfied. Then $x_i(t) = 0$ should be satisfied. Or satisfy the following conditions:

$$B_i = \sum_{j=1, j \neq i}^{n} A_{ij}x_j - A_{ii}x_i, \ \ i,j = 1,2,\cdots,n \tag{6}$$

Lemma 1 in the system $Ax = b$ of inhomogeneous linear equations. If A is a square matrix of order n, then the necessary and sufficient condition for the system to have a solution is: $r(A) = r(Ab)$. When $r(A) = r(Ab) = r = n$, the system has a unique solution. When $r(A) = r(Ab) = r < n$, the equations have multiple solutions. When $r(A) \neq r(Ab)$, the system of equations has no solution [10]. Corollary 1: In formula (6) has $x_i(t) = 0 (i = 1, 2, \cdots, n)$ when $r(A_{n\times n}) \neq r(A_{n\times n} \vdots B_n)$. When $r(A_{n\times n}) = r(A_{n\times n} \vdots B_n) = r = n$, $x_i(t)$ has a unique solution. When $r(A_{n\times n}) = r(A_{n\times n} \vdots B_n) = r < n$, the solution of $x_i(t)$ has at least a value of r not equal to 0.

Corollary 1 shows the number of enterprises in the enterprise may reach a stable state. It shows that the competition of enterprises in the market is very fierce and there is no stable state, and the competition of enterprises in the market will continue. When there are multiple solutions to the non-homogeneous equations, firms in the market will reach a stable state through free competition [11].

If an enterprise satisfies three conditions: (1) the growth rate of one enterprise is greater than the sum of the internal growth rate of all competing enterprises and the ratio of the two enterprises' interaction index to the internal impact index of competing enterprises. (2) The inherent growth rate of some other enterprises is less than the sum of the interaction rate of all competing enterprises and the product of a certain stable state obtained by inference 1. (3) According to the inherent growth rate, mutual influence index and internal influence index of enterprises. The enterprise population that satisfies the second condition will eventually die out in the market. Then the enterprise population satisfying the conditions (1) will eliminate the enterprise population satisfying the conditions (2) in order to achieve enterprise monopoly.

6 Conclusion

The ecological characteristics of enterprise evolution are analyzed. The interaction and influence among multi-group enterprises in the market are considered cross-sectionally. An improved Lotka-Volterra ecological model of enterprise evolutionary competition is proposed under the given hypothesis. The results show that: (1) there are competition, predation and cooperation among enterprises. By changing the interaction between enterprises, it has a significant impact on the number of users. (2) Under the corresponding conditions, some enterprises may be eliminated, and some enterprises may become the monopoly of the market. Therefore, the government should comprehensively analyze the market environment and optimize the market by intervening in the competitive process of enterprises.

Acknowledgments. This research reported herein was supported by the NSFC of China under Grants No. 71571091 and 71771112.

References

1. Cournot, A.A.: Recherches sur les principes mzthematiques de la theorie des richesses. Chez L. Hachette, Paris (1838)
2. Hu, H.: Deduction of natural monopoly theory. Master's thesis of Shanghai Academy of Social Sciences (2006)
3. Sosnick, S.H.: A critique of concepts of workable competition. Q. J. Econ. **72**, 380–423 (1958)
4. Xu, X., Gao, S., Zhou, Q., Liu, J.: Study on evolution of the port system along "maritime silk road" based on logistics and Lotka-Volterra model. Oper. Res. Manage. Sci. **7**(8), 172–181 (2018)
5. Chen, H., Luan, W., Wang, Y.: Evolution model and empirical analysis of port growth. Port Eng. Technol. **46**(2), 41–44 (2009)
6. Cai, M.: Research on evolution of large civil aircraft industry based on logistics model. J. Harbin Univ. Commer. (Nat. Sci. Edn.) **31**(3), 374–377 (2015)
7. Zhou, Z.: Study on competition-cooperation relations of container ports based on Lotka-Volterra model. Chang'an University (2015)
8. Guo, Y., Wu, Z.: A study on knowledge creation of technological innovative network based on Lotka-Volterra model. J. Intell. **31**(6), 139–143 (2012)
9. Li, K., An, S., Wang, C., Zhu, X.: Competition mechanism of enterprises' knowledge transmission based on Lotka-Volterra ecological model. R&D Manage. **30**(3), 75–84 (2018)
10. Kun, W., Yan, Z.: Linear Algebra and Its Applications, pp. 60–65. Machinery Industry Press, Beijing (2007)
11. Li, B., Yu, S.: Some new results of N-population autonomous competition system in Lotka-Volterra model. Applied Mathematics **27**(3), 556–564 (2004)
12. Zeeman, M.L.: Extinction in competitive Lotka-Volterra systems. Proc. Am. Math. Soc. **123** (1), 87–96 (1995)
13. Wang, Y., Xie, W.: A study on dynamic competition between the imitator and the innovator in the ecommerce model. Sci. Sci. Manage. S&T **34**(6), 44–51 (2013)
14. Li, S., Li, S.: Analysis of the competition situation and evolution of domestic civil aviation market based on a nonautonomous Lotka-Volterra model. J. Xi'an Univ. Financ. Econ. **30** (30), 35–42 (2018)

Quality Management Model Focusing on Good Agricultural Practices to Increase Productivity of Pomegranate Producing SMEs in Peru

Mayra Cárdenas[1(✉)], Mayra Rodriguez[1(✉)], Edgar Ramos[1(✉)], Edgardo Carvallo[1(✉)], and Carlos Raymundo[2(✉)]

[1] Ingeniería Industrial, Universidad Peruana de Ciencias Aplicadas (UPC), Lima, Peru
{u201416043,u201314900,pcineram,pcinecar}@upc.edu.pe
[2] Dirección de Investigación, Universidad Peruana de Ciencias Aplicadas (UPC), Lima, Peru
carlos.raymundo@upc.edu.pe

Abstract. Peru's pomegranate productivity is 10.3 tons, which is almost a third of that produced by Turkey (27.5 tons) and almost half of that produced by Spain (20 tons). In this sense, it is necessary that the organizations know the importance of quality management to monitor and maintain product quality. It is important to consider the appropriate methods to adopt and implement quality management so that MSEs may develop a better understanding of how quality management systems should be implemented and the significant, positive impact they generate. This document proposes a model based on quality management and good practices, which frames a series of activities and criteria to provide an objective diagnosis. This model was developed based on information gathered from SMEs in the district of Santiago de Ica.

Keywords: Quality mangement · Good agricultural practices · Food industry · Pomegranate · Sme's, Perú

1 Introduction

Currently, according to the Association of Pomegranate Producers in Peru (Progranada), the exports of this fruit in 2017 were 25,574 tons, which shows a considerable increase when compared to 2011, when just over 4,000 tons were exported. Although there has been an increase in exports, there is also a deficit in demand fulfillment as the country on average only produced 10 tons/ha, which means that fruits not meeting the minimum requirements for export are being produced; therefore, they are sold on the domestic market.

Companies must develop a more thorough understanding of how quality management systems should be implemented to realize their benefits, namely, performance improvement, productivity increase, and customer satisfaction [1]. This need has been widely addressed in previous studies, and there have been numerous researches conducted on these issues, but these are mainly focused on large enterprises because small

T. Ahram et al. (Eds.): IHSED 2019, AISC 1026, pp. 1023–1029, 2020.
https://doi.org/10.1007/978-3-030-27928-8_152

enterprises are believed to lack the resources to adopt a model using the concepts of quality management and good agricultural practices [2].

This article shows how quality management systems can be adopted to improve and ensure product quality and describe the potential benefits after implementing a model based on quality management and good agricultural practices in micro and small businesses in Santiago de Ica. In particular, the research focused on the benefits of quality management systems along with good agricultural practices and the link between them. In this sense, this model seeks to control and maintain product quality through implementing control methods and tools [3].

2 Literature Review

Quality management system (QMS) is relatively well explored among large enterprises but not in small ones [4]. Therefore, a small Swedish enterprise decided to implement a QMS, and in a short time, managed to triple its production capacity and double its profits, establishing itself as a market leader in its sector. Moreover, to achieve these benefits, the enterprise had to reorganize its operations according to quality practices [5].

Many initiatives in small businesses fail due to poor implementation [6]. Hence, the authors developed a model that considers quality management practices and total employee involvement, the latter being a common theme in the literature on quality management in MSEs. The model is based on five stages: knowledge, persuasion, decision, implementation, and confirmation. By applying this model, companies will be able to better understand the role of quality management practices and be aware of the critical factors and implement them more effectively [5]. In addition to acquiring the initial knowledge of implementing and integrating QMSs, small enterprises will be able to implement them gradually and in a controlled manner [6].

In a study conducted by [7], the authors developed a model that considers various factors, such as strategic planning, customer focused-strategy, human resource management, processes, suppliers, and leadership [2]. This model was implemented in small enterprises in Ghana and made it possible to determine how useful and to what extent quality management is practiced in that country. The study found that the model was able to improve quality and sales performance in organizations. The improvement of product quality was reflected through waste reduction, error reduction, and increased capability to prevent defects.

3 Contribution

The pomegranate producing micro and small enterprises in the district of Santiago, Ica, have lower productivity levels compared to other countries that have the same type of land and climate to the Peruvian coast but who produce and export more than the sector under study. For this reason, a basic model is designed that considers the following processes: (see Fig. 1).

As part of the proposal of the basic quality management model for SMEs, the following treatments have been determined for each of the identified processes (Fig. 2).

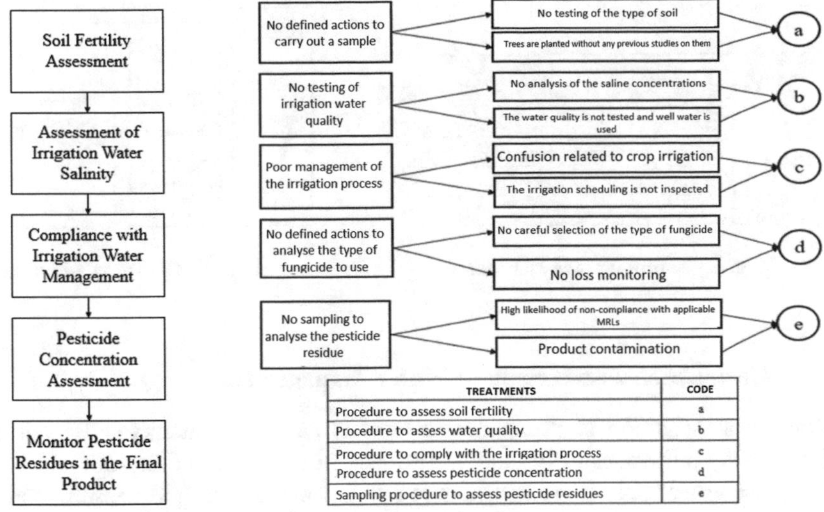

TREATMENTS	CODE
Procedure to assess soil fertility	a
Procedure to assess water quality	b
Procedure to comply with the irrigation process	c
Procedure to assess pesticide concentration	d
Sampling procedure to assess pesticide residues	e

Fig. 1. Model

Fig. 2. Proposal scheme

3.1 Soil Fertility Assessment

In accordance with good agricultural practices, soil testing is crucial to evaluate and guarantee fertility. Therefore, soil sampling is essential to determine nutrient availability and thus ensure the proper root growth.

Proper fertilizer application for plant growth is extremely important as an incorrect application can affect soil productivity.

3.2 Assessment of Irrigation Water Salinity

The assessment of the irrigation water salinity is essential as high concentrations of salt are toxic for crops and constitute a risk for soil salinization.

Likewise, the assessment of irrigation water allows us to optimize crop production as there is a direct relation between the quality of the irrigation water and crop yield. This means that analyzing and supplying the necessary components results in production improvement. Thus, the objective of this study will be to assess the quality of irrigation water, determine the concentration of total dissolved solids, and prevent the use of irrigation water classified as "doubtful/unsafe" or "unsuitable" [8]. (See Fig. 3)

Conductividad Eléctrica (uohms)		pH		Boro (ppm)	
Rango de CE (uohms)		Rango de pH		Rango de concentración de Boro	
Menor a 250	Excelente	Menor a 6.5	No óptimo	Menor a 0.67	Excelente
250-750	Bueno	Entre 6.5 - 7.5	Óptimo	0.67 - menor a 1.33	Bueno
750-2000	Permisible			1.33 - menor a 2.0	Permisible
2000-3000	Dudoso	Mayor a 7.5	No óptimo	2.0 - menor a 2.5	Dudoso
Mayor a 3000	Inadecuado			Mayor a 2.5	Inadecuado
Conductividad Eléctrica (uohms)		pH		Boro (ppm)	

Fig. 3. Sampling analysis record

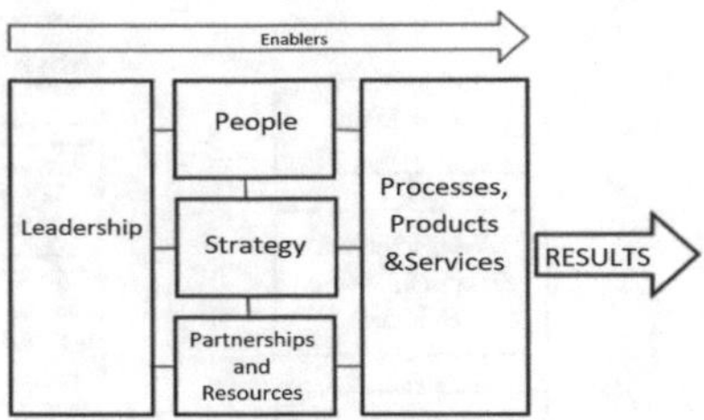

Fig. 4. EFQM adapted model

3.3 Compliance with Irrigation Water Management

Monitoring irrigation of pomegranate trees is an important factor to consider as it significantly influences fruit quality. The lack of irrigation monitoring and scheduling can cause water stress in trees, which results in cracking, turgor pressure, and changes in their properties [9]. A study conducted by [10] indicates that water stress should be avoided as excessive irrigation in pomegranate trees can cause fruit cracking. In this context, this process is carried out to prevent further cracking or changes in the pomegranate fruit.

There are studies regarding the management of crop irrigation. They indicate that the management and monitoring of the volume of water applied per hectare to be used is important to satisfy the total crop water requirement. The authors conclude that the daily irrigation volume applied along with the irrigated area must be monitored to assess and control the irrigated surface and to properly use the irrigation applications which will be recorded in the irrigation schedule to avoid crop abnormalities.

3.4 Pesticide Concentration Assessment

The control of diseases and pests in pomegranate fruits is based on the use of pesticides; however, to be effective, it requires the assessment of the chemical components present in each pesticide. Therefore, to ensure the proper use of these pesticides, the concentration of these chemical components must be inspected and any type of disease and pests that may occur in the crop must be monitored [11, 12]. Each active substance (the active component against pests/plant diseases contained in the plant protection product) must be safe in terms of human health, animal health, and environmental impact. The European Food Safety Authority states that to comply with the aforementioned aspects, the maximum residue levels of pesticides permitted in marketed products of plant or animal origin must be analyzed. This type of procedure is also recommended and regulated by the Government of Canada, which, under the Pest Control Products Act, regulates the sale and use of pesticides that pose an unacceptable risk to human health. In this regard, a procedure containing clearly defined steps to control safe pesticide use is proposed.

3.5 Monitoring Pesticide Residue in the Final Product

The monitoring of pesticide residues is important as this will ensure the quality of the final product and protect consumers from exposure to unacceptable levels of pesticide residues [10, 13]. Moreover, the foreign market can be better accessed, providing a quality product that complies with the maximum residue limits for pesticides established by the Codex Alimentarius Commission [11, 13]. Sampling is important to determine the unacceptable levels of pesticide residue in crops [14]. In their study, the authors mention that to monitor compliance with the maximum residue limits, the minimum sample size to be taken into account for the analysis must be defined, to avoid contributing to 80%–90% of the sampling uncertainty. Therefore, sampling uncertainty will be taken into account as an indicator to monitor the number of minimum samples per hectare to be analyzed and reduce the variability in results and, consequently, maintain the reference value of 10% for sampling uncertainty [10, 14].

4 Validation

To validate our proposal, we have considered several studies in which the European Foundation for *Quality* Management (EFQM) model is taken as a tool to ratify the proposed model [15–17].

The EFQM Excellence Model is best known for ensuring that the quality practices implemented consistently coexist [18–20]. It is also a tool that allows us to diagnose the company's current situation and then identify which aspects need improvement and implement actions that help us achieve our goal [12]. Hence, it is an ideal method to validate the proposals mentioned above (See Fig. 4).

4.1 Leadership

The proposal is aimed at joint participation between management and the company [12, 19]. To this end, the SME's manager or owner is empowered to assign key roles and activities to guarantee the development, implementation, and improvement of the proposed model.

4.2 Strategy

In terms of strategy, the planning of activities, goals, and identification of relevant processes have been clearly defined. We have considered the activities within each process, which upon being performed entail the achievement of the process goal. For example, compliance with irrigation water management (Weekly): 1. Receive the established irrigation schedule, 2. Identify the irrigation method and frequency, 3. Calculate the crop water requirement, 4. Perform irrigation according to the schedule, 5. Record irrigation compliance, 6. Fill the deficit or surplus matrix, and 7. Confirm execution of activities to the caretaker.

4.3 Partnerships and Resources

In the process of pesticide concentration assessment, the resource acquired is disintegrated by the concentration of its components according to SENASA guidelines, and thus, an adequate material management is achieved. By incorporating the filling of records, these would be analyzed and then stored to correctly manage information and knowledge.

4.4 Processes

The necessary process improvements are made through innovation, with the goal of achieving customer satisfaction and adding value to the process. For this purpose, pesticide residues are sampled to ensure the quality of the fruit.

5 Conclusions

This study is focused on increasing the productivity of pomegranate producing SMEs in Santiago, Ica, where we identified that the personnel have a lack of knowledge and experience related to recent QMSs. Further, because it is a small sector of study, we found that there is no QMS that can be applied directly to a micro and small enterprise; hence, conditions must adhere to the scenario of an MSE.

It should be noted that a quality management model represents an opportunity for improvement in the management of SMEs (they constitute more than 97% of companies in the country), considering that currently they do not have QMSs, or if they have, they are totally empirical. Likewise, to obtain favorable results, the participation and commitment of company's management and personnel is required as the way in which businesses conduct their daily activities is changing and each stakeholder must take on more tasks and responsibilities.

References

1. Yang, C.-C.: The effectiveness analysis of the practices in five quality management stages for SMEs. Total Qual. Manage. Bus. Excell. (2018)
2. Arévalo, J., Quispe, G., Raymundo, C.: Sustainable energy model for the production of biomass briquettes based on rice husk in low-income agricultural areas in Peru. Energy Proc. **141**, 138–145 (2017)
3. Zairi, M.: The TQM legacy – Gurus' contributions and theoretical impact. TQM J. **25**(6), 659–676 (2013)
4. Beheshti, H.M., Lollar, J.G.: An empirical study of US SMEs using TQM. Total Qual. Manage. Bus. Excell. **14**(8), 839–847 (2003)
5. Assarlind, M., Gremyr, I.: Critical factors for quality management initiatives in small- and medium-sized enterprises. Total Qual. Manage. Bus. Excell. **25**, 397–411 (2014)
6. Assarlind, M., Gremyr, I.: Iniciando la gestión de la calidad en una pequeña empresa. El diario de TQM **28**(2), 166–179 (2016)

7. Osei Mensah, J., Copuroglu, G., Appiah Fening, F.: El estado de la gestión total de la calidad (TQM) en Ghana. Revista internacional de gestión de calidad y fiabilidad (2012)
8. Calvo, A., Picón, A., Ruiz, L., Cauzo, L.: Análisis contextual y de mediación entre los factores críticos de TQM y los resultados organizativos en el marco del Modelo de Excelencia EFQM. Int. J. Prod. Res. **53**(7), 2186–2201 (2015)
9. Galindo, A., Rodríguez, P., Collado-González, J., Cruz, Z.N., Torrecillas, E., Ondono, S., Corell, M., Moriana, A., Torrecillas, A.: Rainfall intensifies fruit peel cracking in water stressed pomegranate trees. Agric. For. Meteorol. **194**, 29–35 (2014)
10. Munhuweyi, K., Lennox, C., Meitz, J., Opara, L.: Major diseases of pomegranate (Punica granatum L.), their causes and management – a review. Sci. Hortic. **211**, 126–139 (2016)
11. Mphahlele, R., Fawole, O., Stander, M., Opara, U.: Preharvest and postharvest factors influencing bioactive compounds in pomegranate (Punica granatum L.)—a review. Sci. Hortic. **178**, 114–123 (2014)
12. Türkoz, G., Bengü, D., Bakirci, F., Ötles, S.: Pesticide residues in fruits and vegetables from the Aegean región Turkey. Food Chem. **160**, 379–392 (2014)
13. FAO: Norma para la Granada. Codex Stan 310–2013 (2013)
14. Omeroglu, P., Boyacioglu, D., Ambrus, A., Karaali, A., Saner, S.: An overview of steps of pesticide residue analysis and contribution of the individual steps to the measurement ucertainty. Food Anal. Methods **5**, 1469–1480 (2012)
15. Bou-Llusar, J.C., Escrig, A.B., Roca, V., Beltrán, I.: An empirical assessment of the EFQM excellence model: evaluation (2009)
16. Suárez, E., Roldán, J., Calvo, A.: Un análisis estructural del modelo EFQM: una evaluación del papel mediador de la gestión de procesos. J. Bus. Econ. Manage. **15**(5), 862885 (2014)
17. Yousaf, M., Li, J., Lu, J., Ren, T., Cong, R., Fahad, S., Li, X.: Effects of fertilization on crop production and nutrient-supplying capacity under rice-oilseed rape rotation system. Sci. Rep. **7**, 1270 (2017)
18. Talib, F., Rahman, Z., Qureshi, M.N.: Analysis of interaction among the barriers to total quality management implementation using interpretive structural modeling approach. Benchmarking: Int. J. **18**(4), 563–587 (2011)
19. Talib, F., Rahman, Z., Qureshi, M.N.: An empirical investigation of relationship between total quality management practices and quality performance in Indian service companies. Int. J. Qual. Reliab. Manage. **30**(3), 280–318 (2013)
20. Talib, F., Rahman, Z., Qureshi, M.N., Siddiqui, J.: Total quality management and service quality: an exploratory study of management practices and barriers in service industries. Int. J. Serv. Oper. Manage. **10**(1), 94–118 (2011)

Twitter Mining for Multiclass Classification Events of Traffic and Pollution

Verónica Chamorro[1], Richard Rivera[2(✉)], José Varela-Aldás[3], David Castillo-Salazar[3,6], Carlos Borja-Galeas[4,9], Cesar Guevara[5], Hugo Arias-Flores[5], Washington Fierro-Saltos[6,7], Jairo Hidalgo-Guijarro[8], and Marco Yandún-Velasteguí[8]

[1] Facultad de Informática, Universidad Complutense de Madrid, Madrid, Spain
verocham@ucm.es

[2] Escuela de Formación de Tecnólogos, Escuela Politécnica Nacional, Quito, Ecuador
richard.rivera01@epn.edu.ec

[3] SISAu Research Group, Universidad Indoamérica, Ambato, Ecuador
{josevarela,davidcastillo}@uti.edu.ec

[4] Facultad de Arquitectura, Artes y Diseño, Universidad Indoamérica, Quito, Ecuador
carlosborja@uti.edu.ec

[5] Mechatronics and Interactive Systems - MIST Research Center, Universidad Indoamérica, Quito, Ecuador
{cesarguevara,hugoarias}@uti.edu.ec

[6] Facultad de Informática, Universidad Nacional de la Plata, Buenos Aires, Argentina
washington.fierros@info.unlp.edu.ar

[7] Facultad de Ciencias de la Educación, Universidad Estatal de Bolívar, Guanujo, Ecuador

[8] Grupo de Investigación GISAT, Universidad Politécnica Estatal del Carchi, Tulcan, Ecuador
{jairo.hidalgo,marco.yandun}@upec.edu.ec

[9] Facultad de Diseño y Comunicación, Universidad de Palermo, Buenos Aires, Argentina

Abstract. During the last decade social media have generated tons of data, that is the primal information resource for multiple applications. Analyzing this information let us to discover almost immediately unusual situations, such as traffic jumps, traffic accidents, state of the roads, etc.. This research proposes an approach for classifying pollution and traffic tweets automatically. Taking advantage of the information in tweets, it evaluates several machine learning supervised algorithms for text classification, where it determines that the support vector machine (SVM) algorithm achieves the highest accuracy value of 85,8% classifying events of traffic and not traffic. Furthermore, to determine the events that correspond to traffic or pollution we perform a multiclass classification. Where we obtain an accuracy of 78.9%.

Keywords: Twitter event detection · Pollution detection · Traffic detection · Twitter mining · Algorithms of classification · SVM

T. Ahram et al. (Eds.): IHSED 2019, AISC 1026, pp. 1030–1036, 2020.
https://doi.org/10.1007/978-3-030-27928-8_153

1 Introduction

The development of social networks has created new communication and expression paths. In the past few years social networks, specifically Twitter that has incremented its popularity, becoming one of the most influential microblog websites, and a useful information resource. Its shorts messages or tweets contains only 140 characters, which obligates to users to express in concise way. The text of the tweets may present an extensive variety of information about specific situations, also opinions of the users about products or services. These contents can be analyzed to respond appropriately to the feelings expressed by the tweets; for example, to expose new services or products. Another example is when an event is mentioned continuously in several tweets in a specific place, due to a parade or a concert, that is provoking traffic. So, users can take advantage of these useful information in real time, to plan their routes in order to avoid traffic jumps.

Motivated by a previous work by D'Andrea et al. [1] where they present an approach to detect in real time traffic events by analyzing twitter steam analysis, using a dataset of Italian tweets. In contrast, this research, presents an approach to classify automatically tweets related with pollution and traffic. The dataset to evaluate our approach were collected from the generated tweets of Madrid and in Spanish language. Our approach evaluates four machine learning supervised algorithms for text classification, and it demonstrates the superiority of the SVM algorithm, due to it obtains the highest accuracy with a value of 85.8% in a binary classification of events of traffic or not traffic. Likewise, in a multiclass classification the same algorithm, achieves an accuracy value of 78.9%.

2 Classification Approach

This section describes the approach of our classification model. It presents the architecture of the system, then the description of the process in the classification system.

2.1 System Architecture

The architecture of the classification system presented in this work comprises three modules. It starts by collecting and preprocessing tweets, then the core of the system the processing module, and it finishes by classifying tweets. The goal of the proposed system as shown in Fig. 1, is to obtain tweets from the social network Twitter, to process them by applying text mining, in order to classify them by assigning class labels defined, in our case traffic, pollution, traffic and pollution or none of them. Finally, by analysis, preprocessing and classification of the tweets, the system will be able to classify them in the previously mentioned classes.

Tweets Collecting and preprocessing. The tweets are obtained using one or several Finding criteria (e.g., geographical coordinates), in our study we collect tweets published in Madrid. Each of the tweets contain: the text of the tweet, Twitter ID, user ID, date and time, and geographic coordinates. The text of the tweet may contain very

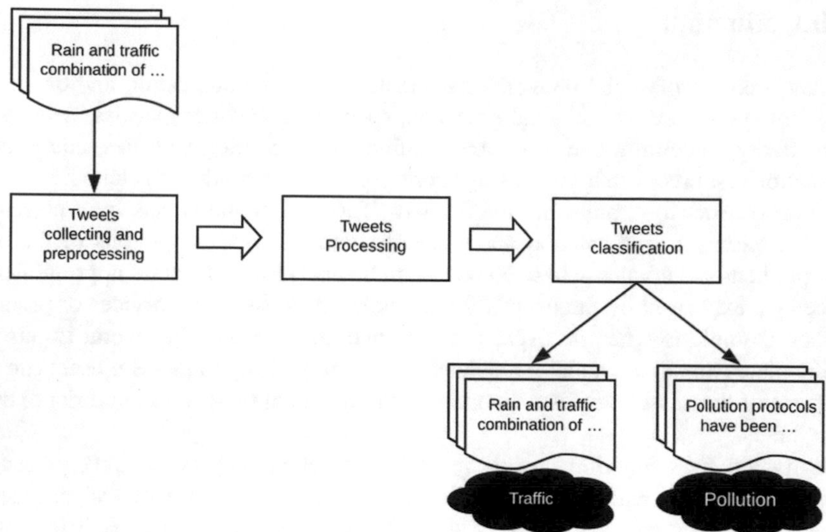

Fig. 1. High level overview of the events detection system based on the analysis of Twitter information.

diverse information that includes, hashtags, links, mentions, emojis, and special characters. In this research, we focus only in tweets in Spanish language. However, previous works performed in other languages [1, 4] demonstrate that these systems can be easily adapted to different languages.

Then, the tweets are pre-processed to extract the text of each tweet and remove all its metadata, we use Python to obtain only the date, time creation and the text of the Tweet. Additional we create a column "words" that contains the text of the Tweet changing all the characters to lowercase and separating them in words. At the end of this elaboration, each tweet appears as a string, that is, a sequence of words.

Tweets Processing. The second module is responsible to transform the set of pre-processed tweets, which is, a set of strings, into a set of numeric vectors to be processed by the Tweets classification module. Figure 2 shows the five text mining techniques applied to a sample tweet.

The first technique is the *tokenization*, it receives tweet samples, then it cleans them, for instance it removes accents, apostrophes, tabulations and spaces, this process also removes all punctuation marks and divides each tweet into tokens. The tokens correspond to tweet words.

The second technique *stop-word filtering* it consists of the elimination of words that provide little or no information to the analysis. Typically, empty words are articles, conjunctions, prepositions, pronouns, or words that are specific to the language or the domain, which are considered as noise [2, 3].

The third technique *stemming* is the process of reducing each word to its root [1], to group semantic related words. When this step finishes each tweet is represented as a sequence of stems.

The fourth technique *stem filtering*, it consists in reducing the number of stems [1], filtering the stems that do not belong to the set of relevant stems. For this, we apply TF-IDF [4, 5] to remove words that consistently appears in most of the tweets, since they may not permit distinguish different tweets. At the end of this step each tweet is represented as a sequence of relevant stems.

The last technique applied is *feature representation* it consist in the construction of vectors with numeric characteristics for each tweet, to classify them, all tweets should have the same number of characteristics, where each characteristic corresponds to an stem and is represented by the weight, zero if the stem does not appear in the tweet or the calculated weight according to the set of relevant stems.

Tweets Classification. The third module is responsible to assign a label of class related with the events of traffic or pollution or none of those. At the end of this module we receive a set of tweets with their labels. To labeling tweets it applies the Support Vector Machine algorithm. Where the parameters of the classification model are identified during the supervised learning phase.

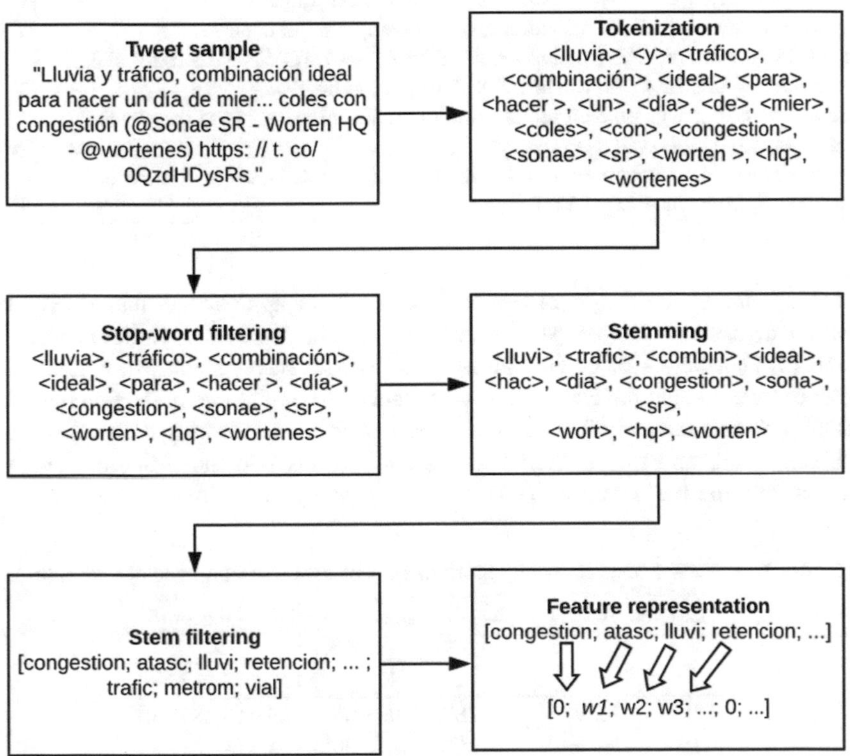

Fig. 2. Text mining techniques applied to a sample tweet

3 Results

The dataset was collected from tweets in the Spanish language of the city of Madrid by establishing two search criteria first the geographic coordinates of Madrid, and the ones that were posted during 2016. In overall, our main dataset comprises around 3M tweets. To train the supervised classification algorithms we manually built four datasets of 1000 tweets each one, which are a set of data T (traffic class), N (class not traffic or pollution, P (pollution class) and PT (pollution and traffic class).Those datasets were tagged manually, by assigning the label of the correct class. Table 1 shows a selection of tweets obtained for each dataset with their respective label added manually.

Table 1. Manually classified tweets in classes P (pollution), PT (pollution and traffic) and N (neither pollution nor traffic) tweet

Tweet	Class
- Lo de madrid no es contaminación, es la aurora boreal autóctona	P
- #Madrid Por contaminación mañana viernes 30 no se podrá aparcar en zona de parquímetros https://t.co/5euvYCSsZA https://t.co/YsuxHkaoxa	P
- Madrid! Sí, ya septiembre, sus atascos, como siempre en el centro…	T
- #RETENCIÓN nivel AMARILLO en AUTOPISTA/AUTOVÍA #A3 (pk 9 al 11 decreciente) MADRID #DGTMadrid #DGT https://t.co/INzPwWwyRe	T
- Así nos vemos actualmente en reposo sin congestión ni nada, eso sí tirando de las luces de la… https://t.co/CQL9VmEwyE	N
- Así nos vemos actualmente en reposo sin congestión ni nada, eso sí tirando de las luces de la… https://t.co/CQL9VmEwyE	N

Table 2 shows the results of the evaluation of our approach, obtained by 3 classification algorithms support vector machine [6, 9], k-nearest neighbors algorithm (k-NN) [3, 7], where k takes the values (1, 3, 5), and Bayes naïve [8], applied to the dataset of tweets using the classes T (tweets related to traffic) and N (tweets not related to traffic). More concretely for each classifier it shows the following metrics: accuracy, precision, recall, and the harmonic average of the las two the F-Score. Where the classifier with the best result is SVM with an accuracy of 85.75%.

Table 2. Results of classification algorithms for classes T (traffic) and N (no traffic).

Classifier	Accuracy (%)	Precision (%)		Recall (%)		F-score (%)	
		T	N	T	N	T	N
SVM	**85.8**	**85.3**	**86.3**	**86.5**	**85.0**	**85.8**	**85.7**
1NN	81.9	83.2	81.2	80.8	83.3	82.0	82.2
3NN	81.7	83.3	80.3	88.8	83.5	81.5	81.8
5NN	81.6	83.7	88.8	88.3	88.8	81.4	81.8
NB	80.6	83.2	88.4	87.7	83.4	80.3	80.8

On the other hand, Table 3, shows the results applied to the dataset of tweets using the classes P (pollution) and T (traffic). Where is ratified SVM as the classifier with the best result, with an accuracy of 78.89%.

Table 3. Results of classification model for classes: P (pollution) and T (traffic).

Classifier	Accuracy (%)	Precision (%)		Recall (%)		F-score (%)	
		P	PT	P	PT	P	PT
SVM	**78.9**	**78.5**	**79.3**	**79.5**	**78.2**	**79.0**	**78.8**
1NN	74.1	75.2	73.6	73.4	75.3	74.27	74.4
3NN	74.0	75.3	72.9	72.6	75.5	73.9	74.1
5NN	73.9	75.6	72.6	72.1	75.8	73.8	74.1
NB	73.1	75.2	71.3	70.8	75.4	72.93	73.3

4 Conclusions

This research, presents an approach for detecting events that involve pollution and traffic, applying an analysis over the data flow of Twitter. This approach focuses on analyzing tweets datasets to classify them in four classes: pollution, traffic, pollution and traffic or none of them.

The resulting system was built with Python, and different open source libraries, to evaluate machine learning algorithms and text mining techniques in order to analyze and classify tweet events. The results show that the SVM algorithm reaches the highest accuracy with a value of 85.8% in a binary classification of events related to traffic or no traffic. Furthermore, this algorithm also reaches the highest accuracy with a value of 78.89% in a multiclass classification of events that involves three classes pollution, traffic and pollution or none of them.

References

1. D'Andrea, E., Ducange, P., Lazzerini, B., Marcelloni, F.: Real-time detection of traffic from twitter stream analysis. IEEE Trans. Intell. Transp. Syst. **16**(4), 2269–2283 (2015)
2. Quinlan, J.R.: C4.5: Programs for Machine Learning. Elsevier, Amsterdam (2014)
3. Cover, T., Hart, P.: Nearest neighbor pattern classification. IEEE Trans. Inf. Theory **13**(1), 21–27 (1967)
4. Aiello, L.M., et al.: Sensing trending topics in Twitter. IEEE Trans. Multimed. **15**(6), 1268–1282 (2013)
5. Patil, L.H., Atique, M.: A novel feature selection based on information gain using WordNet. In: Proceedings of SAI Conference, London, UK, pp. 625–629 (2013)
6. Cortes, C., Vapnik, V.: Support-vector networks. Mach. Learn. **20**(3), 273–297 (1995)
7. Aha, D.W., Kibler, D., Albert, M.K.: Instance-based learning algorithms. Mach. Learn. **6**(1), 37–66 (1991)

8. Kohavi, R., et al.: A study of cross-validation and bootstrap for accuracy estimation and model selection. In: Ijcai, Montreal, Canada, vol. 14, pp. 1137–1145 (1995)
9. Zeng, Z.-Q., Yu, H.-B., Xu, H.-R., Xie, Y.-Q., Gao, J.: Fast training support vector machines using parallel sequential minimal optimization. In: 3rd International Conference on Intelligent System and Knowledge Engineering, ISKE 2008, vol. 1, pp. 997–1001. IEEE (2008)

Impressions of Japanese Character Katakana Strings

Yuta Hiraide(✉) and Masashi Yamada

Graduate School of Engineering, Kanazawa Institute of Technology, 7-1 Ohgigaoka, Kanazawa, Ishikawa 921-8812, Japan
b1519802@planet.kanazawa-it.ac.jp, m-yamada@neptune.kanazawa-it.ac.jp

Abstract. Japanese words are written with three types of characters: kanji (Japanese-Chinese characters), hiragana and katakana. Katakana are frequently used for names of goods and people appearing in anime and video games. Impressions of names are affected by meaning, shape of the characters and sounds when they are pronounced. In the present study, meaningless character strings which consisted of three katakana characters were presented on a computer display. Participants looked at each character string and rated their impressions for it using semantic differential method. They were not allowed to pronounce the character strings. The results of factor analysis showed that the impression space for the character strings was spanned by two dimensions, pleasantness and sharpness. The pleasantness was deeply affected by whether the strings included the voiced consonants, /b/, /d/, /g/ or /z/. The sharpness was deeply affected by the place of articulation of the consonant of the first mora in the strings. These results show that the impressions of the character strings are strongly determined by the sounds when the strings are pronounced, despite the fact that the participants were not allowed to pronounce them.

Keywords: Character strings · Japanese katakana · Impression · Semantic differential method · Factor analysis · Consonant

1 Introduction

Japanese words are written with three types of characters: kanji (Chinese-Japanese characters), hiragana, and katakana. Each character of hiragana and katakana corresponds to each mora of Japanese. One mora of Japanese language is consisted of one consonant plus one vowel (cv), or one vowel (v). Each mora takes almost the same time length in Japanese language. Therefore, hiragana or katakana character strings have isochronous rhythm when they are pronounced. Katakana character strings are frequently used for words from foreign languages. Moreover, katakana character strings are frequently used for names of products and characters who appear in anime and video games. Impressions of katakana strings may be affected by meaning of the words, shape of the characters, and sounds when they are pronounced. Kurokawa authored her book named "Why do names of monsters include /ga/, /gi/, /gu/, /ge/ and /go/" [1]. She summarized the impressions of Japanese character strings and insisted

T. Ahram et al. (Eds.): IHSED 2019, AISC 1026, pp. 1037–1043, 2020.
https://doi.org/10.1007/978-3-030-27928-8_154

that the impressions of character strings are deeply affected by the sounds when they are pronounced. However, her theory was based on her intuition and experience and formal experimental data were not shown in the book. Therefore, the perceptual impressions of Japanese katakana character strings are determined experimentally, in the present study.

2 Experiment

Character strings which consisted of three katakana characters were constructed and used as stimuli. If a string had meaning, it was eliminated from the set of stimuli to avoid the effects of the meaning. There are 71 katakana characters in the Japanese language. Most of the katakana characters are consists of cv or v. One exception is /n/ which is consists of one consonant. In the present experiment, 70 katakana characters, except for /n/, were used to construct the three katakana character strings. Figure 1 shows examples of three katakana characterstrings.

タニヌ	ダフズ	ハチネ	ザゼナ	ナホダ
チテザ	ヂノヅ	ヒズジ	ジナデ	ニジヘ
ツザト	ヅデタ	フヒジ	ズツゼ	ヌドニ
テハヒ	デゾフ	ヘネハ	ゼヂホ	ネヌチ
トヅヂ	ドタゾ	ホダツ	ゾヘノ	ノトテ

Fig. 1. Examples of three katakana character strings

The experiment was divided into four sections. In the first section, katakana characters limited to seven which include consonants /k/, /g/, /p/, /m/, /b/, /r/ and /s/, and were used to construct 63 character strings. In the second section, 35 strings were constructed. In this section, one of the characters with five new consonants, /t/, /d/, /h/, /z/ and /n/ was included at least once in the strings, and the other characters were with the previous seven consonants. In the third section, a character of solo five semivowels, /wa/, /wo/, /ya/, /yu/ and /yo/ was included at least once in the strings, then 40 strings were constructed. In the last section, a character of solo vowels, /a/, /i/, /u/, /e/ and /o/ was included at least once in the strings, then 52 strings were constructed. In total, 198 three-katakana character strings were used as stimuli in the present experiment.

In the experiment, each katakana character string was presented on the computer display. Ten Japanese native students, ranging from 21 to 24 years old, sat on a chair in a dark soundproof room. They looked at each string, then were requested to rate their impressions for each string using semantic differential method [2, 3]. Twenty-three seven-step bipolar scales shown in Table 1 were used as semantic differential scales, in the present experiment. The participants were not allowed to pronounce the strings.

3 Results and Discussion

The rated scores were averaged over participants for each string and for each scale. The average scores were analyzed using factor analysis [4]. The results of the analysis showed a two-dimensional solution accounted for 79% of data variance. The two dimensions are labeled pleasantness and sharpness, after the semantic scales which showed large factor loadings (Table 1). Russell revealed that our emotions can be illustrated by a two-dimensional model, which was spanned by valence (pleasant - unpleasant) and arousal (arousing - calm) axes, in the simplest way [5]. The two factors found in the present study coincided with the two dimensions Russell showed: The pleasantness corresponded to the valence in Russell, and the sharpness corresponded to the arousal.

Table 1. Semantic differential scales and their factor loadings

SD Scale			Factor	
			Pleasantness	**Sharpness**
Dark	-	Bright	0.878	-0.231
Quiet	-	Noisy	-0.871	0.116
Powerless	-	Powerful	-0.866	0.260
Masculine	-	Feminine	0.838	-0.362
Ugly	-	Beautiful	0.972	-0.083
Unpleasant	-	Pleasant	0.942	-0.123
Dirty	-	Beautiful	0.975	-0.035
Hateful	-	Cute	0.901	-0.334
Thin	-	Thick	-0.865	-0.163
Complex	-	Simple	0.908	-0.162
Weak	-	Strong	-0.874	0.276
Heavy	-	Light	0.947	0.013
Small	-	Big	-0.898	-0.003
Safe	-	Dangerous	-0.875	0.393
Dull	-	Vivid	0.484	0.771
Cold	-	Warm	0.229	-0.770
Slow	-	Speedy	0.151	0.903
Round	-	Sharp	-0.036	0.931
Loose	-	Tight	-0.239	0.877
Hard	-	Soft	0.624	-0.697
Dry	-	Wet	0.598	-0.611
Cheap	-	Luxurious	0.421	0.087
Banal	-	Unique	-0.629	-0.107
Cumulative Contribution Rate			0.568	0.794

Each stimulus of 198 strings were plotted on the pleasantness-sharpness plane using factor scores of the stimulus. Then, variance of factor scores was calculated for each moraposition for the pleasantness and sharpness factors, respectively, to clarify which moraposition in the strings significantly affected the impressions. Figure 2 shows the average of variances for eachmora. Figure 2 shows that the variance for the first mora is the smallest compared to the others, both in the pleasantness and sharpness, despite the fact that there are no significant differences, statistically. This implies that the firstmora character affects the impressions of the strings most strongly.

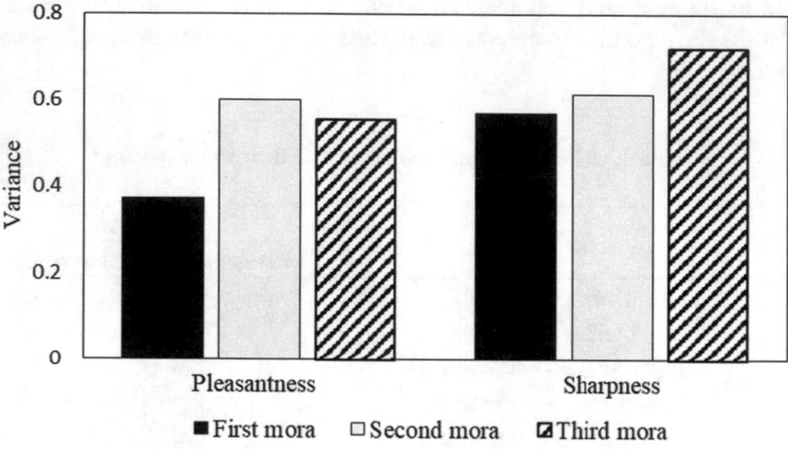

Fig. 2. Averaged variance for each mora position in each factor

Figure 3 shows the plots of the centroid for each first consonant in two dimensional impression plane, spanned by pleasantness and sharpness factors. When the first consonant was, /s/, /t/ or /k/, the strings were felt to be pleasant and sharp. When it was /h/, /p/ or /m/, the strings were feltto be pleasant and dull. When it was /g/ or /z/, the strings were feltto be unpleasant and sharp. When it was /b/, the strings were feltto be unpleasant and dull.

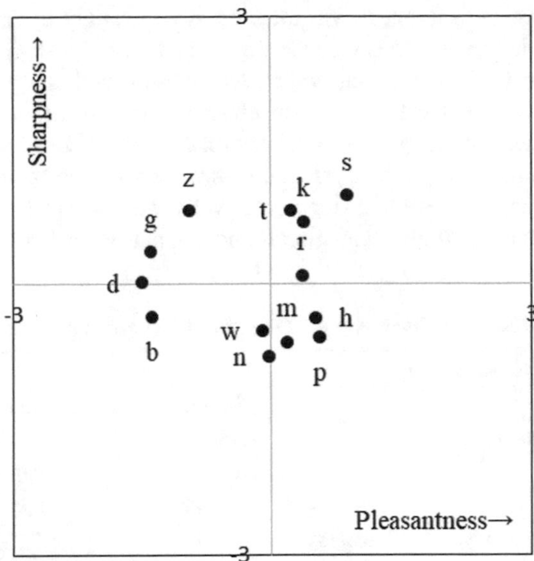

Fig. 3. Centroid of the impressions for each consonant in the first mora.

Figure 4 shows the plots of the strings which include any character with voiced consonants, /b/, /d/, /g/ or /z/, in the three characters and the strings which did not includethem. Figure 4 shows that the pleasantness is deeply affected by whether the strings included the voiced consonants, /b/, /d/, /g/ or /z/; The pleasantness largely declines if these consonants are included in the strings.

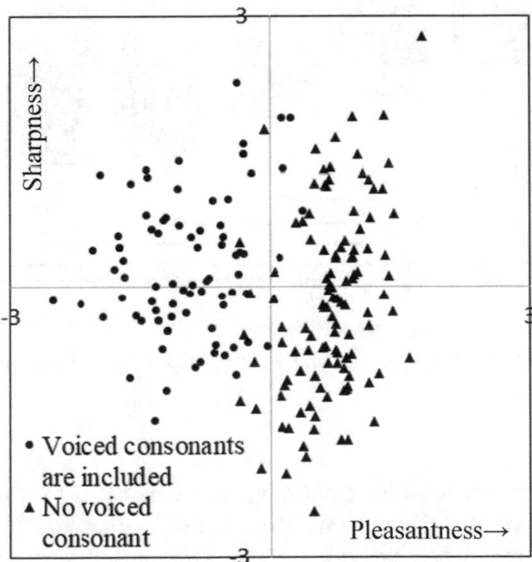

Fig. 4. Impressions of character strings with or without voiced consonants

Figure 2 showed the sharpness was affected most strongly by the first character in the strings. Then, three-way ANOVA was performed with the sharpness as dependent variable, and manner of articulation, voiced /voiceless sound and point of articulation as main effects. The F value for the main effect of point of articulation was high in Table 2. This implies that the point of articulation of the first consonant strongly affects the sharpness. Figure 5 shows the averaged sharpness for each point of articulation in the first mora. Figure 5 shows that dental and velar were felt to be sharp, palatal was felt to be intermediate, and alveolar, glottal and bilabial were felt to be dull.

Table 2. Results of three-way ANOVA for the consonant in the first mora

Dependent variable: sharpness				
Source	Sum of squares	df	Mean square	F
Manner of articulation	8.562	4	2.140	3.316
Voiced/voiceless	0.003	1	0.003	0.004
Point of articulation	24.327	5	4.865	7.537
Manner of articulation * voiced/voiceless	1.131	1	1.131	1.753
Manner of articulation * point of articulation	0	0	–	–
voiced/voiceless * point of articulation	3.103	2	1.552	2.404

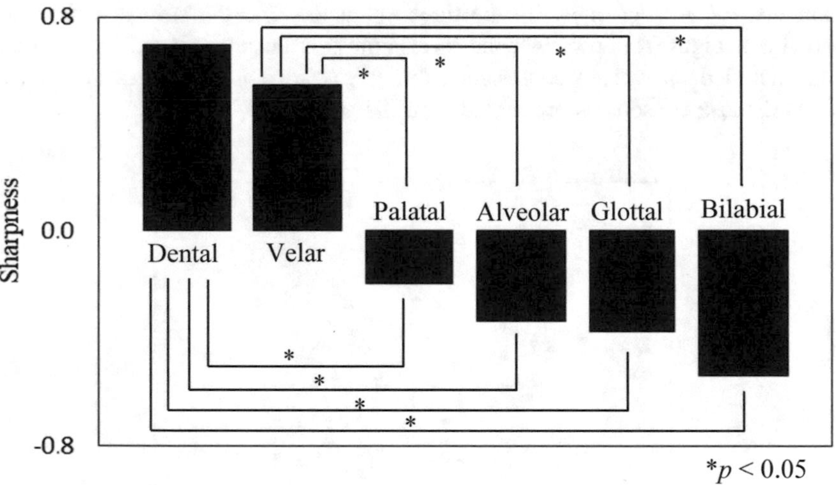

Fig. 5. Averaged sharpness for each point of articulation

4 Conclusion

In the present study, a perceptual experiment was conducted for three-katakana character strings. The results showed that the impressions of the strings were deeply affected by the sounds when the strings were pronounced, despite the fact that the participants were not allowed to pronounce them.

The participants were Japanese native speakers and knew the pronunciations for the katakana strings. In the case of non-native speakers, the shapes of the characters may deeply affect the impressions. Using the results of the present study, products or characters appearing in anime or video games can be named with congruent impressions for the products or characters.

References

1. Kurokawa, I.: Why do names of monsters include /ga/, /gi/, /gu/, /ge/ and /go/? (Kaijyu no nawanaze /ga/, /gi/, /gu/, /ge/ and /go/ nanoka?, in Japanese), Shincho-sya, Tokyo (2004)
2. Osgood, C.E.: The nature and measurement of meaning. Psychol. Bull. **49**(3), 197–237 (1952)
3. Osgood, C.E., Succi, G.J., Tannenbaum, P.H.: The Measurement of Meaning, p. 955. University of Illinois Press, Illinois (1957)
4. Osgood, C.E., Succi, G.J.: Factor analysis of meaning. J. Exp. Psychol. **50**(5), 235–338 (1955)
5. Russell, J.A.: A circumplex model of affect. J. Pers. Soc. Psychol. **39**, 1161–1178 (1980)

Self-cleaning Smart City Street Lighting Design Research Based on Internet of Things Technology

Jian Xu[1] and Jianli Wang[2](✉)

[1] School of Mechanical Engineering, Southeast University, Nanjing 211189, China
jayhsu1993@163.com

[2] Jiangsu Key Laboratory for Design and Manufacture of Micro-Nano Biomedical Instruments, School of Mechanical Engineering, Southeast University, Nanjing 211189, China
wangjianli@seu.edu.cn

Abstract. With the wave of smart city construction sweeping down, as one of the most extensive public infrastructure, streetlights should also be upgraded in the direction of intellectualization, greening and multifunction. Smart city street lighting with wide coverage and high density has tremendous potential value for the Internet of Things. Updating and upgrading the existing street light can easily establish a wide coverage of information perception network, thus establishing the basic network platform of Internet of Things to realize smart city. This paper intends to provide several ideas and schemes for the design of self-cleaning streetlights. Guided by the theory of green design and sustainable design, and based on modern technologies such as Internet of Things, this paper attempts to develop and excavate other advantages that city street lighting system may have.

Keywords: Street lighting system · Smart city · IOT · Green design theory · Human factors · Optimization

1 Introduction

Internet of Things (IoT) technology has broad application prospects in many fields, such as intelligent transportation, environmental protection, information collection and public governance. Based on the IoT technology, smart city is a new form of city that applies new information and communication technology, optimizes the traditional mode of urban planning, construction and management, promotes economic development and improves the happiness index of citizens [8]. The smart city market is very large. According to the Forward-looking Industry Research Institute's "China Ping'an City Construction Trend Forward-looking and Investment Strategic Planning Analysis Report 2018–2023", it is estimated that the smart city market scale in China is about 500 billion US dollars [5].

Smart street lighting is one of the key application fields of the IoT in modern cities. Most countries regard smart street lighting as the necessary infrastructure of IoT

T. Ahram et al. (Eds.): IHSED 2019, AISC 1026, pp. 1044–1050, 2020.
https://doi.org/10.1007/978-3-030-27928-8_155

because it has the characteristics of power, network and wide distribution. The value-added application that can be carried out is full of infinite imagination: Smart street lighting integrates hardware functions such as surveillance camera, 4G micro-base station, multimedia information screen and charging pile of new energy automobile, and transmits collected data and information to "intelligent lighting software system platform" through advanced information sensing technology [1].

At present, the traditional urban street lighting does not have the function of automatic cleaning. The main way to clean the streetlight is to patrol by the municipal departments at a fixed time and clean it by using lifting ladder or high-pressure water gun [7]. Because of the large number and wide distribution of streetlight, the financial pressure of manual maintenance and supervision is enormous. At the same time, once upon the cleaning is not timely, the surface of city street light shade often has a large amount of dust and other garbage accumulation. This paper will try to put forward several design ideas and design schemes in the next chapter for future related research.

This paper is structured as follows: In Sect. 2, the quantitative research method is explained. Section 3 introduces the structure of self-cleaning part. In Sect. 4, mobile phone control interface was designed to show how to control the work of smart streetlights remotely. Finally, Sect. 5 draws some conclusions from the research.

2 Kansei Engineering Research Method

This part focuses on the quantitative study and user preferences of the overall design style of smart city streetlights. Kansei Engineering is to quantify people's subjective perception, exploring the relationship between user perception and design characteristics of objects. This method enables designers to meet the user's aesthetics in the most efficient way.

Perceptual Adjective Pairs. Using the method of category distinction, we first determine the perceptual concept of order 0—useful and then push it down in turn to get enough perceptual adjectives. The perceptual adjectives generated by categorization is paired with antonyms to facilitate perceptual measurement based on semantic difference method. The results are shown in the Table 1 below.

Table 1. Perceptual adjective pairs based on semantic difference method.

01 Safe—Dangerous	02 Modern—Traditional	03 Colorful—Dull
04 Simple—Fussy	05 Clever—Silly	06 Concrete—Abstract
07 Rational—Emotional	08 Serious—Relaxed	

Sample Images. Totally 129 sample images of streetlights are collected from the internet. Then 129 sample pictures were screened by 3 college teachers. 55 unclear and duplicate pictures were removed. Then 74 remaining pictures were classified based on morphology and color. Finally, 20 typical sample pictures were selected for perceptual measurement.

Subjects. The main task of the participants in the experiment is the evaluation of perceptual adjective pairs. Therefore, the selected participants need to get frequent exposure to streetlights and understand perceptual adjective pairs. A total of 40 participants were invited in the experiment, including 20 students majoring in design, 10 university teachers and 10 employees working in the city center. The ratio of men to women is 1 to 1.

Perceptual Measurement Based on Semantic Difference Method. In order to facilitate data collection, collation and statistics, this experiment uses VB to compile an operation program, 40 subjects have used this software to score, the scoring range is 1–7. The data is calculated by the software directly, and the average of 9 perceptual words of each sample is obtained. The whole scoring process is carried out in a quiet laboratory. A total of 38 valid evaluations were retrieved in this experiment, and the corresponding perceptual scores of each sample were obtained by means of calculation, as shown in Table 2.

Table 2. Quantitative analysis results of perceptual measurement.

01	Safe	Modern	Colorful	Simple	Clever	Concrete	Rational	Serious
02	5.675	6.883	1.224	6.009	3.979	4.152	5.221	2.468
03	4.998	1.288	1.102	2.101	2.035	5.566	4.237	3.043
04	5.335	5.775	2.035	6.142	2.235	1.423	4.597	1.798
05	6.829	4.307	1.458	1.021	5.974	5.251	6.002	1.989
06	3.471	3.488	1.964	3.255	2.453	3.252	6.193	3.146
07	5.455	5.845	2.886	4.239	2.436	2.131	3.436	1.189
08	1.523	2.253	2.664	5.873	5.644	2.148	4.592	2.355
09	5.015	4.306	2.336	5.433	3.462	2.466	1.574	2.572
10	3.252	1.334	6.332	4.134	4.322	2.118	1.365	1.665
11	5.578	2.033	1.633	4.251	5.578	6.124	5.324	3.462
12	4.233	1.195	4.754	5.983	3.647	3.125	4.734	2.431
13	4.905	2.236	3.013	4.643	3.814	5.321	4.587	3.436
14	2.476	3.667	3.593	5.658	3.566	2.363	4.216	3.98
15	2.082	2.536	3.422	5.507	3.458	4.324	5.574	3.772
16	3.838	5.698	2.702	5.995	3.973	4.193	2.763	2.296
17	2.466	6.085	5.945	5.809	4.214	3.875	3.454	5.035
18	5.453	4.555	1.887	4.663	4.325	3.985	4.547	1.679
19	6.366	2.143	1.612	4.232	3.426	3.686	6.235	1.875
20	4.992	4.915	3.479	3.674	3.563	3.997	4.452	2.014

Through analysis, we know that the shape characteristics and design elements of street lamps have an impact on the subjective feelings of users. These may be considered as reference and restriction information for the researchers and designers. According to the results, “Modern” and “Clever” will serve as a guide for the next design work.

3 Structure of Self-cleaning Part

In order to realize self-cleaning function of streetlights, the structure of lampshade needs to be redesigned. This section mainly introduces two design ideas of self-cleaning parts' structures.

Smart Wiper. The first design idea comes from windshield wipers of cars. Applying mechanical components similar to automotive wipers to streetlight shades can ideally achieve remote unified control of the working path, strength and time of cleaning components. Compared with traditional manual cleaning method, this scheme eliminates the manual patrol and exempt from using the lifting ladder cleaning process.

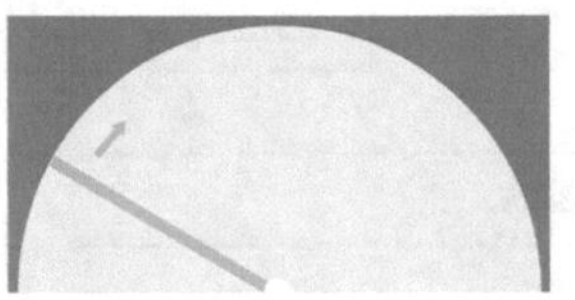
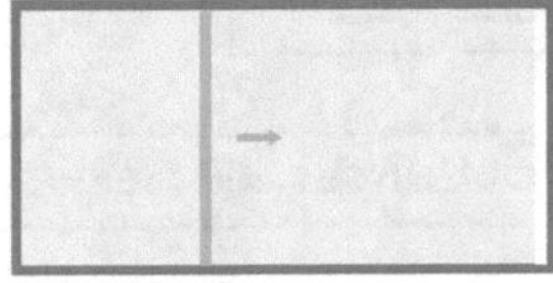
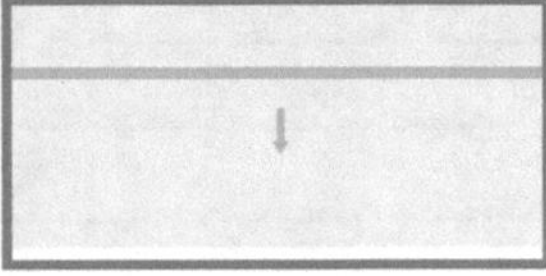

Fig. 1. Different self-cleaning paths of wiper's mechanical components

Figure 1 shows different self-cleaning paths of wiper's mechanical components, supposing the lightshade is a rectangle with a aspect ratio of 5:3. Their work efficiency and cost need further experiments to verify. In general, it can greatly reduce the human and financial expenditure of urban streetlights' cleaning work. However, one of the drawbacks of this method is that the mechanical components used for cleaning may fail or damage after a period of time, so this problem should be taken into account in the design process.

High Pressure Jet Module. The second design idea is to build a high-pressure jet module inside the lightshade. The dirt inside the lampshade can be cleared out by using the air pressure ejected. The remote control system is also used in this case to clean the light shade uniformly, which has the same advantages as the traditional manual cleaning method.

4 Remote Control System of Smart City Street Light

The remote control system is used to realize the functions of remote controlling the brightness, working time and self-cleaning mechanical components of streetlights. The smart streetlight is equipped with light sensor, dust detector and other sensing units to test and control the environmental factors around the streetlight, including ambient temperature, humidity and illumination. The overall structure of the control system is shown in Fig. 2.

Figure 3 shows the architecture design of smart city streetlight remote control system based on Internet of Things. All kinds of sensor units constitute a systematic perception layer, which is responsible for collecting relevant information and data. The sensing units are set up according to the predefined goal of streetlights. At the same

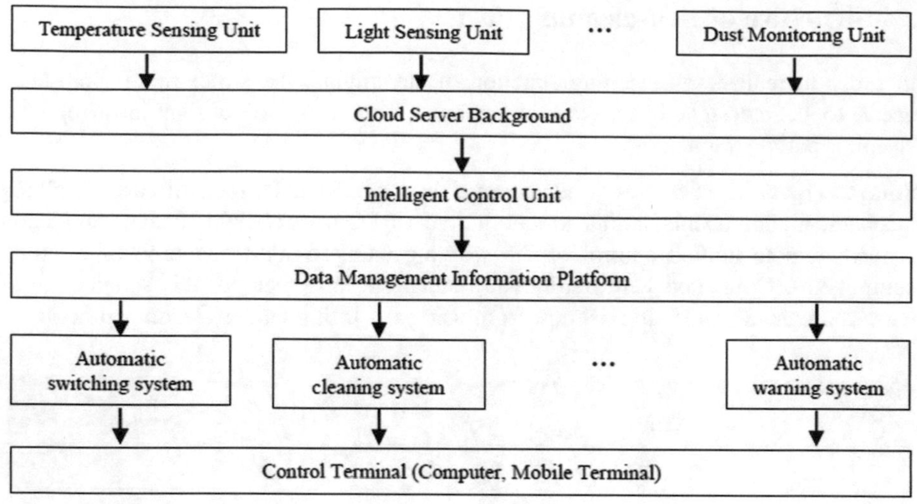

Fig. 2. Overall structure of the remote control system based on IOT

time, the surrounding environment parameters are collected in real time. Then the data are transmitted to the application layer for analysis and processing, and the automation and intellectualization of the control system are realized.

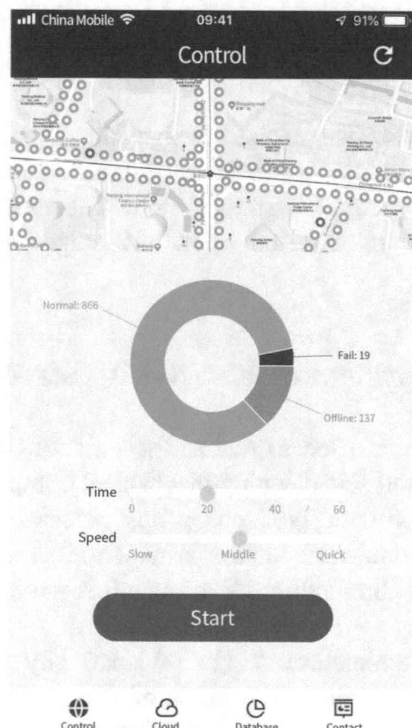

Fig. 3. One of the control interfaces of the mobile terminal

The system supports almost all the terminal devices, and accesses kinds of big data for data query, parameter setting and remote control. It provides the core function of remote control system, integrates the data information from sensor unit and intelligently analyses, and provides reference for the final decision-making of users.

The control software of remote control system has been designed and developed, which has a PC version and a mobile phone version. One of the control interfaces of the mobile terminal is shown as below.

The real-time running data of streetlights will be displayed on the interface, including running time, number and location. This software can switch brightness of streetlights remotely, and start the self-cleaning module to adjust the cleaning time and intensity. If one of the streetlights fails, it can be quickly located from the terminal control interface. The remote control system based on IoT technology can manage thousands of city streetlights. The data collected by various sensors on streetlights will also be displayed on PC and mobile terminals in real time. Managers can easily learn about the operation of city streetlights at any time and anywhere.

5 Conclusion

This paper mainly discusses the design research of self-cleaning smart city streetlights based on Internet of Things technology. Firstly, this paper introduces the current research status of intelligent city streetlights and Internet of Things technology, and finds out the shortcomings. The following section quantitatively analyzes the user's subjective preference for urban streetlights by using the theory of Kansei Engineering. Then two design ideas of streetlight's cleaning modules are put forward. Finally, the design of the whole remote control system and the control interface of the terminal are introduced. This smart city street lighting system saves time and improves work efficiency, saving labor costs and management costs, providing a reliable reference for related research in the future.

Acknowledgments. I would like to take this chance to express my sincere gratitude to those who helped me and assisted me during the writing of my paper, especially my supervisor Professor Jianli Wang. In addition, this work was supported by National Natural Science Foundation of China (No. 51876041).

References

1. Paskaleva, K.: Enabling the smart city: the progress of e-city governance in Europe. Int. J. Innov. Reg. Dev. **1**, 405–422 (2009)
2. Archana, G., Aishwarya N., Anitha, J.: Intelligent street light system. Int. J. Recent Adv. Eng. Technol. (2015)
3. Senkou shoujyo: Kansei engineering, ESJ Information N72. Japan Kansei Engineering Committee, Tokyo, Japan (1998)
4. Zhao, Q.: Kansei engineering and its application in product design. Shandong University, Jinan (2007)

5. Feng, J., Ye, C.A.: Street lamp control system based on wireless chain network design. Wirel. Commun. Technol. 54–57 (2013)
6. Deepanshu Khandelwal, B.M.T., Kritika Mehndiratta, N.K.: Sensor based automatic street lighting system. Int. J. Educ. Sci. Res. Rev. (2015)
7. Akshay, B., Siva, M., Parthasarathi, V., Vasudevan, S.K.: An Innovation in the Field of Street Lighting System with Cost and Energy Efficiency. Indian J. Sci. Technol. (2015)
8. Ye, W., Lu, W., Hong, K., et al.: Road lighting intelligent control system based on NB-IoT technology. J. Illuminat. Eng. **28**, 20–23 (2017)
9. Du, G.H., Wu, Y.X.: Research on Tunnel lighting automatic control system based on environment and traffic volume. Wirel. Netw. Technol. 64–65 (2016)

The Shopping Centre - Architectural Characterization and Evolution

Helen Morais(✉), Amílcar Pires, and Rui Duarte

Lisbon School of Architecture, CIAUD, University of Lisbon, Lisbon, Portugal
helen.morais@campus.ul.pt

Abstract. The analysis model that is presented relates to the Place referenced to the shopping centre. The method used is comparative and occur from the phenomenological analysis. Will be shown and describe the spatial structure that defines these architectural objects, their signs, elements of spatial and functional program and the way it is used. It is expected to describe the physical form, the visual form and spatial form and, consequently, the elements that create the space and provide the aesthetics, experience and recognition of the shopping centre as a Place.

Keywords: Strategy · Place · Project · Shopping center · Architectural program

1 Introduction

In the commercial centre, there is a set of observed phenomena that are experienced through an individual and embodied experience. Phenomenology was the method applied to identify the impact of the interactions of the phenomena with the users and evaluate the receptivity of the structure of the architectural program by the users.

The case studies are two shopping centres inaugurated at the beginning of the development process of this new concept of doing business: the Conjunto Nacional Brasília Shopping Centre in Brasilia (1971) and the Fonte Nova Shopping Centre in Lisbon (1985). The method used is comparative and derives from the phenomenological analysis. An analysis that will be directed by observation for the architectural characterization of these shopping centres.

The volume and space will be displayed from the "geometry", like the one that defines the concepts of mass, surface, amplitude and routes. In other words, concepts will be used as essential elements for the structuring of the commercial center's architectural space, the "interior" and the "outside" (the identification of the public space and the private space), "the main entrance" main facade", the "physical integration of commercial spaces" and "interior routes".

We will elucidate how the architectural program influences materiality and commercial strategy, what are the attributes and elements of the location of space and geometry that impact the formation of the Place in the commercial centre.

T. Ahram et al. (Eds.): IHSED 2019, AISC 1026, pp. 1051–1057, 2020.
https://doi.org/10.1007/978-3-030-27928-8_156

2 The Shopping Centre and the Place

Perceived images embody and produce interactions with the Place. The experience made through the senses of the body experience. According to Maurice Merleau-Ponty, experience is fundamentally a result of the integration of the senses, "My perception is not a sum of visual, tactile and auditory gifts: I perceive totally way with my existence life - I perceive a unique structure of the thing, a unique way of being, that speaks to all my senses at the same time" [1].

From this body interaction described by Merleau-Ponty are formed the elements of the perception of an embodied image. In a shopping centre, the Place has limits between the public and the private, where the community space is delimited, and the exterior and interior are also delimited.

Because it is a complex building, the analysis of shopping centres takes into consideration other dimensions that go beyond body experience, such as the temporal dimension. In the encounter of both arises the desire for the novelty, the fantasy, the ludic and the spectacle. Therefore, the experiences and phenomena in their evolutionary process naturally change over time.

We analyze two distinct shopping centres: the Conjunto Nacional Brasília in Brasilia and the Fonte Nova in Lisbon, with images formed within each context, considering the temporal dimension.

3 The Embodied Image

The Conjunto Nacional Brasilia Shopping conceived in the late 1960s, under a strong influence of the imaginary of the Las Vegas Strip. It located in the centre of the road that connects the centre to the north side and the south side of the city. Its facade was all designed with publicity panels, in a direct allusion to the Strip. It was initially 28 panels in neon that glowed at dusk.

According to Venturi, Bown and Izenour, the Strip based on advertising communication. It has commercial and advertising interests predominating: the sign is more important than architecture, and this reflected in the owner's budget. The sign in the front is a vulgar extravagance, the building in the back, a modest need. What is cheap here is the architecture. Sometimes the construction itself is the advertisement (…) [2].

Although the signs of the Conjunto Nacional do not serve directly to identify the building, it takes the observer to the imaginary of the Strip. Thereby with that the construction of an image embodied from the allusion to the Strip.

The Fonte Nova in Lisbon, conceived in the early 1980s, on the other hand, is the base of the "podium" that receives the towers of dwelling and elevates them so that they are the main elements seen from the outside. The podium, according to Vitruvius Pollio, is a prominent platform or large pedestal on which the temple rests [3] (Fig. 1).

In Fonte Nova, similarly, what highlighted and elevated are the two housing towers and are seen from several points of the city, including from the Second Circular, the arterial road adjacent to the commercial centre that connects the peripheral areas to the city centre of Lisbon (Fig. 2).

Fig. 1. On the left a view of the street and the entrance of the East Facade Fonte Nova in 1985, the centre and the right the same raised in 2019. Source: the author.

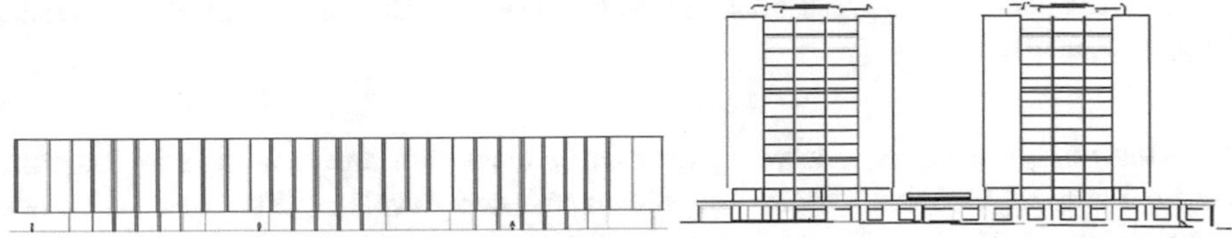

Fig. 2. On the left is the Conjunto Brasilia and on the right is the Fonte Nova. Source: the author.

The Fonte Nova, in addition to from the podium as the base for housing, refers to an interpretation that suggests being the 'modernist version' of Haussmann's Paris, whose apartment blocks had six storeys and commerce on the ground floor. However, on the scale of the city of the modern era, it has two fourteen-story towers, seen at the time, as landmarks in the landscape.

The Conjunto Nacional and Fonte Nova define a structure embodied beyond the enclosure. The embodied experience begins abroad from the identification of its volumes.

However, it is in the inner experience with the other senses that this embodied image is consolidated and fixed in the memory of the attentive observer. When entering the shopping centre, the user moves and then experiences other phenomena.

4 The Spatial Shape

The boundary between the public and the private shopping centre delimited in the entrances. They are the entrance doors that usually delimit the interior space of the exterior. According to Merleau-Ponty, space is not the environment (real or logical) in which things disposed of, but the way by which the position of things becomes possible [2]. In other words, the presence or absence of a door in a space is a positioned element with a specific function of delimiting the interior from the exterior.

In the Conjunto Nacional Brasília, which is a commercial centre without doors, the exterior and interior mainly oriented by the ceiling, floor and pillars. Besides that, this is a modernist building set on 'Pilotis'. It has an orthogonal concrete structure that models the internal circulation in labyrinth rhythm (Fig. 3).

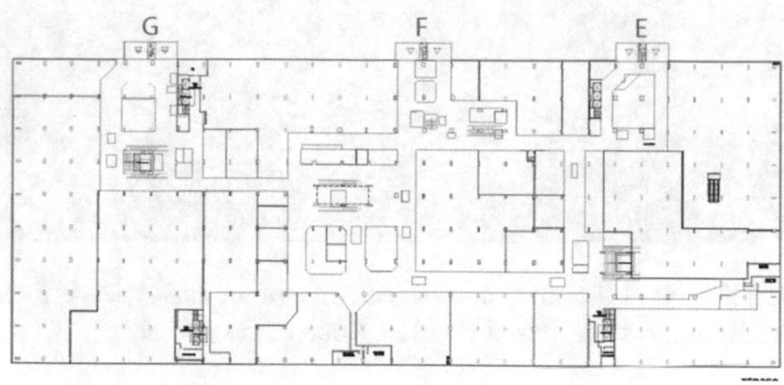

Fig. 3. Floor plan of the ground floor of Conjunto Nacional Brasilia. Source: Ancar Ivanhoe Shopping Centers.

It has many corridors, twelve squares and a restaurant area, and it is in this complexity of interconnected spaces that the experience develops. The user succumbs, therefore, to the rhythm of the labyrinth of the corridors of circulation and experience the relationship between the architecture and the spectacle, as Pallasmaa describes: "*Los escenarios tematizados y los ficticios simulacros arquitectónicos actuales, como los centros comerciales e las plazas urbanas, ejemplifican esta perdida de inocencia y sinceridade cultural*" [4].

The images stimulate and capture, either by the grandeur, or by the excess of the colours, or by the superposition of the sent messages. It is a fictitious scene space, in which it is intended to hide the essential problems of daily life and lead to a new and idyllic world of fantasy and pleasure.

Fonte Nova, unlike the Conjunto Nacional Brasilia, has doors in its three entrances. All located on the ground floor, where there is also a central square that connects to the first basement. The central square is the place of convergence, orientation and dynamism. To some extent, it is the point of equilibrium to establish the order of space (Fig. 4).

When crossing the doors of the shopping centre, there is a concrete delimitation of the interior and the exterior. There is conviction about where the boundaries of the public and the private begin and end. The building has a hermetic and intimate character, being the central square that distributes the flows.

The "show" is discreet, intimate and directed to the senses, translating into a real experience of the materiality of the architectural elements (floor, lighting, wall coverings and skylight). Although in an intimate way, or even as a spectacle, in the Fonte Nova "architecture as experience" comes from the perceptions provided by the senses that interact with memory and imagination.

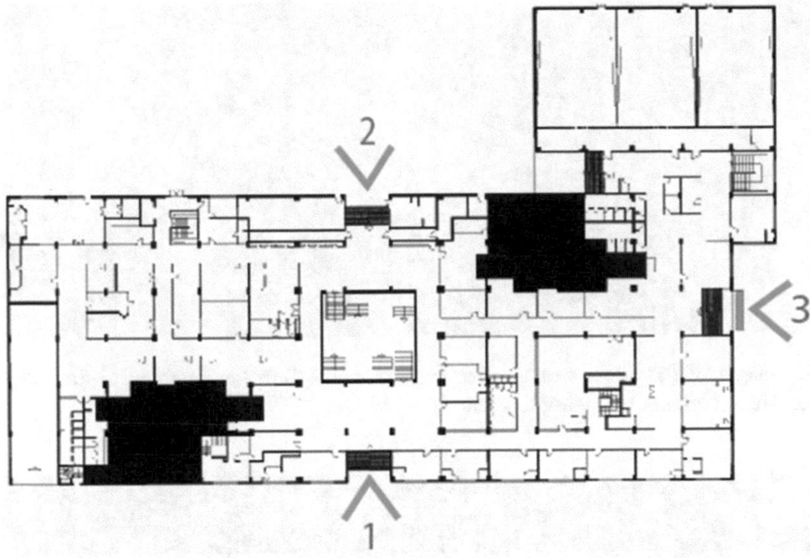

Fig. 4. To the left the floor plan of the ground floor with the respective entrances of the Fonte Nova: 1. East; 2. West; 3. North, on the right the first basement floor plan. Source: Alessia Allegri [5].

5 The Visual Shape

The perception of the visual form is a product of an experience that is corporeal through the sense of sight. However, this embodied experience is also temporal. It is related to a specific time when consciousness builds it. This awareness becomes evident when the senses capture every aspect of experience and register it in the brain and memory. Memory produces imagination, intuition, but also needs to develop empathy and a broad set of sensations that go beyond the five senses. According to Pallasmaa, "even the blind and the dumb can experience their entire embodied existences. However, Steiner's philosophy categorizes twelve senses, one of which is the ego, the sense of self-movement, and together, these three non-Aristotelian senses constitute the existential sense" [6].

For Pallasmaa in architecture, there are relational phenomena in the embodied experience that broadens perception and goes beyond the sense of sight. Similarly, in both shopping centres analyzed, there is an intangible dimension to the senses perceived and fixed in memory.

In the Conjunto Nacional Brasilia and the Fonte Nova, there is an architectural structure that repeated in both. In practice, this structure has been replaced by the images that are seen everywhere: in the corridors of the shopping centres, in the signs and shop windows. These images interact with the psyche of the wearer and a new record forms in their memory. Now it is the experience of marketing, advertising and commercial communication (Figs. 5 and 6).

Fig. 5. Images of the interior of the Conjunto Nacional Brasília shopping centre, May/2019. Source: Ancar Ivanhoe Shopping Centers

Fig. 6. Images of the interior of the Fonte Nova shopping centre, December/2018. Source: the author

In this ‘cultural anthropological’ relation consumption defines the value of both shopping centres. The desire to possess the object poses only one issue in evidence: ‘monumentality’, which is associated with the tendencies that each culture has built. This relation to the object, based on cultural anthropology, determines to a great extent the geometric relations of each one and the elements that construct and delineate the materiality to the Place.

6 Conclusion

The architectural program influences the materiality and strategy of the shopping centre. It is a building that results from contemporary urban forces and culture. Perceiving these forces and their meanings is a pertinent and current task, which is translated, mainly, in an attempt to systematize and update a “language” of an architecture project, that is the shopping centres so that it corresponds to a post-typological phase.

The shopping centres analysed are located in different contexts. However, both contain a strong influence of the cut-out and overlapping images, where the form and especially the meanings, transcend the sensory experience Aristotelian. They are, in fact, an existential experience in which there is an appeal to the emotions, to the ego and to the act of possessing the object, which enchants and dominates the Self.

This point is the where, in both cases, the embodied, spatial, and visual shapes converge on the same question, almost 'anthropophagic': they succumb to the psychoanalytic experience of marketing, 'neuromarketing'. The experience is synthetic, generated by mass culture that produces with one hand individualism and customization and, in the other, a collective cultural identity: the object can be as each culture wants, as long as they are multiple and profitable. It is not the most expensive or the cheapest that is the best, but the one that accommodated in the cultural context.

In the dimension of psychoanalysis, several meanings can grasp the collective sense and cultural identity of each context analysed, such as sustainability, conscious consumption, food wanting to appear healthier, the origin of the product to be sustainable, biological. The Being desires and possessed by the "appearance of the 'Good'" and not necessarily by the 'Beautiful'.

The formation of the Place in the commercial centre is, therefore, the result of these contemporary forces. Characterization is associated with what is shocking. Scale and geometry have a strong aspect of monumentality and consumable now. And the Self is at the centre of all questions, though not perceived as such.

References

1. Merleau-Ponty, M.: Sense Non-Sense. Northwestern University Press, Evanston, Illinois (1964)
2. Venturi, R., Brown, D.S., Izenour, S.: Aprendendo com Las Vegas. Cosac Naify, São Paulo (2003)
3. Pollio, V.: Tratado de Arquitetura Vitrúvio. Martins Fontes, São Paulo (2007)
4. Pallasmaa, J.: La imagen corpórea. Imaginação e imaginário en la arquitectura. Gustavo Gili, Barcelona (2014)
5. Allegri, A.: Mientras Apolo 70 viaja hacia a Alvalaxia XXI , Colombo y Vasco da Gama nos descubren otra ciudad. La dimensión urbana del comercio. Lisboa 1970-2010 (Doctoral dissertation). UPC-ETSAB, Barcelona (2012)
6. Pallasmaa, J.: Essências. Gustavo Gili, São Paulo (2018)

Empirical Assessment of Cyber-physical Systems Influence on Industrial Service Sector: The Manufacturing Industry as a Case Study

Ifetayo Oluwafemi[1(✉)] and Timothy Laseinde[2]

[1] Postgraduate School of Engineering Management, University of Johannesburg, Johannesburg, Republic of South Africa
ijoluwafemi@uj.ac.za

[2] Mechanical and Industrial Engineering Tech Dept, University of Johannesburg, Johannesburg, Republic of South Africa
otlaseinde@uj.ac.za

Abstract. This study examines service innovation in the manufacturing sector which is largely due to the adoption of digitalization and cyber-physical systems. It assesses their influence on the service environment and the long-term effect, based on unprecedented rapid automation witnessed since the dawn of the fourth industrial revolution. Besides the normal business of selling manufactured goods, manufacturing firms are becoming more competitive in their services by engaging in product-related after-market services all through their product lifecycle. The study was carried out using an exploratory method and the findings were primarily qualitative. The study findings showed that the emergence of cyber-physical systems and smart technologies have completely revolutionized and catalyzed more opportunities in the service business rendering throughout product lifecycle. The opportunities created are not limited to maintenance, repair and overhaul service business, but other emerging services, enhanced by cyber-physical systems. Based on the case studies examined in this study, the affordance that could realistically utilize the emerging services were identified.

Keywords: Industrialization · Product-service systems · Cyber-physical · Lifecycle · Automation · Transformation · Manufacturing

1 Introduction

The Industrial sector, popularly known as the "goods-producing section of an economy", is vulnerable and highly prone to paradigm shifts in business processes and manufacturing systems due to continuous inevitable changes. The need for higher quality services, increasing customer expectations, higher profitability, global competitiveness, and changing society needs are major factors that influence the embrace of Cyber-Physical Systems (CPS) in the manufacturing sector. From a technical and business point of view, industries will only remain competitive when internal systems and structures evolve with changing trends. Manufacturing firms are known for producing and selling products. Besides these two functions, many manufacturing firms are also adopting the rule of competitive advantage in their operations by creating

T. Ahram et al. (Eds.): IHSED 2019, AISC 1026, pp. 1058–1065, 2020.
https://doi.org/10.1007/978-3-030-27928-8_157

product-related services all through the product lifecycle. For decades, many firms have experienced great increase in their incomes from the service businesses and with emphasis on product-related services in a bid to expand their businesses [1–4]. Manufacturing firms consider value additional service, by offering scheduled maintenance, emergency/unplanned repairs, and overhauling services, with noticeable technical support assistance for products which signifies the importance of service business. This is often termed the servitization phase in the transformation journey of manufacturing industries [5–11]. Industrial products as well as services is said to be completely interlaced with one another and is referred to as the Industrial Product-Service Systems (IPSS) [12–16]. The usual goods dominant (G-D) pattern experience few inadequacies in respect to interpreting the substitute of value, fresh service-dominant (S-D) logic which is currently adopted amongst investigators in a bid to comprehend cutting-edge business frameworks in the orthodox products-driven business environment. While firms in manufacturing sectors find it difficult to promote servitization, the service business is becoming prominent, because additional stable revenue streams are generated, unlike the cyclic product business.

Designs of electromechanical products are evolving by the day. For high-end equipment's, breakdowns may be capital intensive resulting in various losses such as man-hour loss, customer loss, time loss, revenue loss, and other unquantifiable losses. Technology has greatly changed the phase of equipment breakdowns due to autonomous predictive maintenance systems triggered by artificial intelligence. Equipment's are now interconnected through the Internet of Things (IoT) using signal responses generated from sensors which are conveyed through packets and thereafter decrypted in low-level languages and readable formats. Principally, from an industrial perspective, ease of utilization of equipment, and reliability are major requisites and many equipment manufacturers have and are continually introducing functionalities that addresses reliability, maintainability, usability, and functionalities that supersedes traditional product offerings [26]. These innovations are contributory factor to the emergence of new business models thereby progressively changing the service business landscape. Considering condition monitoring of products as an example, preventive and predictive maintenance minimizes the downtime of equipment's. Manufacturers struggle to take advantage of evolving technological capabilities largely due to cost constraints, and due to limited envisaged cost benefits on a short term bases. The general argument by manufacturers is that same product qualities and production speeds are still achievable with older models of equipment's without considering CPS and other technological solutions. Some manufacturers only consider newer equipment's with CPS features only when significant value additions have been substantiated in other industries utilizing such technologies.

In this study, suitable service delivery frameworks for integrating CPS into product rendering manufacturing industries was explored. From the preliminary study, it was observed that increase in competition among manufacturing industry's service businesses stimulates growth in adoption of newer technologies due to competitiveness.

As service businesses gain prominence, evolving technological potentials are increasingly becoming inevitable in the manufacturing sector. As such, manufacturer's embrace systems that improve service offerings and service process delivery. For this reason, utilization of CPS as a means of improving manufacturing sector service business processes has continually gained prominence.

A typical scenario observed across manufacturing industries is the embrace of autonomous maintenance systems. When service firms detects breakdowns or service requirement through CPS embedded in equipment's, a field technician may then be sent to carry out a physical analytical examination to rectify such issues, and the stored trend in such occurrences may be further analyzed. The captured feedback is useful for machine learning systems. Also, it is highly imperative to highlight the possibilities such as providing distant service using CPS. Trivial glitches can be identified remotely and could be repaired.

The purpose of this study is to examine how the cyber-physical systems affects services business and, also, to identify the service business models. As such, the research questions are as summarized:

- What specific effects do cyber-physical systems have on manufacturing services
- Which business framework or models conform to the manufacturing sector?

The remaining structure of the study is discussed as follows: Firstly, important concepts related to the study were well-defined. Secondly, results in respect to the research question were displayed, which is built on conclusions from several case studies. Finally, a critical discussion and future research possibilities were highlighted for future research.

2 CPS' Effects on Industrial Services

Service business affordances in the manufacturing sector was sort into defined categories from the case studies. The recognized service business affordances were seven (7) in total, as presented in Table 1.

Table 1. Industrialized service business CPSs affordances

Service affordance	Explanation
Use data on operation performance to bring about better equipment	Information from the modern equipment can be utilized to improve the technical futures of the equipment while developing a new version of the equipment
Optimizing operations of the equipment	Data on past operations can be used to improve the operation of the equipment can. Failures can be avoided. Operations can be optimized by considering past usage patterns
Remotely control and manage equipment	Remote service centers can be used to manually manage specific operations of the equipment, with the knowledge that CPSs can obtain control information. CPSs can be rebooted to remotely fix faults

(*continued*)

Table 1. (*continued*)

Service affordance	Explanation
Predict and trigger service activities	Service activities can be generated and forecasted using CPSs Continuous data collection. For instance, routine maintenance activities can be done according to the level of the equipment used as well as its wear and tear of the equipment. By conducting service activities as well fixing service activities at appropriate times, improves the efficiency of the equipment
Diagnose and fix field service activities remotely	It is mostly possible to manage and fix faults of the equipment remotely. Trained staff in the set up comprehensive service centers detects and resolves faults remotely. This increases the efficiency of experienced service technician, having solved the faults remotely, reduces the stress of traveling to the spot where the service is needed initially
Give power and optimize field service	The effectiveness of prevailing service methods, especially the field service tasks can be maximized by industrial CPSs. The Field service tasks can be done quickly and, also, the value of the service could be improved as a result of CPSs. Remote experts and field service technicians will work together to quickly and more effectively solve problems
Services characterized by information and data	Unexpected information and data-driven service opportunities can be achieved as a result of the data and insights gotten from CPSs. For example, where the data belongs to the manufacturer, it can be sold to other stakeholders via standardized interfaces. The service business can take full advantage of this data

It is seen from the case studies that CPSs can immensely influence the service business. The first result will be that the equipment will be effectively monitored, thereby improving its condition. Secondly, remotes can be manually used to control how the equipment functions. Subsequently, the customer can be provided with individual services and solutions can be provided for operational issues. Thirdly, the efficiency of equipment can be maximized using sensor data and industry-specific algorithms, due to the advent of automated processes. If external sensors from other products are also considered, the efficiency of the entire system can be maximized [20]. Fourthly, Predictive and pre-emptive maintenance and repair which are usually carried out in the field can become possible. Customer satisfaction can ultimately go higher as equipment downtime reduces. Being able to remotely detect and diagnose errors and faults is the fifth result. Field agents can be immediately sent to the sites where the faults have been detected. Another point is that CPS data can serve as backing for field service agents. Important information can be provided by strong mobile work support systems like wearable devices, thereby improving efficiency of the service [39].

Spare parts can automatically be requested for by the products to improve first-time fix rate. Lastly, information and data-driven services can be exchanged between data owners and other interested parties. The manufacturers can sell the data they have computed to other service companies to ensure they improve customer service.

As seen from the study findings, CPSs provides huge possibilities of improving effectiveness and industrially drive service innovation. The case adopted in the study confirms that CPSs offer enormous possibilities for the industrial products operational phase.

3 Illustration of How CPSs Influences the Service Environment

It has been deduced from the case studies that increased presence of cyber-physical systems changes the market. One major factor in the service ecosystem can be attributed to technical architecture of CPSs. There is increased complexity in the service environment, caused by the rise in interdisciplinary, complexity and individuals product and service offerings [31]. Product manufacturers, product operators, and service organizations are three key stakeholders that may be recognized. These three stakeholders are influenced when industrial products digitalization ultimately influences the entire environment.

4 Influence on Manufacturers of Industrial Products

"Initial life (IL), conceptualization, description, and realization of equipment, which are the initial stages of the product lifecycle, are essentially in the hands of the manufacturers" [34]. In traditional sales and service businesses, the product operator is the customer which receives the products from the manufacturer of the industrial product. The individual then becomes the custodian of the product. Equipping industrial products with sensors and connectivity will avail the manufacturer the opportunity to retrieve field data from a sold equipment to provide Maintenance, Repairs, and Operations (MRO) service. MRO will be enhanced by utilizing analyzed data from previous services for future improvement of equipment's. Information Technology (IT) and data analysis are two essentials that influence industrial products. Retrieved data are becoming vital components of products as traditional products are developed into CPSs based products, which are essentially interconnected products which receive input and send feedbacks for analysis in a product cloud [20]. The type of industry determines the scenario which allows the stakeholders group the product cloud to easily access to the retrieved data.

The situations in the environment determines the decision of the equipment manufacturers in regards to adequate IT architecture that will ensure smart products and CPSs. Closed system approach and open system approach are the potential approaches [20] as confirmed by the case study reports in this study.

There are copyright mandates on the entire system which includes sensors, analytical abilities, connectivity and interfaces for retrieving information in a closed system approach. Appropriate requirements are put in place in the event of the manufacturing organization having dominance in the market. The manufacturing organizations can profit by having a higher competitive edge if appropriate prerequisites are set and if the strategies for implementation are efficient in a closed CPSs system, as this approach recommends more initial investment. The organization can control and maximize the construction of both the physical and non-physical parts of the system in this approach. This means that the company will be in charge of technology, data, and future systems development. Producers of system components have to be licensed in order to have access and integrate their products into the closed system. A specialist external IT service can supply the analytical prowess of the system. Conversely, in open systems, the end customers can gather as well as customize certain parts of the solution. Several stakeholders such as the equipment manufacturers, third party service organizations and even competing firms in the market can be allowed to interface with the system and add to the transformation of the system based on requirements. The components of the CPSs can be contributed by various parties. For example, the manufacturer can produce the product while a third-party entity incorporates the sensor network and connectivity.

The industrial products are required to collaborate with parties with experience about equipping products with sensor technology and connectivity. The cyber-physical systems are efficiently exploited when tactical partnerships are formed with software companies.

To summarize, a vital factor to efficiently take advantage of is the vast possibilities of cyber-physical systems is effective value co-creation.

5 Impact of CPS on Product Machinists

Product operators require perfect product operations. For production plans to align with set timelines and targets, equipment and operators must function at optimal efficiency. To achieve this level of performance where there is no lag in production plans, digitalizing the traditional production systems is important, and CPSs are considered valuable prospects.

Operators using an open system approach can technically incorporate the equipment or machinery in their database so that the sensor data can be harnessed and analyzed, which will provide profitable information for system improvement. A typical example is the airport operator used in this case study, which focuses on the connection of all equipment such as air-conditioning systems, lighting systems, the elevators, and the escalators used in passenger transport to central facility management system. This leads to the airport operator managing the equipment based on flight plans. This situation is typically applied to industrial robots and factory automation systems currently existing in the automotive industry.

6 The Impact of CPSs on Industrial Sector Service Organizations

Industrial CPSs are paramount improving service effectiveness and promoting service innovation in service organizations. Products operators inform the service organizations about the failure of traditional products. As a result of communication gaps, the service organizations are only informed about a failure and not the full extent of the failure. The products can directly pass information about failures as a result of the sensors embedded in industrial products.

Service organizations can ensure adequate delivery of services, due to the advent of CPSs. Such as when a service organization is in control of supplying and ensuring that industrial machines work effectively, in situation known as performance contracting. However, there is a shift with servitization where the manufacturers have contracted third-party service organizations with the responsibility of proving services to their clients. These service organizations now have an increased competitive edge in the market as they have access to the product's operational data as well as their expertise. This data is increasingly proving to be vital in knowing the operational effectiveness of the service business. Ultimately, manufacturers providing services have a higher market value due to increased competitive edge emerging from CPSs.

7 Conclusion

The service business has witnessed remarkable changes and is still undergoing continuous transformation stimulated by the influence of Cyber-Physical Systems (CPSs). The transformation caused by CPSs on the equipment manufacturing industry service business has been examined by the eleven (11) case studies of manufacturers, service organizations and equipment providers. The service affordances were recognized and the influence of CPSs on several stakeholders in the industrial service environment were equally examined. The study largely focused on the innovative prospects for manufacturers in the serviced manufacturing industry. Despite applying a well-defined research approach, there were some limitations in the research. The case study approach was limited to empirical findings from various studies. However, the results can be more generalized if more cases are examined across various manufacturing industries. This nature of study will require huge resources beyond what was provided for the current study. On the positive side, this study bridges the gap in understanding the role CPSs play in industrial services. This study reveals the affordances of CPSs on the service businesses, by drawing out techniques experts can take advantage of as indicated by emerging technological capabilities. This study provides a leverage for scholars interested in further research relating to challenges and affordances of cyber-physical systems, for service rendering models. Future works in respect to this study can be considered by examining classifications of service setups of CPSs in focal sector based manufacturing industries.

References

1. Cavalieri, S., Pezzotta, G.: Product–service systems engineering: state of the art and research challenges. Comput. Ind. **63**(4), 278–288 (2012)
2. Mont, O.K.: Clarifying the concept of product–service system. J. Clean. Prod. **10**(3), 237–245 (2002)
3. Tukker, A.: Product services for a resource-efficient and circular economy – a review. J. Clean. Prod. **97**, 76–91 (2015). https://0-doi-org.ujlink.uj.ac.za/10.1016/j.jclepro.2013.11.049
4. Windahl, C., Lakemond, N.: Developing integrated solutions: the importance of relationships within the network. Ind. Mark. Manage. **35**(7), 806–818 (2006)
5. Doni, F., Corvino, A., Martini, S.B.: Servitization and sustainability actions. Evidence from European manufacturing companies. J. Environ. Manage. **234**, 367–378 (2019). https://0-doi-org.ujlink.uj.ac.za/10.1016/j.jenvman.2019.01.004
6. Huxtable, J., Schaefer, D.: On servitization of the manufacturing industry in the UK. Proc. CIRP, **52**, 46–51 (2016). https://0-doi-org.ujlink.uj.ac.za/10.1016/j.procir.2016.07.042
7. Kastalli, I.V., Van Looy, B.: Servitization: disentangling the impact of service business model innovation on manufacturing firm performance. J. Oper. Manage. **31**(4), 169–180 (2013)
8. Mastrogiacomo, L., Barravecchia, F., Franceschini, F.: A general overview of manufacturing servitization in Italy. Proc. CIRP, **64**, 121–126 (2017). https://0-doi-org.ujlink.uj.ac.za/10.1016/j.procir.2017.03.010
9. Li, J.H., Lin, L., Chen, D.P., Ma, L.Y.: An empirical study of servitization paradox in China. J. High Technol. Manage. Res. **26**(1), 66–76 (2015). https://0-doi-org.ujlink.uj.ac.za/10.1016/j.hitech.2015.04.007
10. Rymaszewska, A., Helo, P., Gunasekaran, A.: IoT powered servitization of manufacturing – an exploratory case study. Int. J. Prod. Econ. **192**, 92–105 (2017). https://0-doi-org.ujlink.uj.ac.za/10.1016/j.ijpe.2017.02.016
11. Vandermerwe, S., Rada, J.: Servitization of business: adding value by adding services. Eur. Manag. J. **6**(4), 314–324 (1988)
12. Guidat, T., Barquet, A., Widera, H., Rozenfeld, H., Seliger, G.: Guidelines for the definition of innovative industrial product-service systems (PSS) business models for remanufacturing. Proc. CIRP **16**, 193–198 (2014)
13. Everhartz, J., Maiwald, K., Wieseke, J.: Identifying and analyzing the customer situation: drivers for purchasing industrial product service systems. Proc. CIRP **16**, 308–313 (2014)
14. Vargo, S.L., Lusch, R.F.: From goods to service(s): divergences and convergences of logics. Ind. Mark. Manage. **37**(3), 254–259 (2008)
15. Maiwald, K., Wieseke, J., Everhartz, J.: The dark side of providing industrial product-service systems–perceived risk as a key challenge from a customer-centric point of view. Proc. Cirp **16**, 241–246 (2014)
16. Mikusz, M.: Towards an understanding of cyber-physical systems as industrial software-product-service systems. Proc. CIRP **16**, 385–389 (2014)
17. Annunziata, M., Evans, P.C.: Industrial internet: pushing the boundaries of minds and machines. General Electric (2012)

Useful Total Quality Management Critical Success Fundamentals in Higher Education Institution

Ifetayo Oluwafemi(✉) and Timothy Laseinde

Postgraduate School of Engineering Management,
University of Johannesburg, Johannesburg, Republic of South Africa
{ijoluwafemi,otlaseinde}@uj.ac.za

Abstract. Higher education performs highly essential roles or responsibilities in both the economic development and cultural reconstruction of every nation of the world. However, the quality of higher education is crucial in this respect. Therefore, Total Quality Management (TQM) in higher education institutions (HEI) is in perspective in this paper. This high quality can only be achieved by these institutions by having long term planning, formulating yearly quality programs and executing such quality programs to the letter annually so that the vision of the higher education institutions can be accomplished. Applying TQM concepts is an important thing that can go a long way to revive higher education institutions through entrenching awareness about quality into everything or every procedure undertaken by the higher institution. This particular study makes efforts at analyzing TQM concepts in higher education institutions. This is done by outlining some available literature on Critical Success Factors (CSF) of TQM as well as the execution of TQM in all areas to improve the total quality management practices as well as the outcomes of such practices. As a concept or an approach, TQM contributes immensely towards making sure that projects are properly delivered to the target stakeholders based on their requirements. In this paper, five Critical Success Factors have been identified, and these are communication as a means of improving quality, organizational management, employee involvement, training and development, and recognition, and culture. All these must be given proper consideration by HEI before total quality can be achieved or become successful.

Keywords: Total Quality Management · Quality awareness · Higher education institutions · Critical Success Factors

1 Introduction

In recent times, there have been growing concerns about the quality of institutions of higher learning and their quality management in many countries of the world. There is also a serious concern about the improvement devices like the number of accreditations, performance indicators, institutional assessments, enrollment for programs and graduation, quality audits and many more. Therefore, it has become important to delve

T. Ahram et al. (Eds.): IHSED 2019, AISC 1026, pp. 1066–1074, 2020.
https://doi.org/10.1007/978-3-030-27928-8_158

deeply into quality management principles and methodologies, as well as the development of tools that can be applied to improve quality of higher education institutions.

Excellence and quality should be the focus of higher education [1–3]. Therefore, increment in students' enrolment rate and increment in the number of students, changing but practical societal structures and delivery of different programs, as well as job descriptions that produce multifaceted quality of questions should be the hallmark of higher institutions of learning. This is because the core of education, especially higher education should be quality. Quality does not only influence what students learn but also the procedure of learning. Both the procedure and the quality are important because they go hand in hand as both will permit the students to understand why they are learning. The pursuit of ensuring every student achieves the predetermined learning objective outcomes and also acquire values and skills that can help them to play positive roles in their various societies or communities is an issue that should take the front lead on the policy agenda of every progressive country. This will help the country to produce highly qualified graduates to handle or man every sector of the country.

Total quality management (TQM), which has been well explicitly described by different scholars, has been referred to as an approach of management with the purpose of improving both effectiveness and efficiency, cohesiveness and flexibility as well as competitiveness of any organization or business in its entirety. The British Standard Institution has affirmed the core objectives of TQM to be a management philosophy as well as company organizational which its primary intention is to connect the human as well as material resources of any organization adopting its principle in the most efficient way to accomplish the goals of the organization [4].

Based on the definition above, it has become obvious that Quality improvement remains a significant force globally. Regardless of the fact that there are different methods of improving and managing, it has been realized that TQM remains a significant determinant or very important success factor for both manufacturing organizations and Higher Education Institutions across the globe. Strengthened global competition, as well as increasing demand for improved quality in education, has caused significant increase in the number of HEIs. Therefore, it is essential to invest significant resources in the process of adjusting and implementing the core concept of TQM, tools as well as techniques.

TQM has been well illustrated as extensive managerial philosophical approach, which embodies a series of regulatory principles that represent the background of continuously improvement of an institution. TQM comprises the usage of a sound management methods, quantitative quantification procedures as well as human resources having an underlying aim to improve every department within an institution, surpassing the current as well as forthcoming needs of the shareholders as well as increasing the general routine.

In recent past, dominant facts relating to the discipline of quality management have been established by Deming as well as Juran, and these scholars have advocated for having abundant prescriptions in Total Quality Management. The insight from Deming and Juran into the discipline of TQM has afforded very useful and sound understanding of the various principles underlying the idea of TQM. Therefore, TQM has instituted the foundation on which ensuing scholars have been formulating as well as verifying critical success factors on Total Quality Management.

The term Quality Management has been defined by several scholars as a phrase used in encompassing great array of instruments, procedures as well as methodologies for continual improvement in an organization. This paper, therefore, considers the different critical success factors (CSFs) of TQM laying more emphasis on its relationship with Higher Education Institutions. Thus, it has been observed that the quality together with the social relevance of higher education in developing countries globally is deteriorating as a result of several unfavorable factors. This has made TQM a significant feature that shapes the policies of higher education institution in guaranteeing improved quality as well as constant improvement. Therefore, this study recognizes critical success factors of TQM, as such factors aid the improved performance of HEIs, in that way, satisfying the stakeholders' expectancies as much as possible. This paper will identify the useful critical features that are essential for the success of HEI's and that are necessary for understanding TQM Principles that are important for both quality sustenance and improvement. The paper also analyses the relevance of total quality management in the HEIs context in the economic expansion of any nation.

2 Literature Review

Critical Success Factors may be referred to as the restricted number of areas where satisfactory results can guarantee for an organization the successful competitive performance. Critical Success Factors have also been defined as the significant key areas where everything must go right before an organization or business can flourish. therefore, inability of the organization to maintain such areas will lead to non-attainment of the organization's goals for a particular period of time. Therefore, results in the areas of Critical Success Factors must be satisfied before a business' or an organization's performance will not fall short of expectations of both the organization and its customers.

Conducting a research on TQM, Saraph et al. [5] have established a very dependable model for measuring the practice of quality management. The model developed was founded on eight important elements. The eight critical elements include Supplier quality management, Role of quality department, Role of divisional top management and quality policy, Training, Product/service design, data and reporting, Employee relations and Process management operating Quality.

In a similar dimension, the practices highlighted by Saraph et al. [5] were expanded by Ahire et al. [6]. In the expansion, 12 critical factors were identified for the implementation of TQM by Ahire et al. [6]. The factors are as follows: use of statistical process control, Supplier performance, Top management commitment, Customer focus, Supplier quality management, Design quality management, internal quality information, Employee involvement, Benchmarking, Employee empowerment, Employee training, and Product Quality. A surveyed of three hundred and seventy (370) Greek companies was conducted by Fotopoulos et al. [7]. The survey discovered that the issue of leadership coupled with process management and service design are very important. The survey also found that certain factors can be referred to as critical success factors in the implementation of TQM. The factors as these: supplier quality management, customer focus, human resource management, as well as Education and Training.

Despite the fact that a lot of studies have been conducted on TQM, it should be noted that certain scholars occupy an important position in TQM literature. These scholars often advocate for making far more reaching concerns for the implementation of TQM. The scholars are Sila *et al.* [8], Idris *et al.* [9], Singh *et* al. [10], Vouzas *et al.* [11], and Karuppusami *et al.* [12], as well as Prajogo *et al.* [13], among others. These scholars have recommended that there should be further concern for the evaluation of critical success factor. Other scholars include Chapman *et al.* [14], Al-Khalifa *et al.* [15], Alomaim *et al.* [16] and Baidoun [17] just to mention a few. This group of scholars is of the opinion that there are seven common critical success factors of TQM in the Arabian countries. Most of the studies above identified the following factors: customer focus, top management commitment; Recognition and Reward, education and training, vision, employee involvement, supplier quality management and plan statement. The scholars have submitted that such factors are not only applicable to the Arabian countries but also to most developing countries around the world. In their submission, they opine that the factors are common to most developing countries of the world because most of these countries face similar or the same challenges.

Therefore, this study has considered these factors for the evaluation of TQM implementation in Nigerian Iron and Steel Company. There are major factors posing as impediments from implementing TQM in HEIs, viz as viz: Lack of management commitment, poor vision, poor plan statement as well as Government Influence.

3 Importance of TQM in HEI

In an environment where there is competition, HEI's are usually compelled to develop and implement strategies operational in global context. Thus, Total Quality Management (TQM) can be referred to first as a management philosophy and secondly as a thinking process with the capacity to assist institutions of higher learning to move towards achieving excellence in the provision of education. TQM helps organizations, especially institutions of higher learning to create a culture of trust and participation; teamwork and quality-mindedness; zeal for continuous improvement and continuous learning. In the long run, TQM helps to create a working culture that supports the success of HEI.

In many HEI's, TQM has not only become extensively recognized but also successfully implemented. This success has given organizations practicing TQM the edge in competitiveness, both locally and internationally. TQM has also ensured very high-quality services that have satisfied the needs of the stakeholders in different areas. Be this as it may, organizations, especially HEI's must possess detailed knowledge as regards key factors connected with quality performance practices which have been considered highly imperative to increase efficiency as well as enhance growth.

4 Critical Success Factors of TQM in HEI's

4.1 Management Commitment and Leadership

When the Leadership of an institution is directly involved in TQM, this will make the decision-making on any sensitive issues to be prompt, and also helps TQM agenda. This implies that the support of the top management staff, together with its leadership involvement is an indispensable factor required for an effective implementation of TQM in HEI's. In addition, TQM requires adequate financial assistance from the top management so that funding many research projects can be made available.

According to Crosby [18], the commitment of top management is a vital feature necessary for safeguarding implementation of TQM in any organization. For an organization to communicate quality strategy throughout its workforce, it is essential for the leadership of the institution to make an institutional environment that emphases on constant improvement. The commitment from the employees has the capacity to promote the creation of apparent and noticeable quality values. In addition, the management is expected to guide all activities of the employees of the organization towards quality excellence.

4.2 Continuous Improvement

Every unit in an educational institution is expected to be trained towards improvement. For instance, the top management, the colleges, the faculties, and departments should have the same vision so that the entire process of TQM will yield the expected result. Bearing this in mind, it has to be noted that constant improvement must become the order of the day before TQM can be achieved. It is only when this has been done that institutions will maintain control on quality and meet the target or goals of the institution.

4.3 Overall Satisfaction

As a process, Total Quality Management (TQM) focuses on achieving overall satisfaction through the notion of "continuous improvement". The idea of "continuous improvement" came forward after the 1980 s, with the aim of expanding and increasing quality management strategy by adding more aspects related to quality. The concept of Total Quality Management (TQM) in HEIs is a modern management concept, and it has helped in enhancing the competitiveness of organizations so that a good brand name can be achieved for the institutions at all times, especially in the long run.

4.4 Involvement of Faculties and Non-teaching Staff

To achieve both individual and institutional goals, faculties and non-teaching staff members must be involved in the managerial decision-making process. This involvement helps employees to become more focused on their individuals, group and institutional goals. For instance, this has always been recognized by the Japanese and this has accounted for their success in various endeavors across the global markets. This is so because they place remarkable value on the integration of the employees with organizational objectives, process, and equipment. A good combination of all these often lead to very high productivity and great success.

From the perspective of Lawler, if "Employee Involvement" is properly implemented, there will be changes in the fundamental relationship between employees and whatever organization they are working for. This accounts for why Chapman [14] submits that employee engagement has the capacity to increase the in-depth knowledge of organizational policies. It also involves processes like decision making at the lower levels, adopting the experience, knowledge and the ideas for the advancement of the organization. This implies that employees will be accorded proper recognition for their contributions, innovations, ideas, etc. to their various units and the organization as a whole. This also serves as a process of developing confidence in the employees by encouraging them to make decisions and solve problems among themselves without involving the top management of the organization.

4.5 Training

Before an employee can be deeply involved in organizational matters, such an employee needs more training because being deeply involved implies having more responsibility. Therefore, training has been considered as an important factor that can help in quality enhancement or development. It should be noted that not just training but quality training is essential to improve quality of the workforce. Quality training consists of educating, informing and training every employee. It helps employees to increase knowledge. This is done by provide information about the mission and vision of the organization; it also deals with the direction or organizational structure to enable employees to gain skills in an attempt to advance the quality and as a consequence solve the problem.

In his opinion, Johnson [20] has submitted that an organization that requires quality will determine and allocate the benefits of such training to its workforce. This is essential because training or strategic planning does not only focus on the present training needs of the employees but more on training in the future. The result of strategic planning is often a strategic training plan. This can be adopted in predicting the future or subsequent training needs with the focus on the needs of employees as well as consumers' demands.

4.6 Teamwork

Teamwork is one of the important elements of TQM that brings about quality and performance improvements in any institution where TQM is being implemented. Teamwork can bring about peace and harmony among staff members of the institution in order to succeed in the pursuit of quality improvement. In the spirit of togetherness observed via cooperation, commitment as well as participation of all staff member with the motive of accomplishing quality improvement agenda instigated by the institution. Therefore, this involves the collaborative efforts of all staff members, via their contribution in affording skills and assessment as well as the needed experience to a certain task.

4.7 Emphasis on Measurement and Review

In TQM periodical meetings are to be held to review the improvement work that has been taken up in the HEI's. Without reviews, corrective and preventive action taken up cannot be effective. Measurement is also an important aspect in today's world to have a check on quality because it enables us to understand what the current quality levels are and what quality level we aspire.

4.8 Universal Responsibility of Quality

Another basic TQM perception is that the responsibility for quality is not restricted to only one department of the HEI but is the responsibility of all employees and all departments.

4.9 Benchmarking

Benchmarking is a continuous, systematic process of evaluating and comparing the capability and performance of one institution with others, normally recognized as HEI's offering standard and quality education. Benchmarking is often used for improving the quality and communication, to professionalize the institution for improving faster in the competitive era.

4.10 Value Improvement

The essence of value improvement is the ability to meet or exceed the expectations of the stakeholders while removing unnecessary costs. TQM removes unnecessary costs, while simultaneously the expectations and requirements of the stakeholders are fulfilled.

4.11 Initiativeness

Higher education institutions must take up the initiative in influencing events for achieving the Objectives. They must take action for setting high-performance goals and originate action rather than responding to the action of others.

5 Conclusion

To achieve success in the higher education institutions, the management must be very strategic on their managerial approach, especially on how they identify, analyze and react to sensitive issues. In this study, it has been established that the implementation of TQM in the higher education institutions will not only have a positive impact on the institutional goals, missions as well its vision but will also, have an excellent impact on the institutional quality culture, which will, in turn, create more value and proper the institutional quality, thereby assuring growth of the institution. Despite the potential benefits associated with the implementation of TQM, which is bringing a wide range of changes in the higher education sector, many HEIs are not giving precedence its implementation.

Finally, TQM is a considered as a concept and not a process that can be implemented overnight. For any higher education institution to remain competitive and achieve continuous quality and performance improvement, it becomes highly imperative for such HEIs to critically analyses the critical success elements of whatever TQM model to be adopted before its implementation in order to attain a sustainable quality culture.

References

1. Akareem, H.S., Hossain, S.S.: Determinants of education quality: what makes students' perception different? Open Rev. Educ. Res. **3**(1), 52–67 (2016)
2. Al Shobaki, M. J., Naser, S.S.A.: The role of the practice of excellence strategies in education to achieve sustainable competitive advantage to institutions of higher education-faculty of engineering and information technology at Al-Azhar University in Gaza a model (2017)
3. Kumpulainen, K., Lankinen, T.: Striving for educational equity and excellence. In: Miracle of Education Anonymous, pp. 71–82. Springer (2016)
4. Middlehurst, R.: Quality: an organising principle for higher education? High. Educ. Q. **46**(1), 20–38 (1992)
5. Saraph, J.V., Benson, P.G., Schroeder, R.G.: An instrument for measuring the critical factors of quality management. Decis. Sci. **20**(4), 810–829 (1989)
6. Ahire, S.L., Golhar, D.Y., Waller, M.A.: Development and validation of TQM implementation constructs. Decis. Sci. **27**(1), 23–56 (1996)
7. Fotopoulos, C.B., Psomas, E.L.: The impact of "soft" and "hard" TQM elements on quality management results. Int. J. Qual. Reliab. Manage. **26**(2), 150–163 (2009)
8. Sila, I., Ebrahimpour, M.: An investigation of the total quality management survey based research published between 1989 and 2000: a literature review. Int. J. Qual. Reliab. Manage. **19**(7), 902–970 (2002)
9. Idris, M.A., Zairi, M.: Sustaining TQM: a synthesis of literature and proposed research framework. Total Qual. Manage. Bus. Excell. **17**(9), 1245–1260 (2006)
10. Singh, P.J., Smith, A.: An empirically validated quality management measurement instrument. Benchmarking: Int. J. **13**(4), 493–522 (2006)
11. Vouzas, F.K., Gotzamani, K.D.: Best practices of selected Greek organizations on their road to business excellence: the contribution of the new ISO 9000: 2000 series of standards. TQM Mag. **17**(3), 259–266 (2005)
12. Karuppusami, G., Gandhinathan, R.: Pareto analysis of critical success factors of total quality management: a literature review and analysis. TQM Mag. **18**(4), 372–385 (2006)
13. Prajogo, D.I., McDermott, C.M.: The relationship between total quality management practices and organizational culture. Int. J. Oper. Prod. Manage. **25**(11), 1101–1122 (2005)
14. Chapman, R., Al-Khawaldeh, K.: TQM and labour productivity in Jordanian industrial companies. TQM Mag. **14**(4), 248–262 (2002)
15. Al-Khalifa, K.N., Aspinwall, E.M.: The development of total quality management in Qatar. TQM Mag. **12**(3), 194–204 (2000)
16. Alomaim, N., Zihni Tunca, M., Zairi, M.: Customer satisfaction@ virtual organizations. Manage. Decis. **41**(7), 666–670 (2003)

17. Baidoun, S.: The implementation of TQM philosophy in Palestinian organization: a proposed non-prescriptive generic framework. TQM Mag. **16**(3), 174–185 (2004)
18. Crosby, P.B.: Quality is Still Free: Making Quality Certain in Uncertain Times. McGraw-Hill Companies, New York (1996)
19. Schmidt, W.H., Finnigan, J.P.: The Race without A Finish Line: America's Quest for Total Quality. Jossey-Bass Management Series. ERIC, Kingswood (1992)
20. Johnson, H.T.: Relevance regained: total quality management and the role of management accounting. Crit. Perspect. Acc. **5**(3), 259–267 (1994)

Perception of Quality in Higher Education Institutions: A Logical View from the Literature

Ifetayo Oluwafemi[1(✉)] and Timothy Laseinde[2]

[1] Postgraduate School of Engineering Management, University of Johannesburg, Johannesburg, Republic of South Africa
ijoluwafemi@uj.ac.za

[2] Mechanical and Industrial Engineering Tech Dept, University of Johannesburg, Johannesburg, Republic of South Africa
otlaseinde@uj.ac.za

Abstract. The hearty increase in population globally has impelled the educational sector to ensure they remained uncensored in the globalization trend, and to remain competitive. This study surveyed how quality is being viewed from higher education institutions context and examined deeply the operational meaning of quality in higher education institutions. The paper examined various total quality management concepts in some selected Sub-Sahara African countries and discussed the performance management techniques adopted to enhance their educational standard. Quality in higher education institutions was conceptualized in his study and various relevant and commonly used quality constructs were studied.

Keywords: Quality · Quality assessment · Higher education · Quality improvement

1 Introduction

Dynamic increase in population worldwide has enacted competition in all sector of the economy across the globe. This has impelled the educational sector to ensure they're not expurgated in the trend of globalization, this has made them to remain focused and competitive. In the service sectors, manufacturing sector, etc. a lot of efforts are put in place to ensure quality concept is given optimum priority so that they can remain competitive. In a bid for the higher education institution to remain competitive and relevant, quality concept are being integrated to ensure quality of learning [1], and to attain stakeholder satisfaction [2–4], and improve learning processes and competitiveness [5–7]. Quality in higher education institution (HEIs) is context-specific and is conceptualized via multi-level, multifaceted and dynamic approaches [8]. Quality can be defined from the perspective of HEIs as the fitness for purpose, transformation, distinctiveness, perfection and to attain value for money. Quality in the educational contexts is classified into two major categories, administrative quality as well as educational quality [9].

T. Ahram et al. (Eds.): IHSED 2019, AISC 1026, pp. 1075–1083, 2020.
https://doi.org/10.1007/978-3-030-27928-8_159

Educational quality in HEIs contexts refer to teaching processes and the perception of the stakeholders about the quality culture of the institutions, while administrative quality pertains to both infrastructural and administrative activities of the institutions. In order for both educational and administrative quality to be accomplished in any HEIs, the HEIs itself must provide a unique, valuable, affordable and excellent education. Also, ensuring stakeholders, such as employees' and students satisfaction increases loyalty and retention [10–12]. Consequently, describing, evaluating, guaranteeing quality are crucial to HEIs [13]. Insights from the literature have investigated themes like effective leadership [14], quality assurance [15], student engagement [16], teaching quality [17, 18], and peer review teaching [19].

This paper seeks to survey the literature on different perception of quality in HEIs. To this effect, a research questions are being drawn to give an insight on the nitty-gritty of the study. The questions are illustrated below:

1. How quality is being seen and operationalized in HEIS?
2. What has been done regarding quality in HEIs by researchers and how the literature on quality in HEIs is growing?

2 Concept of Total Quality Management in Higher Education Institutions

The idea of TQM has been introduced to the educational sector and mostly to the HEIs system. This is also with the motive to enhance quality improvement in the educational sector. Owing to the high rate of dilapidation in the quality of the educational sector, this brought about introduction and implementation of the concept of TQM in HEIs.

2.1 Comparative Analysis of TQM in South African

Quality issues have been of great concerns to many higher institutions in South African, both private and government-owned Universities. Ramlagan [20] reported in his studies, the challenges been faced in the South African educational systems, he enumerated the challenges encountered by the South African government in implementing and maintaining a TQM framework. He further stated that there exists a policy, where most of the privately owned HEIs in South African were instructed to merge together to form a formidable citadel of learning. However, it was observed that issues arise in the management of this institutions, they were faced with a lot of challenges in the accreditations exercise of their academic programs. From Ramalagan findings, he emphasized the need to adopt and implement a TQM system for proper management of the private higher education institutions in South African. He also made it known in his findings that for the newly merged private institutions to function properly, meet the audit criteria of the South African Department of Education, which ensures adequate delivery of standard education, they must ensure speedy implementation of a TQM system in their internal quality mechanisms.

The Republic of South African understands the need for Quality enhancements and maintenance. Abubakar [21] in his study, stated that maintaining and ensuring quality standard in all sectors, not limited to the educational sector alone should be collective responsibilities, either by individuals or by a group of organization constituted purposely for ensuring quality. The Republic of South African has several quality maintenance groups which form an organization, their specific objective is to promote quality at both secondary and higher educational level. The following are organizations which were established by the Republic of South African for quality performance improvements in their educational sector.

The South African Qualification Authority (SAQA)

Saqa is a quality enhancement figure which was founded by the South African Ministry of Education and labor. The organization is made of 29 members committee, nominated by refined educational stakeholders with reputable track records in education and training. This body is saddled with the responsibility of formulating quality policies and registration requirements for other bodies who are responsible for monitoring and auditing education standards and qualifications. The major responsibility of the SAQA is to supervise the progress and improvement of the National Qualification Framework (NQF). Another task in which SAQA is obliged to carry out is to ensure the implementation of the NQF by ensuring prompt registration as well as accreditation and assign tasks to the NQF. They also ensure registration of national standards and qualifications are done as at when due, which must be in total compliance with international standards.

The National Qualification Framework (NQF)

The National Qualification Framework (NQF), is a substructure of the South African Ministry of Education and labor, which is governed by few principles and guidelines that present a vision and set boundaries, it can also be regarded as a philosophical base substructure which was established for qualification assessment. Comprehensive development, as well as policies implementations, are being ensured within the set boundaries of their scope of operations. The NQF is been referred to as a national simply because it's been serviced by national resources and it involves national efforts in incorporating and enforcing education and training in the quality framework to ensure the NQF becomes an internationally recognized qualification structure.

The South African Council on Higher Education (CHE)

The South African Council on Higher education (CHE), which was purposely set-up for quality enhancement in HEIs in South African, was instituted as an self-regulating statutory body in the year 1998. The body is mainly constituted as a quality advisory committee to the office of the South African Education minister. The committee is obliged to give advice to the minister on all subjects involving higher education policies and issues. The minister holds regular meetings with the members of the CHE periodically to discuss the best way to foster and ensure quality assurance in the South African HEIs.Beyond the strategic function of the CHE, other existing quality management infrastructures exist within Higher Educational institutions, government funding agencies, professional regulatory bodies and the Department of Higher Education (DHET).

2.2 A Comparative Analysis of TQM in Kenya

It is evident that most of the Universities, either privately owned or public universities places priority on the number of students who enrolled in their respective schools rather than focusing on the quality of education impacts the school have on the students. Majority of the privately-owned institutions centered their attention on how the population of their students increases geometrically. Nyaoga [22] emphasized that the educational managers, who are the top management staff of the higher institutions should focus their priority on improving the quality culture of the school rather than placing their main priority on what income the school generates.

He enumerated most of the challenges encountered in most Higher Education Institutions in Kenya, which are an inadequacy in the use of ICT, inadequacy in the funding, particularly in research development and technological innovations, the inclusion of irrelevant courses in their academic curriculum, inability to subscribe to a unified accreditation system.

Despite all these challenges, the educational sectors in the country are obliged to ensure that a quality education is delivered to the masses. The University of Nairobi in Kenya was able to adopt TQM and ensured proper implementation of it. Abubakar [21] mentioned in his study, the benefits in which the University of Nairobi derived by adopting TQM and implementing TQM. He stated how TQM has assisted the University of Nairobi in becoming a reference point of excellence in higher education institutions in Nairobi and its sub-region. He also emphasized how the implementation of TQM has assisted the University to continuously satisfies the accreditations requirement of the Higher Education Regulatory body in the country. It helped in achieving the basics needs and expectations of the customers, that is, the students as well as the staff members.

3 The Concept of Quality in the Educational Sector

Perception of quality in education is quite new in the academic research works. In the eighteenth century, there have been some significant improvements in education, but continuity has been a major issue, no concept or model was in place to sustain the improvement experienced. The changes experienced in education does not only covers the quantitative measures alone but also the qualitative aspects. In enumerating the quantitative criteria, we are talking about the numeric rise in the figure of students who are enrolling in HEIs globally, high increase in the numbers of higher education institutions and many more. All these necessitated the need for quality implementation in education. In the vein of the qualitative aspect, owing to the increase in the population on yearly basis, there comes a need for quality implementation in the country's educational system.

Since the inception of the World war II, most developed countries enjoyed good infrastructure as well as other amenities in their educational sector, and some moves were made to encourage qualitative enhancement in education due to the new emerging challenges faced by the entire society due to the war. This brings about the adoption of quality in the educational sector, that gave room for better planning of what the standard of education should look like.

However, even though the term quality has been devised, it does not have a consolidated concept. What does the word quality mean in education? The question categorizes the core problem, and these are the issues addressed nowadays.

In pursuant of excellence, there is a need for a high level of quality to be integrated into the educational sector, in which its interest will be for maximum results. Quality is also regarded as one of the key intent of education. Everyone is interested in attending quality educational institutions, but it is somewhat impossible to identify what constitutes quality in education.

In developing countries like Nigeria, several institutions still embrace the old-fashioned belief of quality in education. They believed the inclusion of extensive content in the academic curriculum and adopting the best teaching method signifies quality [23, 24]. This type of approach is not suitable in achieving quality in education, this approach will only support academic performance, meanwhile, education is considered as knowledge formally acquired while knowledge itself is considered as education acquired through experience. However, placing priority on the curriculum alone while trying to achieve quality in education is wrong, there are other important factors to be considered such as few others features of human knowledge related to knowledge to value, knowledge which deals on ethics which also directly or indirectly affect the quality of education.

Loreman [25] stressed that to have an inclusive education, there are seven pillars of support that needs to be considered. Each of the important factors enumerated to have an inclusive education is quite evident in the past literature. The factors which are considered as the seven pillars of support to have an inclusive education which implies quality education are illustrated in the chart below.

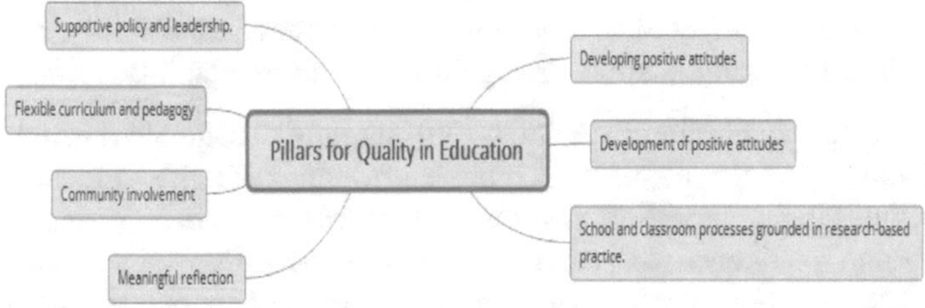

4 TQM in Education

The concept of total quality which invariably means Total Quality Management as adopted by the Americans has been adopted by many schools in the developed countries. Narasimhan [26], in his study, identified Fox Valley Technical College (FVTC) as the first higher education institution in the United State of America that adopted and implemented TQM concept into their education. This made the school more efficient in terms of the placement of graduates, and there was more enhancement of employer's satisfaction, this made their credits more acceptable and a significant improvement was recorded in the learning environment. Due to the success recorded by

FVTC because of the adoption of TQM, many other institutions began to adopt TQM as well. The University of Wisconsin-Maison as well as North Dakota University, Delaware Community College and Oregon State University [27], were the ones who adopted TQM shortly after the TQM implementation at Fox Valley Technical College [28]. In the United State, at 1996, over 160 universities adopted the concept of TQM, which is aimed at quality improvements and over half of this Universities have developed an organizational structure aimed at enhancing quality. There are several reports on adoption of TQM in the United State which can be found in the studies done by Kanji and Malek [29].

Higher education institutions can be regarded as a servicing company that its end motive is providing educational service to students through the application of teaching-learning procedures and some specific products are obtained such as academic performance. In achieving high-quality academic performance, some adequate facilities need to be provided such as good academic curriculum, good lecture rooms, suitable laboratories and many more. Process control and process improvements are very key in achieving a suitable educational system.

A lot of frameworks has been developed in TQM for quality enhancement in the educational sector, the central focus on most of this framework which has been developed, priority was basically placed on the recipients, which is classified as the customer or the beneficiary.

5 Conceptualizing Quality in Higher Education Institutions

In the higher education context, most relevant quality constructs from the literature are total quality management, student learning, quality assurance, service quality, student satisfaction, benchmarking, student engagement and accountability.

Student Learning: According to Ashwin [30], student's academic transformation is a subject of some presage elements which has to do with the student's preceding knowledge, learning approaches, institutional structures, pedagog, and intelligence. Student learning is likened to student's various experiences and could as well be associated with adoption of economics as well as social experiences. Also, academic support in form of bursary to student is another factor that positively contributes towards impartation of knowledge as well as interpersonal skills development among students.

Student Engagement: In HEIs context, student engagement can be viewed as the commitment of students to their studies as well as the way by which the institution structures their academic programs and other social activities that can encourage student's participation. Mandernach [16], enumerated student engagement constructs to be effective, academic, behavioral, cognitive and peer-to-peer academic learning experiences within the faculty and institutional environment. Therefore, student engagement can be correlated to learning results, student retention as well as deep learning.

Student Satisfaction: In his student satisfaction can be defined as the end-result of compounded multifaceted issues faced in HEIs such as physical infrastructure, social climate, feedback, learning, support facilities, supervision, leisure activities and curriculum [31]. Van Kemenade [32], enlisted age, background and gender, value systems like control, commitment, re-invention, and continuous improvement affects student satisfaction. Another factor which is considered as a means of effective service delivery and a way of attaining student satisfaction is student's voice [33, 34]. As affirmed by Heng [35], HEIs can enjoy from the internalization schemes of higher education, especially when they adhere to the voices of the international students.

6 Better Approaches of Introducing Quality

Higher education institutions are considered as the most valuable transformative process institutions where knowledge can be directly acquired. HEIs engages in the development of stakeholders such as students and even the staff members who, in turn, contribute meaningfully to communities, the wider society as a whole and also, to the economy at large. Quality in HEIs can be achieved via the listed approaches which is supported with literatures.

Total Quality Management (TQM): TQM emanated in the United State of America as far back as 1980, this was as a result of the shortfall observed by Hewlett-Packard in some of the manufacturers of US chips for poor quality of products supplied as compared to their Japanese competitors [36]. TQM can be defined as sets of management as well as control processes targeted at ensuring quality of products and service delivery, infusing the entire organization in meeting the customers' needs and expectations. TQM approaches in ensuring quality improvement is an inclusive approach that produce positive impact [37]. TQM in HEIs encourages a disciplined coordination which has facilitated quality improvement in terms of administrative processes, meeting students and staff expectation and course quality.

According to Antony [38], TQM constructs which brings about improvement are student focus, top management commitment, employee involvement, process management, and quality planning.

Quality Assurance: The concept of quality assurance entails approaches such as accreditation assessments, continuous process improvement, and audits which is driven by the top management or leadership of an organization. In HEIs, learning outcomes are improved as a result of external audits conducted with the support of internal staff members initiatives. These learning outcomes are triggered by internal motivation and employee individual readiness for change [39].

Benchmarking: In HEIs, benchmarking brings about improvements in quality of teaching, curriculum design, enhances employability ratios, increases industrial collaborations, international rankings and research dynamics, thereby ensuring academic excellence [6]. Bovill *et al.* [40], identified student views and timely design decisions as a substantial factor for co-creating curricular. Efficacy of teaching is appraised based on proof such as self-evaluation, student ratings, and peer ratings. Madriaga [41], suggested that excellent performance of teaching should be recognized and awards in form of grant for both student and teaching staff should be awarded.

Accountability: In HEIs, both academic independence and self-empowerment are seen as the major tenets of HEIs. Though accountability and sense of commitment has been observed to be lacking.

References

1. Arjomandi, M., Kestell, C., Grimshaw, P.: An EFQM excellence model for higher education quality assessment. In: 20th Annual Conference for the Australasian Association for Engineering Education, 6–9 December 2009: Engineering the Curriculum, p. 1015 (2009)
2. Annamdevula, S., Bellamkonda, R.S.: Effect of student perceived service quality on student satisfaction, loyalty and motivation in Indian universities: development of HiEduQual. J. Model. Manage. **11**(2), 488–517 (2016)
3. Ahmad, S.Z.: Evaluating student satisfaction of quality at international branch campuses. Assess. Eval. High. Educ. **40**(4), 488–507 (2015)
4. Annamdevula, S., Bellamkonda, R.S.: The effects of service quality on student loyalty: the mediating role of student satisfaction. J. Model. Manage. **11**(2), 446–462 (2016)
5. Antunes, M.G., Mucharreira, P.R., Justino, M.T., Quirós, J.T.: Total quality management implementation in portuguese higher education institutions. In: Multidisciplinary Digital Publishing Institute Proceedings, p. 1342 (2018)
6. Tasopoulou, K., Tsiotras, G.: Benchmarking towards excellence in higher education. Benchmarking: Int. J. **24**(3), 617–634 (2017)
7. Ruben, B.D.: Quality in Higher Education. Routledge, London (2018)
8. Schindler, L.A., Puls-Elvidge, S., Welzant, H., Crawford, L.: Definitions of quality in higher education: a synthesis of the literature (2015)
9. Duque, L.C.: A framework for analyzing performance in higher education (2013)
10. Asrar-ul-Haq, M., Kuchinke, K.P., Iqbal, A.: The relationship between corporate social responsibility, job satisfaction, and organizational commitment: case of Pakistani higher education. J. Clean. Prod. **142**, 2352–2363 (2017)
11. Calitz, A., Bosire, S., Cullen, M.: The role of business intelligence in sustainability reporting for South African higher education institutions. Int. J. Sustain. High. Educ. **19**(7), 1185–1203 (2018)
12. Serpa, S., Ferreira, C.M., Santos, A.I., Teixeira, R.: Participatory action research in higher education training. Int. J. Soc. Sci. Stud. **6**, 1 (2018)
13. Harvey, L., Williams, J.: Fifteen years of quality in higher education. Qual. High. Educ. **16** (1), 3–36 (2010)
14. Harris, A., Day, C., Hopkins, D., Hadfield, M., Hargreaves, A., Chapman, C.: Effective Leadership for School Improvement. Routledge, London (2013)
15. Skolnik, M.L.: Quality assurance in higher education as a political process. High. Educ. Manag. Policy **22**(1), 1–20 (2010)
16. Mandernach, B.J.: Assessment of student engagement in higher education: a synthesis of literature and assessment tools. Int. J. Learn. Teach. Educ. Res., vol. **12**(2) (2015)
17. Greatbatch, D., Holland, J.: Teaching quality in higher education: literature review and qualitative research. Department for Business, Innovation and Skills, vol. 21 (2016). https://www.Gov.Uk/Government/Uploads/System/Uploads/Attachment_data/File/524495/He-Teaching-Quality-Literature-Review-Qualitative-Research.Pdf
18. Prakash, G.: Quality in higher education institutions: insights from the literature. TQM J. **30** (6), 732–748 (2018)

19. Thomas, S., Chie, Q.T., Abraham, M., Jalarajan Raj, S., Beh, L.: A qualitative review of literature on peer review of teaching in higher education: an application of the swot framework. Rev. Educ. Res. **84**(1), 112–159 (2014). 03/01; 2019/05
20. Ramlagan, R.: An investigation into quality practices at private higher and further education institutions in the Durban Central Business District (2009)
21. Abubakar, N., Luki, B.N.: A review on total quality management in higher education
22. Nyaoga, R.B., Nyamwange, O., Onger, R., Ombati, T.: Quality management practices in Kenyan educational institutions: the case of the University of Nairobi. Afr. J. Bus. Manage. **1** (2010), 14–28 (2010)
23. Dolton, P.J., Vignoles, A.: Is a broader curriculum better? Econ. Educ. Rev. **21**(5), 415–429 (2002)
24. Harris, J., Mishra, P., Koehler, M.: Teachers' technological pedagogical content knowledge and learning activity types: curriculum-based technology integration reframed. J. Res. Technol. Educ. **41**(4), 393–416 (2009)
25. Loreman, T.: Seven Pillars of Support for Inclusive Education: moving from. Int. J. Whole School. **3**(2), 22–38 (2007)
26. Narasimhan, K.: Organizational climate at the University of Braunton in 1996. Total Qual. Manage. **8**(2–3), 233–237 (1997)
27. Sunder, V.M.: Constructs of quality in higher education services. Int. J. Prod. Perf. Manage. **65**(8), 1091–1111 (2016). 11/14; 2018/12
28. Seymour, D.T.: On Q: Causing Quality in Higher Education. ERIC, Kingswood (1992)
29. Kanji, G.K., Malek, A., Tambi, B.A.: Total quality management in UK higher education institutions. Total Qual. Manage. **10**(1), 129–153 (1999)
30. Ashwin, P., Abbas, A., McLean, M.: How do students' accounts of sociology change over the course of their undergraduate degrees? High. Educ. **67**(2), 219–234 (2014)
31. Wiers-Jenssen, J., Stensaker, B.R., Gr⊘gaard, J.B.: Student satisfaction: towards an empirical deconstruction of the concept. Qual. High. Educ. **8**(2), 183–195 (2002)
32. Van Kemenade, E., Pupius, M., Hardjono, T.W.: More value to defining quality. Qual. High. Educ. **14**(2), 175–185 (2008)
33. Blair, E., Valdez Noel, K.: Improving higher education practice through student evaluation systems: is the student voice being heard? Assess. Eval. High. Educ. **39**(7), 879–894 (2014)
34. Teeroovengadum, V., Kamalanabhan, T., Seebaluck, A.K.: Measuring service quality in higher education: development of a hierarchical model (HESQUAL). Qual. Assur. Educ. **24** (2), 244–258 (2016)
35. Heng, T.T.: Voices of Chinese international students in USA colleges: 'I want to tell them that…'. Stud. High. Educ. **42**(5), 833–850 (2017)
36. Talha, M.: Total quality management (TQM): an overview. Bottom Line **17**(1), 15–19 (2004)
37. Thakkar, J., Deshmukh, S., Shastree, A.: Total quality management (TQM) in self-financed technical institutions: a quality function deployment (QFD) and force field analysis approach. Qual. Assur. Educ. **14**(1), 54–74 (2006)
38. Psomas, E., Antony, J.: Total quality management elements and results in higher education institutions: the Greek case. Qual. Assur. Educ. **25**(2), 206–223 (2017)
39. Liu, S.: Quality assessment of undergraduate education in China: impact on different universities. High. Educ. **66**(4), 391–407 (2013)
40. Bovill, C.: An investigation of co-created curricula within higher education in the UK, Ireland and the USA. Innov. Educ. Teach. Int. **51**(1), 15–25 (2014)
41. Madriaga, M., Morley, K.: Awarding teaching excellence: 'what is it supposed to achieve?' Teacher perceptions of student-led awards. Teach. High. Educ. **21**(2), 166–174 (2016)

Study on Eye Movement Behavior of Interface Complexity

Kaili Yin(✉), Yingwei Zhou, Ning Li, Ziang Chen, and Jinshou Shi

China Institute of Marine Technology and Economy, No. 70, XueYuan South Road, HaiDian District, Beijing 100081, China
yinkaili1012@163.com

Abstract. Eye movement tracking technology is widely used in user research and user experience testing. In order to further study the feasibility of using the first gaze time to evaluate the interface complexity, this paper designed experiments to explore the relationship between the first gaze time and the interface complexity. In this study, interfaces with three levels of complexity were designed, and an appropriate number of users were selected to carry out visual search tasks, and the rule of the first fixation duration changing with the interface complexity was analyzed. The results show that the complexity of the interface is negatively correlated with the duration of the first gaze, which can be used to evaluate the complexity of the interface.

Keywords: User experience testing · Interface complexity · First gaze time

1 Introduction

Complex product interface (referred to as complex interface) has been widely used in aircraft, ships, automobiles and other industries. Due to more and more information of complex product control interface and more and more complex structural relationship, interface complexity affects the operator's recognition, understanding and judgment of interface information. According to statistics, from 2007 to 2010, about 50% of civil aviation accidents in China were caused by pilot misunderstanding and misjudgment caused by interface usability problems, so the study and evaluation of interface complexity is an important aspect of usability research and evaluation [1].

Due to the limitations of traditional interface usability evaluation in practical applications, for example, many test methods are based on the interaction between the subject and the test subject, the words, behaviors and even expressions of the subjects will have an impact on the subjects and their activities [2]. Therefore, eye tracker is one of the commonly used auxiliary tools in the evaluation of interface usability.

Fengpei Hu et al. analyzed the possible problems in the application of eye tracking technology in usability testing based on the human-computer interaction eye tracking research, and proposed the problems that need to be further studied [3]. In order to reduce the subjective influence and error in the traditional usability evaluation method, Shiwei Cheng et al. applied eye movement tracking technology to analyze the cognitive situation in the human-computer interaction process, and proposed the usability

T. Ahram et al. (Eds.): IHSED 2019, AISC 1026, pp. 1084–1089, 2020.
https://doi.org/10.1007/978-3-030-27928-8_160

evaluation method for mobile computing user interface, so as to provide guidance for interface design optimization [4].

When studying the preference of page complexity, Wei Jiang defined the page complexity for the first time and tested four different styles of commodity display pages. At last, he built the relationship model between cognitive style and page complexity preference, and verified the model with the eye-tracking method and questionnaire method based on eye tracker [5]. In 2012, Meijuan Hu et al. studied the Chinese reading strategies of college students and their relationship with eye trajectory, and concluded that the first gaze time could detect the complexity of the interface or task [6].

However, none of the above studies studied the specific characterization relationship between interface complexity and first gaze time. Based on the above studies, this paper focused on the correlation between interface complexity and first gaze time.

2 Eye Movement Experiment Scheme with Interface Complexity

2.1 Experimental Methods

In this experiment, the search task was designed to complete the same task under different interface complexity, and the first gaze time was taken as the interface complexity perceived by the subjects in the search process.

Currently, there are 6 commonly used interface layout presentation methods, and 2 to 3 layout structures are selected for each layout. Among the 15 different layout structures, 3 interfaces of complexity are selected as the three interface layout structures in the experiment [7], as shown in Table 1 and Fig. 1.

Table 1. Interface layout and complexity relation.

Layout structure type	Degree of complexity
Top and bottom structure (below TB instead)	Complexity maximum
Left and right layout (below LR instead)	Medium complexity
T-shaped layout (T is used instead)	Least complexity

2.2 The Experiment Task

Search task: find the shape "⬭" in the interface and judge its position on the screen, if the ellipse on the left side of the screen press "←" on the keyboard, on the right side of the screen press "→" on the keyboard.

2.3 The Experiment Design

(1) **The independent variables: layout structure**
Select the above three layout structures (Fig. 1). Conduct a search task for each layout structure in one experiment, find the target object, and conduct selection operation. Repeat 10 times for each layout experiment.

The size of the experimental image is about 32 cm * 18 cm, filling the entire screen. In the search task, the content is filled with square, circle, diamond and ellipse, among which the ellipse appears once and the square, circle and diamond appear repeatedly. Search the shape ellipse to determine its position (Press on the "←" if left and on the"→" if right) (Fig. 2).

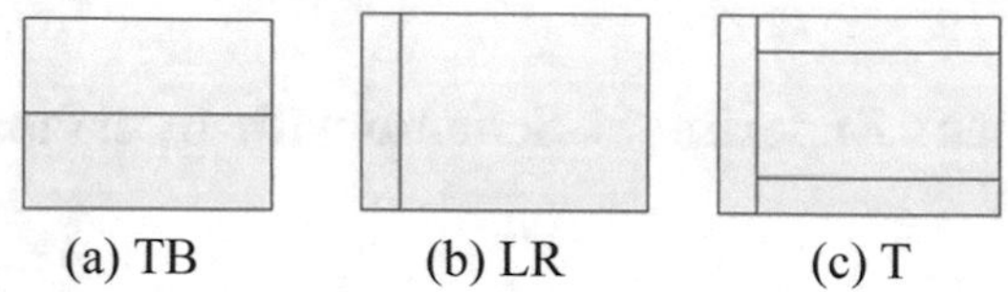

Fig. 1. Schematic diagram of layout structural materials

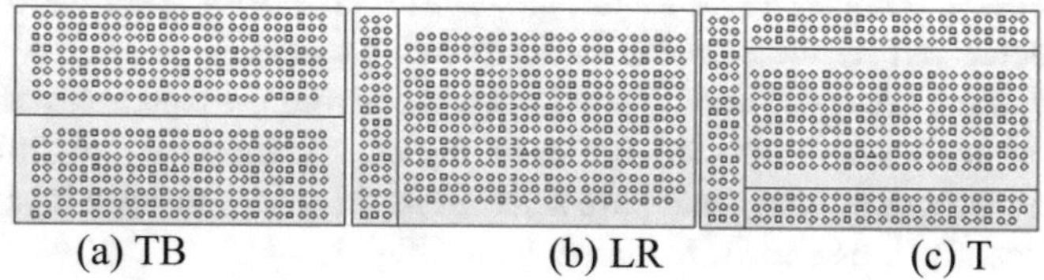

Fig. 2. Schematic diagram of search task materials

(2) **The dependent variable**
In the search task, the response time, accuracy rate and the first fixation time in the eye movement indicators of the participants in each layout structure were collected.

(3) **Interference variables and controls**
In order to avoid the influence of content prominence on visual cognitive order, symmetrical objects were selected for filling. Secondly, shape size, color and structure spacing were controlled to be consistent in each layout structure.

2.4 The Experimental Process

Experimental instructions: inform participants of the experimental process, operation and matters needing attention, click the space bar to enter the next task interface. Fill in the information of participants, click the space bar to enter the experiment. The center of the screen presents "+" 1000 ms to remind participants to pay attention. Rendering white screen to clear visual memory 1000 ms. When presented with a search task, the subject searches for the "ellipse" and determines its position ("ellipse" on the left press "←" on the keyboard, otherwise, press "→" on the keyboard).

2.5 Experimental Environment and Equipment

(1) **The experiment environment**
During the experiment, the amount of interference was strictly controlled. The illuminance in the light environment was 100 lx, and the color temperature was 5000 K, which was kept constant during the experiment.
(2) **Eye tracker**
The eye movement device is German SMI glasses eye movement device, the highest sampling rate of the equipment is 120 Hz. During the experiment, it can be selected between 30, 60, and 120 Hz as required. In this experiment, the sampling rate is 60 Hz.

2.6 Participants

The participants were 16, aged 28 ± 2, left-handed corrected visual acuity 1.0, achromatopsia and other visual symptoms of color discrimination, with ship background knowledge, not experimental design and program developers.

3 The Data Analysis

Traditional mathematical statistics method was used to compare the dependent variables with different complexity, and significance analysis was performed.

3.1 Descriptive Statistical Analysis

Data of 16 people were collected for the experiment, and the original data to be analyzed for the experiment were sorted and statistically analyzed, as shown in Table 2.

Table 2. Raw data statistics

Layout structure	The complexity of the layout structure	First gaze time (/ms)
TB	Maximum	221.67
LR	Medium	232.75
T	Minimum	279.43

It can be found that this is no obvious relationship between the time of first fixation and the complexity of the original statistical data, and the volatility of the original data is large, which needs to be analyzed again after processing, we should cut out the abnormal data in every member's data group and in accordance with the 3σ standard.

Statistical analysis was conducted on the data after pretreatment, and the statistical results were obtained as shown in Fig. 3.

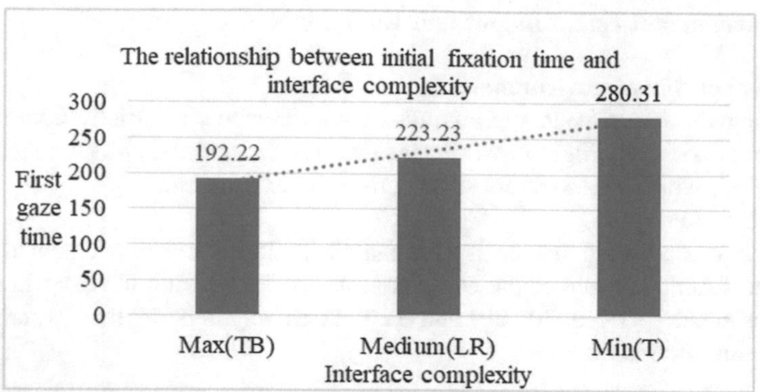

Fig. 3. Relationship between first fixation duration and interface complexity

According to the relationship between the first gaze time and the interface complexity in the data in the table and the figure, in the search task, the more complex the interface layout is, the smaller the first gaze time will be and the negative correlation will be shown.

3.2 Significance Analysis

Variance analysis was performed on the first gaze duration under different complexity to analyze whether the main effect of complexity on the first gaze was significant.

Univariate anova was performed for the first gaze duration under different complexity, and the significant main effect of complexity on the first gaze duration was obtained ($F = 11.47$, $P = 0.00009 < 0.05$).

4 Conclusion

In the search task, there was a significant difference in the duration of the first gaze under different complexity. The more complex the interface layout was, the smaller the duration of the first gaze was, presenting a negative correlation.

Acknowledgments. This study was supports with the Grant No. 41412040304 and No. 6141B03020602.

References

1. Zongbo, W.: Study on the Usability Evaluation of Digital Interface of Aircraft Avionics System (2010). (in Chinese)
2. Chen, Y., Zhang, H.: Situational awareness analysis of general aviation pilots. Traffic Enterp. Manag. **26**, 57–58 (2011). (in Chinese)
3. Hu, F., Han, J., Ge, L.: Review of studies on eye tracking and usability testing. Ergonomics **11** (2005). (in Chinese)

4. Cheng, S., Wu, S., Sun, S.: Eye movement method for mobile computing user interface usability evaluation. Acta Electronica Sinica **37** (2009). (in Chinese)
5. Jiang, W.: Research on WEB page complexity preference of consumers with different cognitive styles, Hefei University of Technology
6. Hu, M.: Chinese reading strategies of college students and their relationship with eye movement trajectory, Suzhou University (2012). (in Chinese)
7. Li, J.: Human-machine interface information coding method for balanced cognitive load (2015). (in Chinese)

Study on the Interactive Mode of Eye Control Mode in Human–Computer Interface

Yingwei Zhou(✉), Ning Li, Bei Zhang, Tuoyang Zhou, Kaili Yin, and Jinshou Shi

China Institute of Marine Technology and Economy,
Xueyuan South Road No. 70, Haidian District, Beijing 100081, China
zhouyingwei_bh@126.com

Abstract. The ergonomic experiment of human-computer interaction mode of eye movement control is carried out for efficient and precise interaction requirements of the human-machine interface. Based on eye movement behavior, operational performance and subjective feeling data, the advantages and disadvantages of eye movement control behaviors such as eye gaze and blink in human-machine interface design are studied. The results show that blink is relatively quicker to pick up the target, and the advantage of eye gaze is that the target control is more targeted.

Keywords: Eye control · Human-machine interface · Interactive mode

1 Introduction

Studies have shown that when the user operates on the human-machine interface, the line of sight will gaze and move in the relevant information as the operation. For example, the user usually shifts the line of sight to an area, moves the mouse, and makes related operations through the mouse. The user can directly make a selection operation when the line of sight shifts his attention to the region of interest, and the interaction efficiency will be effectively improved. Therefore, human-machine interface information interaction through eye-motion control is a more efficient and natural way of information interaction.

There are three basic modes of eye movement: eye gaze, eye movement and visual following [1]. These three modes can not only provide the data of eye movement in the process of psychological activities, such as the operation time required to trigger a specific task, but also optimize and design the human-computer interface through eye movement data. The early application of eye movement tracking technology was mainly limited to psychological research [2]. In terms of user interface operation, the interface constructed based on eye gaze tracking technology is called "gaze user interface" [3]. In 1985, Bolt from MIT developed the first gaze user interface – dynamic window for orchestral music compilation with eye movement [4]. In 2013, Samsung launched its Galaxy S4 smart phone, adding "eye-controlled scrolling" based on eye-gaze tracking and "intelligent pause" based on face recognition. Min Lin from Shanghai University studied the key technology of human-computer interaction based

T. Ahram et al. (Eds.): IHSED 2019, AISC 1026, pp. 1090–1094, 2020.
https://doi.org/10.1007/978-3-030-27928-8_161

on eye movement information, aiming to promote the new interaction mode to engineering application as soon as possible [5, 6].

With the maturity and productization of eye-control technology, the application prospect of interaction in human-computer interface is becoming clear. Based on the data of eye movement behavior, operation performance and subjective perception, the human-machine interaction mode conforming to the eye-control mode was proposed to provide reference for the design of new human-machine interface.

2 Experimental Design

Human-machine ergonomic experiments on various visual behaviors were carried out. Data such as pick time, drag time, drag times, length of eye movement track during the operation of typical tasks were collected. The effects of two kinds of eye movement control modes on eye movement sensitivity, reaction speed and eye movement trajectory were analyzed under different tasks. According to the analysis results, human visual perception and information processing mode are studied to explore the human-computer interactive modes suitable for eye-control interaction.

3 Experimental Variables and Levels

3.1 Independent Variables and Levels

In the experiment, two different eye movement behaviors were set as the control modes, and the interaction experiment was completed by blinking and eye gaze respectively. Considering the normal blink frequency (8–12/min), habitual blink interval (3–8 s), and single blinking time (0.3–0.4 s), the blinking control mode was set as continuous twice, and the eye gaze control mode was set as continuous 2 s.

The main content of the experiment is to operate the control through eye movement behavior as the independent variable of the experiment. The variables and levels of this experiment are shown in Table 1.

Table 1. Experimental variables and levels.

The independent variables	Levels	
Eye movement control	Blinking	twice in a row
	Eye gaze	lasts 2 s

3.2 Dependent Variable

The operation behavior and subjective evaluation of the subjects were selected as the experimental response (dependent variable), and the details are as follows:

(1) Operation behavior: the correct completion time and operation times of the operation task can be used to evaluate the convenience of eye movement behavior control.
(2) Subjective evaluation: the convenience and comfort of eye movement control were evaluated by questionnaire survey.

3.3 Experimental Tasks

Subjects. The subjects were 16 college students, aged between 20 and 26 years old, with naked eye vision no less than 1.0, and right-handed.

Experimental Environment. The experimental environment was simulated fluorescent lighting environment, with the illumination level of 150 lx and the color temperature of 5500 K.

Experimental Tasks. The experimental tasks were two types of simulated operations: command activation and target movement. The operator needed to use the eye movement control prototype system to successfully pick up and drag the target into the specified circular range and successfully drop the target. The correct completion of the above operations was regarded as a natural interactive operation on the human-computer interface.

In the experiment, two different eye movement operation modes, blinking and eye gaze, were used to complete target pickup and drop. In each operation mode, the subjects were required to repeat four sets of experiments in each operation mode, that is a total of 40 trails were required in the whole experiment.

The experimental design recorded the target pickup time, target drag time (the time that the target was successfully dragged into the specified range and put down), the target drag times, the eye movement track and the length of the line of sight during the dragging process and other data under two different operation modes.

4 Data Processing and Analysis

4.1 Descriptive Statistics

The statistics of the mean and standard deviation (SD) of the two different eye movement control modes are shown in Table 2.

Table 2. The Mean and SD of the eye control modes

Control modes	Target pickup time (s)		Target drag time (s)		Target drag times (time)		Length of the line of sight (pixel)	
	Mean	SD	Mean	SD	Mean	SD	Mean	SD
Blank	4.79	7.32	12.44	25.09	1.60	1.31	2095.49	2861.42
Eye gaze	5.77	8.00	7.85	10.63	1.37	0.72	957.03	735.10

P value can be obtained by significance analysis of the above four parameters, as shown in the Table 3.

Table 3. Parameter significance analysis of the eye movement control modes

	Target pickup time	Target drag time	Target drag times	Length of the line of sight
Blank × Eye gaze	0.088	0.003	0.006	0.000

According to the analysis results, for the target pickup time, the blinking mode is 0.98 s shorter than the eye gaze mode, indicating that the blinking mode is faster. For the target drag time, the blinking mode is 4.59 s longer than the eye gaze mode, and the speed is obviously slower. For target drag times, there are more times of blinking, indicating that there are many wrong operations in this mode.

In conclusion, the SD of the target pickup time, target drag times and length of the line of sight of the blinking mode are significantly higher than eye gaze mode, which indicates that the blinking mode has stronger individual differences and relatively more unstable performance.

4.2 Subjective Evaluation

The subjects were asked to evaluate two eye movement control modes of blinking and eye gaze by means of questionnaire, and the evaluation included convenience and comfort, which are rated on a scale of 1–5, with 1 being very poor and 5 being very good.

The subjective evaluation data are shown in Table 4.

Table 4. Mean of subjective evaluation

Control modes	Convenience		Comfort	
	Mean	SD	Mean	SD
Blank	3.125	1.053	2.813	1.184
Eye gaze	3.688	0.982	3.313	1.044

According to the analysis results of subjective evaluation scores, the convenience and comfort of eye gaze are better than that of blink, but neither of the two methods has reached the relatively ideal degree (the mean is below 4), that is, the two eye movement control methods have a general sense of user experience, and the eye gaze method is better.

5 Conclusion

In this study, ergonomic experiments were carried out to compare the differences between the two eye-control modes of blinking and eye gaze in the human-computer interface. According to the experimental results, the conclusions are as follows.

(1) Compared with the two eye-control methods of blinking and eye gaze, blinking is relatively faster for target pickup, but it is not suitable for dynamic operation of designated direction or position, because the behavior of blink will significantly interfere with eye gaze point capture. The advantage of eye gaze is more accurate positioning of the target control, less load of blink behavior, and better interactive experience. However, the interactive process lacks clear sense of timing and effective action feedback.

(2) The two methods of eye movement control have moderate convenience and comfort. Based on the current eye movement technology, good user experience has not been achieved yet.

Acknowledgments. This work was supported by the projects with the Grant NO. 41412040304, NO. 3020102010303, and NO. 6141B03020602.

References

1. Pan, D.: Impact of individual characteristics on space mission performance, Tsinghua University
2. Wang, Y.: Cognitive psychology analysis of user information ability difference. In: Modern Intelligence, vol. 10, pp. 110–114 (2005)
3. Schiffman, H.R.: Sensation and Perception: An Integrated Approach. Wiley, Hoboken (2014)
4. Sternberg, R.J., Sternberg, K.: Cognitive Psychology. Wadsworth/Cengage Learning, Boston (2016)
5. Raash, T., Bailey, I.L.: Repeatability of visual acuity measurement. Optom. Vis. Sci. **75**, 342–348 (1998)
6. Yu, R.-F., Chan, A.H.S.: Visual search time in detection tasks with multiple targets: considering change of the effective stimulus field area. Int. J. Ind. Ergon. **43**, 328–334 (2013)
7. Yang, L., Song, F., Zhang, Y., Yu, R.: Effects of target occurrence probability, target presence and time pressure on visual search performance. Ind. Eng. **19**(6), 83–88 (2016)

Total Quality Management Fundamentals and Evolving Outcomes in Higher Education Institutions

Ifetayo Oluwafemi[1(✉)] and Timothy Laseinde[2]

[1] Postgraduate School of Engineering Management, University of Johannesburg, Johannesburg, RSA
ijoluwafemi@uj.ac.za

[2] Mechanical and Industrial Engineering Technology Department, University of Johannesburg, Johannesburg, RSA
otlaseinde@uj.ac.za

Abstract. Findings from the literature has identified various barriers preventing most higher education institutions from implementing TQM in their respective institutions, this could be attributed to the fact that most TQM associated concepts may not have a substantial continuous impact on the majority of concepts and customs of educational quality. However, this study seeks to identify and discuss various fundamentals of TQM in HEIs, identify various accomplishments of TQM in HEIs. The study contributes to the body of knowledge by analyzing the key TQM fundamentals adopted by various researchers. It was also seen that the most treasured and adopted TQM fundamentals in the HEIs are the ones that concerns the human aspect of the institutions, in other words, the major stakeholders; students, both academic and non-academic staff members and the top management.

Keywords: Quality · Quality assessment · Higher education · Quality improvement

1 Introduction

A better understanding of what a higher education institution (HEIs) is can be likened to complex, complicated and most importantly, a challenging environment. It is an academic environment where competition has become the order of the day, and previously expected government financial support has become irregular [1]. Some of the developing countries are facing financial challenges, especially Nigeria where there is continuous economic crisis where one dollar ($1) is equal to three hundred and sixty naira (N 360). In such situation, it becomes highly imperative for HEIs to thrive in the phase of financial sense, if not they will be wiped out of business [2]. In an attempt to survive and stay competitive, HEIs are looking for ways to meet their customers' expectations, ensure drastic reduction in cost of operation with the goal of attaining an improved efficiency. The HEIs are also coping with the ongoing competitiveness in the world through continuous improvement of processes and via provision of standard quality education [3]. Most of these adopted management applications by HEIs exhibit

T. Ahram et al. (Eds.): IHSED 2019, AISC 1026, pp. 1095–1100, 2020.
https://doi.org/10.1007/978-3-030-27928-8_162

the philosophies of total quality management (TQM). From the perspective of a dynamic environment, in the context of Education, TQM has been identified as a management techniques tool for dealing with the challenges of quality in education as well as its stakeholders challenges, which could be either internal or external [4].

The economic importance of TQM is highly inevitable, though its roots was originally from manufacturing, moving into the service as well as the health sectors, then into the government and educational sector [5–7]. Following the quality challenges observed in the industrial and government sector which prompted the adoption of TQM, a large number of higher institutions are also implementing TQM to salvage the deteriorating quality level of higher institutions across the globe [8]. The adoption of TQM has also gained more popularity in the developed countries such as the United State of America and in the UK [9], several international polytechnics, colleges and universities have implemented the use of TQM as a quality measure tool to enhance quality improvement in their various higher institutions. Also, several international educational institutions, for instance, in the USA, Singapore, Switzerland, Romania, Canada, Australia, Malaysia and in the UK have adopted quality management system, which is in line with ISO 9001 standard [1], the quality philosophies which is similar to total quality management philosophy [10].

2 Barriers to TQM Implementation in HEIs

There is a lot of barriers preventing most HEIs from implementing TQM philosophies in their various HEIs across the globe [3]. This is attributed to the fact that most of the TQM associated concepts do not continuously have a substantial impact on the majority of concepts and customs of educational quality [9]. This could be as a result of infidelity in the universal consensus on the best approach of managing quality in an HEIs environment [11], perhaps, the adoption of TQM concepts from the industrial sector for its use in the educational sector is not always successful [3]. Likewise, the deployment of TQM in the educational sector is not an easy procedure, according to Rosa [12], numerous barriers needs to be overwhelmed to actualize a successful implementation of TQM.

One of the identified barriers encountered by several HEIs in implementing TQM philosophy in their various institutions is likened to unwillingness of both staff and students of the HEIs to use the TQM tools that was borrowed from the industrial sector, and in adopting its approach in their quality management scheme [13]. Another notable barrier is the fact that several researchers from HEIs are still doubtful about adopting it in education [3], most especially about its aptness and fitness into education [14], its applicability is another factor considered as a stumbling block towards its implementation. According to Konidari [15], adaptability and efficiency is another barrier restraining most researchers and the stakeholders of HEIs from adopting TQM philosophy in their various educational institutions.

3 Applicability of TQM Philosophies and Theories in HELs

Several theorists, researchers and practitioners have conducted numerous studies in evaluating theories surrounding the applicability of TQM philosophies in education, which includes both theoretical as well as empirical [16–18]. Nevertheless, there exist a gap in the literature in the discipline of TQM in the educational sector. Mehta *et al.* [4], asserted that much studies has been done in the past decades about TQM in the manufacturing sector; though, full consideration has not been given to the educational sector. However, Soria *et al.* [17], also enumerated that only few empirical studies with regards to expansion of multiple item measurement scales of TQM in education has been considered.

Based on gap identified from the literature, numerous future research suggestions were raised from several scholars. Dumond *et al.* [1], stated that further research is necessary to collect empirical data which could be used in the investigation of TQM in education. According to Soria *et al.* [17], the TQM empirical investigation should take into consideration, the holistic genuineness of the educational sector. A good quality education requires an effective organization as well as a good management system in place, it will make a good sense if researchers can examine numerous techniques used in managing quality in the industrial sectors to see how it fits into the educational sector. As enumerated by Saiti [19], following the differences between the educational sector and the industrial sector, she state that it becomes highly imperative that further analysis and studies should be conducted focusing on implementing the values of total quality management in the educational sector. According to Ardi *et al.* [8], new developments in studies of TQM in HEIs gives avenue for further investigations on impacts of diverse patterns of quality management in different countries. Likewise, Bayraktar *et al.* [20], stated that additional research is quite needed with respect to assessing the critical success factors or elements of TQM in the educational sector in numerous countries. Another scholar, Calvo [21], advocated the use of longitudinal approach for further and future studies in the field of TQM in HEIs, which implies that quality assessment as well as improvement is a process that develops over time. In lieu of the above, it is evident that future research will give more insights on the applicability and implementation of TQM in education.

4 Total Quality Management Fundamentals in HEIs

The implementation of TQM in HEIs requires that the top management of the institutions understand the nitty-gritty of the TQM elements, which are the quality dimensions that entails the formation of the concept of TQM [8]. A previous study have been undertaken empirically to explore the quality fundamentals that institutes the TQM constructs in HEIs [8]. Though, there seems to be some discrepancy as regards dimensions which constitutes these constructs [17]. Due to this, numerous features and models for TQM have been suggested [4], which entails symbiotic elements, such as principles, critical factors, components, values and dimension [22].

As identified by Venkatraman [3], some of the fundamental values founding the construction of a TQM model for HEIs are listed as follows: Leadership, quality culture, employment involvement, continuous improvement as well as innovation in educational activities, fast response as well as management of information, partnership development and customer-driven quality. According to O'Mahony *et al.* [23], after an extant review of the literature, he made know that the most pertinent factors needed to been given attention in the implementation of quality management systems in the HEIs includes: top management commitment and sponsorship (ensuring a proper communication of the mission, via empowerment manner and the use of appropriate information, data, and adequate knowledge of best practices), creating and ensuring culture of continuous improvement as well as focusing on process issues (internal audits, managing by process, self-assessment, benchmarking, measurement, accreditation of the quality management system as well as an evidence-based method to decision making, information and analysis), stakeholder involvement (empowerment of employees, and middle management involvement). Likewise, Svensson *et al.* [14], leveraging on the literature, stated that there is a consensus in the fundamental values that institutes the model for a TQM built for self-assessment in the educational sector. However, these values are continuous improvement, customers focus, fact-based decision-making, both employee and leadership commitment. According to Ali *et al.* [24] in his study, he enumerated ten critical success factors of TQM in the HEIs context, which are relevant to quality initiatives. These were effective communication, visionary leadership, congruent objectives, customer focus, staff selection and deployment, teamwork spirit, recognition and motivation, competent staff, innovation and creativity. A TQM model for quality improvement in Engineering education developed by Mehta [4], suggested institutional resource management, excellence in human resource management, top management commitment as well as visionary leadership, long-term strategy and planning, continuous assessment and improvement, student focus, alumni focus, a quality mission as well as vision statement, employee focus, an information management system, service quality, industry as well as institutional collaboration and innovative academic philosophy.

5 Achievements of Total Quality Management in HEIs

The adoption of TQM philosophy in the HEIs has yielded numerous significant outcomes, especially on both internal as well as external institutional environment. One of the noteworthy outcomes identified through the literatures as stated by Psomas [25], which is also evidence in his work is seen as the improvement of service quality, operational performance and satisfaction attained by both teaching staff and employees are all classified as internal achievement of TQM adoption in HEIs. Also, the external environment benefit of TQM application in HEIs, which implies the society, students as well as the market recorded a great achievement though can't be compared to the achievement derived by the internal environment. This is in accordance with the results of the study carried out by Santo *et al.* [26], enumerating how resourceful TQM can be in terms of attaining suitable market targeted outcomes that disdained any form of fears of likely tendency of quality management to look inwards, thus, increasing core processes. Nevertheless, quality improvement still remains a process whose impacts are treasured mainly at the long run [27], TQM outcomes relating to external environment of HEIs in most of the recent studies are not forthcoming in the short term. Though, in terms of the financial performance, the adoption of TQM in HEIs has aided the financial performance growth of many HEIs across the globe.

According to Psomas [25] study, the achievement of TQM adoption in HEIs, most especially in the internal environment outpaced the outcomes of the external environment.

This is in line with the study of Kanji et al. [28] which was conducted in HEIs in the USA and Malaysia, it was observed that the HEIs experienced a significant performance improvement due to the implementation of TQM elements. Evidences from Calvo *et al.* [21] shows that the Spanish HEIs adoption of EFQM enablers brings about satisfaction of all the stakeholders; students, employee, society and the institution.

6 Conclusion

This study has shield light on the barriers encountered in the implementation of TQM in HEIs, gives an expository insight on the philosophical and theoretical perception of TQM in HEIs, discussed various fundamentals of TQM in HEIs and cited various accomplishments of TQM in HEIs. More precisely, this study contributes to the body of knowledge by analyzing the key TQM fundamentals adopted various researchers. Findings shows that the most appreciated and adopted TQM fundamentals in the HEIs are the ones that concerns the human aspect of the institutions, in other words, the major stakeholders; students, both academic and non-academic staff members and the top management.

References

1. Dumond, E.J., Johnson, T.W.: Managing university business educational quality: ISO or AACSB? Qual. Assur. Educ. **21**(2), 127–144 (2013)
2. Juhl, H.J., Christensen, M.: Quality management in a Danish business school–a head of department perspective. Total Qual. Manag. **19**(7–8), 719–732 (2008)
3. Venkatraman, S.: A framework for implementing TQM in higher education programs. Qual. Assur. Educ. **15**(1), 92–112 (2007)
4. Mehta, N., Verma, P., Seth, N.: Total quality management implementation in engineering education in India: an interpretive structural modelling approach. Total Qual. Manag. Bus. Excell. **25**(1–2), 124–140 (2014)
5. Ajmal, M.M., Tuomi, V., Helo, P.T., Sandhu, M.A.: TQM practices in public sector: case of Finnish healthcare organizations. Int. J. Inf. Syst. Serv. Sect. (IJISSS) **8**(1), 34–44 (2016)
6. Chitra, S., Sanjiv, K.: Road plan to enterprise TQM from manufacturing to library services. Int. J. Inf. Dissem. Technol. **6**(3) (2016)
7. Mosadeghrad, A.M.: Developing and validating a total quality management model for healthcare organisations. TQM J. **27**(5), 544–564 (2015)
8. Ardi, R., Hidayatno, A., Yuri, T., Zagloel, M.: Investigating relationships among quality dimensions in higher education. Qual. Assur. Educ. **20**(4), 408–428 (2012)
9. Chung, D.C.S.: Quality assurance in post-secondary education: some common approaches. Qual. Assur. Educ. **18**(1), 64–77 (2010)

10. Fotopoulos, C.V., Psomas, E.L.: The structural relationships between TQM factors and organizational performance. TQM J. **22**(5), 539–552 (2010)
11. Becket, N., Brookes, M.: Evaluating quality management in university departments. Qual. Assur. Educ. **14**(2), 123–142 (2006)
12. Pires da Rosa, M.J., Saraiva, P.M., Diz, H.: The development of an excellence model for Portuguese higher education institutions. Total Qual. Manag. **12**(7–8), 1010–1017 (2001)
13. Hodgkinson, M., Kelly, M.: Quality management and enhancement processes in UK business schools: a review. Qual. Assur. Educ. **15**(1), 77–91 (2007)
14. Svensson, M., Klefsjö, B.: TQM-based self-assessment in the education sector: experiences from a Swedish upper secondary school project. Qual. Assur. Educ. **14**(4), 299–323 (2006)
15. Konidari, V., Abernot, Y.: From TQM to learning organisation: another way for quality management in educational institutions. Int. J. Qual. Reliab. Manag. **23**(1), 8–26 (2006)
16. Sahney, S., Banwet, D.K., Karunes, S.: Conceptualizing total quality management in higher education. TQM Mag. **16**(2), 145–159 (2004)
17. Soria-García, J., Martínez-Lorente, Á.R.: Development and validation of a measure of the quality management practices in education. Total Qual. Manag. Bus. Excell. **25**(1–2), 57–79 (2014)
18. Thakkar, J., Deshmukh, S., Shastree, A.: Total quality management (TQM) in self-financed technical institutions: a quality function deployment (QFD) and force field analysis approach. Qual. Assur. Educ. **14**(1), 54–74 (2006)
19. Saiti, A.: Leadership and quality management: an analysis of three key features of the Greek education system. Qual. Assur. Educ. **20**(2), 110–138 (2012)
20. Bayraktar, E., Tatoglu, E., Zaim, S.: An instrument for measuring the critical factors of TQM in Turkish higher education. Total Qual. Manag. **19**(6), 551–574 (2008)
21. Calvo-Mora, A., Leal, A., Roldán, J.L.: Relationships between the EFQM model criteria: a study in Spanish universities. Total Qual. Manag. Bus. Excell. **16**(6), 741–770 (2005)
22. Fotopoulos, C.B., Psomas, E.L.: The impact of "soft" and "hard" TQM elements on quality management results. Int. J. Qual. Reliab. Manag. **26**(2), 150–163 (2009)
23. O'Mahony, K., Garavan, T.N.: Implementing a quality management framework in a higher education organisation: a case study. Qual. Assur. Educ. **20**(2), 184–200 (2012)
24. Ali, N.A., Mahat, F., Zairi, M.: Testing the criticality of HR-TQM factors in the Malaysian higher education context. Total Qual. Manag. **21**(11), 1177–1188 (2010)
25. Psomas, E., Antony, J.: Total quality management elements and results in higher education institutions: the Greek case. Qual. Assur. Educ. **25**(2), 206–223 (2017)
26. Santos-Vijande, M.L., Álvarez-González, L.I.: TQM's contribution to marketing implementation and firm's competitiveness. Total Qual. Manag. **20**(2), 171–196 (2009)
27. Calvo-Mora, A., Leal, A., Roldán, J.L.: Using enablers of the EFQM model to manage institutions of higher education. Qual. Assur. Educ. **14**(2), 99–122 (2006)
28. Kanji, G.K., Tambi, A.M.B.A., Wallace, W.: A comparative study of quality practices in higher education institutions in the US and Malaysia. Total Qual. Manag. **10**(3), 357–371 (1999)

Author Index

T. Ahram et al. (Eds.): IHSED 2019, AISC 1026, pp. 1101–1105, 2020.
https://doi.org/10.1007/978-3-030-27928-8

Printed in the United States
By Bookmasters